ALCOHOL AND THE NERVOUS SYSTEM

HANDBOOK OF CLINICAL NEUROLOGY

Series Editors

MICHAEL J. AMINOFF, FRANÇOIS BOLLER, AND DICK F. SWAAB

VOLUME 125

ELSEVIER

EDINBURGH LONDON NEW YORK OXFORD PHILADELPHIA
ST LOUIS SYDNEY TORONTO 2014

ALCOHOL AND THE NERVOUS SYSTEM

Series Editors

MICHAEL J. AMINOFF, FRANÇOIS BOLLER, AND DICK F. SWAAB

Volume Editors

EDITH V. SULLIVAN AND ADOLF PFEFFERBAUM

VOLUME 125

3rd Series

ELSEVIER

EDINBURGH LONDON NEW YORK OXFORD PHILADELPHIA
ST LOUIS SYDNEY TORONTO 2014

ELSEVIER B.V.
Radarweg 29, PO Box 211, 1000 AE Amsterdam, Netherlands
The Boulevard, Langford Lane, Kidlington, Oxford OX5 1GB, UK
225 Wyman Street, Waltham, MA 02451, USA

ISBN: 9780444626196

British Library Cataloguing in Publication Data
A catalogue record for this book is available from the British Library

Library of Congress Cataloging in Publication Data
A catalog record for this book is available from the Library of Congress

Commissioning Editor: Mica Haley
Development Editor: Michael Parkinson
Project Manager: Sujatha Thirugnana Sambandam
Designer/Design Direction: Alan Studholme

The Publisher's policy is to use **paper manufactured from sustainable forests**

Handbook of Clinical Neurology 3rd Series

Available titles

Vol. 79, The human hypothalamus: basic and clinical aspects, Part I, D.F. Swaab ISBN 9780444513571
Vol. 80, The human hypothalamus: basic and clinical aspects, Part II, D.F. Swaab ISBN 9780444514905
Vol. 81, Pain, F. Cervero and T.S. Jensen, eds. ISBN 9780444519016
Vol. 82, Motor neurone disorders and related diseases, A.A. Eisen and P.J. Shaw, eds. ISBN 9780444518941
Vol. 83, Parkinson's disease and related disorders, Part I, W.C. Koller and E. Melamed, eds. ISBN 9780444519009
Vol. 84, Parkinson's disease and related disorders, Part II, W.C. Koller and E. Melamed, eds. ISBN 9780444528933
Vol. 85, HIV/AIDS and the nervous system, P. Portegies and J. Berger, eds. ISBN 9780444520104
Vol. 86, Myopathies, F.L. Mastaglia and D. Hilton Jones, eds. ISBN 9780444518996
Vol. 87, Malformations of the nervous system, H.B. Sarnat and P. Curatolo, eds. ISBN 9780444518965
Vol. 88, Neuropsychology and behavioural neurology, G. Goldenberg and B.C. Miller, eds. ISBN 9780444518972
Vol. 89, Dementias, C. Duyckaerts and I. Litvan, eds. ISBN 9780444518989
Vol. 90, Disorders of consciousness, G.B. Young and E.F.M. Wijdicks, eds. ISBN 9780444518958
Vol. 91, Neuromuscular junction disorders, A.G. Engel, ed. ISBN 9780444520081
Vol. 92, Stroke – Part I: Basic and epidemiological aspects, M. Fisher, ed. ISBN 9780444520036
Vol. 93, Stroke – Part II: Clinical manifestations and pathogenesis, M. Fisher, ed. ISBN 9780444520043
Vol. 94, Stroke – Part III: Investigations and management, M. Fisher, ed. ISBN 9780444520050
Vol. 95, History of neurology, S. Finger, F. Boller and K.L. Tyler, eds. ISBN 9780444520081
Vol. 96, Bacterial infections of the central nervous system, K.L. Roos and A.R. Tunkel, eds. ISBN 9780444520159
Vol. 97, Headache, G. Nappi and M.A. Moskowitz, eds. ISBN 9780444521392
Vol. 98, Sleep disorders Part I, P. Montagna and S. Chokroverty, eds. ISBN 9780444520067
Vol. 99, Sleep disorders Part II, P. Montagna and S. Chokroverty, eds. ISBN 9780444520074
Vol. 100, Hyperkinetic movement disorders, W.J. Weiner and E. Tolosa, eds. ISBN 9780444520142
Vol. 101, Muscular dystrophies, A. Amato and R.C. Griggs, eds. ISBN 9780080450315
Vol. 102, Neuro-ophthalmology, C. Kennard and R.J. Leigh, eds. ISBN 9780444529039
Vol. 103, Ataxic disorders, S.H. Subramony and A. Durr, eds. ISBN 9780444518927
Vol. 104, Neuro-oncology Part I, W. Grisold and R. Sofietti, eds. ISBN 9780444521385
Vol. 105, Neuro-oncology Part II, W. Grisold and R. Sofietti, eds. ISBN 9780444535023
Vol. 106, Neurobiology of psychiatric disorders, T. Schlaepfer and C.B. Nemeroff, eds. ISBN 9780444520029
Vol. 107, Epilepsy Part I, H. Stefan and W.H. Theodore, eds. ISBN 9780444528988
Vol. 108, Epilepsy Part II, H. Stefan and W.H. Theodore, eds. ISBN 9780444528995
Vol. 109, Spinal cord injury, J. Verhaagen and J.W. McDonald III, eds. ISBN 9780444521378
Vol. 110, Neurological rehabilitation, M. Barnes and D.C. Good, eds. ISBN 9780444529015
Vol. 111, Pediatric neurology Part I, O. Dulac, M. Lassonde and H.B. Sarnat, eds. ISBN 9780444528919
Vol. 112, Pediatric neurology Part II, O. Dulac, M. Lassonde and H.B. Sarnat, eds. ISBN 9780444529107
Vol. 113, Pediatric neurology Part III, O. Dulac, M. Lassonde and H.B. Sarnat, eds. ISBN 9780444595652
Vol. 114, Neuroparasitology and tropical neurology, H.H. Garcia, H.B. Tanowitz and O.H. Del Brutto, eds. ISBN 9780444534903
Vol. 115, Peripheral nerve disorders, G. Said and C. Krarup, eds. ISBN 9780444529022
Vol. 116, Brain stimulation, A.M. Lozano and M. Hallett, eds. ISBN 9780444534972
Vol. 117, Autonomic nervous system, R.M. Buijs and D.F. Swaab, eds. ISBN 9780444534910
Vol. 118, Ethical and legal issues in neurology, J.L. Bernat and H.R. Beresford, eds. ISBN 9780444535016
Vol. 119, Neurologic aspects of systemic disease Part I, J. Biller and J.M. Ferro, eds. ISBN 9780702040863
Vol. 120, Neurologic aspects of systemic disease Part II, J. Biller and J.M. Ferro, eds. ISBN 9780702040870
Vol. 121, Neurologic aspects of systemic disease Part III, J. Biller and J.M. Ferro, eds. ISBN 9780702040887
Vol. 122, Multiple sclerosis and related disorders, D.S. Goodin, ed. ISBN 9780444520012
Vol. 123, Neurovirology, A.C. Tselis and J. Booss, eds. ISBN 9780444534880
Vol. 124, Clinical neuroendocrinology, E. Fliers, M. Korbonits and J.A. Romijn, eds. ISBN 9780444596024

Foreword

Artifacts from ancient excavations attest to the use of alcohol at least as early as the Neolithic age (over 10 000 years BC) and it has been suggested that alcoholic beverages, particularly beer, may have preceded bread as a basic food. Furthermore, there are documented instances in more recent history where beer replaced water when the latter was deemed to be undrinkable.

There are clearly two differing and coexisting viewpoints about alcohol. On the one hand, alcoholic beverages must be credited as playing an important role not only in enhancing the enjoyment and quality of life, but also in medicine and religion. On the other hand, alcohol has been responsible for disease, specifically alcoholism and various other disorders relating to alcohol consumption. While more than half of all Americans aged 12 years or older, i.e., an estimated 130.6 million people, report that they consume alcohol, alcohol abuse affects as many as 7-10% of the US population.

In the previous two series of the *Handbook of Clinical Neurology*, founded and edited by Pierre Vinken and George Bruyn, alcoholism was discussed in the context of intoxications of the nervous system. We are very pleased to introduce the present volume of the *Handbook* dealing with the consequences of alcohol abuse on the nervous system, the first time an entire volume has been dedicated specifically to this important problem.

The volume deals with a variety of alcohol-related topics. It includes animal models, genetics, and an indepth presentation of the epidemiology, neuropathology, neurophysiology, and molecular basis of alcoholism. Subsequent chapters deal with acute intoxication and the sequelae of chronic alcohol abuse, including the fetal alcohol disorder. Most chapters deal with the central nervous system, including neuropsychologic and psychiatric implications, but the effect of alcohol on the peripheral nervous system as well as on the liver is also discussed. One chapter deals with alcohol and the law and there are several chapters discussing the pharmacologic treatment of alcoholism as well as non-pharmacologic interventions and prevention.

We have been fortunate in having two distinguished scholars serve as volume editors. Dr. Edith V. Sullivan is Professor of Psychiatry and Behavioral Sciences at the Stanford University School of Medicine and Dr. Adolf Pfefferbaum is Director of the Neuroscience Program in the Center for Health Sciences at SRI International (a non-profit, independent research and innovation center serving government and industry) as well as Professor (Emeritus) of Psychiatry and Behavioral Sciences at the Stanford University School of Medicine. They have assembled a truly international group of contributing authors with acknowledged expertise and have produced with them an authoritative, comprehensive, and up-to-date volume on alcohol and the nervous system. Its availability electronically on Elsevier's Science Direct site as well as in print format should ensure its ready accessibility and facilitate searches for specific information. We are grateful to the volume editors and contributors for their efforts in creating such an invaluable resource. As series editors we read and commented on each of the chapters with great interest. We are therefore confident that both clinicians and researchers in many different medical disciplines will find much in this volume to appeal to them.

We are grateful to Dr. Ellen Witt, Deputy Director, Division of Neuroscience and Behavior, National Institute on Alcohol Abuse and Alcoholism. She and her colleagues provided invaluable advice in the early preparation of the volume. As always, it is a pleasure to thank Elsevier, our publisher - and in particular Michael Parkinson in Lochcarron, and Mica Haley and Kristi Anderson in San Diego – for their unfailing and expert assistance in the development and production of this volume.

Michael J. Aminoff
François Boller
Dick F. Swaab

Preface

For millennia, men and women enjoyed the pleasures of alcohol and have suffered the throes of its misuse. The lifetime prevalence of alcohol use disorders is estimated at 10%. Upwards of one-third of university hospital admissions are related to some form of alcohol abuse, and estimates are even higher for facilities caring for veterans. The neurologic substrates and sequelae of excessive and chronic drinking are salient and can be life-threatening; yet, many are quelled or reversed with sobriety and requisite medical supervision. No known remedy other than one's own metabolism reverses acute alcohol intoxication. Despite its prevalence and devastating effects on the afflicted, family, and society, alcoholism remains an undertreated and underrecognized disease even within the medical profession.

The first volume of the *Handbook of Clinical Neurology* was published nearly 50 years ago. In the 1970s, the *Handbook* published "Intoxications of the Nervous System," a two-volume series (volumes 36 and 37), with updates in the mid-1990s presented in volumes 64 and 65. The four volumes presented extensive reviews of the toxicology of numerous agents, poisons, and venoms, including alcohol. Now in series 3, volume 125, "Alcohol and the Nervous System" is the first of the entire handbook series devoted solely to the clinical and basic science of acute and chronic alcohol abuse and alcoholism. Although the most salient signs and symptoms of alcoholism have a neurologic basis resulting in brain changes considered both neurotoxic and neuroadaptative, their detection is often elusive for several reasons, including lack of awareness and difficulties in confronting the patient with the problem. Principal goals of this volume are to update knowledge on the basic neuroscience of alcoholism, recognize that alcoholism is a disease, and provide the clinician with guidance in identifying alcohol use disorders and in dealing with the difficulties in educating the patient about its untoward medical, psychologic, and social outcomes.

To fulfill these goals, several dozen leading scientists in clinical and basic alcoholism research contributed 39 chapters, making a compendium of evidence-based concepts and thoughtful models on the neurology and neuroscience of alcoholism. The volume presents alcoholism and alcohol use disorders as a disease and a lifespan problem, extending from fetal alcohol spectrum disorder to teenage drinking to chronic problems in adulthood and the age-alcoholism interaction, where older alcoholics exhibit accelerated brain structural insult. The basic neuroscience of alcoholism describes the molecular basis of alcoholism, gene-environment interactions, alcohol metabolism, pharmacokinetics, and interactions with other licit and illicit agents. Human studies using the various *in vivo* neuroimaging modalities to examine brain tissue macrostructure, microstructure, circuitry, chemistry, and neurotransmitter systems reveal damage incurred with excessive drinking but also at least partial recovery with sustained sobriety. The dynamic course of alcoholism captured with clinical observation and rigorous neuroimaging, electrophysiologic, and neuropsychologic study indicates not only the negative side with the demise of brain structure and function but also the hopeful side of recovery of selective brain function and even its structure with abstinence. Systematic studies of human alcoholism indicate that, despite the widespread untoward effects of alcoholism on the brain and its functions, the effects, while widespread, are not merely diffuse but rather show a selective pattern of sparing and impairment, potentially useful in directing efforts in behavioral and pharmacologic treatment development. These studies also provide evidence for the vast heterogeneity of expression of the sequelae and indicate major factors contributing to this variance that need to be considered in treatment and basic science. Salient factors include age, sex, ethnicity, psychiatric and other medical comorbidities, and genetic predisposition.

Why should a psychiatrist and a psychologist edit a volume on alcoholism for the *Handbook of Clinical Neurology*? Reasoning might include recognition that the formal diagnosis of alcohol use disorders is typically based on criteria listed in the *Diagnostic and Statistical Manual of Mental Disorders* (DSM) or *International Classification of Diseases* (ICD), diagnostic schedules focused on psychiatric diseases and disorders, and that the demonstrative behavioral outcomes are well characterized with neuropsychologic testing. Given that alcoholism - whether considered alcohol abuse, alcohol dependence, or alcohol use disorder - targets neurologic systems, alcoholism is an exquisite example of a multidisciplinary disease, requiring consideration from an interaction of disciplines. Its initiation requires consideration from neurology, psychiatry, and psychology and, as the disease advances, morbidities from many facets of internal medicine, hepatology, oncology, and cardiology come into play. Surely, the reader will come away with the understanding that treatment of the disease requires a team effort.

We thank the chapter authors, who represent these multiple medical and scientific disciplines, for contributing chapters of outstanding quality, thereby making our task more an educational experience than an editorial chore. We would also like to acknowledge the series editors for inviting us to assemble and edit this volume and Mike Parkinson, the Development Editor, for his invaluable oversight throughout the publication process.

Edith V. Sullivan
Adolf Pfefferbaum

Contributors

A. Abi-Dargham
Department of Psychiatry and Department of Radiology, Columbia University College of Physicians and Surgeons, New York, NY, USA

R.F. Anton
Center for Drug and Alcohol Programs, Alcohol Research Center, Department of Psychiatry and Behavioral Sciences, Medical University of South Carolina, Charleston, SC, USA

F.C. Baker
Center for Health Sciences, SRI International, Menlo Park, CA, USA and Brain Function Research Group, School of Physiology, University of the Witwatersrand, Johannesburg, South Africa

H. Beaunieux
Inserm, EPHE, Université de Caen/Basse-Normandie, Unité U1077, GIP Cyceron, CHU de Caen, France

H.C. Becker
Charleston Alcohol Research Center, Department of Psychiatry and Behavioral Sciences; Department of Neurosciences, Medical University of South Carolina and Department of Veterans Affairs RHJ Medical Center, Charleston, SC, USA

T.P. Beresford
Department of Psychiatry, School of Medicine, University of Colorado and Laboratory for Clinical and Translational Research in Psychiatry, Department of Veterans Affairs, Denver, CO, USA

S.W. Book
Center for Drug and Alcohol Programs, Alcohol Research Center, Department of Psychiatry and Behavioral Sciences, Medical University of South Carolina, Charleston, SC, USA

J.C.M. Brust
Department of Neurology, Columbia University College of Physicians & Surgeons, New York Neurological Institute, New York, NY, USA

R.F. Butterworth
Neuroscience Research Unit, Hôpital St-Luc (CHUM), University of Montreal, Montreal, Qc., Canada

R. Caetano
School of Public Health, Dallas Regional Campus, University of Texas Health Science Center at Houston, Dallas, TX, USA

J. Camchong
Neurobehavioral Research Inc., Kihei, HI, USA

S. Chanraud
EPHE; INCIA, CNRS UMR 5287, Université Victor Segalen Bordeaux, Bordeaux, France

K.G. Chartier
School of Public Health, Dallas Regional Campus, University of Texas Health Science Center at Houston, Dallas, TX, USA

I.M. Colrain
Center for Health Sciences, SRI International, Menlo Park, CA, USA and Melbourne School of Psychological Sciences, The University of Melbourne, Parkville, Vic., Australia

B.N. Costin
Virginia Commonwealth University Alcohol Research Center and Department of Pharmacology and Toxicology, Virginia Commonwealth University, Richmond, VA, USA

D.W. Crabb
Department of Medicine, Indiana University School of Medicine, Indianapolis, IN, USA

J.C. Crabbe
Portland Alcohol Research Center and Department of Behavioral Neuroscience, Oregon Health & Science University, Portland, OR, USA

F.T. Crews
Bowles Center for Alcohol Studies, Department of Pharmacology and Psychiatry, University of North Carolina, Chapel Hill, NC, USA

H.J. Edenberg
Department of Biochemistry and Molecular Biology, Indiana University School of Medicine, Indianapolis, IN, USA

M. Endres
Neurobehavioral Research Inc., Kihei, HI, USA

F. Eustache
Inserm, EPHE, Université de Caen/Basse-Normandie, Unité U1077, GIP Cyceron, CHU de Caen, France

R. Fama
Department of Psychiatry and Behavioral Sciences, Stanford University School of Medicine, Stanford and Neuroscience Program, SRI International, Menlo Park, CA, USA

G. Fein
Neurobehavioral Research Inc., Kihei, HI, USA

L. Ferguson
Waggoner Center for Alcohol and Addiction Research, University of Texas, Austin, TX, USA

T. Foroud
Department of Medical and Molecular Genetics, Indiana University School of Medicine, Indianapolis, IN, USA

L. Glass
Center for Behavioral Teratology, San Diego State University, San Diego, CA, USA

Z.R. Gravitz
Boston University School of Medicine and VA Boston Healthcare System, Boston, MA, USA

R.A. Harris
Waggoner Center for Alcohol and Addiction Research, University of Texas, Austin, TX, USA

S. Higuchi
National Hospital Organization, Kurihama Medical and Addiction Center, Kanagawa, Japan

J. Jacobus
Psychology Service, VA San Diego Healthcare System and Department of Psychiatry, University of California, San Diego, CA, USA

B.A. Johnson
Department of Psychiatry and Brain Science Research Consortium at the University of Maryland School of Medicine, Baltimore, MD, USA

Y-C. Jung
Department of Psychiatry, Yonsei University College of Medicine, Seoul, Korea

A.O. Karasov
Department of Psychiatry and Behavioral Sciences, Stanford University School of Medicine, Palo Alto, CA, USA

G.F. Koob
Committee on the Neurobiology of Addictive Disorders, The Scripps Research Institute, La Jolla, CA, USA

J.J. Kril
Departments of Pathology and Medicine, Sydney Medical School, University of Sydney, Sydney, Australia

A. Lembke
Department of Psychiatry and Behavioral Sciences, Stanford University, Stanford, CA, USA

B. Lewis
Department of Psychiatry, University of Florida, Gainesville, FL, USA

R.B. Luhar
Boston University School of Medicine and VA Boston Healthcare System, Boston, MA, USA

R.P. Maiya
College of Pharmacy, University of Texas, Austin, TX, USA

P.R. Martin
Department of Psychiatry, Vanderbilt University Medical Center, Nashville, TN, USA

D. Martinez
Department of Psychiatry, Columbia University College of Physicians and Surgeons, New York, NY, USA

S. Matsushita
National Hospital Organization, Kurihama Medical and Addiction Center, Kanagawa, Japan

S.N. Mattson
Center for Behavioral Teratology, San Diego State University, San Diego, CA, USA

R.O. Messing
College of Pharmacy, University of Texas, Austin, TX, USA

D.J. Meyerhoff
Center for Imaging of Neurodegenerative Diseases, San Francisco Veterans Affairs Medical Center and Department of Radiology and Biomedical Imaging, University of California San Francisco, San Francisco, CA, USA

M.F. Miles
Virginia Commonwealth University Alcohol Research Center, Department of Pharmacology and Toxicology, Center for Study of Biological Complexity and Department of Neurology, Virginia Commonwealth University, Richmond, VA, USA

B.A. Mills
School of Public Health, Dallas Regional Campus, University of Texas Health Science Center at Houston, Dallas, TX, USA

S.M. Ruiz
Boston University School of Medicine and VA Boston Healthcare System, Boston, MA, USA

D. Most
Waggoner Center for Alcohol and Addiction Research, University of Texas, Austin, TX, USA

P.J. Mulholland
Charleston Alcohol Research Center, Department of Psychiatry and Behavioral Sciences and Department of Neurosciences, Medical University of South Carolina, Charleston, SC, USA

E.M. Müller-Oehring
Department of Psychiatry and Behavioral Sciences, Stanford University School of Medicine, Stanford and Neuroscience Program, Center for Health Sciences, Biosciences, SRI International, Menlo Park, CA, USA

K. Namkoong
Department of Psychiatry, Yonsei University College of Medicine, Seoul, Korea

C.L. Nicholas
Melbourne School of Psychological Sciences, The University of Melbourne, Parkville, Vic., Australia

S.J. Nixon
Departments of Psychiatry and Psychology, University of Florida, Gainesville, FL, USA

A. Noronha
Division of Neuroscience and Behavior, National Institute on Alcohol Abuse and Alcoholism, Bethesda, MD, USA

M. Oscar-Berman
Boston University School of Medicine and VA Boston Healthcare System, Boston, MA, USA

M.J. Ostacher
Department of Psychiatry and Behavioral Sciences, Stanford University School of Medicine and Veterans Affairs Palo Alto Health Care System, Palo Alto, CA, USA

A-L. Pitel
Inserm, EPHE, Université de Caen/Basse-Normandie, Unité U1077, GIP Cyceron, CHU de Caen, France

M.H. Plawecki
Department of Psychiatry, Indiana University School of Medicine, Indianapolis, IN, USA

B. Porjesz
Henri Begleiter Neurodynamics Laboratory, Department of Psychiatry and Behavioral Sciences, SUNY Downstate Medical Center, Brooklyn, NY, USA

R. Prather
Department of Psychiatry, University of Florida, Gainesville, FL, USA

M. Rangaswamy
Henri Begleiter Neurodynamics Laboratory, Department of Psychiatry and Behavioral Sciences, SUNY Downstate Medical Center, Brooklyn, NY, USA

S. Ravan
Department of Psychiatry, Columbia University College of Physicians and Surgeons, New York, NY, USA

M.T. Reilly
Division of Neuroscience and Behavior, National Institute on Alcohol Abuse and Alcoholism, Bethesda, MD, USA

J.S. Rich
Department of Psychiatry, Vanderbilt University Medical Center, Nashville, TN, USA

J. Rose
Department of Orthopaedic Surgery, Stanford University School of Medicine, Stanford, CA, USA

K.S. Sawyer
Boston University School of Medicine and VA Boston Healthcare System, Boston, MA, USA

J.P. Schacht
Center for Drug and Alcohol Programs, Alcohol Research Center, Department of Psychiatry and Behavioral Sciences, Medical University of South Carolina, Charleston, SC, USA

T. Schulte
Neuroscience Program, Center for Health Sciences, Biosciences, SRI International, Menlo Park, CA, USA

C. Seneviratne
Department of Psychiatry, Institute for Genome Sciences, University of Maryland School of Medicine, Baltimore, MD, USA

D. Seo
Department of Psychiatry, Yale University School of Medicine, New Haven, CT, USA

D. Sheedy
Department of Pathology, Sydney Medical School, University of Sydney, Sydney, Australia

R. Sinha
Department of Psychiatry, Yale University School of Medicine, New Haven, CT, USA

M. Slifstein
Department of Psychiatry, Columbia University College of Physicians and Surgeons, New York, NY, USA

L.M. Squeglia
Department of Psychiatry, University of California, San Diego, CA, USA

M. Stanford
Addiction Medicine and Therapy Services, Santa Clara Valley Health and Hospital System, Santa Clara County, CA, USA

K.K. Sulik
Department of Cell Biology and Physiology and Bowles Center for Alcohol Studies, University of North Carolina, Chapel Hill, NC, USA

E.V. Sullivan
Department of Psychiatry and Behavioral Sciences, Stanford University School of Medicine, Stanford, CA, USA

G.T. Sutherland
Department of Pathology, Sydney Medical School, University of Sydney, Sydney, Australia

S.F. Tapert
Psychology Service, VA San Diego Healthcare System and Department of Psychiatry, University of California, San Diego, CA, USA

B.A. Temple
Department of Psychiatry, School of Medicine, University of Colorado, Denver, CO, USA

P.A.C. Vaeth
School of Public Health, Dallas Regional Campus, University of Texas Health Science Center at Houston, Dallas, TX, USA

M.M. Valmas
Boston University School of Medicine and VA Boston Healthcare System, Boston, MA, USA

R.L. Vassar
Stanford University School of Medicine, Stanford, CA, USA

R.P. Vetreno
Bowles Center for Alcohol Studies, Department of Pharmacology and Psychiatry, University of North Carolina, Chapel Hill, NC, USA

A.L. Ware
Center for Behavioral Teratology, San Diego State University, San Diego, CA, USA

K. Warren
Division of Neuroscience and Behavior, National Institute on Alcohol Abuse and Alcoholism, Bethesda, MD, USA

N. Wongngamnit
Substance Abuse Treatment Program, Department of Veterans Affairs and Department of Psychiatry, School of Medicine, University of Colorado, Denver, CO, USA

N.M. Zahr
Department of Psychiatry and Behavioral Sciences, Stanford University School of Medicine, Stanford and Neuroscience Program, SRI International, Menlo Park, CA, USA

Contents

SECTION 5 Neuropsychology

SECTION 6 Neuroimaging of brain macrostructure and microstructure

SECTION 7 Neuroimaging of neurochemical markers

SECTION 8 Neuroimaging of brain function

SECTION 9 Neuroelectrophysiology

SECTION 10 Fetal alcohol spectrum disorder

SECTION 11 Adolescent drinking

SECTION 12 Other topics

Section 1

Introduction

Handbook of Clinical Neurology, Vol. 125 (3rd series)
Alcohol and the Nervous System
E.V. Sullivan and A. Pfefferbaum, Editors

Chapter 1

Alcoholism: diagnosis, prognosis, epidemiology, and burden of the disease

THOMAS P. BERESFORD[1,2]*, NARIN WONGNGAMNIT[1,3], AND BENJAMIN A. TEMPLE[1]
[1]*Department of Psychiatry, School of Medicine, University of Colorado, Denver, CO, USA*
[2]*Laboratory for Clinical and Translational Research in Psychiatry, Department of Veterans Affairs, Denver, CO, USA*
[3]*Substance Abuse Treatment Program, Department of Veterans Affairs, Denver, CO, USA*

NATURAL HISTORY AND CLINICAL ASSESSMENT

Alcoholism frequency

Alcoholism, known clinically in its most severe form as alcohol dependence (ALD), and in early course forms as alcohol abuse, affects as many as 7–10% of persons in the United States (Vaillant, 1995). It directly affects nearly one in every three families in the United States and carries with it considerable negative social stigma. Those suffering from it often appear to reject any help from others, or even to ignore the presence of the condition itself, in their struggle to give up alcohol use.

Population studies offer some useful background insights to the practicing clinician but, as often characteristic of large epidemiologic studies, may not translate well into the care of specific patients. For example, the National Institute on Alcohol Abuse and Alcoholism (NIAAA) conducted a two-stage prevalence study, the National Epidemiologic Survey on Alcohol Related Conditions (NESARC), in 2001–2002 and 2004–2005 (Hasin et al., 2007). Criteria for ALD and alcohol abuse were borrowed from the *Diagnostic and Statistical Manual of Mental Disorders*, fourth edition (DSM-IV, 1994: American Psychiatric Association, 1994) and its fourth edition text revision (DSM-IV-TR: American Psychiatric Association, 2000) published in 2000, with little difference between the two. The same study group retrospectively compared the results with an NESARC predecessor conducted in 1991–1992 before the publication of the DSM-IV, the National Longitudinal Alcohol Epidemiologic Survey (NLAES). Owing to differing methods used in the two studies by the same researchers, they cannot be considered strictly equivalent with respect to population estimates of alcoholism, but they approximate each other with sufficient similarity to offer prevalence comparison over time. For example, in the NLAES, the percentage of ALD Americans was 4.38 while alcohol abuse was 3.03. In the NESARC, the corresponding figures were 3.81 and 4.65. The combined percent point prevalence totals for ALD plus alcohol abuse are NLAES 7.41 and NESARC 8.46, suggesting an estimated increase in prevalence of problematic alcohol use in 1.5% of the population between the two surveys, from 1992 and 2005. According to the author's (TB) calculations from these frequency data, a prevalence increase of 1.5% translates to some 4.4 million Americans, allowing for the estimated increase in the US population itself over those years. If the same rate persists today, the estimated increase alone would be 5.9 million Americans. Added to the 2005 NESARC base of 25 million, this results in a total of nearly 31 million ALD or alcohol-abusing persons in the United States, or about 10% of the present population. The figures speak to the magnitude of the problem and we will discuss its economic implications further below.

From another viewpoint, the Centers for Disease Control and Prevention (CDC) 2008–2010 National Health Interview Survey (NHIS) is another recent source of detailed alcohol epidemiology figures (Schoenborn et al., 2013). That survey found that rates of use, abuse, and dependence varied by age, gender, location, socioeconomic status, marital status, and ethnicity. The NHIS reports that only one-fifth of American adults are

*Correspondence to: Thomas P. Beresford, M.D., 1055 Clermont Street (116), Denver, CO 80220, USA. Tel: +1-303-399-8020; x 3732, E-mail: thomas.beresford@ucdenver.edu

Table 1.1

Characteristic differences between primary alcohol and polydrug dependence

Primary alcohol dependence	Polydrug dependence
7–10% US population	0.5% US population
Alcohol primary dependence	Polysubstance dependence
Normal childhood	Deprivation/abuse in childhood
No conduct disorder	Conduct disorder symptoms before age 15
Regular use: late teens, 20s	Polydrug dependence: teens
No characteristic personality diagnosis before addictive use	Adult personality disorder diagnoses concurrent with use
Natural remission: 30%/year	Natural remission: 10%/year
With treatment: 45%/year	With treatment: 10%/year

lifetime abstainers from alcohol use. However, many more women abstain than men (26.5% against 14.7%). Young Americans are more likely to drink than others. Also of note is the higher likelihood (35.1%) for adults aged 18–24 years than any other group to have had more than five drinks in one sitting some time in a year. White adults were more likely to drink than other groups: 67.8% versus 52.5% of Black adults and 48% of Asians. Interestingly, level of education was positively correlated with drinking. While 77.3% of those with graduate degrees were current drinkers, only 46.8% without high school diplomas were. Finally, the Midwest had the highest number of current drinkers at 68.6%, while the South had the fewest, with 61%.

Practical typology

For the neurologist, psychiatrist, or other health professional, more practical, disease course-based characteristics of ALD appear more pertinent. The vast majority of ALD cases, for example, involve alcohol as the primary, and often only, drug of sustained, uncontrolled use. Research reports refer to this as primary ALD; it has accrued other synonyms such as type 1 (Cloninger et al., 1981), type A (Brown et al., 1994), or late-onset (Wetterling et al., 2003) alcoholism. By contrast, polydrug dependence – that is, addiction to two or more substances, not including nicotine, that may include alcohol as one of the dependence substances – affects about 0.5% of adults in the United States. Viewed from the perspective of alcoholism research, this group refers to type 2, type B, or early-onset alcoholism. Although the distinction between the two clinical groups with respect to their abstinence prognoses has been known for many years, many clinical programs fail to take this distinction into account. For example, liver transplant programs that separate the two types in follow-up studies demonstrate high rates of abstinence in the primary ALD group after transplant while those that do not make this distinction often report much lower abstinence rates (Lucey, 2007). Prior research on non-transplant samples demonstrates that the much more common primary ALD persons will enjoy a better prognosis than the polydrug-dependent group (Vaillant, 1995). Table 1.1 summarizes some of the characteristic differences between the two in brief form.

It is important to keep in perspective that alcoholism has been viewed in many model systems over many centuries, ranging from moral failing to medical illness, with many other models between. (See the lead author's *Science Update* installments on the website of the National Council on Alcoholism and Drug Dependence, beginning in November 2013, for a brief historic overview.) The present discussion refers to alcoholism in the disease model and as an illness that courses over decades, and rather than a few weeks, months, or even years. This occurs for two practical reasons: (1) it allows application of the methods of clinical science to a ubiquitous, and deadly, condition; and (2) it provides patients with an all-important sense of hope. Quoting one, "It is much easier to think of myself as an ill person trying to get better than as a bad person trying to become good." Nonetheless, it is important to recall that other models provide useful discussions from other points of view with helpful contributions from time to time. Physicians have a particularly critical role to play in applying the principles of clinical science with respect to diagnosis and prognosis, both of which inform specific treatments.

DIAGNOSIS

Alcohol use history

Clinically, two general concepts guide the assessment of alcohol use, with or without other drugs of abuse. First, the interviewer is assessing an addictive disorder and ALD is a form of this phenomenon. Alcohol addiction is second in prevalence only to nicotine addiction in the general population.

Table 1.2

***International Classification of Diseases and Health Problems* (ICD-10) guidelines for a dependence diagnosis (WHO, 2010)**

Physiologic withdrawal state when alcohol use has ceased
Evidence of tolerance, such that increased doses of alcohol are required to achieve effects originally produced by lower doses
A strong desire or sense of compulsion to drink alcohol, during abstinence periods by implication
Difficulties in controlling alcohol-drinking behavior, by implication when once begun, whether in onset, termination, or levels of use
Progressive neglect of alternative healthy activities because of alcohol use
Persisting with alcohol drinking despite clear evidence and prior knowledge of harmful consequences

Second, ALD and other forms of substance dependence are clinical diagnoses based on specific behaviors, rather than quantifications of amounts and frequencies of alcohol use. Table 1.2 lists the World Health Organization's current diagnostic guidelines from the Tenth Revision of the *International Classification of Diseases and Health Problems* (ICD-10) (WHO, 2010), edited from the general guidelines to focus on ALD. The ICD-10 criteria begin with the stipulation that diagnosis of ALD dependence requires evidence that three or more of the six criteria must have occurred together at some time during the previous year.

The ICD-10 criteria describe ALD as a diagnostic syndrome, that is, as a grouping of concurrent, associated symptoms. Very recently, the American Psychiatric Association abandoned the syndrome approach used in previous editions of its *Diagnostic and Statistical Manuals* (American Psychiatric Association, 1987, 2000; Schuckit, 1994). The new fifth edition (American Psychiatric Association, 2013) in this same series abandons the ALD syndrome, preferring a more generally stated alcohol use disorder (AUD) construed on a continuum of mild, moderate, and severe; this is discussed further on in this chapter. For clarity when referring to medical diagnosis, this discussion will retain the ICD-10 version of ALD diagnosis as the most commonly used in clinical settings.

Making a diagnosis

Viewed in its most basic form, the diagnosis of ALD requires specific evidence of phenomena in three clinical domains: (1) physiologic dependence, including tolerance and withdrawal; (2) Loss of Control of alcohol use, often erroneously referred to as "craving"; and (3) a resultant decline in either physical or social functioning, or both. For practical purposes, this discussion will leave questions of precision and "caseness" for the judgments taken from further independent empirical evidence. Instead, the workaday neurologist and other medical specialists may be best served to focus on the three large symptom domains as offering a brief, practical way of remembering and assessing relevant symptoms in clinic settings.

Physiologic dependence: tolerance and withdrawal

Tolerance

Tolerance, more specifically acquired tolerance, refers to the ability of the central nervous system (CNS) to approximate normal functioning in the presence of ever-increasing doses of ethyl alcohol. Clinically, the person reports an acquired need for more alcohol to get the same effect, once noticed at a much lower dose earlier in the natural history of drinking. To assess this, the clinician must establish a baseline drinking effect that has changed over time. This necessitates careful attention to the details of the drinking history.

One useful approach may be to ask what effect the alcohol had when a person first began drinking on his or her own. Results include reports such as nausea, feeling high, or other unique descriptors that the patient can provide: what the person noticed after one or two standard alcohol drinks. A standard drink, each containing roughly ¾ oz. of ethyl alcohol, may be approximately defined as a 12 oz. can or bottle of beer, 6 oz. glass of table wine, or 1.5 oz. shot of whiskey or other spirits.

After establishing a baseline, for example, "In high school I got high after one or two cans of beer," the interviewer may then ask how many standard drinks the person noticed achieved the same effect at the time when the drinking was at its greatest. Some formal criteria require a 50% increase. In the case of alcohol, most heavy drinkers will describe a doubling or more of the amount used for an initial effect. Many ALD patients will describe amounts several times greater than those drunk in the state naïve to alcohol, such as, "Back then I got high after two beers. Now I don't notice anything until after I drink a six pack and a case [24 12-oz. cans or bottles of beer] to feel drunk." This signals that the CNS has adapted to heavy alcohol use, that is, developed a tolerance.

While most forms of tolerance are acquired over time with repeated drinking, research has demonstrated that a minority of individuals have a very high innate tolerance to ethyl alcohol as part of their neural make-up. In research paradigms this is currently referred to as low response (LR) to alcohol (Schuckit et al., 2004).

Clinically, LR can be approached through questions such as: "Some people tell me that they didn't notice an effect of the alcohol when they first began to drink. What was your experience?" This same subjective insensitivity to alcohol in a drinking-naive state is mirrored in reduced neuroendocrine responses and in a lack of neurologic impairment on such tasks as maintaining balance (Schuckit et al., 2004). More ominously, follow-up studies of innately alcohol-tolerant persons indicate a much higher ALD prevalence rate, as high as 29%, far higher than the control group estimated frequency of 11% (Schuckit and Smith, 1996). In such cases, the body appears to lack a "built-in" negative-feedback mechanism for ethanol's effects. This tendency runs in families and serves as a focus of ongoing genetic explorations as to its cause. Whether LR carries the same diagnostic weight as an acquired tolerance may be argued. Clinically, however, it serves as a risk factor for ALD.

Withdrawal

Defined clinically as an excessive activity of the sympathetic nervous system triggered by a rapid decline in alcohol blood level, withdrawal symptoms accompany tolerance in the great majority of individuals. In those who report no withdrawal symptoms despite a history of clear tolerance to alcohol, the clinician must ask whether the patient is drinking in the morning before withdrawal symptoms manifest themselves, or is regularly taking some other CNS depressant, such as a benzodiazepine or an anticholinergic agent, that covers withdrawal symptoms, especially in the morning. On occasion, the clinician will encounter patients who have few or no withdrawal symptoms despite a clear alcohol tolerance. Although relatively infrequent, this recognized clinical phenomenon appears most likely due to a genetically based constitutional invulnerability to withdrawal (Crabbe et al., 2013).

Since ethanol is a short-acting CNS depressant, its quick removal after constant and heavy exposure triggers CNS hyperactivity both centrally, as for example expressed by a subjective sense of jitteriness or impending disaster (anxiety), and peripherally through the symptoms of sympathetic nervous system discharge. Table 1.3 lists acute withdrawal symptoms that patients commonly report.

Of note, neither headache nor blackout is a symptom of alcohol withdrawal. The former, a component of "hangover," often may have to do with fluid shifts in the brain that are thought to activate pain sensors in the arachnoid layer of the brain's coverings. The latter is a state-dependent amnesia that occurs when reaching acutely high levels of ethanol. While these can indicate acute heavy drinking with or without ALD, they do not serve as diagnostic of ALD.

Table 1.3

Symptoms and signs of early alcohol withdrawal

Symptom	Sign
Anxiety	Palpitations (tachycardia >110 beats/min)
Tremor	Rapid breathing
Nausea and vomiting	Low-grade fever
Sweating	Hypertension (>90 mmHg diastolic)

Later and more severe occurrences, however, such as generalized grand mal seizures on or after the first day of withdrawal, and delirium tremens (DTs) occurring after the third day, make up the most severe and most dangerous withdrawal phenomena. Both are often preceded by either or both of two clinical signs: hyperreflexia and ankle clonus, both indications of the severely irritated CNS. The DTs – characterized by the clinical triad of: (1) confusion; (2) usually visual or tactile hallucinations; and (3) extreme rises in vital signs, such as malignant hypertension or tachycardia – presents a true medical emergency and indicates in hospitalization. Because of the patient's confusion during the DTs, he or she may not appreciate the significance of a hospitalization for withdrawal. A corroborating third party can provide clinically crucial information in that instance. In epidemiologically vulnerable populations, such as those admitted to general hospitals, withdrawal seizures and DTs may occur as frequently as 7% or 5% respectively (Kaim et al., 1969). Both should be regarded as emergencies. Active seizures indicate attention to assuring open airway status via standard mandibular extension or, if needed, intubation when continued airway compromise occurs. Both isolated convulsions and status epilepticus respond to parenteral benzodiazepine administration of agents like diazepam or lorazepam, the same agents used to treat alcohol withdrawal. Most regard admission to an intensive care unit as the best site of treatment for the DTs, a volatile condition with reported death rates in the range of 10–15% when left untreated (Clark et al., 2013). Adequate doses of benzodiazepine agents to relieve targeted symptoms, followed by a tapering of those agents over the next few days, constitute the standard treatment of withdrawal symptoms in the United States.

Loss of control phenomenon

The essence of any addiction, and certainly of ALD, the Loss of Control phenomenon, refers to the inability of the affected person to predict with any degree of certainty how much alcohol he or she will drink from one episode to the next (Vaillant, 1995). Clinically, once the

Table 1.4

Assessing Loss of Control

First state: Some people tell me they notice these things when they drink and I would like to know if you've ever noticed them.

1. Some find it very hard to stop drinking, once they start. They feel that they are "off to the races" where drinking is concerned.
2. Some people find themselves drinking more than they wanted to drink or had planned to drink.
3. Some notice that they try to make rules to control their drinking, such as by drinking only certain types of beverages and avoiding others or by restricting drinking to a specific time of day.

drinking episode starts, the ALD person will be unable to stop in the middle of the drinking bout without a very great struggle. Useful questions at interview include the phrasing presented in Table 1.4.

Indirect phrasing of this kind, with the reference to "some people," displaces the patient's concern at being accused of some form of vice and allies both interviewer and patient in looking at one phenomenon - the patient's experience. This lessens the likelihood of an implied adversarial setting in which the interviewer examines the patient as if under a microscope. Experience suggests that an allying approach gains much more detailed, and more useful, information than an adversarial confrontation.

At the same time, it is important to distinguish the Loss of Control phenomenon from "craving." The former has to do with the inability to stop drinking once started. Craving refers to the episodic and often intense desire or compulsion to drink while in an alcohol- or drug-free state, as when between bouts of use. It may also be confused with the patient's intense search for alcohol to relieve withdrawal symptoms, a separate phenomenon as well. The distinction can be made accurately in the clinical setting by paying attention to what occurs during episodes of use and what occurs between them. Loss of Control during use is a hallmark of addiction while craving between use episodes appears to be a much more complex phenomenon involving brain-environment interactions. The ICD-10 diagnostic guidelines include both.

Social or physical decline

The combination of physical addiction and Loss of Control usually results in a patient's spending a large portion of time and effort sustaining heavy drinking or managing or recovering from its effects. To assess this, the clinician asks whether drinking has become a problem with respect to family relationships, legal status, work, friendships, or physical health. Neurologists and other physicians will be especially attuned to the last of these since the physical illness sequelae of ALD result in frequent clinic visits and hospitalizations. The patient's family relationships, however, usually offer the most sensitive indications of social difficulties due to alcohol or drug use. A change in work status is usually one of the last indicators of social problems since income provides the basis for continued use.

ALD, then, may be reliably diagnosed when evidence in all three domains presents. None of the current or immediately past sets of DSM diagnostic criteria require evidence in all three domains and therefore may cast a somewhat wider diagnostic net. The ICD-10 guidelines add the dimension of craving use among their six symptom categories. While either approach is defensible, the physician will do well to choose one or more of the standard definitions and use it consistently over time (Lucey et al., 1993; Lucey, 2007).

GAUGING PROGNOSIS

As noted in Table 1.1, drinking remission rates for primary ALD persons range from about 30% per year without clinical treatment to about 45% per year when clinical treatment is added. This figure is complicated, however, by high relapse rates. Some studies note relapse rates as high as 70% in the first year, with a gradual reduction in relapse frequency to the single-digit range after 6 years (Vaillant, 1995). Clearly, primary ALD patients require treatment efforts aimed at relapse prevention and alcohol-free living. Achieving this has to do with patients resolving a sense of underlying ambivalence about their alcohol use, as discussed further below. For polysubstance-dependent ALD persons, however, outcome studies are much less sanguine. Annual abstinence rates approximate 10% and show little in the way of an incremental response to treatment. For this group, some argue that harm reduction – maintaining life and health until the person "matures out" of their polydrug addictions that include alcohol – offers the best treatment goal. The natural courses of both variations of ALD invite us to consider the specific phenomena that predict recovery and abstinence.

Positive versus negative

How can the physician go about the task of assessing the prognosis for alcohol-free living (Beresford, 2006)? While each individual patient and his or her social network provide unique considerations, a series of general factors, best gleaned from prospective, longitudinal research on populations of alcohol- or drug-dependent persons, offer useful windows on prognosis. In the great majority of cases these include: (1) the alcohol use diagnosis; (2) an understanding of the patient's ambivalence towards continuing use; (3) measures of personal social stability; and (4) a series of four empirically based factors that Vaillant (1995) found to predict stable abstinence in prospective research studies. In addition, those with a co-occurring diagnosis of a major psychiatric disorder require first, an evaluation of psychoactive medication response to the primary disorder and second, a medication adherence history in addition to the factors already mentioned.

Some clinical programs focus more on what they see as negative prognostic predictors – those that predict a return to drinking – rather than positive factors such as those that predict continued abstinence. The negative factors often include mention of first, alcohol or other drug use in the past 6 months – the infamous "6-month rule" that has not been borne out in prospective research (Dimartini et al., 2010), and second, counting previous "recidivist" episodes, a term that calls to mind criminal incarceration and punishment, rather than a remitting/relapsing medical condition. In most cases, a focus on the positive prognostic factors that predict abstinence offers a more sound approach.

Diagnosis is prognosis

The diagnostic procedure described in detail above provides the basis for important prognostic distinctions bearing on continued abstinence. Practically speaking, health professionals must evaluate ALD type as a clinical guide to prognosis. While the 6-month rule has little justification for use in primary ALD (DiMartini et al., 2006), a history of recent use of any dependent drug, including alcohol, in a polydrug-dependent ALD person predicts the limits of targeting only abstinence in treatment. Empiric data on the polydrug group suggests that continuous, stable abstinence comes only after the emotional growth of many years, often not until the person's late 30s or early 40s (Vaillant, 1995). In a person with polydrug-dependent ALD, demonstration of a corroborated, sustained, stable abstinence from both drugs and alcohol for a significant period of time - often measured by one or more years rather than by months - affords confidence in providing medical treatments, including that of a scarce resource such as liver grafts, for which clinical need far exceeds the graft supply.

Abuse rather than dependence

When diagnosing primary ALD, it is important to establish whether the Loss of Control phenomenon exists in each case. Its absence, even in the face of a clear history of tolerance or alcohol-related social or physical problems, for example, strongly suggests a lesser diagnosis of alcohol abuse or, in DSM-5 terms, mild or moderate AUD. As an example of specific importance, in our clinical series (Lucey et al., 1994), about 10% of those applying for a liver transplant for alcoholic cirrhosis did not meet ALD criteria but could clearly be considered in the alcohol abuse category. Abuse generally suggests that behavioral processes moving towards Loss of Control and ALD appear to be in place but that the clinical line into ALD has not yet been crossed. Even though a lesser diagnosis than ALD (Schuckit et al., 2005), therefore, alcohol abuse may still indicate asking the patient to cease alcohol use. In that case, two phenomena often occur: (1) the person achieves and maintains abstinence without a struggle; and (2) the risk of relapse over the long term appears considerably lower. Generally speaking, alcohol abuse offers a much better prognosis with respect to natural course than does ALD. While the lesser diagnosis is sometimes more difficult to define, it is a generally more optimistic diagnosis to make.

Behavioral versus hepatic diagnosis

A significant portion of those candidates referred to as "alcoholic" on the basis of liver disease or other alcohol-related conditions will not merit the ALD diagnosis. In our previous series (Lucey et al., 1994), only about 75% of those with a tissue diagnosis of alcoholic liver disease fit the behavioral ALD diagnosis. Conversely, only about 80% of those referred as "alcoholics" met the ALD diagnosis. Of the remaining 20%, half fit the alcohol abuse diagnosis and half did not merit either. This was especially true of women who, because of a well-known gender-specific vulnerability to alcoholic liver disease, may injure the liver without ever reaching ALD. This limited overlap emphasizes the need for diagnostic precision in establishing the presence of ALD.

Assessing ambivalence and continuing Loss of Control risk

Clinically, longitudinal studies of abstinence make it clear that, once the control of drinking behavior departs, it does not return in the great majority of cases (Vaillant, 1995). Once lost, controlled alcohol or drug use cannot be relearned or reconstituted. In this sense a diagnosis of ALD signals a permanent condition, including a permanent risk of uncontrolled drinking or use. One component of the interview, therefore, requires review of this risk with the patient and a third party. Poor understanding, or ambivalent acceptance, of drinking relapse risk in a patient with normal cognition may indicate an unresolved ambivalence toward ALD itself and therefore the need to refer for alcohol- or other drug-focused behavioral treatment.

Social stability

This is the foundation upon which ongoing medical adherence and favorable prognosis are built. While several social stability scales exist, the simplest comes from the work of Straus and Bacon (1951), who found that the presence of any two of four social factors predicted clinic appointment compliance among ALD persons. These included: (1) being married; (2) not living alone; (3) stable employment for 3 years; and (4) a stable residence for 2 years. Viewed teleologically, falling below the cutoff point suggests, but does not establish, the likelihood of insufficient social resources to support treatment efforts. These may include social isolation and homelessness, two conditions that raise the likelihood of both drinking relapse and medical non-compliance. When present, the specialized involvement of a treatment clinic program aimed at increasing social resources offers the next step towards ALD recovery.

Vaillant's four prognostic factors

In the best available prospective, longitudinal study of ALD prognosis, Vaillant's 8-year study noted four operating factors, any two or more of which, when present, predicted long-term sobriety, defined as 2 years or more in his study. One or none indicated a return to drinking within 1 year (Vaillant, 1995). A proper prognosis evaluation includes them.

Structured time

Because ALD-style drinking requires a lot of time and effort, newly abstinent persons often encounter considerable "dead," or unstructured, time that can itself foster a relapse. Structuring that time corrects this concern and is a frequent ingredient in specialized clinic and self-help group (such as Alcoholics Anonymous) formats. Any setting that fills time with productive and engaging activity - that the person's "heart is in" – fits the need. At interview, a brief assessment of how the patient spends his/her day addresses whether he or she engages in active projects that involve other people, or lives an isolated existence such as filling the time with endless television viewing.

A rehabilitation relationship

Sustained, heavy drinking or drug use serves to isolate the person from others and to create the need in those who come into social contact, such as family members, to "fix" the drinking or use. Direct intervention, such as telling the person not to drink or use, generally results in an interpersonal struggle that works against continued abstinence. Rather, abstinence sets the stage for proper boundaries between person and social network. In brief, the family, a clinic, and a self-help network communicate the same approach to the person: "You can stay but the drinking has to go." Clinic personnel, other professionals, seasoned self-help group members, and family members who have learned to set boundaries on drinking behaviors may qualify as providing these kinds of healthy relationships to ALD patients.

Sources of hope or self-esteem

Regretful thoughts about having hurt others during drinking days frequently return to most ALD people during abstinence periods. Ruminating on them has the effect of lowering hope and self-confidence and thereby promoting a return to drinking: "I am a terrible person, I may as well drink." In assessing prognosis, the interviewer inquires after mechanisms that help a person through such periods. These include hope for the future ("What keeps you going in life? What do you look forward to?") and sources of accomplishment or self-esteem ("What helps you to feel good about yourself? What gives you a sense of usefulness or accomplishment?"). The evaluator may also note: "Most people run into regrets about their drinking days. Does this happen to you?" If the answer is positive, "What helps you get past regretful thoughts without drinking?" For some it may be religious beliefs, for others the "higher power" without religion, as Alcoholics Anonymous describes it, and for others a concentration on items of hope and accomplishment, such as "another day behind me without drinking." Clinically, the absence of forces that counterbalance regretful ruminations presents an ominous sign toward relapse and indicates further specialized treatment to focus on that.

A negative behavioral reinforcer

In other words, "Is there something very painful that will happen to you, without doubt, the very next time you drink alcohol?" Very few circumstances meet these criteria; severe pancreatic pain and the disulfiram-ethanol reaction qualify and are the most frequently encountered. Neuropathies, seizures, cardiomyopathy, liver failure, and most other alcohol-related medical conditions do not fulfill the criteria because of their subtlety and lack of predictability on each occasion. While the others of Vaillant's factors can be considered positive changes in the path to sustained abstinence, this is the only negative one and is generally the least frequently encountered except in the clinic setting. Disulfiram treatment often results in low adherence rates in most instances unless it is involuntary, as when ordered by a court. Cooperation between court and clinic results in much higher disulfiram-assisted abstinence rates

Table 1.5

Addressing prognostic factors systematically

Physical diagnosis
Cognitive impairment: add lactulose or other treatment and re-evaluate
Substance use diagnosis
1. Primary alcohol dependence: proceed to further prognostic evaluation
2. Polydrug dependence with alcohol dependence: verify either lengthy sustained abstinence and the "maturing out" process or verify continued effective treatment as, for example, in methadone maintenance for opiate dependence
3. Abuse: verify/corroborate cessation and abstinence
4. No diagnosis: corroborate and proceed to other treatments as medically or surgically indicated

Substance use prognosis
1. Unresolved ambivalence toward use: focused treatment on this aspect of abstinence; re-evaluate in 3–6 months
2. Unstable social adjustment: establish viable social resources; 3–6 month follow-up reassessment
3. One or none of Vaillant's factors: focused treatment referral; 3–6 month follow-up assessment

(Martin et al., 2003, 2004). This medicine is not without risk, however, and must be given with careful monitoring, especially during the first 3 months of its prescription.

Summing up prognosis

While none of the factors listed under the heading Gauging prognosis, above, may individually offer absolute certainty in respect to sustained abstinence in each individual case, their combination of variables offers a prognostic pattern that appears useful. Our early algorithm (Beresford et al., 1990), for example, predicted high rates of abstinence in ALD liver transplant recipients who present a series of good prognostic factors and this has been borne out in subsequent studies (Lucey, 2007). This approach can be applied to any concurrent disease entity, such as in a clinic specializing in alcoholic neuropathies, with similar results. The individual factors summed in Table 1.5 can be addressed in the initial evaluation and re-evaluated longitudinally.

OTHER PSYCHIATRIC ASSESSMENT

The foregoing discussion relates to the great majority of persons with alcohol or substance use histories who do not have other pre-existing psychiatric conditions. One critically important feature of this patient group recalls that ALD can mimic all of the major psychiatric disorders, including major depressive disorder, bipolar disorder, psychosis, several anxiety disorders, and obsessive compulsive disorder, as well as other entities, such as sleep discontinuity and anorexia. These and others can be assessed through a series of standard screening questions. When addressing them, the evaluator must keep one truly important factor in mind: other psychiatric disorders cannot be established with certainty during periods of heavy drinking and, to a large extent, during periods of other drug use. Diagnostic clarity for the listed psychiatric disorders requires evidence during abstinence periods generally exceeding 1 month. Failing to observe this may lead to misdiagnosis and misutilization of treatments, most often seen in psychoactive medications given gratuitously.

DSM-5 and alcoholism diagnosis: progress or old wine in a new bottle?

With the release of the new DSM-5 manual (American Psychiatric Association, 2013), no discussion of AUDs will be complete without reference to this published approach. For perspective, the DSMs of mental disorders in their various editions were intended as a classification guide. For many, however, they became less a guide than a cookbook, even though never intended to supplant clinical judgment in diagnosis. Since 1952, the DSM has been previously updated four times, the last in 2000, DSM-IV-TR (American Psychiatric Association, 2000). In 2007, the American Psychiatric Association formed a workgroup whose findings led to changes found in DSM-5 in 2013.

The most substantial change replaces alcohol abuse and ALD with a spectrum disorder, "alcohol use disorder." This appears driven by review of alcohol abuse criteria that found the validity of the alcohol abuse diagnosis as considerably lower than that for ALD. Clinical observation also suggested cases in which alcohol abuse appeared more severe than ALD. Driven by these observations, the workgroup looked at combining abuse and dependence criteria in favor of a severity rating (Hasin et al., 2013).

The DSM-5 also added craving, defined as a strong desire or urge to drink, although the workgroup data revealed limited utility to this addition in respect to validity or reliability. This criterion was added since craving served as a target for some attempts at pharmacotherapy and may relate to a subgroup of ALD persons. Interestingly, the ICD-10 guidelines had already incorporated the concept in its syndrome approach.

DSM-5 outlines that, in order to meet a mild severity level, two of 11 criteria must be met. This best matched the new spectrum disorder with current prevalence of DSM-IV alcohol abuse and ALD. This was acknowledged to be a low threshold, with the claim that this would not add to the diagnosed population. Rather, it would allow for identification of mild cases of AUD in the primary care setting, or other settings, where perhaps Screening, Brief Intervention, Referral to Treatment (SBIRT) may be implemented. To define moderate and severe cases, weighing of individual criteria appeared to add no benefit and a numeric strategy was implemented: moderate, four to five criteria; severe, six or more criteria.

The DSM-5 diagnostic model hopes to recognize the range of persons with AUDs more efficiently than the previous models and to align criteria with some newer conceptualizations of the disorder. As always in science, independent verification studies will assess whether the new model is useful. In the meantime, alcoholism, formally defined as ALD, continues with astounding levels of cost to society.

ECONOMIC COSTS OF PROBLEMATIC ALCOHOL USE

Cost of illness: clarifying the concept

The human costs of ALD, here included with the other AUDs, are numerous and diverse. Consequently, any estimated economic price of alcohol abuse appears as a collage of different types of costs. Some are fully or directly attributable to alcohol, others only partially or indirectly so. Some of those costs are borne by the alcohol user, others by someone else. Some of the costs appear quickly, others manifest themselves either chronically or after a long period without previous indication. Some of the costs are attributable to death, others to disease, injury, or incapacitation. Because the price of AUD can depend on some or all of these several factors and others not listed, it also depends, to some degree, on which of these categories are taken into account when calculating cost.

Similarly, the meaning and significance of the cost of AUD rest, in part, on the way that the price is expressed. Cost can be stated in total or in per capita terms. It can be stated as a cost incurred over different durations of time. By taking into account total alcohol consumption, it can be represented in terms of cost per drink. It can also be expressed as a ratio of tax revenues from alcohol to cost of overconsumption. Or it can be stated in qualified, per-person amounts: for example, dollars spent per non-alcohol-abusing person. It can even be expressed in negative terms, as in dollars not contributed to the economy as a result of premature deaths.

Clearly, calculating the cost of AUD involves flexible borders, chosen by researchers with different interests and orientations. Yet, for general purposes, at least domestically, standards exist: the US Public Health Service provides guidelines for cost of illness studies. Using these standards, one study found that, in 2006 (the most recent year with adequate data), AUD cost the American economy \$223.5 billion or \$746 per capita (Bouchery et al., 2011). In 2006, that amounted to 1.6% of gross domestic product (GDP). For perspective, that is a cost close to 40% of the total growth in GDP for that year.

Cost: contribution by category

Following the Public Health Service guidelines, the same study took into account several broad expenditure categories: healthcare costs, productivity losses, and a miscellaneous group mostly involving crime-related expenses. Each group was divided into smaller categories and researched piecemeal. For instance, healthcare costs were calculated by adding together the price of specialty care for AUD, hospitalization, treatment of fetal alcohol syndrome, health insurance, medication, prevention and research expenses, ambulatory care, nursing homes, care for crime victims, and training for medical personnel. Each subcategory tells a story. Consider those costs related to medical personnel. Alcohol abuse diverts the resources of psychiatric, surgical, and emergency departments away from other less-preventable disease types.

For each subcategory, expense figures were culled from sources like the NESARC. Where the specific contribution of alcohol to cost in a category was unclear, as in homicide for example, the researchers consulted CDC-developed software, or Alcohol-Related Disease Impact, to provide clarification. Additionally, the study kept track of which costs were attributable to four costly types of alcohol use: (1) binge drinking; (2) underage drinking; (3) drinking while pregnant; and (4) drinking related to a crime. In places, these groups overlapped with the healthcare, lost productivity, and miscellaneous groups. Sometimes this was illuminating, as when binge drinking was found to account for 57% of all healthcare spending related to excessive drinking.

The reported groups, cost categories, and cost subcategories (Bouchery et al., 2011) provide helpful reference points for more intratotal analysis. Perhaps surprisingly, significantly more of the cost of alcohol abuse came from lost productivity (\$161.3 billion or 72.2%) rather than from medical expenses (\$24.6 billion or 11%). Similarly, a much larger price was incurred by binge drinking (\$170.7 billion or 74.6%) than by any other type.

Interpretation: cost consequences

Alcoholism is costly. But who pays? Overall, in 2006 the share accounted for by alcohol users and relatives (\$92.9 billion or 41.5%) was nearly the same as the part paid for by government (\$94.2 billion or 42.1%). However, the answer also depends on the type of cost under consideration. In the healthcare category of AUD-related expenditures, government spending, at 60.9%, greatly outstripped alcohol abuser contribution to cost. Regardless of exact share, in both the general and the specific case, alcohol abusers pay only part of the cost that they incur.

Because excessive alcohol use transfers costs from alcohol users to non-users, responsible users, and the state, the economic price of alcohol abuse has public policy intersections and implications, particularly in tax law. Some research suggests that demand for alcohol, perhaps unlike demand for other addictive goods, declines in response to increases in price (Cook and Moore, 2002). This in turn suggests that raising alcohol excise taxes could not only increase tax revenues, but also reduce demand for alcohol and the attendant economic harm caused by its overuse.

One report notes that, while alcohol taxes at one time accounted for 40% of federal revenues, they now make up closer to 0.5% (Naimi, 2011). This decline is partly caused by the way federal alcohol taxes are designed: they tax by volume of alcohol rather than as a proportion of the price of alcohol and consequently, when not adjusted, decline over time in real terms, as a result of inflation. Some of the decline in tax revenue also has to do with a drop in American consumption over the last three decades.

ECONOMICS AND ALCOHOL

While alcohol exacts a high economic cost, the causal relationship does not run one way: economic conditions can also influence patterns of AUD. Research conducted regarding the recent economic crisis of 2008-2009 suggests that sharp economic downturns can push drinking patterns in two directions (Bor et al., 2013). While the percentage of the population that drinks declined slightly overall (from 52% to 51.6%), overall consumption increased (0.27 additional drinks per month per person) and binge drinking jumped markedly (by 7.2%). Because binge drinking is such a major contributor to AUD-related costs, particularly in healthcare, the increase is concerning. Rates in all three categories normalized by 2010 as the most acute period of economic distress passed.

ALCOHOL-RELATED SPENDING AND OTHER DISEASE CATEGORIES

How does healthcare spending caused by AUD compare with healthcare spending caused by other similar avoidable public health problems, like smoking and obesity? RAND research from the early 2000s associated problem drinking with an average increase of $150 per person for ambulatory and inpatient care (Sturm, 2002). By contrast, smoking was associated with a $230 increase and obesity with a $395 average climb. While alcohol abuse accounted for less of an increase in healthcare prices per person for those needing care, it nevertheless contributed to increased healthcare costs.

Healthcare spending involves not just clinical work, but also research costs. And National Institutes of Health (NIH) figures on annual spending show that alcohol-related research is funded at much lower levels, on both a total and per-person basis, than is research on other major disease categories. In 2011, roughly 31 million Americans abused alcohol and 13 million suffered from cancer. However, alcohol-related research received funding of $452 million, compared to $5.4 billion for cancer research. Per person suffering from disease, alcohol research received $15 while cancer research received $419, according to our own group's most recent review of published figures (Beresford et al., 2013). And yet the NIH estimates the economic impact of cancer in 2011 to have been $201.5 billion. This is a smaller amount than the $223.5 billion figure for the alcohol-related impact in 2006, which is again the most recent estimate. Coupled with the imbalanced per-patient research funding figures from 2011, it could mean that the high economic impact of AUD is not reflected in a commensurately high financial commitment to research.

CONCLUSION

Alluding to an ancient Chinese curse: alcoholism clinicians, scientists, and patients, and their communities, large and small, all live in interesting times. On the one hand there is much debate on formulations, reformulations, and re-reformulations of diagnosis. On the other, our understanding of prognosis is much better than ever before. At the same time, medications to treat ALD are generally regarded as largely ineffective - a complicated perception that this chapter has not discussed. Meanwhile, the US economy takes a significant beating at the hands of alcoholism. And, every day, many thousands of people fall ill, die, or lose their families owing to this preventable, treatable illness. There is much work to be done, much more to learn, and many more lives and families to be saved.

References

American Psychiatric Association (1987). Diagnostic and statistical manual of mental disorders : DSM-III-R. 3rd edn. American Psychiatric Association, Washington, DC.

American Psychiatric Association (1994). Diagnostic and Statistical Manual of Mental Disorders. fourth edition American Psychiatric Association, Washington, DC.

American Psychiatric Association (2000). Diagnostic and statistical manual of mental disorders: DSM-IV-TR. American Psychiatric Association, Washington, D.C.

American Psychiatric Association (2013). Diagnostic and statistical manual of mental disorders: DSM-5. American Psychiatric Association, Washington, D.C.

Beresford TP (2006 May). Probabilities of relapse and abstinence among liver transplant recipients. Liver Transpl 12 (5): 705–706.

Beresford TP, Turcotte JG, Merion R et al. (1990 Summer). A rational approach to liver transplantation for the alcoholic patient. Psychosomatics 31 (3): 241–254.

Beresford T, Temple B, Maravilla F (2013). NIH Research Alcoholism Funding. Alcohol Clin Exp Res 37 (S 2): 1A–332A.

Bor J, Basu S, Coutts A et al. (2013 May–Jun). Alcohol use during the great recession of 2008–2009. Alcohol Alcohol 48 (3): 343–348.

Bouchery EE, Harwood HJ, Sacks JJ et al. (2011 Nov). Economic costs of excessive alcohol consumption in the U.S., 2006. Am J Prev Med 41 (5): 516–524.

Brown J, Babor TF, Litt MD et al. (1994 Feb 28). The type A/type B distinction. Subtyping alcoholics according to indicators of vulnerability and severity. Ann N Y Acad Sci 708: 23–33.

Clark BJ, Keniston A, Douglas IS et al. (2013 Sep). Healthcare utilization in medical intensive care unit survivors with alcohol withdrawal. Alcohol Clin Exp Res 37 (9): 1536–1543.

Cloninger CR, Bohman M, Sigvardsson S (1981 Aug). Inheritance of alcohol abuse. Cross-fostering analysis of adopted men. Arch Gen Psychiatry 38 (8): 861–868.

Cook PJ, Moore MJ (2002 Mar–Apr). The economics of alcohol abuse and alcohol-control policies. Health Aff (Millwood) 21 (2): 120–133.

Crabbe JC, Kendler KS, Hitzemann RJ (2013). Modeling the diagnostic criteria for alcohol dependence with genetic animal models. Curr Top Behav Neurosci 13: 187–221.

DiMartini A, Day N, Dew MA et al. (2006 May). Alcohol consumption patterns and predictors of use following liver transplantation for alcoholic liver disease. Liver Transpl 12 (5): 813–820.

Dimartini A, Dew MA, Day N et al. (2010 Oct). Trajectories of alcohol consumption following liver transplantation. Am J Transplant 10 (10): 2305–2312.

Hasin DS, Stinson FS, Ogburn E et al. (2007 Jul). Prevalence, correlates, disability, and comorbidity of DSM-IV alcohol abuse and dependence in the United States: results from the National Epidemiologic Survey on Alcohol and Related Conditions. Arch Gen Psychiatry 64 (7): 830–842.

Hasin DS, O'Brien CP, Auriacombe M et al. (2013; Aug 1). DSM-5 criteria for substance use disorders: recommendations and rationale. Am J Psychiatry 170 (8): 834–851.

Kaim SC, Klett CJ, Rothfeld B (1969 Jun). Treatment of the acute alcohol withdrawal state: a comparison of four drugs. Am J Psychiatry 125 (12): 1640–1646.

Lucey MR (2007 Feb). Liver transplantation for alcoholic liver disease: past, present, and future. Liver Transpl 13 (2): 190–192.

Lucey MR, Merion R, Beresford T (1993). Liver Transplantation and the Alcoholic Patient. Cambridge University Press, Cambridge, UK.

Lucey M, Merion RM, Beresford TP (1994). Liver Transplantation and the Alcoholic Patient. Cambridge University Press, Cambridge.

Martin B, Clapp L, Bialkowski D et al. (2003 Mar–Apr). Compliance to supervised disulfiram therapy: a comparison of voluntary and court-ordered patients. Am J Addict 12 (2): 137–143.

Martin BK, Clapp L, Alfers J et al. (2004 Apr). Adherence to court-ordered disulfiram at fifteen months: a naturalistic study. J Subst Abuse Treat 26 (3): 233–236.

Naimi TS (2011 Nov). The cost of alcohol and its corresponding taxes in the U.S.: a massive public subsidy of excessive drinking and alcohol industries. Am J Prev Med 41 (5): 546–547.

Schoenborn CA, Adams PF, Peregoy JA (2013). Health behaviors of adults: United States, 2008–2010. National Center for Health Statistics. Vital Health Stat 10 (257): 5–20.

Schuckit MA (1994). DSM-IV: was it worth all the fuss? Alcohol Alcohol Suppl 2: 459–469.

Schuckit MA, Smith TL (1996 Mar). An 8-year follow-up of 450 sons of alcoholic and control subjects. Arch Gen Psychiatry 53 (3): 202–210.

Schuckit MA, Smith TL, Anderson KG et al. (2004 Dec). Testing the level of response to alcohol: social information processing model of alcoholism risk–a 20-year prospective study. Alcohol Clin Exp Res 28 (12): 1881–1889.

Schuckit MA, Smith TL, Danko GP et al. (2005 Feb). Prospective evaluation of the four DSM-IV criteria for alcohol abuse in a large population. Am J Psychiatry 162 (2): 350–360.

Straus R, Bacon SD (1951 Jun). Alcoholism and social stability; a study of occupational integration in 2,023 male clinic patients. Q J Stud Alcohol 12 (2): 231–260.

Sturm R (2002 Mar–Apr). The effects of obesity, smoking, and drinking on medical problems and costs. Health Aff (Millwood) 21 (2): 245–253.

Vaillant GE (1995). The Natural History of Alcoholism, Revisited, Harvard University Press, Cambridge, MA.

Wetterling T, Veltrup C, John U et al. (2003 May). Late onset alcoholism. Eur Psychiatry 18 (3): 112–118.

WHO (2010). ICD-10 Classifications of Mental and Behavioural Disorder: Clinical Descriptions and Diagnostic Guidelines, World Health Organization, Geneva, 2010.

Handbook of Clinical Neurology, Vol. 125 (3rd series)
Alcohol and the Nervous System
E.V. Sullivan and A. Pfefferbaum, Editors

Chapter 2

Perspectives on the neuroscience of alcohol from the National Institute on Alcohol Abuse and Alcoholism

MATTHEW T. REILLY*, ANTONIO NORONHA, AND KENNETH WARREN
Division of Neuroscience and Behavior, National Institute on Alcohol Abuse and Alcoholism, Bethesda, MD, USA

INTRODUCTION

Alcohol (ethanol), a simple two-carbon psychoactive chemical, has long been a part of the history of human civilization. Indeed, archeologic evidence exists for the production of fermented beverages by humans more than 10 000 years ago (Devor and Cloninger, 1989). Although the chemical structure of alcohol is simple, its effects are complex, ranging from societal and cultural to the structural, cellular, molecular, and behavioral levels. In the United States, along with most western countries, alcohol is consumed as part of social gatherings, weddings, and religious ceremonies and is enjoyed in moderation by the vast majority of the population. However, alcohol consumption also causes a large economic and disease burden. Alcohol abuse can lead to physical dependence with associated end-organ damage, particularly the brain and liver (Li, 2008). Eighteen million Americans (8.5% of the population 18 years of age and older) suffer from alcohol abuse or dependence (Grant et al., 2004; Mokdad et al., 2004). Alcohol problems cost US society an estimated $235 billion annually (Rehm et al., 2009). Therefore, there is a clear need to understand the causes and consequences of this enormous public health concern.

1. The foundation for research on the neurobiological consequences of alcoholism was due, in part, to the creation of the National Institute on Alcohol Abuse and Alcoholism (NIAAA), the lead agency for US research on alcoholism. NIAAA was established under the authority of the Comprehensive Alcohol Abuse and Alcoholism Prevention, Treatment, and Rehabilitation Act of 1970 (P.L. 91-616) with authority to develop and conduct comprehensive health, education, training, research, and planning programs for the prevention and treatment of alcohol abuse and alcoholism. During its more than 40 years of existence, NIAAA has established a large research portfolio including all aspects of the neuroscience of alcohol abuse and dependence. This chapter will briefly describe some of the major areas from this portfolio and a vision for the future for the neuroscience of alcohol abuse and alcoholism research.

THE DIVISION OF NEUROSCIENCE AND BEHAVIOR (DNB) AND ITS MISSION

The DNB was formed within NIAAA in 2002 under the directorship of Dr. Ting-Kai Li. Until this time neuroscience research on alcohol existed as a small branch within the Division of Basic Research. The establishment of a separate division for neuroscience and behavior testifies to the importance of this area for understanding the etiology and treatment of alcoholism as a brain disease. DNB has grown to become the largest division within NIAAA in terms of extramurally funded research grants. The neuroscience research portfolio can be broadly classified into several major themes: genetics, neuroadaptation/neuroplasticity (including human imaging and neuropsychologic studies), behavior, and preclinical medication development. These broad research domains are applied within NIAAA's "alcohol in a lifespan perspective" because alcohol affects all stages of the lifespan, from prenatal to adult (http://www.niaaa.nih.gov/about-niaaa/our-work/strategic-plan). Thus, the long-term mission of DNB is to identify neurobiologic mechanisms in

*Correspondence to: Matthew T. Reilly, Ph.D., National Institutes of Health, National Institute on Alcohol Abuse and Alcoholism, Division of Neuroscience and Behavior, 5635 Fishers Lane, Suite 2065, Bethesda, MD 20892-9304, USA. Tel: +1-301-594-6228, Fax: +1-301-443-1650, E-mail: reillymt@mail.nih.gov

sufficient detail to permit development of effective preventive and therapeutic interventions across the lifespan.

GENETICS

Because alcoholism is a behavioral disorder, genes that are expressed in the brain are obvious candidates that underlie genetic susceptibility to alcoholism. NIAAA and DNB have a long history of supporting studies that examine the genetic vulnerability to alcoholism. Family, twin, and adoption studies over the last 40 years have convincingly demonstrated that part of the susceptibility to alcoholism is due to genetic factors (Mayfield et al., 2008). One of the first studies to demonstrate a genetic involvement was that of Goodwin et al. (1973), who studied the drinking history of 55 adopted-out sons of alcoholics and 78 adopted-out sons of non-alcoholics. Biologic sons of alcoholics who had been adopted by non-related foster families were four times as likely to become alcoholics compared to adopted sons of non-alcoholics. In addition, twin and family studies have determined that almost 60% of the vulnerability to alcoholism is due to genetic factors (McGue, 1999; Prescott and Kendler, 1999). These studies demonstrate that genetic factors are important in determining vulnerability, but do not identify which specific gene variants contribute to this vulnerability. In the late 1980s, it became feasible to use DNA markers to map genotype–phenotype correlations and begin to pinpoint specific genomic loci that influence alcohol phenotypes. Under the leadership of NIAAA and support from Congress, the Collaborative Study on the Genetics of Alcoholism (COGA) was formed in 1989 to map and characterize the genetic variants associated with alcoholism (Edenberg and Foroud, 2006).

2. COGA is a family-based genetic study with nine research sites throughout the USA. Nuclear and multigenerational families were ascertained through a proband in treatment for alcohol dependence. The initial sample consisted of 105 families with 987 individuals. Results from the first genomewide scan provided evidence of linkage for loci on chromosomes 1 and 7 and suggestive linkage on chromosome 2 with alcohol dependence; linkage on chromosome 2, conferring protection against alcohol dependence, near the alcohol dehydrogenase genes was also found (Reich et al., 1998). A follow-up study in 157 independent families densely affected with alcoholism was conducted and provided further evidence of linkage to chromosomes 1 and 7, with new loci identified on chromosomes 2 and 3 (Foroud et al., 2000).
3. The COGA sample now consists of more than 17 000 individuals from over 2000 families with extensive phenotypic and genomic information. In addition to genomewide studies of alcohol dependence, COGA has conducted several candidate gene association studies and found association with more than 20 genes affecting both alcohol dependence and endophenotypes associated with alcoholism (Edenberg and Foroud, 2006; Gunzerath et al., 2011). Finally, in recent years COGA has conducted several genomewide association studies. Family-based and case-control genomewide association studies for alcohol dependence or related phenotypes identified several promising new genetic variants, including C15orf53 (Wang et al., 2013), KCNJ6 (Kang et al., 2012), and GABRA2 (Bierut et al., 2010).
4. Animal models have also been useful in determining the genes associated with alcoholism and related phenotypes (Crabbe et al., 1999). Although alcoholism is a uniquely human disease, animal models that capture specific characteristics of alcoholism have provided a wealth of information on the neuroscience of alcohol. Animal models, particularly rodent models, display many of the traits associated with alcoholism, such as tolerance, dependence, escalation of consumption, and conditioned responses to alcohol. Animal models also provide many experimental advantages over human studies, particularly when examining the genetic underpinnings of alcohol. The environment and genotype can be tightly controlled in animal studies, allowing for more solid conclusions to be drawn on the role of genes versus the environment in vulnerability towards alcoholism (Buck et al., 2012).
5. Quantitative trait loci (QTL) mapping in animal models for alcoholism began in the early 1990s. By 1998, 24 QTLs had been definitively mapped for alcohol responses in rodents (Crabbe et al., 1999). Many of the genes for alcohol responses are syntenic with those in humans, making the translation of these findings feasible to humans. For example, using crosses between C57BL/6 and DBA/2 progenitor mouse strains, which display mild versus severe withdrawal severity, respectively, QTLs were identified on chromosomes 1, 4, and 11 (Buck et al., 1997). Interval specific congenic strains (Fig. 2.1) were used to narrow the chromosome 4 QTL interval down to a region which only contained four expressed genes (Fehr et al., 2002). Sequence and gene expression analysis identified *Mpdz* as the most likely candidate underlying the QTL effect, where 10 coding differences were found in the MPDZ protein between the progenitor strains as well as gene expression differences. The other genes within this QTL interval showed no

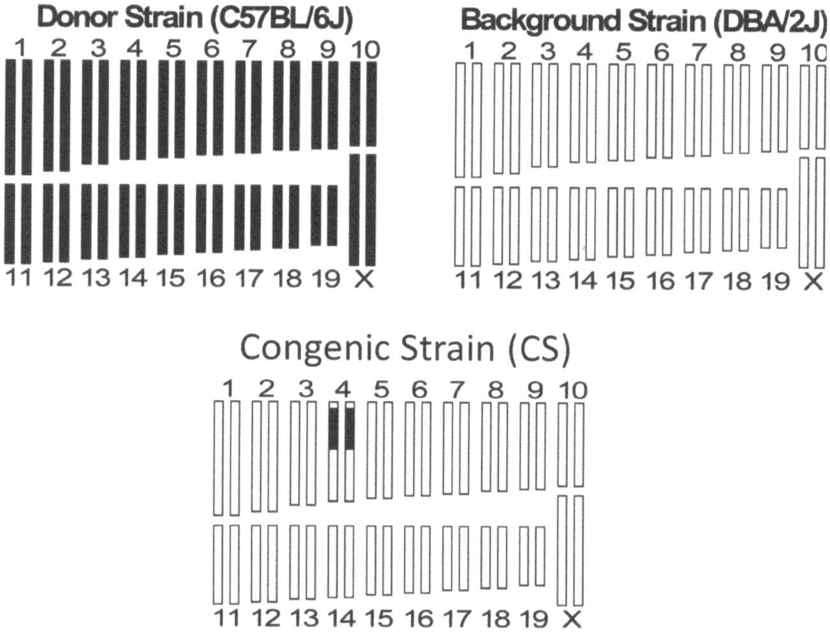

Fig. 2.1. Development of congenic strain. A congenic mouse strain is genetically engineered to isolate a quantitative trait locus (QTL) on a pure (isogenic) genetic background. A region on chromosome 4 that contains the QTL is transferred from a donor strain (C57BL/6J) on to a background strain (DBA/2J). If the congenic strain differs from the background strain in alcohol withdrawal severity, then it can be concluded that genes within the donor strain segment of the congenic strain contribute to the phenotype. By continually testing congenic strains with increasingly smaller donor region segments, the gene (or, in some cases, multiple genes), referred to as quantitative trait genes (QTGs) can be physically mapped. Shown are the 19 mouse chromosomes (and the X chromosome) for the donor, background, and congenic strains.

sequence or expression differences, indicating that *Mpdz* is a quantitative trait gene for alcohol withdrawal severity (Shirley et al., 2004). Human candidate gene association studies provide some evidence that MPDZ is associated with alcohol dependence (Karpyak et al., 2009, 2012).

6. Animal and human genetic studies on alcoholism and related phenotypes have provided a better understanding of the genetic factors that influence vulnerability. Although much more work is needed in this area, the new tools of genomics should enable more rapid identification of all the genetic variants that contribute to alcohol dependence.

NEUROADAPTATION

The second broad theme in DNB's research portfolio is neuroadaptation. Homeostatic mechanisms exist in the brain to counteract the effects of alcohol, leading to tolerance. Upon removal of alcohol (withdrawal) these compensatory or neuroadapative mechanisms are uncovered and produce a withdrawal syndrome that consists of symptoms of hyperactivity, including anxiety, tremor, and seizures (Fig. 2.2) (Becker, 2000). C.K. Himmelsbach first proposed this homeostatic model of drug tolerance and physical dependence in the 1940s (Himmelsbach, 1941). The neuroadaptations underlying tolerance and dependence are thought to drive excessive alcohol use (Becker, 2013). Therefore, a major emphasis of research that NIAAA supports has been to define specific alcohol neuroadaptations, with the ultimate goal of providing new targets for medications development.

7. Alcohol neuroadaptation is a broad area that contains studies that use many different techniques to examine the adaptations that occur in the brain. These studies include techniques ranging from cellular and molecular physiology (Diamond and Gordon, 1997) to genetics and genomics (Rhodes and Crabbe, 2005). Neuroadaptation also involves neuroplasticity. In fact, alcohol dependence has been characterized as a learned, plasticity-dependent process (Walker, 2012). A large literature has amassed over the last 30 years, making it clear that acute and chronic alcohol exposure changes the molecular and cellular components of the brain in significant ways. It is beyond the scope of this article to review all of these neuroadaptations; however, two major NIAAA initiatives on alcohol neuroadaptation will be described as examples.

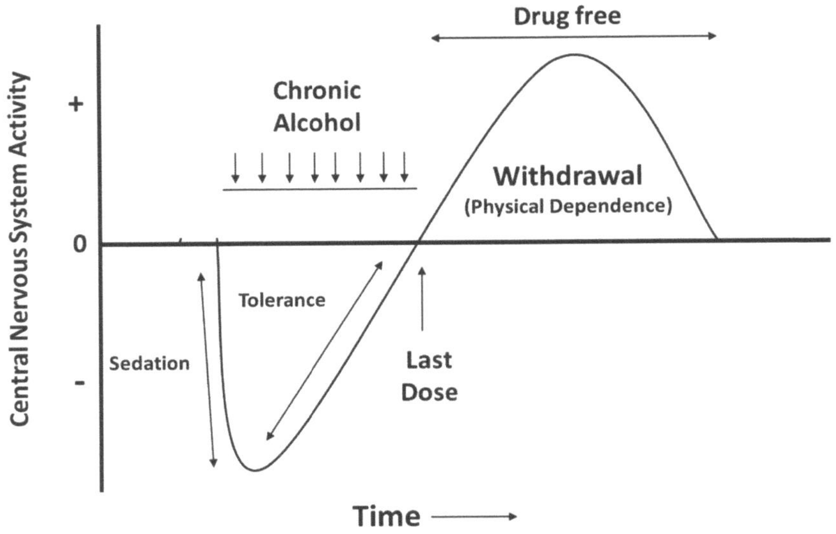

Fig. 2.2. Homeostatic model of alcohol tolerance and withdrawal. Shown is a simplified representation of central nervous system (CNS) activity following chronic alcohol administration and withdrawal. Initial alcohol exposure produces CNS depression (sedation). After multiple exposures to alcohol, the sedative effects of alcohol decrease (tolerance). After the last dose of alcohol, a homeostatic, compensatory response to chronic alcohol exposure occurs that is characterized by an increase in CNS hyperexcitability (withdrawal). (Reproduced fFrom Metten and Crabbe, 1996, with permission from Taylor and Francis.)

8. In 2001 the Integrative Neuroscience Initiative on Alcoholism (INIA) was formed: it consists of 16 interrelated projects that focus on the neuroadaptations that lead to excessive alcohol consumption. Now in its 12th year of funding, the main hypothesis of this consortium is that genetic differences and neuroadaptations in reward circuitry are responsible for individual differences in vulnerability to excessive alcohol consumption. The overall goals of INIA are: (1) to confirm gene targets nominated by gene expression assays or other methods by use of transgenic, knockout, inducible knockout, site-specific knockout, RNAi, in situ hybridization, in vivo electrophysiology, *in vivo* imaging, and next-generation sequencing; and (2) to identify druggable targets that are most promising for medication development for the treatment of alcoholism by use of novel molecules in concert with molecules with existing Food and Drug Administration (FDA) approval in animal models with the most predictive ability (http://www.scripps.edu/california/research/inia/).
9. INIA has made significant progress in identifying novel candidate genes through gene expression analysis. Using a meta-analysis of several large datasets of brain gene expression from rodent models of high and low alcohol preference, almost 4000 unique genes were identified that show differential expression (Mulligan et al., 2006). Among these differentially expressed genes, several overrepresented functional groups were identified, including mitogen-activated protein kinase signaling and transcription regulation pathways. This study highlights the importance of using large genomics datasets to converge on potentially important targets for novel medication development that could combat alcohol abuse and dependence.
10. Finally, this consortium has begun to explore neuroinflammatory adaptations that may drive excessive alcohol consumption. In rodent models of excessive alcohol consumption and human alcoholic postmortem brain, several genes involved with neuroinflammatory processes were identified (Mulligan et al., 2006; Kimpel et al., 2007; Flatscher-Bader et al., 2008; Tabakoff et al., 2008). To validate these gene expression changes, knockout mice for six of these candidate genes were tested for alcohol consumption, including: β_2-microglobulin (B2m), cathepsin S (Ctss), cathepsin F (Ctsf), interleukin-1 receptor antagonist (IL1rn), CD14 molecule (Cd14) and interleukin 6 (IL6). Alcohol consumption and preference were reduced in all six knockout animals, indicating a causal role of these neuroimmune genes in excessive alcohol consumption (Blednov et al., 2012). Several FDA-approved and novel drugs that target these neuroimmune factors are being tested in animal models to further support their role in excessive alcohol consumption.
11. In 2002 a separate but related INIA (INIAstress) consortium was initiated with a focus on stress, neuroadaptation, and excessive alcohol consumption.

Since its inception, INIAstress has had the primary goal of coordinating and facilitating translational, multidisciplinary, and integrative research aimed at elucidating genetic and environmental influences on brain mechanisms that mediate excessive alcohol consumption, the response to stress, and the reciprocal relationship between excessive drinking, the physiologic state of stress, and the subjective state of anxiety (http://www.iniastress.org/). This multisite consortium is examining the unique adaptations in brain circuitry following chronic intermittent alcohol exposure (alcohol allostasis) that impact the subsequent interaction of stress and alcohol to promote further excessive drinking.

12. Acute and chronic stress has long been suspected to lead to excessive alcohol consumption (Anthenelli and Grandison, 2012). However, there is a paucity of direct evidence to support the causal role of stress in excessive alcohol consumption. A comprehensive review of the effects of acute/subchronic stress on alcohol consumption in animals showed that most studies find no change or decreased alcohol consumption. However, studies examining chronic stress, usually early in development, produce enhanced alcohol consumption, particularly in adulthood (Becker et al., 2011). The differences in the effects of stress on alcohol consumption between studies are likely due to a variety of genetic and environmental factors.
13. In contrast to the mostly equivocal findings on the acute effects of stress on subsequent alcohol consumption, chronic intermittent alcohol exposure, acting as a stressor, reliably increases voluntary alcohol consumption. For example, when mice are made dependent on alcohol through repeated cycles of chronic alcohol exposure delivered by inhalation, escalation of voluntary alcohol drinking is observed (Becker and Lopez, 2004; Lopez and Becker, 2005). The neuroadaptations that underlie the dependent state presumably drive the escalation of drinking and enhanced stress sensitivity (Breese et al., 2005). Neuroadaptations in neuroendocrine and extrahypothalamic stress systems are two major brain systems thought to underlie these effects. Corticotropin-releasing factor (extrahypothalamic) (Thorsell et al., 2005; Finn et al., 2007; Gilpin et al., 2008; Sommer et al., 2008; Roberto et al., 2010), neuropeptide Y (Thorsell et al., 2005; Gilpin et al., 2008, 2011), and opioid (Gilpin et al., 2008; Walker and Koob, 2008; Walker et al., 2011) systems have been implicated in excessive drinking that follows chronic intermittent alcohol exposure. The hypothalamic–pituitary–adrenal axis also shows alterations, with a blunting of this axis, as measured by reduced corticosterone after chronic intermittent alcohol exposure (Lopez et al., 2010; Griffin et al., 2011).
14. Collectively, these consortia have identified promising neuroadaptations underlying alcohol dependence and excessive alcohol consumption. Many of these neuroadaptations could serve as novel targets for medication development, which will be a major emphasis by these consortia in the next several years.

BEHAVIOR

Alcoholism is a behavioral disorder and any attempts to understand its etiology must also include the study of behavior. At its core alcoholism is a compulsive behavior, simply defined as an overwhelming desire (or compulsion) for use despite serious adverse consequences (Koob, 2013). Alcohol impacts many neuronal circuits, resulting in neuroadaptations (as described in the preceding section) in areas of the brain that regulate motivation, interoception, emotion, memory, decision making, and cognitive control (Koob and Volkow, 2010). Alcohol negatively impacts all of these behavioral processes, leaving the alcoholic with the inability to control consumption. Understanding aberrant alcohol-induced behaviors is the third major theme in the portfolio of DNB.

15. Koob and colleagues (2014) have proposed a conceptual behavioral framework for alcoholism and the addiction cycle. This cycle involves three stages: binge/intoxication, withdrawal/negative affect, and preoccupation/anticipation. During these three stages positive and negative reinforcement are thought to differentially reinforce behavior. Positive reinforcement is a behavioral mechanism whereby the presentation of a stimulus increases the probability of a behavior, whereas negative reinforcement is a behavioral mechanism whereby a behavior is strengthened by the removal of an aversive stimulus. Further, impulse control disorders are largely characterized by positive reinforcement mechanisms, while compulsive disorders are mediated by negative reinforcement mechanisms (American Psychiatric Association, 1994). During the initial stage of this cycle – binge/intoxication – positive reinforcement mechanisms are thought to control what is considered largely impulsive behavior. However, after repeated binge/intoxication and withdrawal/negative affect episodes, negative reinforcement mechanisms drive what has now become compulsive behavior, such that the negative emotional state induced by withdrawal is alleviated by

continued alcohol consumption. The preoccupation/anticipation stage of this cycle, referred to as "craving" in colloquial parlance, is an interoceptive state characterized by enhanced sensitivity to drug-associated conditioned cues or stress that strengthens the chronic relapsing nature of the addiction cycle.

16. Binge drinking as defined by NIAAA is a pattern of drinking alcohol that brings blood alcohol concentration (BAC) to 0.08 g% or above in 2 hours. In adult men this usually occurs after five drinks and in women after four drinks (http://www.niaaa.nih.gov/alcohol-health/overview-alcohol-consumption/moderate-binge-drinking). The official definition of binge drinking was formally adopted by the National Advisory Council to the NIAAA in 2004, and this provided a much-needed benchmark for researchers and clinicians to help quantify this type of behavior. Binge drinking is endemic in the USA. Eight and a half percent of 13 000 adults surveyed through the National Epidemiological Survey on Alcohol and Related Conditions exceed this definition of binge drinking one to two times per week (Dawson et al., 2008). Binge drinking is also prevalent in adolescence and is a risk factor for the later development of alcohol dependence (Silveri, 2012).
17. Preclinical studies of alcohol-binge drinking behavior in rodents include: two-bottle preference, chronic intermittent access, scheduled high alcohol consumption, schedule-induced polydipsia, intragastric self-administration, the sweetened solution procedure, and the drinking in the dark (DID) procedure (for review, see Crabbe et al., 2011). The majority of these paradigms do not produce BACs greater than 0.08 g/dL, the NIAAA threshold for binge drinking. However, one of these procedures, DID, has been used as a successful model of binge intoxication, where mice exposed to this procedure reliably achieve BACs greater than 0.08 g/dL (Rhodes et al., 2005). The DID procedure capitalizes on rodents' natural tendency to be most active, and consume most of their daily food and fluid, during the dark phase of their circadian cycle (Armstrong, 1980). This model has been proposed as a useful screen for medications that could reduce binge drinking, because acamprosate and a metabotropic glutamate receptor antagonist reduce alcohol binge drinking in this model (Gupta et al., 2008).
18. Preclinical animal models of the withdrawal/negative affect stage of the addiction cycle also exist (Heilig et al., 2010). These behavioral models include repeated gastric intubation (Majchrowicz, 1975), liquid diet (Moy et al., 1997), and chronic alcohol exposure by vapor inhalation (Goldstein and Pal, 1971; Rogers et al., 1979). All of these models produce physical signs of dependence characterized by central nervous system hyperexcitability that are similar to that observed in humans (Finn and Crabbe, 1997; Becker, 2000). For example, withdrawal-related seizures are observed in both animals and humans (Victor, 1970; Deitrich and Harris, 1996; Becker, 2000). Other signs of physical dependence are also observed during withdrawal, such as autonomic nervous system hyperactivity (Crawshaw et al., 1994), and motor abnormalities such as tremor (Frye et al., 1983; Meert et al., 1992; Macey et al., 1996). Psychologic indices (e.g., anxiety and negative affect) are also observed during alcohol withdrawal. In particular, anxiety that occurs upon withdrawal has been proposed to be a major factor that increases risk for relapse (Roelofs, 1985; Driessen et al., 2001; Koob, 2003). Animal and human studies provide evidence that withdrawal-related anxiety increases alcohol consumption (Spanagel et al., 1995; Holter et al., 1998; Henniger et al., 2002; Willinger et al., 2002; Ciccocioppo et al., 2006; Barrenha and Chester, 2007).
19. The preoccupation/anticipation (craving) stage can also be modeled in animals. Craving is somewhat of a nebulous clinical construct, in that it is typically difficult to measure clinically (Tiffany et al., 2000). However, attempts to model craving in animals have resulted in several useful experimental behavioral paradigms. These are broadly classified as "drug-seeking" models, where reinstatement of alcohol consumption is induced by alcohol-associated cues, alcohol priming, stress, or a protracted negative emotional state resulting from alcohol abstinence (Koob, 2000). In addition, although not considered a model of craving per se, drinking induced by chronic intermittent alcohol vapor exposure has been useful in modeling relapse-type consumption after dependence (Becker and Lopez, 2004; Lopez and Becker, 2005; Finn et al., 2007).
20. Over the last 40 years, great progress has been made in animal models of behavior related to alcohol. As discussed above, these models are grounded in Koob's conceptual framework of the addiction cycle and have in many cases shown face, construct, and predictive validity (i.e., in the case of screens for medication development). Finally, many of these behavioral models have proven useful for identifying underlying neural substrates (see section on neuroadaptation, above) and are useful for screening and validation of novel neurobiologic drug targets (see following section on preclinical medications development).

NEUROBEHAVIORAL, STRUCTURAL, AND FUNCTIONAL CONSEQUENCES OF HUMAN ALCOHOLISM

Human imaging studies, including structural magnetic resonance imaging (sMRI), diffusion tensor imaging (DTI), and functional MRI (fMRI), have been instrumental in defining the structural and functional deficits and neuroadaptations resulting from chronic alcohol use (Sullivan and Pfefferbaum, 2005). Alcohol selectively damages specific brain regions, while sparing others. In addition, chronic alcoholism is associated with a range of neuropsychologic impairments that can be linked to brain structural and functional abnormalities resulting from years of alcohol abuse (Sullivan et al., 2010). Some of these neuropsychologic impairments and associated brain alterations may be ameliorated with abstinence from alcohol. NIAAA has a large portfolio of research grants that use a variety of different imaging modalities. Continued discoveries in this field will have important implications for understanding the etiology and possibly the treatment of alcoholism as a brain disease.

21. The frontal lobes of the brain have been consistently shown to be selectively damaged by alcohol abuse (Gilman et al., 1990; Pfefferbaum et al., 1997; Gansler et al., 2000; Oscar-Berman et al., 2004; Dirksen et al., 2006). For example, sMRI studies demonstrated volume shrinkage in the frontal lobe of alcoholics (Pfefferbaum et al., 1997). fMRI studies also indicate functional alterations in the frontal lobes (Tapert et al., 2001). In addition, postmortem brain tissue from alcoholics shows reduced neuron density in the frontal cortex (Harper and Matsumoto, 2005). The frontal lobe of the brain (especially the prefrontal cortex) is essential for executive function. Executive functions require many cognitive processes such as attention, perception, and memory (Eslinger et al., 2013) to enable our ability for judgment, problem solving, decision making, planning, and social conduct. The frontal cortex is also associated with emotion and personality. Alcoholics with damage to the frontal cortex show deficits in affective processing (Kornreich et al., 2002; Townshend and Duka, 2003; Clark et al., 2007), which may result from coordinated abnormalities in visuospatial processes, poor inhibitory control, and interpersonal stress (Oscar-Berman and Marinkovic, 2007). Indeed, consideration of stress factors has led to hypotheses and support for disruption of ventromedial prefrontal circuitry as contributing to alcohol craving and maintenance of alcohol dependence (Sinha, 2012; Seo et al., 2013). Therefore, it is not surprising that many of these executive functions are impaired in alcoholics due to selective damage of the frontal lobes and frontally based circuitry (Crews and Boettiger, 2009).
22. Limbic regions of the brain are also affected by chronic alcohol abuse. One of the key limbic regions affected is the amygdala, which has connections with the prefrontal cortex as well as other brain regions, including the mesolimbic dopamine system (Koob and Volkow, 2010). One of the primary roles of the amygdala is to process the emotional and motivational significance of stimuli (Everitt et al., 2003; LeDoux, 2003). An fMRI study showed that long-term abstinent alcoholics compared to control subjects show a blunting of amygdala activation after presentation of emotional words or facial expressions (Marinkovic et al., 2009). The hippocampus, a region classically associated with memory, is another limbic region, which shows abnormalities after alcohol abuse.
23. Structural imaging studies demonstrate that hippocampal volume is smaller in adolescent (De Bellis et al., 2000) and adult (Sullivan et al., 1995; Agartz et al., 1999; Beresford et al., 2006) alcoholics. This reduced hippocampal volume may be associated with memory impairment in alcoholics (Vetreno et al., 2011). Midbrain mesolimbic dopamine regions also show abnormalities with alcohol abuse. sMRI and fMRI showed that alcoholics display enhanced midbrain activation after presentation of alcohol-related words, and may indicate an automatic attention bias to alcohol-related stimuli in alcoholics (Muller-Oehring et al., 2013). Another sMRI study showed minimal differences in subcortical volumes in long-term abstinent alcoholics, suggesting recovery of the brain with sustained abstinence from alcohol; however, smaller subcortical volumes in long-term alcoholics with comorbid psychiatric disorders were noted, consistent with the possibility of a complex interaction with co-occurring psychiatric conditions (Sameti et al., 2011).
24. The cerebellum is a brain structure that controls voluntary movement and balance, but it also plays a key role in cognition and emotion, which has expanded knowledge of hitherto unknown, but potentially important, brain functionality (Krienen and Buckner, 2009). Chronic alcohol abuse is associated with notable shrinkage of cerebellar tissue (Sullivan et al., 2000a), which may explain the significant movement and balance deficits seen in alcoholics (Sullivan et al., 2000b, 2006). *In vivo* imaging techniques demonstrate significant damage to the cerebellum (Sullivan et al., 2003). Several innovative studies have proposed a role for a frontocerebellar circuit in executive function, a role not traditionally ascribed to the cerebellum per se. In one study,

alcoholics and controls performed a dual-task working memory procedure and underwent sMRI to examine this frontocerebellar circuit. Alcoholics differed from non-alcoholics on spatial working memory tasks and these differences between groups on the memory tasks were associated with differential utilization of specific nodes of the frontocerebellar circuit (Chanraud et al., 2010). In addition, limbic–cerebellar connectivity may be differentially utilized in alcoholics, which may represent a compensatory process to preserve associative memory (Pitel et al., 2013). Further, it has been proposed that the brains of alcoholics may "remap" functional connectivity circuitry to compensate for impairment during recovery (Schulte et al., 2012; Chanraud et al., 2013). These latter findings may indicate novel strategies to help facilitate normal brain functioning in recovering alcoholics. Importantly, further studies should be pursued that examine brain changes during recovery to inform new treatments for cognitive impairments associated with alcoholism.

25. A new area in studying the brain's functional connectivity and the alterations in this connectivity resulting from alcoholism is applying imaging techniques to understand the default mode network (DMN) of the brain. The DMN is an endogenous non-task activity network in the brain that is present at rest or a baseline state of activity, i.e., when the subject is not engaged in any externally goal-directed activity (Raichle et al., 2001; Greicius et al., 2003). These seminal studies showed that brain regions in the DMN most reliably correlated (or connected) are: posterior parietal cortex, precuneus and posterior cingulate cortex, medial prefrontal cortex, cerebellar Crus I and Crus II, and cerebellar lobules VI and IX (Greicius et al., 2003; Habas et al., 2009; Andrews-Hanna et al., 2010). Dysfunction in the DMN and its brain regional connectivity may lead to cognitive impairments (Nomura et al., 2010).
26. In the first study to examine the DMN and functional connectivity in alcoholics (Chanraud et al., 2011), alcoholics and controls underwent fMRI during a spatial working memory task. During rest, the spontaneous slow fluctuations of fMRI signals in the posterior cingulate and cerebellar regions in alcoholics were less synchronized compared with controls, indicative of compromised functional connectivity of the DMN in alcoholics. In addition, the efficiency of the DMN in alcoholics was substantially lower than controls between the posterior cingulate and multiple cerebellar sites. In contrast, alcoholics with sobriety had greater efficiency of the DMN in several brain regions (posterior cingulate cortex and inferior cerebellar lobule VIII). Finally, during the working memory task, alcoholics performed similar to controls, which was associated with greater connectivity between the left posterior cingulate cortex and left cerebellar regions. This was interpreted as a way alcoholics achieve normal performance on a task by using a compensatory brain network that is different from healthy controls (Chanraud et al., 2013). Other studies have continued this line of research, which holds promise for a novel understanding of functional brain connectivity alterations in alcoholism and recovery from alcoholism (Camchong et al., 2013a, b).
27. There has been substantial progress in understanding the structural and functional changes occurring in the alcoholic brain. It is clear that specific regions of the brain are damaged in treatment-seeking alcoholics (e.g., frontal cortex) (Pfefferbaum et al., 1997; Fein et al., 2009), while others are relatively less affected (Rogers et al., 2012); regional volume deficits show similar volume deficit patterns in treatment-naive alcoholics but are less severe (Fein et al., 2002). Regarding functional abnormalities related to alcoholism, functional connectivity in frontocerebellar neural circuits is reduced in alcohol-dependent patients, while connectivity between prefrontal and premotor cortex and several non-cerebellar regions was not different between alcoholics and control subjects (Rogers et al., 2012). It is also clear that certain regions of the brain may recover function during abstinence in alcoholics. Finally, the new area of studying the DMN in alcoholics is deepening our understanding of the functional connectivity of the brain and how alcoholics may use different or compensatory networks to perform cognitive task, although at a potential detrimental cost. These new studies on functional connectivity should benefit from the discoveries being made by the National Institutes of Health Human Connectome Project (http://www.neuroscienceblueprint.nih.gov/connectome/) which has the overarching purpose to acquire and share data about the structural and functional connectivity of the human brain using advanced imaging techniques.

PRECLINICAL MEDICATIONS DEVELOPMENT

As mentioned above, a long-term goal of DNB is to identify the neurobiologic substrates of alcohol abuse and dependence to aid in medications development. This is a lofty goal but one that is critical for ameliorating the enormous suffering imposed by alcohol dependence.

The preclinical medications development program benefits from the scientific advances in the previously described domains of genetics, neuroadaptations, and behavior.

28. There are currently four US FDA-approved medications for the treatment of alcohol dependence: disulfiram, oral naltrexone, acamprosate, and an extended-release injectable form of naltrexone (Litten et al., 2005; Johnson, 2008; O'Malley and O'Connor, 2011). In addition, topiramate is used off-label to treat alcoholism (Johnson et al., 2004) and nalmefene was recently approved in the European Union to treat alcohol dependence. Disulfiram was approved in 1949, while extended-release naltrexone was approved in 2006. Thus, medications development for alcohol dependence has a long history, spanning over 60 years. However, because of the complex, heterogeneous nature of alcohol dependence, these medications work in some, but not all, populations of alcoholics. In an effort to provide a framework for medications development in the next decade, NIAAA has identified seven key long-term objectives (Litten et al., 2012), two of which align with the overall mission of DNB: (1) identify and validate new molecular targets for the treatment of alcohol dependence; and (2) develop and implement screening models using animal and human laboratory paradigms.
29. Up until the late 1980s the main neurobiologic targets for medications development were limited to several key neurotransmitter/neuropeptide systems, including serotonin, dopamine, and endogenous opioids (Gorelick, 1989; Litten and Allen, 1991; Gianoulakis, 1993). Since this time new targets have been explored, sometimes based on a specific hypothesis regarding disease etiology. Several of these novel targets include ionotropic serotonin receptors, adenosine receptors, glutamate receptors, neuropeptide Y, vasopressin, glycine transporters, histone deacetylases, and neuroimmune-related molecules (Litten et al., 2012). However, over the last 10 years the list of potential targets has greatly expanded due in part to the advent of whole-genome discovery-based approaches (Roukos et al., 2014). The use of whole-genome approaches, such as gene expression microarrays (Mulligan et al., 2006) and RNA sequencing (Bottomly et al., 2011), has allowed an overall agnostic approach to target discovery. For example, a genomewide gene expression analysis identified K-ras, a small G protein, as a regulator of excessive alcohol consumption (Repunte-Canonigo et al., 2010b), as well as adiponectin receptor 2 (Repunte-Canonigo et al., 2010a). In addition, one of the main goals of INIA, as mentioned above, is identifying new drug targets by use of genomewide approaches in animal models of excessive alcohol consumption. To this end, INIA is pursuing neuroinflammatory system targets (TLR4, PPAR alpha, IKK beta, PDE4), ion channel targets (sodium channel 4b, BK potassium channels, GIRK2), signal transduction targets (HYOU1, NR4a1, GPR88, DGK zeta), and epigenetic targets (HDAC5, Gnb1, MECP2) (Contet et al., 2011). Finally, it is worth noting that in some cases focusing on a single target for medications development may not be optimal, but rather identifying a combination of targets to alter simultaneously might be more effective. This could be implemented through combined pharmacotherapy or using compounds that act on multiple targets (Johnson et al., 2003; Roth et al., 2004; Johnson, 2008).
30. Another major area in the preclinical medications development program is using animal models to evaluate medications with the ultimate aim of informing decisions to carry out costly clinical trials (Egli, 2005; Koob et al., 2009). An important first step in this process is evaluating the predictive capacity of an animal model by testing drugs that have been used clinically for alcohol dependence in animal models (Johnson et al., 2005; Koob et al., 2009). Establishing this type of "positive control" has great predictive power for the model being used. In addition, "negative controls," drugs that have proven ineffective clinically, such as bromocriptine and ritanserin, should also be tested in the animal model to enhance predictive power (Powell et al., 1995; Johnson et al., 1996).
31. The selection of the appropriate animal model is also an important consideration when testing novel compounds because different models encompass different phases of the addiction cycle (see section on behavior, above). However, evidence exists that demonstrates the usefulness of animal models to evaluate medications for alcohol dependence, particularly those that model biologic factors underlying excessive drinking (Egli, 2005). To this end, NIAAA and DNB initiated a Preclinical Medications Efficacy Testing Program to test medications in three different animal models of alcohol dependence. The models used in this program are: limited-access drinking in P (alcohol-preferring rats) and HAD1 rats (high-alcohol-drinking rats) (Li et al., 1987, 1993), and excessive alcohol drinking in dependent mice (Becker and Lopez, 2004; Lopez and Becker, 2005). In this program a reference compound (positive and/or negative control) is compared against a novel test compound for reduction in drinking. For example,

naltrexone and topiramate, two drugs that showed the most consistent effects in clinical studies (Litten et al., 2005; Johnson, 2008), displayed robust dose-dependent effects on reduction of drinking in these animal models. These data on the positive controls are then used to compare against novel, proprietary drugs submitted to NIAAA by pharmaceutical companies. If one of these novel drugs shows promise in the animal model, then this provides strong evidence for moving forward with a clinical trial to test the compound in humans. In addition, industry's interest in the compound is further enhanced and may provide additional resources for moving the testing of the drug forward.

32. In summary, NIAAA and DNB have laid a solid foundation for preclinical medications development through the identification of novel neurobiologic targets and implementing an effective screening pipeline in animal models. This program should prove useful in identifying novel drug targets and encourage testing in human populations through preliminary efficacy testing in animal models.

FUTURE DIRECTIONS

Since its inception within NIAAA, DNB has played a major role in supporting all contemporary research areas in the understanding of the neuroscience of alcohol abuse and dependence. Several of these key areas were briefly summarized in the preceding sections and provide the ground work for continued research discoveries. The field is making progress in all areas; new genetic variants have been discovered, novel drug targets have been identified, animal models that capture specific stages of the addiction cycle have been developed, and preclinical medications development is enabling new avenues for testing novel compounds for the treatment of alcohol dependence.

33. Several key areas for future neuroscience research are recommended. First, the field of the genetics of alcoholism has begun to appreciate the complex interaction between genes and the environment and tools for assessing these interactions are now available. This is particularly relevant to the area of epigenetics, the molecular modifications of DNA and histones that regulate when and where genes are expressed (Robison and Nestler, 2011). We now have appreciable evidence that alcohol alters epigenetic processes, namely DNA methylation (Zhang et al., 2013) and histone modifications (Pandey et al., 2008; Warnault et al., 2013) (Fig. 2.3). However, whether these epigenetic modifications induced by alcohol are heritable is an unanswered question. To this end, NIAAA

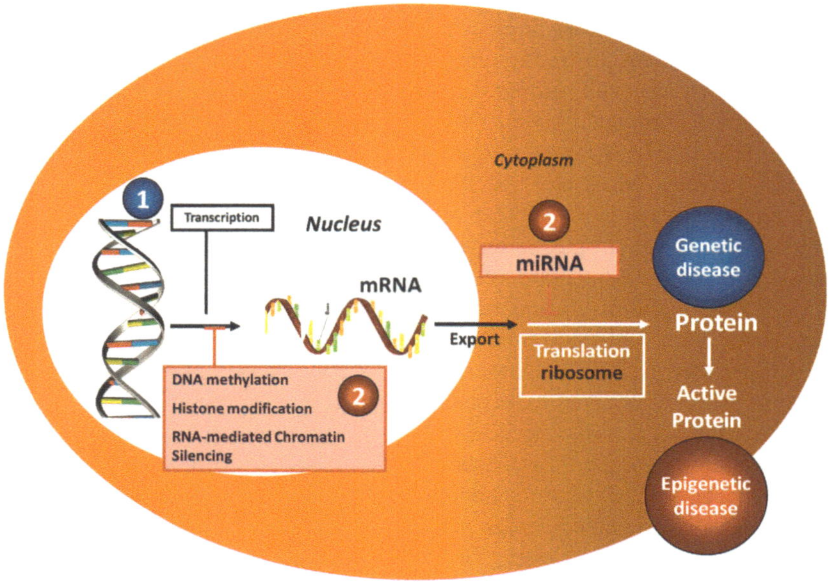

Fig. 2.3. Genetic and epigenetic contributions to disease etiology. Both genetic and epigenetic factors contribute to the etiology of alcoholism. Shown is a cell (neuron), where DNA is transcribed in the nucleus to mRNA, exported to the cytoplasm, and translated into protein. (1) Variation in the DNA sequence can result in altered structure and function of the corresponding protein (genetic disease). Gene expression is regulated by various mechanisms to control the spatiotemporal pattern of expression along with the abundance of the corresponding protein. (2) Epigenetic mechanisms, including DNA methylation, histone modification, RNA-mediated chromatin silencing, and miRNA expression, are critical regulatory processes that influence gene expression and can contribute to disease (epigenetic disease).

released a program announcement (PA) to encourage studies on alcohol-induced transgenerational epigenetic inheritance (cf. PA-13-003/004 "Epigenetic Inheritance and Transgenerational Effects of Alcohol": http://grants.nih.gov/grants/guide/pa-files/PA-13-003.html). If epigenetic modifications are heritable across generations this could have an enormous public health impact for the treatment and prevention of alcohol dependence. In addition, understanding the role of genetic variation in synaptic plasticity resulting from alcohol is understudied, but an important area for future advances (Reilly and Lovinger, 2010). Second, another area of inquiry is centered on the idea of alcohol dependence as a chronic pain disorder and NIAAA issued a PA to promote research in this area (cf. PA-11-267/268 "Mechanistic Studies of Pain and Alcohol Dependence": http://grants.nih.gov/grants/guide/pa-files/PA-11-267.html). There is growing evidence to support the hypothesis that alcohol dependence is among the pathologies arising from aberrant neurobiologic substrates of pain (Egli et al., 2012), which has implications for understanding disease etiology and treatment. Third, alcoholism is often comorbid with other substance abuse disorders and psychiatric disorders (Hasin and Kilcoyne, 2012). Two of the most commonly co-abused drugs are alcohol and nicotine and DNB has encouraged research into this area with the publication of a PA (cf. PA-13-193/194 "Mechanisms of Alcohol and Nicotine Co-Addiction": http://grants.nih.gov/grants/guide/pa-files/PA-13-194.html). Fourth, novel techniques for genetically engineering animal models have allowed sophisticated dissection of the role of specific genes in alcohol phenotypes (Reilly et al., 2012). These new genetically engineered animal models are crucial for understanding gene function and validating novel drug targets for future medications development. Finally, as new imaging techniques become available, substantial progress in understanding the structural and functional alterations in alcoholics and the changes associated with recovery from alcoholism will significantly impact treatment and diagnostic approaches. Namely, the pioneering work of several NIAAA funded investigators in understanding brain functional connectivity by examining the DMN in alcoholics is an excellent example of where significant discoveries will be made (Chanraud et al., 2011). These are just a few of the novel areas of exploration in the neuroscience of alcohol for the coming years. Many more relevant areas of major impact will certainly become apparent with the generation of new knowledge. These areas hold promise for a deeper understanding of the etiology, treatment, and prevention of alcohol abuse and dependence.

References

Agartz I, Momenan R, Rawlings RR et al. (1999). Hippocampal volume in patients with alcohol dependence. Arch Gen Psychiatry 56: 356–363.

American Psychiatric Association (1994). Diagnostic and Statistical Manual of Mental Disorders. 4th edn. American Psychiatric Press, Washington, DC.

Andrews-Hanna JR, Reidler JS, Sepulcre J et al. (2010). Functional-anatomic fractionation of the brain's default network. Neuron 65: 550–562.

Anthenelli R, Grandison L (2012). Effects of stress on alcohol consumption. Alcohol Res 34: 381–382.

Armstrong S (1980). A chronometric approach to the study of feeding behavior. Neurosci Biobehav Rev 4: 27–53.

Barrenha GD, Chester JA (2007). Genetic correlation between innate alcohol preference and fear-potentiated startle in selected mouse lines. Alcohol Clin Exp Res 31: 1081–1088.

Becker HC (2000). Animal models of alcohol withdrawal. Alcohol Res Health 24: 105–113.

Becker HC (2013). Animal models of excessive alcohol consumption in rodents. Curr Top Behav Neurosci 13: 355–377.

Becker HC, Lopez MF (2004). Increased ethanol drinking after repeated chronic ethanol exposure and withdrawal experience in C57BL/6 mice. Alcohol Clin Exp Res 28: 1829–1838.

Becker HC, Lopez MF, Doremus-Fitzwater TL (2011). Effects of stress on alcohol drinking: a review of animal studies. Psychopharmacology (Berl) 218: 131–156.

Beresford TP, Arciniegas DB, Alfers J et al. (2006). Hippocampus volume loss due to chronic heavy drinking. Alcohol Clin Exp Res 30: 1866–1870.

Bierut LJ, Agrawal A, Bucholz KK et al. (2010). A genome-wide association study of alcohol dependence. Proc Natl Acad Sci U S A 107: 5082–5087.

Blednov YA, Ponomarev I, Geil C et al. (2012). Neuroimmune regulation of alcohol consumption: behavioral validation of genes obtained from genomic studies. Addict Biol 17: 108–120.

Bottomly D, Walter NA, Hunter JE et al. (2011). Evaluating gene expression in C57BL/6 J and DBA/2 J mouse striatum using RNA-Seq and microarrays. PLoS One 6: e17820.

Breese GR, Overstreet DH, Knapp DJ et al. (2005). Prior multiple ethanol withdrawals enhance stress-induced anxiety-like behavior: inhibition by CRF1- and benzodiazepine-receptor antagonists and a 5-HT1a-receptor agonist. Neuropsychopharmacology 30: 1662–1669.

Buck KJ, Metten P, Belknap JK et al. (1997). Quantitative trait loci involved in genetic predisposition to acute alcohol withdrawal in mice. J Neurosci 17: 3946–3955.

Buck KJ, Milner LC, Denmark DL et al. (2012). Discovering genes involved in alcohol dependence and other alcohol responses: role of animal models. Alcohol Res 34: 367–374.

Camchong J, Stenger A, Fein G (2013a). Resting-state synchrony during early alcohol abstinence can predict subsequent relapse. Cereb Cortex 23: 2086–2099.

Camchong J, Stenger A, Fein G (2013b). Resting-state synchrony in long-term abstinent alcoholics. Alcohol Clin Exp Res 37: 75–85.

Chanraud S, Pitel AL, Rohlfing T et al. (2010). Dual tasking and working memory in alcoholism: relation to frontocerebellar circuitry. Neuropsychopharmacology 35: 1868–1878.

Chanraud S, Pitel AL, Pfefferbaum A et al. (2011). Disruption of functional connectivity of the default-mode network in alcoholism. Cereb Cortex 21: 2272–2281.

Chanraud S, Pitel AL, Muller-Oehring EM et al. (2013). Remapping the brain to compensate for impairment in recovering alcoholics. Cereb Cortex 23: 97–104.

Ciccocioppo R, Economidou D, Cippitelli A et al. (2006). Genetically selected Marchigian Sardinian alcohol-preferring (msP) rats: an animal model to study the neurobiology of alcoholism. Addict Biol 11: 339–355.

Clark US, Oscar-Berman M, Shagrin B et al. (2007). Alcoholism and judgments of affective stimuli. Neuropsychology 21: 346–362.

Contet C, Gardon O, Filliol D et al. (2011). Identification of genes regulated in the mouse extended amygdala by excessive ethanol drinking associated with dependence. Addict Biol 16: 615–619.

Crabbe JC, Phillips TJ, Buck KJ et al. (1999). Identifying genes for alcohol and drug sensitivity: recent progress and future directions. Trends Neurosci 22: 173–179.

Crabbe JC, Harris RA, Koob GF (2011). Preclinical studies of alcohol binge drinking. Ann N Y Acad Sci 1216: 24–40.

Crawshaw LI, O'Connor CS, Crabbe JC et al. (1994). Temperature regulation in mice during withdrawal from ethanol dependence. Am J Physiol 267: R929–R934.

Crews FT, Boettiger CA (2009). Impulsivity, frontal lobes and risk for addiction. Pharmacol Biochem Behav 93: 237–247.

Dawson DA, Li TK, Grant BF (2008). A prospective study of risk drinking: at risk for what? Drug Alcohol Depend 95: 62–72.

De Bellis MD, Clark DB, Beers SR et al. (2000). Hippocampal volume in adolescent-onset alcohol use disorders. Am J Psychiatry 157: 737–744.

Deitrich RA, Harris RA (1996). How much alcohol should I use in my experiments? Alcohol Clin Exp Res 20: 1–2.

Devor EJ, Cloninger CR (1989). Genetics of alcoholism. Annu Rev Genet 23: 19–36.

Diamond I, Gordon AS (1997). Cellular and molecular neuroscience of alcoholism. Physiol Rev 77: 1–20.

Dirksen CL, Howard JA, Cronin-Golomb A et al. (2006). Patterns of prefrontal dysfunction in alcoholics with and without Korsakoff's syndrome, patients with Parkinson's disease, and patients with rupture and repair of the anterior communicating artery. Neuropsychiatr Dis Treat 2: 327–339.

Driessen M, Meier S, Hill A et al. (2001). The course of anxiety, depression and drinking behaviours after completed detoxification in alcoholics with and without comorbid anxiety and depressive disorders. Alcohol Alcohol 36: 249–255.

Edenberg HJ, Foroud T (2006). The genetics of alcoholism: identifying specific genes through family studies. Addict Biol 11: 386–396.

Egli M (2005). Can experimental paradigms and animal models be used to discover clinically effective medications for alcoholism? Addict Biol 10: 309–319.

Egli M, Koob GF, Edwards S (2012). Alcohol dependence as a chronic pain disorder. Neurosci Biobehav Rev 36: 2179–2192.

Eslinger PJ, Flaherty-Craig CV, Chakara FM (2013). Rehabilitation and management of executive function disorders. Handb Clin Neurol 110: 365–376.

Everitt BJ, Cardinal RN, Parkinson JA et al. (2003). Appetitive behavior: impact of amygdala-dependent mechanisms of emotional learning. Ann N Y Acad Sci 985: 233–250.

Fehr C, Shirley RL, Belknap JK et al. (2002). Congenic mapping of alcohol and pentobarbital withdrawal liability loci to a <1 centimorgan interval of murine chromosome 4: identification of Mpdz as a candidate gene. J Neurosci 22: 3730–3738.

Fein G, Di Sclafani V, Cardenas VA et al. (2002). Cortical gray matter loss in treatment-naive alcohol dependent individuals. Alcohol Clin Exp Res 26: 558–564.

Fein G, Shimotsu R, Chu R et al. (2009). Parietal gray matter volume loss is related to spatial processing deficits in long-term abstinent alcoholic men. Alcohol Clin Exp Res 33: 1806–1814.

Finn DA, Crabbe JC (1997). Exploring alcohol withdrawal syndrome. Alcohol Health Res World 21: 149–156.

Finn DA, Snelling C, Fretwell AM et al. (2007). Increased drinking during withdrawal from intermittent ethanol exposure is blocked by the CRF receptor antagonist D-Phe-CRF(12-41). Alcohol Clin Exp Res 31: 939–949.

Flatscher-Bader T, Zuvela N, Landis N et al. (2008). Smoking and alcoholism target genes associated with plasticity and glutamate transmission in the human ventral tegmental area. Hum Mol Genet 17: 38–51.

Foroud T, Edenberg HJ, Goate A et al. (2000). Alcoholism susceptibility loci: confirmation studies in a replicate sample and further mapping. Alcohol Clin Exp Res 24: 933–945.

Frye GD, Mccown TJ, Breese GR (1983). Characterization of susceptibility to audiogenic seizures in ethanol-dependent rats after microinjection of gamma-aminobutyric acid (GABA) agonists into the inferior colliculus, substantia nigra or medial septum. J Pharmacol Exp Ther 227: 663–670.

Gansler DA, Harris GJ, Oscar-Berman M et al. (2000). Hypoperfusion of inferior frontal brain regions in abstinent alcoholics: a pilot SPECT study. J Stud Alcohol 61: 32–37.

Gianoulakis C (1993). Endogenous opioids and excessive alcohol consumption. J Psychiatry Neurosci 18: 148–156.

Gilman S, Adams K, Koeppe RA et al. (1990). Cerebellar and frontal hypometabolism in alcoholic cerebellar degeneration studied with positron emission tomography. Ann Neurol 28: 775–785.

Gilpin NW, Richardson HN, Koob GF (2008). Effects of CRF1-receptor and opioid-receptor antagonists on dependence-induced increases in alcohol drinking by alcohol-preferring (P) rats. Alcohol Clin Exp Res 32: 1535–1542.

Gilpin NW, Misra K, Herman MA et al. (2011). Neuropeptide Y opposes alcohol effects on gamma-aminobutyric acid release in amygdala and blocks the transition to alcohol dependence. Biol Psychiatry 69: 1091–1099.

Goldstein DB, Pal N (1971). Alcohol dependence produced in mice by inhalation of ethanol: grading the withdrawal reaction. Science 172: 288–290.

Goodwin DW, Schulsinger F, Hermansen L et al. (1973). Alcohol problems in adoptees raised apart from alcoholic biological parents. Arch Gen Psychiatry 28: 238–243.

Gorelick DA (1989). Serotonin uptake blockers and the treatment of alcoholism. Recent Dev Alcohol 7: 267–281.

Grant BF, Dawson DA, Stinson FS et al. (2004). The 12-month prevalence and trends in DSM-IV alcohol abuse and dependence: United States, 1991–1992 and 2001–2002. Drug Alcohol Depend 74: 223–234.

Greicius MD, Krasnow B, Reiss AL et al. (2003). Functional connectivity in the resting brain: a network analysis of the default mode hypothesis. Proc Natl Acad Sci U S A 100: 253–258.

Griffin WC, Overstreet MP, Becker HC (2011). Chronic intermittent ethanol exposure alters CRF release in the amygdala and bed nucleus of the stria terminalis in C57BL/6 J mice. Alcohol Clin Exp Res 35: 69A.

Gunzerath L, Hewitt BG, Li TK et al. (2011). Alcohol research: past, present, and future. Ann N Y Acad Sci 1216: 1–23.

Gupta T, Syed YM, Revis AA et al. (2008). Acute effects of acamprosate and MPEP on ethanol Drinking-in-the-Dark in male C57BL/6 J mice. Alcohol Clin Exp Res 32: 1992–1998.

Habas C, Kamdar N, Nguyen D et al. (2009). Distinct cerebellar contributions to intrinsic connectivity networks. J Neurosci 29: 8586–8594.

Harper C, Matsumoto I (2005). Ethanol and brain damage. Curr Opin Pharmacol 5: 73–78.

Hasin D, Kilcoyne B (2012). Comorbidity of psychiatric and substance use disorders in the United States: current issues and findings from the NESARC. Curr Opin Psychiatry 25: 165–171.

Heilig M, Egli M, Crabbe JC et al. (2010). Acute withdrawal, protracted abstinence and negative affect in alcoholism: are they linked? Addict Biol 15: 169–184.

Henniger MS, Spanagel R, Wigger A et al. (2002). Alcohol self-administration in two rat lines selectively bred for extremes in anxiety-related behavior. Neuropsychopharmacology 26: 729–736.

Himmelsbach CK (1941). The morphine abstinence syndrome, its nature and treatment. Ann Intern Med 15: 829–843.

Holter SM, Engelmann M, Kirschke C et al. (1998). Long-term ethanol self-administration with repeated ethanol deprivation episodes changes ethanol drinking pattern and increases anxiety-related behaviour during ethanol deprivation in rats. Behav Pharmacol 9: 41–48.

Johnson BA (2008). Update on neuropharmacological treatments for alcoholism: scientific basis and clinical findings. Biochem Pharmacol 75: 34–56.

Johnson BA, Jasinski DR, Galloway GP et al. (1996). Ritanserin in the treatment of alcohol dependence – a multi-center clinical trial. Ritanserin Study Group. Psychopharmacology (Berl) 128: 206–215.

Johnson BA, Ait-Daoud N, Bowden CL et al. (2003). Oral topiramate for treatment of alcohol dependence: a randomised controlled trial. Lancet 361: 1677–1685.

Johnson BA, Swift RM, Ait-Daoud N et al. (2004). Development of novel pharmacotherapies for the treatment of alcohol dependence: focus on antiepileptics. Alcohol Clin Exp Res 28: 295–301.

Johnson BA, Mann K, Willenbring ML et al. (2005). Challenges and opportunities for medications development in alcoholism: an international perspective on collaborations between academia and industry. Alcohol Clin Exp Res 29: 1528–1540.

Kang SJ, Rangaswamy M, Manz N et al. (2012). Family-based genome-wide association study of frontal theta oscillations identifies potassium channel gene KCNJ6. Genes Brain Behav 11: 712–719.

Karpyak VM, Kim JH, Biernacka JM et al. (2009). Sequence variations of the human MPDZ gene and association with alcoholism in subjects with European ancestry. Alcohol Clin Exp Res 33: 712–721.

Karpyak VM, Geske JR, Colby CL et al. (2012). Genetic variability in the NMDA-dependent AMPA trafficking cascade is associated with alcohol dependence. Addict Biol 17: 798–806.

Kimpel MW, Strother WN, Mcclintick JN et al. (2007). Functional gene expression differences between inbred alcohol-preferring and -non-preferring rats in five brain regions. Alcohol 41: 95–132.

Koob GF (2000). Animal models of craving for ethanol. Addiction 95 (Suppl 2): S73–S81.

Koob GF (2003). Neuroadaptive mechanisms of addiction: studies on the extended amygdala. Eur Neuropsychopharmacol 13: 442–452.

Koob GF (2013). Theoretical frameworks and mechanistic aspects of alcohol addiction: alcohol addiction as a reward deficit disorder. Curr Top Behav Neurosci 13: 3–30.

Koob GF, Volkow ND (2010). Neurocircuitry of addiction. Neuropsychopharmacology 35: 217–238.

Koob GF, Kenneth Lloyd G, Mason BJ (2009). Development of pharmacotherapies for drug addiction: a Rosetta stone approach. Nat Rev Drug Discov 8: 500–515.

Koob GF, Buck CL, Cohen A et al. (2014). Addiction as a stress surfeit disorder. Neuropharmacology 76: 370–382.

Kornreich C, Philippot P, Foisy ML et al. (2002). Impaired emotional facial expression recognition is associated with interpersonal problems in alcoholism. Alcohol Alcohol 37: 394–400.

Krienen FM, Buckner RL (2009). Segregated fronto-cerebellar circuits revealed by intrinsic functional connectivity. Cereb Cortex 19: 2485–2497.

LeDoux J (2003). The emotional brain, fear, and the amygdala. Cell Mol Neurobiol 23: 727–738.

Li TK (2008). Quantifying the risk for alcohol-use and alcohol-attributable health disorders: present findings and future research needs. J Gastroenterol Hepatol 23 (Suppl 1): S2–S8.

Li TK, Lumeng L, McBride WJ et al. (1987). Rodent lines selected for factors affecting alcohol consumption. Alcohol Alcohol Suppl 1: 91–96.

Li TK, Lumeng L, Doolittle DP (1993). Selective breeding for alcohol preference and associated responses. Behav Genet 23: 163–170.

Litten RZ, Allen JP (1991). Pharmacotherapies for alcoholism: promising agents and clinical issues. Alcohol Clin Exp Res 15: 620–633.

Litten RZ, Fertig J, Mattson M et al. (2005). Development of medications for alcohol use disorders: recent advances and ongoing challenges. Expert Opin Emerg Drugs 10: 323–343.

Litten RZ, Egli M, Heilig M et al. (2012). Medications development to treat alcohol dependence: a vision for the next decade. Addict Biol 17: 513–527.

Lopez MF, Becker HC (2005). Effect of pattern and number of chronic ethanol exposures on subsequent voluntary ethanol intake in C57BL/6 J mice. Psychopharmacology (Berl) 181: 688–696.

Lopez MF, Griffin WC, Becker HC (2010). Ethanol intake, plasma corticosterone levels and brain region CRF levels in ethanol dependent C57BL/6 J mice. Alcohol Clin Exp Res 34: 200A.

Macey DJ, Schulteis G, Heinrichs SC et al. (1996). Time-dependent quantifiable withdrawal from ethanol in the rat: effect of method of dependence induction. Alcohol 13: 163–170.

Majchrowicz E (1975). Induction of physical dependence upon ethanol and the associated behavioral changes in rats. Psychopharmacologia 43: 245–254.

Marinkovic K, Oscar-Berman M, Urban T et al. (2009). Alcoholism and dampened temporal limbic activation to emotional faces. Alcohol Clin Exp Res 33: 1880–1892.

Mayfield RD, Harris RA, Schuckit MA (2008). Genetic factors influencing alcohol dependence. Br J Pharmacol 154: 275–287.

McGue M (1999). Phenotyping alcoholism. Alcohol Clin Exp Res 23: 757–758.

Meert TF, Rassnick S, Huysmans H et al. (1992). Quantification of tremor sensitivity and inhibition of exploratory behaviour during alcohol withdrawal in rats. Behav Pharmacol 3: 601–607.

Metten PM, Crabbe JC (1996). Dependence and withdrawal. In: RA Deitrich, VG Erwin (Eds.), Pharmacological Effects of Alcohol on the Nervous System. CRC Press, Boca Raton, pp. 269–290.

Mokdad AH, Marks JS, Stroup DF et al. (2004). Actual causes of death in the United States, 2000. JAMA 291: 1238–1245.

Moy SS, Knapp DJ, Criswell HE et al. (1997). Flumazenil blockade of anxiety following ethanol withdrawal in rats. Psychopharmacology (Berl) 131: 354–360.

Muller-Oehring EM, Jung YC, Sullivan EV et al. (2013). Midbrain-driven emotion and reward processing in alcoholism. Neuropsychopharmacology 38: 1844–1853.

Mulligan MK, Ponomarev I, Hitzemann RJ et al. (2006). Toward understanding the genetics of alcohol drinking through transcriptome meta-analysis. Proc Natl Acad Sci U S A 103: 6368–6373.

Nomura EM, Gratton C, Visser RM et al. (2010). Double dissociation of two cognitive control networks in patients with focal brain lesions. Proc Natl Acad Sci U S A 107: 12017–12022.

O'Malley SS, O'Connor PG (2011). Medications for unhealthy alcohol use: across the spectrum. Alcohol Res Health 33: 300–312.

Oscar-Berman M, Marinkovic K (2007). Alcohol: effects on neurobehavioral functions and the brain. Neuropsychol Rev 17: 239–257.

Oscar-Berman M, Kirkley SM, Gansler DA et al. (2004). Comparisons of Korsakoff and non-Korsakoff alcoholics on neuropsychological tests of prefrontal brain functioning. Alcohol Clin Exp Res 28: 667–675.

Pandey SC, Ugale R, Zhang H et al. (2008). Brain chromatin remodeling: a novel mechanism of alcoholism. J Neurosci 28: 3729–3737.

Pfefferbaum A, Sullivan EV, Mathalon DH et al. (1997). Frontal lobe volume loss observed with magnetic resonance imaging in older chronic alcoholics. Alcohol Clin Exp Res 21: 521–529.

Pitel AL, Chanraud S, Muller-Oehring EM et al. (2013). Modulation of limbic-cerebellar functional connectivity enables alcoholics to recognize who is who. Brain Struct Funct 218: 683–695.

Powell BJ, Campbell JL, Landon JF et al. (1995). A double-blind, placebo-controlled study of nortriptyline and bromocriptine in male alcoholics subtyped by comorbid psychiatric disorders. Alcohol Clin Exp Res 19: 462–468.

Prescott CA, Kendler KS (1999). Genetic and environmental contributions to alcohol abuse and dependence in a population-based sample of male twins. Am J Psychiatry 156: 34–40.

Raichle ME, Macleod AM, Snyder AZ et al. (2001). A default mode of brain function. Proc Natl Acad Sci U S A 98: 676–682.

Rehm J, Mathers C, Popova S et al. (2009). Global burden of disease and injury and economic cost attributable to alcohol use and alcohol-use disorders. Lancet 373: 2223–2233.

Reich T, Edenberg HJ, Goate A et al. (1998). Genome-wide search for genes affecting the risk for alcohol dependence. Am J Med Genet 81: 207–215.

Reilly MT, Lovinger DM (2010). International Review of Neurobiology. Functional plasticity and genetic variation: insights into the neurobiology of alcoholism. Preface. Int Rev Neurobiol 91: xi–xii.

Reilly MT, Harris RA, Noronha A (2012). Using genetically engineered animal models in the postgenomic era to understand gene function in alcoholism. Alcohol Res 34: 282–291.

Repunte-Canonigo V, Berton F, Cottone P et al. (2010a). A potential role for adiponectin receptor 2 (AdipoR2) in the regulation of alcohol intake. Brain Res 1339: 11–17.

Repunte-Canonigo V, Van Der Stap LD, Chen J et al. (2010b). Genome-wide gene expression analysis identifies K-ras as a regulator of alcohol intake. Brain Res 1339: 1–10.

Rhodes JS, Crabbe JC (2005). Gene expression induced by drugs of abuse. Curr Opin Pharmacol 5: 26–33.

Rhodes JS, Best K, Belknap JK et al. (2005). Evaluation of a simple model of ethanol drinking to intoxication in C57BL/6 J mice. Physiol Behav 84: 53–63.

Roberto M, Cruz MT, Gilpin NW et al. (2010). Corticotropin releasing factor-induced amygdala gamma-aminobutyric acid release plays a key role in alcohol dependence. Biol Psychiatry 67: 831–839.

Robison AJ, Nestler EJ (2011). Transcriptional and epigenetic mechanisms of addiction. Nat Rev Neurosci 12: 623–637.

Roelofs SM (1985). Hyperventilation, anxiety, craving for alcohol: a subacute alcohol withdrawal syndrome. Alcohol 2: 501–505.

Rogers J, Wiener SG, Bloom FE (1979). Long-term ethanol administration methods for rats: advantages of inhalation over intubation or liquid diets. Behav Neural Biol 27: 466–486.

Rogers BP, Parks MH, Nickel MK et al. (2012). Reduced fronto-cerebellar functional connectivity in chronic alcoholic patients. Alcohol Clin Exp Res 36: 294–301.

Roth BL, Sheffler DJ, Kroeze WK (2004). Magic shotguns versus magic bullets: selectively non-selective drugs for mood disorders and schizophrenia. Nat Rev Drug Discov 3: 353–359.

Roukos DH, Ziogas DE, Baltogiannis GG et al. (2014). Novel next-generation sequencing and networks-based therapeutic targets: realistic and more effective drug design and discovery. Curr Pharm Des 20: 11–22.

Sameti M, Smith S, Patenaude B et al. (2011). Subcortical volumes in long-term abstinent alcoholics: associations with psychiatric comorbidity. Alcohol Clin Exp Res 35: 1067–1080.

Schulte T, Muller-Oehring EM, Sullivan EV et al. (2012). Synchrony of corticostriatal-midbrain activation enables normal inhibitory control and conflict processing in recovering alcoholic men. Biol Psychiatry 71: 269–278.

Seo D, Lacadie CM, Tuit K et al. (2013). Disrupted ventromedial prefrontal function, alcohol craving, and subsequent relapse risk. JAMA Psychiatry 70: 727–739.

Shirley RL, Walter NA, Reilly MT et al. (2004). Mpdz is a quantitative trait gene for drug withdrawal seizures. Nat Neurosci 7: 699–700.

Silveri MM (2012). Adolescent brain development and underage drinking in the United States: identifying risks of alcohol use in college populations. Harv Rev Psychiatry 20: 189–200.

Sinha R (2012). How does stress lead to risk of alcohol relapse? Alcohol Res 34: 432–440.

Sommer WH, Rimondini R, Hansson AC et al. (2008). Upregulation of voluntary alcohol intake, behavioral sensitivity to stress, and amygdala crhr1 expression following a history of dependence. Biol Psychiatry 63: 139–145.

Spanagel R, Montkowski A, Allingham K et al. (1995). Anxiety: a potential predictor of vulnerability to the initiation of ethanol self-administration in rats. Psychopharmacology (Berl) 122: 369–373.

Sullivan EV, Pfefferbaum A (2005). Neurocircuitry in alcoholism: a substrate of disruption and repair. Psychopharmacology (Berl) 180: 583–594.

Sullivan EV, Marsh L, Mathalon DH et al. (1995). Anterior hippocampal volume deficits in nonamnesic, aging chronic alcoholics. Alcohol Clin Exp Res 19: 110–122.

Sullivan EV, Deshmukh A, Desmond JE et al. (2000a). Cerebellar volume decline in normal aging, alcoholism, and Korsakoff's syndrome: relation to ataxia. Neuropsychology 14: 341–352.

Sullivan EV, Rosenbloom MJ, Lim KO et al. (2000b). Longitudinal changes in cognition, gait, and balance in abstinent and relapsed alcoholic men: relationships to changes in brain structure. Neuropsychology 14: 178–188.

Sullivan EV, Harding AJ, Pentney R et al. (2003). Disruption of frontocerebellar circuitry and function in alcoholism. Alcohol Clin Exp Res 27: 301–309.

Sullivan EV, Rose J, Pfefferbaum A (2006). Effect of vision, touch and stance on cerebellar vermian-related sway and tremor: a quantitative physiological and MRI study. Cereb Cortex 16: 1077–1086.

Sullivan EV, Harris RA, Pfefferbaum A (2010). Alcohol's effects on brain and behavior. Alcohol Res Health 33: 127–143.

Tabakoff B, Saba L, Kechris K et al. (2008). The genomic determinants of alcohol preference in mice. Mamm Genome 19: 352–365.

Tapert SF, Brown GG, Kindermann SS et al. (2001). fMRI measurement of brain dysfunction in alcohol-dependent young women. Alcohol Clin Exp Res 25: 236–245.

Thorsell A, Slawecki CJ, Ehlers CL (2005). Effects of neuropeptide Y and corticotropin-releasing factor on ethanol intake in Wistar rats: interaction with chronic ethanol exposure. Behav Brain Res 161: 133–140.

Tiffany ST, Carter BL, Singleton EG (2000). Challenges in the manipulation, assessment and interpretation of craving relevant variables. Addiction 95 (Suppl 2): S177–S187.

Townshend JM, Duka T (2003). Mixed emotions: alcoholics' impairments in the recognition of specific emotional facial expressions. Neuropsychologia 41: 773–782.

Vetreno RP, Hall JM, Savage LM (2011). Alcohol-related amnesia and dementia: animal models have revealed the contributions of different etiological factors on neuropathology, neurochemical dysfunction and cognitive impairment. Neurobiol Learn Mem 96: 596–608.

Victor M (1970). The alcohol withdrawal syndrome: theory and practice. Postgrad Med 47: 68–72.

Walker BM (2012). Conceptualizing withdrawal-induced escalation of alcohol self-administration as a learned, plasticity-dependent process. Alcohol 46: 339–348.

Walker BM, Koob GF (2008). Pharmacological evidence for a motivational role of kappa-opioid systems in ethanol dependence. Neuropsychopharmacology 33: 643–652.

Walker BM, Zorrilla EP, Koob GF (2011). Systemic κ-opioid receptor antagonism by nor-binaltorphimine reduces dependence-induced excessive alcohol self-administration in rats. Addict Biol 16: 116–119.

Wang JC, Foroud T, Hinrichs AL et al. (2013). A genome-wide association study of alcohol-dependence symptom counts in extended pedigrees identifies C15orf53. Mol Psychiatry 18: 1218–1224.

Warnault V, Darcq E, Levine A et al. (2013). Chromatin remodeling – a novel strategy to control excessive alcohol drinking. Transl Psychiatry 3: e231.

Willinger U, Lenzinger E, Hornik K et al. (2002). Anxiety as a predictor of relapse in detoxified alcohol-dependent patients. Alcohol Alcohol 37: 609–612.

Zhang R, Miao Q, Wang C et al. (2013). Genome-wide DNA methylation analysis in alcohol dependence. Addict Biol 18: 392–403.

Section 2

Animal models: neurochemistry and metabolism of alcohol

Handbook of Clinical Neurology, Vol. 125 (3rd series)
Alcohol and the Nervous System
E.V. Sullivan and A. Pfefferbaum, Editors

Chapter 3

Neurocircuitry of alcohol addiction: synthesis from animal models

GEORGE F. KOOB*

Committee on the Neurobiology of Addictive Disorders, The Scripps Research Institute, La Jolla, CA, USA

DEFINITIONS AND CONCEPTUAL FRAMEWORK FOR NEUROCIRCUITRY OF ALCOHOLISM

Alcoholism has many definitions that vary from social frameworks to a psychiatric framework. Alcoholism, more generically drug addiction, can be defined as a chronically relapsing disorder characterized by: (1) compulsion to seek and take the drug (alcohol); (2) loss of control in limiting (alcohol) intake; and (3) emergence of a negative emotional state (e.g., dysphoria, anxiety, irritability), reflecting a motivational withdrawal syndrome, when access to the drug (alcohol) is prevented (defined here as dependence; Koob and Le Moal, 1997). These key elements embrace most of the symptoms of substance dependence on alcohol in the *Diagnostic and Statistical Manual of the American Psychiatric Association*, 5th edition (DSM-5; American Psychiatric Association, 2013). Clinically and in animal models, the occasional but limited use of alcohol with the potential for abuse or dependence is distinct from escalated alcohol intake and the emergence of a chronic alcohol-dependent state. Alcoholism and drug addiction in general have been hypothesized to be composed of multiple stages that feed into each other in a three-stage cycle: binge/intoxication, withdrawal/negative affect, and preoccupation/anticipation ("craving": Fig. 3.1). The three stages are conceptualized as interacting with each other, becoming more intense, and ultimately leading to the pathologic state known as addiction (Koob and Le Moal, 1997).

The thesis argued in the present overview, derived largely from work with animal models, is that alcoholism, similar to drug addiction, is a neurocircuitry disorder and that neuroadaptations within specific motivational circuits play an important role in defining and perpetuating alcoholism. The argument herein is that excessive engagement of reward circuitry sets up a core deficit of decreased reward function and increased activation of brain stress systems and more generically drug addiction. Alcoholism also involves substantial neuroadaptations that persist beyond acute withdrawal and trigger relapse and deficits in cognitive function that can also fuel compulsive drinking.

The argument herein is that alcoholism is a reward deficit, stress surfeit, and executive function (self-regulation) disorder. A negative emotional state is a common presentation in most alcoholics during withdrawal and protracted abstinence. Decreases in brain reward function and brain reward neurotransmission in part mediate such reward deficits. Parallel to the decreased function of the reward system is a sensitized brain stress system mediated by the activation of both the hypothalamic–pituitary–adrenal axis and brain stress neurotransmitter systems. Deficits in frontal cortex function combined with powerful neurochemical adaptations and a residual sensitization of brain stress systems are hypothesized to drive both residual impulsivity and compulsivity during protracted abstinence.

The overall guiding hypothesis is that the compulsivity observed in alcoholism has an important negative reinforcement component that perpetuates alcoholism. Such negative emotional states become sensitized over time and set up an allostatic state that perpetuates dependence. Negative emotional states also set up a powerful motivational state for relapse. The neurobiologic substrates that underlie the motivation to seek alcohol will be reviewed, and an argument will be presented that it

*Correspondence to: George F. Koob, Committee on the Neurobiology of Addictive Disorders, The Scripps Research Institute, 10550 North Torrey Pines Road, SP30-2400, La Jolla, CA 92037, USA. Tel: +1-858-784-7062, Fax: +1-858-784-7405, E-mail: gkoob@scripps.edu

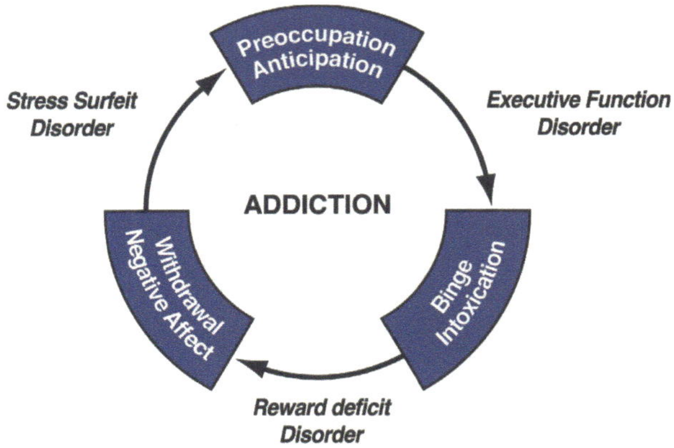

Fig. 3.1. The three stages of the addiction cycle – binge/intoxication, withdrawal/negative affect, and preoccupation/anticipation – from the perspective of functional neurocircuitry deficits that correspond to each stage, respectively: reward deficit disorder, stress surfeit disorder, and executive function disorder.

is the loss of reward function and gain of brain stress function that mediate the negative emotional state associated with alcoholism.

Drug addiction has generally been conceptualized as a disorder that involves elements of both impulsivity and compulsivity, in which impulsivity can be defined behaviorally as "a predisposition toward rapid, unplanned reactions to internal and external stimuli without regard for the negative consequences of these reactions to themselves or others" (Moeller et al., 2001). Impulsivity is measured in two domains: the choice of a smaller, immediate reward over a larger, delayed reward (Rachlin and Green, 1972) or the inability to inhibit behavior by changing the course of action or to stop a response once it is initiated (Logan et al., 1997). Impulsivity is a core deficit in substance abuse disorders (Allen et al., 1998) and neuropsychiatric disorders, such as attention-deficit/hyperactivity disorder. Operationally, delay-to-gratification tasks (e.g., delayed discounting tasks, impulsive choice) and the stop signal or go/no-go task (behavioral impulsivity) have been used as measures of impulsivity (Green et al., 1994; Fillmore and Rush, 2002).

Compulsivity can be defined as elements of behavior that result in perseveration of responding in the face of adverse consequences or perseveration in the face of incorrect responses in choice situations (e.g., operationally, responding for a drug or alcohol in the face of adverse consequences [Wolffgramm and Heyne, 1995] or responding for a drug or alcohol on a progressive-ratio schedule of reinforcement [Walker et al., 2008]). Compulsivity can be considered analogous to some of the key symptoms of substance use disorder outlined by the American Psychiatric Association: continued substance use despite knowledge of having had a persistent or recurrent physical or psychologic problem and a great deal of time spent in activities necessary to obtain the substance (American Psychiatric Association, 2013).

Collapsing the cycles of impulsivity and compulsivity yields a composite addiction cycle that consists of three stages – preoccupation/anticipation, binge/intoxication, and withdrawal/negative affect – in which impulsivity often dominates at the early stages and compulsivity dominates at terminal stages (Fig. 3.2). As an individual moves from impulsivity to compulsivity, a shift occurs from positive reinforcement that drives the motivated behavior to negative reinforcement that drives the motivated behavior (Koob, 2004). Negative reinforcement can be defined as the process by which removal of an aversive stimulus (e.g., negative emotional state of drug withdrawal) increases the probability of a response (e.g., dependence-induced drug intake to relieve the negative emotional state). Note that negative reinforcement is not punishment, although both involve an aversive stimulus. In punishment, the aversive stimulus suppresses behavior, including drug taking (e.g., disulfiram [Antabuse]). Negative reinforcement can be perhaps described in lay terms as reward via relief (i.e., relief reward), such as removal of pain or, in the case of alcoholism, removal of the negative emotional state of acute withdrawal or protracted abstinence (Fig. 3.2).

In alcohol addiction, or alcoholism, a pattern of oral drug taking evolves that is often characterized by binges of alcohol intake that can be daily episodes or prolonged days of heavy drinking and can be characterized by a severe emotional and somatic withdrawal syndrome. Many alcoholics continue with such a binge/withdrawal pattern for extended periods of time, but some individuals can evolve into an opioid-like situation in which they must have alcohol available at all times to avoid the negative consequences of abstinence. Here, intense preoccupation with obtaining alcohol (craving) develops that is linked not only to stimuli associated with obtaining the drug but also to stimuli associated with withdrawal and the aversive motivational state. A pattern develops in which the drug must be obtained to avoid the severe dysphoria and discomfort of abstinence.

ANIMAL MODELS FOR COMPULSIVE ALCOHOL SEEKING

Methods of inducing binge-like drinking with alcohol range from having animals drink alcohol solutions that are made extra palatable with the addition of a sweetener (Ji et al., 2008) to intermittent overnight access (Wise, 1973; Simms et al., 2010), to restricting intake to specific periods of the dark cycle (i.e., drinking-in-the-dark; Rhodes et al., 2005), to models that involve alcohol

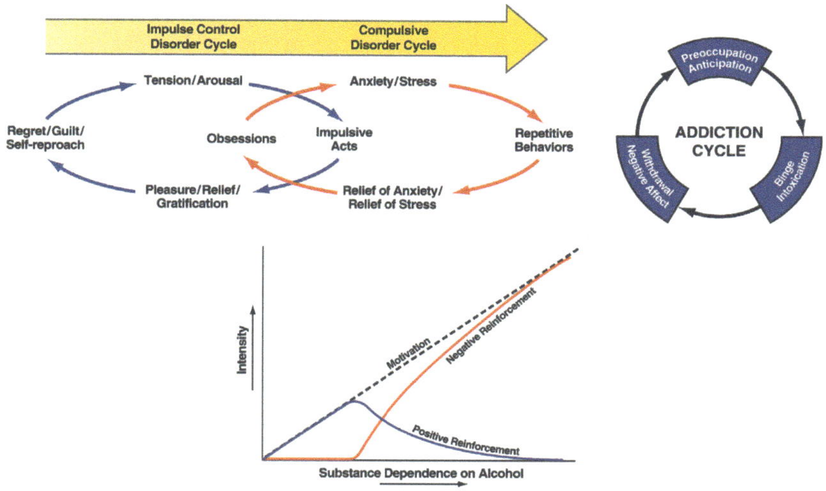

Fig. 3.2. (Top left) The stages of impulse control disorder and compulsive disorder cycles related to the sources of reinforcement. In impulse control disorders, an increasing tension and arousal occur before the impulsive act, with pleasure, gratification, or relief during the act. Following the act, there may or may not be regret or guilt. In compulsive disorders, there are recurrent and persistent thoughts (obsessions) that cause marked anxiety and stress followed by repetitive behaviors (compulsions) that are aimed at preventing or reducing distress. (Reproduced with permission from Koob, 2008b.) Positive reinforcement (pleasure/gratification) is more closely associated with impulse control disorders. Negative reinforcement (relief of anxiety or relief of stress) is more closely associated with compulsive disorders. (Reproduced with permission from Koob, 2013, with permission from Springer Science and Business Media.) (Top right) Collapsing the cycles of impulsivity and compulsivity results in the addiction cycle, conceptualized as three major components: preoccupation/anticipation, binge/intoxication, and withdrawal/negative affect. (Reproduced with permission from Koob, 2008b.) (Bottom) Change in the relative contribution of positive and negative reinforcement constructs during the development of substance dependence on alcohol. (Reproduced with permission from Koob, 2013, with permission from Springer Science and Business Media.)

dependence in animals, such as alcohol vapor inhalation, intragastric alcohol infusion, and alcohol liquid diet. The compulsive use of alcohol derives from multiple sources of reinforcement, and animal models have been developed not only for the acute positive reinforcing effects of ethanol but also for the negative reinforcing effects associated with removal of the aversive effects of ethanol withdrawal. A major early breakthrough was the development of a training procedure that involves access to a sweetened solution and subsequent fading in of ethanol to avoid the aversiveness of the ethanol taste (for review, see Samson, 1987). Subsequent work extended these procedures to measures of self-administration in dependent rats and postdependent rats (Roberts et al., 1996; O'Dell et al., 2004).

High doses of alcohol solution will be self-administered intragastrically after animals are made dependent via passive intragastric infusion, and rats will self-infuse 4–7 g/kg per day of ethanol (Fidler et al., 2006). Here, blood alcohol levels average 0.12 g%, measured 30 minutes after the start of a bout in which rats infuse 1.5 g/kg per 30 minutes.

In an alcohol-liquid diet procedure, the diet is typically the sole source of calories available to rats (e.g., Moy et al., 1997), thereby forcing rats to consume the alcohol. Typically, rats are provided a palatable liquid diet that contains 5–8.7% v/v ethanol as their sole source of calories sufficient to produce dependence and maintain blood alcohol levels of 100–130 mg% during the dark (active drinking) cycle (Schulteis et al., 1996; Brown et al., 1998; Valdez et al., 2004). High responders during withdrawal from liquid diet will reach blood alcohol levels of approximately 80–100 mg% (Schulteis et al., 1996; Gilpin et al., 2009).

Reliable self-administration of ethanol in dependent animals using ethanol vapor exposure has been extensively characterized in rats, in which animals reach blood alcohol levels in the 100–150 mg% range (Roberts et al., 1999, 2000). Similarly, rats with a history of alcohol dependence show increased self-administration of ethanol, even weeks after acute withdrawal (Roberts et al., 2000). In a variant of alcohol vapor exposure with more face validity, intermittent exposure to chronic ethanol using alcohol vapor chambers (14 hours on/10 hours off) produces more rapid escalation to increased ethanol intake and higher amounts of intake (Rimondini et al., 2002; O'Dell et al., 2004), and blood alcohol levels are reliably above 140 mg% after a 30-minute session of self-administration in dependent animals (Richardson et al., 2008). In both the liquid diet and ethanol vapor procedures, alcohol intake is directly related to the blood alcohol range and the pattern of intermittent high-dose alcohol exposure (Gilpin et al., 2009). Although the alcohol vapor model may have limited face validity, considering that alcohol is passively administered to animals, numerous studies have demonstrated that it also has robust predictive validity for alcohol addiction (Heilig and Koob, 2007; Koob et al., 2009).

Similar procedures have been developed for mice and produce reliable increases in bouts of binge-like drinking using the drinking-in-the-dark paradigm or withdrawal-induced drinking in dependent animals. Drinking-in-the-dark involves mice that are allowed to drink 20% ethanol 3 hours after lights out for 2 hours for 3 consecutive days and on the fourth day are allowed 4 hours access, also 3 hours after lights out (Rhodes et al., 2005). Withdrawal-induced drinking usually involves C57BL/6 mice which are exposed to intermittent durations of ethanol vapor (three cycles of 16 hours of vapor and 8 hours of air) and then tested in a 2-hour limited-access ethanol preference drinking test during the circadian dark period (Becker and Lopez, 2004; Lopez and Becker, 2005; Finn et al., 2007). Intermittent ethanol vapor exposure significantly increased 15% (v/v) ethanol intake by 30–50% in the postvapor period, usually after multiple cycles and usually after 24 hours of withdrawal (Finn et al., 2007). Similar results have been reported using an operant response in mice in 60-minute test sessions for 10% (w/v) ethanol with intermittent vapor exposure of 14 hours on/10 hours off (Chu et al., 2007).

ANIMAL MODELS OF MOTIVATION, WITHDRAWAL, AND OPPONENT PROCESS

The concept of motivation in addiction was inextricably linked with hedonic, affective, or emotional states in the context of temporal dynamics by Solomon's opponent process theory of motivation. Solomon and Corbit (1974) postulated that hedonic, affective, or emotional states, once initiated by drugs, are automatically modulated by the central nervous system with mechanisms that reduce the intensity of hedonic feelings. The a-process in drug use consists of positive hedonic responses, occurs shortly after presentation of a stimulus, correlates closely with the intensity, quality, and duration of the reinforcer, and shows tolerance. In contrast, the b-process in drug use appears after the a-process has terminated, consists of negative hedonic responses, and is sluggish in onset, slow to build up to an asymptote, and slow to decay, and gets larger with repeated exposure. The a-process includes affective or hedonic habituation (or tolerance), and the b-process includes affective or hedonic withdrawal (abstinence). The thesis here is that opponent processes begin early in drug taking, reflect changes in the brain reward and stress systems, and later form one of the major motivations for compulsivity in drug taking.

Thus, dependence or the manifestation of a withdrawal syndrome after the removal of chronic drug administration is defined in terms of motivational aspects of dependence, such as the emergence of a negative emotional state (e.g., dysphoria, anxiety, irritability) when access to the drug is prevented (Koob and Le Moal, 2001), rather than on the physical signs of dependence. Indeed, some have argued that the development of such a negative affective state can define dependence as it relates to addiction (Russell, 1976; Koob et al., 1989; Baker et al., 2004).

Alcoholics show dramatic evidence of dysphoric states during acute withdrawal that persist into protracted abstinence. Alcohol withdrawal in humans produces well-documented physical (somatic) symptoms, such as tremor, autonomic hyperactivity, nausea, vomiting, and seizures, but, more importantly, produces significant affective symptoms of anxiety, dysphoria, and depression-like symptoms.

Animal models can also be used to test the hypothesis that there are opponent process-like motivational changes associated with the development of alcohol dependence. Electrical brain stimulation reward or intracranial self-stimulation has a long history as a measure of activity of the brain reward system and of the acute reinforcing effects of drugs of abuse. All drugs of abuse, when administered acutely, decrease brain stimulation reward thresholds (Kornetsky and Esposito, 1979) and when administered chronically increase reward thresholds during withdrawal. Brain stimulation reward involves widespread neurocircuitry in the brain, but the most sensitive sites defined by the lowest thresholds involve the trajectory of the medial forebrain bundle that connects the ventral tegmental area with the basal forebrain (Olds and Milner, 1954; Koob et al., 1977). Although much emphasis was focused initially on the role of the ascending monoamine systems in the medial forebrain bundle in brain stimulation reward, other non-dopaminergic systems in the medial forebrain bundle clearly play a key role (Hernandez et al., 2006).

Rats made dependent using chronic ethanol vapor exposure at blood alcohol levels sufficient to drive excessive drinking showed an increase in brain reward thresholds during withdrawal that lasted up to 3 days postwithdrawal (Schulteis et al., 1995). However, data suggest, similar to other drugs of abuse, that such opponent-like processes can begin with a single dosing (Fig. 3.3). An acute elevation in brain reward thresholds was observed during repeated acute withdrawal from ethanol, bearing a striking resemblance to human

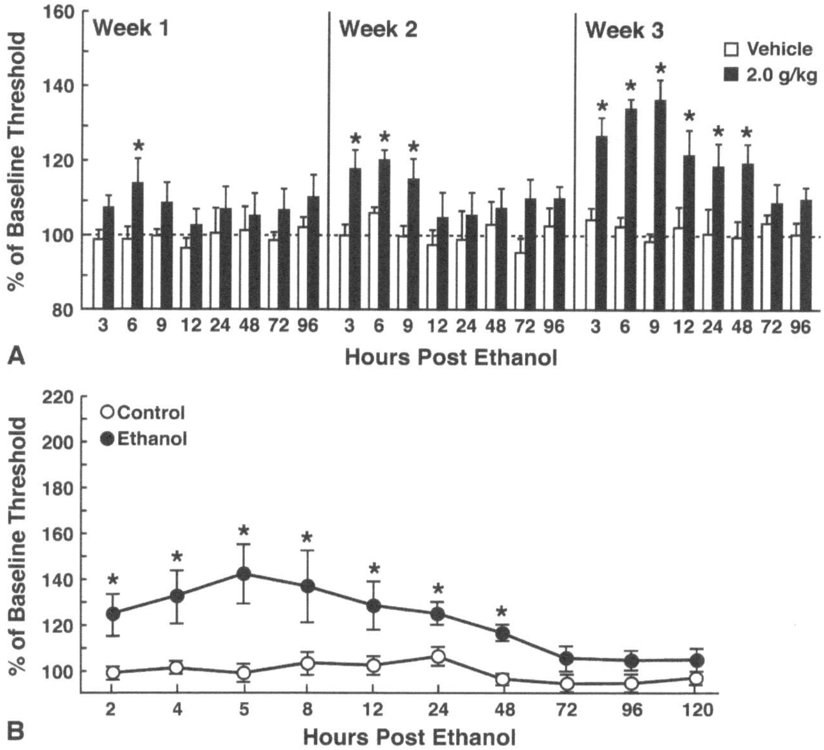

Fig. 3.3. (**A**) Withdrawal from a single bout of acute ethanol intoxication (week 1) resulted in a significant but transient increase in brain reward threshold only with the highest dose of ethanol tested (2.0 g/kg, bottom panel, $*p < 0.05$, compared with vehicle controls at given timepoint postinjection). The effect was significant at 6 hours, a time when blood alcohol levels had declined to virtually undetectable levels following this dose of ethanol. Repeated treatment with this dose for 2 additional weeks resulted in a progressive broadening of the duration of significant threshold elevations. The data are expressed as the mean ± SEM percentage of baseline threshold. $n = 8–10$ per dose group. (Reproduced with permission from Schulteis and Liu, 2006.) (**B**) Time-dependent elevation of intracranial self-stimulation thresholds during ethanol withdrawal. Mean blood alcohol levels were 197.29 mg%. The data are expressed as the mean ± SEM percentage of baseline threshold. $*p < 0.05$, thresholds that were significantly elevated above control levels at 2–48 hours postethanol. Open circles indicate the control condition. Closed circles indicate the ethanol withdrawal condition. (Reproduced with permission from Schulteis et al., 1995. Copyright (1995) National Academy of Sciences, USA.)

subjective reports (Schulteis and Liu, 2006; Fig. 3.3). These results demonstrate that the elevation in brain reward thresholds following prolonged access to alcohol may fail to return to baseline levels between repeated, prolonged exposure to alcohol self-administration (i.e., a residual reward deficit), thus creating the greater elevation in reward thresholds observed during withdrawal from chronic ethanol. Rapid acute tolerance (i.e., tolerance within a given bout of intoxication) and opponent process-like effects in response to the hedonic effects of alcohol have been reported in human studies using the alcohol clamp procedure (Morzorati et al., 2002). These data provide compelling evidence for brain reward dysfunction with chronic alcohol, which provides strong support for a hedonic allostasis model of alcoholism (Koob, 2003).

The dysregulation of brain reward function associated with withdrawal from chronic administration of drugs of abuse is a common element of all drugs of abuse. Withdrawal from chronic cocaine (Markou and Koob, 1991), amphetamine (Paterson et al., 2000), opioids (Schulteis et al., 1994), cannabinoids (Gardner and Vorel, 1998), nicotine (Epping-Jordan et al., 1998), and ethanol (Schulteis et al., 1995) leads to increases in reward thresholds during acute abstinence, and some of these elevations in threshold can last for up to 1 week. These observations lend credence to the hypothesis that opponent processes can set the stage for one aspect of compulsivity in which negative reinforcement mechanisms are engaged.

More recently, the opponent process theory has been expanded into the domains of the neurobiology of drug addiction from a neurocircuitry perspective. An allostatic model of the brain motivational systems has been proposed to explain the persistent changes in motivation that are associated with dependence in addiction (Koob and Le Moal, 2001, 2008). In this formulation, addiction is conceptualized as a cycle of increasing dysregulation of brain reward/antireward mechanisms that results in a negative emotional state that contributes to the compulsive use of drugs. Counteradaptive processes that are part of the normal homeostatic limitation of reward function fail to return within the normal homeostatic range. These counteradaptive processes are hypothesized to be mediated by two mechanisms: within-system neuroadaptations and between-system neuroadaptations (Koob and Bloom, 1988).

In a within-system neuroadaptation, "the primary cellular response element to the drug would itself adapt to neutralize the drug's effects; persistence of the opposing effects after the drug disappears would produce the withdrawal response" (Koob and Bloom, 1988). Thus, a within-system neuroadaptation is a molecular or cellular change within a given reward circuit to accommodate overactivity of hedonic processing associated with addiction, resulting in a decrease in reward function.

The emotional dysregulation associated with the withdrawal/negative affect stage also may involve between-system neuroadaptations in which neurochemical systems other than those involved in the positive rewarding effects of drugs of abuse are recruited or dysregulated by chronic activation of the reward system. "In the between-systems opposing process, a different cellular system and separable molecular apparatus would be triggered by the changes in the primary drug response neurons and would produce the adaptation and tolerance" (Koob and Bloom, 1988). Thus, a between-system neuroadaptation is a circuitry change in which another different circuit (antireward circuit) is activated by the reward circuit and has opposing actions, again limiting reward function. The remainder of this review explores the neuroadaptational changes that occur in the brain emotional systems to account for the neurocircuitry changes that produce opponent processes and are hypothesized to play a key role in the compulsivity of addiction.

NEUROCIRCUITS FOR THE BINGE/ INTOXICATION STAGE ASSOCIATED WITH ALCOHOLISM

Animal models for the acute reinforcing effects of alcohol and the use of selective antagonists for specific neurochemical systems have supported the hypothesis that alcohol at intoxicating doses has a wide but selective action on neurotransmitter systems in the brain reward systems. Neuropharmacologic studies have implicated multiple neurochemical systems in the acute reinforcing effects of alcohol, including γ-aminobutyric acid (GABA), opioid peptides, dopamine, serotonin, and glutamate (Fig. 3.4).

There has been significant work that shows, at least at the pharmacologic level, a role for GABA in the intoxicating effects of alcohol (Martz et al., 1983). Systemic injections of $GABA_A$ receptor antagonists or inverse agonists reverse the motor-impairing effects of alcohol (Hellevuo et al., 1989), the anxiolytic-like effects of alcohol (Liljequist and Engel, 1984), and alcohol drinking (McBride et al., 1988; Balakleevsky et al., 1990; June et al., 1991; Wegelius et al., 1994). $GABA_A$ antagonists or inverse agonists also decrease operant alcohol self-administration (Samson et al., 1987; Rassnick et al., 1993a; Samson and Chappell, 2001) and block alcohol stimulus effects in drug discrimination (Gatto and Grant, 1997; Grant et al., 2000).

Other mechanisms by which alcohol may modulate the GABA system may be via the $GABA_B$ receptor. For example, presynaptic inhibition of GABA release via the selective $GABA_B$ receptor agonist baclofen decreases alcohol self-administration in non-dependent rats (Janak and Gill, 2003) and decreases the alcohol deprivation effect in alcohol-preferring rats (Colombo et al., 2003a, b). Some clinical studies also have shown potential efficacy of the $GABA_B$ agonist baclofen in reducing alcohol craving and alcohol withdrawal (Addolorato et al., 2002a, b).

The endogenous opioid peptide systems have long been hypothesized to play a role in the reinforcing effects of alcohol (Reid and Hunter, 1984; Herz, 1997). Naltrexone decreases alcohol drinking and self-administration in a variety of animal models, and these results led to the clinical use of naltrexone in reducing alcohol consumption and preventing relapse (Altshuler et al., 1980; Reid and Hunter, 1984; Samson and Doyle, 1985; Froehlich et al., 1990; Weiss et al., 1990; Hubbell et al., 1991; Kornet et al., 1991; O'Malley et al., 1992; Volpicelli et al., 1992; Hyytia and Sinclair, 1993; Davidson and Amit, 1997).

Consistent with the overlap in the neuropharmacologic actions of cannabinoid CB_1 receptors and μ opioid receptors, CB_1 receptors modulate alcohol preference in mice. The high alcohol preference of young mice is reduced by a competitive CB_1 receptor antagonist, and young CB_1 receptor knockout mice show a decrease in alcohol preference (Wang et al., 2003).

Significant evidence also supports a role for the mesolimbic dopamine system in alcohol reinforcement. Systemic injections of dopamine receptor antagonists decrease responding for alcohol (Pfeffer and Samson, 1986; Rassnick et al., 1992). Alcohol self-administration increases extracellular levels of dopamine in the nucleus accumbens in non-dependent rats (Weiss et al., 1993). Such increases not only occur during the actual self-administration session but precede the self-administration session, possibly reflecting the incentive

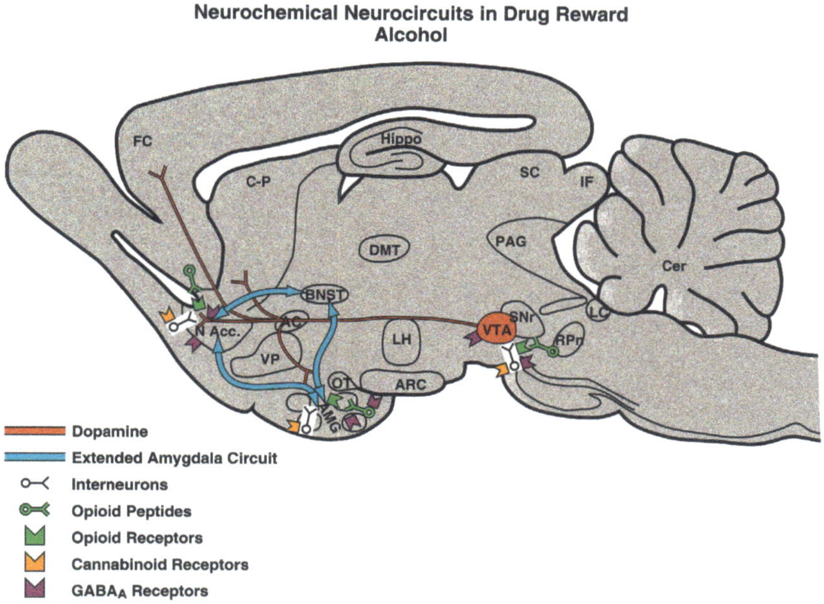

Fig. 3.4. Sagittal section through a representative rodent brain illustrating the pathways and receptor systems implicated in the acute reinforcing actions of alcohol. Alcohol activates γ-aminobutyric acid-A ($GABA_A$) receptors in the ventral tegmental area, nucleus accumbens, and amygdala via either direct actions at the $GABA_A$ receptor or through indirect release of GABA. Alcohol is hypothesized to facilitate the release of opioid peptides in the ventral tegmental area, nucleus accumbens, and central nucleus of the amygdala. Alcohol facilitates the release of dopamine (red) in the nucleus accumbens via an action in either the ventral tegmental area or nucleus accumbens. Endogenous cannabinoids and adenosine may interact with postsynaptic elements in the nucleus accumbens involving dopamine and/or opioid peptide systems. The blue arrows represent the interactions within the extended amygdala system hypothesized to play a key role in alcohol reinforcement. AC, anterior commissure; AMG, amygdala; ARC, arcuate nucleus; BNST, bed nucleus of the stria terminalis; Cer, cerebellum; C-P, caudate-putamen; FC, frontal cortex; Hippo, hippocampus; IF, inferior colliculus; LC, locus coeruleus; LH, lateral hypothalamus; N Acc, nucleus accumbens; OT, olfactory tract; PAG, periaqueductal gray; RPn, reticular pontine nucleus; SC, superior colliculus; SNr, substantia nigra pars reticulata; VP, ventral pallidum; VTA, ventral tegmental area. (Figure based on Koob, 1992. Reproduced with permission from Koob and Le Moal, 2006.)

motivational properties of cues associated with alcohol (Weiss et al., 1993). Also, while the increases in dopamine release in rats are modest, there are more robust increases (50%) observed in animals that carry the OPRM1 118G variant (Ramchandani et al., 2011). However, mesolimbic dopamine does not appear to be critical for the acute reinforcing effects of alcohol, in which lesions of the mesolimbic dopamine system fail to block operant self-administration of alcohol (Myers, 1990; Lyness and Smith, 1992; Rassnick et al., 1993c), suggesting multiple redundant sources of alcohol action but all converging on the nucleus accumbens and amygdala.

Modulation of various aspects of serotonergic transmission, including increases in the synaptic availability of serotonin with precursor loading, and blockade of serotonin reuptake can decrease alcohol intake (Sellers et al., 1992). Antagonists of several serotonin receptor subtypes can decrease alcohol self-administration. Serotonin-3 ($5\text{-}HT_3$) receptor antagonists decrease alcohol self-administration under 24-hour access conditions but are less effective in limited-access conditions (Fadda et al., 1991; Knapp and Pohorecky, 1992; Kostowski et al., 1993; Beardsley et al., 1994; McKinzie et al., 1998; but see Hodge et al., 1993) and have been effective at decreasing relapse in clinical trials (Johnson et al., 2000). $5\text{-}HT_3$ receptor antagonists block the alcohol-stimulated release of dopamine in the mesolimbic dopamine system (Campbell and McBride, 1995). $5\text{-}HT_2$ receptor antagonists, including some with both $5\text{-}HT_{1A}$ agonist and $5\text{-}HT_2$ antagonist action, can selectively decrease acute alcohol reinforcement (Roberts et al., 1998).

Thus, the neuroanatomic substrates for the reinforcing actions of alcohol may also involve neurocircuitry not only of the ventral striatum but also the extended amygdala, which forms a separate entity within the basal forebrain (Alheid and Heimer, 1988). The extended amygdala has been conceptualized to be composed of several basal forebrain structures, including the bed nucleus of the stria terminalis, central nucleus of the amygdala, sublenticular substantia innominata, and a transition zone in the medial part of the nucleus accumbens (e.g., shell; Heimer and Alheid, 1991). The very potent $GABA_A$ receptor antagonist SR 95531, when microinjected into the basal forebrain, significantly decreased responding for alcohol but not water in rats trained to lever press for alcohol (Hyytia and Koob, 1995). When the GABA antagonist was injected bilaterally into the nucleus accumbens, bed nucleus of the stria terminalis, and central nucleus of the amygdala, the most sensitive site for decreasing alcohol responding was the central nucleus of the amygdala (Hyytia and Koob, 1995). Very low doses of the dopamine antagonist fluphenazine injected into the nucleus accumbens will block alcohol self-administration at low doses (Rassnick et al., 1992; Samson et al., 1993). Injections of an opioid antagonist into the central nucleus of the amygdala also significantly reduced alcohol consumption at lower doses than for other sites, such as the nucleus accumbens or lateral ventricle (Heyser et al., 1999), suggesting a role for opioid peptides in the extended amygdala in the acute reinforcing actions of alcohol (Heyser et al., 1999). Microinjection of 5-HT_3 antagonists into the amygdala in non-dependent Wistar rats also significantly attenuated alcohol drinking in limited-access paradigms (2-hour daily sessions; Dyr and Kostowski, 1995).

NEURAL SUBSTRATES FOR THE WITHDRAWAL/NEGATIVE AFFECT STAGE ASSOCIATED WITH ALCOHOLISM

Within-system neuroadaptations that contribute to the compulsivity associated with the dark side of alcoholism

Within-system neuroadaptations to chronic drug exposure include decreases in the function of the same neurotransmitter systems in the same neurocircuits implicated in the acute reinforcing effects of drugs of abuse. One prominent hypothesis is that dopamine systems are compromised in crucial phases of the addiction cycle, such as withdrawal and protracted abstinence. This decrease in dopamine function is hypothesized to lead to decreased motivation for non-drug-related stimuli and increased sensitivity to the abused drug (Melis et al., 2005). Activation of the mesolimbic dopamine system has long been known to be critical for the acute rewarding properties of psychostimulant drugs and to be associated with the acute reinforcing effects of alcohol (Koob, 1992; McBride and Li, 1998; Nestler, 2005). However, the magnitude of the increase in dopaminergic activity produced by alcohol typically pales in comparison to psychostimulant "intoxication." For example, intravenous cocaine self-administration produces a 300% increase in extracellular dopamine (Weiss et al., 1992b) compared with ethanol (which produces a 20–50% increase in extracellular dopamine in the nucleus accumbens; Doyon et al., 2003) and heroin (which does not increase extracellular dopamine in the nucleus accumbens; Table 3.1; but see above for potential genetic differences).

However, the impact of dopamine activation may change with the development of dependence. Basal levels of dopamine activity are known to decrease during withdrawal. Decreases in activity of the mesolimbic dopamine system and decreases in serotonergic neurotransmission in the nucleus accumbens have been observed during acute alcohol withdrawal in animal studies (Rossetti et al., 1992; Weiss et al., 1992a, 1996). In dependent male Wistar rats trained to self-administer ethanol during withdrawal, the release of dopamine and serotonin was monitored by microdialysis in the nucleus accumbens at the end of a 3–5-week ethanol (8.7% w/v) liquid diet regimen, during 8 hours of withdrawal, and during renewed availability of ethanol that involved the opportunity to operantly self-administer ethanol (10% w/v) for 60 minutes, followed by unlimited access to the ethanol liquid diet. In non-dependent rats, operant ethanol self-administration increased both dopamine and serotonin release in the nucleus accumbens. Withdrawal from the chronic ethanol diet produced a progressive suppression of the release of these transmitters over the 8-hour withdrawal period. Self-administration of ethanol reinstated and maintained dopamine release at pre-withdrawal levels but failed to

Table 3.1

Effects of intravenous self-administration of D-amphetamine, cocaine, and heroin and oral self-administration of alcohol on extracellular dopamine levels in the nucleus accumbens using *in vivo* microdialysis

Drug	Increase in dopamine over baseline	References
D-Amphetamine	700%	Di Ciano et al., 1995
Cocaine	200–500%	Weiss et al., 1992b; Di Ciano et al., 1995
Alcohol	25–50%	Weiss et al., 1992a, 1996; Ramchandani et al., 2011
Heroin	<20%	Hemby et al., 1995

completely restore serotonin efflux. These findings suggested that deficits in nucleus accumbens monoamine release may contribute to the negative affective consequences of ethanol withdrawal and thereby motivate ethanol-seeking behavior in dependent subjects (Weiss et al., 1996). Similar dramatic decreases in extracellular dopamine in the nucleus accumbens, measured by microdialysis, were found in a study in which animals were tested 8 hours into ethanol withdrawal produced by chronic repeated ethanol injections of up to 5 g/kg every 6 hours for 6 consecutive days using the Majchrowicz procedure (Majchrowicz, 1975; Rossetti et al., 1999). Thus, as a result, ethanol-dependent animals may show a much greater percentage increase in dopamine release in the nucleus accumbens during ethanol self-administration during withdrawal because baseline levels of dopamine are so low during withdrawal (Weiss et al., 1996). Parallel to these studies is strong evidence that the firing of ventral tegmental area dopamine neurons is dramatically decreased during acute withdrawal from alcohol (Diana et al., 1993).

Imaging studies in drug-addicted humans have consistently shown long-lasting decreases in the numbers of dopamine D_2 receptors in alcoholics compared with controls (Volkow et al., 2002). Additionally, alcohol-dependent subjects had dramatically reduced dopamine release in the striatum response to a pharmacologic challenge with the stimulant drug methylphenidate (Volkow et al., 2007). Decreases in the number of dopamine D_2 receptors, coupled with the decrease in dopaminergic activity, in cocaine, nicotine, and alcohol abusers are hypothesized to produce a decreased sensitivity of reward circuits to stimulation by natural reinforcers (Volkow and Fowler, 2000; Martin-Solch et al., 2001). These findings suggest an overall reduction of the sensitivity of the dopamine component of reward circuitry to natural reinforcers and other drugs in drug-addicted individuals (Fig. 3.5).

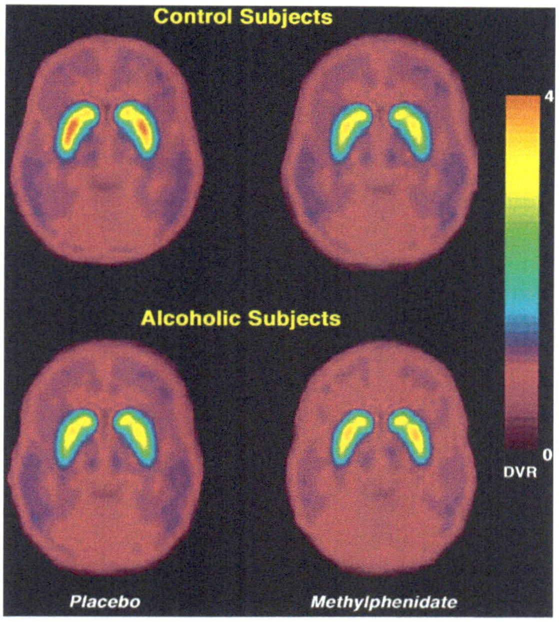

Fig. 3.5. Profound decrease in dopamine release in the striatum in detoxified alcoholics. Average distribution volume ratio (DVR) images for [^{11}C]raclopride in controls ($n = 20$) and alcoholics ($n = 20$) at the level of the striatum after placebo and after methylphenidate. Notice the decrease in specific binding (DVR) with methylphenidate and the attenuated response to methylphenidate in alcoholic subjects compared with controls. (Reproduced with permission from Volkow et al., 2007.)

Other within-system neuroadaptations under this conceptual framework could include increased sensitivity of receptor transduction mechanisms in the nucleus accumbens. Drugs of abuse have acute receptor actions that are linked to intracellular signaling pathways that may undergo adaptations with chronic treatment. In the context of chronic alcohol administration, multiple molecular mechanisms have been hypothesized to counteract the acute effects of ethanol that could be considered within-system neuroadaptations. For example, chronic ethanol decreases GABA receptor function, possibly through downregulation of the α_1 subunit (Mhatre et al., 1993; Devaud et al., 1997). Chronic ethanol also decreases the acute inhibition of adenosine reuptake (i.e., tolerance develops to the inhibition of adenosine by ethanol; Sapru et al., 1994). Perhaps more relevant to the present treatise, whereas acute ethanol activates adenylate cyclase, withdrawal from chronic ethanol decreases cyclic adenosine monophosphate response element-binding protein phosphorylation in the amygdala and is linked to decreases in the function of neuropeptide Y and anxiety-like responses observed during acute ethanol withdrawal (Chance et al., 2000; Pandey, 2004).

Between-system neuroadaptations that contribute to compulsivity associated with the dark side of alcoholism

Brain neurochemical systems involved in arousal-stress modulation may also be engaged within the neurocircuitry of the brain stress systems in an attempt to overcome the chronic presence of the perturbing drug (alcohol) and to restore normal function despite the presence of drug. The neuroanatomic entity termed the extended amygdala (Heimer and Alheid, 1991) may represent a common anatomic substrate that integrates brain arousal-stress systems with hedonic processing systems to produce the between-system opponent process elaborated above. The extended amygdala receives numerous afferents from limbic structures, such as the basolateral amygdala and hippocampus, and sends efferents to the medial part of the ventral pallidum and a large projection to the lateral hypothalamus, thus further defining the specific brain areas that interface classical limbic (emotional) structures with the extrapyramidal motor system (Alheid et al., 1995). The extended amygdala has long been hypothesized to play a key role not only in fear conditioning (Le Doux, 2000) but also in the emotional component of pain processing (Neugebauer et al., 2004).

The brain stress system mediated by CRF systems in both the extended amygdala and hypothalamic–pituitary–adrenal axis is dysregulated by chronic administration of all major drugs with dependence or abuse potential, with a common response of elevated adrenocorticotropic hormone, corticosterone, and amygdala CRF during acute withdrawal from chronic drug administration (Rivier et al., 1984; Koob et al., 1994; Merlo-Pich et al., 1995; Delfs et al., 2000; Rasmussen et al., 2000; Olive et al., 2002; Koob, 2008a).

More specifically, alcohol withdrawal reliably produces anxiety-like responses in animal models that can be reversed by CRF receptor antagonists (Koob, 2008a). Ethanol withdrawal produces anxiety-like behavior that is reversed by intracerebroventricular administration of CRF_1/CRF_2 peptidergic antagonists (Baldwin et al., 1991), small-molecule CRF_1 antagonists (Knapp et al., 2004; Overstreet et al., 2004; Funk et al., 2007), and intracerebral administration of a peptidergic CRF_1/CRF_2 antagonist into the central nucleus of the amygdala (Rassnick et al., 1993b). CRF antagonists injected intracerebroventricularly or systemically also block the potentiated anxiety-like responses to stressors observed during protracted abstinence from chronic ethanol (Valdez et al., 2003; Breese et al., 2005; Sommer et al., 2008).

Perhaps more relevant to the present thesis are studies that show that intermittent alcohol exposure facilitates withdrawal anxiety-like responses and that administration of drug treatments during withdrawal from the first and second alcohol cycles blocked this sensitization of withdrawal (Knapp et al., 2004). Diazepam, flumazenil (a $GABA_A$ receptor partial agonist), and baclofen (a $GABA_B$ receptor agonist) blocked the sensitization of withdrawal, consistent with a within-system neuroadaptation (Knapp et al., 2004, 2005, 2007). However, a CRF_1 antagonist also prevented the sensitization of withdrawal-induced anxiety, consistent with a between-system neuroadaptation (Overstreet et al., 2004, 2005). These results are consistent with the observation that a prolonged history of alcohol exposure produces persistent upregulation of both CRF and CRF_1 receptors in the brain (Zorrilla et al., 2001; Sommer et al., 2008; Roberto et al., 2010).

The ability of CRF antagonists to block the anxiogenic-like and aversive-like motivational effects of drug withdrawal would predict motivational effects of CRF antagonists in animal models of extended access to drugs. A particularly dramatic example of the motivational effects of CRF in dependence can be observed in animal models of ethanol self-administration in dependent animals. During ethanol withdrawal, extrahypothalamic CRF systems become hyperactive, with an increase in extracellular CRF within the central nucleus of the amygdala and bed nucleus of the stria terminalis in dependent rats (Merlo-Pich et al., 1995; Olive et al., 2002; Funk et al., 2006). The dysregulation of brain CRF systems is hypothesized to underlie not only the enhanced anxiety-like behaviors but also the enhanced ethanol self-administration associated with ethanol withdrawal. Supporting this hypothesis, the subtype-non-selective CRF receptor antagonists α-helical $CRF_{9\text{-}41}$ and D-Phe $CRF_{12\text{-}41}$ (intracerebroventricular administration) reduced ethanol self-administration in dependent animals (Valdez et al., 2004). Exposure to repeated cycles of chronic ethanol vapor produced substantial increases in ethanol intake in rats, during both acute withdrawal and protracted abstinence (2 weeks post acute withdrawal; Rimondini et al., 2002; O'Dell et al., 2004). Intracerebroventricular administration of a CRF_1/CRF_2 antagonist blocked the dependence-induced increase in ethanol self-administration during both acute withdrawal and protracted abstinence (Valdez et al., 2004). When administered directly into the central nucleus of the amygdala, a CRF_1/CRF_2 antagonist blocked ethanol self-administration in ethanol-dependent rats (Funk et al., 2006). Systemic injections of small-molecule CRF_1 antagonists also blocked the increased ethanol intake associated with acute withdrawal (Funk et al., 2007; Gehlert et al., 2007). These data suggest an important role for CRF, primarily within the central nucleus of the amygdala, in mediating the increased self-administration associated with dependence. Consistent with the sensitization of the withdrawal response associated with repeated alcohol exposure, a CRF antagonist administered during repeated withdrawal also blocked the development of excessive drinking during withdrawal (Roberto et al., 2010).

Although less well developed, evidence supports a role for noradrenergic systems in the extended amygdala in the negative motivational state and increased self-administration associated with dependence. Substantial evidence has accumulated that suggests that, in animals and humans, central noradrenergic systems are activated during acute withdrawal from ethanol. Alcohol withdrawal in humans is associated with activation of noradrenergic function, and the signs and symptoms of alcohol withdrawal in humans are blocked by postsynaptic β-adrenergic blockade (Romach and Sellers, 1991). Alcohol withdrawal signs are also blocked in animals by administration of α_1 antagonists and β-adrenergic antagonists and selective blockade of norepinephrine synthesis (Trzaskowska and Kostowski, 1983). In dependent rats, the α_1 antagonist prazosin selectively blocked the increased drinking associated with acute withdrawal (Walker et al., 2008). Thus, converging data suggest that noradrenergic neurotransmission is enhanced during ethanol withdrawal and that noradrenergic functional antagonists can block aspects of ethanol withdrawal.

Dynorphin, an opioid peptide that binds to κ opioid receptors, has long been known to show activation with chronic administration of psychostimulants and opioids (Nestler, 2004; Koob, 2008a), and κ opioid receptor agonists produce aversive effects in animals and humans (Mucha and Herz, 1985; Pfeiffer et al., 1986). Although κ agonists suppress non-dependent drinking, possibly via aversive stimulus effects (Wee and Koob, 2010), κ opioid antagonists block the excessive drinking associated with ethanol withdrawal and dependence (Holter et al., 2000; Walker and Koob, 2008), and this effect may be mediated by the shell of the nucleus accumbens (Nealey et al., 2011). Recently, some have argued that the effects of CRF in producing negative emotional states are mediated by activation of κ opioid systems (Land et al., 2008). However, κ receptor activation can activate CRF systems in the spinal cord (Song and Takemori, 1992), and there is pharmacologic evidence that dynorphin systems can activate the CRF system. A CRF_1 antagonist blocked κ agonist-induced reinstatement of cocaine seeking in squirrel monkeys (Valdez et al., 2007).

The dynamic nature of the brain stress system response to challenge is illustrated by the pronounced interaction between central nervous system CRF systems and central nervous system norepinephrine systems. Conceptualized as a feedforward system at multiple levels of the pons and basal forebrain, CRF activates norepinephrine, and norepinephrine in turn activates CRF (Koob, 1999). Much pharmacologic, physiologic, and anatomic evidence supports an important role for a CRF–norepinephrine interaction in the region of the locus coeruleus in response to stressors (Valentino et al., 1991, 1993; Van Bockstaele et al., 1998). However, norepinephrine also stimulates CRF release in the paraventricular nucleus of the hypothalamus (Alonso et al., 1986), bed nucleus of the stria terminalis, and central nucleus of the amygdala. Such feedforward systems were further hypothesized to have powerful functional significance for mobilizing an organism's response to environmental challenge, but such a mechanism may be particularly vulnerable to pathology (Koob, 1999).

Neuropeptide Y is a neuropeptide with dramatic anxiolytic-like properties localized to the amygdala and has been hypothesized to have effects opposite to CRF in the negative motivational state of withdrawal from drugs of abuse (Heilig and Koob, 2007). Significant evidence suggests that activation of neuropeptide Y in the central nucleus of the amygdala can block the motivational aspects of dependence associated with chronic ethanol administration. Neuropeptide Y administered intracerebroventricularly blocked the increased drug intake associated with ethanol dependence (Thorsell et al., 2005a, b). Injection of neuropeptide Y directly into the central nucleus of the amygdala (Gilpin et al., 2008) and viral vector-enhanced expression of neuropeptide Y in the central nucleus of the amygdala also blocked the increased drug intake associated with ethanol dependence (Thorsell et al., 2007). At the cellular level, neuropeptide Y, like CRF_1 antagonists, blocks the increase in GABA release in the central nucleus of the amygdala produced by ethanol and also, when administered chronically, blocks the transition to excessive drinking with the development of dependence (Gilpin et al., 2011).

Thus, acute withdrawal from drugs increases CRF in the central nucleus of the amygdala, which has motivational significance for the anxiety-like effects of acute withdrawal from alcohol and the increased drug intake associated with dependence. Acute withdrawal may also increase the release of norepinephrine in the bed nucleus of the stria terminalis and dynorphin in the nucleus accumbens, both of which may contribute to the negative emotional state associated with dependence. Decreased activity of neuropeptide Y in the central nucleus of the amygdala may contribute to the anxiety-like state associated with ethanol dependence. Activation of brain stress systems (CRF, norepinephrine, dynorphin) combined with inactivation of brain antistress systems (neuropeptide Y) elicits powerful emotional dysregulation in the extended amygdala. Such dysregulation of emotional processing may be a significant contribution to the between-system opponent processes that help maintain dependence and set the stage for more prolonged state changes in emotionality such as in protracted abstinence.

Notably, multiple detoxifications and reinstatement to dependence in humans can "kindle" withdrawal and produce less responsiveness to treatments (Malcolm et al., 2000). Blockade of the brain stress systems, however, can block the development of "kindling" of anxiety-like responses in animal models, lending support to the key role of the brain stress systems in the kindling phenomenon (Breese et al., 2005).

In summary, two processes are hypothesized to form the neurobiologic basis for the withdrawal/negative

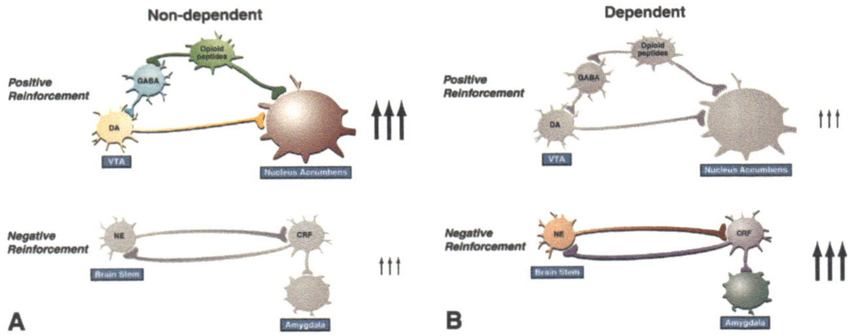

Fig. 3.6. Neurocircuitry associated with the acute positive reinforcing effects of drugs of abuse (**A**) and the negative reinforcement of dependence (**B**) and how it changes in the transition from non-dependent drug taking to dependent drug taking. Key elements of the reward circuit are dopamine (DA) and opioid peptide neurons that intersect at both the ventral tegmental area (VTA) and the nucleus accumbens and are activated during initial use and the early binge/intoxication stage. Key elements of the stress circuit are corticotropin-releasing factor (CRF) and noradrenergic (norepinephrine [NE]) neurons that converge on γ-aminobutyric acid (GABA) interneurons in the central nucleus of the amygdala that are activated during the development of dependence. (Reproduced with permission from Koob and Le Moal, 2008.)

affect stage: loss of function in the reward systems (within-system neuroadaptation) and recruitment of the brain stress systems (between-system neuroadaptation; Koob and Bloom, 1988; Koob and Le Moal, 1997; Fig. 3.6). As dependence and withdrawal develop, brain stress systems, such as CRF, norepinephrine, and dynorphin, are recruited, producing aversive or stress-like states (Aston-Jones et al., 1999; Nestler, 2001; Koob, 2003). The combination of decreases in reward neurotransmitter function and recruitment of brain stress systems provides a powerful motivation for re-engaging in drug taking and drug seeking.

NEURAL SUBSTRATES FOR EXECUTIVE FUNCTION DEFICITS ASSOCIATED WITH ALCOHOLISM

The preoccupation/anticipation stage of the addiction cycle has long been hypothesized to be a key element of relapse in humans and defines addiction as a chronic relapsing disorder. Although often linked to the construct of craving, craving per se has been difficult to measure in human clinical studies (Tiffany et al., 2000) and often does not correlate with relapse. Nevertheless, the stage of the addiction cycle where the individual reinstates drug-seeking behavior after abstinence remains a challenging focus for neurobiologic mechanisms and medications development for treatment.

Animal models of craving can be divided into two domains: alcohol seeking induced by stimuli paired with drug taking (reward craving) and alcohol seeking induced by an acute stressor or a state of stress (relief craving; Heinz et al., 2003; Table 3.2). Animal models of reward craving involve the use of alcohol-primed reinstatement and cue-induced reinstatement (Martin-Fardon and Weiss, 2013). Animal models of relief craving involve stress-induced reinstatement in animals that have acquired drug self-administration and then

Table 3.2

Drug craving

Drug craving	"Drug craving is the desire for the previously experienced effects of a psychoactive substance. This desire can become compelling and can increase in the presence of both internal and external cues, particularly with perceived substance availability. It is characterized by an increased likelihood of drug-seeking behaviour and, in humans, or drug-related thoughts" (United Nations International Drug Control Programme, 1992)
Reward craving	• Induced by stimuli that have been paired with drug self-administration such as environmental cues • Termed conditioned positive reinforcement in experimental psychology • Animal model: cue-induced reinstatement where a cue previously paired with access to a drug reinstates responding for a lever that has been extinguished
Relief craving	• State of protracted abstinence in drug-dependent individuals weeks after acute withdrawal • Conceptualized as a state change characterized by anxiety and dysphoria • Animal model: residual hypersensitivity to states of stress and environmental stressors that lead to reinstatement to drug-seeking behavior

have been subjected to extinction of responding for the drug (Shippenberg and Koob, 2002).

Most evidence from animal studies suggests that drug-induced reinstatement is localized to the medial prefrontal cortex/ventral striatum circuit mediated by the neurotransmitter glutamate (McFarland and Kalivas, 2001). In contrast, neuropharmacologic and neurobiologic studies that use animal models for cue-induced reinstatement involve the basolateral amygdala as a critical substrate with a possible feedforward mechanism possibly through the same prefrontal cortex system involved in drug-induced reinstatement (Weiss et al., 2001; Everitt and Wolf, 2002). Neurotransmitter systems involved in drug-induced reinstatement involve a glutamatergic projection from the frontal cortex to the nucleus accumbens that is modulated by dopamine activity in the frontal cortex. Cue-induced reinstatement involves dopamine modulation in the basolateral amygdala and a glutamatergic projection to the nucleus accumbens from both the basolateral amygdala and ventral subiculum (Vorel et al., 2001; Everitt and Wolf, 2002). In contrast, stress-induced reinstatement of drug-related responding in animal models appears to depend on the activation of both CRF and norepinephrine in elements of the extended amygdala (Shalev et al., 2002; Shaham et al., 2003).

Protracted abstinence from alcohol, as defined in the rat, spans a period when acute physical withdrawal has disappeared, but changes in behavior persist. Elevations in alcohol intake over baseline and increased stress responsivity persists 2–8 weeks postwithdrawal from chronic alcohol (Roberts et al., 2000; Valdez et al., 2002). Rats tested on the elevated plus maze 3–5 weeks postwithdrawal did not show an anxiogenic-like response at baseline, but an anxiogenic-like response was provoked by mild restraint stress only in rats with a history of alcohol dependence. Protracted abstinence, largely described in alcohol dependence models, appears to involve overactive glutamatergic and CRF systems (Valdez et al., 2002; De Witte et al., 2005). This stress-induced anxiogenic-like response was reversed by a competitive CRF antagonist (Valdez et al., 2003). The increased self-administration of alcohol observed during protracted abstinence also was blocked by a competitive CRF antagonist. In rats trained to self-administer alcohol and made dependent and subsequently retested postwithdrawal, a CRF antagonist dose-dependently reduced self-administration only in rats with a history of dependence (Valdez et al., 2002). Rats similarly made dependent with chronic continuous exposure to ethanol showed anxiety-like behavior on the elevated plus maze at 4 weeks postwithdrawal (Valdez et al., 2002). These results suggest that brain CRF systems remain hyperactive during protracted abstinence, and this hyperactivity has motivational significance for excessive alcohol drinking.

Executive control over incentive salience is essential to maintain goal-directed behavior and the flexibility of stimulus–response associations. The prefrontal cortex sends glutamatergic projections directly to mesocortical dopamine neurons in the ventral tegmental area, exerting excitatory control on dopamine in the prefrontal cortex. Thus, the prefrontal cortex is in a good position to inhibit incentive salience and suppress conditioned behavior when a salient cue is presented to the subject. Consistent with this hypothesis, appetitive stimuli activate the prefrontal cortex, and lesions of the prefrontal cortex induce impulsivity (Jentsch and Taylor, 1999). Withdrawal from alcohol is associated with increased glutamate release in the nucleus accumbens (Dahchour et al., 1998). In studies with psychostimulants, cue-induced reinstatement of drug seeking-behavior induces a reliable increase in prefrontal activity and glutamate release in the nucleus accumbens (Reissner and Kalivas, 2010). Increased activity of the prefrontal-glutamatergic system during relapse may be hypothesized to elicit a dramatic glutamatergic response that may mediate craving-like responses during the preoccupation-anticipation stage.

Behavioral procedures have also been developed to reinstate alcohol self-administration with previously neutral stimuli that have been paired with alcohol self-administration or that predict alcohol self-administration (Katner and Weiss, 1999; Katner et al., 1999; Martin-Fardon et al., 2000; Ciccocioppo et al., 2001). Rats come to associate a specific olfactory stimulus with the availability of alcohol, and that stimulus then will reinstate responding in animals subjected to extinction. Consistent with the well-established conditioned cue reactivity in human alcoholics, the motivating effects of alcohol-related stimuli are highly resistant to extinction and retain their efficacy to elicit alcohol-seeking behavior after more than 1 month of repeated testing (Katner et al., 1999; Ciccocioppo et al., 2001). This reinstatement was blocked by systemic administration of naltrexone and a selective μ and δ opioid receptor antagonist (Ciccocioppo et al., 2002), results consistent with human studies that showed that opioid antagonists may blunt the urge to drink elicited by presentation of alcohol-related cues in alcoholics (Monti et al., 1999). Dopamine D_1 and D_2 receptor antagonists also blocked cue-induced reinstatement (Liu and Weiss, 2002a). Nitric oxide synthesis inhibition by *N*-ω-nitro-L-arginine methyl ester also blocks the conditioned reinstatement by alcohol cues, suggesting a glutamatergic component to cue-induced reinstatement to alcohol, as hypothesized above (Liu and Weiss, 2004).

Stress exposure also can reinstate responding for alcohol in rats previously extinguished, and this stress-induced reinstatement is blocked by the competitive CRF antagonist D-Phe $CRF_{12\text{-}41}$. In fact, the CRF antagonist blocks stress-induced reinstatement but not cue-induced reinstatement, and naltrexone blocks cue-induced reinstatement but not stress-induced reinstatement, suggesting two independent neuropharmacologic routes to reinstatement and by extrapolation relapse (Ciccocioppo et al., 2002, 2003; Liu and Weiss, 2002b; Fig. 3.7).

Human imaging studies reveal similar circuit dysregulation during the preoccupation/anticipation stage in alcoholism, as delineated in animal models. Decreased frontal cortex activity parallels deficits in executive function in neuropsychologically challenging tasks in alcoholics with and without Wernicke–Korsakoff's syndrome (for review, see Sullivan and Pfefferbaum, 2005; Goldstein and Volkow, 2011; Oscar-Berman, 2012). In alcoholics, there are impairments in the maintenance of spatial information, disruption of decision making, and impairments in behavioral inhibition. Such frontal cortex-derived executive function disorders in addiction have been linked to deficits in the ability of behavioral treatments to effect recovery from alcoholism postdetoxification (DiClemente, 2007; Crews and Boettiger, 2009). As elegantly outlined by Sullivan and Pfefferbaum (2005), these changes in frontal cortex executive function can sometimes be long-term, sometimes transient, and sometimes compensatory (Sullivan and Pfefferbaum, 2005). Thus, deficits in the prefrontal cortical control of incentive salience may represent a key mechanism to explain individual differences in the vulnerability to addiction, and excessive attribution of incentive salience to drug-related cues as well as a residual hypersensitivity of the brain stress systems may perpetuate excessive drug intake, compulsive behavior, and relapse.

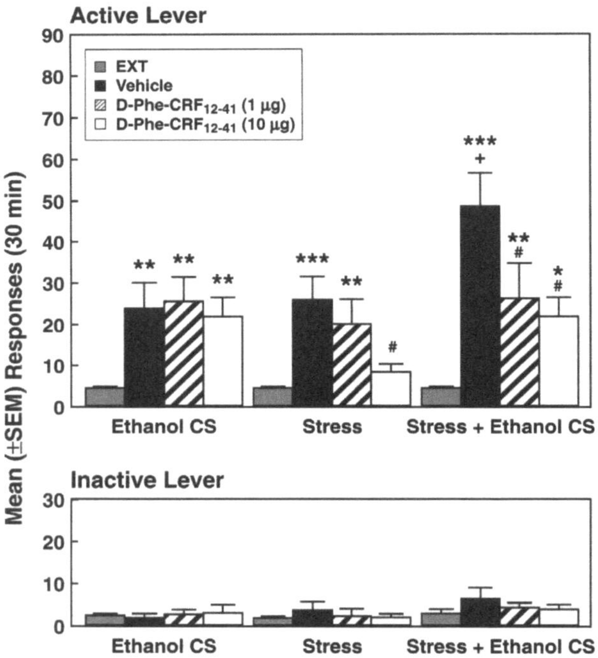

Fig. 3.7. Effects of the corticotropin-releasing factor antagonist D-Phe $CRF_{12\text{-}41}$ on reinstatement after extinction (EXT) in stress, ethanol-conditioned stimulus (CS), and stress + ethanol CS conditions in rats. D-Phe $CRF_{12\text{-}41}$ antagonized footshock- but not ethanol CS-induced responding and partially reversed reinstatement in the stress + ethanol CS condition, but responding remained significantly above extinction levels. For clarity, EXT data have been collapsed across the three groups in each reinstatement condition. $^{*}p<0.05$, $^{**}p<0.01$, $^{***}p<0.001$, significant differences from respective baseline performance. The plus sign (+) indicates statistically significant difference from stress and ethanol CS vehicle control conditions. Hash signs (#) indicate statistically significant difference from vehicle. (Reproduced with permission from Liu and Weiss, 2002b.)

COMPULSIVITY IN ALCOHOLISM: AN ALLOSTATIC VIEW

Compulsivity in alcoholism can derive from neurocircuitry changes at all three stages of the addiction cycle (Fig. 3.8). In the binge/intoxication stage, such changes involve neurocircuits that include enhanced incentive salience (Robinson and Berridge, 1993) and engagement of habit function (Everitt et al., 2008). In the withdrawal/negative affect stage, such changes involve neurocircuits involved in negative emotional states. In the preoccupation/anticipation stage, such changes involve impairment in executive function (Jentsch and Taylor, 1999). However, negative emotional states may strongly impact compulsivity not only by directly driving negative reinforcement but also by enhancing the value of incentive salience, enhancing the value of habit expression, or exacerbating impairments in executive function.

Thus, the overall conceptual theme argued herein is that drug addiction represents a break with homeostatic brain-regulatory mechanisms that regulate the emotional state of the animal. The dysregulation of emotion begins with the binge and subsequent acute withdrawal but leaves a residual neuroadaptive trace that allows rapid "re-addiction" even months and years after detoxification and abstinence. Thus, the emotional dysregulation of alcohol addiction represents more than simple homeostatic dysregulation of hedonic function – it also represents a dynamic break with the homeostasis of this system that has been termed allostasis (Koob, 2003).

Allostasis, originally conceptualized to explain persistent morbidity of arousal and autonomic function, can be defined simply as "stability through change"

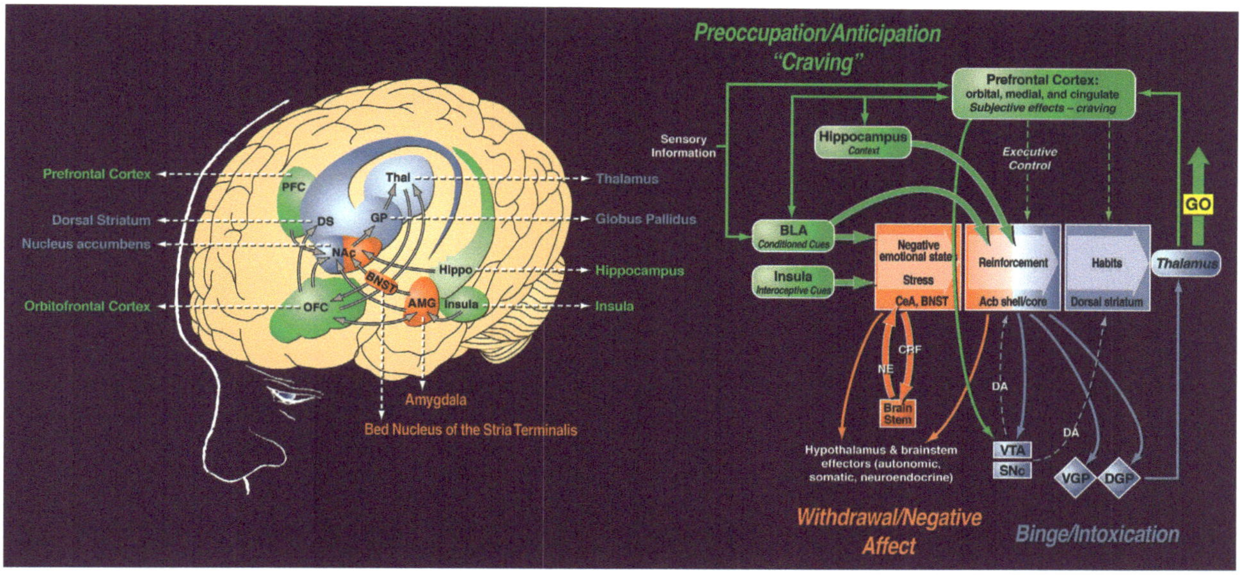

Fig. 3.8. Neurocircuitry schematic illustrating the combination of neuroadaptations in the brain circuitry for the three stages of the addiction cycle that drive drug-seeking behavior in the addicted state. Notice the activation of the ventral striatum/dorsal striatum/extended amygdala driven by cues via hippocampal and basolateral amygdala and stress via the insula. The frontal cortex system is compromised, producing deficits in executive function and contributing to the incentive salience of drugs compared to natural reinforcers. Dopamine systems are compromised, and brain stress systems, such as corticotropin-releasing factor (CRF), are activated to further reset the salience of drugs and drug-related stimuli in the context of an aversive dysphoric state. BLA, basolateral amygdala; Acb, nucleus accumbens; DA, dopamine; VTA, ventral tegmental area; SNc, substantia nigra pars compacta; VGP, ventromedial globus pallidus; NE, norepinephrine. (Modified with permission from Koob et al., 2008.)

(Sterling and Eyer, 1988). Allostasis involves a changed set point with continuous reevaluation of need and continuous readjustment of all parameters toward new set points. The set point in question here is a decreased negative emotional state threshold. An allostatic state can be defined as a state of chronic deviation of the emotional equilibrium from its normal (homeostatic) operating level. The argument is that alcoholism reflects largely a movement to an allostatic state, often before sufficient physical pathology to meet the criteria of allostatic load (McEwen, 2000).

The allostatic state can be envisioned to involve all three stages of the addiction cycle. Increased reactivity of the stress/hypothalamic–pituitary–adrenal axis module may be critical in the initiation of alcohol intake and for the maintenance of drug intake. Activation of the hypothalamic–pituitary–adrenal axis can potentiate the reinforcing effects of drugs (Piazza and Le Moal, 1998). However, this activation can in turn drive amygdala CRF, further exacerbating the development of negative emotional states (Koob and Kreek, 2007). Although the initial deficit in a specific functional circuit that drives excessive drinking might be specific to one stage of the addiction cycle, as the transition to addiction progresses, an individual is ultimately likely to show a progressive and generalized loss of control over many, if not all, systems.

Indeed, it is hypothesized that repeated challenges, such as with repeated alcohol binges, will lead to attempts of the brain via molecular, cellular, and neurocircuitry changes to maintain stability but at a cost (i.e., the allostatic state). This state represents a combination of chronic elevation of the reward set point fueled by decreased function of reward circuits and recruitment of brain stress systems, both of which lead to the compulsivity of alcohol seeking and alcohol taking. How these systems are modulated by other known brain emotional systems localized to the basal forebrain, where the ventral striatum and extended amygdala project to convey emotional valence, how the dysregulation of brain emotional systems impacts on the cognitive domain linked to impairments in executive function, and how individuals differ at the molecular-genetic level of analysis to convey loading on these circuits remain challenges for future research (George and Koob, 2010; Koob, 2013).

ACKNOWLEDGMENTS

The author would like to thank Michael Arends and Mellany Santos for assistance with manuscript preparation and editing. Research was supported by National Institutes of Health grants AA006420, AA020608, and AA008459 from the National Institute on Alcohol

Abuse and Alcoholism and the Pearson Center for Alcoholism and Addiction Research. This is publication number 24030 from The Scripps Research Institute.

References

Addolorato G, Caputo F, Capristo E et al. (2002a). Baclofen efficacy in reducing alcohol craving and intake: a preliminary double-blind randomized controlled study. Alcohol Alcohol 37: 504–508.

Addolorato G, Caputo F, Capristo E et al. (2002b). Rapid suppression of alcohol withdrawal syndrome by baclofen. Am J Med 112: 226–229.

Alheid GF, Heimer L (1988). New perspectives in basal forebrain organization of special relevance for neuropsychiatric disorders: the striatopallidal, amygdaloid, and corticopetal components of substantia innominata. Neuroscience 27: 1–39.

Alheid GF, De Olmos JS, Beltramino CA (1995). Amygdala and extended amygdala. In: G Paxinos (Ed.), The Rat Nervous System. Academic Press, San Diego, pp. 495–578.

Allen TJ, Moeller FG, Rhoades HM et al. (1998). Impulsivity and history of drug dependence. Drug Alcohol Depend 50: 137–145.

Alonso G, Szafarczyk A, Balmefrezol M et al. (1986). Immunocytochemical evidence for stimulatory control by the ventral noradrenergic bundle of parvocellular neurons of the paraventricular nucleus secreting corticotropin-releasing hormone and vasopressin in rats. Brain Res 397: 297–307.

Altshuler HL, Phillips PE, Feinhandler DA (1980). Alteration of ethanol self-administration by naltrexone. Life Sci 26: 679–688.

American Psychiatric Association (2013). Diagnostic and Statistical Manual of Mental Disorders. 5th edn. American Psychiatric Publishing, Washington, DC.

Aston-Jones G, Delfs JM, Druhan J et al. (1999). The bed nucleus of the stria terminalis: a target site for noradrenergic actions in opiate withdrawal. In: JF McGinty (Ed.), Advancing from the Ventral Striatum to the Extended Amygdala: Implications for Neuropsychiatry and Drug Abuse. Annals of the New York Academy of Sciences. Vol. 877. New York Academy of Sciences, New York, pp. 486–498.

Baker TB, Piper ME, McCarthy DE et al. (2004). Addiction motivation reformulated: an affective processing model of negative reinforcement. Psychol Rev 111: 33–51.

Balakleevsky A, Colombo G, Fadda F et al. (1990). Ro 19-4603, a benzodiazepine receptor inverse agonist, attenuates voluntary ethanol consumption in rats selectively bred for high ethanol preference. Alcohol Alcohol 25: 449–452.

Baldwin HA, Rassnick S, Rivier J et al. (1991). CRF antagonist reverses the "anxiogenic" response to ethanol withdrawal in the rat. Psychopharmacology (Berl) 103: 227–232.

Beardsley PM, Lopez OT, Gullikson G (1994). Serotonin $5\text{-}HT_3$ antagonists fail to affect ethanol self-administration of rats. Alcohol 11: 389–395.

Becker HC, Lopez MF (2004). Increased ethanol drinking after repeated chronic ethanol exposure and withdrawal experience in C57BL/6 mice. Alcohol Clin Exp Res 28: 1829–1838.

Breese GR, Overstreet DH, Knapp DJ et al. (2005). Prior multiple ethanol withdrawals enhance stress-induced anxiety-like behavior: inhibition by CRF_1- and benzodiazepine-receptor antagonists and a $5\text{-}HT_{1a}$-receptor agonist. Neuropsychopharmacology 30: 1662–1669.

Brown G, Jackson A, Stephens DN (1998). Effects of repeated withdrawal from chronic ethanol on oral self-administration of ethanol on a progressive ratio schedule. Behav Pharmacol 9: 149–161.

Campbell AD, McBride WJ (1995). Serotonin-3 receptor and ethanol-stimulated dopamine release in the nucleus accumbens. Pharmacol Biochem Behav 51: 835–842.

Chance WT, Sheriff S, Peng F et al. (2000). Antagonism of NPY-induced feeding by pretreatment with cyclic AMP response element binding protein antisense oligonucleotide. Neuropeptides 34: 167–172.

Chu K, Koob GF, Cole M et al. (2007). Dependence-induced increases in ethanol self-administration in mice are blocked by the CRF_1 receptor antagonist antalarmin and by CRF_1 receptor knockout. Pharmacol Biochem Behav 86: 813–821.

Ciccocioppo R, Angeletti S, Weiss F (2001). Long-lasting resistance to extinction of response reinstatement induced by ethanol-related stimuli: role of genetic ethanol preference. Alcohol Clin Exp Res 25: 1414–1419.

Ciccocioppo R, Martin-Fardon R, Weiss F (2002). Effect of selective blockade of μ or δ opioid receptors on reinstatement of alcohol-seeking behavior by drug-associated stimuli in rats. Neuropsychopharmacology 27: 391–399.

Ciccocioppo R, Lin D, Martin-Fardon R et al. (2003). Reinstatement of ethanol-seeking behavior by drug cues following single versus multiple ethanol intoxication in the rat: effects of naltrexone. Psychopharmacology (Berl) 168: 208–215.

Colombo G, Serra S, Brunetti G et al. (2003a). Suppression by baclofen of alcohol deprivation effect in Sardinian alcohol-preferring (sP) rats. Drug Alcohol Depend 70: 105–108.

Colombo G, Vacca G, Serra S et al. (2003b). Baclofen suppresses motivation to consume alcohol in rats. Psychopharmacology (Berl) 167: 221–224.

Crews FT, Boettiger CA (2009). Impulsivity, frontal lobes and risk for addiction. Pharmacol Biochem Behav 93: 237–247.

Dahchour A, de Witte P, Bolo N et al. (1998). Central effects of acamprosate: Part 1. Acamprosate blocks the glutamate increase in the nucleus accumbens microdialysate in ethanol withdrawn rats. Psychiatry Res 82: 107–114.

Davidson D, Amit Z (1997). Naltrexone blocks acquisition of voluntary ethanol intake in rats. Alcohol Clin Exp Res 21: 677–683.

de Witte P, Littleton J, Parot P et al. (2005). Neuroprotective and abstinence-promoting effects of acamprosate: elucidating the mechanism of action. CNS Drugs 19: 517–537.

Delfs JM, Zhu Y, Druhan JP et al. (2000). Noradrenaline in the ventral forebrain is critical for opiate withdrawal-induced aversion. Nature 403: 430–434.

Devaud LL, Fritschy JM, Sieghart W et al. (1997). Bidirectional alterations of $GABA_A$ receptor subunit pepetide levels in rat cortex during chronic ethanol consumption and withdrawal. J Neurochem 69: 126–130.

Diana M, Pistis M, Carboni S et al. (1993). Profound decrement of mesolimbic dopaminergic neuronal activity during ethanol withdrawal syndrome in rats: electrophysiological and biochemical evidence. Proc Natl Acad Sci U S A 90: 7966–7969.

Di Ciano P, Coury A, Depoortere RY et al. (1995). Comparison of changes in extracellular dopamine concentrations in the nucleus accumbens during intravenous self-administration of cocaine or D-amphetamine. Behav Pharmacol 6: 311–322.

DiClemente CC (2007). Mechanisms, determinants and processes of change in the modification of drinking behavior. Alcohol Clin Exp Res 31: 13s–20s.

Doyon WM, York JL, Diaz LM et al. (2003). Dopamine activity in the nucleus accumbens during consummatory phases of oral ethanol self-administration. Alcohol Clin Exp Res 27: 1573–1582.

Dyr W, Kostowski W (1995). Evidence that the amygdala is involved in the inhibitory effects of 5-HT_3 receptor antagonists on alcohol drinking in rats. Alcohol 12: 387–391.

Epping-Jordan MP, Watkins SS, Koob GF et al. (1998). Dramatic decreases in brain reward function during nicotine withdrawal. Nature 393: 76–79.

Everitt BJ, Wolf ME (2002). Psychomotor stimulant addiction: a neural systems perspective. J Neurosci 22: 3312–3320.

Everitt BJ, Belin D, Economidou D et al. (2008). Neural mechanisms underlying the vulnerability to develop compulsive drug-seeking habits and addiction. Philos Trans R Soc Lond [Biol] 363: 3125–3135.

Fadda F, Garau B, Marchei F et al. (1991). MDL 72222, a selective 5-HT_3 receptor antagonist, suppresses voluntary ethanol consumption in alcohol-preferring rats. Alcohol Alcohol 26: 107–110.

Fidler TL, Clews TW, Cunningham CL (2006). Reestablishing an intragastric ethanol self-infusion model in rats. Alcohol Clin Exp Res 30: 414–428.

Fillmore MT, Rush CR (2002). Impaired inhibitory control of behavior in chronic cocaine users. Drug Alcohol Depend 66: 265–273.

Finn DA, Snelling C, Fretwell AM et al. (2007). Increased drinking during withdrawal from intermittent ethanol exposure is blocked by the CRF receptor antagonist D-Phe-CRF(12-41). Alcohol Clin Exp Res 31: 939–949.

Froehlich JC, Harts J, Lumeng L et al. (1990). Naloxone attenuates voluntary ethanol intake in rats selectively bred for high ethanol preference. Pharmacol Biochem Behav 35: 385–390.

Funk CK, O'Dell LE, Crawford EF et al. (2006). Corticotropin-releasing factor within the central nucleus of the amygdala mediates enhanced ethanol self-administration in withdrawn, ethanol-dependent rats. J Neurosci 26: 11324–11332.

Funk CK, Zorrilla EP, Lee M-J et al. (2007). Corticotropin-releasing factor 1 antagonists selectively reduce ethanol self-administration in ethanol-dependent rats. Biol Psychiatry 61: 78–86.

Gardner EL, Vorel SR (1998). Cannabinoid transmission and reward-related events. Neurobiol Dis 5: 502–533.

Gatto GJ, Grant KA (1997). Attenuation of the discriminative stimulus effects of ethanol by the benzodiazepine partial inverse agonist Ro 15-4513. Behav Pharmacol 8: 139–146.

Gehlert DR, Cippitelli A, Thorsell A et al. (2007). 3-(4-Chloro-2-morpholin-4-yl-thiazol-5-yl)-8-(1-ethylpropyl)-2,6-dimethyl-imidazo[1,2-*b*]pyridazine: a novel brain-penetrant, orally available corticotropin-releasing factor receptor 1 antagonist with efficacy in animal models of alcoholism. J Neurosci 27: 2718–2726.

George O, Koob GF (2010). Individual differences in prefrontal cortex function and the transition from drug use to drug dependence. Neurosci Biobehav Rev 35: 232–247.

Gilpin NW, Misra K, Koob GF (2008). Neuropeptide Y in the central nucleus of the amygdala suppresses dependence-induced increases in alcohol drinking. Pharmacol Biochem Behav 90: 475–480.

Gilpin NW, Smith AD, Cole M et al. (2009). Operant behavior and alcohol levels in blood and brain of alcohol-dependent rats. Alcohol Clin Exp Res 33: 2113–2123.

Gilpin NW, Misra K, Herman MA et al. (2011). Neuropeptide Y opposes alcohol effects on gamma-aminobutyric acid release in amygdala and blocks the transition to alcohol dependence. Biol Psychiatry 69: 1091–1099.

Goldstein RZ, Volkow ND (2011). Dysfunction of the prefrontal cortex in addiction: neuroimaging findings and clinical implications. Nat Rev Neurosci 12: 652–669.

Grant KA, Waters CA, Green-Jordan K et al. (2000). Characterization of the discriminative stimulus effects of $GABA_A$ receptor ligands in *Macaca fascicularis* monkeys under different ethanol training conditions. Psychopharmacology (Berl) 152: 181–188.

Green L, Fristoe N, Myerson J (1994). Temporal discounting and preference reversals in choice between delayed outcomes. Psychon Bull Rev 1: 383–389.

Heilig M, Koob GF (2007). A key role for corticotropin-releasing factor in alcohol dependence. Trends Neurosci 30: 399–406.

Heimer L, Alheid G (1991). Piecing together the puzzle of basal forebrain anatomy. In: TC Napier, PW Kalivas, I Hanin (Eds.), The Basal Forebrain: Anatomy to Function, In: Advances in Experimental Medicine and Biology. Vol. 295. Plenum Press, New York, pp. 1–42.

Heinz A, Lober S, Georgi A et al. (2003). Reward craving and withdrawal relief craving: assessment of different motivational pathways to alcohol intake. Alcohol Alcohol 38: 35–39.

Hellevuo K, Kiianmaa K, Korpi ER (1989). Effect of GABAergic drugs on motor impairment from ethanol, barbital and lorazepam in rat lines selected for differential sensitivity to ethanol. Pharmacol Biochem Behav 34: 399–404.

Hemby SE, Martin TJ, Co C et al. (1995). The effects of intravenous heroin administration on extracellular nucleus accumbens dopamine concentrations as determined by *in vivo* microdialysis. J Pharmacol Exp Ther 273: 591–598.

Hernandez G, Hamdani S, Rajabi H et al. (2006). Prolonged rewarding stimulation of the rat medial forebrain bundle:

neurochemical and behavioral consequences. Behav Neurosci 120: 888–904.

Herz A (1997). Endogenous opioid systems and alcohol addiction. Psychopharmacology (Berl) 129: 99–111.

Heyser CJ, Roberts AJ, Schulteis G et al. (1999). Central administration of an opiate antagonist decreases oral ethanol self-administration in rats. Alcohol Clin Exp Res 23: 1468–1476.

Hodge CW, Samson HH, Lewis RS et al. (1993). Specific decreases in ethanol- but not water-reinforced responding produced by the 5-HT_3 antagonist ICS 205-930. Alcohol 10: 191–196.

Holter SM, Henniger MS, Lipkowski AW et al. (2000). Kappa-opioid receptors and relapse-like drinking in long-term ethanol-experienced rats. Psychopharmacology (Berl) 153: 93–102.

Hubbell CL, Marglin SH, Spitalnic SJ et al. (1991). Opioidergic, serotonergic, and dopaminergic manipulations and rats' intake of a sweetened alcoholic beverage. Alcohol 8: 355–367.

Hyytia P, Koob GF (1995). $GABA_A$ receptor antagonism in the extended amygdala decreases ethanol self-administration in rats. Eur J Pharmacol 283: 151–159.

Hyytia P, Sinclair JD (1993). Responding for oral ethanol after naloxone treatment by alcohol-preferring AA rats. Alcohol Clin Exp Res 17: 631–636.

Janak PH, Gill MT (2003). Comparison of the effects of allopregnanolone with direct GABAergic agonists on ethanol self-administration with and without concurrently available sucrose. Alcohol 30: 1–7.

Jentsch JD, Taylor JR (1999). Impulsivity resulting from frontostriatal dysfunction in drug abuse: implications for the control of behavior by reward-related stimuli. Psychopharmacology (Berl) 146: 373–390.

Ji D, Gilpin NW, Richardson HN et al. (2008). Effects of naltrexone, duloxetine, and a corticotropin-releasing factor type 1 receptor antagonist on binge-like alcohol drinking in rats. Behav Pharmacol 19: 1–12.

Johnson BA, Roache JD, Javors MA et al. (2000). Ondansetron for reduction of drinking among biologically predisposed alcoholic patients: a randomized controlled trial. JAMA 284: 963–971.

June HL, Lummis GH, Colker RE et al. (1991). Ro15-4513 attenuates the consumption of ethanol in deprived rats. Alcohol Clin Exp Res 15: 406–411.

Katner SN, Weiss F (1999). Ethanol-associated olfactory stimuli reinstate ethanol-seeking behavior after extinction and modify extracellular dopamine levels in the nucleus accumbens. Alcohol Clin Exp Res 23: 1751–1760.

Katner SN, Magalong JG, Weiss F (1999). Reinstatement of alcohol-seeking behavior by drug-associated discriminative stimuli after prolonged extinction in the rat. Neuropsychopharmacology 20: 471–479.

Knapp DJ, Pohorecky LA (1992). Zacopride, a 5-HT_3 receptor antagonist, reduces voluntary ethanol consumption in rats. Pharmacol Biochem Behav 41: 847–850.

Knapp DJ, Overstreet DH, Moy SS et al. (2004). SB242084, flumazenil, and CRA1000 block ethanol withdrawal-induced anxiety in rats. Alcohol 32: 101–111.

Knapp DJ, Overstreet DH, Breese GR (2005). Modulation of ethanol withdrawal-induced anxiety-like behavior during later withdrawals by treatment of early withdrawals with benzodiazepine/γ-aminobutyric acid ligands. Alcohol Clin Exp Res 29: 553–563.

Knapp DJ, Overstreet DH, Breese GR (2007). Baclofen blocks expression and sensitization of anxiety-like behavior in an animal model of repeated stress and ethanol withdrawal. Alcohol Clin Exp Res 31: 582–595.

Koob GF (1992). Drugs of abuse: anatomy, pharmacology, and function of reward pathways. Trends Pharmacol Sci 13: 177–184.

Koob GF (1999). Corticotropin-releasing factor, norepinephrine and stress. Biol Psychiatry 46: 1167–1180.

Koob GF (2003). Alcoholism: allostasis and beyond. Alcohol Clin Exp Res 27: 232–243.

Koob GF (2004). Allostatic view of motivation: implications for psychopathology. In: RA Bevins, MT Bardo (Eds.), Motivational Factors in the Etiology of Drug Abuse. Nebraska Symposium on Motivation. Vol. 50. University of Nebraska Press, Lincoln, NE, pp. 1–18.

Koob GF (2008a). A role for brain stress systems in addiction. Neuron 59: 11–34.

Koob GF (2008b). Neurobiology of addiction. In: M Galanter, HD Kleber (Eds.), Textbook of Substance Abuse Treatment. 4th edn. American Psychiatric Publishing, Washington, DC, pp. 3–16.

Koob GF (2013). Theoretical frameworks and mechanistic aspects of alcohol addiction: alcohol addiction: alcohol addiction as a reward deficit disorder. In: WH Sommer, R Spanagel (Eds.), Behavioral Neurobiology of Alcohol Addiction, In: Current Topics in Behavioral Neuroscience. Vol. 13. Springer-Verlag, Berlin, pp. 3–30.

Koob GF, Bloom FE (1988). Cellular and molecular mechanisms of drug dependence. Science 242: 715–723.

Koob GF, Kreek MJ (2007). Stress, dysregulation of drug reward pathways, and the transition to drug dependence. Am J Psychiatry 164: 1149–1159.

Koob GF, Le Moal M (1997). Drug abuse: hedonic homeostatic dysregulation. Science 278: 52–58.

Koob GF, Le Moal M (2001). Drug addiction, dysregulation of reward, and allostasis. Neuropsychopharmacology 24: 97–129.

Koob GF, Le Moal M (2006). Neurobiology of Addiction. Academic Press, London.

Koob GF, Le Moal M (2008). Addiction and the brain antireward system. Annu Rev Psychol 59: 29–53.

Koob GF, Winger GD, Meyerhoff JL et al. (1977). Effects of D-amphetamine on concurrent self-stimulation of forebrain and brain stem loci. Brain Res 137: 109–126.

Koob GF, Stinus L, Le Moal M (1989). Opponent process theory of motivation: neurobiological evidence from studies of opiate dependence. Neurosci Biobehav Rev 13: 135–140.

Koob GF, Heinrichs SC, Menzaghi F et al. (1994). Corticotropin releasing factor, stress and behavior. Semin Neurosci 6: 221–229.

Koob GF, Everitt BJ, Robbins TW (2008). Reward, motivation, and addiction. In: LG Squire, D Berg, FE Bloom, S Du Lac, A Ghosh, N Spitzer (Eds.), Fundamental Neuroscience. 3rd edn. Academic Press, Amsterdam, pp. 987–1016.

Koob GF, Lloyd GK, Mason BJ (2009). Development of pharmacotherapies for drug addiction: a Rosetta stone approach. Nat Rev Drug Discov 8: 500–515.

Kornet M, Goosen C, Van Ree JM (1991). Effect of naltrexone on alcohol consumption during chronic alcohol drinking and after a period of imposed abstinence in free-choice drinking rhesus monkeys. Psychopharmacology (Berl) 104: 367–376.

Kornetsky C, Esposito RU (1979). Euphorigenic drugs: effects on the reward pathways of the brain. Fed Proc 38: 2473–2476.

Kostowski W, Dyr W, Krzascik P (1993). The abilities of 5-HT_3 receptor antagonist ICS 205-930 to inhibit alcohol preference and withdrawal seizures in rats. Alcohol 10: 369–373.

Land BB, Bruchas MR, Lemos JC et al. (2008). The dysphoric component of stress is encoded by activation of the dynorphin κ-opioid system. J Neurosci 28: 407–414.

Le Doux JE (2000). Emotion circuits in the brain. Annu Rev Neurosci 23: 155–184.

Liljequist S, Engel JA (1984). The effects of GABA and benzodiazepine receptor antagonists on the anti-conflict actions of diazepam or ethanol. Pharmacol Biochem Behav 21: 521–525.

Liu X, Weiss F (2002a). Reversal of ethanol-seeking behavior by D1 and D2 antagonists in an animal model of relapse: differences in antagonist potency in previously ethanol-dependent versus nondependent rats. J Pharmacol Exp Ther 300: 882–889.

Liu X, Weiss F (2002b). Additive effect of stress and drug cues on reinstatement of ethanol seeking: exacerbation by history of dependence and role of concurrent activation of corticotropin-releasing factor and opioid mechanisms. J Pharmacol Exp Ther 22: 7856–7861.

Liu X, Weiss F (2004). Nitric oxide synthesis inhibition attenuates conditioned reinstatement of ethanol-seeking, but not the primary reinforcing effects of ethanol. Alcohol Clin Exp Res 28: 1194–1199.

Logan GD, Schachar RJ, Tannock R (1997). Impulsivity and inhibitory control. Psychol Sci 8: 60–64.

Lopez MF, Becker HC (2005). Effect of pattern and number of chronic ethanol exposures on subsequent voluntary ethanol intake in C57BL/6 J mice. Psychopharmacology (Berl) 181: 688–696.

Lyness WH, Smith FL (1992). Influence of dopaminergic and serotonergic neurons on intravenous ethanol self-administration in the rat. Pharmacol Biochem Behav 42: 187–192.

Majchrowicz E (1975). Induction of physical dependence upon ethanol and the associated behavioral changes in rats. Psychopharmacologia 43: 245–254.

Malcolm R, Roberts JS, Wang W et al. (2000). Multiple previous detoxifications are associated with less responsive treatment and heavier drinking during an index outpatient detoxification. Alcohol 22: 159–164.

Markou A, Koob GF (1991). Post-cocaine anhedonia: an animal model of cocaine withdrawal. Neuropsychopharmacology 4: 17–26.

Martin-Fardon R, Weiss F (2013). Modeling relapse in animals. Curr Top Behav Neurosci 13: 403–432.

Martin-Fardon R, Ciccocioppo R, Massi M et al. (2000). Nociceptin prevents stress-induced ethanol- but not cocaine-seeking behavior in rats. Neuroreport 11: 1939–1943.

Martin-Solch C, Magyar S, Kunig G et al. (2001). Changes in brain activation associated with reward processing in smokers and nonsmokers: a positron emission tomography study. Exp Brain Res 139: 278–286.

Martz A, Deitrich RA, Harris RA (1983). Behavioral evidence for the involvement of gamma-aminobutyric acid in the actions of ethanol. Eur J Pharmacol 89: 53–62.

McBride WJ, Li TK (1998). Animal models of alcoholism: neurobiology of high alcohol-drinking behavior in rodents. Crit Rev Neurobiol 12: 339–369.

McBride WJ, Murphy JM, Lumeng L et al. (1988). Effects of Ro 15-4513, fluoxetine and desipramine on the intake of ethanol, water and food by the alcohol-preferring (P) and -nonpreferring (NP) lines of rats. Pharmacol Biochem Behav 30: 1045–1050.

McEwen BS (2000). Allostasis and allostatic load: implications for neuropsychopharmacology. Neuropsychopharmacology 22: 108–124.

McFarland K, Kalivas PW (2001). The circuitry mediating cocaine-induced reinstatement of drug-seeking behavior. J Neurosci 21: 8655–8663.

McKinzie DL, Eha R, Cox R et al. (1998). $Serotonin_3$ receptor antagonism of alcohol intake: effects of drinking conditions. Alcohol 15: 291–298.

Melis M, Spiga S, Diana M (2005). The dopamine hypothesis of drug addiction: hypodopaminergic state. Int Rev Neurobiol 63: 101–154.

Merlo-Pich E, Lorang M, Yeganeh M et al. (1995). Increase of extracellular corticotropin-releasing factor-like immunoreactivity levels in the amygdala of awake rats during restraint stress and ethanol withdrawal as measured by microdialysis. J Neurosci 15: 5439–5447.

Mhatre MC, Pena G, Sieghart W et al. (1993). Antibodies specific for GABAA receptor alpha subunits reveal that chronic alcohol treatment down-regulates α-subunit expression in rat brain regions. J Neurosci 61: 1620–1625.

Moeller FG, Barratt ES, Dougherty DM et al. (2001). Psychiatric aspects of impulsivity. Am J Psychiatry 158: 1783–1793.

Monti PM, Rohsenow DJ, Hutchison KE et al. (1999). Naltrexone's effect on cue-elicited craving among alcoholics in treatment. Alcohol Clin Exp Res 23: 1386–1394.

Morzorati SL, Ramchandani VA, Flury L et al. (2002). Self-reported subjective perception of intoxication reflects family history of alcoholism when breath alcohol levels are constant. Alcohol Clin Exp Res 26: 1299–1306.

Moy SS, Knapp DJ, Criswell HE et al. (1997). Flumazenil blockade of anxiety following ethanol withdrawal in rats. Psychopharmacology (Berl) 131: 354–360.

Mucha RF, Herz A (1985). Motivational properties of kappa and mu opioid receptor agonists studied with place and taste preference conditioning. Psychopharmacology (Berl) 86: 274–280.

Myers RD (1990). Anatomical "circuitry" in the brain mediating alcohol drinking revealed by THP-reactive sites in the limbic system. Alcohol 7: 449–459.

Nealey KA, Smith AW, Davis SM et al. (2011). κ-opioid receptors are implicated in the increased potency of intra-accumbens nalmefene in ethanol-dependent rats. Neuropharmacology 61: 35–42.

Nestler EJ (2001). Molecular basis of long-term plasticity underlying addiction. Nat Rev Neurosci 2: 119–128.

Nestler EJ (2004). Historical review: molecular and cellular mechanisms of opiate and cocaine addiction. Trends Pharmacol Sci 25: 210–218.

Nestler EJ (2005). Is there a common molecular pathway for addiction? Nat Neurosci 8: 1445–1449.

Neugebauer V, Li W, Bird GC et al. (2004). The amygdala and persistent pain. Neuroscientist 10: 221–234.

O'Dell LE, Roberts AJ, Smith RT et al. (2004). Enhanced alcohol self-administration after intermittent versus continuous alcohol vapor exposure. Alcohol Clin Exp Res 28: 1676–1682.

O'Malley SS, Jaffe AJ, Chang G et al. (1992). Naltrexone and coping skills therapy for alcohol dependence: a controlled study. Arch Gen Psychiatry 49: 881–887.

Olds J, Milner P (1954). Positive reinforcement produced by electrical stimulation of septal area and other regions of rat brain. J Comp Physiol Psychol 47: 419–427.

Olive MF, Koenig HN, Nannini MA et al. (2002). Elevated extracellular CRF levels in the bed nucleus of the stria terminalis during ethanol withdrawal and reduction by subsequent ethanol intake. Pharmacol Biochem Behav 72: 213–220.

Oscar-Berman M (2012). Function and dysfunction of prefrontal brain circuitry in alcoholic Korsakoff's syndrome. Neuropsychol Rev 22: 154–169.

Overstreet DH, Knapp DJ, Breese GR (2004). Modulation of multiple ethanol withdrawal-induced anxiety-like behavior by CRF and CRF_1 receptors. Pharmacol Biochem Behav 77: 405–413.

Overstreet DH, Knapp DJ, Breese GR (2005). Pharmacological modulation of repeated ethanol withdrawal-induced anxiety-like behavior differs in alcohol-preferring P and Sprague-Dawley rats. Pharmacol Biochem Behav 81: 122–130.

Pandey SC (2004). The gene transcription factor cyclic AMP-responsive element binding protein: role in positive and negative affective states of alcohol addiction. Pharmacol Ther 104: 47–58.

Paterson NE, Myers C, Markou A (2000). Effects of repeated withdrawal from continuous amphetamine administration on brain reward function in rats. Psychopharmacology (Berl) 152: 440–446.

Pfeffer AO, Samson HH (1986). Effect of pimozide on home cage ethanol drinking in the rat: dependence on drinking session length. Drug Alcohol Depend 17: 47–55.

Pfeiffer A, Brantl V, Herz A et al. (1986). Psychotomimesis mediated by κ opiate receptors. Science 233: 774–776.

Piazza PV, Le Moal M (1998). The role of stress in drug self-administration. Trends Pharmacol Sci 19: 67–74.

Rachlin H, Green L (1972). Commitment, choice and self-control. J Exp Anal Behav 17: 15–22.

Ramchandani VA, Umhau J, Pavon FJ et al. (2011). A genetic determinant of the striatal dopamine response to alcohol in men. Mol Psychiatry 16: 809–817.

Rasmussen DD, Boldt BM, Bryant CA et al. (2000). Chronic daily ethanol and withdrawal: 1. Long-term changes in the hypothalamo-pituitary-adrenal axis. Alcohol Clin Exp Res 24: 1836–1849.

Rassnick S, Pulvirenti L, Koob GF (1992). Oral ethanol self-administration in rats is reduced by the administration of dopamine and glutamate receptor antagonists into the nucleus accumbens. Psychopharmacology (Berl) 109: 92–98.

Rassnick S, D'Amico E, Riley E et al. (1993a). GABA antagonist and benzodiazepine partial inverse agonist reduce motivated responding for ethanol. Alcohol Clin Exp Res 17: 124–130.

Rassnick S, Heinrichs SC, Britton KT et al. (1993b). Microinjection of a corticotropin-releasing factor antagonist into the central nucleus of the amygdala reverses anxiogenic-like effects of ethanol withdrawal. Brain Res 605: 25–32.

Rassnick S, Stinus L, Koob GF (1993c). The effects of 6-hydroxydopamine lesions of the nucleus accumbens and the mesolimbic dopamine system on oral self-administration of ethanol in the rat. Brain Res 623: 16–24.

Reid LD, Hunter GA (1984). Morphine and naloxone modulate intake of ethanol. Alcohol 1: 33–37.

Reissner KJ, Kalivas PW (2010). Using glutamate homeostasis as a target for treating addictive disorders. Behav Pharmacol 21: 514–522.

Rhodes JS, Best K, Belknap JK et al. (2005). Evaluation of a simple model of ethanol drinking to intoxication in C57BL/6 J mice. Physiol Behav 84: 53–63.

Richardson HN, Zhao Y, Fekete EM et al. (2008). MPZP: a novel small molecule corticotropin-releasing factor type 1 receptor (CRF_1) antagonist. Pharmacol Biochem Behav 88: 497–510.

Rimondini R, Arlinde C, Sommer W et al. (2002). Long-lasting increase in voluntary ethanol consumption and transcriptional regulation in the rat brain after intermittent exposure to alcohol. FASEB J 16: 27–35.

Rivier C, Bruhn T, Vale W (1984). Effect of ethanol on the hypothalamic–pituitary–adrenal axis in the rat: role of corticotropin-releasing factor (CRF). J Pharmacol Exp Ther 229: 127–131.

Roberto M, Cruz MT, Gilpin NW et al. (2010). Corticotropin releasing factor-induced amygdala gamma-aminobutyric Acid release plays a key role in alcohol dependence. Biol Psychiatry 67: 831–839.

Roberts AJ, Cole M, Koob GF (1996). Intra-amygdala muscimol decreases operant ethanol self-administration in dependent rats. Alcohol Clin Exp Res 20: 1289–1298.

Roberts AJ, McArthur RA, Hull EE et al. (1998). Effects of amperozide, 8-OH-DPAT, and FG 5974 on operant responding for ethanol. Psychopharmacology (Berl) 137: 25–32.

Roberts AJ, Heyser CJ, Koob GF (1999). Operant self-administration of sweetened versus unsweetened ethanol: effects on blood alcohol levels. Alcohol Clin Exp Res 23: 1151–1157.

Roberts AJ, Heyser CJ, Cole M et al. (2000). Excessive ethanol drinking following a history of dependence: animal model of allostasis. Neuropsychopharmacology 22: 581–594.

Robinson TE, Berridge KC (1993). The neural basis of drug craving: an incentive-sensitization theory of addiction. Brain Res Rev 18: 247–291.

Romach MK, Sellers EM (1991). Management of the alcohol withdrawal syndrome. Annu Rev Med 42: 323–340.

Rossetti ZL, Hmaidan Y, Gessa GL (1992). Marked inhibition of mesolimbic dopamine release: a common feature of ethanol, morphine, cocaine and amphetamine abstinence in rats. Eur J Pharmacol 221: 227–234.

Rossetti ZL, Isola D, De Vry J et al. (1999). Effects of nimodipine on extracellular dopamine levels in the rat nucleus accumbens in ethanol withdrawal. Neuropharmacology 38: 1361–1369.

Russell MAH (1976). What is dependence? In: G Edwards (Ed.), Drugs and Drug Dependence. Lexington Books, Lexington, MA, pp. 182–187.

Samson HH (1987). Initiation of ethanol-maintained behavior: a comparison of animal models and their implication to human drinking. In: T Thompson, PB Dews, JE Barrett (Eds.), Neurobehavioral Pharmacology, In: Advances in Behavioral Pharmacology. Vol. 6. Lawrence Erlbaum, Hillsdale, NJ, pp. 221–248.

Samson HH, Chappell A (2001). Muscimol injected into the medial prefrontal cortex of the rat alters ethanol self-administration. Physiol Behav 74: 581–587.

Samson HH, Doyle TF (1985). Oral ethanol self-administration in the rat: effect of naloxone. Pharmacol Biochem Behav 22: 91–99.

Samson HH, Tolliver GA, Pfeffer AO et al. (1987). Oral ethanol reinforcement in the rat: effect of the partial inverse benzodiazepine agonist RO15-4513. Pharmacol Biochem Behav 27: 517–519.

Samson HH, Hodge CW, Tolliver GA et al. (1993). Effect of dopamine agonists and antagonists on ethanol-reinforced behavior: the involvement of the nucleus accumbens. Brain Res Bull 30: 133–141.

Sapru MK, Diamond I, Gordon AS (1994). Adenosine receptors mediate cellular adaptation to ethanol in NG108-15 cells. J Pharmacol Exp Ther 271: 542–548.

Schulteis G, Liu J (2006). Brain reward deficits accompany withdrawal (hangover) from acute ethanol in rats. Alcohol 39: 21–28.

Schulteis G, Markou A, Gold LH et al. (1994). Relative sensitivity to naloxone of multiple indices of opiate withdrawal: a quantitative dose–response analysis. J Pharmacol Exp Ther 271: 1391–1398.

Schulteis G, Markou A, Cole M et al. (1995). Decreased brain reward produced by ethanol withdrawal. Proc Natl Acad Sci U S A 92: 5880–5884.

Schulteis G, Hyytia P, Heinrichs SC et al. (1996). Effects of chronic ethanol exposure on oral self-administration of ethanol or saccharin by Wistar rats. Alcohol Clin Exp Res 20: 164–171.

Sellers EM, Higgins GA, Sobell MB (1992). 5-HT and alcohol abuse. Trends Pharmacol Sci 13: 69–75.

Shaham Y, Shalev U, Lu L et al. (2003). The reinstatement model of drug relapse: history, methodology and major findings. Psychopharmacology (Berl) 168: 3–20.

Shalev U, Grimm JW, Shaham Y (2002). Neurobiology of relapse to heroin and cocaine seeking: a review. Pharmacol Rev 54: 1–42.

Shippenberg TS, Koob GF (2002). Recent advances in animal models of drug addiction and alcoholism. In: KL Davis, D Charney, JT Coyle, C Nemeroff (Eds.), Neuropsychopharmacology: The Fifth Generation of Progress. Lippincott Williams and Wilkins, Philadelphia, pp. 1381–1397.

Simms JA, Bito-Onon JJ, Chatterjee S et al. (2010). Long-Evans rats acquire operant self-administration of 20% ethanol without sucrose fading. Neuropsychopharmacology 35: 1453–1463.

Solomon RL, Corbit JD (1974). An opponent-process theory of motivation: 1. Temporal dynamics of affect. Psychol Rev 81: 119–145.

Sommer WH, Rimondini R, Hansson AC et al. (2008). Upregulation of voluntary alcohol intake, behavioral sensitivity to stress, and amygdala *crhr1* expression following a history of dependence. Biol Psychiatry 63: 139–145.

Song ZH, Takemori AE (1992). Stimulation by corticotropin-releasing factor of the release of immunoreactive dynorphin A from mouse spinal cords in vitro. Eur J Pharmacol 222: 27–32.

Sterling P, Eyer J (1988). Allostasis: a new paradigm to explain arousal pathology. In: S Fisher, J Reason (Eds.), Handbook of Life Stress, Cognition and Health. John Wiley, Chichester, pp. 629–649.

Sullivan EV, Pfefferbaum A (2005). Neurocircuitry in alcoholism: a substrate of disruption and repair. Psychopharmacology (Berl) 180: 583–594.

Thorsell A, Slawecki CJ, Ehlers CL (2005a). Effects of neuropeptide Y and corticotropin-releasing factor on ethanol intake in Wistar rats: interaction with chronic ethanol exposure. Behav Brain Res 161: 133–140.

Thorsell A, Slawecki CJ, Ehlers CL (2005b). Effects of neuropeptide Y on appetitive and consummatory behaviors associated with alcohol drinking in Wistar rats with a history of ethanol exposure. Alcohol Clin Exp Res 29: 584–590.

Thorsell A, Rapunte-Canonigo V, O'Dell L et al. (2007). Viral vector-induced amygdala NPY overexpression reverses increased alcohol intake caused by repeated deprivations in Wistar rats. Brain 130: 1330–1337.

Tiffany ST, Carter BL, Singleton EG (2000). Challenges in the manipulation, assessment and interpretation of craving relevant variables. Addiction 95: s177–s187.

Trzaskowska E, Kostowski W (1983). Further studies on the role of noradrenergic mechanisms in ethanol withdrawal syndrome in rats. Pol J Pharmacol 35: 351–358.

United Nations International Drug Control Programme (1992). Informal Expert Group Meeting on the Craving Mechanism (report no. V92-54439 T). United Nations International Drug Control Programme and World Health Organization, Geneva.

Valdez GR, Roberts AJ, Chan K et al. (2002). Increased ethanol self-administration and anxiety-like behavior during acute withdrawal and protracted abstinence: regulation by corticotropin-releasing factor. Alcohol Clin Exp Res 26: 1494–1501.

Valdez GR, Zorrilla EP, Roberts AJ et al. (2003). Antagonism of corticotropin-releasing factor attenuates the enhanced responsiveness to stress observed during protracted ethanol abstinence. Alcohol 29: 55–60.

Valdez GR, Sabino V, Koob GF (2004). Increased anxiety-like behavior and ethanol self-administration in dependent rats: reversal via corticotropin-releasing factor-2 receptor activation. Alcohol Clin Exp Res 28: 865–872.

Valdez GR, Platt DM, Rowlett JK et al. (2007). κ Agonist-induced reinstatement of cocaine seeking in squirrel monkeys: a role for opioid and stress-related mechanisms. J Pharmacol Exp Ther 323: 525–533.

Valentino RJ, Page ME, Curtis AL (1991). Activation of noradrenergic locus coeruleus neurons by hemodynamic stress is due to local release of corticotropin-releasing factor. Brain Res 555: 25–34.

Valentino RJ, Foote SL, Page ME (1993). The locus coeruleus as a site for integrating corticotropin-releasing factor and noradrenergic mediation of stress responses. In: Y Tache, C Rivier (Eds.), Corticotropin-Releasing Factor and Cytokines: Role in the Stress Response, In: Annals of the New York Academy of Sciences. Vol. 697. New York Academy of Sciences, New York, pp. 173–188.

Van Bockstaele EJ, Colago EE, Valentino RJ (1998). Amygdaloid corticotropin-releasing factor targets locus coeruleus dendrites: substrate for the co-ordination of emotional and cognitive limbs of the stress response. J Neuroendocrinol 10: 743–757.

Volkow ND, Fowler JS (2000). Addiction, a disease of compulsion and drive: involvement of the orbitofrontal cortex. Cereb Cortex 10: 318–325.

Volkow ND, Fowler JS, Wang GJ (2002). Role of dopamine in drug reinforcement and addiction in humans: results from imaging studies. Behav Pharmacol 13: 355–366.

Volkow ND, Wang GJ, Telang F et al. (2007). Profound decreases in dopamine release in striatum in detoxified alcoholics: possible orbitofrontal involvement. J Neurosci 27: 12700–12706.

Volpicelli JR, Alterman AI, Hayashida M et al. (1992). Naltrexone in the treatment of alcohol dependence. Arch Gen Psychiatry 49: 876–880.

Vorel SR, Liu X, Hayes RJ et al. (2001). Relapse to cocaine-seeking after hippocampal theta burst stimulation. Science 292: 1175–1178.

Walker BM, Koob GF (2008). Pharmacological evidence for a motivational role of κ-opioid systems in ethanol dependence. Neuropsychopharmacology 33: 643–652.

Walker BM, Rasmussen DD, Raskind MA et al. (2008). α_1-Noradrenergic receptor antagonism blocks dependence-induced increases in responding for ethanol. Alcohol 42: 91–97.

Wang L, Liu J, Harvey-White J et al. (2003). Endocannabinoid signaling via cannabinoid receptor 1 is involved in ethanol preference and its age-dependent decline in mice. Proc Natl Acad Sci U S A 100: 1393–1398.

Wee S, Koob GF (2010). The role of the dynorphin-κ opioid system in the reinforcing effects of drugs of abuse. Psychopharmacology (Berl) 210: 121–135.

Wegelius K, Honkanen A, Korpi ER (1994). Benzodiazepine receptor ligands modulate ethanol drinking in alcohol-preferring rats. Eur J Pharmacol 263: 141–147.

Weiss F, Mitchiner M, Bloom FE et al. (1990). Free-choice responding for ethanol versus water in alcohol preferring (P) and unselected Wistar rats is differentially modified by naloxone, bromocriptine, and methysergide. Psychopharmacology (Berl) 101: 178–186.

Weiss F, Hurd YL, Ungerstedt U et al. (1992a). Neurochemical correlates of cocaine and ethanol self-administration. In: PW Kalivas, HH Samson (Eds.), The Neurobiology of Drug and Alcohol Addiction, In: Annals of the New York Academy of Sciences. Vol. 654. New York Academy of Sciences, New York, pp. 220–241.

Weiss F, Markou A, Lorang MT et al. (1992b). Basal extracellular dopamine levels in the nucleus accumbens are decreased during cocaine withdrawal after unlimited-access self-administration. Brain Res 593: 314–318.

Weiss F, Lorang MT, Bloom FE et al. (1993). Oral alcohol self-administration stimulates dopamine release in the rat nucleus accumbens: genetic and motivational determinants. J Pharmacol Exp Ther 267: 250–258.

Weiss F, Parsons LH, Schulteis G et al. (1996). Ethanol self-administration restores withdrawal-associated deficiencies in accumbal dopamine and 5-hydroxytryptamine release in dependent rats. J Neurosci 16: 3474–3485.

Weiss F, Ciccocioppo R, Parsons LH et al. (2001). Compulsive drug-seeking behavior and relapse: neuroadaptation, stress, and conditioning factors. In: V Quinones-Jenab (Ed.), The Biological Basis of Cocaine, In: Annals of the New York Academy of Sciences. Vol. 937. New York Academy of Sciences, New York, pp. 1–26.

Wise RA (1973). Voluntary ethanol intake in rats following exposure to ethanol on various schedules. Psychopharmacologia 29: 203–210.

Wolffgramm J, Heyne A (1995). From controlled drug intake to loss of control: the irreversible development of drug addiction in the rat. Behav Brain Res 70: 77–94.

Zorrilla EP, Valdez GR, Weiss F (2001). Changes in levels of regional CRF-like-immunoreactivity and plasma corticosterone during protracted drug withdrawal in dependent rats. Psychopharmacology (Berl) 158: 374–381.

Handbook of Clinical Neurology, Vol. 125 (3rd series)
Alcohol and the Nervous System
E.V. Sullivan and A. Pfefferbaum, Editors

Chapter 4

Metabolism

MARTIN H. PLAWECKI[1] AND DAVID W. CRABB[2]*

[1]*Department of Psychiatry, Indiana University School of Medicine, Indianapolis, IN, USA*

[2]*Department of Medicine, Indiana University School of Medicine, Indianapolis, IN, USA*

OVERVIEW

The physiologic effects of ethanol on the nervous system and other organs are highly dependent upon the systemic concentration over time – in turn, the pharmacodynamic effects of ethanol are heavily influenced by systemic pharmacokinetics. Thus, in order to understand what predisposes some individuals to develop alcohol use disorders, an understanding of this variance is necessary.

Ethanol is most efficiently absorbed within the small intestine and thus, the absorption rate and systemic effect are highly dependent upon gastric emptying. In addition to gastric emptying, the very process of digestion may alter ethanol pharmacokinetics through a redistribution of the circulation to the gut, thereby increasing substrate delivery to the liver. Blood flow to the liver delivers ethanol at a rate that generally exceeds the liver's capacity to metabolize it all. Because both the gastrointestinal tract and liver express drug-metabolizing enzymes such as cytochrome P450s and alcohol dehydrogenases (ADHs) which are genetically variable, metabolic rate may affect the amount of ethanol ultimately reaching the systemic circulation. This chapter will review hepatic and gastrointestinal tract alcohol metabolism and the factors that can alter systemic pharmacokinetics. We conclude with the description of modeling and related technology, which had taught us much about ethanol pharmacokinetics and provides new ways to study its effects upon the brain.

ETHANOL METABOLISM

Hepatic ethanol metabolism

The activities of alcohol and aldehyde dehydrogenases are important determinants of metabolism and responses to ethanol. The strongest evidence for this assertion is the finding that the inheritance of ADH alleles encoding high-activity enzymes or the dominant negative allele (*ALDH1B*2*) for ALDH2 are associated with lower prevalence of alcoholism (Agarwal, 2001). It is natural to ask if non-genetic influences on the activity of these enzymes might also affect responses to ethanol. We review these issues below.

ALCOHOL DEHYDROGENASES

The ADH gene family encodes enzymes that metabolize a wide variety of substrates, including ethanol, retinal, other aliphatic alcohols, hydroxysteroids, and lipid peroxidation products (Table 4.1). Human ADH is a dimeric protein consisting of two 40 kD subunits. The proteins contain catalytic and structural zinc atoms. So far, seven ADH genes and subunits have been identified in humans (Edenberg and Bosron, 1997). The ADH forms can be divided into five major classes (I–V); those that are important for ethanol metabolism are the class I, class II, and class IV isozymes.

Class I isozymes are formed by random dimeric association of the three types of polypeptide subunits, α, β, and γ, encoded by three separate genes, *ADH1A*, *ADH1B*, and *ADH1C3* (Table 4.2), respectively (Edenberg, 2000). These isoenzymes have low K_m (<4 mM) for ethanol, are expressed at highest levels in liver, and play a major role in ethanol metabolism. The *ADH1B* and *ADH1C* genes have allelic variants: *ADH1B*1*, *ADH1B*2*, and *ADH1B*3*, encoding β1, 2, and 3 isozymes, respectively. *ADH1C*1* and *ADH1C*2* encode γ1 and 2 peptides. Class II ADH, encoded by the *ADH2* gene, exists as a single homodimeric enzyme of π subunits, and is also expressed in liver. The class IV enzyme is a homodimeric enzyme of σ subunits encoded by *ADH4* gene; it is expressed primarily in stomach (Seitz and Oneta, 1998).

*Correspondence to: David W. Crabb, M.D., Professor of Medicine and of Biochemistry and Molecular Biology, Emerson Hall Room 317, 545 Barnhill Drive, Indianapolis, IN 46202, USA. Tel: +1-317-274-8438, E-mail: dcrabb@iupui.edu

Table 4.1

Properties of alcohol dehydrogenases (ADH) in humans

Gene locus	New nomenclature	Subunit type	K_m (ethanol)	V_{max}	Tissue distribution
Class I					
ADH1	*ADH1A*	α	4	54	Liver
ADH2	*ADH1B*	ß	0.05–34†	–	Liver, lung
ADH3	*ADH1C*	γ	0.6–1†	–	Liver, stomach
Class II					
ADH4	*ADH2*	π	34	40	Liver, cornea
Class III					
ADH5	*ADH3*	χ	1000	–	Most tissues
Class IV*					
ADH7	*ADH4*	σ, μ	20	1510	Stomach, esophagus, other mucosae
Class V*					
ADH6	*ADH5*	–	30	?	Liver, stomach

The nomenclature used in older literature is shown in the first column, with the current nomenclature in the second column.
*Tentative assignments based upon sequence homologies. K_m values are given in mM and V_{max} values are given in terms of turnover number (per min).
†Kinetic constants vary with isozyme; see Table 4.2.

Table 4.2

Properties of polymorphic forms of human alcohol dehydrogenase (ADH)

Gene locus	Subunit type	K_m (ethanol)	V_{max}	Population
*ADH2*1 (ADH1B*1)*	ß1	0.05	9	Caucasians, African Americans
*ADH2*2 (ADH1B*2)*	ß2	0.9	400	Asians, Israelis
*ADH2*3 (ADH1B*3)*	ß3	34	300	African Americans
*ADH3*1 (ADH1C*1)*	γ1	1.0	87′	All groups
*ADH3*2 (ADH1C*2)*	γ2	0.63	35	Caucasians

The kinetic constants are noted for the homodimers of the subunits listed (Bosron and Li, 1987). Heterodimers behave as if the active sites were independent. The K_m values are in mM and the V_{max} values are given in terms of turnover numbers (per min), as in Table 4.1. The column labeled population indicates which populations have high allele frequencies for these variants. The alleles are not limited to those populations. The older and current nomenclatures for the alleles are shown.

ALDEHYDE DEHYDROGENASES

Aldehyde dehydrogenases (ALDHs) are a group of NAD(P)+-dependent enzymes that catalyze the oxidation of a wide variety of endogenous and exogenous aliphatic and aromatic aldehydes (Lindahl, 1992). Of the 16 recognized ALDH genes, only class I (ALDH1) and class II (ALDH2) isozymes, encoded by *ALDH1A* and *ALDH1B*, respectively, are important for acetaldehyde metabolism (Yoshida et al., 1998). Human ALDH1 and ALDH2 are homotetrameric enzymes with subunits of 54 kD. They are expressed constitutively in many tissues, with substantial variation in expression between tissues. ALDH1 is a ubiquitous cytosolic enzyme expressed at highest levels in liver. It has relatively low catalytic efficiency for acetaldehyde oxidation, having high affinities for all-*trans*- and 9-*cis* retinal (Yoshida et al., 1998). ALDH2 is the mitochondrial enzyme expressed in most tissues, with the highest level in the liver, kidney, muscle, and heart (Stewart et al., 1996). It exhibits high catalytic efficiency for acetaldehyde oxidation (K_m for acetaldehyde $<1\ \mu M$). The fact that individuals with the dominant negative *ALDH1B*2* allele are deficient in enzyme activity and exhibit the "alcohol flushing" syndrome due to elevated blood acetaldehyde after drinking (Yoshida et al., 1998) indicates that it plays a major role for acetaldehyde oxidation *in vivo*.

CYP2E1

Cytochrome P4502E1 (CYP2E1) was originally designated the microsomal ethanol-oxidizing system. It associates with NADPH-cytochrome P450 reductase in the microsomal membrane and reduces molecular oxygen to water as ethanol is oxidized to acetaldehyde. CYP2E1 has a high K_m for ethanol; thus it has a greater role when blood ethanol levels are high. CYP2E1 is unusually "leaky" and generates reactive oxygen species, including hydroxyl radical ($OH \cdot^-$), superoxide anion ($O_2 \cdot^-$), hydrogen peroxide (H_2O_2), and hydroxy ethyl radical (HER·). Thus, CYP2E1 is a major source of oxidative stress. This P450 enzyme also has capacity to activate many xenobiotics to toxic metabolites, most notably acetaminophen (Lieber, 1999). Other P450 enzymes such as CYP1A2 and 3A4 may also play a small role in ethanol oxidation.

VARIATION IN THE ACTIVITY OF ETHANOL-METABOLIZING SYSTEMS

Genetic factors: Genetic variants and gender

Genetic variation in ADH, ALDH2, CYP2E1 and ethanol metabolism. People of different continental origin inherit different sets of ADH and ALDH2 genes (Tables 4.1 and 4.2), and the enzymes encoded by these alleles vary markedly in K_m for ethanol and enzyme activity (V_{max}). ß1ADH is most common in Caucasians, has a relatively low V_{max} and a very low K_m for ethanol. ß2ADH is found commonly in Asians as well as in about 20% of Israelis. It has a substantially higher V_{max} and somewhat higher K_m compared with ß1. The ß3 isozyme was first detected in African Americans. It has a high K_m for ethanol, and high V_{max}. Smaller differences in enzymatic properties are observed between the products of the *ADH1C* alleles. The γ1 isozyme has about twice the V_{max} of the γ2 isozyme, while the K_ms for ethanol are similar. γ1 ADH is found at high frequency in Asians and African Americans; Caucasians have about equal frequency of γ1 and γ2ADH alleles (Burnell and Bosron, 1989). The other ADH loci are not polymorphic in the coding regions.

The widely varying V_{max} and Michaelis constants of the isozymes suggested that individuals inheriting different isozymes would have different alcohol elimination rates (AERs). To date, different *ADH1B*2* genotypes have been correlated at most to only a small portion of the between-individual differences in AERs. The *ADH1B*3* polymorphism has been shown to be associated with a 10% increase in ethanol metabolism (Thomasson et al., 1995). The *ADH1C* polymorphism did not affect ethanol elimination (Couzigou et al., 1991).

The importance of ALDH2 in ethanol oxidation is demonstrated by the alcohol flushing reaction, observed in a large proportion of East Asians. The flushing was correlated with the accumulation of acetaldehyde to high levels (Zeiner et al., 1979; Goedde et al., 1983); in flushers the levels are variable, but may exceed 100 μM (Enomoto et al., 1991). About 40% of Japanese lack ALDH2 activity (Harada et al., 1981).

The mutation responsible for the deficiency results in replacement of glutamate with lysine in the ALDH2 polypeptide (Hempel et al., 1984). The normal allele is called *ALDH1B*1* and the dominant mutant allele is designated *ALDH1B*2*. The dominance is partial, as the homozygotes have far higher acetaldehyde levels after they drink than the heterozygotes, and heterozygotes have residual low K_m ALDH activity (Enomoto et al., 1991). The dominance is explained by a reduction in activity of the ALDH2 enzyme tetramer if even a single mutated polypeptide is present, and decreased stability of the *ALDH1B*2* polypeptides (Xiao et al., 1996).

Studies on the effect of ALDH2 deficiency on alcohol metabolic rates are limited by the flush reaction. Initial studies failed to show a difference in AERs between flushers and non-flushers (Mizoi et al., 1979), but a larger study detected reduced rates of alcohol elimination in ALDH2-deficient individuals when the data were controlled for ADH genotype (Mizoi et al., 1994). A mutation in the promoter of the *ALDH2* was simultaneously reported (Chou et al., 1999; Harada et al., 1999). The A allele is less active than the G allele. These variants were found in all ethnic groups examined. Harada's group showed that the A allele was less common in alcoholics with active ALDH2, and Chou's group found an effect of the variant on drinking behavior among American Jews.

Polymorphisms in CYP2E1 have not been associated with altered alcohol metabolism. A review (Neafsey et al., 2009) concluded that the known genetic variants in the CYP2E1 gene are not well correlated with expression or activity.

Genetic predisposition to alcoholism

Individuals having the *ADH1B*2* allele are at reduced risk of alcoholism. Individuals without alcohol drinking problems have a higher frequency of the *ADH1B*2* allele compared with alcoholics; this has been confirmed in the Atayal natives of Taiwan (Thomasson et al., 1994), and in Spanish patients (Parés et al., 1994) as well as in Israelis (Neumark et al., 1998). These genes also influence drinking behavior among non-alcoholics; for instance, *ADH1B*1* was more common in heavy drinkers than in moderate drinkers (Tanaka et al., 1997). Moreover, the *ADH1C*1* allele (encoding the more active γ ADH isozyme) is now implicated as moderating risk of alcoholism: a recent meta-analysis of over

50 population studies concluded that the *ADH1C*1* was protective (Li et al., 2012). The effect of these variants is proposed to be due to increased rates of formation of acetaldehyde, causing aversive reactions similar to the flush reaction; however, there is little direct experimental evidence for this hypothesis.

Elegant studies from Japan have demonstrated that ALDH2 deficiency reduced quantity and frequency of ethanol consumption and the risk of alcoholism (Harada et al., 1983; Thomasson et al., 1991; Higuchi et al., 1992). Interestingly, among Japanese, the protective effect of being heterozygous for *ALDH1B*2* appears to be decreasing over time (Higuchi et al., 1994), i.e., the frequency of finding *ALDH1B*2* heterozygotes among alcoholics is increasing, presumably because of environmental and cultural changes; however, homozygotes for *ALDH1B*2* appear to be absolutely protected against alcoholism.

Effects of gender on alcohol metabolism. Studies evaluating AERs between men and women are difficult to compare because of lack of standardization of pharmacokinetic measurements, different concentrations of administered ethanol, and different times of administration. Men have greater body water and less body fat than women of similar weight (Marshall et al., 1983; Sutker et al., 1983; Thomasson, 1995); thus oral dosing results in higher blood alcohol concentration (BAC) in women than in men. However, an alternative explanation is less first-pass metabolism in women (Baraona et al., 2001). Some studies concluded that women have faster AERs than men (Cole-Harding and Wilson, 1987; Thomasson et al., 1995), while others have concluded that there are no differences (Marshall et al., 1983; Sutker et al., 1983; Goist and Sutker, 1985; Martin et al., 1985).

The most convincing studies examining gender differences used intravenous administration of ethanol. In one study, ethanol was administered intravenously to nine brother–sister sibling pairs, reducing genetic variation between the subjects. There was a significantly higher rate of ethanol elimination in the women (Mishra et al., 1989). However, the dose of ethanol was not adjusted for differences in lean body mass. No differences in AERs were found by Arthur et al. (1984). Kwo et al. (1998) used intravenous alcohol clamping (see below) to determine AERs while liver volume was determined by computed tomography. Mean AER and mean computed liver volume were not significantly different. These observations led to a conclusion that, while there were no differences in AER per unit of liver volume, women have greater ethanol clearance per unit lean body mass than men (Kwo et al., 1998).

ENVIRONMENTAL FACTORS: EFFECTS OF NUTRITION AND ENDOCRINE FACTORS

Because of the influence of genotypes of ADH and ALDH2 on ethanol metabolism and risk of alcoholism, much work has sought environmental effects on these enzymes. Although much has been learned about the promoters of these genes, it has been difficult to translate these findings to understand variation in human ethanol metabolism.

Fasting

Fasting reduces the liver content of ADH (Crabb et al., 1983), which is paralleled by a reduction in AER (Lumeng and Crabb, 1984). ADH synthetic rate falls during fasting, and the degradation rate increases (Bosron et al., 1984). This may reflect the influence of insulin (Lakshman et al., 1988). Fasting also reduces the activity of gastric ADH (Mezey et al., 1992). Food restriction mimics the effect of fasting on ADH activity and ethanol elimination (Lumeng et al., 1979). Fasting significantly reduced ALDH2 protein levels in liver and kidney, possibly by an effect of elevated cAMP on the transcription factor HNF4 (You et al., 2002).

CYP2E1 is induced in fasting, diabetes, obesity, and by a diet high in fat. These effects are mediated in part by increased levels of ketone bodies (particularly acetone), which, as a substrate for the enzyme, stabilizes it (Yun et al., 1992). In diabetes, CYP2E1 mRNA is stabilized by an unknown mechanism that is reversed by insulin (Gonzalez, 2007). Low levels of insulin during fasting and diabetes, and insulin resistance in obesity and in response to a high-fat diet may contribute to induced CYP2E1.

Protein malnutrition

ADH activity is reduced in protein-restricted rats (Lindros et al., 1979; Wilson et al., 1986; Yang et al., 1994) and blood ethanol and acetaldehyde levels were elevated in animals on low-protein diets (Yang et al., 1994). Feeding rats chronically a low-protein, high-carbohydrate diet caused a reduction of both ADH and ALDH2 activities (Lindros et al., 1979; Yang et al., 1994). The rate of ethanol elimination in these animals was decreased.

Effect of fat content or composition of diet

Cha and Sachan (1994) showed that high-saturated-fat diets reduced ethanol elimination rates and raised plasma acylcarnitines, without affecting liver ADH activity. Acetylcarnitine inhibited ADH activity (Sachan and Cha, 1994), raising the possibility that high-fat diets may reversibly inhibit ADH activity *in situ*. High-fat diets induce CYP2E1 in rodents (Yun et al., 1992).

Effect of dietary ethanol

Ethanol is not generally considered to induce hepatic ADH activity (Kishimoto et al., 1995). However, higher ethanol doses, such as those achieved by intragastric infusion (the Tsukamoto–French model), induced liver ADH activity. This resulted from induction of the transcription factor C/EBPβ and suppression of C/EBPγ and a truncated form of C/EBPβ (He et al., 2002). An additional mechanism could be portal endotoxinemia. Mezey's group (Potter et al., 2003) showed that endotoxin can induce ADH mRNA, protein, and activity, mediated by upstream stimulatory factor. However, we know little about the effect of heavy ethanol use on human ADH expression.

Tomita et al. (1992) reported that 1 week of ethanol feeding of mice reduced mitochondrial ALDH activity dramatically, with activity returning to normal after 4 weeks of feeding; while others have reported no effect of ethanol on mitochondrial ALDH2 (Kishimoto et al., 1995), or increased specific activities of high and low K_m ALDH (Aoki and Itoh, 1989).

Heavy ethanol use induces CYP2E1 predominantly by stabilizing the protein against degradation. It has been estimated for forensic work that heavy drinkers and alcoholics may have an increase in AER of as much as 30%, attributed to the induction of CYP2E1 (Jones, 2010).

Miscellaneous dietary effects

Animal studies suggest that there may be effects of thiamine and vitamin A deficiency on ADH activity and AER, but there are no studies to indicate if these effects are relevant to humans. Similarly, dietary zinc deficiency reduced the activity of hepatic ADH (Das et al., 1984); however, a study examining different degrees of zinc deficiency showed no difference in ADH activity between zinc-deficient animals and the pair-fed controls (Indo et al., 1985). Zinc supplementation restored gastric ADH activity in animals fed a liquid, ethanol-containing diet (Caballeria et al., 1997). It is possible that other minor dietary components, such as isoflavones, daidzein, genistein, and their derivatives may have an effect on ADH activity. In parallel, ALDH2 is inhibited by compounds that can react with reactive cysteines such as disulfiram; this might extend to isothiocyanates found in cruciferous vegetables, citral (3,7-dimethyl-2,6-octadienal), a flavoring and fragrance agent, and the isoflavone prunetin. Whether one could ingest sufficient amounts of these vegetables or compounds to affect ethanol-metabolizing enzyme activities *in vivo* has not been determined.

Endocrine effects

Response elements for thyroid hormone, retinoic acid (Harding and Duester, 1992), and glucocorticoid receptors (Winter et al., 1990) are present in the ADH promoters. However, the effects of these hormones *in vivo* are modest (Qulali and Crabb, 1992; Dipple et al., 1993). Growth hormone increased ADH activity (Mezey and Potter, 1979; Potter et al., 1993), while androgens and thyroid hormone (Mezey and Potter, 1981; Dipple et al., 1993) decreased it. By reducing blood levels of testosterone, ethanol increased ADH activity in male rats (Rachamin et al., 1980). Orchiectomy increased AERs in humans (Mezey et al., 1988).

Gastrointestinal and other tissue metabolism of ethanol

Gut metabolism: first-pass metabolism

Ethanol is rapidly absorbed from the gastrointestinal tract, and most of it is metabolized in the liver. When moderate amounts (e.g., 10–30 g) of ethanol are consumed by normal adult human males in the fed state, a sizeable first-pass metabolism of ethanol can be easily detected (that is, the increase in BAC is less when it is taken orally than when the same dose is given intravenously) (Julkunen et al., 1985). First-pass metabolism is important because it reduces the amount of ethanol reaching the blood stream and target organs like the brain, but whether the stomach or liver, or both, is responsible for this phenomenon is debated. The gastrointestinal tract, just like the liver, expresses alcohol-metabolizing enzymes such as cytochrome P450s, ADH, and ALDH. Gastric mucosa contains several ADHs (β-, γ-, and σ- ADH) that can oxidize ethanol (Moreno and Pares, 1991; Seitz and Oneta, 1998). Ingested ethanol is absorbed slowly from the stomach and very rapidly from the upper small intestine. The fraction absorbed transgastrically may be subject to metabolism. Ethanol leaving the gut is then carried to the liver, where a fraction is metabolized before leaving that organ. In favor of gastric metabolism are the facts that peak BACs are higher in individuals with decreased gastric ADH activity, e.g., women (Frezza et al., 1990; Seitz et al., 1993), individuals with atrophic gastritis, alcoholics (DiPadova et al., 1987), and individuals taking certain medications (Roine et al., 1990; Caballería et al., 1991), and individuals who have undergone gastric bypass surgery for obesity (Woodard et al., 2011). σ-ADH, a major gastric ADH isozyme, is absent in the stomach of about 30% of Asian subjects, and as a group, those lacking this enzyme had lower first-pass metabolism of ethanol (Dohmen et al., 1996), suggesting that σ-ADH is important.

However, the assertion that gastric ADH (Yin et al., 1997) or first-pass metabolism (Ammon et al., 1996) is lower in women than men is contested, and other studies have found no correlation between gastric ADH activity and first-pass metabolism (Brown et al., 1995). Calculation of the total ADH activity in the stomach does not account for the amount of metabolized ethanol needed for differences between oral and intravenous area under the curves (AUCs) (Yin et al., 1997). Human and rat σ-ADHs have markedly different kinetic properties; yet, first-pass metabolism in the two species is similar. Hence, first-pass metabolism of ethanol may also occur in the liver. At low rates of gastric emptying (low portal venous ethanol concentrations) ethanol is efficiently extracted, whereas, at higher rates, the relatively low K_m hepatic ADH isozymes are saturated and ethanol reaches the systemic circulation more easily (Levitt and Levitt, 1994).

Ammon et al. (1996) gave ethanol intravenously and deuterated ethanol by mouth or into the duodenum. This permitted an estimate of both gastric and hepatic first-pass metabolism within subjects. They found that first-pass metabolism was about 8–9% of the oral dose, that it did not differ between men and women, and that the gastric contribution to first-pass metabolism was about 6% of the oral dose. It seems safe to conclude that first-pass metabolism of oral ethanol is usually a small fraction (perhaps 10%) of total body ethanol elimination and, when gastric emptying is rapid or the ethanol dose consumed is high, it is quantitatively less important. Gender differences in first-pass metabolism are probably not major (Ammon et al., 1996).

Alcohol-metabolizing enzymes in the nervous system

While high levels of class I ADH mRNA were found in kidney, stomach, duodenum, colon, and uterus of rats, very low levels were found in brain, thymus, muscle, or heart (Estonius et al., 1993). Subsequent analysis using oligonucleotide hybridization showed only class 3 ADH (χ ADH, which does not oxidize ethanol) in brain (Galter et al., 2003). ALDH1 and ALDH2 mRNAs are expressed in a variety of human tissues besides liver (Stewart et al., 1996); low levels of these mRNAs were found in brain. Blood vessels express ALDH, providing a blood–brain barrier to the entry of acetaldehyde from the blood stream. Of interest, CYP2E1 is expressed in the brain and is inducible by ethanol consumption. This has been demonstrated in rat brains (expression in frontal cortex, olfactory bulb, hippocampus, and cerebellum (Howard et al., 2003) and brainstem (Zhong et al., 2012)). This inducible CYP2E1 has been suggested to play a role in the chronic neurotoxicity of ethanol. Zimatkin et al. (2006) directly examined the ability of rat brain homogenates to convert ethanol to acetaldehyde and acetate and concluded, based on the effects of enzyme inhibitors and the behavior of extracts from brains of genetic knockout animals, that both CYP2E1 and catalase contribute to the oxidation of ethanol in the brain.

SYSTEMIC PHARMACOKINETICS AND PHARMACOKINETIC MODELING

Overview – absorption, first-pass metabolism, distribution, and elimination

Systemic ethanol metabolism is almost entirely hepatic, and until recently was exclusively quantified by BAC after oral or intravenous bolus dosing. After the concentration peaks, there is a pseudolinear decline in BAC. This pseudolinear phase results from the major alcohol-oxidizing enzymes having relatively low K_m for ethanol; thus, they are saturated with ethanol until its concentration falls below 40–50 mg%. At this point, the rate of ethanol disappearance begins to slow in a manner consistent with Michaelis–Menten enzyme kinetics. AERs are typically expressed as the amount of ethanol cleared from the blood per unit time (ß60, mg/dL/hour) or as AER (mg/kg body weight/hour). The physiologic effects of ethanol are highly dependent upon the systemic concentration over time – the pharmacodynamic effects of ethanol are heavily influenced by systemic pharmacokinetics – and thus ethanol pharmacokinetics have been examined extensively.

As noted above, ethanol is most efficiently absorbed within the small intestine and thus, the absorption rate and systemic effects are highly dependent upon gastric emptying. Even with dosage normalization, great variability in absorption exists between individuals upon oral ingestion, e.g., a 1.5–2-fold range in peak concentrations (Fig. 4.1) (Ramchandani et al., 2009), thereby creating a challenge for those attempting to investigate the effects of ethanol on various physiologic processes.

For most of the body, the blood leaving the heart passes through only one capillary bed before returning. However, much of the blood supplied to the gut is collected in the portal venous system, which then supplies the liver, before returning to the heart. Thus, the liver receives both an arterial supply, the hepatic artery, and a venous supply originating from the digestive system, which serves to maximize the metabolic processing of absorbed materials prior to systemic delivery. Digestion results in a redistribution of the circulation to the gut, thereby increasing substrate delivery to the liver for

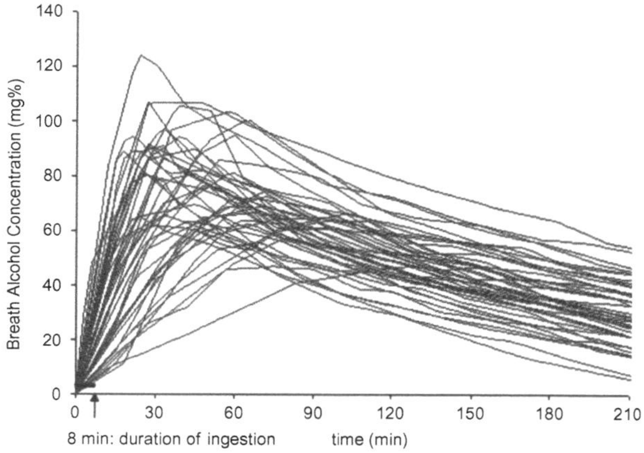

Fig. 4.1. Breath alcohol concentration (BrAC) versus time profiles for 44 subjects following oral alcohol administration of 1 g ethanol per liter total body water. (Data from Ramchandani et al., 2009.) Data represent linear piecewise approximation of BrAC between recorded measurements. Highlighted time section demarcates 8-minute time period of standardized oral alcohol ingestion.

potential metabolism. While the liver begins metabolic processing immediately, it does not have the capacity to metabolize all the ethanol ingested. As previously discussed, the net loss of delivered drug, as measured in the systemic circulation due to the hepatic and gastrointestinal tract influences amongst others, is termed first-pass metabolism.

Virtually all the ethanol that escapes first-pass metabolism is diluted by and distributed in the body's vasculature and peripheral body water compartments (ethanol is a small, highly polarized and water-soluble molecule) until hepatic metabolism is complete, since other remaining routes of elimination (e.g., losses in breath or urine) are negligible. Systemic pharmacokinetic modeling is fundamentally the estimation and subsequent combination of these individual processes – absorption, distribution, and metabolism/elimination, and serves to clarify the contributions of each with changes in BAC over time.

Pharmacokinetic modeling

Pharmacokinetic modeling for ethanol began in the 1930s with Widmark (1932) and continues today in predominantly two forms: phenomenologic and physiologic. Phenomenologic models describe the time series of ethanol concentration in terms of generic compartments, where the number of compartments is the order of the linear dynamical system (i.e., the number of time constants) required to fit the time series (Whitmire et al., 2002). Physiologic models describe the distribution of ethanol in the body in terms of compartments based upon functional anatomic areas/structures (e.g., a liver compartment, including aspects of the portal venous system) and physiologic principles (e.g., conservation of mass) (Whitmire et al., 2002). Depending upon the model, each compartment may have separately defined distribution and elimination kinetics. By contrast, non-physiologic models typically lump many or all of these compartments together and may only attempt to describe the net elimination of ethanol. Thus, physiologic modeling has the potential to correlate genetic, dietary, or environmental changes in alcohol-metabolizing enzymes with changes in pharmacokinetics.

Despite the differences in approach, elimination of ethanol occurs almost exclusively via enzyme systems that obey Michaelis–Menten (MM) kinetics. The MM equation, which is

$$\frac{dC}{dt} = \frac{V_{max} C(t)}{k_m + C(t)}$$

describes the rate of change of concentration as a function of time [$C(t)$] of a particular compound due to the action of an enzyme system, where $V_{\max}$ ([concentration]/[time]) is the maximal metabolism rate, and K_m ([concentration]) is the Michaelis–Menten constant or concentration of the drug at which metabolism is one-half the maximal rate. Note that this definition of K_m is somewhat different from the definition used in enzyme kinetics determined with purified enzymes, in that it assumes that only the concentration of the substrate of interest is subsaturating and changing (and does not take into account concentrations of other substrates or the effects of product inhibition), but it is sufficiently accurate for most models of ethanol metabolism.

As the volume of distribution of ethanol is thought to be the total body water (TBW) space (Jones et al., 1992; Norberg et al., 2001), distribution kinetics are usually described as a diffusion process, where the rate of change of concentration is proportional to the concentration gradient between compartments. Thus, variability in blood flow rates, body compartment water volumes, and hepatic elimination kinetics is also manifest as variability in the time course of ethanol concentration across individuals. Nonetheless, such variability is small in comparison to that due to absorption.

Most physiologically based pharmacokinetic (PBPK) models of ethanol contain one or more compartments, representing functional anatomic areas typically including the portal venous system, and therefore attempt to describe oral and/or intravenous administration of ethanol.

Compartmental/phenomenologic modeling

Numerous investigators have constructed compartmental mass-flow models for ethanol, appropriately tailoring their model architecture to the scientific task at hand, hence only a minority will be discussed. Shoaf (2000) used the human data from previously reported data obtained with constant venous infusion of ethanol to describe a two-compartment model where ethanol is removed from both compartments by two different enzyme systems, where each has MM reaction kinetics. Norberg et al. (2000) also used a two-compartment model with MM elimination to estimate the pharmacokinetic parameters of human subjects who were provided ethanol intravenously. Norberg et al. (2001) then expanded this model to include two MM elimination pathways and compared it to a traditional one-compartment model in the estimation of TBW. Each of these particular models shares several characteristics; they all include MM elimination kinetics, they all assume distribution into the entire TBW space, and they all were used to explain previously recorded breath alcohol concentration data. In contrast, Mumenthaler et al. (2000) provided Caucasian females with weight-standardized drinks and compared several models of elimination. However, each of the models contained some combination of first-order and MM elimination kinetics and was evaluated based upon their fit to experimental data.

Physiologically based modeling

Many investigators have constructed physiologically based pharmacokinetic models for ethanol for a variety of purposes, from observation to attempts at controlling ethanol brain exposure, and in both animal and human subjects. Whitmire et al. (2002) used a five-compartment physiologically based model to describe absorption, distribution, and elimination of ethanol within an animal model. Dogs were given ethanol as a function of body mass; the ethanol was delivered by a feeding tube directly into the stomach. Their study included TBW, stomach mucosa, stomach lumen, duodenal villi, and duodenal lumen compartments. Elimination was modeled as a parallel MM process metabolizing ethanol from both the TBW compartment and the stomach mucosa compartments. In a series of papers described below, Pastino, Sultatos, and colleagues presented extensive results concerning ethanol exposure modeling in the mouse, rat, hamster, and human. These models evolved from a flow-limited PBPK model that included six compartments – lung, rapidly and slowly perfused tissues, fat, brain, liver – which was adapted to include the stomach (Pastino et al., 1996). Pastino et al. (1997) subsequently explored the pharmacokinetics of inhaled ethanol with experimental animal data, with later comparison of their model output with previously published human data. Pastino and Conolly (2000) used this model to investigate first-pass metabolism and bioavailability of orally administered ethanol in male Sprague–Dawley rats. More recently, Martin et al. (2012) presented the development of physiologically based models of the adult, pregnant, and neonatal rat with oral, inhalation, and intravenous routes of ethanol exposure.

Other experimenters have focused on modeling human ethanol metabolism. Pastino et al. (2000) revised their model to include human oral administration data, including the incorporation of ADH polymorphism kinetics, with specific interest in human populations in which the particular isoform might have a clinically significant effect. They included kinetic parameters of ADH1, ADH2, and ADH3 homodimers into single and multiple metabolic equations describing elimination from the gastric and liver compartments and estimated model fitting based upon available oral administration data from Caucasian and African American males. While the data used to generate their simulations were limited to that available in the literature, prediction of blood ethanol concentration and systemic ethanol exposure appeared to be sensitive to the inclusion of multiple isoforms (Pastino et al., 2000).

Sultatos et al. (2004) later augmented their prior modeling efforts with the inclusion of ADH isoform submodels representing hereditary control of these enzymes. They used Monte Carlo sampling methods to quantify the variability in blood ethanol levels output from their model (now consisting of eight compartments: lung, brain, rapidly and slowly perfused tissues, fat, liver, small intestine, and stomach) and compared it to previously published human data sets. In their simulations, peak BAC and time-to-peak BAC were marginally affected (individuals with a β_2 subunit showed decreases in both compared to those without this subunit) while the AUC was very sensitive to ADH genotype (reduced AUC in those with versus without an *ADH1B2* allele, and those with *ADH1C*1/*1* genotype had moderately less AUC than those with *ADH1C*2/*2* status). Furthermore, they reported that the degree of interindividual variability in ethanol pharmacokinetics was not only dependent upon ADH genotype, but also was a function of the ethanol dosage.

At a substantially higher level of detail, Umulis et al. (2005) used a five-compartment structure and modeled both ethanol and its metabolite acetaldehyde, the build-up of which may induce nausea and vomiting and which is the mechanism of protection from alcohol use disorders provided by *ALDH1B* polymorphisms and proposed for ADH variants, discussed above. This model includes four well-mixed compartments (stomach, gastrointestinal, central, and muscle) and a fifth compartment (liver), modeled as a continuous-flowing system with multiple inputs (i.e., a tubular flow reactor). The goal of the investigation by Umulis et al. (2005) was to understand the interaction of ethanol and acetaldehyde kinetics and how changes in enzymatic activity due to genetic variants changed their concentrations. Using previously reported oral administration data, they were able to model the increased acetaldehyde concentrations observed with ALDH2 deficiency and noted that some increases in ethanol concentration could be attributed to the reverse reaction of ADH. Further, they observed that this reverse reaction significantly lowered estimated acetaldehyde concentrations (Umulis et al., 2005).

More recently, Mustavich et al. (2010) adapted aspects of several prior models, including that of Umulis et al. (2005), and data sources to construct a three-compartment model for the purpose of investigating the effect of ADH and ALDH2 kinetic and bitter-taste gene variations on modeled ethanol exposures and subsequent development of alcohol dependence. Exposures were simulated as discrete drinking intervals, with time of consumption determined by acetaldehyde concentration and dosage dependent on TAS2R38 bitter-taste genotype. They then related the eventual development of alcohol dependence to the ethanol AUC as a function of genotype, with the AUC representing a measure of degree of ethanol usage. They reported good agreement between observed and predicted genotype/haplotype frequencies among those with and without alcohol dependence.

Applications

In contrast to studies utilizing models to explore previously recorded data or reevaluate elimination trajectories, O'Connor, Ramchandani, and colleagues have pursued a research program in applied ethanol PBPK modeling. Ramchandani et al. (1999) described a three-compartment PBPK-like model for humans, where distribution kinetics were based in part on physiologically relevant cardiac outflow apportionment. This model is similar in many ways to the PBPK models discussed above, but the motivation for its development was experimental control of the time course of brain ethanol exposure in individual human subjects to facilitate research on the effects of ethanol blood levels or rate of change. The original application was to use the model as part of a system to prescribe an arterial ethanol concentration trajectory, thus a breath alcohol concentration time course, which is identical for all subjects, by determining an appropriate ethanol infusion profile for each. The initially chosen waveform was a linear rise to a target concentration in a specified time period, the maintenance of that concentration for a specified interval, and a measured, but uncontrolled, elimination phase. Beyond exposure control, this waveform provided for a simplified measurement of the AER independent of TBW. If ethanol is infused at a steady-state infusion rate that maintains a constant BAC, a so-called "alcohol clamp," that infusion rate is equal to the rate at which ethanol is removed from the vascular space.

This paradigm has now been used in many investigations at multiple institutions (Morzorati et al., 2002; Subramanian et al., 2002; Neumark et al., 2004; Ramchandani et al., 2006; Yoder et al., 2007; Kareken et al., 2010; Roh et al., 2011; Gilman et al., 2012; and Ralevski et al., 2012, amongst others). Amongst other applications, this work eventually included exploration of gender differences in ethanol metabolism, as described above (Kwo et al., 1998), as well as elucidation of the minimal effect of food composition versus consumption on AER independently of the confound of oral administration (Ramchandani et al., 2001). Subsequent development of the technique has increased the complexity of achievable ethanol exposures (Plawecki et al., 2012; Wetherill et al., 2012) and then intravenous ethanol self-administration paradigms (Zimmermann et al., 2009; Plawecki et al., 2013).

SUMMARY

Generally, independently of their use for ethanol exposure analysis or control, these modeling efforts have provided evidence to support many of the phenomena experienced by ethanol drinkers throughout the world – many factors impact the pharmacokinetics and therefore the pharmacodynamics of ethanol. As ethanol distributes into the TBW space, larger individuals may be less affected by identical doses of ethanol than smaller persons; which is to say they can "hold" more liquor. The rate of ethanol emptying from the stomach into the duodenum can dramatically affect blood and therefore brain ethanol concentration, and may explain some of the wide variation in ethanol pharmacokinetics (Fig. 4.1) and subjective effects after oral dosing. Moreover, gastric emptying may be affected by emotional state (e.g., Geeraerts et al., 2005), food (Jones and Jonsson, 1994), and surgical bypass (Woodard et al., 2011). Many people advocate eating before consumption of ethanol to avoid "drinking on an empty stomach" or can relate differential effects of the same drink on different days, reflecting the variability in brain exposure. Evidence has shown that the rate of change of ethanol brain exposure may also affect subjective response and be indicative of risk (King et al., 2011; Quinn and Fromme, 2011; Wetherill et al., 2012, amongst others). Hence, the same ethanol dose consumed differently may alter response, which relates to the development of many varied drinking styles (e.g. "shots," "chugging," etc.).

The second important aspect of recent studies of ethanol elimination is the variability observed between individuals. The rate of ethanol clearance from the blood in the pseudolinear segment of the elimination curve varies by two- to threefold between individuals (Martin et al., 1985). The test–retest reliability in the oral ethanol challenge method of determining AER is open to criticism, but significant between-individual variation was recently confirmed using intravenous administration techniques (O'Connor et al., 1998, 2000; Ramchandani et al., 2009). The reasons for this variation are incompletely understood, and physiologic modeling has not explained this entirely.

The explanation for these differences is likely to be a complex interplay between all the factors that influence AER (Fig. 4.2). These include differences in ethanol concentration and non-ethanol components of beverages (which would cause variability outside that observed in the laboratory setting), differences in rates of gastric emptying and possibly the complement of ethanol metabolizing enzymes in the gastric mucosa, physiology of the small bowel, and the rate of portal blood flow. Variation in the activity of enzymes catalyzing ethanol oxidation, including genetic, endocrine, and dietary differences, alter expression of the enzymes or the ability of the liver to dispose of acetaldehyde and regenerate NAD +from NADH. The ultimate mathematic model would need to incorporate the actual *in situ* activity of the enzymes (including an assessment of total hepatic enzyme activity, i.e., liver mass and enzyme activity per gram of liver) plus the effects of alterations of substrate (ethanol and NAD+) and products (NADH and acetaldehyde) on the steady-state activity of each step in the pathway. Such an approach was successfully applied to rats (Crabb et al., 1983).

In addition to gastrointestinal and hepatic disposition of ethanol, pharmacokinetics could be altered by changes in ethanol distribution (changes in TBW, body fat, and lean body mass, such as occur between men and women and with aging). Finally, there may be small effects of tissue metabolism of ethanol, and perhaps quantitatively more importantly, the effect of local generation of acetaldehyde or alteration in tissue metabolism caused by redox shifts occurring with ethanol oxidation. If all of these factors could be assessed and modeled in humans, the resulting differences in alcohol pharmacokinetics (conceptually similar to the effects of major allelic variation in ADH and ALDHs) might illuminate some of the between-individual differences which affect the risk of developing alcohol use disorders.

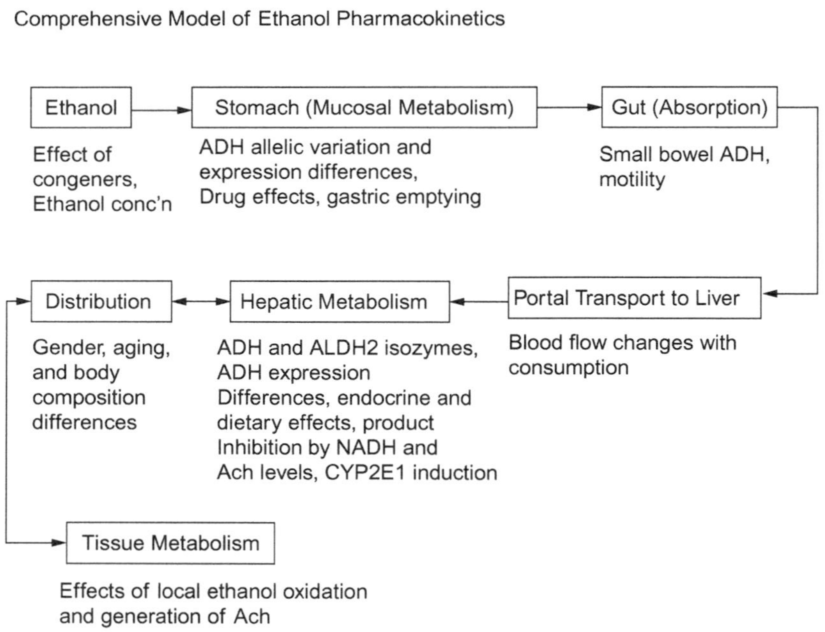

Fig. 4.2. Contributors to ethanol pharmacokinetic variability. To explore the pharmacodynamic effect of ethanol adequately, and thus the mechanisms by which individuals abuse or become dependent upon the substance, investigators must compensate or control for the pharmacokinetic variability of ethanol. This is typically accomplished with modeling techniques. Accurate modeling requires biologic, and therefore mathematical, description of absorption, distribution (and redistribution) and metabolism/elimination kinetics, each of which is independently subject to significant within and between subject variability. A summary of these factors and their interdependence is depicted. ADH, alcohol dehydrogenase; Ach, acetylcholine; Cyp2E1, Cytochrome P4502E1; ALDH2, aldehyde dehydrogenase class II.

ACKNOWLEDGMENTS

MHP and DWC are supported by the Indiana Alcohol Research Center (P60 AA07611, Center on Genetic Determinants of Alcohol Ingestion), funded by the National Institute on Alcohol Abuse and Alcoholism. DWC is also supported by a grant from the Substance Abuse and Mental Health Services Administration (SAMHSA 1U79TI020281, SBIRT Implementation in the Indiana University Educational Innovations Project).

References

Agarwal DP (2001). Genetic polymorphisms of alcohol metabolizing enzymes. Pathol Biol (Paris) 49: 703–709.

Ammon E, Schafer C, Hoffman U et al. (1996). Disposition and first pass metabolism of ethanol in humans: is it gastric or hepatic and does it depend on gender? Clin Pharmacol Toxicol 59: 503–513.

Aoki Y, Itoh H (1989). Effects of alcohol consumption on mitochondrial aldehyde dehydrogenase isoenzymes in rat liver. Enzyme 41: 151–158.

Arthur MJ, Lee A, Wright R (1984). Sex differences in the metabolism of ethanol and acetaldehyde in normal subjects. Clin Sci 67: 397–401.

Baraona E, Abittan CS, Dohmen K et al. (2001). Gender differences in pharmacokinetics of alcohol. Alcohol Clin Exp Res 25: 502–507.

Bosron WF, Li T-K (1987). Catalytic properties of human liver alcohol dehydrogenase isoenzymes. Enzyme 37: 19–28.

Bosron WF, Crabb DW, Housinger TA et al. (1984). Effect of fasting on the activity and turnover of rat liver alcohol dehydrogenase. Alcohol Clin Exp Res 8: 196–200.

Brown AStJM, Flatatrone JR, Wood P et al. (1995). The effect of gastritis on human gastric alcohol dehydrogenase activity and ethanol metabolism. Aliment Pharmacol Ther 9: 57–61.

Burnell JC, Bosron WF (1989). Genetic polymorphism of human liver alcohol dehydrogenase and kinetic properties of the isoenzymes. In: KE Crow, RD Batt (Eds.), Human metabolism of alcohol. 11th edn. CRC Press, Boca Raton, FL, pp. 65–75.

Caballería J, Baraona E, Deulafeu R et al. (1991). Effects of H2-receptor antagonists on gastric alcohol dehydrogenase activity. Dig Dis Sci 36: 1673–1679.

Caballeria J, Gimenez A, Andreu H et al. (1997). Zinc administration improves gastric alcohol dehydrogenase activity and first-pass metabolism of ethanol in alcohol-fed rats. Alcohol Clin Exp Res 21: 1619–1622.

Cha YS, Sachan DS (1994). Opposite effects of dietary saturated and unsaturated fatty acids on ethanol-pharmacokinetics, triglycerides and carnitines. J Am Coll Nutr 13: 338–343.

Chou WY, Stewart MJ, Carr LG et al. (1999). An A/G polymorphism in the promoter of mitochondrial aldehyde dehydrogenase (ALDH2): effects of the sequence variant on transcription factor binding and promoter strength. Alcohol Clin Exp Res 23: 963–968.

Cole-Harding S, Wilson JR (1987). Ethanol metabolism in men and women. J Stud Alcohol 48: 380–387.

Couzigou P, Fleury B, Groppi A et al. (1991). Role of alcohol dehydrogenase polymorphism in ethanol metabolism and alcohol-related diseases. Adv Exp Med Biol 284: 263–270.

Crabb DW, Bosron WF, Li TK (1983). Steady-state kinetic properties of purified rat liver alcohol dehydrogenase: application to predicting alcohol elimination rates in vivo. Arch Biochem Biophys 224: 299–309.

Das I, Burch RE, Hahn HK (1984). Effects of zinc deficiency on ethanol metabolism and alcohol and aldehyde dehydrogenase activities. J Lab Clin Med 104: 610–617.

DiPadova D, Worner TM, Julkunen RJK et al. (1987). Effects of fasting and chronic alcohol consumption on the first-pass metabolism of ethanol. Gastroenterology 92: 1169–1173.

Dipple KM, Qulali M, Ross RA et al. (1993). Effects of thyroxine on the expression of alcohol dehydrogenase in rat liver and kidney. Hepatology 17: 701–706.

Dohmen K, Baraona E, Ishibashi H et al. (1996). Ethnic differences in gastric sigma alcohol dehydrogenase and ethanol first pass metabolism. Alcohol Clin Exp Res 20: 1569–1576.

Edenberg HJ (2000). Regulation of the mammalian alcohol dehydrogenase genes. Prog Nucleic Acid Res Mol Biol 64: 295–341.

Edenberg HJ, Bosron WF (1997). Alcohol dehydrogenase. In: FP Guengerich (Ed.), Biotransformation. Pergamon, New York, pp. 119–131.

Enomoto N, Takase S, Yasuhara M et al. (1991). Acetaldehyde metabolism in different aldehyde dehydrogenase 2 genotypes. Alcohol Clin Exp Res 15: 141–144.

Estonius M, Danielsson O, Karlsson C et al. (1993). Distribution of alcohol and sorbitol dehydrogenases. Assessment of mRNA species in mammalian tissues. Eur J Biochem 215: 497–503.

Frezza M, DiPadova C, Pozzato G et al. (1990). High blood alcohol levels in women. The role of decreased gastric alcohol dehydrogenase activity and first-pass metabolism. N Engl J Med 322: 95–99.

Galter D, Carmine A, Buervenich S et al. (2003). Distribution of class I, III and IV alcohol dehydrogenase mRNAs in the adult rat, mouse and human brain. Eur J Biochem 270: 1316–1326.

Geeraerts B, Vandenberghe J, Van Oudenhove L et al. (2005). Influence of experimentally induced anxiety on gastric sensorimotor function in humans. Gastroenterology 129: 1437–1444.

Gilman JM, Ramchandani VA, Crouss T et al. (2012). Subjective and neural responses to intravenous alcohol in young adults with light and heavy drinking patterns. Neuropsychopharmacology 37: 467–477.

Goedde HW, Agarwal DP, Harada S et al. (1983). Population genetic studies of aldehyde dehydrogenase isozyme deficiency and alcohol sensitivity. Am J Hum Genet 35: 769–772.

Goist Jr KC, Sutker PB (1985). Acute alcohol intoxication and body composition in women and men. Pharmacol Biochem Behav 22: 811–814.

Gonzalez FJ (2007). The 2006 Bernard B. Brodie Award Lecture. Cyp2e1. Drug Metab Dispos 35: 1–8.

Harada S, Agarwal DP, Goedde HW (1981). Aldehyde dehydrogenase deficiency as cause of facial flushing reaction to alcohol in Japanese. Lancet 2: 982.

Harada S, Agarwal DP, Goedde HW et al. (1983). Aldehyde dehydrogenase isozyme variation and alcoholism in Japan. Pharmacol Biochem Behav 18: 151–153.

Harada S, Okubo T, Nakamura T et al. (1999). A novel polymorphism (-357 G/A) of the ALDH2 gene: linkage disequilibrium and an association with alcoholism. Alcohol Clin Exp Res 23: 958–962.

Harding PP, Duester G (1992). Retinoic acid activation and thyroid hormone repression of the human alcohol dehydrogenase gene ADH3. J Biol Chem 267: 14145–14150.

He L, Ronis MJ, Badger TM (2002). Ethanol induction of class I alcohol dehydrogenase expression in the rat occurs through alterations in CCAAT/enhancer binding proteins beta and gamma. J Biol Chem 277: 43572–43577.

Hempel J, Kaiser R, Jörnvall H (1984). Human liver mitochondrial aldehyde dehydrogenase: a C-terminal segment positions and defines the structure corresponding to the one reported to differ in the Oriental enzyme variant. FEBS Lett 173: 367–373.

Higuchi S, Muramatsu T, Shigemori K et al. (1992). The relationship between low Km aldehyde dehydrogenase phenotype and drinking behavior in Japanese. J Stud Alcohol 53: 170–175.

Higuchi S, Matsushita S, Imazeki H et al. (1994). Aldehyde dehydrogenase genotypes in Japanese alcoholics. Lancet 343: 741–742.

Howard LA, Miksys S, Hoffmann E et al. (2003). Brain CYP2E1 is induced by nicotine and ethanol in rat and is higher in smokers and alcoholics. Br J Pharmacol 138: 1376–1386.

Indo Y, Nagata N, Higashi A et al. (1985). Effects of dietary zinc deficiency on hepatic ornithine carbamoyltransferase and alcohol dehydrogenase activities in rats. J Pediatr Gastroenterol Nutr 4: 268–273.

Jones AW (2010). Evidence-based survey of the elimination rates of ethanol from blood with applications in forensic casework. Forensic Sci Int 200: 1–20.

Jones AW, Jonsson KA (1994). Food-induced lowering of blood-ethanol profiles and increased rate of elimination immediately after a meal. J Forensic Sci 39: 1084–1093.

Jones AW, Hahn RG, Stalberg HP (1992). Pharmacokinetics of ethanol in plasma and whole blood: estimation of total body water by the dilution principle. Eur J Clin Pharmacol 42: 445–448.

Julkunen RJK, DiPadova C, Lieber CS (1985). First pass metabolism of ethanol – a gastrointestinal barrier against the systemic toxicity of ethanol. Life Sci 37: 567–573.

Kareken DA, Bragulat V, Dzemidzic M et al. (2010). Family history of alcoholism mediates the frontal response to alcoholic drink odors and alcohol in at-risk drinkers. Neuroimage 50: 267–276.

King AC, deWit H, McNamara PJ et al. (2011). Rewarding, stimulant, and sedative alcohol responses and relationship to future binge drinking. Arch Gen Psychiatry 68: 389–399.

Kishimoto R, Fujiwara I, Kitayama S et al. (1995). Changes in hepatic enzyme activities related to ethanol metabolism in mice following chronic ethanol administration. J Nutr Sci Vitaminol (Tokyo) 41: 527–543.

Kwo PY, Ramchandani VA, O'Connor S et al. (1998). Gender differences in alcohol metabolism: relationship to liver volume and effect of adjusting for body mass. Gastroenterology 115: 1552–1557.

Lakshman MR, Chambers LL, Chirtel SJ et al. (1988). Roles of hormonal and nutritional factors in the regulation of rat liver alcohol dehydrogenase activity and ethanol elimination rate in vivo. Alcohol Clin Exp Res 12: 407–411.

Levitt MD, Levitt DG (1994). The critical role of the rate of ethanol absorption in the interpretation of studies purporting to demonstrate gastric metabolism of ethanol. J Pharmacol Exp Ther 269: 297–304.

Li D, Zhao H, Gelernter J (2012). Further clarification of the contribution of the ADH1C gene to vulnerability of alcoholism and selected liver diseases. Hum Genet 131: 1361–1374.

Lieber CS (1999). Microsomal ethanol-oxidizing system (MEOS): the first 30 years (1968–1998) – a review. Alcohol Clin Exp Res 23: 991–1007.

Lindahl R (1992). Aldehyde dehydrogenases and their role in carcinogenesis. Crit Rev Biochem Mol Biol 27: 283–335.

Lindros KO, Pekkanen L, Koivula T (1979). Enzymatic and metabolic modification of hepatic ethanol and acetaldehyde oxidation by the dietary protein level. Biochem Pharmacol 28: 2313–2320.

Lumeng L, Crabb DW (1984). Rate-determining factors for ethanol metabolism in fasted and castrated male rats. Biochem Pharmacol 33: 2623–2628.

Lumeng L, Bosron WF, Li TK (1979). Quantitative correlation of ethanol elimination rates in vivo with liver alcohol dehydrogenase activities in fed, fasted and food-restricted rats. Biochem Pharmacol 28: 1547–1551.

Marshall AW, Kingstone D, Boss M et al. (1983). Ethanol elimination in males and females: relationship to menstrual cycle and body composition. Hepatology 3: 701–706.

Martin NG, Perl J, Oakeshott JG et al. (1985). A twin study of ethanol metabolism. Behav Genet 15: 93–109.

Martin SA, McLanahan ED, El-Masri H et al. (2012). Development of multi-route physiologically-based pharmacokinetic models for ethanol in the adult, pregnant, and neonatal rat. Inhal Toxicol 24: 698–722.

Mezey E, Potter JJ (1979). Rat liver alcohol dehydrogenase activity: effects of growth hormone and hypophysectomy. Endocrinology 104: 1667–1673.

Mezey E, Potter JJ (1981). Effects of thyroidectomy and triiodothyronine administration on rat liver alcohol dehydrogenase. Gastroenterology 80: 566–574.

Mezey E, Oesterling JE, Potter JJ (1988). Influence of male hormones on rates of ethanol elimination in man. Hepatology 8: 742–744.

Mezey E, Sharma S, Rennie L et al. (1992). Sex differences in gastric alcohol dehydrogenase activity in Sprague–Dawley rats. Gastroenterology 103: 1804–1810.

Mishra L, Sharma S, Potter JJ et al. (1989). More rapid elimination of alcohol in women as compared with their male siblings. Alcohol Clin Exp Res 13: 552–554.

Mizoi Y, Ijiri I, Tatsuno Y et al. (1979). Relationship between facial flushing and blood acetaldehyde levels after alcohol intake. Pharmacol Biochem Behav 10: 303–311.

Mizoi Y, Yamamoto K, Ueno Y et al. (1994). Involvement of genetic polymorphism of alcohol and aldehyde dehydrogenases in individual variation of alcohol metabolism. Alcohol 29: 707–710.

Moreno A, Pares X (1991). Purification and characterization of a new alcohol dehydrogenase from human stomach. J Biol Chem 266: 1128–1133.

Morzorati SL, Ramchandani VA, Flury L et al. (2002). Self-reported subjective perception of intoxication reflects family history of alcoholism when breath alcohol levels are constant. Alcohol Clin Exp Res 26: 1299–1306.

Mumenthaler MS, Taylor JL, Yesavage JA (2000). Ethanol pharmacokinetics in white women: nonlinear model fitting versus zero-order elimination analyses. Alcohol Clin Exp Res 24: 1353–1362.

Mustavich LF, Miller P, Kidd KK et al. (2010). Using a pharmacokinetic model to relate an individual's susceptibility to alcohol dependence to genotypes. Hum Hered 70: 177–193.

Neafsey P, Ginsberg G, Hattis D et al. (2009). Genetic polymorphism in CYP2E1: population distribution of CYP2E1 activity. J Toxicol Environ Health B Crit Rev 12: 362–388.

Neumark YD, Friedlander Y, Thomasson HR et al. (1998). Association of the ADH2*2 allele with reduced ethanol consumption in Jewish men in Israel: a pilot study. J Stud Alcohol 59: 133–139.

Neumark YD, Friedlander Y, Durst R et al. (2004). Alcohol dehydrogenase polymorphisms influence alcohol-elimination rates in a male Jewish population. Alcohol Clin Exp Res 28: 10–14.

Norberg A, Gabrielsson J, Jones AW et al. (2000). Within- and between-subject variations in pharmacokinetic parameters of ethanol by analysis of breath, venous blood, and urine. Br J Clin Pharmacol 49: 399–408.

Norberg A, Sandhagen B, Bratteby LE et al. (2001). Do ethanol and deuterium oxide distribute into the same water space in healthy volunteers? Alcohol Clin Exp Res 25: 1423–1430.

O'Connor S, Morzorati S, Christian J et al. (1998). Clamping BrAC reduces experimental variance: application to the study of acute tolerance to alcohol and alcohol elimination rate. Alcohol Clin Exp Res 22: 202–210.

O'Connor S, Ramchandani VA, Li TK (2000). PBPK modeling as a basis for achieving a steady BrAC of 60 +/– 5 mg% within ten minutes. Alcohol Clin Exp Res 24: 426–427.

Parés X, Farrés J, Parés A et al. (1994). Genetic polymorphism of liver alcohol dehydrogenase in Spanish subjects: significance of alcohol consumption and liver disease. Alcohol Alcohol 29: 701–705.

Pastino GM, Conolly RB (2000). Application of a physiologically based pharmacokinetic model to estimate the bioavailability of ethanol in male rats: distinction between gastric and hepatic pathways of metabolic clearance. Toxicol Sci 55: 256–265.

Pastino GM, Sultatos LG, Flynn EJ (1996). Development and application of a physiologically based pharmacokinetic model for ethanol in the mouse. Alcohol Alcohol 31: 365–374.

Pastino GM, Sultatos LG, Flynn EJ (1997). A comparison of physiologically based pharmacokinetic model predictions and experimental data for inhaled ethanol in male and female B6C3F1 mice, F344 rats, and humans. Toxicol Appl Pharmacol 145: 147–157.

Pastino GM, Flynn EF, Sultatos LG (2000). Genetic polymorphisms in ethanol metabolism: issues and goals for physiologically based pharmacokinetic modeling. Drug Chem Toxicol 23: 179–201.

Plawecki MH, Zimmermann US, Vitvitskiy V et al. (2012). Alcohol exposure rate control through physiologically-based pharmacokinetic (PBPK) modeling. Alcohol Clin Exp Res 36: 1042–1049.

Plawecki MH, Wetherill L, Vitvitskiy V et al. (2013). Voluntary intravenous self-administration of alcohol detects an interaction between GABAergic manipulation and GABRG1 polymorphism genotype: a pilot study. Alcohol Clin Exp Res 37 (suppl 1): E152–E160.

Potter JJ, Yang VW, Mezey E (1993). Regulation of the rat class I alcohol dehydrogenase gene by growth hormone. Biochem Biophys Res Commun 191: 1040–1045.

Potter JJ, Rennie-Tankersley L, Mezey E (2003). Endotoxin enhances liver alcohol dehydrogenase by action through upstream stimulatory factor but not by nuclear factor-kB. J Biol Chem 278: 4353–4357.

Quinn PD, Fromme K (2011). Subjective response to alcohol challenge: a quantitative review. Alcohol Clin Exp Res 35: 1759–1770.

Qulali M, Crabb DW (1992). Corticosterone induces rat liver alcohol dehydrogenase mRNA but not enzyme protein or activity. Alcohol Clin Exp Res 16: 427–431.

Rachamin G, MacDonald JA, Wahid S et al. (1980). Modulation of alcohol dehydrogenase and ethanol metabolism by sex hormones in the spontaneously hypertensive rat. Effect of chronic ethanol administration. Biochem J 186: 489–490.

Ralevski E, Perry Jr EB, D'Souza DC et al. (2012). Preliminary findings on the interactive effects of IV ethanol and IV nicotine on human behavior and cognition: a laboratory study. Nicotine Tob Res 14: 596–606.

Ramchandani VA, Bolane J, Li TK (1999). A physiologically based pharmacokinetic (PBPK) model for alcohol facilitates rapid BrAC clamping. Alcohol Clin Exp Res 23: 617–623.

Ramchandani VA, Kwo PY, Li TK (2001). Effect of food and food composition on alcohol elimination rates in healthy men and women. J Clin Pharmacol 41: 1345–1350.

Ramchandani VA, O'Connor S, Neumark Y et al. (2006). The alcohol clamp: Applications, challenges, and new directions – An RSA 2004 symposium summary. Alcohol Clin Exp Res 30: 155–164.

Ramchandani VA, Plawecki M, Li TK et al. (2009). Intravenous ethanol infusions can mimic the time course of breath alcohol concentrations following oral alcohol administration in healthy volunteers. Alcohol Clin Exp Res 33: 938–944.

Roh S, Matsushita S, Hara S et al. (2011). Role of GABRA2 in moderating subjective responses to alcohol. Alcohol Clin Exp Res 35: 400–407.

Roine R, Gentry RT, Hernández-Munõz R et al. (1990). Aspirin increases blood alcohol concentrations in humans after ingestion of ethanol. JAMA 264: 2406–2408.

Sachan DS, Cha YS (1994). Acetylcarnitine inhibits alcohol dehydrogenase. Biochem Biophys Res Commun 203: 1496–1501.

Seitz HK, Oneta CM (1998). Gastrointestinal alcohol dehydrogenase. Nutr Rev 56: 52–60.

Seitz HK, Egerer G, Simanowski UA et al. (1993). Human gastric alcohol dehydrogenase activity: effect of age, sex, and alcoholism. Gut 34: 1433–1437.

Shoaf S (2000). Pharmacokinetics of intravenous alcohol: two compartment, dual Michaelis–Menten elimination. Alcohol Clin Exp Res 24: 424–425.

Stewart MJ, Malek K, Crabb DW (1996). Distribution of messenger RNAs for aldehyde dehydrogenase 1, aldehyde dehydrogenase 2, and aldehyde dehydrogenase 5 in human tissues. J Invest Med 44: 42–46.

Subramanian MG, Heil SH, Kruger ML et al. (2002). A three-stage alcohol clamp procedure in human subjects. Alcohol Clin Exp Res 26: 1479–1483.

Sultatos LG, Pastino GM, Rosenfeld CA et al. (2004). Incorporation of the genetic control of alcohol dehydrogenase into a physiologically based pharmacokinetic model for ethanol in humans. Toxicol Sci 78: 20–31.

Sutker PB, Tabakoff B, Goist Jr KC et al. (1983). Acute alcohol intoxication, mood states and alcohol metabolism in women and men. Pharmacol Biochem Behav 18 (Suppl 1): 349–354.

Tanaka F, Shiratori Y, Yokosuka O et al. (1997). Polymorphism of alcohol-metabolizing genes affects drinking behavior and alcoholic liver disease in Japanese men. Alcohol Clin Exp Res 21: 596–601.

Thomasson HR (1995). Gender differences in alcohol metabolism. Physiological responses to ethanol. Recent Dev Alcohol 12: 163–179.

Thomasson HR, Edenberg HJ, Crabb DW et al. (1991). Alcohol and aldehyde dehydrogenase genotypes and alcoholism in Chinese men. Am J Hum Genet 48: 677–681.

Thomasson HR, Crabb DW, Edenberg HJ et al. (1994). Low frequency of the ADH2*2 allele among Atayal natives of Taiwan with alcohol use disorders. Alcohol Clin Exp Res 18: 640–643.

Thomasson HR, Beard J, Li T-K (1995). ADH gene polymorphisms are determinants of alcohol pharmacokinetics. Alcohol Clin Exp Res 19: 1494–1499.

Tomita Y, Haseba T, Kurosu M et al. (1992). Effects of chronic ethanol intoxication on aldehyde dehydrogenase in mouse liver. Alcohol Alcohol 27: 171–180.

Umulis DM, Gurmen NM, Singh P et al. (2005). SHSA physiologically based model for ethanol and acetaldehyde metabolism in human beings. Alcohol 335: 3–12.

Wetherill L, Morzorati SL, Foroud T et al. (2012). Subjective perceptions associated with the ascending and descending slopes of breath alcohol exposure vary with recent drinking history. Alcohol Clin Exp Res 36: 1050–1057.

Whitmire D, Cornelius L, Whitmire P (2002). Monte Carlo simulation of an ethanol pharmacokinetic model. Alcohol Clin Exp Res 26: 1484–1493.

Widmark E (1932). Verteilung und Unwandtung des Aethyl Alcohols im Organismus des Hundes. Biochem Z 267: 128–134.

Wilson JS, Korsten MA, Lieber CS (1986). The combined effects of protein deficiency and chronic ethanol administration on rat ethanol metabolism. Hepatology 6: 823–829.

Winter LA, Stewart MJ, Shean ML et al. (1990). A hormone response element upstream from the human alcohol dehydrogenase gene ADH2 consists of three tandem glucocorticoid receptor binding sites. Gene 91: 233–240.

Woodard GA, Downey J, Hernandez-Boussard T et al. (2011). Impaired alcohol metabolism after gastric bypass surgery: a case-crossover trial. J Am Coll Surg 212: 209–214.

Xiao Q, Weiner H, Crabb DW (1996). The mutation in the mitochondrial aldehyde dehydrogenase (ALDH2) gene responsible for alcohol-induced flushing increases turnover of the enzyme tetramers in a dominant fashion. J Clin Invest 98: 2027–2032.

Yang SC, Ito M, Furukawa Y et al. (1994). Comparative study of alcohol metabolism in stroke-prone spontaneously hypertensive rats and Wistar-Kyoto rats fed normal or low levels of dietary protein. J Nutr Sci Vitaminol (Tokyo) 40: 547–555.

Yin S-J, Liao C-S, Wu C-W et al. (1997). Human stomach alcohol and aldehyde dehydrogenases: comparison of expression pattern and activities in alimentary tract. Gastroenterology 112: 766–775.

Yoder KK, Constantinescu CC, Kareken DA et al. (2007). Heterogeneous effects of alcohol on dopamine release in the striatum: a PET study. Alcohol Clin Exp Res 31: 965–973.

Yoshida A, Rzhetsky A, Hsu LC et al. (1998). Human aldehyde dehydrogenase gene family. Eur J Biochem 251: 549–557.

You M, Fischer M, Cho WK et al. (2002). Transcriptional control of the human aldehyde dehydrogenase 2 promoter by hepatocyte nuclear factor 4: inhibition by cyclic AMP and COUP transcription factors. Arch Biochem Biophys 398: 79–86.

Yun YP, Casazza JP, Sohn DH et al. (1992). Pretranslational activation of cytochrome P450IIE during ketosis induced by a high fat diet. Mol Pharmacol 41: 474–479.

Zeiner AR, Paredes A, Christiansen DH (1979). The role of acetaldehyde in mediating reactivity to an acute dose of ethanol among different racial groups. Alcohol Clin Exp Res 3: 11–18.

Zhong Y, Dong G, Luo H et al. (2012). Induction of brain CYP2E1 by chronic ethanol treatment and related oxidative stress in hippocampus, cerebellum, and brainstem. Toxicology 302: 275–284.

Zimatkin SM, Pronko SP, Vasiliou V et al. (2006). Enzymatic mechanisms of ethanol oxidation in the brain. Alcohol Clin Exp Res 30: 1500–1505.

Zimmermann US, Mick I, Laucht M et al. (2009). Offspring of parents with an alcohol use disorder prefer higher levels of brain alcohol exposure in experiments involving computer-assisted self-infusion of ethanol (CASE). Psychopharmacology (Berl) 202: 689–697.

Handbook of Clinical Neurology, Vol. 125 (3rd series)
Alcohol and the Nervous System
E.V. Sullivan and A. Pfefferbaum, Editors

Chapter 5

Use of animal models of alcohol-related behavior

JOHN C. CRABBE*

Portland Alcohol Research Center and Department of Behavioral Neuroscience, Oregon Health & Science University, Portland, OR, USA

INTRODUCTION

Alcoholism (alcohol dependence and alcohol use disorder, henceforth AUD) is quintessentially behavioral in nature. Because alcohol dependence depends for its development and maintenance on the self-administration of alcohol, both the public and, to a lesser extent, the medical community have been slow to recognize their inherently biologic basis. Many lay persons are still surprised to find that AUD is now considered a disease much like any other, and that successful treatment does not often simply consist of a decision to "just say no." Hence the continued search for novel therapeutic compounds to supplement the three drugs currently approved by the Food and Drug Administration to treat AUD in the United States (http://www.niaaa.nih.gov/research/major-initiatives/medications-development-program). The neural mechanisms by which disulfiram (Antabuse), naltrexone (Revia, Vivitrol, and Naltrel) and acamprosate (Campral) help control drinking in some patients are discussed in Chapter 30. Although the benzodiazepines diazepam (Valium) and alprazolam (Xanax) have been approved and are regularly used to treat acute alcohol withdrawal symptoms, many patients never receive continued therapy with any of the three agents that essentially seek to prevent relapse (Saxon and McCarty, 2005).

PHARMACOLOGIC CONSIDERATIONS

Because the criteria for a *Diagnostic and Statistical Manual of Mental Disorders*, 4th edition, text revision (DSM-IV-TR: American Psychiatric Association, 2000) diagnosis of alcohol dependence are based on a maladaptive pattern of alcohol use (see Chapter 1), understanding the behavioral aspects of AUD requires consideration of the pharmacology of the drug, ethyl alcohol. For studies with animal models, the experimenter can arrange access to and delivery of specific doses of drugs to the subject. Most drugs of abuse have specific receptor proteins that gate their initial actions in the brain. Because receptors are not distributed uniformly in brain regions, drug effects are often local to specific brain regions, thereby providing targets for an antidote (e.g., antagonist drugs). Alcohol, however, is a highly unusual drug of abuse. It is relatively impotent, with effective doses of g/kg body weight, as compared, for example, to cocaine (mg/kg) or LSD-25 (μg/kg). Thus, the doses ingested orally require huge investment of liver enzymes (see Chapters 4 and 34). In addition, it is not particularly safe, as the therapeutic index is about 10:1 (i.e., the lethal dose is only an order of magnitude greater than the effective dose). It distributes preferentially into water relative to fat and reaches the entire brain rapidly after ingestion. Although ethanol appears to bind specific pockets in some ion channels, there are no obviously specific antagonist drugs (see Chapter 6). At effective doses, it affects virtually all neurotransmitter systems (Vengeliene et al., 2008). The behavioral responses to alcohol are dose-dependent, and the most prominent feature is the inverted-U shape of alcohol's dose–response effects on general arousal. Low doses are behaviorally stimulating and disinhibitory (the first-drink-at-the-party effect). Moderate doses reveal the characteristics that lead to the drug's classification as a sedative-hypnotic, and as doses become progressively higher, general anesthesia, stupor, or coma ensues and the individual may die for any of several reasons. All these characteristics make alcohol a difficult target for studies attempting to map neuropharmacologic effects and specific neural circuits on to specific behavioral endpoints.

*Correspondence to: John C. Crabbe, Ph.D., Portland Alcohol Research Center, Professor, Department of Behavioral Neuroscience, Oregon Health & Science University, Senior Research Career Scientist, VA Medical Center (R&D 12), 3710 SW US Veterans Hospital Road, Portland, OR 97239, USA. Tel: +1-503-273-5298, Fax: +1-503-721-1029, E-mail: crabbe@ohsu.edu

AUD IS BEHAVIORALLY AND GENETICALLY COMPLEX

Earning a diagnosis of alcohol dependence requires presentation of any three of seven symptom clusters during a 12-month period (see Chapter 1). Five of these seven are clearly behavioral in nature, or describe interoceptive states (e.g., "persistent desire or unsuccessful efforts to cut down or control use"). Because an individual can earn an alcohol dependence diagnosis with any three symptoms, the DSM-IV-TR distinguishes cases with or without physiologic dependence, where the former requires presentation of either tolerance or withdrawal (or, most commonly, both). Revisions in DSM-V are expected to place even more emphasis on psychosocial aspects of AUD (Ahmed, 2012). This categoric system ensures diagnostic heterogeneity within any group of alcohol-dependent subjects, and we presume that this reflects biologic heterogeneity as well. Abundant evidence from twin, family, and adoption studies supports a strong contribution (heritabilities of about 0.4–0.6) of genetics to AUD risk (Cloninger et al., 1981; Kendler et al., 1994; Enoch and Goldman, 2001; Goldman and Ducci, 2007). Most human studies of genetic risk presume explicitly or implicitly that any underlying genetic risk factors in the population of alcohol-dependent subjects reflect a single dimension of genetic risk, but this may not be the case (Kendler et al., 2012).

WHAT CAN BE MODELED?

A basic distinction can be drawn between the two general targets of animal modeling efforts. First, each of the many neurologically relevant behaviors affected by alcohol must be modeled in the laboratory. For example, consider the motor systems and postural instability discussed in Chapter 15. Myriad behavioral assays are commonly used to model these human behaviors in rats and mice (e.g., fixed-speed rotarod, accelerating rotarod, rotating drum, treadmill, with or without barriers, balance beam, dowel balancing, platform test of ambulatory ataxia). Each behavioral domain encompassed within AUD discussed in the chapters of this volume yields a similarly diverse list of rodent assays. Most of these phenotypic assays can be employed to study response to acute ethanol: many can also be used to follow the development of tolerance and to provide an index of behavioral disruption during withdrawal following alcohol dependence. It is well beyond the scope of this review to consider the voluminous literature on alcohol's effects across all of human and rodent behavior.

Second, the use of animal models to portray genetic risk is an even more challenging target. This effort uses many of the same basic behavioral assays, but faces an additional problem. Individual differences in alcohol seeking or alcohol response behavior may arise from genetic or from non-genetic contributions, or their interaction. Because partitioning genetic from non-genetic sources requires genetically informative animal models, this aspect of modeling will be treated separately. For historic reasons, this has been one of the most fruitful areas in research related to alcohol.

CHOICE OF NON-HUMAN ANIMAL SPECIES

Many behaviors are exquisitely sensitive to acute ethanol and show systematic changes with the chronic administration that lead to tolerance and withdrawal. As for other drugs of abuse, there are substantial individual differences in sensitivity to ethanol. These basic characteristics of drug response have been demonstrated in many mammalian species as well as in fish (e.g., goldfish, zebrafish: see Echevarria et al., 2011), birds (e.g., chicks (Smith et al., 2011), pigeons (Pinkston et al., 2009)) and invertebrate species (e.g. *Drosophila* and their larvae, *Caenorhabditis elegans*; see Scholz and Mustard, 2013). A review of research using non-mammalian models identified both strengths and weaknesses of their use (Lovinger and Crabbe, 2005). While there is generally good cross-species face validity for many behaviors, because of substantial dissimilarities in neuropharmacology and brain circuitry underlying non-mammalian model systems vs mammals, I do not consider them further here. Neither do I consider research with non-human primates. Non-human primates unquestionably provide the closest analogs of humans with AUD and have provided many fascinating insights (Helms et al., 2012; Barr, 2013), and the reader is referred to this literature. Thus, the remainder of this review will focus on studies where rats and mice have been used to model phenotypes related to alcohol.

ALCOHOL-RELATED PHENOTYPES

The complexity of the categoric diagnosis and its strong dependence on behavioral criteria suggest that it is not possible to generate a laboratory animal model that recapitulates all the essential features of an AUD diagnosis (McClearn, 1979; Crabbe, 2012). Nonetheless, others have argued that alcohol dependence (Rodd et al., 2004) and cocaine addiction (Ahmed, 2012) are successfully fully modeled in rats. In my view, a more realistic goal is to adopt a reductionistic approach and explore different features of human AUD with specific behavioral assays designed to represent those features in laboratory rodents. These features may be conceptualized as representing at least the following categories: degree of response to an initial challenge with alcohol (sensitivity); brain-based adaptations that ensue following chronic drug administration (these are of two sorts, progressively reduced response to continuation of a standard dose, or tolerance, and progressively increased response, called sensitization); dependence, which must be inferred from the presence of withdrawal symptoms when alcohol is discontinued; and sensitivity to the reinforcing effects of the drug, which may be positive (drug seeking, which in humans is often described as "craving") or negative (avoiding drug withdrawal). Researchers have also often sought to model other neurobiologic factors thought to be comorbid with or to predispose to AUD, such as various aspects of stress responses (Koob, 2008; Vengeliene et al., 2008), personality traits such as impulsivity/disinhibition/loss of control (Dick et al., 2010), or psychiatric conditions, including depression and anxiety disorders.

I believe that a major issue faced by the animal modeler is that there can be no laboratory assays in rodents that fully capture such constructs as "anxiety," "depression," "impulsivity," "craving," or many other behavioral/psychologic constructs. Rather, these are interoceptive states that must be inferred from specific behaviors exhibited by our animal subjects. Not all researchers agree, and some argue that the existing rodent behavioral assays represent faithful representations of these constructs. Most researchers would agree that, although imperfect, the existing assays are essentially sound (for depression and anxiety, see Cryan and Holmes, 2005; for impulsivity, see Dick et al., 2010). Neurologically relevant behaviors affected by acute or chronic alcohol, or withdrawal include somatomotor activity, balance and ataxia-related behaviors, muscle strength, pain sensitivity, tremor, convulsions and seizures, cognition (including attention, decision making, learning, and memory), sleep, impulsive and affective behavior, and electrophysiologic and brain-imaging measures. Although modeling each of these in a rodent may seem more straightforward than modeling depression, there are in fact multiple laboratory assays for each of the above-listed behaviors. The assumption that each is modeling essentially the same human behavior appears not to be warranted. A clear example of the difficulty in selecting a particular rodent behavioral assay to model a specific human disorder is provided in a chapter discussing the variety of human epilepsies and the rodent assays used to model them (Coppola and Moshe, 2012). How genetic studies reveal the complexities inherent in mapping rodent assays on to the human traits they presumably represent is discussed below.

ASSESSING ALCOHOL SENSITIVITY

Alcohol sensitivity in rodents is generally studied after administration of dilute alcohol via intraperitoneal injection. Alcohol rapidly reaches the brain, and behavioral responses can become evident within the first 1–5 minutes after injection. Brain levels reach peak values as early as 2–3 minutes after injection (Ponomarev and Crabbe, 2002) but may be delayed for many minutes, depending on the dose and dilution used: higher and more concentrated doses reach peak levels more quickly. Arterial blood ethanol concentration and brain concentrations equilibrate rapidly, but blood ethanol concentrations estimated from peripheral vessels (e.g., tail vein blood) can be significantly lower for 30–60 minutes before equilibrating with brain concentrations (Goldstein, 1983). The concentration of alcohol employed in rodent studies is generally 10–20%, administered in saline: higher concentrations are thought to produce tissue irritation. Alcohol is sometimes administered by intragastric gavage, where higher concentrations can be employed (Fidler et al., 2012). Absorption, distribution, and duration of exposure to alcohol are substantially slower and longer by this route of administration. Alcohol is generally not administered by subcutaneous or intravenous routes due to its irritant effects.

ASSESSING ALCOHOL TOLERANCE OR SENSITIZATION

When alcohol is repeatedly administered at a fixed dose, response to alcohol generally becomes progressively attenuated. The degree of reduction in response is one way to assess the development of tolerance; the other is to index the amount that dose must be raised to achieve the same degree of response (Kalant et al., 1971). Tolerance develops to most sedative effects of ethanol (e.g., ataxia, hypothermia, cognitive decrements). For the locomotor stimulant effects of low to moderate doses of alcohol in mice, the effect of chronic administration is the opposite of tolerance. As is the case with many psychostimulants (e.g., cocaine, amphetamines), repeated dosing leads to progressively enhanced responses, or sensitization (Phillips, 1997). An unexplained feature of locomotor responses to alcohol is that mice respond with stimulation followed by sedation (i.e., an inverted-U dose–effect curve), whereas rats generally respond only with monophasic reductions in motor activity (but see exceptions, e.g., Correa et al., 2003).

Tolerance is generally seen whether alcohol is repeatedly injected or chronically administered in some other way. Many early studies incorporated alcohol into a liquid diet, fortified with vitamins and other supplements to allow the diet to provide all nutrition (e.g., Kovacs, 2000). Mice or rats exposed to such diets for 1–2 weeks become tolerant. When alcohol is allowed to clear, they show approximately 50% attenuation in their sensitivity to many responses to an acute ethanol challenge (Ritzmann and Tabakoff, 1976; Crabbe et al., 1982;

Smith and Little, 2000). Thus, a naive mouse given a 2 g/kg dose of ethanol by intraperitoneal injection might show a reduction in core body temperature of 1.50°C, whereas a mouse tested several hours after the discontinuation of a week-long exposure to ethanol via liquid diet might only show a reduction of 0.75°C. (Tolerance could also be demonstrated, and quantified, by finding that it required a 3 g/kg injection to reduce temperature by 1.50°C.) Similar degrees of tolerance can be elicited by repeated injections or by exposure to vapor inhalation. The degree of tolerance to ethanol is markedly less than that to opioid and psychostimulant drugs, where effective doses can escalate more than 10-fold in tolerant individuals.

ASSESSING ALCOHOL DEPENDENCE

In addition to the liquid diet used to study tolerance, dependence on alcohol can also be induced by chronic inhalation of alcohol vapor (Goldstein and Pal, 1971; Le Bourhis, 1977) or chronic intragastic infusion (Hardy and Deutsch, 1977; Cunningham et al., 2012). The intragastric route has the disadvantage of being technically quite demanding, but the advantage of largely bypassing issues of taste and smell. Once rendered dependent, even animals that do not prefer to drink alcohol solutions will self-administer substantial doses of alcohol intragastrically (Fidler et al., 2011). The inhalation route is being used more and more frequently in both rats (Roberto et al., 2010) and mice (Lopez et al., 2012). It offers the distinct benefit of good control of blood alcohol levels during dependence induction. However, a disadvantage is that only a single, not very convincing, study suggests that rodents (mice) will self-administer ethanol via this route (Rijk et al., 1982), presumably because of the irritating effects of alcohol vapor. Nonetheless, it has been reported that rats will elect to drink intoxicating amounts of alcohol for many days after withdrawal from a prolonged period of forced exposure by inhalation (Le Bourhis, 1977). Older methods of inducing dependence (multiple daily injections, offering animals ethanol in water solutions as their only source of fluid) have essentially been discontinued in favor of vapor and liquid diets.

Withdrawal signs are myriad. An early review cited parallel evidence for human vs rat and/or mouse withdrawal symptoms in multiple domains, including tremor and muscle fasciculation; changes in activity; increased startle reactivity; irritability; fright; sleep disorders; spontaneous vocalizations; temperature regulation; sweating; and convulsions (Friedman, 1980). Regardless of the method for inducing physical dependence, severity and duration of withdrawal are a function of dose and duration of alcohol treatment.

ASSESSING ALCOHOL REINFORCEMENT

Methods for inferring that alcohol is a reinforcer have been discussed in detail elsewhere (Samson and Hodge, 1996; Samson and Czachowski, 2002; Stephens et al., 2010). For most drugs of abuse, the "gold standard" in the animal models literature is the willingness of the animal to self-administer the drug by the chronic intravenous route. Attempts to achieve intravenous self-administration of alcohol in rodents have met with mixed results but this can be achieved in both rats and mice (Numan et al., 1984; Grahame and Cunningham, 1997; Gass and Olive, 2007; for review, see Green and Grahame, 2008). Intravenous self-administration can be facilitated by a history of dependence induced by forced infusions of ethanol, and animals will subsequently self-administer doses sufficient to prevent withdrawal, but not doses sufficient to produce gross signs of intoxication (Numan, 1981). Most investigators employ variants of the two-bottle preference paradigm (Richter and Campbell, 1940), where bottles of (typically 10%) alcohol and water are continuously available. The percentage of total daily fluid drunk from the alcohol bottle is taken to represent the animal's preference for alcohol. A better measure is the dose of alcohol ingested, because preference ratios vary with the ethanol concentration offered (Richter and Campbell, 1940; McClearn, 1968). Others offer ethanol on an operant schedule of access requiring work such as a lever press (rats) or nose poke (mice) to gain access for oral self-administration (Samson and Hodge, 1996; Leeman et al., 2010; Stephens et al., 2010).

Pavlovian conditioning has also been frequently used to assess the positive hedonic and aversive effects of ethanol. A conditioned place preference can be achieved in rats or mice by administering intraperitoneal doses of alcohol in an environment made distinctive by sensory cues (e.g., patterned walls, floor texture, odor). Alternating injections of saline are given in a second environment with different cues. After several such pairings, the animal is tested drug-free in a chamber that combines the alcohol-paired cues on one side and the saline-paired cues on the other. If the animal spends more time during this choice test on the alcohol-paired side, a positively reinforcing effect of the subjective state elicited by alcohol is inferred. The other Pavlovian test employed assesses alcohol's aversive effects. A water-deprived animal is presented with a short period of access to a novel solution, such as physiologic saline, which it drinks. At the end of the session, the animal is given an intraperitoneal injection of saline or alcohol. The novel solution is offered again at 24- or 48-hour intervals, and the postdrinking injection is repeated. Rats and mice given ethanol injections drink progressively less of the novel solution each test day, which is taken to indicate that the subjective effects of the ethanol injection have a punishing quality. While this may sound contradictory, there are numerous theories for why drugs of abuse can simultaneously have both rewarding and aversive qualities, and the reader is referred to excellent discussions of how both place preference and taste aversion can be conditioned by the complex stimulus properties of alcohol (Cunningham et al., 2009; Verendeev and Riley, 2012). An excellent review concluded that, on balance, genetic differences obtained from preference drinking data were generally consistent with data from other methods used to assess ethanol's reinforcing effects (Green and Grahame, 2008).

One curious feature of ethanol as a reinforcer is that rats and mice typically will not voluntarily ingest (or administer by intravenous or intragastric routes) doses sufficient to produce overt signs of intoxication. This is of course quite different from alcohol-dependent individuals, who typically repeatedly become intoxicated: non-human primates will also drink excessive amounts of alcohol (Grant et al., 2008). Many procedural manipulations (e.g., scheduling access, altering the taste of alcohol-containing solutions) can elevate alcohol drinking in rodents, but these tend to be laborious and require a long time: they also detract from the face validity of the model (for discussion, see Rhodes et al., 2005; Becker, 2013). This limitation led to the development of a murine genetic model of binge-like drinking, Drinking in the Dark (DID), that will be discussed below (Crabbe, 2012).

MODELING GENETIC RISK IN NON-HUMAN ANIMALS

Three genetic methods have been employed to identify specific genes conferring risk or protection, inferred from the model's enhanced or attenuated response to (or seeking of) alcohol. Since 1988, it has been possible to target specific genes in the mouse genome and interrupt their function by inducing transgenic overexpression or targeted deletion of the gene. Recently, this has also become possible in rats. Either of these approaches starts with a candidate gene and explores its effects on alcohol-related behavior. More than 100 genes have individually been targeted in this way in mice. While relatively few instances of overexpression transgenics have been studied, many dozens of null mutations and partial-expression knockdown mice have been tested. While most knockouts/knockdowns are induced in early development, siRNA methods now allow gene silencing in adult rats and mice as well as in specific brain regions (e.g., Jeanblanc et al., 2009). A review of the literature then available on gene-targeted mutant mice in November, 2006, identified 141 papers presenting alcohol-related data for 93 targeted genes. The alcohol-related phenotype that was most frequently studied was two-bottle preference drinking. Of the 76 targeted genes that were studied for preference drinking, manipulations of 20 (26%) of them increased preference, 31 (41%) had no effect, and 25 (33%) decreased drinking (Crabbe et al., 2006). By February, 2008, 175 papers on 107 targeted genes yielded outcomes on preference for 86 targeted genes of 23 genes increasing, 33 genes without effect, and 30 genes decreasing preference (i.e., the proportions had not changed: unpublished data). Thus, turning on or off a gene's function is more likely than not to affect alcohol preference drinking. Numerous other behavioral outcomes were also tabulated in the early review. The utility of gene targeting for specific behaviors of neurologic interest may depend upon the understanding of each neurologic syndrome's underlying human genetic architecture. Single-gene mutants may be most relevant for modeling single-gene autosomal-dominant disorders (e.g., Subramony, 2012), but are unlikely to stand in as complete models for disorders such as AUD. This is not to belittle their value for understanding the role of specific genes in specific behaviors.

The second common genetic animal modeling method is selective breeding. As dogs are bred for show points or dairy cattle for milk composition and yield, rats and mice can be bred to display high vs low sensitivity to an alcohol response. In contrast to gene targeting, this approach begins with a phenotype and pursues the genetic contributions to individual differences in that trait. This method was initially employed in the 1940s at the University of Chile to produce the UChB line of rats selected for their high two-bottle preference for alcohol solutions (Mardones and Segovia-Riquelme, 1983). Typically, an alternative line is selected for the opposite phenotype, in this case low-preference drinking UChA selected line. As mating pairs are picked from each generation because of their extreme scores, generation after generation, the genetic composition of the selected line is enriched for allelic forms that lead to the selected response. This change in alleles occurs at all genes affecting the selected trait. Oppositely selected lines come to diverge very markedly in the selected trait. In a well-controlled selection experiment, gene (allele) frequencies at all other genes do not vary: that is, the original population polymorphisms are maintained across generations. When selected lines are found to differ for other traits (e.g., in neurochemical parameters), this must necessarily have occurred because of effects of the same genes driving the selected trait and provides evidence for the role of the correlated response on the selected trait.

A third genetic approach is the study of inbred strains. More than 100 standard inbred mouse strains are available from the Jackson Laboratory, and several inbred rats strains are also commercially available. An inbred strain is produced by systematically mating brothers and sisters for more than 20 generations. At the end of this process, two identical alleles at each gene have been fixed in all animals: same-sex animals within an inbred strain thus resemble clones or monozygotic twins and have the added feature of a total lack of heterozygosity (Falconer and Mackay, 1996).

Use of both inbred strains and selected lines to assess the mechanisms underlying alcohol-related behaviors is discussed elsewhere (Crabbe et al., 1990).

Rodent lines selectively bred for alcohol responses

Many rodent lines have been selected for sensitivity to or propensity to drink ethanol. These are listed in Table 5.1 and described below.

Alcohol preference selections

Most experiments using the powerful technique of selective breeding have used nearly the identical paradigm. Rats or mice are offered 10% alcohol in tap water vs tap water continuously for about 3 weeks. The animals drinking the highest percentage of their daily fluid from the alcohol bottle are then mated to produce the next generation. A parallel, reproductively independent line is developed by breeding together those individuals with the least alcohol preference. As noted earlier, rat lines have been generated for high vs low drinking since the late 1940s. A recent review listed all the known lines selected for high- vs low-preference drinking, described their genetic characteristics, and referred to then-recent comprehensive reviews of the correlated responses documented in these lines (Crabbe et al., 2010). Updated reviews of certain of these selected lines have since appeared (Bjork et al., 2010; Bell et al., 2012; McBride et al., 2012; Ciccocioppo, 2013). One such review (Ciccocioppo, 2013) gives a concise history of the development of all six pairs of these selected rat lines and of a seventh case where a single line was selected for high preference: this review does not consider the more recent mouse lines. One review considers both the rat and mouse selected lines (Green and Grahame, 2008).

Congruent findings in rat preference selections

Because all these lines were selected for the same trait, it seems reasonable to declare that the most convincing data for believing that a correlated response is relevant to the behavior to be modeled is that it should be detected in all, or nearly all, comparisons of the pairs of high-preference vs low-preference lines. Indeed, when at all possible, a laboratory creating a pair of selectively bred lines should also create a second, replicate pair. The reasoning is, if a correlated response differentiating the two lines cannot be generated in the same laboratory in two iterations of the same selection experiment, it is unlikely to be of crucial importance in explaining the selected trait. Of course, relatively small population sizes for such experiments are dictated by practical (cost and space) considerations. With small population sizes, some genetic differences between lines unavoidably arise by chance (termed genetic drift). Thus, differences between lines, or failures to detect differences between lines, due to genetic drift unrelated to the selected trait cannot be entirely avoided. (For a discussion, see Crabbe et al., 1990, and for a more technical treatment of the genetics, see Falconer and Mackay, 1996.) Due to their larger size, only a single rat selection for preference was strictly replicated, that for High Alcohol Drinking (HAD-1 and HAD-2) vs Low Alcohol Drinking (LAD-1 and LAD-2) rats at Indiana University (Li et al., 1987). Mouse selections for alcohol-related traits have generally been replicated (see later sections).

Most of these rat lines have been fully selected for more than 20 years. Bell and colleagues (2012) tabulate innate neurochemical characteristics reported in comparisons between the five of these six pairs of lines for which more data are available (their Table 3), as well as the results of studies where pharmacologic agents were tested for effects on alcohol-preference drinking (their Table 4). Given that the line pairs are in different laboratories and that methods are often rather laboratory-specific, it is perhaps not surprising that there are many interesting findings in one pair of lines that have not been systematically tested in the others. I could find no case where data from the same brain area were available across all five pairs of lines. Urocortin 1 content in lateral septum and Edinger–Westphal nucleus was compared across four pairs of lines: however, results depended on the specific pair compared. This set of comparisons' face validity is high because all data came from a single laboratory (Turek et al., 2005). 3,4-Dihydroxyphenylacetic acid content of nucleus accumbens was assessed in four pairs of lines: again, the results depended on the line. Three of four lines showed lower dopamine D2 receptor levels in caudate putamen in high drinkers, but there was no difference in the fourth pair, and in a fifth study, the line pair difference in AA vs ANA rats was not replicated (for references, see Table 3 in Bell et al., 2012.)

Comparison of the efficacy of pharmacologic treatments to reduce preference drinking or alcohol self-administration in the high-drinking selected lines also

Table 5.1

Lines of rats or mice selectively bred for alcohol-related behaviors

	Lines	Acronym	Selection phenotype	Relevant human responses	Approximate blood alcohol levels attained	References
Rats	University of Chile Alcohol Drinker and non-Drinker	UChB/UChA	High/low consumption 10% ethanol vs water	Drinking	0.05–0.10%	Quintanilla et al., 2006; Bell et al., 2012; Ciccocioppo, 2013
	ALKO Alcohol and Non-Alcohol	AA/ANA	High/low consumption and preference 10% ethanol vs water	Drinking	0.05–0.10%	Sommer et al., 2006; Bell et al., 2012; Ciccocioppo, 2013
	Alcohol Preferring and Non-preferring; inbred P, inbred NP	P/NP iP/iNP	High/low consumption and preference 10% ethanol vs water	Drinking	0.05–0.10%	Bell et al., 2006, 2012; Ciccocioppo, 2013
	Sardinian Alcohol Preferring and Non-preferring	sP/sNP	High/low consumption and preference 10% ethanol vs water	Drinking	0.05–0.10%	Colombo et al., 2006; Bell et al., 2012; Ciccocioppo, 2013
	Marchigian Sardinian Alcohol Preferring	msP	High consumption and preference 10% ethanol vs water	Drinking	0.05–0.10%	Ciccocioppo et al., 2006; Bell et al., 2012; Ciccocioppo, 2013
	High/Low Alcohol Drinking	HAD-1,2/LAD-1,2	High/low consumption and preference for 10% ethanol vs water	Drinking	0.05–0.10%	Murphy et al., 2002; Bell et al., 2012; Ciccocioppo, 2013
	Alcohol-Tolerant/Non-tolerant	AT/ANT	High/low sensitivity to ataxia on the tilting plane test	Ataxia	0.15–0.25%	Radcliffe et al., 2004, 2009; Eriksson and Sarviharju, 1984
	High/Low Alcohol-Sensitive; inbred High/Low Alcohol-Sensitive	HAS1,2/LAS1,2 iHAS1,2/iLASA1,2	High/low sensitivity to loss of righting reflex	Anesthesia; suppression of reflexes	0.40–0.50%.	Radcliffe et al., 2005
Mice	High/Low Alcohol Preference; crossed High Alcohol Preference	HAP-1,2,3/LAP-1,2,3 cHAP (HAP-1 is extinct)	High/low consumption 10% ethanol vs water	Drinking	0.05–0.10% (cHAP has reached 0.20–0.30% in some tests	Green and Grahame, 2008; Matson and Grahame, 2011
	High Drinking in the Dark	HDID-1/HDID-2	High blood alcohol levels after drinking 20% ethanol in a single-bottle, limited-access exposure	Binge-like drinking	0.08–0.30%	Crabbe et al., 2009, 2011a
	Long Sleep/Short Sleep; inbred Long Sleep/Short Sleep	LS/SS iLS/iSS	Loss of righting reflex	Anesthesia; suppression of reflexes	0.40–0.50%.	Bennett et al., 2006, 2008
	FAST/SLOW	FAST-1,2/SLOW-1,2	Stimulated activity	Euphoria	0.10–0.20%	Phillips, 1997; Fish et al., 2012
	Withdrawal Seizure-Prone/ Seizure-Resistant; inbred WSP and inbred WSR	WSP-1,2/WSR-1,2 iWSP-1,2/iWSR-1,2	Handling induced convulsion severity after chronic ethanol vapor inhalation	Withdrawal seizures; tremor; central nervous system hyperexcitation	0.25–0.40%	Metten and Crabbe, 1996; Finn et al., 2004; Ford et al., 2010

does not reveal any instance where any drug has been tested in all five selected lines. Mostly, this is because the HAD-2 replicate line has apparently not been used for such studies. However, only naltrexone has been tested in all four other lines, where it reliably reduces drinking in AA, HAD-1, P, and sP rats. The benzodiazepine inverse agonist Ro 19-4603, as well as the GABA-B receptor antagonist baclofen and positive modulator compound GS39783, reduced limited-access preference drinking in AA, P, and sP rats, but none has been tested in the HAD lines (for references, see Table 4 in Bell et al., 2012). An earlier review found little evidence that two-bottle preference drinking models provided a valuable route to discovery of novel therapeutic agents, largely because of the high frequency of false-positive findings: that is, many compounds reduce rodent drinking but subsequently do not prove effective in the clinic (Egli, 2005). The recent comprehensive review just cited makes clear that few agents have yet been thoroughly tested in these genetically enriched rat models (Bell et al., 2012).

The preference lines have been tested for a wide variety of behavioral responses to alcohol, ranging from sensitivity through withdrawal and reinforcement. In addition, they have been tested for other behavioral characteristics. Thorough summaries of the relevant findings have regularly appeared. For example, a special issue of *Addiction Biology* (vol. 11 (3–4), 2006) included several such reviews. Some consistent findings are summarized here (and see Table 1 in Bell et al., 2012). All these rat lines voluntarily drink enough alcohol when it is freely available to reach modest blood ethanol concentrations. However, under most circumstances, these levels do not exceed 0.8 mg ethanol/mL blood (0.08%). The National Institute on Alcohol Abuse and Alcoholism defines a human binge as a pattern of intake that achieves this level, typically requiring five or more drinks for men and four or more for women within about 2 hours (NIAAA, 2004). The high-drinking lines will generally work to gain access to alcohol, indicating that they find it rewarding. Several of the lines become tolerant to alcohol, although for the AA rats, that tolerance is largely due to the induction of more rapid metabolism. Only P rats have been shown to drink sufficient alcohol to display withdrawal signs, but this required 6 weeks of drinking (Kampov-Polevoy et al., 2000). However, P rats will self-administer alcohol intragastrically, whereas NP rats will not (Waller et al., 1984). Other pairs of preference lines have not been tested in this paradigm. sP rats will escalate their drinking when alcohol is offered intermittently, and display some alterations in drinking that suggest that they might be dependent, but this study did not directly evaluate withdrawal (Loi et al., 2010). All these lines show a transient escalation of intake when access to alcohol is restored after a period of abstinence, a phenomenon known as the alcohol deprivation effect, but these studies are not entirely consistent across different selected lines (Ciccocioppo, 2013). Several studies have reported that P and sP rats are less sensitive to an ethanol-conditioned taste and place aversion than the low-preferring NP and sNP, respectively; however, AA and ANA rats did not differ in either response in two studies. ALKO AA and ANA rats did not differ in sensitivity to an ethanol-conditioned taste aversion (for references, see Green and Grahame, 2008).

Findings from high (HAP) and low (LAP) alcohol-preferring mice

Starting from a genetically heterogeneous mouse stock, lines were selected for preference drinking using a paradigm nearly identical to that employed for all the rat selections just discussed (Grahame et al., 1999). The only procedural difference was that, during selection, the HAP and LAP mice were not initially exposed to a 4-day period of forced access to 10% alcohol as their only source of fluid. Mice apparently have a somewhat less pronounced initial aversion to 10% ethanol solutions than rats, and selection for preference did not require initial forced exposure. Substantial response to selection was seen: however, breeding difficulties with the HAP1 line led to the decision to select two additional replicate pairs of lines (HAP2/LAP2 and HAP3/LAP3: Oberlin et al., 2011). The HAP1 line is now extinct (Crabbe et al., 2010), but the investigator took an interesting extra step. Before the termination of the HAP1 line, he intercrossed it with the HAP2 line to produce a “crossed HAP” (cHAP) line, which he also continued to select directionally for high drinking. After 11 selected generations, the cHAP line was drinking more than either HAP1 or HAP2 were at that point in their selection (Oberlin et al., 2011), and cHAP mice now drink sufficient alcohol to achieve blood alcohol levels of greater than 2.50 mg/mL (0.25%), or three times the legal driving limit in the United States. HAP2 and HAP3 lines drink less than cHAP, but a little more than the unselected inbred strain, C57BL/6 J (Matson and Grahame, 2011).

All HAP lines show greater preference for saccharin solutions than the LAP lines, consistent with much other data suggesting common genetic influences on preference for ethanol and sweet solutions, in both rodents and humans. These studies, and the less common contradictory findings, are discussed elsewhere (Kampov-Polevoy et al., 1999; Mennella et al., 2010; Oberlin et al., 2011). It is not entirely clear whether saccharin and sucrose preference tests in rodents are indicative of the individual’s sensitivity to reward in general, or whether such preferences are more related to taste, and deficiencies in sucrose preference have been taken as a sign of anhedonia or a depressed state (Cryan and Holmes, 2005). Both HAP1 and HAP2 lines show attenuated development of an ethanol-conditioned taste aversion vs their respective LAP1 and LAP2 lines (Chester et al., 2003). HAP1 mice also failed to show a conditioned place preference for ethanol, while LAP1 mice selected an environment paired with ethanol injections (Grahame et al., 2001).

Selection for drinking in the dark

In an attempt to overcome the reluctance of rats and mice to ingest sufficient alcohol to become intoxicated, we undertook to develop a model of binge-like drinking. We started with C57BL/6J mice because they were known to prefer to drink even relatively high concentrations of alcohol in a variety of paradigms. Mice are offered limited (2–4 hours) access to a single bottle of ethanol early during their circadian dark period, when they normally ingest the majority of their daily food and water. C57BL/6J mice will ingest sufficient alcohol to exceed blood ethanol concentrations of 1.00 mg/mL (0.10%) (Rhodes et al., 2005) and are behaviorally intoxicated (Rhodes et al., 2007). We subsequently developed two lines of mice by selecting for blood ethanol concentrations at the end of the DID session. High Drinking in the Dark (HDID-1 and HDID-2) selected lines reach higher blood ethanol concentrations (average = 1.40 mg/mL, or 0.14%; maximum = 3.00 mg/mL, or 0.30%; unpublished data) than C57BL/6J mice, become intoxicated, and drink to relatively high blood ethanol concentrations even when water is concurrently available (Crabbe et al., 2009, 2010). The DID trait has proven useful for many studies addressing the underlying neurobiologic mechanisms and to test compounds for efficacy in reducing binge-like drinking (for reviews, see Crabbe et al., 2011a; Sprow and Thiele, 2012).

HDID mice achieve their high blood ethanol concentrations by ingesting alcohol in larger bouts, and tend to show more drinking bouts as well (Barkley-Levenson and Crabbe, 2012). The HDID lines are more sensitive than the unselected HS control stock to some effects of acute ethanol (e.g., stimulated activity, ethanol-induced ambulatory ataxia in the parallel rod floor test), and were less sensitive than HS to others (e.g., ethanol-induced ataxia on a balance beam or rotarod). They may develop greater tolerance, but they do not develop more severe acute or chronic withdrawal (Crabbe et al., 2012a, c). Their preference for ethanol solutions is slightly greater when alcohol is continuously available (Crabbe et al., 2011b) and their intake escalates during repeated cycles of dependence and withdrawal; however, the escalation is no greater than that seen in the HS control stock (Crabbe et al., 2013). Exposure to alcohol solutions intermittently did not lead to elevated consumption in these mice, whether the intermittent access was limited to 4 hours/day or was continuous (every other day), despite the effectiveness of the latter manipulation in the C57BL/6J genotype (Crabbe et al., 2012b).

Selection for locomotor response to ethanol

The stimulant response to low-dose alcohol in mice has been taken as an index of alcohol's reinforcing effects. Two pairs of mouse lines have been bred for high (FAST) and low (SLOW) sensitivity to motor stimulant effects (Phillips et al., 1991). Responses to other effects of ethanol have been characterized and FAST mice showed significantly greater ethanol withdrawal severity than SLOW (Shen et al., 1996; Palmer and Phillips, 2002). Studies of the lines' response to other abused drugs have been reviewed (Bergstrom et al., 2003). Greater sensitivity to ethanol and cocaine stimulation appears to be partially due to mesolimbic dopaminergic systems (Meyer et al., 2009) but GABA-B receptor systems also appear likely to play a role (Holstein et al., 2013). Many other responses have been documented in these lines (Phillips and Shen, 1996).

Lines selected to be ethanol Withdrawal Seizure-Prone (WSP) or -Resistant (WSR)

Starting 30 years ago, we bred mice to display severe (WSP-1, WSP-2) or mild (WSR-1, WSR-2) withdrawal convulsions after being made physically dependent on alcohol through forced inhalation. These animals differ in the severity of a number of other alcohol withdrawal signs (Kosobud and Crabbe, 1986). WSP mice also show more severe withdrawal from benzodiazepines, barbiturates, and other central nervous system depressants than WSR mice. WSP mice show an acute withdrawal response after a single high dose of such depressants as well, indicative of their sensitivity. These studies, and early explorations of the basis for the withdrawal convulsions between WSP and WSR mice, have been reviewed (Crabbe and Phillips, 1993; Metten and Crabbe, 1996; Crabbe, 2002). WSP and WSR mice have been used extensively to study the role of GABAergic neuroactive steroids in modulating alcohol withdrawal (for reviews, see Finn et al., 2004, 2010). WSP mice show less ethanol conditioned taste avoidance than WSR, but greater ethanol place preference (Chester et al., 1998; for review, see Ford et al., 2010). WSP and WSR lines have also been used to map genetic loci contributing to differences in alcohol withdrawal severity (Bergeson et al., 2003; Ehlers et al., 2010; Buck et al., 2012).

Long sleep (LS) and short sleep (SS) mice

In one of the earliest selection studies, mice were selectively bred for the duration of loss of righting reflex after a high dose of alcohol. Many dozens of studies of these lines' behavioral and neurochemical differences have been published (for reviews, see Allan and Harris, 1991; Phillips and Crabbe, 1991; Bennett et al., 2006). These animals differ in many respects, and have been most extensively used to map the loci underlying this ethanol response (Owens et al., 2002; Bennett et al., 2008). Sets of rat lines have also been developed for the same response and these lines and their crosses have been used to map quantitative trait loci for this response to high alcohol doses (Radcliffe et al., 2005).

Alcohol-tolerant (AT) and non-tolerant (ANT) rats

These animals were selected not for tolerance, as their name implies, but rather for differential sensitivity to the intoxicating effects of a moderate ethanol dose, using the tilting plane test of ataxic response. They have been used in many studies, including quantitative trait locus mapping studies (Radcliffe et al., 2004, 2009).

Studies with inbred strains

The availability of many inbred mouse strains has been exploited to explore the topography of alcohol-related behaviors. Because inbred strains crystallize a stable genotype (allelic composition at each gene is invariant except for rare, random mutations), strain difference data collected in one laboratory can be compared with data for the same or a different trait, collected in a different laboratory in new mice of the same strains, even if the projects are separated by many years. The data for each trait must be collected with careful attention to maintaining environmental conditions as nearly the same as possible for all strains: when this is accomplished, if the strain mean differences significantly exceed the within-strain individual differences, the presumption is that they reflect predominantly genetic influences. Thus, if two traits reveal patterns of strain mean values that are significantly correlated, the assumption is that some genes must necessarily influence both traits, which is further presumed to reflect similarities in the underlying neurobiologic mechanisms. One of the clearest examples of this characteristic of inbred strains in behavioral genetics is the finding that inbred strain differences in preference for 10% ethanol solutions have been quite replicable. For example, a comparison of data published across multiple laboratories over a time span of more than 50 years revealed that mean inbred strain preferences between five data sets correlated with r values between 0.79 and 0.98 (Wahlsten et al., 2006), indicating high reliability of this trait's genetic underpinnings.

Of more interest is how different traits' genetic influences correlate. We spent several years characterizing eight inbred mouse strains for their sensitivity to acute alcohol intoxication. We employed 11 behavioral assays (e.g., rotarod, balance beam, observer-rated ambulatory ataxia) using multiple doses of ethanol, from which we extracted 18 variables taken to be indices of intoxication. Strains differed significantly in all responses assessed (see references in Crabbe et al., 2005). One goal in this project was to identify clusters of tasks that were essentially revealing the same genetic influences, that is, showed strong strain mean correlations. We thought that some tasks would likely be assessing "balance," others "ataxia," and still others disordered sensorimotor processing. A second goal was to identify any inbred strains that were generally sensitive to alcohol, or resistant to its effects.

What we found was far more revealing, albeit a bit discouraging. We identified no strains that were either generally sensitive or resistant: rather, each task affected some specific strains more and others less. While there were significant correlations across some pairs or clusters of tasks, we saw modest correlations in general. Some hints that tasks could be mapped on to constructs were seen. For example, splayed limbs in the observer-rated ataxia task were significantly correlated with errors on the balance beam and foot slip errors detected in a task called the grid test. All three of these tasks record hind limb positioning errors as the index of intoxication. But our assumptions about what these tasks were measuring were not reflected in the patterns of correlation. if the "balance beam" and the "dowel balancing" task were both indeed indicating ethanol's effects on balance mechanisms, why were the strain mean correlations on these tasks less than $r=0.30$? Each of these tasks was significantly influenced by genetic factors. We concluded that the laboratory tasks commonly employed in mice to indicate behavioral constructs such as "balance," "ataxia," and "coordination" are actually much more complex, and each probably reflects a nearly unique composite contribution of multiple underlying physiologic capabilities, including at least balance, coordination, patterned gait, muscle strength, intact proprioceptive feedback, intact visual capacities, learning, and motivation to perform (Crabbe et al., 2005). Perhaps this should not have surprised us, because an earlier analysis of the effects of a null mutation of the serotonin 1B receptor gene using many of these same behaviors had revealed that the mutant mice were less sensitive than wild types on only two of eight tests employed (Boehm et al., 2000).

On a more positive note, a multivariate analysis of several published and unpublished data sets showed substantial relationships between 15 inbred strain means for consumption of and withdrawal from four drugs. Strains that preferred ethanol solutions also preferred to drink morphine, while a different set of strains preferred both pentobarbital and diazepam solutions. Those strains exhibiting severe withdrawal from ethanol also showed similarly strong withdrawal from diazepam and barbiturate injections, but the pattern of strain means for precipitated morphine withdrawal differed from the other drugs. This study also revealed a pattern of negative genetic correlation between withdrawal from and consumption of ethanol and diazepam, while no such pattern existed for morphine or pentobarbital (for references, see Belknap et al., 2008). That genotypes showing strong ethanol withdrawal tended not to prefer to drink ethanol solutions corroborated earlier data comparing many populations derived from C57BL/6J and DBA/2J inbred strains (Metten et al., 1998).

CAN BEHAVIOR GENETICS REVEAL THE STRUCTURE OF ETHANOL-RELATED BEHAVIOR?

Behavioral data seem deceptively simple. This review has discussed how promiscuously alcohol induces behavioral effects in laboratory rodents. Eliciting a behavioral response to acute or chronic alcohol in rats or mice is easy, and recording the behavioral change has been rendered simple by automation. Interpretation of behavioral data, however, cannot be automated. Because a piece of software automatically records an increase in entries into the open arms of an elevated plus maze after alcohol in a mouse does not demonstrate that the mouse is "less anxious." In studies of lab animals, there is no easy way to validate such behavioral assays. Sometimes there is no way to assess the reliability of the animal's response (for example, retesting behavior on an elevated plus maze leads to response patterns that differ from those seen in the initial test: File et al., 1990). Careful attention to procedural variables and methods is required to perform interpretable behavioral assays with rodents (Wahlsten and Crabbe, 2007), but both general and specific guidance is available for most tests (Wahlsten, 2011).

The importance of test reliability and validity is acute when attempting to ascertain relationships between specific genes and specific behaviors. As manipulation of specific genes and assessment of gene allelic structure and expression have become more commonplace, the wealth of genetic data has begun to inform the interpretation of specific laboratory behavioral assays. Alcohol-related behavior, the subject of this review, offers the most varied and complete set of behavior-genetic data for any drug of abuse. Two principal methods have provided such data: the analysis of inbred strains, and the use of intentionally genetically selected rodent lines that display large differences for a single trait. Ideally, any two traits genetically correlated across pairs of selected lines should also be correlated if examined with a panel of inbred strains. In reality, such a clear result is rarely seen, for a variety of technical genetic reasons, discussed elsewhere (Crabbe et al., 1990). I have discussed several examples drawn from this large literature: what have these studies told us about alcohol-related behavior?

All alcohol related behaviors that can be assessed reliably reveal substantial genetic influences in rats and mice. These behaviors, however, do not map simply on to constructs. Sensitivity to one effect of alcohol does not necessarily predict sensitivity to another. For example, HDID-1 mice were more sensitive than HS controls in one of three measures of ethanol-induced "ataxia," but less sensitive on the other two (Crabbe et al., 2012a). This has been revealed most clearly by studies with inbred mouse strains, where neurologically relevant indices of intoxication reveal more genetic individuality than commonality of influences. The existence of lines of mice selected for sensitivity to one effect of ethanol (locomotor stimulant (Palmer and Phillips, 2002; Beckstead and Phillips, 2009) or hypnotic (Radcliffe et al., 2005)) also shows that such lines differ on some, but not all, other responses to alcohol. Although the data are sparse, we know that tolerance to one effect of alcohol is similarly response-specific, both phenotypically and genetically (Kalant, 1998; Radcliffe et al., 2006). Genetic propensity to develop withdrawal from alcohol is shared with that for pentobarbital and diazepam, but not the pharmacologically dissimilar drug, morphine. These studies are generally consistent across inbred strains and lines selectively bred for alcohol withdrawal severity (Metten and Crabbe, 1996; Belknap et al., 2008). The various assays of ethanol's reward-related effects are generally consistent across inbred strains and selected lines: high-preference drinking, low ethanol-conditioned taste avoidance and high place preference tend to be associated in more comparisons than not (Green and Grahame, 2008; Cunningham et al., 2009), but these relationships are not always consistent, and there are few data for other indices of ethanol reward, such as locomotor stimulation (Fish et al., 2012). The most reasonable conclusion is that these various assays of ethanol reward share some, but not all, genetic determinants. It would be expected that the genetic differences remaining among them are also reflected in distinct neurobiologic substrates. Similarly, the generally consistent genetic data negatively linking measures of withdrawal with those of preference drinking also are not totally in agreement.

An interesting area of application of behavior genetic analyses resulted from a recent finding from a human twin study (Kendler et al., 2012). An analysis of the specific symptoms that underlie an alcohol dependence diagnosis in this sample revealed that genetic risk was substantial, ranging from 36% to 59%; aggregate genetic risk had been repeatedly demonstrated over the years (Goldman and Ducci, 2007). More interestingly, structural equation modeling revealed that genetic risk was not monolithic: rather, three largely independent paths that channeled genetic risk were identified. The seven criteria used to diagnose alcohol dependence according to DSM-IV-TR loaded differentially on these genetic risk factors. Tolerance, one of the seven symptoms, loaded on one factor, while withdrawal loaded on another. We therefore turned to the genetic animal model literature from selected lines and inbred strains to see whether genetic contributions to tolerance and withdrawal were independent, or correlated. The animal data tended to confirm the finding in humans of genetic independence of these aspects of alcohol responsiveness (Crabbe et al., 2011b; Kendler et al., 2012).

In conclusion, alcohol-related behaviors are influenced by genetics. The existence of numerous genetic techniques and animal models reflecting different aspects of those gene–behavior links has helped to clarify what some of the behavioral assays used in laboratory rodents are assessing. We have learned a great deal over the 50+ years they have been employed, and as genomics strategies begin to provide integrative data, the behavioral data will increase in value. While there are no perfect animal models recapitulating all the aspects of genetic risk, I admit to partiality toward the utility of two recent selective breeding projects with mice. One is the cHAP mouse line selected for high levels of drinking under conditions of continuous alcohol availability. The other is our own recent selection of HDID mouse lines for binge-like drinking to intoxication. Together, these models mirror two important aspects of human AUDs, and further explorations of their genetic attributes are likely to be informative.

ACKNOWLEDGMENTS

The author is supported by the US Department of Veterans Affairs, and by the National Institute on Alcohol Abuse and Alcoholism through grants AA10760, AA13519, and AA020245.

REFERENCES

Ahmed SH (2012). The science of making drug-addicted animals. Neuroscience 211: 107–125.

Allan AM, Harris RA (1991). Neurochemical studies of genetic differences in alcohol action. In: JC Crabbe, RA Harris (Eds.), The Genetic Basis for Alcohol and Drug Actions. Plenum, New York, pp. 105–152.

American Psychiatric Association (2000). Diagnostic and statistical manual of mental disorders. 4th edn, text revision American Psychiatric Press, Washington, DC.

Barkley-Levenson AM, Crabbe JC (2012). Ethanol drinking microstructure of a High Drinking in the Dark selected mouse line. Alcohol Clin Exp Res 36: 1330–1339.

Barr CS (2013). Non-human primate models of alcohol-related phenotypes: the influence of genetic and environmental factors. Curr Top Behav Neurosci 13: 223–249.

Becker HC (2013). Animal models of excessive alcohol consumption in rodents. Curr Top Behav Neurosci 13: 355–377.

Beckstead MJ, Phillips TJ (2009). Mice selectively bred for high- or low-alcohol-induced locomotion exhibit differences in dopamine neuron function. J Pharmacol Exp Ther 329: 342–349.

Belknap JK, Metten P, Beckley EH et al. (2008). Multivariate analyses reveal common and drug-specific genetic influences on responses to four drugs of abuse. Trends Pharmacol Sci 29: 537–543.

Bell RL, Rodd ZA, Lumeng L et al. (2006). The alcohol-preferring P rat and animal models of excessive drinking. Addict Biol 11: 270–288.

Bell RL, Sable HJ, Colombo G et al. (2012). Animal models for medications development targeting alcohol abuse using selectively bred rat lines: neurobiological and pharmacological validity. Pharmacol Biochem Behav 103: 119–155.

Bennett B, Downing C, Parker C et al. (2006). Mouse genetic models in alcohol research. Trends Genet 22: 367–374.

Bennett B, Carosone-Link P, Beeson M et al. (2008). Genetic dissection of quantitative trait locus for ethanol sensitivity in long- and short-sleep mice. Genes Brain Behav 7: 659–668.

Bergeson SE, Warren RK, Crabbe JC et al. (2003). Chromosomal loci influencing chronic alcohol withdrawal severity. Mamm Genome 14: 454–463.

Bergstrom HC, Palmer AA, Wood RD et al. (2003). Reverse selection for differential response to the locomotor stimulant effects of ethanol provides evidence for pleiotropic genetic influence on locomotor response to other drugs of abuse. Alcohol Clin Exp Res 27: 1535–1547.

Bjork K, Hansson AC, Sommer WH (2010). Genetic variation and brain gene expression in rodent models of alcoholism implications for medication development. Int Rev Neurobiol 91: 129–171.

Boehm SL, Schafer GL, Phillips TJ et al. (2000). Sensitivity to ethanol-induced motor incoordination in 5-HT(1B) receptor null mutant mice is task-dependent: implications for behavioral assessment of genetically altered mice. Behav Neurosci 114: 401–409.

Buck KJ, Milner LC, Denmark DL et al. (2012). Discovering genes involved in alcohol dependence and other alcohol responses: role of animal models. Alcohol Res 34: 367–374.

Chester JA, Risinger FO, Cunningham CL (1998). Ethanol reward and aversion in mice bred for sensitivity to ethanol withdrawal. Alcohol Clin Exp Res 22: 468–473.

Chester JA, Lumeng L, Li TK et al. (2003). High- and low-alcohol-preferring mice show differences in conditioned taste aversion to alcohol. Alcohol Clin Exp Res 27: 12–18.

Ciccocioppo R (2013). Genetically selected alcohol preferring rats to model human alcoholism. Curr Top Behav Neurosci 13: 251–269.

Ciccocioppo R, Economidou D, Cippitelli A et al. (2006). Genetically selected Marchigian Sardinian alcohol-preferring (msP) rats: an animal model to study the neurobiology of alcoholism. Addict Biol 11: 339–355.

Cloninger CR, Bohman M, Sigvardsson S (1981). Inheritance of alcohol abuse. Cross-fostering analysis of adopted men. Arch Gen Psychiatry 38: 861–868.

Colombo G, Lobina C, Carai MA et al. (2006). Phenotypic characterization of genetically selected Sardinian alcohol-preferring (sP) and -non-preferring (sNP) rats. Addict Biol 11: 324–338.

Coppola A, Moshe SL (2012). Animal models. In: H Stefan, WH Theodore (Eds.), Epilepsy, Part I. Elsevier, pp. 63–98.

Correa M, Arizzi MN, Betz A et al. (2003). Locomotor stimulant effects of intraventricular injections of low doses of ethanol in rats: acute and repeated administration. Psychopharmacology (Berl) 170: 368–375.

Crabbe JC (2002). Genetic contributions to addiction. Ann Rev Psychol 53: 435–462.

Crabbe JC (2012). Translational behaviour-genetic studies of alcohol: are we there yet? Genes Brain Behav 11: 375–386.

Crabbe JC, Phillips TJ (1993). Selective breeding for alcohol withdrawal severity. Behav Genet 23: 171–177.

Crabbe JC, Janowsky JS, Young ER et al. (1982). Tolerance to ethanol hypothermia in inbred mice: genotypic correlations with behavioral responses. Alcohol Clin Exp Res 6: 446–458.

Crabbe JC, Phillips TJ, Kosobud A et al. (1990). Estimation of genetic correlation: interpretation of experiments using selectively bred and inbred animals. Alcohol Clin Exp Res 14: 141–151.

Crabbe JC, Metten P, Cameron AJ et al. (2005). An analysis of the genetics of alcohol intoxication in inbred mice. Neurosci Biobehav Rev 28: 785–802.

Crabbe JC, Phillips TJ, Harris RA et al. (2006). Alcohol-related genes: contributions from studies with genetically engineered mice. Addict Biol 11: 195–269.

Crabbe JC, Metten P, Rhodes JS et al. (2009). A line of mice selected for high blood ethanol concentrations shows drinking in the dark to intoxication. Biol Psychiatry 5: 662–670.

Crabbe JC, Phillips TJ, Belknap JK (2010). The complexity of alcohol drinking: studies in rodent genetic models. Behav Genet 40: 737–750.

Crabbe JC, Harris RA, Koob GF (2011a). Preclinical studies of alcohol binge drinking. Ann N Y Acad Sci 1216: 24–40.

Crabbe JC, Spence SE, Brown LL et al. (2011b). Alcohol preference drinking in a mouse line selectively bred for High Drinking in the Dark. Alcohol 45: 427–440.

Crabbe JC, Colville AM, Kruse LC et al. (2012a). Ethanol tolerance and withdrawal severity in high drinking in the dark selectively bred mice. Alcohol Clin Exp Res 36: 1152–1161.

Crabbe JC, Harkness JH, Spence SE et al. (2012b). Intermittent availability of ethanol does not always lead to elevated drinking in mice. Alcohol Alcohol 47: 509–517.

Crabbe JC, Kruse LC, Colville AM et al. (2012c). Ethanol sensitivity in High Drinking in the Dark selectively bred mice. Alcohol Clin Exp Res 36: 1162–1170.

Crabbe JC, Kendler KS, Hitzemann RJ (2013). Modeling the diagnostic criteria for alcohol dependence with genetic animal models. Curr Top Behav Neurosci 13: 187–221.

Cryan JF, Holmes A (2005). The ascent of mouse: advances in modeling human depression and anxiety. Nat Rev Drug Discov 4: 775–790.

Cunningham CL, Gremel CM, Groblewski PA (2009). Genetic influences on conditioned taste aversion. In: S Reilly, TR Schachtman (Eds.), Conditioned Taste Aversion: Behavioral and Neural Processes. Oxford University Press, New York, pp. 387–421.

Cunningham CL, Fidler TL, Murphy KV et al. (2012). Time-dependent negative reinforcement of ethanol intake by alleviation of acute withdrawal. Biol Psychiatry 73: 249–255.

Dick DM, Smith G, Olausson P et al. (2010). Understanding the construct of impulsivity and its relationship to alcohol use disorders. Addict Biol 15: 217–226.

Echevarria DJ, Toms CN, Jouandot DJ (2011). Alcohol-induced behavior change in zebrafish models. Rev Neurosci 22: 85–93.

Egli M (2005). Can experimental paradigms and animal models be used to discover clinically effective medications for alcoholism? Addict Biol 10: 309–319.

Ehlers CL, Walter NAR, Dick DM et al. (2010). A comparison of selected quantitative trait loci associated with alcohol use phenotypes in humans and mouse models. Addict Biol 15: 185–199.

Enoch MA, Goldman D (2001). The genetics of alcoholism and alcohol abuse. Curr Psychiatry Rep 3: 144–151.

Eriksson CJ, Sarviharju M (1984). Motor impairment, narcosis and hypothermia by ethanol: separate genetic mechanisms. Alcohol 1: 59–62.

Falconer DS, Mackay TFC (1996). Introduction to Quantitative Genetics. 4th edn. Longman, Harlow, England.

Fidler TL, Dion AM, Powers MS et al. (2011). Intragastric self-infusion of ethanol in high- and low-drinking mouse genotypes after passive ethanol exposure. Genes Brain Behav 10: 264–275.

Fidler TL, Powers MS, Ramirez JJ et al. (2012). Dependence induced increases in intragastric alcohol consumption in mice. Addict Biol 17: 13–32.

File SE, Mabbutt PS, Hitchcott PK (1990). Characterisation of the phenomenon of "one-trial tolerance" to the anxiolytic effect of chlordiazepoxide in the elevated plus-maze. Psychopharmacology (Berl) 102: 98–101.

Finn DA, Ford MM, Wiren KM et al. (2004). The role of pregnane neurosteroids in ethanol withdrawal: behavioral genetic approaches. Pharmacol Ther 101: 91–112.

Finn DA, Beckley EH, Kaufman KR et al. (2010). Manipulation of GABAergic steroids: sex differences in the effects on alcohol drinking- and withdrawal-related behaviors. Horm Behav 57: 12–22.

Fish EW, Robinson JE, Krouse MC et al. (2012). Intracranial self-stimulation in FAST and SLOW mice: effects of alcohol and cocaine. Psychopharmacology (Berl) 220: 719–730.

Ford MM, Fretwell AM, Anacker AM et al. (2010). The influence of selection for ethanol withdrawal severity on traits associated with ethanol self-administration and reinforcement. Alcohol Clin Exp Res 35: 326–337.

Friedman HJ (1980). Assessment of physical dependence on and withdrawal from ethanol in animals. In: H Rigter, JC Crabbe (Eds.), Alcohol Tolerance and Dependence. Elsevier/North-Holland Biomedical Press, Amsterdam, pp. 93–121.

Gass JT, Olive MF (2007). Reinstatement of ethanol-seeking behavior following intravenous self-administration in Wistar rats. Alcohol Clin Exp Res 31: 1441–1445.

Goldman D, Ducci F (2007). Deconstruction of vulnerability to complex diseases: enhanced effect sizes and power of intermediate phenotypes. Sci World J 7: 124–130.

Goldstein DB (1983). Pharmacology of Alcohol. Oxford University Press, Oxford.

Goldstein DB, Pal N (1971). Alcohol dependence produced in mice by inhalation of ethanol: grading the withdrawal reaction. Science 172: 288–290.

Grahame NJ, Cunningham CL (1997). Intravenous ethanol self-administration in C57BL/6J and DBA/2J mice. Alcohol Clin Exp Res 21: 56–62.

Grahame NJ, Li TK, Lumeng L (1999). Selective breeding for high and low alcohol preference in mice. Behav Genet 29: 47–57.

Grahame NJ, Chester JA, Rodd-Henricks K et al. (2001). Alcohol place preference conditioning in high- and low-alcohol preferring selected lines of mice. Pharmacol Biochem Behav 68: 805–814.

Grant KA, Leng X, Green HL et al. (2008). Drinking typography established by scheduled induction predicts chronic heavy drinking in a monkey model of ethanol self-administration. Alcohol Clin Exp Res 32: 1824–1838.

Green AS, Grahame NJ (2008). Ethanol drinking in rodents: is free-choice drinking related to the reinforcing effects of ethanol? Alcohol 42: 1–11.

Hardy WT, Deutsch JA (1977). Preference to ethanol in dependent rats. Behav Biol 20: 482–492.

Helms CM, Messaoudi I, Jeng S et al. (2012). A longitudinal analysis of circulating stress-related proteins and chronic ethanol self-administration in cynomolgus macaques. Alcohol Clin Exp Res 36: 995–1003.

Holstein SE, Li N, Eshleman AJ et al. (2013). GABA(B) receptor activation attenuates the stimulant but not mesolimbic dopamine response to ethanol in FAST mice. Behav Brain Res 237: 49–58.

Jeanblanc J, He DY, Carnicella S et al. (2009). Endogenous BDNF in the dorsolateral striatum gates alcohol drinking. J Neurosci 29: 13494–13502.

Kalant H (1998). Research on tolerance: What can we learn from history? Alcohol Clin Exp Res 22: 67–76.

Kalant H, LeBlanc AE, Gibbins RJ (1971). Tolerance to, and dependence on, some non-opiate psychotropic drugs. Pharmacol Rev 23: 135–191.

Kampov-Polevoy AB, Garbutt JC, Janowsky DS (1999). Association between preference for sweets and excessive alcohol intake: a review of animal and human studies. Alcohol Alcohol 34: 386–395.

Kampov-Polevoy AB, Matthews DB, Gause L et al. (2000). P rats develop physical dependence on alcohol via voluntary drinking: changes in seizure thresholds, anxiety, and patterns of alcohol drinking. Alcohol Clin Exp Res 24: 278–284.

Kendler KS, Neale MC, Heath AC et al. (1994). A twin-family study of alcoholism in women. Am J Psychiatry 151: 707–715.

Kendler KS, Aggen SH, Prescott CA et al. (2012). Evidence for multiple genetic factors underlying the DSM-IV criteria for alcohol dependence. Mol Psychiatry 17: 1306–1315.

Koob GF (2008). A role for brain stress systems in addiction. Neuron 59: 11–34.

Kosobud A, Crabbe JC (1986). Ethanol withdrawal in mice bred to be genetically prone or resistant to ethanol withdrawal seizures. J Pharmacol Exp Ther 238 (1): 170–327.

Kovacs GL (2000). The role of atrial natriuretic peptide in alcohol withdrawal: a peripheral indicator and central modulator? Eur J Pharmacol 405: 103–112.

Le Bourhis B (1977). Sur l'établissement de la dépendence des rats à l'égard de l'alcool. Physiol Behav 18: 475–478.

Leeman RF, Heilig M, Cunningham CL et al. (2010). Ethanol consumption: how should we measure it? Achieving concilience between human and animal phenotypes. Addict Biol 15: 109–124.

Li TK, Lumeng L, McBride WJ et al. (1987). Rodent lines selected for factors affecting alcohol consumption. Alcohol Alcohol Suppl 1: 91–96.

Loi B, Lobina C, Maccioni P et al. (2010). Increase in alcohol intake, reduced flexibility of alcohol drinking, and evidence of signs of alcohol intoxication in Sardinian alcohol-preferring rats exposed to intermittent access to 20% alcohol. Alcohol Clin Exp Res 34: 2147–2154.

Lopez MF, Griffin III WC, Melendez RI et al. (2012). Repeated cycles of chronic intermittent ethanol exposure leads to the development of tolerance to aversive effects of ethanol in C57BL/6 J mice. Alcohol Clin Exp Res 36: 1180–1187.

Lovinger DM, Crabbe JC (2005). Laboratory models of alcoholism: treatment target identification and insight into mechanisms. Nat Neurosci 8: 1471–1480.

Mardones J, Segovia-Riquelme N (1983). Thirty-two years of selection of rats by ethanol preference: UChA and UChB strains. Neurobehav Toxicol Teratol 5: 171–178.

Matson LM, Grahame NJ (2013). Pharmacologically relevant intake during chronic, free-choice drinking rhythms in selectively bred high alcohol-preferring mice. Addict Biol 13: 921–929.

McBride WJ, Kimpel MW, McClintick JN et al. (2012). Gene expression in the ventral tegmental area of 5 pairs of rat lines selectively bred for high or low ethanol consumption. Pharmacol Biochem Behav 102: 275–285.

McClearn GE (1968). Genetics and motivation of the mouse. In: WJ Arnold (Ed.), Nebraska Symposium on Motivation. University of Nebraska Press, Lincoln, Nebraska, pp. 47–83.

McClearn GE (1979). Genetics and alcoholism simulacra. Alcohol Clin Exp Res 3: 255–258.

Mennella JA, Pepino MY, Lehmann-Castor SM et al. (2010). Sweet preferences and analgesia during childhood: effects of family history of alcoholism and depression. Addiction 105: 666–675.

Metten P, Crabbe JC (1996). Dependence and withdrawal. In: RA Deitrich, VG Erwin (Eds.), Pharmacological Effects of Ethanol on the Nervous System. CRC Press, Boca Raton, FL, pp. 269–290.

Metten P, Phillips TJ, Crabbe JC et al. (1998). High genetic susceptibility to ethanol withdrawal predicts low ethanol consumption. Mamm Genome 9: 983–990.

Meyer PJ, Meshul CK, Phillips TJ (2009). Ethanol- and cocaine-induced locomotion are genetically related to increases in accumbal dopamine. Genes Brain Behav 8: 346–355.

Murphy JM, Stewart RB, Bell RL et al. (2002). Phenotypic and genotypic characterization of the Indiana University rat lines selectively bred for high and low alcohol preference. Behav Genet 32: 363–388.

NIAAA Newsletter (2004). NIAAA Council Approves Definition of Binge Drinking, Vol. 3, DHHS-NIH, Bethesda, MD, p.3.

Numan R (1981). Multiple exposures to ethanol facilitate intravenous self-administration of ethanol by rats. Pharmacol Biochem Behav 15: 101–108.

Numan R, Naparzewska AM, Adler CM (1984). Absence of reinforcement with low dose intravenous ethanol self-administration in rats. Pharmacol Biochem Behav 21: 609–615.

Oberlin B, Best C, Matson L et al. (2011). Derivation and characterization of replicate high- and low-alcohol preferring lines of mice and a high-drinking crossed HAP line. Behav Genet 41: 288–302.

Owens JC, Bennett B, Johnson TE (2002). Possible pleiotropic effects of genes specifying sedative/hypnotic sensitivity to ethanol on other alcohol-related traits. Alcohol Clin Exp Res 26: 1461–1467.

Palmer AA, Phillips TJ (2002). Effect of forward and reverse selection for ethanol-induced locomotor response on other measures of ethanol sensitivity. Alcohol Clin Exp Res 26: 1322–1329.

Phillips TJ (1997). Behavior genetics of drug sensitization. Crit Rev Neurobiol 11: 21–33.

Phillips TJ, Crabbe JC (1991). Behavioral studies of genetic differences in alcohol action. In: JC Crabbe, RA Harris (Eds.), The Genetic Basis of Alcohol and Drug Actions. Plenum Press, New York, pp. 25–104.

Phillips TJ, Shen EH (1996). Neurochemical bases of locomotion and ethanol stimulant effects. Int Rev Neurobiol 39: 243–282.

Phillips TJ, Burkhart-Kasch S, Terdal ES et al. (1991). Response to selection for ethanol-induced locomotor activation: genetic analyses and selection response characterization. Psychopharmacology (Berl) 103: 557–566.

Pinkston JW, Ginsburg BC, Lamb RJ (2009). Examination of reinforcement magnitude on the pharmacological disruption of fixed-ratio performance. Exp Clin Psychopharmacol 17: 237–246.

Ponomarev I, Crabbe JC (2002). A novel method to assess initial sensitivity and acute functional tolerance to hypnotic effects of ethanol. J Pharmacol Exp Ther 302: 257–263.

Quintanilla ME, Israel Y, Sapag A et al. (2006). The UChA and UChB rat lines: metabolic and genetic differences influencing ethanol intake. Addict Biol 11: 310–323.

Radcliffe RA, Hoffmann SE, Deng XS et al. (2004). Behavioral characterization of alcohol-tolerant and alcohol-nontolerant rat lines and an F(2) generation. Behav Genet 34: 453–463.

Radcliffe RA, Floyd KL, Drahnak JA et al. (2005). Genetic dissociation between ethanol sensitivity and rapid tolerance in mouse and rat strains selectively bred for differential ethanol sensitivity. Alcohol Clin Exp Res 29: 1580–1589.

Radcliffe RA, Floyd KL, Lee MJ (2006). Rapid ethanol tolerance mediated by adaptations in acute tolerance in inbred mouse strains. Pharmacol Biochem Behav 84: 524–534.

Radcliffe RA, Erwin VG, Bludeau P et al. (2009). A major QTL for acute ethanol sensitivity in the alcohol tolerant and non-tolerant selected rat lines. Genes Brain Behav 8: 611–625.

Rhodes JS, Best K, Belknap JK et al. (2005). Evaluation of a simple model of ethanol drinking to intoxication in C57BL/6J mice. Physiol Behav 84: 53–63.

Rhodes JS, Ford MM, Yu C-H et al. (2007). Mouse inbred strain differences in ethanol drinking to intoxication. Genes Brain Behav 6: 1–18.

Richter CP, Campbell KH (1940). Alcohol taste thresholds and concentrations of solution preferred by rats. Science 91: 507–508.

Rijk H, Crabbe JC, Rigter H (1982). A mouse model of alcoholism. Physiol Behav 29: 833–839.

Ritzmann RF, Tabakoff B (1976). Ethanol, serotonin metabolism, and body temperature. Ann N Y Acad Sci 273: 247–255.

Roberto M, Cruz MT, Gilpin NW et al. (2010). Corticotropin releasing factor-induced amygdala gamma-aminobutyric acid release plays a key role in alcohol dependence. Biol Psychiatry 67: 831–839.

Rodd ZA, Bell RL, Sable HJ et al. (2004). Recent advances in animal models of alcohol craving and relapse. Pharmacol Biochem Behav 79: 439–450.

Samson HH, Czachowski CL (2002). Behavioral measures of alcohol self-administration and intake control: rodent models. Int Rev Neurobiol 54: 107–143.

Samson HH, Hodge CW (1996). Neurobehavioral regulation of ethanol intake. In: RA Deitrich, VG Erwin (Eds.), Pharmacological Effects of Ethanol on the Nervous System. CRC Press, Boca Raton, FL, pp. 203–226.

Saxon AJ, McCarty D (2005). Challenges in the adoption of new pharmacotherapeutics for addiction to alcohol and other drugs. Pharmacol Ther 108: 119–128.

Scholz H, Mustard JA (2013). Invertebrate models of alcoholism. Curr Top Behav Neurosci 13: 433–457.

Shen EH, Dorow JD, Huson M et al. (1996). Correlated responses to selection in FAST and SLOW mice: effects of ethanol on ataxia, temperature, sedation, and withdrawal. Alcohol Clin Exp Res 20: 688–696.

Smith JW, Little HJ (2000). Differential effects of a dihydropyridine calcium channel antagonist on the components of ethanol tolerance. Brain Res 863: 9–19.

Smith SM, Flentke GR, Kragtorp KA et al. (2011). Ethanol exposure during the early first trimester equivalent impairs reflexive motor activity and heightens fearfulness in an avian model. Alcohol 45: 57–63.

Sommer W, Hyytia P, Kiianmaa K (2006). The alcohol-preferring AA and alcohol-avoiding ANA rats: neurobiology of the regulation of alcohol drinking. Addict Biol 11: 289–309.

Sprow GM, Thiele TE (2012). The neurobiology of binge-like ethanol drinking: Evidence from rodent models. Physiol Behav 106: 325–331.

Stephens DN, Duka T, Crombag HS et al. (2010). Reward sensitivity: issues of measurement, and achieving consilience between human and animal phenotypes. Addict Biol 15: 145–168.

Subramony SH (2012). Approach to ataxic diseases. In: SH Subramony, A Durr (Eds.), Ataxic Disorders, Elsevier, pp.127–134.

What is place of publication? Turek VF, Tsivkovskaia NO, Hyytia P et al. (2005). Urocortin 1 expression in five pairs of rat lines selectively bred for differences in alcohol drinking. Psychopharmacology (Berl) 181: 511–517.

Vengeliene V, Bilbao A, Molander A et al. (2008). Neuropharmacology of alcohol addiction. Br J Pharmacol 154: 299–315.

Verendeev A, Riley AL (2012). Conditioned taste aversion and drugs of abuse: history and interpretation. Neurosci Biobehav Rev 36: 2193–2205.

Wahlsten D (2011). Mouse Behavioral Testing. Academic Press, London.

Wahlsten D, Crabbe JC (2007). Behavioral testing. In: JG Fox, C Newcomer, A Smith et al. (Eds.), The Mouse in Biomedical Research, Vol. 3. Normative Biology, Immunology and Husbandry. Elsevier, Amsterdam.

Wahlsten D, Bachmanov A, Finn DA et al. (2006). Stability of inbred mouse strain differences in behavior and brain size between laboratories and across decades. Proc Natl Acad Sci U S A 103: 16364–16369.

Waller MB, McBride WJ, Gatto GJ et al. (1984). Intragastric self-infusion of ethanol by ethanol-preferring and -nonpreferring lines of rats. Science 225: 78–80.

Section 3

Molecular basis of alcoholism

Handbook of Clinical Neurology, Vol. 125 (3rd series)
Alcohol and the Nervous System
E.V. Sullivan and A. Pfefferbaum, Editors

Chapter 6

Molecular basis of alcoholism

DANA MOST, LAURA FERGUSON, AND R. ADRON HARRIS*
Waggoner Center for Alcohol and Addiction Research, University of Texas, Austin, TX, USA

MOLECULAR MECHANISMS UNDERLYING ACUTE AND CHRONIC ALCOHOLISM

Introduction

Alcohols are organic compounds containing a hydroxyl (-OH) group attached to a carbon atom and the aliphatic alcohols are described by the general formula $C_nH_{2n+1}OH$. Ethanol, the psychoactive constituent of alcohol, has been used recreationally for tens of thousands of years (Hanson, 1995) and is one of the largest health burdens on society (Cargiulo, 2007). For the remainder of this chapter alcohol will refer to ethanol, and alcoholism is the functional equivalent of alcohol use disorder, as outlined by the *Diagnostic and Statistical Manual of Mental Disorders*, fifth edition (DSM-V) (American Psychiatric Association, 2013).

The maladaptive behaviors of alcoholism rely on changes in the brain that lead to compulsive and excessive drinking, afflicting all organs, with damage as a secondary consequence of alcoholism. While acute use of alcohol, such as binge drinking and intoxication, causes cellular changes in the brain that last for hours, chronic alcohol use induces widespread neuroadaptations in the nervous system that can last a lifetime. This involves the remodeling of synapses that are dependent upon changes in gene expression in the presence of chronic alcohol use (Wilke et al., 1994), and is illustrated in Figure 6.1.

In order to fully understand the effects of alcohol on behavior and thus enable the development of efficacious treatments, it is necessary to understand the actions of alcohol at the molecular level. It is remarkable how little is known about alcohol's molecular targets, in view of alcohol's burden on public health and its long-term and widespread use. One reason for this might be alcohol's low binding affinity to proteins, reflected by the fact that clinically relevant intoxication levels of alcohol are measured in millimolar concentrations, whereas most other drugs of abuse are measured in nanomolar concentrations. A consequence of the high concentrations in the body is that there are a large number of potential molecular targets.

In the early days of alcohol research it was thought that, because of the small size and non-specific nature of the alcohol molecule, it would likely not have a specific binding domain on proteins, but would instead interfere with the lipid membranes of the central nervous system (CNS). However, alcohol has only been shown to interact with the lipid bilayer at concentrations much higher than clinically relevant (Pang et al., 1980; Goldstein, 1984; Peoples et al., 1996). For this reason, researchers have focused their search on other molecular targets of alcohol. Most of the examples below come from animal models and postmortem tissue analysis of alcoholics, animal models, and cell cultures. Many of the molecular pathways that are sensitive to alcohol are highly conserved across species such as humans, mice, rats, worms, and fruit flies (Dick and Foroud, 2002, Lewohl et al., 2011). In addition, considerable progress has been made in defining binding cavities for alcohol in several proteins, including ion channels (Harris et al., 2008; Howard et al., 2011, Sauguet et al., 2013).

This chapter will discuss the known and proposed molecular targets of alcohol in the brain that may be important for behaviors meeting the DSM-V criteria for alcohol use disorder. Additionally, the molecular mechanisms of approved and prospective treatments for alcoholism will be discussed.

Primary targets

Alcohol's effects on the brain are diverse and include changes in levels and function of neurotransmitters,

*Correspondence to: Dr. R. Adron Harris, The University of Texas at Austin, 1 University Station A4800, Austin, TX 78712, USA. Tel: +1-512-232-2514, E-mail: harris@austin.utexas.edu

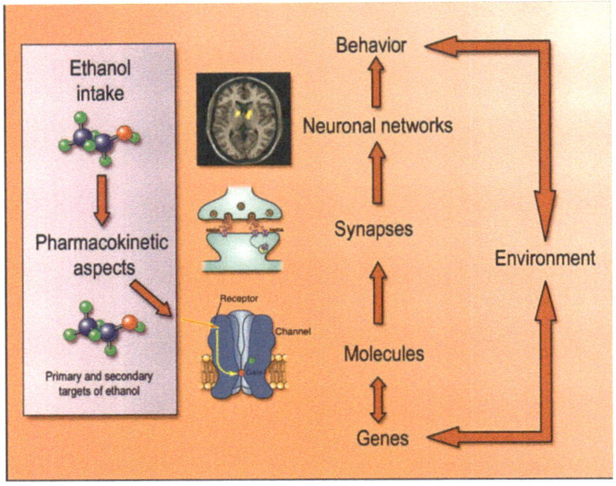

Fig. 6.1. A schematic illustrating the dynamic interactions between alcohol and the nervous system. (Adapted from Spanagel, 2009.)

receptors, enzymes, and other molecules, culminating in synaptic changes in brain circuitry regulating compulsivity and inhibition. Changes in these molecular systems lead to tolerance and withdrawal when alcohol is removed from the system.

Alcohol metabolism and distribution, which are discussed in depth in Chapter 4, can be summarized as follows: After consumption, alcohol is absorbed into the blood stream through the stomach and intestines, and it readily crosses the blood–brain barrier. Alcohol is distributed with body water and is found at approximately the same concentration in all tissues, including the nervous system. Most of the metabolism of alcohol occurs in the liver, and the brain has limited metabolism by the mitochondrial cytochrome P450 (CYP2E1), catalase, and other pathways (Zakhari, 2006).

CYP2E1 metabolism of alcohol produces acetaldehyde and is a source of acetaldehyde in brain. Additional acetaldehyde may enter the brain from peripheral conversion of alcohol to acetaldehyde by an enzyme found in the liver called alcohol dehydrogenase (ADH). In fact, ADH can be considered a "target" of alcohol since it binds alcohol through a zinc atom on ADH and the hydroxyl group of alcohol. Details on the proposed receptor-binding sites of alcohol are beyond the scope of this chapter. Interested readers may wish to explore this subject further in these reviews (Harris et al., 2008, Howard et al., 2011).

In the periphery, acetaldehyde is the primary metabolite of alcohol and is responsible for the flushing effect that encompasses face flushing, nausea, vomiting, headache, tachycardia, and sweating. Disulfiram (Antabuse), the first of three approved treatments for alcoholism, employs the aversive nature of acetaldehyde by inhibiting aldehyde dehydrogenase enzyme, thereby allowing for an accumulation of acetaldehyde. Disulfiram is a rather non-specific enzyme inhibitor and it may be useful in the treatment of cocaine dependence and even cancer chemotherapy due to actions on sites other than aldehyde dehydrogenase (Shorter and Kosten, 2011, Schmitt et al., 2012). Although disulfiram decreases drinking when taken regularly, it has low patient compliance because of the aversive effects (Moriarty, 1950; Barth and Malcolm, 2010). In addition to the peripheral flushing effect, acetaldehyde may have actions in the brain relevant to the acute effects of alcohol (Quertemont et al., 2005) and even the development of alcoholism (Deng and Deitrich, 2008). Acetaldehyde is self-administered by rodents through intravenous and intracerebral ventricular routes (Amit et al., 1977, Brown et al., 1979; Myers et al., 1984). Rats show preference for the physical place where they received central or peripheral administration of acetaldehyde over places where they received saline (conditioned place preference) (Quintanilla and Tampier, 2003, Spina et al., 2010). Although more research is necessary to fully understand these mechanisms, the neurobiological – and consequently the behavioral – actions of alcohol most likely depend on central contributions from both alcohol and its metabolites, acetaldehyde and acetate (Jiang et al., 2013). The development of alcohol tolerance and dependence comes from alterations in brain structure and function over time, and is long-lasting. This involves the remodeling of synapses that are dependent upon changes in gene expression in the presence of chronic alcohol (Rhodes and Crabbe, 2005). We will review the molecular adaptations at the acute and chronic stages of alcohol use that may underlie the hallmarks of alcoholism: withdrawal, tolerance, relapse, and craving.

It is important to note that these long-lasting drug-responsive alterations are unlikely to be encoded in RNAs or proteins due to the fast turnover rate of those molecules. This suggests that DNA is in charge of cellular memory, and that epigenetic mechanisms may be critical components of long-term learning and memory processes (Levenson and Sweatt, 2005), as well as chronic dependence on drugs (Renthal and Nestler, 2008).

DNA

Both the sequence and structure of DNA molecules control all downstream processes such as RNA transcription and protein translation, and both can contribute to the development, progression, and persistence of alcoholism. A great deal of work has been done in the field of genes and molecular pathways involved in fetal alcohol syndrome and alcohol-related developmental disorders (Goodlett et al., 2005; Sulik, 2005) and will be covered in other chapters of this book. Also covered in another chapter of this book are results from genome-wide association and quantitative trait loci studies that reveal mutations that may be important for predicting future alcoholism in certain groups or individuals.

Recent research has emphasized the notion that epigenetic mechanisms (which exert lasting control over gene expression through structural modifications of the DNA without altering the sequence) could mediate stable changes in brain function associated with alcoholism (Waddington, 2012). DNA methylation, histone acetylation, and phosphorylation are three common epigenetic-enzymatic modifications to chromatin structure that make the DNA less or more accessible to transcription factors and enzymes, thus changing the transcriptional activity of the target genes. Chronic exposure to alcohol was found to induce changes in the chromatin structure, specifically on gene promoters causing changes in gene expression in alcoholics (Guerri and Pascual, 2010; Ponomarev et al., 2012). This suggests that epigenetic mechanisms are involved in both biochemical and behavioral responses to alcohol.

Histone deacetylase (HDAC) is an enzyme that removes the acetyl group from histone proteins on DNA, making the DNA less accessible to transcription factors. Acute alcohol was found to decrease HDAC activity and increase acetylation of histones (H3 and H4) in the rat amygdala. Conversely, withdrawal from chronic alcohol in rats was found to increase HDAC activity and decrease H3 and H4 acetylation in the rat amygdala (Pandey et al., 2008). An HDAC inhibitor (trichostatin A) blocks the increase in HDAC activity and rescues the deficits in H3 and H4 acetylation in the amygdala. This change prevents the development of withdrawal-related anxiety in rats, suggesting a potential role for HDAC inhibitors as therapeutic agents in treating alcohol withdrawal symptoms (Pandey et al., 2008). CREB-binding protein (CBP) is a histone acetylase (HAT) that acetylates nearby histones, and allows for subsequent transcriptional activation. Acute alcohol exposure increases levels of CBP and neuropeptide Y (NPY), while withdrawal produces the opposite effect. Moreover, the withdrawal-induced anxiety behavior was found to correlate with the levels of CBP and NPY in the amygdala (Pandey et al., 2008). CBP was found to be important for normal learning and memory mechanisms in the brain since mutations of CBP cause a form of mental retardation in humans (Tsankova et al., 2007). This finding suggests an important mechanism in alcoholism, since addiction is thought to be an aberrant learning and memory process (Hyman et al., 2006). Figure 6.2 summarizes the epigenetic changes associated with alcohol.

Mice lacking a specific HDAC, HDAC5, display normal rewarding responses to acute cocaine exposure and become hypersensitive when chronically exposed to

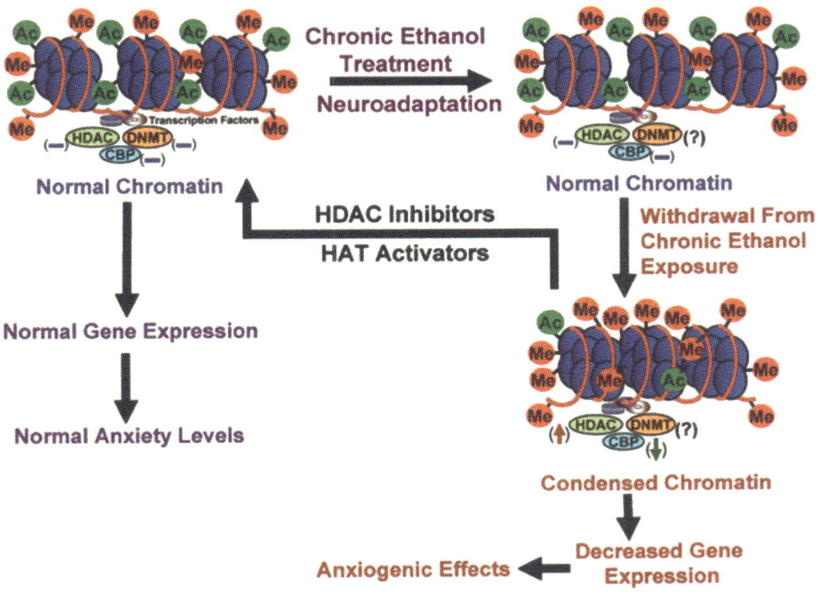

Fig. 6.2. Epigenetic modifications associated with alcoholism. CREB-binding protein (CBP) and histone deacetylase (HDAC) activity are altered by acute alcohol exposure. Chronic alcohol exposure results in neuroadaptations opposing the acute effects of alcohol in order to maintain a homeostatic state. Withdrawal after chronic alcohol exposure is associated with increased HDAC activity and decreased levels of CBP and the associated histone acetylation (Starkman et al., 2012). DNMT, DNA methyltransferase; Ac, acetyl; Me, methyl; HAT, histone acetylase.

cocaine (Renthal et al., 2007). Chronic cocaine administration inactivates HDAC5 by exporting it out of the nucleus, resulting in histone hyperacetylation and increased mRNA expression of HDAC5 target genes. An example of a target gene of HDAC5 that is found to be increased during both chronic cocaine and alcohol exposure is neurokinin 1 receptor (NK1R). This receptor is also known as the substance P receptor, due to its importance in pain transmission. Silencing the translation of the NK1R using a technique known as RNA interference was found to reduce alcohol drinking in mice, emphasizing the role that NK1R plays in alcoholism (Baek et al., 2010). In fact, an NK1R antagonist (L822429) decreased voluntary alcohol consumption, suppressed stress-induced reinstatement of alcohol seeking, and increased sensitivity to the sedative effects of alcohol in rats (Schank et al., 2011). Similar effects were also seen in detoxified alcoholic inpatients, as an NK1R antagonist suppressed spontaneous alcohol cravings, improved overall well-being, blunted cravings induced by a challenge procedure, and attenuated concomitant cortisol responses. An analysis of brain function in human alcoholics compared with controls while performing behavioral emotional tasks was performed using functional magnetic resonance imaging and corroborated the above results, suggesting NK1R antagonism as a potential therapeutic target for the treatment of alcoholism (George et al., 2008).

DNA is the master regulator in the cell and is possibly a molecular implementation of the persistence of drug effects (or even memories) in the brain. DNA exerts its control through gene expression, which includes: RNA transcription, protein translation, and regulatory processes like microRNA and RNA splicing. In addition to alcohol's indirect effect on gene expression via epigenetic actions on chromatin, alcohol can directly target the gene expression machinery, which will be explored below and is detailed in Figure 6.3.

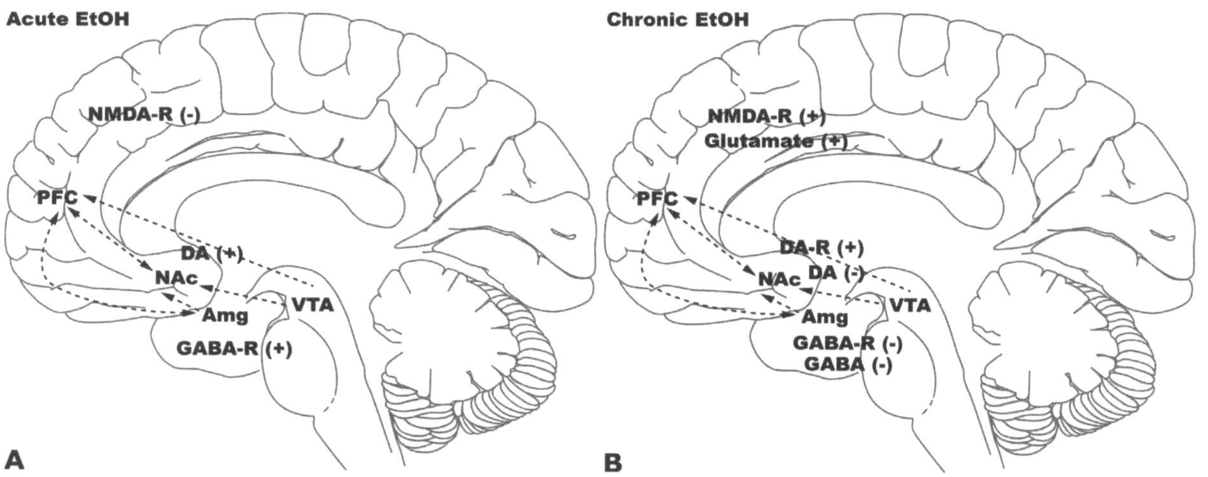

Fig. 6.3. Synaptic transmission mechanism and the synaptic neuroadaptations occurring after chronic alcohol use.

(**A**) Synaptic transmission is the process by which the brain communicates information. Depending on the inputs to the presynaptic neuron's dendrites, an action potential (i.e., nerve impulse) will be generated at the axon hillock and propagate down the axon through the movement of charged particles (i.e., the ions – Na^+ and K^+). The synapse is the gap between the axon of the presynaptic neuron and the dendrite of the postsynaptic neuron. In order for the information to cross the synapse, the electrical signal of the action potential must be converted into a chemical signal. This is achieved by releasing chemicals (neurotransmitters) from the presynaptic nerve terminal in a voltage-dependent (and therefore calcium-dependent) manner. The presynaptic nerve terminal contains the neurotransmitter release machinery needed for this to occur. Once released the neurotransmitter will diffuse across the synapse and bind to receptors on the postsynaptic nerve's dendrites. The two major types of receptors of concern here are ligand-gated ion channels and G-protein-coupled receptors. Ligand-gated ion channels undergo a conformational change when a ligand (i.e., neurotransmitter) is bound and allow a particular ion (e.g. Cl^-, Na^+, Ca^{2+}, etc.) to flow into or out of the cell. When G-protein-coupled receptors are activated they affect secondary messengers and molecular cascades, resulting in changes in the postsynaptic neuron. It is necessary to understand the basics of synaptic transmission because acute and chronic alcohol exposure causes changes in many molecules important for transmission and changes transmission properties. (Adapted from Clapp et al., 2008.)

(**B**) An example of a synapse stimulated by alcohol use and the molecular cellular cascade associated with alcohol use. GABA, γ-aminobutyric acid; NMR, NMDA (*N*-methyl-D-asparatate) receptor; VGCC, voltage-gated calcium channel; Camp, cyclic adenosine monophosphate; PKA, protein kinase A; PKC, protein kinase C; caMK, calmoldulin kinase; MAPk, mitogen-activated protein kinase; CREB, cyclic AMP response element-binding protein. (Adapted from Koob et al., 2005.)

Neurotransmitter systems

Glutamate

Glutamate is the primary excitatory neurotransmitter in the brain. Glutamatergic (glutamate-using) systems contain both ionotropic (ligand-gated receptors coupled to the flow of charged particles – *N*-methyl-D-aspartate (NMDA), α-amino-3-hydroxy-5-methyl-4-isoxazolepropionic acid (AMPA), kainate) and metabotropic (receptors coupled to G proteins – mGluR) receptors. The glutamatergic system plays an important role in long-term potentiation and long-term depression and therefore has physiologic implications that are important for learning and memory. Generally, all glutamate receptors are inhibited by acute alcohol treatment, although some subtypes are only affected by very high concentrations. Alcohol's acute actions on the glutamatergic system have been implicated in tolerance, withdrawal, craving, relapse, and dependence.

Alcohol acts as a non-competitive inhibitor of the AMPA/kainate receptors at high concentrations (Dildy-Mayfield and Harris, 1992, 1995, Akinshola et al., 2003). Out of all the glutamatergic receptors, NMDA receptors are the most sensitive to the effects of alcohol. Alcohol's binding site on the NMDA receptor is not known, and there has been some evidence to suggest that alcohol exerts its effect on the NMDA receptor through protein kinase C (Li et al., 2005). Acute application of alcohol to hippocampal neuronal slices reduces NMDA activity at a concentration that produces intoxication in humans and is linearly related to alcohol's intoxicating potency (Lovinger et al., 1989).

In a drug discrimination test, MK-801, an NMDA antagonist, elicited the same response that animals were trained to give in the presence of alcohol, indicating that alcohol's NMDA-antagonizing effects are important for mediating the subjective effects of alcohol (Butelman et al., 1993). In recently detoxified alcoholics, ketamine, an NMDA antagonist, was able to mimic the behavioral effects of alcohol (Krystal et al., 1998). The production of alcohol-like effects by ketamine supports the clinical role of NMDA receptor function facilitating alcohol's effects on humans.

The response to alcohol depends on the subunit composition of the glutamatergic receptor. A study using rat NMDA receptor subunits expressed in *Xenopus* oocytes suggests that NMDA receptors containing certain subunits are more sensitive to alcohol than others (Raeder et al., 2008). Alterations in subunit composition of glutamate receptors were seen after chronic use of alcohol in mice (Ortiz et al., 1995). Chronic alcohol was also found to increase the expression of AMPA1 receptor subunit in the brains of alcoholics (Lewohl et al., 2000).

"Acute tolerance" is the term used to describe the decreased inhibitory effects of alcohol over a short period of time. NMDA receptor activity in cultured neurons returns to baseline levels after less than 1 hour of alcohol exposure (Lai et al., 2004). Once alcohol is removed from the system, the hyperglutamatergic state can produce a severe withdrawal syndrome characterized by agitation, anxiety, and disorientation and is associated with a susceptibility to seizures and excitotoxic cell death. NMDA receptor antagonists can be applied to protect cells from this type of death (Hoffman et al., 1995; al Qatari et al., 2001). The acute withdrawal syndrome is clinically relevant and well characterized. The question remains as to whether the neuroadaptations leading to acute withdrawal syndrome contribute to the propensity of relapse or lead to alcoholism.

Acute alcohol inhibits the function of the NMDA receptor, while chronic use of alcohol seems to upregulate NMDA receptor expression in the brain (Qiang and Ticku, 2005). This upregulation is thought to counteract the acute inhibition of the glutamatergic system and also the sedative effects of increased γ-aminobutyric acid receptor A (GABA(A)) activity, as discussed below. Figure 6.4 presents an illustration of the main neurotransmitter systems and brain regions involved in the neuroadaptations associated with acute and chronic alcohol use.

Acamprosate, one of the three approved treatments for alcoholism, is proposed to exert at least part of its effect by altering glutamatergic function (De Witte et al., 2005), though the exact mechanism by which it interacts is unclear. Acamprosate was found to reduce relapse rate, increase abstinence rate, and decrease excessive drinking in alcohol-dependent rats, and had no effect in non-dependent rats (Spanagel et al., 1996a, b, c). In dozens of clinical trials conducted in Europe, about half of the alcoholics treated with acamprosate maintained sobriety compared to the placebo group (Mason and Ownby, 2000; Mann et al., 2004), but results of US clinical trials have shown less beneficial effects of acamprosate (Anton et al., 2006; Mason et al., 2006). Several studies seeking to identify acamprosate's mechanism of action have failed to show direct modulation of NMDA-Rs at clinically relevant concentrations (Reilly et al., 2008) and a recent study suggests that calcium is the active moiety of acamprosate (Spanagel et al., 2013). Application of acamprosate on GABA(A), glycine, vanilloid-1 receptors and voltage-gated Na^+ channels did not exert any effect (Reilly et al., 2008). On the other hand, knocking out some of the mGluR subunits of the metabotropic glutamate receptors in mice or by using an mGluR antagonist reduces the ability of acamprosate to affect behavior (Blednov and Harris, 2008). Taken together, the evidence indicates that

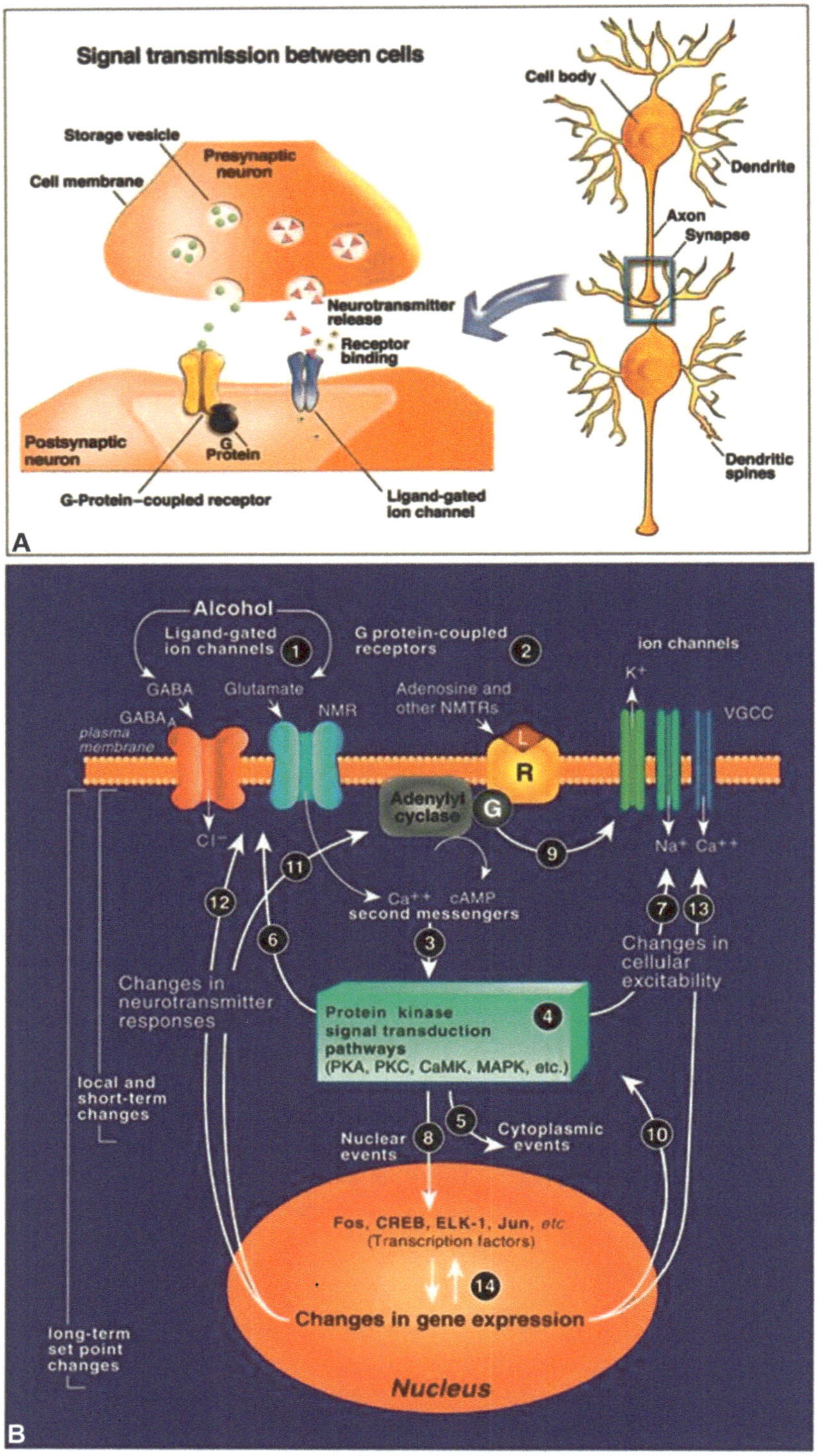

Fig. 6.4. The molecular adaptations in the mesocorticolimbic pathway occurring with alcoholism.

(**A**) The mesocorticolimbic system includes several brain areas: prefrontal cortex (PFC), nucleus accumbens (NAc), ventral tegmental area (VTA), and amygdala (Amg). Alcohol consumption has a major effect on the dopaminergic, glutamatergic, and GABAergic systems in this pathway. Acute effects of alcohol include: increased dopamine release (associated with reward), increased GABA receptor activity (associated with anxiolysis, sedation, and motor incoordination), and decreased glutamate receptor activity. DA, dopamine; EtOH, ethanol; NMDA-R, *N*-methyl-D-aspartate receptor; GABA-R, γ-aminobutyric acid receptor.

(**B**) Chronic alcohol consumption causes neuroadaptations to oppose the effects of acute alcohol and include: decreased dopamine release and increased dopamine receptor expression, decreased GABA receptor expression, increased glutamate release and increased NMDA receptor expression.

alcohol's modulation of the glutamatergic system is an important molecular mechanism by which alcohol exerts its behavioral effects and a potential target for the treatment of alcoholism.

Dopamine

Dopamine is a neurotransmitter in the CNS and binds to several different types of receptors (D1 and D2 families). Dopamine is thought to contribute to alcoholism by signaling in the midbrain dopaminergic system, a brain circuit involved in associative learning, incentive salience, and reward prediction (Gonzales et al., 2004). The midbrain dopaminergic system originates in the ventral tegmental area and projects to regions of the brain such as the striatum, nucleus accumbens, and prefrontal cortex (PFC). Alcohol and other drugs of abuse increase dopaminergic activity in the midbrain region of rodents and humans (Boileau et al., 2003). Dopamine release in the midbrain partially mediates the positive-reinforcing properties of acute alcohol exposure necessary for the development of alcoholism (Raeder et al., 2008). Moreover, preference for alcohol has been directly correlated with alcohol-induced dopamine release in the midbrain. This is illustrated by the fact that rats which are bred to prefer alcohol release more dopamine than wild-type rats in an alcohol self-administration study (Weiss et al., 1993). Also, mice lacking different dopamine receptors and transporters show modified alcohol preference compared with controls, further illustrating dopamine's involvement in alcohol-related behaviors (Crabbe et al., 2006).

Acute alcohol exposure activates dopamine reward pathways, whereas chronic treatment produces a hypodopaminergic state associated with dysphoria, which can lead to craving and relapse (Koob and Volkow, 2010). A 1-year-long chronic alcohol treatment in rats decreases dopamine and its metabolite in the striatum, decreases tyrosine hydroxylase protein levels, and increases dopamine transporter protein levels compared to controls (Rothblat et al., 2001). Positron emission tomography scans show that chronic alcoholics have fewer D2 receptors when compared with non-alcoholics. D2 receptors in the striatum are mainly localized on GABA-synthesizing cells. These results provide evidence of GABAergic involvement in the dopaminergic abnormalities seen in alcoholics (Volkow et al., 2002). These results also reflect the brain's homeostatic mechanism to adapt to initial increases in dopamine levels after chronic exposure to alcohol.

The dopamine receptor antagonist fluphenazine will block alcohol self-administration when injected into the nucleus accumbens (Rassnick et al., 1992). Additionally, several clinical trials have shown that aripiprazole, an atypical antipsychotic, reduces craving and increases positive subjective feelings in alcoholics (Martinotti et al., 2007; Brunetti et al., 2012). Although these clinical trials are promising, the mechanism by which dopamine malfunction caused by alcohol contributes to clinical alcoholism is currently unknown.

Gaba(A)

GABA is the primary inhibitory neurotransmitter, and the GABA(A) receptor, a ligand-gated chloride channel, is the most abundant inhibitory receptor in the mammalian brain. Acute alcohol exposure enhances GABA(A) function. The molecular actions by which alcohol may exert its effects on GABAergic activity are by directly binding to the receptor, increasing presynaptic release of GABA or releasing GABAergic steroids (Lobo and Harris, 2008). This acute action of alcohol on the GABAergic system, along with the effects of chronic alcohol use, is illustrated in Figure 6.5. Alcohol intoxication is characterized by anxiolysis, sedation, impaired cognitive function, hypnosis, and impaired motor function. Many of these behaviors are mimicked by the administration of pharmacologic GABA(A) agonists like muscimol and benzodiazepines (Grobin et al., 1998). Moreover, the effects of alcohol can be diminished by using pharmacologic GABA antagonists like bucucilline and picrotoxin (Hyytia and Koob, 1995), suggesting that GABA(A) signaling is directly involved in the acute actions of alcohol. Electrophysiologic data support alcohol's actions on GABA(A) receptors, showing potentiation of GABA-mediated chloride influx following alcohol administration in a variety of preparations. Mihic et al. (1997) identified a 45-amino-acid residue necessary and sufficient for the enhancement of GABA(A) receptor function by alcohol, suggesting that alcohol's binding site is between the transmembrane 2 and the transmembrane 3 regions of the receptor (Mihic et al., 1997). This is supported by crystallographic analysis of alcohol binding to a related ligand-gated ion channel, GLIC (Sauget et al., 2013). Chronic alcohol use causes changes in GABA receptor subunit composition in human alcoholics (Lewohl et al., 1997), and it should be noted that alcohol's pharmacologic action on GABA(A) receptors (and consequently behavioral responses) strongly depends on the subunit composition of the receptor. Alcohol increases GABAergic neurotransmission over a wide range of concentrations, with some studies showing that delta-containing receptors are more sensitive than other GABA(A) receptors (Wallner et al., 2003). Use of mutant mice either lacking a subunit or containing an ethanol-resistant subunit demonstrated that the alpha-2 and alpha-3 subunits take part in mediating the motor-impairing effects of alcohol and

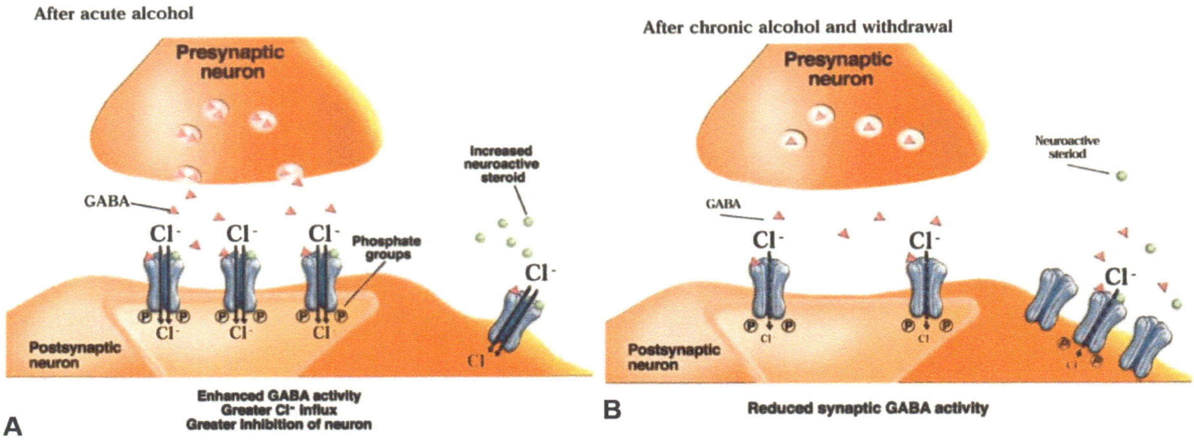

Fig. 6.5. The γ-aminobutyric acid (GABA) system and the effects of acute (**A**) and chronic (**B**) alcohol exposure.
(**A**) GABA acts in part through the activation of GABA(A) receptors, which are ligand-gated ion channels that allow chloride ions (Cl^-) to flow into the neuron after GABA binds to it. Greater influx of Cl^- into the neuron makes it more difficult for the cell to generate a new action potential (i.e., nerve impulse). This is why GABA is considered to be an inhibitory neurotransmitter. The molecular actions by which alcohol may exert its effects on GABAergic activity is by directly binding to the receptor, increasing presynaptic release of GABA, or releasing GABAergic steroids that activate extrasynaptic receptors (Clapp et al., 2008).
(**B**) After chronic alcohol exposure and during withdrawal, GABA activity at the synapse is reduced, leading to reduced inhibition of the postsynaptic neuron. This results in development of anxiety and hyperexcitability. (Adapted from Clapp et al., 2008.)

that the alpha-2 subunit takes part in the aversive properties of alcohol (Blednov et al., 2011b, 2013).

Chronic alcohol use causes the downregulation of GABA(A) receptors due to the initial overstimulation by the alcohol. In turn, glutamate receptors are upregulated to counteract the sedative effects of increased GABA(A). During withdrawal from alcohol, the excess of glutamate receptors combined with the lack of GABA(A), causes anxiety, dysphoria, and in severe cases seizures. Topiramate, an anticonvulsant seizure medication, was found to reduce voluntary alcohol intake in humans with alcohol dependence in recent clinical trials (Johnson et al., 2008). Alcoholic patients achieved a rate of continuous abstinence much higher than those in a placebo group. They also reported fewer cravings compared to a placebo group but complained of other side-effects. Topiramate is associated with glutamate system inhibition, although the exact mechanism by which topiramate promotes abstinence is not yet understood.

GABA(B)

GABA(B) receptors are metabotropic receptors responsible for mediating slow inhibitory responses in the brain and have been implicated in alcoholism. Acute alcohol was seen to increase inhibitory GABA(B) signaling in a concentration-dependent manner in rat midbrain dopaminergic neurons in a current and single-electrode voltage clamp electrophysiologic experiment (Federici et al., 2009). Silencing the GABA(B) receptor in fruit flies using RNA interference decreases acute motor impairment after alcohol exposure (Dzitoyeva et al., 2003). Similarly, in mice, cerebellar injections of GABA(B) agonists accentuated and antagonists attenuated motor impairment following acute alcohol (Dar, 1996). Conversely, recordings in CA1 pyramidal neurons have indicated that GABA(B) signaling was immune to the acute effects of alcohol but became modulated after chronic exposure, perhaps because of neuroadaptations from dependence (Frye et al., 1991). Baclofen (b-parachlorophenol GABA), a GABA(B) agonist prescribed to treat muscle spasticity, reduces relapse in dependent humans and decreases alcohol consumption in rats (Addolorato et al., 2002a b; Maccioni and Colombo, 2009, Agabio et al., 2012). Dr. Olivier Ameisen describes attending over 5000 AA meetings over the course of several years in an attempt to reduce his alcohol craving and abstain from drinking. Despite all efforts he continued drinking and eventually treated himself with high doses of baclofen; he was able to "cure" his disease (Ameisen, 2005, 2012). Preclinical studies show that baclofen is able to suppress withdrawal symptoms in alcoholics, promote alcohol abstinence, and reduce withdrawal-related anxiety and alcohol craving (Addolorato et al., 2002a). There is also evidence for baclofen having no significant effect on alcoholics (Garbutt et al., 2010), suggesting that baclofen may be effective in treating only certain subtypes of alcoholics. There is still research to be done on baclofen and its side-effects, and therefore it is not yet approved by the Food and Drug Administration (FDA) for the treatment of alcoholism (Leggio et al., 2010).

RNA splicing is a process that removes the intervening, non-coding sequences of genes (introns) from pre-mRNA and joins the protein-coding sequences (exons) together in order to enable translation of mRNA into a protein. Sometimes exons are extended or skipped, or introns are retained during the splicing process, leading to "alternative splicing." Alternative splicing of the pre-mRNA creates different splice variants for the same gene, which results in a functional diversity of proteins (Matlin et al., 2005). Eventually, the mRNA transcripts are exported to the cytoplasm and translated into different isoforms of the same protein that vary in their functional properties (Coetzee et al., 1999; Dredge et al., 2001). Drugs of abuse such as alcohol can affect alternative splicing. For example, chronic alcohol consumption enhances the complexity of GABA(B) receptor splicing (Lee et al., 2010, 2013). Human alcoholics show differences in GABA(B) receptor splice variants when compared to non-alcoholics, which may contribute to the pharmacologic effects of GABA(B) agonists such as baclofen in the treatment of alcoholism (Lee et al., 2013).

In summary, pharmacologic, behavioral, and electrophysiologic evidence supports a role for GABA(B) in mediating the effects of alcohol, and that many of the behaviors associated with alcoholism result from the direct alcohol-induced modulation of GABAergic neurotransmission. It should be noted that the specific behaviors elicited by alcohol depend highly on receptor subtype expression and cellular location. These behaviors will hopefully become evident in the future through behavioral characterization of genetically manipulated mice.

Serotonin (5-HT)

Serotonin is a neurotransmitter that serves many functions, including regulating mood, sleep, appetite, learning, memory, and other phenomena. The supply of neuronal serotonin originates mostly from neurons in the raphe nuclei whose axons innervate almost the entire brain, reflecting serotonin's vast physiologic roles. There are many types of receptors for serotonin in the CNS and all are metabotropic, except for 5-HT3, which is a ligand-gated ion channel similar to nicotinic acetylcholine receptors. Evidence indicates that serotonin is involved in mediating both acute and chronic alcohol action and perhaps the development and maintenance of alcoholism.

Alcohol facilitates serotonergic transmission in part by increasing the potency with which the agonist can activate the 5-HT3 receptor (Sung et al., 2000) and the time spent in the open state (Zhou et al., 1998). In general an inverse relationship between serotonin transmission and alcohol drinking has been established. Increased serotonergic transmission is associated with less alcohol consumption and less serotonergic transmission is linked to more alcohol consumption, in animal models of drinking and in human alcoholics (for review, see Lovinger, 1997).

Electrophysiologic recordings in neuroblastoma cells show that acute alcohol potentiates 5-HT3 receptor-mediated ion currents (Lovinger, 1991). Acute alcohol was also found to activate the 5-HT3 receptor in oocytes (Harris et al., 1995), ganglion neurons (Lovinger and White, 1991), frontal cortex neurons (Sung et al., 2000), and human embryonic kidney 293 cells (Lovinger and Zhou, 1994).

Serotonin levels in animal brains are elevated after acute alcohol exposure (Murphy et al., 1982; LeMarquand et al., 1994). Mice lacking the 5-HT1B serotonin receptor consume larger amounts of alcohol compared to wild-type. Mice lacking the receptor show much higher incoordination compared to wild-type mice (after just one injection of alcohol), suggesting this receptor is involved in the intoxication process (Crabbe et al., 1996). The 5-HT2 receptor seems to be important for the reinforcing properties of alcohol because antagonists selectively decrease acute alcohol reinforcement (Roberts et al., 1998). Blockade of the serotonin transporter (5-HTT) with fluoxetine, a selective serotonin reuptake inhibitor (SSRI) (a widely used antidepressant), or by genetic knockout, decreases alcohol consumption in rodents (Kelai et al., 2003). Decreased consumption of alcohol following SSRI treatment has been observed in almost every rat model of alcoholism (Brown et al., 1979; Sellers et al., 1992; Ciccocioppo et al., 2006). Human studies have given less consistent results, but it appears that SSRIs may be effective in some subpopulations of alcoholics, such as late-onset individuals (Kranzler et al., 2012).

In addition to enhancing 5-HT activity in the brain using SSRIs, another approach widely used to understand serotonin's effect on alcohol consumption has been to selectively block the 5-HT3 receptor. 5-HT3 antagonists decrease alcohol self-administration (Fadda et al., 1991; Hodge et al., 1993) and consumption in rodents (Sellers et al., 1994).

In humans, the drug ondansetron, a 5-HT3 antagonist that is FDA-approved for the treatment of chemotherapy-induced nausea, is a promising treatment for alcoholics. In a clinical trial, alcoholics were randomly selected to receive either ondansetron or a placebo for 11 weeks. The ondansetron patients with early-onset alcoholism had fewer drinks per day and reported more days of total abstinence than the placebo group (Johnson et al., 2000). In another clinical trial with 71 alcoholic men, almost four times as many people receiving a 6-week treatment with a low dose of ondansetron (0.25 mg) showed significant levels of decreased drinking when compared to placebo (Sellers et al., 1994). In a double-blind placebo-controlled, Latin square cross-over design study analyzing 5-HT3 receptor's function in mediating the reinforcing properties of alcohol (Johnson et al., 1993), 16 healthy social drinkers received ondansetron treatment and the resulting subjective psychologic effects to a small amount of alcohol were recorded. Ondansetron treatment decreased the subjective pleasurable effects of alcohol and the desire to drink (Johnson et al., 1993). Since the serotonin transporter is important for regulating the serotonergic system, alleles at the gene encoding 5-HTT might predict the severity of the alcoholism and the therapeutic response to treatment with ondansetron. A clinical trial with 283 alcoholics found that ondansetron recipients had fewer drinks per day and more days spent totally abstinent than those who received placebo. The effect was greater in alcoholics with the LL genotype than the SS or LS genotype of the 5'-regulatory region of the serotonin transporter gene (Johnson et al., 2011). Moreover, animal studies have demonstrated that ondansetron might also be useful for treating opioid withdrawal symptoms (Pinelli et al., 1997).

Big potassium (BK) channels

The BK channel is a high-conductance calcium- and voltage-dependent potassium channel (Atkinson et al., 1991) and plays a dominant role in shaping neuronal activity and, unlike other voltage-gated ion channels, is strongly affected by alcohol (Treistman and Martin, 2009). Tolerance, which is the loss of drug effectiveness over time, is an important component of addiction. Acute tolerance can have a rapid onset within minutes of alcohol exposure, whereas tolerance after prolonged alcohol exposure develops over the course of days or weeks. The degree of acute behavioral tolerance to alcohol exhibited by a naïve subject can predict the likelihood of alcoholism (Treistman and Martin, 2009). Thus, the determinants of acute tolerance are important to understand. The BK channel is a key target in the development of tolerance in invertebrates and mammals (Davies et al., 2003; Martin et al., 2008). There are several variables that influence the response of the BK channel to alcohol, including subunit composition, splice variant, and post-translational mechanisms (see section on microRNA, below). BK channels from neurons of wild-type mice (in which the β4 subunit is well represented) exhibit little tolerance. By contrast, neuronal BK channels from β4 knockout mice do display acute tolerance. In addition to displaying tolerance, the β4 knockout mice drink more than their wild-type counterparts in an alcohol self-administration assay (Martin et al., 2008).

Drugs of abuse such as alcohol affect alternative splicing. The development of acute tolerance to alcohol provides another example that demonstrates the effects of alternative splicing of mRNAs on alcohol-related behavior (seen previously with the GABA(B) receptors). Tolerance was found to involve the splicing of the mRNA of the alpha subunit of the BK channel. The BK channel was found to have a sensitive and non-sensitive splice variant to alcohol. Within minutes of exposure to alcohol, the expression of the more alcohol-sensitive variant was selectively decreased, therefore producing tolerance (Treistman and Martin, 2009).

Transcription factors

Transcription factors serve as a key mechanism by which distinct gene programs are controlled, because they bind to highly specific DNA-regulatory sequences (control elements) (Renthal and Nestler, 2008; Rahman, 2012). Activation of alcohol-responsive transcription factors is also likely to result in changes in the expression of those genes with the corresponding control elements.

Heat shock factor protein 1 (HSF1) is a transcription factor which was found to be involved in the effects of acute alcohol application. Acute alcohol was found to increase GABA-R a4 subunit mRNA by increasing the binding of HSF1 to the promoter regions of GABA, increasing its transcription. A small interfering RNA (siRNA), used to block expression of HSF1, reduced HSF1 protein and reduced the alcohol-induced increases in the GABA-R a4 subunit, while introducing an active form of HSF1-induced GABA-R a4 transcription in the absence of alcohol (Pignataro et al., 2007). Acute alcohol also facilitates activation of HSF1 to the promoter region of synaptotagmin-1, a protein involved in synaptic transmission and release, suggesting that alcohol has a direct control of neurotransmitter release in an acute fashion (Pignataro et al., 2007, 2009). HSF family of mRNA was found to be different between alcoholic and control postmortem frontal cortices (Lewohl et al., 2000), as well as in cultured cortical neurons exposed to chronic alcohol treatment (Wang et al., 2007). These results indicate that these transcription factors are also involved in alcohol dependence.

The activation of several signaling pathways involving cAMP, Ca^{2+} and extracellular signal regulated kinase lead to the phosphorylation of CREB, a transcription factor cyclic AMP (cAMP) response element-binding protein (CREB) (Winstanley et al., 2007). Some chromatin-remodeling enzymes target chromatin by interacting with specific transcription factors by guiding them to a specific locus on the DNA. When phosphorylated, the transcription factor CREB interacts with CBP, a HAT that helps facilitate target gene activation by acetylating neighboring histones. This process was found to play a major role in behavioral responses to cocaine. Long-term treatment with alcohol was found to increase CREB transcription factor activation, which in turn increases the binding of CREB to the promoter of the glutamate receptor, which then increases the transcription of a specific glutamate receptor subunit NR2B. Site-directed mutation in the sequence where CREB binds abolished the stimulatory effect by alcohol, suggesting that CREB is involved in mediating alcohol-induced upregulation of the NR2B gene (Rani et al., 2005). This is a great example of an alcohol-responsive transcription factor affecting downstream expression of receptors which are also involved in mediating the response to alcohol.

Another example suggesting that the downstream transcription of genes is selective to either the acute or chronic stages of addiction is that of the immediate-early genes, such as *c-Fos* and *FosB*. These immediate-early genes are induced rapidly in the brain by acute use of drugs such as cocaine and alcohol, whereas the transcription of genes such as *Cdk5* and *Bdnf* is induced by chronic use of cocaine (Taylor et al., 2007). Chronic and compulsive use of drugs also causes the accumulation of delta fosB, a member of the Fos family of transcription factors. Delta fosB accumulates in specific regions of the brain in response to many chronic stimuli and persists for several weeks after the end of the stimulus (Watanabe et al., 2009; Damez-Werno et al., 2012). Delta fosB represents a molecular mechanism that can initiate and then sustain changes in gene expression that persist long after drug exposure ceases (Nestler et al., 2001).

microRNAs

MicroRNAs are short, non-coding RNA strands that regulate the translation of target mRNA into protein posttranscriptionally by binding to complementary sequences of the mRNA 3'-untranslated region (Schratt et al., 2006). MicroRNAs are highly abundant in the brain and have also been found to mediate the neuroadaptations induced by exposure to drugs of abuse, such as alcohol (Lewohl et al., 2011; Mayfield and Nunez, 2012). Chronic alcohol use causes global changes in microRNA expression in human alcoholic brains as well as in a variety of animal models (Li and van der Vaart, 2011).

The mechanism by which alcohol produces changes in microRNA expression is unknown. As discussed previously in this chapter, alcohol has the ability to affect receptor tyrosine kinases and mitogen-activated protein kinases, which in turn regulate transcription factor activity and thus microRNA gene expression. Alcohol can also alter the epigenetics on the DNA sites for microRNA, thereby altering their expression.

The mechanism contributing to the development of molecular tolerance to alcohol specifically involving the BK channel (discussed above) was found to be controlled by a microRNA (Pietrzykowski et al., 2008). In rodent neurons alcohol upregulates a particular microRNA 9 (miR9) which subsequently affects the expression of specific splice variants of the BK channel, some of which are less or more sensitive to alcohol. This mechanism suggests that alcohol-sensitive microRNAs are regulatory master-switches for the development of tolerance, a crucial component of alcohol addiction.

Many studies use computational approaches to identify potential interactions of alcohol with gene expression. Networks of highly correlated RNAs enable the drawing of potential cellular pathways of alcohol-responsive RNAs (Wolstenholme et al., 2011; Ponomarev et al., 2012). In this study *Dicer* was found to be a downregulated alcohol-responsive gene with the greatest overrepresentation of microRNA-targeting effects in human alcoholics. This suggests a negative-feedback loop between microRNAs and *Dicer* to control their own availability (Lewohl et al., 2011).

Synaptic mRNA translation and microRNAs

While most mRNAs are restricted to the neuronal soma, significant amounts of mRNA are found in synaptic compartments of the cell and translated into protein locally (Steward and Levy, 1982; Steward and Banker, 1992). RNAs – specifically in the synaptic compartments of the cell – are found to play a key role in the different states of addiction such as dependence, tolerance, withdrawal and craving (Russo et al., 2010; Mayfield and Nunez, 2012). Similar to mRNA, microRNAs have also been found in synaptic compartments of the cell, suggesting a local form of mRNA regulation by microRNA (Schuman et al., 2006; Schratt, 2009). Some studies examined the composition of microRNAs in the synapse and revealed many synaptically enriched and depleted microRNA families (Lugli et al., 2005; Eipper-Mains et al., 2011; Pichardo-Casas et al., 2012). Lugli et al. (2005) have elegantly shown that the microRNA-processing enzymes such as Dicer and elf2c are also enriched in the synapse and that their activity is modulated by neuronal activity. Other studies from this group show the synaptic enrichment of microRNA precursor forms, suggesting that neuronal activity might activate the Dicer enzyme to process the synaptic microRNA precursor into the mature microRNA form, thus rendering it active in the synapse. Recent findings in *Aplysia californica* indicate that synaptic microRNA-124 (miR-124) takes part in regulating long-term facilitation, suggesting a mechanism by which microRNAs serve as a dynamic brake on synaptic mRNA translation and emphasizing the importance of this highly conserved mechanism (Rajasethupathy et al., 2009).

A single microRNA has the potential to target many genes, and multiple microRNAs can cooperate to target the same genes (Grimson et al., 2007; Lewohl et al., 2011). microRNA-7 (miR-7) and microRNA-153 (miR-153) were found to be differentially expressed between human alcoholic brain tissue and control group brain tissue. Interestingly, miR-7 and miR-153 were found to both regulate the expression of α-synuclein (Doxakis, 2010). α-synuclein is a protein that plays a major role in neurotransmitter release in presynaptic terminals (Liu et al., 2004; Greten-Harrison et al., 2010) and is involved in dopaminergic neurotransmission and neurodegenerative disorders (Doxakis, 2010). Studies show that alcohol dependence in humans, as well as in rodents, is related to levels of α-synuclein (Bonsch et al., 2005a, b). Interestingly, overexpression of mir-7 and mir-153 significantly reduces endogenous alpha-synuclein levels, whereas inhibition of mir-7 and mir-153 enhances translation of alpha-synuclein. These findings illustrate a mechanism by which alcohol changes the expression of two microRNAs and how they can cooperate to target an mRNA that is known to be involved in alcoholism as well. This cooperation between the microRNAs allows regulation of gene expression with a reduced number of active microRNAs. Since alcoholism is a complex trait with global changes in gene expression, microRNAs serve as good targets for treatment, since they control global cellular changes in gene expression.

Neuroimmune

Substantial evidence has accumulated that implicates an unlikely target, the neuroimmune system, in the development, progression, and persistence of alcoholism. The term neuroimmune refers to signaling molecules that were first associated with innate immunity but are also commonly found in the brain. The brain uses immune signaling systems as neuromodulators that have functions distinct from their role in the immune system and are critical for normal brain functions like neuronal plasticity (Boulanger et al., 2001). One example of this is the major histocompatibility complex class I molecule, which serves as a primary mediator of immune response in the periphery but modulates activity-dependent refinement and plasticity in cortical synapses and the developing visual system in the CNS (Glynn et al., 2011; Elmer and McAllister, 2012). Alcohol-related behavior can be affected by immune signaling that originates in the brain or immune signaling derived from the periphery that crosses the blood–brain barrier to act on the brain.

Ingesting alcohol activates the neuroimmune system (Crews et al., 2011), which is proposed to lead to further increases in alcohol consumption, producing an escalating feedforward loop not conducive to a homeostatic state. Alcohol consumption compromises the tight junctions in gut epithelium, allowing lipopolysaccharide (LPS), a Gram-negative bacterial endotoxin that is normally confined to the gut, to leak into the blood stream, where it binds to and activates Toll-like receptors (TLRs) expressed on liver Kupffer cells and other tissues. This initiates a signaling cascade that culminates in the release of proinflammatory cytokines in the blood stream that can then cross the blood–brain barrier and interact with the brain, thus affecting behavior. Researchers have found that alcohol increases TLR expression in the brain and increases its sensitivity to LPS (Crews et al., 2012; Vetreno and Crews, 2012). To test how peripheral release of LPS would affect drinking behavior, Blednov et al. (2011a) injected mice with LPS and found that a single injection produces long-lasting increases in alcohol consumption consistent with neuroimmune signaling mediating the reinforcing properties of alcohol. The single injection of LPS also increased the firing rate of dopamine neurons in the ventral tegmental area, providing an example of how neuroimmune activation following peripheral LPS administration modulates brain reward circuitry (Blednov et al., 2011a). Indeed, peripheral immune function has been found to be important for other mental illnesses such as schizophrenia, depression, and autism (Dantzer et al., 2008; Kelley and Dantzer, 2011; Derecki et al., 2012; Jones and Thomsen, 2013; Takao et al., 2013); however, the mechanism by which peripheral immune activation exerts its effects on the brain is unknown and presents a field to be explored.

Immune signaling molecules are also found within the brain, where they can be used for normal, non-immune signaling, but can also reflect pathology and lead to neurodegeneration if proinflammation is left unchecked (El Khoury, 2010). Neurodegenerative effects of immune signaling have been demonstrated in Alzheimer's and Parkinson's disease (Carta and Pisanu, 2013; Hickman and El Khoury, 2013). Neurodegeneration has been observed in alcoholics and is especially prominent in the PFC (Fadda and Rossetti, 1998). Neurodegeneration in the PFC can affect judgment and reasoning capabilities and further exacerbate chronic alcohol consumption. It is possible that neuroimmune activation from alcohol consumption could contribute to the neurodegeneration seen in alcoholics (Blanco et al., 2005; Hua et al., 2007; Alfonso-Loeches et al., 2010; Qin and Crews, 2012).

Other evidence linking neuroimmune function with alcoholism is that immune gene expression in the brain was found to be altered in human alcoholics (Liu et al., 2006; Crews et al., 2012; Ponomarev et al., 2012), mice, and fruit flies after alcohol exposure (Qin et al., 2008; Kong et al., 2010), and rodent genetic models of high alcohol consumption (Mulligan et al., 2006; Saba et al., 2006). Furthermore, mice lacking genes related to immune function show decreased alcohol consumption compared to littermate controls (Blednov et al., 2012). Acute and chronic alcohol use activates microglia, the resident macrophages of the brain, and increases proinflammatory cytokines via nuclear factor κ B (NFκB) in the brain (Crews et al., 2011). Behavioral responses to acute alcohol were altered in mice that are lacking TLR receptors, TLR2 or TLR4 (Wu et al., 2011). In a key publication, the knockdown of TLR4 in the rat amygdala using siRNA decreased alcohol self-administration, demonstrating that neuroimmune signaling independent of input from peripheral cytokines was sufficient in regulating alcohol behavior (Liu et al., 2011).

Pioglitazone and rosiglitazone, agonists of the peroxisome proliferator activator receptor (PPAR) type gamma, have anti-inflammatory properties largely mediated through their ability to inhibit the transcription factor NFκB and thus decrease proinflammatory cytokine production (Daynes and Jones, 2002). Pioglitazone was found to reduce alcohol drinking, abolish reinstatement of alcohol seeking, reduce alcohol self-administration, and decrease the severity of physical withdrawal symptoms in rats (Stopponi et al., 2011). Another PPAR agonist, clofibrate, prevents the acquisition of nicotine self-administration in naïve rats and monkeys and decreases nicotine self-administration in nicotine-dependent rats and monkeys, suggesting that the PPAR agonist might be viable options for treating the neuroimmune pathologies in alcoholism and other forms of addiction. The pharmacologic blockade of neuroimmune activation reduces alcohol reward and decreases consumption using many types of anti-inflammatory drugs like minocycline, doxycycline, topiramate, anakinra, indomethacin, and CAPE (Agarwal, 2001; Pascual et al., 2007; Breslin et al., 2010; Wu et al., 2011; McIver et al., 2012; Blednov et al., 2013; Zalewska-Kaszubska et al., 2013). Taken together, these findings have built substantial evidence for neuroimmune modulation of acute and chronic alcohol consumption and offer unique unexplored targets for therapeutic intervention.

SUMMARY AND FUTURE DIRECTIONS

Excessive alcohol consumption causes widespread persistent changes throughout the brain. These molecular and cellular adaptations are thought to be the mechanisms by which neurons adapt to chronic alcohol use. These changes can eventually lead to alcoholism, depending on several factors, such as genetic predisposition, sex and environmental factors such as stress, age of drinking onset, and access to alcohol (Fig. 6.6). Once a person becomes an alcoholic, drinking becomes increasingly compulsive and seems to escapes voluntary control. The heterogeneous nature of alcoholism reflects the complex and multifaceted nature of alcohol's molecular effects on the CNS. The fact that multiple neurotransmitter systems are affected by alcohol makes it

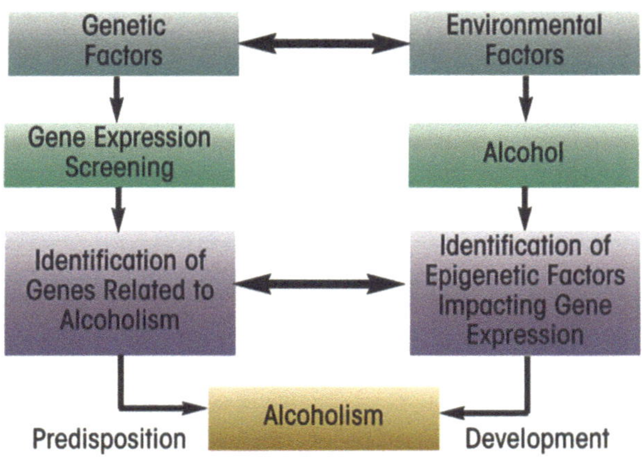

Fig. 6.6. Alcoholism is a complex disease caused by genetic and environmental factors. The neural changes characterized by alcoholism result from a complex dynamic system with a plethora of contributing factors. These factors are both environmental and internal. These internal factors depend on complex genetic states and molecular interactions. Since the normal brain is one of recursive feedback loops and regulatory controls, a drug like alcohol (whose effects are global) demonstrates radical alterations with a variety of entry points. Articulating the causes of alcoholism thus relies on research at a variety of scales and modalities. (Adapted from Starkman et al., 2012.)

Table 6.1

Summary table of all pharmaceutic treatments discussed in this chapter

Drug	Description	Advantages	Disadvantages
Disulfiram (Antabuse)*	Inhibitor of aldehyde dehydrogenase enzyme	Prevents alcohol consumption	Aversive therapy and does not reduce alcohol cravings, leading to poor patient compliance
Acamprosate (Campral)*	Mechanism of action is unclear. Proposed to be altering glutamatergic system	Few side-effects and high patient compliance	Only for patients who had already been withdrawn from alcohol. Abstinence rates are lower than naltrexone and the combination of naltrexone and acamprosate
Naltrexone*	Mu-opioid receptor antagonist	Reduces craving. Highest abstinent rates among FDA-approved drugs for alcoholism	Anxiety, trouble sleeping, nausea. A better response to naltrexone has been associated with possession of the G allele of the A118G polymorphism of the μ-opioid receptor gene (*OPRM1*)
Benzodiazepines†	GABA(A)-positive modulators	Treats withdrawal symptoms	Has abuse potential. Causes sedation. Cross-tolerance with alcohol
Ondansetron (Zofran)†	5-HT3 receptor antagonist	May reduce alcohol consumption	
Fluoxetine (Prozac)†	Selective serotonin reuptake inhibitor (SSRI)	May reduce alcohol consumption in some patients (late-onset individuals)	Human studies show inconsistent results. SSRI side-effects
Fluphenazine†	Dopamine receptor antagonist	Reduced drinking in rodents	No human clinical trial. Typical antipsychotic side-effects
Aripiprazole (Abilify)†	Dopamine receptor antagonist	Aripiprazole reduces craving and is well tolerated	Atypical antipsychotic side-effects
Topiramate (Topamax)†	Mechanism of action is unclear. Associated with glutamatergic system inhibition	Successful in treating volunteers who were drinking heavily	May effect memory/thinking and cause sedation
Baclofen (Kemstro, Lioresal, Liofen, Gablofen, Beklo and Baclosan)†	GABA(B) agonist		Effectiveness is controversial
Pioglitazone (Actos)† Rosiglitazone (Avandia)† Clofibrate (Atromid-S)†	Peroxisome proliferator activator receptor (PPAR) gamma agonists (pioglitazone and rosiglitazone) PPAR alpha agonist (clofibrate)	Drugs have been found to be hepatoprotective and neuroprotective in rodents	No published human data
Minocycline (Minocin)† Doxycycline (Vibramycin, Monodox, Oracea, Doryx, Vibrox, Adoxa, Doxyhexal, Doxylin, Doxoral, Doxy-1 and Atridox)† Anakinra (Kineret)† Indomethacin (Indocin)† CAPE	Anti-inflammatory antibiotics	Well tolerated	No human data. Alters metabolism of oral contraceptives
Gabapentin†	GABA analog, calcium channel and GABA modulator	Increases sleep quality in alcoholics. Analgesic and anxiolytic	Causes sedation. Does not affect mood or craving

*Food and Drug Administration (FDA)-approved for treatments of alcoholism.

†FDA-approved for the treatment of other diseases.

GABA, γ-aminobutyric acid.

Table 6.2

Summary table of genes discussed in this chapter

Gene symbol	Gene name	Function	Reference
CYP2E1	Cytochrome P450	Involved in ethanol metabolism	Zakhari, 2006
ADH	Alcohol dehydrogenase	Involved in ethanol metabolism. A primary target of alcohol. Binds alcohol through a zinc atom on ADH and the hydroxyl group of alcohol	Zakhari, 2006
CBP	CREB-binding protein	Allows for transcriptional activation through histone acetylase activity. Acute alcohol exposure increases levels of CBP. Withdrawal produces the opposite effect	Pandey et al., 2008
NPY	Neuropeptide Y	Acute alcohol exposure increases levels of NPY. Withdrawal produces the opposite effect	Pandey et al., 2008
HDAC5	Histone deacetylase 5	Mice lacking HDAC5 become hypersensitive when chronically exposed to cocaine. Chronic cocaine administration inactivates HDAC5 by exporting it out of the nucleus, resulting in histone hyperacetylation and increased mRNA expression of HDAC5 target genes	Renthal et al., 2007
NK1R	Neurokinin 1 receptor	NK1R is increased during both chronic cocaine and alcohol exposure. Silencing the translation of the NK1R using a technique known as RNA interference was found to reduce alcohol drinking in mice	Baek et al., 2010
		NK1R antagonist (L822429) decreased voluntary alcohol consumption, suppressed stress-induced reinstatement of alcohol seeking, and increased sensitivity to the sedative effects of alcohol in rats	Schank et al., 2011
		An NK1R antagonist suppressed spontaneous alcohol cravings, improved overall well-being, blunted cravings induced by a challenge procedure, and attenuated concomitant cortisol responses in detoxified alcoholic inpatients. An analysis of brain function in human alcoholics compared with controls, while performing behavioral emotional tasks, was performed using functional magnetic resonance imaging and corroborated the above results, suggesting NK1R antagonism as a potential therapeutic target for the treatment of alcoholism	George et al., 2008

difficult to pinpoint the molecular mechanisms that are primarily responsible for the disease. Indeed, alcohol and its metabolites' molecular targets are many and varied, and include modifications of the genome, transcriptome, and proteome. Alcoholism is most likely the result of the cumulative interactions within and between all three systems. Nevertheless, identifying the molecular targets is a crucial step in understanding alcoholism and developing new treatments. Tables 6.1–6.3 summarize all the pharmaceutic treatments, genes, and neurotransmitter systems discussed in this chapter.

An outstanding question that remains is: what could be responsible for the persistent drug memory that is able to linger for years and causes relapse despite all the negative consequences of alcoholism? Molecular turnover (except for DNA) is relatively rapid. DNA sequence is not changed by drug exposure, but DNA protein (chromatin) structure can be changed by environmental events, including drugs, and these epigenetic mechanisms provide promising approaches to the long-lasting effects of drug dependence. Genomics, proteomics, and transriptomics are promising fields that will help elucidate the molecular mechanisms of alcoholism.

It is clear today that the GABA and glutamate systems play a major role in alcohol's acute anxiolytic effects, as well as withdrawal-induced hyperexcitability that leads to seizures, neurotoxicity, and cellular death. It is hypothesized that dopamine partly mediates the acute rewarding effects of alcohol during intoxication,

Table 6.3

Summary of neurotransmitter systems discussed in this chapter

Glutamate	
Acute alcohol inhibits NMDA receptor function, while chronic use of alcohol upregulates NMDA receptor expression in the brain	Lovinger et al., 1989; Qiang and Ticku, 2005
Alcohol acts as a non-competitive inhibitor of the AMPA/kainate receptors at high concentrations	Dildy-Mayfield and Harris, 1992, 1995; Akinshola et al., 2003
MK-801, an NMDA antagonist, mimics the subjective effects of alcohol in animals	Butelman et al., 1993
Ketamine, an NMDA antagonist, mimics the behavioral effects of alcohol in detoxified alcoholics	Krystal et al., 1998
Chronic alcohol increases the expression of AMPA1 receptor subunit in the brains of alcoholics	Lewohl et al., 2000
NMDA receptors containing specific subunits are more sensitive to alcohol than others	Raeder et al., 2008
Alterations in subunit composition of glutamate receptors were seen after chronic use of alcohol in mice	Ortiz et al., 1995
Dopamine	
Acute alcohol exposure activates dopamine reward pathways, whereas chronic treatment produces a hypodopaminergic state associated with dysphoria, which could lead to craving and relapse	Koob and Volkow, 2010
Alcohol increases dopaminergic activity in the midbrain region of rodents and humans	Boileau et al., 2003
Dopamine release in the midbrain partially mediates the positive reinforcing properties of acute alcohol exposure	Raeder et al., 2008
Alcohol-preferring rats release more DA than wild-type rats in an alcohol self-administration study	Weiss et al., 1993
Mice lacking different DA receptors and transporters show modified alcohol preference compared with controls	Crabbe et al., 2006
Chronic alcohol treatment decreases DA and its metabolite in the striatum, decreases tyrosine hydroxylase protein levels, and increases DA transporter protein levels in rats compared to controls	Rothblat et al., 2001
PET scans show that chronic alcoholics have fewer D2 receptors when compared with non-alcoholics	Volkow et al., 2002
The DA receptor antagonist fluphenazine will block alcohol self-administration when injected into the nucleus accumbens	Rassnick et al., 1992
Aripiprazole, an atypical antipsychotic, reduces craving and increases positive subjective feelings in alcoholics	Martinotti et al., 2007, Brunetti et al., 2012
GABA(A)	
Behavioral intoxication is mimicked by the administration of pharmacologic GABA(A) agonists like muscimol and benzodiazepines	Grobin et al., 1998
The effects of alcohol can be diminished by using pharmacologic GABA antagonists like bicuculline and picrotoxin	Hyytia and Koob, 1995
Chronic alcohol use causes changes in GABA receptor subunit composition in human alcoholics	Lewohl et al., 1997
Alcohol increases GABAergic neurotransmission over a wide range of concentrations, with some studies showing that delta-containing receptors are more sensitive than other GABA(A) receptors	Wallner et al., 2003
Alpha-2 and alpha-3 subunits take part in mediating the motor-impairing effects of alcohol	Blednov et al., 2011b
Alpha-2 subunit takes part in the aversive properties of alcohol	Blednov et al., 2013
GABA(B)	
Acute alcohol increases inhibitory GABA(B) signaling in a concentration-dependent manner in rat midbrain dopaminergic neurons in a current and single-electrode voltage clamp electrophysiologic experiment	Federici et al., 2009
Silencing the GABA(B) receptor in fruit flies using RNA interference decreases acute motor impairment after alcohol exposure	Dzitoyeva et al., 2003
Cerebellar injections in mice of GABA(B) agonists accentuated and antagonists attenuated motor impairment following acute alcohol	Dar, 1996

Table 6.3

Continued

CA1 pyramidal neuron recordings indicated that GABA(B) signaling was immune to the acute effects of alcohol but became modulated after chronic exposure	Frye et al., 1991
Baclofen, a GABA(B) agonist, reduces relapse in dependent humans and decreases alcohol consumption in rats	Addolorato et al., 2002a, b; Maccioni and Colombo, 2009; Agabio et al., 2012
GABA(B) has complex alternative splicing leading to several splice variants in the alcoholic brain. Alternative splicing might contribute to the pharmacologic effects of GABA(B) agonists such as baclofen in the treatment of alcoholism	Lee et al., 2013
Serotonin (5-HT)	
Acute alcohol potentiates 5-HT3 receptor	Lovinger, 1991; Lovinger and White, 1991; Lovinger and Zhou, 1994; Harris et al., 1995; Sung et al., 2000
Serotonin levels in animal brains are elevated after acute alcohol exposure	Murphy et al., 1982; LeMarquand et al., 1994
Mice lacking the 5-HT1B serotonin receptor consume larger amounts of alcohol compared to wild type	Crabbe et al., 1996
5-HT2 antagonists selectively decrease acute alcohol reinforcement	Roberts et al., 1998
Blockade of the serotonin transporter (5-HTT) with selective serotonin reuptake inhibitors (SSRIs), or by genetic knockout, decreases alcohol consumption in rats and partially in humans	Brown et al., 1979; Sellers et al., 1992; Kelai et al., 2003; Ciccocioppo, Economidou et al., 2006; Kranzler et al., 2012
Selective antagonists of the 5-HT3 receptor decrease alcohol self-administration and consumption in rodents	Fadda et al., 1991; Hodge et al., 1993; Sellers et al., 1994
Selective antagonists of the 5-HT3 receptor decrease the amount of drinks per day and increase the amount of abstinence time in alcoholics. Also, decreases the subjective pleasurable effects of alcohol and the desire to drink	Johnson et al., 1993, 2000; Sellers et al., 1994
The effect was greater in alcoholics with the LL genotype than the SS or LS genotype of the 5'-regulatory region of the serotonin transporter gene	Johnson et al., 2011

NMDA, *N*-methyl-D-aspartate; AMPA, α-amino-3-hydroxy-5-methyl-4-isoxazolepropionic acid; DA, dopamine; PET, positron emission tomography; GABA, γ-aminobutyric acid.

the anhedonic state during the development of alcoholism, and the fear of anhedonia during withdrawal, potentially leading to relapse. Cross-tolerance of alcohol with barbiturates suggests overlapping pharmacologic actions involving GABA(A) receptors, but the exact binding domain for alcohol on its primary targets remains to be discovered.

Alcohol's propensity for modifying the neural landscape is reflected by its many responsive molecules and signaling pathways, such as receptors, kinases, scaffolding proteins, neurotransmitters, hormones, chaperones, transcription factors, and cytokines. Establishing causality between alcohol and the following cellular and behavioral adaptation has proven elusive, and identifying the small effects on each system that culminate in alcoholism is a complex challenge. The availability of a large variety of animal models (ranging from flies and worms to primates), as well as the advancement of neurobiologic techniques (including gene expression microarray and whole-genome RNA sequencing proteomics) provides hope for discovering the molecular mechanisms underpinning the neurobiology of alcoholism and revealing new therapeutic targets for this societal health burden.

References

Addolorato G, Caputo F, Capristo E et al. (2002a). Baclofen efficacy in reducing alcohol craving and intake: a preliminary double-blind randomized controlled study. Alcohol Alcohol 37: 504–508.

Addolorato G, Caputo F, Capristo E et al. (2002b). Rapid suppression of alcohol withdrawal syndrome by baclofen. Am J Med 112: 226–229.

Agabio R, Maccioni P, Carai MA et al. (2012). The development of medications for alcohol-use disorders targeting the GABAB receptor system. Recent Pat CNS Drug Discov 7: 113–128.

Agarwal DP (2001). Genetic polymorphisms of alcohol metabolizing enzymes. Pathol Biol (Paris) 49: 703–709.

Akinshola BE, Yasuda RP, Peoples RW et al. (2003). Ethanol sensitivity of recombinant homomeric and heteromeric

AMPA receptor subunits expressed in *Xenopus* oocytes. Alcohol Clin Exp Res 27: 1876–1883.

al Qatari M, Khan S, Harris B et al. (2001). Acamprosate is neuroprotective against glutamate-induced excitotoxicity when enhanced by ethanol withdrawal in neocortical cultures of fetal rat brain. Alcohol Clin Exp Res 25: 1276–1283.

Alfonso-Loeches S, Pascual-Lucas M, Blanco AM et al. (2010). Pivotal role of TLR4 receptors in alcohol-induced neuroinflammation and brain damage. J Neurosci 30: 8285–8295.

Ameisen O (2005). Complete and prolonged suppression of symptoms and consequences of alcohol-dependence using high-dose baclofen: a self-case report of a physician. Alcohol Alcohol 40: 147–150.

Ameisen O (2012). Suppressing addiction using high-dose baclofen, rather than perpetuating it using substitution therapy. J Psychopharmacol 26: 1042–1043, author reply 1044.

American Psychiatric Association (2013). Diagnostic and statistical manual of mental disorders, 5th edn. American Psychiatric Publishing, Arlington, VA.

Amit Z, Brown ZW, Rockman GE (1977). Possible involvement of acetaldehyde, norepinephrine and their tetrahydroisoquinoline derivatives in the regulation of ethanol seld-administration. Drug Alcohol Depend 2: 495–500.

Anton RF, O'Malley SS, Ciraulo DA et al. (2006). Combined pharmacotherapies and behavioral interventions for alcohol dependence: the COMBINE study: a randomized controlled trial. JAMA 295: 2003–2017.

Atkinson NS, Robertson GA, Ganetzky B (1991). A component of calcium-activated potassium channels encoded by the *Drosophila* slo locus. Science 253: 551–555.

Baek MN, Jung KH, Halder D et al. (2010). Artificial microRNA-based neurokinin-1 receptor gene silencing reduces alcohol consumption in mice. Neurosci Lett 475: 124–128.

Barth KS, Malcolm RJ (2010). Disulfiram: an old therapeutic with new applications. CNS Neurol Disord Drug Targets 9: 5–12.

Blanco AM, Valles SL, Pascual M et al. (2005). Involvement of TLR4/type I IL-1 receptor signaling in the induction of inflammatory mediators and cell death induced by ethanol in cultured astrocytes. J Immunol 175: 6893–6899.

Blednov YA, Harris RA (2008). Metabotropic glutamate receptor 5 (mGluR5) regulation of ethanol sedation, dependence and consumption: relationship to acamprosate actions. Int J Neuropsychopharmacol 11: 775–793.

Blednov YA, Benavidez JM, Geil C et al. (2011a). Activation of inflammatory signaling by lipopolysaccharide produces a prolonged increase of voluntary alcohol intake in mice. Brain Behav Immun 25 (Suppl. 1): S92–S105.

Blednov YA, Borghese CM, McCracken ML et al. (2011b). Loss of ethanol conditioned taste aversion and motor stimulation in knockin mice with ethanol-insensitive alpha2-containing GABA(A) receptors. J Pharmacol Exp Ther 336: 145–154.

Blednov YA, Ponomarev I, Geil C et al. (2012). Neuroimmune regulation of alcohol consumption: behavioral validation of genes obtained from genomic studies. Addict Biol 17: 108–120.

Blednov YA, Benavidez JM, Black M et al. (2013). Linking GABA(A) receptor subunits to alcohol-induced conditioned taste aversion and recovery from acute alcohol intoxication. Neuropharmacology 67: 46–56.

Boileau I, Assaad JM, Pihl RO et al. (2003). Alcohol promotes dopamine release in the human nucleus accumbens. Synapse 49: 226–231.

Bonsch D, Greifenberg V, Bayerlein K et al. (2005a). Alpha-synuclein protein levels are increased in alcoholic patients and are linked to craving. Alcohol Clin Exp Res 29: 763–765.

Bonsch D, Lenz B, Kornhuber J et al. (2005b). DNA hypermethylation of the alpha synuclein promoter in patients with alcoholism. Neuroreport 16: 167–170.

Boulanger LM, Huh GS, Shatz CJ (2001). Neuronal plasticity and cellular immunity: shared molecular mechanisms. Curr Opin Neurobiol 11: 568–578.

Breslin FJ, Johnson BA, Lynch WJ (2010). Effect of topiramate treatment on ethanol consumption in rats. Psychopharmacology (Berl) 207: 529–534.

Brown ZW, Amit Z, Rockman GE (1979). Intraventricular self-administration of acetaldehyde, but not ethanol, in naive laboratory rats. Psychopharmacology (Berl) 64: 271–276.

Brunetti M, Di Tizio L, Dezi S et al. (2012). Aripiprazole, alcohol and substance abuse: a review. Eur Rev Med Pharmacol Sci 16: 1346–1354.

Butelman ER, Baron SP, Woods JH (1993). Ethanol effects in pigeons trained to discriminate MK-801, PCP or CGS-19755. Behav Pharmacol 4: 57–60.

Cargiulo T (2007). Understanding the health impact of alcohol dependence. Am J Health Syst Pharm 64 (Suppl. 3): S5–S11.

Carta AR, Pisanu A (2013). Modulating microglia activity with PPAR-gamma agonists: a promising therapy for Parkinson's disease? Neurotox Res 23: 112–123.

Ciccocioppo R, Economidou D, Cippitelli A et al. (2006). Genetically selected Marchigian Sardinian alcohol-preferring (msP) rats: an animal model to study the neurobiology of alcoholism. Addict Biol 11: 339–355.

Clapp P, Bhave SV, Hoffman PL (2008). How adaptation of the brain to alcohol leads to dependence: a pharmacological perspective. Alcohol Res Health 31: 310–339.

Coetzee WA, Amarillo Y, Chiu J et al. (1999). Molecular diversity of K+channels. Ann N Y Acad Sci 868: 233–285.

Crabbe JC, Phillips TJ, Feller DJ et al. (1996). Elevated alcohol consumption in null mutant mice lacking 5-HT1B serotonin receptors. Nat Genet 14: 98–101.

Crabbe JC, Phillips TJ, Harris RA et al. (2006). Alcohol-related genes: contributions from studies with genetically engineered mice. Addict Biol 11: 195–269.

Crews FT, Zou J, Qin L (2011). Induction of innate immune genes in brain create the neurobiology of addiction. Brain Behav Immun 25 (Suppl. 1): S4–S12.

Crews FT, Qin L, Sheedy D et al. (2012). High mobility group box 1/Toll-like receptor danger signaling increases brain neuroimmune activation in alcohol dependence. Biol Psychiatry 73: 602–612.

Damez-Werno D, LaPlant Q, Sun H et al. (2012). Drug experience epigenetically primes Fosb gene inducibility in rat nucleus accumbens. J Neurosci 32: 10267–10272.

Dantzer R, O'Connor JC, Freund GG et al. (2008). From inflammation to sickness and depression: when the immune system subjugates the brain. Nat Rev Neurosci 9: 46–56.

Dar MS (1996). Mouse cerebellar GABAB participation in the expression of acute ethanol-induced ataxia and in its modulation by the cerebellar adenosinergic A1 system. Brain Res Bull 41: 53–59.

Davies AG, Pierce-Shimomura JT, Kim H et al. (2003). A central role of the BK potassium channel in behavioral responses to ethanol in *C. elegans*. Cell 115: 655–666.

Daynes RA, Jones DC (2002). Emerging roles of PPARs in inflammation and immunity. Nat Rev Immunol 2: 748–759.

De Witte P, Littleton J, Parot P et al. (2005). Neuroprotective and abstinence-promoting effects of acamprosate: elucidating the mechanism of action. CNS Drugs 19: 517–537.

Deng XS, Deitrich RA (2008). Putative role of brain acetaldehyde in ethanol addiction. Curr Drug Abuse Rev 1: 3–8.

Derecki NC, Cronk JC, Lu Z et al. (2012). Wild-type microglia arrest pathology in a mouse model of Rett syndrome. Nature 484: 105–109.

Dick DM, Foroud T (2002). Genetic strategies to detect genes involved in alcoholism and alcohol-related traits. Alcohol Res Health 26: 172–180.

Dildy-Mayfield JE, Harris RA (1992). Acute and chronic ethanol exposure alters the function of hippocampal kainate receptors expressed in *Xenopus* oocytes. J Neurochem 58: 1569–1572.

Dildy-Mayfield JE, Harris RA (1995). Ethanol inhibits kainate responses of glutamate receptors expressed in *Xenopus* oocytes: role of calcium and protein kinase C. J Neurosci 15: 3162–3171.

Doxakis E (2010). Post-transcriptional regulation of alpha-synuclein expression by mir-7 and mir-153. J Biol Chem 285: 12726–12734.

Dredge BK, Polydorides AD, Darnell RB (2001). The splice of life: alternative splicing and neurological disease. Nat Rev Neurosci 2: 43–50.

Dzitoyeva S, Dimitrijevic N, Manev H (2003). Gamma-aminobutyric acid B receptor 1 mediates behavior-impairing actions of alcohol in *Drosophila*: adult RNA interference and pharmacological evidence. Proc Natl Acad Sci U S A 100: 5485–5490.

Eipper-Mains JE, Kiraly DD, Palakodeti D et al. (2011). microRNA-Seq reveals cocaine-regulated expression of striatal microRNAs. RNA 17: 1529–1543.

El Khoury J (2010). Neurodegeneration and the neuroimmune system. Nat Med 16: 1369–1370.

Elmer BM, McAllister AK (2012). Major histocompatibility complex class I proteins in brain development and plasticity. Trends Neurosci 35: 660–670.

Fadda F, Rossetti ZL (1998). Chronic ethanol consumption: from neuroadaptation to neurodegeneration. Prog Neurobiol 56: 385–431.

Fadda F, Garau B, Marchei F et al. (1991). MDL 72222, a selective 5-HT3 receptor antagonist, suppresses voluntary ethanol consumption in alcohol-preferring rats. Alcohol Alcohol 26: 107–110.

Federici M, Nistico R, Giustizieri M et al. (2009). Ethanol enhances GABAB-mediated inhibitory postsynaptic transmission on rat midbrain dopaminergic neurons by facilitating GIRK currents. Eur J Neurosci 29: 1369–1377.

Frye GD, Taylor L, Trzeciakowski JP et al. (1991). Effects of acute and chronic ethanol treatment on pre- and postsynaptic responses to baclofen in rat hippocampus. Brain Res 560: 84–91.

Garbutt JC, Kampov-Polevoy AB, Gallop R et al. (2010). Efficacy and safety of baclofen for alcohol dependence: a randomized, double-blind, placebo-controlled trial. Alcohol Clin Exp Res 34: 1849–1857.

George DT, Gilman J, Hersh A et al. (2008). Neurokinin 1 receptor antagonism as a possible therapy for alcoholism. Science 319: 1536–1539.

Glynn MW, Elmer BM, Garay PA et al. (2011). MHCI negatively regulates synapse density during the establishment of cortical connections. Nat Neurosci 14: 442–451.

Goldstein DB (1984). The effects of drugs on membrane fluidity. Annu Rev Pharmacol Toxicol 24: 43–64.

Gonzales RA, Job MO, Doyon WM (2004). The role of mesolimbic dopamine in the development and maintenance of ethanol reinforcement. Pharmacol Ther 103: 121–146.

Goodlett CR, Horn KH, Zhou FC (2005). Alcohol teratogenesis: mechanisms of damage and strategies for intervention. Exp Biol Med (Maywood) 230: 394–406.

Greten-Harrison B, Polydoro M, Morimoto-Tomita M et al. (2010). alphabetagamma-Synuclein triple knockout mice reveal age-dependent neuronal dysfunction. Proc Natl Acad Sci U S A 107: 19573–19578.

Grimson A, Farh KK, Johnston WK et al. (2007). MicroRNA targeting specificity in mammals: determinants beyond seed pairing. Mol Cell 27: 91–105.

Grobin AC, Matthews DB, Devaud LL et al. (1998). The role of GABA(A) receptors in the acute and chronic effects of ethanol. Psychopharmacology (Berl) 139: 2–19.

Guerri C, Pascual M (2010). Mechanisms involved in the neurotoxic, cognitive, and neurobehavioral effects of alcohol consumption during adolescence. Alcohol 44: 15–26.

Hanson DJ (1995). Preventing alcohol abuse : alcohol, culture, and control. Praeger, Westport, CT.

Harris RA, Mihic SJ, Dildy-Mayfield JE et al. (1995). Actions of anesthetics on ligand-gated ion channels: role of receptor subunit composition. FASEB J 9: 1454–1462.

Harris RA, Trudell JR, Mihic SJ (2008). Ethanol's molecular targets. Sci Signal 1: re7.

Hickman SE, El Khoury J (2013). The neuroimmune system in Alzheimer's disease: the glass is half full. J Alzheimers Dis 33 (Suppl. 1): S295–S302.

Hodge CW, Samson HH, Lewis RS et al. (1993). Specific decreases in ethanol- but not water-reinforced responding produced by the 5-HT3 antagonist ICS 205-930. Alcohol 10: 191–196.

Hoffman PL, Iorio KR, Snell LD et al. (1995). Attenuation of glutamate-induced neurotoxicity in chronically ethanol-exposed cerebellar granule cells by NMDA receptor antagonists and ganglioside GM1. Alcohol Clin Exp Res 19: 721–726.

Howard RJ, Slesinger PA, Davies DL et al. (2011). Alcohol-binding sites in distinct brain proteins: the quest for atomic level resolution. Alcohol Clin Exp Res 35: 1561–1573.

Hua F, Ma J, Ha T et al. (2007). Activation of Toll-like receptor 4 signaling contributes to hippocampal neuronal death following global cerebral ischemia/reperfusion. J Neuroimmunol 190: 101–111.

Hyman SE, Malenka RC, Nestler EJ (2006). Neural mechanisms of addiction: the role of reward-related learning and memory. Annu Rev Neurosci 29: 565–598.

Hyytia P, Koob GF (1995). GABAA receptor antagonism in the extended amygdala decreases ethanol self-administration in rats. Eur J Pharmacol 283: 151–159.

Jiang L, Gulanski BI, De Feyter HM et al. (2013). Increased brain uptake and oxidation of acetate in heavy drinkers. J Clin Invest 123: 1605–1614.

Johnson BA, Campling GM, Griffiths P et al. (1993). Attenuation of some alcohol-induced mood changes and the desire to drink by 5-HT3 receptor blockade: a preliminary study in healthy male volunteers. Psychopharmacology (Berl) 112: 142–144.

Johnson BA, Roache JD, Javors MA et al. (2000). Ondansetron for reduction of drinking among biologically predisposed alcoholic patients: a randomized controlled trial. JAMA 284: 963–971.

Johnson BA, Rosenthal N, Capece JA et al. (2008). Improvement of physical health and quality of life of alcohol-dependent individuals with topiramate treatment: US multisite randomized controlled trial. Arch Intern Med 168: 1188–1199.

Johnson BA, Ait-Daoud N, Seneviratne C et al. (2011). Pharmacogenetic approach at the serotonin transporter gene as a method of reducing the severity of alcohol drinking. Am J Psychiatry 168: 265–275.

Jones KA, Thomsen C (2013). The role of the innate immune system in psychiatric disorders. Mol Cell Neurosci 53: 52–62.

Kelai S, Aissi F, Lesch KP et al. (2003). Alcohol intake after serotonin transporter inactivation in mice. Alcohol Alcohol 38: 386–389.

Kelley KW, Dantzer R (2011). Alcoholism and inflammation: neuroimmunology of behavioral and mood disorders. Brain Behav Immun 25 (Suppl. 1): S13–S20.

Kong EC, Allouche L, Chapot PA et al. (2010). Ethanol-regulated genes that contribute to ethanol sensitivity and rapid tolerance in *Drosophila*. Alcohol Clin Exp Res 34: 302–316.

Koob GF, Volkow ND (2010). Neurocircuitry of addiction. Neuropsychopharmacology 35: 217–238.

Koob GF, Le Moal M, Koob G et al. (2005). Neurobiology of Addiction. Elsevier, Burlington.

Kranzler HR, Feinn R, Armeli S et al. (2012). Comparison of alcoholism subtypes as moderators of the response to sertraline treatment. Alcohol Clin Exp Res 36: 509–516.

Krystal JH, Petrakis IL, Webb E et al. (1998). Dose-related ethanol-like effects of the NMDA antagonist, ketamine, in recently detoxified alcoholics. Arch Gen Psychiatry 55: 354–360.

Lai CC, Chang MC, Lin HH (2004). Acute tolerance to ethanol inhibition of NMDA-induced responses in rat rostral ventrolateral medulla neurons. J Biomed Sci 11: 482–492.

Lee C, Mayfield RD, Harris RA (2010). Intron 4 containing novel GABAB1 isoforms impair GABAB receptor function. PLoS One 5: e 14044.

Lee C, Mayfield RD, Harris RA (2013). Altered GABAB1 receptor splicing in alcoholics. Biol Psychiatry. In peer review (in press).

Leggio L, Garbutt JC, Addolorato G (2010). Effectiveness and safety of baclofen in the treatment of alcohol dependent patients. CNS Neurol Disord Drug Targets 9: 33–44.

LeMarquand D, Pihl RO, Benkelfat C (1994). Serotonin and alcohol intake, abuse, and dependence: findings of animal studies. Biol Psychiatry 36: 395–421.

Levenson JM, Sweatt JD (2005). Epigenetic mechanisms in memory formation. Nat Rev Neurosci 6: 108–118.

Lewohl JM, Crane DI, Dodd PR (1997). Expression of the alpha 1, alpha 2 and alpha 3 isoforms of the GABAA receptor in human alcoholic brain. Brain Res 751: 102–112.

Lewohl JM, Wang L, Miles MF et al. (2000). Gene expression in human alcoholism: microarray analysis of frontal cortex. Alcohol Clin Exp Res 24: 1873–1882.

Lewohl JM, Nunez YO, Dodd PR et al. (2011). Up-regulation of microRNAs in brain of human alcoholics. Alcohol Clin Exp Res 35: 1928–1937.

Li MD, van der Vaart AD (2011). MicroRNAs in addiction: adaptation's middlemen? Mol Psychiatry 16: 1159–1168.

Li HF, Mochly-Rosen D, Kendig JJ (2005). Protein kinase Cgamma mediates ethanol withdrawal hyper-responsiveness of NMDA receptor currents in spinal cord motor neurons. Br J Pharmacol 144: 301–307.

Liu S, Ninan I, Antonova I et al. (2004). alpha-Synuclein produces a long-lasting increase in neurotransmitter release. EMBO J 23: 4506–4516.

Liu J, Lewohl JM, Harris RA et al. (2006). Patterns of gene expression in the frontal cortex discriminate alcoholic from nonalcoholic individuals. Neuropsychopharmacology 31: 1574–1582.

Liu J, Yang AR, Kelly T et al. (2011). Binge alcohol drinking is associated with GABAA alpha2-regulated Toll-like receptor 4 (TLR4) expression in the central amygdala. Proc Natl Acad Sci U S A 108: 4465–4470.

Lobo IA, Harris RA (2008). GABA(A) receptors and alcohol. Pharmacol Biochem Behav 90: 90–94.

Lovinger DM (1991). Ethanol potentiation of 5-HT3 receptor-mediated ion current in NCB-20 neuroblastoma cells. Neurosci Lett 122: 57–60.

Lovinger DM (1997). Serotonin's role in alcohol's effects on the brain. Alcohol Health Res World 21: 114–120.

Lovinger DM, White G (1991). Ethanol potentiation of 5-hydroxytryptamine3 receptor-mediated ion current in neuroblastoma cells and isolated adult mammalian neurons. Mol Pharmacol 40: 263–270.

Lovinger DM, Zhou Q (1994). Alcohols potentiate ion current mediated by recombinant 5-HT3RA receptors expressed in a mammalian cell line. Neuropharmacology 33: 1567–1572.

Lovinger DM, White G, Weight FF (1989). Ethanol inhibits NMDA-activated ion current in hippocampal neurons. Science 243: 1721–1724.

Lugli G, Larson J, Martone ME et al. (2005). Dicer and eIF2c are enriched at postsynaptic densities in adult mouse brain and are modified by neuronal activity in a calpain-dependent manner. J Neurochem 94: 896–905.

Maccioni P, Colombo G (2009). Role of the GABA(B) receptor in alcohol-seeking and drinking behavior. Alcohol 43: 555–558.

Mann K, Lehert P, Morgan MY (2004). The efficacy of acamprosate in the maintenance of abstinence in alcohol-dependent individuals: results of a meta-analysis. Alcohol Clin Exp Res 28: 51–63.

Martin GE, Hendrickson LM, Penta KL et al. (2008). Identification of a BK channel auxiliary protein controlling molecular and behavioral tolerance to alcohol. Proc Natl Acad Sci U S A 105: 17543–17548.

Martinotti G, Di Nicola M, Janiri L (2007). Efficacy and safety of aripiprazole in alcohol dependence. Am J Drug Alcohol Abuse 33: 393–401.

Mason BJ, Ownby RL (2000). Acamprosate for the treatment of alcohol dependence: a review of double-blind, placebo-controlled trials. CNS Spectr 5: 58–69.

Mason BJ, Goodman AM, Chabac S et al. (2006). Effect of oral acamprosate on abstinence in patients with alcohol dependence in a double-blind, placebo-controlled trial: the role of patient motivation. J Psychiatr Res 40: 383–393.

Matlin AJ, Clark F, Smith CW (2005). Understanding alternative splicing: towards a cellular code. Nat Rev Mol Cell Biol 6: 386–398.

Mayfield RD, Nunez YO (2012). Understanding alcoholism through microRNA signatures in brains of human alcoholics. Frontiers in Genetics 3: 17–29.

McIver SR, Muccigrosso MM, Haydon PG (2012). The effect of doxycycline on alcohol consumption and sensitivity: consideration for inducible transgenic mouse models. Exp Biol Med (Maywood) 237: 1129–1133.

Mihic SJ, Ye Q, Wick MJ et al. (1997). Sites of alcohol and volatile anaesthetic action on GABA-A and glycine receptors. Nature 389: 385–389.

Moriarty JD (1950). The use of antabus in the therapy of alcoholic patients. Calif Med 73: 144–147.

Mulligan MK, Ponomarev I, Hitzemann RJ et al. (2006). Toward understanding the genetics of alcohol drinking through transcriptome meta-analysis. Proc Natl Acad Sci U S A 103: 6368–6373.

Murphy JM, McBride WJ, Lumeng L et al. (1982). Regional brain levels of monoamines in alcohol-preferring and -nonpreferring lines of rats. Pharmacol Biochem Behav 16: 145–149.

Myers WD, Ng KT, Marzuki S et al. (1984). Alteration of alcohol drinking in the rat by peripherally self-administered acetaldehyde. Alcohol 1: 229–236.

Nestler EJ, Barrot M, Self DW (2001). DeltaFosB: a sustained molecular switch for addiction. Proc Natl Acad Sci U S A 98: 11042–11046.

Ortiz J, Fitzgerald LW, Charlton M et al. (1995). Biochemical actions of chronic ethanol exposure in the mesolimbic dopamine system. Synapse 21: 289–298.

Pandey SC, Ugale R, Zhang H et al. (2008). Brain chromatin remodeling: a novel mechanism of alcoholism. J Neurosci 28: 3729–3737.

Pang KY, Braswell LM, Chang L et al. (1980). The perturbation of lipid bilayers by general anesthetics: a quantitative test of the disordered lipid hypothesis. Mol Pharmacol 18: 84–90.

Pascual M, Blanco AM, Cauli O et al. (2007). Intermittent ethanol exposure induces inflammatory brain damage and causes long-term behavioural alterations in adolescent rats. Eur J Neurosci 25: 541–550.

Peoples RW, Li C, Weight FF (1996). Lipid vs protein theories of alcohol action in the nervous system. Annu Rev Pharmacol Toxicol 36: 185–201.

Pichardo-Casas I, Goff LA, Swerdel MR et al. (2012). Expression profiling of synaptic microRNAs from the adult rat brain identifies regional differences and seizure-induced dynamic modulation. Brain Res 1436: 20–33.

Pietrzykowski AZ, Friesen RM, Martin GE et al. (2008). Posttranscriptional regulation of BK channel splice variant stability by miR-9 underlies neuroadaptation to alcohol. Neuron 59: 274–287.

Pignataro L, Miller AN, Ma L et al. (2007). Alcohol regulates gene expression in neurons via activation of heat shock factor 1. J Neurosci 27: 12957–12966.

Pignataro L, Varodayan FP, Tannenholz LE et al. (2009). The regulation of neuronal gene expression by alcohol. Pharmacol Ther 124: 324–335.

Pinelli A, Trivulzio S, Tomasoni L (1997). Effects of ondansetron administration on opioid withdrawal syndrome observed in rats. Eur J Pharmacol 340: 111–119.

Ponomarev I, Wang S, Zhang L et al. (2012). Gene coexpression networks in human brain identify epigenetic modifications in alcohol dependence. J Neurosci 32: 1884–1897.

Qiang M, Ticku MK (2005). Role of AP-1 in ethanol-induced N-methyl-D-aspartate receptor 2B subunit gene up-regulation in mouse cortical neurons. J Neurochem 95: 1332–1341.

Qin L, Crews FT (2012). NADPH oxidase and reactive oxygen species contribute to alcohol-induced microglial activation and neurodegeneration. J Neuroinflammation 9: 5.

Qin L, He J, Hanes RN et al. (2008). Increased systemic and brain cytokine production and neuroinflammation by endotoxin following ethanol treatment. J Neuroinflammation 5: 10.

Quertemont E, Tambour S, Tirelli E (2005). The role of acetaldehyde in the neurobehavioral effects of ethanol: a comprehensive review of animal studies. Prog Neurobiol 75: 247–274.

Quintanilla ME, Tampier L (2003). Acetaldehyde-reinforcing effects: differences in low-alcohol-drinking (UChA) and high-alcohol-drinking (UChB) rats. Alcohol 31: 63–69.

Raeder H, Holter SM, Hartmann AM et al. (2008). Expression of *N*-methyl-D-aspartate (NMDA) receptor subunits and splice variants in an animal model of long-term voluntary alcohol self-administration. Drug Alcohol Depend 96: 16–21.

Rahman S (2012). Epigenetic mechanisms: targets for treatment of alcohol dependence and drug addiction. CNS Neurol Disord Drug Targets 11: 101.

Rajasethupathy P, Fiumara F, Sheridan R et al. (2009). Characterization of small RNAs in Aplysia reveals a role for miR-124 in constraining synaptic plasticity through CREB. Neuron 63: 803–817.

Rani CS, Qiang M, Ticku MK (2005). Potential role of cAMP response element-binding protein in ethanol-induced *N*-methyl-D-aspartate receptor 2B subunit gene transcription in fetal mouse cortical cells. Mol Pharmacol 67: 2126–2136.

Rassnick S, Pulvirenti L, Koob GF (1992). Oral ethanol self-administration in rats is reduced by the administration of dopamine and glutamate receptor antagonists into the nucleus accumbens. Psychopharmacology (Berl) 109: 92–98.

Reilly MT, Lobo IA, McCracken LM et al. (2008). Effects of acamprosate on neuronal receptors and ion channels expressed in *Xenopus* oocytes. Alcohol Clin Exp Res 32: 188–196.

Renthal W, Nestler EJ (2008). Epigenetic mechanisms in drug addiction. Trends Mol Med 14: 341–350.

Renthal W, Maze I, Krishnan V et al. (2007). Histone deacetylase 5 epigenetically controls behavioral adaptations to chronic emotional stimuli. Neuron 56: 517–529.

Rhodes JS, Crabbe JC (2005). Gene expression induced by drugs of abuse. Curr Opin Pharmacol 5: 26–33.

Roberts AJ, McArthur RA, Hull EE et al. (1998). Effects of amperozide, 8-OH-DPAT, and FG 5974 on operant responding for ethanol. Psychopharmacology (Berl) 137: 25–32.

Rothblat DS, Rubin E, Schneider JS (2001). Effects of chronic alcohol ingestion on the mesostriatal dopamine system in the rat. Neurosci Lett 300: 63–66.

Russo SJ, Dietz DM, Dumitriu D et al. (2010). The addicted synapse: mechanisms of synaptic and structural plasticity in nucleus accumbens. Trends Neurosci 33: 267–276.

Saba L, Bhave SV, Grahame N et al. (2006). Candidate genes and their regulatory elements: alcohol preference and tolerance. Mamm Genome 17: 669–688.

Sauguet L, Howard RJ, Malherbe L et al. (2013). Structural basis for potentiation by alcohols and anaesthetics in a ligand-gated ion channel. Nat Commun 4: 1697.

Schank JR, Pickens CL, Rowe KE et al. (2011). Stress-induced reinstatement of alcohol-seeking in rats is selectively suppressed by the neurokinin 1 (NK1) antagonist L822429. Psychopharmacology (Berl) 218: 111–119.

Schmitt SM, Frezza M, Dou QP (2012). New applications of old metal-binding drugs in the treatment of human cancer. Front Biosci (Schol Ed) 4: 375–391.

Schratt G (2009). microRNAs at the synapse. Nat Rev Neurosci 10: 842–849.

Schratt GM, Tuebing F, Nigh EA et al. (2006). A brain-specific microRNA regulates dendritic spine development. Nature 439: 283–289.

Schuman EM, Dynes JL, Steward O (2006). Synaptic regulation of translation of dendritic mRNAs. J Neurosci 26: 7143–7146.

Sellers EM, Higgins GA, Sobell MB (1992). 5-HT and alcohol abuse. Trends Pharmacol Sci 13: 69–75.

Sellers EM, Toneatto T, Romach MK et al. (1994). Clinical efficacy of the 5-HT3 antagonist ondansetron in alcohol abuse and dependence. Alcohol Clin Exp Res 18: 879–885.

Shorter D, Kosten TR (2011). Novel pharmacotherapeutic treatments for cocaine addiction. BMC Med 9: 119.

Spanagel R (2009). Alcoholism: a systems approach from molecular physiology to addictive behavior. Physiol Rev 89: 649–705.

Spanagel R, Holter SM, Allingham K et al. (1996a). Acamprosate and alcohol: I. Effects on alcohol intake following alcohol deprivation in the rat. Eur J Pharmacol 305: 39–44.

Spanagel R, Putzke J, Stefferl A et al. (1996b). Acamprosate and alcohol: II. Effects on alcohol withdrawal in the rat. Eur J Pharmacol 305: 45–50.

Spanagel R, Zieglgansberger W, Hundt W (1996c). Acamprosate and alcohol: III. Effects on alcohol discrimination in the rat. Eur J Pharmacol 305: 51–56.

Spanagel R, Vengeliene V, Jandeleit B et al. (2013). Acamprosate Produces its Anti-Relapse Effects via Calcium. Neuropsychopharmacology 39: 783–791.

Spina L, Longoni R, Vinci S et al. (2010). Role of dopamine D1 receptors and extracellular signal regulated kinase in the motivational properties of acetaldehyde as assessed by place preference conditioning. Alcohol Clin Exp Res 34: 607–616.

Starkman BG, Sakharkar AJ, Pandey SC (2012). Epigenetics-beyond the genome in alcoholism. Alcohol Res Health 34: 293–305.

Steward O, Banker GA (1992). Getting the message from the gene to the synapse: sorting and intracellular transport of RNA in neurons. Trends Neurosci 15: 180–186.

Steward O, Levy WB (1982). Preferential localization of polyribosomes under the base of dendritic spines in granule cells of the dentate gyrus. J Neurosci 2: 284–291.

Stopponi S, Somaini L, Cippitelli A et al. (2011). Activation of nuclear PPARgamma receptors by the antidiabetic agent pioglitazone suppresses alcohol drinking and relapse to alcohol seeking. Biol Psychiatry 69: 642–649.

Sulik KK (2005). Genesis of alcohol-induced craniofacial dysmorphism. Exp Biol Med (Maywood) 230: 366–375.

Sung KW, Engel SR, Allan AM et al. (2000). 5-HT(3) receptor function and potentiation by alcohols in frontal cortex neurons from transgenic mice overexpressing the receptor. Neuropharmacology 39: 2346–2351.

Takao K, Kobayashi K, Hagihara H et al. (2013). Deficiency of Schnurri-2, an MHC enhancer binding protein, induces mild chronic inflammation in the brain and confers molecular, neuronal, and behavioral phenotypes related to schizophrenia. Neuropsychopharmacology 38: 1409–1425.

Taylor JR, Lynch WJ, Sanchez H et al. (2007). Inhibition of Cdk5 in the nucleus accumbens enhances the locomotor-activating and incentive-motivational effects of cocaine. Proc Natl Acad Sci U S A 104: 4147–4152.

Treistman SN, Martin GE (2009). BK Channels: mediators and models for alcohol tolerance. Trends Neurosci 32: 629–637.

Tsankova N, Renthal W, Kumar A et al. (2007). Epigenetic regulation in psychiatric disorders. Nat Rev Neurosci 8: 355–367.

Vetreno RP, Crews FT (2012). Adolescent binge drinking increases expression of the danger signal receptor agonist HMGB1 and Toll-like receptors in the adult prefrontal cortex. Neuroscience 226: 475–488.

Volkow ND, Wang GJ, Maynard L et al. (2002). Effects of alcohol detoxification on dopamine D2 receptors in alcoholics: a preliminary study. Psychiatry Res 116: 163–172.

Waddington CH (2012). The epigenotype. 1942. Int J Epidemiol 41: 10–13.

Wallner M, Hanchar HJ, Olsen RW (2003). Ethanol enhances alpha 4 beta 3 delta and alpha 6 beta 3 delta gamma-aminobutyric acid type A receptors at low concentrations known to affect humans. Proc Natl Acad Sci U S A 100: 15218–15223.

Wang Y, Krishnan HR, Ghezzi A et al. (2007). Drug-induced epigenetic changes produce drug tolerance. PLoS Biol 5: e265.

Watanabe H, Henriksson R, Ohnishi YN et al. (2009). FOSB proteins in the orbitofrontal and dorsolateral prefrontal cortices of human alcoholics. Addict Biol 14: 294–297.

Weiss F, Lorang MT, Bloom FE et al. (1993). Oral alcohol self-administration stimulates dopamine release in the rat nucleus accumbens: genetic and motivational determinants. J Pharmacol Exp Ther 267: 250–258.

Wilke N, Sganga M, Barhite S et al. (1994). Effects of alcohol on gene expression in neural cells. EXS 71: 49–59.

Winstanley CA, LaPlant Q, Theobald DE et al. (2007). DeltaFosB induction in orbitofrontal cortex mediates tolerance to cocaine-induced cognitive dysfunction. J Neurosci 27: 10497–10507.

Wolstenholme JT, Warner JA, Capparuccini MI et al. (2011). Genomic analysis of individual differences in ethanol drinking: evidence for non-genetic factors in C57BL/6 mice. PLoS One 6: e21100.

Wu Y, Lousberg EL, Moldenhauer LM et al. (2011). Attenuation of microglial and IL-1 signaling protects mice from acute alcohol-induced sedation and/or motor impairment. Brain Behav Immun 25 (Suppl. 1): S155–S164.

Zakhari S (2006). Overview: how is alcohol metabolized by the body? Alcohol Res Health 29: 245–254.

Zalewska-Kaszubska J, Bajer B, Gorska D et al. (2013). Effect of repeated treatment with topiramate on voluntary alcohol intake and beta-endorphin plasma level in Warsaw alcohol high-preferring rats. Psychopharmacology (Berl) 225: 275–281.

Zhou Q, Verdoorn TA, Lovinger DM (1998). Alcohols potentiate the function of 5-HT3 receptor-channels on NCB-20 neuroblastoma cells by favouring and stabilizing the open channel state. J Physiol 507: 335–352.

Section 4

Neurologic signs and consequences

Handbook of Clinical Neurology, Vol. 125 (3rd series)
Alcohol and the Nervous System
E.V. Sullivan and A. Pfefferbaum, Editors

Chapter 7

Alcohol: intoxication and poisoning – diagnosis and treatment

YOUNG-CHUL JUNG* AND KEE NAMKOONG
Department of Psychiatry, Yonsei University College of Medicine, Seoul, Korea

INTRODUCTION

Alcohol intoxication refers to a clinically harmful condition induced by recent ingestion of alcohol, when alcohol and its metabolites accumulate in the blood stream faster than they can be metabolized by the liver. Due to the long history and widespread use of alcohol as a recreational beverage, the clinical manifestations of alcohol intoxication are usually not taken seriously and considered to subside spontaneously with time; however, the adverse effects of alcohol at sufficiently high levels can cause coma and respiratory depression. In addition, individuals who seek medical treatment for acute alcohol intoxication likely have additional medical problems related to chronic alcohol consumption or alcohol dependence.

The aims of this chapter are to delineate and discuss: (1) the acute and chronic effects of ethanol on organ systems; (2) diagnostic criteria of alcohol intoxication; and (3) clinical conditions, which can accompany acute alcohol intoxication, especially those that are associated with chronic alcohol consumption (e.g., hepatic encephalopathy and Wernicke's encephalopathy); and (4) treatment.

THE EFFECTS OF ETHANOL ON ORGAN SYSTEMS

Alcohol is absorbed primarily through the small intestine. Only 10% of consumed alcohol is absorbed from the stomach. Initial mucosal absorption begins within 10 minutes of consumption and serum peak blood alcohol concentration is reached between 30 and 90 minutes (Marco and Kelen, 1990). A standard single drink contains about 10–12 mg of ethanol, which is estimated to increase the blood alcohol concentration of a 70-kg (155-lb) man by 15–20 mg/dL. On the other hand, 90% of absorbed alcohol is metabolized in the liver through oxidation at an average rate of 15 mg/dL per hour. Therefore, alcohol consumption exceeding this rate results in accumulation in the body and induces intoxication symptoms. The signs and symptoms of acute intoxication are greater when the blood alcohol concentration is rising than when it is declining. This implies that some adaptation process during exposure contributes to acute tolerance during a single dose (Vogel-Sprott and Fillmore, 1993; Kalant, 1996).

Although alcohol has the potential to affect almost every organ, the major adverse effects of alcohol that gain clinical attention are the neurologic, gastrointestinal, cardiovascular and respiratory problems.

Nervous system

The first major problem of the nervous system related to alcohol intoxication is temporary anterograde amnesia, usually referred to as blackout, in which the person is unable to recall all ("en bloc") or part ("fragmentary") of what occurred during an intoxicated drinking episode (Goodwin, 1995). In contrast with passing out (i.e., losing consciousness), persons experiencing blackout may appear normal, make conversations, and even perform complicated tasks. Blackout results from a rapid increase in alcohol blood concentration blocking the consolidation of short-term memories into long-term memories, a process, which involves the hippocampus and medial temporal structures (White et al., 2000). Intoxicated persons may have relatively intact remote memory and be unaware of suffering from a specific recent memory impairment and being unable to recall events that happened several minutes before (Squire and Alvarez, 1995; Miller and Matzel, 2000). Temporary blackouts should be distinguished from global amnesia, which is a chronic condition associated with Korsakoff's syndrome (Kopelman, 1995).

*Correspondence to: Young-Chul Jung, M.D., Ph.D., Department of Psychiatry, Yonsei University College of Medicine, 50 Yonsei-ro, Seodaemun-gu, Seoul 120-752, Korea. Tel: +82-2-2228-1620, E-mail: eugenejung@yuhs.ac

Another common consequence of alcohol intoxication is impaired judgment and insight (awareness of illness or impairment). Intoxicated patients usually deny, or understate, their impaired motor and cognitive functions, which increases the risk of accidents such as driving while intoxicated. Alcohol-induced cognitive impairment is observed in various domains such as verbal and visuospatial memory, information processing, inhibitory control, and working memory (Schweizer and Vogel-Sprott, 2008).

Alcohol is believed to improve insomnia and promote sleep; however the potential benefits are limited to the initial step in falling asleep. Virtually, alcohol disrupts the stages of sleep throughout the night (i.e., shortens the duration of deep slow-wave sleep and rapid eye movement sleep), resulting in sleep fragmentation and non-restorative sleep. Alcohol exacerbates obstructive sleep apnea through relaxing the pharynx muscles, which narrow of the upper airway (Issa and Sullivan, 1982).

After an intoxication episode, chronic alcoholics may express signs of peripheral neuropathy, including limb numbness, tingling and burning sensations, and paresthesias. Signs are usually bilateral and are more pronounced distally than proximally (Peters et al., 2006).

Gastrointestinal system

Alcohol intake can irritate the esophagus and stomach mucosa causing esophagitis, gastritis, and gastric ulcers. High concentrations of alcohol can cause spasm of the stomach pyloric valve, resulting in nausea and vomiting. Vigorous retching or vomiting can induce hematemesis (vomiting of blood) due to a Mallory–Weiss lesion, which is a longitudinal laceration in the mucosa at the gastroesophageal junction.

Heavy drinking can result in inflammation of the pancreas that can lead to acute pancreatitis, which accompanies an acute episode of abdominal pain with nausea and vomiting. Alcohol impairs gluconeogenesis and fatty acid oxidation in the liver, resulting in fat accumulation in liver cells, which might be found as an enlarged liver on physical examination. Fatty liver is reversible if abstinence is maintained and does not necessarily predispose the victim to a serious live damage; however, repeated episodes of heavy drinking will commonly lead to alcoholic hepatitis and hepatic cirrhosis (Maddrey, 2000).

Cardiovascular system

The relation between alcohol intake, especially moderate drinking, and cardiovascular diseases is complex, involving both protective and harmful effects (Russell et al., 2009). Daily heavy drinking results in a linear dose-dependent increase in blood pressure, which returns to normal within weeks of abstinence (Taylor et al., 2009). Chronic heavy drinkers carry an increased risk for dilated cardiomyopathy. Cardiomyopathy contributes to unexplained arrhythmias with left ventricular impairment, mitral valve regurgitation with associated mural thrombi and congestive heart failure. Malignant ventricular arrhythmias can occur temporarily after heavy episodic drinking in individuals showing no other evidence of heart disease (Uyarel et al., 2005). Activation of aldehyde dehydrogenase-2, which catalyzes the oxidation of cytotoxic aldehyde, reduces ischemic heart damage in rodent models (Chen et al., 2008).

Other effects

The main life-threatening consequence of high alcohol blood concentration is respiratory depression. Alcohol intoxication decreases the airway sensitivity and gag reflex to foreign bodies, which might be related to higher risk of aspiration. Alcohol also interferes with mucociliary clearance (Vonghia et al., 2008). Chronic alcohol consumption with nutritional deficiencies suppresses the bone marrow hematopoietic system, resulting in decreased production of red blood cells (anemia), white blood cells (neutropenia) and platelets (thrombocytopenia). Enlarged red blood cells (increased mean corpuscular volume) are observed in chronic alcoholics (Palmieri et al., 2006). Thrombocytopenia increases the risk of excessive bleeding (e.g., hemorrhagic stroke), which usually resolves within a week of abstinence from alcohol unless accompanied with hepatic cirrhosis or splenomegaly. Independent of peripheral neuropathy, chronic alcoholics can experience muscle weakness or cramps caused by acute alcoholic myopathy (Preedy et al., 2001). Heavy drinking interferes with calcium absorption and metabolism in the skeletal system. Chronic alcoholics may manifest lower bone density, and decreased growth in the epiphyses, leading to an increased risk for falling-related fractures and osteonecrosis of the femoral head (Kaukonen et al., 2006).

Repeated heavy drinking in adolescence can disturb normal sexual development and alter the onset and progress of puberty. Modest drinking can increase sexual drive but might also cause erectile dysfunction in men. Even in the absence of liver cirrhosis, chronic alcoholic men may show testicular atrophy, gynecomastia, decreases in ejaculate volume, and changes in body hair (Van Thiel et al., 1975). Chronic excessive alcohol consumption by women can influence sex hormone levels and alter ovarian function, resulting in amenorrhea, infertility, and an increased risk of spontaneous abortion (Gill, 2000). Hormonal changes related with acute alcohol intoxication include increase in cortisol levels (Badrick et al., 2008), diuresis due to inhibition of vasopressin secretion, and decrease in serum thyroxine and triiodothyronine (Kolakowska and Swigar, 1977).

DIAGNOSIS

The smell of alcohol on the patient's breath is the first cue that leads to the impression of alcohol intoxication. The diagnosis can usually be made based on history taking and physical examination. Intoxicated patients are likely to deny or understate their maladaptive pattern of alcohol use. Therefore, a thorough history of the individual's alcohol use should be taken from other informants if possible. The information about the time of the last drink is critical in order to prevent and manage symptoms of alcohol withdrawal, which can develop as soon as 6–8 hours after the cessation of drinking. However, various conditions can cause symptoms that mimic alcohol intoxication, while the altered mental status of the patient makes interviewing difficult and puts the clinician in a situation depending more on physical examination and laboratory analysis.

Physical examination must start with analyzing vital signs. Signs and symptoms that imply chronic alcohol abuse or dependence, including capillary prominence, spider nevi, telangiectasias, and palmar erythema, must be checked. Elevated mean corpuscular volume, serum gamma-glutamyl transpeptidase, and carbohydrate-deficient transferring are sensitive indicators of chronic alcohol abuse (Lumeng, 1986; Sillanaukee, 1996). Cardiac and chest examination with electrocardiography and chest radiography must be performed. Computed tomography (CT) of the brain should be ordered when neurologic signs are present or a head trauma event is suspected. Brain atrophy, evident as ventricular enlargement and widened cortical sulci on magnetic resonance imaging and CT scans, occurs in chronic alcoholics; these changes are greatest in the prefrontal cortex and are partially reversible if abstinence is maintained (Sullivan and Pfefferbaum, 2005).

Clinical features

According to the American Psychiatric Association's (2013) *Diagnostic and Statistical Manual of Mental Disorders* (DSM-5) and the World Health Organization's (1992) *International Classification of Diseases* (ICD-10) criteria, alcohol intoxication is a clinical diagnosis based on presence of clinically significant problematic behavioral or psychological changes, which are accompanied by physiologic signs of intoxication, including slurred speech, incoordination, unsteady gait, nystagmus, conjunctival injection, impairment in attention or memory, stupor, or coma.

Blood alcohol concentration

The effects of alcohol are related to blood alcohol concentration (Table 7.1); however, the extent of acute alcohol intoxication also depends on several factors, including the individual body weight, the volume and period of alcohol ingestion, the percentage of alcohol in the beverage, and whether alcohol was taken on an empty stomach or with food, which delays the absorption. Tolerance is a sign of alcohol dependence, but can also develop to a certain level in non-dependent regular drinkers. In addition, the individual differences in signs and symptoms may vary according to genetic susceptibility (Macgregor et al., 2009).

The legal definition of alcohol intoxication has been developed in reference to the major effects on driving abilities. The legal intoxication level that imposes penalties for driving in most states in the USA is 80 mg ethanol per deciliter of blood (80 mg/dL or 0.08 g/dL), a level when evidence of slowed motor performance and decreased thinking ability becomes apparent; however, signs indicating loosening of judgments and inhibitions starts to appear at below 50 mg/dL, i.e., after one to two drinks. At blood concentrations between 100 and 200 mg/dL (0.10–0.20%), mood lability and over-expression (e.g., anger or sadness), impairment in coordination (ataxia) become obvious (Yost, 2002; Schuckit, 2005). Although individual differences vary according to various factors, anyone who does not show significant levels of motor and cognitive impairment at this blood alcohol concentration probably has significant tolerance to alcohol, which strongly indicates a history of chronic alcohol abuse. At blood alcohol concentrations between 200–300 mg/dL (0.20–0.30%), marked slurring of speech, nausea and vomiting due to stomach pyloric valve spam, memory impairments including blackouts are manifested. The risk of aspiration increases at this blood alcohol concentration. At higher blood alcohol concentrations, alcohol affects breathing and heart rate leading to a high risk of respiratory depression, coma, and death (Adinoff et al., 1988).

Alcohol levels can also be determined by breath analysis (James et al., 1984) or with a saliva dipstick, although these methods are less reliable (Schwartz et al., 1989). Due to the short half-life of ethanol, blood or breath alcohol analyses provide no information about the chronic effects of alcohol abuse or heavy drinking. Therefore, clinicians should always keep in mind that these examinations have limitations because they do not necessarily correlate with clinical presentations and do not predict clinical severity or outcome (Marco and Kelen, 1990).

Alcohol flush reaction

Alcohol is metabolized by two enzymes: alcohol dehydrogenase and aldehyde dehydrogenase. Alcohol dehydrogenase catalyzes the conversion of alcohol into acetaldehyde, which is a toxic compound, whereas aldehyde dehydrogenase catalyzes the conversion of

Table 7.1

Clinical manifestations of alcohol intoxication according to blood alcohol concentration (BAC)

BAC	Approximate alcohol consumption*	Clinical manifestation
< 50 mg/dL	1–2 drinks	Mild euphoria, slowed motor performance,
> 50 mg/dL (10.9 mmol/L)	3–5 drinks	Impaired sensation, incoordination
> 100 mg/dL (21.7 mmol/L)	6–10 drinks	Mood lability, cognitive and memory difficulties, marked incoordination, ataxia
> 200 mg/dL (43.4 mmol/L)	> 10 drinks	Nausea, vomiting, nystagmus, alcoholic blackouts, marked slurring of speech, aspiration risk
> 300 mg/dL (65.1 mmol/L)		Hypoventilation, hypothermia, cardiac arrhythmia
> 400 mg/dL (86.8 mmol/L)		Coma, respiratory arrest, death

*A standard single drink contains about 10–12 mg of ethanol, which is estimated to increase the blood alcohol concentration of a 70-kg (155-lb) man by 15–20 mg/dL.

acetaldehyde into acetic acid. Alcohol flush reaction is a condition in which an individual experiences flushes, headache, palpitation, and drowsiness due to accumulation of acetaldehyde. Flushing is associated with the erythema (reddening caused by dilation of capillaries) of the face, neck, shoulders, and in some cases, the entire body after as few as one or two drinks. It is also referred as the Asian flush syndrome, as it is known to be experienced more by people of Asian descent. Alcohol flush reaction is highly correlated with variants of the genes that are responsible of coding for alcohol dehydrogenase and aldehyde dehydrogenase (Yoshida et al., 1984; Eng et al., 2007; Macgregor et al., 2009).

Idiosyncratic alcohol intoxication

Idiosyncratic alcohol intoxication is proposed as a behavioral syndrome that develops rapidly after a small amount of alcohol that would have minimal behavioral effects on most persons. However, the validity of idiosyncratic alcohol intoxication as a diagnostic entity is still under debate (Maletzky, 1976; Hollender, 1979). The diagnosis is especially important in forensic cases because idiosyncratic alcohol intoxication can be used in a person's defense as an unexpected, pathologic reaction to alcohol. The clinical features are confusion, disorientation, transitory delusions, illusion and visual hallucinations, increased psychomotor activity, and impulsive behaviors like suicide attempts.

DIFFERENTIAL DIAGNOSIS

In many cases, the intoxicated patient has comorbid medical conditions, which are usually related to chronic alcohol abuse. For this reason, additional investigations to identify potential problems needing particular attention should be considered, depending on the clinical features of the patient. Special attention should be paid to the changes of mental status, which might vary across a spectrum, comprising mild euphoria, disinhibition, lethargy, and coma. Mental statuses that are not consistent with the information of history taking (i.e., the amount and period of drinking) should immediately require clinical attention and further evaluation.

Several different conditions can mimic or be masked by the altered mental status changes of acute alcohol intoxication (Table 7.2). These conditions include other substance-related intoxication, metabolic alterations, infectious diseases, neurologic causes (such as seizure and trauma), and respiratory depression (Vonghia et al., 2008). These clinical conditions, potentially life-threatening, resemble the non-specific signs of alcohol intoxication and, therefore, must be ruled out while evaluating a patient with alcohol intoxication. Hepatic encephalopathy and Wernicke encephalopathy are two representative conditions with clinical presentations that are difficult to differentiate from those related to alcohol intoxication, but if not treated appropriately, might rapidly progress to coma and death or chronically to irreversible cognitive compromise.

Hepatic encephalopathy

Hepatic encephalopathy is defined as a disturbance in central nervous system function because of hepatic insufficiency. Hepatic encephalopathy is caused by accumulation of ammonia and toxic substances in the blood stream that are normally detoxified in the liver. A precipitating event is often identified, such as gastrointestinal bleeding or injudicious use of central nervous system-depressing drugs. Patients may be confused or exhibit a change in personality and mood that mimics alcohol intoxication. Asterixis is often present and can be examined by having the patient extend the arms and bend the wrists back, which elicits a flapping tremor, i.e., a sudden forward movement of the wrist.

These signs are present in both acute and chronic liver failure and potentially reversible upon correction of precipitating factors and improvement of liver function (Blei et al., 2001). In the advanced stages, however, it may ultimately lead to hepatic coma and death. The diagnosis of hepatic encephalopathy requires laboratory analysis of liver function. Increased ammonia level is a typical sign. Nonetheless, clinicians should keep in mind that the correlation between ammonia levels and the severity of liver disease is not always consistent. Ammonia levels can be normal in 10% of hepatic encephalopathy patients (Ong et al., 2003). Alcohol intoxication can precipitate hepatic encephalopathy by exacerbating underlying problems such as electrolyte disturbances, infection, and dehydration. Thus, hepatic encephalopathy should always be suspected when an intoxicated patient exhibits mental status changes in line with these potential precipitating events (Cash et al., 2010). Severe encephalopathy can progress to cerebral edema, elevated intracranial pressure, and brain herniation in acute liver failure (Desjardins et al., 2012).

Table 7.2

Differential diagnosis of alcohol intoxication

Substance-related	Intoxication with other alcohol
	Methanol
	Isopropyl alcohol
	Psychoactive drugs
	Cocaine
	Opiates
	Benzodiazepine/barbiturates
	Disulfiram
Metabolic	Hepatic encephalopathy
	Hypoglycemia
	Electrolyte abnormalities
	Hyponatremia
	Alcoholic ketoacidosis
	Diabetic ketoacidosis
Infectious	Sepsis
	Meningitis
	Encephalitis
Neurologic	Alcohol withdrawal
	Wernicke–Korsakoff syndrome
	Cerebrovascular accidents
	Seizure disorders
Trauma	Closed head injuries
Respiratory	Hypoxia secondary to aspiration
	Respiratory depression
Others	Hypotension
	Hypothermia
	Dehydration
	Hyper-/hypothyroidism

Wernicke–Korsakoff syndrome

Alcoholism still remains the most common cause of thiamine deficiency in industrialized countries. Alcoholics are at special risk for thiamine deficiency because of poor nutritional intake, compromised absorption from the gastrointestinal tract, reduced thiamine storage, and impaired thiamine utilization (Sechi and Serra, 2007). Wernicke's encephalopathy is an acute neurologic disorder characterized by mental status changes, ataxia (affecting primarily the gait), and a variety of ocular motility abnormalities due to thiamine deficiency. Mental status changes, which are the most common symptom of Wernicke's encephalopathy, are non-specific and can range from apathy to inability to concentrate, confusion and, if not treated appropriately, coma. Nystagmus (involuntary eye movements) and ophthalmoplegia (paralyses of conjugate gaze) are common signs, whereas sluggish reactions of the pupils to light or anisocoria (unequal size of pupils) are relatively uncommon. Ataxia (loss of coordinated movement of upper or lower limbs) is usually reported as dyscoordinated gait (Harper et al., 1986; Thomson, 2000). A quantitative test detected ataxia in 36% of uncomplicated alcoholics as a sign of preclinical Wernicke's encephalopathy or residual of subclinical Wernicke's encephalopathy, indicative of the sensitivity of vigorous assessment in detecting deficits (Pitel et al., 2010). The diagnosis is clinical and is mainly supported by the dramatic improvement in neurologic signs with parenteral thiamine. In patients suspected of having Wernicke's encephalopathy, parenteral thiamine (500 mg three times per day) should be initiated immediately to be effective in treatment of Wernicke's encephalopathy and prevention of Korsakoff's syndrome (Thomson et al., 2002). If Wernicke's encephalopathy is left untreated, 80% of patients may progress into Korsakoff's syndrome. The cardinal feature of Korsakoff's syndrome is a disproportionate impairment in declarative memory relative to other components of cognitive functioning, which becomes more obvious as the confusional stage of Wernicke encephalopathy resolves (Kopelman, 1995).

TREATMENT

The treatment of an intoxicated patient entails supportive and symptomatic care. Management starts with airway assessment and examination of the cardiac and respiratory function. The immediate steps include a complete assessment of the patient's medical status, including drinking history of alcohol or other possible substances that might interfere with clinical condition. The rapid absorption of alcohol through intestinal mucosa, which starts within 10 minutes, makes gastric lavage or activated charcoal of limited value. Prevention of aspiration is mandatory; therefore, placement of the patient in a lateral position may be helpful. Intravenous fluid solution should be administered to hydrate the patient and to correct electrolyte imbalances and hypoglycemia. Hyponatremia must be managed carefully because rapid correction can induce central pontine myelinolysis (Uchino et al., 2003). To prevent Wernicke's encephalopathy, parenteral thiamine must be administered prior to any administration of glucose (Maldonado, 2010). Parenteral thiamine therapy is recommended as malnutrition and alcohol can impair the intestinal thiamine absorption by ~70% (Thomson et al., 2002). Antiemetic drugs may be useful in patients with nausea and vomiting. Prolonged vomiting might indicated an unappreciated head injury. Slurred speech, ataxia, altered mental status, and other neurologic signs of alcohol intoxication can mimic or mask a comorbid medical condition; therefore, repeated assessments of the mental status and neurologic signs are critical.

There is no antidote for alcohol intoxication. In addition, there is no established treatment effective for preventing or treating alcohol hangover (Pittler et al., 2005). Metadoxine (pyridoxol L-2-pyrrolidone-5-carbohydrate) is proposed to accelerate ethanol metabolism due to several mechanisms, including an increase in acetaldehyde dehydrogenase activity, ethanol and acetaldehyde plasma clearance, and urinary elimination of ketones (Shpilenia et al., 2002; Addolorato et al., 2003). Dihydromyricetin, a flavonoid component of herbal medicine, is reported to be effective in counteracting acute alcohol intoxication and alcohol withdrawal symptoms (Shen et al., 2012). Recently, there have been efforts to use biomimetic nanocomplexes containing alcohol oxidase and catalase, which reduce blood alcohol levels, as an alternative antidote and prophylactic for alcohol intoxication (Liu et al., 2013).

References

Addolorato G, Ancona C, Capristo E et al. (2003). Metadoxine in the treatment of acute and chronic alcoholism: a review. Int J Immunopathol Pharmacol 16: 207–214.

Adinoff B, Bone GH, Linnolia M (1988). Acute ethanol poisoning and the ethanol withdrawal syndrome. Med Toxicol Adverse Drug Exp 3: 172–196.

American Psychiatric Association (2013). Diagnostic and statistical manual of mental disorders. 5th ed. American Psychiatric Publishing, Arlington, VA.

Badrick E, Bobak M, Britton A et al. (2008). The relationship between alcohol consumption and cortisol secretion in an aging cohort. J Clin Endocrinol Metab 93: 750–757.

Blei AT, Cordoba J, Practice Parameters Committee of the American College of Gastroenterology (2001). Hepatic encephalopathy. Am J Gastroenterol 96: 1968–1976.

Cash WJ, McConville P, McDermott E et al. (2010). Current concepts in the assessment and treatment of hepatic encephalopathy. Q J Med 103: 9–16.

Chen C-H, Budas GR, Churchill EN et al. (2008). Activation of aldehyde dehydrogenase-2 reduces ischemic damage to the heart. Science 321: 1493–1495.

Desjardins P, Du T, Jiang W et al. (2012). Pathogenesis of hepatic encephalopathy and brain edema in acute liver failure: role of glutamine redefined. Neurochem Int 60: 690–696.

Eng MY, Luczak SE, Wall TL (2007). ALDH2, ADH1B, and ADH1C genotypes in Asians: a literature review. Alcohol Res Health 30: 22–27.

Gill J (2000). The effects of moderate alcohol consumption on female hormone levels and reproductive function. Alcohol Alcohol 35: 417–423.

Goodwin DW (1995). Alcohol amnesia. Addiction 90: 315–317.

Harper CG, Giles M, Finlay-Jones R (1986). Clinical signs in the Wernicke–Korsakoff complex: a retrospective analysis of 131 cases diagnosed at necropsy. J Neurol Neurosurg Psychiatry 49: 341–345.

Hollender MH (1979). Pathological intoxication – is there such an entity? J Clin Psychiatry 40: 424–426.

Issa FG, Sullivan CE (1982). Alcohol, snoring and sleep apnea. J Neurol Neurosurg Psychiatry 45: 353–359.

James JJ, Dargon D, Day RG (1984). Serum vs. breath alcohol levels and accidental injury: analysis among US Army personnel in an emergency room setting. Mil Med 149: 369–374.

Kalant H (1996). Current state of knowledge about the mechanisms of alcohol tolerance. Addict Biol 1: 133–141.

Kaukonen JP, Nurmi-Luthje I, Luthje P et al. (2006). Acute alcohol use among patients with acute hip fractures: a descriptive incidence study in southeastern Finland. Alcohol Alcohol 41: 345–348.

Kolakowska T, Swigar ME (1977). Thyroid function in depression and alcohol abuse: a retrospective study. Arch Gen Psychiatry 34: 984–988.

Kopelman MD (1995). The Korsakoff syndrome. Br J Psychiatry 166: 154–173.

Liu Y, Du J, Yan M (2013). Biomimetic enzyme nanocomplexes and their use as antidotes and preventive measures for alcohol intoxication. Nat Nanotechnol 8: 187–192.

Lumeng L (1986). New diagnostic markers of alcohol abuse. Hepatology 6: 742–745.

Macgregor S, Lind PA, Bucholz KK et al. (2009). Associations of ADH and ALDH2 gene variation with self report alcohol reactions, consumption and dependence: an integrated analysis. Hum Mol Genet 18: 580–593.

Maddrey WC (2000). Alcohol-induced liver disease. Clin Liver Dis 4: 115–131, vii.

Maldonado JR (2010). An approach to the patient with substance use and abuse. Med Clin North Am 94: 1169–1205, x-i.

Maletzky BM (1976). The diagnosis of pathological intoxication. J Stud Alcohol 37: 1215–1228.

Marco CA, Kelen GD (1990). Acute intoxication. Emerg Med Clin North Am 8: 731–748.

Miller RR, Matzel LD (2000). Memory involves far more than 'consolidation'. Nat Rev Neurosci 1: 214–216.

Ong JP, Aggarwal A, Krieger D et al. (2003). Correlation between ammonia levels and the severity of hepatic encephalopathy. Am J Med 114: 188–193.

Palmieri VO, Cicco G, Minerva F et al. (2006). Red blood cells (RBC) deformability and aggregability: alterations in alcoholism. Adv Exp Med Biol 578: 125–131.

Peters TJ, Kotowicz J, Nyka W et al. (2006). Treatment of alcoholic polyneuropathy with vitamin B complex: a randomized controlled trial. Alcohol Alcohol 41: 636–642.

Pitel AL, Zahr NM, Jackson K et al. (2010). Signs of preclinical Wernicke's encephalopathy and thiamine levels as predictors of neuropsychological deficits in alcoholism without Korsakoff's syndrome. Neuropsychopharmacology 36: 580–588.

Pittler MH, Verster JC, Ernst E (2005). Interventions for preventing or treating alcohol hangover: systematic review of randomised controlled trials. BMJ 331: 1515–1518.

Preedy VR, Adachi J, Ueno Y et al. (2001). Alcoholic skeletal muscle myopathy: definitions, features, contribution of neuropathy, impact and diagnosis. Eur J Neurol 8: 677–687.

Russell M, Chu BC, Banerjee A et al. (2009). Drinking patterns and myocardial infarction: a linear dose–response model. Alcohol Clin Exp Res 33: 324–331.

Schuckit MA (2005). Alcohol-related disorders. In: BJ Sadock, VA Sadock (Eds.), Comprehensive Textbook of Psychiatry. Lippincott Williams and Wilkins, Philadelphia PA, pp. 1168–1188.

Schwartz RH, O'Donnell RM, Thorne MM et al. (1989). Evaluation of colorimetric dipstick test to detect alcohol in saliva: a pilot study. Ann Emerg Med 18: 1001–1003.

Schweizer TA, Vogel-Sprott M (2008). Alcohol-impaired speed and accuracy of cognitive functions: a review of acute tolerance and recovery of cognitive performance. Exp Clin Psychopharmacol 16: 240–250.

Sechi G, Serra A (2007). Wernicke's encephalopathy: new clinical settings and recent advances in diagnosis and management. Lancet Neurol 6: 442–455.

Shen Y1, Lindemeyer AK, Gonzalez C et al. (2012). Dihydromyricetin as a novel anti-alcohol intoxication medication. J Neurosci 32: 390–401.

Shpilenia LS, Muzychenko AP, Casbarrini G et al. (2002). Metadoxine in acute alcohol intoxication: a double blind, randomized, placebo-controlled study. Alcohol Clin Exp Res 26: 340–346.

Sillanaukee P (1996). Laboratory markers of alcohol abuse. Alcohol Alcohol 31: 613–616.

Squire LR, Alvarez P (1995). Retrograde amnesia and memory consolidation: a neurobiological perspective. Curr Opin Neurobiol 5: 169–177.

Sullivan EV, Pfefferbaum A (2005). Neurocircuitry in alcoholism: a substrate of disruption and repair. Psychopharmacology (Berl) 180: 583–594.

Taylor B, Irving HM, Baliunas D et al. (2009). Alcohol and hypertension: gender differences in dose-response relationships determined through systematic review and meta-analysis. Addiction 104: 1981–1990.

Thomson AD (2000). Mechanisms of vitamin deficiency in chronic alcohol misusers and the development of the Wernicke–Korsakoff syndrome. Alcohol Alcohol Suppl 35: 2–7.

Thomson AD, Cook CC, Touquet R et al. (2002). The Royal College of Physicians report on alcohol: guidelines for managing Wernicke's encephalopathy in the accident and emergency department. Alcohol Alcohol Suppl 37: 513–521.

Uchino A, Yuzuriha T, Murakami M et al. (2003). Magnetic resonance imaging of sequelae of central pontine myelinolysis in chronic alcohol abusers. Neuroradiology 45: 877–880.

Uyarel H, Ozdol C, Gencer AM et al. (2005). Acute alcohol intake and QT dispersion in healthy subjects. J Stud Alcohol 66: 555–558.

Van Thiel DH, Gavaler JS, Lester R et al. (1975). Alcohol-induced testicular atrophy. An experimental model for hypogonadism occurring in chronic alcoholic men. Gastroenterology 69: 326–332.

Vogel-Sprott M, Fillmore MT (1993). Impairment and recovery under repeated doses of alcohol: effects of response-outcomes. Pharmacol Biochem Behav 45: 59–63.

Vonghia L, Leggio L, Ferrulli A et al. (2008). Acute alcohol intoxication. Eur J Intern Med 19: 561–567.

White AM, Matthews DB, Best PJ (2000). Ethanol, memory, and hippocampal function: a review of recent findings. Hippocampus 10: 88–93.

World Health Organization (1992). International Classification of Diseases, Tenth Revision. World Health Organization, Geneva, Switzerland.

Yoshida A, Huang IY, Ikawa M (1984). Molecular abnormality of an inactive aldehyde dehydrogenase variant commonly found in Orientals. Proc Natl Acad Sci U S A 81: 258–261.

Yost DA (2002). Acute care for alcohol intoxication. Postgrad Med 112: 25–26.

Handbook of Clinical Neurology, Vol. 125 (3rd series)
Alcohol and the Nervous System
E.V. Sullivan and A. Pfefferbaum, Editors

Chapter 8

Acute withdrawal: diagnosis and treatment

JOHN C.M. BRUST*

Department of Neurology, Columbia University College of Physicians & Surgeons, New York Neurological Institute, New York, NY, USA

INTRODUCTION

As with other drugs used recreationally, ethanol dependence is of two types. "Psychic dependence" is compulsive drug-seeking behavior in which procurement of a drug to achieve pleasure or avoid discomfort becomes a daily preoccupation. "Physical dependence" is a maladaptive state in which abrupt cessation of prolonged heavy drug use results in physiologic and cognitive disturbances. "Addiction" refers to psychic dependence. "Alcohol withdrawal" refers to the symptoms and signs produced by abrupt abstinence in physically dependent drinkers (Table 8.1). Such symptoms cover a spectrum from mild "hangover" to life-threatening seizures and delirium tremens (DTs) (Brust, 2004).

Chronic heavy drinking produces more severe and prolonged withdrawal symptoms (Victor and Adams, 1953), and repeated prior bouts of alcohol withdrawal are a risk for a more severe course ("allostasis," "kindling") (Monte et al., 2010).

SYMPTOMS AND SIGNS

Hangover

"Hangover," consisting of headache, malaise, fatigue, nausea, sweating, and tremulousness, can follow a single brief period of heavy drinking. Motor incoordination is often a feature, relevant to driving or piloting. Factors in addition to alcohol probably contribute to symptoms, including dehydration, lack of food, and insufficient sleep. Acetaldehyde is implicated, as well as congeners in dark liquors such as whiskey or brandy (Wiese et al., 2000).

Tremor

The most common withdrawal symptom is tremor, which usually appears within a few hours of the last drink and is promptly relieved by alcohol. With continued abstinence, tremor becomes more intense and is accompanied by anxiety, insomnia, easy startling, hyperactivity, nystagmus, flushing, sweating, anorexia, nausea, vomiting, weakness, tachypnea, tachycardia, and systolic hypertension. Tremor is distal, coarse, 6–8 Hz, irregular, and worse with movement, interfering with eating or even with standing. During this early period of withdrawal, mentation is usually normal except for inattentiveness, agitation, or pre-existing cognitive impairment; more severely altered mental status should raise consideration of seizures, Wernicke–Korsakoff syndrome, meningitis, or intracranial hemorrhage. Without treatment, tremor usually subsides within a few days but can be more prolonged, and some patients feel "shaky" inside for several weeks (Victor and Adams, 1953; Neiman et al., 1990).

Hallucinosis

Another common feature of early withdrawal is perceptual disturbance, including vivid dreams, nightmares, illusions, and hallucinations, which can be auditory, visual, tactile, olfactory, or a combination ("alcoholic hallucinosis"). Visual hallucinations are more common, with imagery of insects, animals, people, or disembodied heads. They may occur only with eye closure and tend to be fragmentary, lasting minutes at a time over several days. Paranoid delusions are sometimes an accompaniment. Of uncertain status are hallucinations that occur

*Correspondence to: John C.M. Brust, M.D., Department of Neurology, Columbia University College of Physicians & Surgeons, New York Neurological Institute, Room 1404, 710 West 168th Street, New York, NY, 10025, USA. Tel: +1-212-342-1516, E-mail: jcb2@cumc.columbia.edu

Table 8.1

Ethanol withdrawal syndromes

Early
Tremor: distal, coarse, worse with movement, usually promptly relieved by ethanol
Hallucinations: visual, auditory, tactile, or olfactory in the presence of normal alertness and attentiveness and variable insight
Seizures: usually generalized and single or clustered
Late
Delirium: marked inattentiveness, often with agitation, hallucinations, and fluctuating level of alertness; coarse tremor usually present
Autonomic instability: tachycardia, hypertension, fever, hypothermia, sweating, flushing
Abnormal movements: parkinsonism, chorea, dystonia, myoclonus, catatonia

in some patients during active drinking or after more than a week of abstinence. Some workers have proposed that alcoholic hallucinosis is of two types, one of which is independent of withdrawal. Of 70 alcoholic patients with hallucinations (visual, auditory, or both), 24 had no other symptoms, and the rest had tremor or seizures; four later developed DTs (Victor and Hope, 1958). Eight continued to have hallucinations for months or years, sometimes with ideas of reference, flat affect, and loose associations. None had evidence of schizophrenia prior to the first bout of hallucinosis. A study using controls similarly found no history of schizophrenia among patients before their hallucinosis or among their relatives (Schuckit and Winokur, 1971).

A withdrawing alcoholic patient, blind from ocular injury, had formed visual hallucinations, which he recognized as unreal and at the same time developed Anton syndrome, believing his vision had returned and confabulating descriptions (Schwarz and Brust, 1984). A patient with left hemineglect had withdrawal visual hallucinations restricted to the right visual hemifield (Chamorro et al., 1990).

Abnormal movements

Transient parkinsonism is described in alcoholics older than 50 years of age, beginning within a few days of the last drink or sometimes during drinking (Shandling et al., 1990). Symptoms tend to clear over days or weeks without treatment, and such patients have been followed for years without developing Parkinson's disease. Transient chorea has also occurred during alcohol withdrawal. Usually oral-lingual and sometimes spreading to neck or arm muscles, it typically affects younger patients during the second week of abstinence (Fornazzari and Carlen, 1982). Ethanol acutely decreases striatal dopamine release, perhaps accounting for parkinsonism, with receptor upregulation resulting in chorea (Neiman et al., 1990).

Also described during alcohol withdrawal are dystonia variably affecting arms, neck, trunk, or tongue (Bijjal et al., 2012) and myoclonus unresponsive to anticonvulsants (Drake, 1983).

Catatonia with mutism and posturing can occur during alcohol withdrawal, either as the presenting feature or appearing during benzodiazepine therapy and clearing with dosage increase (Geoffroy et al., 2012).

Seizures

Alcohol can precipitate seizures in patients with diagnosed epilepsy, but the amount required is disputed. So are recommendations as to whether social drinking is safe among epileptics. A study in which epileptic subjects drank one to three glasses of vodka over a 2-hour period daily for 16 weeks found neither increased seizure frequency nor electroencephalographic worsening (Hoppener et al., 1983).

Seizures occur in alcoholics whether or not they are epileptic (Hillbom et al., 2003). Indirect mechanisms include trauma, central nervous system infection, stroke, hypoglycemia, and other alcohol-related systemic disease. The term "alcohol-related seizure" refers to seizures in heavy drinkers in the absence of epilepsy or indirect cause. Such seizures are often encountered during early alcohol withdrawal.

A widely cited study reviewed 241 patients with alcoholism and seizures, either incident (new onset) or prevalent (recurrent) (Victor and Brausch, 1967). Three seizures occurred more than 2 weeks after the last drink; the rest occurred within 72 hours of abstinence. Other studies have confirmed the role of withdrawal. Typically, a single generalized tonic-clonic seizure occurs 6–48 hours after cessation of chronic drinking. There may be a cluster of seizures over a few hours. Status epilepticus occurs in fewer than 10%, but, among patients with status epilepticus, ethanol is a major contributor (Lowenstein and Alldredge, 1993). Focal features, not always attributable to previous head injury or other brain lesions, are observed in up to 25% (Earnest and Yarnell, 1976). Alcohol-related seizures can occur in otherwise asymptomatic patients, or they can accompany tremor or hallucinosis. Up to a third of patients progress to DTs if untreated (Victor and Brausch, 1967).

Most studies of alcohol-related seizures have concluded that they usually occur after months or years of excessive drinking (Victor and Adams, 1953; Rathlev et al., 1994; Bartolomei, 2006). In a study of 10 opioid-dependent volunteers, subjects became abruptly abstinent after several weeks of round-the-clock drinking; two had seizures upon withdrawal, but both had been heavy drinkers in the past (Isbell et al., 1955). In another report of detoxification without pharmacotherapy, seizures occurred in only 1% of 1024 subjects (Whitfield et al., 1978). A case-control study of new-onset unprovoked seizures found that daily ingestion of 50 grams absolute alcohol raised the odds ratio above one, and at 200 grams daily it was 20; the minimal duration of drinking that conferred risk of seizures could not be determined (Ng et al., 1988). In that study statistical analysis failed to identify a clear-cut temporal relationship between seizures and early abstinence; many seizures occurred during active drinking or more than a week after the last drink. Another case-control study found similar risk for new-onset seizures; odds ratios rose above one at 50 grams daily absolute alcohol in men and 25 grams daily in women (Leone et al., 1997). That study did not address the relationship between seizures and active drinking.

A meta-analysis of 18 studies which "quantified the association between alcohol consumption and epilepsy" concluded that four drinks daily were associated with a risk ratio (RR) of 1.81, six drinks daily with a RR of 2.44, and eight drinks daily with a RR of 3.27 (Samokhvalov et al., 2010). The results suggested that alcohol conferred risk for "spontaneous seizures not immediately related to alcohol intake." It is therefore possible that alcohol causes seizures by more than one mechanism, one of which is independent of withdrawal per se (Hauser et al., 1988). The risk of seizures increases with repeated alcohol detoxification, consistent with a kindling phenomenon and permanent synaptic alteration. Such a mechanism might also lead to long-term cognitive and emotional deficits which contribute to loss of control over drinking (Stephens and Duka, 2008).

The diagnosis of alcohol-related seizures requires exclusion of other lesions, and computed tomography scan or magnetic resonance imaging is indicated when seizures are of new onset. Of 259 patients with presumed alcohol-related seizures and normal neurologic examinations, 16 (6.2%) had unsuspected intracranial lesions, including subdural hematoma, subdural hygroma, vascular malformation, cysticercosis, aneurysm, neoplasm, skull fracture with subarachnoid hemorrhage, and cerebral infarction; in 10 of the 16, the findings altered management (Earnest et al., 1988). Lumbar puncture is indicated if meningitis or subarachnoid hemorrhage is suspected and imaging does not show intracranial blood. Although periodic lateralizing epileptiform discharges are sometimes observed during alcohol withdrawal (Gorelick and Wilkins, 1986), the electroencephalogram is usually normal (Fisch et al., 1989).

Delirium tremens

In contrast to withdrawal tremor, hallucinosis, and seizures, DTs usually appears 48–72 hours after the last drink. Symptoms include not only tremor and disordered sensory perception, but also delirium (extreme inattentiveness and non-awareness of the environment, usually with agitation and sometimes with decreased or fluctuating alertness), autonomic overactivity, and frequently a fatal outcome. DTs tends to be overdiagnosed in subjects with tremor and hallucinosis; in one series the syndrome developed in fewer than 5% of hospitalized patients with symptomatic alcohol withdrawal (Victor and Adams, 1953). Seizures may precede DTs, but seizures during DTs suggest an additional diagnosis such as meningitis or cerebral trauma.

DTs tends to begin and end abruptly, lasting hours to a few days. Symptoms can wax and wane in severity or subside gradually. Concurrent disease such as liver failure, pancreatitis, or pneumonia increases the severity of symptoms, and relapses can occur for several weeks (Gorelick and Wilkins, 1986). Elderly patients are more severely affected. A typical patient is agitated and grossly tremulous, with tachycardia, fever, and profuse sweating. Tremor may include the face, tongue, and pharynx, and the patient may shout at hallucinated objects. Fluid loss can be marked; heat stroke and myoglobinuria are described. Calmer patients may say they are drinking in a bar or, with extreme suggestibility, claim to see objects described by the examiner but not actually present. In some patients delirium persists for up to a few weeks after autonomic hyperactivity has cleared (Hersh et al., 1997).

A clear distinction between early withdrawal (tremor, hallucinosis, or seizures, with a clear sensorium) and later DTs is not always apparent, and some clinicians view symptoms along a continuum of severity, determined by the amount of recent drinking, the total duration of drinking, and, consistent with a kindling model, the number of previous withdrawal episodes.

Mortality for DTs is reportedly as high as 15%, usually attributable to associated disease such as pneumonia or sepsis. Some patients, however, develop unexplained shock or fatal cardiac arrhythmia. An increase in the QT interval is a frequent occurrence during alcohol withdrawal, and Takotsubo cardiomyopathy is described (Hughes, 2009; Omar et al., 2012).

COMORBID DISORDERS

Symptoms of alcohol withdrawal may appear in ambulatory drinkers or in patients hospitalized for other medical or surgical reasons (especially trauma). In either setting, management begins with consideration of comorbid conditions, which include pancreatitis, alcoholic liver disease, gastrointestinal hemorrhage, rhabdomyolysis, co-intoxication, alcoholic ketoacidosis, cardiovascular disease, including arrhythmia, and infection, especially pneumonia and sepsis. Admission studies for patients with moderate to severe alcohol withdrawal should include complete blood count; serum electrolytes, including magnesium; serum glucose, calcium, phosphate, lipase, and creatine kinase; renal and liver function tests; blood alcohol and blood and urine toxicology studies; chest radiograph; electrocardiogram; urinalysis; arterial gas analysis; and blood, urine, and sputum cultures (Carlson et al., 2012).

Table 8.2

Treatment of ethanol withdrawal

Prevention or reduction of early mild symptoms
Chlordiazepoxide 25–100 mg or diazepam 5–20 mg or lorazepam 2–6 mg orally every hour or two until patient is mildly intoxicated, then carefully titrated doses, tapering over 3–6 days
For non-benzodiazepine alternatives, see text
Thiamine and multivitamins
Seizures
Single seizure: lorazepam 2–4 mg intravenously; status epilepticus: treat conventionally with intravenous benzodiazepine, other anticonvulsants as needed
Delirium tremens
Diazepam 10 mg intravenously or lorazepam 4 mg intravenously or intramuscularly, repeated every 5–15 minutes until calming or normalization of vital signs. Maintenance doses every 1–4 hours as needed
If refractory to benzodiazepines, endotracheal intubation and a barbiturate or propofol in doses producing general anesthesia
Careful attention to fluid and electrolyte balance; several liters of saline per day, or even pressors, may be needed
Cooling for high fever
Prevent or correct hypoglycemia
Thiamine and multivitamins
Consider coexisting illness, e.g., liver failure, pancreatitis, sepsis, meningitis, or subdural hematoma

TREATMENT

Non-pharmacologic therapy

Some clinicians recommend that mildly symptomatic patients be managed non-pharmacologically with reassurance, reduced sensory stimuli, rest, hydration, and nutrition (Whitfield et al., 1978; Gorelick and Wilkins, 1986; Craig et al., 2011). Such an approach would avoid unnecessary medication as well as allow negatively reinforcing symptoms, which reduce the likelihood of relapse. The sedative and respiratory depressant effects of benzodiazepines are particularly problematic in patients with coexisting illness such as liver failure or chronic obstructive pulmonary disease. In an ambulatory setting, moreover, such agents have addictive liability.

Pharmacotherapy

Benzodiazepines

Dozens of drugs have been recommended for treating symptoms of alcohol withdrawal. Comparisons are difficult; studies often lack controls or they address different treatment goals, e.g., relief of tremor, prevention of seizures, prevention of DTs, or management of DTs (Amato et al., 2010). In most North American centers, benzodiazepines have been the agents of choice for several decades (Table 8.2). Clomethiazole, a sedative widely used in Europe, is unavailable in the United States (Bonnet et al., 2011). Randomized controlled trials have demonstrated superiority of benzodiazepines to placebo. Although data are insufficient to demonstrate superiority to other pharmacotherapies, no other agent has demonstrated superiority to benzodiazepine (Holbrook et al., 1999; Amato et al., 2011). Benzodiazepines are cross-tolerant with alcohol and can be used orally or parenterally.

Opinions differ as to which benzodiazepine is preferable (Bird and Makela, 1994). Benzodiazepines are metabolized in the liver, but whereas diazepam is oxidized to pharmacologically active metabolites which prolong its effective half-life to 30–60 hours, lorazepam undergoes glucuronidation with inactive metabolites and a half-life of 10–20 hours (Carlson et al., 2012). Diazepam can therefore be given infrequently once a patient is stabilized. The shorter half-life of lorazepam allows a more rapid steady state without extensive accumulation. Midazolam, with a half-life of 2 hours, has been used in critically ill patients. Older patients and heavy smokers may require unusually high doses of benzodiazepines.

Symptom-triggered pharmacotherapy has largely supplanted fixed-dose schedules in both outpatient and inpatient settings (Jaeger et al., 2001; Cassidy et al., 2012). Symptom-triggered therapy protocols rely on a number of standardized instruments (Williams et al., 2001). For example, the Clinical Institute for Withdrawal Assessment for Alcohol (CIWA-Ar) scale assesses at regular intervals vital signs, nausea/vomiting, tremor, paroxysmal sweating, anxiety, agitation, tactile, auditory, or visual disturbances, headache, and orientation/clouding of consciousness. (In using CIWA-Ar, one must keep in mind that the symptoms being measured are non-specific (Hecksel et al., 2008).) Symptom-triggered therapy is superior to fixed-dose schedules in reducing the likelihood of progression to DTs and is less likely to result in sedative accumulation.

For either ambulatory or hospitalized patients, tremor, anxiety, and other mild symptoms can be managed with a long-acting benzodiazepine. In a loading-dose regimen a bolus (e.g., diazepam 5–20 mg or chlordiazepoxide 100 mg) is given orally every hour or two until the patient is mildly intoxicated (calming, dysarthria, ataxia, fine nystagmus); subsequent administration is based on the reappearance of withdrawal symptoms. Because long-acting benzodiazepines accumulate, doses must be carefully titrated to avoid oversedation (including precipitation of hepatic encephalopathy). More severe symptoms may call for parenteral administration of diazepam or lorazepam or a shorter-acting benzodiazepine repeated every few minutes until calming and then titrated to symptoms.

Phenobarbital

Phenobarbital is often used in conjunction with a benzodiazepine to treat alcohol withdrawal. Unlike benzodiazepines, which are GABAergic by increasing the frequency of chloride channel opening, phenobarbital increases the duration of chloride channel opening. Phenobarbital also inhibits glutamate receptors. These different mechanisms of action might be the basis of reports that add-on phenobarbital reduces dosage requirements for benzodiazepine and need for intensive care unit admission or mechanical ventilation (Gold et al., 2007; Rosenson et al., 2012). A prospective trial of phenobarbital vs lorazepam for acute alcohol withdrawal found equal efficacy (Hendey et al., 2011).

Anticonvulsants

In addition to treating immediate symptoms and preventing complications, a goal of treatment for alcohol withdrawal is initiation of long-term preventive therapy. Although cross-tolerance with alcohol provides biologic rationale to benzodiazepine therapy, such pharmacologic interaction might confer risk for benzodiazepine withdrawal symptoms (which are similar to those of alcohol withdrawal) and resumption of drinking. Moreover, benzodiazepines have their own abuse potential, and they cause sedation and cognitive impairment. Such considerations have led to the use of alternative pharmacotherapies, most notably anticonvulsant drugs with GABAergic actions (Ait-Dauod et al., 2006; Leggio et al., 2008).

Anticonvulsants recommended on the basis of randomized double-blind studies comparing their efficacy to either a benzodiazepine or placebo include carbamazepine (Malcolm et al., 2002), gabapentin (Myrick et al., 2009), and valproate (Reoux et al., 2001). Anticonvulsants recommended on the basis of anecdotal and open-label reports include topiramate (Choi et al., 2005), lamotrigine (Djokic et al., 2011), vigabatrin (Ait-Dauod et al., 2006), zonisamide (Rubio et al., 2010), and oxcarbazepine (Leggio et al., 2008). Randomized placebo-controlled trials failed to show evidence in favor of either levetiracetam (Richter et al., 2010) or pregabalin (Forg et al., 2012) compared with placebo.

A Cochrane review of 56 randomized controlled trials comparing anticonvulsants to either a placebo or other pharmacologic treatment concluded that evidence was insufficient to favor anticonvulsants (Minozzi et al., 2011). There were "suggestions that carbamazepine may actually be more effective in treating some aspects of alcohol withdrawal when compared to benzodiazepines." Although anticonvulsants "seem to have limited side-effects. . .adverse effects are not rigorously reported in the analyzed trials."

Baclofen

In rats the GABA-B receptor agonist baclofen suppressed symptoms of alcohol withdrawal, and, conversely, a GABA-B receptor antagonist triggered alcohol withdrawal seizures (Columbo et al., 2004). Anecdotal reports describe baclofen suppression of alcohol withdrawal symptoms in alcoholic patients (Addolarato et al., 2002), and a randomized, double-blind trial showed that, compared to placebo, baclofen reduced the need for high doses of benzodiazepines in withdrawing alcoholics (Lyon et al., 2011). A Cochrane review, however, found no other controlled studies and concluded, "The evidence of recommending baclofen for alcohol withdrawal syndrome is insufficient" (Liu and Wang, 2011).

Gamma-hydroxybutyric acid

Gamma-hydroxybutyric acid (GHB) is a direct agonist at GHB and GABA-B receptors and indirectly, by being metabolized to GABA, at GABA-A receptors (Caputo et al., 2009). A number of randomized controlled trials compared GHB either to other drugs or to placebo in treating alcohol withdrawal symptoms (Elsing et al., 2009). A Cochrane review of six studies concluded, "GHB is effective compared to placebo in the treatment of alcohol withdrawal syndrome. . .but the results do not provide sufficient evidence in favor of GHB compared to benzodiazepines and chlormethiazole" (Leone et al., 2010).

Neuroleptics

Although effective in some animal models (Uzbay, 2012), neuroleptics are not cross-tolerant with alcohol, and they lower seizure threshold, prolong the QT interval, and cause hypotension, liver damage, acute dystonia, skin rash, bone marrow suppression, and impaired thermoregulation. In several studies, neuroleptic treatment was associated with an increased frequency of seizures, DTs, and death (Kaim and Klett, 1972). Neuroleptics might be considered in patients whose only symptoms are hallucinations accompanied by delusions or in whom hallucinations have outlasted other withdrawal symptoms.

Beta-blockers and alpha-2-agonists

Ethanol withdrawal symptoms are associated with elevated catecholamine levels (Linnoila et al., 1987); symptoms of sympathetic overdrive include hypertension, tachycardia, sweating, and tremor. Beta-adrenergic blockers such as propranolol or atenolol and alpha-2 agonists such as clonidine or dexmedetomidine have been used to treat such symptoms. In one study, however, propranolol exacerbated hallucinations (Jacob et al., 1983), and in another study, patients receiving clonidine had higher frequencies of hallucinations, seizures, and orthostatic hypotension (Robinson et al., 1989). Beta-blockers and alpha-2 agonists are not cross-tolerant with alcohol. They are unlikely to control seizures or progression to DTs, and they should be considered adjunctive treatment to benzodiazepines (Muzyk et al., 2011; Carlson et al., 2012).

Ethanol

Ethyl alcohol, the treatment non-hospitalized alcoholics choose for themselves, has a low therapeutic index as well as direct organ toxicity (including the brain). Even though most patients resume drinking on discharge, oral or parenteral ethyl alcohol is an inappropriate agent for treating withdrawal.

N-methyl-d-aspartate receptor blockers

Increased glutamatergic neurotransmission during withdrawal led to promising studies in animals using *N*-methyl-d-aspartate (NMDA) receptor blockers (especially NR2B subunit-selective) (Kotlinska and Bochenski, 2008). Unlike GABAergic drugs such as benzodiazepines, glutamatergic blockers would theoretically prevent excitotoxicity and kindled withdrawal symptoms. Promising results have been anecdotally described in patients with mild withdrawal symptoms who received lamotrigine, topiramate, or memantine (Krupitsky et al., 2007; Carlson et al., 2012).

Other pharmacotherapies

Other treatment approaches include nitric oxide synthase inhibitors (Adams and Cicero, 1998), nitrous oxide (Lichtigfeld and Gillman, 1996), and oxytocin (Pedersen et al., 2012).

TREATMENT OF SEVERE SYMPTOMS

Patients with severe symptoms – seizures or DTs – are appropriately managed in an intensive care unit, as are those with critical comorbid disorders, persistent fever, or need for escalating sedation therapy (Carlson et al., 2012).

Seizures

In animals and humans, valproate and carbamazepine prevented alcohol-related seizures, but phenytoin did not (Rathlev et al., 1994). Because alcohol-related seizures usually occur singly or in a cluster over a few hours, the risk of recurrence has often passed by the time treatment decisions are made. In a randomized double-blind study, intravenous lorazepam (2 mg) given in an emergency department after a generalized seizure significantly reduced the risk of recurrence over the next 6 hours (D'Onofrio et al., 1999).

Treatment of status epilepticus during alcohol withdrawal is conventional. Intravenous benzodiazepine and phenobarbital have the advantage, compared to phenytoin, of preventing or reducing other withdrawal symptoms.

Long-term anticonvulsants are generally not indicated in patients with alcohol-related seizures. Abstainers do not need them, and drinkers do not take them. Patients with epilepsy whose seizures are aggravated by alcohol but also occur independently of drinking usually do require anticonvulsant treatment even though compliance is unlikely.

Delirium tremens

Unlike early ethanol withdrawal, DTs, once present, is not readily reversible by any agent, including sedatives cross-tolerant with alcohol. Treatment of DTs is a medical emergency requiring sedation in an intensive care unit (Brust, 2004; Carlson et al., 2012). Patients vary in the amount of benzodiazepine necessary for effective sedation; some require huge doses, for example, more than 2500 mg diazepam in the first 48 hours (McMicken, 1990). In some patients benzodiazepines are ineffective, possibly related to benzodiazepine receptor–chloride channel "uncoupling" at the GABA receptor complex. Continued symptoms and abnormal vital signs after 3 hours of high-dose benzodiazepine administration might identify such patients and call for additional therapy with phenobarbital or propofol (which, like phenobarbital, not only facilitates GABA-A activity but also inhibits glutamate NMDA activity) (Hack et al., 2006; Carlson et al., 2012).

OTHER MANAGEMENT CONSIDERATIONS

Brains of patients with liver failure are hypersensitive to any sedative, and coma persisting as sedatives are tapered may be hepatic in origin.

Most patients with DTs are dehydrated, some severely so, and many require up to 10 liters of intravenous saline daily. Patients with liver disease, however, retain salt and water. Hyponatremia must be treated cautiously to avoid central pontine myelinolysis. Hypokalemia can cause cardiac arrhythmia. Fever, with or without infection, is often marked, requiring cooling. Coexisting disorders, including hypoglycemia, liver failure, pancreatitis, sepsis, meningitis, and intracranial hemorrhage, may go unrecognized.

Hypomagnesemia occurs early during alcohol withdrawal and returns to normal before the appearance of DTs. Probably both intracellular shift and body loss contribute to hypomagnesemia, which should be corrected (Flink, 1986). Hypokalemia and hypocalcemia also require correction, and hypocalcemia sometimes responds to treatment only after hypomagnesemia is corrected.

Glucose can precipitate Wernicke–Korsakoff syndrome, and because there is often impaired gastrointestinal absorption, thiamine and multivitamins are given parenterally.

SUMMARY

Symptoms of alcohol withdrawal range in severity from mild "hangover" to fatal DTs, and treatment depends on the particular syndrome encountered. It is perhaps ironic that alcohol withdrawal is potentially more dangerous than withdrawal from illicit drugs such as heroin, cannabis, or cocaine.

References

Adams ML, Cicero TJ (1998). Alcohol intoxication and withdrawal: the role of nitric oxide. Alcohol 16: 153–158.

Addolarato G, Caputo F, Capristo E et al. (2002). Rapid suppression of alcohol withdrawal syndrome by baclofen. Am J Med 112: 226–229.

Ait-Dauod N, Malcolm RA, Johnson BA (2006). An overview of medications for the treatment of alcohol withdrawal and alcohol dependence with an emphasis on the use of older and newer anticonvulsants. Addict Behav 31: 1628–1649.

Amato L, Minozzi S, Vecchi S et al. (2010). Benzodiazepines for alcohol withdrawal (review). The Cochrane Collaboration. http://www.thecochranelibrary.com.

Amato L, Minozzi S, Davoli M (2011). Efficacy and safety of pharmacological interventions for the treatment of the alcohol withdrawal syndrome. Cochrane Database Syst Rev (6): CD008537.

Bartolomei F (2006). Epilepsy and alcohol. Epileptic Disord 8: S72–S78.

Bijjal S, Subodh BN, Janardhanan C et al. (2012). Dystonia as a presenting feature of alcohol withdrawal. J Neuropsychiatry Clin Neurosci 24: E15–E16.

Bird RD, Makela EH (1994). Alcohol withdrawal: what is the benzodiazepine of choice? Ann Pharmacother 28: 67–71.

Bonnet U, Lensing M, Specka M et al. (2011). Comparison of two oral symptom-triggered pharmacological inpatient treatments of acute alcohol withdrawal: clomethiazole vs. clonazepam. Alcohol Alcohol 46: 68–73.

Brust JCM (2004). Neurological Aspects of Substance Abuse, 2nd edn, Elsevier, Philadelphia, pp. 317–425.

Caputo F, Vignoli T, Maremmani I et al. (2009). Gamma hydroxylbutyric acid (GHB) for the treatment of alcohol dependence: a review. Int J Environ Res Public Health 6: 1917–1929.

Carlson RW, Kumar NN, Wong-Mckinstry E et al. (2012). Alcohol withdrawal syndrome. Crit Care Clin 28: 549–585.

Cassidy EM, O'Sullivan I, Bradshaw P et al. (2012). Symptom-triggered benzodiazepine therapy for alcohol withdrawal syndrome in the emergency department: a comparison with the standard fixed dose benzodiazepine regimen. Emerg Med J 29: 802–804.

Chamorro AA, Sacco RL, Ciecierski K et al. (1990). Visual hemineglect and hemihallucinations in a patient with a subcortical infarction. Neurology 40: 1463–1464.

Choi EA, Ki SW, Kim SE et al. (2005). The efficacy and safety of topiramate in the treatment of alcohol withdrawal. J Korean Neuropsychiatric Assoc 44: 328–333.

Columbo G, Addolorato G, Agabio R et al. (2004). Role of GABA B receptor in alcohol dependence: reducing effect of baclofen on alcohol intake and alcohol motivational properties in rats and amelioration of alcohol withdrawal syndrome and alcohol craving in human subjects. Neurotox Res 6: 403–414.

Craig M, Pennacchia A, Wright NR et al. (2011). Evaluation of un-medicated, self-paced alcohol withdrawal. PLoS One 6: e22994.

D'Onofrio G, Rathlev NK, Ulrich AS et al. (1999). Lorazepam for the prevention of recurrent seizures related to alcohol. N Engl J Med 340: 915–919.

Djokic G, Lazic D, Nenadovic M et al. (2011). Lamotrigine augmentation in delirium tremens. Srp Arh Celok Lek 139 (Suppl. 1): 41–45.

Drake ME (1983). Recurrent spontaneous myoclonus in alcohol withdrawal. South Med J 76: 1040–1042.

Earnest MP, Yarnell PR (1976). Seizure admissions to a city hospital: the role of alcohol. Epilepsia 17: 387–393.

Earnest MP, Feldman H, Marx JA et al. (1988). Intracranial lesions shown by CT scans in 259 cases of first alcohol-related seizures. Neurology 38: 1561–1565.

Elsing C, Stremmel W, Grenda U et al. (2009). Gamma-hydroxybutyric acid versus clomethiazole for the treatment of alcohol withdrawal syndrome in a medical intensive care unit: an open, single-center randomized study. Am J Drug Alcohol Abuse 35: 189–192.

Fisch BJ, Hauser WA, Brust JCM et al. (1989). The EEG response to diffuse and patterned photic stimulation during acute untreated alcohol withdrawal. Neurology 39: 434–436.

Flink EB (1986). Magnesium deficiency in alcoholism. Alcohol Clin Exp Res 10: 590–594.

Forg A, Hein J, Volkmar K et al. (2012). Efficacy and safety of pregabalin in the treatment of alcohol withdrawal syndrome: a randomized placebo-controlled trial. Alcohol Alcohol 47: 149–155.

Fornazzari L, Carlen PL (1982). Transient choreiform dyskinesias during alcohol withdrawal. Can J Neurol Sci 9: 89–90.

Geoffroy PA, Rolland B, Cottencin O (2012). Catatonia and alcohol withdrawal: a complex and undetermined syndrome. Alcohol Alcohol 47: 288–290.

Gold JA, Rimal B, Nolan A et al. (2007). A strategy of escalating doses of benzodiazepines and phenobarbital administration reduces the need for mechanical ventilation in delirium tremens. Crit Care Med 35: 724–730.

Gorelick DA, Wilkins JN (1986). Special aspects of human alcohol withdrawal. Recent Dev Alcohol 4: 283.

Hack JB, Hoffman RS, Nelson LS (2006). Resistant alcohol withdrawal: does an unexpectedly large sedative requirement identify these patients early? J Med Toxicol 2: 55–60.

Hauser WE, Ng SKC, Brust JCM (1988). Alcohol, seizures, and epilepsy. Epilepsia 29 (Suppl. 2): S66–S78.

Hecksel KA, Bostwick JM, Jaeger TM et al. (2008). Inappropriate use of symptom-triggered therapy for alcohol withdrawal in the general hospital. Mayo Clin Proc 83: 274–279.

Hendey GW, Dery RA, Barnes RL et al. (2011). A prospective, randomized, trial of phenobarbital versus benzodiazepines for acute alcohol withdrawal. Am J Emerg Med 29: 382–385.

Hersh D, Kranzler HR, Meyer RE (1997). Persistent delirium following cessation of heavy alcohol consumption: diagnostic and treatment implications. Am J Psychiatry 154: 846–851.

Hillbom M, Pieninkeroinen I, Leone M (2003). Seizures in alcohol-dependent patients. Epidemiology, pathophysiology and management. CNS Drugs 17: 1013–1030.

Holbrook AM, Crowther R, Lotter A et al. (1999). Meta-analysis of benzodiazepine use in the treatment of acute alcohol withdrawal. CMAJ 160: 649–655.

Hoppener RJ, Kuyer A, van der Lugt PJM (1983). Epilepsy and alcohol: the influence of social alcohol intake on seizures and treatment in epilepsy. Epilepsia 24: 459–471.

Hughes JR (2009). Alcohol withdrawal seizures. Epilepsy Behav 15: 92–97.

Isbell H, Fraser HF, Wikler A (1955). An experimental study of the etiology of "rum fits" and delirium tremens. Q J Stud Alcohol 16: 1–33.

Jacob MS, Zilm DH, Macleod SM et al. (1983). Propranolol-associated confused states during alcohol withdrawal. J Clin Psychopharmacol 3: 185–187.

Jaeger TM, Lohr RH, Pankratz VS (2001). Symptom-triggered therapy for alcohol withdrawal syndrome in medical inpatients. Mayo Clin Proc 76: 695–701.

Kaim SC, Klett CJ (1972). Treatment of delirium tremens: a comparison of four drugs. Q J Stud Alcohol 33: 1065–1072.

Kotlinska J, Bochenski M (2008). The influence of various glutamate receptors antagonists on anxiety-like effect of ethanol withdrawal in a plus-maze test in rats. Eur J Pharmacol 598: 57–63.

Krupitsky EM, Rudenko AA, Burakov AM et al. (2007). Antiglutamatergic strategies for ethanol detoxification: comparison with placebo and diazepam. Alcohol Clin Exp Res 31: 604–611.

Leggio L, Kenna GA, Swift RM (2008). New developments for pharmacological treatment of alcohol withdrawal syndrome. A focus on non-benzodiazepine GABAergic medications. Prog Neuropsychopharmacol Biol Psychiatry 32: 1106–1117.

Leone M, Bottacchi E, Morgando E et al. (1997). Alcohol use is a risk factor for a first generalized tonic-clonic seizure. Neurology 48: 614–620.

Leone MA, Vigna-Taglianti F, Avanzi G et al. (2010). Gamma-hydroxybutyrate (GHB) for treatment of alcohol withdrawal and prevention of relapses. Cochrane Database Syst Rev 17 (2): CD006266.

Lichtigfeld FJ, Gillman MA (1996). Role of dopamine mesolimbic system in opioid action of psychotropic analgesic nitrous oxide (PAN) in alcohol and drug withdrawal. Clin Neuropharmacol 19: 246–251.

Linnoila M, Mefford I, Nutt D et al. (1987). Alcohol withdrawal and noradrenergic function. Ann Intern Med 107: 875–889.

Liu J, Wang L (2011). Baclofen for alcohol withdrawal. Cochrane Database Syst Rev 19 (1): CD008502.

Lowenstein DH, Alldredge BK (1993). Status epilepticus at an urban public hospital in the 1980s. Neurology 43: 483–488.

Lyon JE, Khan RA, Gessert CE et al. (2011). Treating alcohol withdrawal with oral baclofen: a randomized, double-blind, placebo-controlled trial. J Hosp Med 6: 469–474.

Malcolm R, Myrick H, Roberts J et al. (2002). The effects of carbamazepine and lorazepam on single versus multiple previous alcohol withdrawals in an outpatient randomized trial. J Gen Intern Med 17: 349–355.

McMicken DB (1990). Alcohol withdrawal syndromes. Emerg Med Clin North Am 8: 805–819.

Minozzi S, Amato L, Vecchi S et al. (2011). Anticonvulsants for alcohol withdrawal. Cochrane Database Syst Rev 17(3): CD005064.

Monte R, Rabunal R, Casariego E et al. (2010). Clinical aspects: analysis of the factors determining survival of alcohol withdrawal patients in a general hospital. Alcohol Alcohol 45: 151–158.

Muzyk AJ, Fowler JA, Norwood DK et al. (2011). Role of alpha-2-agonists in the treatment of acute alcohol withdrawal. Ann Pharmacother 45: 649–657.

Myrick H, Malcolm R, Randall PK et al. (2009). A double-blind trial of gabapentin versus lorazepam in the treatment of alcohol withdrawal. Alcohol Clin Exp Res 33: 1582–1588.

Neiman J, Lang AE, Fornazzari L et al. (1990). Movement disorders in alcoholism. A review. Neurology 40: 741–746.

Ng SKC, Hauser WA, Brust JCM et al. (1988). Alcohol consumption and withdrawal in new onset seizures. N Engl J Med 319: 666–673.

Omar HR, Abdelmalak HD, Komorova I et al. (2012). Alcohol withdrawal-induced Takotsubo. Intern Emerg Med 7 (Suppl. 2): S107–S108.

Pedersen CA, Smedley KL, Leserman J et al. (2012). Intranasal oxytocin blocks alcohol withdrawal in human subjects. Alcohol Clin Exp Res: 1–6.

Rathlev NK, D'Onofrio G, Fish SS et al. (1994). The lack of efficacy of phenytoin in the prevention of recurrent alcohol-related seizures. Ann Emerg Med 23: 513–518.

Reoux JP, Saxon AJ, Malte CA et al. (2001). Divalproex sodium in alcohol withdrawal: a randomized double-blind placebo-controlled clinical trial. Alcohol Clin Exp Res 25: 1324–1329.

Richter C, Hinzpeter A, Schmidt F et al. (2010). Levetiracetam for the treatment of alcohol withdrawal syndrome: a multicenter, prospective, randomized, placebo-controlled trial. J Clin Psychopharmacol 30: 720–725.

Robinson BJ, Robinson GM, Maling TJ et al. (1989). Is clonidine useful in the treatment of alcohol withdrawal? Alcohol Clin Exp Res 13: 95–98.

Rosenson J, Clements C, Simon B et al. (2012). Phenobarbital for acute alcohol withdrawal: a prospective randomized double-blind placebo-controlled study. J Emerg Med 07 (056): 1–7.

Rubio G, Lopez-Munoz F, Ponce G et al. (2010). Zonisamide versus diazepam in the treatment of alcohol withdrawal syndrome. Pharmacopsychiatry 43: 257–262.

Samokhvalov AV, Irving H, Mohapatra S et al. (2010). Alcohol consumption, unprovoked seizures, and epilepsy: a systematic review and meta-analysis. Epilepsia 51: 1177–1184.

Schuckit MA, Winokur G (1971). Alcoholic hallucinosis and schizophrenia. A negative study. Br J Psychiatry 119: 549–550.

Schwarz BE, Brust JCM (1984). Anton's syndrome accompanying withdrawal hallucinosis in a blind alcoholic. Neurology 34: 969–973.

Shandling M, Carlen PL, Lang AE (1990). Parkinsonism in alcohol withdrawal: a follow-up study. Mov Disord 5: 36–39.

Stephens DN, Duka T (2008). Cognitive and emotional consequences of binge drinking: role of amygdala and prefrontal cortex. Philos Trans R Soc Lond B Biol Sci 363: 3169–3179.

Uzbay TI (2012). Atypical antipsychotic drugs and ethanol withdrawal syndrome: a review. Alcohol Alcohol 47: 33–41.

Victor M, Adams RD (1953). The effect of alcohol on the nervous system. Res Publ Assoc Res Nerv Ment Dis 32: 526–573.

Victor M, Brausch CC (1967). The role of abstinence in the genesis of alcoholic epilepsy. Epilepsia 8: 1–20.

Victor M, Hope JM (1958). The phenomenon of auditory hallucinations in chronic alcoholism. J Nerv Ment Dis 126: 451–481.

Whitfield CL, Thompson G, Lamb A et al. (1978). Detoxification of 1024 alcoholic patients without psychoactive drugs. JAMA 293: 1409–1410.

Wiese JG, Shlipak MG, Browner WS (2000). The alcohol hangover. Ann Intern Med 132: 897–902.

Williams D, Lewis J, McBride A (2001). A comparison of rating scales for the alcohol-withdrawal syndrome. Alcohol Alcohol 36: 104–108.

Handbook of Clinical Neurology, Vol. 125 (3rd series)
Alcohol and the Nervous System
E.V. Sullivan and A. Pfefferbaum, Editors

Chapter 9

Neurochemical mechanisms of alcohol withdrawal

HOWARD C. BECKER[1,2,3]* AND PATRICK J. MULHOLLAND[1,2]

[1]*Charleston Alcohol Research Center, Department of Psychiatry and Behavioral Sciences, Charleston, SC, USA*

[2]*Department of Neurosciences, Medical University of South Carolina, Charleston, SC, USA*

[3]*Department of Veterans Affairs RHJ Medical Center, Charleston, SC, USA*

INTRODUCTION

Alcohol abuse and alcoholism are significant public health concerns that continue to exact enormous burdens on the healthcare industry as well as producing a broad range of societal problems, including judicial (e.g., crime), economic (e.g., damage/loss of property, reduced productivity in the workplace) and, most tragically, collateral damage to personal/family relationships. While many individuals abuse alcohol without being dependent on the drug, continued excessive alcohol consumption can lead to the development of dependence.

A serious consequence of dependence is the emergence of symptoms associated with the alcohol withdrawal syndrome. When drinking is abruptly terminated or substantially reduced in the dependent individual, a characteristic withdrawal syndrome ensues. Clinical features of alcohol withdrawal include signs of central nervous system (CNS) hyperexcitability, heightened autonomic nervous system activation and, in its most severe form, hallucinosis and delirium tremens (Saitz, 1998; Becker, 2000). In addition to physical signs of withdrawal, a constellation of symptoms contributing to psychologic discomfort and negative affect constitutes a prominent component of the withdrawal syndrome (Becker, 2008; Koob, 2013). While many of the signs and symptoms of withdrawal abate within 5–7 days, some symptoms (primarily those related to mood and emotional disturbances) have been reported to linger on for a protracted period of time (Heilig et al., 2010). Moreover, it has been suggested that the negative affective or emotional components of withdrawal, while more subtle in nature, may constitute significant motivational factors that lead to resumption of alcohol-seeking behavior (relapse) (Becker, 2012, 2013; Koob, 2013). The high rate of recidivism in alcoholism underscores the relapsing nature of the disease. Thus, it is not uncommon for alcohol-dependent individuals to experience multiple episodes of withdrawal, with repeated attempts at abstinence tragically failing and individuals reverting back to excessive, unhealthy drinking habits.

The development of alcohol dependence is a complex and dynamic process that ultimately reflects a maladaptive neurophysiologic state. Perturbations in a wide range of neurochemical systems produced by the chronic presence of alcohol contribute to significant changes in neural activity (neuroadaptations) that ultimately compromise the functional integrity of the brain. Indeed, the development of alcohol dependence is thought to reflect an allostatic state fueled by progressive dysregulation of neurophysiologic systems beyond normal homeostatic limits (Koob and Le Moal, 2001). Many of these neuroadaptive changes associated with dependence are integral to brain reward and stress systems (e.g., Hansson et al., 2008; Koob and Le Moal, 2008). Indeed, manifestations of this allostatic state not only underlie the emergence and expression of many alcohol withdrawal symptoms, but also contribute to persistent vulnerability to relapse. Additionally, the potential for alcohol to alleviate discomfort associated with withdrawal-related stress/dysphoria may serve as a powerful motivational force that not only enhances relapse vulnerability, but also favors escalation of alcohol drinking to even higher levels. This chapter highlights the hallmark features of the alcohol withdrawal syndrome, and describes neuroadaptations in a wide array of neurotransmitter and neuromodulator systems as they relate

*Correspondence to: Howard C. Becker, Ph.D., Charleston Alcohol Research Center, Departments of Psychiatry and Neurosciences, Medical University of South Carolina, 67 President Street, MSC-861, Charleston, SC 29425, USA. Tel: +1-843-792-5207, E-mail: beckerh@musc.edu

to the expression of various signs and symptoms of alcohol withdrawal, as well as their relationship to the significant clinical problem of relapse and uncontrolled dangerous drinking.

SIGNS AND SYMPTOMS OF THE ALCOHOL WITHDRAWAL SYNDROME

CNS hyperexcitability

A hallmark feature of the alcohol withdrawal syndrome is general CNS hyperexcitability. This is thought to reflect compensatory neural activity, induced by the depressant effects of alcohol, that is unmasked when the drug is withdrawn from the brain (Finn and Crabbe, 1997; Littleton, 1998; Becker, 2000). Both electrographic and behavioral measures of withdrawal-related seizure activity have been extensively documented in animals and humans (Victor, 1970; Deitrich et al., 1996; Porjesz and Begleiter, 1996; Becker, 2000). Electrographic measures include increased frequency of spontaneous as well as evoked perturbations in electroencephalogram (EEG) activity that include spike and sharp-wave epileptiform activity and more global synchronized high-voltage spindling activity. In animal studies, motor convulsions may occur spontaneously, but can be more readily elicited by exposure to sensory stimuli (e.g., audiogenic), handling manipulation, electroconvulsive stimulation, and various chemoconvulsant agents. Both electrographic and behavioral measures of withdrawal-related seizure activity in animals are highly sensitive to clinically effective anticonvulsants (Crabbe, 1992; Becker, 1996; Watson et al., 1997; Becker and Veatch, 2002; Veatch and Becker, 2005). In humans, sophisticated quantitative frequency analysis of EEG (power spectral analysis) has revealed more subtle and long-lasting functional CNS alterations resulting from chronic alcohol exposure and withdrawal (Sand et al., 2010). Detecting subtle (subclinical) indices of seizure-like activity may be particularly important in light of evidence indicating that repeated withdrawal experience (even relatively mild such episodes) may result in sensitization or a "kindling" effect of withdrawal (Ballenger and Post, 1978; Becker, 1996, 1998).

Kindling of alcohol withdrawal is thought to reflect a process whereby seizure activity (and other withdrawal symptoms) progressively worsens with repeated detoxification experiences. It has been postulated that such a process may underlie the commonly observed progression of withdrawal symptoms from relatively mild responses (e.g., irritability, tremors) characteristic of initial withdrawal episodes to more severe symptoms (e.g., seizures) associated with subsequent withdrawal episodes (Ballenger and Post, 1978). Indeed, a number of clinical studies involving retrospective analyses have reported a positive relationship between the likelihood of seizures occurring during a given withdrawal episode and a history of previous detoxifications (Brown et al., 1988; Lechtenberg and Worner, 1991; Moak and Anton, 1996). An increased risk of seizures in alcoholics with a history of multiple detoxifications is of clinical significance since poorer prognosis and a higher mortality rate have been reported for patients presenting with alcohol withdrawal-related seizures in contrast to individuals with seizures of unknown etiology (Pieninkeroinen et al., 1992). Further, a history of multiple previous detoxifications was reported to be associated with more severe and medically complicated withdrawal syndromes, as well as an increased likelihood of hospital readmission for alcohol-related problems (Booth and Blow, 1993). Preclinical studies have provided critical support and validation of these findings, demonstrating progressive intensification of withdrawal symptoms over repeated withdrawal experiences (Becker, 1998, 1999). From a treatment perspective, detecting withdrawal-related CNS perturbations may be important, as early intervention may be significant in quelling a kindling-like process and providing preventive care for alcoholics (Malcolm et al., 2000, 2002; Veatch and Becker, 2005).

Enhanced sensory reactivity is also thought to reflect heightened CNS excitability during alcohol withdrawal. In humans, this is typically measured by assessing a startle response (eye blink) to an auditory stimulus. In contrast to the dampening effect of alcohol on acoustic startle amplitude in healthy controls (Grillon et al., 1994), exaggerated startle responses have been reported in detoxified alcohol-dependent subjects (Krystal et al., 1997; Schellekens et al., 2012). In several animal studies the startle response has been measured in response to auditory or tactile (air puff) stimuli following withdrawal from chronic alcohol exposure, but these studies have generally yielded mixed results, i.e., startle reactivity has been reported to be increased, decreased, or unchanged during alcohol withdrawal (Rassnick et al., 1992; Macey et al., 1996; Ponomarev and Crabbe, 1999; Vandergriff et al., 2000; Chester et al., 2005; Chester and Barrenha, 2007; Reilly et al., 2009). Deficits in prepulse inhibition of the startle response have been suggested to be related to increased risk for alcoholism (Grillon et al., 2000), and a study in rats indicated that this phenotype may be predictive of more severe withdrawal symptoms following subsequent chronic alcohol exposure (Kayir et al., 2010).

Autonomic nervous system hyperactivity

An array of physiologic symptoms commonly experienced during withdrawal reflect manifestations of heightened autonomic nervous system activity, more specifically, increased sympathetic activity (e.g., Hawley et al., 1994; King et al., 1994; Patkar et al., 2003; Kahkonen, 2004; Rasmussen et al., 2006). These include tachycardia, increased blood pressure, diaphoresis (heavy sweating), body temperature dysregulation, and gastrointestinal disturbances (nausea, vomiting). Animal models have demonstrated many of these classic sympathomimetic withdrawal symptoms, including altered cardiovascular function (Kashkin et al., 2008; Shirafuji et al., 2010), central and behavioral thermal dysregulation (Crawshaw et al., 1994), and gastrointestinal-related symptoms, such as diarrhea and reduced food and water intake (Friedman, 1980; Kliethermes, 2005). Most of these symptoms resolve within the acute phase of withdrawal. Given evidence indicating elevated sympathetic activity associated with alcohol withdrawal, it is not surprising that drugs that reduce noradrenergic tone, such as adrenergic beta-blockers and alpha-2 agonists, have been shown to be useful as adjuncts for treatment of alcohol withdrawal symptoms related to sympathetic overdrive (Mayo-Smith, 1997; Riihioja et al., 1997; Muzyk et al., 2011).

Tremor is another frequent symptom of alcohol withdrawal, and it is thought to emerge as a manifestation of sympathetic hyperactivity (Koller et al., 1985; Charles et al., 1999). Animal studies have used subjective rating scores (Frye et al., 1983; Bone et al., 1989) as well as more quantitative measures (Macey et al., 1996) to demonstrate increased tremor during withdrawal. Interestingly, while mostly anecdotal, human alcoholics often report resumption of drinking linked to a desire to self-medicate the "shakes" (tremor) during early abstinence. In this vein, it is interesting that alcohol withdrawal has been reported to potentiate the tremorogenic effects of nicotine (Gothoni and Ikola, 1985). Thus, the high prevalence/comorbidity of alcohol and nicotine dependence may relate, at least in part, to alcohol's ability to moderate tremor resulting from alcohol withdrawal as well as nicotine use.

Sleep disturbances

A common complaint among alcoholics is disruption in sleep, which typically emerges during early periods of abstinence and often extends to the more protracted phase of abstinence (Brower et al., 2001; Landolt and Gillin, 2001; Cohn et al., 2003; Colrain et al., 2009; Brower and Perron, 2010). Insomnia during abstinence is characterized by fragmentation of sleep architecture that manifests as increased sleep latency, reduced total sleep, compromised sleep efficiency, and a transient increase (rebound) in rapid-eye movement sleep. Animal studies employing techniques to measure EEG in freely moving rats (Ehlers and Slawecki, 2000; Kubota et al., 2002) and mice (Veatch, 2006; Wiggins et al., 2013) undergoing alcohol withdrawal have demonstrated significant and long-lasting alterations in sleep architecture that are similar to those observed in human alcoholics during abstinence. Chronic alcohol exposure and withdrawal also have been shown to alter circadian clock function (which is critical for regulation of the wake–sleep cycle) in rats (Rosenwasser et al., 2005; Sharma et al., 2010) and mice (Seggio et al., 2009; Brager et al., 2010; Logan et al., 2012). Likewise, disruptions in circadian clock function have been noted in human alcoholics during abstinence (Brower et al., 2001; Rosenwasser, 2001), and there is recent evidence for decreased expression of several circadian clock genes (Huang et al., 2010a) as well as a polymorphism in one of the genes (*Per3*) that is associated with insomnia severity in alcohol-dependent subjects (Brower et al., 2012). Clinical studies have noted the self-reported link between disrupted sleep during abstinence and increased risk of relapse (Clark et al., 1998; Drummond et al., 1998; Feige et al., 2007; Malcolm et al., 2007; Steinig et al., 2011; Smith et al., 2013). It is noteworthy that a number of clinical studies have recently begun to evaluate treatment strategies for addressing this aspect of the abstinence syndrome, especially in the context of relapse prevention (Le Bon et al., 2003; Staner et al., 2006; Malcolm et al., 2007; Brower et al., 2008).

Measures of psychologic discomfort and negative affect

Anxiety

Increased anxiety represents a significant component of the alcohol withdrawal syndrome. In humans, anxiety emerges during the early abstinence phase and in many cases lingers for an extended period of time. Of significance, the psychologic discomfort associated with anxiety experienced during abstinence, even after most acute physical symptoms have subsided, has been suggested to play a prominent role in increasing risk for relapse as well as perpetuating continued use/abuse of alcohol (Roelofs, 1985; Becker, 1999; Koob and Le Moal, 2001; Schneider et al., 2001). Indeed, both preclinical and clinical studies suggest a link between anxiety and propensity to self-administer alcohol (Spanagel et al., 1995; Willinger et al., 2002; Barrenha and Chester, 2007), although this relationship is by no means simple in animals or human alcoholics (Schuckit and Hesselbrock, 1994; Henniger et al., 2002; Correia et al., 2009; Heilig et al., 2010).

A number of experimental procedures have been used to demonstrate increased behavioral anxiety in animal models of alcohol dependence and withdrawal

(Becker, 2000; Kliethermes, 2005). Many of these models involve procedures that exploit the natural tendency for rodents to avoid environments that may be considered dangerous or threatening and, thereby, elicit an internal state of "fear" or "anxiety" (e.g., bright open spaces). For example, behavioral measures of elevated anxiety during withdrawal have been demonstrated in animals, as indexed by reduced activity in the central portions of an open field arena or in open quadrants of a maze (e.g., elevated plus/zero mazes), reduced entries (increased avoidance) of a brightly illuminated portion of a two-compartment chamber (light–dark box test), and reduced social interaction behavior in a novel environment (Becker, 2000; Kliethermes, 2005). While these models have been effectively used to document alcohol withdrawal-related behavioral measures of anxiety, results have sometimes been variable (especially in studies using mice) due to a number of confounds, most notably, non-specific reductions in general activity (Kliethermes, 2005). Nevertheless, anxiety and other symptoms reflective of a state of psychologic discomfort (e.g., irritability, agitation) are firmly established as clinically significant components of the alcohol withdrawal syndrome. In fact, many treatments commonly used for managing detoxification (e.g., benzodiazepines) not only serve to reduce life-threatening aspects of the syndrome (e.g., grand mal seizures), but also target symptoms such as anxiety that contribute to negative affect and increased relapse vulnerability.

Heightened stress responsiveness

Increased stress reactivity is another feature of alcohol withdrawal that has been studied in humans and animal models. In many instances, heightened stress responsiveness persists long after physical signs and even many overt psychologic symptoms of withdrawal have dissipated. Although complaints about irritability and increased sensitivity to everyday stressors have long been recognized clinically in alcoholic patients, corresponding human data are just beginning to emerge. For example, an increased stress response (elevated heart rate and cortisol levels) was reported in abstinent alcoholics following exposure to a social stressor (Trier Social Stress test), with the effect modified by the degree of drinking history and duration of abstinence prior to testing (Starcke et al., 2013). Recent neuroimaging (functional magnetic resonance imaging) studies have demonstrated exaggerated brain responses to visual stimuli that evoked negative emotions, with results suggesting altered functional neural connectivity indicative of aberrant cortical modulation of emotional processing (George et al., 2008; Gilman and Hommer, 2008; O'Daly et al., 2012). There is also evidence indicating that alcohol-dependent patients abstinent for 4–8 months exhibit altered brain activation in response to stress- or alcohol-relevant cues (as opposed to neutral cues), and this altered cortical activation was predictive of greater relapse susceptibility and severity (Seo et al., 2013). Collectively, these results add to a growing body of literature that indicates abstinent alcohol-dependent individuals display stress-induced emotional and physiologic dysregulation along with dysfunctional brain activity, which all contribute to increased alcohol craving and relapse vulnerability (Sinha, 2013).

Similarly, increased behavioral reactivity to stress following chronic alcohol exposure and withdrawal has also been demonstrated in animals. For example, enhanced response to stress- induced anxiety has been reported in alcohol-dependent rats using a number of testing protocols, with the effects often persisting over a long period of time (months) following withdrawal (Valdez et al., 2003; Breese et al., 2005; Rylkova et al., 2009; Huang et al., 2010b; Gillett et al., 2013). Of clinical significance, a history of chronic alcohol exposure and withdrawal has been shown to enhance the ability of stress to trigger relapse-like behavior in animal models (Le et al., 2000; Liu and Weiss, 2002a; Gehlert et al., 2007; Marinelli et al., 2007; Sommer et al., 2008). Clinical and experimental evidence indicates that increased sensitivity to stress during abstinence reflects, in large part, adaptations in neuroendocrine and brain stress systems induced by chronic alcohol exposure (see below). Taken together, enhanced stress reactivity in dependent subjects has both physiologic implications as well as cognitive/behavioral potential for influencing relapse vulnerability.

Anhedonia/dysphoria

Another common feature of the alcohol withdrawal syndrome relates to anhedonia, i.e., a reduced ability to derive pleasure from events/stimuli typically perceived as rewarding. Although anecdotal reports have long noted the significance of this dysphoric aspect of alcohol withdrawal, recent clinical studies have used a number of assessment instruments to better quantify the subjective anhedonic/dysphoric state that has been shown, in many cases, to endure for a protracted period of time during abstinence (Janiri et al., 2005; Martinotti et al., 2008; Pozzi et al., 2008; Hatzigiakoumis et al., 2011). Withdrawal-related anhedonia has typically been modeled in animal studies using an intracranial self-stimulation (ICSS) procedure. For example, during withdrawal from chronic alcohol exposure, rats have been shown to exhibit significant increases in ICSS threshold (i.e., the minimal amount of electrical stimulation delivered to reward pathways in the brain that is perceived as rewarding) (Schulteis et al., 1995; Chester et al., 2006; Rylkova et al., 2009). It has been suggested that a dopamine hypofunctional state may underlie anhedonia/dysphoria associated with protracted withdrawal (Heinz et al., 1995; Diana et al., 1996; Weiss et al., 1996; Bailey et al., 2001; Martinez et al., 2005), although other neuroadaptive changes (e.g., corticotropin-releasing factor (CRF), dynorphin/kappa opiate receptors) undoubtedly play a contributory role as well (Heilig and Koob, 2007; Shippenberg et al., 2007). Recently, some clinical studies have targeted withdrawal-related anhedonia in efforts to treat alcohol-dependent subjects (Martinotti et al., 2010, 2011).

NEUROCHEMICAL ADAPTATIONS PRODUCED BY CHRONIC ALCOHOL AND WITHDRAWAL

Prolonged excessive alcohol consumption sets in motion a host of neuroadaptive changes initially triggered to compensate for, and mitigate effects of, continued presence of alcohol in the brain. With continued drinking, adaptations in other systems not initially affected by alcohol emerge that contribute to a state of CNS disequilibrium and dysfunction. When abstinence is attempted and alcohol is eliminated from the CNS, manifestations of these adaptations are unveiled, as evidenced by the expression of myriad signs and symptoms that constitute the alcohol withdrawal syndrome (Fig. 9.1). As previously noted, many of these adaptations also have significant motivational implications with regard to increased relapse vulnerability and excessive drinking, both characteristic of dependence. This section will provide an overview of adaptations in a wide array of neurochemical and neuromodulatory systems associated with alcohol dependence, with particular emphasis on their relationship to various signs and symptoms of the alcohol withdrawal syndrome, as well as their postulated role in underlying enhanced relapse susceptibility and perpetuation of excessive, unhealthy alcohol consumption.

Adaptations in amino acid neurotransmitter systems

The amino acids glutamate and γ-aminobutyric acid (GABA) are the major excitatory and inhibitory neurotransmitters, respectively, in the CNS. While alcohol initially facilitates the inhibitory actions of GABA and inhibits excitatory effects mediated by glutamate transmission, chronic alcohol exposure results in compensatory changes in these amino acid transmitter systems that are opposite in nature and revealed upon withdrawal. Manifestations of this resultant imbalance in GABA-mediated inhibition and glutamate-mediated

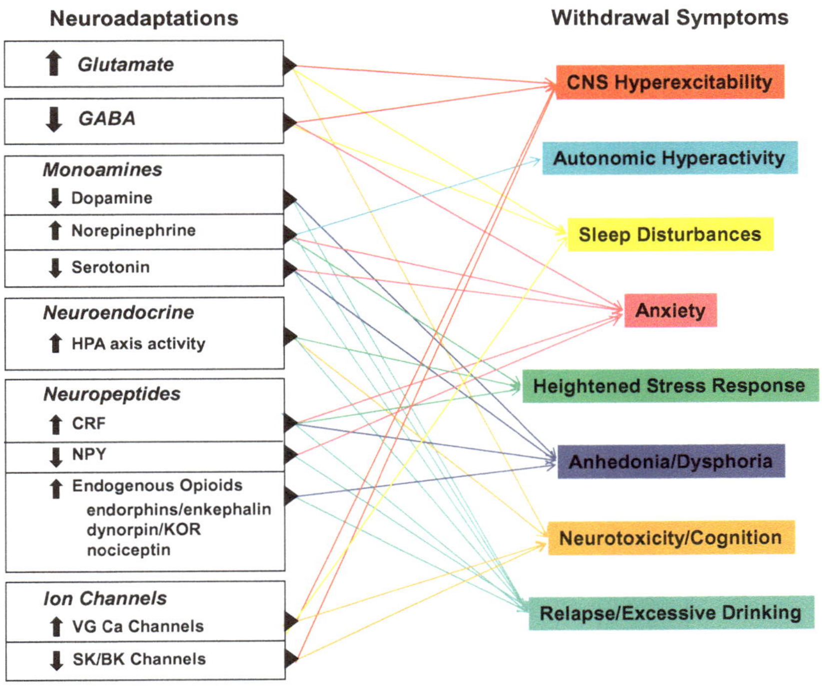

Fig. 9.1. Neurochemical adaptations resulting from chronic alcohol exposure and withdrawal and their relationship to various features of the alcohol withdrawal syndrome. GABA, γ-aminobutyric acid; HPA, hypothalamic–pituitary–adrenocortical; CRF, corticitropin-releasing factor; NPY, neuropeptide Y; CNS, central nervous system; KOR, kappa opiate receptor; VG, voltage-gated; Ca, calcium; SK, small conductance; BK, large conductance.

excitation in the CNS are known to underlie expression of various withdrawal symptoms, most notably CNS hyperexcitability (Becker, 1998; Littleton, 1998; Hillbom et al., 2003). Indeed, it is well established that neuroadaptations in glutamatergic and GABAergic signaling systems following chronic alcohol exposure play a prominent role in mediating a variety of dependence and withdrawal-related sequlae (Fadda and Rossetti, 1998; Lovinger and Roberto, 2013).

Glutamate

The α-amino-3-hydroxy-5-methyl-4-isoxazolepropionic acid (AMPA) and *N*-methyl-D-aspartate (NMDA) subtypes of glutamate receptors are among the most widely distributed and abundant receptors in the brain. AMPA receptors are heterotetramers comprising GluR1–4 subunits and mediate most fast synaptic neurotransmission. NMDA receptors are also heterotetramers and are composed of an obligatory GluN1 subunit co-assembled with a least one type of regulatory GluN2A–D subunit. The majority of NMDA receptors in the adult brain are composed of GluN2A and GluN2B subunits. NMDA receptors mediate the slow component of excitatory postsynaptic potentials. Glutamate also operates at two metabotropic receptor subtypes: mGluR1-like (mGlur1 and mGluR5) and mGluR2-like (mGluR2, mGluR3, and mGluR4). Both ionotropic (AMPA and NMDA) and metabotropic (mGluRs) receptors have been implicated in a wide array of alcohol-associated phenotypes, including those related to dependence and withdrawal.

Compensatory changes produced by chronic alcohol exposure produce a hyperglutamatergic state in the brain that underlies increased susceptibility to chronic alcohol exposure/withdrawal-induced neurotoxicity (Stepanyan et al., 2008; Prendergast and Mulholland, 2012) and altered synaptic plasticity (Chandler et al., 2006; McCool et al., 2010; Holmes et al., 2012; Zorumski et al., 2014). Also resulting is behavioral expression of several withdrawal symptoms, particularly those associated with CNS hyperexcitability (e.g., seizures) (Hillbom et al., 2003; Gass and Olive, 2008).

Studies using a variety of preparations have demonstrated thatincreased NMDA receptor-mediated excitatory transmission following chronic alcohol exposure are highly complex, involving changes in trafficking and phosphorylation of NMDA receptor subunits. For example, *in vitro* and *in vivo* studies have shown that chronic alcohol exposure results in compensatory trafficking of NMDA receptors containing GluN1 and GluN2B subunits to synaptic sites without alterations in expression of NMDA receptors in extrasynaptic domains (Mulholland and Chandler, 2007; Kroener et al., 2012). Further, evidence suggests that increased expression of GluN1 subunits leading to activity-dependent enhanced targeting of NMDA receptors to the synapse involves alternative splicing of the carboxyl-terminal (C2' cassette) of GluN1 subunits (Mu et al., 2003; Clapp et al., 2010). At the same time, there is evidence for internalization of GluN2A subunits via clathrin-dependent endocytosis (Suvarna et al., 2005), with the result being an increased proportion of NMDA receptors reflecting a GluN1/GluN2B conformation.

Chronic alcohol exposure also results in phosphorylation of the GluN2B subunits by the Src family tyrosine kinase Fyn (via dissociation from the scaffolding protein RACK1), resulting in increased NMDA receptor channel activity (Miyakawa et al., 1997; Yaka et al., 2003). Interestingly, downregulation of protein tyrosine phosphatase in the dorsomedial striatum prevented phosphorylation of GluN2B, reduced withdrawal-associated NMDA receptor activity, and excessive alcohol drinking in mice (Ben Hamida et al., 2013). These adaptive changes in NMDA expression and function are suggested to contribute to withdrawal-related CNS hyperexcitability and escalation of drinking and cognitive impairments associated with dependence (Wang et al., 2010; Kroener et al., 2012; Ben Hamida et al., 2013).

Similarly, chronic alcohol-induced enhancement of AMPA receptor expression and function has been reported in cortex (Haugbol et al., 2005), hippocampus (Bruckner et al., 1997), basolateral amygdala (Lack et al., 2007), and dorsomedial striatum (Wang et al., 2012). Selective pharmacologic blockade of AMPA receptors in the basolateral amygdala attenuated withdrawal-related anxiety-like behaviors in rats, suggesting a role for adaptations in AMPA receptors in expression of withdrawal symptoms (Lack et al., 2007).

Recent clinical studies have employed imaging technology involving magnetic resonance spectroscopy procedures to measure glutamate levels in brain. Some studies have reported reduced cortical glutamate levels associated with dependence and acute withdrawal (Mon et al., 2012; Ende et al., 2013), while other studies have shown elevated glutamate activity in anterior cingulate and striatum in alcohol-dependent subjects (Hermann et al., 2012; Bauer et al., 2013). Congruent with these latter findings, studies in rats utilizing a dependence model that involves chronic exposure to alcohol vapor have shown increased glutamate levels in prefrontal cortex (Hermann et al., 2012) and basal ganglia (Zahr et al., 2009; Gu et al., 2013). Likewise, mouse and rat studies employing microdialysis procedures have demonstrated chronic alcohol and acute withdrawal-induced elevation in extracellular glutamate levels in several brain regions, including hippocampus and dorsal and ventral subregions of the striatum

(Rossetti and Carboni, 1995; Dahchour et al., 2000; Dahchour and De Witte, 2003; Kapasova and Szumlinski, 2008). Repeated cycles of chronic intermittent exposure to alcohol vapor were shown to increase basal levels of glutamate in the nucleus accumbens (NAc) at time points beyond acute withdrawal. Pharmacologic studies demonstrated that increased glutamate tone in this brain region plays an important role in promoting excessive drinking associated with dependence (Griffin et al., 2014).

Finally, a history of alcohol dependence in rats also has been shown to alter functional activity and sensitivity of mGluR2/3 and mGluR5 receptors to drugs that modulate alcohol-seeking behavior and reinstatement of alcohol responding following presentation of cues previously associated with alcohol reinforcement (Sidhpura et al., 2010; Kufahl et al., 2011; Meinhardt et al., 2013). These results have clinical implications because a number of drugs known to alter central glutamate activity have been investigated as potential treatments for reducing alcohol craving and consumption in human alcoholics. Unfortunately, mixed results have been generated from some of these studies with acamprosate (Umhau et al., 2010; Mann et al., 2012; Mason and Lehert, 2012; Berger et al., 2013b; Yahn et al., 2013) and anticonvulsants such as topiramate, gabapentin, pregabalin, and levetiracetam (De Sousa, 2010; Johnson and Ait-Daoud, 2010; Guglielmo et al., 2012; Mason et al., 2014). Clearly, this is an area that warrants more exploration and examination.

GABA

GABA operates at two types of receptors: ionotropic $GABA_A$ receptors and metabotropic $GABA_B$ receptors. $GABA_A$ receptors are composed of at least 19 subunit variants (α1–6, β1–3, γ1–3, δ, ϵ, π, θ, and ρ1–3), and the assembly of these different subunits confers unique pharmacologic sensitivity and receptor function. In comparison, there are two subtypes of $GABA_B$ receptors that form functional heterodimers in brain. Chronic exposure to alcohol is well documented to induce neuroadaptive changes in pre- and postsynaptic GABAergic transmission and expression of receptor subunit transcript/peptide levels that are temporally, subunit, and brain region-dependent (Kumar et al., 2009; Lovinger and Roberto, 2013). For example, studies in hippocampal and cortical pyramidal neurons have shown that chronic alcohol exposure reduces the frequency, amplitude, and decay time of mini inhibitory postsynaptic currents, effects that all contribute to increased neural excitability during withdrawal (Fleming et al., 2009). Chronic alcohol also reduces α1 subunits and increases α4 subunits of synaptic $GABA_A$ receptors, and this bidirectional effect on α1 and α4 subunit expression and trafficking can be modulated by protein kinase C phosphorylation (Kumar et al., 2002). Evidence also suggests that the enhanced internalization of the $GABA_A$ α1 subunit by chronic alcohol exposure is regulated by adaptor complex-2 and clathrin-mediated endocytosis. Additionally, chronic alcohol exposure decreases extrasynaptic $GABA_A$-mediated tonic current recorded from neurons in the hippocampus and cortex (Liang et al., 2004; Fleming et al., 2011), and this corresponds with a decrease in extrasynaptic $GABA_A$ receptors containing the δ subunit in hippocampus (Cagetti et al., 2003). Collectively, these $GABA_A$ receptor adaptations play a prominent role in mediating expression of various withdrawal signs and symptoms, including general CNS hyperexcitability and affective components, including anxiety.

Positive allosteric modulators of $GABA_A$ receptors (i.e., benzodiazepines) have been the gold standard for treatment of alcohol detoxification in the United States, owing to their anticonvulsant and anxiolytic pharmacologic profile. However, concerns about their abuse and dependence liability, amnestic effects, and potential for untoward ethanol interactions have spurred attention and greater investigation of other anticonvulsant agents (e.g., gabapentin, topiramate) as alternatives (Johnson and Ait-Daoud, 2000; Litten et al., 2005; Malcolm et al., 2007; Myrick et al., 2009; Addolorato et al., 2012; Cooper and Vernon, 2013).

Although the effects of chronic alcohol exposure on $GABA_B$ receptors have not been extensively studied, alcohol dependence also appears to regulate presynaptic $GABA_B$ receptor function. For example, alcohol-dependent rats showed decreased sensitivity to $GABA_B$ receptor agonists and antagonists on evoked inhibitory postsynaptic currents recorded from central amygdala neurons (Roberto et al., 2008). Studies in animals and humans have suggested that direct $GABA_B$ receptor agonists, such as baclofen, are effective in reducing alcohol self-administration (Colombo et al., 2000; Addolorato et al., 2006, 2012). In one study, rats with a history of dependence exhibited a leftward shift in the dose–effect curve for baclofen to reduce alcohol consumption (greater sensitivity) compared to non-dependent rats (Walker and Koob, 2007).

Adaptations in monoamine systems

Dopamine

The role of dopamine in various facets of alcohol (and other drug) addiction has been extensively studied over several decades. Dopamine transmission is mediated via two groups of G-protein-linked receptors: D1-like (D1 and D5 receptors) and D2-like (D2, D3, and D4 receptors) that are classified on the basis of opposing transduction mechanisms that involve stimulating versus inhibiting adenylate cyclase activity, respectively. Dopamine, through its actions via mesolimbic and mesocortical pathways, has long been viewed as playing a central role in mediating alcohol reward (Gonzales et al., 2004; Spanagel, 2009). Chronic alcohol exposure has been reported to produce persistent neuroadaptive changes in dopamine transmission within the mesoaccumbens reward circuitry, including increased firing rate in ventral tegmental area (VTA) dopamine neurons (Brodie, 2002), increased basal extracellular levels of dopamine in the NAc (Smith and Weiss, 1999; Sahr et al., 2004; Thielen et al., 2004), and alterations in dopamine receptor function (Liu and Weiss, 2002b; Engleman et al., 2003). By contrast, withdrawal from chronic alcohol exposure has been reported to result in decreased VTA dopamine neuronal activity (Diana et al., 1996; Bailey et al., 2001), and reduced basal levels of dopamine in ventral and dorsal subregions of the striatum, possibly due to enhanced dopamine uptake (Weiss et al., 1996; Budygin et al., 2007; Barak et al., 2011).

Many of these effects have been shown to persist beyond acute withdrawal, and this dopamine hypofunctional state has been suggested to underlie, in part, the negative affect and dysphoria associated with protracted abstinence, as well as the motivation to re-engage in alcohol-drinking behavior (Diana, 2011; Charlet et al., 2013). While there is some evidence that dopamine antagonists reduce alcohol craving and consumption, concern about their side-effects has hampered their general use in treating alcohol dependence (Swift, 2010). The D2 dopamine receptor partial agonist aripiprazole has shown some efficacy in treating alcohol dependence (Anton et al., 2008a; Kranzler et al., 2008; Martinotti et al., 2009). The mixed agonist/antagonist mechanism of this drug may enable restoration of dopamine function during acute withdrawal and then blunt dopamine stimulation produced by drinking (relapse) (Myrick et al., 2010).

Altered cortical dopamine activity has been suggested to contribute to severe alcohol withdrawal symptoms that typically constitute delirium tremens (e.g., extreme agitation, hallucinations, psychosis). Antipychotics (both typical and atypical) have been used to manage such severe withdrawal, but they are typically used as an adjunct treatment (often administered in combination with benzodiazepines or anticonvulsants) because alone, they offer no protection against seizures. There are also concerns about severe sedation and potential respiratory depression (Mayo-Smith, 1997; Mainerova et al., 2013).

Norepinephrine

Norepinephrine is widely distributed throughout the brain, arising principally from neurons within the locus coeruleus, and noradrenergic activity has been shown to play a key role in regulating behaviors related to attention, arousal, and stress. The norepinephrine system also has been implicated in alcohol dependence, and alcohol withdrawal in particular. Chronic alcohol exposure results in increased peripheral and central noradrenergic activity, which underlies both somatic and affective manifestations of alcohol withdrawal. For example, drugs that temper noradrenergic activity either by stimulating presynaptic autoreceptors (alpha-2 adrenergic agonists: e.g., clonidine, dexmedetomidine) or blocking postsynaptic receptors (beta-adrenergic antagonists: e.g., propranolol) have been shown in preclinical and clinical studies to ameliorate various withdrawal symptoms related to heightened autonomic (sympathetic) nervous activity. Accordingly, this pharmacologic approach has proven to be useful as an adjunct in the management of alcohol detoxification (Muzyk et al., 2011).

Additionally, there is evidence that alcohol dependence-related adaptations in brain norepinephrine activity might play a role in influencing motivation to drink. For example, blocking postsynaptic alpha-1 adrenergic receptors with the antagonist prazosin reduced alcohol consumption in both dependent rats (Walker et al., 2008) and alcohol-dependent humans (Simpson et al., 2009). Likewise, treatment with beta-adrenoceptor antagonists (e.g., propranolol) also reduced drinking in dependent rats (Gilpin and Koob, 2010).

Serotonin

Serotonin predominantly arises from neurons within the raphe nuclei of the hindbrain, which send broad projections that innervate all levels of the brain. Serotonin exerts its known role in modulating various regulatory behaviors (e.g., feeding, sleep/arousal, aggression), mood, and emotional aspects of motivational behavior via several metabotropic (5-HT1 and 5-HT2 subtypes) and ionotropic (5-HT3) receptor systems throughout the brain. While changes in serotonin activity are not thought to play a significant role in mediating somatic signs of alcohol withdrawal, its general role in regulating mood and affect suggests serotonin may contribute to withdrawal-related dysphoria, as well as motivational effects of alcohol. Indeed, chronic alcohol exposure reduces serotonin levels in several brain regions, and there is a large body of evidence indicating a negative relationship between serotonin levels in brain and propensity to self-administer alcohol (Murphy et al., 2002; Casu et al., 2004; Petrakis, 2006). As is the case with dopamine, alcohol self-administration following withdrawal restored reduced extracellular levels of serotonin in the NAc produced by chronic alcohol exposure in rats (Weiss et al., 1996).

Given the role of serotonin in affective illness, it is not surprising that selective serotonin reuptake inhibitors (SSRIs) have been the focus of treatment for alcohol dependence and comorbid depression (Pettinati et al., 2010), anxiety (Book et al., 2008), and posttraumatic stress disorder (Petrakis et al., 2012). While there is some evidence indicating that SSRIs reduce drinking in animal studies (Maurel et al., 1999; Naranjo and Knoke, 2001), results are more mixed in clinical studies, possibly related to a unique polymorphism in the serotonin transporter (Kranzler et al., 2012), the presence and nature of comorbid illness (Johnson and Ait-Daoud, 2000), and other alcoholism-related endophenotypes (Pettinati et al., 2000). Interestingly, a drug that selectively inhibits serotonin reuptake (fluoxetine) and one that blocks both serotonin and norepinephrine reuptake (milnacipran) were both found to be effective in reducing alcohol self-administration in dependent rats (Simon O'Brien et al., 2011), while a novel triple monoamine uptake inhibitor was reported to reduce alcohol intake in rats with a genetic predisposition for high alcohol preference (Yang et al., 2012). Finally, the 5-HT3 receptor antagonist ondansetron has been shown to reduce alcohol consumption in animals (Hodge et al., 2004) and human alcohol-dependent subjects (Johnson et al., 2000; Kranzler et al., 2003; Johnson, 2004), with recent evidence suggesting that polymorphisms in the 5-HT3 receptor and serotonin transporter moderate treatment efficacy (Johnson et al., 2013).

Adaptations in neuropeptide systems

CORTICOTROPIN-RELEASING FACTOR

Chronic alcohol exposure engages a number of neuropeptide systems in the brain, with CRF most extensively studied in animal models of dependence (Heilig and Koob, 2007; Koob and Zorrilla, 2010; Lowery and Thiele, 2010). CRF is a 41-amino-acid neuropeptide that is widely distributed throughout the brain and, along with related peptides (urocortin; Ucn1, Ucn2, Ucn3), interacts with two G-protein-coupled receptor subtypes (CRF1 and CRF2) to produce its physiologic and behavioral effects (Bale and Vale, 2004). While there is some evidence indicating a role for urocortin peptides (especially Ucn2 and Ucn3) and CRF2 receptors in alcohol dependence and withdrawal (e.g., Valdez et al., 2004; Funk and Koob, 2007; Ryabinin et al., 2012), most attention has focused on adaptations in CRF–CRF1 receptor activity within brain and neuroendocrine systems in relation to chronic alcohol exposure and withdrawal-related consequences.

CRF emanating from the hypothalamus plays a key role in regulating the neuroendocrine function of the hypothalamic–pituitary–adrenocortical (HPA) axis. Specifically, CRF neurons residing in the paraventricular nucleus of the hypothalamus regulate HPA axis activity via control of glucocorticoid production and release, which is critical for orchestrating behavioral and physiologic responses to stress. A large preclinical and clinical literature has demonstrated profound disturbances in HPA axis function following chronic alcohol exposure (Rivier, 2000; Wand, 2000; Stephens and Wand, 2012). Elevated glucocorticoid levels resulting from dependence-related HPA axis activation not only underlie altered stress responsiveness in dependent subjects, but it also may contribute to amplified motivation to drink as well as other ramifications of the dependence state. Studies in mice and rats also have shown that withdrawal following chronic alcohol consumption produced elevated corticosterone levels in the prefrontal cortex and hippocampus that persisted long after plasma corticosterone levels returned to baseline levels (Little et al., 2008). There are also significant changes in expression of receptors in brain that bind glucocorticoids (glucocorticoid but not mineralocorticoid receptors) during abstinence, and there is recent evidence that activity at these receptors may play a role in relapse and perpetuation of excessive alcohol consumption (Vendruscolo et al., 2012). These results are congruent with findings showing that elevated corticosteroids enhance the propensity to consume alcohol through an interaction with mesolimbic and mesocortical reward pathways (Fahlke et al., 1995, 1996; Piazza and Le Moal, 1997; Koenig and Olive, 2004; Uhart and Wand, 2009). Further, elevations in brain glucocorticoid concentrations following chronic alcohol exposure and withdrawal may contribute to the cognitive deficits and neurotoxic damage that are commonly associated with alcohol dependence (Rose et al., 2010).

Beyond alterations in neuroendocrine (HPA axis) function, chronic alcohol exposure has been shown to alter CRF activity independently of the HPA axis (Heilig and Koob, 2007; Heilig et al., 2010; Koob and Zorrilla, 2010). Increased CRF activity in several brain structures following chronic alcohol exposure is thought to play a key role in the emergence of negative affective withdrawal symptoms (e.g., anxiety, dysphoria) that may be especially relevant in enhancing susceptibility to relapse and promoting return to excessive levels of drinking (Koob and Kreek, 2007; Heilig et al., 2010; Lowery and Thiele, 2010; Becker, 2012). For example, studies in rats involving pharmacologic manipulation of CRF activity has implicated a role of CRF1 receptors in mediating increased anxiety associated with withdrawal from chronic alcohol exposure delivered in a liquid diet (Baldwin et al., 1991; Rassnick et al., 1993; Valdez et al., 2003; Breese et al., 2005; Huang et al., 2010b), by intragastric administration (Gehlert et al.,

2007), and via vaporized alcohol in inhalation chambers (Sommer et al., 2008), although a role for CRF2 receptors cannot be ruled out (Valdez et al., 2004). Additionally, increased CRF activity in brain structures integral to reward and stress pathways has been shown to play an important role in mediating the ability of stress to trigger relapse-like behavior (Le et al., 2000; Liu and Weiss, 2002a; Gehlert et al., 2007; Marinelli et al., 2007; Sommer et al., 2008). A number of studies have shown that pharmacologic blockade or genetic deletion of CRF1 receptors impedes escalation of alcohol consumption in dependent animals (Chu et al., 2007; Funk et al., 2007; Gehlert et al., 2007; Gilpin et al., 2008b; Sommer et al., 2008; Roberto et al., 2010; Molander et al., 2012). Finally, studies in humans (Chen et al., 2010; Schmid et al., 2010; Blomeyer et al., 2011), non-human primates (Barr et al., 2008, 2009), and rats (Hansson et al., 2006; Ayanwuyi et al., 2013) have indicated a strong genetic influence on the role of CRF in mediating stress responsiveness as well as alcohol drinking and risk for dependence. Taken together, there is a large body of evidence that indicates that dependence-related alterations in brain CRF activity both within and outside the HPA axis play a significant role in mediating various symptoms of alcohol withdrawal as well as excessive drinking associated with dependence.

Neuropeptide Y

Neuropeptide Y (NPY) is a 36-amino acid peptide that is widely distributed in the CNS and operates at five distinct receptors (Y1–5). NPY is commonly viewed as an "antistress," anxiolytic signaling molecule (Heilig et al., 1994; Valdez and Koob, 2004), and it has been implicated in various facets of alcohol dependence and withdrawal (Thorsell, 2007; Gilpin, 2012). For example, reduced NPY mRNA and protein levels in the central and medial (but not basolateral) subregions of the amygdala were shown to be associated with withdrawal-related anxiety behavior following chronic alcohol treatment in rats (Roy and Pandey, 2002; Zhang and Pandey, 2003; Zhang et al., 2010). Intraventricular administration of NPY following chronic exposure to alcohol vapor reduced alcohol self-administration (Thorsell et al., 2005a, b). Additionally, central administration of NPY during repeated withdrawal cycles from chronic alcohol vapor exposure blocked the progressive increase in alcohol consumption in rats (Gilpin et al., 2011). Similar results were obtained in another study with direct infusion of NPY into the central nucleus of the amygdala following chronic alcohol treatment in a liquid diet (Gilpin et al., 2008a). While studies have demonstrated a role for Y1 (Sparrow et al., 2012), Y2 (Thorsell et al., 2002), and Y5 (Schroeder et al., 2005) receptors in modulating alcohol consumption, few studies have dissected the relative role of these NPY receptor subtypes in mediating alcohol withdrawal anxiety and drinking. Nevertheless, evidence to date suggests that adaptive changes in NPY function in dependence contributes to withdrawal-related anxiety and increased propensity to self-administer alcohol.

Opioid polypeptides

Alcohol interactions with endogenous opioid systems in the brain are well established in the animal and human literature. The endogenous opioid polypeptides endorphin, enkephalin, and dynorphin are distributed throughout the brain and principally operate at mu (MOR), delta (DOR), and kappa (KOR) opiate receptors, respectively, to produce their behavioral and physiologic actions (Bodnar, 2012). Activity at mu and delta receptors mediates classic opioid effects (e.g., analgesia, euphoria, sedation) while drugs that mimic the endogenous ligand for kappa receptors exert an opposite pharmacologic profile (e.g., increased pain sensitivity, dysphoria). Given the role of endogenous opioids in reward processes and motivational behaviors, most attention has focused on pharmacologic agents that alter endogenous opioid activity as treatments for alcohol dependence. A host of studies have shown that opiate antagonists (primarily targeting mu receptors) reduce alcohol craving and consumption, ultimately leading to naltrexone being Food and Drug Administration-approved as a medication for treatment of alcoholism (O'Malley et al., 1992; Volpicelli et al., 1992). Recent studies have shown that a polymorphism in mu opiate receptors confers differential sensitivity to the treatment efficacy of naltrexone (Ray and Hutchison, 2007; Anton et al., 2008b). Also, there is some evidence that antagonists with greater selectivity for delta receptors (e.g., nalmefene) may have promise as effective treatments (Mason et al., 1999; Mann et al., 2013). While there is ample preclinical evidence indicating that antagonism of both MOR and DOR is effective in reducing alcohol consumption (e.g., Ciccocioppo et al., 2002), some studies have shown that sensitivity to DOR antagonism is greater in dependent compared to non-dependent animals (Walker and Koob, 2008; Nealey et al., 2011).

There is growing interest in chronic alcohol-induced changes in the dynorphin/KOR system, particularly as such changes bear on somatic as well as the emergence of dysphoria/anxiety components of alcohol withdrawal (Walker et al., 2012). For example, there is some evidence to indicate that dynorphin (prodynorphin mRNA) expression in brain relates to genetic propensity for withdrawal seizures following chronic alcohol treatment (Beadles-Bohling et al., 2000). Additionally, KOR antagonists were shown to attenuate acute alcohol withdrawal ("hangover") anxiety (Schank et al., 2012; Valdez and Harshberger, 2012); KOR activity has been implicated in negative affect/emotional states produced by chronic exposure to alcohol via a liquid diet (Gillett et al., 2013) or vapor inhalation (Berger et al., 2013a; Kissler et al., 2013). Also, KOR antagonism has been demonstrated to reduce escalated alcohol consumption in dependent rats without altering responding/intake in non-dependent rats (Walker and Koob, 2008; Nealey et al., 2011; Walker et al., 2012). Interestingly, a polymorphism in the gene encoding KOR (*OPRK1*) was shown to be associated with human alcohol dependence (Edenberg et al., 2008). Collectively, these studies indicate that increased dynorphin/KOR activity resulting from chronic alcohol exposure significantly contributes to dysphoric and negative emotional aspects of withdrawal, and blockade of KOR effectively suppresses alcohol self-administration that would otherwise be elevated as a function of dependence.

Nociceptin

Nociceptin (orphanin FQ) is a 17-amino-acid peptide that is structurally similar to dynorphin, but it possesses unique pharmacologic actions through binding with high affinity to opioid receptor-like 1 (ORL-1), also known as NOP (Reinscheid et al., 1995; Lambert, 2008). Devoid of mu, delta, and kappa receptor activity, nociceptin exerts antianxiety and antistress effects at NOP receptors. Recent studies have suggested that this peptide may be involved in the expression of various alcohol withdrawal symptoms, as well as alcohol self-administration behavior. For example, the gene that encodes the NOP receptor (*Oprl1*) was shown to be robustly upregulated in prefrontal cortex following chronic alcohol exposure in a mouse model of dependence (Melendez et al., 2012). Further, central administration of nociceptin reduced expression of somatic signs of withdrawal as well as increased anxiety following chronic alcohol treatment (Economidou et al., 2011; Aujla et al., 2013). While several studies have shown that nociceptin (and NOP agonists) decreases alcohol consumption, particularly in rats with a genetic predisposition for high alcohol intake (Ciccocioppo et al., 2004; Economidou et al., 2008), dependent rats have been shown to exhibit greater sensitivity to this effect in comparison with non-dependent animals (Martin-Fardon et al., 2010). The potential for drugs that target NOP receptors for treatment of alcohol dependence awaits further investigation.

Other neuropeptides

Other neuropeptides recently implicated in alcohol dependence include orexin/hypocretin (Bayerlein et al., 2011; von der Goltz et al., 2011; Kim et al., 2012), substance P/neurokinin1 receptors (George et al., 2008; Thorsell et al., 2010; Schank et al., 2011), and neuropeptide S (Ruggeri et al., 2010; Enquist et al., 2012). While all these neuropeptide system have been implicated in motivational effects of alcohol, their role in mediating withdrawal symptoms as well as excessive alcohol consumption associated with dependence remains to be determined.

Adaptations in ion channels

CNS neurons express a wide range of calcium- or voltage-gated potassium (K^+) and calcium (Ca^{2+}) ion channels. These ion channels play a critical role in modulating many aspects of neuronal physiology (i.e., intrinsic excitability, dendritic integration, tonic firing frequency, spike frequency adaptation, and action potential repolarization). The distribution and density of K^+ and Ca^{2+} ion channels can vary by cell type and subcellular localization within the dendritic field. For example, some channels are enriched in dendritic spines of pyramidal neurons in the hippocampus and cortex, while others are highly expressed on interneurons or distal apical dendrites of CA1 pyramidal neurons. Findings from recent studies have demonstrated that chronic alcohol exposure can produce adaptive changes in the expression and function of these dendritic ion channels. Below, we describe evidence that neuroadaptations in voltage-gated Ca^{2+} channels (VGCCs), small-conductance (SK) and large-conductance (BK) K^+ channels, and voltage-gated K^+ (Kv) channels critically regulate hyperexcitability, local inhibitory neurocircuitry, and seizure activity associated with withdrawal from chronic alcohol exposure.

Voltage-gated Ca^{2+} channels

Activation of VGCCs at depolarized membrane potentials allows Ca^{2+} entry into neurons through the channel pore where it can then influence Ca^{2+}-activated K^+ channels, gene expression, neuronal excitability and firing patterns, and neurotransmitter release. VGCCs can be classified as low- or high-threshold voltage-activated or by their sensitivity to selective toxins. Chronic alcohol exposure has been shown to increase high-voltage activated L-type VGCCs in brain (Littleton, 1998; Walter and Messing, 1999; Katsura et al., 2005). Blocking L-type channels significantly reduces the severity of seizure activity during withdrawal and prevented c-Fos induction in the cortex, hippocampus, and striatum (Bouchenafa and Littleton, 1998). Prolonged alcohol exposure also increases high-voltage activated N- and P-type VGCCs in the frontal cortex, hippocampus, and inferior colliculus (McMahon et al., 2000). More recently, chronic intermittent alcohol exposure or prolonged consumption of an alcohol liquid diet produced increases in low-threshold, transient T-type VGCCs in thalamic neurons (Nordskog et al., 2006; Graef et al., 2011). Like the hippocampus, the thalamus contributes to the generation of brain rhythms (e.g., theta oscillations, sleep spindles) (Buzsaki, 2002; Steriade, 2006). Thalamic neurons in the midline reuniens nucleus have strong recurrent connections with the frontal cortex and hippocampus and drive corticothalamic-hippocampal network activity (Bertram et al., 2008). Increased T-type channel function during alcohol withdrawal is associated with enhanced burst firing of thalamic neurons and disruptions in EEG theta power (Graef et al., 2011). Further, alcohol withdrawal-related disruption in theta power was prevented in mice by treatment with ethosuximide, a T-type channel blocker (Graef et al., 2011). Collectively, these data suggest that the neuroadaptations in VGCCs produced by alcohol dependence contribute to CNS hyperexcitability during withdrawal.

Small-conductance Ca^{2+}-activated K^+ channels

SK channels regulate membrane excitability by shaping excitatory postsynaptic potentials (EPSP) and controlling intrinsic activity, dendritic integration, and pacemaker firing (Bond et al., 2005; Fakler and Adelman, 2008). SK channels are solely activated by transient elevations of intracellular Ca^{2+} and form functional heteromeric complexes with calmodulin, that acts as a high-affinity Ca^{2+} sensor (Lee et al., 2003; Maylie et al., 2004; Allen et al., 2007). In pyramidal neurons of the hippocampus, amygdala and prefrontal cortex, SK2 channels are enriched in the postsynaptic density of dendritic spines, where they form a Ca^{2+}-mediated negative-feedback loop with synaptic NMDA receptors (Faber et al., 2005; Ngo-Anh et al., 2005; Bloodgood and Sabatini, 2007; Lin et al., 2008; Faber, 2010; Mulholland et al., 2011). Activation of this feedback loop is thought to act as a postsynaptic shunt to decrease spine Ca^{2+} transients and synaptic depolarization. Consistent with this idea are studies showing that blocking SK channels in pyramidal neurons increases EPSP amplitude and facilitates NMDA receptor-dependent synaptic plasticity (Faber et al., 2005, 2008; Lin et al., 2008; Faber, 2010). SK channels also can modulate the ability of NMDA receptors to induce burst firing in midbrain dopamine neurons (Seutin et al., 1993; Johnson and Seutin, 1997).

Recent evidence has demonstrated that chronic alcohol exposure reduced SK-mediated currents recorded from cultured CA1 pyramidal neurons (Mulholland et al., 2011). Chronic alcohol exposure also decreased expression of SK2 channels while at the same time increasing expression of NMDA receptor GluN1 and GluN2B subunits in hippocampus. This bidirectional regulation of SK2 channels and NMDA receptors leads to a disruption of the Ca^{2+}-mediated negative-feedback loop and enhanced hyperexcitability during acute alcohol withdrawal (Mulholland et al., 2011). Positive modulation of SK channels attenuated acute withdrawal-related epileptiform burst firing and behavioral signs of seizure activity (Mulholland et al., 2011; Mulholland, 2012). Thus, these data demonstrate that downregulation of SK2 channels contributes to alcohol-associated plasticity of hippocampal glutamatergic synapses, and that reduced expression of SK2 channels in dendritic spines plays a role in the alcohol withdrawal syndrome.

In addition to alcohol-induced adaptations in hippocampal SK channels, recent studies have shown that acute and chronic alcohol exposure alters firing properties of VTA dopamine and NAc core medium spiny neurons through adaptations in SK channel function and expression (Brodie et al., 1999; Hopf et al., 2007, 2010). For example, SK-mediated currents were significantly reduced in rats 7 days following withdrawal from a chronic ethanol treatment regimen (Hopf et al., 2007). In another study, rats self-administering alcohol over a 5-week period exhibited increased intrinsic excitability of neurons in the NAc core (Hopf et al., 2010). Taken together, these data suggest that withdrawal from chronic alcohol exposure is associated with reduced SK channel function in the VTA and NAc core, and these adaptations are important for regulating intrinsic excitability and NMDA receptor-dependent burst firing during alcohol withdrawal.

Large-conductance voltage- and Ca^{2+}-activated K^+ channels

A number of studies have implicated an important role for somatodendritic BK channels in ethanol tolerance, adaptive plasticity, and withdrawal-associated seizures. BK channels are a product of the *KCNMA1* (*Slo1*) gene and its orthologs (e.g., d*Slo* in *Drosophila*), and are ubiquitously expressed in the CNS. BK channels are comprised of α and β auxiliary subunits in a 1:1 stoichiometry, and these channels can be activated by increases in intracellular Ca^{2+} or transmembrane positive voltage shifts (Latorre and Brauchi, 2006). BK channels serve to shape dendritic Ca^{2+} spikes and modulate the release of growth hormones, neuropeptides (vasopressin and oxytocin), and various neurotransmitters (Storm, 1990; Dopico et al., 1996; Jakab et al., 1997). Of relevance to alcohol, BK channels are known to influence neurotransmitter release, nociception, intrinsic excitability of NAc medium spiny neurons, motor incoordination, and alcohol withdrawal-induced seizures (Brodie et al., 2007; Treistman and Martin, 2009; Ghezzi et al., 2012).

Unlike other members of the voltage-gated, TM6 family of K^+ channels, BK channels are very sensitive to low (10–50 mM) alcohol concentrations in a number of structures (Treistman and Martin, 2009). Alcohol activation of BK channels has been reported in a wide variety of neuronal and non-neuronal preparations and model systems (i.e., oocytes, HEK 293 cells, *Caenorhabditis elegans*, *Drosophila*). Exposure of neuronal terminals to alcohol for 24 hours declusters and internalizes surface BK channels and renders them less sensitive to the potentiating effects of alcohol on channel function (Pietrzykowski et al., 2004). While the β_4 subunit has been shown to play a critical role in mediating tolerance to alcohol-induced modulation of BK channel function (Martin et al., 2008), CaMKII phosphorylation of Thr107 of the alpha subunit of BK channels increased channel activity and mediated the switch from alcohol-induced activation of the channel-to-channel inhibition (Treistman and Martin, 2009). In *Drosophila*, alcohol exposure reduced seizure threshold at 1 and 7 days into withdrawal, but not in flies where *slo* expression was genetically eliminated (Ghezzi et al., 2012). Further, heat shock induction of *slo* mimicked the lowering effect of alcohol on seizure threshold, suggesting a role for *slo* induction in the susceptibility to withdrawal seizures. Thus, altered function of Ca^{2+}-activated K^+ channels represents a developing area that is of particular importance in understanding alcohol-associated molecular and behavioral tolerance, adaptive plasticity, and withdrawal severity.

A-TYPE K^+ CHANNELS

Voltage-dependent potassium (Kv) channels composed of Kv1 and Kv4 subunits underlie the transient A-type K^+ current (I_A) in hippocampal, cortical, and NAc medium spiny neurons, where they influence a wide range of functions, including the shaping and propagation characteristics of action potentials, and complex cognitive processes (Hoffman et al., 1997; Chen et al., 2006; Lockridge and Yuan, 2011). Emerging evidence also suggests a role for Kv4 channels in ethanol dependence and withdrawal. For example, chronic intermittent alcohol exposure significantly increased I_A in NAc medium spiny neurons (Marty and Spigelman, 2012) and decreased I_A in hippocampal CA1 interneurons (Li et al., 2013). Chronic alcohol exposure and withdrawal also significantly reduce surface expression and function of Kv4.2 channels in CA1 pyramidal neurons (Mulholland et al., 2010). The chronic intermittent alcohol exposure-associated bidirectional changes between I_A in NAc medium spiny neuronss and hippocampus suggests divergent cell-specific regulation of Kv4 channel subunits or posttranslational modifications (Tkatch et al., 2000). Together, these data indicate divergent cell-specific regulation of Kv4 channel subunits, and suggest that neuroadaptations in Kv4 channels may be an important factor underlying altered synaptic plasticity, poor cognitive performance, and aberrant excitability associated with prolonged alcohol exposure and withdrawal.

CONCLUSIONS

Alcoholism is a chronic relapsing disease and, thus, it is not uncommon for many dependent individuals to attempt abstinence on numerous occasions, only to find themselves progressing to unhealthy excessive drinking once a "slip" (relapse) occurs. When dependent individuals completely stop or significantly reduce their alcohol drinking, a characteristic withdrawal syndrome ensues. The alcohol withdrawal syndrome includes signs of CNS hyperexcitability (i.e., seizure activity, sometimes culminating in behavioral convulsions and sensory hyperreactivity), physiologic manifestations of heightened autonomic nervous system activity (e.g., tachycardia, hypertension, diaphoresis, body temperature dysregulation, and gastrointestinal disturbances), motor problems (e.g., tremor) and, in its most severe form, hallucinosis and delirium tremens. In addition to these physiologic signs of withdrawal, a constellation of symptoms contributing to psychologic discomfort and negative affect constitutes a prominent component of the withdrawal syndrome. Withdrawal symptoms that fall within the domain of psychologic discomfort and negative affect (mood) include sleep disturbances, anxiety and irritability, increased stress reactivity, and a general state of anhedonia/dysphoria. While most physiologic symptoms of withdrawal typically abate within a few days, many of the psychologic symptoms of withdrawal linger for protracted periods of time, and these latter symptoms are thought to play an important role in underlying increased susceptibility to relapse.

The development of alcohol dependence is a complex and dynamic process that involves profound perturbations in neuroendocrine systems (i.e., HPA axis activity) as well as neuroadaptations in a host of neurochemical systems, including classic neurotransmitter systems (e.g., glutamate, GABA, monoamines), neuropeptides (e.g., CRF, NPY, endogenous opioids), and numerous ion channels (e.g., VGCCs, SK channels, BK channels). The maladaptive nature of these changes induced by chronic alcohol exposure is revealed when alcohol is eliminated from the brain during withdrawal. Manifestations of these neuroadaptations are seen in the expression of various symptoms related to physiologic and psychologic components of the alcohol withdrawal syndrome. Further, many of these adaptive changes set in motion as a result of chronic alcohol exposure and withdrawal experience occur within brain reward and stress systems. The progressive dysregulation of brain reward and stress systems beyond normal homeostatic limits is thought to fuel a state of allostasis that not only underlies expression of withdrawal symptoms, but also drives enhanced relapse vulnerability perpetuation of excessive harmful levels of drinking. The challenge ahead is to better understand the pathophysiologic mechanisms of alcohol dependence so that more effective and targeted therapeutics can be developed that not only effectively treat symptoms of withdrawal, but also reduce risk of relapse and temper motivation to engage in unhealthy drinking.

ACKNOWLEDGMENTS

This work was supported by National Institutes of Health grants P50 AA010761 (HCB), U01 AA014095 (HCB), U01 AA020929 (HCB), R00 AA017922 (PJM), U01 AA020930 (PJM), and U01 AA019967 (PJM) from the National Institute on Alcohol Abuse and Alcoholism, and the Department of Veterans Affairs Medical Research (HCB).

References

Addolorato G, Leggio L, Agabio R et al. (2006). Baclofen: a new drug for the treatment of alcohol dependence. Int J Clin Pract 60: 1003–1008.

Addolorato G, Leggio L, Hopf FW et al. (2012). Novel therapeutic strategies for alcohol and drug addiction: focus on GABA, ion channels and transcranial magnetic stimulation. Neuropsychopharmacology 37: 163–177.

Allen D, Fakler B, Maylie J et al. (2007). Organization and regulation of small conductance Ca^{2+}-activated K^+ channel multiprotein complexes. J Neurosci 27: 2369–2376.

Anton RF, Kranzler H, Breder C et al. (2008a). A randomized, multicenter, double-blind, placebo-controlled study of the efficacy and safety of aripiprazole for the treatment of alcohol dependence. J Clin Psychopharmacol 28: 5–12.

Anton RF, Oroszi G, O'Malley S et al. (2008b). An evaluation of mu-opioid receptor (OPRM1) as a predictor of naltrexone response in the treatment of alcohol dependence: results from the Combined Pharmacotherapies and Behavioral Interventions for Alcohol Dependence (COMBINE) study. Arch Gen Psychiatry 65: 135–144.

Aujla H, Cannarsa R, Romualdi P et al. (2013). Modification of anxiety-like behaviors by nociceptin/orphanin FQ (N/OFQ) and time-dependent changes in N/OFQ-NOP gene expression following ethanol withdrawal. Addict Biol 18: 467–479.

Ayanwuyi LO, Carvajal F, Lerma-Cabrera JM et al. (2013). Role of a genetic polymorphism in the corticotropin-releasing factor receptor 1 gene in alcohol drinking and seeking behaviors of Marchigian Sardinian alcohol-preferring rats. Front Psychiatry 4: 23.

Bailey CP, O'Callaghan MJ, Croft AP et al. (2001). Alterations in mesolimbic dopamine function during the abstinence period following chronic ethanol consumption. Neuropharmacology 41: 989–999.

Baldwin HA, Rassnick S, Rivier J et al. (1991). CRF antagonist reverses the "anxiogenic" response to ethanol withdrawal in the rat. Psychopharmacology (Berl) 103: 227–232.

Bale TL, Vale WW (2004). CRF and CRF receptors: role in stress responsivity and other behaviors. Annu Rev Pharmacol Toxicol 44: 525–557.

Ballenger JC, Post RM (1978). Kindling as a model for alcohol withdrawal syndromes. Br J Psychiatry 133: 1–14.

Barak S, Carnicella S, Yowell QV et al. (2011). Glial cell line-derived neurotrophic factor reverses alcohol-induced allostasis of the mesolimbic dopaminergic system: implications for alcohol reward and seeking. J Neurosci 31: 9885–9894.

Barr CS, Dvoskin RL, Yuan Q et al. (2008). CRH haplotype as a factor influencing cerebrospinal fluid levels of corticotropin-releasing hormone, hypothalamic-pituitary-adrenal axis activity, temperament, and alcohol consumption in rhesus macaques. Arch Gen Psychiatry 65: 934–944.

Barr CS, Dvoskin RL, Gupte M et al. (2009). Functional CRH variation increases stress-induced alcohol consumption in primates. Proc Natl Acad Sci U S A 106: 14593–14598.

Barrenha GD, Chester JA (2007). Genetic correlation between innate alcohol preference and fear-potentiated startle in selected mouse lines. Alcohol Clin Exp Res 31: 1081–1088.

Bauer J, Pedersen A, Scherbaum N et al. (2013). Craving in alcohol-dependent patients after detoxification is related to glutamatergic dysfunction in the nucleus accumbens and the anterior cingulate cortex. Neuropsychopharmacology 38: 1401–1408.

Bayerlein K, Kraus T, Leinonen I et al. (2011). Orexin A expression and promoter methylation in patients with alcohol dependence comparing acute and protracted withdrawal. Alcohol 45: 541–547.

Beadles-Bohling AS, Crabbe JC, Wiren KM (2000). Elevated prodynorphin expression associated with ethanol withdrawal convulsions. Neurochem Int 37: 463–472.

Becker HC (1996). The alcohol withdrawal "kindling" phenomenon: clinical and experimental findings. Alcohol Clin Exp Res 20: 121A–124A.

Becker HC (1998). Kindling in alcohol withdrawal. Alcohol Health Res World 22: 25–33.

Becker HC (1999). Alcohol withdrawal: neuroadaptation and sensitization. CNS Spectr 4: 38–65.

Becker HC (2000). Animal models of alcohol withdrawal. Alcohol Res Health 24: 105–113.

Becker HC (2008). Alcohol dependence, withdrawal, and relapse. Alcohol Res Health 31: 348–361.

Becker HC (2012). Effects of alcohol dependence and withdrawal on stress responsiveness and alcohol consumption. Alcohol Res 34: 448–458.

Becker HC (2013). Animal models of excessive alcohol consumption in rodents. Curr Topics Behav Neurosci 13: 355–377.

Becker HC, Veatch LM (2002). Effects of lorazepam treatment for multiple ethanol withdrawals in mice. Alcohol Clin Exp Res 26: 371–380.

Ben Hamida S, Darcq E, Wang J et al. (2013). Protein tyrosine phosphatase alpha in the dorsomedial striatum promotes excessive ethanol-drinking behaviors. J Neurosci 33: 14369–14378.

Berger AL, Williams AM, McGinnis MM et al. (2013a). Affective cue-induced escalation of alcohol self-administration and increased 22-kHz ultrasonic vocalizations during alcohol withdrawal: role of kappa-opioid receptors. Neuropsychopharmacology 38: 647–654.

Berger L, Fisher M, Brondino M et al. (2013b). Efficacy of acamprosate for alcohol dependence in a family medicine setting in the United States: a randomized, double-blind, placebo-controlled study. Alcohol Clin Exp Res 37: 668–674.

Bertram EH, Zhang D, Williamson JM (2008). Multiple roles of midline dorsal thalamic nuclei in induction and spread of limbic seizures. Epilepsia 49: 256–268.

Blomeyer D, Buchmann AF, Schmid B et al. (2011). Age at first drink moderates the impact of current stressful life events on drinking behavior in young adults. Alcohol Clin Exp Res 35: 1142–1148.

Bloodgood BL, Sabatini BL (2007). Nonlinear regulation of unitary synaptic signals by CaV(2.3) voltage-sensitive calcium channels located in dendritic spines. Neuron 53: 249–260.

Bodnar RJ (2012). Endogenous opiates and behavior: 2011. Peptides 38: 463–522.

Bond CT, Maylie J, Adelman JP (2005). SK channels in excitability, pacemaking and synaptic integration. Curr Opin Neurobiol 15: 305–311.

Bone GH, Majchrowicz E, Martin PR et al. (1989). A comparison of calcium antagonists and diazepam in reducing ethanol withdrawal tremors. Psychopharmacology (Berl) 99: 386–388.

Book SW, Thomas SE, Randall PK et al. (2008). Paroxetine reduces social anxiety in individuals with a co-occurring alcohol use disorder. J Anxiety Disord 22: 310–318.

Booth BM, Blow FC (1993). The kindling hypothesis: further evidence from a U.S. national study of alcoholic men. Alcohol Alcohol 28: 593–598.

Bouchenafa O, Littleton JM (1998). Expression of c-Fos protein immunoreactivity in rat brain during ethanol withdrawal is prevented by nifedipine. Alcohol 15: 71–76.

Brager AJ, Ruby CL, Prosser RA et al. (2010). Chronic ethanol disrupts circadian photic entrainment and daily locomotor activity in the mouse. Alcohol Clin Exp Res 34: 1266–1273.

Breese GR, Overstreet DH, Knapp DJ et al. (2005). Prior multiple ethanol withdrawals enhance stress-induced anxiety-like behavior: inhibition by CRF1- and benzodiazepine-receptor antagonists and a 5-HT1a-receptor agonist. Neuropsychopharmacology 30: 1662–1669.

Brodie MS (2002). Increased ethanol excitation of dopaminergic neurons of the ventral tegmental area after chronic ethanol treatment. Alcohol Clin Exp Res 26: 1024–1030.

Brodie MS, McElvain MA, Bunney EB et al. (1999). Pharmacological reduction of small conductance calcium-activated potassium current (SK) potentiates the excitatory effect of ethanol on ventral tegmental area dopamine neurons. J Pharmacol Exp Ther 290: 325–333.

Brodie MS, Scholz A, Weiger TM et al. (2007). Ethanol interactions with calcium-dependent potassium channels. Alcohol Clin Exp Res 31: 1625–1632.

Brower KJ, Perron BE (2010). Prevalence and correlates of withdrawal-related insomnia among adults with alcohol dependence: results from a national survey. Am J Addict 19: 238–244.

Brower KJ, Aldrich MS, Robinson EA et al. (2001). Insomnia, self-medication, and relapse to alcoholism. Am J Psychiatry 158: 399–404.

Brower KJ, Myra Kim H, Strobbe S et al. (2008). A randomized double-blind pilot trial of gabapentin versus placebo to treat alcohol dependence and comorbid insomnia. Alcohol Clin Exp Res 32: 1429–1438.

Brower KJ, Wojnar M, Sliwerska E et al. (2012). PER3 polymorphism and insomnia severity in alcohol dependence. Sleep 35: 571–577.

Brown ME, Anton RF, Malcolm R et al. (1988). Alcohol detoxification and withdrawal seizures: clinical support for a kindling hypothesis. Biol Psychiatry 23: 507–514.

Bruckner MK, Rossner S, Arendt T (1997). Differential changes in the expression of AMPA receptors genes in rat brain after chronic exposure to ethanol: an in situ hybridization study. J Hirnforsch 38: 369–376.

Budygin EA, Oleson EB, Mathews TA et al. (2007). Effects of chronic alcohol exposure on dopamine uptake in rat nucleus accumbens and caudate putamen. Psychopharmacology (Berl) 193: 495–501.

Buzsaki G (2002). Theta oscillations in the hippocampus. Neuron 33: 325–340.

Cagetti E, Liang J, Spigelman I et al. (2003). Withdrawal from chronic intermittent ethanol treatment changes subunit composition, reduces synaptic function, and decreases behavioral responses to positive allosteric modulators of GABAA receptors. Mol Pharmacol 63: 53–64.

Casu MA, Pisu C, Lobina C et al. (2004). Immunocytochemical study of the forebrain serotonergic innervation in Sardinian alcohol-preferring rats. Psychopharmacology (Berl) 172: 341–351.

Chandler LJ, Carpenter-Hyland E, Hendricson AW et al. (2006). Structural and functional modifications in glutamateric synapses following prolonged ethanol exposure. Alcohol Clin Exp Res 30: 368–376.

Charles PD, Esper GJ, Davis TL et al. (1999). Classification of tremor and update on treatment. Am Fam Physician 59: 1565–1572.

Charlet K, Beck A, Heinz A (2013). The dopamine system in mediating alcohol effects in humans. Curr Top Behav Neurosci 13: 461–488.

Chen X, Yuan LL, Zhao C et al. (2006). Deletion of Kv4.2 gene eliminates dendritic A-type K^+ current and enhances induction of long-term potentiation in hippocampal CA1 pyramidal neurons. J Neurosci 26: 12143–12151.

Chen AC, Manz N, Tang Y et al. (2010). Single-nucleotide polymorphisms in corticotropin releasing hormone receptor 1 gene (CRHR1) are associated with quantitative trait of event-related potential and alcohol dependence. Alcohol Clin Exp Res 34: 988–996.

Chester JA, Barrenha GD (2007). Acoustic startle at baseline and during acute alcohol withdrawal in replicate mouse lines selectively bred for high or low alcohol preference. Alcohol Clin Exp Res 31: 1633–1644.

Chester JA, Blose AM, Froehlich JC (2005). Effects of chronic alcohol treatment on acoustic startle reactivity during withdrawal and subsequent alcohol intake in high and low alcohol drinking rats. Alcohol Alcohol 40: 379–387.

Chester JA, Rausch EJ, June HL et al. (2006). Decreased reward during acute alcohol withdrawal in rats selectively bred for low alcohol drinking. Alcohol 38: 165–172.

Chu K, Koob GF, Cole M et al. (2007). Dependence-induced increases in ethanol self-administration in mice are blocked by the CRF1 receptor antagonist antalarmin and by CRF1 receptor knockout. Pharmacol Biochem Behav 86: 813–821.

Ciccocioppo R, Martin-Fardon R, Weiss F (2002). Effect of selective blockade of mu(1) or delta opioid receptors on reinstatement of alcohol-seeking behavior by drug-associated stimuli in rats. Neuropsychopharmacology 27: 391–399.

Ciccocioppo R, Economidou D, Fedeli A et al. (2004). Attenuation of ethanol self-administration and of conditioned reinstatement of alcohol-seeking behaviour by the antiopioid peptide nociceptin/orphanin FQ in alcohol-preferring rats. Psychopharmacology (Berl) 172: 170–178.

Clapp P, Gibson ES, Dell'acqua ML et al. (2010). Phosphorylation regulates removal of synaptic *N*-methyl-D-aspartate receptors after withdrawal from chronic ethanol exposure. J Pharmacol Exp Ther 332: 720–729.

Clark CP, Gillin JC, Golshan S et al. (1998). Increased REM sleep density at admission predicts relapse by three months in primary alcoholics with a lifetime diagnosis of secondary depression. Biol Psychiatry 43: 601–607.

Cohn TJ, Foster JH, Peters TJ (2003). Sequential studies of sleep disturbance and quality of life in abstaining alcoholics. Addict Biol 8: 455–462.

Colombo G, Agabio R, Carai MA et al. (2000). Ability of baclofen in reducing alcohol intake and withdrawal severity: I Preclinical evidence. Alcohol Clin Exp Res 24: 58–66.

Colrain IM, Turlington S, Baker FC (2009). Impact of alcoholism on sleep architecture and EEG power spectra in men and women. Sleep 32: 1341–1352.

Cooper E, Vernon J (2013). The effectiveness of pharmacological approaches in the treatment of alcohol withdrawal syndrome (AWS): a literature review. J Psychiatr Ment Health Nurs 20: 601–612.

Correia D, Ribeiro AF, Brunialti Godard AL et al. (2009). Trait anxiety and ethanol: anxiolysis in high-anxiety mice and no relation to intake behavior in an addiction model. Prog Neuropsychopharmacol Biol Psychiatry 33: 880–888.

Crabbe JC (1992). Antagonism of ethanol withdrawal convulsions in withdrawal seizure prone mice by diazepam and abecarnil. Eur J Pharmacol 221: 85–90.

Crawshaw LI, O'Connor CS, Crabbe JC et al. (1994). Temperature regulation in mice during withdrawal from ethanol dependence. Am J Physiol 267: R929–R934.

Dahchour A, De Witte P (2003). Excitatory and inhibitory amino acid changes during repeated episodes of ethanol withdrawal: an in vivo microdialysis study. Eur J Pharmacol 459: 171–178.

Dahchour A, Hoffman A, Deitrich R et al. (2000). Effects of ethanol on extracellular amino acid levels in high-and low-alcohol sensitive rats: a microdialysis study. Alcohol Alcohol 35: 548–553.

De Sousa A (2010). The role of topiramate and other anticonvulsants in the treatment of alcohol dependence: a clinical review. CNS Neurol Disord Drug Targets 9: 45–49.

Deitrich RA, Radcliffe R, Erwin VG (1996). Pharmacological effects in the development of physiological tolerance and physical dependence. In: H Begleiter, B Kissin (Eds.), The pharmacology of alcohol and alcohol dependence. Oxford University Press, New York, pp. 431–476.

Diana M (2011). The dopamine hypothesis of drug addiction and its potential therapeutic value. Front Psychiatry 2: 64.

Diana M, Pistis M, Muntoni A et al. (1996). Mesolimbic dopaminergic reduction outlasts ethanol withdrawal syndrome: evidence of protracted abstinence. Neuroscience 71: 411–415.

Dopico AM, Lemos JR, Treistman SN (1996). Ethanol increases the activity of large conductance, Ca(2+)-activated K+channels in isolated neurohypophysial terminals. Mol Pharmacol 49: 40–48.

Drummond SP, Gillin JC, Smith TL et al. (1998). The sleep of abstinent pure primary alcoholic patients: natural course and relationship to relapse. Alcohol Clin Exp Res 22: 1796–1802.

Economidou D, Hansson AC, Weiss F et al. (2008). Dysregulation of nociceptin/orphanin FQ activity in the amygdala is linked to excessive alcohol drinking in the rat. Biol Psychiatry 64: 211–218.

Economidou D, Cippitelli A, Stopponi S et al. (2011). Activation of brain NOP receptors attenuates acute and protracted alcohol withdrawal symptoms in the rat. Alcohol Clin Exp Res 35: 747–755.

Edenberg HJ, Wang J, Tian H et al. (2008). A regulatory variation in OPRK1, the gene encoding the kappa-opioid receptor, is associated with alcohol dependence. Hum Mol Genet 17: 1783–1789.

Ehlers CL, Slawecki CJ (2000). Effects of chronic ethanol exposure on sleep in rats. Alcohol 20: 173–179.

Ende G, Hermann D, Demirakca T et al. (2013). Loss of control of alcohol use and severity of alcohol dependence in non-treatment-seeking heavy drinkers are related to lower glutamate in frontal white matter. Alcohol Clin Exp Res 37: 1643–1649.

Engleman EA, McBride WJ, Li TK et al. (2003). Ethanol drinking experience attenuates (−)sulpiride-induced increases in extracellular dopamine levels in the nucleus accumbens of alcohol-preferring (P) rats. Alcohol Clin Exp Res 27: 424–431.

Enquist J, Ferwerda M, Madhavan A et al. (2012). Chronic ethanol potentiates the effect of neuropeptide s in the basolateral amygdala and shows increased anxiolytic and anti-depressive effects. Neuropsychopharmacology 37: 2436–2445.

Faber ES (2010). Functional interplay between NMDA receptors, SK channels and voltage-gated Ca^{2+} channels regulates synaptic excitability in the medial prefrontal cortex. J Physiol 588: 1281–1292.

Faber ES, Delaney AJ, Sah P (2005). SK channels regulate excitatory synaptic transmission and plasticity in the lateral amygdala. Nat Neurosci 8: 635–641.

Faber ES, Delaney AJ, Power JM et al. (2008). Modulation of SK channel trafficking by beta adrenoceptors enhances excitatory synaptic transmission and plasticity in the amygdala. J Neurosci 28: 10803–10813.

Fadda F, Rossetti ZL (1998). Chronic ethanol consumption: from neuroadaptation to neurodegeneration. Prog Neurobiol 56: 385–431.

Fahlke C, Hard E, Eriksson CJ et al. (1995). Consequence of long-term exposure to corticosterone or dexamethasone on ethanol consumption in the adrenalectomized rat, and the effect of type I and type II corticosteroid receptor antagonists. Psychopharmacology (Berl) 117: 216–224.

Fahlke C, Hard E, Hansen S (1996). Facilitation of ethanol consumption by intracerebroventricular infusions of corticosterone. Psychopharmacology (Berl) 127: 133–139.

Fakler B, Adelman JP (2008). Control of K(Ca) channels by calcium nano/microdomains. Neuron 59: 873–881.

Feige B, Scaal S, Hornyak M et al. (2007). Sleep electroencephalographic spectral power after withdrawal from alcohol in alcohol-dependent patients. Alcohol Clin Exp Res 31: 19–27.

Finn DA, Crabbe JC (1997). Exploring alcohol withdrawal syndrome. Alcohol Health Res World 21: 149–156.

Fleming RL, Manis PB, Morrow AL (2009). The effects of acute and chronic ethanol exposure on presynaptic and postsynaptic gamma-aminobutyric acid (GABA) neurotransmission in cultured cortical and hippocampal neurons. Alcohol 43: 603–618.

Fleming RL, Acheson SK, Moore SD et al. (2011). GABA transport modulates the ethanol sensitivity of tonic inhibition in the rat dentate gyrus. Alcohol 45: 577–583.

Friedman HJ (1980). Assessment of physical dependence on and withdrawal from ethanol in animals. In: H Rigter, JC Crabbe (Eds.), Alcohol tolerance and dependence. Elsevier/North Holland Biomedical Press, Amsterdam, pp. 93–121.

Frye GD, McCown TJ, Breese GR (1983). Differential sensitivity of ethanol withdrawal signs in the rat to gamma-aminobutyric acid (GABA)mimetics: blockade of audiogenic seizures but not forelimb tremors. J Pharmacol Exp Ther 226: 720–725.

Funk CK, Koob GF (2007). A CRF(2) agonist administered into the central nucleus of the amygdala decreases ethanol self-administration in ethanol-dependent rats. Brain Res 1155: 172–178.

Funk CK, Zorrilla EP, Lee MJ et al. (2007). Corticotropin-releasing factor 1 antagonists selectively reduce ethanol self-administration in ethanol-dependent rats. Biol Psychiatry 61: 78–86.

Gass JT, Olive MF (2008). Glutamatergic substrates of drug addiction and alcoholism. Biochem Pharmacol 75: 218–265.

Gehlert DR, Cippitelli A, Thorsell A et al. (2007). 3-(4-Chloro-2-morpholin-4-yl-thiazol-5-yl)-8-(1-ethylpropyl)-2,6-dimethyl-imidazo [1,2-b]pyridazine: a novel brain-penetrant, orally available corticotropin-releasing factor receptor 1 antagonist with efficacy in animal models of alcoholism. J Neurosci 27: 2718–2726.

George DT, Gilman J, Hersh J et al. (2008). Neurokinin 1 receptor antagonism as a possible therapy for alcoholism. Science 319: 1536–1539.

Ghezzi A, Krishnan HR, Atkinson NS (2012). Susceptibility to ethanol withdrawal seizures is produced by BK channel gene expression. Addict Biol, doi: 10.1111/j.1369-1600.2012.00465.x.

Gillett K, Harshberger E, Valdez GR (2013). Protracted withdrawal from ethanol and enhanced responsiveness stress: regulation via the dynorphin/kappa opioid receptor system. Alcohol 47: 359–365.

Gilman JM, Hommer DW (2008). Modulation of brain response to emotional images by alcohol cues in alcohol-dependent patients. Addict Biol 13: 423–434.

Gilpin NW (2012). Neuropeptide Y (NPY) in the extended amygdala is recruited during the transition to alcohol dependence. Neuropeptides 46: 253–259.

Gilpin NW, Koob GF (2010). Effects of beta-adrenoceptor antagonists on alcohol drinking by alcohol-dependent rats. Psychopharmacology (Berl) 212: 431–439.

Gilpin NW, Misra K, Koob GF (2008a). Neuropeptide Y in the central nucleus of the amygdala suppresses dependence-induced increases in alcohol drinking. Pharmacol Biochem Behav 90: 475–480.

Gilpin NW, Richardson HN, Koob GF (2008b). Effects of CRF1-receptor and opioid-receptor antagonists on dependence-induced increases in alcohol drinking by alcohol-preferring (P) rats. Alcohol Clin Exp Res 32: 1535–1542.

Gilpin NW, Misra K, Herman MA et al. (2011). Neuropeptide Y opposes alcohol effects on gamma-aminobutyric acid release in amygdala and blocks the transition to alcohol dependence. Biol Psychiatry 69: 1091–1099.

Gonzales RA, Job MO, Doyon WM (2004). The role of mesolimbic dopamine in the development and maintenance of ethanol reinforcement. Pharmacol Ther 103: 121–146.

Gothoni P, Ikola W (1985). Ethanol withdrawal tremor potentiates the tremorogenic action of nicotine. Med Biol 63: 131–136.

Graef JD, Huitt TW, Nordskog BK et al. (2011). Disrupted thalamic T-type Ca^{2+} channel expression and function during ethanol exposure and withdrawal. J Neurophysiol 105: 528–540.

Griffin III WC, Haun HL, Hazelbaker CL et al. (2014). Increased extracellular glutamate in the nucleus accumbens promotes excessive ethanol drinking in ethanol dependent mice. Neuropsychopharmacology 39: 707–717.

Grillon C, Sinha R, O'Malley SS (1994). Effects of ethanol on the acoustic startle reflex in humans. Psychopharmacology (Berl) 114: 167–171.

Grillon C, Sinha R, Ameli R et al. (2000). Effects of alcohol on baseline startle and prepulse inhibition in young men at risk for alcoholism and/or anxiety disorders. J Stud Alcohol 61: 46–54.

Gu M, Zahr NM, Spielman DM et al. (2013). Quantification of glutamate and glutamine using constant-time point-resolved spectroscopy at 3 T. NMR Biomed 26: 164–172.

Guglielmo R, Martinotti G, Clerici M et al. (2012). Pregabalin for alcohol dependence: a critical review of the literature. Adv Ther 29: 947–957.

Hansson AC, Cippitelli A, Sommer WH et al. (2006). Variation at the rat Crhr1 locus and sensitivity to relapse into alcohol seeking induced by environmental stress. Proc Natl Acad Sci U S A 103: 15236–15241.

Hansson AC, Rimondini R, Neznanova O et al. (2008). Neuroplasticity in brain reward circuitry following a history of ethanol dependence. Eur J Neurosci 27: 1912–1922.

Hatzigiakoumis DS, Martinotti G, Giannantonio MD et al. (2011). Anhedonia and substance dependence: clinical correlates and treatment options. Front Psychiatry 2: 10.

Haugbol SR, Ebert B, Ulrichsen J (2005). Upregulation of glutamate receptor subtypes during alcohol withdrawal in rats. Alcohol Alcohol 40: 89–95.

Hawley RJ, Nemeroff CB, Bissette G et al. (1994). Neurochemical correlates of sympathetic activation during severe alcohol withdrawal. Alcohol Clin Exp Res 18: 1312–1316.

Heilig M, Koob GF (2007). A key role for corticotropin-releasing factor in alcohol dependence. Trends Neurosci 30: 399–406.

Heilig M, Koob GF, Ekman R et al. (1994). Corticotropin-releasing factor and neuropeptide Y: role in emotional integration. Trends Neurosci 17: 80–85.

Heilig M, Egli M, Crabbe JC et al. (2010). Acute withdrawal, protracted abstinence and negative affect in alcoholism: are they linked? Addict Biol 15: 169–184.

Heinz A, Lichtenberg-Kraag B, Baum SS et al. (1995). Evidence for prolonged recovery of dopaminergic transmission after detoxification in alcoholics with poor treatment outcome. J Neural Transm Gen Sect 102: 149–157.

Henniger MS, Spanagel R, Wigger A et al. (2002). Alcohol self-administration in two rat lines selectively bred for extremes in anxiety-related behavior. Neuropsychopharmacology 26: 729–736.

Hermann D, Weber-Fahr W, Sartorius A et al. (2012). Translational magnetic resonance spectroscopy reveals excessive central glutamate levels during alcohol withdrawal in humans and rats. Biol Psychiatry 71: 1015–1021.

Hillbom M, Pieninkeroinen I, Leone M (2003). Seizures in alcohol-dependent patients: epidemiology, pathophysiology and management. CNS Drugs 17: 1013–1030.

Hodge CW, Kelley SP, Bratt AM et al. (2004). 5-HT(3A) receptor subunit is required for 5-HT3 antagonist-induced reductions in alcohol drinking. Neuropsychopharmacology 29: 1807–1813.

Hoffman DA, Magee JC, Colbert CM et al. (1997). K^+ channel regulation of signal propagation in dendrites of hippocampal pyramidal neurons. Nature 387: 869–875.

Holmes A, Fitzgerald PJ, MacPherson KP et al. (2012). Chronic alcohol remodels prefrontal neurons and disrupts NMDAR-mediated fear extinction encoding. Nat Neurosci 15: 1359–1361.

Hopf FW, Martin M, Chen BT et al. (2007). Withdrawal from intermittent ethanol exposure increases probability of burst firing in VTA neurons in vitro. J Neurophysiol 98: 2297–2310.

Hopf FW, Bowers MS, Chang SJ et al. (2010). Reduced nucleus accumbens SK channel activity enhances alcohol seeking during abstinence. Neuron 65: 682–694.

Huang MC, Ho CW, Chen CH et al. (2010a). Reduced expression of circadian clock genes in male alcoholic patients. Alcohol Clin Exp Res 34: 1899–1904.

Huang MM, Overstreet DH, Knapp DJ et al. (2010b). Corticotropin-releasing factor (CRF) sensitization of ethanol withdrawal-induced anxiety-like behavior is brain site specific and mediated by CRF-1 receptors: relation to stress-induced sensitization. J Pharmacol Exp Ther 332: 298–307.

Jakab M, Weiger TM, Hermann A (1997). Ethanol activates maxi Ca^{2+}-activated K^+ channels of clonal pituitary (GH3) cells. J Membr Biol 157: 237–245.

Janiri L, Martinotti G, Dario T et al. (2005). Anhedonia and substance-related symptoms in detoxified substance-dependent subjects: a correlation study. Neuropsychobiology 52: 37–44.

Johnson BA (2004). Role of the serotonergic system in the neurobiology of alcoholism: implications for treatment. CNS Drugs 18: 1105–1118.

Johnson BA, Ait-Daoud N (2000). Neuropharmacological treatments for alcoholism: scientific basis and clinical findings. Psychopharmacology (Berl) 149: 327–344.

Johnson BA, Ait-Daoud N (2010). Topiramate in the new generation of drugs: efficacy in the treatment of alcoholic patients. Curr Pharm Des 16: 2103–2112.

Johnson SW, Seutin V (1997). Bicuculline methiodide potentiates NMDA-dependent burst firing in rat dopamine neurons by blocking apamin-sensitive Ca^{2+}-activated K^+ currents. Neurosci Lett 231: 13–16.

Johnson BA, Ait-Daoud N, Prihoda TJ (2000). Combining ondansetron and naltrexone effectively treats biologically predisposed alcoholics: from hypotheses to preliminary clinical evidence. Alcohol Clin Exp Res 24: 737–742.

Johnson BA, Seneviratne C, Wang XQ et al. (2013). Determination of genotype combinations that can predict the outcome of the treatment of alcohol dependence using the 5-HT(3) antagonist ondansetron. Am J Psychiatry 170: 1020–1031.

Kahkonen S (2004). Mechanisms of cardiovascular dysregulation during alcohol withdrawal. Prog Neuropsychopharmacol Biol Psychiatry 28: 937–941.

Kapasova Z, Szumlinski KK (2008). Strain differences in alcohol-induced neurochemical plasticity: a role for accumbens glutamate in alcohol intake. Alcohol Clin Exp Res 32: 617–631.

Kashkin VA, Zvartau EE, Fedorova OV et al. (2008). Endogenous bufadienolide mediates pressor response to ethanol withdrawal in rats. Eur Neuropsychopharmacol 18: 74–77.

Katsura M, Torigoe F, Hayashida S et al. (2005). Ethanol physical dependence is accompanied by up-regulated expression of L-type high voltage-gated calcium channel alpha1 subunits in mouse brain. Brain Res 1039: 211–215.

Kayir H, Yavuz O, Goktalay G et al. (2010). The relationship between baseline prepulse inhibition levels and ethanol withdrawal severity in rats. Prog Neuropsychopharmacol Biol Psychiatry 34: 1507–1514.

Kim AK, Brown RM, Lawrence AJ (2012). The role of orexins/hypocretins in alcohol use and abuse: an appetitive-reward relationship. Frontiers Behavior Neurosci 6: 78.

King AC, Parsons OA, Bernardy NC et al. (1994). Drinking history is related to persistent blood pressure dysregulation in postwithdrawal alcoholics. Alcohol Clin Exp Res 18: 1172–1176.

Kissler JL, Sirohi S, Reis DJ et al. (2013). The one-two punch of alcoholism: role of central amygdala dynorphins/kappa-opioid receptors. Biol Psychiatry, doi: 10.1016/j.biopsych.2013.03.014.

Kliethermes CL (2005). Anxiety-like behaviors following chronic ethanol exposure. Neurosci Biobehav Rev 28: 837–850.

Koenig HN, Olive MF (2004). The glucocorticoid receptor antagonist mifepristone reduces ethanol intake in rats under limited access conditions. Psychoneuroendocrinology 29: 999–1003.

Koller W, O'Hara R, Dorus W et al. (1985). Tremor in chronic alcoholism. Neurology 35: 1660–1662.

Koob GF (2013). Theoretical frameworks and mechanistic aspects of alcohol addiction: alcohol addiction as a reward deficit disorder. Curr Top Behav Neurosci 13: 3–30.

Koob G, Kreek MJ (2007). Stress, dysregulation of drug reward pathways, and the transition to drug dependence. Am J Psychiatry 164: 1149–1159.

Koob GF, Le Moal M (2001). Drug addiction, dysregulation of reward, and allostasis. Neuropsychopharmacology 24: 97–129.

Koob GF, Le Moal M (2008). Addiction and the brain antireward system. Annu Rev Psychol 59: 29–53.

Koob GF, Zorrilla EP (2010). Neurobiological mechanisms of addiction: focus on corticotropin-releasing factor. Curr Opin Investig Drugs 11: 63–71.

Kranzler HR, Pierucci-Lagha A, Feinn R et al. (2003). Effects of ondansetron in early- versus late-onset alcoholics: a prospective, open-label study. Alcohol Clin Exp Res 27: 1150–1155.

Kranzler HR, Covault J, Pierucci-Lagha A et al. (2008). Effects of aripiprazole on subjective and physiological responses to alcohol. Alcohol Clin Exp Res 32: 573–579.

Kranzler HR, Armeli S, Tennen H (2012). Post-treatment outcomes in a double-blind, randomized trial of sertraline for alcohol dependence. Alcohol Clin Exp Res 36: 739–744.

Kroener S, Mulholland PJ, New NN et al. (2012). Chronic alcohol exposure alters behavioral and synaptic plasticity of the rodent prefrontal cortex. PLoS One 7: e37541.

Krystal JH, Webb E, Grillon C et al. (1997). Evidence of acoustic startle hyperreflexia in recently detoxified early onset male alcoholics: modulation by yohimbine and m-chlorophenylpiperazine (mCPP). Psychopharmacology (Berl) 131: 207–215.

Kubota T, De A, Brown RA et al. (2002). Diurnal effects of acute and chronic administration of ethanol on sleep in rats. Alcohol Clin Exp Res 26: 1153–1161.

Kufahl PR, Martin-Fardon R, Weiss F (2011). Enhanced sensitivity to attenuation of conditioned reinstatement by the mGluR 2/3 agonist LY379268 and increased functional activity of mGluR 2/3 in rats with a history of ethanol dependence. Neuropsychopharmacology 36: 2762–2773.

Kumar S, Sieghart W, Morrow AL (2002). Association of protein kinase C with GABA(A) receptors containing alpha1 and alpha4 subunits in the cerebral cortex: selective effects of chronic ethanol consumption. J Neurochem 82: 110–117.

Kumar S, Porcu P, Werner DF et al. (2009). The role of GABA(A) receptors in the acute and chronic effects of ethanol: a decade of progress. Psychopharmacology (Berl) 205: 529–564.

Lack AK, Diaz MR, Chappell A et al. (2007). Chronic ethanol and withdrawal differentially modulate pre- and postsynaptic function at glutamatergic synapses in rat basolateral amygdala. J Neurophysiol 98: 3185–3196.

Lambert DG (2008). The nociceptin/orphanin FQ receptor: a target with broad therapeutic potential. Nat Rev Drug Discov 7: 694–710.

Landolt HP, Gillin JC (2001). Sleep abnormalities during abstinence in alcohol-dependent patients. Aetiology and management. CNS Drugs 15: 413–425.

Latorre R, Brauchi S (2006). Large conductance Ca^{2+}-activated K^+ (BK) channel: activation by Ca^{2+} and voltage. Biol Res 39: 385–401.

Le Bon O, Murphy JR, Staner L et al. (2003). Double-blind, placebo-controlled study of the efficacy of trazodone in alcohol post-withdrawal syndrome: polysomnographic and clinical evaluations. J Clin Psychopharmacol 23: 377–383.

Le AD, Harding S, Juzytsch W et al. (2000). The role of corticotrophin-releasing factor in stress-induced relapse to alcohol-seeking behavior in rats. Psychopharmacology (Berl) 150: 317–324.

Lechtenberg R, Worner TM (1991). Relative kindling effects of detoxification and non-detoxification admissions in alcoholics. Alcohol Alcohol 26: 221–225.

Lee WS, Ngo-Anh TJ, Bruening-Wright A et al. (2003). Small conductance Ca^{2+}-activated K^+ channels and calmodulin: cell surface expression and gating. J Biol Chem 278: 25940–25946.

Li Q, Fleming RL, Acheson SK et al. (2013). Long-term modulation of A-type K(+) conductances in hippocampal CA1 interneurons in rats after chronic intermittent ethanol exposure during adolescence or adulthood. Alcohol Clin Exp Res 37: 2074–2085.

Liang J, Cagetti E, Olsen RW et al. (2004). Altered pharmacology of synaptic and extrasynaptic GABAA receptors on CA1 hippocampal neurons is consistent with subunit changes in a model of alcohol withdrawal and dependence. J Pharmacol Exp Ther 310: 1234–1245.

Lin MT, Lujan R, Watanabe M et al. (2008). SK2 channel plasticity contributes to LTP at Schaffer collateral-CA1 synapses. Nat Neurosci 11: 170–177.

Litten RZ, Fertig J, Mattson M et al. (2005). Development of medications for alcohol use disorders: recent advances and ongoing challenges. Expert Opin Emerg Drugs 10: 323–343.

Little HJ, Croft AP, O'Callaghan MJ et al. (2008). Selective increases in regional brain glucocorticoid: a novel effect of chronic alcohol. Neuroscience 156: 1017–1027.

Littleton J (1998). Neurochemical mechanisms underlying alcohol withdrawal. Alcohol Health Res World 22: 13–24.

Liu X, Weiss F (2002a). Additive effect of stress and drug cues on reinstatement of ethanol seeking: exacerbation by history of dependence and role of concurrent activation of corticotropin-releasing factor and opioid mechanisms. J Neurosci 22: 7856–7861.

Liu X, Weiss F (2002b). Reversal of ethanol-seeking behavior by D1 and D2 antagonists in an animal model of relapse: differences in antagonist potency in previously ethanol-dependent versus nondependent rats. J Pharmacol Exp Ther 300: 882–889.

Lockridge A, Yuan LL (2011). Spatial learning deficits in mice lacking A-type K(+) channel subunits. Hippocampus 21: 1152–1156.

Logan RW, McCulley 3rd WD, Seggio JA et al. (2012). Effects of withdrawal from chronic intermittent ethanol vapor on the level and circadian periodicity of running-wheel activity in C57BL/6 J and C3H/HeJ mice. Alcohol Clin Exp Res 36: 467–476.

Lovinger DM, Roberto M (2013). Synaptic effects induced by alcohol. Curr Top Behav Neurosci 13: 31–86.

Lowery EG, Thiele TE (2010). Pre-clinical evidence that corticotropin-releasing factor (CRF) receptor antagonists are promising targets for pharmacological treatment of alcoholism. CNS Neurol Disord Drug Targets 9: 77–86.

Macey DJ, Schulteis G, Heinrichs SC et al. (1996). Time-dependent quantifiable withdrawal from ethanol in the rat: effect of method of dependence induction. Alcohol 13: 163–170.

Mainerova B, Prasko J, Latalova K et al. (2013). Alcohol withdrawal delirium – diagnosis, course and treatment. Biomedical papers of the Medical Faculty of the University Palacky. Olomouc, Czechoslovakia.

Malcolm R, Roberts J, Wang W et al. (2000). Multiple previous detoxifications are associated with less responsive treatment and heavier drinking during an index outpatient detoxification. Alcohol 22: 159–164.

Malcolm R, Myrick LH, Roberts J et al. (2002). The effects of carbamazepine and lorazepam on single versus multiple previous alcohol withdrawals in an outpatient randomized trial. J Gen Intern Med 17: 349–355.

Malcolm R, Myrick LH, Veatch LM et al. (2007). Self-reported sleep, sleepiness, and repeated alcohol withdrawals: a randomized, double blind, controlled comparison of lorazepam vs gabapentin. J Clin Sleep Med 3: 24–32.

Mann K, Lemenager T, Hoffmann S et al. (2012). Results of a double-blind, placebo-controlled pharmacotherapy trial in alcoholism conducted in Germany and comparison with the US COMBINE study. Addict Biol 18: 937–946.

Mann K, Bladstrom A, Torup L et al. (2013). Extending the treatment options in alcohol dependence: a randomized controlled study of as-needed nalmefene. Biol Psychiatry 73: 706–713.

Marinelli PW, Funk D, Juzytsch W et al. (2007). The CRF1 receptor antagonist antalarmin attenuates yohimbine-induced increases in operant alcohol self-administration and reinstatement of alcohol seeking in rats. Psychopharmacology (Berl) 195: 345–355.

Martin GE, Hendrickson LM, Penta KL et al. (2008). Identification of a BK channel auxiliary protein controlling molecular and behavioral tolerance to alcohol. Proc Natl Acad Sci U S A 105: 17543–17548.

Martinez D, Gil R, Slifstein M et al. (2005). Alcohol dependence is associated with blunted dopamine transmission in the ventral striatum. Biol Psychiatry 58: 779–786.

Martin-Fardon R, Zorrilla EP, Ciccocioppo R et al. (2010). Role of innate and drug-induced dysregulation of brain stress and arousal systems in addiction: focus on corticotropin-releasing factor, nociceptin/orphanin FQ, and orexin/hypocretin. Brain Res 1314: 145–161.

Martinotti G, Nicola MD, Reina D et al. (2008). Alcohol protracted withdrawal syndrome: the role of anhedonia. Subst Use Misuse 43: 271–284.

Martinotti G, Di Nicola M, Di Giannantonio M et al. (2009). Aripiprazole in the treatment of patients with alcohol dependence: a double-blind, comparison trial vs. naltrexone. J Psychopharmacol 23: 123–129.

Martinotti G, Reina D, Di Nicola M et al. (2010). Acetyl-L-carnitine for alcohol craving and relapse prevention in anhedonic alcoholics: a randomized, double-blind, placebo-controlled pilot trial. Alcohol Alcohol 45: 449–455.

Martinotti G, Andreoli S, Reina D et al. (2011). Acetyl-L-Carnitine in the treatment of anhedonia, melancholic and negative symptoms in alcohol dependent subjects. Prog Neuropsychopharmacol Biol Psychiatry 35: 953–958.

Marty VN, Spigelman I (2012). Long-lasting alterations in membrane properties, K(+) currents, and glutamatergic synaptic currents of nucleus accumbens medium spiny neurons in a rat model of alcohol dependence. Front Neurosci 6: 86.

Mason BJ, Lehert P (2012). Acamprosate for alcohol dependence: a sex-specific meta-analysis based on individual patient data. Alcohol Clin Exp Res 36: 497–508.

Mason BJ, Salvato FR, Williams LD et al. (1999). A double-blind, placebo-controlled study of oral nalmefene for alcohol dependence. Arch Gen Psychiatry 56: 719–724.

Mason BJ, Quello S, Goodell V et al. (2014). Gabapentin treatment for alcohol dependence: a randomized clinical trial. JAMA Intern Med 174: 70–77.

Maurel S, De Vry J, De Beun R et al. (1999). 5-HT2A and 5-HT2C/5-HT1B receptors are differentially involved in alcohol preference and consummatory behavior in cAA rats. Pharmacol Biochem Behav 62: 89–96.

Maylie J, Bond CT, Herson PS et al. (2004). Small conductance Ca^{2+}-activated K^+ channels and calmodulin. J Physiol 554: 255–261.

Mayo-Smith MF (1997). Pharmacological management of alcohol withdrawal. A meta-analysis and evidence-based practice guideline American Society of Addiction Medicine Working Group on Pharmacological Management of Alcohol Withdrawal. JAMA 278: 144–151.

McCool BA, Christian DT, Diaz MR et al. (2010). Glutamate plasticity in the drunken amygdala: the making of an anxious synapse. Int Rev Neurobiol 91: 205–233.

McMahon T, Andersen R, Metten P et al. (2000). Protein kinase C epsilon mediates up-regulation of N-type calcium channels by ethanol. Mol Pharmacol 57: 53–58.

Meinhardt MW, Hansson AC, Perreau-Lenz S et al. (2013). Rescue of infralimbic mGluR2 deficit restores control over drug-seeking behavior in alcohol dependence. J Neurosci 33: 2794–2806.

Melendez RI, McGinty JF, Kalivas PW et al. (2012). Brain region-specific gene expression changes after chronic intermittent ethanol exposure and early withdrawal in C57BL/6 J mice. Addict Biol 17: 351–364.

Miyakawa T, Yagi T, Kitazawa H et al. (1997). Fyn-kinase as a determinant of ethanol sensitivity: relation to NMDA-receptor function. Science 278: 698–701.

Moak DH, Anton RF (1996). Alcohol-related seizures and the kindling effect of repeated detoxifications: the influence of cocaine. Alcohol Alcohol 31: 135–143.

Molander A, Vengeliene V, Heilig M et al. (2012). Brain-specific inactivation of the Crhr1 gene inhibits post-dependent and stress-induced alcohol intake, but does not affect relapse-like drinking. Neuropsychopharmacology 37: 1047–1056.

Mon A, Durazzo TC, Meyerhoff DJ (2012). Glutamate, GABA, and other cortical metabolite concentrations during early abstinence from alcohol and their associations with neurocognitive changes. Drug Alcohol Depend 125: 27–36.

Mu Y, Otsuka T, Horton AC et al. (2003). Activity-dependent mRNA splicing controls ER export and synaptic delivery of NMDA receptors. Neuron 40: 581–594.

Mulholland PJ (2012). K(Ca)2 channels: novel therapeutic targets for treating alcohol withdrawal and escalation of alcohol consumption. Alcohol 46: 309–315.

Mulholland PJ, Chandler LJ (2007). The thorny side of addiction: adaptive plasticity and dendritic spines. Scientific World Journal 7: 9–21.

Mulholland PJ, Kroener S, Chandler LJ (2010). Chronic ethanol exposure enhances backpropagating action potential-induced calcium transients in distal apical dendrites of ca1 pyramidal neurons. Alcohol Clin Exp Res 34: 140A.

Mulholland PJ, Becker HC, Woodward JJ et al. (2011). Small conductance calcium-activated potassium type 2 channels regulate alcohol-associated plasticity of glutamatergic synapses. Biol Psychiatry 69: 625–632.

Murphy JM, Stewart RB, Bell RL et al. (2002). Phenotypic and genotypic characterization of the Indiana University rat lines selectively bred for high and low alcohol preference. Behav Genet 32: 363–388.

Muzyk AJ, Fowler JA, Norwood DK et al. (2011). Role of alpha2-agonists in the treatment of acute alcohol withdrawal. Ann Pharmacother 45: 649–657.

Myrick H, Malcolm R, Randall PK et al. (2009). A double-blind trial of gabapentin versus lorazepam in the treatment of alcohol withdrawal. Alcohol Clin Exp Res 33: 1582–1588.

Myrick H, Li X, Randall PK et al. (2010). The effect of aripiprazole on cue-induced brain activation and drinking parameters in alcoholics. J Clin Psychopharmacol 30: 365–372.

Naranjo CA, Knoke DM (2001). The role of selective serotonin reuptake inhibitors in reducing alcohol consumption. J Clin Psychiatry 62 (Suppl. 20): 18–25.

Nealey KA, Smith AW, Davis SM et al. (2011). Kappa-opioid receptors are implicated in the increased potency of intra-accumbens nalmefene in ethanol-dependent rats. Neuropharmacology 61: 35–42.

Ngo-Anh TJ, Bloodgood BL, Lin M et al. (2005). SK channels and NMDA receptors form a Ca^{2+}-mediated feedback loop in dendritic spines. Nat Neurosci 8: 642–649.

Nordskog BK, Hammarback JA, Godwin DW (2006). Diurnal gene expression patterns of T-type calcium channels and their modulation by ethanol. Neuroscience 141: 1365–1373.

O'Daly OG, Trick L, Scaife J et al. (2012). Withdrawal-associated increases and decreases in functional neural connectivity associated with altered emotional regulation in alcoholism. Neuropsychopharmacology 37: 2267–2276.

O'Malley SS, Jaffe AJ, Chang G et al. (1992). Naltrexone and coping skills therapy for alcohol dependence. A controlled study. Arch Gen Psychiatry 49: 881–887.

Patkar AA, Gopalakrishnan R, Naik PC et al. (2003). Changes in plasma noradrenaline and serotonin levels and craving during alcohol withdrawal. Alcohol Alcohol 38: 224–231.

Petrakis IL (2006). A rational approach to the pharmacotherapy of alcohol dependence. J Clin Psychopharmacol 26 (Suppl. 1): S3–S12.

Petrakis IL, Ralevski E, Desai N et al. (2012). Noradrenergic vs serotonergic antidepressant with or without naltrexone for veterans with PTSD and comorbid alcohol dependence. Neuropsychopharmacology 37: 996–1004.

Pettinati HM, Rukstalis MR, Luck GJ et al. (2000). Gender and psychiatric comorbidity: impact on clinical presentation of alcohol dependence. Am J Addict 9: 242–252.

Pettinati HM, Oslin DW, Kampman KM et al. (2010). A double-blind, placebo-controlled trial combining sertraline and naltrexone for treating co-occurring depression and alcohol dependence. Am J Psychiatry 167: 668–675.

Piazza PV, Le Moal M (1997). Glucocorticoids as a biological substrate of reward: physiological and pathophysiological implications. Brain Res Brain Res Rev 25: 359–372.

Pieninkeroinen IP, Telakivi TM, Hillbom ME (1992). Outcome in subjects with alcohol-provoked seizures. Alcohol Clin Exp Res 16: 955–959.

Pietrzykowski AZ, Martin GE, Puig SI et al. (2004). Alcohol tolerance in large-conductance, calcium-activated potassium channels of CNS terminals is intrinsic and includes two components: decreased ethanol potentiation and decreased channel density. J Neurosci 24: 8322–8332.

Ponomarev I, Crabbe JC (1999). Genetic association between chronic ethanol withdrawal severity and acoustic startle parameters WSP and WSR mice. Alcohol Clin Exp Res 23: 1730–1735.

Porjesz B, Begleiter H (1996). Effects of alcohol on electrophysiological activity of the brain. In: H Begleiter, B Kissin (Eds.), The Pharmacology of Alcohol and Alcohol Dependence. Oxford University Press, New York, pp. 207–247.

Pozzi G, Martinotti G, Reina D et al. (2008). The assessment of post-detoxification anhedonia: influence of clinical and psychosocial variables. Subst Use Misuse 43: 722–732.

Prendergast MA, Mulholland PJ (2012). Glucocorticoid and polyamine interactions in the plasticity of glutamatergic synapses that contribute to ethanol-associated dependence and neuronal injury. Addict Biol 17: 209–223.

Rasmussen DD, Wilkinson CW, Raskind MA (2006). Chronic daily ethanol and withdrawal: 6. Effects on rat sympathoadrenal activity during “abstinence”. Alcohol 38: 173–177.

Rassnick S, Koob GF, Geyer MA (1992). Responding to acoustic startle during chronic ethanol intoxication and withdrawal. Psychopharmacology (Berl) 106: 351–358.

Rassnick S, Heinrichs SC, Britton KT et al. (1993). Microinjection of a corticotropin-releasing factor antagonist into the central nucleus of the amygdala reverses anxiogenic-like effects of ethanol withdrawal. Brain Res 605: 25–32.

Ray LA, Hutchison KE (2007). Effects of naltrexone on alcohol sensitivity and genetic moderators of medication response: a double-blind placebo-controlled study. Arch Gen Psychiatry 64: 1069–1077.

Reilly W, Koirala B, Devaud LL (2009). Sex differences in acoustic startle responses and seizure thresholds between ethanol-withdrawn male and female rats. Alcohol Alcohol 44: 561–566.

Reinscheid RK, Nothacker HP, Bourson A et al. (1995). Orphanin FQ: a neuropeptide that activates an opioidlike G protein-coupled receptor. Science 270: 792–794.

Riihioja P, Jaatinen P, Oksanen H et al. (1997). Dexmedetomidine, diazepam, and propranolol in the treatment of ethanol withdrawal symptoms in the rat. Alcohol Clin Exp Res 21: 804–808.

Rivier C (2000). Effects of alcohol on the neuroendocrine system. In: A Noronha, M Eckardt, K Warren (Eds.), Review of NIAAA’s neuroscience and behavioral research portfolio: NIAAA Research Monograph No 34. National Institute on Alcohol Abuse and Alcoholism, Bethesda, MD, pp. 61–81.

Roberto M, Gilpin NW, O’Dell LE et al. (2008). Cellular and behavioral interactions of gabapentin with alcohol dependence. J Neurosci 28: 5762–5771.

Roberto M, Cruz MT, Gilpin NW et al. (2010). Corticotropin releasing factor-induced amygdala gamma-aminobutyric acid release plays a key role in alcohol dependence. Biol Psychiatry 67: 831–839.

Roelofs SM (1985). Hyperventilation, anxiety, craving for alcohol: a subacute alcohol withdrawal syndrome. Alcohol 2: 501–505.

Rose AK, Shaw SG, Prendergast MA et al. (2010). The importance of glucocorticoids in alcohol dependence and neurotoxicity. Alcohol Clin Exp Res 34: 2011–2018.

Rosenwasser AM (2001). Alcohol, antidepressants, and circadian rhythms. Human and animal models. Alcohol Res Health 25: 126–135.

Rosenwasser AM, Fecteau ME, Logan RW (2005). Effects of ethanol intake and ethanol withdrawal on free-running circadian activity rhythms in rats. Physiol Behav 84: 537–542.

Rossetti ZL, Carboni S (1995). Ethanol withdrawal is associated with increased extracellular glutamate in the rat striatum. Eur J Pharmacol 283: 177–183.

Roy A, Pandey SC (2002). The decreased cellular expression of neuropeptide Y protein in rat brain structures during ethanol withdrawal after chronic ethanol exposure. Alcohol Clin Exp Res 26: 796–803.

Ruggeri B, Braconi S, Cannella N et al. (2010). Neuropeptide S receptor gene expression in alcohol withdrawal and protracted abstinence in postdependent rats. Alcohol Clin Exp Res 34: 90–97.

Ryabinin AE, Tsoory MM, Kozicz T et al. (2012). Urocortins: CRF’s siblings and their potential role in anxiety, depression and alcohol drinking behavior. Alcohol 46: 349–357.

Rylkova D, Shah HP, Small E et al. (2009). Deficit in brain reward function and acute and protracted anxiety-like behavior after discontinuation of a chronic alcohol liquid diet in rats. Psychopharmacology (Berl) 203: 629–640.

Sahr AE, Thielen RJ, Lumeng L et al. (2004). Long-lasting alterations of the mesolimbic dopamine system after periadolescent ethanol drinking by alcohol-preferring rats. Alcohol Clin Exp Res 28: 702–711.

Saitz R (1998). Introduction to alcohol withdrawal. Alcohol Health Res World 22: 5–12.

Sand T, Bjork M, Brathen G et al. (2010). Quantitative EEG in patients with alcohol-related seizures. Alcohol Clin Exp Res 34: 1751–1758.

Schank JR, Pickens CL, Rowe KE et al. (2011). Stress-induced reinstatement of alcohol-seeking in rats is selectively suppressed by the neurokinin 1 (NK1) antagonist L822429. Psychopharmacology (Berl) 218: 111–119.

Schank JR, Goldstein AL, Rowe KE et al. (2012). The kappa opioid receptor antagonist JDTic attenuates alcohol seeking and withdrawal anxiety. Addict Biol 17: 634–647.

Schellekens AF, Mulders PC, Ellenbroek B et al. (2012). Early-onset alcohol dependence increases the acoustic startle reflex. Alcohol Clin Exp Res 36: 1075–1083.

Schmid B, Blomeyer D, Treutlein J et al. (2010). Interacting effects of CRHR1 gene and stressful life events on drinking initiation and progression among 19-year-olds. Int J Neuropsychopharmacol 13: 703–714.

Schneider U, Altmann A, Baumann M et al. (2001). Comorbid anxiety and affective disorder in alcohol-dependent patients seeking treatment: the first Multicentre Study in Germany. Alcohol Alcohol 36: 219–223.

Schroeder JP, Overstreet DH, Hodge CW (2005). The neuropeptide-Y Y5 receptor antagonist L-152,804 decreases alcohol self-administration in inbred alcohol-preferring (iP) rats. Alcohol 36: 179–186.

Schuckit MA, Hesselbrock V (1994). Alcohol dependence and anxiety disorders: what is the relationship? Am J Psychiatry 151: 1723–1734.

Schulteis G, Markou A, Cole M et al. (1995). Decreased brain reward produced by ethanol withdrawal. Proc Natl Acad Sci U S A 92: 5880–5884.

Seggio JA, Fixaris MC, Reed JD et al. (2009). Chronic ethanol intake alters circadian phase shifting and free-running period in mice. J Biol Rhythms 24: 304–312.

Seo D, Lacadie CM, Tuit K et al. (2013). Disrupted ventromedial prefrontal function, alcohol craving, and subsequent relapse risk. JAMA Psychiatry 70: 727–739.

Seutin V, Johnson SW, North RA (1993). Apamin increases NMDA-induced burst-firing of rat mesencephalic dopamine neurons. Brain Res 630: 341–344.

Sharma R, Engemann SC, Sahota P et al. (2010). Effects of ethanol on extracellular levels of adenosine in the basal forebrain: an in vivo microdialysis study in freely behaving rats. Alcohol Clin Exp Res 34: 813–818.

Shippenberg TS, Zapata A, Chefer VI (2007). Dynorphin and the pathophysiology of drug addiction. Pharmacol Ther 116: 306–321.

Shirafuji S, Liu J, Okamura N et al. (2010). QT interval dispersion and cardiac sympathovagal balance shift in rats with acute ethanol withdrawal. Alcohol Clin Exp Res 34: 223–230.

Sidhpura N, Weiss F, Martin-Fardon R (2010). Effects of the mGlu2/3 agonist LY379268 and the mGlu5 antagonist MTEP on ethanol seeking and reinforcement are differentially altered in rats with a history of ethanol dependence. Biol Psychiatry 67: 804–811.

Simon O'Brien E, Legastelois R, Houchi H et al. (2011). Fluoxetine, desipramine, and the dual antidepressant milnacipran reduce alcohol self-administration and/or relapse in dependent rats. Neuropsychopharmacology 36: 1518–1530.

Simpson TL, Saxon AJ, Meredith CW et al. (2009). A pilot trial of the alpha-1 adrenergic antagonist, prazosin, for alcohol dependence. Alcohol Clin Exp Res 33: 255–263.

Sinha R (2013). Modeling relapse situations in the human laboratory. Curr Top Behav Neurosci 13: 379–402.

Smith AD, Weiss F (1999). Ethanol exposure differentially alters central monoamine neurotransmission in alcohol-preferring versus -nonpreferring rats. J Pharmacol Exp Ther 288: 1223–1228.

Smith N, Hill R, Marshall J et al. (2013). Sleep related beliefs and their association with alcohol relapse following residential alcohol detoxification treatment. Behav Cogn Psychother: 1–12.

Sommer WH, Rimondini R, Hansson AC et al. (2008). Upregulation of voluntary alcohol intake, behavioral sensitivity to stress, and amygdala crhr1 expression following a history of dependence. Biol Psychiatry 63: 139–145.

Spanagel R (2009). Alcoholism: a systems approach from molecular physiology to addictive behavior. Physiol Rev 89: 649–705.

Spanagel R, Montkowski A, Allingham K et al. (1995). Anxiety: a potential predictor of vulnerability to the initiation of ethanol self-administration in rats. Psychopharmacology (Berl) 122: 369–373.

Sparrow AM, Lowery-Gionta EG, Pleil KE et al. (2012). Central neuropeptide Y modulates binge-like ethanol drinking in C57BL/6 J mice via Y1 and Y2 receptors. Neuropsychopharmacology 37: 1409–1421.

Staner L, Boeijinga P, Danel T et al. (2006). Effects of acamprosate on sleep during alcohol withdrawal: a double-blind placebo-controlled polysomnographic study in alcohol-dependent subjects. Alcohol Clin Exp Res 30: 1492–1499.

Starcke K, van Holst RJ, van den Brink W et al. (2013). Physiological and endocrine reactions to psychosocial stress in alcohol use disorders: duration of abstinence matters. Alcohol Clin Exp Res 37: 1343–1350.

Steinig J, Foraita R, Happe S et al. (2011). Perception of sleep and dreams in alcohol-dependent patients during detoxication and abstinence. Alcohol Alcohol 46: 143–147.

Stepanyan TD, Farook JM, Kowalski A et al. (2008). Alcohol withdrawal-induced hippocampal neurotoxicity in vitro and seizures in vivo are both reduced by memantine. Alcohol Clin Exp Res 32: 2128–2135.

Stephens MA, Wand G (2012). Stress and the HPA axis: role of glucocorticoids in alcohol dependence. Alcohol Res 34: 468–483.

Steriade M (2006). Grouping of brain rhythms in corticothalamic systems. Neuroscience 137: 1087–1106.

Storm JF (1990). Potassium currents in hippocampal pyramidal cells. Prog Brain Res 83: 161–187.

Suvarna N, Borgland SL, Wang J et al. (2005). Ethanol alters trafficking and functional *N*-methyl-D-aspartate receptor NR2 subunit ratio via H-Ras. J Biol Chem 280: 31450–31459.

Swift R (2010). Medications acting on the dopaminergic system in the treatment of alcoholic patients. Curr Pharm Des 16: 2136–2140.

Thielen RJ, Engleman EA, Rodd ZA et al. (2004). Ethanol drinking and deprivation alter dopaminergic and serotonergic function in the nucleus accumbens of alcohol-preferring rats. J Pharmacol Exp Ther 309: 216–225.

Thorsell A (2007). Neuropeptide Y (NPY) in alcohol intake and dependence. Peptides 28: 480–483.

Thorsell A, Rimondini R, Heilig M (2002). Blockade of central neuropeptide Y (NPY) Y2 receptors reduces ethanol self-administration in rats. Neurosci Lett 332: 1–4.

Thorsell A, Slawecki CJ, Ehlers CL (2005a). Effects of neuropeptide Y and corticotropin-releasing factor on ethanol intake in Wistar rats: interaction with chronic ethanol exposure. Behav Brain Res 161: 133–140.

Thorsell A, Slawecki CJ, Ehlers CL (2005b). Effects of neuropeptide Y on appetitive and consummatory behaviors associated with alcohol drinking in wistar rats with a history of ethanol exposure. Alcohol Clin Exp Res 29: 584–590.

Thorsell A, Schank JR, Singley E et al. (2010). Neurokinin-1 receptors (NK1R:s), alcohol consumption, and alcohol reward in mice. Psychopharmacology (Berl) 209: 103–111.

Tkatch T, Baranauskas G, Surmeier DJ (2000). Kv4.2 mRNA abundance and A-type K(+) current amplitude are linearly related in basal ganglia and basal forebrain neurons. J Neurosci 20: 579–588.

Treistman SN, Martin GE (2009). BK Channels: mediators and models for alcohol tolerance. Trends Neurosci 32: 629–637.

Uhart M, Wand GS (2009). Stress, alcohol and drug interaction: an update of human research. Addict Biol 14: 43–64.

Umhau JC, Momenan R, Schwandt ML et al. (2010). Effect of acamprosate on magnetic resonance spectroscopy measures of central glutamate in detoxified alcohol-dependent individuals: a randomized controlled experimental medicine study. Arch Gen Psychiatry 67: 1069–1077.

Valdez GR, Harshberger E (2012). Kappa opioid regulation of anxiety-like behavior during acute ethanol withdrawal. Pharmacol Biochem Behav 102: 44–47.

Valdez GR, Koob GF (2004). Allostasis and dysregulation of corticotropin-releasing factor and neuropeptide Y systems: implications for the development of alcoholism. Pharmacol Biochem Behav 79: 671–689.

Valdez GR, Zorrilla EP, Roberts AJ et al. (2003). Antagonism of corticotropin-releasing factor attenuates the enhanced responsiveness to stress observed during protracted ethanol abstinence. Alcohol 29: 55–60.

Valdez GR, Sabino V, Koob GF (2004). Increased anxiety-like behavior and ethanol self-administration in dependent rats: reversal via corticotropin-releasing factor-2 receptor activation. Alcohol Clin Exp Res 28: 865–872.

Vandergriff J, Kallman MJ, Rasmussen K (2000). Moxonidine, a selective imidazoline-1 receptor agonist, suppresses the effects of ethanol withdrawal on the acoustic startle response in rats. Biol Psychiatry 47: 874–879.

Veatch LM (2006). Disruptions in sleep time and sleep architecture in a mouse model of repeated ethanol withdrawal. Alcohol Clin Exp Res 30: 1214–1222.

Veatch LM, Becker HC (2005). Lorazepam and MK-801 effects on behavioral and electrographic indices of alcohol withdrawal sensitization. Brain Res 1065: 92–106.

Vendruscolo LF, Barbier E, Schlosburg JE et al. (2012). Corticosteroid-dependent plasticity mediates compulsive alcohol drinking in rats. J Neurosci 32: 7563–7571.

Victor M (1970). The alcohol withdrawal syndrome: theory and practice. Postgrad Med 47: 68–72.

Volpicelli JR, Alterman AI, Hayashida M et al. (1992). Naltrexone in the treatment of alcohol dependence. Arch Gen Psychiatry 49: 876–880.

von der Goltz C, Koopmann A, Dinter C et al. (2011). Involvement of orexin in the regulation of stress, depression and reward in alcohol dependence. Horm Behav 60: 644–650.

Walker BM, Koob GF (2007). The gamma-aminobutyric acid-B receptor agonist baclofen attenuates responding for ethanol in ethanol-dependent rats. Alcohol Clin Exp Res 31: 11–18.

Walker BM, Koob GF (2008). Pharmacological evidence for a motivational role of kappa-opioid systems in ethanol dependence. Neuropsychopharmacology 33: 643–652.

Walker BM, Rasmussen DD, Raskind MA et al. (2008). alpha1-noradrenergic receptor antagonism blocks dependence-induced increases in responding for ethanol. Alcohol 42: 91–97.

Walker BM, Valdez GR, McLaughlin JP et al. (2012). Targeting dynorphin/kappa opioid receptor systems to treat alcohol abuse and dependence. Alcohol 46: 359–370.

Walter HJ, Messing RO (1999). Regulation of neuronal voltage-gated calcium channels by ethanol. Neurochem Int 35: 95–101.

Wand G (2000). Hypothalamic-pituitary-adrenal axis: Changes and risk for alcoholism. In: A Noronha, M Eckardt, K Warren (Eds.), Review of NIAAA's neuroscience and behavioral research portfolio: NIAAA Research Monograph No 34, National Institute on Alcohol Abuse and Alcoholism, Bethesda, MD, pp. 397–415.

Wang J, Lanfranco MF, Gibb SL et al. (2010). Long-lasting adaptations of the NR2B-containing NMDA receptors in the dorsomedial striatum play a crucial role in alcohol consumption and relapse. J Neurosci 30: 10187–10198.

Wang J, Ben Hamida S, Darcq E et al. (2012). Ethanol-mediated facilitation of AMPA receptor function in the dorsomedial striatum: implications for alcohol drinking behavior. J Neurosci 32: 15124–15132.

Watson WP, Robinson E, Little HJ (1997). The novel anticonvulsant, gabapentin, protects against both convulsant and anxiogenic aspects of the ethanol withdrawal syndrome. Neuropharmacology 36: 1369–1375.

Weiss F, Parsons LH, Schulteis G et al. (1996). Ethanol self-administration restores withdrawal-associated deficiencies in accumbal dopamine and 5-hydroxytryptamine release in dependent rats. J Neurosci 16: 3474–3485.

Wiggins WF, Graef JD, Huitt TW et al. (2013). Ethosuximide reduces ethanol withdrawal-mediated disruptions in sleep-related EEG patterns. Alcohol Clin Exp Res 37: 372–382.

Willinger U, Lenzinger E, Hornik K et al. (2002). Anxiety as a predictor of relapse in detoxified alcohol-dependent patients. Alcohol Alcohol 37: 609–612.

Yahn SL, Watterson LR, Olive MF (2013). Safety and efficacy of acamprosate for the treatment of alcohol dependence. Subst Abus 6: 1–12.

Yaka R, Phamluong K, Ron D (2003). Scaffolding of Fyn kinase to the NMDA receptor determines brain region sensitivity to ethanol. J Neurosci 23: 3623–3632.

Yang AR, Yi HS, Warnock KT et al. (2012). Effects of the triple monoamine uptake inhibitor DOV 102,677 on alcohol-motivated responding and antidepressant activity in alcohol-preferring (P) rats. Alcohol Clin Exp Res 36: 863–873.

Zahr NM, Crawford EL, Hsu O et al. (2009). In vivo glutamate decline associated with kainic acid-induced status epilepticus. Brain Res 1300: 65–78.

Zhang H, Pandey SC (2003). Effects of PKA modulation on the expression of neuropeptide Y in rat amygdaloid structures during ethanol withdrawal. Peptides 24: 1397–1402.

Zhang H, Sakharkar AJ, Shi G et al. (2010). Neuropeptide Y signaling in the central nucleus of amygdala regulates alcohol-drinking and anxiety-like behaviors of alcohol-preferring rats. Alcohol Clin Exp Res 34: 451–461.

Zorumski CF, Mennerick S, Izumi Y (2014). Acute and chronic effects of ethanol on learning-related synaptic plasticity. Alcohol 48: 1–17.

Handbook of Clinical Neurology, Vol. 125 (3rd series)
Alcohol and the Nervous System
E.V. Sullivan and A. Pfefferbaum, Editors

Chapter 10

Molecular and neurologic responses to chronic alcohol use

B.N. COSTIN[1] AND M.F. MILES[2]*

[1]*Virginia Commonwealth University Alcohol Research Center and Department of Pharmacology and Toxicology, Virginia Commonwealth University, Richmond, VA, USA*

[2]*Virginia Commonwealth University Alcohol Research Center, Department of Pharmacology and Toxicology, Center for Study of Biological Complexity and Department of Neurology, Virginia Commonwealth University, Richmond, VA, USA*

INTRODUCTION

According to the National Institutes of Health as of April, 2010, 28% of adults aged 18 and older consume alcohol at levels that put them at risk for developing alcoholism, liver disease, and other medical and psychosocial problems (NIAAA, 2010). Alcohol use disorders (AUD) include a spectrum extending from episodic binge drinking to full-blown alcoholism. There is no perfect definition of alcoholism, but most diagnoses require individuals to have been drinking heavily over an extended period of time and to have subsequently suffered multiple major life problems due to their alcohol consumption (Schuckit, 1987; First and Tasman, 2004). At-risk drinking increases one's chances of injuries; health problems, including liver disease, heart disease, sleep disorders, depression, stroke, bleeding from the stomach, sexually transmitted diseases and several types of cancer; and drinking during pregnancy increases the risk of birth defects for the unborn child (NIAAA, 2010). However, chronic alcohol exposure or severe abuse can cause more direct effects on the brain and is the subject of this chapter.

Alcohol is a central nervous system (CNS) depressant that produces euphoria, decreased anxiety, and behavioral excitation at low blood concentrations, and a spectrum of behavioral responses comprising acute intoxication (drowsiness, ataxia, slurred speech, stupor, and coma) at greater concentrations. Much progress has been made in understanding cellular and molecular mechanisms underlying alcohol actions in the nervous system (Vengeliene et al., 2008; Spanagel, 2009). Ligand-gated ion channels have been identified as specific targets for alcohol, and specific brain regions and pathways involved in addictive behavior have been described. Molecular and genetic studies in both humans and animal models promise new advances in understanding the molecular pathophysiology and genetics of alcoholism. Alcohol is used by a large segment of the population in social situations with minimal risk (e.g. complications of intoxication, fetal alcohol spectrum disorder), but a subset of alcohol consumers develop problems due to AUD. The development of AUD is caused by brain plasticity events that produce craving and habitual alcohol-seeking behavior. In addition, such chronic or high-dose alcohol use causes other toxic or adaptive responses within the CNS, as well as in virtually every organ system. This widespread effect on multiple organ systems makes alcohol a unique drug of abuse. Other chapters of this volume describe the more acute molecular and clinical brain responses to alcohol, alcohol withdrawal, and the neurobiology of addiction in alcoholism. Therefore, in this chapter, we will review current concepts on the neurologic complications of chronic, excessive exposure to alcohol and molecular mechanisms underlying these disorders.

MOLECULAR MECHANISMS OF CHRONIC ALCOHOL ACTION ON THE BRAIN

Alcohol (ethanol) is a simple two-carbon molecule that rapidly diffuses into virtually every biologic compartment in the body upon ingestion. Brain levels of alcohol rise within minutes of consuming an alcoholic beverage and signs of intoxication can be observed within minutes of high-dose alcohol administration. Thus it is largely assumed that acute alcohol actions occur at the level

*Correspondence to: Michael F. Miles, M.D., Ph.D., Department of Pharmacology and Toxicology, Virginia Commonwealth University, PO Box 980599, Richmond, VA 23298, USA. Tel: +1-804-827-4054, Fax: +1-804-828-6432, E-mail: mfmiles@vcu.edu

of direct or indirect effects on ion channels or neurotransmission. Such actions of alcohol are discussed in other chapters of this volume. However, the development of alcohol craving and addiction and other neurologic sequelae of chronic alcohol exposure (discussed below) clearly entail additional mechanisms.

Genetic contributions to alcoholism

Elegant studies in both humans and animal models over the last several decades have clearly shown that "genes" influence the development of alcoholism and other behavioral or molecular responses to alcohol. Genetic predisposition contributes to approximately 50% of the vulnerability for developing alcohol dependence (Goodwin et al., 1974; Prescott and Kendler, 1999). However, there has been only limited success in actually identifying candidate genes that contribute to the variable occurrence of alcohol dependence, much less other sequelae such as alcoholic dementia. Success stories include the identification of polymorphisms in isoforms of alcohol dehydrogenase and aldehyde dehydrogenase genes, which influence alcohol metabolism and the risk for alcoholism (Park et al., 2013). Additionally, recent studies on polymorphisms of the mu-opioid receptor gene show promise as a pharmacogenetic risk factor altering responses to naltrexone, a mu-opioid receptor antagonist that is approved by the Food and Drug Administatration for treatment of alcoholism (Ray and Hutchison, 2007; Ramchandani et al., 2011). Recent genomewide association studies suggest additional candidate loci of genes, but the majority of the genetic contribution to alcohol dependence still has not been identified (Kapoor et al., 2013; Kos et al., 2013).

The difficulty with existing human genetic studies is likely due to the occurrence of rare polymorphisms of small effect size in a large number of genes being causal elements in complex traits such as alcoholism. Moreover, even in cases where candidate genes for alcoholism have been implicated, the detection of regulatory network-wide systems is usually beyond the power of these approaches. Additionally, complex interactions between environmental factors (e.g., social stress) and genetic background may further complicate the identification of genetic risk factors. This leaves the question unanswered as to how to identify mechanisms underlying the development of alcohol dependence and alcohol-related neurologic disease.

Modulation of gene expression with chronic alcohol

Although identifying genes associated with genetic risk for alcoholism in humans has been difficult, there has been robust demonstration that acute or chronic alcohol exposure can alter expression of the genome both in cultured neural cells and *in vivo* with brain tissue from animals or humans (Miles et al., 1991; Miles, 1995; Kerns et al., 2005; Contet, 2012). Alterations in mRNA transcript abundance evoked by alcohol or substance abuse have been proposed as mechanisms underlying enduring neuroadaptations leading to abuse and addiction (Miles, 1995; Nestler and Aghajanian, 1997). Disrupted homeostatic control of gene networks is also a possible mechanism underlying CNS toxicity from compulsive alcohol or drug use. Furthermore, genetic differences in gene expression responses to alcohol are also thought to be an important mechanism underlying a predisposition to alcoholism or other complex traits (Schadt et al., 2003; Chesler et al., 2005). Thus, studies in animal models or humans on gene expression networks associated with alcoholism or alcohol-related behaviors have been an area of intense investigation. Such work has, over the last decade, generated large collections of alcohol-responsive or alcohol consumption-related genes through use of whole-genome expression profiling techniques such as microarrays or, more recently, RNA-seq (Kerns et al., 2005; Mulligan et al., 2006; Farris and Miles, 2012).

Identifying high-throughput lists of genes relevant to alcohol brain responses has not automatically resulted in improved understanding of mechanisms underlying alcoholism or new therapeutics for the disease. Recent work has used sophisticated statistical approaches at deriving "gene networks" from the genomewide expression data (Ponomarev et al., 2012; Wolen and Miles, 2012; Iancu et al., 2013). This results in the identification of so-called "hub genes" that may serve as master regulators for entire networks of alcohol-responsive genes, and thus be optimal targets for future drug development efforts (Mehrabian et al., 2005). Some investigators have also employed either integration of such networks across species and/or combining them with expression and behavioral genetic studies so as to more accurately identify networks relevant to alcohol behaviors or other brain responses (Tabakoff et al., 2009; Zhao et al., 2012; Vanderlinden et al., 2013). Using such approaches, some investigators have identified novel genes that show cross-species modulation of alcohol behaviors and hold promise for the development of future innovative therapeutic approaches (Bhandari et al., 2012). Taken together, genomic studies on alcohol suggest that alteration in the regional and cell-specific expression of specific gene networks in brain is a major mechanism underlying the neuroplasticity constituting alcohol dependence and alcohol-related toxicity (Fig. 10.1).

The explosion of genomic information about alcohol and the brain that has been produced since the turn of the century is almost bewildering in magnitude and scope (Contet, 2012; Farris and Miles, 2012). How can such a

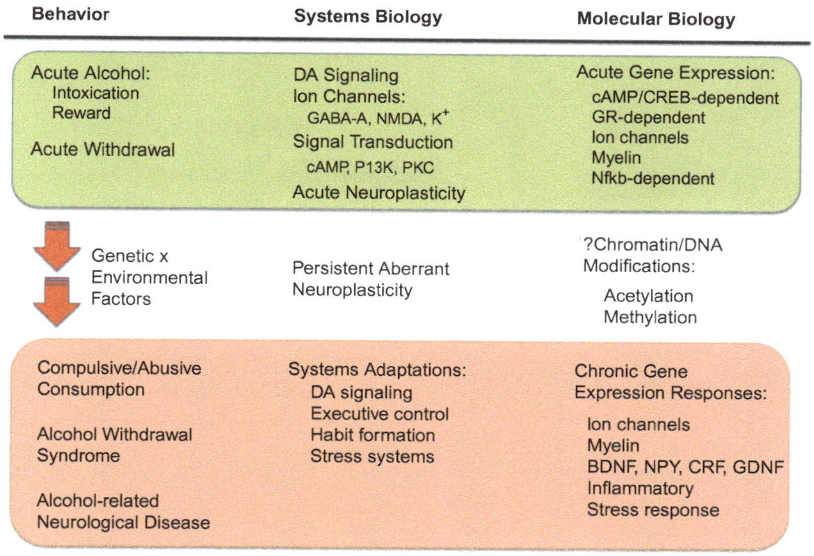

Fig. 10.1. Overview of molecular adaptive events leading to alcoholism and neurologic complications of alcohol use disorders (AUD). The figure documents major aspects of current knowledge regarding behavioral, systems biology, and molecular mechanisms leading from acute alcohol exposure to chronic consequences such as AUD and neurologic complications. Schema are illustrative only and not meant to be all-inclusive. Acute alcohol has major direct effects on ion channels (γ-aminobutyric acid A (GABA-A), *N*-methyl-D-aspartate (NMDA), and K^+). Alcohol-induced release of dopamine (DA) is a major factor in the rewarding properties of the drug. Alteration in numerous signaling cascades (e.g., cAMP, PI-3-kinase (PI3K), and protein kinase C (PKC)) and transcription factors (GR, glucocorticoid receptor; CREB; and Nfkb, Nf-kappa-B) may function in acute and chronic neuroplasticity by altering gene expression at the transcriptional, translational, or posttranslational level. Gene expression alterations are thought to play a central role in both acute and chronic neuroadaptation to alcohol. Epigenetic changes (DNA or histone modifications) are proposed as a mechanism underlying persistence of gene expression/neuroadaptive events, leading to clinical sequelae such as compulsive drinking, recidivism, and alcohol-related neurologic disease. BDNF, brain-derived neurotropic factor; NPY, neuropeptide Y; CRF, corticotropin-releasing factor; GDNF, glial cell-derived neurotropic factor.

simple molecule produce such a scale of changes in brain gene expression and what are their functional implications? Despite the complexity, several major conclusions are beginning to become evident about alcohol and the brain transcriptome:

1. Many ligand-gated ion channels (e.g., γ-aminobutyric acid-A (GABA) receptor subunits) or other neuromodulatory molecules (such as brain-derived neurotrophic factor or neuropeptide Y) that have been implicated in alcohol actions by behavioral pharmacology or physiology studies are also regulated at the level of gene expression by acute or chronic alcohol exposure (Spanagel, 2009; Contet, 2012; Farris and Miles, 2012). This reinforces the crucial role of these molecules in alcohol behaviors and also validates genomic studies as a tool for identifying novel alcohol targets.
2. Neuroinflammatory signaling genes are prominently regulated by acute alcohol and chronic exposure in both humans and animal models (Blednov et al., 2011, 2012). These suggest that neuroinflammation could function in both chronic behavioral responses to alcohol and alcohol-induced neurologic disorders.
3. Alcohol regulates multiple myelin-related genes both acutely and with chronic exposure and basal myelin expression levels can affect both sensitivity to acute alcohol and alcohol consumption (Lewohl et al., 2000; Kerns et al., 2005; Farris and Miles, 2012; Wolen and Miles, 2012). These data suggest that "myelin tone" may be both a risk factor and a target in alcoholism (Fig. 10.2).
4. Although multiple signaling cascades have been implicated by genomic studies, cascades linked to insulin-like growth factor and signaling by AKT kinase appear throughout multiple genomic and other molecular studies (Kerns et al., 2005; Neasta et al., 2011).

These data have suggested these and related pathways as novel targets for therapeutic development (Barak et al., 2013; Ron and Messing, 2013).

Persistence of gene expression changes – role for epigenetic regulation?

Suggesting that gene networks regulated by acute or chronic alcohol have a functional role in the development

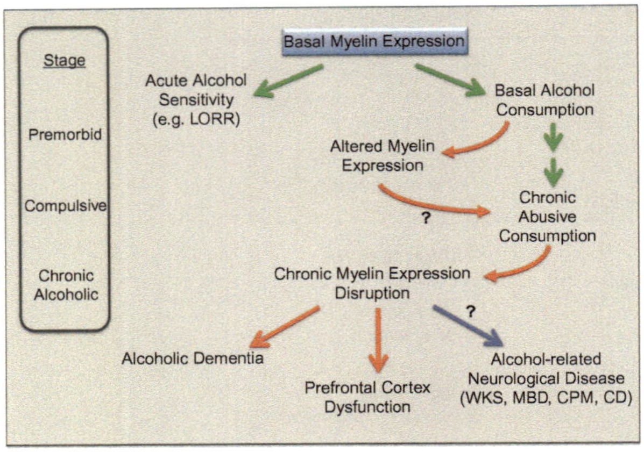

Fig. 10.2. Hypothetic scheme for the role of myelin in alcoholism and alcohol-related neurologic disorders. Basal myelin expression is pictured as affecting initial sensitivity to alcohol, such as with loss-of-righting reflex (LORR), and perhaps relatedly, alcohol consumption. Acute alcohol, however, also alters myelin gene expression, which may contribute to frontal-lobe dysfunction and the consequent development of abusive alcohol consumption. Chronic alcohol exposure produces prominent decreases in myelin gene expression and myelin structural abnormalities. These are envisioned as contributing to frontal-lobe dysfunction in alcoholism and alcoholic dementia. Additionally, alcohol-induced myelin pathology may contribute as a risk factor to other alcohol-related neurologic diseases, such as Wernicke–Korsakoff's syndrome (WKS), Machiafava–Bignami disease (MBD), central pontine myelinolysis (CPM), and cerebellar degeneration (CD).

of alcohol dependence or other alcohol-related neurologic disorders presents one major mechanistic hurtle – persistence. Explaining how such changes in gene expression could persist during periods of even prolonged abstinence, and thus influence the high rate of recidivism seen in alcoholics, is a challenging question. Recently, there has been a large expansion of studies in the area of "epigenetics" as an effort to explain non-genetic, long-lasting alterations in the transcriptome that could be causal in alcohol dependence or neurologic disorders. Epigenetics refers to a modification of the chromatin structure/function (e.g., by methylation or acetylation of histones) or DNA bases (e.g., by methylation), rather than changes in the actual DNA sequence, as a mechanism for altering gene expression in a potentially lifelong or transgenerational fashion (Borrelli et al., 2008; Kaminen-Ahola et al., 2010; Starkman et al., 2012). Expression of genes regulating epigenetic modifications have been found to correlate with individual variations in alcohol consumption in mice and drugs modifying histone acetylation produced alterations in alcohol consumption in such models (Wolstenholme et al., 2011).

NEUROLOGY OF CHRONIC ALCOHOL ACTION IN THE CENTRAL NERVOUS SYSTEM

Overview of clinical syndromes

Neurologic responses to abusive or chronic alcohol intake are often classified as direct versus secondary, but this assumes one knows the underlying mechanisms of the disorder. Alcoholism is a chronic disease characterized by uncontrolled craving for alcohol and physical dependence on the drug. The latter produces a characteristic severe withdrawal syndrome when alcohol drinking is suddenly decreased. In addition to craving and physical dependence, a set of characteristic neurologic disorders may also accompany long-term abusive drinking (Table 10.1). Additional prominent neurologic sequelae of alcohol abuse include fetal alcohol spectrum disorder and peripheral end-organ damage such as neuropathy and myopathy, but these are discussed elsewhere in this volume. Indirect neurologic complications of alcoholism, such as hepatic encephalopathy and traumatic brain injury, are also beyond the scope of this chapter.

Alcohol blackouts

Consumption of large quantities of alcohol, particularly rapidly, can cause episodes of memory impairment, termed "alcohol blackouts." Although alcohol blackouts can be considered to occur as an acute response to alcohol, they were initially described as a risk factor for alcoholism and thus will be discussed here as relevant to neurologic responses to chronic alcohol. Alcohol blackouts occur in the absence of any loss of consciousness or seizure activity and represent a type of anterograde amnesia. Such memory loss can be partial or fragmentary, such as not recalling names or isolated aspects of an experience, or they can be "en bloc," where there is

Table 10.1

Neurologic consequences of chronic exposure to alcohol

Withdrawal-induced	Chronic/abusive exposure
Alcohol-withdrawal seizures	Alcohol blackouts (fragmentary/en bloc)
Alcohol-withdrawal syndrome and delirium tremens	Craving
	Tolerance
	Wernicke–Korsakoff syndrome
	Cerebellar degeneration
	Primary alcoholic dementia
	Central pontine myelinolysis
	Machiafava–Bignami disease

complete lack of memory formation for a defined period of time (Goodwin et al., 1969; White, 2003). En bloc episodes appear to represent deficits of memory transfer from short-term to longer-term storage and the individuals can function with apparent intact short-term memory and conduct complicated tasks such as conversations. Fragmentary alcohol blackouts are more common and there is a suggestion that they occur at lower blood alcohol levels. For both types of blackout, however, there is at least substantial anecdotal evidence to suggest that the events are associated with rapidly rising blood alcohol levels, but a linear relationship to absolute blood alcohol levels has also been described (Perry et al., 2006).

Early work suggested that alcoholics exclusively suffered such memory blackouts. Indeed, "alcoholic blackouts" were considered a premonitory symptom of incipient alcoholism (Jellinek, 1946). However such work suffered from the fact that most studies were indeed conducted only on alcoholics. More recent work has clearly shown that apparently normal individuals undergoing binge social drinking episodes can experience fragmentary or en bloc alcohol blackouts, with the former occurring about three times more frequently in studies on college students (Hartzler and Fromme, 2003; Rose and Grant, 2010). Because of the high prevalence of binge drinking among college students, many recent studies on alcohol-related blackouts have been conducted on that population. The occurrence of such blackouts is apparently not a benign event as there are higher rates of violence, accidents, vandalism, and unprotected sex amongst college students experiencing blackouts (White et al., 2002). Furthermore, although Jellinek's original premise of alcohol blackouts as a predictor of future alcoholism might not have been correct, there have been a variety of studies suggesting that, particularly with individuals having a history of repeated alcohol blackouts, there is a greater incidence of early-onset problem drinking (Rose and Grant, 2010). Twin studies have shown a substantial genetic component to the occurrence of alcohol-related blackouts and shared genetic risk with the frequency of intoxication (Hurley et al., 2011).

It is also apparent that not all individuals, despite high blood alcohol levels, will experience alcohol blackouts. Twin studies, mentioned above, showing a greater than 50% heritability rate for alcohol blackouts, support the concept that some individuals are predisposed to these events. Recent longitudinal neuroimaging studies have suggested that individuals susceptible to future alcohol blackouts have underlying neurobiologic differences in memory-related tasks (Wetherill et al., 2013). Other neuroimaging studies have shown differences in neural activity during contextual learning tasks only in the presence of alcohol for individuals with a history of fragmentary alcohol blackouts (Wetherill et al., 2012).

Neuroimaging and clinical laboratory studies in humans and behavioral pharmacology studies in animals models have contributed greatly of late to our understanding of possible mechanisms underlying different neurobiologic aspects of memory and alcohol-related blackouts (Rose and Grant, 2010). Some investigators initially hypothesized that alcohol-related blackouts represented an example of "state-dependent learning," such that the memories laid down during an alcohol-drinking binge could only be recalled in the presence of alcohol intoxication (Goodwin, 1974). More recent findings, however, have not fully supported such theories (Weissenborn and Duka, 2000). Studies over the last two decades have shown that alcohol impairs explicit memory and spatial memory tasks, which are highly dependent on intact hippocampal function, perhaps by interfering with septohippocampal regulation of hippocampal neuronal circuitry (Givens et al., 2000).

The exact mechanisms for alcohol impairment of hippocampal circuits are not clear, but alcohol does produce dose-dependent inhibition of long-term potentiation, likely by actions at *N*-methyl-D-aspartate (NMDA)-type glutamatergic synaptic currents (Lovinger et al., 1989; Morrisett and Swartzwelder, 1993). Other work has suggested that alcohol effects on hippocampal NMDA receptor function occur at higher alcohol concentrations and that, at lower doses, increases in septohippocampal GABA receptor activity might also function in alcohol-related blackouts. As suggested by the time course, epidemiology, and neuroanatomic localization of alcohol-related blackouts, this disorder appears distinct from the mechanisms underlying Korsakoff's psychosis and primary alcoholic dementia (see below).

Craving

The current definition of craving is "the desire to experience the effect(s) of a previously experienced psychoactive substance" (Vengeliene et al., 2008). This topic is discussed elsewhere in this volume but is also briefly discussed here since craving is a major neuroadaptive event leading to AUD. Despite experiencing adverse neurologic and environmental consequences of alcohol, alcoholics crave alcohol and exhibit habitual, uncontrolled drinking. Thus, these three components typify alcoholism – uncontrolled craving, habitual consumption, and loss of normal judgment regarding adverse consequences. Craving can occur following long-term abstinence from alcohol and can be elicited by exposure to alcohol and alcohol-associated stimuli in abstinent alcoholics (Heinz et al., 2003; Vengeliene et al., 2008). Functional magnetic resonance imaging (MRI) and positron emission tomography studies have identified critical brain regions and neurocircuitry underlying alcohol craving (George et al., 2001; de Greck et al., 2009; Fryer et al., 2013; Jasinska et al., 2013).

Such neuroimaging studies, combined with decades of neurophysiology and behavioral pharmacology work on alcohol in animal models, have led to the development of an elegant model describing the transition from social alcohol use to AUD (Koob and Volkow. 2010), although the model also applies to drugs of abuse. This model describes three stages of drug use: intoxicating/binge, withdrawal/negative affect, and a final stage of preoccupation/anticipation or "craving." Data supporting such a model evoke interplay of different neural pathways and neurotransmitter systems at each stage. Craving is thought to have major influences from hippocampus, basal lateral amygdala, orbital frontal cortex, and medial prefrontal cortex (cingulate). Glutamate projections from medial prefrontal cortex and basal lateral amygdala to the ventral striatum are thought to have a major role in the craving stage, producing changes in dopaminergic reward pathways.

Clinical observations indicate that craving for alcohol persists for very extended periods (if not permanently) in abstinent alcoholics and that recidivism is a major obstacle to treatment of alcoholism. The mechanism underlying such prolonged risk for recidivism is unknown and obviously is an area of intense study. As discussed above, one major hypothesis for the mechanism of AUD and abuse of other drugs concerns alterations in brain gene expression as a direct or indirect consequence of drug exposure (Miles, 1995; Nestler and Aghajanian, 1997). Changes in gene expression evoked by chronic alcohol exposure, perhaps codified by epigenetic alterations of DNA and chromatin, could thus lead to persistent alterations in synaptic plasticity, and ultimately long-lived behavioral consequences (Fig. 10.1).

Tolerance and dependence

Once chronic abusive alcohol intake is established, one prevalent brain adaptive response concerns the development of tolerance to the intoxicating effects of alcohol. Such individuals can appear sober at blood alcohol concentrations of 89–108 mmol/L (Urso et al., 1981), well above the legal limits of intoxication (<22 mmol/L). The highest recorded blood alcohol level to be survived is 328 mmol/L, which was measured in an ambulatory alcoholic who stopped drinking 3 days previously (Johnson et al., 1982). These data highlight early observations by Goldberg (1941), showing that naïve individuals became ataxic at lower blood alcohol levels than heavy drinkers. Tolerance thus represents a CNS pharmacodynamic adaptive response which improves neurologic function at a given blood alcohol level.

Prolonged alcohol consumption leads to the development of tolerance and physical dependence. Physical dependence refers to the existence of spontaneous behavioral disturbances, which are produced by alcohol removal and suppressed by alcohol replacement. Physical dependence underlies the alcohol withdrawal syndrome. Animal models have suggested that there is an inverse genetic relationship between the severity of alcohol withdrawal symptoms and the drive for alcohol consumption (Metten et al., 1998). The mechanisms of physical dependence are less well understood than those responsible for acute intoxication. However, it is known that tolerance and physical dependence occur due to compensatory changes in the function of alcohol's acute targets, including many of the ion channels and second-messenger systems discussed earlier (Spanagel, 2009) and in other chapters of this volume.

Changes in the function or localization of GABA-A and NMDA receptors occur with acute or chronic alcohol exposure in a variety of different *in vivo* and *in vitro* models (Allan and Harris, 1987; Charlton et al., 1997; Anderson et al., 2007; Diaz et al., 2011). Numerous studies have identified changes in GABA-A receptor or NMDA receptor subunit mRNA/protein following chronic alcohol, suggesting that adaptive changes in the expression of these two major targets of acute alcohol action might have a major contribution in the development of tolerance/dependence. Alcohol-induced decreases in GABA-A receptor subunits could explain the cross-tolerance between alcohol and barbiturates or benzodiazepines, which also act on GABA-A receptors. A role for GABA-A receptors in alcohol tolerance is also implied by the usefulness of benzodiazepines or barbiturates in treating alcohol withdrawal.

Whereas acute alcohol exposure inhibits the function of NMDA receptors, chronic alcohol treatment results in increased NMDA receptor expression (Iorio et al., 1992; Allgaier, 2002). This increase in the number of NMDA receptors offsets the acute inhibitory actions of alcohol on the NMDA receptor and may contribute to hyperactivity during withdrawal. Additionally, hyperactivity during withdrawal could result from changes in the subunit composition of the NMDA receptor, as it has been shown that chronic alcohol consumption increases NMDA NR1 and NR2A (Snell et al., 1996) and NR2B (Kalluri et al., 1998) subunit levels in certain brain regions.

Alcohol withdrawal syndrome overview

Approximately 2 million Americans experience the symptoms of alcohol withdrawal yearly (Kosten and O'Connor, 2003; Bayard et al., 2004). As discussed in detail above, prolonged alcohol consumption leads to compensatory functional changes in alcohol's target molecules and alcohol withdrawal syndrome occurs when alcohol use is abruptly discontinued and these changes are unmasked. Alcohol withdrawal syndrome may occur intentionally if the person voluntarily ceases alcohol consumption or unintentionally if abstinence is due to illness or injury (Hall and Zador, 1997). In any case, alcohol withdrawal syndrome indicates the development of a physical dependence on alcohol (Uzbay, 2012). Depending on the amount of alcohol consumed and the duration of an individual's drinking habit, alcohol withdrawal syndrome can range from minor symptoms such as irritability, agitation, anxiety, headache, anorexia, diaphoresis, palpitations, insomnia, nausea and vomiting, and tremors; to moderate symptoms, including confusion and autonomic hyperactivity; to more severe symptoms, including muscle rigidity, seizures, and delirium tremens (Kosten and O'Connor, 2003; Bayard et al., 2004). Minor withdrawal symptoms occur within the first 6–12 hours following the cessation of alcohol use (Hall and Zador, 1997; Bayard et al., 2004). Mild hallucinations can occur 12–14 hours following the cessation of alcohol and usually resolve within 48 hours, but they may last up to 6 days (Bayard et al., 2004; Corfee, 2011). Alcohol withdrawal seizures typically begin 24–48 hours following cessation of drinking (Hall and Zador, 1997; Bayard et al., 2004; Rogawski, 2005; Corfee, 2011). Finally, 48 hours to 5 days following the cessation of alcohol use, delirium tremens can begin and includes marked hallucinations, disorientation, tachycardia, hypertension, fever, agitation, and diaphoresis – symptoms that usually peak 5 days following drinking cessation (Hall and Zador, 1997; Bayard et al., 2004; Corfee, 2011). In most cases, the symptoms of moderate alcohol withdrawal end 2–7 days following consumption of the last drink and do not require medical intervention. However, in more severe cases, individuals may need to be withdrawn from alcohol under medical supervision (Hall and Zador, 1997).

Withdrawal seizures

Alcohol abuse accounts for 20–40% of all new-onset seizures seen in emergency departments, particularly in male patients between 30 and 60 years of age (Tardy et al., 1995; McMicken and Liss, 2011). Additionally, another large clinical study showed that alcohol use was one of the four leading etiologies for status epilepticus (Lowenstein and Alldredge, 1993). Alcohol withdrawal seizures are usually tonic-clonic, although partial seizures can also occur. Because alcohol withdrawal seizures are pharmacologically induced, the pathophysiologic mechanisms of the seizures are different from those of the seizures occurring in genetic and acquired epilepsy (Freedland and McMicken, 1993; Rogawski, 2005). Withdrawal seizures are more common in patients who have a history of recurrent detoxifications (Ballenger and Post, 1978; Bayard et al., 2004; Duka et al., 2004) and the severity and duration of withdrawal seizures increase with repeated withdrawal experiences (Becker et al., 1998). Alternate causes for seizures other than alcohol withdrawal syndrome should be considered if there is no recent abstinence from drinking, if the patient has suffered from a traumatic event or has a fever, if the seizures are focal, and if the seizures occur more than 48 hours following an individual's last drink (Bayard et al., 2004). Treatment and prevention of alcohol withdrawal seizures are remarkably resistant to traditional anticonvulsants such as carbamazepine or phenytoin but benzodiazepines are effective (Rogawski, 2005; McMicken and Liss, 2011).

Rodent models of the human alcohol withdrawal-related tonic-clonic seizures have been used to identify their underlying mechanisms. Rodents chronically exposed to alcohol via intragastric alcohol administration, inhalation, or feeding of a liquid diet for periods of 2–21 days exhibit sound-evoked audiogenic seizures 1–3 days following cessation of alcohol intake (McCown and Breese, 1990; Devaud et al., 1997; Becker et al., 1998; Becker and Veatch, 2002). The audiogenic seizures are the best-studied type of alcohol withdrawal seizure and are mediated largely in the brainstem, which makes the mechanisms mediating these seizures unique from those believed to mediate other clinical seizure types (Rogawski, 2005). The mechanisms of alcohol withdrawal seizures are not fully understood, but it appears that the inferior colliculus plays a major role in alcohol withdrawal-induced audiogenic seizures in rodents. For example, multiple alcohol withdrawal episodes in rats facilitate the development of inferior colliculus kindling (McCown and Breese, 1990). Alterations in GABA-A receptor localization or receptor gene expression, or NMDA receptor expression, likely play a critical mechanistic role in the development of alcohol withdrawal-related seizures (Rogawski, 2005). Rodent models suggest that kindling, a form of neuroplasticity causing increased neuronal excitability, can occur with repeated alcohol withdrawal episodes (McCown and Breese, 1990; Becker et al., 1998). Such neuroplasticity could further contribute to withdrawal seizures, and possibly the genesis of long-standing seizure disorders in the absence of alcohol withdrawal.

Delirium tremens

Delirium tremens is the most severe manifestation of the alcohol withdrawal syndrome. It is characterized by severe disorientation, exaggerated sympathetic activity, psychomotor agitation, and marked hallucinations. Risk factors include concurrent acute medical illness, daily heavy alcohol consumption, history of delirium tremens or withdrawal seizures, old age, abnormal liver function, and severe withdrawal symptoms (Bayard et al., 2004; Corfee, 2011). Although many symptoms of alcohol withdrawal syndrome begin early following cessation of the drug, the manifestations of delirium tremens can develop abruptly several days following alcohol cessation. Delirium tremens can mimic alternate critical illnesses such as sepsis or head injury, may precipitate respiratory and cardiovascular collapse, and, if left untreated, death may occur from respiratory or cardiovascular collapse (Hall and Zador, 1997; Corfee, 2011). In one study, 24% of 200 consecutive alcohol-dependent patients admitted to an inner-city US hospital developed delirium tremens and 8% of those individuals died from complications of delirium tremens (Ferguson et al., 1996). Among other factors, those who developed delirium tremens were more likely to be unemployed, homeless, and to show signs of more severe alcohol-related end-organ disease (Ferguson et al., 1996; Eyer et al., 2011). While high mortality rates (30–50%) from delirium tremens were reported in earlier clinical studies, mortality today has generally been reduced to less than 3–5% (Lukan et al., 2002) and is usually associated with other alcohol-related diseases such as hepatitis, aspiration, and fatal cardiac arrhythmias (Eastes, 2010; Carlson et al., 2012).

Wernicke–Korsakoff syndrome

Wernicke–Korsakoff syndrome (WKS) describes a constellation of neurologic and cognitive problems that comprises both Wernicke's encephalopathy (WE) and Korsakoff syndrome (KS). WE is an acute, potentially reversible, neuropsychiatric disorder caused by thiamine deficiency (Donnino et al., 2007; Isenberg-Grzeda et al., 2012). The altered cognition of WE can progress to KS, a chronic and usually permanent impairment in formation of new memories with relative preservation of other mental functions (Toth et al., 2002). Additionally, KS has additional causes other than WE, including primary CNS lymphoma (Toth et al., 2002) and mammillothalamic tract infarction (Yoneoka et al., 2004). Approximately 80% of patients with acute WE will develop KS (Donnino et al., 2007). With early recognition and treatment, morbidity associated with WKS can be prevented. WE and KS were first described separately by Wernicke and Korsakoff as distinct pathologies (Donnino et al., 2007; Isenberg-Grzeda et al., 2012). It was not until 1956 that it was noted that patients diagnosed with KS revealed neuropathologic lesions in the identical brain regions as patients with WE (Malamud and Skillicorn, 1956).

The classic clinical triad of WE consists of mental status impairment, ophthalmoplegia, and gait ataxia, but the complete triad may only appear in as few as 10% of cases (Harper et al., 1986). Therefore, reliance on the complete triad as the sole criterion for the disease is inadequate and may lead to misdiagnoses. The most common component of the disease determined from retrospective autopsy studies is mental status change (Harper et al., 1986). The degree and nature of cognitive impairment can range from apathy or confusion to, although rare, coma (Donnino et al., 2007). Unfortunately, WE is often recognized only on autopsy (Torvik et al., 1982; Harper et al., 1986). One study at Ullevål Hospital in Oslo showed that, among 8735 autopsies performed over 5 years, there were 70 cases of WE. Examination of the clinical records showed that stupor and coma were the dominating symptoms in active cases of WE (Torvik et al., 1982).

Treatment of WE involves the timely administration of thiamine. Thiamine is a cofactor for several essential enzymes in the Krebs cycle and the pentose phosphate pathway. In the presence of adenosine triphosphate, thiamine is converted to thiamine pyrophosphate, which functions in carbohydrate metabolism as a coenzyme in the decarboxylation of pyruvic and α-ketoglutaric acid. A serious deficiency of thiamine leads to alterations in metabolism, including the accumulation of pyruvate and α-ketoglutarate in the blood, and a decrease in the activity of transketolase, an enzyme in the pentose phosphate pathway (Hoyumpa, 1980). Traditionally in the treatment of WE, 100 mg thiamine is administered once per day intramuscularly or orally, but there is some controversy as to whether this dose is appropriately high to support passage of thiamine across the blood–brain barrier (Chataway and Hardman, 1995; Cook et al., 1998). In one study, Ambrose et al. (2001) compared five groups of subjects, all of whom were detoxifying from alcohol, in a randomized, double-blind, multidose study of thiamine treatment. Subjects were given different doses (range 5–200 mg) of intramuscular thiamine for 2 consecutive days. The posttreatment performance of these groups then was examined on a test of working memory, the delayed alternation task. The subjects taking the higher doses of thiamine performed best on the delayed alternation task, but the results did not show a stepwise dose response (Ambrose et al., 2001). Thus, a common treatment for possible thiamine deficiency consists of intravenous thiamine (10 mg) for several days, followed by daily oral thiamine (100 mg) for at least a month, if not indefinitely (Donnino et al., 2007).

Common misconceptions about WKS persist, including the idea that it is a rare disease and that it is only a disease seen in alcoholics (Donnino et al., 2007; Weathers and Lewis, 2009). Alcoholism appears to be the most common factor predisposing individuals in the United States to WE; however, WE can occur in any patient with nutritional deficiency (Donnino et al., 2007). Older autopsy series reports have revealed WE in 10% of patients with acquired immunodeficiency syndrome (AIDS) (Boldorini et al., 1992), 6% of patients with bone marrow transplants (Bleggi-Torres et al., 2000), and 12.5% of alcoholics (Torvik et al., 1982).

Cerebellar degeneration

The most common degenerative condition affecting the cerebellum is chronic alcoholism, which typically occurs after 10 or more years of alcohol abuse (Baker et al., 1999; Andersen, 2004). Alcoholic cerebellar degeneration is characterized by ataxia of stance and gait. Individuals with cerebellar degeneration may adopt a wide-based gait with shortened steps to compensate for their loss of balance (Andersen, 2004). Additionally, eye movement disorders such as nystagmus, poor handwriting, incoordination of the upper extremities, and mild dysarthria may be present (Diener et al., 1984; Johnson-Greene et al., 1997).

The first comprehensive study of alcoholic cerebellar degeneration was performed by Victor et al. (1959), who established that degeneration of the Purkinje cells in the central lobule and the anterior folia of the vermis resulted in lower-limb incoordination and ataxia. Since this study, additional reports have documented neuropathologic alterations in the structure of the cerebellum following long-term alcohol consumption (Johnson-Greene et al., 1997; Baker et al., 1999; Harper et al., 2003). In one such study, Baker et al. reported a significant decrease in Purkinje cell density and molecular layer volume in the cerebellar vermis in thiamine-deficient chronic alcoholics compared to chronic alcoholics and non-alcoholics. These thiamine-deficient alcoholics also had a significant decrease in the estimated Purkinje cell number in the flocculi (Baker et al., 1999). In this study, chronic alcoholics did not differ significantly from the non-alcoholic controls in all cerebellar measures, suggesting that chronic alcohol consumption per se does not necessarily damage human cerebellar tissue. Along these lines, it has been suggested that alcoholic cerebellar degeneration might be caused by thiamine deficiency and liver disease, and represents the same disease as WE (Butterworth, 1995; Baker et al., 1999). In support of this theory, cerebellar dysfunction frequently improves or stabilizes with abstinence and improved nutrition (Diener et al., 1984). However, other studies have documented cerebellar atrophy in well-nourished alcoholics and noted a relationship with duration and severity of alcohol intake (Nicolas et al., 2000), showing graded volume deficits from controls to uncomplicated alcoholics, to alcoholics with KS (Sullivan et al., 2000; Sullivan and Pfefferbaum, 2009).

Primary alcoholic dementia

In recent decades it has become evident that chronic alcohol use can cause lasting cognitive impairment, or alcoholic dementia, even in the absence of nutritional deficiency or cerebral trauma. Alcohol-related dementia has been estimated to represent approximately 10% of all cases of dementia (Gupta and Warner, 2008). In a review of epidemiologic, neurologic, cognitive, and imaging data by Smith and Atkinson (1995), "heavy alcohol use" (roughly defined) was seen as a contributing factor in 21–24% cases of dementia. Many alcoholics demonstrate gradually progressive multidomain cognitive impairment rather than a more restrictive amnestic disorder which might be more typical of WKS (Brust. 2010). There is some suggestion that alcoholic dementia is at least partially reversible on cessation of alcohol consumption.

Numerous pathologic and neuroimaging studies have documented brain shrinkage with alcoholism, even in the apparent absence of thiamine deficiency (Melgaard et al., 1990; Pfefferbaum et al., 1992; Kril and Halliday, 1999). Such brain tissue loss appears to be primarily due to shrinkage of myelin, although dendritic abnormalities and neuronal loss have also been observed (reviewed in Harper et al., 2003). Myelin and neuronal loss appears to particularly involve the prefrontal and frontal lobe areas, with older alcoholics showing even more severe tissue loss (Pfefferbaum et al., 1992, 1997). Frontocerebellar tracts and interhemispheric white-matter tracts appear to convey aspects of the cognitive disruption seen with alcoholic dementia (Fadda and Rossetti, 1998; Chanraud et al., 2010, 2011; Ridley et al., 2013). Diffusion tensor imaging studies of white-matter abnormalities in alcoholics highlight a disruption of myelin microstructure in prefrontal areas and the corpus callosum (Pfefferbaum and Sullivan, 2005) that appears distinct from callosal atrophy seen with Alzheimer's disease (Pitel et al., 2010). Recent neuroimaging studies on a primate model of chronic oral alcohol consumption clearly document that alcohol itself, rather than just nutritional factors, can cause cerebral cortical gray-matter loss (Kroenke et al., 2013).

There are numerous neuroimaging reports documenting that myelin and brain tissue loss with alcoholism is at least partially reversible on cessation of drinking. In one study by Gazdzinski et al. (2005), the authors showed marked brain tissue volume recovery in alcoholics that were abstinent for 1 month and the tissue volume continued to increase over the next 6–9 months of abstinence, but at a slower rate than the first month following initial abstinence.

The mechanisms underlying alcoholic dementia are currently unknown. There is considerable animal model evidence suggesting that chronic alcohol exposure can cause neurotoxicity, perhaps related to withdrawal-induced glutamate excitotoxicity (Hoffman, 1995) or oxidative stress (Brust, 2010). Additionally, the alcohol metabolite acetaldehyde binds to select proteins, resulting in the formation of adducts. There have been reports suggesting a relationship between acetaldehyde adducts in the brain and brain damage (Nakamura et al., 2000).

Molecular changes evoked by chronic alcohol consumption also alter expression of numerous brain gene expression networks (Figs 10.1 and 10.2), which could contribute to neurotoxicity or functional disruption in alcoholic dementia. As mentioned earlier, alcohol has both acute and chronic effects on the expression of myelin-related genes (Fig. 10.2). Given the prominent disruption of myelin seen with neuropathologic and neuroimaging studies on alcoholic dementia, it is possible that alcohol modulation of myelin gene expression is functionally related to at least aspects of alcoholic dementia. Intriguingly, a recent study demonstrated that variation in myelin gene expression levels correlate with a measure of acute sensitivity to alcohol (loss-of-righting reflex, LORR), in mouse models (Farris and Miles, 2013). Since acute sensitivity to alcohol is a well-established risk factor for development of AUD, myelin expression and structure could have a central mechanistic role in both the genesis of AUD and the development of alcohol-related neurologic disorders such as alcoholic dementia (Fig. 10.2).

Central pontine myelinolysis

Central pontine myelinolysis (CPM) is characterized by a loss of oligodendrocytes and myelin in the central pons (Adams et al., 1959). Extrapontine myelinolysis (EPM) has also been described most often involving the white matter of the cerebellum, lateral geniculate body, putamen, thalamus, hippocampus, and cerebral transition of white and gray matter (Martin, 2004). The most common cause of CPM is an overly rapid correction of hyponatremia in patients with conditions such as alcoholism that lead to nutritional or electrolyte stress. Hyponatremia developing over a 48-hour period is considered acute and causes a more critical patient presentation. Chronic hyponatremia takes greater than 48 hours to develop and can indicate a higher risk for CPM (Lien and Shapiro, 2007).

The central pons may be especially vulnerable to osmotic injury because, unlike most areas of the brain, gray matter and white matter are intermixed. Typically, oligodendrocytes are embedded within white matter and physically isolated from gray matter. It has been proposed that, in CPM, oligodendrocytes in close proximity to gray matter are exposed to a myelin toxic substance following osmotic stress. The symptoms of CPM are varied and include confusion, impaired cognition, dysarthria, dysphasia, gait instability, weakness or paralysis, and generalized seizures (Hurley et al., 2011).

Despite considerable evidence supporting hyponatremia as the proximal cause of CPM, there are both clinical and model organism data suggesting that alcohol itself may perturb myelin structure or that alcoholics may have an underlying "myelin tone" which predisposes them to both alcoholism and other environmental insults that might cause myelinolysis (Fig. 10.2). As discussed elsewhere in this chapter, alcoholics show widespread (but particularly frontal, callosal, and hippocampal) myelin abnormalities, some of which are reversible with abstinence (Alhassoon et al., 2012). Alcoholics show a down-regulation of multiple myelin-related genes, particularly in hippocampus and prefrontal areas (Lewohl et al., 2000; McClintick et al., 2013). Animal models show that acute alcohol regulates myelin gene expression in prefrontal cortex (Kerns et al., 2005) and that a gene network consisting of multiple myelin genes shows strong correlations with both acute alcohol sensitivity and alcohol consumption in mice (Farris and Miles, 2012, 2013). Furthermore, recent MRI studies suggest that adolescents prone to AUD have premorbid abnormalities in myelin structure, as has been predicted by animal studies (Kerns et al., 2005; De Bellis et al., 2008; Farris and Miles, 2012, 2013). Thus, it remains a possibility that alcoholics are predisposed to myelin disorders by environmental insults such as hyponatremia with CPM.

There is a considerable range in prognosis in patients with clinically symptomatic CPM or EPM (Menger and Jorg, 1999; Mochizuki et al., 2003; Odier et al., 2010). Menger and Jorg examined 44 cases, most (42/44) with chronic alcoholism. Of the 32 patients with follow-up data who survived (2 died in the acute phase), 34% (11/32) experienced no significant functional deficits; 34% (11/32) experienced minor neurologic deficits; and 31% (10/32) remained dependent on personal help (Menger and Jorg, 1999). Mochizuki et al. also studied patients with alcoholism and reported that 44% (4/9) of the patients in their study were free of neurologic deficits 6 months following CPM. Additionally, the authors noted that most of their cohort (8/9) did not have an episode of acute correction of hyponatremia (Mochizuki et al., 2003). Treatment algorithms include considerations of current symptoms, whether hyponatremia is acute or chronic, presence of chronic malnourishment, underlying medical illnesses or complications, and other accompanying electrolyte abnormalities (Hurley et al., 2011). Hyponatremia occurs when the serum sodium drops below 135 mmol/L and can be treated by restoring sodium levels to normal. However, this must be done at a rate not exceeding 0.5 mmol/hour (Lien and Shapiro, 2007; Hurley et al., 2011).

Marchiafava–Bignami disease

Another neurologic disorder affecting primarily myelin and occurring with chronic alcohol abuse is Marchiafava–Bignami disease (MBD). Originally described as the "red wine drinker's encephalopathy," MBD is a rare and somewhat poorly understood disease that causes mania, depression, paranoia, dementia, seizures, paresis, and ataxia (Brust, 2010). Corpus callosum demyelination is the hallmark of MBD disease (Tozakidou et al., 2011), but demyelination may extend to the optic chiasm and tracts, cerebellar peduncle, subcortical region, neighboring white matter, and, rarely, cortical gray matter (Kim et al., 2007). The disease may present in two major clinical forms: acute and chronic. Acutely, severe impairment of consciousness, seizures, and muscle rigidity often result in death after several days. In the chronic form, an interhemispheric disconnection syndrome, such as limb apraxia, tactile agraphia, unilateral agraphia, hemialexia, and dementia, can be seen and can last for months to years (Kim et al., 2007). Clinicians can confuse MBD with WE, and some reports have shown that MBD and WE often occur concurrently (Johkura et al., 2005; Kim et al., 2007). Furthermore, there are scattered reports of response to thiamine replacement in cases diagnosed as MBD (Aggarwal et al., 2011). Thus, thiamine replacement should be given in cases of suspected MBD but it is not clear as to whether thiamine disorders are mechanistically involved with most cases of MBD. Furthermore, MBD-like lesions on MRI scans have been infrequently reported with a number of other clinical conditions such as ischemia, multiple sclerosis, lupus, and posterior reversible encephalopathy syndrome. Some of these may be responsible for cases of MBD having milder or reversible courses.

As discussed above for CPM, it remains an open question as to whether myelin abnormalities either underlying or accompanying AUD could contribute to the occurrence of MBD. In particular, fetal alcohol spectrum disorder is known to sometimes produce striking deficits in myelination, particularly involving the corpus callosum. MBD could thus result from the occurrence of a myelin-toxic environmental event, superimposed on a myelin state that is already compromised. Future longitudinal studies on fetal alcohol spectrum disorder patients or alcoholics would be required to confirm such a hypothesis.

CONCLUSIONS

This chapter highlights current knowledge on the molecular and clinical aspects of chronic alcohol effects on the CNS. This drug is almost ubiquitous, widely enjoyed socially, but produces a diverse spectrum of neurologic disease when abused. We have not detailed all current molecular information since the field is moving extremely rapidly and various molecular underpinnings of alcohol action are elaborated in other chapters of this volume. Alcohol interacts acutely predominantly with GABA-A and NMDA receptors, but triggers diverse signaling events within well-defined neural pathways. These result in adaptive changes in gene expression that ultimately produce two major states: addiction and toxicity. Epigenetic modifications of chromatin could lead to long-lived or even transgenerational changes in gene expression, thus producing aspects of the heritability of AUD and long-term behaviors such as recidivism. The diverse clinical syndromes produced by chronic alcohol actions in the CNS reflect the molecular pathology and predominantly involve aspects of tolerance/withdrawal and selective vulnerability (WKS, CPM, MBD). Additionally, deleterious aspects of chronic alcohol on signaling, synaptic transmission, and cell toxicity lead to primary alcoholic dementia. Genetically determined aspects of myelin structure and alcohol actions on myelin gene expression may be a prominent molecular component underlying predisposition to, or causation of, AUD and multiple other neurologic complications of chronic alcohol. The dramatic progress made in understanding molecular actions of alcohol hold great promise for our eventual treatment or prevention of AUD and neurologic complications from chronic abuse of alcohol.

ACKNOWLEDGMENT

This work was supported by National Institute on Alcohol Abuse and Alcoholism grants: U01AA016667, P20AA017828 (MFM), and F31AA020141 (BNC).

References

Adams RD, Victor M, Mancall EL (1959). Central pontine myelinolysis: a hitherto undescribed disease occurring in alcoholic and malnourished patients. AMA Arch Neurol Psychiatry 81: 154–172.

Aggarwal A, Khandelwal A, Jiloha RC (2011). A case of Marchiafava Bignami disease: complete recovery with thiamine. J Neuropsychiatry Clin Neurosci 23: E28.

Alhassoon OM, Sorg SF, Taylor MJ et al. (2012). Callosal white matter microstructural recovery in abstinent alcoholics: a longitudinal diffusion tensor imaging study. Alcohol Clin Exp Res 36: 1922–1931.

Allan AM, Harris RA (1987). Acute and chronic ethanol treatments alter GABA receptor-operated chloride channels. Pharmacol Biochem Behav 27: 665–670.

Allgaier C (2002). Ethanol sensitivity of NMDA receptors. Neurochem Int 41: 377–382.

Ambrose ML, Bowden SC, Whelan G (2001). Thiamin treatment and working memory function of alcohol-dependent people: preliminary findings. Alcohol Clin Exp Res 25: 112–116.

Andersen BB (2004). Reduction of Purkinje cell volume in cerebellum of alcoholics. Brain Res 1007: 10–18.

Anderson NJ, Daunais JB, Friedman DP et al. (2007). Long-term ethanol self-administration by the nonhuman primate, macaca fascicularis, decreases the benzodiazepine sensitivity of amygdala GABA(A) receptors. Alcohol Clin Exp Res 31: 1061–1070.

Baker KG, Harding AJ, Halliday GM et al. (1999). Neuronal loss in functional zones of the cerebellum of chronic alcoholics with and without Wernicke's encephalopathy. Neuroscience 91: 429–438.

Ballenger JC, Post RM (1978). Kindling as a model for alcohol withdrawal syndromes. Br J Psychiatry 133: 1–14.

Barak S, Liu F, Ben Hamida S et al. (2013). Disruption of alcohol-related memories by mtorc1 inhibition prevents relapse. Nat Neurosci 16: 1111–1117.

Bayard M, McIntyre J, Hill KR et al. (2004). Alcohol withdrawal syndrome. Am Fam Physician 69: 1443–1450.

Becker HC, Veatch LM (2002). Effects of lorazepam treatment for multiple ethanol withdrawals in mice. Alcohol Clin Exp Res 26: 371–380.

Becker HC, Veatch LM, Diaz-Granados JL (1998). Repeated ethanol withdrawal experience selectively alters sensitivity to different chemoconvulsant drugs in mice. Psychopharmacology (Berl) 139: 145–153.

Bhandari P, Hill JS, Farris SP et al. (2012). Chloride intracellular channels modulate acute ethanol behaviors in *Drosophila, Caenorhabditis elegans* and mice. Genes Brain Behav 11: 387–397.

Blednov YA, Benavidez JM, Geil C et al. (2011). Activation of inflammatory signaling by lipopolysaccharide produces a prolonged increase of voluntary alcohol intake in mice. Brain Behav Immun 25 (Suppl 1): S92–S105.

Blednov YA, Ponomarev I, Geil C et al. (2012). Neuroimmune regulation of alcohol consumption: behavioral validation of genes obtained from genomic studies. Addict Biol 17: 108–120.

Bleggi-Torres LF, de Medeiros BC, Werner B et al. (2000). Neuropathological findings after bone marrow transplantation: an autopsy study of 180 cases. Bone Marrow Transplant 25: 301–307.

Boldorini R, Vago L, Lechi A et al. (1992). Wernicke's encephalopathy: occurrence and pathological aspects in a series of 400 AIDS patients. Acta Biomed Ateneo Parmense 63: 43–49.

Borrelli E, Nestler EJ, Allis CD et al. (2008). Decoding the epigenetic language of neuronal plasticity. Neuron 60: 961–974.

Brust JC (2010). Ethanol and cognition: indirect effects, neurotoxicity and neuroprotection: a review. Int J Environ Res Public Health 7: 1540–1557.

Butterworth RF (1995). Pathophysiology of alcoholic brain damage: synergistic effects of ethanol, thiamine deficiency and alcoholic liver disease. Metab Brain Dis 10: 1–8.

Carlson RW, Kumar NN, Wong-Mckinstry E et al. (2012). Alcohol withdrawal syndrome. Crit Care Clin 28: 549–585.

Chanraud S, Pitel AL, Rohlfing T et al. (2010). Dual tasking and working memory in alcoholism: relation to frontocerebellar circuitry. Neuropsychopharmacology 35: 1868–1878.

Chanraud S, Pitel AL, Pfefferbaum A et al. (2011). Disruption of functional connectivity of the default-mode network in alcoholism. Cereb Cortex 21: 2272–2281.

Charlton ME, Sweetnam PM, Fitzgerald LW et al. (1997). Chronic ethanol administration regulates the expression of GABAA receptor alpha 1 and alpha 5 subunits in the ventral tegmental area and hippocampus. J Neurochem 68: 121–127.

Chataway J, Hardman E (1995). Thiamine in Wernicke's syndrome – how much and how long? Postgrad Med J 71: 249.

Chesler EJ, Lu L, Shou S et al. (2005). Complex trait analysis of gene expression uncovers polygenic and pleiotropic networks that modulate nervous system function. Nat Genet 37: 233–242.

Contet C (2012). Gene expression under the influence: Transcriptional profiling of ethanol in the brain. Curr Psychopharmacol 1: 301–314.

Cook CC, Hallwood PM, Thomson AD (1998). B vitamin deficiency and neuropsychiatric syndromes in alcohol misuse. Alcohol Alcohol 33: 317–336.

Corfee FA (2011). Alcohol withdrawal in the critical care unit. Aust Crit Care 24: 110–116.

De Bellis MD, Van Voorhees E, Hooper SR et al. (2008). Diffusion tensor measures of the corpus callosum in adolescents with adolescent onset alcohol use disorders. Alcohol Clin Exp Res 32: 395–404.

de Greck M, Supady A, Thiemann R et al. (2009). Decreased neural activity in reward circuitry during personal reference in abstinent alcoholics – a fMRI study. Hum Brain Mapp 30: 1691–1704.

Devaud LL, Fritschy JM, Sieghart W et al. (1997). Bidirectional alterations of GABA(A) receptor subunit peptide levels in rat cortex during chronic ethanol consumption and withdrawal. J Neurochem 69: 126–130.

Diaz MR, Christian DT, Anderson NJ et al. (2011). Chronic ethanol and withdrawal differentially modulate lateral/basolateral amygdala paracapsular and local gabaergic synapses. J Pharmacol Exp Ther 337: 162–170.

Diener HC, Dichgans J, Bacher M et al. (1984). Improvement of ataxia in alcoholic cerebellar atrophy through alcohol abstinence. J Neurol 231: 258–262.

Donnino MW, Vega J, Miller J et al. (2007). Myths and misconceptions of Wernicke's encephalopathy: what every emergency physician should know. Ann Emerg Med 50: 715–721.

Duka T, Gentry J, Malcolm R et al. (2004). Consequences of multiple withdrawals from alcohol. Alcohol Clin Exp Res 28: 233–246.

Eastes LE (2010). Alcohol withdrawal syndrome in trauma patients: a review. J Emerg Nurs 36: 507–509.

Eyer F, Schuster T, Felgenhauer N et al. (2011). Risk assessment of moderate to severe alcohol withdrawal – predictors for seizures and delirium tremens in the course of withdrawal. Alcohol Alcohol 46: 427–433.

Fadda F, Rossetti ZL (1998). Chronic ethanol consumption: from neuroadaptation to neurodegeneration. Prog Neurobiol 56: 385–431.

Farris SP, Miles MF (2012). Ethanol modulation of gene networks: implications for alcoholism. Neurobiol Dis 45: 115–121.

Farris SP, Miles MF (2013). Fyn-dependent gene networks in acute ethanol sensitivity. PLoS One 8: e82435.

Ferguson JA, Suelzer CJ, Eckert GJ et al. (1996). Risk factors for delirium tremens development. J Gen Intern Med 11: 410–414.

First MB, Tasman A (2004). DSM-IV-TR mental disorders: Diagnosis, etiology, and treatment. John Wiley, Chichester, West Sussex.

Freedland ES, McMicken DB (1993). Alcohol-related seizures, part I: Pathophysiology, differential diagnosis, and evaluation. J Emerg Med 11: 463–473.

Fryer SL, Jorgensen KW, Yetter EJ et al. (2013). Differential brain response to alcohol cue distractors across stages of alcohol dependence. Biol Psychol 92: 282–291.

Gazdzinski S, Durazzo TC, Meyerhoff DJ (2005). Temporal dynamics and determinants of whole brain tissue volume changes during recovery from alcohol dependence. Drug Alcohol Depend 78: 263–273.

George MS, Anton RF, Bloomer C et al. (2001). Activation of prefrontal cortex and anterior thalamus in alcoholic subjects on exposure to alcohol-specific cues. Arch Gen Psychiatry 58: 345–352.

Givens B, Williams JM, Gill TM (2000). Septohippocampal pathway as a site for the memory-impairing effects of ethanol. Hippocampus 10: 111–121.

Goldberg L (1941). Criteria of alcohol intoxication in animals in relation to blood alcohol. Acta Physiol Scand 3: 71–81.

Goodwin DW (1974). Alcoholic blackout and state-dependent learning. Fed Proc 33: 1833–1835.

Goodwin DW, Crane JB, Guze SB (1969). Alcoholic "blackouts": A review and clinical study of 100 alcoholics. Am J Psychiatry 126: 191–198.

Goodwin DW, Schulsinger F, Moller N et al. (1974). Drinking problems in adopted and nonadopted sons of alcoholics. Arch Gen Psychiatry 31: 164–169.

Gupta S, Warner J (2008). Alcohol-related dementia: A 21st-century silent epidemic? Br J Psychiatry 193: 351–353.

Hall W, Zador D (1997). The alcohol withdrawal syndrome. Lancet 349: 1897–1900.

Harper CG, Giles M, Finlay-Jones R (1986). Clinical signs in the Wernicke-Korsakoff complex: a retrospective analysis of 131 cases diagnosed at necropsy. J Neurol Neurosurg Psychiatry 49: 341–345.

Harper C, Dixon G, Sheedy D et al. (2003). Neuropathological alterations in alcoholic brains. Studies arising from the new south wales tissue resource centre. Prog Neuropsychopharmacol Biol Psychiatry 27: 951–961.

Hartzler B, Fromme K (2003). Fragmentary and en bloc blackouts: similarity and distinction among episodes of alcohol-induced memory loss. J Stud Alcohol 64: 547–550.

Heinz A, Lober S, Georgi A et al. (2003). Reward craving and withdrawal relief craving: assessment of different motivational pathways to alcohol intake. Alcohol Alcohol 38: 35–39.

Hoffman PL (1995). Glutamate receptors in alcohol withdrawal-induced neurotoxicity. Metab Brain Dis 10: 73–79.

Hoyumpa Jr AM (1980). Mechanisms of thiamin deficiency in chronic alcoholism. Am J Clin Nutr 33: 2750–2761.

Hurley RA, Filley CM, Taber KH (2011). Central pontine myelinolysis: a metabolic disorder of myelin. J Neuropsychiatry Clin Neurosci 23: 369–374.

Iancu OD, Oberbeck D, Darakjian P et al. (2013). Selection for drinking in the dark alters brain gene coexpression networks. Alcohol Clin Exp Res 37: 1295–1303.

Iorio KR, Reinlib L, Tabakoff B et al. (1992). Chronic exposure of cerebellar granule cells to ethanol results in increased *N*-methyl-D-aspartate receptor function. Mol Pharmacol 41: 1142–1148.

Isenberg-Grzeda E, Kutner HE, Nicolson SE (2012). Wernicke-Korsakoff syndrome: under-recognized and under-treated. Psychosomatics 53: 507–516.

Jasinska AJ, Stein EA, Kaiser J et al. (2013). Factors modulating neural reactivity to drug cues in addiction: a survey of human neuroimaging studies. Neurosci Biobehav Rev 38C: 1–16.

Jellinek EM (1946). Phases in the drinking history of alcoholics. Q J Stud Alcohol 7: 1–88.

Johkura K, Naito M, Naka T (2005). Cortical involvement in Marchiafava-Bignami disease. AJNR Am J Neuroradiol 26: 670–673.

Johnson RA, Noll EC, Rodney WM (1982). Survival after a serum ethanol concentration of 1 1/2%. Lancet 2: 1394.

Johnson-Greene D, Adams KM, Gilman S et al. (1997). Impaired upper limb coordination in alcoholic cerebellar degeneration. Arch Neurol 54: 436–439.

Kalluri HS, Mehta AK, Ticku MK (1998). Up-regulation of nmda receptor subunits in rat brain following chronic ethanol treatment. Brain Res Mol Brain Res 58: 221–224.

Kaminen-Ahola N, Ahola A, Maga M et al. (2010). Maternal ethanol consumption alters the epigenotype and the phenotype of offspring in a mouse model. PLoS Genet 6: e1000811.

Kapoor M, Wang JC, Wetherill L et al. (2013). A meta-analysis of two genome-wide association studies to identify novel loci for maximum number of alcoholic drinks. Hum Genet 132: 1141–1151.

Kerns RT, Ravindranathan A, Hassan S et al. (2005). Ethanol-responsive brain region expression networks: Implications for behavioral responses to acute ethanol in dba/2j versus c57bl/6j mice. J Neurosci 25: 2255–2266.

Kim MJ, Kim JK, Yoo BG et al. (2007). Acute Marchiafava-Bignami disease with widespread callosal and cortical lesions. J Korean Med Sci 22: 908–911.

Koob GF, Volkow ND (2010). Neurocircuitry of addiction. Neuropsychopharmacology 35: 217–238.

Kos MZ, Yan J, Dick DM et al. (2013). Common biological networks underlie genetic risk for alcoholism in African- and European-American populations. Genes Brain Behav 12: 532–542.

Kosten TR, O'Connor PG (2003). Management of drug and alcohol withdrawal. N Engl J Med 348: 1786–1795.

Kril JJ, Halliday GM (1999). Brain shrinkage in alcoholics: a decade on and what have we learned? Prog Neurobiol 58: 381–387.

Kroenke CD, Rohlfing T, Park B et al. (2013). Monkeys that voluntarily and chronically drink alcohol damage their brains: a longitudinal MRI study. Neuropsychopharmacology 39: 823–830.

Lewohl JM, Wang L, Miles MF et al. (2000). Gene expression in human alcoholism: microarray analysis of frontal cortex. Alcohol Clin Exp Res 24: 1873–1882.

Lien YH, Shapiro JI (2007). Hyponatremia: clinical diagnosis and management. Am J Med 120: 653–658.

Lovinger DM, White G, Weight FF (1989). Ethanol inhibits NMDA-activated ion current in hippocampal neurons. Science 243: 1721–1724.

Lowenstein DH, Alldredge BK (1993). Status epilepticus at an urban public hospital in the 1980s. Neurology 43: 483–488.

Lukan JK, Reed Jr DN, Looney SW et al. (2002). Risk factors for delirium tremens in trauma patients. J Trauma 53: 901–906.

Malamud N, Skillicorn SA (1956). Relationship between the Wernicke and the Korsakoff syndrome; a clinicopathologic study of seventy cases. AMA Arch Neurol Psychiatry 76: 585–596.

Martin RJ (2004). Central pontine and extrapontine myelinolysis: the osmotic demyelination syndromes. J Neurol Neurosurg Psychiatry 75 (Suppl 3): iii22–iii28.

McClintick JN, Xuei X, Tischfield JA et al. (2013). Stress-response pathways are altered in the hippocampus of chronic alcoholics. Alcohol 47: 505–515.

McCown TJ, Breese GR (1990). Multiple withdrawals from chronic ethanol "kindles" inferior collicular seizure activity: Evidence for kindling of seizures associated with alcoholism. Alcohol Clin Exp Res 14: 394–399.

McMicken D, Liss JL (2011). Alcohol-related seizures. Emerg Med Clin North Am 29: 117–124.

Mehrabian M, Allayee H, Stockton J et al. (2005). Integrating genotypic and expression data in a segregating mouse population to identify 5-lipoxygenase as a susceptibility gene for obesity and bone traits. Nat Genet 37: 1224–1233.

Melgaard B, Henriksen L, Ahlgren P et al. (1990). Regional cerebral blood flow in chronic alcoholics measured by single photon emission computerized tomography. Acta Neurol Scand 82: 87–93.

Menger H, Jorg J (1999). Outcome of central pontine and extrapontine myelinolysis (n = 44). J Neurol 246: 700–705.

Metten P, Phillips TJ, Crabbe JC et al. (1998). High genetic susceptibility to ethanol withdrawal predicts low ethanol consumption. Mamm Genome 9: 983–990.

Miles MF (1995). Alcohol's effects on gene expression. Alcohol Health Res World 19: 237–243.

Miles MF, Diaz JE, DeGuzman VS (1991). Mechanisms of neuronal adaptation to ethanol. Ethanol induces hsc70 gene transcription in ng108-15 neuroblastoma x glioma cells. J Biol Chem 266: 2409–2414.

Mochizuki H, Masaki T, Miyakawa T et al. (2003). Benign type of central pontine myelinolysis in alcoholism – clinical, neuroradiological and electrophysiological findings. J Neurol 250: 1077–1083.

Morrisett RA, Swartzwelder HS (1993). Attenuation of hippocampal long-term potentiation by ethanol: a patch-clamp analysis of glutamatergic and gabaergic mechanisms. J Neurosci 13: 2264–2272.

Mulligan MK, Ponomarev I, Hitzemann RJ et al. (2006). Toward understanding the genetics of alcohol drinking through transcriptome meta-analysis. Proc Natl Acad Sci U S A 103: 6368–6373.

Nakamura K, Iwahashi K, Itoh M et al. (2000). Immunohistochemical study on acetaldehyde adducts in alcohol-fed mice. Alcohol Clin Exp Res 24 (Suppl): 93S–96S.

Neasta J, Ben Hamida S, Yowell QV et al. (2011). Akt signaling pathway in the nucleus accumbens mediates excessive alcohol drinking behaviors. Biol Psychiatry 70: 575–582.

Nestler EJ, Aghajanian GK (1997). Molecular and cellular basis of addiction. Science 278: 58–63.

NIAAA (2010). Rethinking drinking. edited by U.S. Department of Health and Human Services National Institutes of Health http://pubs.niaaa.nih.gov/.

Nicolas JM, Fernandez-Sola J, Robert J et al. (2000). High ethanol intake and malnutrition in alcoholic cerebellar shrinkage. QJM 93: 449–456.

Odier C, Nguyen DK, Panisset M (2010). Central pontine and extrapontine myelinolysis: from epileptic and other manifestations to cognitive prognosis. J Neurol 257: 1176–1180.

Park BL, Kim JW, Cheong HS et al. (2013). Extended genetic effects of ADH cluster genes on the risk of alcohol dependence: from GWAS to replication. Hum Genet 132: 657–668.

Perry PJ, Argo TR, Barnett MJ et al. (2006). The association of alcohol-induced blackouts and grayouts to blood alcohol concentrations. J Forensic Sci 51: 896–899.

Pfefferbaum A, Sullivan EV (2005). Disruption of brain white matter microstructure by excessive intracellular and extracellular fluid in alcoholism: evidence from diffusion tensor imaging. Neuropsychopharmacology 30: 423–432.

Pfefferbaum A, Lim KO, Zipursky RB et al. (1992). Brain gray and white matter volume loss accelerates with aging in chronic alcoholics: a quantitative MRI study. Alcohol Clin Exp Res 16: 1078–1089.

Pfefferbaum A, Sullivan EV, Mathalon DH et al. (1997). Frontal lobe volume loss observed with magnetic resonance imaging in older chronic alcoholics. Alcohol Clin Exp Res 21: 521–529.

Pitel AL, Chanraud S, Sullivan EV et al. (2010). Callosal microstructural abnormalities in alzheimer's disease and alcoholism: same phenotype, different mechanisms. Psychiatry Res 184: 49–56.

Ponomarev I, Wang S, Zhang L et al. (2012). Gene coexpression networks in human brain identify epigenetic modifications in alcohol dependence. J Neurosci 32: 1884–1897.

Prescott CA, Kendler KS (1999). Genetic and environmental contributions to alcohol abuse and dependence in a population-based sample of male twins. Am J Psychiatry 156: 34–40.

Ramchandani VA, Umhau J, Pavon FJ et al. (2011). A genetic determinant of the striatal dopamine response to alcohol in men. Mol Psychiatry 16: 809–817.

Ray LA, Hutchison KE (2007). Effects of naltrexone on alcohol sensitivity and genetic moderators of medication response: a double-blind placebo-controlled study. Arch Gen Psychiatry 64: 1069–1077.

Ridley NJ, Draper B, Withall A (2013). Alcohol-related dementia: an update of the evidence. Alzheimers Res Ther 5: 3.

Rogawski MA (2005). Update on the neurobiology of alcohol withdrawal seizures. Epilepsy Current 5: 225–230.

Ron D, Messing RO (2013). Signaling pathways mediating alcohol effects. Curr Topics Behav Neurosci 13: 87–126.

Rose ME, Grant JE (2010). Alcohol-induced blackout. Phenomenology, biological basis, and gender differences. J Addiction Med 4: 61–73.

Schadt EE, Monks SA, Drake TA et al. (2003). Genetics of gene expression surveyed in maize, mouse and man. Nature 422: 297–302.

Schuckit MA (1987). Biological vulnerability to alcoholism. J Consult Clin Psychol 55: 301–309.

Smith DM, Atkinson RM (1995). Alcoholism and dementia. Int J Addict 30: 1843–1869.

Snell LD, Nunley KR, Lickteig RL et al. (1996). Regional and subunit specific changes in nmda receptor MRNA and immunoreactivity in mouse brain following chronic ethanol ingestion. Brain Res Mol Brain Res 40: 71–78.

Spanagel R (2009). Alcoholism: a systems approach from molecular physiology to addictive behavior. Physiol Rev 89: 649–705.

Starkman BG, Sakharkar AJ, Pandey SC (2012). Epigenetics – beyond the genome in alcoholism. Alcohol Res 34: 293–305.

Sullivan EV, Pfefferbaum A (2009). Neuroimaging of the Wernicke-Korsakoff syndrome. Alcohol Alcohol 44: 155–165.

Sullivan EV, Deshmukh A, Desmond JE et al. (2000). Cerebellar volume decline in normal aging, alcoholism, and Korsakoff's syndrome: relation to ataxia. Neuropsychology 14: 341–352.

Tabakoff B, Saba L, Printz M et al. (2009). Genetical genomic determinants of alcohol consumption in rats and humans. BMC Biol 7: 70.

Tardy B, Lafond P, Convers P et al. (1995). Adult first generalized seizure: etiology, biological tests, EEG, CT scan, in an ED. Am J Emerg Med 13: 1–5.

Torvik A, Lindboe CF, Rogde S (1982). Brain lesions in alcoholics. A neuropathological study with clinical correlations. J Neurol Sci 56: 233–248.

Toth C, Voll C, Macaulay R (2002). Primary CNS lymphoma as a cause of Korsakoff syndrome. Surg Neurol 57: 41–45.

Tozakidou M, Stippich C, Fischmann A (2011). Teaching neuroimages: radiologic findings in Marchiafava-Bignami disease. Neurology 77: e67.

Urso T, Gavaler JS, Van Thiel DH (1981). Blood ethanol levels in sober alcohol users seen in an emergency room. Life Sci 28: 1053–1056.

Uzbay TI (2012). Atypical antipsychotic drugs and ethanol withdrawal syndrome: a review. Alcohol Alcohol 47: 33–41.

Vanderlinden LA, Saba LM, Kechris K et al. (2013). Whole brain and brain regional coexpression network interactions associated with predisposition to alcohol consumption. PLoS One 8: e68878.

Vengeliene V, Bilbao A, Molander A et al. (2008). Neuropharmacology of alcohol addiction. Br J Pharmacol 154: 299–315.

Victor M, Adams RD, Mancall EL (1959). A restricted form of cerebellar cortical degeneration occurring in alcoholic patients. Arch Neurol 1: 579–688.

Weathers AL, Lewis SL (2009). Rare and unusual . . . Or are they? Less commonly diagnosed encephalopathies associated with systemic disease. Semin Neurol 29: 136–153.

Weissenborn R, Duka T (2000). State-dependent effects of alcohol on explicit memory: the role of semantic associations. Psychopharmacology (Berl) 149: 98–106.

Wetherill RR, Schnyer DM, Fromme K (2012). Acute alcohol effects on contextual memory bold response: differences based on fragmentary blackout history. Alcohol Clin Exp Res 36: 1108–1115.

Wetherill RR, Castro N, Squeglia LM et al. (2013). Atypical neural activity during inhibitory processing in substance-naive youth who later experience alcohol-induced blackouts. Drug Alcohol Depend 128: 243–249.

White AM (2003). What happened? Alcohol, memory blackouts, and the brain. Alcohol Res Health 27: 186–196.

White AM, Jamieson-Drake DW, Swartzwelder HS (2002). Prevalence and correlates of alcohol-induced blackouts among college students: results of an e-mail survey. J Am Coll Health 51 (117–119): 122–131.

Wolen AR, Miles MF (2012). Identifying gene networks underlying the neurobiology of ethanol and alcoholism. Alcohol Res 34: 306–317.

Wolstenholme JT, Warner JA, Capparuccini MI et al. (2011). Genomic analysis of individual differences in ethanol drinking: evidence for non-genetic factors in c57bl/6 mice. PLoS One 6: e21100.

Yoneoka Y, Takeda N, Inoue A et al. (2004). Acute Korsakoff syndrome following mammillothalamic tract infarction. AJNR Am J Neuroradiol 25: 964–968.

Zhao Z, Guo AY, van den Oord EJ et al. (2012). Multi-species data integration and gene ranking enrich significant results in an alcoholism genome-wide association study. BMC Genomics 13 (Suppl 8): S16.

Section 5

Neuropsychology

Handbook of Clinical Neurology, Vol. 125 (3rd series)
Alcohol and the Nervous System
E.V. Sullivan and A. Pfefferbaum, Editors

Chapter 11

Methods of association and dissociation for establishing selective brain–behavior relations

ROSEMARY FAMA[1,2]* AND EDITH V. SULLIVAN[1]

[1]*Department of Psychiatry and Behavioral Sciences, Stanford University School of Medicine, Stanford, CA, USA*

[2]*Neuroscience Program, SRI International, Menlo Park, CA, USA*

HISTORIC BACKGROUND

Myriad theories attempting to explain the underlying causes of cognitive, emotional, and behavioral abnormalities have existed throughout history. In ancient Greek civilization the heart was the organ touted as the seat of the soul, responsible for mental functions. A shift from this belief to one professing that the seat of mental functions was the head dates back to Democritus and Plato (Finger, 1994), but it was not until Galen (AD 129–200) that sensory, motor, and cognitive processes were hypothesized as distinct functions of the nervous system and that the brain was the center of this nervous system (see Zola-Morgan, 1995, for a review).

From the time of Galen to approximately the 18th century, debate ensued as to whether the ventricular system or the brain itself subserved cognition, emotion, and behavior. Nemesius (c. 400 AD), for example, postulated that sensation and perception resided in the anterior ventricles, thinking and reasoning in the third ventricle, and memory in the fourth ventricle (Benton, 1988). Modern concepts of functional localization, highlighting where select cognitive processes (e.g., language, memory, visuospatial abilities) are localized and organized in the brain, began in earnest in the 19th century with the observations, experiments, and writings of physicians including Franz Joseph Gall, Johann Spurzheim, Hughlings Jackson, David Ferrier, Paul Broca, Carl Wernicke, and Eduard Hitzig.

Historically, brain–behavior relations were inferred through individual case studies or aggregates of cases with a common lesion locus, where a localized brain lesion was associated with a specific functional deficit. Gall, in the early 1800s, was the first to confidently hypothesize that the brain consisted of different parts or regions, and that these different regions subserved dissociable cognitive functions (Benton, 1988). Gall's theory was strengthened when, in the mid-1800s, Paul Broca observed a series of patients with speech production impairment, who on autopsy were observed to have left frontal lobe damage (Berker et al., 1986). From these observations, Broca hypothesized that the left frontal lobe was responsible for productive speech (Lazar, 2011; Lazar and Mohr, 2011). Broca's discovery of the relationship between non-fluent aphasia and anterior left hemisphere disease was critical to the inception of hemispheric specialization and localization of function, identifying the neural mechanisms underlying behavioral processes (Benton, 1988). In a similar vein, Carl Wernicke observed a series of patients who had difficulty comprehending spoken language (Wilkins and Brody, 1970). Unlike Broca's patients who were unable to produce words, many of Wernicke's patients could produce speech; however, the speech produced was generally unintelligible. Autopsy revealed lesions in the posterior temporal lobe which led Wernicke to postulate that the posterior temporal lobe subserved the ability to understand and produce intelligible language.

Although reports of these brain–behavior relationships were groundbreaking, Jackson (1878) wisely cautioned his fellow scientists that "to locate the lesion which destroys speech and to locate speech are two different things." Kurt Goldstein reiterated Jackson's warning about correlation and specificity in the mid-1900s (Goldstein, 1946):

*Correspondence to: Rosemary Fama, Ph.D., Department of Psychiatry and Behavioral Sciences, Stanford University School of Medicine (MC5723), 401 Quarry Road, Stanford, CA 94305-5723, USA. Tel: +1-650-859-5064, Fax: +1-650-859-2743, E-mail: rfama@stanford.edu

We are by no means justified to infer directly from a correlation between a localized defect and a defect in performance a relationship between the concerned area and a definite performance. The facts allow only localisation of defects, but not a localisation of performances. The latter remains a theoretical interpretation which can be tried only after a careful analysis of the functions corresponding to the performance and defect. Such an analysis induces a concept about brain function and its relation to performance which, as unsatisfactory as it may have been till now, differs in principle from the concept of circumscribed localisation (pp. 25–26).

A simple correlation or relation between anatomical structure and behavioral function provides evidence for an association and can be useful in exploratory analysis, but it is inadequate for demonstrating specificity between a neuroanatomical structure and behavioral function.

SINGLE DISSOCIATION MODEL: LESION STUDIES

Demonstrating a single dissociation between structures and functions extends a simple association by showing that a lesion is related to a specific cognitive, motor, or behavioral function and is not related to a different cognitive, motor, or behavioral function. As Figure 11.1 shows, a dissociation is demonstrated when lesion "1" is associated with impairment in function "1" and lesion "1" is not associated with impairment in function "2." Alternatively, one can demonstrate that function '1' is associated with lesion "1" and not associated with lesion "2." This dissociation model provides stronger evidence of the specificity between structure and function

Single Dissociation

	Speech production	Speech comprehension
Left frontal lesion	+	—

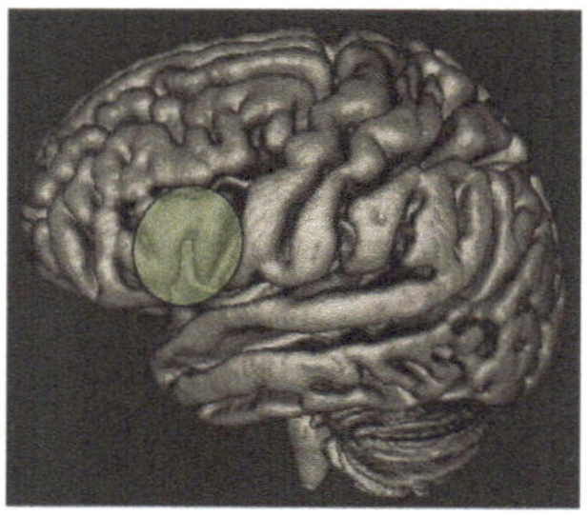

Fig. 11.1. This figure demonstrates the single dissociation model. Individuals with lesions in the left frontal lobe (green-shaded area) often demonstrate speech production but not speech comprehension impairments.

compared with simply demonstrating an association between a single structure and a single function.

The single dissociation model is inadequate for drawing conclusions about the specificity between brain structure and function. Lashley (1952) and Teuber (1955) highlighted a number of these problems, which they cautioned could lead to incorrect conclusions about functional localization. For example, deficits observed in function "1" and not function "2" could be due to differences between the tests used to assess these functions (one test being more or less demanding or sensitive to brain impairment or requiring greater cognitive resources than the other) as opposed to demonstrating brain–behavior specificity (Jones, 1983; Young et al., 2000). Thus, differences in structure–function associations could be due to test demand differences and not to selectivity between brain structure and function.

DOUBLE DISSOCIATION MODEL: LESION STUDIES

Ability to parse component processes of complex behaviors is necessary for understanding structure–function relationships and drawing conclusions about functional localization. Deficits in memory, for example, can result from impairment in a number of different cognitive component processes affecting encoding, storage, or retrieval of learned information. Frontal lobe lesions are often associated with impairments in the ability to organize and plan, and these impairments can contribute to problems with memory; this is not to say, however, that the frontal lobe is the "seat of memory functions." Thus, even when a brain region is associated with a particular cognitive or motor ability and that brain region is also shown to not be associated with a different function, it does not necessarily reflect a selective relationship between brain structure and function. Rather, it could demonstrate that this brain region serves a supportive role in the mediation of that function.

To address the limitations of the single dissociation model, Teuber (1955) championed the model of double dissociation (for historical reviews, see Milner and Teuber, 1968; Bates et al., 2003; Bigler, 2009). In the double dissociation model, structure–function relationships can be inferred between two or more individuals or clinical diagnoses when two functions are disrupted independently from one another: Lesion "1" is associated with impairment in function "1" but not in function "2," and lesion "2" (a lesion in a different area of the brain) is associated with impairment in function "2" but not in function "1." This model lends stronger evidence than does the single dissociation model that a particular brain region is specifically associated with a particular function. Figure 11.2 demonstrates a double dissociation based on the example

Double Dissociation

	Speech production	Speech comprehension
Left frontal lesion	+	—
Left temporoparietal lesion	—	+

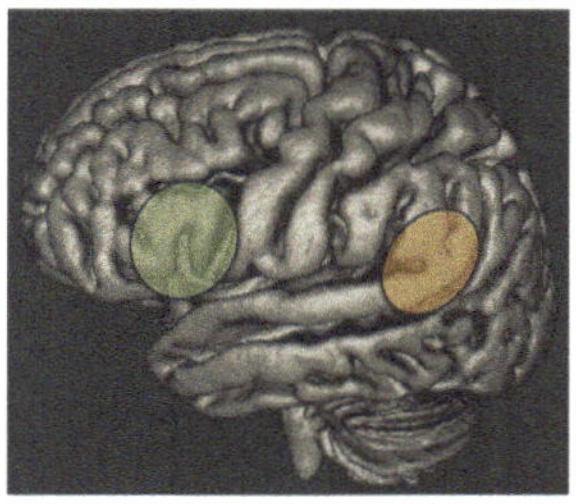

Fig. 11.2. This figure demonstrates the classical double dissociation model. Left frontal lobe lesions (green-shaded area) are associated with impairments in speech production and not speech comprehension, whereas left temporoparietal lobe lesions (orange-shaded area) are associated with the opposite pattern, specifically, impairment in speech comprehension with relative sparing of speech production. This double dissociation lends confidence to the selectivity of brain structure–function relations, where a specific brain region is more relevant to a selective process than to another.

of language production and comprehension. In this example, a left frontal cortical lesion is associated with speech production but not language comprehension, whereas a left temporoparietal cortical lesion is associated with language comprehension but not speech production. This double dissociation supports the hypothesis that the left frontal lobe is selectively related to speech production, whereas the left temporoparietal lobe is selectively related to language comprehension.

DOUBLE DISSOCIATION MODEL IN CONDITIONS AFFECTING MULTIPLE NEURAL SYSTEMS

Modifications of the classic double dissociation model used to establish specificity between brain structure and function in lesion studies have been successfully used in neurologic conditions that affect multiple brain structures, regions, and systems. Models include studies that compare clinical groups with a control group, clinical groups with one another, and heterogeneity within a single clinical group. These modifications to the classic double dissociation model are presented next.

Between-groups model

Many neurodegenerative conditions (e.g., Alzheimer's disease, Parkinson's disease, Huntington's disease (HD)) affect multiple neural systems and cannot be characterized by a single, discrete lesion. In these neurologic diseases, a double dissociation can be demonstrated and brain–behavior relations inferred by showing that one group of patients with a specific disease (group "1") demonstrates one pattern of spared and impaired functions (impaired function "1" and spared function "2"), whereas another group of patients with a different disease (group "2") demonstrates the opposite pattern of spared and impaired functions (impaired function "2" and spared function "1") (Jones, 1983). This pattern of observations suggests that brain abnormalities in these two conditions result in different functional impairments and that these functional impairments, in turn, are related to specific and dissociable neural networks selectively disrupted by the disease.

Dissociability of memory processes (e.g., explicit versus implicit memory processes) has been demonstrated using this model. Individuals, for example, with Korsakoff's syndrome (KS), an amnesia associated with chronic heavy drinking and concomitant thiamine deficiency, show severe anterograde explicit memory impairment with relatively intact implicit, procedural memory compared with control subjects (Cermak et al., 1985; Fama et al., 2006). By contrast, individuals with HD demonstrate the opposite pattern, with relatively intact explicit memory but impaired implicit, procedural memory (Heindel et al., 1988; Gabrieli et al., 1997). These patterns of sparing and impairment lend support to the hypothesis that explicit memory and implicit memory are dissociable processes and that the different neuropathology associated with KS and HD underlies the dissociation: thalamic regions and associated neural networks in KS and striatal regions and associated neural networks in HD. This type of double dissociation can be demonstrated by comparing two neurologic groups to a normal control group (Fig. 11.3, left) or to each other (Fig. 11.3, right) (Jones, 1983). A control group can be helpful in establishing whether clinical groups show impairment relative to controls as well as a difference in cognitive performance pattern. For example, one group demonstrates impairment compared with control subjects while the other group performs comparably to controls, or both clinical groups demonstrate impairment that simply differs in level of severity between the groups.

Within-group model

Another modification of the classic double dissociation model can be used within a single neurologic condition when heterogeneity and variance in brain damage and behavioral impairment are adequate for correlational analyses. Alcohol use disorders modify brain structures and, although not a result of a frank lesion, neurologic dysfunction in alcoholism includes widespread neuronal damage and cognitive, motor, and behavioral deficits. These deficits show a specific pattern of selective

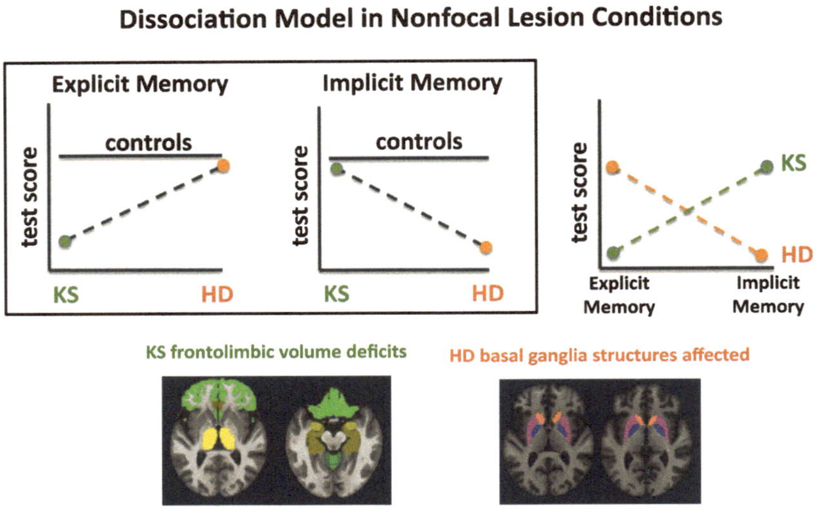

Fig. 11.3. Dissociations can be demonstrated in non-focal lesion conditions. Here, explicit and implicit memory scores in individuals with Korsakoff's syndrome (KS) and others with Huntington's disease (HD) are compared with a control group and between themselves. Studies have generally demonstrated that explicit memory processes are more affected than implicit memory processes in KS, whereas the opposite pattern, greater impairment in implicit than explicit memory processes, has been demonstrated in conditions with primary subcortical dysfunction such as HD. This pattern of deficits can be visually compared with a group of controls (left and middle graphs) or can be shown as a cross-over interaction between the two neurologic groups (right graph). The differences in mnemonic processes observed in these conditions are presumably caused by the different neuroanatomic regions affected in these disorders: frontolimbic deficits in KS and basal ganglia dysfunction in HD.

damage, with particular brain structures (e.g., frontal lobes, corpus callosum, and cerebellum) and selective neural networks more affected than others (Fein et al., 1995; Pfefferbaum et al., 1997, 2006, 2009; Meyerhoff, 2005; Makris et al., 2008). Understanding how alcohol-related brain abnormalities affect cognition, motor, and emotional processes is clinically and theoretically relevant. Identifying patterns of impairment and sparing lends evidence as to the specificity of different neural systems and cognitive processes (Caramazza, 1986) and to the functional and neural separation of abilities (Bates et al., 2003).

Unlike neurodegenerative conditions, brain impairments and behavioral deficits associated with alcoholism can improve or worsen depending on the time assessed during its typical dynamic course. Abstinence, for example, can be associated with increased volumes of gray and white matter and improvement in selective cognitive and motor functioning (Cardenas et al., 2007; Rosenbloom and Pfefferbaum, 2008; Sullivan et al., 2010). Heterogeneity in both brain structure and cognitive processes measured with quantitative measures of brain structure (magnetic resonance imaging (MRI): regional tissue volume) and quantitative measures of behavioral processes (e.g., neuropsychologic tests of cognition, motor, and emotional abilities) allows for brain—behavior associations to be demonstrated with correlational analysis on a continuum. Single and double dissociations in brain structure and function have been observed in alcoholism; examples of how this model can be used in alcoholism are shown below.

A single dissociation involving cerebellar volume and postural balance (Sullivan et al., 2006) was demonstrated when longer sway paths measured on a force platform, indicating imbalance during quiet standing, correlated with smaller volumes of the anterior cerebellar vermis (Fig. 11.4). In addition to this simple association, no relationship between lateral cerebellar volumes and postural balance was observed. This pattern of findings demonstrates the selectivity of the relationship between vermian (but not lateral) cerebellar volumes and postural stability using the single dissociation model.

A double dissociation in alcoholism was identified by Pfefferbaum and colleagues (2006) (Fig. 11.5) using the mean diffusivity metric of MR diffusion tensor imaging measuring local microstructural integrity of the corpus callosum and two cognitive tests. In this study, lower working memory scores were associated with greater diffusivity in the genu but not the splenium, whereas poorer matrix reasoning scores correlated with higher diffusivity in the splenium but not the genu. This double dissociation provides evidence for the specificity of fiber tract integrity of the anterior versus posterior corpus callosum to information processing and reasoning. These examples demonstrate that selective brain regions within chronic alcoholism can differentially affect behavioral processes.

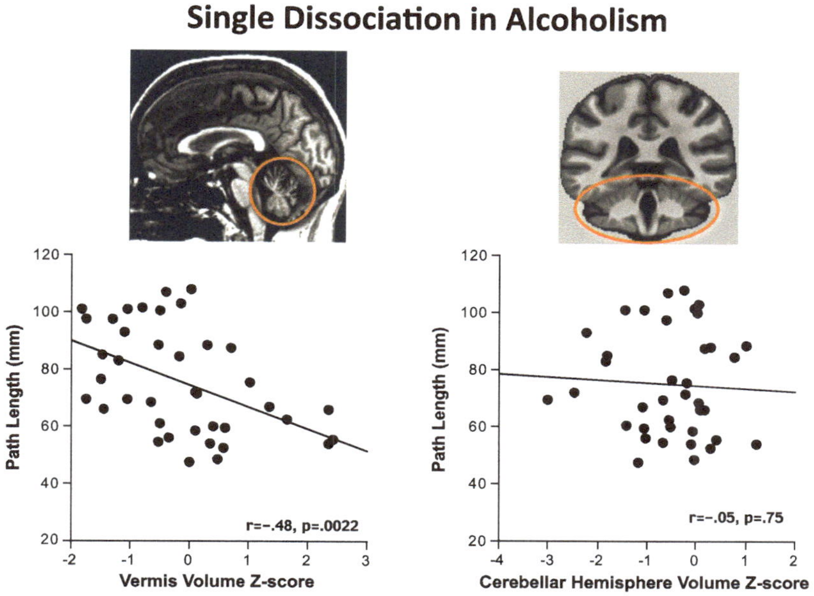

Fig. 11.4. These scatterplots depict a single dissociation within a group of alcoholics. Here it was demonstrated that volumes of the anterior vermis, but not lateral cerebellar hemisphere volumes, were associated with balance stability as assessed by path length sway.

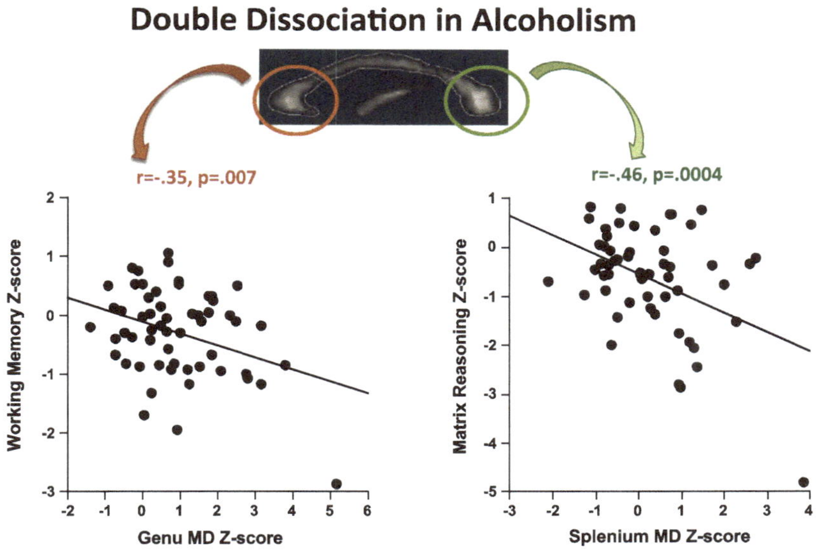

Fig. 11.5. In a group of 57 alcoholics, lower working memory scores were selectively correlated with high mean diffusivity (MD) in the genu but not the splenium, whereas lower visuospatial test scores were selectively correlated with high MD in the splenium but not the genu. Selectivity of these relations was established with multiple regression analysis. As reported in Pfefferbaum et al. (2006, p. 1004), "Genu and splenium MD values of the focal regions were entered as simultaneous predictors of working memory and accounted for 13% of the variance. Genu MD made a significant independent contribution to the variance (partial $F=4.10$, $p<0.05$), whereas splenium MD did not (partial $F=1.18$, $p>0.70$). For matrix reasoning, both genu MD (partial $F=3.55$, $p<0.07$) and splenium MD (partial $F=4.92$, $p=0.03$) were independent contributors to the total variance (26%). When all four regional DTI [diffusion tensor imaging] measures (i.e., genu fractional anisotropy (FA) and MD and splenium FA and MD) were entered as predictors of matrix reasoning, 28% of the variance was accounted for but only splenium MD made a significant independent contribution (partial $F=4.28$, $p<0.05$) to the overall variance. (For a detailed description of DTI methods, see Chapter 17 in this volume.)

Multiple dissociations and relevance to establishing network selectivity

Although the classic double dissociation model and its modifications provide strong evidence for selective brain structure–function relations, there are theoretic limitations to this model (Van Orden et al., 2001; Barr and Goldberg, 2003). A fundamental limitation is the assumption of modularity of function, which assumes that each structure and function can be separate and dissociable (Shallice, 1988), and that there are at most two components to differentiate from one another (Baddeley, 2003). That limitation, however, might be considered a "straw man" argument and be readily supplanted by models involving multiple dissociations. Neural networks and systems underlie complex neuropsychologic functions; a functional and structural network model has supplanted the simple localization model. The network approach posits that neuropsychologic functions (e.g., memory, attention, visuospatial processes) comprise multiple component processes, any of which, if impaired, affects that function.

The fundamental principles of double dissociation can be applied to intrinsic functional networks, which are identified with resting-state functional MRI, are replicable across study sites, are dissociable, and appear to be a network substrate of complex constellations of behavior. For example, nodes of the default mode network (DMN), comprising the medial prefrontal cortex, posterior cingulate cortex, precuneus, and medial temporal lobe (Raichle et al., 2001; Greicius et al., 2003, 2009), are maximally active during rest and considered to serve functions of self-reference. The DMN is distinct from other intrinsic networks, such as the executive control network, which links dorsolateral frontal and parietal structures and is relevant to salience processing functions (Seeley et al., 2007; Tomasi et al., 2009). These networks are dissociable and are differentially vulnerable to disease (Seeley et al., 2009). The distinction of these networks indicates their dissociability and therefore the need for rigorous application of selectivity models to establish their specificity constellation of functions.

Regardless of the model, brain–behavior relationships demonstrate associations (correlations) and do not speak to causation or directionality of effect. Despite these limitations, the double (or multiple) dissociation model has been and continues to be a powerful heuristic and statistical tool to pursue objective and quantitative understanding of complex cognitive systems (Baddeley, 2003). Knowledge of the current limitations of the double dissociation model and understanding of the strength of the inferences obtained from the results of these studies can guide future hypotheses concerning the relationships of brain structure and function in chronic alcoholism and other conditions affecting the brain and nervous system whether characterized by discrete or widespread damage.

ACKNOWLEDGMENT

This research was supported by grants from the National Institute on Alcohol Abuse and Alcoholism AA017347, AA010723, AA017168, and AA021697. We thank Adolf Pfefferbaum, M.D. for his helpful comments on the text and supplying the MR images for the figures throughout the chapter.

References

Baddeley A (2003). Double dissociations: not magic, but still useful. Cortex 39: 129–131.

Barr WB, Goldberg E (2003). Pitfalls in the method of double dissociation: delineating the cognitive functions of the hippocampus. Cortex 39: 153–157.

Bates E, Appelbaum M, Salcedo J et al. (2003). Quantifying dissociations in neuropsychological research. J Clin Exp Neuropsychol 25: 1128–1153.

Benton AL (1988). Neuropsychology: past, present, and future. In: F Boller, J Grafman (Eds.), In: Handbook of Neuropsychology, Vol. 1. Elsevier Science, Amsterdam, pp. 1–29.

Berker EA, Berker AH, Smith A (1986). Translation of Broca's 1865 report. Localization of speech in the third left frontal convolution. Arch Neurol 43: 1065–1072.

Bigler ED (2009). Hans-Lukas Teuber and 'The riddle of frontal lobe frontal in man' as published in The frontal granular cortex and behavior (1964). Neuropsychol Rev 19: 9–24.

Caramazza A (1986). On drawing inferences about the structure of normal cognitive systems from the analysis of patterns of impaired performance: the case for single-patient studies. Brain Cogn 5: 41–66.

Cardenas VA, Studholme C, Gazdzinski S et al. (2007). Deformation-based morphometry of brain changes in alcohol dependence and abstinence. Neuroimage 34: 879–887.

Cermak LS, Talbot N, Chandler K et al. (1985). The perceptual priming phenomenon in amnesia. Neuropsychologia 23: 615–622.

Fama R, Pfefferbaum A, Sullivan EV (2006). Visuoperceptual priming in alcoholic Korsakoff syndrome. Alcohol Clin Exp Res 30: 680–687.

Fein G, Meyerhoff DJ, Weiner MW (1995). Magnetic resonance spectroscopy of the brain in alcohol abuse. Alcohol Health Res World 19: 3056–3314.

Finger S (1994). Origins of Neuroscience: a history of explorations into brain function. Oxford University Press, New York.

Gabrieli JD, Stebbins GT, Singh J et al. (1997). Intact mirror-tracing and impaired rotary-pursuit skill learning in patients with Huntington's disease: evidence for dissociable memory systems in skill learning. Neuropsychology 11: 272–281.

Goldstein K (1946). Remarks on localisation. Confin Neurol 7: 25–34.

Greicius MD, Krasnow B, Reiss AL et al. (2003). Functional connectivity in the resting brain: a network analysis of the default mode hypothesis. Proc Natl Acad Sci U S A 100: 253–258.

Greicius MD, Supekar K, Menon V et al. (2009). Resting-state functional connectivity reflects structural connectivity in the default mode network. Cereb Cortex 19: 72–78.

Heindel WC, Butters M, Salmon DP (1988). Impaired learning of a motor skill in patients with Huntington's disease. Behav Neurosci 102: 141–147.

Jackson HJ (1878). On affections of speech from disease of the brain. Brain 1: 304–330.

Jones GV (1983). On double dissociation of function. Neuropsychologia 21: 397–400.

Lashley KS (1952). Functional interpretation of anatomic patterns. Res Publ Assoc Res Nerv Ment Dis 30: 529–547.

Lazar RM (2011). Broca and the biology of language. Neuropsychol Rev 21: 225–226.

Lazar RM, Mohr JP (2011). Revisiting the contributions of Paul Broca to the study of aphasia. Neuropsychol Rev 21: 236–239.

Makris N, Oscar-Berman M, Jaffin SK et al. (2008). Decreased volume of the brain reward system in alcoholism. Biol Psychiatry 64: 192–202.

Meyerhoff DJ (2005). Brain spectroscopic imaging, morphometry, and cognition in social drinkers and recovering alcoholics. Alcohol Clin Exp Res 29: 153–154.

Milner B, Teuber H-L (1968). Alteration of perception and memory in man: reflections: reflections on methods. In: L Weiskrantz (Ed.), Analysis of Behavioral Change. Harper & Row, New York, pp. 268–375.

Pfefferbaum A, Sullivan EV, Mathalon DH et al. (1997). Frontal lobe volume loss observed with magnetic resonance imaging in older chronic alcoholics. Alcohol Clin Exp Res 21: 521–529.

Pfefferbaum A, Adalsteinsson E, Sullivan EV (2006). Dysmorphology and microstructural degradation of the corpus callosum: interaction of age and alcoholism. Neurobiol Aging 27: 994–1009.

Pfefferbaum A, Rosenbloom MJ, Rohlfing T et al. (2009). Degradation of association and projection white matter systems in alcoholism detected with quantitative fiber tracking. Biol Psychiatry 65: 680–690.

Raichle M, MacLeod A, Snyder A et al. (2001). A default mode of brain function. Proc Natl Acad Sci U S A 98: 676–682.

Rosenbloom MJ, Pfefferbaum A (2008). Magnetic resonance imaging of the living brain: evidence for brain degeneration among alcoholics and recovery with abstinence. Alcohol Res Health 31: 362–376.

Seeley WW, Menon V, Schatzberg AF et al. (2007). Dissociable intrinsic connectivity networks for salience processing and executive control. J Neurosci 27: 2349–2356.

Seeley WW, Crawford RK, Zhou J et al. (2009). Neurodegenerative diseases target large-scale human brain networks. Neuron 62: 42–52.

Shallice T (1988). From Neuropsychology to Mental Structure. Cambridge University Press, Cambridge.

Sullivan EV, Rose J, Pfefferbaum A (2006). Effect of vision, touch, and stance on cerebellar vermian-related sway and tremor: a quantitative MRI and physiological study. Cereb Cortex 16: 1077–1086.

Sullivan EV, Harris RA, Pfefferbaum A (2010). Alcohol's effects on brain and behavior. Alcohol Res Health 33: 127–143.

Teuber H-L (1955). Physiological psychology. Annu Rev Psychol 6: 267–296.

Tomasi D, Volkow ND, Wang R et al. (2009). Dopamine transporters in striatum correlate with deactivation in the default mode network during visuospatial attention. PLoS One 4: e6102.

Van Orden GC, Pennington BF, Stone GO (2001). What do double dissociations prove? Cognit Sci 25: 111–172.

Wilkins RH, Brody IA (1970). Wernicke's sensory aphasia. Arch Neurol 22: 279–282, English translation of C. Wernicke (1874) Der aphasische Symptomencomplex: Eine psychologische Studie auf anatomischer Basis. Breslau: Max Cohn & Weigert.

Young MP, Hilgetag CC, Scannell JW (2000). On imputing function to structure from the behavioural effects of brain lesions. Philos Trans R Soc Lond B Biol Sci 355: 147–161.

Zola-Morgan S (1995). Localization of brain function: the legacy of Franz Joseph Gall (1758–1828). Annu Rev Neurosci 18: 359–383.

Handbook of Clinical Neurology, Vol. 125 (3rd series)
Alcohol and the Nervous System
E.V. Sullivan and A. Pfefferbaum, Editors

Chapter 12

Profiles of impaired, spared, and recovered neuropsychologic processes in alcoholism

MARLENE OSCAR-BERMAN*, MARY M. VALMAS**, KAYLE S. SAWYER**, SUSAN MOSHER RUIZ**, RIYA B. LUHAR**, AND ZOE R. GRAVITZ**
Boston University School of Medicine and VA Boston Healthcare System, Boston, MA, USA

INTRODUCTION

In this chapter, we first survey research related to general cognitive competencies and deficiencies in alcoholism, and the broad areas of brain involvement. This is followed by a review of five principal profiles of alcoholism-related neuropsychologic functions: memory, executive functions, emotion and psychosocial skills, visuospatial cognition, and psychomotor abilities. Our review of the evidence indicates that, while certain features of these functions are impaired, other aspects may be spared or recovered. The research has employed diverse neuropsychologic and behavioral assessment procedures, as well as measures of brain structure and function. We begin our review by addressing some concerns regarding the selection of research participants and the tests used to study them. Attention to methodology is important, because variations and inconsistencies among studies can lead to discordant findings and conclusions. We also consider problems related to assessing impaired, spared, and recovered functions. Ultimately, our objective is to reveal areas for future research and to inform clinical strategies in the treatment of alcoholism.

THE PARTICIPANTS AND THE TESTS

Our review focuses on studies of uncomplicated alcoholism as distinguished from special cases of alcohol-related disorders such as Korsakoff's syndrome (Oscar-Berman, 2012), alcohol dementia (Brust, 2010; Vetreno et al., 2011), and Marchiafava–Bignami disease (Kim et al., 2007). Our review excludes studies of adolescent drinkers, as well as research investigating binge drinking, the episodic heavy consumption of alcohol. Additionally, although studies using acute alcohol administration have direct applicability to driving situations, work-related hazards, and other societally relevant concerns, here we consider residual neuropsychologic functions associated with long-term chronic alcoholism.

Characteristics of the participants

In understanding the consequences of alcoholism, drinking patterns are important considerations, i.e., the type, amount, and frequency of alcohol consumed; the age of onset of drinking; the severity and duration of the abuse or dependency; and the duration of alcohol abstinence. Repeated episodes of withdrawal also may be a risk factor for cognitive impairment and for decreased cognitive recovery due to neurotoxic lesions and excitotoxicity (Loeber et al., 2010). Studies of residual effects typically include participants who have abused alcohol for a long period of time (e.g., 5 years or more), as this permits the measurement of cumulative effects. Additionally, investigators interested in studying alcohol's residual influences generally enroll participants who have been abstinent for a minimum of 3–4 weeks. This duration of abstinence is important for obtaining stable levels of performance after ethanol and its metabolites have been eliminated from the body (Oscar-Berman and Marinkovic, 2007). In this review, we include studies meeting that criterion unless specified otherwise.

When investigating brain–behavior relationships in individuals with a history of alcoholism, it is imperative to keep in mind that, in addition to drinking patterns, many factors influence the expression and

*Correspondence to: Marlene Oscar-Berman, Ph.D., Boston University School of Medicine, L-815, 72 E. Newton St., Boston, MA 02118, USA. Tel: +1-617-638-4803, E-mail: oscar@bu.edu
**These authors contributed equally to the preparation of this chapter

course of the disease, and can interact to aggravate alcoholism's effects (Petrakis et al., 2002). Consequently, methodologic differences with respect to exclusion and inclusion criteria contribute to disparate findings across studies. These differences include the following: family history of alcoholism (Cardenas et al., 2005); treatment history (Di Sclafani et al., 2008); age (Pfefferbaum et al., 2006; Schottenbauer et al., 2007; Westlye et al., 2010); gender (Fattore et al., 2008; Devaud and Prendergast, 2009; Nixon, 2013; Ruiz et al., 2013); the use or abuse of medicines, nicotine, or other drugs (Durazzo et al., 2006; Cosgrove et al., 2011); body mass index (Pfefferbaum et al., 2009; Gazdzinski et al., 2010a); and comorbid medical, neurologic, and psychiatric conditions (Di Sclafani et al., 2007; Fama et al., 2009; Oscar-Berman et al., 2009; Charness, 2009; Sameti et al., 2011; Martin, 2013). Common comorbid medical complications are malnutrition, diseases of the liver and the cardiovascular system, human immunodeficiency virus (HIV)/acquired immunodeficiency syndrome (AIDS), neurologic conditions such as head injury and inflammation of the brain (encephalopathy), and fetal alcohol spectrum disorders. Common comorbid psychiatric conditions are depression, anxiety, posttraumatic stress disorder, antisocial personality disorder, and schizophrenia.

Assessing neuropsychologic functions and the brain

In an attempt to clarify the deficits and intact abilities of individuals with obvious or suspected central nervous system (CNS) pathology, neuropsychologists have relied on a variety of assessment procedures. Test instruments are selected to engage both general and specific abilities that depend upon the integrity of the brain regions involved in those functions. In general, the primary neuropsychologic domains assessed in alcoholics have been memory, executive functions (e.g., attention, abstraction, problem solving, organizing and planning, and inhibition), emotion and psychosocial skills, visuospatial cognition, and psychomotor abilities. Examples of specific assessment instruments are referred to in the sections below, and descriptions of the tests can be found in secondary sources (Strauss et al., 2006; Lezak et al., 2012). Importantly, because these tests tend to measure broad categories of abilities, there is no universally accepted consensus on specific attributes that define the functions they assess.

Neuroimaging techniques have become essential adjunctive tools used to investigate functional, anatomic, and biochemical characteristics of the brain. In the study of alcoholism, commonly used methods include magnetic resonance imaging (MRI), functional MRI (fMRI), diffusion tensor imaging (DTI), positron emission tomography, magnetoencephalography, MR perfusion, and MR spectroscopy (Schulte et al., 2012c). These measurement tools permit researchers to evaluate attributes of the brain associated with particular neuropsychologic processes, and to assess the size and integrity of the underlying neural tissues. Most research endeavors typically employ a single neuroimaging method with individual subjects. However, Schulte et al. (2012c) have suggested that the combined use of several techniques would be highly advantageous in defining brain abnormalities, because each method has unique strengths and weaknesses.

THE IMPAIRED, THE SPARED, AND THE RECOVERED

The various domains of neuropsychologic functioning and their associated brain networks have different patterns of impairment, preservation, and recovery, hence our tripartite categorization. Before reviewing the alcoholism-related functional domains of interest, we first consider separate issues relevant to each of the three categories.

Impairments

Uncomplicated alcoholism has been associated with abnormalities in cognitive, behavioral, and emotional functioning. Although these abnormalities may not be evident through superficial observation, neuropsychologic tests have been able to illuminate the impairments (Leber and Parsons, 1982; Tivis et al., 1995; Clark et al., 2007; Rourke and Grant, 2009). Similarly, neuroimaging techniques have revealed reliable differences between alcoholic and non-alcoholic control groups in brain activation patterns and structural integrity, with the most vulnerable networks having been identified as the frontocerebellar system (see Fig. 12.2, below) and mesocorticolimbic circuitry (see Fig. 12.3, below) (Agartz et al., 1999; Makris et al., 2008; Fama et al., 2012; Chanraud et al., 2013). However, relatively few studies have demonstrated definitive and direct links between specific component neuropsychologic impairments and measures of highly localized brain abnormalities.

Spared functions and compensation

Individuals with a history of long-term uncomplicated alcoholism appear free from impairments in certain basic neuropsychologic functions. Some researchers have suggested that the lack of detectable deficits in these seemingly spared domains could be due to compensatory recruitment of alternative brain regions to facilitate cognitive processes (Sullivan and Pfefferbaum, 2005; Oscar-Berman and Marinkovic, 2007; Chanraud et al., 2013). In support of this notion, fMRI studies have demonstrated altered patterns of neural activation among alcoholics, with reduced activity in some regions (Chanraud et al., 2010, 2011), but simultaneous increases in activity in others (Pfefferbaum et al., 2001a; Desmond et al., 2003; Chanraud-Guillermo et al., 2009; Chanraud et al., 2013). As such, it can be difficult to distinguish between functions that have been truly spared and those that appear unimpaired due to compensatory mechanisms or other recovery of function.

Descriptions of spared CNS structures, while not prone to concerns about compensation in the same way as neuropsychologic performance, are subject to issues of possible tissue growth with sustained sobriety, which could mask effects that might have been observable earlier in abstinence. The picture is further complicated in cross-sectional studies when trying to identify structures that have been truly spared and those that have recovered after cessation of drinking. In other words, brain regions that appear to be spared from damage might contain tissues that have been repaired or regenerated.

Another cautionary note is in order with regard to neuropsychologic processes or brain structures in which there has been a failure to detect alcoholism-related relationships. That is, although negative results from research studies may be due to genuinely preserved abilities and brain tissue, it is possible that the data sets lack the statistical power needed to detect *bona fide* differences between populations. Accordingly, studies that have examined certain functions and structures but have not found alcoholism-related abnormalities cannot characterize confidently the functions or regions as spared, because the results may have been obtained from inadequate sample sizes or because the variability within the measures was too large. In many of those studies, the probability that real associations exist but were not detected (type II error) might be high but not reported. Thus, conclusions regarding spared functions cannot always be drawn with adequate certainty from the current literature. In any case, the identification of preserved functions – whether they are spared, compensatory, or recovered – is of great benefit to the scientific and healthcare communities. Once intact functions are identified, clinicians can work with patients to harness and strengthen these abilities in order to accelerate recovery and improve their quality of life.

Recovery

Although differences in outcomes observed in neuropsychologic and brain abnormalities are complicated by the fact that the course of recovery can be dependent on the extent of impairment at the time of drinking cessation, abstinent alcoholics have shown improvements in neuropsychologic functioning with continued sobriety (Kish et al., 1980; Oscar-Berman et al., 2004; Rosenbloom et al., 2004; Erickson and White, 2009). Likewise, abstinence is accompanied by brain structural changes, such as increased cortical thickness in the brain's extended reward and oversight system (Durazzo et al., 2011). Processes proposed to account for recovery of cognitive functioning in alcoholism include neural repair (e.g., regeneration) and reorganization, such that new or additional neural networks are recruited to accomplish a task (Crews et al., 2005; Sullivan and Pfefferbaum, 2005; Chanraud et al., 2013).

Neuropsychologic findings provided the earliest insights into restored function in abstinent alcoholics (Oscar-Berman and Marinkovic, 2007), and neuroimaging has permitted the quantification of neural tissue repair and recovery with abstinence (Buhler and Mann, 2011). Early neuroimaging evidence for structural brain damage and subsequent improvement among abstinent alcoholics was reported by Carlen et al. (1978), who demonstrated reduced atrophy in recently abstinent alcoholics; improvement was not seen in those who relapsed. Modern MRI studies have confirmed and extended those findings to include demonstrations of recovery in brain volume, microstructure, and neurochemistry (Pfefferbaum et al., 1995; Demirakca et al., 2011; Durazzo et al., 2011; Alhassoon et al., 2012). This body of knowledge has grown steadily to include manifestations of recovery in a wide spectrum of tissue properties across several imaging modalities (Oscar-Berman and Marinkovic, 2007; Schulte et al., 2012c). In particular, improvements in brain tissue volumes following periods of abstinence from alcohol have been observed in gray matter (Durazzo et al., 2011) and in white matter (Ruiz et al., 2013), often with corresponding decreases in the volume of cerebrospinal fluid (CSF). Potential mechanisms that have been suggested to underlie recovery of neural tissue during sobriety include remyelination, neurogenesis, and rehydration (Nixon, 2006; Rosenbloom and Pfefferbaum, 2008). Accordingly, recovery of neuropsychologic functioning has been directly associated with improvements in the structure of neural tissue (Muller-Oehring et al., 2009; Stavro et al., 2012).

PROFILES OF DAMAGE AND REPAIR

In the following sections, we review neuropsychologic functioning in alcoholism. We begin with an overview of general abilities and intelligence in conjunction with findings related to widespread structural brain abnormalities and evidence of recovery. We then review profiles of five functional domains and associated brain regions.

General neuropsychologic abilities

Methods for evaluating general abilities and overall intelligence have been in use for more than a century, and continue to be employed regularly today (Strauss et al., 2006; Lezak et al., 2012). Here we describe examples from research that has looked at general abilities in the study of long-term chronic alcoholism, and we conclude that their utility in characterizing deficits in alcoholism has been limited, in part because of the variety of functions they assess.

Many studies have reported poorer performance by alcoholics as compared to non-alcoholic controls in general cognitive abilities or intelligence (Sullivan et al., 2000b, 2002b; Davies et al., 2005; Rosenbloom et al., 2007), but the deficits have been mild, or they tended to improve within a year of abstinence (Bates et al., 2002). Other studies have reported no alcoholism-related impairments in general intelligence (Oscar-Berman et al., 2004; Fein et al., 2006; Foisy et al., 2007; Marinkovic et al., 2009; Sullivan et al., 2010; Pitel et al., 2012). When general abilities are assessed according to discrete component factors and functional domains, a different picture emerges. That is, investigators who have examined performance on specific IQ subtests or have employed extensive batteries of specialized neuropsychologic tests have observed alcoholism-related impairments in some domains but not in others (Tivis et al., 1995; Beatty et al., 2000; Davies et al., 2005; Fein et al., 2006; Oscar-Berman et al., 2009). For example, Davies et al. (2005) reported no differences between alcoholics and controls on prorated Full Scale IQ scores, but alcoholics performed worse on Vocabulary and Digit Symbol subtests. Stavro et al. (2012) found only a minimal impairment in general intelligence in a meta-analysis of 62 studies examining cognitive abilities in alcoholism. However, they did find moderate impairments across 11 separate cognitive functions: verbal fluency/language, speed of processing, working memory, attention, problem solving/executive functions, inhibition/impulsivity, verbal learning, verbal memory, visual learning, visual memory, and visuospatial abilities. The meta-analysis indicated that the levels of impairment for the separate areas of dysfunction remained impaired after 1 year of sobriety. Importantly, the study stressed the need to examine the distinct component processes that contribute to general abilities and intelligence.

Widespread brain damage in alcoholism

Structural abnormalities have been found in widespread brain regions of alcoholics. Relatively smaller volumes have been noted not only in total brain size (Pfefferbaum et al., 1992; Mann et al., 2005; Chen et al., 2012), including the cerebellum (Chanraud et al., 2007), but also in cerebral subdivisions and regional areas (Pfefferbaum et al., 1992, 2001b; Makris et al., 2008; Fein et al., 2009; Chen et al., 2012) and numerous subcortical structures (Chanraud et al., 2007; Makris et al., 2008). Smaller white-matter volumes likewise have been reported (Harper et al., 1990; Pfefferbaum et al., 1995, 2001b; Chen et al., 2012), including the corpus callosum (Fig. 12.1) (Pfefferbaum et al., 1996; Agartz et al., 1999; Ruiz et al., 2013). Smaller brain volume is accompanied by larger CSF volume (Pfefferbaum et al., 1992, 1998, 2001b; Agartz et al., 1999; Bendszus et al., 2001; Chen et al., 2012), and researchers have reported larger

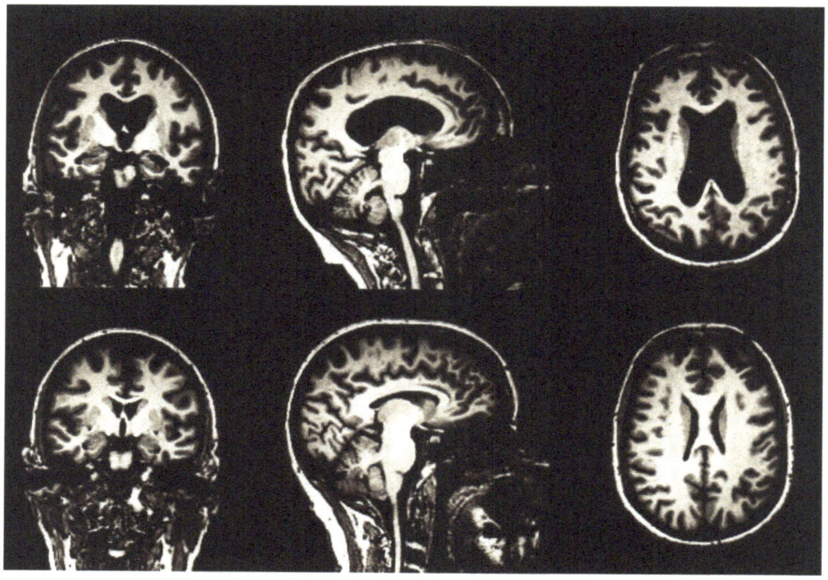

Fig. 12.1. Severe brain atrophy in a 72-year-old alcoholic woman (top row) compared to an age-matched non-alcoholic control woman (bottom row). Most prominent indications of alcoholism-related brain damage include enlarged ventricles, thinning of the corpus callosum, and hippocampal atrophy.

volumes of the lateral ventricles (Zipursky et al., 1989; Pfefferbaum et al., 1998, 2001b; Ruiz et al., 2013) and the third ventricle (Pfefferbaum et al., 2001b).

Structural improvement of gray- and white-matter tissue has been associated with abstinence. Pfefferbaum et al. (1995) found a trend for increases in anterior cortical gray-matter volumes after 20 days of sobriety, and Agartz et al. (2003) reported whole-brain increases in white-matter volumes and corresponding decreases in CSF after 1 month of abstinence. Recovery of white-matter microstructure has been demonstrated in longitudinal studies, with improvements in fractional anisotropy (a measure of microscopic abnormalities in axons) in the corpus callosum after 1 year of abstinence (Alhassoon et al., 2012) and decreases in mean radial diffusivity (a measure related to axonal demyelination) after only 1 month of sobriety, suggesting recovery of axonal integrity (Gazdzinski et al., 2010b). However, Monnig et al. (2012) reported persistent widespread fractional anisotropy deficits in bilateral parietal regions after a year of abstinence. Another marker for detecting reversal of global alcoholism-related brain damage that remains understudied is cerebral perfusion. The few studies examining perfusion improvements using arterial spin labeling MRI have yielded conflicting results. Mon et al. (2009) reported improvement of frontal and parietal gray-matter perfusion after 5 weeks of abstinence, but Durazzo et al. (2010) were unable to detect longitudinal perfusion increases in parietal gray matter after an average of 35 days of abstinence. Thus, findings associated with recovery of overall brain integrity have been inconsistent.

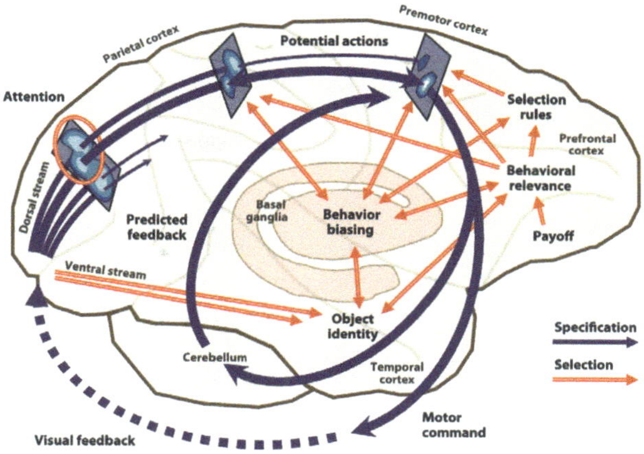

Fig. 12.2. The frontocerebellar system (Cisek and Kalaska, 2010). This figure illustrates interactions among cortical areas, basal ganglia, and the cerebellum. Reproduced with permission of *Annual Reviews*.

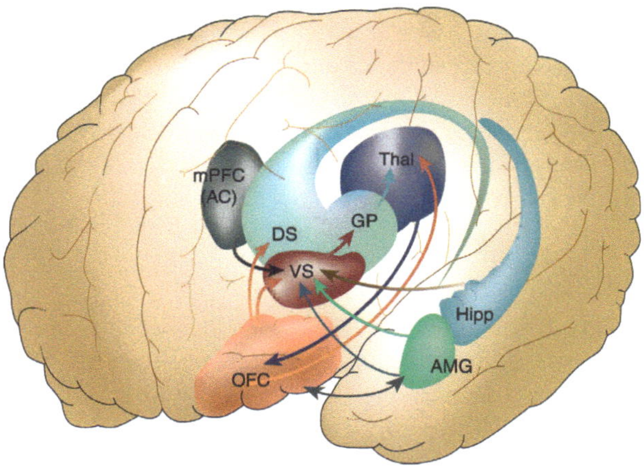

Fig. 12.3. The mesocorticolimbic circuitry (Salloway et al., 2001). This figure highlights the following structures: orbitofrontal cortex (OFC), medial prefrontal cortex (mPFC), anterior cingulate (AC), amygdala (AMG), hippocampus (Hipp), thalamus (Thal), ventral striatum (VS), dorsal striatum (DS), and globus pallidus (GP). Not depicted are the mammillary bodies of the hypothalamus, damage to which has been considered a hallmark of alcoholic Korsakoff's syndrome (Oscar-Berman, 2012). Adapted with permission from Salloway et al., 2001, *The Frontal Lobes and Neuropsychiatric Illness* (copyright 2001), American Psychiatric Association.

Frontocerebellar and mesocorticolimbic structures

It is important to note that, although abnormalities have been found in widespread brain regions of uncomplicated alcoholics, researchers who have investigated networks of neural structures subserving the specific profiles of neuropsychologic domains observed to be impaired (as discussed below) have emphasized the contributions of two distinct but overlapping systems: the frontocerebellar system (Fig. 12.2) and the mesocorticolimbic circuitry (Fig. 12.3) (Sullivan et al., 2003; Fuster, 2008; Harris et al., 2008; Makris et al., 2008; Goldstein and Volkow, 2011). Importantly, both of these extensive systems control functions that could prevent alcoholics from being able to employ good judgment and avoid the negative behaviors and actions associated with addiction.

THE FIVE FUNCTIONAL DOMAINS

In order to refine our understanding of alcoholism-related impaired, spared, and recovered abilities, we focus on five specific functional domains: (1) memory; (2) executive functions; (3) emotion and psychosocial skills; (4) visuospatial cognition; and (5) psychomotor abilities. Each neuropsychologic domain has a unique profile of relative impairment, or lack thereof. For each domain, behavioral deficits are discussed first (summarized in Table 12.1), along with their associated brain abnormalities (Table 12.2), followed by evidence of compensation or recovery.

Memory

Memory refers to the storage of new and old information for later use. Memories can be stored for very short periods of time – seconds or even milliseconds, during which information is encoded and consolidated – or for long periods of time – weeks, years, or a lifetime. Table 12.3 represents a descriptive, atheoretic way of viewing human memory, emphasizing temporal characteristics of the information being stored. Because memory is not unitary, theorists have considered its many different forms, mediated by separate component processes, governed by different principles, and controlled by extensive brain circuitries (Tulving and Craik, 2000). In relation to alcoholism, memory has been explored most frequently with respect to the type of material or event being remembered and the duration of storage. "Working memory," which stores transitory information in the service of future action despite distractions, is considered a component of executive functions (discussed below).

Impairments

Alcoholics have demonstrated some degree of memory impairment, despite evidence that they tend to overestimate their memory capacity (metamemory: Le Berre et al., 2010). The nature and extent of the impairments depend to some degree upon characteristics of the neuropsychologic tests (task parameters such as timing of stimulus presentation and type of materials) and on the integrity of other cognitive functions supporting memory. For example, Marinkovic et al. (2009), using the Wechsler Memory Scale, which measures a number of different types of memory, including verbal, auditory, visual, short-term, and working memory, found that alcoholics' reaction times in a deep word-encoding task were negatively correlated both with working memory scores and performance on a separate test of verbal fluency. These findings suggest that participants with worse executive functioning skills found the encoding task more difficult. The reaction times also showed a tendency to correlate with duration of heavy drinking, indicating that long-term heavy alcohol intake impacted speed of word encoding. Other investigators used extensive batteries of neuropsychologic tests (including the Wechsler Memory Scale, Dementia Rating Scale, Rey-Osterrieth Complex Figures Test, and a word/face recognition test) and found that alcoholics had significantly impaired verbal, non-verbal, and spatial short-term memory (Sullivan et al., 2000b, 2002b).

Alcoholics also have been found to perform below normal on tests that evaluate memory typologies or processes such as episodic memory (Le Berre et al., 2010) and free and delayed recall (Pitel et al., 2007; Noël et al., 2012). The tests of episodic memory used by Le Berre et al. (2010) examined free recall, semantic cued recall, and recognition memory for shallow and deeply encoded words; total recall and recognition scores were below those of the controls. Similarly, Pitel et al. (2007) employed a variety of episodic memory tests and found impairments in several components, including encoding and retrieval processes, contextual memory, autonoetic consciousness (the ability to analyze our own thoughts), and learning abilities. Noël et al. (2012) also detected impairments in episodic memory using the California Verbal Learning Test, and suggested that this defect likely reflected executive dysfunction. That is, rather than a retrieval deficit, alcoholics might have difficulties applying strategies in organizing information for later use. Kopera et al. (2012) studied alcoholic patients who presented with deficits in visual episodic memory, and observed that non-verbal domains such as attention, visual memory, and working memory were impaired.

Neuropsychologic indications of dysfunctional memory acquisition and processing have been investigated using neuroimaging techniques. In an fMRI study, Akine et al. (2007) employed a long-term false-recognition memory paradigm comparing alcoholics to controls on task performance and brain activation. Although the researchers did not find alcoholics to be impaired on the task, they did find differing patterns of brain activity between the groups. Alcoholics exhibited diminished responses in several small and isolated locations of the brain (right prefrontal cortex, anterior cingulate, right pulvinar, and right ventral striatum). Because the alcoholics in this study were relatively young (mean age in their 30s) and had been abstinent for an average of more than 3 years, the authors suggested that the abnormal fMRI results, in the absence of observed behavioral deficits, might indicate latent lesions or subclinical pathology.

In addition to abnormal brain activation, structural compromise has been observed in brain regions important for memory functions. Investigators have reported reduced alcoholism-related volumes in the mesocorticolimbic system, including medial temporal-lobe structures (Sullivan et al., 1995; Harding et al., 1997; Agartz et al., 1999; Gazdzinski et al., 2005a; Cardenas et al., 2007, 2011; Chanraud et al., 2007; Sameti et al., 2011). White-matter abnormalities in memory-associated regions also have been reported. Pfefferbaum et al. (2009), using quantitative fiber tracking derived from

Table 12.1

References to research reviewed in this chapter. The table is organized by neuropsychologic domain and assessment tool, and it indicates findings of (1) residual impairments associated with long-term chronic alcoholism, and (2) compensation or recovery of functions with abstinence

Neuropsychologic measures	Impairments	Compensation and recovery
General ability and intelligence		
Global Assessment of Functioning	Pfefferbaum et al., 2009; Rosenbloom et al., 2009; Schulte et al., 2012b	
Mehrfachwahl-Wortschatz-Intelligenztest MWT-B	Mann et al., 2005	
National Adult Reading Test Score	Pfefferbaum et al., 2001b	
Peabody IQ	Rosenbloom et al., 2009	
SILS Abstraction: mental age	Beatty et al., 1996, 2000	
SILS Vocabulary: mental age	Beatty et al., 1996, 2000	
WAIS Full Scale IQ	Davies et al., 2005	
WAIS Performance IQ	Makris et al., 2008	
WAIS Similarities		Kish et al., 1980 (between 15 and 21 days)
WAIS Vocabulary	Pfefferbaum et al., 1992; Davies et al., 2005	
WASI General IQ Index	Rosenbloom et al., 2007	
Memory		
California Verbal Learning Test	Noël et al., 2012	
Dementia Rating Scale	Sullivan et al., 2002b, 2010; Davies et al., 2005	
Dementia Rating Scale: Memory Scale	Rosenbloom et al., 2004	Rosenbloom et al., 2004 (4-year follow-up)
Ecological Contextual Memory Test	Pitel et al., 2007	
Face-name association learning and recognition fMRI tasks	Pitel et al., 2013	
Feeling of Knowing Task	Le Berre et al., 2010, 2012	
Free and Cued Selective Reminding Test	Pitel et al., 2007	
Free and Cued Selective Reminding Test	Pitel et al., 2007; Le Berre et al., 2012	
Grober and Buschke (free recall)	Chanraud-Guillermo et al., 2009	
Other word recall tasks	Le Berre et al., 2010	
Spondee Test	Pitel et al., 2007; Le Berre et al., 2012	
WMS Delayed Memory	Fama et al., 2012	
WMS Logical Memory I and II (story recall)	Sullivan et al., 2000b; Davies et al., 2005	
WMS Verbal composite	Reed et al., 1992; Fama et al., 2004	
WMS Verbal Recall		Sullivan et al., 2000a (2- to 12-month follow-up)
WMS Visual Memory	Fama et al., 2012	
WMS-R Memory Index	Sullivan et al., 2002b, 2010; Davies et al., 2005	
Executive functions		
Alternate Response Task		Pitel et al., 2009 (6-month follow-up)
Iowa Gambling Task	Fernández-Serrano et al., 2010; Le Berre et al., 2012	
Recency	Sullivan et al., 2000b	
Ruff Figural Fluency Task	Sullivan et al., 2002b; Oscar-Berman et al., 2009	
Self Ordered Pointing	Sullivan et al., 2000b	
Trail Making Test (A and/or B)	Ratti et al., 2002; Sullivan et al., 2002b; Davies et al., 2005; Pfefferbaum et al., 2006 (component of working memory composite score); Chanraud et al., 2007; Oscar-Berman et al., 2009; Loeber et al., 2010; Le Berre et al., 2012	Kish et al., 1980 (between 15 and 21 days); Loeber et al., 2010 (6-month follow-up)

Continued

Table 12.1

Continued

Neuropsychologic measures	Impairments	Compensation and recovery
Verbal Fluency (e.g., COWAT)	Pitel et al., 2007; Oscar-Berman et al., 2009; Fernández-Serrano et al., 2010	
Visual Search	Sullivan et al., 2000b	
Wisconsin Card Sorting Test	Sullivan et al., 1993, 2000b, 2002b; Ratti et al., 2002; Fama et al., 2004; Chanraud et al., 2007; Oscar-Berman et al., 2009	Sullivan et al., 2000a (2- to 12-month follow-up)
Working and short-term memory		
Brown-Peterson Distractor Test	Sullivan et al., 2000b	
STM Nonverbal	Sullivan et al., 2002b; Rosenbloom et al., 2004	Sullivan et al., 2000a (2- to 12-month follow-up); Rosenbloom et al., 2004 (4-year follow-up)
STM Verbal	Sullivan et al., 2002b; Rosenbloom et al., 2004	
WAIS Block Span	Pfefferbaum et al., 2006	Sullivan et al., 2000a (2- to 12-month follow-up)
WAIS Digit Span	Pfefferbaum et al., 2006	Kish et al., 1980 (between 15 and 21 days, and 21 and 110 days of abstinence); Sullivan et al., 2000a (2- to 12-month follow-up); Alhassoon et al., 2012 (1-year follow-up)
WAIS Digit Symbol	Reed et al., 1992; Beatty et al., 2000; Ratti et al., 2002; Oscar-Berman et al., 2004, 2009; Rosenbloom et al., 2004; Davies et al., 2005; Makris et al., 2008	
WAIS Letter-Number Sequencing	Chanraud et al., 2007; Uekermann et al., 2007	
Executive Functions – Selective Attention		
WAIS Digit Cancellation	Ratti et al., 2002	
Inhibition and impulsivity		
Attentional Assessment Test	Pitel et al., 2007	
Childhood Behavior Disorders Checklist	Beatty et al., 1996	
Modified Card Sorting Task	Le Berre et al., 2012	
Modified Stroop	Lusher et al., 2004; Pitel et al., 2007	
N-back task	Pitel et al., 2007	
South Oaks Gambling Screen	Mitchell et al., 2005	
Stroop Color Naming	Ratti et al., 2002; Uekermann et al., 2007	
Stroop Color Word Naming	Ratti et al., 2002; Le Berre et al., 2012	
Stroop Word Reading	Uekermann et al., 2007	
Utah Criteria	Beatty et al., 1996	
Weintraub and Mesulam's Letter/Symbol Cancellation Task	Beatty et al., 1996	
Emotion and psychosocial skills		
Emotions and mood		
Beck Depression Inventory	Beatty et al., 1996; Monnot et al., 2001; Mann et al., 2005; Mitchell et al., 2005; Uekermann et al., 2005; Foisy et al., 2007; Maurage et al., 2007, 2008, 2009, 2011, 2013; Pfefferbaum et al., 2009; Rosenbloom et al., 2009; Witkiewitz and Aracelliz, 2009; Durazzo et al., 2010; Loeber et al., 2010; Kornreich et al., 2011, 2013	
	Di Sclafani et al., 2008	

Table 12.1

Continued

Neuropsychologic measures	Impairments	Compensation and recovery
Depression Scale of the Minnesota Multiphasic Personality Inventory		
Depression, Anxiety, and Stress Scale	Mitchell et al., 2005	
Hamilton Rating Scale for Depression	Makris et al., 2008; Oscar-Berman et al., 2009	
POMS Anger	Oscar-Berman et al., 2009	
POMS Depression	Makris et al., 2008; Oscar-Berman et al., 2009	
POMS Tension	Oscar-Berman et al., 2009	
Reiss–Epstein Anxiety Sensitivity Index	Di Sclafani et al., 2008	
Spielberger Trait Anxiety	Davies et al., 2005	
State-Trait Anxiety Index	Beatty et al., 1996; Di Sclafani et al., 2008; Maurage et al., 2008, 2011, 2013; Townshend and Duka, 2003; O'Daly et al., 2012; Foisy et al., 2007; Kornreich et al., 2011, 2013	
Zung Anxiety Scale	Kornreich et al., 2001, 2002	
Zung Depression Scale	Kornreich et al., 2001, 2002	
Face processing		
Benton Facial Recognition Test	Foisy et al., 2007	
Ekman and Friesen Pictures	Maurage et al., 2007; Kornreich et al., 2013	
Emotional Facial Expression Test	Foisy et al., 2007	
International Affective Picture System	Gilman et al., 2010	
Matsumoto and Ekman Pictures	Kornreich et al., 2002; Philippot et al., 1999	
Personality		
Hypomania Scale of the Minnesota Multiphasic Personality Inventory	Di Sclafani et al., 2008	
NEO Personality Inventory	Salloum et al., 2007	
Psychopathic Deviance Scale of the Minnesota Multiphasic Personality Inventory	Di Sclafani et al., 2008	
Psychosocial skills		
Adaptation Skills Battery	Nixon et al., 1992	
Aprosodia Battery	Monnot et al., 2001, 2002	
Barratt Impulsivity Scale	Mitchell et al., 2005	
Emotion-Regulation Skills Questionnaire	Berking et al., 2011	
Empathy Quotient	Maurage et al., 2011	
Florida Affect Battery-Revised (German Version)	Uekermann et al., 2005	
Interpersonal Reactivity Index (Emotional Empathy Factors)	Maurage et al., 2011; Gizewski et al., 2013	
Inventory of Interpersonal Problems	Kornreich et al., 2002; Maurage et al., 2009, 2011	
Montreal Affective Voices	Kornreich et al., 2013	
Rotter's Locus of Control Scale	Mitchell et al., 2005	
Socialization Scale of the California Psychological Inventory	Di Sclafani et al., 2008	
Toronto Alexithymia Scale	Stasiewicz et al., 2012, Thorberg et al., 2011, Maurage et al., 2011	
Trait Emotional Intelligence Questionnaire	Kornreich et al., 2011	
WAIS Comprehension	Oscar-Berman et al., 2009	
WAIS Object Assembly	Beatty et al., 1996	
WAIS Picture Arrangement	Oscar-Berman et al., 2009	
Wason Selection Task	Kornreich et al., 2011	
	Hibbert and Best, 2011	

Continued

Table 12.1

Continued

Neuropsychologic measures	Impairments	Compensation and recovery
World Health Organization Quality of Life BREF		
Visuospatial cognition		
CANTAB Pattern Recognition Memory	Townshend and Duka, 2003; Kopera et al., 2012	
CANTAB Spatial Working Memory Span	Kopera et al., 2012	
Category Test	Reed et al., 1992	
Gollin Incomplete Pictures	Sullivan et al., 2000b; Fama et al., 2004	
Hidden Figures Test	Sullivan et al., 2000b, 2002b; Fama et al., 2004	
Memory for Designs		Kish et al., 1980 (between 15 and 21 days)
Mental Rotation Test	Beatty et al., 1996	
Rey-Osterrieth Complex Figure Test	Beatty et al., 1996; Sullivan et al., 2000b, 2002b	
Spatial Processing (Mean of Block Design, Microcog Clocks, Microcog Tic Tac)	Fein et al., 2006	
WAIS Block Design	Beatty et al., 1996; Sullivan et al., 2002b; Oscar-Berman et al., 2009	Kish et al., 1980 (between 15 and 21 days)
WAIS Matrix Reasoning	Pfefferbaum et al., 2006	
WMS Drawing	Sullivan et al., 2000b	Sullivan et al., 2000a (2- to 12-month follow-up)
WMS Visual Learning	Reed et al., 1992	
WMS Visual Recall	Reed et al., 1992	
Psychomotor skills		
Alternated Finger Tapping	Sullivan et al., 2002a	
Finger Tapping	Parks et al., 2003	
Fregly-Graybiel Walk-a-Line Test	Pfefferbaum et al., 2006	
Walk-a-Line Ataxia Battery	Sullivan et al., 2000b; Fein and Greenstein, 2013	Rosenbloom et al., 2004 (4-year follow-up)
ICARS Kinetic	Fitzpatrick et al., 2012	
ICARS Oculomotor	Fitzpatrick et al., 2012	
ICARS Posture and Gait	Fitzpatrick et al., 2012	
ICARS Speech	Fitzpatrick et al., 2012	
ICARS Total	Fitzpatrick et al., 2012	
Movement Time	Sullivan et al., 2002a	
Simple Reaction Time Task	York and Biederman, 1991; Ratti et al., 2002; Maurage et al., 2008; Kopera et al., 2012	
Stand Heel-to-Toe	Rosenbloom et al., 2004	
Stand on One Foot	Rosenbloom et al., 2004	
Symbol Digit Modalities	Fama et al., 2012	
Symbol-Digit Substitution Test and Digit Symbol Substitution Test		Rosenbloom et al., 2004 (4-year follow-up)
Task-Based Reaction Time	Maurage et al., 2008; Fein et al., 2010; Pandey et al., 2012	
Walk Heel-to-Toe	Rosenbloom et al., 2004	

BREF, the short form (26 items) of the World Health Organization Quality of Life-100 instrument; CANTAB, Cambridge Neuropsychological Test Automated Battery; COWAT, Controlled Oral Word Association Test; fMRI, functional magnetic resonance imaging; ICARS, International Cooperative Ataxia Rating Scale; IQ, Intelligence Quotient; NEO, Neuroticism-Extroversion-Openness Inventory; POMS, Profile of Mood States; SILS, Shipley Institute of Living Scale; STM, Short-Term Memory; WAIS, Wechsler Adult Intelligence Scale; WASI, Wechsler Abbreviated Scale of Intelligence; WMS, Wechsler Memory Scale. For descriptions of most tests cited, refer to Lezak et al. (2012) and Strauss et al. (2006).

Table 12.2

References to research reviewed in this chapter. The table is organized by anatomic regions, and it indicates neuroimaging findings based upon brain functional and structural methods. The names of the anatomic regions are listed as given by the researchers in the citations. All impairments are compared to healthy controls.

	Volumetric impairments	Volumetric recovery	Functional brain abnormalities
Total brain volume	Pfefferbaum et al., 1992; Mann et al., 2005; Chen et al., 2012	Gazdzinski et al., 2005a (1-month follow-up); Mann et al., 2005 (5-week follow-up)	
Total gray matter	Fein et al., 2009; Chen et al., 2012		
Total cortical gray matter	Pfefferbaum et al., 1992, 2001b; Makris et al., 2008		
Frontal lobes	Pfefferbaum et al., 1992, 1998; Gazdzinski et al., 2005b		
Prefrontal region	Pfefferbaum et al., 1992	Sullivan, 2003	
Dorsolateral prefrontal cortex	Cardenas et al., 2007; Chanraud et al., 2007; Makris et al., 2008		Pfefferbaum et al., 2001a (2-Back Task); Akine et al., 2007 (Modified False Recognition Task); Parks et al., 2010 (Finger-Tapping)
Ventrolateral prefrontal cortex			Maurage et al., 2012 (Cyberball Task)
Orbitofrontal cortex		Demirakca et al., 2011 (3-month follow-up)	O'Daly et al., 2012 (Facial Expression Recognition Task)
Superior frontal gyrus			Chanraud-Guillermo et al., 2009 (Auditory Language Task); Maurage et al., 2013 (Face-Voice Interaction Task)
Middle frontal gyrus			Pfefferbaum et al., 2001a (2-Back Task); Chanraud-Guillermo et al., 2009 (Auditory Language Task); Maurage et al., 2013 (Face-Voice Interaction Task); Maurage et al., 2012 (Cyberball Task); Pitel et al., 2013 (Face-Name Associative Learning Task)
Inferior frontal gyrus			Pfefferbaum et al., 2001a (2-Back Task); Pitel et al., 2013 (Face-Name Associative Learning Task)
Precentral gyrus (primary motor cortex)	Fein et al., 2009		Pfefferbaum et al., 2001a (2-Back Task); Parks et al., 2003 (Finger-Tapping Task); Maurage et al., 2013 (Face-Voice Interaction Task); O'Daly et al., 2012 (Facial Expression Recognition Task); Pitel et al., 2013 (Face-Name Associative Learning Task)
Cingulate gyrus		Demirakca et al., 2011 (3-month follow-up)	Maurage et al., 2013 (Face-Voice Interaction Task)
Anterior cingulate	Chanraud et al., 2007; Fein et al., 2009		Akine et al., 2007 (Modified False Recognition Task); Salloum et al., 2007 (Emotional Facial Expressions)
Middle cingulate gyrus			Pitel et al., 2013 (Face-Name Associative Learning Task)
Posterior cingulate cortex			Maurage et al., 2013 (Face-Voice Interaction Task); Schulte et al., 2012a (Stroop Match-to-Sample Task)
Parietal lobes	Pfefferbaum et al., 1992; Gazdzinski et al., 2005a; Fein et al., 2009		

Continued

Table 12.2

Continued

	Volumetric impairments	Volumetric recovery	Functional brain abnormalities
Superior parietal			Maurage et al., 2013 (Face-Voice Interaction Task); Pitel et al., 2013 (Face-Name Associative Learning Task)
Mesial parietal	Fein et al., 2009		
Lateral parietal	Fein et al., 2009		
Inferior parietal			Pfefferbaum et al., 2001a (2-Back Task); Pitel et al., 2013 (Face-Name Associative Learning Task)
Supramarginal gyrus			
Angular gyrus			Pfefferbaum et al., 2001a (2-Back Task); Pitel et al., 2013 (Face-Name Associative Learning Task)
Postcentral gyrus (primary somatosensory cortex)	Fein et al., 2009		Parks et al., 2010 (Finger-Tapping)
Temporoparietal	Pfefferbaum et al., 1992		
Dorsolateral frontal and parietal	Chanraud et al., 2007		
Precuneus			Maurage et al., 2013 (Face-Voice Interaction Task)
Insular cortex	Chanraud et al., 2007; Makris et al., 2008; Fein et al., 2009	Demirakca et al., 2011 (3-month follow-up)	Gizewski et al., 2013 (modified Reading the Mind In The Eyes Test); Maurage et al., 2012 (Cyberball Task); O'Daly et al., 2012 (Facial Expression Recognition Task); Pitel et al., 2013 (Face-Name Associative Learning Task)
Temporal lobes	Gazdzinski et al., 2005b; Cardenas et al., 2007		
Superior temporal	Chanraud et al., 2007		Maurage et al., 2013 (Face-Voice Interaction Task); Pitel et al., 2013 (Face-Name Associative Learning Task)
Inferior temporal (lingual gyrus)	Chanraud et al., 2007; Fein et al., 2009		
Frontotemporal	Pfefferbaum et al., 1992		
Middle temporal			Maurage et al., 2013 (Face-Voice Interaction Task)
Parahippocampal gyrus			
Fusiform gyrus			Maurage et al., 2013 (Face-Voice Interaction Task)
Occipital lobes	Fein et al., 2009		
Parieto-occipital	Pfefferbaum et al., 1992		
Medial occipital			Maurage et al., 2013 (Face-Voice Interaction Task)
Inferior occipital			Maurage et al., 2013 (Face-Voice Interaction Task)
Visual association (in occipital lobe)	Fein et al., 2009		Pfefferbaum et al., 2001a (2-Back Task)
Primary visual (in occipital lobe)	Fein et al., 2009		
Calcarine gyrus			Pitel et al., 2013 (Face-Name Associative Learning Task)
Total subcortical	Pfefferbaum et al., 1992		
Thalamus	Gazdzinski et al., 2005b; Chanraud et al., 2007		Akine et al., 2007 (Modified False Recognition Task)
Ventral striatum			Akine et al., 2007 (Modified False Recognition Task)
Nucleus accumbens	Makris et al., 2008		
Caudate			Pitel et al., 2013 (Face-Name Associative Learning Task)

Table 12.2

Continued

	Volumetric impairments	Volumetric recovery	Functional brain abnormalities
Amygdala	Makris et al., 2008; Wobrock et al., 2009		Marinkovic et al., 2009 (Face-Encoding Task); O'Daly et al., 2012 (Facial Expression Recognition Task)
Hippocampus	Sullivan et al., 1995; Harding et al., 1997; Agartz et al., 1999		Marinkovic et al., 2009 (Face-Encoding Task)
Midbrain			Schulte et al., 2012a (Stroop Match-to-Sample Task)
Pons	Sullivan, 2003; Chanraud et al., 2007		
Total cerebellum	Chanraud et al., 2007		
Cerebellar other	Chanraud et al., 2007		Parks et al., 2003 (Finger-Tapping Task); Chanraud-Guillermo et al., 2009 (Auditory Language Task); Pitel et al., 2013 (Face-Name Associative Learning Task)

	Volumetric impairments	Volumetric recovery	Diffusion impairments	Diffusion recovery
Total white matter (WM)		Agartz et al., 2003; Demirakca et al., 2011 (3-month follow-up)		
Total cerebral WM	Harper et al., 1990; Pfefferbaum et al., 1992, 1995, 2001b; Chen et al., 2012			
Frontal WM	Gazdzinski et al., 2005b; Chanraud et al., 2007	Gazdzinski et al., 2010b (1-month follow-up)	Gazdzinski et al., 2010b (high MD)	Gazdzinski et al., 2010b (increased MD after 1 week of abstinence, decreased MD and increased FA after 1 month of abstinence)
Frontal forceps			Pfefferbaum et al., 2009 (low FA, high ADC); Rosenbloom et al., 2009 (low FA)	
Parietal WM	Gazdzinski et al., 2005b		Gazdzinski et al., 2010b (high MD)	Gazdzinski et al., 2010b (increased MD after 1 week of abstinence)
Temporal WM	Chanraud et al., 2007		Gazdzinski et al., 2010b (high MD)	Gazdzinski et al., 2010b (decreased MD and increased FA after 1 month of abstinence)
Occipital WM				Gazdzinski et al., 2010b (increased MD after 1 week of abstinence)
Cingulum bundle			Harris et al., 2008 (low FA); Pfefferbaum et al., 2009 (low FA, high ADC); Rosenbloom et al., 2009 (high ADC)	

Continued

Table 12.2

Continued

	Volumetric impairments	Volumetric recovery	Diffusion impairments	Diffusion recovery
Corpus callosum	Agartz et al., 1999; Pfefferbaum et al., 1996, 2006; Chanraud et al., 2007; Ruiz et al., 2013		Muller-Oehring et al., 2009; Alhassoon et al., 2012 (low FA)	
Internal capsule			Pfefferbaum et al., 2009 (high ADC)	
External capsule			Pfefferbaum et al., 2009 (high ADC); Rosenbloom et al., 2009 (high ADC)	
Superior longitudinal fasciculus			Harris et al., 2008 (low FA); Pfefferbaum et al., 2009 (high ADC); Rosenbloom et al., 2009 (high ADC)	
Fornix			Pfefferbaum et al., 2009 (high ADC)	
Superior temporal gyrus WM			Harris et al., 2008 (low FA)	
Arcuate fasciculus			Harris et al., 2008 (low FA)	
Frontotemporal junction, limen insula			Harris et al., 2008 (low FA)	
Anterior paracingulate WM	Chanraud et al., 2007			
Uncinate fasciculus			Schulte et al., 2012b (low FA)	
Total ventricles or cerebrospinal fluid	Pfefferbaum et al., 1992; Agartz et al., 1999; Kubota et al., 2001; Bendszus et al., 2001; Mann et al., 2005; Chanraud et al., 2007; Chen et al., 2012	Carlen et al., 1978; Gazdzinski et al., 2005a (1-month follow-up); Mann et al., 2005 (5-week follow-up), Demirakca et al., 2011 (3-month follow-up)		
Lateral ventricles	Pfefferbaum et al., 1992, 1998, 2001b	Zipursky et al., 1989 (19- to 28-day follow-up); Wobrock et al., 2009 (6- to 9-month follow-up)		
Third ventricle	Pfefferbaum et al., 1992, 2001b			
Sulcal or extracerebral cerebrospinal fluid	Pfefferbaum et al., 1998, 2001b; Mann et al., 2005	Mann et al., 2005 (5-week follow-up)		

ADC, apparent diffusion coefficient; FA, fractional anisotropy; MD, mean diffusivity.

Table 12.3

Forms of human memory*

	Sensory memory	Immediate memory	Short-term memory	Long-term memory
Terms used	Iconic/acoustic storage	Primary/working memory	Secondary memory	Tertiary memory
Durations	Fractions of a second	Up to about 45 seconds	Minutes to weeks	Weeks to years; can be permanent
Examples	Subliminal perception	Digit span	A new song	Your name

*Memories are stored for time periods ranging from milliseconds to years. Working memory, a component of executive functions, also is discussed in the section on executive functions. Long-term memory is divided into declarative (explicit) memory and procedural (implicit) memory. Declarative memory can be further subdivided into episodic memory (memory for our experiences) and semantic memory (for facts). Procedural memory is memory of skills and knowing how to do things.

DTI, observed defects in microstructural fiber integrity in frontal forceps, internal and external capsules, fornix, and superior cingulate and longitudinal fasciculi. Chanraud et al. (2009) examined white-matter integrity in conjunction with measures of verbal episodic memory derived from the Free and Cued Selective Reminding Test. The investigators found that low verbal episodic memory performance in alcoholics was associated with structural compromise of axons in frontal, temporal, hippocampal, and parahippocampal regions.

Compensation and recovery

Although metamemory, short-term memory, episodic memory, and free and delayed recall are impaired in uncomplicated alcoholism, deficits in other memory components have not consistently been identified (Crews et al., 2005). Remote memories, especially those formed before the onset of prolonged heavy drinking, remain relatively preserved compared to recently acquired memories (Mearns and Lees-Haley, 1993). Recovery of some forms of memory functioning has been found to vary in relation to duration of abstinence. On episodic memory tasks, deficits generally have not been found following prolonged abstinence (Reed et al., 1992; Rourke and Grant, 1999; Fein et al., 2006). Pitel et al. (2009) reported that for alcoholics who abstained over a 6-month interval, episodic memory performance (as assessed by a selective reminding test) returned to a level of functioning comparable to that of non-alcoholic controls, while this level of recovery was not observed in alcoholic patients who relapsed. Additionally, Pitel et al. (2013) tested abstinent alcoholics on a face-name learning and recognition-memory task during an fMRI scan. At rest, fMRI activation patterns in the left hippocampal and cerebellar regions were positively synchronized in controls, but negatively synchronized in the alcoholics. During initial learning of the task, the alcoholics did not differ from controls on measures of performance, but the alcoholic group had lower cerebellar activation. During memory-task engagement, both groups exhibited hippocampal-cerebellar fMRI desynchronization. The authors speculated that, because there were no significant group differences in recognition-memory performance, the alcoholics' modulated hippocampal-cerebellar activity was compensatory.

In summary, the consequences of chronic alcoholism for memory functioning are complex and varied (Pitel and Beaunieux, 2013). Uncomplicated alcoholics often experience impairments in working memory, short-term memory, and episodic memory. Some of these deficiencies have been linked to executive dysfunction and damage to associated structures, while others have been connected with damaged mesocorticolimbic regions. Fortunately, long-term memories tend to be preserved, and episodic memory abilities can improve with sobriety.

Executive functions

Executive functions are complex cognitive abilities requiring the synchronization of several subprocesses to achieve conscious and non-conscious goals (Alvarez and Emory, 2006). They control and regulate other abilities and behaviors and involve complicated mental skills that help us to connect past experiences with present actions, plan future behavior, organize, judge, change behavior and strategies, and remember details for our decision making. Relevant skills include sustained and selective attention, working memory, planning, organization, problem solving, abstraction, and the ability to inhibit inappropriate responses. Intact executive functioning has been associated with integrity of frontal brain circuitry, which involves an overlapping and interconnected network. In alcoholics, the most compromised cortical areas of this network are dorsolateral prefrontal, ventral/orbitofrontal, and anterior cingulate regions. Although theorists differ on whether executive functions are unified with respect to process and component cognitive abilities (Lezak et al., 2012), there is abundant evidence that executive functioning is impaired in uncomplicated alcoholics, and that abnormal frontal brain circuitry is involved.

Impairments

Using tasks adopted from non-human animal models (comparative neuropsychology) to test components of executive functioning in alcoholics, Oscar-Berman and colleagues have demonstrated deficits in working memory, problem solving, and susceptibility to interference (see Oscar-Berman and Bardenhagen, 1998, for review). Fernández-Serrano et al. (2010) administered conventional neuropsychologic measures of fluency, analogic reasoning, interference, cognitive flexibility, decision making, self-regulation, and working memory to alcoholics and other drug abusers and found that alcohol abuse was associated with verbal fluency and decision-making deficits. Several groups have reported that alcoholics performed significantly worse than non-alcoholic controls on the Trail Making Test, a measure of visual-conceptual abilities requiring divided attention, mental flexibility, set-shifting proficiency, and motor skills (e.g., Davies et al., 2005; Chanraud et al., 2007; Oscar-Berman et al., 2009; Loeber et al., 2010). Others have reported alcoholism-related impairments on the Wisconsin Card Sorting Test (e.g., Sullivan et al., 1993; Ratti et al., 2002; Oscar-Berman et al., 2009), which measures abstract thinking, cognitive flexibility and shifting, concept identification, hypothesis generation, and sensitivity to feedback. Additional evidence of executive dysfunction in alcoholics has been reported as deficits on the Stroop Color-Word Association test (Ratti et al., 2002) or modifications of the Stroop test (Lusher et al., 2004) that tap into selective attention, perceptual interference, response inhibition, and information-processing speed. Tests of impulsivity have shown alcoholism-related impairments on Go/No-Go tasks (Pandey et al., 2012) that require subjects to respond quickly to a target while inhibiting responses when a stop-signal appears. Additional reports of poor executive functioning in alcoholics come from a variety of other tests, including the Letter-Number Sequencing test, which relies on working memory (Chanraud et al., 2007); the Ruff Figural Frequency Test, requiring subjects to quickly draw as many unique designs as possible by connecting dots (Oscar-Berman et al., 2009); a delay-discounting decision-making test in which subjects choose between a large but delayed reward and a small but more immediate reward (Mitchell et al., 2005); and the Iowa Gambling Test (Fernández-Serrano et al., 2010). The many findings from these combined studies of alcoholics clearly demonstrate executive dysfunction as a characteristic of the disorder. Even though descriptions of the impairments vary across reports, the inconsistencies are plausible because of the multiplicity and complexity of executive functioning components, and because researchers have used diverse tasks to measure them. In any case, executive dysfunction has been found to be associated with low motivation to change drinking behaviors (Le Berre et al., 2012). This creates difficult conditions for recovery from addiction and therefore, likely serves as a cognitive barrier to treatment. Loss of inhibition, poor insight, distractibility, perseverative responding, impaired decision making, and difficulties with impulse control contribute to increased likelihood of relapse (Crews and Boettiger, 2009; Sullivan et al., 2010).

Goldstein and Volkow (2002) have provided an informative review of neuroimaging findings showing dysfunction of prefrontal cortex in addictions. They suggested that prefrontal abnormalities affect limbic reward regions and lead to dysregulation of executive functions, especially the ability to inhibit maladaptive or disadvantageous behaviors. In alcoholism, abnormalities of task-related and resting-state functional connectivity have been documented in frontal brain circuitry supporting executive functions. In an fMRI study designed to assess reward-guided adaptive decision making during a visual reinforcement-learning task, Park et al. (2010) found that, compared to controls, alcoholics had lower functional connectivity in the right dorsolateral prefrontal cortex and ventral striatum. The abnormalities in frontostriatal activity were associated with decision-making difficulties; however, mean sobriety was only 17 days and may not represent persistent effects. Schulte et al. (2012a) used a modified Stroop task to examine inhibitory control and conflict processing. In controls, posterior cingulate cortex was activated during response switching and deactivated during response repetition, whereas alcoholics showed the opposite pattern, and greater deviations from the normal activity correlated with higher amounts of lifetime alcohol consumption. A functional dissociation of brain network connectivity between the groups further showed that the controls exhibited greater cortico-cortical connectivity among medial prefrontal, middle cingulate, and posterior cingulate cortices than the alcoholics. In contrast, the alcoholics exhibited greater orbitofrontal-midbrain network connectivity than the controls.

Impairments in executive functioning in alcoholics have been associated with underlying structural abnormalities in frontocerebellar circuitry (Moselhy et al., 2001; Sullivan, 2003; Sullivan et al., 2003; Chanraud et al., 2007; Oscar-Berman and Marinkovic, 2007; Oscar-Berman et al., 2009). Significant volume reductions have been reported in the frontal lobes

(Pfefferbaum et al., 1992; Agartz et al., 1999; Kubota et al., 2001; Chanraud et al., 2007; Chen et al., 2012), including reduced gray matter in dorsolateral prefrontal cortex (Chanraud et al., 2007; Makris et al., 2008) and anterior cingulate cortex (Chanraud et al., 2007). Analyses of cortical thickness in alcoholics have demonstrated that the frontal lobes are particularly vulnerable to thinning (Fortier et al., 2011). With DTI scans, researchers have found white-matter abnormalities in the cingulum bundle, superior longitudinal fasciculi, and internal and external capsules (Harris et al., 2008; Pfefferbaum et al., 2009; Rosenbloom et al., 2009; Schulte et al., 2012b).

Compensation and recovery

Studies have reported positive associations between executive functioning and maintenance of sobriety. For example, Loeber et al. (2010) reported that improvement of attention and executive functioning, measured by the Trail-Making Test and the Wisconsin Card Sorting Test, occurred within the second of two 3-month follow-up sessions subsequent to initial detoxification. Also, Pitel et al. (2009) reported that alcoholic patients who abstained over a 6-month interval returned to levels of performance comparable to non-alcoholic controls on executive functioning tasks measuring inhibition, flexibility, and updating, while recovery was not observed for relapsers. Additionally, factors such as relapse rates and personality disorders have been associated with different outcomes in recovery of executive functioning. Andó et al. (2012) compared personality traits in recovering alcoholics abstinent for 12 weeks to those abstinent for 3 years or more. Compared to the long-term abstinent patients, the short-term abstinent patients perceived higher levels of stress and used non-adaptive coping strategies. However, both groups showed evidence of persistent decision-making deficits. The researchers suggested that an adaptive personality profile of stress reduction skills and effective coping strategies likely helps to compensate for decision-making deficits and assists in maintaining long-term abstinence.

When component processes of executive functions have not been found to be impaired, there may have been compensatory input through alternate brain pathways. For example, Gilman et al. (2010) found that, despite not showing performance deficits during decision-making fMRI tasks, alcoholics recruited a larger frontal brain network than non-alcoholics, especially in the left inferior and right middle frontal gyri. Additionally, Pfefferbaum et al. (2001a) did not identify differences between alcoholics and controls in a spatial working memory fMRI task, but did find that alcoholics used more inferior and posterior frontal regions than controls, who activated dorsolateral prefrontal cortex primarily. Similarly, Desmond et al. (2003) reported that alcoholics with sobriety periods from 4 weeks to 2 years, while not showing performance impairments (accuracy and reaction time) on a verbal working-memory fMRI task, exhibited greater fMRI activation than a control group in left frontal and right superior cerebellar regions. These findings suggested that increased activation in language and articulatory-control regions may reflect the compensatory recruitment necessary to perform simple decision-making or working-memory tasks.

Recovery of neurocognitive functioning in the executive domain has been associated with improvements in brain neurochemistry. Bartsch et al. (2007) administered an attentional test, requiring prompt detection of three subtly different target stimuli, which also measures aspects of concentration and coordination. They found that performance by alcoholics correlated with increases in choline and *N*-acetylaspartate levels with 6–7 weeks of sobriety, particularly in frontomesial areas. Durazzo et al. (2011) used arterial spin labeling perfusion MRI of frontal and parietal gray matter to study groups of alcoholics with differing durations of abstinence. They found that, compared to alcoholics who abstained during a 12-month period, those who resumed drinking showed significantly lower frontal gray-matter perfusion; there were no group differences for parietal gray matter.

In summary, impairments in the domain of executive functioning consistently have been shown in alcoholics, but some executive impairments may be reversed with sobriety. Because of the role of frontal brain systems in directing cognition and behavior, this finding has important implications for overcoming addiction. Because many alcoholics report that a central problem is an inability to act in concert with their conscious plans and intentions, especially with respect to abstaining from alcohol, it may be encouraging to note that the brain system involved in these cognitive processes can and does recover with continued abstinence.

Emotion and psychosocial skills

Emotional signals, expressed in voices, faces, and body language, are important aspects of social interactions. Deficiencies in expressing and interpreting these signals can interfere with interpersonal relations, which in turn can lead to increased alcohol consumption as a dysfunctional coping mechanism. Affective and social functions are complex and rely on neural substrates of emotion, intrinsic motivation, decision making, and sensitivity to reinforcement. Multiple brain networks are involved, most notably mesocorticolimbic and frontoparietal circuitries (Insel and Fernald, 2004; Frith and Frith, 2007; Schulte et al., 2010; Goldstein and Volkow, 2011). These circuitries are involved in many biobehavioral functions that are impaired in alcoholics, and breakdown within the networks can initiate drug use or relapse after protracted abstinence (Oscar-Berman and Bowirrat, 2005).

IMPAIRMENTS

Alcoholics exhibit deficits in several aspects of emotional functioning. Research on the perception and decoding of emotional expressions has shown that alcoholics are impaired in perceiving emotions in facial expressions (Philippot et al., 1999; Kornreich et al., 2001, 2002; Maurage et al., 2008, 2009) and in recognizing emotional prosody in speech – a non-linguistic aspect of language that conveys the speaker's feelings and attitudes (Oscar-Berman et al., 1990a; Monnot et al., 2001, 2002). For example, Oscar-Berman et al. (1990a) reported that alcoholics had difficulty judging emotional intonations and emotional semantic content of spoken sentences, and Monnot et al. (2001, 2002) found that alcoholics were deficient in the ability to detect emotion and attitude in the voice. These findings indicate a defect in decoding affective prosody. Uekermann et al. (2005) reported that alcoholics were impaired in identifying semantic content when it did not match affective prosody, and in matching affective prosody to facial expressions, deficits that may be related to cross-modal difficulties (Oscar-Berman et al., 1990b; Maurage et al., 2007, 2013). Research has found that alcoholics responded abnormally to negative emotional stimuli (Clark et al., 2007) and overestimated the intensity of negative facial expressions such as fear, sadness, anger, and threat (Oscar-Berman et al., 1990a; Kornreich et al., 2001, 2013; Townshend and Duka, 2003; Clark et al., 2007; Maurage et al., 2009). Using images of morphed facial expressions with varying emotional intensities, Philippot et al. (1999) also reported that alcoholics overestimated the intensity of emotional expressions, and they misinterpreted most of the emotional expressions at all levels of intensity. Moreover, the alcoholics were unaware of their misperceptions.

In an extensive review of studies of social cognition in alcoholics, Uekermann and Daum (2008) summarized behavioral findings of diminished social cognitive functioning in alcoholism such as impairments in facial affect perception (as noted above), humor processing (Uekermann et al., 2007), empathy (Maurage et al., 2007; Uekermann et al., 2007), and theory-of-mind (the ability to ascribe mental states to others) (Uekermann et al., 2007). Kornreich et al. (2011) reported that alcoholics demonstrated impaired reasoning for social contract, precaution, and understanding conditional social rules. The alcoholics also had lower emotional intelligence scores, suggesting that these factors contributed to impaired interpersonal functioning. Further, higher levels of alexithymia, characterized by difficulties identifying, differentiating, and expressing feelings, have been reported in alcoholism (Maurage et al., 2011; Stasiewicz et al., 2012), and positive correlations between alexithymia measures and severity of alcohol use have been reported (Thorberg et al., 2011).

A number of fMRI studies have shown alcoholism-related abnormalities in the prefrontal cortex and the limbic system, which are important in emotion and cognition. For example, one fMRI study examined brain activation patterns in alcoholic and control subjects as they were shown photographs of faces with positive, negative, and neutral expressions (Marinkovic et al., 2009). Non-alcoholic subjects had strong amygdala and hippocampus activation when viewing emotional (compared with neutral) expressions, while alcoholics had similar responses to both neutral and emotional stimuli, which reflected blunted limbic activation to emotional faces. Another fMRI study showed that alcoholic subjects were impaired in recognizing expressions of fearful faces and had lower activation in orbitofrontal and insular cortex than controls when viewing emotional faces (O'Daly et al., 2012). Gilman et al. (2010) found that alcoholics demonstrated greater activation in the inferior frontal gyrus, fusiform gyrus, and amygdala when they passively viewed negative facial expressions, and decreased activation in the precuneus and parahippocampal gyrus when they viewed positive facial expressions. Low activation in the anterior cingulate cortex during decoding of facial expressions of fear, disgust, and sadness also has been reported (Salloum et al., 2007). In a study examining functional brain activity during a task assessing cognitive and affective aspects of empathy, Gizewski et al. (2013) found that alcohol history was associated with low activation in the right anterior insula and deficits in affective empathy. Maurage et al. (2012) studied psychosocial functioning in the context of a game that simulated interpersonal interactions, and found higher insula and lower frontal activation during social exclusion conditions. In another fMRI study, Maurage et al. (2013) found that, during a face-voice interaction task, alcoholic subjects had less activation than controls in the superior temporal sulcus, inferior occipital gyrus, middle frontal gyrus, and superior parietal lobule, as well as poor functional connectivity in frontal regions.

In addition to atypical activation patterns in alcoholics, diminished volumes have been reported for brain structures associated with emotional and social functioning, such as orbitofrontal cortex, rostral anterior cingulate cortex, nucleus accumbens, amygdala, hippocampus, and insula (Makris et al., 2008; Wobrock et al., 2009). These regions, which are part of the "social brain" (Insel and Fernald, 2004) and neural circuitry underlying emotion and reward processing (Schulte et al., 2010), are disproportionately affected in alcoholism (Moselhy et al., 2001; Oscar-Berman and Marinkovic, 2007; Makris et al., 2008). In a study that examined the relationship between DTI-based fiber tracking and fMRI during an emotional Stroop task, Schulte et al. (2012b) found that alcoholics demonstrated poorer Stroop-word performance, suggesting higher emotional interference, and this performance was correlated with lower white-matter integrity in the cingulate and corpus callosum.

Compensation and recovery

Despite persistence of emotional and psychosocial impairments (Kornreich et al., 2001, 2002; Foisy et al., 2007; Maurage et al., 2008), recovery of these functions has been reported in relation to relapse rates. For example, Witkiewitz and Aracelliz (2009) found in a longitudinal study that negative affect such as depression and anger contributed to relapse risk. Also, Berking et al. (2011) found that deficits in emotion regulation skills, especially the ability to tolerate negative emotions, predicted relapse 3 months after cognitive behavioral treatment for alcoholism. Impairments in decoding of emotional facial expressions are related to interpersonal problems (Kornreich et al., 2002) and may contribute to relapse risk. As such, social and interpersonal skills training for individuals in recovery have focused on the overestimation of intensity of emotions, and misattribution of negative emotions such as anger and fear has been developed as a component of cognitive behavioral treatments in relapse prevention (for reviews, see Morgenstern and Longabaugh, 2000; Witkiewitz and Marlatt, 2004). Additionally, social-cognitive and emotional-processing factors may affect the recovery of alcoholic men and alcoholic women differently, as gender differences in interpersonal problem-solving skills have been reported (Nixon et al., 1992). In any case, abstinence has been associated with increases in positive social relationships (Hibbert and Best, 2011), improved self-efficacy and coping abilities (Rose et al., 2012), and higher rates of help-seeking behavior (Dawson et al., 2006).

In summary, alcoholics have been shown to have dysfunctional social and emotional processing abilities. Although those problems continue after abstinence, recovery of emotional and psychosocial functioning is central to positive treatment outcomes, relapse prevention, and maintenance of significant relationships.

Visuospatial cognition

Visuospatial cognition refers to visuoperceptual, visuospatial, and visuoconstruction abilities involving conscious and non-conscious visually guided competencies. These abilities are defined operationally in terms of performance on specific tests. Visuoperceptual skills are reflected in the ability to identify stimuli such as objects, faces, and patterns. Visuospatial skills involve the localization of objects in space, navigation, and conceptualization of distances, areas, and volumes. Visuoconstructive abilities involve organizing elements into correct spatial relationships, such as when drawing a house or doing a jigsaw puzzle. Alcoholism-related impairments in visuospatial cognition are among the most frequently reported.

Impairments

Neuropsychologic tests consistently have shown that alcoholics suffer from persistent deficits in visuospatial cognition (Beatty et al., 1996; Oscar-Berman and Schendan, 2000; Sullivan et al., 2000b, 2002b, 2010; Fama et al., 2004; Fein et al., 2006). Early demonstrations of these deficits in alcoholics were apparent as relatively lower scores on Wechsler Adult Intelligence Scale Performance subtests compared to Verbal subtests (reviewed by Ellis and Oscar-Berman, 1989). This discrepancy was largely accounted for by scores on three of the individual Performance subtests, which relied strongly on visuospatial cognitive capacities: Block Design, Object Assembly, and Digit Symbol. Other examples of tests on which alcoholics have performed poorly are the Wechsler Memory Scale Drawing Test (Sullivan et al., 2000b), a letter/symbol cancellation test (Beatty et al., 1996), the Mental Rotation Test (Beatty et al., 1996), the Hidden Figures Test (Sullivan et al., 2000b; Fama et al., 2004), the Rey-Osterrieth Complex Figure Test (Beatty et al., 1996; Sullivan et al., 2000b), and the Gollin Incomplete Pictures Test (Sullivan et al., 2000b; Fama et al., 2004). These tests clearly do not measure homogeneous abilities. Despite the array of subcomponents that have been examined, however, alcoholics' impairments in visuospatial cognition are evident and endure over time.

In accordance with the neuropsychologic deficits observed, brain abnormalities have been identified in a large network of interconnected areas communicating with each other in feedforward and feedback loops to serve visuospatial cognition. Cortical regions throughout the entire cerebrum are involved, as well as the cerebellum and the thalamus (reviewed by Fitzpatrick et al., 2008, 2012). Morphometry studies have shown volume deficits among alcoholics, compared with controls, within the parietal lobes (Fein et al., 2002; Gazdzinski et al., 2005b; Chanraud et al., 2007), and cortical shrinkage in the parietal lobes has been associated with poor performance in spatial processing (Fein et al., 2009). Although parietal cortical volume (measured with conventional MRI) was not a predictor of visuospatial ability in alcoholics, cerebellar hemispheric white matter was such a predictor (Sullivan, 2003). Rosenbloom et al. (2009) examined associations between regional white-matter microstructural integrity (measured with DTI) and visuoconstruction accuracy (as measured by the Rey-Osterrieth Complex Figure Test) in alcoholics and healthy controls. For both groups, the investigators found an association between poorer copy accuracy and increased radial diffusivity in the occipital forceps and external capsule. In the alcoholics, poor immediate recall of the figure was associated with compromised fiber microstructure (lower fractional anisotropy) in inferior cingulate bundles.

Compensation and recovery

Despite the persistence of deficits in visuospatial cognition after years of abstinence, when intact performance levels are observed, they have been attributed to reorganization of brain functioning (Pfefferbaum et al., 2001a). For example, Fama et al. (2004) reported that, whereas visuospatial abilities predicted performance on visuoperception tasks by non-alcoholic control subjects, for alcoholics it was executive ability, as measured by the Wisconsin Card Sorting Test, that consistently predicted performance on visuoperception tasks. Because completing the Wisconsin Card Sorting Test relies heavily on frontal brain circuitry, the authors suggested that alcoholics likely engage additional frontal systems to perform visuoperceptual tasks at the level of non-alcoholics. In another study, Rosenbloom et al. (2004) compared the cognitive performance of alcoholic and non-alcoholic women across three durations of abstinence (3 months, 1 year, and 4 years) and found better performance with increased lengths of sobriety for visuospatial scores as measured by the Digit Symbol subtest of the Wechsler Adult Intelligence Scale. After 4 years of abstinence, the alcoholic women had achieved a performance level that was not significantly different from that of non-alcoholic women.

In summary, a considerable body of research indicates selective deficiencies of multiple facets of visuospatial cognition in association with long-term alcoholism. There is evidence that some of these deficits may recover with abstinence, and that improvements are associated with reorganized brain functioning. In other words, the abilities that appear to have remained intact or recovered through significant lengths of abstinence may be explained by recourse to the compensatory action of recruitment of other brain regions through a dynamic process of reorganization.

Psychomotor abilities

Psychomotor abilities include skills such as quick reaction time, grip strength, fine-finger movement, postural abilities, gait, and balance. Persistent alcoholism-related psychomotor impairments are characteristic of the disorder.

Impairments

Accounts of impairments in reaction time pervade the literature on alcoholism and range from simple tests of psychomotor speed to complex tasks that incorporate higher-level skills. An example of the former was reported by Pfefferbaum et al. (2009), who tested alcoholics and controls for fine-finger movement. The test was timed and required that study participants turn a knurled post with the index finger and thumb of each hand separately, and then both hands. The alcoholics were slower than controls on both the unimanual and bimanual conditions. In another study, Kopera et al. (2012) found that alcoholics performed poorly on a simple reaction time task that was part of a larger neuropsychologic assessment battery. The task required participants to push a button in response to a stimulus that appeared at random intervals in the same location on a computer screen. In a similar study, Ratti et al. (2002) found that alcoholics were impaired in both simple and choice reaction times. In the simple reaction time task, participants released or pressed a key when a target appeared. In the choice reaction time task, a stimulus could appear in one of three locations, and participants had to press a corresponding key. In both of these studies, the authors suggested that the motor response speeds themselves were not necessarily impaired, but the slower reaction times likely reflected reduced cognitive processing speed. In fact, some researchers have proposed that alcoholics are impaired in cognitive efficiency, that is, the ability to process information both quickly and accurately. In other words, alcoholics demonstrate a speed–accuracy trade-off, such that greater emphasis is placed on responding accurately at the expense of speed, leading to decrements in efficiency (Sullivan et al., 2002a). This conjecture appears confirmed by additional reports of deficits in reaction time, which have ranged from simple target-striking tasks (York and Biederman, 1991) to delays in identifying emotional expressions in facial stimuli (Maurage et al., 2008; Marinkovic et al., 2009; Fein et al., 2010), and performing executive functioning tasks (Pandey et al., 2012).

Although upper-limb strength usually remains intact, alcoholics often tend to demonstrate deficits in lower-limb competencies, such as posture, gait, and balance (Sullivan et al., 2000b). A number of studies have tested alcoholic and control subjects on the Walk-a-Line Ataxia Battery (Sullivan et al., 2000b; Pfefferbaum et al., 2006; Smith and Fein, 2011; Fein and Greenstein, 2013), which consists of three parts, each to be performed with eyes open and then closed. First, subjects stand with feet placed heel-to-toe and arms folded across the chest; next, subjects stand on one foot; finally, subjects walk heel-to-toe. Each of these studies showed impairments in alcoholics; the deficits were persistent and slow to improve (Fein and Greenstein, 2013). Finally, lifetime alcohol consumption, disease duration, and age have been negatively associated with scores on the International Cooperative Ataxia Rating Scale, representing not only poor posture and gait, but also kinetic, speech, and oculomotor abilities (Fitzpatrick et al., 2012).

Neuroimaging has augmented many of these neuropsychologic measures in understanding the psychomotor functioning deficits in alcoholics. Rogers et al. (2012) investigated the functional connectivity between cerebellar and cortical brain regions in alcoholic patients. The connectivity analyses focused on cortical regions that exhibited fMRI responses associated with non-dominant-hand finger tapping in the patients, but not in controls. Functional connectivity in the patients was attenuated in circuits involving: (1) premotor areas and the superior cerebellum; and (2) prefrontal cortex and the inferior cerebellum. This pattern suggested that structural damage within frontocerebellar circuitry might be responsible for the motor deficits. However, because the alcoholics were abstinent for short durations (average of 5 days), the results might not reflect persistent effects. Pfefferbaum et al. (2009) also used DTI to assess white-matter structural integrity in conjunction with psychomotor measures of balance and response speed. Balance was measured by requiring subjects to stand on one foot, and response speed was measured when subjects turned a knurled post between the thumb and forefinger, and when subjects performed the Digit Symbol subtest of the Wechsler Adult Intelligence Scale. Alcoholics had poorer balance, and they were slower on the speed tests. Poorer performance on all tests correlated with DTI measures of regional white-matter degradation in frontal and occipital forceps, internal and external capsules, fornix, and inferior cingulate bundle.

Compensation and recovery

Psychomotor functions, including postural and gait stability, have varying rates of recovery in relation to length of abstinence and gender. Sullivan and colleagues (2000a, b) found marginally improved gait and balance in abstinent alcoholic men tested after 1 month of sobriety and again 2–12 months later. However, Rosenbloom et al. (2004) found that alcoholic women tested at 15 weeks of sobriety, and then retested 1 and 4 years later, still showed persistent deficits in gait and balance, although there were improvements in psychomotor speed. Smith and Fein (2011) found that alcoholic men and women with multiyear abstinence (1.5–22.4 years) showed more normal gait and balance than alcoholics with short-term (6–15 weeks) abstinence. For the short-term abstinent alcoholics only, gender comparisons indicated that women performed worse than men on two static balance measures with eyes closed (standing heel-to-toe and standing on left leg).

Parks and colleagues (2003, 2010) used fMRI to explore brain activation patterns during finger-tapping tasks performed by recently abstinent alcoholics (less than 2 weeks on average) and control participants. In the 2003 study, alcoholics were significantly slower than controls. Although alcoholics' brain activation patterns were more extensive in motor and premotor cortex and in cerebellar regions compared to the controls, a measure of efficiency of brain activation (tapping rate per unit of brain activation) suggested motor inefficiency and compensatory alterations of corticocerebellar circuits. In the 2010 study, Parks and colleagues were able to equate the groups on self-paced and externally paced finger-tapping rates, and still found that brain activation patterns for alcoholics differed from controls. Under both conditions, tapping with the non-dominant hand was associated with more activation by the alcoholics in frontal, parietal, and temporal brain regions. Additionally, the self-paced tapping activated frontocerebellar networks in the controls, but only precuneus in the alcoholics. Thus, rather than enlisting a frontocerebellar network, as did the controls, the alcoholics recruited a higher-order, "more effortful" or "less automatic," parietal planning mode not required by the controls. The authors concluded that their findings reflected different neurocognitive strategies and brain circuitries used by alcoholics and controls to complete the task.

In summary, severe deficits in psychomotor abilities demonstrated by alcoholics are associated with functional and structural brain abnormalities. Improvements in these abilities tend to correspond to duration of abstinence, and may differ for men and women. Additionally, to compensate for behavioral inefficiencies, alcoholics may employ brain circuitry that differs from non-alcoholics.

SUMMARY AND CONCLUSIONS

Alcoholism has heterogeneous origins and manifestations, depending upon variables such as family history, age, gender, and mental or physical health. Consequently, the neuropsychologic profiles associated with alcoholism are not uniform among individuals. Within and across research studies, variability among participants is substantial and contributes to the characteristics associated with differential outcomes. Nevertheless, evidence for impairments of neuropsychologic processes following years of chronic alcoholism is unequivocal. These impairments are accompanied by structural damage to gray and white matter in the brains of alcoholics, and are measurable on both macrostructural and microstructural levels. The profiles of impairments among alcoholics show that some neuropsychologic processes are more permanently disrupted than others. Memory, executive functions, emotion and psychosocial skills, visuospatial cognition, and psychomotor abilities are particularly affected. In accordance with this pattern of deficits, the brain networks that underlie the most impaired functions involve the frontocerebellar system and mesocorticolimbic circuitry.

Over time, usually with abstinence from alcohol, behavioral functions can appear relatively unimpaired. Although it can be difficult to distinguish processes that have been spared, compensated for, or recovered, the brain appears to become reorganized to provide compensation for structural and behavioral deficits. Longitudinal studies have provided clear evidence for recovery of both neuropsychologic functioning and brain tissues. Further support for recovery of brain volumes with abstinence has come from cross-sectional studies that have demonstrated positive associations between prolonged abstinence and larger volumes of cortical and subcortical gray matter, and cortical and callosal white matter. Despite substantial evidence for recovery of neuropsychologic performance and brain structure with sobriety, some abnormalities persist through abstinence, especially problems with executive functioning, visuospatial cognition, and motor functions.

Because of differences in participant characteristics and methodologic variations across studies, aspects of neurobehavioral and brain disruption and repair remain insufficiently documented. However, by relying on a combination of clinical and scientific approaches, including the continued use of longitudinal studies and multiple neuroimaging technologies, future research will help to refine the compensatory roles of healthy brain systems, the degree to which abstinence and treatment facilitate the reversal of brain atrophy and dysfunction, and the importance of individual differences to outcome. Ultimately, an important goal of neuropsychologic and neuroimaging research into profiles of impaired, spared, and recovered functions in alcoholism is to inform successful strategies for prevention and treatment. Commendable advances already have been made.

ACKNOWLEDGMENTS

The writing of this chapter was supported by grants from the National Institutes of Health, NIAAA R01 AA07112 and K05 AA00219, and the Medical Research Service of the US Department of Veterans Affairs.

REFERENCES

Agartz I, Momenan R, Rawlings RR et al. (1999). Hippocampal volume in patients with alcohol dependence. Arch Gen Psychiatry 56: 356–363.

Agartz I, Brag S, Franck J et al. (2003). MR volumetry during acute alcohol withdrawal and abstinence: a descriptive study. Alcohol 38: 71–78.

Akine Y, Kato M, Muramatsu T et al. (2007). Altered brain activation by a false recognition task in young abstinent patients with alcohol dependence. Alcohol Clin Exp Res 31: 1589–1597.

Alhassoon OM, Sorg SF, Taylor MJ et al. (2012). Callosal white matter microstructural recovery in abstinent alcoholics: a longitudinal diffusion tensor imaging study. Alcohol Clin Exp Res 36: 1–10.

Alvarez JA, Emory E (2006). Executive function and the frontal lobes: a meta-analytic review. Neuropsychol Rev 16: 17–42.

Andó B, Must A, Kurgyis E et al. (2012). Personality traits and coping compensate for disadvantageous decision-making in long-term alcohol abstinence. Alcohol Alcohol 47: 18–24.

Bartsch AJ, Homola G, Biller A et al. (2007). Manifestations of early brain recovery associated with abstinence from alcoholism. Brain 130: 36–47.

Bates ME, Bowden SC, Barry D (2002). Neurocognitive impairment associated with alcohol use disorders: implications for treatment. Exp Clin Psychopharmacol 10: 193–212.

Beatty WW, Hames KA, Blanco CR et al. (1996). Visuospatial perception, construction and memory in alcoholism. J Stud Alcohol 57: 136–143.

Beatty WW, Tivis R, Stott HD et al. (2000). Neuropsychological deficits in sober alcoholics: influences of chronicity and recent alcohol consumption. Alcohol Clin Exp Res 24: 149–154.

Bendszus M, Weijers HG, Wiesbeck G et al. (2001). Sequential MR imaging and proton MR spectroscopy in patients who underwent recent detoxification for chronic alcoholism: correlation with clinical and neuropsychological data. AJNR Am J Neuroradiol 22: 1926–1932.

Berking M, Margraf M, Ebert D et al. (2011). Deficits in emotion-regulation skills predict alcohol use during and after cognitive behavioral therapy for alcohol dependence. J Consult Clin Psychol 79: 307–318.

Brust JC (2010). Ethanol and cognition: indirect effects, neurotoxicity and neuroprotection: a review. Int J Environ Res Public Health 7: 1540–1557.

Buhler M, Mann K (2011). Alcohol and the human brain: a systematic review of different neuroimaging methods. Alcohol Clin Exp Res 35: 1771–1793.

Cardenas VA, Studholme C, Meyerhoff DJ et al. (2005). Chronic active heavy drinking and family history of problem drinking modulate regional brain tissue volumes. Psychiatry Res 138: 115–130.

Cardenas VA, Studholme C, Gazdzinski S et al. (2007). Deformation-based morphometry of brain changes in alcohol dependence and abstinence. Neuroimage 34: 879–887.

Cardenas VA, Durazzo TC, Gazdzinski S et al. (2011). Brain morphology at entry into treatment for alcohol dependence is related to relapse propensity. Biol Psychiatry 70: 561–567.

Carlen PL, Wortzman G, Holgate RC et al. (1978). Reversible cerebral atrophy in recently abstinent chronic alcoholics measured by computed tomography scans. Science 200: 1076–1078.

Chanraud S, Martelli C, Delain F et al. (2007). Brain morphometry and cognitive performance in detoxified alcohol-dependents with preserved psychosocial functioning. Neuropsychopharmacology 32: 429–438.

Chanraud S, Leroy C, Martelli C et al. (2009). Episodic memory in detoxified alcoholics: contribution of grey matter microstructure alteration. PLoS One 4: e6786.

Chanraud S, Pitel AL, Rohlfing T et al. (2010). Dual tasking and working memory in alcoholism: relation to frontocerebellar circuitry. Neuropsychopharmacology 35: 1868–1878.

Chanraud S, Pitel AL, Pfefferbaum A et al. (2011). Disruption of functional connectivity of the default-mode network in alcoholism. Cereb Cortex 21: 2272–2281.

Chanraud S, Pitel AL, Muller-Oehring EM et al. (2013). Remapping the brain to compensate for impairment in recovering alcoholics. Cereb Cortex 23: 97–104.

Chanraud-Guillermo S, Andoh J, Martelli C et al. (2009). Imaging of language-related brain regions in detoxified alcoholics. Alcohol Clin Exp Res 33: 977–984.

Charness ME (2009). Overview of the chronic neurologic complications of alcohol. 1-13, http://www.uptodate.com.

Chen CH, Walker J, Momenan R et al. (2012). Relationship between liver function and brain shrinkage in patients with alcohol dependence. Alcohol Clin Exp Res 36: 625–632.

Cisek P, Kalaska JF (2010). Neural mechanisms for interacting with a world full of action choices. Annu Rev Neurosci 33: 269–298.

Clark US, Oscar-Berman M, Shagrin B et al. (2007). Alcoholism and judgments of affective stimuli. Neuropsychology 21: 346–362.

Cosgrove KP, Esterlis I, Mason GF et al. (2011). Neuroimaging insights into the role of cortical GABA systems and the influence of nicotine on the recovery from alcohol dependence. Neuropharmacology 60: 1318–1325.

Crews FT, Boettiger CA (2009). Impulsivity, frontal lobes and risk for addiction. Pharmacol Biochem Behav 93: 237–247.

Crews FT, Buckley T, Dodd PR et al. (2005). Alcoholic neurobiology: changes in dependence and recovery. Alcohol Clin Exp Res 29: 1504–1513.

Davies SJ, Pandit SA, Feeney A et al. (2005). Is there cognitive impairment in clinically 'healthy' abstinent alcohol dependence? Alcohol Alcohol 40: 498–503.

Dawson DA, Grant BF, Stinson FS et al. (2006). Maturing out of alcohol dependence: the impact of transitional life events. J Stud Alcohol 67: 195–203.

Demirakca T, Ende G, Kammerer N et al. (2011). Effects of alcoholism and continued abstinence on brain volumes in both genders. Alcohol Clin Exp Res 35: 1678–1685.

Desmond JE, Chen SH, DeRosa E et al. (2003). Increased frontocerebellar activation in alcoholics during verbal working memory: an fMRI study. Neuroimage 19: 1510–1520.

Devaud LL, Prendergast MA (2009). Introduction to the special issue of Alcohol and Alcoholism on sex/gender differences in responses to alcohol. Alcohol Alcohol 44: 533–534.

Di Sclafani V, Finn P, Fein G (2007). Psychiatric comorbidity in long-term abstinent alcoholic individuals. Alcohol Clin Exp Res 31: 795–803.

Di Sclafani V, Finn P, Fein G (2008). Treatment-naive active alcoholics have greater psychiatric comorbidity than normal controls but less than treated abstinent alcoholics. Drug Alcohol Depend 98: 115–122.

Durazzo TC, Rothlind JC, Gazdzinski S et al. (2006). A comparison of neurocognitive function in nonsmoking and chronically smoking short-term abstinent alcoholics. Alcohol 39: 1–11.

Durazzo TC, Gazdzinski S, Mon A et al. (2010). Cortical perfusion in alcohol-dependent individuals during short-term abstinence: relationships to resumption of hazardous drinking after treatment. Alcohol 44: 201–210.

Durazzo TC, Tosun D, Buckley S et al. (2011). Cortical thickness, surface area, and volume of the brain reward system in alcohol dependence: relationships to relapse and extended abstinence. Alcohol Clin Exp Res 35: 1187–1200.

Ellis RJ, Oscar-Berman M (1989). Alcoholism, aging, and functional cerebral asymmetries. Psychol Bull 106: 128–147.

Erickson CK, White WL (2009). The neurobiology of addiction recovery. Alcohol Treat Q 27: 338–345.

Fama R, Pfefferbaum A, Sullivan EV (2004). Perceptual learning in detoxified alcoholic men: contributions from explicit memory, executive function, and age. Alcohol Clin Exp Res 28: 1657–1665.

Fama R, Rosenbloom MJ, Nichols BN et al. (2009). Working and episodic memory in HIV infection, alcoholism, and their comorbidity: baseline and 1-year follow-up examinations. Alcohol Clin Exp Res 33: 1815–1824.

Fama R, Rosenbloom MJ, Sassoon SA et al. (2012). Differential effect of alcoholism and HIV infection on visuomotor procedural learning and retention. Alcohol Clin Exp Res 36: 1738–1747.

Fattore L, Altea S, Fratta W (2008). Sex differences in drug addiction: a review of animal and human studies. Womens Health 4: 51–65.

Fein G, Greenstein D (2013). Gait and balance deficits in chronic alcoholics: no improvement from 10 weeks through 1 year abstinence. Alcohol Clin Exp Res 37: 86–95.

Fein G, Di Sclafani V, Cardenas VA et al. (2002). Cortical gray matter loss in treatment-naive alcohol dependent individuals. Alcohol Clin Exp Res 26: 558–564.

Fein G, Torres J, Price LJ et al. (2006). Cognitive performance in long-term abstinent alcoholic individuals. Alcohol Clin Exp Res 30: 1538–1544.

Fein G, Shimotsu R, Chu R et al. (2009). Parietal gray matter volume loss is related to spatial processing deficits in long-term abstinent alcoholic men. Alcohol Clin Exp Res 33: 1806–1814.

Fein G, Key K, Szymanski MD (2010). ERP and RT delays in long-term abstinent alcoholics in processing of emotional facial expressions during gender and emotion categorization tasks. Alcohol Clin Exp Res 34: 1127–1139.

Fernández-Serrano MJ, Perez-Garcia M, Rio-Valle JS et al. (2010). Neuropsychological consequences of alcohol and drug abuse on different components of executive functions. J Psychopharmacol 24: 1317–1332.

Fitzpatrick LE, Jackson M, Crowe SF (2008). The relationship between alcoholic cerebellar degeneration and cognitive and emotional functioning. Neurosci Biobehav Rev 32: 466–485.

Fitzpatrick LE, Jackson M, Crowe SF (2012). Characterization of cerebellar ataxia in chronic alcoholics using the International Cooperative Ataxia Rating Scale (ICARS). Alcohol Clin Exp Res 35: 1943–1951.

Foisy ML, Kornreich C, Fobe A et al. (2007). Impaired emotional facial expression recognition in alcohol dependence: do these deficits persist with midterm abstinence? Alcohol Clin Exp Res 31: 404–410.

Fortier CB, Leritz EC, Salat DH et al. (2011). Reduced cortical thickness in abstinent alcoholics and association with alcoholic behavior. Alcohol Clin Exp Res 35: 2193–2201.

Frith CD, Frith U (2007). Social cognition in humans. Curr Biol 17: R724–R732.

Fuster JM (2008). The Prefrontal Cortex. 4th edn. Academic Press, London.

Gazdzinski S, Durazzo TC, Meyerhoff DJ (2005a). Temporal dynamics and determinants of whole brain tissue volume changes during recovery from alcohol dependence. Drug Alcohol Depend 78: 263–273.

Gazdzinski S, Durazzo TC, Studholme C et al. (2005b). Quantitative brain MRI in alcohol dependence: preliminary evidence for effects of concurrent chronic cigarette smoking on regional brain volumes. Alcohol Clin Exp Res 29: 1484–1495.

Gazdzinski S, Durazzo TC, Mon A et al. (2010a). Body mass index is associated with brain metabolite levels in alcohol dependence: a multimodal magnetic resonance study. Alcohol Clin Exp Res 34: 2089–2096.

Gazdzinski S, Durazzo TC, Mon A et al. (2010b). Cerebral white matter recovery in abstinent alcoholics: a multimodality magnetic resonance study. Brain 133: 1043–1053.

Gilman JM, Davis MB, Hommer DW (2010). Greater activation in left hemisphere language-related regions during simple judgment tasks among substance-dependent patients in treatment for alcoholism. Alcohol Clin Exp Res 34: 331–341.

Gizewski ER, Muller BW, Scherbaum N et al. (2013). The impact of alcohol dependence on social brain function. Addict Biol 18: 109–120.

Goldstein RZ, Volkow ND (2002). Drug addiction and its underlying neurobiological basis: neuroimaging evidence for the involvement of the frontal cortex. Am J Psychiatry 159: 1642–1652.

Goldstein RZ, Volkow ND (2011). Dysfunction of the prefrontal cortex in addiction: neuroimaging findings and clinical implications. Nat Rev Neurosci 12: 652–669.

Harding AJ, Wong A, Svoboda M et al. (1997). Chronic alcohol consumption does not cause hippocampal neuron loss in humans. Hippocampus 7: 78–87.

Harper CG, Smith NA, Kril JJ (1990). The effects of alcohol on the female brain: a neuropathological study. Alcohol Alcohol 25: 445–448.

Harris GJ, Jaffin SK, Hodge SM et al. (2008). Frontal white matter and cingulum diffusion tensor imaging deficits in alcoholism. Alcohol Clin Exp Res 32: 1001–1013.

Hibbert LJ, Best DW (2011). Assessing recovery and functioning in former problem drinkers at different stages of their recovery journeys. Drug Alcohol Rev 30: 12–20.

Insel TR, Fernald RD (2004). How the brain processes social information: searching for the social brain. Annu Rev Neurosci 27: 697–722.

Kim MJ, Kim JK, Yoo BG et al. (2007). Acute Marchiafava-Bignami disease with widespread callosal and cortical lesions. J Korean Med Sci 22: 908–911.

Kish GB, Hagen JM, Woody MM et al. (1980). Alcoholics' recovery from cerebral impairment as a function of duration of abstinence. J Clin Psychol 36: 584–589.

Kopera M, Wojnar M, Brower K et al. (2012). Cognitive functions in abstinent alcohol-dependent patients. Alcohol 46: 665–671.

Kornreich C, Blairy S, Philippot P et al. (2001). Deficits in recognition of emotional facial expression are still present in alcoholics after mid- to long-term abstinence. J Stud Alcohol 62: 533–542.

Kornreich C, Philippot P, Foisy ML et al. (2002). Impaired emotional facial expression recognition is associated with interpersonal problems in alcoholism. Alcohol Alcohol 37: 394–400.

Kornreich C, Delle-Vigne D, Knittel J et al. (2011). Impaired conditional reasoning in alcoholics: a negative impact on social interactions and risky behaviors? Addiction 106: 951–959.

Kornreich C, Brevers D, Canivet D et al. (2013). Impaired processing of emotion in music, faces and voices supports a generalized emotional decoding deficit in alcoholism. Addiction 108: 80–88.

Kubota M, Nakazaki S, Hirai S et al. (2001). Alcohol consumption and frontal lobe shrinkage: study of 1432 non-alcoholic subjects. J Neurol Neurosurg Psychiatry 71: 104–106.

Le Berre AP, Pinon K, Vabret F et al. (2010). Study of metamemory in patients with chronic alcoholism using a feeling-of-knowing episodic memory task. Alcohol Clin Exp Res 34: 1888–1898.

Le Berre AP, Vabret F, Cauvin C et al. (2012). Cognitive barriers to readiness to change in alcohol-dependent patients. Alcohol Clin Exp Res 1–8.

Leber WR, Parsons OA (1982). Premature aging and alcoholism. Int J Addict 17: 61–88.

Lezak MD, Howieson DB, Bigler ED et al. (2012). Neuropsychological assessment. 5th edn. Oxford University Press, New York.

Loeber S, Duka T, Welzel Marquez H et al. (2010). Effects of repeated withdrawal from alcohol on recovery of cognitive impairment under abstinence and rate of relapse. Alcohol Alcohol 45: 541–547.

Lusher J, Chandler C, Ball D (2004). Alcohol dependence and the alcohol Stroop paradigm: evidence and issues. Drug Alcohol Depend 75: 225–231.

Makris N, Oscar-Berman M, Jaffin SK et al. (2008). Decreased volume of the brain reward system in alcoholism. Biol Psychiatry 64: 192–202.

Mann K, Ackermann K, Croissant B et al. (2005). Neuroimaging of gender differences in alcohol dependence: are women more vulnerable? Alcohol Clin Exp Res 29: 896–901.

Marinkovic K, Oscar-Berman M, Urban T et al. (2009). Alcoholism and dampened temporal limbic activation to emotional faces. Alcohol Clin Exp Res 33: 1880–1892.

Martin PR (2013). Psychiatric comorbidities of alcoholism. In: A Pfefferbaum, EV Sullivan (Eds.), Alcohol and the nervous system. Elsevier, New York.

Maurage P, Campanella S, Philippot P et al. (2007). The crossmodal facilitation effect is disrupted in alcoholism: a study with emotional stimuli. Alcohol Alcohol 42: 552–559.

Maurage P, Campanella S, Philippot P et al. (2008). Face processing in chronic alcoholism: a specific deficit for emotional features. Alcohol Clin Exp Res 32: 600–606.

Maurage P, Campanella S, Philippot P et al. (2009). Impaired emotional facial expression decoding in alcoholism is also present for emotional prosody and body postures. Alcohol Alcohol 44: 476–485.

Maurage P, Grynberg D, Noel X et al. (2011). Dissociation between affective and cognitive empathy in alcoholism: a specific deficit for the emotional dimension. Alcohol Clin Exp Res 35: 1662–1668.

Maurage P, Joassin F, Philippot P et al. (2012). Disrupted regulation of social exclusion in alcohol-dependence: an FMRI study. Neuropsychopharmacology 37: 2067–2075.

Maurage P, Joassin F, Pesenti M et al. (2013). The neural network sustaining crossmodal integration is impaired in alcohol-dependence: an fMRI study. Cortex 49: 1610–1626.

Mearns J, Lees-Haley PR (1993). Discriminating neuropsychological sequelae of head injury from alcohol-abuse-induced deficits: a review and analysis. J Clin Psychol 49: 714–720.

Mitchell JM, Fields HL, D'Esposito M et al. (2005). Impulsive responding in alcoholics. Alcohol Clin Exp Res 29: 2158–2169.

Mon A, Durazzo TC, Gazdzinski S et al. (2009). The impact of chronic cigarette smoking on recovery from cortical gray matter perfusion deficits in alcohol dependence: longitudinal arterial spin labeling MRI. Alcohol Clin Exp Res 33: 1314–1321.

Monnig MA, Caprihan A, Yeo RA et al. (2012). Diffusion tensor imaging of white matter networks in individuals with current and remitted alcohol use disorders and comorbid conditions. Psychol Addict Behav 27: 455–465.

Monnot M, Nixon S, Lovallo W et al. (2001). Altered emotional perception in alcoholics: deficits in affective prosody comprehension. Alcohol Clin Exp Res 25: 362–369.

Monnot M, Lovallo WR, Nixon SJ et al. (2002). Neurological basis of deficits in affective prosody comprehension among alcoholics and fetal alcohol-exposed adults. J Neuropsychiatry Clin Neurosci 14: 321–328.

Morgenstern J, Longabaugh R (2000). Cognitive-behavioral treatment for alcohol dependence: a review of evidence for its hypothesized mechanisms of action. Addiction 95: 1475–1490.

Moselhy HF, Georgiou G, Kahn A (2001). Frontal lobe changes in alcoholism: a review of the literature. Alcohol Alcohol 36: 357–368.

Muller-Oehring EM, Schulte T, Fama R et al. (2009). Global-local interference is related to callosal compromise in alcoholism: a behavior-DTI association study. Alcohol Clin Exp Res 33: 477–489.

Nixon K (2006). Alcohol and adult neurogenesis: roles in neurodegeneration and recovery in chronic alcoholism. Hippocampus 16: 287–295.

Nixon SJ (2013). Sex differences. In: A Pfefferbaum, EV Sullivan (Eds.), Alcohol and the Nervous System. Elsevier, New York.

Nixon SJ, Tivis R, Parsons OA (1992). Interpersonal problem-solving in male and female alcoholics. Alcohol Clin Exp Res 16: 684–687.

Noël X, Van der Linden M, Brevers D et al. (2012). The contribution of executive functions deficits to impaired episodic memory in individuals with alcoholism. Psychiatry Res 198: 116–122.

O'Daly OG, Trick L, Scaife J et al. (2012). Withdrawal-associated increases and decreases in functional neural connectivity associated with altered emotional regulation in alcoholism. Neuropsychopharmacology 37: 2267–2276.

Oscar-Berman M (2012). Function and dysfunction of prefrontal brain circuitry in alcoholic Korsakoff's syndrome. Neuropsychol Rev 22: 154–169.

Oscar-Berman M, Bardenhagen F (1998). Nonhuman animal models of memory dysfunction in neurodegenerative disease. In: A Troster (Ed.), Memory in Neurodegenerative Disease. Cambridge University Press, New York, pp. 3–20.

Oscar-Berman M, Bowirrat A (2005). Genetic influences in emotional dysfunction and alcoholism-related brain damage. Neuropsychiatr Dis Treat 1: 211–229.

Oscar-Berman M, Marinkovic K (2007). Alcohol: effects on neurobehavioral functions and the brain. Neuropsychol Rev 17: 239–257.

Oscar-Berman M, Schendan E (2000). Assymmetries of brain function in alcoholism: Relationship to aging. In: LT Connor, LK Obler (Eds.), Neurobehavior of

Language and Cognition: Studies of Normal Aging and Brain Damage, Kluwer Academic Publishing, New York, pp. 213–240.
Oscar-Berman M, Hancock M, Mildworf B et al. (1990a). Emotional perception and memory in alcoholism and aging. Alcohol Clin Exp Res 14: 383–393.
Oscar-Berman M, Pulaski JL, Hutner N et al. (1990b). Cross-modal functions in alcoholism and aging. Neuropsychologia 28: 851–869.
Oscar-Berman M, Kirkley SM, Gansler DA et al. (2004). Comparisons of Korsakoff and non-Korsakoff alcoholics on neuropsychological tests of prefrontal brain functioning. Alcohol Clin Exp Res 28: 667–675.
Oscar-Berman M, Valmas MM, Sawyer KS et al. (2009). Frontal brain dysfunction in alcoholism with and without antisocial personality disorder. Neuropsychiatr Dis Treat 5: 309–326.
Pandey AK, Kamarajan C, Tang Y et al. (2012). Neurocognitive deficits in male alcoholics: an ERP/sLORETA analysis of the N2 component in an equal probability Go/NoGo task. Biol Psychol 89: 170–182.
Park SQ, Kahnt T, Beck A et al. (2010). Prefrontal cortex fails to learn from reward prediction errors in alcohol dependence. J Neurosci 30: 7749–7753.
Parks MH, Morgan VL, Pickens DR et al. (2003). Brain fMRI activation associated with self-paced finger tapping in chronic alcohol-dependent patients. Alcohol Clin Exp Res 27: 704–711.
Parks MH, Greenberg DS, Nickel MK et al. (2010). Recruitment of additional brain regions to accomplish simple motor tasks in chronic alcohol-dependent patients. Alcohol Clin Exp Res 34: 1098–1109.
Petrakis IL, Gonzalez G, Rosenheck R et al. (2002). Comorbidity of alcoholism and psychiatric disorders. Alcohol Res Health 26: 81–89.
Pfefferbaum A, Lim KO, Zipursky RB et al. (1992). Brain gray and white matter volume loss accelerates with aging in chronic alcoholics: a quantitative MRI study. Alcohol Clin Exp Res 16: 1078–1089.
Pfefferbaum A, Sullivan EV, Mathalon DH et al. (1995). Longitudinal changes in magnetic resonance imaging brain volumes in abstinent and relapsed alcoholics. Alcohol Clin Exp Res 19: 1177–1191.
Pfefferbaum A, Lim KO, Desmond JE et al. (1996). Thinning of the corpus callosum in older alcoholic men: a magnetic resonance imaging study. Alcohol Clin Exp Res 20: 752–757.
Pfefferbaum A, Sullivan EV, Rosenbloom MJ et al. (1998). A controlled study of cortical gray matter and ventricular changes in alcoholic men over a 5-year interval. Arch Gen Psychiatry 55: 905–912.
Pfefferbaum A, Desmond JE, Galloway C et al. (2001a). Reorganization of frontal systems used by alcoholics for spatial working memory: an fMRI study. Neuroimage 14: 7–20.
Pfefferbaum A, Rosenbloom M, Deshmukh A et al. (2001b). Sex differences in the effects of alcohol on brain structure. Am J Psychiatry 158: 188–197.
Pfefferbaum A, Adalsteinsson E, Sullivan EV (2006). Dysmorphology and microstructural degradation of the corpus callosum: interaction of age and alcoholism. Neurobiol Aging 27: 994–1009.
Pfefferbaum A, Rosenbloom M, Rohlfing T et al. (2009). Degradation of association and projection white matter systems in alcoholism detected with quantitative fiber tracking. Biol Psychiatry 65: 680–690.
Philippot P, Kornreich C, Blairy S et al. (1999). Alcoholics' deficits in the decoding of emotional facial expression. Alcohol Clin Exp Res 23: 1031–1038.
Pitel AL, Beaunieux H (2013). Component processes of memory. In: A Pfefferbaum, EV Sullivan (Eds.), Alcohol and the Nervous System, Elsevier, New York.
Pitel AL, Beaunieux H, Witkowski T et al. (2007). Genuine episodic memory deficits and executive dysfunctions in alcoholic subjects early in abstinence. Alcohol Clin Exp Res 31: 1169–1178.
Pitel AL, Chanraud S, Muller-Oehring EM et al. (2013). Modulation of limbic-cerebellar functional connectivity enables alcoholics to recognize who is who. Brain Struct Funct 218: 683–695.
Pitel AL, Rivier J, Beaunieux H et al. (2009). Changes in the episodic memory and executive functions of abstinent and relapsed alcoholics over a 6-month period. Alcohol Clin Exp Res 33: 490–498.
Pitel AL, Chanraud S, Rohlfing T et al. (2012). Face-name association learning and brain structural substrates in alcoholism. Alcohol Clin Exp Res 36: 1171–1179.
Pitel AL, Chanraud S, Muller-Oehring EM et al. (2013). Modulation of limbic-cerebellar functional connectivity enables alcoholics to recognize who is who. Brain Struct Funct 218: 683–695.
Ratti MT, Bo P, Giardini A et al. (2002). Chronic alcoholism and the frontal lobe: which executive functions are imparied? Acta Neurol Scand 105: 276–281.
Reed RJ, Grant I, Rourke SB (1992). Long-term abstinent alcoholics have normal memory. Alcohol Clin Exp Res 16: 677–683.
Rogers BP, Parks MH, Nickel MK et al. (2012). Reduced fronto-cerebellar functional connectivity in chronic alcoholic patients. Alcohol Clin Exp Res 36: 294–301.
Rose GL, Skelly JM, Badger GJ et al. (2012). Interactive voice response for relapse prevention following cognitive-behavioral therapy for alcohol use disorders: a pilot study. Psychol Serv 9: 174–184.
Rosenbloom M, Pfefferbaum A (2008). Magnetic resonance imaging of the living brain: evidence for brain degeneration among alcoholics and recovery with abstinence. Alcohol Res Health 31: 362–376.
Rosenbloom MJ, Pfefferbaum A, Sullivan EV (2004). Recovery of short-term memory and psychomotor speed but not postural stability with long-term sobriety in alcoholic women. Neuropsychology 18: 589–597.
Rosenbloom MJ, Rohlfing T, O'Reilly AW et al. (2007). Improvement in memory and static balance with abstinence in alcoholic men and women: selective relations with change in brain structure. Psychiatry Res 155: 91–102.

Rosenbloom MJ, Sassoon SA, Pfefferbaum A et al. (2009). Contribution of regional white matter integrity to visuospatial construction accuracy, organizational strategy, and memory for a complex figure in abstinent alcoholics. Brain Imaging Behav 3: 379–390.

Rourke SB, Grant I (1999). The interactive effects of age and length of abstinence on the recovery of neuropsychological functioning in chronic male alcoholics: a 2-year follow-up study. J Int Neuropsychol Soc 5: 234–246.

Rourke SB, Grant I (2009). The neurobehavioral correlates of alcoholism. In: I Grant, K Adams (Eds.), Neuropsychological Assessment of Neuropsychiatric and Neuromedical Disorders, Oxford University Press, Oxford, pp. 398–454.

Ruiz SM, Oscar-Berman M, Sawyer KS et al. (2013). Drinking history associations with regional white matter volumes in alcoholic men and women. Alcohol Clin Exp Res 37: 110–122.

Salloum JB, Ramchandani VA, Bodurka J et al. (2007). Blunted rostral anterior cingulate response during a simplified decoding task of negative emotional facial expressions in alcoholic patients. Alcohol Clin Exp Res 31: 1490–1504.

Salloway SMD, Malloy PF, Duffy JD (2001). The Frontal Lobes and Neuropsychiatric Illness, American Psychiatric Press, Washington, DC.

Sameti M, Smith S, Patenaude B et al. (2011). Subcortical volumes in long-term abstinent alcoholics: associations with psychiatric comorbidity. Alcohol Clin Exp Res 35: 1067–1080.

Schottenbauer MA, Hommer D, Weingartner H (2007). Memory deficits among alcoholics: performance on a selective reminding task. Neuropsychol Dev Cogn B Aging Neuropsychol Cogn 14: 505–516.

Schulte T, Muller-Oehring EM, Pfefferbaum A et al. (2010). Neurocircuitry of emotion and cognition in alcoholism: contributions from white matter fiber tractography. Dialogues Clin Neurosci 12: 554–560.

Schulte T, Muller-Oehring EM, Sullivan EV et al. (2012a). Synchrony of corticostriatal-midbrain activation enables normal inhibitory control and conflict processing in recovering alcoholic men. Biol Psychiatry 71: 269–278.

Schulte T, Muller-Oehring EM, Sullivan EV et al. (2012b). White matter fiber compromise contributes differentially to attention and emotion processing impairment in alcoholism, HIV-infection, and their comorbidity. Neuropsychologia 50: 2812–2822.

Schulte T, Oberlin BG, Kareken DA et al. (2012c). How acute and chronic alcohol consumption affects brain networks: insights into multimodal neuroimaging. Alcohol Clin Exp Res 36: 2017–2027.

Smith S, Fein G (2011). Persistent but less severe ataxia in long-term versus short-term abstinent alcoholic men and women: a cross-sectional analysis. Alcohol Clin Exp Res 35: 2184–2192.

Stasiewicz PR, Bradizza CM, Gudleski GD et al. (2012). The relationship of alexithymia to emotional dysregulation within an alcohol dependent treatment sample. Addict Behav 37: 469–476.

Stavro K, Pelletier J, Potvin S (2012). Widespread and sustained cognitive deficits in alcoholism: a meta-analysis. Addict Biol 18: 203–213.

Strauss E, Sherman EMS, Spreen O (2006). A Compendium of Neuropsychological Tests: Administration, Norms, and Commentary. 3rd edn. Oxford University Press, New York.

Sullivan EV (2003). Compromised pontocerebellar and cerebellothalamocortical systems: speculations on their contributions to cognitive and motor impairment in nonamnesic alcoholism. Alcohol Clin Exp Res 27: 1409–1419.

Sullivan EV, Pfefferbaum A (2005). Neurocircuitry in alcoholism: a substrate of disruption and repair. Psychopharmacology (Berl) 180: 583–594.

Sullivan EV, Mathalon DH, Zipursky RB et al. (1993). Factors of the Wisconsin Card Sorting Test as measures of frontal-lobe function in schizzophrenia and in chronic alcoholism. Psychiatry Res 46: 175–199.

Sullivan EV, Marsh L, Mathalon DH et al. (1995). Anterior hippocampal volume deficits in nonamnesic, aging chronic alcoholics. Alcohol Clin Exp Res 19: 110–122.

Sullivan EV, Rosenbloom MJ, Lim KO et al. (2000a). Longitudinal changes in cognition, gait, and balance in abstinent and relapsed alcoholic men: relationships to changes in brain structure. Neuropsychology 14: 178–188.

Sullivan EV, Rosenbloom MJ, Pfefferbaum A (2000b). Pattern of motor and cognitive deficits in detoxified alcoholic men. Alcohol Clin Exp Res 24: 611–621.

Sullivan EV, Desmond JE, Lim KO et al. (2002a). Speed and efficiency but not accuracy or timing deficits of limb movements in alcoholic men and women. Alcohol Clin Exp Res 26: 705–713.

Sullivan EV, Fama R, Rosenbloom MJ et al. (2002b). A profile of neuropsychological deficits in alcoholic women. Neuropsychology 16: 74–83.

Sullivan EV, Harding AJ, Pentney R et al. (2003). Disruption of frontocerebellar circuitry and function in alcoholism. Alcohol Clin Exp Res 27: 301–309.

Sullivan EV, Harris RA, Pfefferbaum A (2010). Alcohol's effects on brain and behavior. Alcohol Res Health 33: 127–143.

Thorberg FA, Young RM, Sullivan KA et al. (2011). Alexithymia, craving and attachment in a heavy drinking population. Addict Behav 36: 427–430.

Tivis R, Beatty WW, Nixon SJ et al. (1995). Patterns of cognitive impairment among alcoholics: are there subtypes? Alcohol Clin Exp Res 19: 496–500.

Townshend JM, Duka T (2003). Mixed emotions: alcoholics' impairments in the recognition of specific emotional facial expressions. Neuropsychologia 41: 773–782.

Tulving E, Craik FIM (2000). The Oxford Handbook of Memory, Oxford Handbook Series. Oxford University Press, New York.

Uekermann J, Daum I (2008). Social cognition in alcoholism: a link to prefrontal cortex dysfunction? Addiction 103: 726–735.

Uekermann J, Daum I, Schlebusch P et al. (2005). Processing of affective stimuli in alcoholism. Cortex 41: 189–194.

Uekermann J, Channon S, Winkel K et al. (2007). Theory of mind, humour processing and executive functioning in alcoholism. Addiction 102: 232–240.

Vetreno RP, Hall JM, Savage LM (2011). Alcohol-related amnesia and dementia: animal models have revealed the contributions of different etiological factors on neuropathology, neurochemical dysfunction and cognitive impairment. Neurobiol Learn Mem 96: 596–608.

Westlye LT, Walhovd KB, Dale AM et al. (2010). Life-span changes of the human brain white matter: diffusion tensor imaging (DTI) and volumetry. Cereb Cortex 20: 2055–2068.

Witkiewitz K, Aracelliz N (2009). Dynamic association between negative affect and alcohol lapses following alcohol treatment. J Consult Clin Pscychol 77: 633–644.

Witkiewitz K, Marlatt GA (2004). Relapse prevention for alcohol and drug problems: that was Zen, this is Tao. Am Psychol 59: 224–235.

Wobrock T, Falkai P, Schneider-Axmann T et al. (2009). Effects of abstinence on brain morphology in alcoholism: a MRI study. Eur Arch Psychiatry Clin Neurosci 259: 143–150.

York JL, Biederman I (1991). Hand movement speed and accuracy in detoxified alcoholics. Alcohol Clin Exp Res 15: 982–990.

Zipursky RB, Lim KC, Pfefferbaum A (1989). MRI study of brain changes with short-term abstinence from alcohol. Alcohol Clin Exp Res 13: 664–666.

Handbook of Clinical Neurology, Vol. 125 (3rd series)
Alcohol and the Nervous System
E.V. Sullivan and A. Pfefferbaum, Editors

Chapter 13

Component processes of memory in alcoholism: pattern of compromise and neural substrates

ANNE-LISE PITEL*, FRANCIS EUSTACHE, AND HELENE BEAUNIEUX
Inserm, EPHE, Université de Caen/Basse-Normandie, Unité U1077, GIP Cyceron, CHU de Caen, France

Chronic and excessive alcohol consumption results in brain damage and attendant compromised neuropsychologic abilities involving motor, gait, balance, and cognitive deficits. These functional consequences can occur even in uncomplicated alcoholics, i.e., those without clinically detectable nutritional deficiencies such as Korsakoff's syndrome (KS) or liver dysfunction, often associated with chronic alcoholism. These neuropsychologic deficits encompass mainly visuospatial disabilities, executive dysfunctions, and memory impairments. This chapter focuses on memory capacities in adults with alcohol dependence. Other cognitive functions will be treated only to the extent that they may interact with memory functioning. For each of the five memory systems described in humans (Fig. 13.1), sparing and impairments of memory components are described in alcoholics without neurologic complication as well as in alcoholics with KS.

WORKING MEMORY AND EXECUTIVE FUNCTIONS

Definition

Working memory is the memory system that allows temporary storage and manipulation of the information necessary for complex cognitive tasks such as language comprehension, learning, and reasoning. Working memory, which requires the simultaneous storage and processing of information, is composed of slave systems and a central executive (Baddeley and Hitch, 1974; Fig. 13.1). The slave systems are short-term storage systems, comprising the phonologic loop, which processes verbal information and is therefore highly involved in language activities, and the visuospatial sketchpad, which processes non-verbal, i.e., visuospatial information. The third slave system of working memory, known as the episodic buffer, is viewed as responsible for linking information across domains and maintaining such multimodal information (Baddeley, 2000). The central executive is a system heavily involved in handling novel situations outside the "automatic" processes. It is regarded as being similar to the executive functions (Baddeley, 1996), which are necessary for goal-directed behavior. They include the ability to initiate and stop actions, to monitor and change behavior as needed, and to plan future behavior when faced with novel tasks and situations (Norman and Shallice, 1986). Executive functions, therefore, enable us to anticipate outcomes and adapt to changing situations.

Slave systems of working memory

Regarding the nature of the information processed by this memory system, the storage of both verbal and non-verbal components can be impaired in uncomplicated alcoholics (Becker et al., 1983; Beatty et al., 1996; Sullivan et al., 1997, 2000; Pitel et al., 2007a; Kopera et al., 2012), although non-verbal working memory is observed typically as more severely affected (Sullivan et al., 2000). The episodic buffer was also found to be impaired in alcoholics (Pitel et al., 2007a).

Central executive of working memory

The central executive of working memory is classically described as compromised in uncomplicated alcoholic subjects. Using standard tasks of inhibition, flexibility, categorization, deduction of rules, organization, updating, and planning, uncomplicated alcoholic subjects have mostly been found to be impaired (Sullivan et al., 2000; Noël et al., 2001, 2012; Pitel et al., 2007a). Executive

*Correspondence to: A-L. Pitel, Inserm-EPHE-Université de Caen/Basse-Normandie, Unité U1077, GIP Cyceron, Bd H. Becquerel, BP 5229, 14074 Caen Cedex, France. E-mail: pitel@cyceron.fr

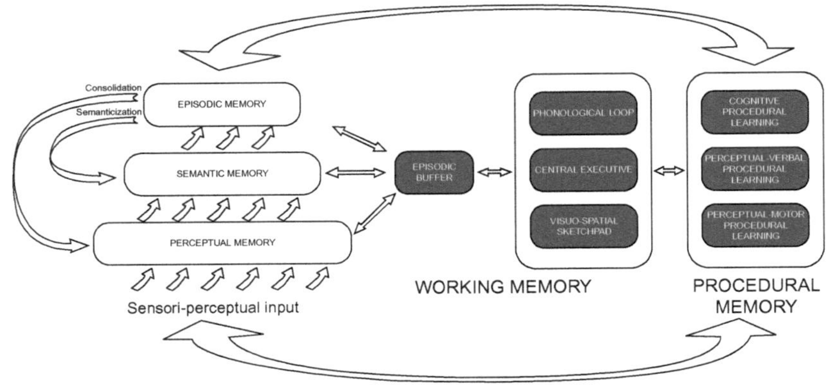

Fig. 13.1. Memory neostructural intersystemic model (MNESIS: reproduced from Eustache and Desgranges, 2008, with permission from Springer Science and Business Media).

functions have also been examined with the Behavioural Assessment of Dysexecutive Syndrome (BADS), a battery of executive tests designed to have ecologic validity. Ihara et al. (2000) showed that alcoholics were impaired on the total score of the BADS and that two of its six subtests, the temporal judgment and the modified six elements tests, were particularly sensitive to alcohol-related executive dysfunction. These findings indicate that, in patients with alcoholism, executive deficits cover elementary aspects of the central executive and extend to everyday executive performance.

Gambling tasks have repeatedly been used to examine decision-making abilities in alcoholics. Impaired decision-making performance related to poor response inhibition (Noël et al., 2007) and high novelty-seeking behavior (Noël et al., 2011). The most affected alcoholics were those with two or more previous detoxifications (Loeber et al., 2009). Low decision-making scores correlated with gray-matter abnormalities in the ventromedial prefrontal cortex, dorsal anterior cingulate cortex, and right hippocampal formation (Le Berre et al., 2012). Taken together, those results indicate that decision-making deficits in alcoholism may result from impairment of both compromised emotional and cognitive networks.

Neural substrates of working memory in alcoholism

Neuroanatomic substrates of the executive dysfunctions are classically associated with compromise of prefrontal structure or metabolism. Additional substrates of dysfunction involve damage to gray-matter structure of frontocerebellar circuitry (Sullivan, 2003; Fig. 13.2), which connects the frontal cortex and the cerebellum, two spatially disparate brain regions, through the pons and thalamus (Schmahmann and Pandya, 1997; Middleton and Strick, 2000). Chanraud et al. (2010) found that both alcoholics and controls relied on the integrity of certain nodes of the corticocerebellar systems to perform verbal and spatial working-memory tasks. However, in the most challenging condition (spatial working-memory task with interference), alcoholics performed lower than controls and exhibited different patterns of brain–behavior relationships. Thus, components of this corticocerebellar system not typically related to normal performance in dual-task conditions may be available to augment otherwise dampened performance by alcoholics.

Smaller white-matter volume in the midbrain of alcohol-dependent subjects was also found to be related to impaired performance in executive tasks involving flexibility capacities (Chanraud et al., 2007). Uncomplicated alcoholic subjects had fewer white-matter fiber tracts coursing between the midbrain and the pons and a smaller pons volume than controls. Furthermore, the number of fibers identified with neuroimaging (diffusion tensor imaging) correlated with the alcoholic's performance in the flexibility task. Using the same neuroimaging technique, Harris et al. (2008) showed a relationship between working-memory performance and the integrity of the superior longitudinal fasciculus in control subjects. No such relationship was seen in alcoholic patients, in whom the microstructure of this white-matter bundle was compromised. Previous investigations also provided evidence of an association between certain components of executive functions (attention skills and inhibition) and white-matter structure in the corpus callosum (Rosenbloom et al., 2003) and between verbal working memory and the medial septal/diagonal band (Sullivan et al., 2005). These later findings suggest a cholinergic mechanism for the working-memory impairments in uncomplicated alcoholics.

Functional magnetic resonance imaging (fMRI) provides further information about the status of working memory in uncomplicated alcoholics. Such fMRI studies

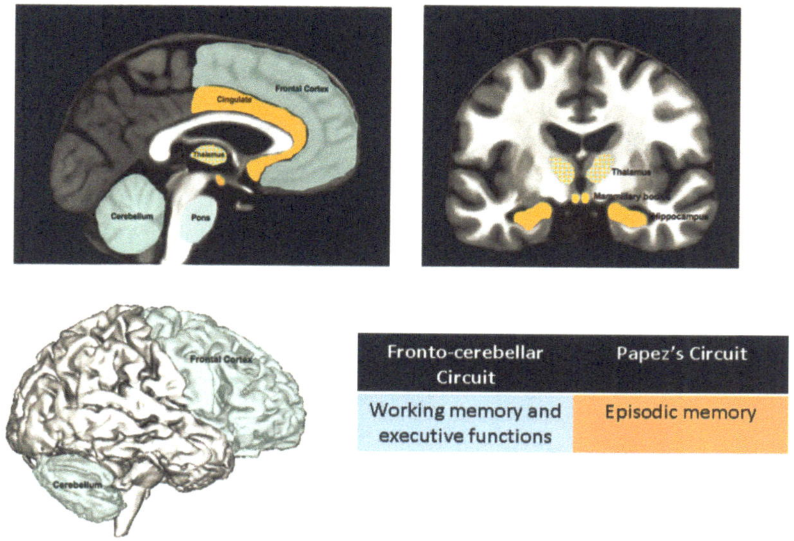

Fig. 13.2. Brain circuits commonly damaged in alcohol-dependent subjects and related to selective cognitive impairments in these subjects, notably executive functions and episodic memory. Sagittal (top left), coronal (top right), and surface-rendered (bottom left) views of brain regions that are part of the frontocerebellar circuit (in blue) and Papez circuit (in yellow) and their relations with cognitive impairments (bottom right). Alterations within one or both of these circuits are usually found in alcohol-dependent subjects (Reproduced from Chanraud et al., 2011b, with permission.)

show that, while performing a verbal working-memory task, alcoholics exhibit greater activation in the left frontal and right cerebellar regions than controls, suggesting that the articulatory control system of verbal working memory requires a compensatory increase in alcoholics in order to maintain the same level of performance (Desmond et al., 2003). Such functional reorganization was also revealed in alcoholics engaged in a visual spatial working-memory task (Pfefferbaum et al., 2001). In this case, despite equivalence in behavioral performance, uncomplicated alcoholics activated the ventral "what" stream usually used for encoding of verbal material, whereas controls activated the dorsal "where" stream appropriately used to encode visuospatial information.

Chanraud et al. (2011a) examined whether performance on a spatial working-memory task was associated with functional connectivity within the default mode network. This network is defined by intrinsic and correlated activity of selective brain regions (Raichle et al., 2001) when the brain is not involved in an externally imposed goal-directed activity. The brain regions with reliably synchronous activity are the posterior parietal cortex, precuneus and posterior cingulate cortex, medial prefrontal cortex, and selective cerebellar lobules (Greicius et al., 2003; Andrews-Hanna et al., 2010). The results showed that functional connectivity within this network is compromised at rest with lower synchronization in alcoholics than in controls. On the contrary, the connectivity was more robust in alcoholics than controls during the spatial working-memory task. Since alcoholics performed on par with controls, the authors interpreted this higher default mode network synchronization as compensatory networking to achieve normal performance (Chanraud et al., 2013).

Working memory in Korsakoff's syndrome

Regarding the slave systems of working memory, alcoholics with and without KS do not differ from each other on span tasks assessing the visuospatial sketchpad, the phonologic loop, and the episodic buffer (Joyce and Robbins, 1991; Pitel et al., 2008). This deficit of the slave systems of working memory was equally severe in the KS and the alcoholic groups; the phonologic loop, the visuospatial sketchpad or the episodic buffer did not seem to be disproportionately affected. Moreover, the slave systems were impaired to the same extent in the KS and non-KS alcoholic groups.

Regarding executive functions, alcoholics with KS have compromised abilities in organization, inhibition, flexibility, updating, and integration (Jacobson et al., 1990; Brokate et al., 2003; Hildebrandt et al., 2004; Oscar-Berman et al., 2004; Pitel et al., 2008; see Oscar-Berman, 2012 for review). Organization (Jacobson et al., 1990; Joyce and Robbins, 1991; Hildebrandt et al., 2004; Oscar-Berman et al., 2004) and updating abilities (Brokate et al., 2003; Hildebrandt et al., 2004) are more severely impaired in alcoholics with KS than in those without KS but the two groups do not appear to differ in terms of inhibition,

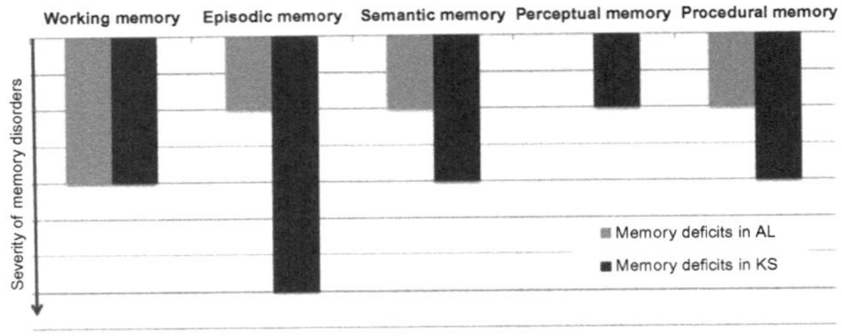

Fig. 13.3. Graded effects of memory disorders in alcoholism. AL, alcoholism; KS, Korsakoff syndrome.

flexibility, and integration abilities (Pitel et al., 2008; Fig. 13.3). Analyses of individual working memory results indicate a total mixture of the two patient groups, with patients' performance ranging from normal to severely impaired.

In short, KS have an overall working-memory disorder affecting both the slave systems and the central executive. Abnormalities in brain regions know to subserve these functions have been found in the frontocerebellar circuit of alcoholics with KS compared with control subjects and notably in the frontal cortex, cerebellum, thalamus, and pons (Christie et al., 1988; Jernigan et al., 1991; Colchester et al., 2001; Reed et al., 2003; Pitel et al., 2009a; see Oscar-Berman, 2012 for review). Using a region-of-interest approach, a neuroimaging study reported graded effects of volume deficits in the cerebellum, thalamus, and pons, from mild damage in uncomplicated alcoholics to moderate or severe damage in KS patients (Sullivan and Pfefferbaum, 2009). However, those differences between alcoholics with and without KS were not significant. A voxel-by-voxel investigation confirmed considerable similarities in the pattern of gray- and white-matter damage within the frontocerebellar circuitry of the two patient groups (Pitel et al., 2012). Thus, alcohol-related working-memory deficits and frontocerebellar damage seem to be shared by alcoholics with and without KS, potentially reflecting the harmful effects of chronic and excessive alcohol consumption on those brain regions. Therefore, those neuropsychologic impairments and the associated brain alterations cannot constitute relevant diagnostic criteria for KS.

EPISODIC MEMORY

Definition

Episodic memory is currently described as the memory system in charge of the encoding, storage, and retrieval of personally experienced events, associated with a precise spatial and temporal context of encoding. Episodic memory allows the conscious recollection of happenings and events from one's personal past and the mental projection of anticipated events into one's subjective future (Wheeler et al., 1997). Recollection of episodic events includes autonoetic awareness, which is the impression of re-experiencing or reliving the past and mentally traveling back in subjective time (Tulving, 2001). Of the many components of memory, episodic memory is hierarchically the highest memory system (Fig. 13.1) – the most sophisticated but also the most sensitive to pathology, trauma, and toxicity.

Encoding and retrieval processes

Most studies investigating episodic memory in chronic alcoholism examined learning abilities, i.e., capacities to improve the level of performance with practice, by means of classic psychometric tasks. These investigations reported a non-specific nature of learning impairments (Salmon et al., 1986; Sullivan et al., 1997) affecting both verbal (mainly lists of words (Sherer et al., 1992)) and non-verbal information (face-name associations (Schaeffer and Parsons, 1987; Everett et al., 1988; Beatty et al., 1995; Tivis et al., 1995), complex figure (Sullivan et al., 1992), symbols, visual location or visual pattern (Kopera et al., 2012). Although sober alcoholics commonly perform below controls, both groups can show evidence of learning over trials (Ryan et al., 1980). Learning difficulties have sometimes been attributed to an impoverished generation of spontaneous learning strategies (Sullivan et al., 1992; Noël et al., 2012), which may also account for the poor performance on free-recall tasks (Weingartner et al., 1996; Noël et al., 2012). Sober alcoholics, however, can show impairment in recognizing test information after spontaneous encoding as well as on a free recall after deep encoding. The timing of these deficits suggests that, at alcohol treatment entry, alcoholics present an impairment of both encoding and retrieval abilities (Pitel et al., 2007a).

Contextual memory, source memory, and prospective memory

The spatiotemporal context of encoding is also impaired in alcoholism, affecting both temporal order (Salmon et al., 1986; Sullivan et al., 1997) and spatial context recognition (Pitel et al., 2007b). Moreover, alcoholic subjects tend not to recall complete episodes, i.e., correct factual information associated with the correct spatiotemporal context of encoding, indicating that episodic memories are not complete (Pitel et al., 2007b). Also affected in alcoholism are source memory, which is the ability to discriminate the origin or source of information (Schwartz et al., 2002), and prospective memory (Heffernan, 2008 for review), which refers to the cognitive ability of remembering to carry out an intended action at some future point in time. Alcoholics report greater prospective memory impairments and do not appear to use sufficient internal or external memory strategies to compensate for these deficits (Heffernan et al., 2002). Lastly, alcoholism leads to a deficit of autonoetic consciousness (Pitel et al., 2007b), which reflects difficulties in re-experiencing or reliving the past.

Metamemory

Taken together, these findings suggest that most components of episodic memory are impaired in uncomplicated alcoholic subjects. Interestingly, alcoholics seem to exhibit a limited consciousness of those episodic memory deficits. In support of this position, Le Berre et al. (2010) showed that alcoholics have a tendency to overestimate their episodic memory performance on a metamemory task, while judging themselves as efficient as controls about memory capacity when evaluated with a questionnaire. This metamemory impairment observed in alcoholics without neurologic complication may be a mild form of anosognosia observed in alcoholics with KS. The association of episodic memory deficits and a limited consciousness of such impairments may result in poor everyday life functioning, as suggested by recent findings showing impaired results on a computerized activity task with ecologic validity (shopping task, Laloyaux et al., 2012). In alcoholics, poor performance on this task correlated with poor performance on the verbal episodic memory, processing speed and executive tasks, everyday life functioning, and extended duration of illness.

Episodic memory and executive functions

Because of the links between episodic memory and working memory, the relationships between executive dysfunctions and episodic memory impairments have been explored in alcoholics to determine whether executive dysfunction could be responsible for the deterioration of certain episodic memory processes (Pitel et al., 2007b; Noël et al., 2012). The episodic memory impairment observed in uncomplicated alcoholic subjects was not linked solely to executive dysfunction (Pitel et al., 2007b), suggesting that chronic alcoholism may cause genuine episodic memory deficits and are not simply an indirect consequence of executive deficits. Further, episodic memory deficits were more related to executive abilities in alcoholics than in controls (Noël et al., 2012), which indicates that alcoholics' episodic memory disorders were characterized by impaired effortful executive processes.

Neural substrates of episodic memory deficits in alcoholism

There may well be a link between the episodic memory disturbance in uncomplicated alcoholic subjects and neuroanatomic abnormalities in regions involved in episodic memory functioning (Fig. 13.2). For example, postmortem investigations have reported atrophy in the mammillary bodies and dorsomedial nucleus of the thalamus even in uncomplicated alcoholics (Sheedy et al., 1999; Harding et al., 2000; Harper et al., 2003). *In vivo* imaging studies also reported volume loss in mammillary body, hippocampus, and thalamus (Sullivan et al., 1995, 1999; Cardenas et al., 2007; Chanraud et al., 2007) but failed to show any correlation between gray-matter macrostructural abnormalities and episodic memory impairments. Using diffusion tensor imaging, Chanraud et al. (2009) confirmed the absence of a relationship between episodic memory performance and regional volumes but found that low verbal episodic memory correlated with altered gray-matter microstructure in parahippocampal areas, frontal cortex, and left temporal cortex. Thus, microstructural damage may be a better predictor of episodic memory disorders than macrostructural damage in alcoholism. Episodic memory deficits may also be associated with damage of white-matter fiber bundles and tracts, including compromised cingulate bundle and fornix (Pfefferbaum et al., 2009; Schulte et al., 2010), which interconnect the hippocampus, thalamus, and mammillary body in the Papez's circuit, thus leading to an interruption of this brain network.

Autobiographical memory

Autobiographical memory, which is formed by different types of representation from specific personal events (episodic components) to general knowledge about oneself (semantic component: Tulving et al., 1988; Conway, 2001), has been reported to be impaired in uncomplicated alcoholics (D'Argembeau et al., 2006). When thoroughly evaluated and compared with non-alcoholic controls, alcoholics were seen to recall specific memories less frequently and general memories more frequently, suggesting difficulties in strategically accessing episodic autobiographic memory. Nevertheless, when a specific past event was accessed, alcoholic subjects subjectively experienced as many sensory and contextual details associated with the event as controls. Depending whether those autobiographic memory disorders concern specifically the drinking period or the whole lifespan, several hypotheses can be formulated: encoding deficits related to the acute effects of alcohol consumption or encoding and/or retrieval deficits related to the chronic effects of alcohol consumption on the frontal cortex (D'Argembeau et al., 2006) and the hippocampus.

Episodic memory in Korsakoff's syndrome

Only a few studies have made direct comparisons of performance between alcoholics with and without KS, showing broadly that alcoholic patients with KS are more severely impaired than non-Korsakoff alcoholics (see Emsley et al., 1996; Krabbendam et al., 2000; Fama et al., 2012a for review). When compared on the different episodic memory components (Pitel et al., 2008), there were similarities in the profiles of episodic memory deficits in the two groups, even though the extent of the deficits was altogether different in KS and alcoholism, given that KS systematically obtained lower scores than alcoholics without KS (Fig. 13.3). In addition to the severity of the impairments, the one feature of episodic memory deficits that distinguished alcoholics with and without KS was the disproportionate impairment of encoding abilities compared with retrieval ones (Cermak and Butters, 1972; Asselen et al., 2005). Analyses of individual episodic memory results revealed an unexpected partial overlap of the two patient groups, involving the uncomplicated alcoholics with the worst performances and the alcoholics KS with the best ones (Pitel et al., 2008). Those uncomplicated alcoholics with equivocal episodic memory performance, similar to those of KS, may be regarded as alcoholics at risk of developing KS and should be identified in order to receive particular attention and preventive action before their memory disorders have harmful repercussions on their daily lives.

The identification of alcoholics at risk for Wernicke encephalopathy (WE) and KS may benefit from the use of the criteria established by Caine et al. (1997) and validated in their neuropathologic study of alcoholic cases. According to these authors, alcoholics with at least two of the four following signs were considered to have "signs of WE": dietary deficiency, oculomotor abnormalities, cerebellar dysfunction, and altered mental state or mild cognitive impairment. Pitel et al. (2011) applied those criteria to a group of ostensibly uncomplicated alcoholics and extended them to classify individuals with only one criterion as "at risk for WE." They found that 27% met no criteria, 57% were at risk for WE, and 16% had signs of WE, which is in agreement with the prevalence rates reported postmortem in this population (Harper et al., 1986). Neuropsychologic performance of these alcoholic subgroups was graded, with those meeting zero criteria not differing from controls, those meeting one criterion presenting mild to moderate deficits on some of the functional domains, and those meeting two or more criteria having the most severe deficits on each of the domains examined. All together these findings indicate that the presence of signs of WE may explain, at least partially, the heterogeneity of alcoholism-related cognitive and motor deficits. This initial study provides promising evidence for the utility of Caine et al.'s criteria in identifying alcoholics at risk for neurologic complications.

All together, these neuropsychologic findings suggest that alcoholics with KS may be considered as uncomplicated alcoholics, whose existing deficits, potentially due to the neurotoxic effects of ethanol on Papez's circuit, have been exacerbated by thiamine deficiency, leading to amnesic syndrome. This hypothesis is corroborated by neuroimaging results indicating graded brain volume shrinkage from uncomplicated alcoholics to alcoholics with KS, notably in regions involved in episodic memory functioning such as hippocampus, mammillary bodies, and thalamus (Sullivan and Pfefferbaum, 2009). A voxel-by-voxel brain comparison of gray- and white-matter volumes in alcoholics with and without KS revealed considerable similarities in the pattern of gray- and white-matter damage in the two groups of alcoholics (Pitel et al., 2012). In this study, only a few key cerebral regions, including the medial thalami, mammillary bodies, and corpus callosum, were more severely damaged in alcoholics with KS than in uncomplicated alcoholics. Statistical cluster analyses indicated that alcoholic patients with medial thalamic volume less than two standard deviations from controls were included within the same cluster as patients with KS. Further studies combining refined neuropsychologic assessment and neuroimaging examinations within the same patients are required to determine whether alcoholics with small thalamic volumes are also those with equivocal episodic memory performances, similar to those of KS.

SEMANTIC MEMORY

Definition

Semantic memory refers to the memory of meaning, understanding, general knowledge about the world, and other concept-based knowledge unrelated to specific experiences. The level of consciousness associated with semantic memory is noetic because it is independent of context encoding and personal relevance (Tulving, 1985, 2001).

Little is known about the effects of chronic and excessive alcohol consumption on the organization or extent of the pre-existing semantic networks. Only Fama et al. (2011) studied remote semantic memory for a specific kind of information (related to presidential elections from 1928 to 2004) and found preserved performance on that task in alcoholics. Impaired remote semantic was observed only in alcoholics who were infected with the human immunodeficiency virus (HIV). The few investigations of semantic memory conducted in alcoholics examined ability to acquire new semantic concepts.

New semantic learning

Early in sobriety, alcoholics have impairment in novel semantic encoding and, more specifically, a deficit in acquiring new labels information (Pitel et al., 2007b). Learning new categories and features can be achieved by alcoholics but at a slower rate than controls. Compared with controls, alcoholics invoke different and inefficient cognitive strategies to attempt to compensate for impaired episodic and working memory. Alcoholics invoked specific cognitive processes, such as working-memory abilities, in addition to those invoked by controls. Compensatory strategies have already been suggested in neuroimaging studies, showing that alcoholics recruit higher-level cognitive systems to perform even simple tasks at control levels (Pfefferbaum et al., 2001; Desmond et al., 2003; Chanraud et al., 2013).

Pitel et al. (2010) tested whether the use of errorless learning, which is a rehabilitation method that limits error production during learning, could compensate for those impaired semantic learning strategies and enable alcoholics to perform on par with controls. They compared new labels learning performance in alcoholics in errorless learning condition with that of another group of alcoholics in a trial-and-error learning condition. The authors found that the alcoholic group in the errorless condition performed significantly better than the alcoholic group in the trial-and-error condition but did not differ from the two control groups (whatever the learning condition). Thus, the use of errorless learning may be relevant for alcoholics when learning of new semantic labels is considered.

Semantic memory in Korsakoff's syndrome

Abilities to acquire new semantic knowledge have been examined in alcoholics with KS compared with alcoholics without KS (Pitel et al., 2009b). Both patient groups exhibited label learning deficits but patients with KS were more severely impaired than uncomplicated alcoholics (Fig. 13.3). Only KS exhibited impaired category and feature learning results compared with control subjects. Episodic memory performance accounted for label learning results in control subjects and uncomplicated alcoholics but not in KS patients, and for category and feature learning results for none of the groups. These findings suggest that the severe deficit of new label learning observed in KS may be considered a consequence of episodic memory disorders, whereas the specific impairment in category and feature learning may reflect a genuine deficit of new learning in semantic memory in KS. As with uncomplicated alcoholics, whether the pre-existing semantic network is preserved or impaired in alcoholics with KS remains to be determined.

PERCEPTUAL MEMORY

Definition

Perceptual memory receives, stores, and makes available to other systems information about perceptual features of physical objects. It is assumed to be involved in perceptual priming, which refers to the effect in which exposure to the form of a stimulus influences a response to a later stimulus. Such experience-based changes in perceptual memory implicitly manifest themselves as enhancement in the perceptual features of objects (Tulving and Schacter, 1990). Perceptual memory includes both conscious and non-conscious processing of sensoriperceptual information. It can be considered the root of the human memory systems (Fig. 13.1), because it is through perceptual memory that information is subsequently and progressively transferred into the different representation memory systems.

Implicit perceptual learning

While deficits in basic visuospatial abilities have been documented in uncomplicated alcoholic subjects (Fox et al., 2000; Sullivan et al., 2000), less is known about implicit visual perceptual learning processes. One of the implicit perceptual learning tasks is the picture fragment completion stimuli, where participants identify a set of fragmented pictures once and then a second time, intermixed with new pictures. The ability to identify the stimuli on initial testing requires intact visuospatial abilities, whereas identification of these stimuli at later testing sessions includes an additional implicit memory component. Implicit memory is reflected in the decrease in identification difficulty for pictures that were previously presented. Given that picture fragment completion tasks can be performed at normal levels by subjects with episodic memory deficits, these tests are assumed to assess implicit memory for perceptual features, i.e., perceptual memory.

When using this implicit perceptual learning paradigm, alcoholics showed the commonly reported impairment in visuoperception but normal levels of perceptual learning (Fama et al., 2004). This pattern of spared and impaired processes suggests that the alcoholics were able to take normal advantage of prior exposure to fragmented line drawings to enhance performance. However, although the alcoholics performed at the same level as the controls on the implicit perceptual learning task, each group used different strategies to perform the task successfully. Whereas controls used basic visuospatial processes to perform the perceptual learning task, alcoholic subjects invoked higher-order executive processes to establish normal learning. This difference in strategy was a further indication that alcoholics recruit more cognitively demanding and less efficient neural systems than controls to perform a task at normal levels, possibly because of alcoholism-related dysfunction in optimal neural systems.

Perceptual memory in Korsakoff's syndrome

Patients with KS show perceptual learning with repeated exposure to fragmented pictures or words, although not necessarily at normal levels (Warrington and Weiskrantz, 1968); however, KS can retain implicit perceptual information over prolonged periods (Warrington and Weiskrantz, 1968; Cave and Squire, 1992; Fahle and Daum, 2002). Verfaellie et al. (1996) reported that KS could identify previously seen fragmented pictures more readily than new pictures; however, the priming effect observed in KS was significantly smaller than that of the uncomplicated alcoholic group (Fig. 13.3). After continued exposure to the stimuli, alcoholics with KS were found to show additional learning, despite non-mnemonic visuospatial deficits and profound explicit memory impairment for the test stimuli (Fama et al., 2006). The persistence of learning beyond the life of the percept suggests that perceptual learning and memory reflect a relatively intact memory process in KS.

PROCEDURAL MEMORY

Definition

Procedural memory is defined as the memory system in charge of the encoding, storage, and retrieval of the procedures (rather than episodes) that underlie motor, visuospatial, or cognitive skills. This memory system is connected to both episodic memory and working memory (Fig. 13.1), which are required during encoding of a new procedure, i.e., during procedural learning (Beaunieux et al., 2006). Procedural learning is a dynamic process involving different phases (cognitive, associative, and autonomous) and resulting in the automation of the procedure (Anderson, 1992). The cognitive stage of such acquisition involves non-procedural functions (Ackerman and Cianciolo, 2000), such as episodic memory and working memory. Impairments of these memory systems in uncomplicated alcoholics may therefore hinder procedural acquisition and hamper its automation.

Cognitive procedural learning

Early in alcohol sobriety, uncomplicated alcoholic subjects solve cognitive procedural problems more slowly than controls and make more moves to achieve a new cognitive procedure (Pitel et al., 2007b). Alcoholics are also slower than controls, sacrificing speed for accuracy during learning. By dint of repeating numerous trials, alcoholic subjects manage to catch up with the controls and achieve similar performance. However, alcoholics are typically not at the same stage of learning as controls would be, being still in a problem-solving mode at the end of the learning phase and implementing higher-order cognitive processes to achieve normal learning levels. Their episodic and working-memory deficits may prevent them from completing the cognitive and associative stages and thus from automating the cognitive procedure. A recent sober alcoholic may be able to acquire a new cognitive procedure, but it would be more difficult and take longer than for non-alcoholic counterparts.

Visuomotor procedural learning

Visuomotor procedural learning and retention have been studied in alcoholics using the rotary pursuit task (Fama et al., 2012b). Despite impairment of explicit memory and psychomotor speed, alcoholics showed visuomotor learning across rotary pursuit testing sessions, as well as retention and consolidation over time.

Procedural memory in Korsakoff's syndrome

While procedures acquired before the development of amnesia are likely to be preserved in alcoholic patients with KS, the ability of KS patients to learn new procedures has been called into question. Only a few studies examined procedural learning in KS, some of them showing preserved learning abilities (Cermak and Butters, 1973; Charness et al., 1988; Beaunieux et al., 1998; Swinnen et al., 2005; Kessels et al., 2007), whereas others reported impaired ones (Butters et al., 1985; Nissen et al., 1989; Schmidtke et al., 1996). When focusing on cognitive procedural learning, previous studies reported deficits of new learning abilities and ascribed them to working-memory (Butters et al., 1985) and episodic memory (Xu and Corkin, 2001) deficits frequently observed in KS. And yet, an investigation of the relationships between non-procedural cognitive functions and cognitive procedural performance in a group of 20 amnesic patients, including seven KS, did not reveal any direct evidence of a role of episodic memory impairment in cognitive procedural learning deficits (Schmidtke et al., 1996). In this study, executive functions were involved, notably at the initial stage of the learning process, when rule-directed behavior, planning, and problem solving are required. These data suggest that prolonged dependence on intellectual abilities in the amnesic group may indicate a delayed transition to more advanced stages of the skill acquisition.

A recent study (Beaunieux et al., 2013) examined cognitive procedural learning abilities in alcoholics with and without KS compared with control subjects. KS patients achieved poorer scores than controls and uncomplicated alcoholics (Fig. 13.3), who presented a slowdown in the cognitive procedural learning but managed to reach a normal level of performance at the end of the learning process (Pitel et al., 2007b). Because of their severe episodic memory deficits, KS were unable to implement such compensatory strategies, and their performance remained lower than both controls and alcoholics until the end of the learning sessions. The findings indicated that amnesia observed in KS may explain the slowdown in the dynamics of learning in this group and the absence of automation of the cognitive procedure within the 40 trials conducted in this study. Those results challenge the traditional description of preserved procedural learning in KS.

FACTORS CONTRIBUTING TO THE HETEROGENEITY OF MEMORY DISORDERS IN ALCOHOLISM

Human alcohol consumption is marked by considerable heterogeneity in its consequences on memory. Disabilities range from mild to moderate in uncomplicated alcoholism to severe and debilitating in KS. Potential sources of variability include age, gender, alcohol use pattern, smoking, nutritional status, and psychiatric comorbidity.

Age

Normal aging disproportionately affects some brain regions, and these compromised brain areas in the healthy elderly are also vulnerable to alcohol neurotoxicity. Age and alcoholism interact, thereby increasing brain vulnerability to alcohol's neurotoxicity in aging (Sullivan and Pfefferbaum, 2003; Pfefferbaum et al., 2006) and the chances of more severely impaired cognitive capacities in the elderly who abuse alcohol.

Gender

Gender may also be a factor influencing the nature and the severity of the memory disorders in uncomplicated alcoholics (Liu et al., 2010). For a comprehensive review, please see Chapter 16.

Alcohol use pattern

The relationships between drinking history and severity of memory disorders is still a matter of debate. Drinking history can be characterized in terms of age at onset (first contact with alcohol or onset of alcoholism), quantity (number of drinks per day), and frequency (number of days drinking per month) of alcohol consumption. Studies yield heterogeneous results. It remains unknown why some alcoholic subjects exhibit severe impairments, whereas others with measurably similar alcohol consumption levels appear to drink with impunity.

Smoking status

A growing body of evidence suggests that chronic cigarette smoking, which occurs with high prevalence in alcoholism, itself results in abnormalities in brain morphology, blood flow, neurochemistry, and neurophysiology, leading to cognitive dysfunction (Durazzo et al., 2007). The association of greater smoking chronicity and poorer performance on multiple measures after control for alcohol consumption suggests that chronic smoking adds an additional burden to neurocognitive function in those with alcohol dependence (Durazzo et al., 2010).

Nutritional status

Sustained heavy drinking frequently occurs at the expense of eating, rendering alcoholics, who also exhibit compromised vitamin absorption (Thomson, 2000), at risk for nutritional depletion. That can result in the development of neurologic complications such as WE and KS. Thiamine deficiency appears even in alcoholics without clinically apparent WE (Thomson et al., 1987) and may be related to their memory deficits (Pitel et al., 2011).

Psychiatric comorbidity

Almost 60% of alcoholics have at least one additional psychiatric (mood or anxiety) or past substance dependence comorbidity. However, psychiatric comorbidity does not necessarily compound poor results on memory and intelligence tests associated with chronic alcohol misuse, suggesting that poorer performance on these tasks may be more a function of alcoholism per se than non-alcoholic comorbidity (Rosenbloom et al., 2005). On the contrary, Finn et al. (2009) found that alcohol-related cognitive impairment can be partly accounted for by externalizing behaviors, including childhood conduct disorder, adult antisocial behavior, and other substance abuse than alcohol-related problems per se (see Chapter 14).

Treatment-naïve versus treatment-seeking patients

According to a recent epidemiologic survey in the United States (NESARC: Moss et al., 2007), only about a quarter of alcoholics ever seek treatment. Some current studies indicate that those alcoholics who remain drinkers are not necessarily affected by alcoholism to a significant extent (Smith and Fein, 2010). These alcoholics may enjoy a special protection, such as relatively safe drinking pattern or safe environment for drinking (such as healthy diet or exercise).

REVERSIBILITY OF ALCOHOL-RELATED MEMORY DISORDERS

Uncomplicated alcoholism

Some cognitive recovery may take place with sustained abstinence from alcohol consumption. The greatest recovery occurs during the first month of drinking cessation (Kish et al., 1980; Bartsch et al., 2007; Manning et al., 2008), but further improvements are detectable with prolonged abstinence. Indeed, alcoholics have been shown to present memory capacities equivalent to those of control subjects after several years of abstinence (Reed et al., 1992; Rourke and Grant, 1999; Fein et al., 2006). However, some investigations report persistent cognitive impairments in alcoholics 6 months after drinking cessation (Munro et al., 2000) and even after more than 5 years' abstinence (Brandt et al., 1983). A recent meta-analysis examined cognition in alcohol abuse and dependence and the duration of abstinence necessary to achieve cognitive recovery (Stavro et al., 2012). Findings revealed moderate impairment across several cognitive domains during short-term abstinence (<1 month) and intermediate-term abstinence (2–12 months). Small-effect-size estimates were found for long-term abstinence (>1 year). These results suggest that significant impairment across multiple cognitive functions remains stable during the first year of abstinence from alcohol. The lack of accuracy as to the length of abstinence required for normalization may be due to variability in the ability of different cognitive functions to recover (Brandt et al., 1983; Rosenbloom et al., 2004), complicating simultaneous analysis of several domains of cognition.

Three main methodologic issues inherent to the investigation of recovery with abstinence from alcohol in alcohol-dependent individuals have to be considered. Firstly, the interaction between age and length of abstinence (Fein et al., 1990; Reed et al., 1992; Rourke and Grant, 1999; Munro et al., 2000) may induce a slowdown in the course of recovery in older alcoholics, notably because of a decrease in brain plasticity in aging. Secondly, the use of cross-sectional comparisons between groups of alcoholic subjects who are abstinent for two different lengths is not recommended. Even when patient groups are matched in demographic and alcoholism characteristics, cross-sectional comparisons yield less reliable results than longitudinal studies, which monitor recovery in a single sample of subjects. The period of abstinence is variable and represents only a mean or a threshold, leading to an inaccurate picture of the differences between groups. The last factor, which has to be considered when evaluating normalization of abilities, is interim drinking during the follow-up (Yohman et al., 1985; Fein et al., 1990; Rourke and Grant, 1999; Rosenbloom et al., 2004). For example, interim drinking is considered differently according to the investigations, in that alcoholic subjects who have resumed drinking to a modest degree (perhaps just one drink) may be included either in the "abstinent" (Johnson-Greene et al., 1997) or the "relapsed" group (Rourke and Grant, 1999; Rosenbloom et al., 2004), whereas some studies consider abstinent subjects as "having not drunk to intoxication" (Munro et al., 2000). Under conditions of strictly monitored long-term alcohol abstinence, memory capacities have been shown to return to normal in small samples of alcoholics (Rosenbloom et al., 2004; Bartels et al., 2007; Pitel et al., 2009c).

Recovery of cognitive function has occasionally been associated with at least partial recovery of alcoholism-related brain shrinkage. This brain recovery may be selective according to the neuropsychologic changes observed (Rosenbloom et al., 2007) and associations were observed between memory improvement and lateral ventricular volume change. In effect, the reduction of alcoholics' brain weight and volume, which is well documented, is largely due to an overall decrease of white-matter volume, including demyelination, axonal and glial loss, and dehydration (Harper et al., 2003). Recent findings suggest that the reversibility of the brain damage reflects not only rehydration but also growth of the white-matter volume (Cardenas et al., 2007) associated with increased levels of *N*-acetyl aspartate (a marker of neuronal integrity) and choline (a marker of cell membrane turnover) (Bartsch et al., 2007) and modest increase in cortical gray-matter volume (Pfefferbaum et al., 2006). Regarding the hippocampus, a principal neural substrate of episodic memory consolidation, experiments using rodents models of binge drinking indicate that adult neurogenesis in the hippocampus is inhibited during alcohol intoxication but returns to normal with cessation of drinking (Nixon, 2006). Whether substantial and functionally viable neurogenesis occurs in the human hippocampus or other brain structures remains unknown.

Korsakoff's syndrome

There are very few studies on cognitive and brain recovery in abstinent KS. Fujiwara et al. (2008) conducted a longitudinal neuropsychologic investigation of 20 patients with KS. At baseline and the 2-year follow-up sessions, the majority of the patients were impaired compared with control subjects. Comparing T1 and T2 revealed no significant decline in any of the investigated functions. Instead, general knowledge, visual long-term memory, and verbal fluency improved slightly after 2 years, though they still remained within pathologic range.

Yet, a longitudinal study with a 9-month follow-up positron emission tomography examination showed increased metabolism in the anterior cingulate and parietal cortex in an alcohol-abstinent and thiamine-replete KS patient (Fellgiebel, 2003). Dissociations between permanent and reversible cognitive deficits and brain damage would enable us to differentiate those resulting from chronic alcohol consumption, common to uncomplicated alcoholism and potentially reversible, from those specific to KS and considered as permanent.

CLINICAL IMPLICATIONS OF MEMORY DISORDERS IN THE TREATMENT OF ALCOHOL DEPENDENCE

Neuropsychologic impairments, and more especially memory disorders, could impede uncomplicated alcoholic subjects from benefiting fully from cognitive and behavioral treatment approaches for alcohol dependence. Indeed, cognitive dysfunctions of alcoholic subjects have been shown to be linked with the outcome of the treatment (Tapert et al., 2004). Successful alcohol treatment and subsequent abstinence may require high-level cognitive processes and consequently may not be feasible by alcoholic subjects with extensive impaired memory capacities. Because the classic treatment of alcoholism, notably based on cognitive behavioral treatment, may rely on the acquisition of novel semantic information or cognitive procedures, not all alcoholic subjects may be cognitively able to acquire such complex novel knowledge. Consequently, current forms of treatment may not be appropriate for alcoholic subjects with neuropsychologic impairments (Tapert et al., 2004). The assessment of neuropsychologic deficits at alcohol treatment entry may prove useful for clinical decision making and the choice of the most appropriate treatment (Bates et al., 2002). Lastly, repetition of material or procedures to be learned would be a useful strategy with alcoholics who are in educational treatment programs, wherein they are expected to acquire new information that may be relevant in later settings (Nixon et al., 1998). The neuropsychologic screening of these cognitive impairments may prove useful at alcohol treatment entry, allowing clinicians to ascertain whether alcoholic subjects are capable of undergoing standard therapy or whether it needs to be adjusted in order to take account of memory disorders. Follow-up testing would document change in memory function that could alter therapeutic programs dynamically.

CONCLUSION

While alcohol-related memory deficits were initially considered only through the prism of KS, it is now clear that chronic alcohol consumption results in memory disorders in alcoholics without ostensible neurologic complications. It was suggested that alcoholic patients with KS, uncomplicated alcoholics, and heavy social drinkers represent separate points along a single scale of cognitive impairments (Ryback, 1971). Recent work showing that the pattern of memory sparing and impairment in uncomplicated alcoholics is broadly consistent with that of alcoholic Korsakoff patients challenges the question of the specificity of the memory profile in KS and revives the continuity hypothesis of disordered memory processes. Most of the memory deficits in alcoholic Korsakoff subjects seem to be already present in uncomplicated alcoholics, either in an equivalent form or a less severe one. Taken together, these findings lend weight to a revised continuity theory between uncomplicated alcoholics and alcoholic Korsakoff subjects.

References

Ackerman PL, Cianciolo AT (2000). Cognitive, perceptual-speed, and psychomotor determinants of individual differences during skill acquisition. J Exp Psychol Appl 6: 259–290.

Anderson JR (1992). Automaticity and the ACT* theory. Am J Psychol 105: 165–180.

Andrews-Hanna JR, Reidler JS, Sepulcre J et al. (2010). Functional-anatomic fractionation of the brain's default network. Neuron 65: 550–562.

Asselen MVan, Kessels RPC, Wester AJ (2005). Spatial working memory and contextual cueing in patients with Korsakoff amnesia. J Clin Exp Neuropsychol 27: 37–41.

Baddeley A (1996). Exploring the central executive. Q J Exp Psychol A 49: 5–28.

Baddeley A (2000). The episodic buffer: a new component of working memory? Trends Cognitive Sci 4: 417–423.

Baddeley A, Hitch G (1974). Working memory. In: G Bower (Ed.), Psychology of Learning and Motivation. vol. 8. Academic Press, New York, pp. 47–87.

Bartels C, Kunert H-J, Stawicki S et al. (2007). Recovery of hippocampus-related functions in chronic alcoholics during monitored long-term abstinence. Alcohol Alcohol 42: 92–102.

Bartsch AJ, Homola G, Biller A et al. (2007). Manifestations of early brain recovery associated with abstinence from alcoholism. Brain 130: 36–47.

Bates ME, Bowden SC, Barry D (2002). Neurocognitive impairment associated with alcohol use disorders: implications for treatment. Exp Clin Psychopharmacol 10: 193–212.

Beatty WW, Katzung VM, Moreland VJ et al. (1995). Neuropsychological performance of recently abstinent alcoholics and cocaine abusers. Drug Alcohol Depend 37: 247–253.

Beatty WW, Hames KA, Blanco CR et al. (1996). Visuospatial perception, construction and memory in alcoholism. J Stud Alcohol 57: 136–143.

Beaunieux H, Desgranges B, Lalevée C et al. (1998). Preservation of cognitive procedural memory in a case of Korsakoff's syndrome: methodological and theoretical insights. Percept Mot Skills 86: 1267–1287.

Beaunieux H, Hubert V, Witkowski T et al. (2006). Which processes are involved in cognitive procedural learning? Memory 14: 521–539.

Beaunieux H, Pitel AL, Witkowski T et al. (2013). Dynamics of the cognitive procedural learning in alcoholics with Korsakoff's syndrome. Alcohol Clin Exp Res 37: 1025–1032. http://dx.doi.org/10.1111/acer.12054. Epub 2012 Dec 27.

Becker JT, Butters N, Hermann A et al. (1983). A comparison of the effects of long-term alcohol abuse and aging on the performance of verbal and nonverbal divided attention tasks. Alcohol Clin Exp Res 7: 213–219.

Brandt J, Butters N, Ryan C et al. (1983). Cognitive loss and recovery in long-term alcohol abusers. Arch Gen Psychiatry 40: 435–442.

Brokate B, Hildebrandt H, Eling P et al. (2003). Frontal lobe dysfunctions in Korsakoff's syndrome and chronic alcoholism: continuity or discontinuity? Neuropsychology 17: 420–428.

Butters N, Wolfe J, Martone M et al. (1985). Memory disorders associated with Huntington's disease: verbal recall, verbal recognition and procedural memory. Neuropsychologia 23: 729–743.

Caine D, Halliday GM, Kril JJ et al. (1997). Operational criteria for the classification of chronic alcoholics: identification of Wernicke's encephalopathy. J Neurol Neurosurg Psychiatry 62: 51–60.

Cardenas VA, Studholme C, Gazdzinski S et al. (2007). Deformation-based morphometry of brain changes in alcohol dependence and abstinence. Neuroimage 34: 879–887.

Cave CB, Squire LR (1992). Intact and long-lasting repetition priming in amnesia. J Exp Psychol 18: 509–520.

Cermak LS, Butters N (1972). The role of interference and encoding in the short-term memory deficits of Korsakoff patients. Neuropsychologia 10: 89–95.

Cermak L, Butters N (1973). Information processing deficits of alcoholic Korsakoff patients. Q J Stud Alcohol 34: 1110–1132.

Chanraud S, Martelli C, Delain F et al. (2007). Brain morphometry and cognitive performance in detoxified alcohol-dependents with preserved psychosocial functioning. Neuropsychopharmacology 32: 429–438.

Chanraud S, Leroy C, Martelli C et al. (2009). Episodic memory in detoxified alcoholics: contribution of grey matter microstructure alteration. PLoS One 4: e6786.

Chanraud S, Pitel A-L, Rohlfing T et al. (2010). Dual tasking and working memory in alcoholism: relation to frontocerebellar circuitry. Neuropsychopharmacology 35: 1868–1878.

Chanraud S, Pitel A-L, Pfefferbaum A et al. (2011a). Disruption of functional connectivity of the default-mode network in alcoholism. Cereb Cortex 21: 2272–2281.

Chanraud S, Pitel A-L, Sullivan EV (2011b). Structural imaging of substance abuse. In: ME Shenton, BI Turetsky (Eds.), Understanding Neuropsychiatric Disorders: Insights from Neuroimaging. Cambridge University Press, Cambridge, pp. 403–428.

Chanraud S, Pitel AL, Müller-Oehring EM et al. (2013). Remapping the brain to compensate for impairment in recovering alcoholics. Cereb Cortex 23: 97–104.

Charness N, Milberg W, Alexander MP (1988). Teaching an amnesic a complex cognitive skill. Brain Cogn 8: 253–272.

Christie JE, Kean DM, Douglas RH et al. (1988). Magnetic resonance imaging in pre-senile dementia of the Alzheimer-type, multi-infarct dementia and Korsakoff's syndrome. Psychol Med 18: 319–329.

Colchester A, Kingsley D, Lasserson D et al. (2001). Structural MRI volumetric analysis in patients with organic amnesia, 1: methods and comparative findings across diagnostic groups. J Neurol Neurosurg Psychiatry 71: 13–22.

Conway MA (2001). Sensory-perceptual episodic memory and its context: autobiographical memory. Philos Transact R Soc Lond Ser B 356: 1375–1384.

D'Argembeau A, Van Der Linden M, Verbanck P et al. (2006). Autobiographical memory in non-amnesic alcohol-dependent patients. Psychol Med 36: 1707–1715.

Desmond JE, Chen SHA, DeRosa E et al. (2003). Increased frontocerebellar activation in alcoholics during verbal working memory: an fMRI study. Neuroimage 19: 1510–1520.

Durazzo TC, Gazdzinski S, Meyerhoff DJ (2007). The neurobiological and neurocognitive consequences of chronic cigarette smoking in alcohol use disorders. Alcohol Alcohol 42: 174–185.

Durazzo TC, Fryer SL, Rothlind JC et al. (2010). Measures of learning, memory and processing speed accurately predict smoking status in short-term abstinent treatment-seeking alcohol-dependent individuals. Alcohol Alcohol 45: 507–513.

Emsley R, Smith R, Roberts M et al. (1996). Magnetic resonance imaging in Alcoholic Korsakoff's syndrome: evidence for an association with alcoholic dementia. Alcohol Alcohol 31: 479–486.

Eustache F, Desgranges B (2008). MNESIS: towards the integration of current multisystem models of memory. Neuropsychol Rev 18: 53–69.

Everett M, Schaeffer KW, Parsons OA (1988). Learning impairment in male and female alcoholics. Arch Clin Neuropsychol 3: 203–211.

Fahle M, Daum I (2002). Perceptual learning in amnesia. Neuropsychologia 40: 1167–1172.

Fama R, Pfefferbaum A, Sullivan EV (2004). Perceptual learning in detoxified alcoholic men: contributions from explicit memory, executive function, and age. Alcohol Clin Exp Res 28: 1657–1665.

Fama R, Pfefferbaum A, Sullivan EV (2006). Visuoperceptual learning in alcoholic Korsakoff syndrome. Alcohol Clin Exp Res 30: 680–687.

Fama R, Rosenbloom MJ, Sassoon SA et al. (2011). Remote semantic memory for public figures in HIV infection, alcoholism, and their comorbidity. Alcohol Clin Exp Res 35: 265–276.

Fama R, Pitel A-L, Sullivan EV (2012a). Anterograde episodic memory in Korsakoff syndrome. Neuropsychol Rev 22: 93–104.

Fama R, Rosenbloom MJ, Sassoon SA et al. (2012b). Differential effect of alcoholism and HIV infection on visuomotor procedural learning and retention. Alcohol Clin Exp Res 36: 1738–1747.

Fein G, Bachman L, Fisher S et al. (1990). Cognitive impairments in abstinent alcoholics. West J Med 152: 531–537.

Fein G, Torres J, Price LJ et al. (2006). Cognitive performance in long-term abstinent alcoholic individuals. Alcohol Clin Exp Res 30: 1538–1544.

Fellgiebel A (2003). Persistence of disturbed thalamic glucose metabolism in a case of Wernicke–Korsakoff syndrome. Psychiatry Res Neuroimaging 124: 105–112.

Finn PR, Rickert ME, Miller MA et al. (2009). Reduced cognitive ability in alcohol dependence: examining the role of covarying externalizing psychopathology. J Abnorm Psychol 118: 100–116.

Fox A, Coltheart M, Solowij N et al. (2000). Dissociable cognitive impairments in problem drinkers. Alcohol Alcohol 35: 52–54.

Fujiwara E, Brand M, Borsutzky S et al. (2008). Cognitive performance of detoxified alcoholic Korsakoff syndrome patients remains stable over two years. J Clin Exp Neuropsychol 30: 576–587.

Greicius MD, Krasnow B, Reiss AL et al. (2003). Functional connectivity in the resting brain: a network analysis of the default mode hypothesis. Proc Natl Acad Sci U S A 100: 253–258.

Harding A, Halliday G, Caine D et al. (2000). Degeneration of anterior thalamic nuclei differentiates alcoholics with amnesia. Brain 123: 141–154.

Harper CG, Giles M, Finlay-Jones R (1986). Clinical signs in the Wernicke-Korsakoff complex: a retrospective analysis of 131 cases diagnosed at necropsy. J Neurol Neurosurg Psychiatry 49: 341–345.

Harper C, Dixon G, Sheedy D et al. (2003). Neuropathological alterations in alcoholic brains. Studies arising from the New South Wales Tissue Resource Centre. Prog Neuropsychopharmacol Biol Psychiatry 27: 951–961.

Harris GJ, Jaffin SK, Hodge SM et al. (2008). Frontal white matter and cingulum diffusion tensor imaging deficits in alcoholism. Alcohol Clin Exp Res 32: 1001–1013.

Heffernan TM (2008). The impact of excessive alcohol use on prospective memory: a brief review. Curr Drug Abuse Rev 1: 36–41.

Heffernan TM, Moss M, Ling J (2002). Subjective ratings of prospective memory deficits in chronic heavy alcohol users. Alcohol Alcohol 37: 269–271.

Hildebrandt H, Brokate B, Eling P et al. (2004). Response shifting and inhibition, but not working memory, are impaired after long-term heavy alcohol consumption. Neuropsychology 18: 203–211.

Ihara H, Berrios GE, London M (2000). Group and case study of the dysexecutive syndrome in alcoholism without amnesia. J Neurol Neurosurg Psychiatry 68: 731–737.

Jacobson RR, Acker CF, Lishman WA (1990). Patterns of neuropsychological deficit in alcoholic Korsakoff's syndrome. Psychol Med 20: 321–334.

Jernigan TL, Schafer K, Butters N et al. (1991). Magnetic resonance imaging of alcoholic Korsakoff patients. Neuropsychopharmacology 4: 175–186.

Johnson-Greene D, Adams KM, Kluin KJ et al. (1997). Effects of abstinence and relapse upon neuropsychological function and cerebral glucose metabolism in severe chronic alcoholism. J Clin Exp Neuropsychol 19: 37–41.

Joyce EM, Robbins TW (1991). Frontal lobe function in Korsakoff and non-Korsakoff alcoholics: planning and spatial working memory. Neuropsychologia 29: 709–723.

Kessels RPC, Van Loon E, Wester AJ (2007). Route learning in amnesia: a comparison of trial-and-error and errorless learning in patients with the Korsakoff syndrome. Clin Rehabil 21: 905–911.

Kish GB, Hagen JM, Woody MM et al. (1980). Alcoholics' recovery from cerebral impairment as a function of duration of abstinence. J Clin Psychol 36: 584–589.

Kopera M, Wojnar M, Brower K et al. (2012). Cognitive functions in abstinent alcohol-dependent patients. Alcohol 46: 1–7.

Krabbendam L, Visser P, Derix M et al. (2000). Normal cognitive performance in patients with chronic alcoholism in contrast to patients with Korsakoff's syndrome. J Neuropsychiatry Clin Neurosci 12: 44–50.

Laloyaux J, Michel C, Mourad H et al. (2012). Performance on an everyday life activity in persons diagnosed with alcohol dependency compared to healthy controls: relations between a computerized shopping task and cognitive and clinical variables. Alcohol Alcohol 47: 240–247.

Le Berre AP, Pinon K, Vabret F et al. (2010). Study of metamemory in patients with chronic alcoholism using a feeling-of-knowing episodic memory task. Alcohol Clin Exp Res 34: 1888–1898.

Le Berre A, Rauchs G, La Joie R et al. (2012). Impaired decision-making and brain shrinkage in alcoholism. Eur Psychiatry 29: 125–133.

Liu I-C, Chiu C-H, Yang T-T (2010). The effects of gender and a co-occurring depressive disorder on neurocognitive functioning in patients with alcohol dependence. Alcohol Alcohol 45: 231–236.

Loeber S, Duka T, Welzel H et al. (2009). Impairment of cognitive abilities and decision making after chronic use of alcohol: the impact of multiple detoxifications. Alcohol Alcohol 44: 372–381.

Manning V, Wanigaratne S, Best D et al. (2008). Changes in neuropsychological functioning during alcohol detoxification. Eur Addict Res 14: 226–233.

Middleton FA, Strick PL (2000). Basal ganglia output and cognition: evidence from anatomical, behavioral, and clinical studies. Brain Cogn 42: 183–200.

Moss HB, Chen CM, Yi H (2007). Subtypes of alcohol dependence in a nationally representative sample. Drug Alcohol Depend 91: 149–158.

Munro CA, Saxton J, Butters MA (2000). The neuropsychological consequences of abstinence among older alcoholics: a cross-sectional study. Alcohol Clin Exp Res 24: 1510–1516.

Nissen MJ, Willingham D, Hartman M (1989). Explicit and implicit remembering: when is learning preserved in amnesia? Neuropsychologia 27: 341–352.

Nixon K (2006). Alcohol and adult neurogenesis: roles in neurodegeneration and recovery in chronic alcoholism. Hippocampus 16: 287–295.

Nixon SJ, Paul R, Phillips M (1998). Cognitive efficiency in alcoholics and polysubstance abusers. Alcohol Clin Exp Res 22: 1414–1420.

Noël X, Van Der Linden M, Schmidt N et al. (2001). Supervisory attentional system in nonamnesic alcoholic men. Arch Gen Psychiatry 58: 1152–1158.

Noël X, Bechara A, Dan B et al. (2007). Response inhibition deficit is involved in poor decision making under risk in nonamnesic individuals with alcoholism. Neuropsychology 21: 778–786.

Noël X, Brevers D, Bechara A et al. (2011). Neurocognitive determinants of novelty and sensation-seeking in individuals with alcoholism. Alcohol Alcohol 46: 407–415.

Noël X, Van der Linden M, Brevers D et al. (2012). The contribution of executive functions deficits to impaired episodic memory in individuals with alcoholism. Psychiatry Res 198: 116–122.

Norman D, Shallice T (1986). Attention to action: Willed and automatic control of behaviour. In: R Davidson, G Schwartz, D Shapiro (Eds.), Consciousness and Self-Regulation: Advances in Research and Theory. Vol. 4. Plenum, New York, pp. 1–18.

Oscar-Berman M (2012). Function and dysfunction of prefrontal brain circuitry in alcoholic Korsakoff's syndrome. Neuropsychol Rev 22: 154–169.

Oscar-Berman M, Kirkley SM, Gansler DA et al. (2004). Comparisons of Korsakoff and non-Korsakoff alcoholics on neuropsychological tests of prefrontal brain functioning. Alcohol Clin Exp Res 28: 667–675.

Pfefferbaum A, Desmond JE, Galloway C et al. (2001). Reorganization of frontal systems used by alcoholics for spatial working memory: an fMRI study. Neuroimage 14: 7–20.

Pfefferbaum A, Adalsteinsson E, Sullivan EV (2006). Dysmorphology and microstructural degradation of the corpus callosum: interaction of age and alcoholism. Neurobiol Aging 27: 994–1009.

Pfefferbaum A, Rosenbloom M, Rohlfing T et al. (2009). Degradation of association and projection white matter systems in alcoholism detected with quantitative fiber tracking. Biol Psychiatry 65: 680–690.

Pitel AL, Beaunieux H, Witkowski T et al. (2007a). Genuine episodic memory deficits and executive dysfunctions in alcoholic subjects early in abstinence. Alcohol Clin Exp Res 31: 1169–1178.

Pitel AL, Witkowski T, Vabret F et al. (2007b). Effect of episodic and working memory impairments on semantic and cognitive procedural learning at alcohol treatment entry. Alcohol Clin Exp Res 31: 238–248.

Pitel AL, Beaunieux H, Witkowski T et al. (2008). Episodic and working memory deficits in alcoholic Korsakoff patients: the continuity theory revisited. Alcohol Clin Exp Res 32: 1229–1241.

Pitel A-L, Aupée A-M, Chételat G et al. (2009a). Morphological and glucose metabolism abnormalities in alcoholic Korsakoff's syndrome: group comparisons and individual analyses. PloS one 4: e7748.

Pitel AL, Beaunieux H, Guillery-Girard B et al. (2009b). How do Korsakoff patients learn new concepts? Neuropsychologia 47: 879–886.

Pitel AL, Rivier J, Beaunieux H et al. (2009c). Changes in the episodic memory and executive functions of abstinent and relapsed alcoholics over a 6-month period. Alcohol Clin Exp Res 33: 490–498.

Pitel AL, Perruchet P, Vabret F et al. (2010). The advantage of errorless learning for the acquisition of new concepts' labels in alcoholics. Psychol Med 40: 497–502.

Pitel A-L, Zahr NM, Jackson K et al. (2011). Signs of preclinical Wernicke's encephalopathy and thiamine levels as predictors of neuropsychological deficits in alcoholism without Korsakoff's syndrome. Neuropsychopharmacology 36: 580–588.

Pitel AL, Chételat G, Le Berre AP et al. (2012). Macrostructural abnormalities in Korsakoff syndrome compared with uncomplicated alcoholism. Neurology 78: 1330–1333.

Raichle ME, MacLeod AM, Snyder AZ et al. (2001). A default mode of brain function. Proc Natl Acad Sci U S A 98: 676–682.

Reed RJ, Grant I, Rourke SB (1992). Long-term abstinent alcoholics have normal memory. Alcohol Clin Exp Res 16: 677–683.

Reed LJ, Lasserson D, Marsden P et al. (2003). FDG-PET findings in the Wernicke-Korsakoff syndrome. Cortex 39: 1027–1045.

Rosenbloom M, Sullivan EV, Pfefferbaum A (2003). Using magnetic resonance imaging and diffusion tensor imaging to assess brain damage in alcoholics. Alcohol Res Health 27: 146–152.

Rosenbloom MJ, Pfefferbaum A, Sullivan EV (2004). Recovery of short-term memory and psychomotor speed but not postural stability with long-term sobriety in alcoholic women. Neuropsychology 18: 589–597.

Rosenbloom MJ, O'Reilly A, Sassoon SA et al. (2005). Persistent cognitive deficits in community-treated alcoholic men and women volunteering for research: limited contribution from psychiatric comorbidity. J Studies Alcohol 66: 254–265.

Rosenbloom MJ, Rohlfing T, O'Reilly AW et al. (2007). Improvement in memory and static balance with abstinence in alcoholic men and women: selective relations with change in brain structure. Psychiatry Res 155: 91–102.

Rourke SB, Grant I (1999). The interactive effects of age and length of abstinence on the recovery of neuropsychological functioning in chronic male alcoholics: a 2-year follow-up study. J Int Neuropsychol Soc 5: 234–246.

Ryan C, Butters N, Montgomery K et al. (1980). Memory deficits in chronic alcoholics: continuities between the "intact" alcoholic and the alcoholic Korsakoff patient. Adv Exp Med Biol 126: 701–718.

Ryback R (1971). The continuum and specificity of the effects of alcohol on memory: a review. Q J Studies Alcohol 32: 995–1016.

Salmon DP, Butters N, Schuckit M (1986). Memory for temporal order and frequency of occurrence in detoxified alcoholics. Alcohol 3: 323–329.

Schaeffer K, Parsons O (1987). Learning impairment in alcoholics using an ecologically relevant test. J Nerv Ment Dis 175: 213–218.

Schmahmann JD, Pandya DN (1997). The cerebrocerebellar system. Int Rev Neurobiol 41: 31–60.

Schmidtke K, Handschu R, Vollmer H (1996). Cognitive procedural learning in amnesia. Brain Cogn 32: 441–467.

Schulte T, Müller-Oehring E, Pfefferbaum A et al. (2010). Neurocircuitry of emotion and cognition in alcoholism: contributions from white matter fiber tractography. Dialogues Clin Neurosci 12: 554–560.

Schwartz BL, Parker ES, Deutsch SI et al. (2002). Source monitoring in alcoholism. J Clin Exp Neuropsychol 24: 806–817.

Sheedy D, Lara A, Garrick T et al. (1999). Size of mamillary bodies in health and disease: useful measurements in neuroradiological diagnosis of Wernicke's encephalopathy. Alcohol Clin Exp Res 23: 1624–1628.

Sherer M, Nixon SJ, Parsons OA et al. (1992). Performance of alcoholic words test and brain-damaged subjects on the Luria memory. Arch Clin Neuropsychol 7: 499–504.

Smith S, Fein G (2010). Cognitive performance in treatment-naïve active alcoholics. Alcohol Clin Exp Res 34: 2097–2105.

Stavro K, Pelletier J, Potvin S (2012). Widespread and sustained cognitive deficits in alcoholism: a meta-analysis. Addict Biol 18: 203–213.

Sullivan EV (2003). Compromised pontocerebellar and cerebellothalamocortical systems: speculations on their contributions to cognitive and motor impairment in nonamnesic alcoholism. Alcohol Clin Exp Res 27: 1409–1419.

Sullivan EV, Pfefferbaum A (2003). Diffusion tensor imaging in normal aging and neuropsychiatric disorders. Eur J Radiol 45: 244–255.

Sullivan EV, Pfefferbaum A (2009). Neuroimaging of the Wernicke-Korsakoff syndrome. Alcohol Alcohol 44: 155–165.

Sullivan EV, Mathalon DH, Ha CN et al. (1992). The contribution of constructional accuracy and organizational strategy to nonverbal recall in schizophrenia and chronic alcoholism. Biol Psychiatry 32: 312–333.

Sullivan EV, Marsh L, Mathalon DH et al. (1995). Anterior hippocampal volume deficits in nonamnesic, aging chronic alcoholics. Alcohol Clin Exp Res 19: 110–122.

Sullivan EV, Shear PK, Zipursky RB et al. (1997). Patterns of content, contextual, and working memory impairments in schizophrenia and nonamnesic alcoholism. Neuropsychology 11: 195–206.

Sullivan EV, Lane B, Deshmukh A et al. (1999). In vivo mammillary body volume deficits in amnesic and nonamnesic alcoholics. Alcohol Clin Exp Res 23: 1629–1636.

Sullivan EV, Rosenbloom MJ, Pfefferbaum A (2000). Pattern of motor and cognitive deficits in detoxified alcoholic men. Alcohol Clin Exp Res 24: 611–621.

Sullivan EV, Deshmukh A, De Rosa E et al. (2005). Striatal and forebrain nuclei volumes: contribution to motor function and working memory deficits in alcoholism. Biol Psychiatry 57: 768–776.

Swinnen SP, Puttemans V, Lamote S (2005). Procedural memory in Korsakoff's disease under different movement feedback conditions. Behav Brain Res 159: 127–133.

Tapert SF, Ozyurt SS, Myers MG et al. (2004). Neurocognitive ability in adults coping with alcohol and drug relapse temptations. Am J Drug Alcohol Abuse 30: 445–460.

Thomson AD (2000). Mechanisms of vitamin deficiency in chronic alcohol misusers and the development of the Wernicke-Korsakoff syndrome. Alcohol Alcohol 35: 2–7.

Thomson AD, Jeyasingham MD, Pratt OE et al. (1987). Nutrition and alcoholic encephalopathies. Acta Med Scand Suppl 717: 55–65.

Tivis R, Beatty WW, Nixon J,S et al. (1995). Patterns of cognitive impairment among alcoholics: are there subtypes? Alcohol Clin Exp Res 19: 496–500.

Tulving E (1985). Memory and consciousness. Can Psychologist 26: 1–12.

Tulving E (2001). Episodic memory and common sense: how far apart? Philos Transact R Soc Lond Ser B 356: 1505–1515.

Tulving E, Schacter DL (1990). Priming and human memory systems phenomena of priming. Science 247: 301–306.

Tulving E, Schacter DL, McLachlan DR et al. (1988). Priming of semantic autobiographical knowledge: a case study of retrograde amnesia. Brain Cogn 8: 3–20.

Verfaellie M, Gabrieli JDE, Vaidya CJ et al. (1996). Implicit memory for pictures in amnesia: role of etiology and priming task. Neuropsychology 10: 517–528.

Warrington E, Weiskrantz L (1968). New method of testing long-term retention with special reference to amnesic patients. Nature 217: 972–974.

Weingartner HJ, Andreason PJ, Hommer DW et al. (1996). Monitoring the source of memory in detoxified alcoholics. Biol Psychiatry 40: 43–53.

Wheeler MA, Stuss DT, Tulving E (1997). Toward a theory of episodic memory: the frontal lobes and autonoetic consciousness. Psychol Bull 121: 331–354.

Xu Y, Corkin S (2001). H.M. revisits the Tower of Hanoi Puzzle. Neuropsychology 15: 69–79.

Yohman JR, Parsons OA, Leber WR (1985). Lack of recovery in male alcoholics' neuropsychological performance one year after treatment. Alcohol Clin Exp Res 9: 114–117.

Handbook of Clinical Neurology, Vol. 125 (3rd series)
Alcohol and the Nervous System
E.V. Sullivan and A. Pfefferbaum, Editors

Chapter 14

Decision making, risky behavior, and alcoholism

JAZMIN CAMCHONG, MICHAEL ENDRES, AND GEORGE FEIN*
Neurobehavioral Research Inc., Kihei, HI, USA

DECISION MAKING, RISKY BEHAVIOR, AND ALCOHOLISM

A hallmark symptom of alcoholism is the continual engagement in dangerous or risky drinking behavior despite recurring social or personal difficulties such as alienation from friends and family, trouble with the law and at work, or problems with physical and emotional well-being (American Psychiatric Association, 1994). It has been suggested that the alcoholic's propensity for such risky behavior is manifested by impulsive decision-making processes, wherein potential short-term appetitive outcomes of drinking (e.g., intoxication) are deemed more important than potential long-term aversive consequences of drinking (e.g., drunk-driving arrests). Dual-process models of alcoholism hold that separate but interrelated neurocognitive pathways to impulsive decision making exist (Volkow et al., 2013) – one reflected by weak "top-down" executive control over impulsive and compulsive urges to consume alcohol, the other reflected by a strong "bottom-up" appetitive drive in impulsive and compulsive urges to consume alcohol (Li et al., 2009). The current chapter surveys evidence that functional and organizational patterns in brain executive control and appetitive drive networks characterize impulsive decision making and risky behavior in alcoholics, and that these patterns differ at different stages of alcoholism dependence and recovery.

POOR EXECUTIVE CONTROL LEADS TO POOR DECISION MAKING

Impairments in executive cognitive ability have been consistently noted in alcoholics at various stages of alcohol dependence and recovery. Actively drinking alcoholics typically perform more poorly than non-substance-abusing controls in a variety of neurocognitive domains related to attention, memory, and contingency learning (Bechara and Martin, 2004; Finn and Hall, 2004; Finn et al., 2009; Bogg and Finn, 2010). Finn et al. (2009) showed that severity of alcoholism and other related externalizing problems predicted poor short-term and working memory capacity, with the greatest impairments exhibited by those with other co-occurring externalizing behavior problems. Alcoholics' poor cognitive capacity also has been linked with a broad constellation of personality traits related to impulsiveness, poor harm avoidance, and antisocial behavior tendencies (Bogg and Finn, 2010). In the same vein, laboratory evidence has suggested that poor executive cognitive capacity is partially responsible for impulsive decision making in alcoholics and related externalizing syndromes (Finn, 2002; Endres et al., 2011; Bogg et al., 2012). Recent evidence also shows that, when combined with highly impulsive trait dispositions, poor executive cognitive capacity predicts more heavy and dangerous drinking behavior in active alcoholics (Gunn and Finn, 2013). In summary, the evidence suggests that impaired or weak executive cognitive control processes are at least partially responsible for the impulsive decision-making tendencies and risky behaviors of actively drinking alcoholics.

Alcoholics who stop drinking, and remain sober, experience recovery in some, but not all, domains of neuropsychologic functioning. Studies have shown that, while neuropsychologic impairments in executive cognitive ability are present in recently detoxified and short-term abstinent alcoholics (STAA), such impairments may resolve with extended abstinence; nonetheless, the major recovery is observed in the first year of abstinence (Nixon and Glenn, 1995; Mann et al., 1999; Sullivan et al., 2000; Fein et al., 2006; Stavro et al., 2013). A recent meta-analysis by Stavro et al. (2013) found that widespread

*Correspondence to: George Fein, Ph.D., President and Senior Scientist, Neurobehavioral Research Inc., 1325 South Kihei Rd, Suite 216, Kihei, HI 96753, USA. Tel: +1-808-250-3725, E-mail: george@nbresearch.com

deficits in executive cognitive ability characterized recently detoxified and STAA; but such impairments were dramatically attenuated in long-term abstinent alcoholics (LTAA). However, Stavro et al. (2013) also noted a dearth of research on the persistence of inhibitory control/impulsive decision-making deficits in those with multiyear abstinence. A longitudinal study by Sullivan et al. (2000) found similar evidence of recovery, demonstrating alcoholics with sustained abstinence improved in short-term memory, visuospatial processing, attention, and gait and balance abilities. Likewise, Fein et al. (2006) showed that LTAA with an average of 6.7 years' abstinence did not differ from non-substance-abusing controls in numerous executive cognitive domains, including cognitive abstraction/flexibility, attention, short-term/working memory, psychomotor function/reaction time, and verbal skills. However, Fein et al. (2006) did find that LTAA had difficulty processing visuospatial information. Thus, evidence suggests that executive cognitive impairments in attention, memory, and motor control to a large extent normalize with extended abstinence. However, as reviewed below, evidence also suggests that cognitive impairments related to the processing of appetitive and/or emotional information remain present in multiyear abstinent alcoholics.

STRONG APPETITIVE DRIVE LEADS TO POOR DECISION MAKING

Psychometric evidence has suggested that impulsive decision making in alcoholism and related externalizing syndromes comprises three independent domains: myopia to the future, unwillingness to delay gratification, and difficulties with passive avoidance (Dom et al., 2007). Myopic decision making reflects insensitivity to the long-term aversive consequences of appetitive goal-directed behavior and has been assessed with simulated gambling tasks such as the Iowa Gambling Task (Bechara et al., 1994). Preference for immediate gratification reflects a tendency to choose smaller immediate appetitive outcomes over larger delayed appetitive outcomes and has been assessed with intertemporal choice paradigms such as the delay discounting task (DDT) (Rachlin et al., 1991). Poor passive avoidance refers to difficulty stopping or withholding proponent or appetitive behavior so as to avoid suboptimal or aversive outcomes and has been assessed with measures of inhibitory control, such as the stop-signal (SST) (Logan et al., 1997), go/no-go (GNG) (Newman et al., 1985), or balloon analog risk task (Lejuez et al., 2002).

Acute alcohol use has disinhibiting effects on behavior control, impairing rational decision making and fueling risky behavior such as gambling, getting a tattoo, or having unprotected sex (Tzambazis and Stough, 2000; Lin et al., 2005; Purdie et al., 2011). Separate studies have shown active alcoholics typically make more impulsive or disinhibited decisions than non-substance-abusing controls, with those presenting other co-occurring externalizing symptoms exhibiting the poorest performances on the Iowa Gambling Task, DDT, and GNG (Mazas et al., 2000; Finn et al., 2002; Cantrell et al., 2008; Bobova et al., 2009; Endres et al., 2011). Evidence also suggests that impaired processing of aversive feedback leads to impulsive decision making in alcoholism and related externalizing syndromes, and this information-processing impairment is associated with factors that predate the onset of clinical problems, such as a family history of alcoholism (Finn et al., 1994; Fein and Chang, 2008).

Impulsive decision-making tendencies also have been observed in abstinent alcoholics, and unlike impairments in executive cognitive ability, evidence has suggested that impulsive decision-making tendencies persist for weeks, months, and even years of abstinence. Research has shown abstinent alcoholics to be less willing to delay gratification in the DDT relative to non-substance-abusing controls, and this effect was still present after controlling for impulsive personality traits (Mitchell et al., 2005). Evidence also has suggested recently detoxified alcoholics make more myopic decisions in the Iowa Gambling Task than controls, with little to no improvement in this capacity at 1-month follow-up (Dom et al., 2007). Similar myopic decision-making tendencies have also been found in STAA and LTAA (Fein et al., 2010b), with short-term and long-term abstainers showing comparable impulsive decision making during the Iowa Gambling Task relative to non-substance abusing controls (Ando et al., 2012). LTAA also are known to have persistent difficulties with emotion perception and affective decision making (Philippot et al., 1999; Kornreich et al., 2001; Foisy et al., 2007; Fein et al., 2010b; Endres and Fein, 2012). Endres and Fein (2012) recently showed that LTAAs' persistent difficulties with passive avoidance in an affective GNG task were largely due to a history of externalizing problems and independent of more general difficulties in lexical decision making. However, research has suggested a dissociation between antisocial symptoms and traits in LTAA, suggesting that antisocial behaviors return to normal levels in the face of elevated antisocial and reward-seeking trait dispositions (Fein and Fein, 2012). In light of the fact that alcoholics can remain sober in spite of elevated levels of impulsive traits and decision-making tendencies, researchers have suggested that other compensatory mechanisms come into play, such as heightened executive control and conscientiousness aid in abstinence maintenance (Ando et al., 2012; Camchong et al., 2013a). When together, research has shown impulsive decision-making tendencies are present in alcoholics throughout the recovery process, but such impulsive personality and decision-making tendencies do not appear to militate against the ability to maintain abstinence.

It is still not completely clear whether poor decision making and risky behavior contribute to the vulnerability of becoming an alcoholic or whether they are a consequence of chronic alcohol use. There is evidence supporting the idea that poor decision making exists before alcohol dependence. A longitudinal study on young adults in college found that poor decision making (as assessed with the Iowa Gambling Task) at second year of college is associated with heavy drinking 2 years later (Goudriaan et al., 2009). Poor decision making (assessed with the Iowa Gambling and Stroop tasks) has also been identified in individuals at risk for alcoholism (with family history of alcoholism) when compared to healthy controls (Lovallo et al., 2006). Heavy-drinking adolescents have shown poor decision making and higher impulsivity than light-drinking adolescents (Field et al., 2007).

There is also evidence suggesting that chronic alcohol consumption during adolescence, a period in which decision-making brain regions are still developing, has a long-term effect on adult decision making. Nasrallah et al. (2009), using a probability-discounting instrumental response task, found that adult rats that had chronic alcohol vs. placebo use during adolescence had poorer decision making (did not properly weigh benefits and costs) and preferred more risky options (chose large but uncertain rewards over small but certain rewards). Further research needs to be conducted to better determine whether poor decision making is present before or is a consequence of alcoholism.

OTHER ASPECTS OF BEHAVIOR RELATED TO POOR DECISION MAKING IN ALCOHOLICS

Even though poor decision making has been identified in alcoholics and individuals at risk of becoming alcoholics, this behavioral deficit is not specific to alcoholism. Poor decision making in alcoholism has been associated with other aspects of behavior such as antisocial tendencies, stress, number of detoxifications, and onset of alcohol dependence.

An individual with antisocial personality disorder (ASPD) does not adhere to accepted societal norms and often engages in criminal and other impulsive behavior (American Psychiatric Association, 1994). ASPD is more prevalent in alcoholics than in healthy controls (Compton et al., 2005; Goldstein et al., 2007). Previous research has proposed that poor decision making in alcohol-dependent individuals is increased in those with a comorbid ASPD. A study by Miranda et al. (2009) compared risky decision making (Iowa Gambling Task performance) between men with alcohol dependence (abstinent >30 days) with and without ASPD comorbidity compared with non-substance-abusing control men. Both alcoholic groups made significantly more risky decisions (manifested as poorer Iowa Gambling Task performance) than the control group. However, while alcoholics without ASPD comorbidity showed a slow and steady improvement in advantageous decision making during the task, alcoholics with ASPD comorbidity showed an initial improvement but were unable to sustain it. Moreover, authors reported evidence that impulsive behavioral traits (e.g., propensity for rash impulsive behavior) mediated disadvantageous decision making in alcoholics with ASPD comorbidity. These findings suggest that poor decision making in alcoholics is closely related to impulsive antisocial tendencies. Because there is evidence that alcoholics' antisocial behavior resolves with extended periods of abstinence, while antisocial disposition does not (Fein and Fein, 2012), findings of poor decision making in LTAA might be related to a persistent disposition to appetitive drive tendencies.

There is evidence that decision making improves when stress is induced in actively heavy drinkers. A study that compared decision making in heavy and light drinkers reported that heavy drinkers made significantly more disadvantageous decisions than light drinkers (with more attention to gains and less attention to losses). When anticipatory stress was induced (by asking them to give a speech), decision making was improved in both heavy and light drinkers (by increasing attention to losses) (Gullo and Stieger, 2011). Evidence from this study suggests that stress is needed to better focus attention on negative consequences and hence better decision making. It should be noted, however, that the effect of stress on decision making may be different in individuals that have been diagnosed with alcohol dependence or substance dependence (Zhang et al., 2011). In addition, the level of stress induced by requesting participants to give a speech may not be severe, chronic, or realistic enough to negatively affect drinkers' decision making.

Another aspect that has been found to affect decision making in individuals with alcohol dependence is the number of detoxifications an individual has been through. A study that compared a group with fewer than two detoxifications (low-detox group) versus a group with more than two detoxifications (high-detox group) reported that the low-detox group had a significantly greater rate of improvement in choices throughout the Iowa Gambling Task than the high-detox group

(Loeber et al., 2009). This evidence suggests that individuals with repeated withdrawal episodes lack the ability to make proper decisions and to adapt behavior to newly learned rules. More research needs to be conducted to find out whether poor decision making leads to repeated relapse episodes or whether the negative effects of repeated withdrawals affect decision making.

Higher levels of impulsive decision making have been associated with early-onset alcoholism (Dom et al., 2006). Early-onset alcoholics have shown higher levels of impulsive decision making during a delayed discounting task than late-onset alcoholics and controls (Dom et al., 2006). This claim, however, needs to be further examined considering other factors related to poor decision making in early-onset alcoholism, such as fewer years of education, potentially higher genetic load, or longer lifetime alcohol use.

NEURAL CORRELATES OF DECISION MAKING AND RISKY BEHAVIOR IN ALCOHOLISM

Behavioral aspects of alcoholism such as poor decision making and risky behavior have been associated with corresponding brain functional differences from non-substance-abusing individuals. The executive control (principally mediated by cortical regions, particularly in frontal cortex) and appetitive drive (principally mediated by striatal regions) brain networks have prominent roles in decision making and risky behavior in alcoholism. During its early stages, alcohol consumption is a goal-directed behavior, initiated and executed by regions within the executive control network (such as dorsolateral prefrontal cortex (DLPFC) and anterior cingulate cortex) with its rewarding effects processed by appetitive drive regions (such as nucleus accumbens). After alcohol is repeatedly consumed, evidence for or consistent with poor decision making, consumption becomes more automatic (with more involvement of appetitive drive regions such as caudate and putamen) and less voluntary (with less involvement of executive control regions) (Everitt and Robbins, 2005). Alcohol consumption shifts to a more habitual mode, particularly as a response to avoid withdrawal symptoms. An individual with alcohol dependence seeks alcohol compulsively, a behavior that has been associated with increased activity of appetitive drive regions when presented with an alcohol cue, and a lack of engagement of prefrontal regions when required to stop a prepotent, maladaptive behavior such as alcohol consumption. The quality of interaction between and within these functional networks (frontal executive and striatal appetitive networks) is becoming an increasingly relevant topic in alcoholism and addiction in general (see . 14.1 for a review of involved networks).

To examine functional brain activity and synchrony associated with poor decision making in alcoholics, participants are required to perform tasks while functional magnetic resonance imaging data are collected. Numerous neuroimaging studies have used specific tasks to tap into functional brain differences (in frontal executive and striatal appetitive networks, shown in Fig. 14.1) associated with poor decision making and risky behavior during alcohol use, abuse, and dependence.

BRAIN FUNCTION ASSOCIATED WITH DECISION MAKING IN BINGE DRINKING

A study by Xiao et al. (2012) compared brain functional activity of 14 binge-drinking adolescents while they made decisions during the Iowa Gambling Task compared with 14 adolescents who had never consumed alcohol. Binge drinkers had both behavioral and brain functional differences manifested as inflexible poor decision making (kept making disadvantageous decision) together with higher activity than non-drinkers in limbic brain regions known to play a role in the emotional aspects of decision making (amygdala and insula). Because they also found that adolescents that had higher limbic activity while performing the decision-making task had more self-reported drinking problems, Xiao et al. (2012) proposed that a hyperactive bottom-up affective network may be a good candidate for a biologic marker that can be used to identify individuals at risk of alcohol dependence. A previous behavioral study used the original Iowa Gambling Task and a variant of the Iowa Gambling Task to dissect whether poor decision making found in binge drinkers is attributable to insensitivity to long-term consequences (lack of executive control, planning) or to hypersensitivity to reward (enhanced appetitive drive) (Johnson et al., 2008). After comparing Iowa Gambling Task performance between adolescent binge drinkers to adolescent never-drinkers, Johnson et al. (2008) found that poor decision making in binge drinkers is associated with hypersensitivity to reward. Results from these studies provide evidence that at early stages of alcohol abuse there is enhanced behavior ruled by appetitive drive together with a hyperactive bottom-up network.

BRAIN FUNCTION ASSOCIATED WITH DECISION MAKING IN ACTIVE DRINKERS WITH ALCOHOL DEPENDENCE

The neural correlates of decision making have been examined in active drinkers with a diagnosis of alcohol dependence (Amlung et al., 2013). Amlung et al. (2013) compared brain activity during delayed reward discounting between active drinkers with an alcohol dependence

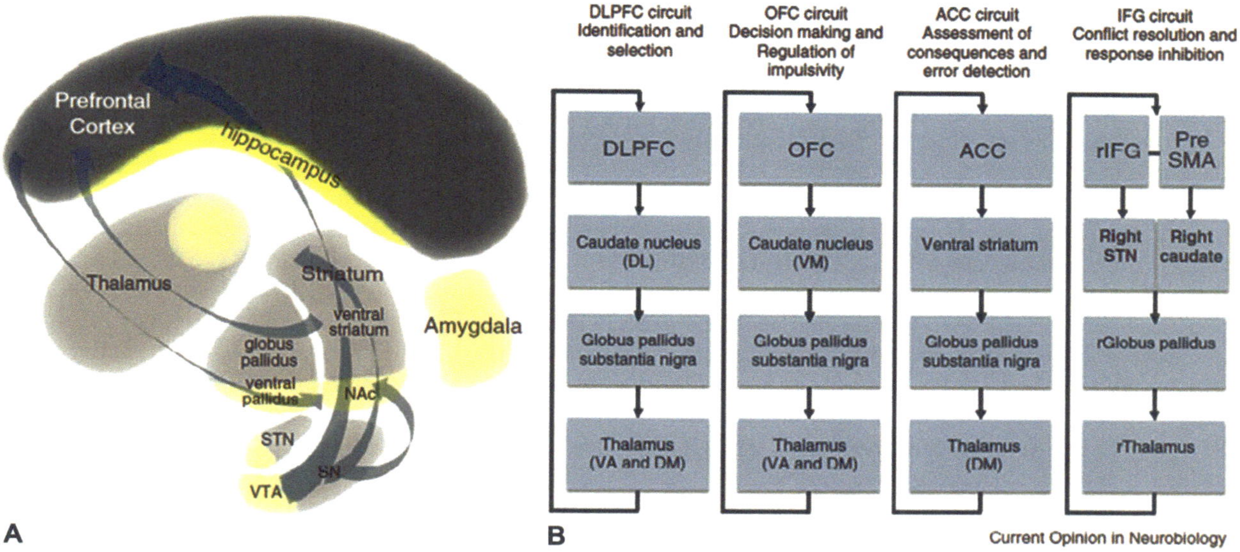

Fig. 14.1. Frontal and striatal regions involved in decision making and stimulus–response habits. (**A**) Schematic anatomic representation of the mesocorticolimbic dopamine system in the human brain, highlighting several key processing stations: ventral tegmental area (VTA) and substantia nigra (SN), nucleus accumbens (NAc) in the ventral striatum, thalamus, and subthalamic nuclei (STN), and prefrontal cortex (PFC), among others. (**B**) Four of the frontostriatal cortical circuits that appear to play major roles in executive function and inhibitory control. DL, dorsolateral; DM, dorsomedial; VA, ventroanterior; VM, ventromedial; r, right; IFG, inferior frontal gyrus; preSMA, presomatic motor area; STN, subthalamic nucleus; ACC, anterior cingulate cortex; OFC, orbitofrontal cortex. (Adapted from Volkow et al., 2013.)

diagnosis (AUD+; $n = 13$) versus active drinkers without an alcohol dependence diagnosis (AUD–; $n = 12$). AUD+ showed more impulsive decision making and higher activity in executive control (DLPFC) and attention (precuneus) regions than AUD– during delayed-reward decisions. Although the study has a small sample size, was limited to men, but did compare drinkers with a control (non-drinking) group, this study provides important evidence of greater brain activity in executive control regions when alcoholics are required to delay gratification, a finding that may reflect alcoholics' increased demand of executive control when required to make decisions on behavior ruled by appetitive drive. These neural differences need to be explored further in larger samples that include a comparison to control groups.

BRAIN FUNCTION ASSOCIATED WITH DECISION MAKING IN SHORT-TERM ABSTINENT ALCOHOLICS

Although it is important to understand the neural network differences in active drinkers, findings from these studies may be affected by the acute effects of alcohol and caution needs to be taken when comparing such results with those in abstinent treated samples. Active drinking samples may comprise a different population from treated samples – they may not have an alcohol use disorder or have less severe alcohol use disorders than treatment samples (Fein and Landman, 2005; Di Sclafani et al., 2008; Fein et al., 2010a). The neural substrates of decision making have been extensively investigated in STAA.

Park et al. (2010) compared brain activity during reward-guided decision making between 20 STAA with at least 7 days of abstinence (mean 16.9 days of abstinence) and 16 healthy controls. Subjects performed a task that required them to make a choice between two abstract stimuli to be allocated a probabilistically predetermined reward. This task measured learning rates and outcome because subjects needed to update the reward values associated with the stimuli to guide their future responses. While STAA did not differ from controls in striatal activity associated with predicting errors during task performance, they did show significantly lower frontostriatal synchrony (i.e., ventral striatum and right DLPFC) when presented with feedback of wins versus losses. Park et al. (2010) suggest that, while STAA do not have impaired brain activity associated with the representation of prediction error in the ventral striatum (nucleus accumbens), they do not seem to relay this information to brain regions that mediate higher executive control processes (DLPFC) and are not able to integrate reward-related information for proper control of behavior following a reward (Park et al., 2010). While this study was limited to only men and to the examination of functional synchrony of the ventral striatum, it provides valuable evidence pointing to differences in functional synchrony between brain regions that mediate appetitive drive and regions that guide decision making in STAA versus healthy controls. The lack of interaction between these executive control and appetitive drive regions may underlie alcoholics' difficulty in guiding future reward-related choices and goal-directed behavior. Furthermore, results from this study suggest that these neural network differences are evident after short periods of abstinence.

A study by Li et al. (2009) investigated differences in brain activity specifically related to impulsive control and risk taking between 24 STAA and 24 healthy controls using the SST. By dissecting the component processes of SST performance (response inhibition, error processing, posterror slowing, risk taking) they could examine neural differences specific to discrete aspects of impulse control and risky behavior. The authors reported that, while groups did not differ in overall SST performance, STAA compared to healthy controls showed significantly: (1) lower left DLPFC activity when inhibiting a response; (2) lower right DLPFC activity during posterror slowing (behavioral adjustment after an error); (3) lower amygdala, striatal, and posterior cingulate cortex activity during posttrial speeding (representing a risk-taking decision); and (4) higher activity in visual cortex and anterior cingulate cortex (ACC) when failing to stop a response. Results from this study showed that, even though overall performance did not differ between groups, STAA showed differences in brain activity during individual trials that reflect specific instances of poor decision making, such as posttrial speeding and failure to stop a response.

Evidence from studies mentioned above suggest that poor decision making and risky behavior in STAA are associated with brain functional differences in regions within the executive control (lower activity) and appetitive drive (higher activity) networks known to be involved in cognitive and emotional control. Whether these brain functional differences predispose to alcohol problems or are a consequence of unhealthy alcohol use is not clear.

BRAIN FUNCTION ASSOCIATED WITH DECISION MAKING IN INDIVIDUALS AT RISK FOR ALCOHOLISM

Neuroimaging studies on adolescents at risk for alcoholism have identified brain functional differences associated with poor decision making in such samples. A study by Acheson et al. (2009) compared brain function between 15 adolescents with a family history of alcoholism (FHP) and 19 with a negative family history of alcoholism (FHN) when performing the Iowa Gambling Task. While groups did not differ in behavioral performance, the FHP group showed more activity in left dorsal ACC and left caudate nucleus than the FHN group. Because activity in dorsal ACC has been associated with risk prediction and activity in caudate nucleus with risky responses, Acheson et al. (2009) suggest that FHP need to recruit additional brain regions for proper decision making involving risk taking when compared to FHN. A study by Cservenka and Nagel (2012) provided a different perspective on predisposing biologic markers of alcoholism. They investigated whether there were differences in functional activity associated with risk taking when FHP versus FHN adolescents performed the "Wheel of Fortune" decision-making task. While there were no behavioral differences in risk taking, FHP showed significantly lower activity in right DLPFC and right cerebellar tonsil than FHN in the risky versus safe decision-making contrast. The authors suggest that attenuated DLPFC response to risk taking in FHP may be associated with poorer cognitive control and hence poorer behavioral regulation in FHP, making them vulnerable to failure in the decision to avoid risk taking and ultimately contributing to out-of-control alcohol consumption.

A study by Norman et al. (2011) investigated the neural correlates of another aspect that contributes to risky behavior in adolescents, the lack of response inhibition. They examined brain function in adolescents while they performed a GNG task that measured response inhibition and response selection. Based on follow-up interviews assessing alcohol use information (after a mean of 4.2 years), the sample was classified as either heavy alcohol users ($n=21$) or healthy controls ($n=17$). While there were no behavioral differences in impulse control during the GNG task at baseline, adolescents that later became heavy alcohol users showed significantly lower activity during no-go trials than healthy controls in prefrontal regions (inferior frontal gyrus, dorsolateral and medial frontal gyri), striatum (putamen), and inferior parietal lobules. The authors suggest that, when required to exert control over a prepotent response, individuals that are at risk of becoming alcoholics show less responsive frontal regions even before the onset of drinking problems. Although these findings are important, they still do not provide decisive evidence for a predetermined predisposition because these may indicate that future problem drinkers have either a pre-existing disadvantage (maybe genetic) due to poor engagement of frontal regions needed to properly exert executive control on impulsive and risky behaviors or a delayed cortical maturity, in which activation in the executive control network is still not fully specialized, but never reaches its full potential because its development is later stunted by alcohol use. More studies need to be conducted to fully address these issues.

BRAIN FUNCTION ASSOCIATED WITH DECISION MAKING IN LONG-TERM ABSTINENT ALCOHOLICS

To investigate whether brain functional differences identified in actively drinking or recently abstinent alcoholics are permanent or can be overcome with long periods of abstinence, it is essential to examine LTAA. The examination of executive and appetitive drive networks during rest in LTAA has yielded results consistent with an ongoing compensatory mechanism in LTAA. A study by Camchong et al. (2013a), that compared resting-state synchrony of executive and appetitive drive networks between 23 LTAA (with more than 18 months of abstinence from alcohol use) and 23 healthy controls found that LTAA have lower synchrony of appetitive drive regions and greater synchrony of executive control regions than healthy controls. Moreover, these resting-state synchrony differences were positively correlated with performance in a task that measured cognitive flexibility in decision making (e.g., higher resting-state synchrony of executive control regions significantly correlated with better performance). Results from this study reflect adaptive mechanisms in brain ongoing functional organization that may support proper executive control of behavior in LTAA and successfully stop behavior that may lead to relapse. A follow-up study by Camchong et al. (2013b) that studied this effect in STAA (6–15 weeks of abstinence) found ordered effects from STAA to LTAA within both the executive control and appetitive drive networks: higher resting-state synchrony of the executive control network (Fig. 14.2) and lower resting-state synchrony of the reward-processing network (Fig. 14.3) than healthy controls. These results provided further evidence that this compensatory mechanism follows an adaptive progression from short- to long-term abstinence. Longitudinal studies of brain functional organization need to be conducted to explore the hypothesis suggesting that synchrony within the executive control network progressively increases and synchrony within the appetitive drive network progressively decreases with length of abstinence.

Differences in brain function and organization in prefrontal executive control regions are not specific to decision making and risky behavior in alcoholics. Alcoholics have shown similar differences, as described above, in neural activity in executive control regions (DLPFC and anterior cingulate cortex) during tasks that hone in on other aspects of executive function, such as spatial working memory (Pfefferbaum et al., 2001) or verbal working memory (Desmond et al., 2003; Cservenka and Nagel, 2012). Moreover, brain activity and functional organization differences associated with executive control differences have been identified in other samples with addiction to other substances such as cocaine and nicotine (Gu et al., 2010; Janes et al., 2010; Camchong et al., 2011).

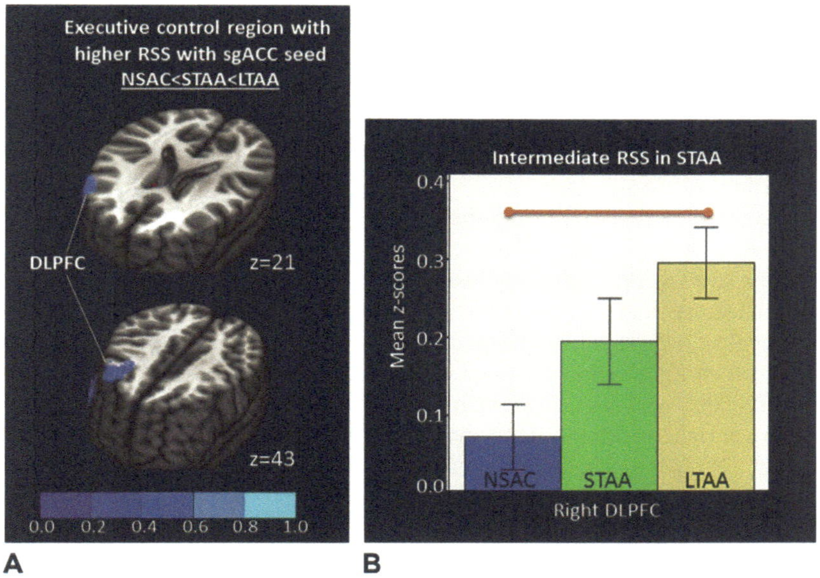

Fig. 14.2. (**A**) Three-dimensional MNI brain in neurologic orientation with slices cut at $z = 21$ and 43 with region (in blue) that showed a linear trend (NSAC < STAA < LTAA) in strength of resting-state synchrony (RSS) in the executive control network. (**B**) Bar graph illustrating significant linear trend between groups in strength of RSS between subgenual anterior cingulate (sgACC) and right dorsolateral prefrontal cortex (DLPFC) in STAA (green bar), LTAA (beige bar), and NSAC (blue bar). Red lines represent significant post hoc differences between groups. MNI, Montreal Neurological Institute; LTAA, long-term abstinent alcoholics; STAA, short-term abstinent alcoholics; NSAC, non-substance-abusing controls.

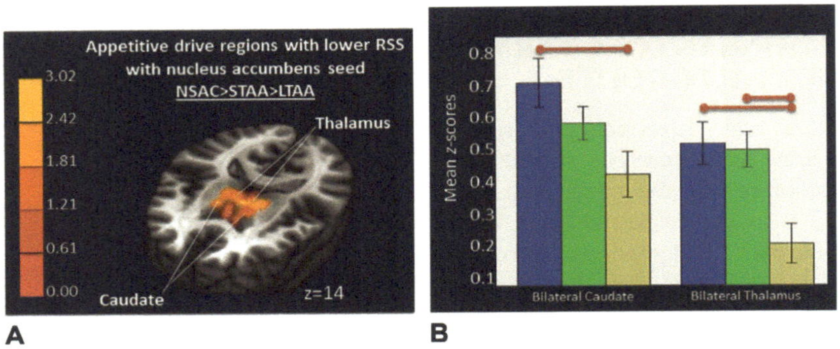

Fig. 14.3. (**A**) Three-dimensional MNI brain in neurologic orientation with slice cut at $z = 14$ with region (in orange/red) that showed a linear trend (NSAC > STAA > LTAA) in strength of resting-state synchrony (RSS) in the appetitive drive network. (**B**) Bar graph illustrating significant linear trends between groups in strength of RSS between nucleus accumbens (NAcc) and appetitive drive regions in STAA (green bars), LTAA (beige bars), and NSAC (blue bars). Red lines represent significant post hoc differences between groups. MNI, Montreal Neurological Institute; LTAA, long-term abstinent alcoholics; STAA, short-term abstinent alcoholics; NSAC, non-substance-abusing controls.

Although neuroimaging studies have identified brain function and organization differences in executive and appetitive drive networks in alcoholics, these differences vary depending on whether the alcoholic is actively drinking or has been abstinent for short or extended periods of time. A number of studies reviewed in the sections above do not show performance differences between alcoholic and comparison groups, but instead find differences between groups in brain activation patterns associated with comparable performance. Such findings are often interpreted as indicating lack of efficiency in alcoholics vs controls of specific brain regions in performing some function. However, it is important to remember that such a statement is a hypothesis, not a finding. If an impairment in some function is present (less efficient processing by a brain region is an impairment), there should be experimental challenges that will result in impaired performance on some task.

A recent paper by Chanraud et al. (2013) found that, while controls and recovering alcoholics had similar performance levels during a spatial working memory task, they recruited different brain networks. While controls recruited prefrontal-cerebellar regions known to mediate working memory, recovering alcoholics recruited two additional frontocerebellar networks, presumably to compensate and achieve normal working memory performance (Chanraud et al., 2013). It is also important to emphasize that the phenomena discussed above are not necessarily reflected in all aspects of behavior. For example, Fein et al. (2004) found impaired performance on the Iowa Gambling Task in LTAA – they were able to achieve multiyear abstinence despite the persistent decision-making impairments on the Iowa Gambling Task. In contrast, Fein et al. (2006) found normal Iowa Gambling Task performance in actively drinking treatment-naïve alcoholics. These individuals drank harmfully (a poor decision) despite having normal Iowa Gambling Task performance. These examples point out that propensities and inherited vulnerabilities do not fully control behavior. The research reviewed above suggests some of the adaptive changes that may take place in the development of alcoholism and in the struggle to achieve long-term abstinence.

References

Acheson A, Robinson JL, Glahn DC et al. (2009). Differential activation of the anterior cingulate cortex and caudate nucleus during a gambling simulation in persons with a family history of alcoholism: studies from the Oklahoma Family Health Patterns Project. Drug Alcohol Depend 100: 17–23.

American Psychiatric Association (1994). Diagnostic and Statistical Manual of Mental Disorders (DSM-IV). 4th edn American Psychiatric Association, Washington, DC.

Amlung M, Sweet LH, Acker J et al. (2013). Dissociable brain signatures of choice conflict and immediate reward preferences in alcohol use disorders. Addict Biol. http://dx.doi.org/10.1111/adb.12017.

Ando B, Must A, Kurgyis E et al. (2012). Personality traits and coping compensate for disadvantageous decision-making in long-term alcohol abstinence. Alcohol Alcohol 47: 18–24.

Bechara A, Martin EM (2004). Impaired decision making related to working memory deficits in individuals with substance addictions. Neuropsychology 18: 152–162.

Bechara A, Damasio AR, Damasio H et al. (1994). Insensitivity to future consequences following damage to human prefrontal cortex. Cognition 50: 7–15.

Bobova L, Finn PR, Rickert ME et al. (2009). Disinhibitory psychopathology and delay discounting in alcohol dependence: personality and cognitive correlates. Exp Clin Psychopharmacol 17: 51–61.

Bogg T, Finn PR (2010). A self-regulatory model of behavioral disinhibition in late adolescence: integrating personality traits, externalizing psychopathology, and cognitive capacity. J Pers 78: 441–470.

Bogg T, Fukunaga R, Finn P et al. (2012). Cognitive control links alcohol use, trait disinhibition, and reduce cognitive capacity: evidence for medial prefrontal cortex dysregulation during reward-seeking behavior. Drug Alcohol Depend 122: 112–118.

Camchong J, Macdonald 3rd AW, Nelson B et al. (2011). Frontal hyperconnectivity related to discounting and reversal learning in cocaine subjects. Biol Psychiatry 69: 1117–1123.

Camchong J, Stenger A, Fein G (2013a). Resting-state synchrony in long-term abstinent alcoholics. Alcohol Clin Exp Res 37: 75–85.

Camchong J, Stenger VA, Fein G (2013b). Resting-state synchrony in short-term versus long-term abstinent alcoholics. Alcohol Clin Exp Res 37: 794–803.

Cantrell H, Finn PR, Rickert ME et al. (2008). Decision making in alcohol dependence: insensitivity to future consequences and comborbid disinhibitory psychopatology. Alcohol Clin Exp Res 32: 1398–1407.

Chanraud S, Pitel AL, Muller-Oehring EM et al. (2013). Remapping the brain to compensate for impairment in recovering alcoholics. Cereb Cortex 23: 97–104.

Compton WM, Conway KP, Stinson FS et al. (2005). Prevalence, correlates, and comorbidity of DSM-IV antisocial personality syndromes and alcohol and specific drug use disorders in the United States: results from the national epidemiologic survey on alcohol and related conditions. J Clin Psychiatry 66: 677–685.

Cservenka A, Nagel BJ (2012). Risky decision-making: an FMRI study of youth at high risk for alcoholism. Alcohol Clin Exp Res 36: 604–615.

Desmond JE, Chen SH, Derosa E et al. (2003). Increased frontocerebellar activation in alcoholics during verbal working memory: an fMRI study. Neuroimage 19: 1510–1520.

Di Sclafani V, Finn P, Fein G (2008). Treatment-naive active alcoholics have greater psychiatric comorbidity than normal controls but less than treated abstinent alcoholics. Drug Alcohol Depend 98: 115–122.

Dom G, de Wilde B, Hulstijn W et al. (2006). Decision-making deficits in alcohol-dependent patients with and without comorbid personality disorder. Alcohol Clin Exp Res 30: 1670–1677.

Dom G, de Wilde B, Hulstijn W et al. (2007). Dimensions of impulsive behaviour in abstinent alcoholics. Pers Indiv Differ 42: 465–476.

Endres MJ, Fein G (2012). Emotion-word processing difficulties in abstinent alcoholics with and without lifetime externalizing disorders. Alcohol Clin Exp Res: 1–8.

Endres MJ, Rickert ME, Bogg T et al. (2011). Externalizing psychopathology and behavioral disinhibition: working memory mediates signal discriminability and reinforcement moderates response bias in approach-avoidance learning. J Abnorm Psychol 120: 336–351.

Everitt BJ, Robbins TW (2005). Neural systems of reinforcement for drug addiction: from actions to habits to compulsion. Nat Neurosci 8: 1481–1489.

Fein G, Chang M (2008). Smaller feedback ERN amplitudes during the BART are associated with a greater family history density of alcohol problems in treatment-naive alcoholics. Drug Alcohol Depend 92: 141–148.

Fein G, Fein D (2012). Antisocial symptoms decrease to normal levels in long-term abstinence. Alcohol Clin Exp Res 37: E271–E280.

Fein G, Landman B (2005). Treated and treatment-naive alcoholics come from different populations. Alcohol Clin Exp Res 35: 19–26.

Fein G, Klein L, Finn P (2004). Impairment on a simulated gambling task in long-term abstinent alcoholics. Alcohol Clin Exp Res 28: 1487–1491.

Fein G, Torres J, Price LJ et al. (2006). Cognitive performance in long-term abstinent alcoholic individuals. Alcohol Clin Exp Res 30: 1538–1544.

Fein G, Di Sclafani V, Finn P (2010a). Sensation seeking in long-term abstinent alcoholics, treatment-naive active alcoholics, and nonalcoholic controls. Alcohol Clin Exp Res 34: 1045–1051.

Fein G, Key K, Szymanski MD (2010b). ERP and RT delays in long-term abstinent alcoholics in processing of emotional facial expressions during gender and emotion categorization tasks. Alcohol Clin Exp Res 34: 1127–1139.

Field M, Christiansen P, Cole J et al. (2007). Delay discounting and the alcohol Stroop in heavy drinking adolescents. Alcohol Clin Exp Res 102: 579–586.

Finn PR (2002). Motivation, working memory, and decision making: a cognitive-motivational theory of personality vulnerability to alcoholism. Behav Cogn Neurosci Rev 1: 183–205.

Finn PR, Hall J (2004). Cognitive ability and risk for alcoholism: short-term memory capacity and intelligence moderate personality risk for alcohol problems. J Abnorm Psychol 113: 569–581.

Finn PR, Kessler DN, Hussong AM (1994). Risk for alcoholism and classical conditioning to signals for punishment: evidence for a weak behavioral inhibition system? J Abnorm Psychol 103: 293–301.

Finn PR, Mazas CA, Justus AN et al. (2002). Early-onset alcoholism with conduct disorder: go/no go learning deficits, working memory capacity, and personality. Alcohol Clin Exp Res 26: 186–206.

Finn PR, Rickert ME, Miller MA et al. (2009). Reduced cognitive ability in alcohol dependence: examining the role of covarying externalizing psychopathology. J Abnorm Psychol 118: 100–116.

Foisy ML, Kornreich C, Fobe A et al. (2007). Impaired emotional facial expression recognition in alcohol dependence: do these deficits persist with midterm abstinence? Alcohol Clin Exp Res 31: 404–410.

Goldstein RB, Compton WM, Pulay AJ et al. (2007). Antisocial behavioral syndromes and DSM-IV drug use disorders in the United States: results from the National Epidemiologic Survey on Alcohol and Related Conditions. Drug Alcohol Depend 90: 145–158.

Goudriaan AE, Slutske WS, Krull JL et al. (2009). Longitudinal patterns of gambling activities and associated risk factors in college students. Alcohol Clin Exp Res 104: 1219–1232.

Gu H, Salmeron BJ, Ross TJ et al. (2010). Mesocorticolimbic circuits are impaired in chronic cocaine users as demonstrated by resting-state functional connectivity. Neuroimage 53: 593–601.

Gullo MJ, Stieger AA (2011). Anticipatory stress restores decision-making deficits in heavy drinkers by increasing sensitivity to losses. Drug Alcohol Depend 117: 204–210.

Gunn RL, Finn PR (2013). Impulsivity partially mediates the association between working memory and alcohol problems. Alcohol 47: 5.

Janes AC, Pizzagalli DA, Richardt S et al. (2010). Brain reactivity to smoking cues prior to smoking cessation predicts ability to maintain tobacco abstinence. Biol Psychiatry 67: 722–729.

Johnson CA, Xiao L, Palmer P et al. (2008). Affective decision-making deficits, linked to a dysfunctional ventromedial prefrontal cortex, revealed in 10th grade Chinese adolescent binge drinkers. Neuropsychologia 46: 714–726.

Kornreich C, Blairy S, Philippot P et al. (2001). Deficits in recognition of emotional facial expression are still present in alcoholics after mid- to long-term abstinence. J Stud Alcohol 62: 533–542.

Lejuez CW, Read JP, Kahler CW et al. (2002). Evaluation of a behavioral measure of risk taking: the Balloon Analogue Risk Task (BART). J Exp Psychol Appl 8: 75–84.

Li CS, Luo X, Yan P et al. (2009). Altered impulse control in alcohol dependence: neural measures of stop signal performance. Alcohol Clin Exp Res 33: 740–750.

Lin D, Li X, Yang H et al. (2005). Alcohol intoxication and sexual risk behaviors among rural-tourban. Drug Alcohol Depend 79: 103–112.

Loeber S, Duka T, Welzel H et al. (2009). Impairment of cognitive abilities and decision making after chronic use of alcohol: the impact of multiple detoxifications. Alcohol Alcohol 44: 372–381.

Logan G, Schachar R, Tannock R (1997). Impulsivity and inhibitory control. Psychol Sci 3: 60–64.

Lovallo WR, Yechiam E, Sorocco KH et al. (2006). Working memory and decision-making biases in young adults with a family history of alcoholism: studies from the Oklahoma family health patterns project. Alcohol Clin Exp Res 30: 763–773.

Mann K, Gunther A, Stetter F et al. (1999). Rapid recovery from cognitive deficits in abstinent alcoholics: a controlled test-retest study. Alcohol Clin Exp Res 34: 567–574.

Mazas CA, Finn PR, Steinmetz JE (2000). Decision-making biases, antisocial personality, and early-onset alcoholism. Alcohol Clin Exp Res 24: 1036–1040.

Miranda Jr R, Mackillop J, Meyerson LA et al. (2009). Influence of antisocial and psychopathic traits on decision-making biases in alcoholics. Alcohol Clin Exp Res 33: 817–825.

Mitchell JM, Fields HL, D'Esposito M et al. (2005). Impulsive responding in alcoholics. Alcohol Clin Exp Res 29: 2158–2169.

Nasrallah NA, Yang TW, Bernstein IL (2009). Long-term risk preference and suboptimal decision making following adolescent alcohol use. Proc Natl Acad Sci U S A 106: 17600–17604.

Newman J, Widom C, Nathan S (1985). Passive avoidance in syndromes of disinhibition: psychopathy and extraversion. J Pers Soc Psychol 48: 1316–1327.

Nixon SJ, Glenn SW (1995). Cognitive psychosocial performance and recovery in female alcoholics. Plenum Press, New York.

Norman AL, Pulido C, Squeglia LM et al. (2011). Neural activation during inhibition predicts initiation of substance use in adolescence. Drug Alcohol Depend 119: 216–223.

Park SQ, Kahnt T, Beck A et al. (2010). Prefrontal cortex fails to learn from reward prediction errors in alcohol dependence. J Neurosci 30: 7749–7753.

Pfefferbaum A, Desmond JE et al. (2001). Reorganization of frontal systems used by alcoholics for spatial working memory: an fMRI study. Neuroimage 14: 7–20.

Philippot P, Kornreich C, Blairy S et al. (1999). Alcoholics' deficits in the decoding of emotional facial expression. Alcohol Clin Exp Res 23: 1031–1038.

Purdie MP, Norris J, Davis KC et al. (2011). The effects of acute alcohol intoxication, partner risk level, and general intention to have unprotected sex on women's sexual decision making with a new partner. Exp Clin Psychopharmacol 19: 378–388.

Rachlin H, Raineri A, Cross D (1991). Subjective probability and delay. J Exp Anal Behav 55: 233–244.

Stavro K, Pelletier J, Potvin S (2013). Widespread and sustained cognitive deficits in alcoholism: a meta-analysis. Addict Biol 18: 203–213.

Sullivan EV, Rosenbloom MJ, Lim KO et al. (2000). Longitudinal changes in cognition, gait, and balance in abstinent and relapsed alcoholic men: relationships to changes in brain structure. Neuropsychology 14: 178–188.

Tzambazis K, Stough C (2000). Alcohol impairs speed of information processing and simple and choice reaction time and differentially impairs higher-order cognitive abilities. Alcohol Alcohol 35: 197–201.

Volkow ND, Wang G-J, Tomasi D et al. (2013). Unbalanced neuronal circuits in addiction. Curr Opin Neurobiol 23: 1–10.

Xiao L, Becgara A, Gong Q et al. (2012). Abnormal affective decision making revealed in adolescent binge drinkers using functional magnetic imaging study. Psychol Addict Behav 27: 443.

Zhang X, Shi J, Zhao L et al. (2011). Effects of stress on decision-making deficits in formerly heroin-dependent patients after different durations of abstinence. Psychiatry 168: 610–616.

Handbook of Clinical Neurology, Vol. 125 (3rd series)
Alcohol and the Nervous System
E.V. Sullivan and A. Pfefferbaum, Editors

Chapter 15

Motor systems and postural instability

RACHEL L. VASSAR[1] AND JESSICA ROSE[2]*
[1]*Stanford University School of Medicine, Stanford, CA, USA*
[2]*Department of Orthopaedic Surgery, Stanford University School of Medicine, Stanford, CA, USA*

INTRODUCTION

Motor systems related to balance, posture, and gait demonstrate temporary and long-term effects from alcohol consumption. Acute alcohol intoxication is associated with postural instability. Chronic alcohol consumption results in damage to the motor and sensory systems related to posture and stability that persists after sobriety is achieved. Alcohol dependence has been found to increase the risk of falling and associated head traumas and orthopedic fractures (Kool et al., 2009). Excessive sway and postural tremor occur in many individuals with a history of alcoholism and are correlated with atrophy of the anterior superior vermis of the cerebellum. Gender, age, and duration of alcoholism may all contribute to the long-term prognosis of motor impairments from alcohol dependence. Understanding the underlying physiologic basis of alcohol-related motor impairments and postural instability may allow for improved identification and treatment.

POSTURAL CONTROL IN HEALTHY INDIVIDUALS

The integration of signals from the visual, vestibular, and somatosensory systems is required for maintaining balance, posture, and precise motor control. Normal postural control relies on these multiple inputs to the central nervous system (CNS) to monitor limb placement and make any necessary postural adjustments.

Fast, automatic signals from the CNS govern most postural adjustments, but posture can also be voluntarily changed and can be influenced by long-term practice and environmental factors. Integration of the sensorimotor and musculoskeletal systems is required for correct orientation and adjustment of posture. Static equilibrium maintains a desired position and dynamic equilibrium is the process of maintaining an upright posture during a controlled movement. In general, for a mass to remain in static equilibrium, the center of mass (CoM) must lie within the base of support, where as in dynamic equilibrium, the CoM constantly changes and is often outside the base of support. Postural control is considered to be a dynamic process, recruiting multiple sensory inputs specific to the task, environment, and desired behavior (Horak and Macpherson, 2011).

Maintaining posture requires not only the stabilization of the CoM but also the coordination of body parts in relation to each other and to the environment. The positioning of the trunk is particularly critical to the placement of the limbs and neck. Head placement can also have a significant impact on overall postural equilibrium as it is the primary receiver and interpreter of sensory inputs. Stability of the head and, by extension, the eyes can allow for simplification of visual and vestibular inputs. Together the stability of the trunk, head, and limbs determines the overall posture of the individual (Horak and Macpherson, 2011).

In a standing human, the support base area is relatively small and the CoM is high, resulting in relative instability. A variety of small muscular forces are needed to constantly regulate the stability of the body and prevent it from toppling. In reality the human body is rarely static and must estimate and predict movements to retain stability of posture. The combination of optimal coordination of muscle results in movement and requires constant adjustment and monitoring (Horak and Macpherson, 2011).

Identification of the vertical axis is critical to the correct positioning of the trunk. The vestibular system receives otolith signals that encode the orientation of the head with respect to gravity. These signals allow

*Correspondence to: Jessica Rose, Ph.D., Associate Professor, Department of Orthopaedic Surgery, Stanford University School of Medicine, 770 Welch Road, Room 423, Stanford, CA 94304, USA. Tel: +1-650-380-3249, E-mail: jessica.rose@stanford.edu

for comparison of the trunk's placement and the gravity vector, and subsequent adjustments can be made. The positioning of the head in relation to gravity also influences perception of vertical, and stabilization of the head (and eyes) is useful for the maintenance of postural dynamic equilibrium, particularly during complex motor activities (Kandel et al., 2000; Horak and Macpherson, 2011).

The upper and lower extremities also contribute to identification of the body's orientation in space and help make adjustments to posture. Limbs receive somatosensory information and can provide a frame of reference to which the trunk can adjust. The limbs also are often involved in disturbances in balance and must transmit signals about the nature of the perturbation to the CNS. Postural adjustments must then be coordinated and often involve changes in the placement of the extremities to alter the CoM to retain equilibrium. In a normal human, sudden destabilizing forces result in fast electromyogram responses (70–100 ms). The response required for stabilization depends on the direction and amplitude of the force as well as the original position. Adaptation is the term used to describe the response mediated by the CNS to correct balance after a perturbation. Healthy, sober individuals have an innate ability to quickly adjust limb and torso placement to minimize the impact of a destabilizing movement. Adaptation is critical to posture correction but signaling from the CNS to perform this response is slowed by alcohol intoxication (Kandel et al., 2000).

Multiple brain regions participate in the identification and coordination of sensory inputs and musculoskeletal outputs. The human employs receptor organs to collect sensory information, and a weighting of the inputs is required to optimize the resulting movements. Humans use the somatosensory system (including mechanoreceptors on the skin), the vestibular system (in the semicircular canals), and the visual system to receive sensory cues. These signals are integrated in the CNS by a variety of brain regions and neural connections. The brainstem, and in particular the medulla and pons, integrates vestibular, somatosensory, and visual information from the peripheral nervous system. Vestibulospinal and reticulospinal pathways extend from the brainstem to motor neurons related to neck and upper torso movement and rotation, as well as interneurons that project to muscles of the extremities (Horak and Macpherson, 2011).

The basal ganglia play a critical role in the maintenance of posture, as individuals with lesions in this region demonstrate significant difficulties with postural stability. The basal ganglia have complex involvement in integration of signals from a variety of sources, including the brainstem and other subcortical nuclei. Connections between the basal ganglia and sensorimotor and supplementary motor cortices allow for coordination of centrally initiated adjustments of posture before and during voluntary motor movements. Based on studies in patients with Parkinson's disease, the basal ganglia likely help to modify posture based on environmental factors such as surface conditions (i.e., incline) or stabilizing aides. Although the basal ganglia participate in a variety of functions related to sensorimotor integration required for postural stability, the region is likely not central to the initiation of adaptive responses (Horak and Macpherson, 2011).

The cerebellum has a vital role in postural control and helps with sensory integration. Damage to the cerebellum can result in a variety of profound difficulties with stability and locomotion. Most significant are lesions within the vermis and anterior lobe of the cerebellum, a region separated from the cerebellar hemispheres by the primary fissure. The anterior lobe is the destination for a variety of somatosensory inputs and has projections to the spinal cord. Severe ataxia, gait difficulties, and anterior–posterior trunk oscillations result from damage to the anterior lobe of the cerebellum, indicating the particular involvement of this region of the cerebellum in posture and motor control. Also significant is the delayed response to automatic postural readjustments in individuals with cerebellar lesions. The cerebellum is believed to be involved in initiation of the response to postural perturbation (Horak and Macpherson, 2011).

Involvement of cortical regions in the direct control of posture is debated. The sensorimotor cortex is primarily involved in voluntary movement but may also participate in anticipation of postural changes required to maintain dynamic equilibrium during planned movements. Cortical tracts have not been found to participate in postural response to unexpected perturbations due to the speed with which these adjustments are made, but fine-tuned responses with longer signal latencies likely employ neural tracts reaching the cortex. For refinements in posture the cortex participates in integration of sensory clues and anticipates necessary postural changes. Studies of patients with posterior frontal lobe (secondary motor cortex) lesions support this theory, that the cortex participates in planning and refinement of movements associated with balance (Gurfinkel and Élner, 1988).

PRENATAL EXPOSURE TO ALCOHOL

Exposure to alcohol in the prenatal environment has significant detrimental effects on brain development and health outcomes. Motor studies on infants with prenatal alcohol exposure (PAE) and fetal alcohol syndrome (FAS) demonstrate impairments in fine and gross motor performance starting as early as the first days of life and persisting through childhood and even into adulthood (Kalberg et al., 2006). In addition, neuroimaging studies have found regional brain volume differences related to PAE, including decreased total brain volume, frontal lobe, hippocampus, cerebellar vermis, caudate, putamen, and corpus callosum. These physiologic changes likely contribute to fetal alcohol-associated functional deficits (Astley et al., 2009).

A large number of studies have demonstrated the lasting motor deficits induced by exposure to alcohol *in utero*. Simmons et al. (2010) studied children (average age 13 years) with FAS and PAE and age-equivalent controls, and found slowed reaction time to stimuli when complex movement patterns were required. Children with diagnosed FAS had substantially longer reaction times compared to both the PAE and control groups, suggesting that high levels of alcohol *in utero* have much more serious consequences on future motor function than low alcohol exposure. Other studies on infants with fetal alcohol exposure have found impairment in isometric force regulation (Simmons et al., 2012) and motor task function, particularly on fine motor tasks (Connor et al., 2006; Bay and Kesmodel, 2011). Roebuck et al. (1998) analyzed balance in subjects (average age 12 years) with PAE and found that PAE subjects relied more heavily on somatosensory cues. When visual or somatosensory external cues were manipulated, the PAE children had more difficulty than the controls with regulation of balance (Roebuck et al., 1998).

Connor et al. (2006) performed a 25-year prospective longitudinal study on individuals exposed *in utero* to varying amounts of alcohol. This study found that only individuals exposed to high fetal alcohol levels with resulting neuropsychologic deficits in childhood had continued motor impairments as adults. Though the mean motor performance between PAE subjects and comparison subjects failed to reach significance, three-quarters of PAE subjects had motor task scores below the "normal" range for the comparison subjects (Connor et al., 2006).

Fetal exposure to alcohol has negative effects on motor function, coordination, and postural stability. Infants with FAS, exposed to high levels of alcohol *in utero*, generally experience the most significant impairments that persist into adulthood. Serious motor-related disabilities, such as cerebral palsy, are known to have a higher incidence in infants exposed to high levels of alcohol in the pre/perinatal period (O'Leary et al., 2012). The dose-dependent response to fetal alcohol exposure requires further study (Bay and Kesmodel, 2011), but some evidence suggests that even small doses of alcohol may have negative effects on brain development in regions that modulate cognitive and motor functions.

ACUTE ALCOHOL INTOXICATION: NEUROLOGIC AND BEHAVIORAL CHANGES IN POSTURAL CONTROL

The long-term damage to postural control is most notable when observed in chronic alcoholics, but short-term alcohol intoxication results in acute and temporary deficits in the control of balance and posture. Acute alcohol intoxication provides a window into the process of integration of multiple sensory inputs required for maintaining upright stance. Common signs of alcohol intoxication include spinocerebellar and vestibulocerebellar ataxia, oculomotor changes, and increased reliance on visuospatial clues. As a result, alcohol induces delays in motor and sensory processing and functioning, and causes deficits in postural control, particularly when faced with balance perturbations. Furthermore, alcohol intoxication delays the ability to detect postural changes and enact the appropriate response (Modig et al., 2012a).

Chemical effects of ethanol on postural control

The CNS is composed of 25–50% GABAergic neurons (Ribak and Yan, 2000). Ethanol interacts selectively with the $GABA_A$ receptor and changes the permeability of nerve cell membranes to the chloride ion, reducing the rate and amplitude of nerve impulses transmitted across the synapse. Thus, ethanol is an indirect GABA agonist, causing neural signals to travel more slowly and take longer to trigger a muscular response. The active monitoring and adjustment of limbs and the torso, involved in adaptation to perturbations, become impaired and delayed. The effect on the CNS is dose-dependent, with higher blood alcohol concentration (BAC) levels corresponding to more severe inhibition of nerve impulse transmission. This relationship appears to be non-linear, however, as higher blood alcohol levels result in substantially higher inhibition of voltage-gated calcium channels and slower synaptic and neuromuscular junction signaling (Hanchar et al., 2005; Sullivan et al., 2009).

Motor neurons within the spinal cord have decreased excitability after exposure to ethanol, as demonstrated by von Dincklage et al. (2007). Measurement of excitability assessed the reflex response of the soleus muscle, which can be achieved by stimulating the femoral nerve. Von Dincklage et al. attributed the decrease in motor neuron excitability to an increase in the presynaptic inhibition of group Ia afferent fibers, mediated by increased $GABA_A$ binding resulting in increased chloride ion conductance.

Ikeda et al. (1980) demonstrated a higher level of sensitivity to alcohol within neurons within the vestibular nucleus of the cat brain, compared to within the trigeminal and medial geniculate nuclei. Though this has not been duplicated in humans, the selective vulnerability of cells and neural tracts within motor regions of the brain (the cerebellum and pons, for example) has been proposed and needs further investigation (Ikeda et al., 1980; von Dincklage et al., 2007).

Instability during acute intoxication

Alcohol intoxication and its effects on motor systems can be measured in laboratory settings using tasks that assess balance and risk of falling when an individual is subjected to challenging stability tasks. Vibrations of the calf muscle in particular have been used as a method for disrupting balance, as well as force platforms, which jolt participants in various directions and test subjects' ability to learn how to readjust to compensate for direction-specific perturbations (Modig et al., 2012a).

Modig et al. (2012a, b) investigated the role of alcohol in the vestibular and oculomotor control of posture. Subjects were given varying amounts of alcohol to create three subject populations with BAC of 0.0%, 0.06%, or 0.10%. Results demonstrated that alcohol intoxication decreased stability both during quiet stance and during balance perturbation tasks (Modig et al., 2012b).

The Modig study also found that subjects at 0.10% BAC had greater instability in the lateral direction compared to the anteroposterior direction, likely due to substantial slowing of nerve impulse transmission to muscles controlling lateral movements. The study also suggests that alcohol intoxication results in a lowered ability to assess movement and posture along multiple axes and results in unidirectional attention to anteroposterior movement, perhaps because those movements are generally larger in amplitude. Subjects with both 0.06% and 0.10% BAC were significantly impaired in their ability to adapt to repeated perturbations, particularly in the lateral direction. Long-term performance on motor tasks was also impaired in alcohol-intoxicated subjects, as their destabilization increased over the course of the study (Fig. 15.1B). This effect, however, could also be the result of decreased attention over time, since attentional resources are critical to performance on longer-duration motor tasks and postural perturbations (Modig et al., 2012a, b).

Hafstrom et al. (2007) also found that acute alcohol intoxication impairs the ability to interpret gravitational vestibular clues for assessment of vertical. Visual assessment of one's environment is critical to the ability to remain upright, and alcohol has been demonstrated to impair perception of horizontal and vertical directionality. Determination of the subjective visual horizontal and vertical requires integration and interpretation of information from the vestibular, visual, and somatosensory receptor organs. The vestibular system (and specifically the otolithic organs, the utriculus and sacculus) allows

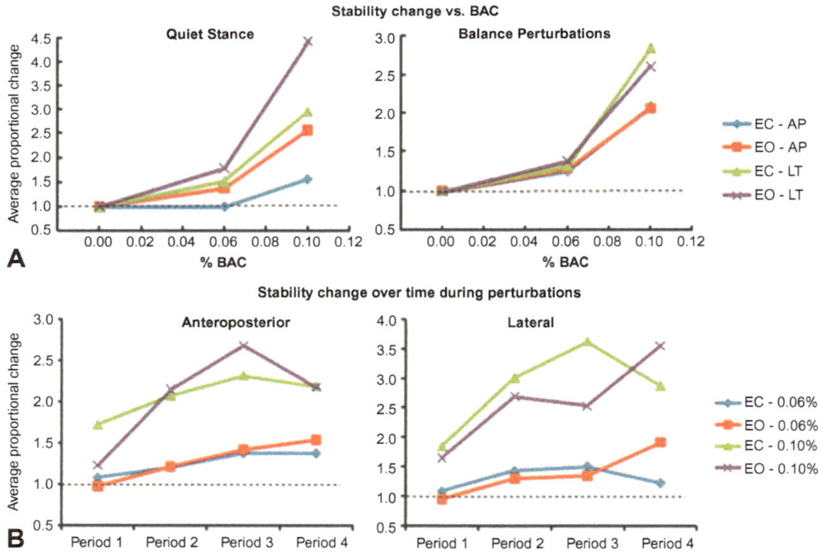

Fig. 15.1. (**A**) The mean proportional changes of total torque variance caused by alcohol intoxication at different blood alcohol concentration (BAC) relative to 0.0% BAC during quiet stance and during balance perturbations. (**B**) Graphical illustration of proportional change of total torque variance at different BAC relative to 0.0% BAC during each of the four 50-second periods of balance perturbations. The recorded changes in anteroposterior (AP) and lateral (LT) directions during tests with eyes closed (EC) and eyes open (EO) are presented. The dotted value of 1 on the *y*-axis represents the baseline value while sober (0.0% BAC) and a value of 2 on the *y*-axis represents that the recorded torque variance has doubled compared with the baseline value. $n=25$ (13 female, 12 male). (Adapted from Modig et al., 2012a.)

for identification of gravity's directionality. Though evidence on the direct effect of ethanol on the utriculus and sacculus is limited, Hafstrom et al. (2007) identified otolith functional deficits resulting from alcohol consumption.

When under the influence of alcohol, the visual system does not effectively compensate for the impaired vestibular system. Alcohol intoxication changes eye movement patterns, including decreased postrotatory nystagmus and impaired saccadic and tracking movements. Thus visual perception changes related to alcohol begin with differential gaze patterns and continue with altered integration of visual signals. Hafstrom et al. (2007) found alcohol dose-dependent increases in variation of the perception of horizontal and vertical beams of light, suggesting impairment in the vestibular otolith function. A higher percentage of errors was made when bars were rotated in the counterclockwise direction, compared to in the clockwise direction. This observation was interpreted to be related to handedness of the subjects (all but one were right-handed) and may be related to hemispheric differences in energy inputs. This study also found an increase in reliance on visual cues, rather than the proprioceptive and vestibular inputs, in subjects with higher BAC levels (Hafstrom et al., 2007).

The cerebellum in acute alcohol intoxication

Alcohol intoxication has been demonstrated to impair nerve signal pathways within the cerebellum, particularly in granule cells (Hanchar et al., 2005). Studies of cognitive deficits during acute alcohol intoxication suggest that alcohol may impair short-term learning, particularly in the cerebellum, thus reducing the ability to adapt to repeated motor challenges. But the cerebellum is not the only brain region involved in maintaining posture. Lowered attention to posture during acute alcohol intoxication may also decrease the ability to adjust and correct balance (Modig et al., 2012a).

CHRONIC ALCOHOLISM: LONG-TERM MOTOR AND NEUROLOGIC EFFECTS

Long-term consumption of alcohol results in damage to motor pathways, cerebellar tracts, and the peripheral nervous system, causing deficits in the reception, integration, and interpretation of sensory cues required for postural control. Ataxia, gait abnormalities, excessive sway, and physiologic tremor have all been well documented as significant motor symptoms associated with chronic alcoholism, even in individuals without complications such as from Wernicke–Korsakoff syndrome. These consequences can persist even in abstinent patients with a history of chronic alcohol dependence, and the magnitude and duration of the neurologic and behavioral effects are debated but functional deficits appear to extend well into sobriety (Sullivan et al., 2005).

Motor symptoms

Motor impairment caused by decreased ability to integrate sensory inputs and coordinate muscular outputs results in a substantial increase in risk of falls and orthopedic fractures in chronic alcoholic patients, even after long periods of sobriety. Kool et al. (2009) assessed unintentional fall prevalence among young and middle-aged adults related to alcohol exposure and found an increase in falls after moderate alcohol consumption, and a possible dose-dependent response. Hartshorne et al. (1997) found that nearly 50% of fatal falls occurring at ground level involve alcohol intoxication. Furthermore, chronic alcoholics experience substantially more falls than controls, as one study found that 60% of alcoholics reported a fall in the previous year compared to 30% of controls (Rosenbloom et al., 2005). The combination of alcohol-related postural instability and natural decline of balance with aging causes particularly high risk of falls in elderly individuals with a history of alcoholism.

Significant motor deficits have been documented in patients with chronic alcoholism compared to control subjects. For example, alcoholic men performed an average of 1.2 SD worse than controls on balance tasks such as the heel-to-toe walk or a one-foot stand (Sullivan et al., 2000). This deficit appears to be even more significant than cognitive deficits caused by alcoholism, suggesting that motor skills are selectively vulnerable to impairment. Balance tasks appear most difficult for subjects with chronic alcoholism, as opposed to upper-limb tasks, in which alcoholic subjects performed similarly to controls. Furthermore, patients without obvious sway or ataxia may still demonstrate significant motor deficits, as measured by one-foot stands or heel-to-toe walking, when compared to control subjects. These findings suggest that chronic alcoholism results in damage to mechanisms required for postural stability.

Excessive sway and postural tremor

Excessive sway during quiet stance has been documented in chronic as well as recovering alcoholics and is characterized by its anterior–posterior directionality. Postural tremor accompanying this characteristic sway is usually observed in subjects with chronic alcoholism and can be quantified using posturography studies and balance platform testing. Tremor of the trunk in alcoholic men is characterized by a frequency of approximately 2–5 Hz (Diener et al., 1983). Subsequent studies have also observed tremors occurring in alcoholic women, with a frequency of 5–7 Hz. Sullivan et al. (2010b) proposed that this higher-frequency tremor may be indicative of damage within the thalamus and deficits in cerebellar-thalamic relays. This finding was corroborated by evidence from a monkey model of Parkinson's disease, in which lesions in thalamic nuclei resulted in a 5–7-Hz postural tremor that was reduced, but not eliminated, with input from visual inputs (Guehl et al., 2003).

Sullivan et al. performed a biomechanic, quantitative analysis of excessive sway behaviors in alcoholics and healthy control subjects, assisted by a force platform to record center of pressure in subjects. Alcoholic subjects had been sober for an average of 75 days but continued to demonstrate significant impairments in their musculoskeletal coordination. Alcoholic men and women had longer sway paths during quiet stance without balancing aids, as demonstrated in Figure 15.2. Sway in both alcoholic men and women occurs predominantly in the anterior–posterior direction, rather than in the medial–lateral direction (Sullivan et al., 2005, 2010b).

On average, the alcoholic subjects demonstrated significantly longer sway pathways and were less stable than controls when attempting to maintain quiet stance on the force platform. Assistance from visual and tactile cues improved balance in all groups (Fig. 15.2B, D).

The sway pattern and length demonstrated by alcoholic subjects were reduced to near-normal levels with assistance from stabilizing aids (Fig. 15.3). These findings suggest that the alcoholic subjects have intact sensorimotor integration processes, demonstrating selective vulnerability. Postural tremor was also reduced in alcoholic subjects when stabilizing cues were provided (Sullivan et al., 2010b).

Visual and tactile cues

The use of external cues to assist with balance has been demonstrated as an effective strategy for diminishing sway. Analyzing the differential effect of various

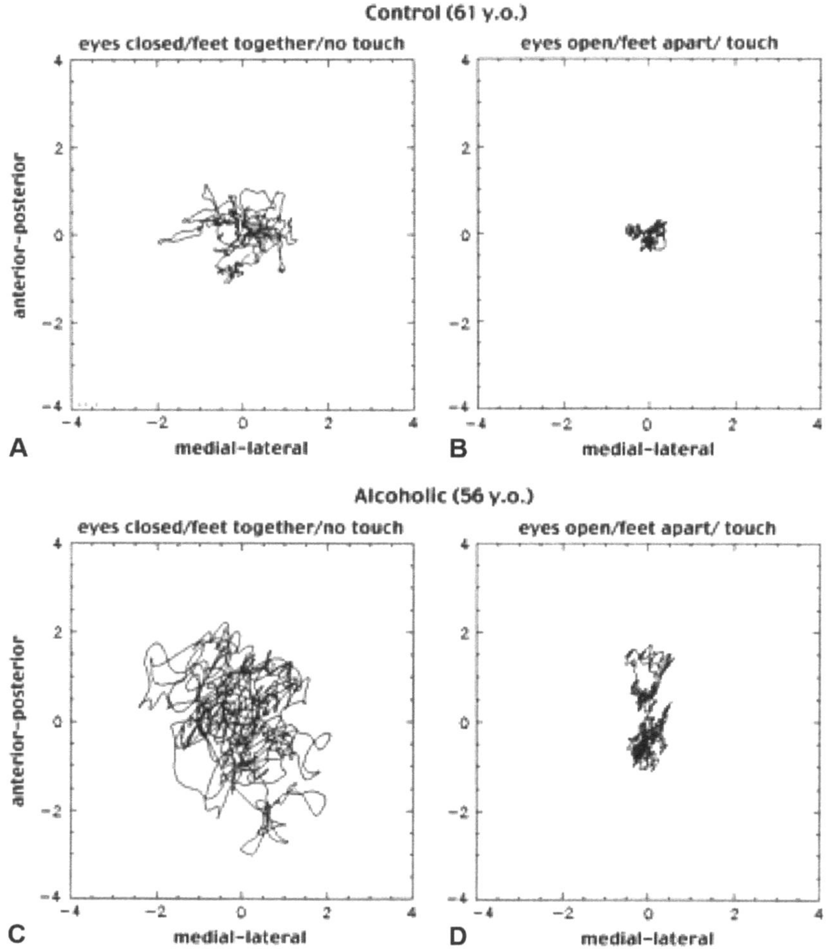

Fig. 15.2. Sample data from a healthy control subject (**A** and **B**) and alcoholic man (**C** and **D**) demonstrating sway patterns in the anterior–posterior and medial–lateral directions while standing on a force platform, with and without sensorimotor cues. (Reproduced from Sullivan et al., 2005.)

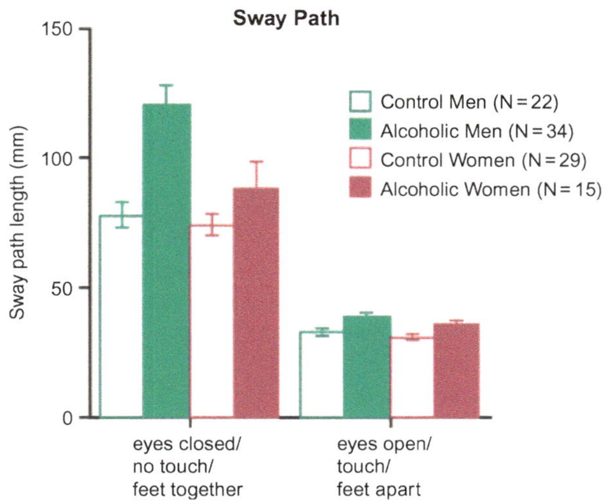

Fig. 15.3. Alcoholic men and women demonstrated longer sway path lengths during quiet stance compared to controls without assistance from visual and tactile aids. Sway decreased for all subjects, but most significantly for the alcoholic men, when stabilizing clues were provided. (Reproduced from Sullivan et al., 2010b.)

external cues may allow for deeper understanding of the roles each sensory input plays in the maintenance of posture. Furthermore, determining the differential advantages these external cues provide to alcoholic subjects versus healthy controls could allow for understanding of which sensory systems are selectively vulnerable in alcoholism. Sullivan et al. proposed that alcoholic men employ a high degree of muscle stiffening and bracing to maintain balance. Though these techniques may be advantageous for postural stability when faced with small perturbations, in the long term these techniques may result in increased instability and likelihood of falls due to inflexibility and muscle stiffness. Furthermore, Sullivan et al. demonstrated that stance with feet together is more difficult for alcoholic subjects than controls, but stance with feet apart reduces, but does not eliminate, deficits in performance by the alcoholic subjects (Sullivan et al., 2005, 2010b).

Sullivan et al. (2010b) found that alcoholic women and healthy controls both benefited from visual cues (eyes open vs eyes closed) and tactile cues (a plastic pole to touch for reference). Men decreased their sway most substantially. Alcoholic women still demonstrated significantly greater sway patterns than controls with eyes open and with the assistance of tactile inputs (Sullivan et al., 2010b). Some studies find that visual clues help subjects with chronic alcoholism adjust their stance (Sullivan et al., 2005), but other studies demonstrate evidence that visual cues can be distracting for acutely intoxicated subjects (Modig et al., 2012a). In summary, studies are inconsistent in their findings about the benefits of visual input for adaptations, and more research is needed to understand the role of the visual system in maintenance of balance in patients with chronic alcoholism. Understanding how sensorimotor feedback can improve postural stability may benefit therapeutic strategies in patients with alcoholism or other pathologies resulting in instability. For example, the simple provision of a cane or a walking stick may allow for significant improvements in gait and stability among alcoholic and/or elderly patients, and may reduce fall risk (Sullivan et al., 2010b).

Studies that assess the role of various stabilizing aids in postural equilibrium help to show how information from multiple sensory inputs can be used both by alcoholic and normal controls. These studies also consistently demonstrate the prolonged deficits in postural control within alcoholic subjects even after long periods of sobriety. The postural impairments associated with long-term alcoholism are linked to underlying neurologic damage responsible for the motor symptoms that occur while under the influence of alcohol and during sobriety.

Neurologic changes in chronic alcoholism

Brain-imaging studies using techniques such as magnetic resonance imaging (MRI) have allowed for quantification of regional brain differences in subjects with chronic alcohol dependence compared to controls. In people with a history of alcoholism, several brain structures are significantly smaller in volume compared to healthy controls, including the frontal lobe, the corpus callosum, and anterior superior vermis of the cerebellum. Imaging studies also allow visualization of moderate potential for recovery in brain tissue volume after a period of abstinence, and alcoholics who resume drinking show continued reduction in regional brain volumes (Cardenas et al., 2007).

A magnetic resonance technique, diffusion tensor imaging (DTI), has more recently been used to study coherence and degradation of neural fibers. DTI allows for quantification of microstructural white matter (WM) abnormalities underlying functional deficits in people with chronic alcoholism. The integrity of WM fibers is most commonly assessed using fractional anisotropy (FA), a measure that is generally higher in WM tracts with homogeneous and cohesive structure. Another measure used in DTI analyses is mean diffusivity (MD), for which higher values generally correspond to increased water diffusion and less cellular density. MD can be further analyzed in terms of its three perpendicular vectors. Axial diffusivity (AD) is a measure of water diffusion along the principal axis of diffusion, and radial diffusivity (RD) comprises the two vectors of diffusion perpendicular to the primary direction of diffusion. RD is thought to decrease as a result of increased axonal myelination. DTI can be used to quantify the integrity of fiber tracks and may allow for more detailed analysis than MRI of WM microstructure (Rosenbloom et al., 2003).

Cortical changes

Neuropathologic studies of human cases with long-term alcohol dependence demonstrate significant overall brain shrinkage, and this loss of brain matter can primarily be attributed to cerebral hemisphere WM shrinkage (Harper and Kril, 1985; Kril et al., 1997; Harper, 1998). WM volume appears to be negatively correlated with self-reported lifetime alcohol consumption (Kril et al., 1997; Kril and Butterworth, 1997). Gray matter (GM) loss has also been noted in the cortices of alcohol-dependent subjects (Pfefferbaum et al., 1992). Specifically related to motor cortices, loss of GM volume has been observed in alcoholic subjects within the precentral gyrus (Mechtcheriakov et al., 2007), a cortical region involved in motor function. Momenan et al. (2012) found more severe damage, indicated by cortical thinning, in the right hemisphere of alcoholic subjects, compared to the left.

Corpus callosum

Studies of WM integrity have used MRI and DTI to analyze changes in tract coherence within brains exposed long-term to alcohol. Decreased fiber integrity and thinning of the corpus callosum, as well as WM hyperintensity signals, are associated with impairment of postural regulation in alcoholic men (Pfefferbaum et al., 1996) and women (Hommer et al., 1996). Pfefferbaum et al. demonstrated a decrease in FA within the corpus callosum of alcoholic men and women compared to control subjects (Fig. 15.4) (Pfefferbaum et al., 2000, 2002, 2010).

The corpus callosum may be particularly vulnerable to injury from alcohol exposure. The corpus callosum allows for interhemispheric processing and damage to the integrity of fibers likely results in slower integration and processing of information. The corpus callosum is implicated in the integration of visual and vestibular

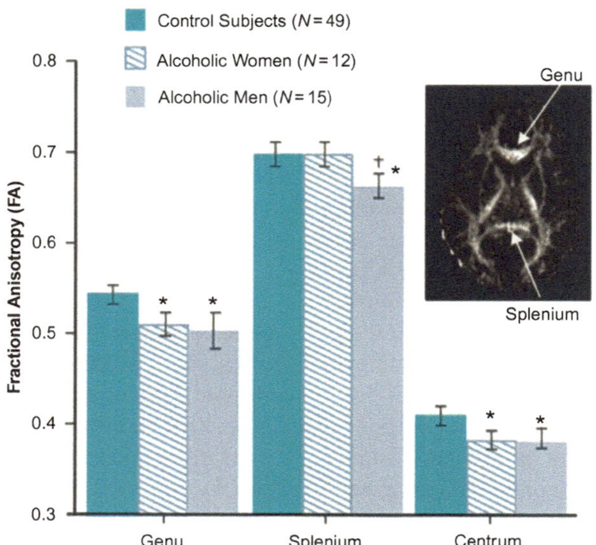

Fig. 15.4. Diffusion tensor imaging was used to characterize diffusion within the corpus callosa of alcoholic men, women, and control subjects. Fractional anisotropy (FA) was decreased in alcoholic men in the genu, splenium, and centrum semiovale. Alcoholic women had decreased FA within the genu and centrum semiovale. (Modified from Pfefferbaum et al., 2000, 2002; Rosenbloom et al., 2003.)

signals required for posture, and degradation of the WM within the corpus callosum has been found to correlate with performance on visuospatial tasks (Müller-Oehring et al., 2009).

Sullivan et al. (2010a) found a deficit in volume (−0.61 SD) of the corpus callosum in alcoholic women compared to healthy controls. More volumetric and connectivity studies in alcoholic women and men will be necessary to develop a more detailed understanding of the microstructural changes that occur in the corpus callosum due to prolonged exposure to alcohol.

Cerebellum

Postmortem autopsy and imaging studies have determined that chronic alcoholism results in abnormally small volume of the superior anterior vermis of the cerebellum. Studies of motor function and brain MRIs have consistently documented the relationship between decreased volume of the cerebellar vermis (Sullivan et al., 2000) and postural instability in alcoholics (Sullivan et al., 2005). Changes in the cerebellum from long-term exposure to alcohol result from a multitude of factors, including thiamine depletion, damage to glial cells, lowered energy production, and general oxidative stress (Zahr et al., 2010). Cerebellar atrophy has also been found in patients with Wernicke–Korsakoff syndrome (thiamine deficiency often secondary to alcoholism) who exhibit ataxia (Zahr et al., 2010).

At a cellular level, the deficits within the cerebellum likely result from the diminished number of Purkinje cells, which appear to be particularly vulnerable to damage from alcohol (Pentney, 1982) and also to thiamine deficiency (Baker et al., 1999;' Mulholland, 2006). It has been suggested that a thiamine deficiency, common among alcohol-dependent subjects, could be an underlying cause of cerebellar atrophy in alcoholism, as a loss of Purkinje cells and shrinkage of the granule layer of the cerebellum have been demonstrated in rats (Mulholland et al., 2005) and humans (Witt, 1985) exposed to a thiamine-deficient diet (Irle and Markowitsch, 1983; Mulholland, 2006; Ke et al., 2009).

Lower FA, likely representing disruptions in WM fiber integrity, and changes in metabolic processing have been reported in alcoholic subjects within cerebellar regions and tracts connecting to the pons and midbrain (Meyerhoff et al., 2004; Pfefferbaum et al., 2009). Though the majority of DTI studies on alcoholism suggest a selective disruption in WM structures within frontal regions of the brain, there is evidence of damage to cerebellar WM and GM structures (Sullivan, 2003; Chanraud et al., 2009; Yeh et al., 2009).

Studies of motor deficits in alcoholics have found longer sway pathways during quiet stance, particularly in the anterior–posterior direction, are correlated inversely with size of the anterior superior vermis of the cerebellum, but not with the volume of the cerebellar hemispheres or the corpus callosum. This finding has been repeated in multiple studies of alcoholic men and women. Anterior superior cerebellar vermis lesions and animal studies suggest that damage to this region results in postural instability similar to that observed in alcoholic subjects (Diener et al., 1983). Positron emission tomography studies in healthy subjects have identified increased cellular activity within the anterior superior cerebellar vermis when subjects performed tasks requiring postural control, further implicating this region in postural control (Ouchi, 1999; Zahr et al., 2010). The volume of the anterior superior vermis of the cerebellum has also been correlated with magnitude of deviations from vertical during upright stance in recovering alcoholics, and its volume appears to increase with the length of sobriety (Sullivan et al., 2010a).

Pons and thalamus

The pons is particularly vulnerable in alcoholism to degradation of its WM fibers, known as central pontine myelinolysis. On a cellular level, neurons within the pons are reduced in number, likely due to liver disease or thiamine deficiency. Patients with pathology of the pons demonstrate motor difficulties in processes including swallowing, talking, or walking. Neural circuitry connecting the pons through the cerebellum up to the motor and sensory regions of the cortex (known as the pontocerebellar tracts) modulates motor function, and thus damage to this tract is likely involved in some of the deficits in alcoholism (Pfefferbaum et al., 2002; Sullivan, 2003; Sullivan et al., 2009).

The thalamus also plays a critical role in the transduction of motor-related information between many brain regions, including the brainstem, cerebellum, basal ganglia, and cortex. The thalamus appears to be diminished in size in alcoholics (Kril and Butterworth, 1997) and this shrinkage is likely an indication of damage to cerebellothalamocortical pathways. Sullivan et al. also found decreased thalamic volume in alcoholics but volume demonstrated no significant correlation with motor impairment (Sullivan et al., 2009, 2010a).

As a unit, the pontocerebellar and cerebellothalamocortical pathways connecting the pons, thalamus, and cerebellum to cortical structures appear to participate in the coordination of posture, and are selectively vulnerable to damage in chronic alcoholism (Chanraud et al., 2008). Both the pons and thalamus show volumetric deficits in alcoholism, but postural instability likely arises from both regional abnormalities and disruptions to the circuitry connecting these regions. Using DTI to collect connectivity data may further elucidate how microstructural changes in these pathways relate to motor deficits observed in alcoholism.

Peripheral nervous system

Peripheral neuropathy is a relatively common complication resulting from chronic alcoholism and can impair sensory awareness and motor function. Studies suggest an incidence rate of 25–66% in chronic alcoholics, though detection is difficult and criteria used to classify neuropathy likely underestimate prevalence (Mellion et al., 2011; Chopra and Tiwari, 2012). Originally ethanol and its metabolic products were believed to be neurotoxins that directly damaged peripheral nerves. Subsequently, when thiamine deficiency was discovered as the underlying condition in Wernicke's encephalopathy and beriberi, researchers shifted towards the belief that nutritional deficits or the inadequate biologic utilization of thiamine played a larger role in alcohol-induced peripheral neuropathy. Research, however, has failed to conclusively demonstrate the direct or indirect association between thiamine deficiency and neuropathy (Mellion et al., 2011).

The neurotoxic effect of ethanol on the brain and liver has been well documented, but the effect of ethanol on the peripheral nervous system is more complex. Though nutritional deficits may play a role in some of the neuropathies experienced by alcoholics, studies have demonstrated normal nutritional status in many patients with peripheral neuropathy (Behse and Buchthal, 1977). Lifetime exposure to ethanol likely plays a more significant role in risk for developing peripheral neuropathy, either due to direct toxic effects of the ethanol and its metabolites, or from toxic impurities such as lead in alcoholic beverages. Peripheral neuropathy in diabetic patients can have deleterious effects on gait and motor abilities (Martinelli et al., 2012) and likely causes deficits in motor function among patients with chronic alcoholism. For example, mechanoreceptors on the soles of the feet allow for assessment of the supporting surface and relay this information to the CNS for interpretation. No known study has investigated the specific role of neuropathy in the motor impairment caused by alcoholism, but likely it plays a non-trivial role in overall sensory inputs critical to the maintenance of postural equilibrium.

INFLUENCE OF LENGTH OF ALCOHOL DEPENDENCE AND SOBRIETY

Severity of motor impairments and other symptoms from alcoholism relates to a variety of factors, including duration of alcoholism, age, and other health determinants and comorbidities.

In a study by Wöber et al. (2009) on recovering alcohol-dependent subjects, estimated lifetime alcohol consumption was correlated with postural imbalance and measurements of peripheral neuropathy, but not evidence of cerebellar damage. Conversely, estimated alcohol consumption in the previous 6 months was not correlated with postural or neurologic measures. The type of alcohol dependence significantly influenced measurements of subjects' postural sway. Subjects with neurologic and psychiatric diagnoses prior to the onset of the alcoholism had the most severe postural impairments but also had the highest estimated lifetime alcohol consumption. Likely the duration and intensity of alcoholism, combined with other environmental and medical factors, contribute to the long-term damage to the neural control of posture (Wöber et al., 2009). Other studies have found that duration of sobriety for recovering alcoholics is negatively correlated with length of sway path when measured on a force plate stabilogram (Sullivan et al., 2005).

A study by Sullivan et al. (2010a) in alcoholic women and controls found no associations between magnitude of sway during quiet stance, severity of alcoholism, and length of sobriety. More likely, degree of impairment and the possibility of recovery are correlated to the severity of damage to the anterior vermis of the cerebellum. These correlations need to be studied further, as current findings report inconsistent results on severity of alcoholism, length of sobriety, and long-term measures of health, posture, and recovery potential.

Influence of age

Postural stability decreases in healthy aging individuals particularly after age 60, as the sensory system begins to decline. Elderly adults experience substantially more falls, likely due to reduced and abnormal signaling from the somatosensory and vestibular systems. During quiet stance, the sway path lengths of elderly individuals increase compared to younger individuals, regardless of assistance from tactile or visual clues. After age 40, individuals tend to rely more heavily on the visual system during locomotion to maintain equilibrium (Choy et al., 2003; Rose and Gamble, 2006), but as vision declines, so does the ability to use vision for orientation and stability. Complete loss of visual input results in a fivefold increase in the loss of balance (Judge et al., 1995). As elderly individuals have decreased sensory input, particularly from the extremities, and rely more heavily on visual cues despite declining vision, aging itself is a critical risk factor for postural instability. Older subjects are particularly vulnerable to brain damage associated with heavy alcohol consumption, likely because motor deficits are confounded by general degradation of brain regions controlling posture and balance. Peripheral neuropathy, a common issue in aging adults, particularly those with diabetes, can also be exacerbated by alcoholism and contribute to further loss of sensory input. Combined with a history of alcoholism, elderly individuals experience high rates of falling and have significant difficulties with postural control (Rose and Gamble, 2006).

Sexual dimorphism

Sexual dimorphism in the neurologic and symptomatic expression of chronic alcoholism has been debated in the literature. Alcohol is metabolized differently in men and women, leading many studies to conclude that women may be particularly vulnerable to brain damage from chronic alcoholism (Baraona et al., 2001; Silva et al., 2009; Ceylan-Isik et al., 2010). Volumetric changes in the brain resulting from alcoholism appear to be sexually dimorphic (Pfefferbaum, 2001). Consumption behaviors and frequency are also different in men and women, and thus research on the effects of sex on the ramifications of alcoholism requires the consideration of many variables. The term "telescoping effect" has been used to describe the observed phenomenon that women may demonstrate similar severity of neurophysiologic markers from alcoholism compared to men, despite overall lower alcohol consumption. These findings have been demonstrated using computed tomography studies, in which women developed alcoholism and the underlying brain pathologies after shorter periods of alcoholism compared to men (Mann et al., 2005). MRI studies have found more significant volumetric changes in alcoholic women compared to men; however, when Sullivan et al. (2009) measured gender effects on cerebellar and pons volume changes in alcoholics, no evidence for the "telescoping effect" was shown. In fact, men appeared to have more significant structural deficits within the brain, even when adjusting for alcohol history and age (Sullivan et al., 2009).

Studies have demonstrated more severe motor deficits in alcoholic men compared to women. Alcoholic men demonstrate ~2 SD deficits in postural control (measured in sway path length magnitude) compared to healthy controls, whereas ~1 SD deficits were observed in alcoholic women. These sex differences could be an exaggerated result of age-related postural control differences observed in healthy aging men vs women (Sullivan et al., 2010b).

Based on posturography analysis in alcoholic men and women, the self-reported lifetime alcohol consumption was positively correlated with length of sway path in men but not women. These disparities must be considered along with the knowledge that the alcoholic men in the Sullivan et al. (2005) study had consumed 2.5 times more alcohol over their lifespan compared to the women in the Sullivan et al. (2010a) study on women. Since severity and duration of alcoholism often differ in women and men, multiple factors must be considered in separating sex from the other factors influencing alcoholic behavior and recovery.

A number of studies have suggested that sex may play a role in potential and rate of functional and neurologic recovery after chronic alcoholism (Smith and Fein, 2011). In the treatment of stroke, for example, a sexually dimorphic response to ischemia and resulting cell death has prompted the development of sex-specific treatment (Manwani and McCullough, 2011). Different treatment strategies based on sex may be necessary to treat the underlying neurologic, psychologic, and postural changes associated with prolonged sobriety.

RECOVERY

Deficits in postural control can extend beyond the period of alcoholism. Evidence is mixed about the possibility of partial or full recovery of stability and normal gait within former alcoholics. Sober individuals with a past history of alcoholism often continue to suffer from mild sway during quiet standing, putting them at continued risk for falls. Neurologic deficits, such as atrophy of the cerebellar vermis, remain even after subjects establish sobriety and likely are the underlying causes of the prolonged postural abnormalities. Improvement in balance after a year or more of sobriety has been demonstrated along with expansion of neural tissues, such as the cerebellum, implicated in balance (Sullivan et al., 2005).

A study in recovering alcoholic women by Rosenbloom et al. (2005) found prolonged selective impairment in postural stability, even after 30 months of sobriety. Three years after initial testing, women with a history of alcoholism had continued impairments in tests of balance during quiet stance compared to age-matched controls, despite normal performance on cognitive and psychomotor tasks. At initial testing (approximately 3 months after sobriety), the group with a history of alcohol dependence demonstrated impairments in all cognitive and motor tests, compared to age-matched controls. The potential for significant improvements in non-verbal short-term memory and speed in psychomotor tasks demonstrated the potential for recovery, but postural instability persisted, suggesting its selective vulnerability (Rosenbloom et al., 2005).

Even during the first 2 weeks of abstinence from alcohol, regional volume changes can be observed in the brain relative to controls, and recovery rates are region-dependent. For example, the cerebellum appears to gain volume faster than the cingulate gyrus (Van Eijk et al., 2013). Imaging studies have demonstrated the potential for improvement in WM integrity with sobriety. A study by Alhassoon et al. (2012) demonstrated decreased fiber integrity and myelination within the corpus callosum of long-term alcoholics compared to controls, but also substantial improvement over the course of a year following detoxification. Along with evidence from Pfefferbaum et al. (2000) demonstrating decreased connectivity within the corpus callosum in chronic alcoholism, the Alhassoon et al. study suggests that the corpus callosum might be a brain region in which neurogenesis and regrowth of myelination may allow for improvement after alcoholism (Pfefferbaum et al., 2000; Alhassoon et al., 2012).

Figure 15.5 demonstrates the potential for WM recovery of the corpus callosum in individuals with a history of alcoholism. Comparing time points 1 and 2, measured 2 weeks and 1 year into abstinence, FA increased and

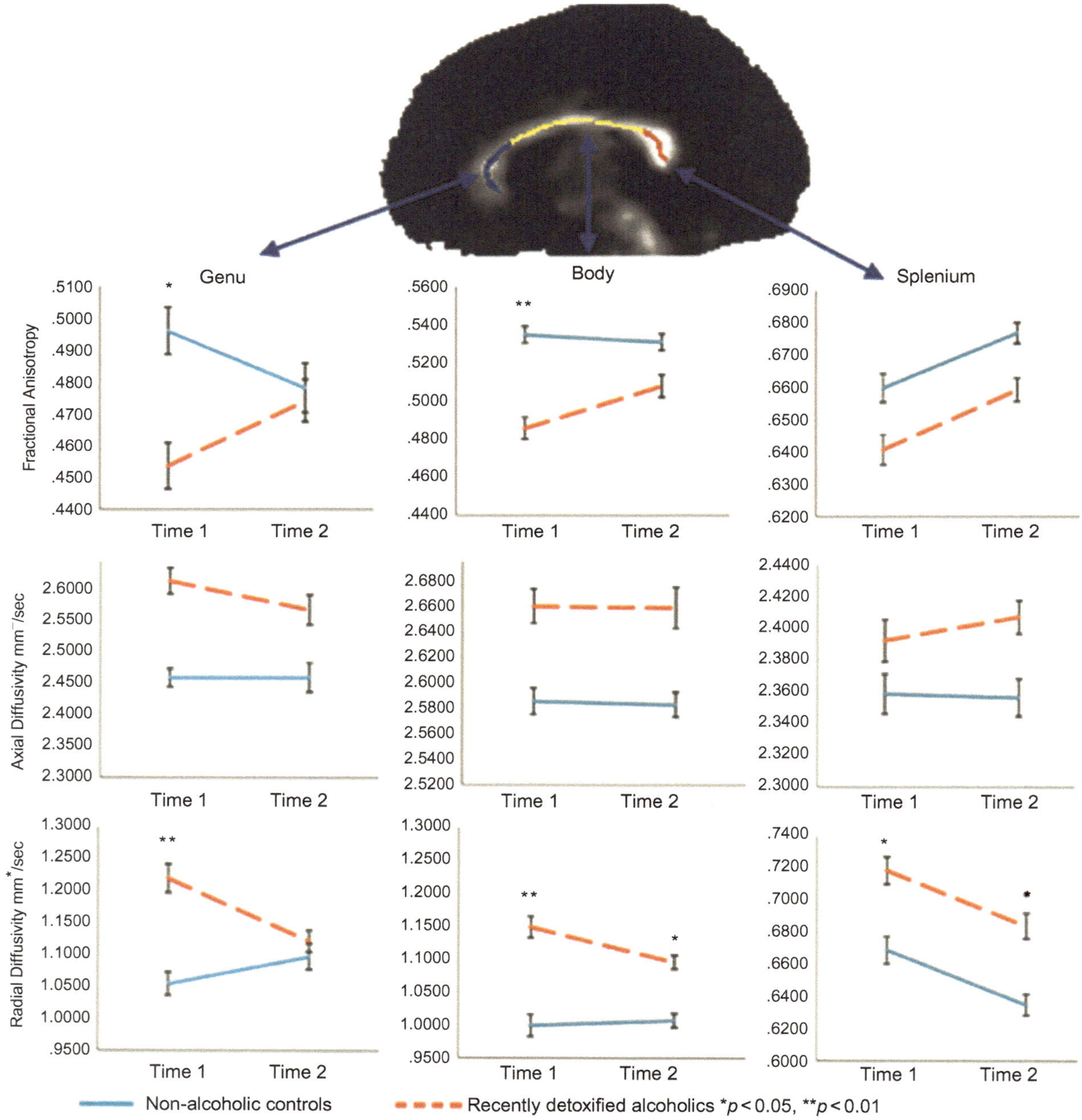

Fig. 15.5. Evidence of WM recovery in the corpus callosum within a subset of patients with alcoholism, as measured using fractional anisotropy, axial diffusivity, and radial diffusivity from diffusion tensor imaging. (Reproduced from Alhassoon et al., 2012.)

RD decreased in the recovering alcoholics, suggestive of improved coherence of neural tracts. The changes observed in the control subjects, however, indicate wide variability in the diffusion measures, thus further research is required (Alhassoon et al., 2012).

Studies of volumetric changes within the brains of recovering alcoholics have been performed using voxel-based and deformation-based morphometry, comparing recovering alcoholics who abstained vs relapsed with controls. Overall cortical atrophy appears to be significantly reversible with abstinence from alcohol, and correlations between sobriety and regional brain volume have been found in the thalamus, corpus callosum, anterior cingulate, insula, and subcortical WM. Increases in cerebellar volume, specifically within the anterior superior vermis, have been associated with duration of abstinence from alcohol, and lack of relapse (Judge et al., 1995; Cardenas et al., 2007). Furthermore, change towards normal levels of metabolism within the cerebellum, assessed with magnetic resonance spectroscopy, occurs simultaneously with improvements in postural stability (Rosenbloom et al., 2007).

CONCLUSION

Acute alcohol intoxication and chronic alcohol dependence alter the neurologic control of posture and motor function. Ethanol delays the conduction of electric signals from the CNS to the muscles controlling posture. Consequently, individuals under the influence of alcohol have impaired stability and delayed motor responses. Chronic alcoholism results in postural tremors and excessive sway during quiet stance that can persist even after sobriety has been achieved. Underlying neurologic changes due to chronic alcoholism have been found to be associated with these characteristic postural changes and include decreased volume of the anterior superior vermis of the cerebellum, decreased connectivity within the corpus callosum, and overall cortical atrophy. Imaging studies highlight the potential for partial recovery from neurologic and motor deficits caused by alcoholism. Emerging evidence on the motor and neurologic changes caused by alcohol dependence may allow for improved treatment and prevention of the morbidities associated with alcoholism.

References

Alhassoon OM, Sorg SF, Taylor MJ et al. (2012). Callosal white matter microstructural recovery in abstinent alcoholics: a longitudinal diffusion tensor imaging study. Alcohol Clin Exp Res 36: 1922–1931.

Astley SJ, Aylward EH, Olson HC et al. (2009). Magnetic resonance imaging outcomes from a comprehensive magnetic resonance study of children with fetal alcohol spectrum disorders. Alcohol Clin Exp Res 33: 1671–1689.

Baker KG, Harding AJ, Halliday GM et al. (1999). Neuronal loss in functional zones of the cerebellum of chronic alcoholics with and without Wernicke's encephalopathy. Neuroscience 91: 429–438.

Baraona E, Abittan CS, Dohmen K et al. (2001). Gender differences in pharmacokinetics of alcohol. Alcohol Clin Exp Res 25: 502–507.

Bay B, Kesmodel US (2011). Prenatal alcohol exposure – a systematic review of the effects on child motor function. Acta Obstet Gynecol Scand 90: 210–226.

Behse F, Buchthal F (1977). Alcoholic neuropathy: clinical, electrophysiological, and biopsy findings. Ann Neurol 2: 95–110.

Cardenas VA, Studholme C, Gazdzinski S et al. (2007). Deformation-based morphometry of brain changes in alcohol dependence and abstinence. Neuroimage 34: 879–887.

Ceylan-Isik AF, McBride SM, Ren J (2010). Sex difference in alcoholism: who is at a greater risk for development of alcoholic complication? Life Sci 87: 133–138.

Chanraud S, Reynaud M, Wessa M et al. (2008). Diffusion tensor tractography in mesencephalic bundles: relation to mental flexibility in detoxified alcohol-dependent subjects. Neuropsychopharmacology 34: 1223–1232.

Chanraud S, Leroy C, Martelli C et al. (2009). Episodic memory in detoxified alcoholics: contribution of grey matter microstructure alteration. PLoS ONE 4: e6786.

Chopra K, Tiwari V (2012). Alcoholic neuropathy: possible mechanisms and future treatment possibilities. Br J Clin Pharmacol 73: 348–362.

Choy NL, Brauer S, Nitz J (2003). Changes in postural stability in women aged 20 to 80 years. J Gerontol A Biol Sci Med Sci 58: M525–M530.

Connor PD, Sampson PD, Streissguth AP et al. (2006). Effects of prenatal alcohol exposure on fine motor coordination and balance: a study of two adult samples. Neuropsychologia 44: 744–751.

Diener HC, Dichgans J, Bacher M et al. (1983). Mechanisms of postural ataxia after intake of alcohol. Z Rechtsmed 90: 159–165.

Guehl D, Pessiglione M, Francois C et al. (2003). Tremor-related activity of neurons in the "motor" thalamus: changes in firing rate and pattern in the MPTP vervet model of parkinsonism. Eur J Neurosci 17: 2388–2400.

Gurfinkel VS, Élner AM (1988). Contribution of the frontal lobe secondary motor area to organization of postural components in human voluntary movement. Neurophysiology 20: 5–11.

Hafstrom A, Modig F, Karlberg M et al. (2007). Increased visual dependence and otolith dysfunction with alcohol intoxication. Neuroreport 18: 391–394.

Hanchar HJ, Dodson PD, Olsen RW et al. (2005). Alcohol-induced motor impairment caused by increased extrasynaptic GABAA receptor activity. Nat Neurosci 8: 339–345.

Harper C (1998). The neuropathology of alcohol-specific brain damage, or does alcohol damage the brain? J Neuropathol Exp Neurol 57: 101–110.

Harper C, Kril J (1985). Brain atrophy in chronic alcoholic patients: a quantitative pathological study. J Neurol Neurosurg Psychiatry 48: 211–217.

Hartshorne N, Harruff R, Alvord Jr E (1997). Fatal head injuries in ground-level falls. Am J Forensic Med Pathol 18: 258–264.

Hommer D, Momenan R, Rawlings R et al. (1996). Decreased corpus callosum size among alcoholic women. Arch Neurol 53: 359–363.

Horak FB, Macpherson JM (2011). Postural orientation and equilibrium. In: R Terjung (Ed.), Comprehensive Physiology, John Wiley, Hoboken, NJ, USA.

Ikeda Y, Sasa M, Takaori S (1980). Selective effect of ethanol on the vestibular nucleus neurons in the cat. Jpn J Pharmacol 30: 665–673.

Irle E, Markowitsch HJ (1983). Widespread neuroanatomical damage and learning deficits following chronic alcohol consumption or vitamin-B1 (thiamine) deficiency in rats. Behav Brain Res 9: 277–294.

Judge JO, King MB, Whipple R et al. (1995). Dynamic balance in older persons: effects of reduced visual and proprioceptive input. J Gerontol A Biol Sci Med Sci 50: M263–M270.

Kalberg WO, Provost B, Tollison SJ et al. (2006). Comparison of motor delays in young children with fetal alcohol syndrome to those with prenatal alcohol exposure and with no prenatal alcohol exposure. Alcohol Clin Exp Res 30: 2037–2045.

Kandel ER, Schwartz JH, Jessell TM (Eds.), (2000). Posture, in: Principles of Neural Science, McGraw-Hill, Health Professions Division, New York.

Ke Z-J, Wang X, Fan Z et al. (2009). Ethanol promotes thiamine deficiency-induced neuronal death: involvement of double-stranded RNA-activated protein kinase. Alcohol Clin Exp Res 33: 1097–1103.

Kool B, Ameratunga S, Jackson R (2009). The role of alcohol in unintentional falls among young and middle-aged adults: a systematic review of epidemiological studies. Injury Prevent 15: 341–347.

Kril JJ, Butterworth RF (1997). Diencephalic and cerebellar pathology in alcoholic and nonalcoholic patients with end-stage liver disease. Hepatology 26: 837–841.

Kril J, Halliday G, Svoboda M et al. (1997). The cerebral cortex is damaged in chronic alcoholics. Neuroscience 79: 983–998.

Mann K, Ackermann K, Croissant B et al. (2005). Neuroimaging of gender differences in alcohol dependence: are women more vulnerable? Alcohol Clin Exp Res 29: 896–901.

Manwani B, McCullough LD (2011). Sexual dimorphism in ischemic stroke: lessons from the laboratory. Womens Health (Lond Engl) 7: 319–339.

Martinelli AR, Mantovani AM, Nozabieli AJL et al. (2012). Muscle strength and ankle mobility for the gait parameters in diabetic neuropathies. Foot (Edinb) 23: 17–21.

Mechtcheriakov S, Brenneis C, Egger K et al. (2007). A widespread distinct pattern of cerebral atrophy in patients with alcohol addiction revealed by voxel-based morphometry. J Neurol Neurosurg Psychiatry 78: 610–614.

Mellion M, Gilchrist JM, De La Monte S (2011). Alcohol-related peripheral neuropathy: nutritional, toxic, or both? Muscle Nerve 43: 309–316.

Meyerhoff DJ, Blumenfeld R, Truran D et al. (2004). Effects of heavy drinking, binge drinking, and family history of alcoholism on regional brain metabolites. Alcohol Clin Exp Res 28: 650–661.

Modig F, Patel M, Magnusson M et al. (2012a). Study I: Effects of 0.06% and 0.10% blood alcohol concentration on human postural control. Gait Posture 35: 410–418.

Modig F, Patel M, Magnusson M et al. (2012b). Study II: Mechanoreceptive sensation is of increased importance for human postural control under alcohol intoxication. Gait Posture 35: 419–427.

Momenan R, Steckler LE, Saad ZS et al. (2012). Effects of alcohol dependence on cortical thickness as determined by magnetic resonance imaging. Psychiatry Res 204: 101–111.

Mulholland PJ (2006). Susceptibility of the cerebellum to thiamine deficiency. Cerebellum 5: 55–63.

Mulholland PJ, Self RL, Stepanyan TD et al. (2005). Thiamine deficiency in the pathogenesis of chronic ethanol-associated cerebellar damage in vitro. Neuroscience 135: 1129–1139.

Müller-Oehring EM, Schulte T, Fama R et al. (2009). Global–local interference is related to callosal compromise in alcoholism: a behavior-DTI association study. Alcohol Clin Exp Res 33: 477–489.

O'Leary CM, Watson L, D'Antoine H et al. (2012). Heavy maternal alcohol consumption and cerebral palsy in the offspring. Dev Med Child Neurol 54: 224–230.

Ouchi Y (1999). Brain activation during maintenance of standing postures in humans. Brain 122: 329–338.

Pentney RJ (1982). Quantitative analysis of ethanol effects on Purkinje cell dendritic tree. Brain Res 249: 397–401.

Pfefferbaum A (2001). Sex differences in the effects of alcohol on brain structure. Am J Psychiatry 158: 188–197.

Pfefferbaum A, Lim KO, Zipursky RB et al. (1992). Brain gray and white matter volume loss accelerates with aging in chronic alcoholics: a quantitative MRI study. Alcohol Clin Exp Res 16: 1078–1089.

Pfefferbaum A, Lim KO, Desmond JE et al. (1996). Thinning of the corpus callosum in older alcoholic men: a magnetic resonance imaging study. Alcohol Clin Exp Res 20: 752–757.

Pfefferbaum A, Sullivan EV, Hedehus M et al. (2000). In vivo detection and functional correlates of white matter microstructural disruption in chronic alcoholism. Alcohol Clin Exp Res 24: 1214–1221.

Pfefferbaum A, Rosenbloom M, Serventi KL et al. (2002). Corpus callosum, pons, and cortical white matter in alcoholic women. Alcohol Clin Exp Res 26: 400–406.

Pfefferbaum A, Rosenbloom M, Rohlfing T et al. (2009). Degradation of association and projection white matter systems in alcoholism detected with quantitative fiber tracking. Biol Psychiatry 65: 680–690.

Pfefferbaum A, Rosenbloom MJ, Fama R et al. (2010). Transcallosal white matter degradation detected with quantitative fiber tracking in alcoholic men and women: selective relations to dissociable functions. Alcohol Clin Exp Res 34: 1201–1211.

Ribak C, Yan X (2000). GABA neurons in the neocortex. In: DL Martin, RW Olsen (Eds.), GABA in the nervous system: the view at fifty years. PA, Lippincott Williams and Wilkins, Philadelphia, pp. 357–368.

Roebuck TM, Simmons RW, Mattson SN et al. (1998). Prenatal exposure to alcohol affects the ability to maintain postural balance. Alcohol Clin Exp Res 22: 252–258.

Rose J, Gamble JG (Eds.), (2006). Human walking, 3rd edn. 3rd edn Lippincott Williams & Wilkins, Philadelphia.

Rosenbloom M, Sullivan EV, Pfefferbaum A (2003). Using magnetic resonance imaging and diffusion tensor imaging to assess brain damage in alcoholics. Alcohol Res Health 27: 146–152.

Rosenbloom M, O'Reilly A, Sassoon S et al. (2005). Persistent cognitive deficits in community-treated alcoholic men and women volunteering for research: limited contribution from psychiatric comorbidity. J Stud Alcohol Drugs 66: 254–256.

Rosenbloom MJ, Rohlfing T, O'Reilly AW et al. (2007). Improvement in memory and static balance with abstinence in alcoholic men and women: selective relations with change in brain structure. Psychiatry Res Neuroimaging 155: 91–102.

Silva SM, Santos-Marques MJ, Madeira MD (2009). Sexually dimorphic response of the hypothalamo-pituitary-adrenal axis to chronic alcohol consumption and withdrawal. Brain Res 1303: 61–73.

Simmons RW, Thomas JD, Levy SS et al. (2010). Motor response programming and movement time in children with heavy prenatal alcohol exposure. Alcohol 44: 371–378.

Simmons RW, Nguyen TT, Levy SS et al. (2012). Children with heavy prenatal alcohol exposure exhibit deficits when regulating isometric force. Alcohol Clin Exp Res 36: 302–309.

Smith S, Fein G (2011). Persistent but less severe ataxia in long-term versus short-term abstinent alcoholic men and women: a cross-sectional analysis. Alcohol Clin Exp Res 35: 2184–2192.

Sullivan EV (2003). Compromised pontocerebellar and cerebellothalamocortical systems: speculations on their contributions to cognitive and motor impairment in nonamnesic alcoholism. Alcohol Clin Exp Res 27: 1409–1419.

Sullivan EV, Deshmukh A, Desmond JE et al. (2000). Cerebellar volume decline in normal aging, alcoholism, and Korsakoff's syndrome: relation to ataxia. Neuropsychology 14: 341–352.

Sullivan EV, Rose J, Pfefferbaum A (2005). Effect of vision, touch and stance on cerebellar vermian-related sway and tremor: a quantitative physiological and MRI study. Cereb Cortex 16: 1077–1086.

Sullivan EV, Rohlfing T, Pfefferbaum A (2009). Pontocerebellar volume deficits and ataxia in alcoholic men and women: no evidence for "telescoping". Psychopharmacology (Berl) 208: 279–290.

Sullivan EV, Rose J, Pfefferbaum A (2010a). Physiological and focal cerebellar substrates of abnormal postural sway and tremor in alcoholic women. Biol Psychiatry 67: 44–51.

Sullivan EV, Rose J, Pfefferbaum A (2010b). Mechanisms of postural control in alcoholic men and women: biomechanical analysis of musculoskeletal coordination during quiet standing. Alcohol Clin Exp Res 34: 528–537.

Van Eijk J, Demirakca T, Frischknecht U et al. (2013). Rapid partial regeneration of brain volume during the first 14 days of abstinence from alcohol. Alcohol Clin Exp Res 37: 67–74.

Von Dincklage F, Benzke M, Rehberg B et al. (2007). Ethanol reduces motoneuronal excitability and increases presynaptic inhibition of Ia afferents in the human spinal cord. Brain Res 1173: 78–83.

Witt ED (1985). Neuroanatomical consequences of thiamine deficiency: a comparative analysis. Alcohol Alcohol 20: 201–221.

Wöber C, Wöber-Bingöl, ç, Karwautz A et al. (2009). Postural control and lifetime alcohol consumption in alcohol-dependent patients. Acta Neurol Scand 99, 48–99, 53.

Yeh P-H, Simpson K, Durazzo TC et al. (2009). Tract-based spatial statistics (TBSS) of diffusion tensor imaging data in alcohol dependence: abnormalities of the motivational neurocircuitry. Psychiatry Res Neuroimaging 173: 22–30.

Zahr NM, Pitel A-L, Chanraud S et al. (2010). Contributions of studies on alcohol use disorders to understanding cerebellar function. Neuropsychol Rev 20: 280–289.

Handbook of Clinical Neurology, Vol. 125 (3rd series)
Alcohol and the Nervous System
E.V. Sullivan and A. Pfefferbaum, Editors

Chapter 16

Sex differences in alcohol-related neurobehavioral consequences

SARA JO NIXON[1,2]*, ROBERT PRATHER[1], AND BEN LEWIS[1]
[1]*Department of Psychiatry, University of Florida, Gainesville, FL, USA*
[2]*Department of Psychology, University of Florida, Gainesville, FL, USA*

INTRODUCTION

Systematic study of the neurobehavioral effects of alcohol consumption in humans has been conducted for nearly a century (Victor and Adams, 1953; Talland, 1965). Despite early (Lisansky, 1957; Drake et al., 1990; Nixon, 1994) and continuing (e.g., Sullivan et al., 2002; Ruiz et al., 2013) discourse regarding the importance of studying both men and women, most published work has included predominantly male samples or failed to report sex-dependent analyses. This research bias is likely due to several factors, including: (1) intentional sample restriction imposed to reduce within-group variance; (2) the underrepresentation of women in treatment settings, disallowing sufficient numbers for statistically meaningful analysis; and (3) heterogeneous published outcomes serving to dissuade the conduct of this work.

Rather than reflecting the irrelevance of sex differences research, these challenges reinforce their importance. In this chapter, we present existing research, organizing it within two broad sections. The first section provides a review of studies examining effects of acute alcohol consumption. The second section addresses the long-term effects of chronic alcohol misuse. From the perspective of scientific development, it is imperative that research be placed within its historical context. Therefore, we included older as well as more recent studies and regret the exclusion of many worthy investigations.

The tasks, methods, and measures applied in this literature can be confusing. To establish a common lexicon and reference, Tables 16.1 through 16.3 were constructed to describe key concepts. Table 16.1 presents a simplified overview of neurobehavioral areas or domains frequently explored in studies of alcohol's effects on neuropsychologic function. Within each domain, the primary brain areas presumed to mediate these functions are also listed. The second column lists the primary cognitive/mental functions associated with each of the domains and the third column lists exemplar tasks. References are provided for each task in the third column; one is a primary reference for the exemplar test; the second references an alcohol study using the task. The complexity of neural system integration and the interdependence of domains cannot be appreciated in this format. We encourage readers interested in a broader perspective to consult Chapter 12 in this volume or other resources, including Lezak and colleagues (2012) and Strauss et al. (2006).

Table 16.2 provides an overview of neurophysiologic measures frequently used in alcohol studies. For the purposes of this chapter, we have limited the outline to those of electroencephalography (EEG) and event-related potentials (ERPs). As discussed in Chapter 23, these methods involve the non-invasive measurement of changes in the brain's electrophysiology using scalp electrodes. These methods are useful in identifying the temporal resolution of underlying processes. EEG is analyzed on the basis of segmented frequencies associated with varying degrees of arousal or engagement. ERPs are time-locked negative (N) or positive (P) shifts in the baseline activity following stimulus presentation. Shifts occurring within approximately 100 ms of the stimulus presentation are typically linked to physical characteristics of the stimulus (e.g., modality, intensity). Later shifts are presumed to be largely influenced by task demands (e.g., counting specific stimulus events, detecting repeated events). Column 2 identifies relevant EEG and ERP components and the neurocognitive process(es) presumably reflected by their occurrence. Column 3 is similar to the corresponding column in Table 16.1, listing general and alcohol-related references. Several excellent references are available to readers

*Correspondence to: Sara Jo Nixon, Professor and Co-Vice Chair, University of Florida, Departments of Psychiatry and Psychology, PO Box 100256, Gainesville, FL 32610, USA. Tel: +1-352-294-4900, E-mail: sjnixon@ufl.edu

Table 16.1

Neuropsychologic domains*†

Neuropsychologic domain/brain areas	Functions of the domain	Exemplar tasks/sources[3]
Executive functions Frontal areas Limbic system Diencephalon Cerebellum‡	Abstraction Problem solving Introspection Motor and strategic planning Set shifting	Wisconsin Card Sorting Task **Heaton and Pendleton, 1981;** Lawton-Craddock et al., 2003; Loeber et al., 2010 Stroop Test **Stroop, 1935; Trenerry, 1989;** Duka and Townshend, 2004; Nixon et al., 2007 Trail Making Test Part B **Reitan, 1955; Reitan and Wolfson, 1993;** Durazzo et al., 2006; Gilbertson et al., 2009; Lewis and Nixon, 2013 Verbal Fluency **Thurstone, 1938; Benton, 1968;** Hewett et al., 1991; Green et al., 2010; Smith and Fein, 2010
Visuospatial Right temporal lobe Right parietal lobe	Spatial orientation of self and others, and objects in space	Block Design Beatty et al., 1997; **Wechsler, 1997b;** Durazzo et al., 2007 Little Man Test **Acker and Acker, 1982;** Glenn and Parsons, 1991; Lawton-Craddock et al., 2003
Learning/memory Visual Right hemisphere Temporal lobe Hippocampus Parietal lobe Visual association cortices Verbal Left hemisphere Temporal lobe Hippocampus Auditory association cortices and language areas	Acquisition, storage, and retrieval of information	Visual Wechsler Memory Scale (Visual Reproduction) **Wechsler, 1987, 1997a**; Sullivan et al., 2000; Errico et al., 2002 Rey-Osterrieth Complex Figure Test Beatty et al., 1997; Sullivan et al., 2000; **Lezak et al., 2012** Verbal California Verbal Learning Test Bates, 1997; **Delis et al., 2000;** Durazzo and Meyerhoff, 2007 Wechsler Memory Scale (Logical Memory) Nixon et al., 1987; **Wechsler, 1987, 1997a;** Gilbertson et al., 2011
Psychomotor Cerebellum Sensory/motor cortices Striatum	Coordination of fine body movements	Finger tapping Løberg, 1980; **Reitan and Wolfson, 1993;** Rogers et al., 2012 Grooved pegboard **Kløve, 1963;** Tivis et al., 1993; Zahr et al., 2009 Trail Making Test Part A **Reitan, 1955; Reitan and Wolfson, 1993;** Durazzo and Meyerhoff, 2007; Gilbertson et al., 2009; Lewis and Nixon, 2013
Gait/balance Cerebellum Sensory/motor cortices Striatum Corpus callosum	Control of postural stability and movement patterns	Sway path analysis **Mauritz et al., 1979;** Sullivan et al., 2009, 2010a Fregly Ataxia Battery **Fregly and Graybiel, 1968;** Sullivan et al., 2000; Durazzo et al., 2006; Fein et al., 2012

*References in **bold** type note those of general interest.

†Attentional processes often defined as executive functions underlie performance across multiple domains. Comprehensive neuropsychologic reviews appropriate detailed discussion to attention in its varied forms, a level of discussion not possible herein. Please consult resources provided elsewhere.

‡The cerebellum has not been traditionally included in this domain. Current literature (e.g., O'Halloran et al, 2012; Stoodley, 2012) suggests, however, that it may play an important role in a variety of processes within this domain.

Table 16.2

Neurophysiologic methods

Method	Description	Exemplar sources*
Electroencephalography (EEG)	Brain activity is segmented into frequencies reflecting arousal and/or cognitive activity: • delta (1–3 Hz): slow-wave sleep • theta (3.5–7.5 Hz): drowsiness • alpha (8–11.5 Hz): alertness • beta (12–28 Hz): concentration • gamma (28.5–50 Hz): sensory processing	**Berger, 1929**; Enoch et al., 1999; Bauer, 2001; Kouri and Lukas, 2001; Rangaswamy et al., 2003
Event-related potentials (ERP)	Time-locked shifts varying in cognitive engagement: • P1 ($\leq$ 100 ms)/N1 ($\sim$100 ms): selective attention • N2 ($\sim$200 ms): physical deviance detection • P3 ($\sim$300 ms): target detection • N4 ($\sim$400 ms): contextual incongruity	**Dawson, 1947; Rugg and Coles, 1996;** Nixon et al., 2002; Porjesz and Begleiter, 2003; Bauer et al., 2006; Ehlers et al., 2011

*References in **bold** type note those of general interest.

interested in additional study, including Rugg and Coles (1996) and Fabiani et al. (2000).

Neuroimaging techniques and their application are rapidly evolving (see Chapters 17 and 22). In contrast to the temporal resolution enabled by electrophysiologic techniques, neuroimaging techniques are applied because of their superior spatial resolution, allowing a more precise study of compromise in specific structures or brain areas. Table 16.3 outlines the neuroimaging techniques reviewed in this chapter. Consistent with the previous tables, the first column identifies the method, the second column lists primary measures, and the third column provides related references. In addition to relevant chapters in the current volume, other related readings include Kaufman (2001), Ciccarelli et al. (2008), Dolan (2008), and Friston (2009).

NEUROBEHAVIORAL CONSEQUENCES OF ACUTE ALCOHOL ADMINISTRATION

There is an extensive and complex literature addressing alcohol's acute effects, a significant portion of which is summarized in an older review by Holloway (1994). Although the majority of studies have focused on doses producing peak blood alcohol concentrations (BACs) of $\sim$0.08–0.10 g/dL, a substantial number have considered the effect of more "moderate" doses (i.e., those producing BACs of $\sim$0.05 g/dL). Most of this work has relied heavily on young drinkers (but see Collins and Mertens, 1988; Gilbertson et al., 2009; Sklar et al., 2012). It is often assumed that women achieve higher blood levels than men when administered equivalent doses (i.e., grams of alcohol per kg weight: Watson et al., 1981; Dubowski, 1985). To enhance the probability of achieving equivalent alcohol levels in male and female participants, some investigators have modified dosing formulas (e.g., Dougherty et al., 2000; Rickenbacher et al., 2011) to women's doses. As illustrated below, heightened peak alcohol levels are not uniformly observed, even if sex-related adjustments are not applied (e.g., Wang et al., 2003; Weissenborn and Duka, 2003; Marczinski et al., 2008). It might be noted that there is a current emphasis on evaluating the effects of lower alcohol doses among healthy older drinkers (Gilbertson et al., 2009; Sklar et al., 2012). However, sex differences have not yet been sufficiently examined to inform the question adequately. Thus, the large majority of cited studies engage young adult drinkers ($\sim$21–35 years of age). In distilling this literature, we have organized it using the three broad categories: neuropsychology (Table 16.1), neurophysiology (Table 16.2), and neuroimaging (Table 16.3).

Neuropsychological/behavioral consequences

In an early contrast of alcohol's effects in men and women, Jones and Jones (1976) administered a common dose (0.66 g/kg) to healthy drinkers. They found that women performed more poorly than did men on tests of delayed, but not immediate, verbal recall. However, women also achieved higher BACs than did men, confounding interpretation. In contrast, using a higher common dose of 0.10 g/kg, Burns and Moskowitz (1978) observed no sex differences on a variety of tests including body sway, visual backward masking, response speed, or hand steadiness. They also reported no differences in alcohol pharmacokinetics between men and women. To investigate a range of alcohol doses, Mills

Table 16.3

Neuroimaging methods

Method	Description/example		Exemplar sources*
Computed tomography (CT)	Multiple X-ray images combined to map brain structure		**Hounsfield, 1973;** Bergman et al., 1980; Carlen et al., 1986; Jacobson, 1986; Mann et al., 2005
Magnetic resonance imaging (MRI)	Structural assessment of gray and white matter		**Lauterbur, 1973;** Makris et al., 2008
Diffusion tensor imaging (DTI)†	Visualization of white-matter fiber tracts based on restricted movement of water molecules		**Le Bihan et al., 1986;** Pfefferbaum et al., 2006; Harris et al., 2008; Colrain et al., 2011
Functional magnetic resonance imaging (fMRI)†	Indirect index of brain activation via blood oxygen level-dependent (BOLD) changes during task performance		**Belliveau et al., 1991;** Parks et al, 2010; Rickenbacher et al., 2011; Squeglia et al, 2011
Positron emission tomography (PET) Single photon emission computed tomography (SPECT)	Maps brain activity and physiology by measuring radio-labeled tracers, such as:		**Phelps et al., 1975;** Adams et al., 1998; Gansler et al., 2000; Lingford-Hughes et al., 2000; Wang et al., 2003; Volkow et al., 2008
	Assesses	Label	
	Cerebral blood flow	^{15}O-water	
	Dopamine receptor density	^{11}C-raclopride	
	GABA receptor density	^{123}I-iomazenil	
	Cerebral metabolic rate	^{18} F-fluorodeoxyglucose	
Magnetic resonance spectroscopy (MRS)	Quantifies metabolic markers of neuronal integrity and signaling, including:		**Bottomley et al., 1983;** Spielman et al., 1993; Durazzo et al., 2008; Gazdzinski et al., 2010
	Assesses	Metabolite	
	Neuronal viability	*N*-acetyl aspartate	
	Myelin/cell membrane integrity	Choline compounds	
	Neuronal/glial bioenergetics	Creatine	

***Bold** references refer to those of general interest.
†DTI and fMRI derive from fundamental MRI methodologies.
GABA, γ-aminobutyric acid.

and Bisgrove (1983) examined performance under three conditions with doses determined by body fat indices. Subjects received placebo (0.05 g/kg), a low dose (i.e., 0.33 (20–30% body fat) vs 0.37 (10–20% body fat) g/kg) or a high dose (i.e., 0.68 g/kg (20–30% body fat) vs 0.76 g/kg (10–20% body fat)). They found no sex differences in performance on the placebo or low doses. At the high dose, women performed more poorly than did men on divided-attention tasks, but there were no sex differences on the measures of body sway.

Extending this work by examining the possible association between menstrual cycle phase and alcohol sensitivity, Niaura and colleagues (1987) examined a 0.65 g/kg alcohol dose in young men and women. As they expected, alcohol concentrations were higher among women. This outcome was largely due to the alcohol levels achieved by women in the middle stage of the menstrual cycle and those not using hormonal birth control. Performance measures were obtained on both the ascending and descending alcohol limbs. After controlling for sex differences in peak alcohol level, men and women performed quite similarly. The only sex difference was that men were less impaired than women on the descending limb, when alcohol concentrations were decreasing, on measures of verbal memory.

In 1999, Mumenthaler and colleagues reviewed much of this earlier literature and concluded that sex differences within acute alcohol studies were dependent on both task and alcohol dose. For example, matching the sexes on target alcohol level produced infrequent differences in reaction time/response latency, psychomotor tasks, or body sway and upright stability. When observed, differences tended to be seen at higher alcohol levels and on higher-order cognitive tasks, such as divided attention, decision making, and memory tasks. They also raised the question of the necessity for controlling for menstrual cycle in alcohol studies. The issue remains unresolved, with recent literature inconsistently controlling for or considering this variable (e.g., Wang et al., 2003 vs Marczinski et al., 2008).

In attempting to clarify the process by which alcohol impairs performance, Dougherty and colleagues (2000) conducted a repeated-measures study with a placebo and two active doses adjusted for anticipated sex differences in pharmacokinetics. Men received a low dose of 0.5 g/kg and high dose of 1.0 g/kg, whereas women's doses were reduced by 8%. This adjustment resulted in equivalent alcohol levels of ~0.04 g/dL and ~0.09 g/dL for the low and high doses, respectively, for men and women. The investigators found that alcohol, relative to placebo, reduced correct target identifications, increased the number of false positives/hits (correct identifications), and decreased discriminability between target and non-target items. Contrary to expectations, it also increased response bias (resulted in more conservative responding). Unfortunately, it is not clear to what extent sex was considered as an independent variable.

Fillmore and colleagues have conducted a number of studies exploring the neurobehavioral effects of acute alcohol administration. Much of this work has been directed to the study of moderately high to high doses of alcohol, e.g., ~0.65 g/kg to ~1.0 g/kg, and their effects on inhibitory behavioral processes (e.g., Fillmore et al., 2008). As illustrated in a recent review of a portion of this body of work, sex differences are frequently observed, but vary substantially in their degree (Miller et al., 2009). Across studies, subjects, ranging in age from 21 to 35, received an alcohol dose of 0.65 g/kg unadjusted for sex. The dose was selected to achieve a blood alcohol level of ~0.08 g/dL. Participants completed behavioral testing emphasizing executive functions such as behavioral set shifting, attentional processes such as vigilance, and psychomotor skills, as reflected in tasks such as the pursuit rotor task. They also completed brief driving scenarios using a single-screen driving simulator. Alcohol reliably impaired performance across most dependent variables for both men and women. Women, as reflected in an analysis of effect sizes and pattern of impairment, appeared to be more sensitive than men, demonstrating greater deficits in rapid information processing, behavioral inhibition, steering rate, and sustaining lane position. Women also indicated higher levels of subjective intoxication (considered themselves to be more intoxicated) than men who received the same levels of alcohol. These subjective and behavioral differences were found despite lack of significant differences between men and women in peak alcohol concentrations.

Weissenborn and Duka (2003) examined the acute effects of alcohol in a different type of battery. Administering either a placebo or 0.8 g/kg dose, they examined performance on a traditional measure of executive functioning, the Tower of London test, as well as tests of visual-spatial memory and pattern recognition. Alcohol impaired performance on the Tower of London test as well as the spatial recognition test. It did not affect performance on the spatial working memory or pattern recognition assessments. Men and women performed similarly and achieved equivalent alcohol concentrations during testing. Regardless of sex, moderate to heavy drinkers who were also binge drinkers performed more poorly than non-binge drinkers on the spatial working memory and pattern recognition tests. Others have noted that binge or heavy episodic drinking interacts with acute alcohol administration. For example, Fillmore and Weafer (2012) found that binge drinkers experienced less compromise in motor responses relative to non-binge drinkers after alcohol administration. In contrast, behavioral inhibition was similarly compromised in the two groups.

There is a growing literature addressing the acute effects of lower doses of alcohol (Gilbertson et al., 2009; Sklar et al., 2012; Lewis et al., 2013). Some of these reports have found that low alcohol levels may improve performance on psychomotor tasks depending on age and BAC limb tested (Gilbertson et al., 2009). Although provocative, these reports are insufficiently powered to explicitly address sex differences. An exception to this limitation is recent work published by Friedman and colleagues (2011), who examined the effects of doses resulting in alcohol concentrations of ~0.05 g/dL on a test of perceptual judgment (judging the shorter of two lines under a series of control manipulations). They found that alcohol increased error rates on both the ascending and descending limbs. Importantly, this result was due almost entirely to the negative effect of alcohol on performance among the men. Although performance for women under the alcohol condition was worse than under the placebo condition, comparisons failed to reach significance.

Overall, the literature suggests similar, although not identical, effects of acute alcohol consumption in men and women. Whether women achieve higher blood levels than men given equivalent alcohol doses appears surprisingly unclear. It is also not known whether variability in controlling for stage of menstrual cycle might contribute to inconsistency in the findings. Regardless, it is evident that even low to moderate doses of alcohol may influence neuropsychologic and behavioral functions in both men and women. However, alcohol's effects are conditional on a number of variables, including limb of the blood alcohol curve, task demands, and individual drinking histories.

Neurophysiologic consequences

There is a large body of work regarding the effects of acute alcohol administration on neurophysiologic measures (Polich and Criado, 2006). However, in conducting our review we found a relative paucity of studies specifically examining sex differences. One study which explicitly addressed sex differences was conducted by Jaaskelainen and colleagues (1995). Using an auditory paradigm, they examined alcohol-related changes in a host of ERP components, including the N1, P2, and the mismatch negativity (MMN) under a low-moderate alcohol dose of 0.5 g/kg. The N1 and P2 are noted in Table 16.2; however, the MMN demands additional attention. The MMN occurs between approximately 100 and 250 ms after stimulus onset and is suggested to represent an early, preconscious recognition of deviation in the stimulus' physical characteristics. This recognition, in turn, advances the process of active or directed attention. Thus, the MMN may be seen to reflect the transition to more cognitively engaged processes. Their analyses revealed that alcohol significantly reduced the amplitudes for all three of the measures. Latencies for the N1 and P2 were not affected, although the latency for MMN onset was significantly delayed. Of particular interest was the finding that the MMN exhibited considerably greater alcohol sensitivity than did other ERP components. This pattern suggests that alcohol might impair earlier, more automatic processes while extracting less of a cost to later ones. Of current relevance was their failure to observe sex differences in the effect of acute alcohol on MMN.

Easdon and colleagues (2005) examined the effects of two alcohol doses, either 0.56 g/kg or 0.80 g/kg, in a sample of men and women. Using a Go/No-Go task to assess behavioral inhibition, they found that both doses increased errors. Both doses also reduced amplitudes in earlier negative and later positive ERP components across all trial types. These authors argue that these findings support the hypothesis that alcohol's acute effects on behavioral inhibition may be driven by impairment in attentional processing. However, the question of sex differences was not addressed.

Although not based on brain electrophysiology, a study by Hutchison and his early colleagues (1997) is appropriate for inclusion. Using electromyography, they examined the effects of a low dose of alcohol designed to achieve a peak alcohol concentration of 0.025 g/dL on prepulse inhibition (PPI) of the startle response (i.e., eye blink) to auditory stimuli. In this paradigm, the magnitude of the eye blink is assessed for conditions in which a startling auditory stimulus is or is not preceded by a non-startling tone. Typically, responsiveness to the second tone is attenuated when preceded by a non-startling stimulus. This reduction is presumed to reflect early, centrally mediated attentional or gating processes. These investigators found that the presence of alcohol affected PPI in a manner contingent on baseline levels. Alcohol reduced PPI among subjects who demonstrated low PPI at baseline, while increasing it among those who had high baseline PPI. Of particular relevance to this chapter is the fact that opposite effects were found for men and women: alcohol reduced the PPI among men while increasing it among women.

Applying another measure of physiologic response, Udo and colleagues (2009) examined sex differences in heart rate variability. They applied sex-adjusted doses to achieve alcohol levels of 0.09 g/dL and measured response to alcohol cues and emotionally negative, positive, and neutral picture cues. Alcohol significantly dampened heart rate variability, but only among men. This effect was observed for both negative and positive pictures relative to neutral pictures. The authors propose this dampened response may reflect a generalized alcohol-related reduction in arousal among men. Using ERP methodology, Franken et al. (2007) examined the effects of a lower alcohol dose (0.5 g/kg) on arousal using a similar picture protocol and found alcohol-related reductions in the very late positive potential (700–1000 ms). This reduction was restricted to negative stimuli. The authors report that there were no sex differences in the ERP measurements or in reported drinking habits. However, it is not clear to what extent the sex by alcohol interaction was examined.

Polich and Kok (1995) noted in their review of P300 ERP literature that sex differences are prevalent, with women producing higher amplitudes than men. A review of the literature found little evidence of sex differences for the P300 in acute alcohol administration studies. Individual differences in specific variables may be more powerful influences than sex. For example, individuals with a low level of response to alcohol are postulated to be at increased risk, relative to those with high levels of response, for developing alcohol-related problems through developing habits of heavier drinking to achieve desired outcomes (e.g., Schuckit et al., 2007; Schuckit, 2009). Bartholow and colleagues (2010) have examined whether sex and/or alcohol sensitivity may impact P3 parameters. Men and women completed a visual oddball paradigm that included five types of pictures: alcoholic beverages, non-alcoholic beverages, adventure-related scenes, erotic scenes, or neutral scenes. P3 was altered only to the alcoholic beverages and only among both men and women with low alcohol sensitivity.

In summary, neurophysiologic studies fail to generate a uniform conclusion. The degree to which sex differences are ascertained appears to be influenced by the same types of variables discussed above, including the task demands, the nature of the stimuli, the alcohol dose, individual differences in sensitivity, and the specific outcome assessed.

Neuroimaging correlates of acute alcohol administration

Several imaging methods have been applied in the study of acute alcohol. As noted elsewhere, few include sufficient numbers of women to allow analysis, and within these, even fewer report analysis of sex effects.

Using positron emission tomography (PET), Urban et al. (2010) examined dopamine release after a moderately high alcohol dose (0.75 g/kg) in men and women. As expected, enhanced dopamine release was greatest in the ventral striatum, with other areas of the striatum showing less effect. Men and women were affected differently, however. The authors estimate that the alcohol challenge relative to placebo challenge increased extracellular dopamine in this area by 138% for men. For women, the increase was 69%. Although the magnitude varied, alcohol's effects were also greater for men in other areas of the striatum. Another relevant finding was that enhanced dopamine release in the ventral striatum was negatively related to the maximum number of drinks per day for men. That is, more frequent heavier drinking was associated with less dopamine release with alcohol challenge among men. The import of a positive family history of alcoholism could not be ascertained due to its relatively low representation among both men (2/11) and women (3/10). Additional research is needed to ascertain if this differential responsiveness correlates with the development of alcohol problems and/or known risk factors for developing alcoholism.

Wang and colleagues (2003) also noted sex differences. Subjects were administered an alcohol dose of 0.75 g/kg, after which they underwent PET scanning, focusing on glucose metabolism. Alcohol, relative to placebo, reduced whole-brain metabolism in both men and women. This reduction was greatest in the occipital cortex and cerebellum for both men and women. The magnitude of this reduction, however, varied significantly by sex. Among the men, the average whole-brain reduction was approximately 25%, whereas for women it was 14%. By contrast, alcohol tended to impair cognitive performance more for women than men, although the difference failed to achieve statistical significance. These differences could not be attributed to differences in alcohol concentrations because men and women did not differ in their peak levels or other characteristics of the metabolic curve.

In another study from this group (Volkow et al., 2006), lower doses of alcohol were administered to men and women. Subjects in this study received one of two active doses (i.e., 0.25 g/kg or 0.5 g/kg). The analyses showed that brain metabolism was significantly reduced even with these lower doses. Whole-brain metabolism was reduced by 10% for low doses and 23% for moderate doses. Specific brain regions were differentially affected by the doses. The low dose affected primarily cortical areas whereas the reduction associated with the moderate dose was more widespread, with both cortical and subcortical regions (i.e., cerebellum, thalamus, basal ganglia, mesencephalon) being affected. Neither dose was associated with cognitive impairment. It is unfortunate given the sex differences observed in their earlier paper that sample sizes prohibited analysis of sex effects.

Volkow and colleagues (2008) conducted a follow-up study of their earlier higher-dose (0.75 g/kg; Wang et al., 2003) study by exploring an alternative analysis strategy. In this analysis, they applied the coefficient of variation as an indicator of cerebral differentiation (i.e., functional heterogeneity/homogeneity). The assumption was that the more heterogeneous a brain area is, the more differentiated it will be in its function. This analysis suggested that, although alcohol increases the brain's homogeneity overall, it affects specific regions differently. For example, it reduced heterogeneity to a greater extent in the right versus the left hemisphere. Regional differences were also observed, with alcohol-related reduction in coefficient of variation (i.e., heterogeneity) being particularly noteworthy in the right frontal and left parietal lobes. A review of subject characteristics suggests that Wang et al. (2003) and the current study share a common subject pool. However, sex differences were not addressed in this later study.

Rickenbacher and colleagues (2011) used magnetic resonance (MR) arterial spin labeling to quantitate alcohol's effects on resting cerebral perfusion. Men and women received sex-adjusted doses (0.6 g/kg (men) vs 0.5 g/kg (women)) in a within-subjects placebo-controlled study. The resting scan occurred on the descending limb, where BACs were ~0.04–0.05 g/dL. Alcohol increased bilateral frontal perfusion among men only. By contrast, under placebo conditions, women exhibited greater perfusion in the left hemisphere frontal, parietal, and temporal areas than did men. Correlational analyses including the entire sample indicated that alcohol-related perfusion was correlated with traits associated with risk for alcoholism, i.e., impulsivity/antisociality.

A recent study by Tolentino and colleagues (2011) also applied arterial spin labeling in a mixed-sex study using higher alcohol doses (0.7 g/kg for women, 0.75 g/kg for men). Their intent was to explore the relations between cerebral perfusion and alcohol response (high response vs low response). They proposed that low-responding subjects would experience greater reduced alcohol-related perfusion than high-responding subjects, consistent with the hypothesis that lower responsivity is associated with reduced central response to alcohol. Overall, alcohol increased perfusion upward of 17% relative to placebo. These increases were most notable in frontal regions. Supporting their hypothesis, they found that these increases were greater for high than low responders. These results could not be attributed to group differences in pharmacokinetics or drinking frequency. Unfortunately, although $\sim$43% of the sample ($n = 88$) were women, sex differences were not addressed.

As we have seen previously, the effects of alcohol are remarkably heterogeneous. Alcohol enhances dopamine release in various areas of the striatum and reduces metabolism at both low and higher doses. There are some studies suggesting that these measures are less affected in women than men. However, because the literature often insufficiently addresses potential interactions and the effects of the confounding individual variables, a general statement is premature.

Section summary

Overall, our review suggests that much work is needed to clarify the nature of sex differences in the neurobehavioral effects of acute alcohol administration. It is telling that even basic presumptions regarding sex differences in the pharmacokinetics of alcohol are not uniformly supported. Women may be more sensitive to alcohol when behavioral outcomes are observed, whereas men may be more sensitive to alcohol's neurophysiologic effect. However, the findings are not sufficiently consistent within or across these domains to merit strong conclusions.

NEUROBEHAVIORAL CONSEQUENCES ASSOCIATED WITH ALCOHOLISM

General notes

Recent nosologic systems distinguish between disorders of alcohol abuse and alcohol dependence (e.g., American Psychiatric Association Work Group to Revise DSM-III, 1987; American Psychiatric Association Task Force on DSM-IV, 1994; but see Edwards et al., 2013). Prior to the mid-1990s, this distinction was infrequently applied in research settings, resulting in diagnostically heterogeneous samples typically defined as "alcoholics" (e.g., Nixon et al., 1987). Since the mid-1990s, more stringent selection criteria have been applied, and current alcohol-related research typically, but not exclusively, reserves the term "alcoholic" for those individuals meeting criteria for alcohol dependence. This refinement in selection criteria has not substantially altered conclusions regarding the pattern of neurobehavioral outcomes (e.g., Parsons, 1987; Oscar-Berman and Marinkovic, 2007). Therefore, in an effort to minimize complexity and enhance readability, we apply the term "alcoholic" to both types of sample.

Additionally, most studies cited in this section include alcoholics who have completed detoxification and achieved minimum abstinence periods (e.g., $\sim$ 21 days), and whose psychiatric and medical histories do not confound conclusions regarding neurobehavioral compromise. Thus, the alcohol-related aberrations we discuss are unlikely attributable to subclinical withdrawal processes or concomitant medical/psychiatric disorders.

Finally, neurobehavioral compromise in detoxified alcoholics is heterogeneous, varying in scope and severity across and within studies. Furthermore, relatively few alcoholics demonstrate deficits consistent with clinical impairment (e.g., Parsons, 1986; Rourke and Grant, 2009). Instead, their performance is statistically inferior to that of age- and education-equated community comparison groups, i.e., relative to persons engaged in common sociocultural/economic communities. This subtle, subclinical compromise can be readily overlooked without appropriate testing.

Neuropsychologic concomitants

Alcohol-related deficits are reported across a wide range of neuropsychologic domains, including executive functions, visuospatial processing, visual and verbal learning and memory, psychomotor processes, and gait and balance (see Chapter 12 and Parsons et al., 1987; Oscar-Berman and Marinkovic, 2003, 2007). Although their differential vulnerability is not fully understood, the domains do not appear to be equally susceptible to chronic alcohol. In typical samples of detoxified alcoholics, those domains most commonly compromised are executive function, visual-spatial processes, and gait/balance.

In one of the earliest available comparisons of cognitive differences between alcoholic men and women, Zelen and colleagues (1966) concluded that the sexes displayed equivalent conceptual quotients. They differed modestly, but significantly, on IQ measures, with men scoring more highly. Subsequent examinations of alcoholic women by Parsons' group (Fabian et al., 1981; Silberstein and Parsons, 1981), employing a more extensive battery of instruments (e.g., digit span, digit symbol, tactual performance, trail making, and numerous other tasks), confirmed neurocognitive deficits relative to healthy controls, and indicated generally equivalent deficits to those noted among alcoholic men. Although some subtle differences suggested a less severe pattern of deficits among women, no direct experimental comparison between men and women was conducted.

Acker (1986) compared alcoholic and control men and women in a variety of domains, including spatial orientation, psychomotor speed, verbal and spatial memory, visuoperceptual ability, and abstracting. After matching participants on age, education, and IQ, alcoholic women performed more poorly than their male counterparts on tasks involving memory, psychomotor skills, abstract reasoning, and visuoperceptual skill. This differential effect of alcohol was in contrast to lower daily consumption (217 vs 300 g/day), more days abstinent prior to testing (23.2 vs 15.5), and fewer years' drinking (6.2 vs 13.1) among women.

Consistently, in a report by Bergman (1987), more severe neurocognitive deficits were observed in women relative to alcoholic men. However, the degree to which these deficits may have reflected baseline sex differences is unclear because gender by alcoholism interactions were not reported. Deficits were observed in visuospatial, abstracting/problem solving, sensorimotor, and selected memory domains. Women reported approximately half the duration of "subjective loss of control" (6.7 vs 12.4 years) and less than two-thirds maximum quantity of alcohol consumption (189 vs 299 g) relative to alcoholic men. Bergman also used computed tomography (CT) to assess aberrations in brain morphology among this group. The relationship between the patterns of consequences was particularly interesting: alcoholic women displayed more severe neurocognitive deficits, whereas alcoholic men displayed greater morphologic aberration (i.e., larger third ventricles). Although some correlations between morphologic variables and neuropsychologic performance were noted, particularly for size of the third ventricle in older alcoholics, morphologic measures appeared to account only modestly (<20%) for variance in performance. This finding highlights the subtlety and complexity of these deficits as well as the importance of multimodal, multimethod assessment.

Further work was generally supportive of sex differences in vulnerability to alcohol-related neurobehavioral deficits. Glenn and colleagues (1988) reported equivalent learning and memory deficits for alcoholic men and women relative to their community controls. Indices of quantity/frequency (average number of ounces of absolute ethanol consumed per day in the 6 months prior to treatment) of consumption were equivalent between men and women (7.75 oz/day for women vs 9.1 oz/day for men). However, women reported shorter durations of heavy drinking (8.40 years vs 11.13 years in men). In a second study, Glenn and Parsons (1992) considered processing efficiency in alcoholic individuals across a comprehensive battery assessing verbal skills, learning and memory, problem solving/abstracting, and perceptual-motor skills. They found equivalent efficiency deficits in alcoholic men and women relative to sex-matched controls. Again, this sample reflected greater years of alcoholism among men (10.4 years for women vs 13.2 years for men), but no difference in the quantity/frequency index (13.3 oz/day for women vs 12.7 oz/day for men).

Additional investigations by this group (Glenn et al., 1993) examined a wide range of neuropsychologic tasks (including, but not limited to, tests of vocabulary, abstracting, immediate and delayed figural and semantic memories, visuospatial skills, and set shifting). This study also found equivalent impairment in alcoholic men and women relative to sex-appropriate controls. Alcoholic women in this sample also reported shorter problem drinking histories (10.33 years vs 13.78 years for men) while displaying equivalent quantity/frequency indices (13.09 oz/day for women vs 12.87 oz/day for men). This type of neuropsychologic performance pattern, illustrated in Figure 16.1, when accompanied by shorter drinking histories in women has often been cited as being consistent with the "telescoping" of women's alcoholism, i.e., a more rapid progression to significant general and mental health consequences (Randall et al., 1999; Keyes et al., 2010). In light of this focus, it is interesting that the men and women in a number of these samples do not differ in their recent drinking histories (Glenn and Parsons, 1992; Glenn et al., 1993; Nixon et al., 2007). As detailed by Keyes et al. (2010) and applied in Lewis and Nixon (2014), the concept of women's telescoping may be largely influenced by the specific milestones being reviewed.

Consistent with a telescoping hypothesis, Liu and colleagues (2010) demonstrated that alcoholic women performed more poorly than alcoholic men on spatial working memory tasks despite initiating regular drinking later. No community comparison groups were used, prohibiting analysis of a sex by group interaction. Using a series of neurocognitive tasks, Flannery and colleagues (2007) observed poorer performance in motor speed, visuoperceptual processing, decision making, and cognitive flexibility (Stroop) among alcoholic women relative to alcoholic men or control participants. Again, these deficits were noted despite reporting fewer years of alcoholism relative to their male counterparts.

Although lacking direct empiric comparison between men and women, two contrasting studies including

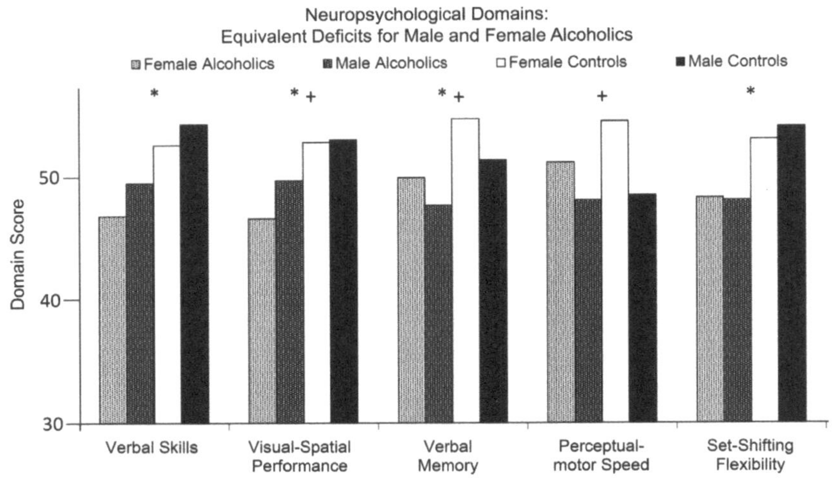

Fig. 16.1. Domain scores have been standardized at a mean score of 50 (±10). No significant sex by group interaction was noted. Significant group (alcoholic vs control) effects are marked with an asterisk ($Fs(1,197) > 13.54$, $p < 0.0005$). Significant sex effects are marked with a plus sign ($Fs(1,197) > 5.69$, $p < 0.05$). (Modified from Glenn et al., 1993.)

alcoholic men (Sullivan et al., 2000) and women (Sullivan et al., 2002) remain notable. Sullivan and colleagues revealed differential consequences of alcohol dependence as a function of sex. Deficits in gait and executive function were observed among alcoholic men, whereas fluency and short-term memory deficits were obtained among alcoholic women. Although limb mobility and declarative memory appeared to be spared, alcohol-associated deficits in visuospatial ability and balance were demonstrated by both sexes. Lifetime alcohol intake among women in these samples appeared to be approximately one-third (522 kg) the level of alcoholic men (1356 kg). Although the general findings are consistent with telescoping, it is pertinent that the specific areas of compromise varied by sex.

Most of the published work focuses on men and women in the first 30–60 days of abstinence. However, a number of groups have explored neuropsychologic consequences persisting into longer-term abstinence. A subset of these works has specifically examined women. Using a cross-sectional design, Fabian and Parsons (1983) observed that women averaging 4 years of abstinence ($M = 48.5$ months, SD $= 42.1$) were inferior to controls in perceptual motor skills, but equivalent in several other measures (e.g., problem solving, abstraction). In a later longitudinal study conducted by Rosenbloom and colleagues (2004), substantial improvements in memory and psychomotor skill were observed among women following an average of 4 years of sobriety ($M = 44.6$, range = 6–72 months), although persistent deficits in standing balance were noted.

A few investigative teams have examined longer-term recovery in men and women. In one report, Fein's group (Fein et al., 2006) observed cognitive deficits in an extended-abstinence group, averaging approximately 7 years abstinence ($M_{males} = 6.9 \pm 6.2$; $M_{females} = 6.5 \pm 5.6$ years). Although some deficits in spatial processing persisted, no sex by group interactions in cognitive processing were noted in cognitive flexibility, attention, psychomotor function, reaction time, spatial processing, verbal skills, or memory (working, immediate, and/or delayed memory). Smith and Fein (2011) described greater ataxic deficits in short-term abstinent (1–3 months) alcoholic women relative to their male counterparts, but noted that this sex difference was not present in longer-term abstinence, although ataxic deficits (i.e., gait, balance) persisted among both men and women.

In summary, the literature reflects more severe neuropsychologic consequences among alcoholic women than men in some cases (e.g., Liu et al., 2010) and equivalent deficits in others (e.g., Glenn and Parsons, 1992). Some studies suggest these differences may be related to telescoping given indications of lower regular consumption, shorter duration of alcoholism, or lower lifetime consumption among women (e.g., Flannery et al., 2007). However, several groups report similar levels of consumption, particularly in the months prior to treatment entry. Thus, sex-contingent effects in the neuropsychologic consequences of chronic alcohol use appear varied, subtle, and only partially explained by telescoping hypotheses.

Neurophysiologic concomitants: EEG

Neurophysiologic investigations of chronic alcohol use have been informed by EEG for some time, with numerous reports suggesting disruption in resting rhythms. An indepth review of alcohol-associated alterations in EEG is discussed elsewhere in this volume (see Chapter 23). Briefly, higher theta power (Rangaswamy et al., 2003), unstable alpha rhythms (Begleiter and Platz, 1972), increased beta power (Winterer et al., 1998; Rangaswamy et al., 2003), and lower gamma power (Porjesz and Begleiter, 2003) have all been noted in alcohol-dependent populations. These aberrations have been associated with behavioral disturbances; for instance, increased theta power is associated with reduced episodic retrieval, recognition memory, and frontal inhibitory control (see Campanella et al., 2009, for review).

Among the relatively few studies specifically investigating sex differences in resting EEG rhythms, reports are inconsistent. For example, Rangaswamy and colleagues (2003) confirmed earlier reports suggesting higher theta power among alcoholic individuals, and additionally detected a sex difference in regional theta prominence. While prominent at parietal regions for both men and women, only men demonstrated prominence in the central region. Propping and colleagues (1981) examined alcoholics of both sexes, noting shifts from the alpha and theta to the beta bands in the resting EEG rhythms of female, but not male, alcoholics. Relatives of the alcoholics, who did not themselves abuse alcohol, showed similar shifts. The authors concluded that, at least among women, a desynchronized resting EEG pattern might enhance risk for developing alcoholism.

Neurophysiologic concomitants: ERP

Electrophysiologic activity during information processing has also received substantial attention. Alcohol-related deficits are noted in a number of ERP components, including the N1, N2, N4, MMN, and P3 (see Campanella et al., 2009 for review). Although many of these studies are underpowered or fail to address sex differences, there are a number of relevant studies.

Among traditional ERP components, the P3 has been most frequently explored. Occurring between 250 and 600 ms following stimulus presentation, the P3 is presumed to reflect processes related to working memory and contextual updating (Porjesz and Begleiter, 1982; Polich, 2004) and is affected by a number of conditions, including alcohol dependence. Several investigations over the last two decades have focused specifically on women. For example, Suresh and colleagues (2003) examined auditory P3s in alcoholic vs non-alcoholic women. They noted lower P3 amplitudes in alcoholic women that appeared uncorrelated with comorbid depression or drug dependence. P3 latencies did not vary between groups. These observations are consistent with the work of Prabhu and colleagues (2001), who examined visual P3 in a sample of women, noting amplitude reductions in the frontal and central regions of female alcoholics compared with controls.

Other studies provide direct comparisons of men and women. For example, as part of the Collaborative Study on the Genetics of Alcoholism (COGA), Porjesz and Begleiter (1998) examined two large groups of families: those affected by alcohol dependence, and those unaffected (controls). In addition to noting aberrant visual P3 amplitudes in these affected families, they also demonstrated that alcoholic men had lower P3 amplitudes than alcoholic women.

Parsons and colleagues (1990, reviewed in Parsons, 1994) conducted an interesting study with alcoholic and control men and women who were further grouped on the basis of whether they had a family history of alcoholism. This ambitious study produced complex, but intriguing, results. Alcoholic women did not differ from their sex-specific controls on any ERP measure. Furthermore, having a positive family history did not influence outcomes for the women. For men, however, alcoholism was associated with lower amplitudes and/or longer latencies for N1, N2, and P3. Furthermore, men with a positive family history, regardless of alcohol status, also showed reduced amplitudes for N1. Although alcoholic women were generally unaffected on ERP measures, they were affected to an equal or greater degree (depending on the test) than alcoholic men on neuropsychologic tests (on which family history was not a significant factor).

Nixon and colleagues (2002) examined alcohol-related aberrations in the N4. The N4 is a negative shift in the brain's electrophysiologic activity occurring between approximately 350 and 650 ms after the presentation of an incongruent stimulus. They used a semantic incongruity task where subjects read sentences that were or were not congruous on the basis of the last word (e.g., "The pizza is too hot to cry/eat"). They found that the N4 amplitude was similarly low for alcoholic men and women relative to their controls, although latency was unaffected. Ceballos et al. (2005) used a different N4 task and also observed alcohol-related reductions in amplitude. In this study, groups were matched for sex distribution, but sex was not examined as an independent variable.

In brief summary, this literature indicates that in investigations of alcoholic women, electrophysiologic disruptions are found, but not consistently. In comparisons of alcoholic men and women, men appear more affected; however, several investigations fail to detect sex differences. These results suggest that electrophysiologic perturbations are evident among alcoholic men and women, although subtle, yet consistent, sex differences (e.g., greater aberrations in the P3 amplitude among men) exist. These differences contrast with neuropsychologic findings suggesting greater sensitivity among women. As mentioned earlier, the discrepancies between sex differences detected in various measurement modalities (i.e., cognitive behavioral, electrophysiologic, neuroimaging) underscore both the subtlety of the differences and the importance of their multimodal examination. These discrepancies may reflect differences in compensatory mechanisms and/or vulnerability between men and women, but this possibility has yet to be thoroughly explored.

Neuroimaging concomitants: structural

Chronic, excessive alcohol consumption disrupts brain structure and function, as evidenced in global reductions in gray matter (e.g., Hommer et al., 2001), as well as white-matter macrostructures (e.g., magnetic resonance imaging (MRI): Pfefferbaum et al., 1997; Demirakca et al., 2011) and microstructures (e.g., diffusion tensor imaging (DTI): Pfefferbaum et al., 2009). Volumetric abnormalities are noted in several areas, including thalamic nuclei, hippocampus, cerebellum, and corpus callosum, as well as frontal, parietal, and cingulate cortices. These reductions are frequently accompanied by increases in cerebrospinal fluid (CSF) volume (e.g., Mechtcheriakov et al., 2007; Demirakca et al., 2011). Several groups have explicitly considered sex differences using various neuroimaging techniques.

Several investigations note greater deficits in alcoholic women than men. For example, Hommer and colleagues (2001) observed a smaller proportion of intracranial content occupied by gray and white matter (relative to increased volumes of sulcal and ventricular CSF) in alcoholics, with the magnitude of difference between alcoholic and control women being greater than that for alcoholic and control men. These observations were consistent with earlier work (Hommer et al., 1996) demonstrating smaller corpora callosa in alcoholic women, relative to alcoholic men or control women (alcoholic men did not differ from controls). Whereas the Hommer team used methods assessing macrostructural compromise (e.g., MRI), Pfefferbaum and colleagues (2009) used a method to assess microstructural integrity (DTI). They observed evidence for greater regional white-matter degradation among alcoholic women relative to alcoholic men. These observations are consistent with a telescoping hypothesis, suggesting that the magnitude of alcohol-associated consequences may be greater among women.

By contrast, other work indicates greater deficits among alcoholic men. For example, using MRI, Pfefferbaum and colleagues (2001) determined that, relative to controls, alcoholic men displayed ventricular enlargement, larger cortical sulci, and reductions in cortical white matter which were not noted among their female counterparts. However, in a later study including MRI and DTI comparisons, alcoholic women were less affected on MRI measures, but equally affected on DTI measures (Pfefferbaum and Sullivan, 2002). These contrasting findings reflect the need for micro- as well as macrostructual assessments and suggest that the absence of macrostructural compromise in alcoholic women should not be interpreted as reflecting a normal or unaffected system. Interestingly, several studies note an association between lifetime alcohol consumption and structural or microstructural deficits among men, but not women (e.g., Pfefferbaum et al., 2009). Further complicating this issue, Ruiz and colleagues (2013) note differential consequences, with greater sensitivity to drinking history noted in frontal, temporal, ventricular, and collosal areas among alcoholic women, and with alcoholic men showing effects primarily in corpus collosum.

In addition to these conflicting reports, several works indicate equivalent brain changes among alcoholic men and women. Analyses based on DTI (Pfefferbaum et al., 2010), CT (Mann et al., 2005), MRI (Pfefferbaum and Sullivan, 2002; Sullivan et al., 2010b), and voxel-based morphometry of MRI (Mechtcheriakov et al., 2007; Demirakca et al., 2011) have all failed to detect sex-contingent morphologic differences in alcoholics. Data from Demirakca and colleagues (2011) are illustrated in Figure 16.2, showing that alcoholics, regardless of sex, in early abstinence display equivalent brain changes relative to appropriate healthy controls.

Among the works reflecting equivalent structural insults, conclusions regarding differences in alcohol consumption/chronicity between men and women remain equivocal. In a recent report, Demirakca and colleagues (2011) employed voxel-based morphology measures to examine gray- and white-matter changes in a population of recently detoxified alcoholics. Although noting decreases in global white and gray matter volumes, increases in global CSF volume, and local GM reductions in cingulate gyrus and insula, no sex-contingent differences were noted, either in very early abstinence or during a 3-month re-examination. Men and women in this sample were also equivalent in their drinking habits, including their lifetime alcohol consumption, drinking in the previous 3 months, and the duration of their illness. Equivalent insults were also noted by Pfefferbaum and Sullivan's group, who observed proportional deficits in pontocerebellar volumes between male and female alcoholics (Sullivan et al., 2009), as well as equivalent evidence for white-matter neuropathology (Pfefferbaum et al., 2010).

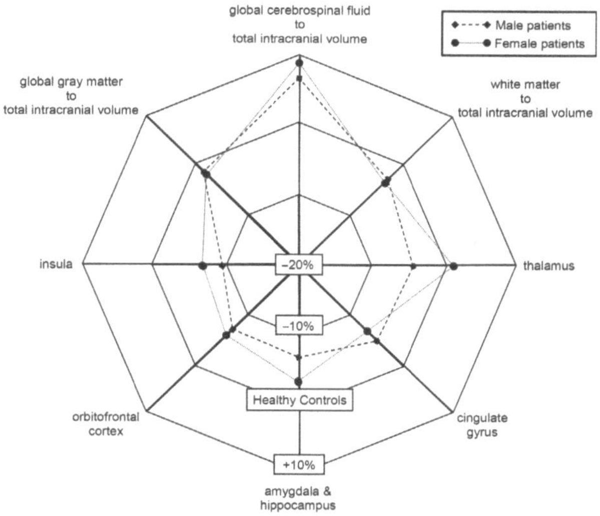

Fig. 16.2. Illustration of gray- and white-matter volume loss, global cerebrospinal fluid (CSF) increase, and regional volume loss in male and female alcoholics relative to healthy controls. No significant sex effects observed for white matter to total intracranial volume (TIV) ratio, gray matter to TIV ratio, or CSF to TIV ratio. Significant sex differences were noted for the insula, cingulate gyrus, orbitofrontal cortex, amygdala and hippocampus, and the thalamus. No significant interaction was noted between sex and group (alcoholics vs healthy controls). (Modified from Demirakca et al., 2011.)

In both of these investigations, men and women were matched for their alcohol consumption, suggesting no evidence for telescoping. This conclusion differed from their earlier examination of male and female alcoholics, who were not matched in this manner (Pfefferbaum and Sullivan, 2002). In this study the microstructural integrity of cortical and collosal white matter was similarly affected across sex (although differential effects in centrum and splenium were noted); however, lifetime alcohol consumption was approximately five times greater among alcoholic men. Studies by Mann and colleagues (1992, 2005) also evidenced sex-contingent differences in drinking patterns. Using CT scans, their group noted equivalent brain atrophy, including smaller brain volumes, larger CSF volumes, and greater global atrophy, albeit following significantly shorter durations of alcohol dependence, among the alcoholic women. These findings are consistent with an early CT examination which suggested equivalent abnormalities despite both shorter drinking histories and lower levels of consumption among women (Jacobson, 1986); however, only indirect comparison to other work was conducted. Thus, among studies detecting equivalent insults, conclusions regarding telescoping are inconsistent, with some evincing sex-dependent differences in drinking behavior and others demonstrating lesser levels of consumption and/or shorter drinking histories.

Several groups have considered sex differences in the recovery process, either by using repeated-measures designs or by examining populations abstinent for extended periods; however, findings are similarly equivocal. Although alcohol-associated morphologic changes appear to begin reversal in as few as 14 days (van Eijk et al., 2013), in at least in some areas these changes appear unlikely to revert completely, even following years of abstinence (Fein et al., 2009). In this latter MRI examination, Fein and colleagues noted gray-matter reductions in the orbital frontal, primary visual, and primary somatosensory cortices of long-term abstinent alcoholic men, but not women, despite an average length of abstinence of 6.25 years for both groups. They speculated that men were either more vulnerable to alcohol-related damage or less able to recover.

Regarding rates of recovery, Ruiz and colleagues (2013) noted stronger association between abstinence duration and white-matter recovery among women; however, this relationship switched after 1 year of abstinence, with a greater association apparent among alcoholic men. Both of these studies are consistent with early work which indicated both more rapid and more complete recovery among women (Jacobson, 1986). Yet, in two studies investigating reversal during early abstinence, no sex differences in recovery progression were noted (Mann et al., 2005; Demirakca et al., 2011).

Overall, this literature suggests that alcohol-related changes in brain structure between men and women are likely to be subtle and involve specific areas. The literature remains mixed in regard to whether sex differences are related to differences in drinking habits and history. Fortunately, at least partial recovery of structural deficits appears possible during abstinence. Although some work is suggestive of a greater and more

Table 16.4

Structural neuroimaging studies contrasting sex differences in alcoholism

Authors	Date	Findings	Method
Mann et al.	1992	ALC_M and ALC_W displayed equivalent ventricular enlargement. ALC_M had greater dependence chronicity	CT
Hommer et al.	1996	ALC_W had reduced callosum volume relative to ALC_M or CON_W. ALC_M and CON_M did not differ	MRI
Hommer et al.	2001	ALC_W had increased CSF volume relative to ALC_M. ALC_W differed from CON_W more markedly than ALC_M from CON_M	MRI
Pfefferbaum et al.	2001	ALC_M displayed enlarged ventricles and cortical sulci and reduced cortical WM vs ALC_W, who did not differ from CON_W	MRI
Pfefferbaum and Sullivan	2002	ALC_M and ALC_W both demonstrated WM deficits, with some sex-specific effects noted (e.g., splenium in ALC_M)	MRI, DTI
Mann et al.	2005	Relative to ALC_M, ALC_W had increased intracranial/ventricular volumes and reduced brain volume. Alcohol consumption and dependence chronicity were less in ALC_W than ALC_M	CT
Pfefferbaum et al.	2006	ALC_M displayed greater WM degradation in left and right hemispheres and a midsagittal sample, relative to ALC_W	DTI
Mechtcheriakov et al.	2007	ALC_M and ALC_W displayed equivalent deficits in WM and GM volumes	MRI
Pfefferbaum et al.	2009	ALCW showed more DTI signs of WM degradation than ALCM in several fiber bundles. Participants were matched on drinking histories	DTI
Sullivan et al.	2010b	ALC_M and ALC_W displayed less infretentorium total area than CON_M and CON_W. If matched on alcohol use, ALC_M displayed a greater decrease than ALC_W in all infratentorial areas except anterior vermis	MRI
Pfefferbaum et al.	2010	Fractional anisotropy was equivalent between ALC_M and ALC_W. Participants were matched on drinking histories	DTI
Demirakca et al.	2011	ALC_M and ALC_W displayed equivalent increases in CSF volume and decreases in WM/GM volume	MRI
Ruiz et al.	2013	ALC_M had reduced callosum WM volume relative to CON_M. ALC_W did not differ statistically from CON_W	MRI

ALC_M, alcoholic men; ALC_W, alcoholic women; CT, computed tomography; CON_M, control men; CON_W, control women; MRI, magnetic resonance imaging; FA, fractional anisotropy; CSF, cerebrospinal fluid; WM, white matter; GM, gray matter; DTI, diffusion tensor imaging. Studies discussed in the text which did not afford direct comparison of sex differences have been omitted (e.g., Jacobson, 1986).

rapid recovery among women, this body of literature remains small. Table 16.4 contains a brief summary of findings from structural imaging studies that include gender analysis.

Neuroimaging concomitants: functional and metabolic contrasts

Functional MRI (fMRI) measures changes in the magnetic field due to alterations in blood oxygen, referred to as the BOLD (blood oxygen level-dependent) response. It allows comparisons between activation patterns across tasks and between subject groups. Diminished frontal and parietal activation is noted among alcoholic relative to control women in a spatial working memory task (Tapert et al., 2001). In a further examination of spatial working memory, Tapert's group (Caldwell et al., 2005) demonstrated altered activation in superior frontal, superior temporal, cingulate, and fusiform regions among young men and women with alcohol use disorders, relative to controls. Interestingly, they also demonstrated an interactive effect, wherein women with alcohol use disorders differed from controls to a greater degree than did men. Other work by this group (Squeglia et al., 2011) also reported different patterns of alcohol compromise for young men and women, with young women being affected to a greater degree (Medina et al., 2008; Hanson et al, 2011; Jacobus and Tapert, 2013). While these findings may thus lack applicability to dependent adult populations, they serve to highlight that functional deficits may appear early in disease progression.

Additional techniques have contributed to understanding neurophysiologic alterations in brain at the molecular and functional levels. Single-photon emission CT (SPECT) and PET techniques, described above and in Table 16.2, have helped elucidate the involvement of specific neurotransmitter systems in alcohol dependence (e.g., Nishikawa et al., 2009), as well as some sex-specific changes in select neurotransmitter systems. For example, Lingford-Hughes and colleagues (2000) noted trends toward reductions in γ-aminobutyric acid (GABA) receptor levels in the cerebellum, occipital lobes, and parietal cortices of alcoholic relative to control women, which contrasted with reductions located primarily in the frontal cortex of alcoholic men related to their controls (Lingford-Hughes et al., 1998).

Magnetic resonance spectroscopy (MRS) involves quantitation of metabolic changes in brain tissue. MRS has been used in the detection of numerous metabolic changes in alcoholics, including lowered levels of *N*-acetylaspartate, choline, creatine, and GABA, as well as heightened myoinositol and glutamate ratios, all considered important markers of activity and integrity of various brain regions (see Buhler and Mann, 2011, for review). These metabolic alterations appear at least partially sex-dependent. Specifically, frontal lobe white-matter concentrations of *N*-acetylaspartate, a marker of neuronal density and function, are lower in alcoholic than control men (~11% difference), whereas alcoholic and control women do not differ (<1% difference; Schweinsburg et al., 2003). These findings underscore the complexity of sex-specific alcohol effects and may help explain the lack of correlation between neuropsychologic and gross morphologic aberrations noted in numerous studies.

OTHER CONSIDERATIONS PERTINENT TO EXPLORING SEX DIFFERENCES AND ALCOHOL EFFECTS

Throughout this chapter, we have focused on sex differences *per se*. We have not addressed the biologic, psychologic, and social milieus that not only differ between men and women, but also influence alcohol effects. We know from a larger literature that many factors may influence the consequences of both acute and chronic alcohol consumption. For example, we mentioned the influence of low versus high levels of alcohol responsivity and noted a similar influence in men and women. Other potential modulating factors of alcohol's effects in men and women include psychiatric comorbidity (Hasin et al., 2007; Crum et al., 2013), chronic cigarette smoking (Durazzo et al., 2012), family histories for alcohol and other drugs of abuse (Parsons et al., 1990), genetics (Kendler et al., 2011), and age (for a review of relevant factors, see Gilbertson et al., 2008). An additional consideration involves the population of individuals who reduce drinking or initiate recovery process without formal treatment. This group has received limited attention in the literature and thus is not addressed here.

Many authors share our experience that recruiting sufficient numbers of men and women to analyze modulating or confounding factors is almost prohibitive. Furthermore, men and women are often regarded as rather homogeneous groups. Little attention has been directed to potentially relevant individual differences within women and/or men. To respond to continuing inconsistencies, programmatic research using multimethod and multimodal designs (e.g., multiple doses, multiple neuropsychologic and/or neurophysiologic measures) is required.

References

Acker C (1986). Neuropsychological deficits in alcoholics: the relative contributions of gender and drinking history. Br J Addict 81: 395–403.

Acker W, Acker C (1982). Bexley Maudsley Automated Processing Screening and Bexley Maudsley Category Sorting Test Manual, NFER-Nelson Publishing, Windsor, UK.

Adams KM, Gilman S, Johnson-Greene D et al. (1998). The significance of family history status in relation to neuropsychological test performance and cerebral glucose metabolism studied with positron emission tomography in older alcoholic patients. Alcohol Clin Exp Res 22: 105–110.

American Psychiatric Association Task Force on DSM-IV (1994). Diagnostic and statistical manual of mental disorders, 4th edn, American Psychiatric Association, Washington, DC.

American Psychiatric Association Work Group to Revise DSM-III (1987). Diagnostic and statistical manual of mental disorders : DSM-III-R, 3rd edn, American Psychiatric Association, Washington, DC.

Bartholow BD, Lust SA, Tragesser SL (2010). Specificity of P3 event-related potential reactivity to alcohol cues in individuals low in alcohol sensitivity. Psychol Addict Behav 24: 220–228.

Bates ME (1997). Stability of neuropsychological assessments early in alcoholism treatment. J Stud Alcohol 58: 617–621.

Bauer LO (2001). Predicting relapse to alcohol and drug abuse via quantitative electroencephalography. Neuropsychopharmacology 25: 332–340.

Bauer LO, Ceballos NA, Taylor RE (2006). Interactive effects of hypertension and alcohol dependence on the P300 event-related potential in African-Americans. Clin EEG Neurosci 37: 210–214.

Beatty WW, Blanco CR, Hames KA et al. (1997). Spatial cognition in alcoholics: influence of concurrent abuse of other drugs. Drug Alcohol Depend 44: 167–174.

Begleiter H, Platz A (1972). The effects of alcohol on the central nervous system in humans. In: B Kissin, H Begleiter (Eds.), The biology of alcoholism: Physiology and behavior, Plenum Press, New York, pp. 293–343.

Belliveau JW, Kennedy DN, McKinstry RC et al. (1991). Functional mapping of the human visual cortex by magnetic resonance imaging. Science 254: 716–719.

Benton AL (1968). Differential behavioral effects in frontal lobe disease. Neuropsychologia 6: 53–60.

Berger H (1929). Über das Elektrenkephalogramm des Menschen. Arch Psychiatrie Nerv 87: 527–570.

Bergman H (1987). Brain dysfunction related to alcoholism: some results from the KARTAD project. In: OA Parsons, PE Nathan (Eds.), Neuropsychology of Alcoholism: Implications for Diagnosis and Treatment, Guilford Press, New York, NY.

Bergman H, Borg S, Hindmarsh T et al. (1980). Computed tomography of the brain and neuropsychological assessment of male alcoholic patients and random sample from the general male population. Acta Psychiatr Scand 62: 47–56.

Bottomley PA, Hart HR, Edelstein WA et al. (1983). NMR imaging/spectroscopy system to study both anatomy and metabolism. Lancet 2: 273–274.

Buhler M, Mann K (2011). Alcohol and the human brain: a systematic review of different neuroimaging methods. Alcohol Clin Exp Res 35: 1771–1793.

Burns M, Moskowitz H (1978). Gender-related differences in impairment of performance by alcohol. In: F Seixas (Ed.), Currents in Alcoholism. Grune & Stratton, New York, pp. 479–492.

Caldwell LC, Schweinsburg AD, Nagel BJ et al. (2005). Gender and adolescent alcohol use disorders on BOLD (blood oxygen level dependent) response to spatial working memory. Alcohol Alcohol 40: 194–200.

Campanella S, Petit G, Maurage P et al. (2009). Chronic alcoholism: insights from neurophysiology. Neurophysiol Clin 39: 191–207.

Carlen PL, Penn RD, Fornazzari L et al. (1986). Computerized tomographic scan assessment of alcoholic brain damage and its potential reversibility. Alcohol Clin Exp Res 10: 226–232.

Ceballos NA, Houston RJ, Smith ND et al. (2005). N400 as an index of semantic expectancies: differential effects of alcohol and cocaine dependence. Alcohol Clin Exp Res 29: 936–943.

Ciccarelli O, Catani M, Johansen-Berg H et al. (2008). Diffusion-based tractography in neurological disorders: concepts, applications, and future developments. Lancet neurology 7: 715–727.

Collins WE, Mertens HW (1988). Age, alcohol, and simulated altitude: effects on performance and breathalyzer scores. Aviat Space Environ Med 59: 1026–1033.

Colrain IM, Sullivan EV, Ford JM et al. (2011). Frontally mediated inhibitory processing and white matter microstructure: age and alcoholism effects. Psychopharmacology (Berl) 213: 669–679.

Crum RM, La Flair L, Storr CL et al. (2013). Reports of drinking to self-medicate anxiety symptoms: longitudinal assessment for subgroups of individuals with alcohol dependence. Depress Anxiety 30: 174–183.

Dawson GD (1947). Cerebral responses to electrical stimulation of the waking human brain. J Neurol Neurosurg Psychiatry 10: 134–140.

Delis D, Kapla E, Kramer J et al. (2000). California Verbal Learning Test-II, The Psychological Corporation, San Antonio, TX.

Demirakca T, Ende G, Kammerer N et al. (2011). Effects of alcoholism and continued abstinence on brain volumes in both genders. Alcohol Clin Exp Res 35: 1678–1685.

Dolan RJ (2008). Neuroimaging of cognition: past, present, and future. Neuron 60: 496–502.

Dougherty DM, Marsh DM, Moeller FG et al. (2000). Effects of moderate and high doses of alcohol on attention, impulsivity, discriminability, and response bias in immediate and delayed memory task performance. Alcohol Clin Exp Res 24: 1702–1711.

Drake AI, Hannay HJ, Gam J (1990). Effects of chronic alcoholism on hemispheric functioning: an examination of gender differences for cognitive and dichotic listening tasks. J Clin Exp Neuropsychol 12: 781–797.

Dubowski KM (1985). Absorption, distribution and elimination of alcohol: highway safety aspects. J Stud Alcohol Supplement 10: 98–108.

Duka T, Townshend JM (2004). The priming effect of alcohol pre-load on attentional bias to alcohol-related stimuli. Psychopharmacology (Berl) 176: 353–361.

Durazzo TC, Meyerhoff DJ (2007). Neurobiological and neurocognitive effects of chronic cigarette smoking and alcoholism. Front Biosci 12: 4079–4100.

Durazzo TC, Rothlind JC, Gazdzinski S et al. (2006). A comparison of neurocognitive function in nonsmoking and chronically smoking short-term abstinent alcoholics. Alcohol 39: 1–11.

Durazzo TC, Rothlind JC, Gazdzinski S et al. (2007). Chronic smoking is associated with differential neurocognitive recovery in abstinent alcoholic patients: a preliminary investigation. Alcohol Clin Exp Res 31: 1114–1127.

Durazzo TC, Gazdzinski S, Yeh PH et al. (2008). Combined neuroimaging, neurocognitive and psychiatric factors to predict alcohol consumption following treatment for alcohol dependence. Alcohol Alcohol 43: 683–691.

Durazzo TC, Meyerhoff DJ, Nixon SJ (2012). A comprehensive assessment of neurocognition in middle-aged chronic cigarette smokers. Drug Alcohol Depend 122: 105–111.

Easdon C, Izenberg A, Armilio ML et al. (2005). Alcohol consumption impairs stimulus- and error-related processing during a Go/No-Go Task. Brain Res Cogn Brain Res 25: 873–883.

Edwards AC, Gillespie NA, Aggen SH et al. (2013). Assessment of a modified DSM-5 diagnosis of alcohol use disorder in a genetically informative population. Alcohol Clin Exp Res 37: 443–451.

Ehlers CL, Phillips E, Criado JR et al. (2011). N4 component responses to pre-pulse startle stimuli in young adults: relationship to alcohol dependence. Psychiatry Res 188: 237–244.

Enoch MA, White KV, Harris CR et al. (1999). Association of low-voltage alpha EEG with a subtype of alcohol use disorders. Alcohol Clin Exp Res 23: 1312–1319.

Errico AL, King AC, Lovallo WR et al. (2002). Cortisol dysregulation and cognitive impairment in abstinent male alcoholics. Alcohol Clin Exp Res 26: 1198–1204.

Fabian MS, Parsons OA (1983). Differential improvement of cognitive functions in recovering alcoholic women. J Abnorm Psychol 92: 87–95.

Fabian MS, Parsons OA, Silberstein JA (1981). Impaired perceptual – cognitive functioning in women alcoholics. Cross-validated findings. J Stud Alcohol 42: 217–229.

Fabiani M, Gratton G, Coles M (2000). Event-related brain potentials: Methods, theory and applications. In: J Cacioppo, L Tassinary, G Bernston (Eds.), Handbook of Psychophysiology. Cambridge University Press, Cambridge, UK, pp. 53–84.

Fein G, Torres J, Price LJ et al. (2006). Cognitive performance in long-term abstinent individuals. Alcohol Clin Exp Res 30: 1538–1544.

Fein G, Shimotsu R, Chu R et al. (2009). Parietal gray matter volume loss is related to spatial processing deficits in long-term abstinent alcoholic men. Alcohol Clin Exp Res 33: 1806–1814.

Fein G, Smith S, Greenstein D (2012). Gait and balance in treatment-naive active alcoholics with and without a lifetime drug codependence. Alcohol Clin Exp Res 36: 1550–1562.

Fillmore MT, Weafer J (2012). Acute tolerance to alcohol in at-risk binge drinkers. Psychol Addict Behav 26: 693–702.

Fillmore MT, Blackburn JS, Harrison EL (2008). Acute disinhibiting effects of alcohol as a factor in risky driving behavior. Drug Alcohol Depend 95: 97–106.

Flannery B, Fishbein D, Krupitsky E et al. (2007). Gender differences in neurocognitive functioning among alcohol-dependent Russian patients. Alcohol Clin Exp Res 31: 745–754.

Franken IH, Nijs IM, Muris P et al. (2007). Alcohol selectively reduces brain activity during the affective processing of negative information. Alcohol Clin Exp Res 31: 919–927.

Fregly AR, Graybiel A (1968). An ataxia test battery not requiring the use of rails. Aerosp Med 39: 277–282.

Friedman TW, Robinson SR, Yelland GW (2011). Impaired perceptual judgment at low blood alcohol concentrations. Alcohol 45: 1–8.

Friston KJ (2009). Modalities, modes, and models in functional neuroimaging. Science 326: 399–403.

Gansler DA, Harris GJ, Oscar-Berman M et al. (2000). Hypoperfusion of inferior frontal brain regions in abstinent alcoholics: a pilot SPECT study. J Stud Alcohol 61: 32–37.

Gazdzinski S, Durazzo TC, Mon A et al. (2010). Body mass index is associated with brain metabolite levels in alcohol dependence – a multimodal magnetic resonance study. Alcohol Clin Exp Res 34: 2089–2096.

Gilbertson RJ, Prather RA, Nixon SJ (2008). The role of selected factors in the development and consequences of alcohol dependence. Alcohol Res Health 31: 389–399.

Gilbertson RJ, Ceballos NA, Prather R et al. (2009). Effects of acute alcohol consumption in older and younger adults: perceived impairment versus psychomotor performance. J Stud Alcohol Drugs 70: 242–252.

Gilbertson R, Boissoneault J, Prather R et al. (2011). Nicotine effects on immediate and delayed verbal memory after substance use detoxification. J Clin Exp Neuropsychol 33: 609–618.

Glenn SW, Parsons OA (1991). Effects of alcoholism and instructional conditions on speed/accuracy tradeoffs. Alcohol Clin Exp Res 15: 612–619.

Glenn SW, Parsons OA (1992). Neuropsychological efficiency measures in male and female alcoholics. J Stud Alcohol 53: 546–552.

Glenn SW, Parsons OA, Sinha R et al. (1988). The effects of repeated withdrawals from alcohol on the memory of male and female alcoholics. Alcohol Alcohol 23: 337–342.

Glenn SW, Errico AL, Parsons OA et al. (1993). The role of antisocial, affective, and childhood behavioral characteristics in alcoholics' neuropsychological performance. Alcohol Clin Exp Res 17: 162–169.

Green A, Garrick T, Sheedy D et al. (2010). The effect of moderate to heavy alcohol consumption on neuropsychological performance as measured by the repeatable battery for the assessment of neuropsychological status. Alcohol Clin Exp Res 34: 443–450.

Hanson KL, Cummins K, Tapert SF et al. (2011). Changes in neuropsychological functioning over 10 years following adolescent substance abuse treatment. Psychol Addict Behav 25: 127–142.

Harris GJ, Jaffin SK, Hodge SM et al. (2008). Frontal white matter and cingulum diffusion tensor imaging deficits in alcoholism. Alcohol Clin Exp Res 32: 1001–1013.

Hasin DS, Stinson FS, Ogburn E et al. (2007). Prevalence, correlates, disability, and comorbidity of DSM-IV alcohol abuse and dependence in the United States: results from the National Epidemiologic Survey on Alcohol and Related Conditions. Arch Gen Psychiatry 64: 830–842.

Heaton RK, Pendleton MG (1981). Use of neuropsychological tests to predict adult patients' everyday functioning. J Consult Clin Psychol 49: 807–821.

Hewett LJ, Nixon SJ, Glenn W et al. (1991). Verbal fluency deficits in female alcoholics. J Clin Psychol 47: 716–720.

Holloway FA (1994). Low-dose alcohol effects on human behavior and performance: A review of post-1984 research. Pub. No. 94–35919, Federal Aviation Administration, Washington, DC.

Hommer D, Momenan R, Rawlings R et al. (1996). Decreased corpus callosum size among alcoholic women. Arch Neurol 53: 359–363.

Hommer D, Momenan R, Kaiser E et al. (2001). Evidence for a gender-related effect of alcoholism on brain volumes. Am J Psychiatry 158: 198–204.

Hounsfield GN (1973). Computerised transverse axial scanning (tomography): Part 1. Description of system. Br J Radiol 46: 1016–1022.

Hutchison KE, Rohsenow D, Monti P et al. (1997). Prepulse inhibition of the startle reflex: preliminary study of the effects of a low dose of alcohol in humans. Alcohol Clin Exp Res 21: 1312–1319.

Jaaskelainen IP, Lehtokoski A, Alho K et al. (1995). Low dose of ethanol suppresses mismatch negativity of auditory event-related potentials. Alcohol Clin Exp Res 19: 607–610.

Jacobson R (1986). Female alcoholics: a controlled CT brain scan and clinical study. Br J Addict 81: 661–669.

Jacobus J, Tapert SF (2013). Neurotoxic effects of alcohol in adolescence. Annu Rev Clin Psychol 9: 703–721.

Jones BM, Jones MK (1976). Alcohol effects in women during the menstrual cycle. Ann N Y Acad Sci 273: 576–587.

Kaufman MJ (2001). Brain imaging in substance abuse: Research, clinical, and forensic applications. In: MJ Kaufman (Ed.), Forensic science and medicine, Humana Press, Totowa, N.J.

Kendler KS, Kalsi G, Holmans PA et al. (2011). Genomewide association analysis of symptoms of alcohol dependence in the molecular genetics of schizophrenia (MGS2) control sample. Alcohol Clin Exp Res 35: 963–975.

Keyes KM, Martins SS, Blanco C et al. (2010). Telescoping and gender differences in alcohol dependence: new evidence from two national surveys. Am J Psychiatry 167: 969–976.

Kløve H (1963). Clinical neuropsychology. Med Clin North Am 47: 1647–1658.

Kouri EM, Lukas SE (2001). Electroencephalography, topographic mapping, and event-related potentials in substance abuse research. In: MJ Kaufman (Ed.), Brain Imaging in Substance Abuse. , Humana Press, Totowa, pp. 1–27.

Lauterbur PC (1973). Image formation by induced local interactions: examples employing nuclear magnetic resonance. Nature 242: 190–191.

Lawton-Craddock A, Nixon SJ, Tivis R (2003). Cognitive efficiency in stimulant abusers with and without alcohol dependence. Alcohol Clin Exp Res 27: 457–464.

Le Bihan D, Breton E, Lallemand D et al. (1986). MR imaging of intravoxel incoherent motions: application to diffusion and perfusion in neurologic disorders. Radiology 161: 401–407.

Lewis B, Nixon SJ (2013). Cognitive flexibility during breath alcohol pleateau is associated with previous drinking measures. Alcohol 47: 333–338.

Lewis B, Nixon SJ (2014). Characterizing gender differences in treatment seekers. Alcohol Clin Exp Res 38: 275–284.

Lewis B, Boissoneault J, Gilbertson R et al. (2013). Neurophysiological correlates of moderate alcohol consumption in older and younger social drinkers. Alcohol Clin Exp Res 37: 941–951.

Lezak MD, Howieson DB, Bigler ED et al. (2012). Neuropsychological assessment, 5th edn, Oxford University Press, Oxford.

Lingford-Hughes AR, Acton PD, Gacinovic S et al. (1998). Reduced levels of GABA-benzodiazepine receptor in alcohol dependency in the absence of grey matter atrophy. Br J Psychiatry 173: 116–122.

Lingford-Hughes AR, Acton PD, Gacinovic S et al. (2000). Levels of gamma-aminobutyric acid-benzodiazepine receptors in abstinent, alcohol-dependent women: preliminary findings from an 123I-iomazenil single photon emission tomography study. Alcohol Clin Exp Res 24: 1449–1455.

Lisansky ES (1957). Alcoholism in women: social and psychological concomitants. I. Social history data. Q J Stud Alcohol 18: 588–623.

Liu IC, Chiu CH, Yang TT (2010). The effects of gender and a co-occurring depressive disorder on neurocognitive functioning in patients with alcohol dependence. Alcohol Alcohol 45: 231–236.

Løberg T (1980). Alcohol misuse and neuropsychological deficits in men. J Stud Alcohol 41: 119–128.

Loeber S, Duka T, Welzel Marquez H et al. (2010). Effects of repeated withdrawal from alcohol on recovery of cognitive impairment under abstinence and rate of relapse. Alcohol Alcohol 45: 541–547.

Makris N, Oscar-Berman M, Jaffin SK et al. (2008). Decreased volume of the brain reward system in alcoholism. Biol Psychiatry 64: 192–202.

Mann K, Batra A, Gunthner A et al. (1992). Do women develop alcoholic brain damage more readily than men? Alcohol Clin Exp Res 16: 1052–1056.

Mann K, Ackermann K, Croissant B et al. (2005). Neuroimaging of gender differences in alcohol dependence: are women more vulnerable? Alcohol Clin Exp Res 29: 896–901.

Marczinski CA, Harrison EL, Fillmore MT (2008). Effects of alcohol on simulated driving and perceived driving impairment in binge drinkers. Alcohol Clin Exp Res 32: 1329–1337.

Mauritz KH, Dichgans J, Hufschmidt A (1979). Quantitative analysis of stance in late cortical cerebellar atrophy of the anterior lobe and other forms of cerebellar ataxia. Brain 102: 461–482.

Mechtcheriakov S, Brenneis C, Egger K et al. (2007). A widespread distinct pattern of cerebral atrophy in patients with alcohol addiction revealed by voxel-based morphometry. J Neurol Neurosurg Psychiatry 78: 610–614.

Medina KL, McQueeny T, Nagel BJ et al. (2008). Prefrontal cortex volumes in adolescents with alcohol use disorders: unique gender effects. Alcohol Clin Exp Res 32: 386–394.

Miller MA, Weafer J, Fillmore MT (2009). Gender differences in alcohol impairment of simulated driving performance and driving-related skills. Alcohol Alcohol 44: 586–593.

Mills KC, Bisgrove EZ (1983). Body sway and divided attention performance under the influence of alcohol: dose–response differences between males and females. Alcohol Clin Exp Res 7: 393–397.

Mumenthaler MS, Taylor JL, O'Hara R et al. (1999). Gender differences in moderate drinking effects. Alcohol Res Health 23: 55–64.

Niaura RS, Nathan PE, Frankenstein W et al. (1987). Gender differences in acute psychomotor, cognitive, and pharmacokinetic response to alcohol. Addict Behav 12: 345–356.

Nishikawa M, Diksic M, Sakai Y et al. (2009). Alterations in brain serotonin synthesis in male alcoholics measured using positron emission tomography. Alcohol Clin Exp Res 33: 233–239.

Nixon SJ (1994). Cognitive deficits in alcoholic women. Alcohol Health Res World 18: 228–232.

Nixon SJ, Kujawski A, Parsons OA et al. (1987). Semantic (verbal) and figural memory impairment in alcoholics. J Clin Exp Neuropsychol 9: 311–322.

Nixon SJ, Tivis R, Ceballos N et al. (2002). Neurophysiological efficiency in male and female alcoholics. Prog Neuropsychopharmacol Biol Psychiatry 26: 919–927.

Nixon SJ, Lawton-Craddock A, Tivis RD et al. (2007). Nicotine's effects on attentional efficiency in alcoholics. Alcohol Clin Exp Res 31: 2083–2091.

O'Halloran CJ, Kinsella GJ, Storey E (2012). The cerebellum and neuropsychological functioning: a critical review. J Clin Exp Neuropsychol 34: 35–56.

Oscar-Berman M, Marinkovic K (2003). Alcoholism and the brain: an overview. Alcohol Res Health 27: 125–133.

Oscar-Berman M, Marinkovic K (2007). Alcohol: effects on neurobehavioral functions and the brain. Neuropsychol Rev 17: 239–257.

Parks MH, Greenberg DS, Nickel MK et al. (2010). Recruitment of additional brain regions to accomplish simple motor tasks in chronic alcohol-dependent patients. Alcohol Clin Exp Res 34: 1098–1109.

Parsons OA (1986). Alcoholics' neuropsychological impairment: current findings and conclusions. Ann Behav Med 8: 13–19.

Parsons OA (1987). Neuropsychological consequences of alcohol abuse: many questions – some answers. In: OA Parsons, N Butters, PE Nathan (Eds.), Neuropsychology of alcoholism: implications for diagnosis and treatment. Guilford Press, New York, pp. 153–175.

Parsons OA (1994). Neuropsychological measures and event-related potentials in alcoholics: interrelationships, long-term reliabilities, and prediction of resumption of drinking. J Clin Psychol 50: 37–46.

Parsons OA, Butters N, Nathan PE (1987). Neuropsychology of alcoholism: implications for diagnosis and treatment, Guilford Press, New York.

Parsons OA, Sinha R, Williams HL (1990). Relationships between neuropsychological test performance and event-related potentials in alcoholic and nonalcoholic samples. Alcohol Clin Exp Res 14: 746–755.

Pfefferbaum A, Sullivan EV (2002). Microstructural but not macrostructural disruption of white matter in women with chronic alcoholism. Neuroimage 15: 708–718.

Pfefferbaum A, Sullivan EV, Mathalon DH et al. (1997). Frontal lobe volume loss observed with magnetic resonance imaging in older chronic alcoholics. Alcohol Clin Exp Res 21: 521–529.

Pfefferbaum A, Rosenbloom M, Deshmukh A et al. (2001). Sex differences in the effects of alcohol on brain structure. Am J Psychiatry 158: 188–197.

Pfefferbaum A, Adalsteinsson E, Sullivan EV (2006). Dysmorphology and microstructural degradation of the corpus callosum: interaction of age and alcoholism. Neurobiol Aging 27: 994–1009.

Pfefferbaum A, Rosenbloom M, Rohlfing T et al. (2009). Degradation of association and projection white matter systems in alcoholism detected with quantitative fiber tracking. Biol Psychiatry 65: 680–690.

Pfefferbaum A, Rosenbloom MJ, Fama R et al. (2010). Transcallosal white matter degradation detected with quantitative fiber tracking in alcoholic men and women: selective relations to dissociable functions. Alcohol Clin Exp Res 34: 1201–1211.

Phelps ME, Hoffman EJ, Mullani NA et al. (1975). Application of annihilation coincidence detection to transaxial reconstruction tomography. J Nucl Med 16: 210–224.

Polich J (2004). Clinical application of the P300 event-related brain potential. Phys Med Rehabil Clin N Am 15: 133–161.

Polich J, Criado JR (2006). Neuropsychology and neuropharmacology of P3a and P3b. Int J Psychophysiol 60: 172–185.

Polich J, Kok A (1995). Cognitive and biological determinants of P300: an integrative review. Biol Psychol 41: 103–146.

Porjesz B, Begleiter H (1982). Evoked brain potential differentiation between geriatric subjects and chronic alcoholics with brain dysfunction. Adv Neurol 32: 117–124.

Porjesz B, Begleiter H (1998). Genetic basis of event-related potentials and their relationship to alcoholism and alcohol use. J Clin Neurophysiol 15: 44–57.

Porjesz B, Begleiter H (2003). Alcoholism and human electrophysiology. Alcohol Res Health 27: 153–160.

Prabhu VR, Porjesz B, Chorlian DB et al. (2001). Visual P3 in female alcoholics. Alcohol Clin Exp Res 25: 531–539.

Propping P, Kruger J, Mark N (1981). Genetic disposition to alcoholism. An EEG study in alcoholics and their relatives. Hum Genet 59: 51–59.

Randall CL, Roberts JS, Del Boca FK et al. (1999). Telescoping of landmark events associated with drinking: a gender comparison. J Stud Alcohol 60: 252–260.

Rangaswamy M, Porjesz B, Chorlian DB et al. (2003). Theta power in the EEG of alcoholics. Alcohol Clin Exp Res 27: 607–615.

Reitan RM (1955). The relation of the trail making test to organic brain damage. J Consult Psychol 19: 393–394.

Reitan RM, Wolfson D (1993). The Halstead-Reitan neuropsychological test battery: theory and clinical interpretation, 2nd edn, Neuropsychology Press, Tucson, AZ.

Rickenbacher E, Greve DN, Azma S et al. (2011). Effects of alcohol intoxication and gender on cerebral perfusion: an arterial spin labeling study. Alcohol 45: 725–737.

Rogers BP, Parks MH, Nickel MK et al. (2012). Reduced fronto-cerebellar functional connectivity in chronic alcoholic patients. Alcohol Clin Exp Res 36: 294–301.

Rosenbloom MJ, Pfefferbaum A, Sullivan EV (2004). Recovery of short-term memory and psychomotor speed but not postural stability with long-term sobriety in alcoholic women. Neuropsychology 18: 589–597.

Rourke SB, Grant I (2009). The neurobehavior correlates of alcoholism. In: I Grant, K Adams (Eds.), Neuropsychological Assessment of Neuropsychiatric and Neuromedical Disorders. Oxford University Press, New York, NY, pp. 398–454.

Rugg MD, Coles MGH (1996). Electrophysiology of mind: event-related brain potentials and cognition, Oxford psychology series; no. 25, Oxford University Press, Oxford.

Ruiz SM, Oscar-Berman M, Sawyer KS et al. (2013). Drinking history associations with regional white matter volumes in alcoholic men and women. Alcohol Clin Exp Res 37: 110–122.

Schuckit MA (2009). An overview of genetic influences in alcoholism. J Subst Abuse Treat 36: S5–S14.

Schuckit MA, Smith TL, Danko GP et al. (2007). The ability of the Self-Rating of the Effects of Alcohol (SRE) Scale to predict alcohol-related outcomes five years later. J Stud Alcohol Drugs 68: 371–378.

Schweinsburg BC, Alhassoon OM, Taylor MJ et al. (2003). Effects of alcoholism and gender on brain metabolism. Am J Psychiatry 160: 1180–1183.

Silberstein JA, Parsons OA (1981). Neuropsychological impairment in female alcoholics: replication and extension. J Abnorm Psychol 90: 179–182.

Sklar AL, Gilbertson R, Boissoneault J et al. (2012). Differential effects of moderate alcohol consumption on performance among older and younger adults. Alcohol Clin Exp Res 36: 2150–2156.

Smith S, Fein G (2010). Cognitive performance in treatment-naive active alcoholics. Alcohol Clin Exp Res 34: 2097–2105.

Smith S, Fein G (2011). Persistent but less severe ataxia in long-term versus short-term abstinent alcoholic men and women: a cross-sectional analysis. Alcohol Clin Exp Res 35: 2184–2192.

Spielman DM, Glover GH, Macovski A et al. (1993). Magnetic resonance spectroscopic imaging of ethanol in the human brain: a feasibility study. Alcohol Clin Exp Res 17: 1072–1077.

Squeglia LM, Schweinsburg AD, Pulido C et al. (2011). Adolescent binge drinking linked to abnormal spatial working memory brain activation: differential gender effects. Alcohol Clin Exp Res 35: 1831–1841.

Stoodley CJ (2012). There cerebellum and cognition: evidence from functional imnaging studies. Cerebellum 11: 352–365.

Strauss E, Sherman EMS, Spreen O (2006). A compendium of neuropsychological tests: administration, norms, and commentary, 3rd edn, Oxford University Press, Oxford.

Stroop JR (1935). Studies of interference in serial verbal reactions. Exp Psychol 18: 643–662.

Sullivan E, Rosenbloom M, Pfefferbaum A (2000). Pattern of motor and cognitive deficits in detoxified alcoholic men. Alcohol Clin Exp Res 24: 611–621.

Sullivan EV, Fama R, Rosenbloom MJ et al. (2002). A profile of neuropsychological deficits in alcoholic women. Neuropsychology 16: 74–83.

Sullivan EV, Rose J, Rohlfing T et al. (2009). Postural sway reduction in aging men and women: relation to brain structure, cognitive status, and stabilizing factors. Neurobiol Aging 30: 793–807.

Sullivan EV, Rose J, Pfefferbaum A (2010a). Mechanisms of postural control in alcoholic men and women: biomechanical analysis of musculoskeletal coordination during quiet standing. Alcohol Clin Exp Res 34: 528–537.

Sullivan EV, Rohlfing T, Pfefferbaum A (2010b). Pontocerebellar volume deficits and ataxia in alcoholic men and women: no evidence for "telescoping". Psychopharmacology (Berl) 208: 279–290.

Suresh S, Porjesz B, Chorlian DB et al. (2003). Auditory P3 in female alcoholics. Alcohol Clin Exp Res 27: 1064–1074.

Talland GA (1965). Deranged Memory, Academic Press, New York.

Tapert SF, Brown GG, Kindermann SS et al. (2001). fMRI measurement of brain dysfunction in alcohol-dependent young women. Alcohol Clin Exp Res 25: 236–245.

Thurstone LL (1938). Primary mental abilities, University of Chicago Press, Chicago.

Tivis LJ, Parsons OA, Glenn SW et al. (1993). Differences in cognitive impairment between VA and community treatment center alcoholics. Psychol Addict Behav 7: 43–51.

Tolentino NJ, Wierenga CE, Hall S et al. (2011). Alcohol effects on cerebral blood flow in subjects with low and high responses to alcohol. Alcohol Clin Exp Res 35: 1034–1040.

Trenerry MR (1989). Stroop Neuropsychological Screening Test manual, Psychological Assessment Resources, Odessa, FL.

Udo T, Bates ME, Mun EY et al. (2009). Gender differences in acute alcohol effects on self-regulation of arousal in response to emotional and alcohol-related picture cues. Psychol Addict Behav 23: 196–204.

Urban NB, Kegeles LS, Slifstein M et al. (2010). Sex differences in striatal dopamine release in young adults after oral alcohol challenge: a positron emission tomography imaging study with [(1)(1)C]raclopride. Biol Psychiatry 68: 689–696.

van Eijk J, Demirakca T, Frischknecht U et al. (2013). Rapid partial regeneration of brain volume during the first 14 days of abstinence from alcohol. Alcohol Clin Exp Res 37: 67–74.

Victor M, Adams RD (1953). The effect of alcohol on the nervous system. Metabolic and Toxic Diseases of the Nervous System: Proceedings of the Association for Research in Nervous and Mental Disease, Vol. 32. 526–573.

Volkow ND, Wang GJ, Franceschi D et al. (2006). Low doses of alcohol substantially decrease glucose metabolism in the human brain. Neuroimage 29: 295–301.

Volkow ND, Ma Y, Zhu W et al. (2008). Moderate doses of alcohol disrupt the functional organization of the human brain. Psychiatry Res 162: 205–213.

Wang GJ, Volkow ND, Fowler JS et al. (2003). Alcohol intoxication induces greater reductions in brain metabolism in male than in female subjects. Alcohol Clin Exp Res 27: 909–917.

Watson PE, Watson ID, Batt RD (1981). Prediction of blood alcohol concentrations in human subjects. Updating the Widmark Equation. J Stud Alcohol 42: 547–556.

Wechsler D (1987). WMS-R: Wechsler Memory Scale – Revised: Manual, The Psychological Corporation, New York.

Wechsler D (1997a). Wechsler Memory Scale. Third Edition manual, Psychological Corporation, San Antonio, TX.

Wechsler D (1997b). Weschler Adult Intelligence Scale-III, Psychological Corporation, San Antonia, TX.

Weissenborn R, Duka T (2003). Acute alcohol effects on cognitive function in social drinkers: their relationship to drinking habits. Psychopharmacology (Berl) 165: 306–312.

Winterer G, Kloppel B, Heinz A et al. (1998). Quantitative EEG (QEEG) predicts relapse in patients with chronic alcoholism and points to a frontally pronounced cerebral disturbance. Psychiatry Res 78: 101–113.

Zahr NM, Rohlfing T, Pfefferbaum A et al. (2009). Problem solving, working memory, and motor correlates of association and commissural fiber bundles in normal aging: a quantitative fiber tracking study. Neuroimage 44: 1050–1062.

Zelen SL, Fox J, Gould E et al. (1966). Sex-contingent differences between male and female alcoholics. J Clin Psychol 22: 160–165.

Section 6

Neuroimaging of brain macrostructure and microstructure

Handbook of Clinical Neurology, Vol. 125 (3rd series)
Alcohol and the Nervous System
E.V. Sullivan and A. Pfefferbaum, Editors

Chapter 17

Structural and microstructral imaging of the brain in alcohol use disorders

NATALIE M. ZAHR[1,2]*

[1]*Department of Psychiatry and Behavioral Sciences, Stanford University School of Medicine, Stanford, CA, USA*

[2]*Neuroscience Program, SRI International, Menlo Park, CA, USA*

INTRODUCTION

Apart from direct effects on the brain, excessive alcohol consumption is associated with increased risk for trauma (i.e., traumatic brain injury) (e.g., Alterman and Tarter, 1985; Chen et al., 2012), seizures (Eyer et al., 2011; Martindale et al., 2011), and stroke (de los Rios et al., 2012; Suzuki and Izumi, 2012), each of which can have effects on brain structure independently of alcohol or each other. Further, by affecting peripheral organs (including the alimentary tract (e.g., Bienia et al., 2002; Duell et al., 2012), liver (e.g., Cederbaum, 2012), heart (e.g., Roerecke et al., 2011), pancreas (e.g., Andersen et al., 2008), kidneys (e.g., Schaeffner and Ritz, 2012), and lungs (e.g., Yeligar et al., 2012)), alcohol can alter the brain. Mechanisms of indirect alcohol effects on brain via peripheral changes are likely mediated via soluble factors (e.g., de la Monte et al., 2012). Particularly well described are the effects of alcohol-related thiamine deficiency on the brain (Zahr et al., 2011). Malnutrition, vomiting, and diarrhea are common in chronic alcoholism and can contribute to thiamine deficiency (Morgan, 1982; Fields et al., 1994; Gloria et al., 1997; Ross et al., 2012). In addition, the ability of the gastrointestinal tract to absorb necessary quantities of thiamine is diminished in alcoholics (Hoyumpa, 1980; Thomson, 2000) and the liver, which stores a large part of the body's supplies of thiamine, if diseased, can have a reduced capacity to store thiamine (Levy et al., 2002; Butterworth, 2009). Consequently, the function of essential thiamine requiring enzymes in the brain (e.g., transketolase, pyruvate dehydrogenase, and α-ketoacid dehydrogenase) is compromised, leading to oxidative stress, cellular energy impairment, and eventually neuronal loss (Thomson et al., 2012).

In order to evaluate its central nervous system (CNS) effects, researchers distinguish "uncomplicated alcoholism" from the various clinically diagnosable consequences of chronic alcohol consumption, including Wernicke's encephalopathy (WE: i.e., the disease associated with thiamine deficiency), Korsakoff's syndrome (KS), hepatic encephalopathy (HE), central pontine myelinolysis (CPM), alcoholic cerebellar degeneration (ACD), alcoholic dementia, and Marchiafava–Bignami disease (MBD). The radiologic evaluation of clinically defined syndromes associated with chronic alcoholism, each with relatively unique radiologic signatures (Table 17.1 and Fig. 17.1), provides guideposts for the interrogation of the brain in uncomplicated alcoholism. The aim of this chapter, then, is to review current macrostructural findings from magnetic resonance imaging (MRI) studies and microstructural findings from diffusion tensor imaging (DTI) studies of alcohol-related CNS disorders as a framework for findings in uncomplicated alcoholism, while additionally covering results in abstinence and relapse.

STRUCTURAL MAGNETIC RESONANCE IMAGING

Since the early 1980s, conventional structural MRI has made possible the visualization of the living human brain. MRI is possible because the nuclei of certain atoms exhibit nuclear magnetic resonance (NMR). When exposed to a strong magnetic field, nuclei with NMR potential spin at a particular rate (the resonant frequency, determined by the strength of the magnetic

*Correspondence to: Natalie M. Zahr, Stanford University School of Medicine, Department of Psychiatry and Behavioral Sciences, 401 Quarry Rd, Stanford, CA 94305, USA. Tel: +1-650-859-5243, Fax: +1-650-859-2743, E-mail: nzahr@stanford.edu

Table 17.1

Primary and secondary brain regions that are targets of alcohol-related syndromes

Alcoholism-related syndrome	Abbreviation	Primary targeted region(s)	Secondary targeted regions
Wernicke's encephalopathy	WE	Mammillary bodies, periaqueductal gray matter, tissue surrounding third ventricle	Dorsal medulla, tectal plates, olivary bodies, pons
Korsakoff's syndrome	KS	Mammillary bodies, hippocampus, thalamus, orbitofrontal cortices	Cerebellum, pons
Hepatic encephalopathy	HE	Globus pallidus, substantia nigra	Corticospinal tract, cortex
Central pontine myelinolysis	CPM	Pons	Basal ganglia, thalamus, cerebral gray–white-matter junctions
Alcoholic cerebellar degeneration	ACD	Cerebellum	
Alcohol-related dementia	ARD	Frontal cortex	
Marchiafava–Bignami disease	MBD	Corpus callosum	Cortex

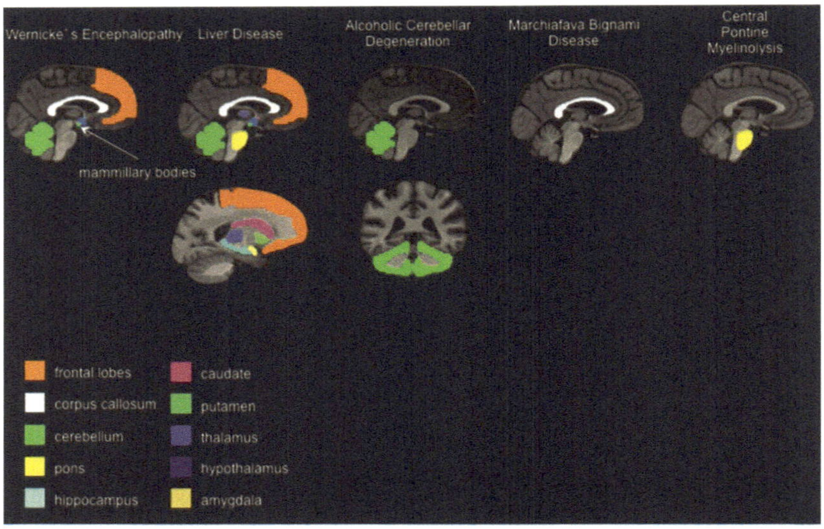

Fig. 17.1. Brain regions targeted by alcohol-related disease. (Adapted from: Zahr et al., 2011.)

field), and align themselves with the field. If nuclei are then presented with radio waves at their resonant frequency, they absorb this energy, and their alignment in the magnetic field is disturbed. When the radio waves are turned off, the relaxation (i.e., free induction decay) of the nuclei as they return to thermodynamic neutrality induces a small current in the receiver coil, an "antenna" used to detect this MR signal. Magnetic field gradients can be used to vary the static magnetic field so that variable precession frequencies become associated with different spatial locations. Hydrogen is the most frequently imaged nucleus in MRI because it is present in great abundance in biologic tissues; detailed images of the brain are possible because the different brain tissue types (i.e., gray matter, white matter, cerebrospinal fluid (CSF)) contain different proportions of water (Rumboldt et al., 2012).

The magnetic resonance image is primarily determined by three variables: proton density, T1, and T2. Proton density reflects the number of hydrogen nuclei stimulated; T1 is an exponential time constant describing the time required for nuclei to realign with the external magnetic field (higher T1 values reflect a longer time required by nuclei to return to equilibrium after perturbation); and T2 is a tissue-specific time constant describing signal loss due to the exchange of energy between hydrogen protons and adjacent nuclei. T1 and T2 tend to reflect the proportion of free to bound water in tissue.

The concentration of mobile hydrogen atoms (i.e., water) and therefore proton density, as well as T1 and T2, differs in various biologic tissue types, thereby providing the contrast necessary to visualize details of brain structure. The biologic significance of these variables, however, is not yet fully understood.

MRI, in contrast to computed tomography (CT), offers flexibility in the plane in which the brain is visualized. Because magnetic gradients can be applied in the three (orthogonal) directions to provide information about the spatial location of signals, the brain can be viewed from bottom to top (axial), as in conventional CT images, from front to back (coronal), from left to right (sagittal), or at any oblique angle to these planes. This flexibility also enables greater accuracy in aligning images with internal landmarks, an essential consideration for ensuring consistency of data from replicate images from the same individual (Rohlfing, 2006).

STRUCTURAL MRI FINDINGS IN SYNDROMES ASSOCIATED WITH ALCOHOLISM

The nutritional deficiency for thiamine in chronic alcoholism results in WE; if untreated, WE patients can develop KS, a severe neurologic disorder characterized by anterograde amnesia (Harper, 2006). In acute WE, MRI can detect symmetric, bilateral hyperintense foci, clearly visible on T2-weighted and fluid attenuation inversion recovery (FLAIR) images, in periaqueductal gray matter, mammillary bodies, and tissue surrounding the third ventricle (Lenz et al., 2002; Sullivan and Pfefferbaum, 2009) (Fig. 17.2). Additionally observed hyperintense areas in WE include the dorsal medulla, tectal plates (Ha et al., 2012), olivary bodies, and dorsal pons (Liou et al., 2012). MRI group analysis of KS patients compared with control subjects reveals substantial volume loss of the mammillary bodies in some, but not all, affected alcoholics (but see Victor et al., 1989; Shear et al., 1996; Sheedy et al., 1999; Sullivan et al., 1999b). In contrast to early MR studies suggesting that KS affects the mammillary bodies while sparing the hippocampi (Squire et al., 1990), more recent work demonstrates hippocampal volume deficits in KS (Sullivan and Marsh, 2003). Other regions affected by KS are the thalamus, orbitofrontal cortex (Jernigan et al., 1991a), cerebellum, and pons (Sullivan and Pfefferbaum, 2009).

HE, occurring in acute or chronic liver disease, including acute liver failure and cirrhosis, is believed to arise, at least partially, from elevations in circulating levels of ammonia. HE presents with confusion, disorientation, and poor coordination (Vaquero et al., 2003; Prakash and Mullen, 2010). T1-weighted imaging in HE reveals bilateral, symmetric, high-intensity signals in basal ganglia structures, particularly the globus pallidus and substantia nigra (Naegele et al., 2000; Cordoba et al., 2002) (Fig. 17.3), probably due to manganese deposition and T1 shortening (Butterworth, 2013), while T2-weighted FLAIR shows hyperintense signals along the corticospinal tract and diffuse increases in white-matter signal intensities in the cerebral hemispheres (Rovira et al., 2002, 2008). Of note, although discriminating features of WE and HE have been outlined, these diseases can be difficult to differentially diagnose and distinguish, as they can present with similar clinical features and comparable changes on MRI, especially among alcoholics (Thorarinsson et al., 2011).

CPM, associated with electrolyte disturbances, specifically with aggressive correction of hyponatraemia (Chua et al., 2002), can manifest with clinical features such as pseudobulbar palsy, ataxia, and acute changes in consciousness (Kumar et al., 2006). Classically, CPM was characterized by the presence of a symmetric triangular or "bat-wing" lesion in the pons, as visualized with hypointense T1 and hyperintense T2-weighted images (Kleinschmidt-Demasters et al., 2006) (Fig. 17.4) and reflecting demyelination (Goldman and Horoupian, 1981). Because extrapontine regions (e.g., basal ganglia, thalami, and cerebral gray–white-matter junctions) have also been reported as affected in CPM, the term "osmotic myelinolysis" has been coined (Chua et al., 2002), despite suggestions that pathology in extrapontine regions may

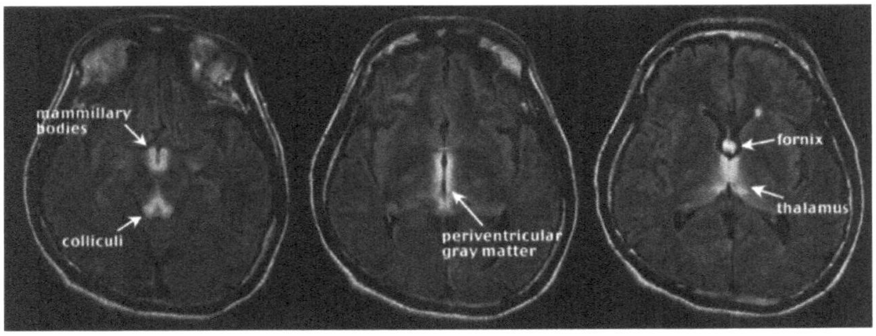

Fig. 17.2. Wernicke's encephalopathy. (Reproduced from Sullivan and Pfefferbaum, 2009.)

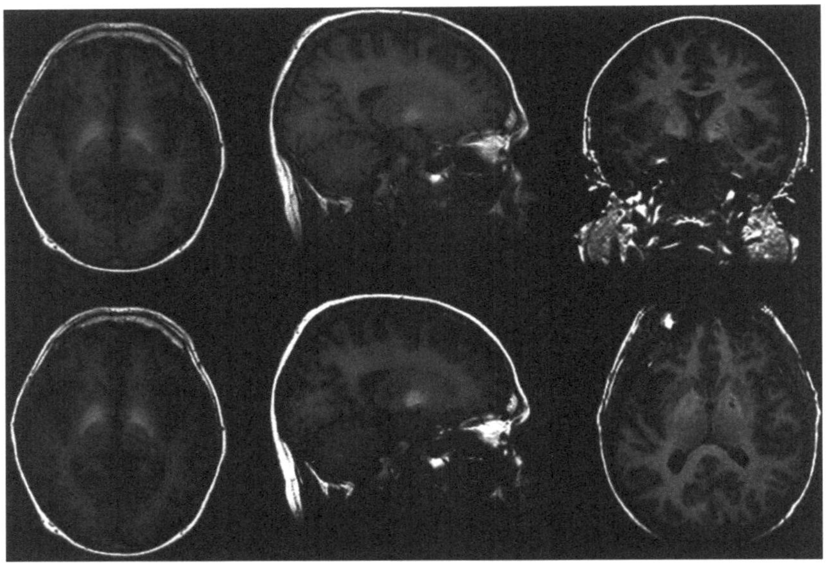

Fig. 17.3. Hepatic encephalopathy. (Reproduced from Rosenbloom et al., 2010.)

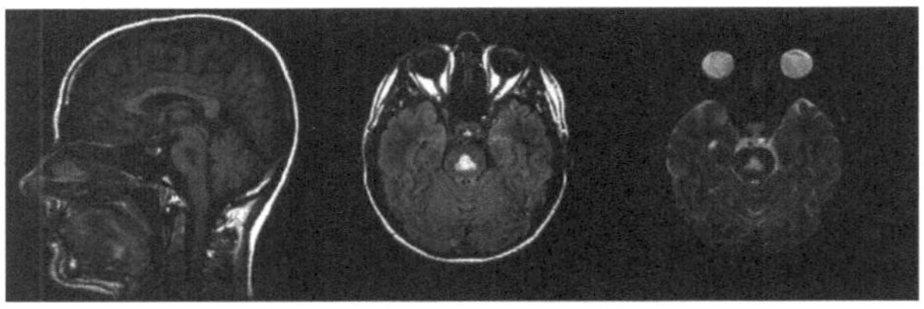

Fig. 17.4. Central pontine myelinolysis.

not strictly represent demyelination (Kleinschmidt-Demasters et al., 2006; Kumar et al., 2006).

ACD presents most frequently with ataxia, although clinical signs can include nystagmus and dysarthria (Fitzpatrick et al., 2012). Neuroimaging in ACD demonstrates damage disproportionately apparent in anterior superior portions of the cerebellar vermis (Sullivan et al., 2000c), with postmortem pathology indicating loss of cerebellar Purkinje cells (Feuerlein, 1977).

Alcoholic dementia, or alcohol-related dementia (ARD), a currently preferred term, remains a controversial diagnosis because of confounding syndromes such as WE and HE. Nevertheless, certain clinically distinguishing features of ARD include its presence in socially isolated men at a younger age of onset than other types of dementia (Draper et al., 2011; Ridley et al., 2013), deficits in visuospatial, executive, and mnemonic functions (Schmidt et al., 2005), slower progress compared with other types of dementia (Gupta and Warner, 2008), and partial reversibility (Oslin and Cary, 2003). ARD is considered a frontal dementia (Stewart, 2006); in support of such categorization, forensic evaluation of a sample of alcoholic brains noted a consistent pattern of synapse loss in the superior laminae of the frontal cortex (i.e., Brodmann's area 10), not related to liver disease (Brun and Andersson, 2001).

The pathogenesis of MBD, a disease marked by mildly impaired mental status (e.g., confusion) and sometimes dysarthria (Lee et al., 2011) or ataxia (Arbelaez et al., 2003), is poorly understood but may be related to nutritional deficiencies in addition to chronic alcohol consumption (Kawamura et al., 1985). Although traditionally characterized by demyelination and necrosis of the corpus callosum, a number of reports identify cortical lesions in so-called MBD (Johkura et al., 2005; Khaw and Heinrich, 2006; Ihn et al., 2007; Yoshizaki et al., 2010; Namekawa et al., 2013). Such data, however, represent single case studies and may reflect inaccurate MBD diagnoses.

Given the aforementioned findings in clinically differential and diagnosable alcohol-related syndromes, the question whether similar brain insults also appear in alcoholics who do not manifest the full spectrum of symptoms present in these clinically recognized conditions is posed; that is, what is observed in brains of uncomplicated alcoholics? Quantitative MRI has shown that relatively mild yet significant structural deficits characteristic of alcoholic syndromes can occur in uncomplicated alcoholics.

STRUCTURAL MRI FINDINGS IN UNCOMPLICATED ALCOHOLISM

Although less severe than observed in KS, mild volume deficits in the mammillary bodies (Shear et al., 1996; Sullivan et al., 1999a), hippocampi, and thalami are reported in uncomplicated alcoholics (Sullivan, 2003; De Bellis et al., 2005; Chanraud et al., 2007; Pitel et al., 2012; van Holst et al., 2012). These structures show a graded effect of volume deficits (KS > uncomplicated alcoholism > normal controls; Fig. 17.5). That atrophic changes are not unique to amnesic KS alcoholics implies that mammillary body damage is not a prerequisite for the development of amnesia in alcoholism (Shear et al., 1996). MR findings also show hippocampal volume deficits in alcoholics compared with controls (Sullivan et al., 1995b; Agartz et al., 1999; Laakso et al., 2000; Kurth et al., 2004; Beresford et al., 2006; Wilhelm et al., 2008). Hippocampal volume deficits in alcoholism are influenced by age (Sullivan et al., 1995a), even though age-related decline is difficult to detect in cross-sectional study (Sullivan et al., 2005a; Raz et al., 2010; Pfefferbaum et al., 2013). While deficits in hippocampal volume are not related to seizure incidence (Sullivan et al., 1996; Bleich et al., 2003), temporal lobe white matter may be sensitive to alcohol withdrawal seizures (Sullivan et al., 1996). Hippocampal volume shrinkage in alcoholism is attributed to loss of white matter and decreased axonal diameter (Harding et al., 1997). Glial cell loss (Korbo, 1999) or reduced incorporation of newly formed neurons to the dentate gyrus (Nixon and Crews, 2004; He et al., 2005), however, could also affect hippocampal volume in alcoholism.

Other regions selectively affected in WE and KS include the orbitofrontal cortices (KS), periaqueductal gray matter, and tissue surrounding the third ventricle

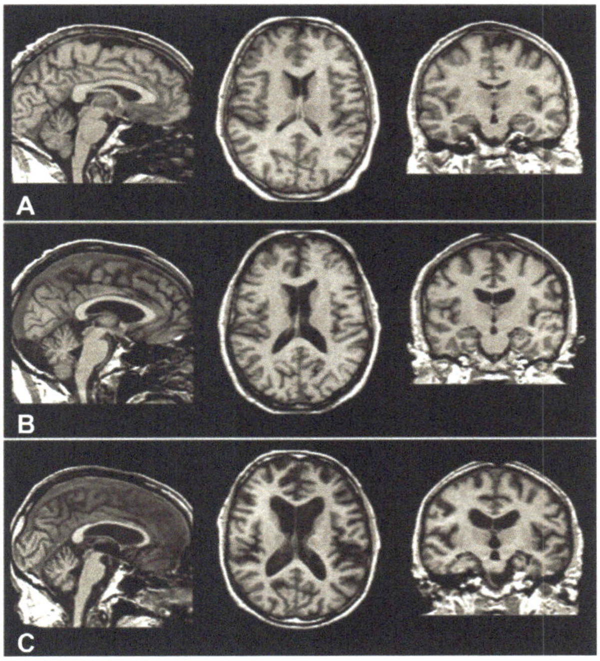

Fig. 17.5. Brain volume deficits in Korsakoff's syndrome compared with uncomplicated alcoholism. (Reproduced from Zahr et al., 2011.)

(WE). With respect to uncomplicated alcoholics, reports suggest that propensity to relapse following sobriety is related to pronounced atrophy in bilateral orbitofrontal cortices (Cardenas et al., 2011; Durazzo et al., 2011; Beck et al., 2012) and the third ventricle can be further enlarged with resumption of chronic alcohol consumption (Sullivan et al., 2000b; Pfefferbaum et al., 2001). There are currently no studies regarding periaqueductal gray-matter volume in uncomplicated alcoholics. Acute alcohol exposure to mice, however, results in robust enhancement of glutamatergic synaptic transmission and increased firing rate of glutamatergic neurons in ventral periaqueductal gray matter (Li et al., 2013). In rats, anxiety-like behavior following alcohol withdrawal also involves glutamatergic transmission in periaqueductal gray matter (Ezequiel Leite and Nobre, 2012). The key regions affected in HE include the globus pallidus and substantia nigra. Volume effects on these two structures have not been reported in uncomplicated alcoholics, but in children with fetal alcohol syndrome, globus pallidus volume is reduced in size compared with unaffected children (Nardelli et al., 2011). By contrast, other basal ganglia nodes of reward circuitry have been described as affected in uncomplicated alcoholism (Makris et al., 2008; Durazzo et al., 2011): MRI studies have revealed smaller volumes of caudate (Boutte et al., 2012), putamen (Jernigan et al., 1991b), amygdala (Fein et al., 2006), and nucleus accumbens, especially in more recently sober alcoholics compared with controls (Sullivan et al., 2005b). Given the role of the amygdala in emotional regulation and behavioral control (for review, see McBride, 2002), however, there is speculation that premorbid amygdala volume deficits put individuals at heightened risk for developing alcohol use disorders (Kamarajan et al., 2006; Benegal et al., 2007; Clarke et al., 2008).

The pons is targeted by CPM and the cerebellum by ACD. Total infratentorial volume (including pons, cerebellar hemispheres, vermis, fissures, cisterns, and fourth ventricle) is significantly smaller in uncomplicated alcoholics than controls. The volume of the pons (Pfefferbaum et al., 2002b; Sullivan, 2003; Chanraud et al., 2009c) and cerebellum (i.e., hemispheres: Sullivan et al., 2000a, c; De Bellis et al., 2005; Chanraud et al., 2007, 2009a; Boutte et al., 2012) (Fig. 17.6). is smaller in uncomplicated alcoholics than controls. Alcoholism-related volume deficits are also prevalent in gray and white matter (Shear et al., 1996; Sullivan et al., 2003) of the cerebellar vermis (Antunez et al., 1998; Piguet et al., 2006; Sullivan et al., 2006, 2010a), predominantly in anterior superior but not posterior inferior regions (Sullivan et al., 2000c).

The frontal cortex is selectively damaged in ARD. With respect to cortical regions in uncomplicated

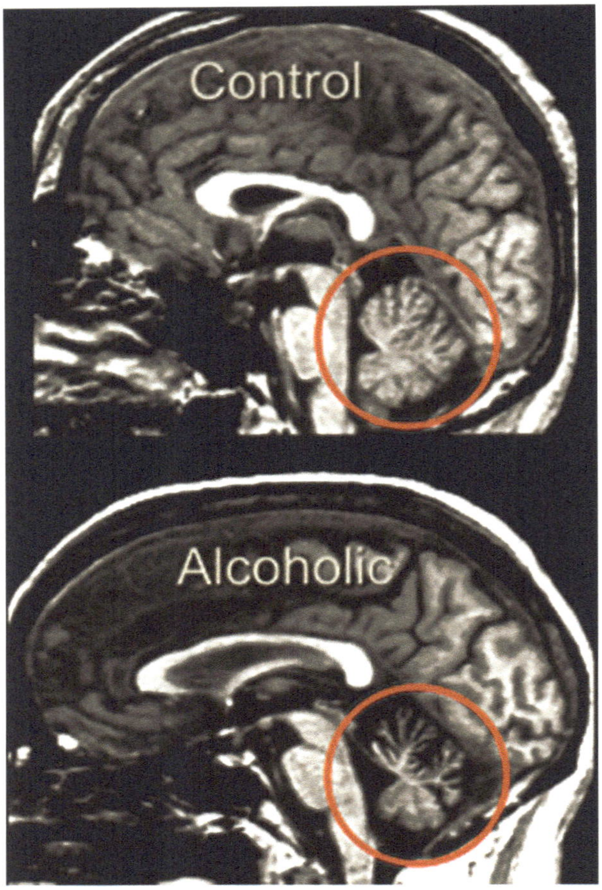

Fig. 17.6. Cerebellar volume deficits in uncomplicated alcoholism. (Reproduced from Sullivan et al., 2010b.)

alcoholism, various methods, including semiautomated procedures for brain tissue segmentation (Pfefferbaum, 1992), voxel-based morphometry (Jang et al., 2007; Mechtcheriakov et al., 2007), and deformation-based morphometry (Cardenas et al., 2007), report significant widespread shrinkage of both cortical gray and white matter with corresponding increases in CSF-filled spaces (Jernigan et al., 1991b; Pfefferbaum et al., 1992). In particular, older, but not younger, adult alcoholics show disproportionate deficits in both gray- and white-matter cortical volume, especially in the frontal lobes when volumes are statistically adjusted for brain tissue decline associated with normal aging (Pfefferbaum et al., 1997; Cardenas et al., 2005, 2007). This is the case even in comparisons made in groups selected on alcohol consumption, where older alcoholics have drunk equivalent amounts over their lifetime as younger alcoholics.

Thinning of the corpus callosum occurs in uncomplicated alcoholics and is more prominent in the anterior than posterior regions (Pfefferbaum et al., 1996; Estruch et al., 1997) (Fig. 17.7). As with WE and KS, evidence for MBD-like pathology in uncomplicated alcoholism raises the possibility of a continuum of graded brain dysmorphology.

STRUCTURAL MRI FINDINGS IN RECOVERY FROM ALCOHOLISM

Longitudinal MRI investigations show reduction of ventricular dilatation following weeks (Schroth et al., 1988; Zipursky et al., 1989) or months (Shear et al., 1994) of

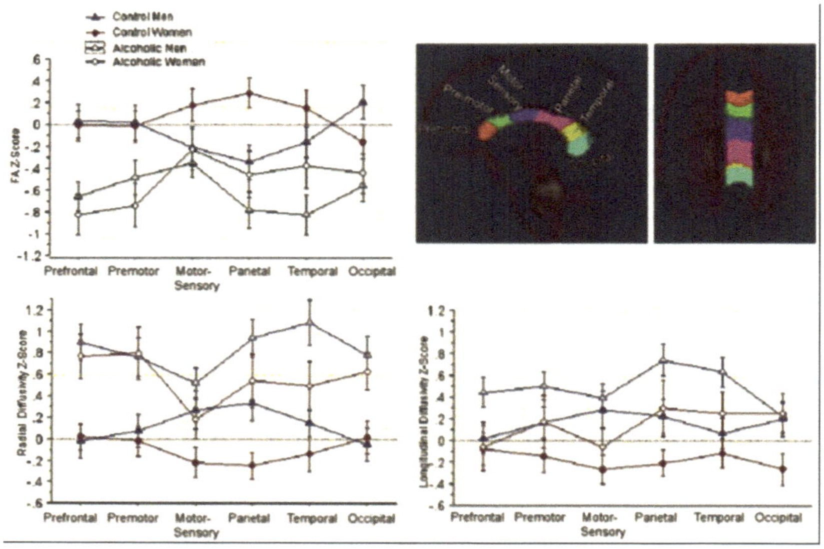

Fig. 17.7. Corpus callosum microstructural compromise in uncomplicated alcoholism. (Reproduced from Pfefferbaum et al., 2010.)

drinking cessation. Reduction of lateral ventricles precedes reduction of third ventricular volume (Pfefferbaum et al., 1995) and may be related to improvement in hematocrit, hemoglobin, and red blood cell counts (Pfefferbaum et al., 2004). Structures showing volume gains with abstinence include the entire cerebral cortex (Liu et al., 2000), temporal, insular, and anterior cingulate cortices (Cardenas et al., 2007), the amygdala (Wrase et al., 2008) (a finding which would argue against a premorbid volume deficit), thalamus (Cardenas et al., 2007), hippocampus (Liu et al., 2000; Wrase et al., 2008), brainstem, and cerebellar cortex (Liu et al., 2000; Cardenas et al., 2007).

Sober alcoholics reveal several associations between brain volume gain as determined by MRI and improvement in neuropsychologic test performance: reduction in the volume of the lateral ventricles is related to improved memory performance (Rosenbloom et al., 2007), third ventricle to improved non-verbal short-term memory performance (Sullivan et al., 2000b), and fourth ventricle to improvement in measures of ataxia (Rosenbloom et al., 2007).

The brain's capacity to return to "normal" following long-term sobriety is unknown. Short-term (6 weeks) abstinence seems sufficient to observe some brain volume recovery, but does not result in equivalent brain volumes between recovering chronic alcoholics and controls (Mann et al., 2005). Whether recovery is complete is difficult to determine. Aging is a factor since older compared with younger alcoholics exhibit reduced capacity for recovery (Fein et al., 1990; Reed et al., 1992; Rourke and Grant, 1999; Munro et al., 2000); longer periods of abstinence may be required for follow-up investigations; some brain damage, such as neuronal loss (Harper, 2007), may be irreversible, even with extended abstinence; and alcohol-dependent subjects may have premorbid brain volume deficits (Schottenbauer et al., 2007).

Despite evidence for recovery of brain volume with abstinence, the mechanisms accounting for recovery remain unclear. The hypothesis of brain rehydration was not supported by evidence for increased T2, which reflects fluctuation in brain free water levels (Schroth et al., 1988). An alternative explanation, neurogenesis (e.g., Mandyam and Koob, 2012), is not likely to be substantial enough to replace the volume loss observed in chronic alcoholism, nor is it clear that new neurons can migrate from neurogenic zones to distant areas of volume loss (Rakic, 2002). On the other hand, adequate volume recovery may be explained by white-matter regeneration since oligodendrocytes have the capacity to repair myelin and remyelinate neurons (Kipp et al., 2012) and oligodendrocyte progenitor cells have the potential to migrate long distances (Tirotta et al., 2010). Indeed, alcoholics who relapse show decrease of white matter (Pfefferbaum et al., 1995), while continued abstinence is associated with increased white matter (Shear et al., 1994), notably in the corpus callosum and subcortical white matter (Cardenas et al., 2007). DTI, by allowing microstructural examination of white matter, has begun to help answer questions such as whether alcohol predominantly affects white more than gray matter and whether brain volume recovery in abstinence can be accounted for by recovery of white-matter volume.

DIFFUSION TENSOR IMAGING

A number of books and papers provide extensive descriptions of the principles of DTI (Chien et al., 1990; Pierpaoli et al., 1996; Poupon et al., 1999; Le Bihan et al., 2001; Basser and Jones, 2002; Le Bihan, 2003; Gerig et al., 2005; Jones, 2005; Sullivan and Pfefferbaum, 2010). Here, only a brief description of DTI and its metrics is provided. DTI takes advantage of the fact that MR images of the brain are predominantly maps of water protons with contrast created by their immediate environment and their motility. In regions with few or no constraints imposed by physical boundaries, such as CSF in the ventricles, water movement is random and uniform in every direction and is therefore isotropic. In contrast to CSF, the path of a water molecule along a white-matter fiber is constrained by physical boundaries such as the axon sheath causing movement to be greater along the long axis of the fiber than across it. This movement is called anisotropic; diffusion along the long axis of a fiber (axial or longitudinal diffusion) is greater than diffusion across the fiber (radial or transverse diffusion) (Song et al., 2002).

The application of several magnetic gradients during image acquisition allows the detection of microscopic water movement. Freely diffusing particles will move more during image acquisition than those with physical restrictions. To characterize the orientation of the diffusion motion in three-dimensional space, observations are made by applying the diffusion gradients in at least six non-collinear orientations. For each voxel, the amount of diffusion is quantified by calculating the ratio of the signals with and without the diffusion gradients for each of the six or more gradient directions, resulting in at least six different diffusion-weighted images, each comprising signal decrease due to the movement of water protons in the orientation of that particular gradient application.

DTI quantification requires computation of a tensor, which is a mathematic description of a three-dimensional ellipsoid depicting the magnitude and orientation of diffusion in individual voxels. The tensor is associated with three corresponding orientational vectors (eigenvectors, λ_1, λ_2, λ_3), describing the diffusion ellipsoid by its major axes. The eigenvalue average, or trace, reflects the magnitude of diffusion, referred to as mean diffusivity (MD) or the apparent diffusion coefficient (ADC). The extent to which one eigenvalue, λ_1, dominates the other two, λ_2 and λ_3, determines the degree of anisotropy, that is, the degree of orientational preference within a voxel, typically measured as fractional anisotropy (FA), ranging between 0 and 1 on a normalized scale (Pierpaoli et al., 1996). The largest eigenvalue, λ_1, is the axial (or longitudinal) diffusivity, λ_L, and reflects axonal integrity, whereas λ_2 and λ_3 quantify radial (or transverse) diffusivity, $\lambda_T = (\lambda_2 + \lambda_3)/2$, and reflect myelin integrity (Song et al., 2002; Sun et al., 2006). Thus, disruption of white-matter microstructure detectable with DTI can reflect compromise of myelin, cytoskeletal structure, or axonal density (Basser, 1995; Basser and Pierpaoli, 1996; Spielman et al., 1996).

FA is the most commonly reported DTI metric, varying in magnitude with the characteristics of the tissue microstructure. For example, FA of the ventricular system, which contains mostly CSF, is near 0, whereas FA of the corpus callosum, where fibers are arranged in a regular and parallel fashion, can approach 0.8–0.9. Lower than expected FA (and the typically associated higher MD/ADC) in a region of fully volumed white matter can be an index of compromised white-matter integrity. FA, however, is quite sensitive to tissue inhomogeneity from crossing fibers within a voxel (Pierpaoli et al., 2001) and partial voluming (Pfefferbaum et al., 2003). Thus, if the fully volumed white-matter voxels are in a region where multiple fiber tracts cross in different directions, such as adjacent to the corpus callosum, FA will be lower, not necessarily because of reduced fiber integrity but because no single orientation predominates within a voxel (Virta et al., 1999; Pierpaoli et al., 2001).

Several approaches have been used to quantify DTI metrics, including identification of regions on FA maps or voxel-by-voxel comparison with statistical parametric mapping (http:/www.fil.ion.ucl.ac.uk/spm/) of brains normalized to a common space or template. One of the more desirable approaches is the use of quantitative fiber tracking to depict selective commissures (e.g., corpus callosum), projection fibers, and association fibers.

Whole-brain analysis

One of the approaches to DTI quantification is voxel-based analysis of whole-brain DTI data, which is useful for identifying regions that differ with respect to diffusion metrics between groups when researchers do not have an *a priori* hypothesis regarding specific brain region. In voxel-by-voxel analysis, each subject's diffusion images are registered into standard space, and then voxel-wise statistics are carried out to detect regional differences between populations or to find areas that correlate with a covariate of interest. The voxel-based approach is less sensitive than other methods of DTI quantification because spatial normalization processes can be imperfect (see Snook et al., 2007 for a further description of whole-brain DTI quantification methodologies), and a certain amount of residual morphometric differences might remain. A new approach named tract-based spatial statistics (TBSS), however, aims to remove such differences completely by using a "mean FA skeleton" (Smith et al., 2006), representing the center of fibers that is common to all subjects. Then, in the individual images, FA values can be assessed in the "heart" of the fiber of interest (Douaud et al., 2007). An advantage of TBSS over region of interest (ROI)-based methods is that it provides an automated, observer-independent method of analyzing DTI data. However, TBSS is prone to partial volume effects, is sensitive to crossing fibers in areas of multiple fiber orientation, and is inappropriately named as it is not a depiction of tracts, but a sample-based method, and the "FA skeleton" does not necessarily represent meaningful anatomic tracts.

Region of interest analysis

When a study is focused on a particular brain region, ROI analysis that involves operator-dependent manual outlining or automated parcellation routines is often employed. ROI analysis is time consuming and requires practice to achieve adequate measurement reliability. Inaccurate ROI segmentation can also result in partial voluming of tissue or CSF unintended for inclusion. Another problem is the choice of images on which to draw the ROIs. One study overcame this problem by using a tissue segmentation procedure for distinguishing different brain tissue types in the tensor images and then applying ROIs based on anatomic images to the segmented DTI data (Pfefferbaum et al., 2000a). Overall, while manual ROI placement has its own challenges, it can provide complementary information and convergent validity to whole-brain analysis and furthermore allows relevant correlational analyses with cognitive measures.

Quantitative fiber tracking and tractography

The degree to which the diffusion orientation of a voxel is similar to its neighbors, that is, shows orientational coherence between voxels (Jones et al., 1999; Pfefferbaum et al., 2000b), serves the conceptual basis for quantitative fiber tracking (e.g., Fillard and Gerig, 2003; Gerig et al., 2005; Mori et al., 2005; Le Bihan, 2007) and provides exquisite visual modeling of fiber systems. Analogous to following the linear trajectory of the longitudinal axis of bricks in a path, intervoxel coherence requires that neighboring eigenvectors do not vary by more than a set criterion and that intravoxel FA reaches a minimum value. An advantage of quantitative fiber tracking over other methods of DTI quantification is its ability to measure the diffusion properties of isolated fiber tracts along their full extent. A number of methods are now available for fiber identification and tracking quantification (e.g., Gerig et al., 2005; Mori et al., 2005; Le Bihan, 2007).

DTI FINDINGS IN SYNDROMES ASSOCIATED WITH ALCOHOLISM

Relatively few DTI studies have been conducted in the syndromes associated with alcoholism. Diffusion-weighted imaging in a 31-year-old alcoholic man with WE revealed high diffusivity specific to tissue in the mammillary bodies that resolved following 2 weeks of thiamine treatment. The authors concluded that the high diffusion was indicative of extracellular edema rather than cellular damage, which would not have resolved at follow-up (Bergui et al., 2001). Another case study indicated that, although diffusion-weighted imaging was useful in visually identifying thalamic signal abnormalities in an alcoholic with WE, diffusivity values were normal (Ducreux et al., 2002). In a final case study of WE resulting from alcohol abuse, diffusivity was low, paralleling the pattern observed in the chronic phase of ischemic stroke (Doherty et al., 2002).

HE due to alcoholism compared to other forms of HE (e.g., as a result of viral infection or primary biliary cirrhosis) appears to have different effects on DTI parameters (Miese et al., 2006). ROI (Kale et al., 2006) and voxel-based (Kumar et al., 2008) analysis in alcohol-related cirrhosis with or without HE showed that MD was elevated, while FA was not affected, in corpus callosum, internal capsule, and frontal white matter; the ROI-based study found only MD effects, while the voxel-based study found effects on both FA and MD of occipital white matter. These findings suggest an increase in the interstitial brain water in alcoholic HE patients compared with controls.

A DTI study of a single case of presumed CPM showed elevated MD in the middle cerebellar peduncles and no effects on the corticospinal tract (Min et al., 2012). FA values of seven segments of the corpus callosum in a single MBD patient compared with a normal control were diminished progressively in an anteroposterior fashion (Sair et al., 2006). Another DTI study of a single case of MBD similarly found that the corpus callosum fibers most affected were closest to the frontal horn of the lateral ventricles, presumably affecting fibers destined for the prefrontal regions of the formal lobe, while fibers to the cingulate region and from anterior corona radiata seem to be relatively spared (Pacheco et al., 2012).

DTI FINDINGS IN UNCOMPLICATED ALCOHOLISM

DTI has revealed microstructural damage related to alcoholism in cerebral areas that appear intact in structural MRI analyses (e.g., Pfefferbaum and Sullivan, 2002; Sullivan et al., 2003). Corpus callosum findings in uncomplicated alcoholics are common and, as observed for MBD, show greater anterior than posterior effects (e.g., Schulte et al., 2005; Arnone et al., 2006; Liu et al., 2010; Pitel et al., 2010; Konrad et al., 2012). A study using both voxel-based and ROI approaches in alcoholics found that FA was diminished in right but not left superior longitudinal fasciculus, orbitofrontal white matter, and cingulum bundles (Harris et al., 2008). Quantitative fiber tracking has demonstrated in alcoholics compared with controls greater FA deficits in anterior than in posterior fibers of supratentorial and infratentorial white-matter bundles as well as diminished FA in tracts of the corpus callosum, centrum semiovale (Pfefferbaum et al., 2000b, 2002a; Pfefferbaum and Sullivan, 2005), internal and external capsules, fornix, superior cingulate, longitudinal fasciculi (Pfefferbaum et al., 2009a), corticopontine bundles (Chanraud et al., 2009c), and frontolimbic fibers (Harris et al., 2008).

Alcoholics demonstrate an age–alcoholism interaction in which older alcoholics have higher diffusivity in callosal genu than would be expected based on age (Pfefferbaum et al., 2006) and women appear to have more severe white-matter fiber damage than men, despite having similar lifetime alcohol consumption (Pfefferbaum et al., 2007). Inverse correlations between FA and diffusivity in genu and centrum semiovale suggest that damage of white matter in alcoholism is attributable, at least in part, to the accumulation of intracellular and extracellular fluid in excess of that occurring in aging, and that the differential influence of these fluid compartments can vary across brain regions (Pfefferbaum and Sullivan, 2005). Exercise might attenuate the effects of heavy alcohol consumption on white-matter damage, as suggested by a report evaluating both heavy drinking and aerobic exercise (Karoly et al., 2013).

Regarding DTI–function relationships in alcoholism, FA in right anterior cingulate and left motor areas correlates with executive and psychomotor performance (Konrad et al., 2012) and splenium correlates with working memory (Pfefferbaum et al., 2000b). A double dissociation was found, showing that higher diffusivity in sensorimotor and parietal bundles was associated with poorer balance but not psychomotor speed, whereas higher diffusivity in prefrontal and temporal bundles was associated with slower psychomotor speed but not balance (Pfefferbaum et al., 2010). DTI changes in multiple supratentorial and infratentorial fiber systems in alcoholics correlate with impairment in speeded performance and postural stability (Pfefferbaum et al., 2009b), frontal fibers connecting left and right hemispheres predict performance on a coordinated psychomotor task (Rosenbloom et al., 2008), and number of reconstructed fibers running between the pons and the midbrain is related to cognitive flexibility performance (Chanraud et al., 2009c). Gray-matter diffusivity in hippocampus, which is lower in alcoholics than in controls, is related to episodic memory impairment (Chanraud et al., 2009b).

DTI FINDINGS IN RECOVERY FROM ALCOHOLISM

Similar to structural MRI findings demonstrating pronounced atrophy of orbitofrontal cortices in abstinent alcoholics who were likely to resume drinking (i.e., Cardenas et al., 2011; Durazzo et al., 2011; Beck et al., 2012), DTI identified alcoholic individuals more likely to resume drinking 6 months following initial evaluation based on lower FA and higher diffusivity in frontal white matter at baseline (Sorg et al., 2012). Increases in FA and decreases in diffusivity have been interpreted as evidence for white-matter recovery with abstinence. Substantiation for recovery has been shown in corpus callosum genu and body at 1 year compared with 2 weeks of abstinence (Alhassoon et al., 2012), and in frontal white matter at 1 month compared with 1 week of abstinence, at least in non-smoking, sober alcoholics (Gazdzinski et al., 2010).

CONCLUSION

Despite evidence for damage, alcoholics who maintain sobriety over extended periods show improvements in both brain volume and white matter integrity, potentially reflecting fiber reorganization and myelin restoration, indicative of a neural mechanism explaining recovery and perhaps enhancing chances for sustained sobriety.

References

Agartz I, Momenan R, Rawlings RR et al. (1999). Hippocampal volume in patients with alcohol dependence. Arch Gen Psychiatry 56: 356–363.

Alhassoon OM, Sorg SF, Taylor MJ et al. (2012). Callosal white matter microstructural recovery in abstinent alcoholics: a longitudinal diffusion tensor imaging study. Alcohol Clin Exp Res 36: 1922–1931.

Alterman AI, Tarter RE (1985). Relationship between familial alcoholism and head injury. J Stud Alcohol 46: 256–258.

Andersen AM, Novovic S, Ersboll AK et al. (2008). Mortality in alcohol and biliary acute pancreatitis. Pancreas 36: 432–434.

Antunez E, Estruch R, Cardenal C et al. (1998). Usefulness of CT and MR imaging in the diagnosis of acute Wernicke's encephalopathy. AJR Am J Roentgenol 171: 1131–1137.

Arbelaez A, Pajon A, Castillo M (2003). Acute Marchiafava-Bignami disease: MR findings in two patients. AJNR Am J Neuroradiol 24: 1955–1957.

Arnone D, Abou-Saleh MT, Barrick TR (2006). Diffusion tensor imaging of the corpus callosum in addiction. Neuropsychobiology 54: 107–113.

Basser PJ (1995). Inferring microstructural features and the physiological state of tissues from diffusion-weighted images. NMR Biomed 8: 333–344.

Basser PJ, Jones DK (2002). Diffusion-tensor MRI: theory, experimental design and data analysis – a technical review. NMR Biomed 15: 456–467.

Basser PJ, Pierpaoli C (1996). Microstructural and physiological features of tissues elucidated by quantitative diffusion tensor MRI. J Magn Reson B 111: 209–219.

Beck A, Wustenberg T, Genauck A et al. (2012). Effect of brain structure, brain function, and brain connectivity on relapse in alcohol-dependent patients. Arch Gen Psychiatry 69: 842–852.

Benegal V, Antony G, Venkatasubramanian G et al. (2007). Gray matter volume abnormalities and externalizing symptoms in subjects at high risk for alcohol dependence. Addict Biol 12: 122–132.

Beresford TP, Arciniegas DB, Alfers J et al. (2006). Hippocampus volume loss due to chronic heavy drinking. Alcohol Clin Exp Res 30: 1866–1870.

Bergui M, Bradac GB, Zhong JJ et al. (2001). Diffusion-weighted MR in reversible Wernicke encephalopathy. Neuroradiology 43: 969–972.

Bienia A, Sodolski W, Luchowska E (2002). The effect of chronic alcohol abuse on gastric and duodenal mucosa. Ann Univ Mariae Curie Sklodowska [Med] 57: 570–582.

Bleich S, Sperling W, Degner D et al. (2003). Lack of association between hippocampal volume reduction and first-onset alcohol withdrawal seizure. A volumetric MRI study. Alcohol Alcohol 38: 40–44.

Boutte D, Calhoun VD, Chen J et al. (2012). Association of genetic copy number variations at 11 q14.2 with brain regional volume differences in an alcohol use disorder population. Alcohol 46: 519–527.

Brun A, Andersson J (2001). Frontal dysfunction and frontal cortical synapse loss in alcoholism – the main cause of alcohol dementia? Dement Geriatr Cogn Disord 12: 289–294.

Butterworth RF (2009). Thiamine deficiency-related brain dysfunction in chronic liver failure. Metab Brain Dis 24: 189–196.

Butterworth RF (2013). Parkinsonism in cirrhosis: pathogenesis and current therapeutic options. Metab Brain Dis 28: 261–267.

Cardenas VA, Studholme C, Meyerhoff DJ et al. (2005). Chronic active heavy drinking and family history of problem drinking modulate regional brain tissue volumes. Psychiatry Res 138: 115–130.

Cardenas VA, Studholme C, Gazdzinski S et al. (2007). Deformation-based morphometry of brain changes in alcohol dependence and abstinence. Neuroimage 34: 879–887.

Cardenas VA, Durazzo TC, Gazdzinski S et al. (2011). Brain morphology at entry into treatment for alcohol dependence is related to relapse propensity. Biol Psychiatry 70: 561–567.

Cederbaum AI (2012). Alcohol metabolism. Clin Liver Dis 16: 667–685.

Chanraud S, Martelli C, Delain F et al. (2007). Brain morphometry and cognitive performance in detoxified alcohol-dependents with preserved psychosocial functioning. Neuropsychopharmacology 32: 429–438.

Chanraud S, Leroy C, Martelli C et al. (2009a). Episodic memory in detoxified alcoholics: contribution of grey matter microstructure alteration. PLoS One 4: e6786.

Chanraud S, Reynaud M, Wessa M et al. (2009b). Diffusion tensor tractography in mesencephalic bundles: relation to mental flexibility in detoxified alcohol-dependent subject. Neuropsychopharmacology 34: 1223–1232.

Chanraud S, Reynaud M, Wessa M et al. (2009c). Diffusion tensor tractography in mesencephalic bundles: relation to mental flexibility in detoxified alcohol-dependent subjects. Neuropsychopharmacology 34: 1223–1232.

Chen CM, Yi HY, Yoon YH et al. (2012). Alcohol use at time of injury and survival following traumatic brain injury: results from the National Trauma Data Bank. J Stud Alcohol Drugs 73: 531–541.

Chien D, Buxton RB, Kwong KK et al. (1990). MR diffusion imaging of the human brain. J Comput Assist Tomogr 14: 514–520.

Chua GC, Sitoh YY, Lim CC et al. (2002). MRI findings in osmotic myelinolysis. Clin Radiol 57: 800–806.

Clarke TK, Treutlein J, Zimmermann US et al. (2008). HPA-axis activity in alcoholism: examples for a gene–environment interaction. Addict Biol 13: 1–14.

Cordoba J, Sanpedro F, Alonso J et al. (2002). 1H magnetic resonance in the study of hepatic encephalopathy in humans. Metab Brain Dis 17: 415–429.

De Bellis MD, Narasimhan A, Thatcher DL et al. (2005). Prefrontal cortex, thalamus, and cerebellar volumes in adolescents and young adults with adolescent-onset alcohol use disorders and comorbid mental disorders. Alcohol Clin Exp Res 29: 1590–1600.

de la Monte S, Derdak Z, Wands JR (2012). Alcohol, insulin resistance and the liver-brain axis. J Gastroenterol Hepatol 27 (Suppl 2): 33–41.

de los Rios F, Kleindorfer DO, Khoury J et al. (2012). Trends in substance abuse preceding stroke among young adults: a population-based study. Stroke 43: 3179–3183.

Doherty MJ, Watson NF, Uchino K et al. (2002). Diffusion abnormalities in patients with Wernicke encephalopathy. Neurology 58: 655–657.

Douaud G, Smith S, Jenkinson M et al. (2007). Anatomically related grey and white matter abnormalities in adolescent-onset schizophrenia. Brain 130: 2375–2386.

Draper B, Karmel R, Gibson D et al. (2011). Alcohol-related cognitive impairment in New South Wales hospital patients aged 50 years and over. Aust N Z J Psychiatry 45: 985–992.

Ducreux D, Petit-Lacour MC, Benoudiba F et al. (2002). Diffusion-weighted imaging in a case of Wernicke encephalopathy. J Neuroradiol 29: 39–42.

Duell EJ, Sala N, Travier N et al. (2012). Genetic variation in alcohol dehydrogenase (ADH1A, ADH1B, ADH1C, ADH7) and aldehyde dehydrogenase (ALDH2), alcohol consumption and gastric cancer risk in the European Prospective Investigation into Cancer and Nutrition (EPIC) cohort. Carcinogenesis 33: 361–367.

Durazzo TC, Tosun D, Buckley S et al. (2011). Cortical thickness, surface area, and volume of the brain reward system in alcohol dependence: relationships to relapse and extended abstinence. Alcohol Clin Exp Res 35: 1187–1200.

Estruch R, Nicolas JM, Salamero M et al. (1997). Atrophy of the corpus callosum in chronic alcoholism. J Neurol Sci 146: 145–151.

Eyer F, Schuster T, Felgenhauer N et al. (2011). Risk assessment of moderate to severe alcohol withdrawal – predictors for seizures and delirium tremens in the course of withdrawal. Alcohol Alcohol 46: 427–433.

Ezequiel Leite L, Nobre MJ (2012). The negative effects of alcohol hangover on high-anxiety phenotype rats are influenced by the glutamate receptors of the dorsal midbrain. Neuroscience 213: 93–105.

Fein G, Bachman L, Fisher S et al. (1990). Cognitive impairments in abstinent alcoholics. West J Med 152: 531–537.

Fein G, Landman B, Tran H et al. (2006). Brain atrophy in long-term abstinent alcoholics who demonstrate impairment on a simulated gambling task. Neuroimage 32: 1465–1471.

Feuerlein W (1977). Neuropsychiatric disorders of alcoholism. Nutr Metab 21: 163–174.

Fields JZ, Turk A, Durkin M et al. (1994). Increased gastrointestinal symptoms in chronic alcoholics. Am J Gastroenterol 89: 382–386.

Fillard P, Gerig G (2003). Analysis tool for diffusion tensor MRI. In: Proceedings of Medical Image Computing and Computer-assisted Intervention, Lecture Notes in Computer Science. Saint-Malo, Springer, France, pp. 967–968.

Fitzpatrick LE, Jackson M, Crowe SF (2012). Characterization of cerebellar ataxia in chronic alcoholics using the International Cooperative Ataxia Rating Scale (ICARS). Alcohol Clin Exp Res 36: 1942–1951.

Gazdzinski S, Durazzo TC, Mon A et al. (2010). Cerebral white matter recovery in abstinent alcoholics – a multimodality magnetic resonance study. Brain 133: 1043–1053.

Gerig G, Corouge I, Vachet C et al. (2005). Quantitative analysis of diffusion properties of white matter fiber tracts: a validation study. 13th Proceedings of the International Society for Magnetic Resonance in Medicine, Abstract no. 1337, International Society for Magnetic Resonance in Medicine, Miami, FL.

Gloria L, Cravo M, Camilo ME et al. (1997). Nutritional deficiencies in chronic alcoholics: relation to dietary intake and alcohol consumption. Am J Gastroenterol 92: 485–489.

Goldman JE, Horoupian DS (1981). Demyelination of the lateral geniculate nucleus in central pontine myelinolysis. Ann Neurol 9: 185–189.

Gupta S, Warner J (2008). Alcohol-related dementia: a 21st-century silent epidemic? Br J Psychiatry 193: 351–353.

Ha ND, Weon YC, Jang JC et al. (2012). Spectrum of MR imaging findings in Wernicke encephalopathy: are atypical areas of involvement only present in nonalcoholic patients? AJNR Am J Neuroradiol 33: 1398–1402.

Harding AJ, Wong A, Svoboda M et al. (1997). Chronic alcohol consumption does not cause hippocampal neuron loss in humans. Hippocampus 7: 78–87.

Harper C (2006). Thiamine (vitamin B1) deficiency and associated brain damage is still common throughout the world and prevention is simple and safe! Eur J Neurol 13: 1078–1082.

Harper C (2007). The neurotoxicity of alcohol. Hum Exp Toxicol 26: 251–257.

Harris GJ, Jaffin SK, Hodge SM et al. (2008). Frontal white matter and cingulum diffusion tensor imaging deficits in alcoholism. Alcohol Clin Exp Res 32: 1001–1013.

He DY, McGough NN, Ravindranathan A et al. (2005). Glial cell line-derived neurotrophic factor mediates the desirable actions of the anti-addiction drug ibogaine against alcohol consumption. J Neurosci 25: 619–628.

Hoyumpa Jr AM (1980). Mechanisms of thiamin deficiency in chronic alcoholism. Am J Clin Nutr 33: 2750–2761.

Ihn YK, Hwang SS, Park YH (2007). Acute Marchiafava-Bignami disease: diffusion-weighted MRI in cortical and callosal involvement. Yonsei Med J 48: 321–324.

Jang DP, Namkoong K, Kim JJ et al. (2007). The relationship between brain morphometry and neuropsychological performance in alcohol dependence. Neurosci Lett 428: 21–26.

Jernigan TL, Schafer K, Butters N et al. (1991a). Magnetic resonance imaging of alcoholic Korsakoff patients. Neuropsychopharmacology 4: 175–186.

Jernigan TL, Butters N, DiTraglia G et al. (1991b). Reduced cerebral grey matter observed in alcoholics using magnetic resonance imaging. Alcohol Clin Exp Res 15: 418–427.

Johkura K, Naito M, Naka T (2005). Cortical involvement in Marchiafava-Bignami disease. AJNR Am J Neuroradiol 26: 670–673.

Jones DK (2005). Fundamentals of diffusion MR imaging. In: J Gillard, AD Waldman, P Barker (Eds.), Clincal MR Neuroimaging, Cambridge University Press, Cambridge.

Jones D, Simmons A, Williams S et al. (1999). Non-invasive assessment of axonal fiber connectivity in the human brain via diffusion tensor MRI. Magn Reson Med 42: 37–41.

Kale RA, Gupta RK, Saraswat VA et al. (2006). Demonstration of interstitial cerebral edema with diffusion tensor MR imaging in type C hepatic encephalopathy. Hepatology 43: 698–706.

Kamarajan C, Porjesz B, Jones K et al. (2006). Event-related oscillations in offspring of alcoholics: neurocognitive disinhibition as a risk for alcoholism. Biol Psychiatry 59: 625–634.

Karoly HC, Stevens CJ, Thayer RE et al. (2013). Aerobic exercise moderates the effect of heavy alcohol consumption on white matter damage. Alcohol Clin Exp Res 37: 1508–1515.

Kawamura M, Shiota J, Yagishita T et al. (1985). Marchiafava-Bignami disease: computed tomographic scan and magnetic resonance imaging. Ann Neurol 18: 103–104.

Khaw AV, Heinrich A (2006). Marchiafava-Bignami disease: diffusion-weighted MRI in corpus callosum and cortical lesions. Neurology 66: 1286, author reply 1286.

Kipp M, Victor M, Martino G et al. (2012). Endogeneous remyelination: findings in human studies. CNS Neurol Disord Drug Targets 11: 598–609.

Kleinschmidt-Demasters BK, Rojiani AM, Filley CM (2006). Central and extrapontine myelinolysis: then. . .and now. J Neuropathol Exp Neurol 65: 1–11.

Konrad A, Vucurevic G, Lorscheider M et al. (2012). Broad disruption of brain white matter microstructure and relationship with neuropsychological performance in male patients with severe alcohol dependence. Alcohol Alcohol 47: 118–126.

Korbo L (1999). Glial cell loss in the hippocampus of alcoholics. Alcohol Clin Exp Res 23: 164–168.

Kumar S, Fowler M, Gonzalez-Toledo E et al. (2006). Central pontine myelinolysis, an update. Neurol Res 28: 360–366.

Kumar R, Gupta RK, Elderkin-Thompson V et al. (2008). Voxel-based diffusion tensor magnetic resonance imaging evaluation of low-grade hepatic encephalopathy. J Magn Reson Imaging 27: 1061–1068.

Kurth C, Wegerer V, Reulbach U et al. (2004). Analysis of hippocampal atrophy in alcoholic patients by a Kohonen feature map. Neuroreport 15: 367–371.

Laakso MP, Vaurio O, Savolainen L et al. (2000). A volumetric MRI study of the hippocampus in type 1 and 2 alcoholism. Behav Brain Res 109: 177–186.

Le Bihan D (2003). Looking into the functional architecture of the brain with diffusion MRI. Nat Rev Neurosci 4: 469–480.

Le Bihan D (2007). The 'wet mind': water and functional neuroimaging. Phys Med Biol 52: R57–R90.

Le Bihan D, Mangin J, Poupon C et al. (2001). Diffusion tensor imaging: concepts and applications. J Magn Reson Imaging 13: 534–546.

Lee SH, Kim SS, Kim SH et al. (2011). Acute Marchiafava-Bignami disease with selective involvement of the precentral cortex and splenium: a serial magnetic resonance imaging study. Neurologist 17: 213–217.

Lenz V, Vargas MI, Bin JF et al. (2002). Value of MRI findings in Gayet-Wernicke encephalopathy. J Neuroradiol 29: 153–160.

Levy S, Herve C, Delacoux E et al. (2002). Thiamine deficiency in hepatitis C virus and alcohol-related liver diseases. Dig Dis Sci 47: 543–548.

Li C, McCall NM, Lopez AJ et al. (2013). Alcohol effects on synaptic transmission in periaqueductal gray dopamine neurons. Alcohol 47: 279–287.

Liou KC, Kuo SF, Chen LA (2012). Wernicke encephalopathy with atypical magnetic resonance imaging. Am J Emerg Med 30 (2086): e2081–e2083.

Liu RSN, Lemieux L, Shorvon SD et al. (2000). Association between brain size and abstinence from alcohol. Lancet 355: 1969–1970.

Liu IC, Chiu CH, Chen CJ et al. (2010). The microstructural integrity of the corpus callosum and associated impulsivity in alcohol dependence: a tractography-based segmentation study using diffusion spectrum imaging. Psychiatry Res 184: 128–134.

Makris N, Oscar-Berman M, Jaffin SK et al. (2008). Decreased volume of the brain reward system in alcoholism. Biol Psychiatry 64: 192–202.

Mandyam CD, Koob GF (2012). The addicted brain craves new neurons: putative role for adult-born progenitors in promoting recovery. Trends Neurosci 35: 250–260.

Mann KF, Ackermann K, Croissan B et al. (2005). Neuroimaging of gender differences in alcoholism: are women more vulnerable? Alcohol Clin Exp Res 29: 896–901.

Martindale JL, Goldstein JN, Pallin DJ (2011). Emergency department seizure epidemiology. Emerg Med Clin North Am 29: 15–27.

McBride WJ (2002). Central nucleus of the amygdala and the effects of alcohol and alcohol-drinking behavior in rodents. Pharmacol Biochem Behav 71: 509–515.

Mechtcheriakov S, Brenneis C, Egger K et al. (2007). A widespread distinct pattern of cerebral atrophy in patients with alcohol addiction revealed by voxel-based morphometry. J Neurol Neurosurg Psychiatry 78: 610–614.

Miese F, Kircheis G, Wittsack HJ et al. (2006). 1H-MR spectroscopy, magnetization transfer, and diffusion-weighted imaging in alcoholic and nonalcoholic patients with cirrhosis with hepatic encephalopathy. AJNR Am J Neuroradiol 27: 1019–1026.

Min Y, Park SH, Hwang SB (2012). Corticospinal tract and pontocerebellar fiber of central pontine myelinolysis. Ann Rehabil Med 36: 887–892.

Morgan MY (1982). Alcohol and nutrition. Br Med Bull 38: 21–29.

Mori S, Wakana S, Nagae-Poetscher LM et al. (2005). An atlas of human white matter, Elsevier, Amsterdam, The Netherlands.

Munro CA, Saxton J, Butters MA (2000). The neuropsychological consequences of abstinence among older alcoholics: a cross-sectional study. Alcohol Clin Exp Res 24: 1510–1516.

Naegele T, Grodd W, Viebahn R et al. (2000). MR imaging and (1)H spectroscopy of brain metabolites in hepatic encephalopathy: time-course of renormalization after liver transplantation. Radiology 216: 683–691.

Namekawa M, Nakamura Y, Nakano I (2013). Cortical involvement in Marchiafava-Bignami disease can be a predictor of a poor prognosis: a case report and review of the literature. Intern Med 52: 811–813.

Nardelli A, Lebel C, Rasmussen C et al. (2011). Extensive deep gray matter volume reductions in children and adolescents with fetal alcohol spectrum disorders. Alcohol Clin Exp Res 35: 1404–1417.

Nixon K, Crews FT (2004). Temporally specific burst in cell proliferation increases hippocampal neurogenesis in protracted abstinence from alcohol. J Neurosci 24: 9714–9722.

Oslin DW, Cary MS (2003). Alcohol-related dementia: validation of diagnostic criteria. Am J Geriatr Psychiatry 11: 441–447.

Pacheco FT, Rego MM, do Rego JI et al. (2012). "Ears of the lynx" sign in a Marchiafava-Bignami patient: structural basis and fiber-tracking DTI contribution to the understanding of this imaging abnormality. J Neuroimaging 24: 205–207.

Pfefferbaum A (1992). Use of MRI to evaluate structural changes in addictive medicine. In: Proceedings of the American Society of Addiction Medicine's 23rd Annual Medical-Scientific Conference.

Pfefferbaum A, Sullivan EV (2002). Microstructural but not macrostructural disruption of white matter in women with chronic alcoholism. Neuroimage 15: 708–718.

Pfefferbaum A, Sullivan EV (2005). Disruption of brain white matter microstructure by excessive intracellular and extracellular fluid in alcoholism: evidence from diffusion tensor imaging. Neuropsychopharmacology 30: 423–432.

Pfefferbaum A, Lim KO, Rosenbloom MJ (1992). Structural imaging of the brain in chronic alcoholism. Imaging in Alcohol Research: NIAAA Research Monograph S Zakhari, E Witt (Eds.), 21: U.S. Dept Health and Human Services, Rockville, MD, pp. 99–120.

Pfefferbaum A, Sullivan EV, Mathalon DH et al. (1995). Longitudinal changes in magnetic resonance imaging brain volumes in abstinent and relapsed alcoholics. Alcohol Clin Exp Res 19: 1177–1191.

Pfefferbaum A, Lim KO, Desmond JE et al. (1996). Thinning of the corpus callosum in older alcoholic men: a magnetic resonance imaging study. Alcohol Clin Exp Res 20: 752–757.

Pfefferbaum A, Sullivan EV, Mathalon DH et al. (1997). Frontal lobe volume loss observed with magnetic resonance imaging in older chronic alcoholics. Alcohol Clin Exp Res 21: 521–529.

Pfefferbaum A, Sullivan EV, Hedehus M et al. (2000a). Age-related decline in brain white matter anisotropy measured with spatially corrected echo-planar diffusion tensor imaging. Magn Reson Med 44: 259–268.

Pfefferbaum A, Sullivan EV, Hedehus M et al. (2000b). In vivo detection and functional correlates of white matter microstructural disruption in chronic alcoholism. Alcohol Clin Exp Res 24: 1214–1221.

Pfefferbaum A, Rosenbloom M, Deshmukh A et al. (2001). Sex differences in the effects of alcohol on brain structure. Am J Psychiatry 158: 188–197.

Pfefferbaum A, Rosenbloom M, Sullivan EV (2002a). Alcoholism and AIDS: magnetic resonance imaging approaches for detecting interactive neuropathology. Alcohol Clin Exp Res 26: 1031–1046.

Pfefferbaum A, Rosenbloom MJ, Serventi K et al. (2002b). Corpus callosum, pons and cortical white matter in alcoholic women. Alcohol Clin Exp Res 26: 400–405.

Pfefferbaum A, Adalsteinsson E, Sullivan EV (2003). Replicability of diffusion tensor imaging measurements of fractional anisotropy and trace in brain. J Magn Reson Imaging 18: 427–433.

Pfefferbaum A, Rosenbloom MJ, Serventi KL et al. (2004). Brain volumes, RBC status, and hepatic function in alcoholics after 1 and 4 weeks of sobriety: predictors of outcome. Am J Psychiatry 161: 1190–1196.

Pfefferbaum A, Adalsteinsson E, Sullivan EV (2006). Dysmorphology and microstructural degradation of the corpus callosum: interaction of age and alcoholism. Neurobiol Aging 27: 994–1009.

Pfefferbaum A, Rosenbloom MJ, Adalsteinsson E et al. (2007). Diffusion tensor imaging with quantitative fiber tracking in HIV infection and alcoholism comorbidity: synergistic white matter damage. Brain 130: 48–64.

Pfefferbaum A, Rosenbloom M, Rohlfing T et al. (2009a). Degradation of association and projection white matter systems in alcoholism detected with quantitative fiber tracking. Biol Psychiatry 65: 680–690.

Pfefferbaum A, Rosenbloom MJ, Rohlfing T et al. (2009b). Degradation of association and projection white matter systems in alcoholism detected with quantitative fiber tracking. Biol Psychiatry 65: 680–690.

Pfefferbaum A, Rosenbloom MJ, Fama R et al. (2010). Transcallosal white matter degradation detected with quantitative fiber tracking in alcoholic men and women: selective relations to dissociable functions. Alcohol Clin Exp Res 34: 1201–1211.

Pfefferbaum A, Rohlfing T, Rosenbloom MJ et al. (2013). Variation in longitudinal trajectories of regional brain volumes of healthy men and women (ages 10 to 85 years) measured with atlas-based parcellation of MRI. Neuroimage 65: 176–193.

Pierpaoli C, Jezzard P, Basser PJ et al. (1996). Diffusion tensor MR imaging of the human brain. Radiology 201: 637–648.

Pierpaoli C, Barnett A, Pajevic S et al. (2001). Water diffusion changes in Wallerian degeneration and their dependence on white matter architecture. Neuroimage 13: 1174–1185.

Piguet O, Cramsie J, Bennett HP et al. (2006). Contributions of age and alcohol consumption to cerebellar integrity, gait and cognition in non-demented very old individuals. Eur Arch Psychiatry Clin Neurosci 256: 504–511.

Pitel A-L, Chanraud S, Sullivan E et al. (2010). Callosal microstructural abnormalities in Alzheimer disease and alcoholism: same phenotype, different mechanisms. Psychiatry Research: Neuroimaging 184: 49–56.

Pitel AL, Chetelat G, Le Berre AP et al. (2012). Macrostructural abnormalities in Korsakoff syndrome compared with uncomplicated alcoholism. Neurology 78: 1330–1333.

Poupon C, Clark CA, Frouin V et al. (1999). Tracking white matter fascicles with diffusion tensory imaging. Proceedings of the Seventh International Society for Magnetic Resonance in Medicine; 325.

Prakash R, Mullen KD (2010). Mechanisms, diagnosis and management of hepatic encephalopathy. Nat Rev Gastroenterol Hepatol 7: 515–525.

Rakic P (2002). Neurogenesis in adult primates. Prog Brain Res 138: 3–14.

Raz N, Ghisletta P, Rodrigue KM et al. (2010). Trajectories of brain aging in middle-aged and older adults: regional and individual differences. Neuroimage 51: 501–511.

Reed RJ, Grant I, Rourke SB (1992). Long-term abstinent alcoholics have normal memory. Alcohol Clin Exp Res 16: 677–683.

Ridley NJ, Draper B, Withall A (2013). Alcohol-related dementia: an update of the evidence. Alzheimers Res Ther 5: 3.

Roerecke M, Greenfield TK, Kerr WC et al. (2011). Heavy drinking occasions in relation to ischaemic heart disease mortality – an 11-22 year follow-up of the 1984 and 1995 US National Alcohol Surveys. Int J Epidemiol 40: 1401–1410.

Rohlfing T (2006). Transformation model and constraints cause bias in statistics on deformation fields. Med Image Comput Comput Assist Interv Int Conf Med Image Comput Comput Assist Interv 9: 207–214.

Rosenbloom MJ, Rohlfing T, O'Reilly A et al. (2007). Improvement in memory and static balance with abstinence in alcoholic men and women: selective relations with changes in regional ventricular volumes. Psychiatry Res Neuroimaging 155: 91–102.

Rosenbloom MJ, Sassoon SA, Fama R et al. (2008). Frontal callosal fiber integrity selectively predicts coordinated psychomotor performance in alcoholism. Brain Imaging Behav 2: 74–83.

Rosenbloom MJ, Sullivan EV, Pfefferbaum A (2010). Effect of neurotoxins on brain tissue and function revealed through in vivo MR imaging and spectroscopy. In: GJ Harry, HA Tilson (Eds.), Neurotoxicology, Informa Healthcare, New York.

Ross LJ, Wilson M, Banks M et al. (2012). Prevalence of malnutrition and nutritional risk factors in patients undergoing alcohol and drug treatment. Nutrition 28: 738–743.

Rourke SB, Grant I (1999). The interactive effects of age and length of abstinence on the recovery of neuropsychological functioning in chronic male alcoholics: a 2-year follow-up study. J Int Neuropsychol Soc 5: 234–246.

Rovira A, Cordoba J, Sanpedro F et al. (2002). Normalization of T2 signal abnormalities in hemispheric white matter with liver transplant. Neurology 59: 335–341.

Rovira A, Alonso J, Cordoba J (2008). MR imaging findings in hepatic encephalopathy. AJNR Am J Neuroradiol 29: 1612–1621.

Rumboldt Z, Castillo M, Huang B et al. (2012). Brain Imaging with MRI and CT: An Image Pattern Approach, Cambridge University Press, Cambridge.

Sair HI, Mohamed FB, Patel S et al. (2006). Diffusion tensor imaging and fiber-tracking in Marchiafava-Bignami disease. J Neuroimaging 16: 281–285.

Schaeffner E, Ritz E (2012). Alcohol and kidney damage: a Janus-faced relationship. Kidney Int 81: 816–818.

Schmidt KS, Gallo JL, Ferri C et al. (2005). The neuropsychological profile of alcohol-related dementia suggests cortical and subcortical pathology. Dement Geriatr Cogn Disord 20: 286–291.

Schottenbauer MA, Momenan R, Kerick M et al. (2007). Relationships among aging, IQ, and intracranial volume in alcoholics and control subjects. Neuropsychology 21: 337–345.

Schroth G, Naegele T, Klose U et al. (1988). Reversible brain shrinkage in abstinent alcoholics, measured by MRI. Neuroradiology 30: 385–389.

Schulte T, Sullivan EV, Muller-Oehring EM et al. (2005). Corpus callosal microstructural integrity influences interhemispheric processing: a diffusion tensor imaging study. Cereb Cortex 15: 1384–1392.

Shear PK, Jernigan TL, Butters N (1994). Volumetric magnetic resonance imaging quantification of longitudinal brain changes in abstinent alcoholics. Alcohol Clin Exp Res 18: 172–176.

Shear PK, Sullivan EV, Lane B et al. (1996). Mammillary body and cerebellar shrinkage in chronic alcoholics with and without amnesia. Alcohol Clin Exp Res 20: 1489–1495.

Sheedy D, Lara A, Garrick T et al. (1999). Size of mamillary bodies in health and disease: useful measurements in neuroradiological diagnosis of Wernicke's encephalopathy. Alcohol Clin Exp Res 23: 1624–1628.

Smith SM, Jenkinson M, Johansen-Berg H et al. (2006). Tract-based spatial statistics: voxelwise analysis of multi-subject diffusion data. Neuroimage 31: 1487–1505.

Snook L, Plewes C, Beaulieu C (2007). Voxel based versus region of interest analysis in diffusion tensor imaging of neurodevelopment. Neuroimage 34: 243–252.

Song SK, Sun SW, Ramsbottom MJ et al. (2002). Dysmyelination revealed through MRI as increased radial (but unchanged axial) diffusion of water. Neuroimage 17: 1429–1436.

Sorg SF, Taylor MJ, Alhassoon OM et al. (2012). Frontal white matter integrity predictors of adult alcohol treatment outcome. Biol Psychiatry 71: 262–268.

Spielman D, Butts K, de Crespigny A et al. (1996). Diffusion-weighted imaging of clinical stroke. Intl J Neuroradiol 1: 44–55.

Squire LR, Amaral DG, Press GA (1990). Magnetic resonance imaging of the hippocampal formation and mammillary nuclei distinguish medial temporal lobe and diencephalic amnesia. J Neurosci 10: 3106–3117.

Stewart JT (2006). The frontal/subcortical dementias: common dementing illnesses associated with prominent and disturbing behavioral changes. Geriatrics 61: 23–27.

Sullivan EV (2003). Compromised pontocerebellar and cerebellothalamocortical systems: speculations on their contributions to cognitive and motor impairment in nonamnesic alcoholism. Alcohol Clin Exp Res 27: 1409–1419.

Sullivan EV, Marsh L (2003). Hippocampal volume deficits in alcoholic Korsakoff's syndrome. Neurology 61: 1716–1719.

Sullivan EV, Pfefferbaum A (2009). Neuroimaging of the Wernicke-Korsakoff syndrome. Alcohol Alcohol 44: 155–165.

Sullivan EV, Pfefferbaum A (2010). Diffusion tensor imaging in aging and age-related disorders. In: DK Jones (Ed.), Diffusion MRI: Theory, Methods and Applications, Oxford University Press, Oxford, pp. 624–643.

Sullivan EV, Marsh L, Mathalon DH et al. (1995a). Anterior hippocampal volume deficits in nonamnesic, aging chronic alcoholics. Alcohol Clin Exp Res 19: 110–122.

Sullivan EV, Marsh L, Mathalon DH et al. (1995b). Anterior hippocampal volume deficits in nonamnesic, aging chronic alcoholics. Alcohol Clin Exp Res 19: 110–122.

Sullivan EV, Marsh L, Mathalon DH et al. (1996). Relationship between alcohol withdrawal seizures and temporal lobe white matter volume deficits. Alcohol Clin Exp Res 20: 348–354.

Sullivan EV, Lane B, Deshmukh A et al. (1999a). In vivo mammillary body volume deficits in amnesic and nonamnesic alcoholics. Alcohol Clin Exp Res 23: 1629–1636.

Sullivan EV, Lane B, Rosenbloom MJ et al. (1999b). In vivo mammillary body volume deficits in amnesic and nonamnesic alcoholics. Alcohol Clin Exp Res 23: 1629–1636.

Sullivan EV, Rosenbloom MJ, Pfefferbaum A (2000a). Pattern of motor and cognitive deficits in detoxified alcoholic men. Alcohol Clin Exp Res 24: 611–621.

Sullivan EV, Rosenbloom MJ, Lim KO et al. (2000b). Longitudinal changes in cognition, gait, and balance in abstinent and relapsed alcoholic men: relationships to changes in brain structure. Neuropsychology 14: 178–188.

Sullivan EV, Deshmukh A, Desmond JE et al. (2000c). Cerebellar volume decline in normal aging, alcoholism, and Korsakoff's syndrome: relation to ataxia. Neuropsychology 14: 341–352.

Sullivan EV, Harding AJ, Pentney R et al. (2003). Disruption of frontocerebellar circuitry and function in alcoholism. Alcohol Clin Exp Res 27: 301–309.

Sullivan EV, Marsh L, Pfefferbaum A (2005a). Preservation of hippocampal volume throughout adulthood in healthy men and women. Neurobiol Aging 26: 1093–1098.

Sullivan EV, Deshmukh A, De Rosa E et al. (2005b). Striatal and forebrain nuclei volumes: contribution to motor function and working memory deficits in alcoholism. Biol Psychiatry 57: 768–776.

Sullivan EV, Rose J, Pfefferbaum A (2006). Effect of vision, touch and stance on cerebellar vermian-related sway and tremor: a quantitative physiological and MRI study. Cereb Cortex 16: 1077–1086.

Sullivan EV, Rose J, Pfefferbaum A (2010a). Physiological and focal cerebellar substrates of abnormal postural sway and tremor in alcoholic women. Biol Psychiatry 67: 44–51.

Sullivan EV, Harris A, Pfefferbaum A (2010b). Alcohol's effects on brain and behavior. Alcohol Res Health 33: 1–2.

Sun SW, Liang HF, Trinkaus K et al. (2006). Noninvasive detection of cuprizone induced axonal damage and demyelination in the mouse corpus callosum. Magn Reson Med 55: 302–308.

Suzuki K, Izumi M (2012). Alcohol is a risk factor not for thalamic but for putaminal hemorrhage: the Akita Stroke Registry. J Stroke Cerebrovasc Dis 22: 1064–1069.

Thomson AD (2000). Mechanisms of vitamin deficiency in chronic alcohol misusers and the development of the Wernicke-Korsakoff syndrome. Alcohol Alcohol Suppl 35: 2–7.

Thomson AD, Guerrini I, Marshall EJ (2012). The evolution and treatment of Korsakoff's syndrome: out of sight, out of mind? Neuropsychol Rev 22: 81–92.

Thorarinsson BL, Olafsson E, Kjartansson O et al. (2011). Wernicke's encephalopathy in chronic alcoholics. Laeknabladid 97: 21–29.

Tirotta E, Carbajal KS, Schaumburg CS et al. (2010). Cell replacement therapies to promote remyelination in a viral model of demyelination. J Neuroimmunol 224: 101–107.

van Holst RJ, de Ruiter MB, van den Brink W et al. (2012). A voxel-based morphometry study comparing problem gamblers, alcohol abusers, and healthy controls. Drug Alcohol Depend 124: 142–148.
Vaquero J, Chung C, Cahill ME et al. (2003). Pathogenesis of hepatic encephalopathy in acute liver failure. Semin Liver Dis 23: 259–269.
Victor M, Adams RD, Collins GH (1989). The Wernicke–Korsakoff syndrome and related neurologic disorders due to alcoholism and malnutrition. 2nd edn F.A. Davis, Philadelphia.
Virta A, Barnett A, Pierpaoli C (1999). Visualizing and characterizing white matter fiber structure and architecture in the human pyramidal tract using diffusion tensor MR. Magn Reson Imaging 17: 1121–1133.
Wilhelm J, Frieling H, Hillemacher T et al. (2008). Hippocampal volume loss in patients with alcoholism is influenced by the consumed type of alcoholic beverage. Alcohol Alcohol 43: 296–299.
Wrase J, Makris N, Braus DF et al. (2008). Amygdala volume associated with alcohol abuse relapse and craving. Am J Psychiatry 165: 1179–1184.
Yeligar SM, Harris FL, Hart CM et al. (2012). Ethanol induces oxidative stress in alveolar macrophag via upregulation of NADPH oxidases. J Immunol 188: 3648–3657.
Yoshizaki T, Hashimoto T, Fujimoto K et al. (2010). Evolution of callosal and cortical lesions on MRI in Marchiafava-Bignami disease. Case Rep Neurol 2: 19–23.
Zahr NM, Kaufman KL, Harper CG (2011). Clinical and pathological features of alcohol-related brain damage. Nat Rev Neurol 7: 284–294.
Zipursky RB, Lim KO, Pfefferbaum A (1989). MRI study of brain changes with short term abstinence from alcohol. Alcohol Clin Exp Res 13: 664–666.

Section 7

Neuroimaging of neurochemical markers

Handbook of Clinical Neurology, Vol. 125 (3rd series)
Alcohol and the Nervous System
E.V. Sullivan and A. Pfefferbaum, Editors

Chapter 18

Molecular imaging in alcohol dependence

SHERVIN RAVAN[2], DIANA MARTINEZ[2], MARK SLIFSTEIN[2], AND ANISSA ABI-DARGHAM[1,2]*
[1]*Department of Radiology, Columbia University College of Physicians and Surgeons, New York, NY, USA*
[2]*Department of Psychiatry, Columbia University College of Physicians and Surgeons, New York, NY, USA*

INTRODUCTION

Alcohol dependence is a chronic relapsing condition manifested by compulsive alcohol use and alcohol-seeking behavior, leading to impaired social and occupational functioning (Koob et al., 2004). Approximately 9% of the adult population in the United States meets criteria for an alcohol use disorder, defined as continued drinking despite adverse consequences (Friedmann, 2013), and as such, there has been extensive research into the genetic, epigenetic, psychosocial, and neurobiologic mechanisms underlying alcohol dependence. Over the past three decades, our understanding of the neurobiology of drug addiction and alcohol dependence has advanced greatly as a result of ground-breaking developments in brain-imaging technology. Specifically, structural and functional brain-imaging techniques have revolutionized addiction research through their ability to provide unprecedented access to the brain, anatomically, physiologically, and longitudinally.

Molecular imaging modalities, specifically positron emission tomography (PET) and single-photon emission computed tomography (SPECT), are commonly used in studies of alcohol dependence (Mena et al., 2005). These techniques are unique in their ability to quantify both biochemical and physiologic processes in living organisms (Parvaz et al., 2011). Furthermore, measurements can be localized to specific brain regions, allowing for correlation between neurochemical processes and anatomic findings (Cosgrove, 2010). Ultimately, molecular imaging techniques provide non-invasive tools for investigating the links between alcohol dependence, biology, behavior, and environment (Mena et al., 2005).

In preclinical research, molecular imaging enables researchers to conduct longitudinal studies on individual animals, allowing for evaluation of disease progression, treatment response, vulnerability, and outcome prediction. It is currently difficult to conduct longitudinal studies of alcohol dependence in humans – a task which would require baseline imaging in alcohol-naïve subjects followed by repeat imaging when, or if, a subject becomes alcohol-dependent (Cosgrove, 2010). Currently, molecular imaging data from alcohol-dependent individuals are usually compared to that of age-matched controls (Urban and Martinez, 2012). In some studies, alcohol-dependent individuals are imaged in early abstinence (usually 1–4 weeks postdetoxification) or during late abstinence (usually greater than 4 weeks postdetoxification). Thus, at this time, molecular imaging studies of alcohol dependence cannot identify whether measured differences between alcohol-dependent individuals and healthy controls are pre-existing and predisposing to alcohol dependence or reflect consequences of excessive alcohol consumption (Cosgrove, 2010).

Through incompletely understood processes, alcohol also modulates the activity of multiple neurotransmitter systems, including the dopaminergic, serotonergic, GABAergic, opioid, and cannabinoid systems (Hillemacher, 2011). Dopamine is the most studied of these given its role in reward and reinforcement, and the wider availability of radiotracers for dopaminergic parameters. Using PET and SPECT, researchers have been able to measure the binding of radioligands to these various receptor systems, within both healthy controls and alcohol-dependent subjects, helping to elucidate further the neurochemical mechanisms involved

*Correspondence to: Anissa Abi-Dargham, M.D., Department of Psychiatry, Columbia University, 1051 Riverside Drive, Unit 31, New York, NY 10032, USA. Tel: +1-212-543-5066, Fax: +1-212-568-6171, E-mail: aa324@columbia.edu

in the development and maintenance of alcohol dependence (Martinez and Narendran, 2010).

REVIEW OF IMAGING METHODS

A basic understanding of molecular imaging data is essential for accurate interpretation of the studies discussed in this chapter. The most commonly used outcome measure in *in vivo* studies of neuroreceptors is binding potential relative to the non-displaceable compartment (BP_{ND}). BP_{ND} is the equilibrium ratio of the concentration of specifically bound radioligand relative to the sum of the free and non-specifically bound radioligand, estimated from activity in a reference region devoid of receptors (Innis et al., 2007). The equation for BP_{ND} is:

$$BP_{ND} = f_{ND} * \frac{B_{MAX}}{K_D}$$

where B_{MAX} is the concentration of D_2 receptors, K_D is the inverse of the affinity of the radiotracer for the receptor, and f_{ND} is the free fraction of ligand in the non-displaceable tissue compartment in the brain (Slifstein and Laruelle, 2001). BP_{ND} is also equal to the ratio at equilibrium of specifically bound radioligand to that of non-displaceable radioligand in tissue. Alternatively the equilibrium ratio relative to either the total or free (unbound) concentration of radioligand in plasma can be used, as opposed to a reference region; these are denoted as BP_P or BP_F respectively (Innis et al., 2007).

In this chapter, we will cover the major contributions of molecular imaging research to our current understanding of alcohol dependence. This review will focus on neurotransmitter systems for which existing radiotracers have been employed successfully in human studies of alcohol dependence: (1) dopamine; (2) γ-aminobutyric acid (GABA); (3) opioid; (4) serotonin; and (5) cannabinoid.

DOPAMINE AND ALCOHOL DEPENDENCE

Dopamine is a catecholamine neurotransmitter involved in myriad brain functions, including voluntary movement, behavior, cognition, motivation, reward, reinforcement, sleep, mood, attention, learning, and working memory (Hou et al., 2012). Originating within the substantia nigra (SN) pars compacta, ventral tegmental area (VTA), and hypothalamus, dopaminergic axons project to various brain regions through four major pathways: (1) mesocortical; (2) mesolimbic; (3) nigrostriatal; and (4) tubuloinfundibular pathway. There is a large body of evidence implicating altered dopaminergic neurotransmission in the development of alcohol dependence (Martinez and Narendran, 2010), and the following section provides an overview and discussion of neurochemical imaging studies investigating this link.

The dopamine receptor in alcohol dependence

The mesolimbic dopamine system plays an integral role in mediating the rewarding and reinforcing properties of addictive substances, including alcohol (Gilpin and Koob, 2008). Dopamine-mediated signaling in addiction occurs primarily through interactions with postsynaptic dopamine receptors in the striatum and limbic cortex (Haber, 2011). There are five subtypes of dopamine receptors (D_1–D_5), which are classified into two groups or families: D_1-like (D_1 and D_5) and D_2-like (D_2, D_3, D_4) (Hou et al., 2012). Within the complex neurobiologic meshwork that underlies reward and reinforcement, the D_2-like receptors (found predominantly within the extended striatum) are thought to be critical in neurobiologic mechanisms leading to alcohol dependence (Rominger et al., 2012). For the purposes of this text we will be referring to the D_2-like family of receptors simply as D_2 receptors.

Neurochemical imaging of dopamine receptors in alcohol dependence

A number of PET and SPECT studies have investigated the relationship between the D_2 family of receptors and alcoholism. PET imaging studies of alcohol dependence and the D_2 receptor have exclusively used three radioligands: (1) [^{11}C]raclopride; (2) [^{18}F]desmethoxyfallypride (DMFP); and (3) [^{18}F]fallypride. It is important to note that [^{11}C]raclopride and [^{18}F]DMFP label striatal D_2 receptors while [^{18}F]fallypride binds both striatal and extrastriatal D_2 receptors, such as those of the amygdala and hippocampus (Rominger et al., 2012). SPECT imaging studies have used [^{123}I]IBZM and [^{123}I]epidepride (Martinez and Narendran, 2010).

The majority of PET/SPECT imaging studies have shown 10–20% lower striatal D_2 binding potential in alcoholics compared to healthy controls (Fig. 18.1) (Hietala et al., 1994; Volkow et al., 1996b, 2002a; Heinz et al., 2004, 2005b; Martinez et al., 2005). Decreases of similar magnitude have been observed in PET studies of heroin dependence, cocaine dependence, methamphetamine abuse, and even obesity (Martinez et al., 2005). Regions of significantly lower D_2 binding potential in alcoholism have been shown across the striatum, including the bilateral ventral striatum, which contains the nucleus accumbens, and receives dopaminergic neurons originating in the VTA (Martinez et al., 2005). This pathway constitutes the mesolimbic corticostriatal loop, which receives input from the amygdala and hippocampus, and is involved in mediating emotional behavior and reward processing (Haber, 2011). Human autoradiography studies have reported similar alterations, with

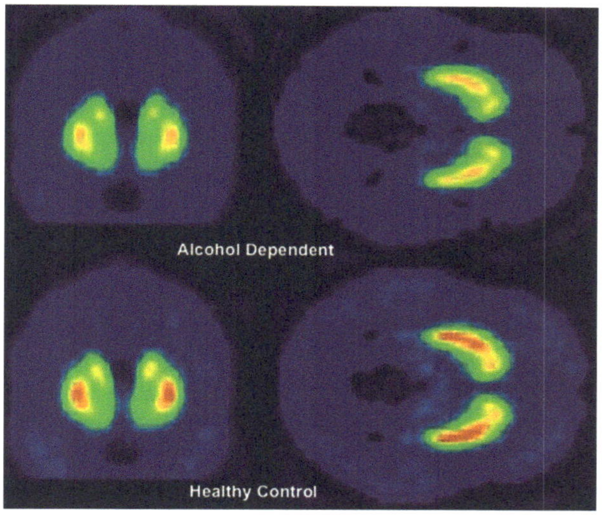

Fig. 18.1. Positron emission tomography images showing mean D_2 receptor binding (with [^{11}C]raclopride) in alcohol-dependent subjects ($n = 15$) and healthy controls ($n = 15$). D_2 receptor binding potential relative to the non-displaceable compartment (BP_{ND}) was significantly lower in the striatum of alcohol-dependent subjects.

20–32% lower D_2 receptor binding in the dorsal striatum (caudate and putamen) and nucleus accumbens when compared to control subjects (Tupala et al., 2001, 2003a, b). A recent study by Rominger et al. (2012), however, failed to show any significant differences in striatal D_2 binding in alcoholics compared to controls using [^{18}F] fallypride, although differences were observed in the caudate with voxel-based analysis. It should be noted that, while the scanning parameters in this study were appropriate for measuring binding potential in extra-striatal regions (to be discussed later in this section), a longer scan time is optimal for obtaining measures of striatal D_2 receptors (Gorelick et al., 2008).

PET imaging has also provided addiction researchers with a tool for identifying associations between receptor binding and behavioral outcomes, such as alcohol craving and relapse, as well as enabled correlation of binding with brain activation, assessed with functional magnetic resonance imaging (fMRI). Alcohol-dependent subjects with decreased D_2 receptor binding within the ventral striatum exhibit increased cue-induced activation within the medial prefrontal cortex and anterior cingulate gyrus (George et al., 2001). The anterior cingulate gyrus is thought to be involved in attention, conditioned drug seeking, and alcohol craving (Volkow and Fowler, 2000; Spinks, 2000; George et al., 2001). Thus, a dysfunction in dopaminergic transmission, manifested partly by reduced D_2 receptor levels, may result in enhanced attentional processing of alcohol-related signals (Heinz et al., 2004). There also appears to be an inverse relationship between striatal D_2 receptor binding and severity of alcohol craving, such that low D_2 binding predicts increased alcohol craving (assessed by the Alcohol Craving Scale) (Heinz et al., 2004). Furthermore, the severity of alcohol craving is positively correlated with alcohol relapse, so that as craving increases so does the risk for relapse (Heinz et al., 2005b). These findings suggest a possible mechanism by which low D_2 receptor concentration results in altered frontal cortical activation, which, in turn, may drive behavioral changes, lending to continued alcohol intake (Heinz et al., 2004).

The identification of associations between alcohol dependence and alterations of neurotransmitter systems raises the question of whether these changes predate alcohol dependence and represent vulnerability to addiction or, rather, are acquired consequences of long-term exposure. The lower D_2 binding potential in alcoholics, compared to control subjects, may be a consequence of long-term alcohol use, possibly as a result of counteradaptive receptor downregulation, alcohol-related neurotoxicity, or a combination of the two (Volkow et al., 2006).

Thus far, no single study has been able to definitively identify a direct causal relationship between excessive alcohol use and decreased D_2 receptor availability. Volkow et al. (2002b) measured D_2 receptor availability in 14 alcohol-dependent individuals during both early (within 6 weeks) and late (1–4 months) abstinence, failing to observe significant differences in binding between these time points (although striatal D_2 receptor availability was reduced in alcoholics when compared to healthy controls). Furthermore, no significant relationship was noted between D_2 receptor availability and days since last alcohol use, suggesting that D_2 receptor availability is unaffected by withdrawal or long-term abstinence, at least within the time frame of 4 months. The study of Rominger et al. (2012), discussed above, imaged 17 alcoholic patients using [^{18}F]fallypride during immediate withdrawal (within 24 hours), early abstinence (1–2 weeks), and, at 1 year (when 4 of the 17 reported either marked reduction in alcohol intake or complete abstinence), failed to show differences in striatal D_2 binding between immediate withdrawal and early abstinence in alcoholics. However, in the group scanned at 1 year there was a 30% increase in striatal and thalamic D_2 receptor binding when compared to scans from immediate withdrawal and early abstinence. Although this finding suggests recovery of D_2 receptors during abstinence or marked reduction in alcohol intake, the sample size is small ($n = 4$) and further long-term prospective investigations are needed before definitive conclusions regarding D_2 receptor recovery during abstinence can be made.

Alternatively, decreased D_2 binding potential may predate and predispose to alcohol dependence, representing a possible gene–environment-based vulnerability to alcoholism (Jalan et al., 2000; Volkow et al., 2006; Haber, 2011). This model has been further explored by a number of animal and human PET studies which have linked D_2 receptor expression and binding to genetic and environmental factors as well as behavioral outcomes, such as social hierarchy, family history of alcoholism, and perception of reward or "high." We describe some of these studies below.

Investigating the relationship between social dominance and D_2 receptor expression in rhesus monkeys, Grant et al. (1998) reported a significantly lower distribution volume of [^{18}F]FCP (D_2 receptor ligand) in subordinate female monkeys compared to dominant female monkeys. In a novel follow-up study, Morgan et al. (2002) imaged 20 male monkeys while housed individually, prior to social group formation, and then again after 3 months, once dominance structures had been established. Intriguingly, when compared to baseline scans, dominant monkeys demonstrated a 20% higher average distribution volume of [^{18}F]FCP compared to subordinate monkeys, whereas the distribution volume of [^{18}F]FCP remained unchanged in subordinate monkeys. Furthermore, subordinate monkeys demonstrated a significantly higher propensity to self-administer cocaine when given the choice between cocaine and saline, while dominant monkeys responded to cocaine and saline similarly. Correspondingly, overexpression of D_2 receptors in alcohol-preferring P-rats results in a marked reduction of alcohol self-administration (Thanos et al., 2001). This provides strong evidence that the expression of D_2 receptors is, at least in part, influenced by environmental factors (such as dominance and stress), and that neurochemical changes in response can occur in relatively short periods of time. In combination with the fact that cocaine serves as a greater reward for subordinate, and not dominant, monkeys, this suggests that environmental conditions and their biologic and physiologic consequences (such as stress) produce alterations in receptor expression that affect an organism's vulnerability to addiction (Haber, 2011).

There are also several lines of evidence supporting the role of family history of alcoholism in the expression of D_2 receptors (Yoder et al., 2005; Volkow et al., 2006). Non-alcoholic members of alcoholic families demonstrate higher D_2 receptor binding within the striatum, suggesting a possible adaptation conferring protection against alcohol dependence (Volkow et al., 2006). The mechanisms by which D_2 receptor expression relates to risk and vulnerability are not well understood. D_2 receptor levels have been associated with the reinforcing properties of the stimulant methylphenidate in non-drug abusers, such that subjects with low D_2 receptor-binding potential report acute intravenous (IV) administration as "pleasant," whereas individuals with high D_2 receptor binding report acute IV administration as "unpleasant" (Volkow et al., 2002a). Furthermore, healthy non-drug-abusing humans with high D_2 receptor levels experience higher levels of intoxication after IV alcohol when compared to subjects with low D_2 receptor levels, possibly coding the experience as "unpleasant," similar to how subjects with high D_2 receptor levels do with methylphenidate use (Yoder et al., 2005). This suggests that initial drug reinforcement and reward may be dictated by the subjective experience of alcohol intake and intoxication, outcomes that appear to be influenced by D_2 receptor expression (Volkow et al., 2006).

Although dopamine receptors are most ubiquitous within the striatum, extrastriatal dopamine receptors are also thought to be involved in alcohol dependence (Rominger et al., 2012). Currently, there are few molecular imaging data on extrastriatal D_2 receptors in alcohol dependence. In addition to their study of striatal D_2 receptors, Rominger et al. (2012) investigated extrastriatal D_2 receptors in alcoholics, reporting decreases in D_2 receptor binding within the thalamus, insular cortex, temporal cortex, and hippocampus ($n = 17$) during early abstinence (1–2 weeks) when compared to healthy controls. These findings extend the presence of decreased D_2 receptor availability to extrastriatal regions and further support the association of attenuated dopaminergic transmission in alcohol dependence.

Alcohol dependence and presynaptic dopamine

In addition to the dopamine receptor, presynaptic dopamine (within intraneuronal vesicles), dopamine synthesis, and the dopamine transporter have been measured in alcohol dependence using molecular imaging techniques.

PET imaging studies have measured presynaptic dopamine function using three methods: (1) imaging the neuronal uptake of the radiotracer [^{18}F]DOPA, which is taken up by presynaptic dopamine neurons and provides an index of presynaptic dopamine synthesis capacity; (2) (+)[^{18}F]dihydrotetrabenazine, which labels the type 2 vesicular monoamine transporters (VMAT2) of the dopamine vesicles and provides an estimate of presynaptic dopamine vesicles in the striatum; and (3) imaging with [^{11}C]raclopride before and after a psychostimulant challenge, which increases intrasynaptic dopamine and reduces the number of D_2 receptors available for binding to the radiotracer. Thus, a comparison of [^{11}C]raclopride binding before and after a stimulant challenge results in a decrease in radiotracer binding, which is linearly correlated with the increase in endogenous dopamine, and provides an indirect measure of presynaptic dopamine release (Gould et al., 2012).

Dopamine synthesis in alcohol dependence

The relationship between alcoholism and presynaptic dopamine synthesis is unclear. To date, there have only been two PET imaging studies evaluating presynaptic dopamine synthesis in alcohol-dependent subjects (Gould et al., 2012). The first study by Tiihonen et al. (1998) demonstrated increases in [^{18}F]DOPA uptake in the left putamen and right caudate of alcohol-dependent subjects when compared to healthy controls. This finding may represent a compensatory increase in dopamine synthesis in response to decreased postsynaptic dopaminergic activity (decreased D_2 receptor concentration) (Tiihonen et al., 1998). The second, and more recent, study revealed no statistically significant differences in net striatal [^{18}F] DOPA uptake between alcoholics and healthy controls. However, there was a significant negative correlation between [^{18}F]DOPA uptake and alcohol craving, where low uptake, specifically in the ventral striatum, was associated with greater craving for alcohol (Heinz et al., 2005b). In conjunction with the fact that craving is positively correlated with relapse (discussed above), this suggests that alcoholics with low dopamine synthesis capacity are at increased risk of relapse (Heinz et al., 2005b).

Alternatively, using the VMAT2 radioligand (+)[^{18}F] DTBZ, Gilman et al. (1998) reported lower radioligand uptake (Ki) within the caudate nucleus (8.6%) and putamen (6.2%) as well as lower (+)[^{18}F]DTBZ distribution volume within the caudate nucleus (4.5%) and putamen (10.4%) of alcoholics compared to control subjects, a finding consistent with loss of presynaptic dopamine stores. Similarly, studies using [^{11}C]raclopride displacement with stimulant challenge report lower endogenous dopamine release in alcohol-dependent subjects compared to healthy controls (Martinez and Narendran, 2010). The earliest of these studies demonstrated regional alterations in [^{11}C] raclopride displacement in the ventral limbic striatum, suggesting blunted endogenous dopamine release in alcohol dependence (Martinez et al., 2005). A subsequent study showed similar results, but reported lower presynaptic dopamine release in the putamen as well as the ventral striatum (Volkow et al., 2007). It is important to note that levels of VMAT2 may influence the magnitude of stimulant effect on extracellular dopamine, such that a loss of VMAT2 would be consistent with blunted amphetamine-induced [^{11}C]raclopride displacement (Patel et al., 2003).

The dopamine transporter in alcohol dependence

In addition to the D_2 receptor and dopamine release, PET imaging of the dopamine system has included studies of the dopamine transporter (DAT), a monoamine transporter localized within the presynaptic terminal of dopaminergic neurons (Volkow et al., 1996a). The DAT serves as a marker of dopamine neurons and regulates dopaminergic neurotransmission by mediating synaptic dopamine reuptake (Bhaskar et al., 2012). These studies have utilized a variety of methods (PET/SPECT/ autoradiography) and radiotracers, including [^{11}C] raclopride, [^{123}I]beta-CIT and [^{11}C]D-threomethylphenidate, but provide inconsistent results (Gould et al., 2012). There is some evidence that DAT concentrations may be decreased during early withdrawal (<3 weeks) (Laine et al., 1999a, b) and prolonged sobriety (Tiihonen et al., 1995; Repo et al., 1999), when compared to healthy controls. Other studies, however, have failed to show significant changes in DAT availability beyond early withdrawal (Volkow et al., 1996b; Laine et al., 1999a, b Heinz et al., 2000a). Laine et al. (1999a, b) reported decreased DAT binding during early withdrawal, while SPECT data from the same population at 4 weeks of abstinence demonstrated interim increases in DAT availability (at levels similar to those of healthy controls). Thus, although there may be a short-lived decrease in DAT availability during immediate abstinence, there does not appear to be a lasting alteration of DAT availability in alcoholism.

Acute alcohol intoxication and dopamine imaging

As discussed earlier, dopamine is involved in the acquisition and maintenance of addictive behaviors. The mechanisms underlying alcohol-induced reward are currently not well understood. In contrast to other drugs of abuse, such as cocaine, alcohol interacts with multiple neurotransmitter systems within the brain and thus identification of discrete reward pathways has proven complex (Urban et al., 2010). It is hypothesized that alcohol-induced dopamine release within the ventral striatum (involved in reward and motivation) contributes to the reinforcing properties of alcohol (Urban et al., 2010).

PET imaging techniques have contributed to elucidating the neurochemical effects of acute alcohol consumption. Alcohol-induced dopamine release in the ventral striatum has been observed in rodents (Di Chiara and Imperato, 1988), while the effect of alcohol administration on [^{11}C]raclopride displacement in healthy non-alcoholic humans is less well understood. Unlike psychostimulant drugs such as amphetamine and cocaine, it is unclear how alcohol stimulates dopamine release (Urban et al., 2010). Both direct and indirect mechanisms have been proposed, including excitation of dopamine neurons in the VTA and removal of GABAergic inhibition of dopamine neurons (Gianoulakis, 2004, 2009).

PET imaging studies of acute alcohol intoxication and dopamine transmission have utilized both IV and oral routes of alcohol administration. In addition to the pharmacologic effects of alcohol, oral alcohol supplies sensory stimuli, including taste and smell, as well as the anticipation of reward, all of which have been shown to stimulate dopamine release within the striatum (Boileau et al., 2003). This complicates interpretation of data from human studies using oral alcohol, as conditioned cues and anticipation may play a role in striatal dopamine release (Ramchandani et al., 2011). Interestingly, in rats habituated to alcohol administration, researchers have shown little increase in dopamine after alcohol intake and no increase in dopamine after alcohol-related cues, suggesting that, while dopamine may play a key role, it is not the sole substrate mediating reinforcement (Nurmi et al., 1998).

The risk of alcoholism in men is double that of women, and PET imaging studies have helped identify neurochemical responses that may be contributory. When exposed to similar levels of alcohol, striatal dopamine release in response to alcohol in men is approximately double that of women (Urban et al., 2010). If we assume that alcohol-induced dopamine release within the striatum mediates reward and drives reinforcement, these data suggest a possible gender difference in the responsivity of dopamine release to alcohol that may explain behavioral sex differences. Similar sex differences in striatal dopamine release have also been observed with amphetamine in stimulant-naive healthy subjects (Tretter et al., 1997). As discussed above, it is difficult to separate contributions of the pharmacologic effects of alcohol on to dopamine release from sensory and prediction or expectation of reward effects. All these may play a role in the differential gender responsiveness to alcohol-induced dopamine release.

In summary, results from molecular imaging studies in alcohol dependence point towards a hypofunction of mesolimbic dopamine signaling in alcohol dependence (Fig. 18.2). Similar to other drugs of abuse, longitudinal studies are needed to determine whether this is a result of use or a predisposing factor.

GABA AND ALCOHOL DEPENDENCE

Many of alcohol's central effects are mediated through interactions with GABA receptors in the human brain. Of the two GABA receptor types in the brain, GABA-A and GABA-B, GABA-A receptors are thought to play the most important role in the development of alcohol dependence (Cosgrove et al., 2011). GABA-A receptors in the human are pentameric ligand-gated chloride ion channels potentially composed of 19 subunits, the most prevalent being the alpha, beta, and gamma subunits (Krystal et al., 2006).

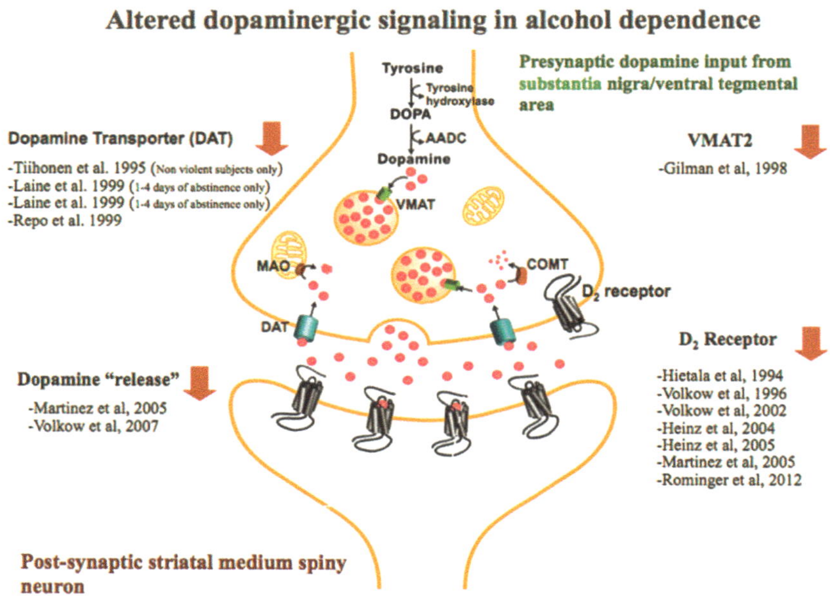

Fig. 18.2. Schematic showing the elements of dopamine signaling in the striatum that are affected in alcohol dependence. These include decreases in the dopamine transporter DAT, presynaptic dopamine release (measured as displacement of the radiotracer [^{11}C]raclopride and a stimulant challenge), the type 2 vesicular monoamine transporter (VMAT2), and the dopamine type 2 receptor. Together, the majority of positron emission tomography studies show that dopamine signaling is reduced in alcohol dependence.

Thus far, PET and SPECT studies have utilized three radiotracers in imaging GABA-A receptor availability in alcoholism: the non-selective GABA receptor radiotracers [^{11}C]flumazenil and [^{123}I]iomazenil, which bind to the benzodiazepine site on the GABA-A receptor, and the alpha-5 subunit specific radiotracer [^{11}C] Ro15 4513. Similar to dopamine transmission in alcohol dependence, it has been proposed that alcohol dependence is associated with attenuated GABAergic function (Abi-Dargham et al., 1998).

To date, GABA-A receptor imaging studies with tracers for the benzodiazepine site in alcoholism have yielded discrepant results. GABA-A receptor binding has been observed to be lower at 1 month, 3 months, and 7 months of abstinence in the medial frontal cortex and cerebellum of alcoholics compared to healthy controls (Abi-Dargham et al., 1998; Lingford-Hughes et al., 1998, 2005). Conversely, a longitudinal study that controlled for smoking reported an increase in GABA-A receptor concentration, observed at 1 week of withdrawal in alcohol-dependent non-smokers ($n = 8$) compared to alcohol-dependent smokers ($n = 15$) and controls ($n = 15$). Interestingly, at 4 weeks of abstinence this effect had normalized to control levels (Staley et al., 2005). Two studies have reported no difference in GABA-A receptor binding in alcoholics when compared to healthy controls (Litton et al., 1993; Lingford-Hughes et al., 2000). In a more recent PET study, a significantly lower distribution volume of [^{11}C]Ro15 4513 was observed within the nucleus accumbens and right hippocampus of alcohol-dependent subjects ($n = 8$) abstinent for at least 6 weeks compared to healthy controls, with similar trends in the parahippocampal gyri and amygdala (Lingford-Hughes et al., 2012). A possible interpretation of this data is that, while there may be a relative decrease in GABA-A receptor binding in the later stages of abstinence (greater than 3 weeks), a transient increase may occur during the immediate withdrawal period. The mechanisms responsible for these findings are unclear, and given the variability of data and small sample sizes, a definitive conclusion would require further investigations with larger study populations and more reproducible results.

In addition to demographic differences and small study populations, brain atrophy can influence PET and SPECT results. This is especially true with imaging targets located within the cortical gray matter, like the GABA-A receptor. In the presence of gray-matter atrophy, an observation of decreased receptor binding may actually represent partial volume averaging from smaller brain volumes and not lower receptor concentration (Cosgrove et al., 2011). In one study, Lingford-Hughes et al. (1998) observed decreased GABA-A receptor binding within multiple brain regions in the absence of gray-matter atrophy in alcoholics compared to controls, suggesting these changes may truly represent neurochemical alterations at the receptor level rather than reflecting global or regional brain atrophy.

If alterations in GABA-A receptor binding are, in fact, associated with alcoholism, we are again left with the question of causality. Whether altered GABAergic neurotransmission, manifested by diminished GABA-A receptor concentration, predates alcohol use, is a result of alcohol-related neurotoxicity, or a combination of both, is unclear. One hypothesis is that vulnerability to alcoholism is conferred via pre-existing reductions in GABA-A concentration, through either genetic or environmental factors, resulting in increased alcohol consumption as a means of restoring attenuated GABAergic function. Excessive alcohol intake may also act as a neurotoxin, either in the absence of pre-existing GABAergic dysfunction or synergistically with it, to facilitate continued alcohol consumption by producing a GABAergic deficiency or by magnifying a pre-existing GABAergic deficit (Gilman et al., 1996).

As discussed earlier, focally diminished GABA-A receptor concentration has been documented within the superomedial frontal lobes bilaterally, regions thought to be integral in the modulation of affect, motivation, inhibition of behavior, selective attention and higher cortical functions, such as abstraction, rule, and concept formation. This presents a potential mechanism by which pre-existing GABAergic deficiency within the frontal lobes, along with direct alcohol-mediated neurotoxicity, may result in decreased inhibition of behavior and an impaired ability to identify high-risk situations, in addition to difficulties modulating anxiety (Gilman et al., 1996).

OPIOID RECEPTORS AND ALCOHOL DEPENDENCE

It is generally accepted that the reinforcing effects of alcohol are, in part, modulated by opioid peptides and receptors (Weerts et al., 2011). The three most studied endogenous opioid peptides are β-endorphins, enkephalins, and dynorphins. These peptides interact with three subtypes of opioid receptors: mu-opioid receptors (MOR), delta-opioid receptors (DOR), and kappa-opioid receptors (KOR). The binding affinity of endogenous opioid peptides to opioid receptors is differential, with β-endorphins binding equally with MORs and DORs while enkephalins bind to DORs with greater affinity than MORs, and dynorphins bind selectively to KORs (Koob et al., 1998; Gianoulakis, 2009).

The major functions mediated by opioid receptors include modulation of responses to painful stimuli and stressors, reward and reinforcement, and homeostatic regulatory functions (Gianoulakis, 2004). The MORs and DORs are thought to function in mediating reward and reinforcement via interactions with endogenous opioid peptides and agonists such as alcohol and morphine. This is supported by data showing that β-endorphins and enkephalins increase dopamine release within the ventral striatum and nucleus accumbens (thought to modulate reinforcement) via interactions with MORs, and DORs to a lesser extent (Weerts et al., 2011). Activation of KORs, however, appears to decrease dopamine release in the ventral striatum, suggesting that agonistic action at the KOR may prevent reinforcement (Gianoulakis, 2009).

Several PET studies have investigated the opioid system in alcohol dependence (Bencherif et al., 2004; Heinz et al., 2005a; Henriksen and Willoch, 2008; Williams et al., 2009; Weerts et al., 2011). The radiotracers used in studies of opioid receptors in alcoholism include, [^{11}C]carfentanil (MOR-selective tracer), [^{11}C]diprenorphine (non-selective opioid reeptor tracer), and [^{11}C] methylnaltrindole (DOR tracer) (Henriksen and Willoch, 2008). Results from these studies have been variable (Bencherif et al., 2004; Heinz et al., 2005a; Henriksen and Willoch, 2008; Williams et al., 2009; Weerts et al., 2011). Using [^{11}C]carfentanil, Bencherif et al. (2004) showed decreases in MOR binding within the right dorsal lateral prefrontal cortex, right anterior frontal cortex, and right parietal cortex of recently abstinent alcoholics when compared to healthy controls ($n=8$). However, Heinz et al. (2005a) demonstrated increases in MOR binding in the ventral striatum (no cortical results) of recently abstinent alcoholics when compared to healthy controls ($n=25$), with no decreases in any brain regions. Williams et al. (2009) also reported a non-significant increase (~8%) in the global volume of distribution (reflection of opioid receptor availability) of [^{11}C]diprenorphine in alcoholics when compared to healthy controls ($n=11$). Most recently, Weerts et al. (2011) used both [^{11}C]carfentanil and [^{11}C]methylnaltrindole, a DOR-selective tracer, to show increases in MOR binding within multiple brain regions, including the ventral striatum, amygdala, insula, and thalamus in alcoholics when compared to healthy controls, while reporting similar, but insignificant, trends in DOR binding in alcoholics compared to healthy controls (~5 days' abstinence). Although results of increased DOR binding did not reach statistical significance, the data provide some evidence that, in addition to MORs, DORs may also play a role in alcohol dependence.

It remains unclear whether increased MOR binding predates alcohol dependence or reflects the neurochemical consequences of excessive alcohol intake. MOR knockout mice have been shown to self-administer alcohol at reduced levels when compared to controls (Hall et al., 2001), and in humans, genetic variants for MORs, DORs, and KORs have been noted as risk factors for drug and alcohol dependence (Zhang et al., 2008). Furthermore, Heinz et al. (2005a) and Williams et al. (2009) have shown that opioid receptor binding and availability remain largely unchanged in late abstinence (~1–3 months) when compared to early abstinence, even in cases of interim alcohol relapse. This is consistent with data in cocaine-dependent subjects showing increased MOR binding within the striatum, both in early and late abstinence, suggesting that altered opioid receptor binding may predate excessive alcohol intake (Zubieta et al., 1996).

Alternatively, increases in MOR binding may be due to a reduction in competition from endogenous opioid peptides (Williams et al., 2009). In humans, low plasma β-endorphin levels have been observed in alcohol-dependent subjects, both in early and late abstinence. Thus, as opioid receptor binding to both [^{11}C]carfentanil and [^{11}C]diprenorphine (non-selective opioid receptor tracer) has been shown to be sensitive to endogenous opioid levels, reduced levels of endogenous endorphins may manifest as increased MOR or DOR binding (Williams et al., 2009). This is supported by genetic data linking polymorphisms resulting in low β-endorphin levels to the development of drug or alcohol dependence (Del Arbol et al., 2007).

Alcohol craving is thought to be one manifestation of the many internal conflicts alcoholics face during early abstinence (Williams et al., 2009). Given the correlation between severity of alcohol craving and relapse rate (O'Brien et al., 1992), PET researchers have included measures of alcohol craving in their studies, hoping to identify correlative relationships with receptor availability. Several studies have indeed reported associations between opioid receptor binding and alcohol craving, although results thus far have been discrepant. Significant negative associations between MOR and alcohol craving have been observed within multiple cortical regions, especially the right dorsolateral prefrontal cortex (thought to play a role in visual and verbal memory as well as the representation, manipulation, and active maintenance of the attentional demands of a task: Bencherif et al., 2004; Goldstein et al., 2004; Weerts et al., 2011). Intriguingly, negative associations between MOR binding and depressive symptoms have been reported in similar brain regions, suggesting a possible shared neural circuitry that may underlie both states (Bencherif et al., 2004). Studies have also shown increased activation within these brain regions in response to drug-related cues in alcohol and cocaine dependence using fMRI (Maas et al., 1998; George et al., 2001).

Positive associations between opioid receptor binding and alcohol craving have also been observed (Heinz et al., 2005a; Williams et al., 2009). These correlations reach statistical significance within the nucleus accumbens and the anterior cingulate cortex in early abstinence, and within the anterior cingulate cortex in late abstinence (although data in late abstinence were from four subjects only) (Williams et al., 2009). There is also evidence that the opioid receptor antagonist naltrexone produces significant reductions in cue-induced alcohol craving in alcohol-dependent subjects when compared with placebo. Furthermore, naltrexone-induced reductions in cue-induced alcohol craving appear to be associated with decreased blood oxygen level-dependent magnetic resonance activation within the ventral striatum (during real-time cue presentation) (Myrick et al., 2008). It should be noted that meta-analyses of randomized clinical trials report only small to moderate effect size of naltrexone in reducing drinking and relapse in alcohol-dependent subjects (Weerts et al., 2011). Overall, these data suggest a role for the anterior cingulate cortex and ventral striatum in alcohol craving and the maintenance of alcohol dependence.

SEROTONIN AND ALCOHOL DEPENDENCE

The central serotonergic system is integral in mediating mood and emotion, and disturbances of serotonergic neurotransmission have been implicated in diminished inhibitory control of behavior and pathologic alcohol consumption (Nishikawa et al., 2009). Preclinical studies have consistently reported associations between low endogenous serotonergic neurotransmission and increased alcohol preference (McBride et al., 1991, 1993; LeMarquand et al., 1994a, b; Tomkins et al., 1994). Furthermore, pharmacologic agents that increase synaptic levels of serotonin, such as 5-HT uptake blockers, releasers, and agonists, have been shown to decrease alcohol consumption in animals and human alcoholics (LeMarquand et al., 1994a, b).

The multiple effects of serotonin are mediated through binding with a large group of receptors, classified according to mechanism of action, structural diversity, and pharmacologic specifications (Savli et al., 2012). To date, researchers have identified seven distinct families of serotonin receptors (5-HT1 through 5-HT7), at least 16 subtypes of 5-HT receptors, and a serotonin transporter (SERT) (Kitson, 2007). Of these, only the 5-HT1A, 5-HT1B, SERT, and 5-HT synthesis have been imaged *in vivo* in human alcohol dependence (Takano et al., 2011). The serotonin transporter mediates the reuptake of serotonin from the synaptic cleft and modulates its extracellular concentration, and is the primary target for the most commonly prescribed antidepressants, the selective serotonin reuptake inhibitors (Meyer, 2007). The 5-HT1A receptors are the most ubiquitous serotonergic receptors in the mammalian brain and function as autoreceptors in the raphe nuclei and postsynaptic receptors in cortical/limbic serotonergic terminals. The 5-HT1A autoreceptor in the dorsal raphe nuclei is of special importance as it modulates the activity of serotonergic neurons and serves as the target for agonist effects of anxiolytics such as buspirone and tandospirone (Takano et al., 2011). The 5-HT1B receptors are found presynaptically on both serotonergic neurons and non-serotonergic neurons, functioning as autoreceptors and heteroreceptors, and mediate the release of multiple neurotransmitters, including serotonin and dopamine (Savli et al., 2012).

PET and SPECT studies imaging the serotonin transporter in alcohol dependence have yielded conflicting results. Multiple radiotracers have been utilized, including [^{123}-I]B-CIT, [^{123}-I]I-ADAM, [^{11}C]McN5652, and [^{11}C] DASB, while others are currently in development (Cosgrove, 2010; Takano et al., 2011). Heinz et al. (1998) reported a 30% reduction in [^{123}-I]B-CIT binding within the midbrain of alcoholics during early abstinence, which correlated inversely with anxiety and depression. This association is consistent with well-documented data establishing anxiety and depression as highly comorbid with alcohol dependence (Cosgrove, 2010). In a follow-up study, Heinz et al. (2000b) again showed decreases in midbrain SERT binding, but only in alcoholics homozygous for the long allele of the SERT gene promoter region, when compared with healthy control subjects with the same genotype. Szabo et al. (2004) also observed a significant decrease in the distribution volume of [^{11}C]McN5652 in multiple brain regions (midbrain, thalamus, amygdala, pons, cingulate, orbitofrontal cortex, and cerebellum) in alcoholics when compared to healthy controls.

By contrast, two later studies using [$^{11-}$C]DASB, which binds SERT with high affinity and selectivity, reported no differences in serotonin transporter binding between alcoholics and healthy controls (Brown et al., 2007; Martinez et al., 2009b). In the most recent study imaging SERT in alcoholism, Ho et al. (2011) used [^{123}-I] I-ADAM to demonstrate lower SERT binding within the midbrain of alcoholics without comorbid anxiety and/or depression.

Given these conflicting findings, the association between SERT availability and alcohol dependence is still incompletely understood. Potential sources for discrepant results include differences in radiotracer characteristics, stage of withdrawal, data analysis (specifically of the midbrain regions), and patient heterogeneity (Martinez et al., 2009b). [^{123}I]B-CIT binding has a very low ratio of specific to non-specific binding in brain regions outside the midbrain, which limits quantification of radiotracer signal (Cosgrove, 2010). By contrast, PET radioligands used in studies of serotonin transporters in alcoholism provide better signal-to-noise ratio in brain regions outside the midbrain, such as the thalamus, amygdala, cingulate, and orbitofrontal cortex (Urban and Martinez, 2012). Furthermore, the majority of these studies did not match for tobacco smoking, and one study suggests that smoking may alter SERT binding in alcoholics (Cosgrove et al., 2009).

In addition to changes in SERT binding, preclinical studies show alterations of central serotonin receptors with prolonged alcohol use (Naranjo and Bremner, 1994; Czachowski, 2005). Alcohol-preferring rats exhibit decreases in total brain serotonin content, loss of serotonergic cells, and compensatory increases in 5-HT1A receptors in the cerebral cortex and hippocampus (Wong et al., 1990; Zhou et al., 1994). 5-HT1A receptor knockout and heterozygous mice demonstrate high levels of anxiety-like behaviors, whereas mice overexpressing 5-HT1A receptors exhibit reduced levels of anxiety-like behaviors. It has also been shown that expression of 5-HT1A receptors during early development, but not in adulthood, is vital for rescuing the anxiety-related phenotype in 5-HT1A receptor knockout mice (Gross et al., 2002). Here it is again important to note that, in humans, anxiety and depression are highly comorbid with alcohol dependence, and that anxiety is an independent risk factor for alcoholism (Buckner et al., 2008).

Until recently, imaging of 5-HT receptors in alcoholism has been limited by a lack of suitable serotonin receptor radioligands. At present, there have only been two PET studies imaging serotonin receptors in alcoholism. The first imaged 5-HT1A receptors using the radiotracer [^{11}C]WAY-100635, a 5-HT1A receptor antagonist, and the second, more recent investigation, imaged 5-HT1B receptors using [^{11}C] P943, a 5-HT1B antagonist (Martinez et al., 2009b; Hu et al., 2010). No differences in 5-HT1A receptor binding were observed in alcoholics relative to healthy controls, whereas significant increases in 5-HT1B receptor binding were detected in the ventral striatum of alcoholics during early abstinence when compared to healthy controls. The molecular basis of increased 5-HT1B receptor binding is not yet understood. Possible mechanisms include overexpression of 5-HT1B receptors, higher affinity of 5-HT1B receptors to the radiotracer, and low intrasynaptic serotonin concentration with subsequently higher levels of unoccupied 5-HT1B receptors (Hu et al., 2010).

In contrast to PET imaging studies of dopamine neurotransmission with amphetamine challenge, there are no validated methods for *in vivo* measurement of intrasynaptic serotonin neurotransmission at this time. Investigators have measured the cerebrospinal fluid (CSF) metabolite of serotonin, 5-HIAA, in alcoholics as an indirect signifier of serotonergic activity and have shown an inverse relationships between 5-HIAA concentration in the CSF and alcohol use as well as with B-CIT binding in the raphe nucleus (LeMarquand et al., 1994a, b; Heinz et al., 2002). In a recent study, the radiotracer α-[^{11}C]-methyl-L-tryptophan was used to reveal alterations in regional brain serotonin synthesis in non-depressed chronic alcoholics. Specifically, alcoholics showed significantly lower serotonin synthesis in Brodmann areas 9, 10, and 32 and higher serotonin synthesis in Brodmann areas 19 and 41 (located within the occipital lobe and left superior temporal gyrus, respectively) when compared to controls (Nishikawa et al., 2009). These areas are involved in mediation of planning, self-control, and social behaviors (Knoch and Fehr, 2007). Researchers have also shown that, in response to m-chlorophenylpiperazine, a partial 5HT2C agonist, alcoholics exhibit diminished regional glucose utilization (measured with fludeoxyglucose-PET) in the orbital and prefrontal cortices when compared to healthy controls (Hommer et al., 1997). Interestingly, no differences in serotonin synthesis were noted within the raphe nucleus. This should be interpreted with caution as scanner resolution along with the relatively small size of the dorsal raphe may have resulted in small variations of serotonin synthesis being missed (Nishikawa et al., 2009). Combined with the previously discussed finding of increased striatal 5-HT1B binding in alcoholics, these data suggest altered serotoninergic modulation of corticostriatal circuits in alcohol dependence (Storvik et al., 2006).

There is increasing evidence that alcoholism, depression, and anxiety share serotonergic dysfunction as a point of pathogenic convergence. Traumatic experiences during childhood are consistently correlated with increased risk for developing anxiety-related neuropsychiatric diseases, many of which are highly comorbid with alcoholism (Teicher et al., 2003). Moreover, peer-reared monkeys exhibit dysfunctions of the hypothalamic–pituitary–adrenal axis, 5-HT system, and high levels of anxiety-like behaviors when compared with animals reared by their mothers (Higley et al., 1992; Shannon et al., 2005; Ichise et al., 2006). In a recent study using [^{18}F]FPWAY, an analogue of [^{11}C] WAY100635 and a selective 5-HT1A receptor radiotracer, Spinelli et al. (2010) showed that, in both males and females, peer-reared monkeys exhibit decreased overall 5-HT1A receptor binding when compared to mother-reared monkeys. One exception to this finding was within the dorsomedial prefrontal cortex of peer-reared females, where receptor availability and regional volumes were higher (Spinelli et al., 2009, 2010). Taken together, these results suggest a possible state-dependent mechanism by which serotonergic dysfunction may develop in early childhood, predisposing the individual to anxiety, depression, and alcohol dependence later in life.

CANNABINOID RECEPTORS AND ALCOHOL DEPENDENCE

There is mounting evidence implicating the endogenous cannabinoid system in the development and maintenance of alcohol dependence (Serrano and Parsons, 2011). The central nervous system effects of endogenous cannabinoids appear to be mediated through the CB1 receptor, located within the cerebral cortex, hippocampus, striatum, cerebellum, and midbrain (Vinod and Hungund, 2006). Midbrain CB1 receptors are located on GABAergic interneurons (specifically in the VTA and SN), where they modulate inhibitory input on to dopamine neurons projecting into the striatum and prefrontal cortex (Luscher and Malenka, 2011). CB1 receptor knockout mice and rodents administered CB1 receptor antagonists demonstrate reduced voluntary alcohol consumption while administration of a CB1 receptor agonist results in increased alcohol intake in rodents (Thanos et al., 2005; Vinod and Hungund, 2006).

At present, two studies have evaluated CB1 receptors in alcoholism and have yielded conflicting results. Neumeister et al. (2012) reported an approximate 20% increase in the distribution volume of [^{11}C]OMAR in multiple brain regions (amygdala, hippocampus, putamen, insula, anterior and posterior cingulate cortices, and orbitofrontal cortex) of male alcoholics ($n=8$) during early abstinence (2–4 weeks) when compared to healthy controls. Using [^{18}F]FMPEP-d2, Hirvonen et al. (2012) imaged 18 male alcoholics during immediate abstinence (3–7 days) and early abstinence (2–4 weeks), revealing a 20–30% reduction in whole-brain distribution volume which persisted across both time points and was negatively correlated with years of alcohol abuse. Given the relatively small number of subjects imaged and the discrepancy of current findings, it is unclear, at this time, whether alcohol dependence is truly associated with changes in the cannabinoid system.

CONCLUSIONS

In summary, D_2 receptor binding and dopamine release have consistently been observed to be lower in alcohol-dependent subjects compared to controls, predominantly affecting the ventral striatum. A deficit in GABAergic parameters has been described, at least in some studies of abstinent patients, but not consistently through all studies and not in early withdrawal. Brain atrophy may have contributed to some of these measured decreases in GABA receptor indices. Studies of the opiate receptors have suggested increases in the MOR subtype, although again not consistently across all studies. Serotonin receptors and transporters have been examined and most studies have failed to detect alterations in alcoholics. And most recently, studies of the CB1 receptor have yielded discrepant results (see Table 18.1 for a comprehensive list of studies and findings discussed in this chapter).

There appears to be a convergence of molecular imaging findings, across some of the neurotransmitters systems discussed, on to the ventral striatum, a region that is central for processing reward function in the brain. These include decreases in dopaminergic parameters (Hietala et al., 1994; Volkow et al., 1996b, 2002b, 2007; Heinz et al., 2004, 2005b; Martinez et al., 2005), decreases in the alpha-5 subtype of GABA-A receptors (Lingford-Hughes et al., 2012), and increases in MOR receptor binding (Heinz et al., 2005a; Weerts et al., 2011). Another site of convergence seems to be the prefrontal cortex, where studies have shown decreases in indices for GABA-A, and opiate receptors (Abi-Dargham et al., 1998; Bencherif et al., 2004). More work is needed to clarify the role of these systems in the various stages of alcoholism. It is also likely that dopaminergic alterations that are robustly measured in alcoholism represent aberrant regulation of the dopaminergic system by one or more of these systems. Better tools and larger studies are needed to understand the role of these interactions.

While alcohol has many differential effects not shared by other drugs of abuse, the effects on the dopamine system appear to be common across addictions. PET imaging studies show decreased D_2 receptor binding in cocaine dependence (Volkow et al., 1990, 1993, 1997; Martinez et al., 2004, 2009a, 2011), methamphetamine abuse (Volkow et al., 2001; Lee et al., 2009; Boileau et al., 2012; Wang et al., 2012), opiate dependence (Wang et al., 1997; Zijlstra et al., 2008; Martinez et al., 2012), and tobacco dependence (Brody et al., 2004; Fehr et al., 2008; Brown et al., 2012; Stokes et al., 2012), although not cannabis abuse (Sevy et al., 2008; Stokes et al., 2012; Urban et al., 2012; Albrecht et al., 2013). Similarly, stimulant-induced dopamine release is blunted across substance use disorders. This has been reported in cocaine abuse (Volkow et al., 1997; Malison et al., 1999; Martinez et al., 2007, 2011), alcohol dependence (Martinez et al., 2005; Volkow et al., 2007), opiate dependence (Martinez et al., 2012), and methamphetamine abuse (Wang et al., 2012). Together, these studies suggest that low D_2 receptor binding and low dopamine release serve as common biomarkers for addiction in general. While alcohol's effects on dopamine are indirect, probably mediated by its action on multiple targets, which regulate dopamine cell

Table 18.1

Radioligand binding in alcohol dependence compared with healthy controls

Target	Reference	AD subjects	Withdrawal period	Radioligand	Outcome measure	Results (AD vs HC)
Dopamine						
D_2 receptor	Hietala et al., 1994	9 M	23 ± 62 weeks	$[^{11}C]$raclopride (PET)	BP_{ND}	↓ in striatum (20%)
	Volkow et al., 1996b	10 M (9 M, 1 F)	52 ± 48 days	$[^{11}C]$raclopride (PET)	BP_{ND}	↓ in striatum (20%)
	Volkow et al., 2002b	8 (7 M, 1 F)	6–20 weeks	$[^{11}C]$raclopride (PET)	BP_{ND}	↓ in CAD (14%) and PUT (18%) in early detox. (17 ± 10 days); NS late compared to early detox (1–4 months)
	Heinz et al., 2004, 2005b	11 M	2–4 weeks	$[^{18}F]$ desmethoxyfallypride (PET)	BP_{ND}	↓ in the PUT and VST
	Martinez et al., 2005	15 (13 M, 2 F)	14 days	$[^{11}C]$raclopride (PET)	BP_{ND}	↓ in the AST (16%), SMST (19%), and VST (13%)
	Rominger et al., 2012	17 M, (17 for first scan, 14 for second, 4 for third)	Within 1 day of detox ($n = 17$), at 1–2 weeks ($n = 14$), at 1 year ($n = 4$)	$[^{18}F]$fallypride (PET)	BP_{ND}	↓ in THA, HIPP, insula and temporal cortex (10–20%). NS after short-term follow-up, ↑ in striatum after 1 year's abstinence
Dopamine synthesis	Tiihonen et al., 1998	10 M	19 ± 57 days	6-$[^{18}F]$-fluorodopa (FDOPA) (PET)	Ki	↑ mean striatal uptake (9–36%)
	Heinz et al., 2005b	12 M	36 ± 22 days	6-$[^{18}F]$-fluorodopa (FDOPA) (PET)	Ki	NS
Presynaptic dopamine storage (VMAT2)	Gilman et al., 1998	7 M	4–36 weeks	$[(+)[^{11}C]DTBZ]$ (PET)	K_1, V_T	↓ K_1 in CAD (8.6%) and PUT (6.2%); ↓ V_T in the CAD (4.5%) and PUT (10.4%)
Dopamine release	Martinez et al., 2005	15 (13 M, 2 F)	14 days	$[^{11}C]$raclopride (PET), amphetamine challenge	ΔBP_{ND}	↓ ΔBP_{ND} in VST, AD vs HC
	Volkow et al., 2007	20 M	1 week	$[^{11}C]$raclopride (PET), methylphenidate challenge	ΔBP_{ND}	↓ in PUT ($\Delta BP_{ND} = 11\%$ in AD, 21% in HC) and VST (8% in AD, 27% in HC)
Dopamine transporter (DAT)	Tiihonen et al., 1995	29 (27, 2)	8 weeks	$[^{123}I]\beta$-CIT (SPECT)	BP_{ND}	↓ in striatum of non-violent AD ($n = 10$), slightly ↑in striatum of violent AD ($n = 19$)

	Volkow et al., 1996b	5 M	52 ± 48 days	[11C]*d-threo* methylphenidate (PET)	BP_{ND}	NS
	Laine et al., 1999a, b	28 (23, 5)	Scan 1 1–4 days, scan 2 after 4 weeks	[123I]β-CIT (SPECT)	BP_{ND}	↓ during immediate withdrawal (1–4 days). NS at 4 weeks
	Repo et al., 1999	9 M	43 ± 62 days (range 7–165 days)	[123I]-PE2I	BP_{ND}	↓ in striatum
	Heinz et al., 2000a	14 (11 M, 3 F)	3–5 weeks	[123I]β-CIT (SPECT)	BP_{ND}	NS (effect of VNTR genotype in PUT)
GABA						
GABA-A receptor	Litton et al., 1993	5 M	Mean 17 days (range 10–22 days)	[11C]flumazenil (PET)	Bmax, Kd	NS
	Abi-Dargham et al., 1998	11 M	1–6 months	[123I]iomazenil (SPECT)	V_T	↓ in PFC, ACC, and cerebellum
	Lingford-Hughes et al., 1998	12 M	23 ± 50 months	[123I]iomazenil (SPECT)	V_T	↓ in frontal, parietal, and temporal cortices
	Lingford-Hughes et al., 2000	9 F	30 ± 53 months	[123I]iomazenil (SPECT)	V_T	↓ trend level in cerebellum, occipital lobes, and parietal cortex (left > right)
	Lingford-Hughes et al., 2005	11 M	7 ± 2 months	[123C]flumazenil (PET)	V_T	NS, but trend ↓ in OFC and ACC (6–8%)
	Staley et al., 2005	27 M	1 scan at 5 ± 3 days (*n* = 23) and 1 scan at 30 ± 8 (*n* = 20)	[123I]Iomazenil (SPECT)	V_T	↑ GABA-A at 1 week detox in AD non-smokers compared to AD smokers and controls, No differences between groups at 4 weeks detox
	Lingford-Hughes et al., 2012	8 M	33 ± 45 months	[11C]Ro15 4513 (PET)	V_T	↓ in whole brain, specifically in Nacc, parahippocampal gyri, right HIPP, and right AMY
Opioid receptors						
μ-opioid receptor	Bencherif et al., 2004	8 M	5 days	[11C]carfentanil (PET)	BP_{ND}	↓ in the right DLPFC, right anterior frontal cortex and right parietal cortex
	Heinz et al., 2005a	25 M	First scan at 1–3 weeks (range 6–22 days), second scan at 4–6 weeks (range 37–51 days)	[11C]carfentanil (PET)	BP_{ND}	↑ in bilateral VST both in early and late abstinence but NS between scans
	Weerts et al., 2011	25 (18 M 7 F)	5 days	[11C]carfentanil (PET)	BP_{ND}	↑ in VST, AMY, CAD, GP, insula, PUT, and THA
∂-opioid receptor	Weerts et al., 2011	20 (15 M 5 F)	5 days	[11C]MeNTL (PET)	BP_{ND}	NS
General opioid receptor	Williams et al., 2009	11 M	Scan 1 at 14–30 days (*n* = 11), scan 2 at 89–132 days (*n* = 4)	[11C]diprenorphine (PET)	V_T	NS

Continued

Table 18.1

Continued

Target	Reference	AD subjects	Withdrawal period	Radioligand	Outcome measure	Results (AD vs HC)
Serotonin						
Serotonin transporter (SERT)	Heinz et al., 1998	22 M	3–5 weeks	$[^{123}I]\beta$-CIT (SPECT)	BP_{ND}	↓ in RN
	Heinz et al., 2000b	14 M	24–32 days	$[^{123}I]\beta$-CIT (SPECT)	BP_{ND}	↓ in RN of long allele 5-HTT AD compared with long allele 5-HTT HC
	Szabo et al., 2004	15	17 ± 10 (minimum 2; maximum 37) years	$[^{11}C]$(+)McN5652 (PET)	V_T ,BP_{ND}	↓ V_T in midbrain, THA, AMY, pons, cingulate, OFC, and cerebellum; ↓ BP_{ND} in midbrain only
	Brown et al., 2007	30 (24 M 6 F)	14 ± 2 days	$[^{11}C]$DASB (PET)	BP_{ND}	NS
	Martinez et al., 2009b	12 (10 M/2 F)	14 days	$[^{11}C]$DASB (PET)	BP_P, BP_{ND}	NS
	Ho et al., 2011	28	1 week	$[^{123}I]$-ADAM (SPECT)	BP_{ND}	↓ in the midbrain of AD without comorbid anxiety and/or depression
5-HT1A receptor	Martinez et al., 2009b	13 (11 M 2 F)	14 days	$[^{11}C]$WAY 100635 (PET)	BP_P, BP_{ND}	NS
5-HT1B receptor	Hu et al., 2010	12 (7 M 5 F)	At least 4 weeks	$[^{11}C]$ P943 (PET)	BP_{ND}	↑ in VST
Cannabinoid						
CB1 receptor	Neumeister et al., 2012	8 M	4 weeks	$[^{11}C]$OMAR (PET)	V_T	↑ 20% in AMY, HIPP, PUT, insula, ACC and PCC, and OFC
	Hirvonen et al., 2012	18 M	First scan at 3–7 days, second scan at 2–4 weeks	$[^{18}F]$FMPEP-d_2 (PET)	V_T	↓ 20–30% at both time points in all regions examined

Subjects: AD, alcohol-dependent; HC, healthy control; M, male; F, female. Outcome measures: V_T, total distribution volume (Innis et al., 2007); BP_{ND}, binding potential with respect to non-displaceable compartment (Innis et al., 2007); BP_P, binding potential with respect to arterial plasma concentration of tracer (Innis et al., 2007); ΔBP_{ND}, percent change in BP_{ND} across conditions; K_1, blood to brain delivery rate constant (Innis et al., 2007); Ki, steady-state uptake rate (Patlak et al., 1983). Results: ↑, AD increased compared to controls; ↓, AD decreased compared to controls; NS, no significant difference between AD and HC. Results (brain regions): ACC, anterior cingulate cortex; AMY, amygdala; AST, associative striatum; CAD, caudate nucleus; DLPFC, dorsolateral prefrontal cortex; GP, globus pallidus; HIPP, hippocampus; Nacc, nucleus accumbens; OFC, orbitofrontal cortex; PCC, posterior cingulate cortex; PFC, prefrontal cortex; PUT, putamen; SMST, sensorimotor striatum; THA, thalamus.

firing in the midbrain, other drugs, such as cocaine, have more direct effects on the dopamine system. Nevertheless, regardless of mechanisms, these drugs all result in similar profound deficits in dopaminergic indices after chronic use.

Imaging studies of other neuroreceptor systems have shown fewer similarities with other drugs of abuse. While PET imaging studies of the MOR in alcohol dependence, in general, report higher MOR binding in striatal and cortical brain regions, studies in cocaine abuse show higher binding in cortical brain regions (prefrontal and temporal cortex) only (Zubieta et al., 1996; Gorelick et al., 2005). However, given the discrepancies discussed above, more work is needed to derive definitive conclusions in this regard.

While many transmitter systems are involved in the mechanisms of action of alcohol, from the initial use through the transition to compulsive use, few, outside of dopamine, have been well characterized. Radiotracers for these targets, with good signal-to-noise properties, are needed for use in systematic studies encompassing all stages of the addictive process. Among those, radiotracers that allow investigation of the glutamatergic system are particularly few and necessary, as many lines of evidence indicate a major role for glutamate in the transition from acute early use to chronic dependence (Kalivas and Volkow, 2011). While few tracers have recently become available, such as [^{11}C]ABP688 for the mGlur5 receptor (Miyake et al., 2011), many more are needed. In particular, the *N*-methyl-D-asparate (NMDA) receptor is a relevant focus for studies of alcoholism. The NMDA receptor is targeted by alcohol at high doses (Hoffman et al., 1990), is modulated by dopaminergic transmission in opposite direction via D_1 and D_2 receptors (Laruelle et al., 2005; Narendran et al., 2005), and mediates long-term synaptic plasticity associated with addictive processes (Kalivas and Volkow, 2011). Furthermore, understanding the molecular mechanism of vulnerability, prior to initiation of heavy use, is crucial for developing better prevention strategies. This is a particularly important and promising application of PET imaging in alcohol-related disorders. As new radiotracers become available, we expect the next few years will show progress in the ability of molecular imaging tools to unravel the complex interplay of neural circuits involved in the development of alcohol use disorders.

References

Abi-Dargham A, Krystal JH, Anjilvel S et al. (1998). Alterations of benzodiazepine receptors in type II alcoholic subjects measured with SPECT and [^{123}I]iomazenil. Am J Psychiatry 155: 1550–1555.

Albrecht DS, Skosnik PD, Vollmer JM et al. (2013). Striatal D(2)/D(3) receptor availability is inversely correlated with cannabis consumption in chronic marijuana users. Drug Alcohol Depend 128: 52–57.

Bencherif B, Wand GS, Mccaul ME et al. (2004). Mu-opioid receptor binding measured by [^{11}C]carfentanil positron emission tomography is related to craving and mood in alcohol dependence. Biol Psychiatry 55: 255–262.

Bhaskar LV, Thangaraj K, Wasnik S et al. (2012). Dopamine transporter (DAT1) VNTR polymorphism and alcoholism in two culturally different populations of south India. Am J Addict 21: 343–347.

Boileau I, Assaad J-M, Pihl RO et al. (2003). Alcohol promotes dopamine release in the human nucleus accumbens. Synapse 49: 226–231.

Boileau I, Payer D, Houle S et al. (2012). Higher binding of the dopamine D3 receptor-preferring ligand [^{11}C]-(+)-propyl-hexahydro-naphtho-oxazin in methamphetamine polydrug users: a positron emission tomography study. J Neurosci 32: 1353–1359.

Brody AL, Olmstead RE, London ED et al. (2004). Smoking-induced ventral striatum dopamine release. Am J Psychiatry 161: 1211–1218.

Brown AK, George DT, Fujita M et al. (2007). PET [^{11}C] DASB imaging of serotonin transporters in patients with alcoholism. Alcohol Clin Exp Res 31: 28–32.

Brown AK, Mandelkern MA, Farahi J et al. (2012). Sex differences in striatal dopamine D_2/D_3 receptor availability in smokers and non-smokers. Int J Neuropsychopharmacol 15: 989–994.

Buckner JD, Schmidt NB, Lang AR et al. (2008). Specificity of social anxiety disorder as a risk factor for alcohol and cannabis dependence. J Psychiatr Res 42: 230–239.

Cosgrove KP (2010). Imaging receptor changes in human drug abusers. Curr Top Behav Neurosci 3: 199–217.

Cosgrove KP, Krantzler E, Frohlich EB et al. (2009). Dopamine and serotonin transporter availability during acute alcohol withdrawal: effects of comorbid tobacco smoking. Neuropsychopharmacology 34: 2218–2226.

Cosgrove KP, Esterlis I, Mason GF et al. (2011). Neuroimaging insights into the role of cortical GABA systems and the influence of nicotine on the recovery from alcohol dependence. Neuropharmacology 60: 1318–1325.

Czachowski CL (2005). Manipulations of serotonin function in the nucleus accumbens core produce differential effects on ethanol and sucrose seeking and intake. Alcohol Clin Exp Res 29: 1146–1155.

Del Arbol JL, Rico Irles J, Contreras I et al. (2007). Plasma concentrations of beta-endorphins in the children of alcoholic patients. An Med Interna 24: 273–277.

Di Chiara G, Imperato A (1988). Drugs abused by humans preferentially increase synaptic dopamine concentrations in the mesolimbic system of freely moving rats. Proc Natl Acad Sci U S A 85: 5274–5278.

Fehr C, Yakushev I, Hohmann N et al. (2008). Association of low striatal dopamine D_2 receptor availability with nicotine dependence similar to that seen with other drugs of abuse. Am J Psychiatry 165: 507–514.

Friedmann PD (2013). Alcohol use in adults. N Engl J Med 368: 365–373.

George MS, Anton RF, Bloomer C et al. (2001). Activation of prefrontal cortex and anterior thalamus in alcoholic subjects on exposure to alcohol-specific cues. Arch Gen Psychiatry 58: 345–352.

Gianoulakis C (2004). Endogenous opioids and addiction to alcohol and other drugs of abuse. Curr Top Med Chem 4: 39–50.

Gianoulakis C (2009). Endogenous opioids and addiction to alcohol and other drugs of abuse. Curr Top Med Chem 9: 999–1015.

Gilman S, Junck L, Martorello S et al. (1996). Positron emission tomographic studies of cerebral benzodiazepine-receptor binding in chronic alcoholics. Ann Neurol 40: 163–171.

Gilman S, Koeppe RA, Adams KM et al. (1998). Decreased striatal monoaminergic terminals in severe chronic alcoholism demonstrated with (+)[^{11}C]dihydrotetrabenazine and positron emission tomography. Ann Neurol 44: 326–333.

Gilpin NW, Koob GF (2008). Neurobiology of alcohol dependence: focus on motivational mechanisms. Alcohol Res Health 31: 185–195.

Goldstein RZ, Leskovjan AC, Hoff AL et al. (2004). Severity of neuropsychological impairment in cocaine and alcohol addiction: association with metabolism in the prefrontal cortex. Neuropsychologia 42: 1447–1458.

Gorelick DA, Kim YK, Bencherif B et al. (2005). Imaging brain mu-opioid receptors in abstinent cocaine users: time course and relation to cocaine craving. Biol Psychiatry 57: 1573–1582.

Gorelick DA, Kim YK, Bencherif B et al. (2008). Brain mu-opioid receptor binding: relationship to relapse to cocaine use after monitored abstinence. Psychopharmacology (Berl) 200: 475–486.

Gould RW, Porrino LJ, Nader MA (2012). Nonhuman primate models of addiction and PET imaging: dopamine system dysregulation. Curr Top Behav Neurosci 11: 25–44.

Grant KA, Shively CA, Nader MA et al. (1998). Effect of social status on striatal dopamine D_2 receptor binding characteristics in cynomolgus monkeys assessed with positron emission tomography. Synapse 29: 80–83.

Gross C, Zhuang X, Stark K et al. (2002). Serotonin1A receptor acts during development to establish normal anxiety-like behaviour in the adult. Nature 416: 396–400.

Haber SN (2011). Neuroanatomy of reward: a view from the ventral striatum. In: JA Gottfried (Ed.), Neurobiology of Sensation and Reward. CRC Press, Boca Raton, FL.

Hall FS, Sora I, Uhl GR (2001). Ethanol consumption and reward are decreased in mu-opiate receptor knockout mice. Psychopharmacology (Berl) 154: 43–49.

Heinz A, Ragan P, Jones DW et al. (1998). Reduced central serotonin transporters in alcoholism. Am J Psychiatry 155: 1544–1549.

Heinz A, Goldman D, Jones DW et al. (2000a). Genotype influences in vivo dopamine transporter availability in human striatum. Neuropsychopharmacology 22: 133–139.

Heinz A, Jones DW, Mazzanti C et al. (2000b). A relationship between serotonin transporter genotype and in vivo protein expression and alcohol neurotoxicity. Biol Psychiatry 47: 643–649.

Heinz A, Jones DW, Bissette G et al. (2002). Relationship between cortisol and serotonin metabolites and transporters in alcolholism. Pharmacopsychiatry 35: 127–134.

Heinz A, Siessmeier T, Wrase J et al. (2004). Correlation between dopamine D_2 receptors in the ventral striatum and central processing of alcohol cues and craving. Am J Psychiatry 161: 1783–1789.

Heinz A, Reimold M, Wrase J et al. (2005a). Correlation of stable elevations in striatal mu-opioid receptor availability in detoxified alcoholic patients with alcohol craving: a positron emission tomography study using carbon 11-labeled carfentanil. Arch Gen Psychiatry 62: 57–64.

Heinz A, Siessmeier T, Wrase J et al. (2005b). Correlation of alcohol craving with striatal dopamine synthesis capacity and $D_2/_3$ receptor availability: a combined [^{18}F]DOPA and [^{18}F]DMFP PET study in detoxified alcoholic patients. Am J Psychiatry 162: 1515–1520.

Henriksen G, Willoch F (2008). Imaging of opioid receptors in the central nervous system. Brain 131: 1171–1196.

Hietala J, West C, Syvalahti E et al. (1994). Striatal D_2 dopamine receptor binding characteristics in vivo in patients with alcohol dependence. Psychopharmacology (Berl) 116: 285–290.

Higley JD, Suomi SJ, Linnoila M (1992). A longitudinal assessment of CSF monoamine metabolite and plasma cortisol concentrations in young rhesus monkeys. Biol Psychiatry 32: 127–145.

Hillemacher T (2011). Biological mechanisms in alcohol dependence – new perspectives. Alcohol Alcohol 46: 224–230.

Hirvonen J, Zanotti-Fregonara P, Umhau JC et al. (2012). Reduced cannabinoid CB1 receptor binding in alcohol dependence measured with positron emission tomography. Mol Psychiatry 18: 916–921.

Ho PS, Shih MC, Ma KH et al. (2011). Availability of the serotonin transporter in patients with alcohol dependence. World J Biol Psychiatry 12: 134–142.

Hoffman PL, Rabe CS, Grant KA et al. (1990). Ethanol and the NMDA receptor. Alcohol 7: 229–231.

Hommer D, Andreasen P, Rio D et al. (1997). Effects of m-chlorophenylpiperazine on regional brain glucose utilization: a positron emission tomographic comparison of alcoholic and control subjects. J Neurosci 17: 2796–2806.

Hou H, Tian M, Zhang H (2012). Positron emission tomography molecular imaging of dopaminergic system in drug addiction. Anat Rec 295: 722–733.

Hu J, Henry S, Gallezot J-D et al. (2010). Serotonin 1B receptor imaging in alcohol dependence. Biol Psychiatry 67: 800–803.

Ichise M, Vines DC, Gura T et al. (2006). Effects of early life stress on [^{11}C]DASB positron emission tomography imaging of serotonin transporters in adolescent peer- and mother-reared rhesus monkeys. J Neurosci 26: 4638–4643.

Innis RB, Cunningham VJ, Delforge J et al. (2007). Consensus nomenclature for in vivo imaging of reversibly binding radioligands. J Cereb Blood Flow Metab 27: 1533–1539.

Jalan R, Turjanski N, Taylor-Robinson SD et al. (2000). Increased availability of central benzodiazepine receptors in patients with chronic hepatic encephalopathy and alcohol related cirrhosis. Gut 46: 546–552.

Kalivas PW, Volkow ND (2011). New medications for drug addiction hiding in glutamatergic neuroplasticity. Mol Psychiatry 16: 974–986.

Kitson SL (2007). 5-Hydroxytryptamine (5-HT) receptor ligands. Curr Pharm Des 13: 2621–2637.

Knoch D, Fehr E (2007). Resisting the power of temptations: the right prefrontal cortex and self-control. Ann N Y Acad Sci 1104: 123–134.

Koob GF, Roberts AJ, Schulteis G et al. (1998). Neurocircuitry targets in ethanol reward and dependence. Alcohol Clin Exp Res 22: 3–9.

Koob GF, Ahmed SH, Boutrel B et al. (2004). Neurobiological mechanisms in the transition from drug use to drug dependence. Neurosci Biobehav Rev 27: 739–749.

Krystal JH, Staley J, Mason G et al. (2006). γ-aminobutyric acid type a receptors and alcoholism: intoxication, dependence, vulnerability, and treatment. Arch Gen Psychiatry 63: 957–968.

Laine TP, Ahonen A, Rasanen P et al. (1999a). Dopamine transporter availability and depressive symptoms during alcohol withdrawal. Psychiatry Res 90: 153–157.

Laine TP, Ahonen A, Torniainen P et al. (1999b). Dopamine transporters increase in human brain after alcohol withdrawal. Mol Psychiatry 4 (104–105): 189–191.

Laruelle M, Frankle WG, Narendran R et al. (2005). Mechanism of action of antipsychotic drugs: from dopamine D(2) receptor antagonism to glutamate NMDA facilitation. Clin Ther 27 (Suppl A): S16–S24.

Lee B, London ED, Poldrack RA et al. (2009). Striatal dopamine D_2/D_3 receptor availability is reduced in methamphetamine dependence and is linked to impulsivity. J Neurosci 29: 14734–14740.

Lemarquand D, Pihl RO, Benkelfat C (1994a). Serotonin and alcohol intake, abuse, and dependence: clinical evidence. Biol Psychiatry 36: 326–337.

Lemarquand D, Pihl RO, Benkelfat C (1994b). Serotonin and alcohol intake, abuse, and dependence: findings of animal studies. Biol Psychiatry 36: 395–421.

Lingford-Hughes AR, Acton PD, Gacinovic S et al. (1998). Reduced levels of GABA-benzodiazepine receptor in alcohol dependency in the absence of grey matter atrophy. Br J Psychiatry 173: 116–122.

Lingford-Hughes AR, Acton PD, Gacinovic S et al. (2000). Levels of γ-aminobutyric acid-benzodiazepine receptors in abstinent, alcohol-dependent women: preliminary findings from an ^{123}I-iomazenil single photon emission tomography study. Alcohol Clin Exp Res 24: 1449–1455.

Lingford-Hughes AR, Wilson SJ, Cunningham VJ et al. (2005). GABA-benzodiazepine receptor function in alcohol dependence: a combined ^{11}C-flumazenil PET and pharmacodynamic study. Psychopharmacology (Berl) 180: 595–606.

Lingford-Hughes A, Reid AG, Myers J et al. (2012). A [^{11}C] Ro15 4513 PET study suggests that alcohol dependence in man is associated with reduced α5 benzodiazepine receptors in limbic regions. J Psychopharmacol 26: 273–281.

Litton JE, Neiman J, Pauli S et al. (1993). PET analysis of [^{11}C] flumazenil binding to benzodiazepine receptors in chronic alcohol-dependent men and healthy controls. Psychiatry Res 50: 1–13.

Luscher C, Malenka RC (2011). Drug-evoked synaptic plasticity in addiction: from molecular changes to circuit remodeling. Neuron 69: 650–663.

Maas LC, Lukas SE, Kaufman MJ et al. (1998). Functional magnetic resonance imaging of human brain activation during cue-induced cocaine craving. Am J Psychiatry 155: 124–126.

Malison RT, Mechanic KY, Klummp H et al. (1999). Reduced amphetamine-stimulated dopamine release in cocaine addicts as measured by [^{123}I]IBZM SPECT. J Nucl Med 40: 110P.

Martinez D, Narendran R (2010). Imaging neurotransmitter release by drugs of abuse. Curr Top Behav Neurosci 3: 219–245.

Martinez D, Broft A, Foltin RW et al. (2004). Cocaine dependence and D_2 receptor availability in the functional subdivisions of the striatum: relationship with cocaine-seeking behavior. Neuropsychopharmacology 29: 1190–1202.

Martinez D, Gil R, Slifstein M et al. (2005). Alcohol dependence is associated with blunted dopamine transmission in the ventral striatum. Biol Psychiatry 58: 779–786.

Martinez D, Narendran R, Foltin RW et al. (2007). Amphetamine-induced dopamine release: markedly blunted in cocaine dependence and predictive of the choice to self-administer cocaine. Am J Psychiatry 164: 622–629.

Martinez D, Greene K, Broft A et al. (2009a). Lower level of endogenous dopamine in patients with cocaine dependence: findings from PET imaging of D(2)/D(3) receptors following acute dopamine depletion. Am J Psychiatry 166: 1170–1177.

Martinez D, Slifstein M, Gil R et al. (2009b). Positron emission tomography imaging of the serotonin transporter and 5-HT1A receptor in alcohol dependence. Biol Psychiatry 65: 175–180.

Martinez D, Carpenter KM, Liu F et al. (2011). Imaging dopamine transmission in cocaine dependence: link between neurochemistry and response to treatment. Am J Psychiatry 168: 634–641.

Martinez D, Saccone PA, Liu F et al. (2012). Deficits in dopamine D(2) receptors and presynaptic dopamine in heroin dependence: commonalities and differences with other types of addiction. Biol Psychiatry 71: 192–198.

Mcbride WJ, Murphy JM, Gatto GJ et al. (1991). Serotonin and dopamine systems regulating alcohol intake. Alcohol Alcohol Suppl 1: 411–416.

Mcbride WJ, Chernet E, Dyr W et al. (1993). Densities of dopamine D2 receptors are reduced in CNS regions of alcohol-preferring P rats. Alcohol 10: 387–390.

Mena JC, Cuellar H, Vargas D et al. (2005). PET and SPECT in drug and substance abuse. Top Magn Reson Imaging 16: 253–256.

Meyer JH (2007). Imaging the serotonin transporter during major depressive disorder and antidepressant treatment. J Psychiatry Neurosci 32: 86–102.

Miyake N, Skinbjerg M, Easwaramoorthy B et al. (2011). Imaging changes in glutamate transmission in vivo with the metabotropic glutamate receptor 5 tracer [^{11}C] ABP688 and *N*-acetylcysteine challenge. Biol Psychiatry 69: 822–824.

Morgan D, Grant KA, Gage HD et al. (2002). Social dominance in monkeys: dopamine D_2 receptors and cocaine self-administration. Nat Neurosci 5: 169–174.

Myrick H, Anton RF, Li X et al. (2008). Effect of naltrexone and ondansetron on alcohol cue-induced activation of the ventral striatum in alcohol-dependent people. Arch Gen Psychiatry 65: 466–475.

Naranjo CA, Bremner KE (1994). Serotonin-altering medications and desire, consumption and effects of alcohol-treatment implications. EXS 71: 209–219.

Narendran R, Frankle WG, Keefe R et al. (2005). Altered prefrontal dopaminergic function in chronic recreational ketamine users. Am J Psychiatry 162: 2352–2359.

Neumeister A, Normandin MD, Murrough JW et al. (2012). Positron emission tomography shows elevated cannabinoid CB1 receptor binding in men with alcohol dependence. Alcohol Clin Exp Res 36: 2104–2109.

Nishikawa M, Diksic M, Sakai Y et al. (2009). Alterations in brain serotonin synthesis in male alcoholics measured using positron emission tomography. Alcohol Clin Exp Res 33: 233–239.

Nurmi M, Sinclair JD, Kiianmaa K (1998). Dopamine release during ethanol drinking in AA rats. Alcohol Clin Exp Res 22: 1628–1633.

O'brien CP, Childress AR, Mclellan AT et al. (1992). Classical conditioning in drug-dependent humans. Ann N Y Acad Sci 654: 400–415.

Parvaz MA, Alia-Klein N, Woicik PA et al. (2011). Neuroimaging for drug addiction and related behaviors. Rev Neurosci 22: 609–624.

Patel J, Mooslehner KA, Chan PM et al. (2003). Presynaptic control of striatal dopamine neurotransmission in adult vesicular monoamine transporter 2 (VMAT2) mutant mice. J Neurochem 85: 898–910.

Patlak CS, Blasberg RG, Fenstermacher JD (1983). Graphical evaluation of blood-to-brain transfer constants from multiple-time uptake data. J Cereb Blood Flow Metab 3: 1–7.

Ramchandani VA, Umhau J, Pavon FJ et al. (2011). A genetic determinant of the striatal dopamine response to alcohol in men. Mol Psychiatry 16: 809–817.

Repo E, Kuikka JT, Bergström KA et al. (1999). Dopamine transporter and D_2-receptor density in late-onset alcoholism. Psychopharmacology (Berl) 147: 314–318.

Rominger A, Cumming P, Xiong G et al. (2012). [^{18}F]fallypride PET measurement of striatal and extrastriatal dopamine $D_2/_3$ receptor availability in recently abstinent alcoholics. Addict Biol 17: 490–503.

Savli M, Bauer A, Mitterhauser M et al. (2012). Normative database of the serotonergic system in healthy subjects using multi-tracer PET. Neuroimage 63: 447–459.

Serrano A, Parsons LH (2011). Endocannabinoid influence in drug reinforcement, dependence and addiction-related behaviors. Pharmacol Ther 132: 215–241.

Sevy S, Smith GS, Ma Y et al. (2008). Cerebral glucose metabolism and D_2/D_3 receptor availability in young adults with cannabis dependence measured with positron emission tomography. Psychopharmacology (Berl) 197: 549–556.

Shannon C, Schwandt ML, Champoux M et al. (2005). Maternal absence and stability of individual differences in CSF 5-HIAA concentrations in rhesus monkey infants. Am J Psychiatry 162: 1658–1664.

Slifstein M, Laruelle M (2001). Models and methods for derivation of in vivo neuroreceptor parameters with PET and SPECT reversible radiotracers. Nucl Med Biol 28: 595–608.

Spinelli S, Chefer S, Suomi SJ et al. (2009). EArly-life stress induces long-term morphologic changes in primate brain. Arch Gen Psychiatry 66: 658–665.

Spinelli S, Chefer S, Carson RE et al. (2010). Effects of early-life stress on serotonin(1A) receptors in juvenile Rhesus monkeys measured by positron emission tomography. Biol Psychiatry 67: 1146–1153.

Spinks R (2000). The prefrontal cortex: Anatomy, physiology, and neuropsychology of the frontal lobe, 3rd edition. Am J Psychiatry 157: 1038.

Staley JK, Gottschalk C, Petrakis IL et al. (2005). Cortical γ-aminobutyric acid type a-benzodiazepine receptors in recovery from alcohol dependence: relationship to features of alcohol dependence and cigarette smoking. Arch Gen Psychiatry 62: 877–888.

Stokes PR, Egerton A, Watson B et al. (2012). History of cannabis use is not associated with alterations in striatal dopamine D_2/D_3 receptor availability. J Psychopharmacol 26: 144–149.

Storvik M, Tiihonen J, Haukijarvi T et al. (2006). Lower serotonin transporter binding in caudate in alcoholics. Synapse 59: 144–151.

Szabo Z, Owonikoko T, Peyrot M et al. (2004). Positron emission tomography imaging of the serotonin transporter in subjects with a history of alcoholism. Biol Psychiatry 55: 766–771.

Takano H, Ito H, Takahashi H et al. (2011). Serotonergic neurotransmission in the living human brain: A positron emission tomography study using [^{11}C]dasb and [^{11}C] WAY100635 in young healthy men. Synapse 65: 624–633.

Teicher MH, Andersen SL, Polcari A et al. (2003). The neurobiological consequences of early stress and childhood maltreatment. Neurosci Biobehav Rev 27: 33–44.

Thanos PK, Volkow ND, Freimuth P et al. (2001). Overexpression of dopamine D_2 receptors reduces alcohol self-administration. J Neurochem 78: 1094–1103.

Thanos PK, Dimitrakakis ES, Rice O et al. (2005). Ethanol self-administration and ethanol conditioned place preference are reduced in mice lacking cannabinoid CB1 receptors. Behav Brain Res 164: 206–213.

Tiihonen J, Kuikka J, Bergstrom K et al. (1995). Altered striatal dopamine re-uptake site densities in habitually violent and non-violent alcoholics. Nat Med 1: 654–657.

Tiihonen J, Vilkman H, Rasanen P et al. (1998). Striatal presynaptic dopamine function in type 1 alcoholics measured with positron emission tomography. Mol Psychiatry 3: 156–161.

Tomkins DM, Higgins GA, Sellers EM (1994). Low doses of the 5-HT1A agonist 8-hydroxy-2-(di-n-propylamino)-tetralin (8-OH DPAT) increase ethanol intake. Psychopharmacology (Berl) 115: 173–179.

Tretter V, Ehya N, Fuchs K et al. (1997). Stoichiometry and assembly of a recombinant GABAA receptor subtype. J Neurosci 17: 2728–2737.

Tupala E, Hall H, Bergstrom K et al. (2001). Dopamine D(2)/D (3)-receptor and transporter densities in nucleus accumbens and amygdala of type 1 and 2 alcoholics. Mol Psychiatry 6: 261–267.

Tupala E, Hall H, Bergstrom K et al. (2003a). Different effect of age on dopamine transporters in the dorsal and ventral striatum of controls and alcoholics. Synapse 48: 205–211.

Tupala E, Hall H, Bergstrom K et al. (2003b). Dopamine D_2 receptors and transporters in type 1 and 2 alcoholics measured with human whole hemisphere autoradiography. Hum Brain Mapp 20: 91–102.

Urban NB, Martinez D (2012). Neurobiology of addiction: insight from neurochemical imaging. Psychiatr Clin North Am 35: 521–541.

Urban NB, Kegeles LS, Slifstein M et al. (2010). Sex differences in striatal dopamine release in young adults after oral alcohol challenge: a positron emission tomography imaging study with [(1)(1)C]raclopride. Biol Psychiatry 68: 689–696.

Urban NB, Slifstein M, Thompson JL et al. (2012). Dopamine release in chronic cannabis users: a $[^{11}C]$raclopride positron emission tomography study. Biol Psychiatry 71: 677–683.

Vinod KY, Hungund BL (2006). Cannabinoid-1 receptor: a novel target for the treatment of neuropsychiatric disorders. Expert Opin Ther Targets 10: 203–210.

Volkow ND, Fowler JS (2000). Addiction, a disease of compulsion and drive: involvement of the orbitofrontal cortex. Cereb Cortex 10: 318–325.

Volkow ND, Fowler JS, Wolf AP et al. (1990). Effects of chronic cocaine abuse on postsynaptic dopamine receptors. Am J Psychiatry 147: 719–724.

Volkow ND, Fowler JS, Wang GJ et al. (1993). Decreased dopamine D_2 receptor availability is associated with reduced frontal metabolism in cocaine abusers. Synapse 14: 169–177.

Volkow ND, Fowler JS, Gatley SJ et al. (1996a). PET evaluation of the dopamine system of the human brain. J Nucl Med 37: 1242–1256.

Volkow ND, Wang G-J, Fowler JS et al. (1996b). Decreases in dopamine receptors but not in dopamine transporters in alcoholics. Alcohol Clin Exp Res 20: 1594–1598.

Volkow ND, Wang GJ, Fowler JS et al. (1997). Decreased striatal dopaminergic responsiveness in detoxified cocaine-dependent subjects. Nature 386: 830–833.

Volkow ND, Chang L, Wang GJ et al. (2001). Low level of brain dopamine D_2 receptors in methamphetamine abusers: association with metabolism in the orbitofrontal cortex. Am J Psychiatry 158: 2015–2021.

Volkow ND, Wang GJ, Fowler JS et al. (2002a). Brain DA D_2 receptors predict reinforcing effects of stimulants in humans: replication study. Synapse 46: 79–82.

Volkow ND, Wang GJ, Maynard L et al. (2002b). Effects of alcohol detoxification on dopamine D_2 receptors in alcoholics: a preliminary study. Psychiatry Res 116: 163–172.

Volkow ND, Wang GJ, Begleiter H et al. (2006). High levels of dopamine D_2 receptors in unaffected members of alcoholic families: Possible protective factors. Arch Gen Psychiatry 63: 999–1008.

Volkow ND, Wang GJ, Telang F et al. (2007). Profound decreases in dopamine release in striatum in detoxified alcoholics: possible orbitofrontal involvement. J Neurosci 27: 12700–12706.

Wang GJ, Volkow ND, Fowler JS et al. (1997). Dopamine D_2 receptor availability in opiate-dependent subjects before and after naloxone-precipitated withdrawal. Neuropsychopharmacology 16: 174–182.

Wang GJ, Smith L, Volkow ND et al. (2012). Decreased dopamine activity predicts relapse in methamphetamine abusers. Mol Psychiatry 17: 918–925.

Weerts EM, Wand GS, Kuwabara H et al. (2011). Positron emission tomography imaging of mu- and delta-opioid receptor binding in alcohol-dependent and healthy control subjects. Alcohol Clin Exp Res 35: 2162–2173.

Williams TM, Davies SJC, Taylor LG et al. (2009). Brain opioid receptor binding in early abstinence from alcohol dependence and relationship to craving: an $[^{11}C]$diprenorphine PET study. Eur Neuropsychopharmacol 19: 740–748.

Wong DT, Threlkeld PG, Lumeng L et al. (1990). Higher density of serotonin-1A receptors in the hippocampus and cerebral cortex of alcohol-preferring P rats. Life Sci 46: 231–235.

Yoder KK, Kareken DA, Seyoum RA et al. (2005). Dopamine d_2 receptor availability is associated with subjective responses to alcohol. Alcohol Clin Exp Res 29: 965–970.

Zhang H, Kranzler HR, Yang BZ et al. (2008). The OPRD1 and OPRK1 loci in alcohol or drug dependence: OPRD1 variation modulates substance dependence risk. Mol Psychiatry 13: 531–543.

Zhou FC, Pu CF, Murphy J et al. (1994). Serotonergic neurons in the alcohol preferring rats. Alcohol 11: 397–403.

Zijlstra F, Booij J, Van Den Brink W et al. (2008). Striatal dopamine D_2 receptor binding and dopamine release during cue-elicited craving in recently abstinent opiate-dependent males. Eur Neuropsychopharmacol 18: 262–270.

Zubieta JK, Gorelick DA, Stauffer R et al. (1996). Increased mu opioid receptor binding detected by PET in cocaine-dependent men is associated with cocaine craving. Nat Med 2: 1225–1229.

Handbook of Clinical Neurology, Vol. 125 (3rd series)
Alcohol and the Nervous System
E.V. Sullivan and A. Pfefferbaum, Editors

Chapter 19

Brain proton magnetic resonance spectroscopy of alcohol use disorders

DIETER J. MEYERHOFF*
Center for Imaging of Neurodegenerative Diseases, San Francisco Veterans Affairs Medical Center, Department of Radiology and Biomedical Imaging, University of California San Francisco, San Francisco, CA, USA

INTRODUCTION

Magnetic resonance imaging (MRI) is a powerful noninvasive tool in the hands of clinicians to diagnose specific medical conditions affecting the structure of the brain that is usually displayed in an image format. MRI is also a valuable tool in the hands of clinical researchers trying to understand morphologic changes associated with disease. MRI takes advantage of the most abundant chemical in the brain, water, and its two hydrogen nuclei per water molecule. Because of its high concentration in the molar range, modern structural MR images are clear and crisp, showing exquisite structural detail down into the submillimeter range. Manufacturers of MR equipment have focused on optimizing image quality since the first images of a human were obtained (in this initial attempt, a finger) in 1977 by Sir Peter Mansfield and Andrew A. Maudsley (Mansfield, 1977; Mansfield and Maudsley, 1977). Many different MR methodologies were derived from the principles and lessons learned while developing structural MRI into the invaluable diagnostic tool it is today: perfusion imaging allows cerebral blood flow to be measured without the use of an external contrast agent; diffusion imaging allows measurement and visualization of water diffusion along the fibers connecting cortical brain regions; and functional MRI allows mapping cortical brain activity based on blood flow needs at a location of neural activity in response to a cognitive task.

Recent years have shown an explosion of research activity around these MR methodologies that all take advantage of different physical properties of the water molecules in the brain. As powerful as these methods are (and all of them as applied to brain changes associated with alcohol use disorders (AUD) are described in other chapters), they do not tell us about anything else than water in different structural and molecular environments and how these environments are potentially affected by the specific substance use behaviors of an individual with AUD, including, but not limited to, the chronic use of alcohol. What happens in and to the brain associated with these chronic behaviors should be evaluated with methods that go beyond their sequelae or influence on brain water.

Many brain chemicals are of critical importance in alcohol-related neuropathology and in fact are targeted by different measurement methodologies: molecular imaging methods such as positron emission tomography (PET) and single-photon emission computed tomography (SPECT) study various receptors present in the brain at very low concentrations (in the nanomolar range). In order to create a measurable signal, however, it is necessary to inject specifically synthesized short-lived radioactive tracers into the blood stream of the individual undergoing such studies.

Magnetic resonance spectroscopy (MRS) on the other hand does not require the use of an exogenous agent to yield a signal from chemicals in the brain that are present in millimolar concentrations and that have been shown to correlate to brain function and behavior. This methodology can be applied to yield images of metabolite distributions from throughout the brain or to yield quantitative information on different chemicals within a preselected brain region. MRS has the additional advantage that it can be performed on the very same equipment used for routine MRI, even in the same session as the MRI exam. Depending on what specific type of MRS equipment and experiment is used, up to

*Correspondence to: Dieter J. Meyerhoff, VA Medical Center/CIND, 4150 Clement St. #114M, San Francisco, CA 94121, USA. Tel: +1-415-221-4810 x4803, E-mail: dieter.meyerhoff@ucsf.edu

18 different chemicals can be measured from selectable regions of almost the entire brain by analyzing their respective signals within a known frequency range.

In this chapter, the first section describes neurochemicals, that is, hydrogen-bearing metabolites that are most commonly measured by *in vivo* MRS (therefore called proton MRS or ^{1}H MRS), continues to summarize our knowledge on their general functions in the brain, and ends with a brief description of basic human ^{1}H MRS methodology. The second section discusses brain ^{1}H MRS findings, both cross-sectional and longitudinal, published in the biomedical research and clinical literature of individuals in treatment and of treatment-naïve individuals with AUD; the neuropathology of uncomplicated alcohol dependence is briefly summarized as a basis for interpreting the *in vivo* findings. The third section critically evaluates the influence of the most common substance use comorbidities in AUD – chronic cigarette smoking and illicit substance use – on regional brain metabolite levels. The fourth section describes two recent research directions: the search for neurobiologic correlates of relapse, and neuroimaging genetics, which investigates genetic influences on brain metabolite levels; the section finishes with ideas about future directions for ^{1}H MRS in AUD, based on current research, knowledge gaps, and perceived insufficiencies in the reviewed literature.

NEUROCHEMICALS MEASURED BY ^{1}H MRS AND BASIC MRS METHODS

MRS yields spectra with several "peaks" or "resonances" that represent signals from different neurochemicals spread out across a frequency range that neighbors the water resonance imaged in MRI. The frequency range is commonly expressed as parts per million (ppm), a frequency scale that is independent of the magnetic field strength of the MR scanner used for the spectral acquisition. The intensities of these peaks (i.e., their integrals or fitted curve areas) are directly proportional to the concentration of the signal-generating metabolites, and are modulated by some physical parameters of the measured metabolites (relaxation times) and experimental parameters. As such, "curve fitting" and proper calibration of the signal intensities allow quantitation of metabolite levels in absolute concentrations (mmol/kg or mmol/L), in absolute "institutional units," or as levels relative to a reference compound in the same brain region presumed to be stable and unaffected by disease status (metabolite ratio). There are advantages and disadvantages associated with either quantitation approach. Only the first quantitation approach properly accounts for potential differences in relaxation times between individuals or groups of individuals, but these metabolite- and region-specific parameters are very time consuming to measure. Absolute quantitation also requires careful accounting for measured amounts of tissue, cerebrospinal fluid, and water signal in the spectroscopy volumes, as well as for differences of potential receiver and transmitter gains between individual measurements. Quantitation in "institutional units" usually does not correct for potential relaxation time differences, but calculates absolute regional metabolite concentrations relative to the strong water signal obtained from the same region. Calculating metabolite ratios is procedurally and computationally least demanding as it does not require the above-mentioned corrections to individual metabolite signals; however, interpretation of findings can be problematic when both nominator and denominator change as a function of the disease under investigation. (For further technical details, see Jansen et al., 2006.)

In vivo ^{1}H MRS can measure the intracellular levels of as many as 18 metabolites in healthy brain parenchyma at ultra-high magnetic field strengths of 7 or 9.4 T (Tkác et al., 2001; Mekle et al., 2009; Marjanska et al., 2012). However, some of these are not observable at the lower field strengths typically used for clinical research MRS (between 1.5 and 4 T). Thus, ^{1}H MRS studies in the alcohol literature have focused on a subset of metabolites including the neuronal metabolite *N*-acetylaspartate (NAA), the glial metabolite myo-inositol (mIno), choline-containing compounds (Cho) such as glycerophosphorylcholine and phosphorylcholine, creatine metabolites (Cr) (the sum of creatine and phosphocreatine), and more recently, glutamate (Glu) and γ-aminobutyric acid (GABA), which are critical for neurotransmission (Fig. 19.1). Alcohol itself

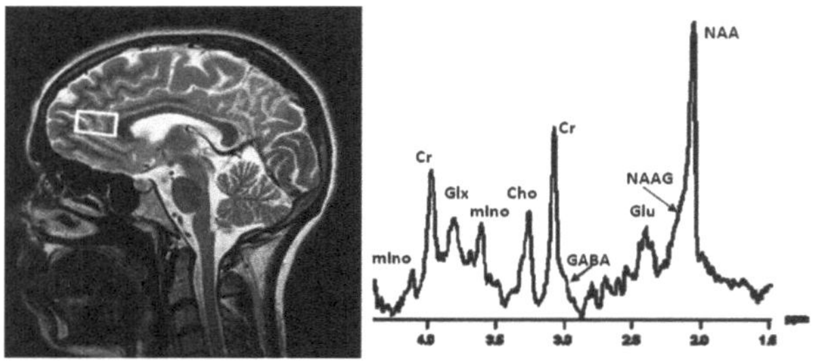

Fig. 19.1. Single-volume ^{1}H magnetic resonance (MR) spectrum (right) obtained from the anterior cingulate cortex (see volume of interest on corresponding MR image, left) at 4 T with a 25-ms echo time. The main spectral resonances are labeled and their peak areas can be quantitated via curve fitting; they are proportional to the absolute concentration of the metabolites from which they originate. NAA, *N*-acetylaspartate; NAAG, *N*-acetylaspartyl glutamate; Glu, glutamate; Glx, glutamate + glutamine; GABA, gamma-aminobutyric acid (which is usually quantitated via J-editing from different spectra); Cr, creatine + phosphocreatine; Cho, choline-containing metabolites; mIno, myo-inositol.

can be detected in the brain after alcohol consumption (Fein and Meyerhoff, 2000) and there is a body of work that describes the "visibility" of the alcohol signal in the brain as a function of membrane rigidity and chronic tolerance to alcohol; however, these studies are not reviewed here and the interested reader is referred to Kroenke et al. (2013). Here we give a brief description of the commonly detected proton metabolites, their role in alcohol dependence, and how they have furthered our understanding of the neurobiology of alcohol dependence and addiction in general. For an extensive description of MRS methodology, the biochemical roles of these metabolites, and a consensus paper on the use of ^{1}H MRS in the clinical setting, see Govindaraju et al. (2000), Vion-Dury et al. (1994), and Öz et al. (2014).

NAA is an amino acid found in high concentrations in all structural elements of neurons and it is virtually absent in mature glial cells (Moffett et al., 1991; Simmons et al., 1991; Benarroch, 2008). The "NAA" peak, the strongest peak in an *in vivo* spectrum of a healthy brain, is composed of overlapping signals from NAA and the much less concentrated *N*-acetylaspartylglutamate (NAAG, which acts to regulate glutamate and dopamine release). NAA is synthesized in neuronal mitochondria and exported to the cytosol; it flows out into the extracellular space and diffuses to oligodendrocytes, where it is rapidly hydrolyzed (Baslow, 2003a). NAA acts as an organic osmolyte controlling cellular water distribution (Baslow, 2003b), provides acetate for myelin lipid synthesis in oligodendrocytes, and facilitates energy metabolism in neuronal mitochondria. Abnormalities in NAA synthesis, transport, or breakdown (i.e., mitochondrial and neuronal dysfunction) as well as overall loss of neuronal soma size may contribute to a compromised steady-state concentration that is measurable by ^{1}H MRS. Because of these functions in the brain, the NAA signal is generally considered a biomarker of neuronal viability or integrity (Miller, 1991; Vion-Dury et al., 1994; De Stefano et al., 1995; Hugg et al., 1996; Sullivan, 2000; Schuff et al., 2001; Baslow and Guilfoyle, 2007; Rigotti et al., 2007). Studies have shown that NAA reduction is reversible in various pathologies, which suggests that NAA does not necessarily reflect the density or concentration of neurons *per se* but more the plasticity of neuronal components and the pathologic processes affecting the metabolic functioning of neurons (Bertolino et al., 2001), particularly neuronal bioenergetics (Pan and Takahashi, 2005; Baslow and Guilfoyle, 2007). NAA is distributed fairly homogeneously throughout the brain, at a concentration of approximately 10 mmol/L, very similar in gray matter and white matter; NAAG (around 2 mmo/L) exhibits higher concentrations in white matter than in gray matter (Pouwels and Frahm, 1998; Haga et al., 2009). NAA tends to decrease throughout the brain with normal aging (Ende et al., 2000; Haga et al., 2009; Maudsley et al., 2012), even after correcting for age-related gray-matter loss (Maudsley et al., 2012).

The "Cho" peak in a ^{1}H MRS spectrum represents the trimethyl ammonium resonance from phosphocholine and glycerophosphocholine, with free choline and the neurotransmitter acetylcholine contributing less than 5% (Khiat et al., 2000). Phosphatidylcholine, the main constituent of myelin and cell membranes, is invisible to *in vivo* MRS due to its restricted molecular mobility. The detectable Cho are intermediates of membrane synthesis and breakdown (Pettegrew et al., 1987, 1990) with a marked regional variability in the brain (highest concentrations in the cerebellum and in rostral cerebrum) (Pouwels and Frahm, 1998). The Cho concentration is between 1.5 and 2.5 mmol/L and increases with age (Haga et al., 2009; Maudsley et al., 2012; Zahr et al., 2013). The Cho signal reflects cell membrane turnover and density (Miller et al., 1996), myelin content (Ross and Bluml, 2001) and/or cell types associated with tissue inflammation (Brenner et al., 1993; Venkatesh et al., 2001).

The "Cr" peak in a spectrum represents creatine and phosphocreatine. The Cr signal is a marker of brain energy metabolism, reflecting overall bioenergetics of neuronal and glial tissues (Balestrino et al., 2002; Ferguson et al., 2002). The metabolite concentration calculated from the Cr MRS signal is about 9 mmol/L, with the highest levels seen in cerebellum, paralleling the distribution of creatine kinase and energy-requiring processes in the brain (Pouwels and Frahm, 1998). As Cr has an important buffer capacity in cellular energy metabolism, its brain concentration is not necessarily stable or invariant under pathologic conditions. Cr levels are dependent on brain activity (Ke et al., 2002) and age (Haga et al., 2009; Maudsley et al., 2012; Zahr et al., 2013) and vary across diseases or pathologic states (Ross and Bluml, 2001; Rosen and Lenkinski, 2007) including alcohol dependence (Durazzo et al., 2010a; Mon et al., 2012; Abé et al., 2013). Therefore, Cr cannot be considered a useful internal concentration reference (Ross and Michaelis, 1994; Sartorius et al., 2008) and the common practice of using the Cr peak signal as an internal concentration reference for other metabolite levels from the same region of interest (e.g., NAA/Cr, Cho/Cr) is problematic (Lee et al., 2007), although experimentally straightforward and methodologically appealing when comparing metabolite measures across individuals. The Cr, NAA, and Cho resonances are single peaks and can therefore be detected rather readily at all field strengths and any practical echo time, an experimental parameter that is critical for optimal detection of other metabolites.

mIno, a carbohydrate that structurally resembles glucose, is a constituent of phosphatidylinositol, an important component of the phospholipid bilayer of cell membranes. mIno is synthesized by and resides predominantly in brain glia, and it is incapable of crossing the blood–brain barrier (Brand et al., 1993). This simple cyclic carbohydrate has been suggested to be a glial marker, to regulate the cellular osmotic environment and maintain cell volume (Brand et al., 1993; Schweinsburg et al., 2000; Ross and Bluml, 2001), and to be involved in intracellular second-messenger system functioning (Fisher et al., 2002). A larger than normal "mIno" peak in an *in vivo* spectrum therefore can reflect glial activation, gliosis, neuroinflammation, an osmotic response to cell shrinkage, or a combination of such processes (Rosen and Lenkinski, 2007). The mIno signal has a complex appearance (i.e., it is not a single resonance, but a coupled multiplet) and its MRS detection is therefore optimal at the shortest echo times practically possible.

Glu, the major excitatory neurotransmitter in the human brain and an important metabolite in all cells, is a mediator of synaptic plasticity and as such implicated in the initiation and maintenance of all addictive disorders (Kalivas and O'Brien, 2008). In a neurotransmitter cycle between neurons and glia, Glu is released from synaptic vesicles of neurons (where it is highly concentrated) into the cytoplasm, actively transported into surrounding glial cells, where it is converted to glutamine (Gln), and then released into the extracellular space to be taken up again by neurons, where Glu is resynthesized. The MR-detectable Glu signal derives mainly from intracellular (including vesicular) and cytosolic pools (Kam and Nicoll, 2007) and is thought to be linked to neurotransmitter activity. Although Glu is more highly concentrated in the brain than NAA (about 11 mmol/L in gray matter, less in white matter) (Pouwels and Frahm, 1998), its MRS detection sensitivity is relatively low: the Glu signal is quite complex, being dispersed over a wide frequency range and overlapping with equally complicated Gln and other resonances. However, with the advent of static magnetic field strengths of 3 T and higher, with optimal echo times (Schubert et al., 2004), specialized acquisition schemes that simplify the Glu signal (Hurd et al., 2004), and with spectral fitting routines that model the complex signal, the measurement of regionally specific levels of Glu (and in some cases also Gln) has become less exceptional.

GABA is the main inhibitory neurotransmitter in the mammalian brain. The entire metabolically active pool of GABA has a concentration of approximately 1.2 mmol/L in human cortex (Rothman et al., 1993; Hetherington et al., 1998), very near the detection limit of *in vivo* MRS. The neurotransmitter pool is a relatively small fraction of the entire pool and is in dynamic equilibrium with it. Abnormalities of GABAergic neurotransmission are implicated in many disorders, including AUD, and in neurologic and psychiatric diseases such as epilepsy, mood and anxiety disorders, and schizophrenia. Due to its relatively low concentration and small MRS signal in close proximity to strong singlet resonances, *in vivo* GABA quantification is technically challenging. Special MRS methods are used that require the subtraction of two spectra from the same brain region acquired with slightly different experimental parameters (Rothman et al., 1993; Kaiser et al., 2008). Although large cerebral molecules (lipids and peptides) may contribute to the commonly detected GABA signal obtained by difference spectroscopy (Choi et al., 2007) and although chronic heavy drinking is known to be associated with brain lipid alterations, changes of the intensity of the detected MRS signal are largely ascribed to changes in GABA (Xin et al., 2010).

The metabolite signal intensities are proportional to metabolite concentrations and their quantification is either in absolute terms or relative to a cerebral concentration standard. Relative or absolute metabolite concentrations can be derived from spectra that are obtained either from one preselected brain region at a time in the case of single-voxel spectroscopy (SVS) or from multiple brain regions simultaneously in the case of MR spectroscopic imaging (MRSI). During a typical MRS examination, SVS spectra are obtained from rectilinear (block-shaped) volumes of interest (VOI) placed in one to four different brain regions (e.g., anterior cingulate cortex (ACC), parietal white matter, cerebellum, thalamus) that each have to be selected before the patient examination begins. The typical VOI size is 4–16 cc and each spectrum requires 2–12 minutes of acquisition time, excluding experimental optimization. The 10 000 times stronger water signal (used to reconstruct high-resolution structural MRIs) is suppressed in MRS by frequency-selective saturation, so that the weak signals from metabolites can be detected from volumes that are much larger than the small voxels that make up an MRI. SVS spectra typically have relatively high signal-to-noise ratios, which facilitates the integration (curve fitting) of metabolite peaks needed for metabolite quantitation. MRSI, conceptually a combination of MRI and MRS, allows imaging of metabolite distributions from either several slices or a three-dimensional volume covering most of the brain. MRSI obtains many hundreds of spectra simultaneously from VOIs that are approximately 1 cc in size within about 15–45 minute. MRSI spectra are typically of lower signal-to-noise ratio than SVS spectra, but they allow evaluation of the distribution of metabolites throughout the imaged regions. In addition, MRSI spectra can be reconstructed from brain regions after the patient has left the MR scanner, allowing greater flexibility for regional data analyses and postacquisition explorations. A major challenge of MRSI is the methodologically and computationally more demanding processing of the MRSI data set compared to that for SVS data. Most scanner manufacturers provide good SVS-processing capabilities, but MRSI-processing methods are still rudimentary and often require offline processing with non-commercial software.

^{1}H MRS OF ALCOHOL USE DISORDERS

Most ^{1}H MRS studies of AUD were performed in alcohol-dependent individuals recruited after acute withdrawal from substance abuse treatment centers (often referred to as "treatment seekers"). The study participants were usually receiving standard care within the rigid structures of inpatient or outpatient treatment that did not include the use of pharmacotherapies (except perhaps disulfiram, which is not centrally active). These studies were therefore purely observational (except for that of Umhau et al., 2010), which describes neuropharmacologic research), providing a snapshot of the brain after withdrawal at various durations of abstinence from alcohol. This research has shown that duration of abstinence has a significant effect on brain metabolite concentrations, as the brain adapts to the sudden absence of alcohol from the system. This realization spurred longitudinal studies aimed at understanding brain plasticity associated with abstinence. Across published studies, the age of the alcohol-dependent study participants varies – though most subjects were in their late 40s – and so does the duration of chronic alcohol consumption over lifetime before study. Duration of abstinence and age, which is usually highly correlated with duration of drinking, are important factors that influence brain metabolite concentrations, and they need to be carefully considered when comparing ^{1}H MRS results across different studies. Table 19.1 lists ^{1}H MRS studies published in peer-reviewed journals on the effects of chronic alcohol consumption in patients with AUD. The study participants are referred to as "alcoholics" in case of treatment seekers and "treatment-naïve heavy drinkers" when referring to individuals with AUD who were recruited from the community. To the best of our knowledge, Table 19.1 contains a complete list of all 34 original ^{1}H MRS studies of AUD published in the English research literature as of chapter completion (February 2013); reviews or symposia proceedings are not included.

Cross-sectional ^{1}H MRS

^{1}H MRS studies have generally described lower NAA and Cho levels in several brain regions of detoxified alcohol-dependent individuals compared to non-alcoholic controls, suggesting relatively widespread neuronal, glial, and membrane abnormalities. The first MRS study to measure NAA levels in abstinent alcohol-dependent individuals employed ^{1}H MRSI (Fein et al., 1994) and suggested neuronal injury in the frontal cortex. Most subsequent studies employed SVS in the frontal lobes and cerebellum of alcoholics after 3–40 days of sobriety. Several groups reported low NAA levels in mostly frontal cortices and white matter (Jagannathan et al., 1996; Seitz et al., 1999; Bendszus et al., 2001; Parks et al., 2002; Durazzo et al., 2004, 2013; Ende et al., 2005; Wang et al., 2009; Mon et al., 2012), suggesting neuronal injury, atrophied dendrites and/or axons, or derangement of neuronal metabolism in brain regions most vulnerable to chronic alcohol abuse as indicated by neuropathology (see below). However, not all studies report lower mesial frontal cortical NAA in recently detoxified alcohol-dependent individuals (Bartsch et al., 2007; Yeo et al., 2013), and this has been related to lower severity of dependence and "uncomplicated, not heavily nicotine abusing alcoholics" (Bartsch et al., 2007), younger age and shorter duration of heavy drinking, and sex (Schweinsburg et al., 2003). In the latter study, alcohol-dependent men and women of similar age, with similar length of "alcoholism" and with comparable alcohol consumption quantities had equally low NAA levels in the frontal white matter. However, only women had significantly lower NAA in an ACC gray-matter region. This careful study suggests a greater vulnerability of frontal gray-matter neurons to chronic alcohol consumption in women than men.

Studies also revealed lower Cho in the cerebellum (Bendszus et al., 2001; Parks et al., 2002; Ende et al., 2005; Bartsch et al., 2007), frontal gray matter (Lee et al., 2007), thalamus, and parietal lobe (Durazzo et al., 2004) relative to light-drinking controls. Lower than normal Cho may reflect less membrane turnover, lower glial cell activity including demyelination, and/or undetected or subclinical pathologies such as thiamine deficiency and liver cirrhosis (Zahr et al., 2010). Only one of several groups who measured mIno in treatment seekers reported elevated mIno in the frontal lobe and in thalamic gray matter of 1-month-abstinent alcohol-dependent individuals using single-volume MRS (Schweinsburg et al., 2000, 2001); all others cited above showed normal or low mIno levels. High mIno levels in early abstinence suggest astrogliosis and/or osmotic changes. Cortical mIno concentrations are correlated positively with measures of learning and memory in treatment seekers (Mon et al., 2012), whereas NAA deficits in frontal white matter and levels of NAA, Cho, and mIno in the cerebellum have been related to lower neurocognitive and motor functioning (Bendszus et al., 2001;

Table 19.1

Original reports of human *in vivo* ^{1}H magnetic resonance spectroscopy in alcohol use disorder (up to February 2013)

Authors, year of publication	Study groups	Duration of abstinence	Age (mean ± SD)	Major findings (all findings statistically significant unless otherwise stated)
Fein et al., 1994	11 alcoholics, 9 controls	3–24 months	65 ± 9	Greater frontal than parietal cortical NAA loss in alcoholics vs controls
Martin et al., 1995	10 alcoholics, 9 controls	0–28 days	44 ± 9	Increase of Cho/NAA in cerebellar vermis with abstinence
Jagannathan et al., 1996	10 alcoholics, 27 controls	1 month	45 ± 4	Low NAA/Cr and NAA/Cho in cerebellum, frontal lobe, and thalamus
Seitz et al., 1999	11 alcoholics, 12 controls	4 days	41 ± 11	Low NAA/Cr at TE = 135 ms and low Cho/Cr at TE = 5 ms
Behar et al., 1999	5 alcoholics, 10 controls	5 weeks	35 ± 7	Low GABA + homocarnosine in occipital lobe
Schweinsburg et al., 2000	9 alcoholics, 5 controls	5 weeks ($n = 4$) and ≥ 1.5 years ($n = 5$)	46 ± 7	High mI in anterior cingulate cortex and thalamus at 5 weeks. NAA, Cho, and Cr not altered. mI normal at 6 years
Schweinsburg et al., 2001	37 alcoholics, 15 controls	1 month	39 ± 9	Low NAA in frontal WM, high mI in WM
Bendszus et al., 2001	17 alcoholics, 12 controls	2 days and 5 weeks	40 ± 8	Low NAA/Cr in cerebellum and frontal lobe, low Cho/Cr in cerebellum (days 3–6) and normalized values after 5 weeks
O'Neill et al., 2001	8 treatment-naïve heavy drinkers, 12 alcoholics	2 years (alcoholics)	43 ± 7	NAA, Cr, and Cho throughout brain similar between groups
Parks et al., 2002	31 alcoholics, 12 controls	4 days and 3 months	41 ± 9	Low cerebellar NAA and Cho. NAA increase over 3 months ($n = 11$)
Schweinsburg et al., 2003	25 alcoholics (11 F, 17 M), 25 controls (12 F, 13 M)	40 days (F), 26 days (M)	43 ± 8 (F) 36 ± 6 (M)	Low frontal white-matter NAA in female and male alcoholics. Low frontal gray-matter NAA only in female alcoholics
Meyerhoff et al., 2004	46 treatment-naïve heavy drinkers, 52 controls	12 hours	41 ± 9	Low NAA in frontal WM of heavy drinkers vs controls (driven by those without a positive family history of alcoholism, associated with lower executive and working memory functions and lower frontal P3b amplitudes). High Cr, ml, and Cho in parietal GM (but not WM) of heavy drinkers vs controls
Durazzo et al., 2004	24 alcoholics, 26 controls	1 week	50 ± 6	Low NAA and Cho in frontal lobe, low Cho in thalamus and parietal lobe. Low NAA throughout brain of smoking vs non-smoking alcoholics. Normal Cr and mI
Ende et al., 2005	33 alcoholics (12 female), 30 controls (10 female)	3 weeks, 3 months and 6 months	45 ± 9	Low Cho in cerebellum and frontal lobe, trend to low NAA in frontal WM; Cho increases over 3 months ($n = 14$); no longitudinal NAA changes over 3 and 6 months ($n = 11$)
Ende et al., 2006	42 social drinkers		~40 ± 10	Greater alcohol consumption correlates with higher frontal Cho
Mason et al., 2006	12 alcoholics, 8 controls	1 week and 1 month	39 ± 8	No GABA group differences in occipital gray matter at 1 week. Higher GABA in non-smoking vs smoking alcoholics. GABA decrease in non-smoking alcoholics over 1 month

Durazzo et al., 2006a	25 alcoholics, 29 controls	1 week and 4 weeks	49 ± 7	NAA and Cho increase in frontal and parietal lobes over 4 weeks. mI and Cr in frontal white matter increase over 4 weeks. Changes more pronounced in non-smoking alcoholics
Lee et al., 2007	13 alcoholics (smokers), 18 controls (non-smokers)	16 days	33 ± 3	Low Cho and Cr, high Glu/Cr, normal NAA in the anterior cingulate cortex (high Glu/Cr due to low Cr?). High Glu measures correlate with better short-term memory and attention, but greater past alcohol consumption. No metabolic differences in insula
Bartsch et al., 2007	15 alcoholics, 10 controls	5 days and 6 weeks	44 ± 8	Low frontomesial NAA and low cerebellar Cho. Both increase over 6–7 weeks
Durazzo et al., 2008	70 alcoholics	1 month	51 ± 9	Low NAA in temporal gray matter and frontal white matter and low Cho in frontal gray matter predict relapse within 6–12 months of treatment
Gazdzinski et al., 2008a	35 alcoholics, 32 heavy drinkers	1 week (alcoholics)	44 ± 8	Low NAA, Cho, and mI throughout the brain of treatment-seeking alcoholics vs non-treated heavy drinkers
Gazdzinski et al., 2008b	24 alcoholics, 14 controls	1 week and 4 weeks	49 ± 9	Low NAA and Cho in medial temporal lobe. NAA and Cho increase in non-smoking alcoholics over 4 weeks
Wang et al., 2009	48 alcoholics (26 smoking, 22 non-smoking), 26 controls	1 month	49 ± 8	Only in smoking alcoholics, NAA lower in frontal WM region of low fractional anisotropy (FA) than in controls. No NAA differences in frontal WM with normal FA
Durazzo et al., 2010b	51 alcoholics 26 controls	1 week	48 ± 9	Low NAA and Cr in brain reward system of those who relapse within 6–12 months of treatment vs abstainers and controls
Gazdzinski et al., 2010	54 alcoholics	Approx. 1 month	50 ± 8	Higher BMI associated with lower NAA, Cho, Cr, and m-Ino in frontal lobe, subcortical nuclei, and vermis. Increased alcohol intake related to elevated m-Ino in multiple brain regions and to smaller volumes of frontal and temporal GM
Umhau et al., 2010	33 alcoholics (15 acamprosate, 18 placebo)	Days 4 and 25 of study medication	33 ± 1	Longitudinal decrease of Glu/Cr in anterior cingulate cortex of those treated with acamprosate. No significant change in any metabolite concentrations in those treated with placebo.
Modi et al., 2011	9 alcoholics, 13 controls	1 week	44 ± 7	High Cho/Cr in occipital lobe (high Cho? low Cr?)
Yeo et al., 2011	146 treatment-naïve heavy drinkers	≤3 weeks	32 ± 9	A specific type of genetic mutation (copy number variation) correlated with NAA, Glx, and Cr in anterior cingulate cortex
Thoma et al., 2011	7 treatment-naïve heavy drinkers, 6 alcoholics, 17 controls	>1 year (alcoholics)	34 ± 7	Significant substance use and psychiatric comorbidity. In anterior cingulate cortex low Glu and high Gln in both drinking groups. Related to drinking severity. No differences between drinking groups
Durazzo et al., 2013	76 alcoholics (43 smoking, 33 non-smoking), 42 controls (non-smoking)	1 week	51 ± 9 (alcoholics), 45 ± 9 (controls)	Lower NAA in smoking vs non-smoking alcoholics and controls in the DLPFC, insula, superior corona radiata, and the total brain reward system. Differences between alcoholic groups not mediated by alcohol consumption, hypertension, psychiatric comorbidities
Abé et al., 2013	40 alcoholics, 28 polysubstance users, 16 controls	1 month	49 ± 9	No significant metabolic differences between abstinent alcoholics and controls. In the DLPFC of alcohol-dependent polysubstance users, significantly lower concentrations of NAA, Cho, and mI than both alcoholics and controls. Cr was significantly lower in polysubstance users compared to alcoholics

Continued

Table 19.1

Continued

Authors, year of publication	Study groups	Duration of abstinence	Age (mean ± SD)	Major findings (all findings statistically significant unless otherwise stated)
Hermann et al., 2012	47 alcoholics, 57 controls	1 day and 14 days	46 ± 1	High Glu at 1 day in prefrontal cortex of alcoholics vs controls. Normal Glu at 14 days. Confirmed in parallel rat studies
Mon et al., 2012	44 alcoholics, 16 controls	9 days and 5 weeks	51 ± 9	Low NAA, Cho, Cr, and Glu in anterior cingulate cortex at 1 week recover to normal levels at 5 weeks. GABA and mI normal at both times. No abnormalities or changes in DLPFC and POC. Higher cortical mI at 1 week correlates with worse cognition and greater alcohol consumption
Xia et al., 2012	49 treatment-naïve heavy drinkers, 45 controls (all non-smoking)	0	46 ± 6	Low NAA/Cr in prefrontal GM and WM and worse executive function. NAA levels and executive skills influenced by different SNPs of the metabotropic glutamate receptor 3 (GMR3) gene. Some of these genotypic differences also found in controls

NAA, *N*-acetylaspartate; Cho, choline-containing metabolites; Cr, creatine; TE, echo time; GABA, γ-aminobutyric acid; mI, WM, white matter; BMI, body mass index; m-Ino, myo-inositol; GM, gray matter; Glx, glutamate; Gln, glutamine; DLPFC, dorsolateral prefrontal cortex; POC, parieto-occipital cortex; SNP, single nucleotide polymorphism.

Parks et al., 2002). One recent study reported higher Cho/Cr in the occipital lobe of alcohol-dependent patients (Modi et al., 2011), but it is unclear if this finding related to high Cho and/or low Cr concentrations. Two groups using absolute quantitation approaches recently reported significantly lower Cr levels in frontal gray matter of 1-week-abstinent treatment seekers (Mon et al., 2012), who relapsed within the year after treatment (Durazzo et al., 2010a), and in smoking alcoholics compared to non-smoking controls (Lee et al., 2007). These studies demonstrate that Cr, an indicator of general cellular high-energy metabolism and buffer of high-energy metabolism, may be decreased in alcohol dependence, a finding not necessarily surprising considering parallel findings of reduced membrane and glial turnover, neuronal injury, and reduced glucose metabolism/consumption in alcohol dependence. The overall pattern of low NAA, low Cho, and low Cr observed in abstinent alcohol-dependent individuals signals a general compromise of bioenergetics and metabolism of brain tissue with potential loss of cells (both neurons and glia), lower neuronal viability, and reduced membrane turnover. This spectral pattern is distinctly different from common neuropathologies studied by ^{1}H MRS (e.g., brain tumors, dementia, epilepsy, human immunodeficiency virus (HIV) infection, multiple sclerosis), which all show low NAA together with high Cho, high mIno, or both. Such patterns are generally interpreted to reflect compromised neuronal viability in the presence of gliosis or pronounced neuroinflammation (Öz et al., 2014). Because the observed brain metabolites have more than one function in the brain (depending on the pathology under investigation, drinking status – active or sober – as well as length of alcohol abstinence), their functional selectivity is not yet fully understood, making data interpretation indefinite.

For the interpretation of MRS findings in AUD, researchers lean heavily on neuropathologic reports from individuals with uncomplicated AUD. Some (Kril et al., 1996; Harding et al., 1997; Korbo, 1999; Kril and Halliday, 1999), but not all (Jensen and Pakkenberg, 1993; Fabricius et al., 2007), postmortem studies reveal neuronal loss mainly in the dorsolateral frontal cortex and the cerebellum as well as glial loss in the hippocampi. Others demonstrate reduced density and size of glial cells in the dorsolateral prefrontal cortex (Miguel-Hidalgo et al., 2002) and lower neuronal and glial cell density in the orbitofrontal cortex (Miguel-Hidalgo et al., 2006). Loss of dendritic arbor and shrinkage of neuronal cell body volume or neuronal death and Wallerian degeneration of myelinated axons (Harper and Kril, 1989; Schwab and Bartholdi, 1996) occur primarily in the frontal lobes, especially the superior frontal cortex (Harper, 2009). Anterior white-matter loss may involve disturbances in both myelin and axonal integrity (Harper, 2009). The mechanisms of chronic alcohol-associated brain injury and neurocognitive dysfunction are hypothesized to involve glutamate and homocysteine-induced excitotoxicity, reduced levels of brain-derived neurotrophic factors, increased oxidative stress and free radical levels, neuroinflammation, thiamine and other nutritional deficiencies, increased aldehyde levels, hepatic dysfunction, and genetic vulnerability (for review, see Durazzo and Meyerhoff, 2007). These mechanisms may work independently or together in AUD to affect cellular oxidative phosphorylation and alter cerebral cellular structures or organelles, membrane phospholipids, myelin, DNA, gene expression, protein synthesis, and cellular metabolism. Associations between MRS-determined metabolite concentrations and neurocognitive deficits are frequently observed (Meyerhoff et al., 2004; Durazzo and Meyerhoff, 2007; Meyerhoff and Durazzo, 2008), demonstrating the functional relevance and clinical significance of their measurements. Thus, *in vivo* ^{1}H MRS, through its unique sensitivity to functionally significant and specific cellular changes and causative as well as mediating agents of brain injury, contributes considerably to our understanding of mechanistic brain alterations in individuals with AUD and of metabolic neuroplasticity during sustained abstinence from alcohol.

Longitudinal ^{1}H MRS during alcohol abstinence

The neuropathologic correlates of extended abstinence in neocortical and subcortical gray matter are increases of dendritic arbor and neuronal cell body volume as well as myelin structure changes and tissue density increases (Dlugos and Pentney, 1997). These changes are associated with partial reversal of brain atrophy on MRI (Pfefferbaum et al., 1995; Durazzo and Meyerhoff, 2007). Underlying these structural changes are basic metabolic and cellular processes during abstinence from alcohol that may be elucidated by longitudinal ^{1}H MRS. Neuroplasticity in the form of metabolic recovery occurs in the frontal lobes and cerebellum within just a few weeks of abstinence. Importantly, these metabolic improvements have been associated with functional improvements. Three early studies (Martin et al., 1995; Bendszus et al., 2001; Parks et al., 2002) reported increases in frontal and cerebellar lobar NAA/Cr, cerebellar lobar Cho/Cr, and Cho/NAA in the cerebellar vermis after varying durations of abstinence (3–9 weeks). After a few weeks, a higher frontal NAA/Cr ratio was related to better auditory-verbal learning and memory and increased cerebellar vermis NAA/Cr correlated positively with attention/concentration. Cho/NAA in the cerebellar vermis did not recover completely after 3 months, suggesting that this might indicate continued compromise of cerebellar vermis tissue (Parks et al., 2002), consistent with neuropathologic findings (Harper, 1998). Several studies using absolute molar quantitation showed significant increases of NAA and Cho (Ende et al., 2005; Durazzo et al., 2006a; Bartsch et al., 2007; Gazdzinski et al., 2008b; Mon et al., 2012) in cortical, subcortical, and white-matter regions over 4–9 weeks of abstinence, which correlated significantly with cognitive improvements (Durazzo et al., 2006a; Bartsch et al., 2007). The longitudinal metabolic improvements and their correlations with cognitive recovery were most prominent in non-smoking alcoholics or in alcoholics consuming fewer than 10 cigarettes per day and suggest increasing neuronal viability (recovery of neuronal size, dendritic sprouting, regeneration, or synaptogenesis) during sustained abstinence from alcohol and membrane changes, including potentially remyelination.

One study reported persistently low NAA in frontal white matter between 3 weeks and 6 months of abstinence, together with significant Cho increases (Ende et al., 2005). Although in this study NAA may have recovered somewhat before the first assessment, this suggests low-level but persistent neuronal injury in abstinent alcohol-dependent individuals. Significant increases over 4 weeks of abstinence were also reported for mIno and Cr in the frontal white matter, primarily in non-smoking alcoholics (Durazzo et al., 2006a), and for Cho and Cr in the ACC (Mon et al., 2012). Finally, and although not a true longitudinal study, the only group reporting high regional mIno concentrations in 1-month-abstinent alcoholics reported normal mIno levels in another group of alcoholics abstinent for an average of 6 years (Schweinsburg et al., 2000, 2001).

Taken together, these longitudinal ^{1}H MRS studies reflect brain plasticity after cessation of chronic alcohol consumption. Converging evidence suggests that the brain of an alcohol-dependent individual in abstinence recovers from neuronal injury, membrane and osmotic abnormalities toward the normal control range over a period of 1–3 months of sustained abstinence, even after 20–30 years of chronic alcohol consumption. These metabolic improvements are paralleled by recovery of cognitive abilities. Correlations between changes in metabolite levels and changes of cognitive performance during abstinence have been observed, but corresponding analyses are uncommon as they require complete data sets from longitudinal assessments. (For a review of research on cognitive recovery and MRS measures during alcohol abstinence, see Durazzo and Meyerhoff, 2007.)

GABA and Glu concentrations in alcohol use disorders

Although quantitative GABA and Glu measures give limited insight into molecular mechanisms, measuring their basal concentrations in psychiatric disorders promises to help with psychiatric diagnosis, may inform pharmacotherapy, and can monitor patient response to drug therapy (Agarwal and Renshaw, 2012; Maddock and Buonocore, 2012). GABA and Glu levels demonstrate functional significance relevant to AUD: extracellular Glu levels measured by microdialysis relate to drug seeking and drug taking in animals, whereas MRS-determined cortical Glu (Zahr et al., 2008) and GABA relate to cognition (Abé et al., 2013), depressive states (Sanacora et al., 2004), and extroversion (Goto et al., 2010) in humans. In non-alcoholic controls, medial prefrontal cortex Glu concentrations correlated negatively with sensation seeking and positively with impulsivity scores (Gallinat et al., 2007a; Hoerst et al., 2010), and ACC Glu/Cr levels related to cognitive control (Falkenberg et al., 2012). In alcohol dependence, a small early study demonstrated low GABA in the occipital lobe at 5 weeks of abstinence (Behar et al., 1999). A larger follow-up study by the same research group (Mason et al., 2006) showed no significant occipital cortex GABA deficits at 1 week of abstinence compared with non-alcoholic controls, but further decreases over 1 month of abstinence in non-smoking alcoholics only. In addition, GABA levels were higher in non-smoking vs smoking alcoholics at 1 week of abstinence. These studies demonstrate changing GABA levels with duration of abstinence and modulation by smoking status. A recent larger high-field MRS study found largely normal GABA levels in frontal and parieto-occipital cortices (POCs) of 50-year-old alcoholics after both 9 days and 5 weeks of abstinence relative to non-smoking healthy controls, with little, if any, modulating effects of cigarette smoking on GABA levels (Mon et al., 2012).

The concentration of Glu in the ACC was significantly higher relative to light or non-drinking controls at 1 day after last alcoholic drink during withdrawal, when alcohol was still detectable in the brain of many study participants (Hermann et al., 2012) and significantly lower than in controls at 9 days of abstinence (Mon et al., 2012). ACC Glu levels on average decreased significantly between day 1 and 14 of abstinence (Hermann et al., 2012) but behaved quite differently across individuals, decreasing in 15/27 cases, increasing in 4/27 and staying unchanged in 8/27 cases. Between 9 and 32 days of abstinence (Mon et al., 2012), initially low Glu levels in ACC and POC increased significantly on average, but also showed variability across individuals in the ACC, increasing in 6/11 cases, decreasing in 1, and staying unchanged in 4/11 cases. The ACC Glu increases in this study were accompanied by significant NAA and Cr recovery, signaling reversal of compromised prefrontal gray-matter bioenergetics or metabolism. Similarly, ACC Glu/Cr tended to increase in 18 alcoholics between 4 and 25 days of abstinence, with no significant NAA ratio changes (Umhau et al., 2010). In another report of Glu levels in 16-day abstinent alcoholics, ACC Glu concentration was normal (Glu/Cr was elevated, but this was due to significantly decreased Cr concentration) (Lee et al., 2007). Taken together, these reports confirm the view that Glu homeostasis is restored during early sobriety, with Glu levels fluctuating dynamically around levels found in controls and normalizing within a few weeks of abstinence. Similar cortical Glu dynamics have been observed in rat alcohol administration protocols (Hermann et al., 2012).

While prefrontal Glu levels are transiently increased during withdrawal, excitotoxic injury may occur (Prendergast et al., 2000; Harris et al., 2003; De Witte, 2004). As studies show that Glu levels decrease rapidly and homeostasis is restored within a few weeks after withdrawal, a single short period of elevated Glu may not promote significant neuronal injury. However, typical alcoholics cycle through relapse/remit phases for much of their drinking lives, so that prefrontal neuronal injury from withdrawal-induced Glu elevations may accumulate over the lifespan. Such an assertion is consistent with the observation that repeated detoxifications are associated with greater loss of control over future drinking (Duka et al., 2011) and with delayed, if not reduced, recovery from executive dysfunction (Durazzo and Meyerhoff, 2007; Loeber et al., 2010).

High Glu levels during withdrawal were hypothesized to be related to future relapse (Mann and Hermann, 2010). If indeed so, acamprosate, one of the three pharmaceuticals approved by the Food and Drug Administration to treat alcoholism and thought to facilitate Glu homeostasis, could be used to reduce relapse most effectively in those with particularly high Glu levels during withdrawal. This hypothesis, however, was not confirmed (Hermann, unpublished observations), in part probably because prefrontal Glu levels early in abstinence are transient and state-dependent and do not reflect the long-term challenges to sobriety faced during recovery from alcoholic drinking. Nevertheless, this idea is an example of how ^{1}H MRS can be used to identify potential subgroups of patients in whom certain medications may be more effective than others or to monitor the efficacy of treatment that mitigates some metabolite imbalance. Case in point is a study that found decreasing Glu/Cr in the ACC of alcohol-dependent individuals treated with acamprosate between 4 and 25 days of abstinence (Umhau et al., 2010). Placebo treatment on the other hand was associated with a trend to a Glu/Cr increase over the same interval. Prefrontal NAA/Cr and NAA/Cho did not change significantly during therapy in either medication group of this study. The MRS measure of Glu was not correlated with cerebrospinal fluid levels of glutamate, suggesting that these glutamate measures reflect different pools. As MRS measures of Glu (and GABA) have been shown to correlate with behavior and cognition, they may in fact better reflect the neurotransmitter pool). Constituting neuropharmacologic research, this report is the only ^{1}H MRS study of AUD to date that was not mainly observational. *In vivo* Glu and GABA measurements will improve our understanding of substance use behavior and relapse, expanding the role of ^{1}H MRS in neuropsychiatric disease (Maddock and Buonocore, 2012).

Treatment-seeking vs treatment-naïve alcohol-dependent individuals

Less than 10% of individuals with AUD seek treatment at some time in their lives (Fein and Landman, 2005; Hasin et al., 2007; Moss et al., 2007). Nevertheless, most of what we know about the effects of AUD on the human brain has been gleaned from studies of these treatment seekers recruited from substance use treatment programs (Fein et al., 2002); moreover, only a diminishingly small fraction of these individuals volunteer to participate in MRS research. Thus, it can be easily understood that the research data obtained from these convenience samples are heavily biased and may not realistically represent the state of the brain in the much larger population of individuals with AUD who never seek treatment in their lives. Studying treatment seekers thus can only generate conclusions with questionable relevance to the majority of individuals with AUD. We therefore review in this section the limited research that has been conducted in treatment-naïve individuals recruited from the community. As opposed to abstinent treatment seekers, they actively consume alcohol at the time of study (but are usually studied with basically 0% blood alcohol concentration several hours after their last alcoholic drink), and have not had a chance to recover from any potential brain abnormalities during the abstinence period before MRS study. Furthermore, treatment-seeking vs treatment-naïve heavy drinkers generally have more severe medical, psychiatric, and substance use comorbidities that may confound brain MRS measures in AUD widely thought to reflect effects from chronic drinking; they also show greater alcohol consumption over lifetime (>50%) and more periods of abstinence than their treatment-naive counterparts (Fein and Landman, 2005). Therefore, it may be reasonable to expect that treatment seekers in early abstinence have generally greater brain abnormalities than active heavy drinkers and that both do not simply represent a continuum of AUD on a common trajectory or drinking scale.

A case in point is an early ^{1}H MRSI study that demonstrated that NAA, Cho, and Cr concentrations throughout the brain of treatment-naïve heavy drinkers were statistically equivalent to those in alcohol-dependent individuals, who had been abstinent – and thus had had time to recover from their (presumed) metabolite deficits – for an average of 2 years (O'Neill et al., 2001). Relative to light-drinking controls, metabolite abnormalities in community-dwelling, treatment-naive heavy drinkers (most of them alcohol-dependent) were less pronounced and demonstrated a pattern different from that reported in recently detoxified treatment seekers: we reported lower NAA in frontal white and parietal gray matter, as well as higher parietal gray matter Cr, Cho, and mIno relative to healthy controls (Meyerhoff et al., 2004; Gazdzinski et al., 2008a), while another recent study found lower-than-normal NAA/Cr in prefrontal gray and white matter (Xia et al., 2012). As weak as the metabolite level abnormalities were, they correlated with poorer performances on executive skills, working memory and processing speed (Fig. 19.2) as well as with lower frontal P300b amplitudes (Meyerhoff et al., 2004; Xia et al., 2012). Furthermore, family history of alcohol problems (FHA) and drinking pattern (binge vs non-binge) as well as age and sex modulated the brain MRS abnormalities in treatment-naïve heavy drinkers (Meyerhoff et al., 2004). Relative to treatment seekers in their mid-40s, NAA, Cho, and mIno concentrations in multiple brain regions of treatment-naïve individuals of similar age were higher and parietal and thalamic atrophy less pronounced after correcting for differences in drinking severity (Gazdzinski et al., 2008a). The only study that assessed Glu levels in treatment-naïve heavy drinkers showed low Glu and high Gln in the ACC compared to light-drinking controls (Thoma et al., 2011). However, the study included only 7 heavy drinkers, their Glu levels were similar to those in 6 1-year-abstinent alcoholics, and significant substance use and psychiatric comorbidity were present in this relatively young cohort (mean age 35 years) that either or both conditions could have affected metabolite measures. In a larger follow-up study of these heavy drinkers (most of them dependent, with a mean age of 31 years, and studied 3 days after their last alcohol intake), metabolite levels in the ACC overall (i.e., a multivariate general linear model including NAA, Cr, Cho, mIno, and Glx) were not different between treatment-naïve and treatment-seeking heavy drinkers after correction for greater and longer alcohol consumption in the treatment seekers (Yeo et al., 2013). More surprisingly though, in the model comparing all metabolite levels between the combined groups of heavy drinkers and age-matched controls, all metabolite levels were weakly elevated in the heavy drinkers and related to more years of heavy drinking.

In a study of social drinkers, greater alcohol consumption, i.e., up to 2 drinks/day, was correlated with

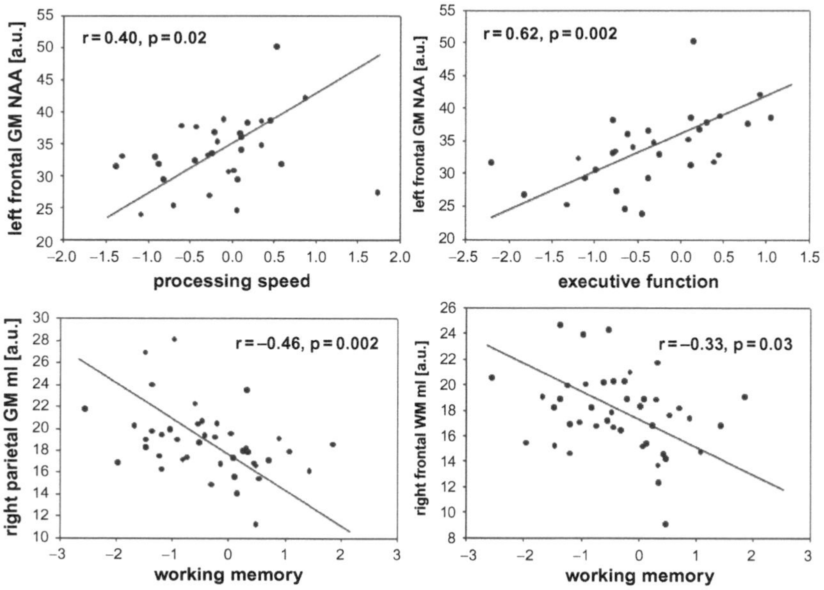

Fig. 19.2. Relationships of regional gray matter (GM) and white matter (WM) metabolite concentrations (arbitrary units) with measures of cognitive functioning in treatment-naïve heavy drinkers recruited from the community (here, mI = myo-inositol). NAA, *N*-acetylaspartate. (Adapted from Meyerhoff et al., 2004.)

higher Cho measures (Ende et al., 2006). Although no interpretation was provided, the finding suggests that Cho levels, at least in the light-to-moderate drinking range typical for control groups in the alcohol MR research literature, are influenced by drinking severity and altered physical properties of the metabolites contributing to the Cho signal (i.e., relaxation times). This observation may explain some of the discrepant Cho findings in the AUD literature.

Altogether, most studies suggest that treatment-naïve (i.e., currently drinking) heavy drinkers have measurable neuronal injury (low NAA) in some brain regions with functional (cognitive) ramifications, and some suggest higher-than-normal energy metabolism (high Cr) and osmotic changes (mIno). Importantly, the regional pattern and magnitude of metabolite levels are different from those of treatment seekers found especially after several decades of drinking, suggesting that we cannot generalize our knowledge of brain abnormalities in abstinent alcoholics to the bulk of individuals with AUD who are not seeking treatment (Fein and Landman, 2005; Durazzo and Meyerhoff, 2007). Alcohol use patterns describing frequency and quantity of use over lifetime are different between these AUD populations and treatment seekers present with greater neuropsychiatric (Di Sclafani et al., 2008) and substance use comorbidities and potentially with greater subclinical liver disease and neuropathology. These factors, individually and in concert, have measurable effects on MRS-observable brain metabolite levels and should influence the interpretation of any MRS data obtained from individuals with AUD.

EFFECTS OF COMMON SUBSTANCE USE COMORBIDITIES ON ^{1}H MRS MEASURES IN ALCOHOL USE DISORDERS

Most ^{1}H MRS studies of AUD described in Table 19.1 ostensibly include participants with dependence on or abuse of alcohol only (unless otherwise specified in the patient characterization section). In most of these studies, screening procedures excluded individuals with other substance dependence and medical and psychiatric comorbidities that can also affect the outcome measures (see Chapter 33 on psychiatric comorbidities in AUD). However, AUD researchers in general have allowed abuse of other substances as defined by standard diagnostic criteria (*Diagnostic and Statistical Manual of Mental Disorders* or *International Classification of Diseases*), pathology that is common in alcoholics such as hypertension and depressive symptomatology, as well as other more subclinical pathology. The reader of these research articles is advised to inquisitively study the participant characterization section to appreciate the comorbid conditions that may have affected the studies' outcome measures. Such examination reveals that some research participants described in the reports listed in Table 19.1 may have been better characterized as polysubstance abusers rather than simply alcohol-dependent individuals.

More recent attempts have been undertaken to more formally understand the brain changes in alcohol-dependent individuals as a function of the most common substance use comorbidities, the chronic misuse of tobacco and illicit substances. This research has been spurred by treatment providers, who have pointed out that the "pure" alcoholic has become less common over the years and that comorbid substance abuse is the clinical reality today. In fact, the US-based National Epidemiologic Survey on Alcohol and Related Conditions (NESARC) finds a number of premorbid and comorbid disorders associated with AUD (Mertens et al., 2003; Stinson et al., 2005; Mansell et al., 2006; Hasin et al., 2007). These contribute considerable variability to the pattern and magnitude of the observed neurobiologic and neurocognitive abnormalities. One of the most common substance use comorbidities in AUD is chronic cigarette smoking (Daeppen et al., 2000; John et al., 2003; Room, 2004). It is recognized that today, more than half of individuals in substance abuse treatment are not only alcohol-dependent, but are also abusing or dependent on illicit drugs, such as cannabis, psychostimulants, heroin, and prescription drugs (Medina et al., 2004; Kedia et al., 2007). For a summary of ^{1}H MRS data on "pure" psychostimulant and polydrug users we refer the reader to an excellent review (Licata and Renshaw, 2010). Below, we review the published literature on the differential effects of AUD with and without tobacco or substance use comorbidities on brain ^{1}H MRS measures.

Smoking comorbidity

An estimated 60–80% of individuals with AUD smoke cigarettes chronically (Romberger and Grant, 2004; Durazzo et al., 2007). Emerging research suggests that chronic smoking, independent of AUD, is associated with abnormalities in brain morphology, cerebral blood flow, neurochemistry, and neurocognition that are similar to those reported in AUD (for review, see Durazzo and Meyerhoff, 2007). We investigated the effects of concurrent chronic cigarette smoking on lobar and subcortical metabolite concentrations in alcohol-dependent individuals (Durazzo et al., 2004) and on longitudinal brain metabolite changes during short-term abstinence from alcohol (Durazzo et al., 2006a). Smoking alcoholics compared with their non-smoking counterparts at 1 week of abstinence had lower NAA in frontal white matter, parietal gray matter, and lenticular nuclei and lower NAA and Cho in the midbrain. Alcohol dependence, independent of smoking, was associated with lower Cho in the thalami and parietal lobes and lower frontal lobe NAA and Cho, consistent with other reports (see above). The regional Cr or mIno levels were not influenced by drinking or smoking status. Greater nicotine dependence and more cigarettes smoked per day were correlated with lower NAA in thalamic and lenticular nuclei. Cerebellar vermis NAA was positively correlated with visuomotor scanning speed (smokers) and visuospatial learning and memory (non-smokers). Largely consistent with metabolic findings in non-alcoholic chronic smokers (Gallinat et al., 2007b), our ^{1}H MRSI findings suggest that chronic smoking compounds alcohol-induced neuronal injury and cell membrane injury in the frontal lobes of persons with AUD and that it has independent adverse effects on neuronal viability and cell membrane turnover/synthesis in the cerebellar vermis and midbrain.

In a related study of a similar patient population (Durazzo et al., 2013), we evaluated if the widespread smoking-related metabolite abnormalities described above (Durazzo et al., 2004) were specific to the cortical and subcortical components of the extended brain reward/executive oversight system (BREOS), a network of overlapping circuitry that is critically involved in the development and maintenance of all forms of addictive disorders. Smoking alcoholics had significantly lower NAA concentrations than either non-smoking alcoholics or controls in the dorsolateral prefrontal cortex (DLPFC), insula, white matter of the superior corona radiata, and the BREOS overall (Fig. 19.3); NAA deficits were non-significant when averaged over spectroscopy volumes from the entire frontal cortex or entire frontal white-matter lobe. These focal metabolic differences, similar to observed cortical thinning (Durazzo et al., 2013), were not mediated by alcohol consumption and common medical or psychiatric comorbidities. Also of note are the normal NAA levels in non-smoking alcoholics that were independent of the lower drinking severity in the non-smokers; unaltered NAA levels were also noted in the ACC of alcoholics who smoked fewer than 10 cigarettes per day (Bartsch et al., 2007).

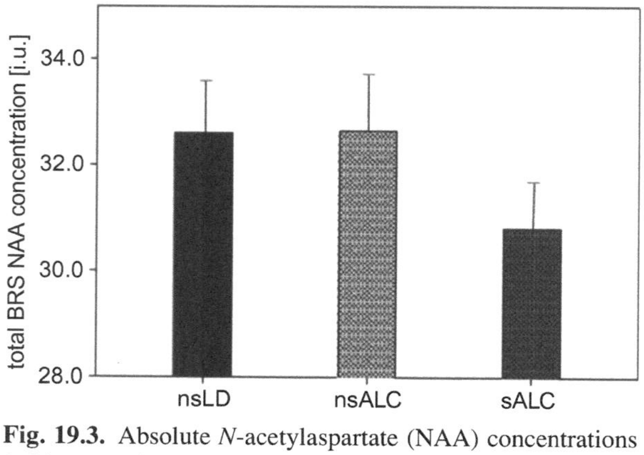

Fig. 19.3. Absolute *N*-acetylaspartate (NAA) concentrations (arbitrary units) averaged over cortical brain regions of the brain reward system (BRS) in non-smoking light/non-drinking controls (nsLD), non-smoking (nsALC) and smoking recovering alcohol-dependent individuals (sALC) at 5 weeks of abstinence from alcohol. (Adapted from Durazzo et al., 2013.)

Thus, chronic smoking in alcohol dependence specifically is associated with significant NAA abnormalities in critical anterior brain regions that are implicated in the development and maintenance of addictive disorders. The smoking-associated metabolic brain abnormalities may be clinically significant as they are accompanied by worse performance on cognitive tests that require fast and flexible processing, including set shifting, processing speed, and cognitive efficiency (see Friend et al., 2005; Durazzo et al., 2006b; Glass et al., 2006 and references cited therein). The metabolite–cognition relationships are often different in pattern or magnitude between smoking and non-smoking alcoholics (Durazzo et al., 2004, 2006a; Gazdzinski et al., 2008b), thereby giving clues to the detrimental impact of chronic smoking on functional neurocircuitry.

Over approximately 1 month of abstinence from alcohol, frontal and parietal NAA and Cho increased significantly (Durazzo et al., 2006a) (see section on longitudinal ^{1}H MRS during alcohol abstinence). Although not strongly decreased at baseline, mIno and Cr levels in frontal white matter also increased during abstinence. Specifically, non-smokers showed widespread increases of all metabolite levels, whereas increases in smokers were much less pronounced and seen in fewer brain regions. At 1 month of abstinence, white-matter NAA in the area of the anterior corona radiata was lower than normal in smoking but not in non-smoking alcoholics (Wang et al., 2009). In non-smokers, metabolite gains were related to improvements in cognition, while smoking alcoholics showed significantly fewer relationships of this kind (Durazzo et al., 2006a). Furthermore, the longer the abstainers had smoked, the smaller the longitudinal increases in regional NAA and Cho concentrations during recovery. Similarly, in a study of medial temporal lobe (including hippocampus), NAA and Cho levels increased and mIno tended to increase over 1 month of abstinence in non-smokers but not in smokers (Gazdzinski et al., 2008b).

^{1}H MRS showed that brain GABA concentrations are also influenced by cigarette smoking in healthy controls (Zhu and Chiappinelli, 1999; Epperson et al., 2005) and during abstinence from alcohol in alcohol-dependent individuals (Mason et al., 2006). At 1 week of abstinence, GABA levels in the occipital cortex were higher in alcohol-dependent non-smokers than smokers. Two weeks later, GABA levels had decreased to similar levels in both groups. Higher GABA in early alcohol abstinence may reflect compensation for reduced cortical benzodiazepine-$GABA_A$ receptor function thought to contribute to alcohol tolerance and withdrawal, while the subsequent GABA decrease may reflect "normalization" of $GABA_A$ receptor function with sobriety (see review in Cosgrove et al., 2011). A recent larger high-field MRS study found normal GABA levels in dorsolateral, medial prefrontal, and POCs at both 9 days and 5 weeks of abstinence, with little, if any, effects of cigarette smoking on regional cortical GABA concentrations (Mon et al., 2012).

Together, the reviewed material suggests that chronic cigarette smoking affects both neurobiology and neurocognition. If confirmed in additional human studies and animal models, we need to consider that smoking and non-smoking individuals with AUD may differ in the nature or extent of their response to pharmacologic and behavioral interventions designed to promote abstinence from alcohol. Thus, effective treatment may have to consider the smoking status of the treatment-seeking individual.

Other substance use comorbidity

In short-term abstinent individuals with comorbid AUD and cocaine dependence, frontal lobe NAA/Cr and Cho/Cr were normal, but GABA/Cr was lower than in controls and similar to levels found in "pure" cocaine dependence (Ke et al., 2004). The study did not include "pure" alcohol-dependent individuals. Using high magnetic field strength (4 T) ^{1}H MRS in three cortical brain regions (ACC, DLPFC, POC), we compared 40 alcohol-dependent individuals and 28 age-matched alcohol-dependent polysubstance users at 1 month of abstinence from alcohol, psychostimulants, and cannabis to 16 drug-free controls. As reported earlier (Mon et al., 2012), after this period of abstinence metabolite levels in "pure" alcoholics and controls were statistically equivalent. However, the polysubstance users at 1 month of abstinence had lower concentrations of NAA, Cho, and mIno in the DLPFC compared to "pure" alcoholics and controls, while Cr concentration was lower compared to alcoholics only; these metabolite levels were normal in the ACC and POC. GABA, largely normal in alcohol dependence throughout abstinence (Mon et al., 2012), tended to be selectively lower in the ACC of polysubstance users compared with both alcoholics and controls (mean difference of 17–20%), and GABA levels in the DLPFC related negatively to greater cocaine consumption. The metabolic group differences in our study were unrelated to age, alcohol consumption, body mass index, smoking status, depression, and anxiety symptomatologies. It appears that, of the cortical regions examined in these polysubstance users, the DLPFC is most strongly compromised. This cortical region is in charge of self-control, -monitoring, and -censoring (i.e., critical for response inhibition, working memory, reasoning, problem solving, planning, organization, and set shifting) and several cortical metabolite concentrations were related to corresponding measures of cognitive performance. By contrast, "pure" alcohol-dependent individuals at 1 week of abstinence show NAA, Cho, and Cr reductions only in the ACC (Mon et al., 2012), a region important for decision making, organizing and integrating information. Together, these findings demonstrate regionally and functionally different as well as longer-lasting metabolic deficits in alcohol-dependent individuals with than without comorbid illicit substance dependence. This metabolic signature may therefore serve as MRS biomarker in substance abusers and suggests targets for pharmacologic and behavioral treatment specific to polysubstance abuse. For example, the GABA data support the view that efforts to increase GABA levels in substance users through topiramate treatment should be a promising treatment possibility (Johnson, 2005).

All but one (Glu) of the measured metabolite levels were lower in the DLPFC of polysubstance users relative to individuals with alcohol dependence only (Abé et al., 2013). Thus, in addition to dose-related cellular injury to neurons, we have to consider alternative interpretations of the reductions, such as presumed associated neuronal or glial edema related to substance use; the associated higher tissue water content might lead to overall lower metabolite concentrations. In addition, osmotic imbalance and/or cell swelling may contribute to the observed MRS differences in the DLPFC of polysubstance users. Finally, DLPFC mIno correlated positively with days of abstinence, ranging from 10 to 45 days. This suggests that at least some brain metabolic abnormalities in alcohol-dependent polysubstance users recover with extended abstinence, even if slower than in "pure" alcohol dependence.

These studies demonstrate that chronic comorbid tobacco and illicit drug use each have significant and unique effects on metabolite concentrations and related cognitive performance in alcohol dependence. While studies in "pure" alcohol-dependent individuals have contributed to our understanding of the detrimental brain changes associated with an alcohol dependence diagnosis, the knowledge gained from these studies cannot necessarily be generalized to brain changes in the majority of alcohol-dependent individuals who have comorbid substance dependence. For MR methods, including ^{1}H MRS, to become generally useful clinical tools, we have to study the clinical reality of polysubstance use and adopt our analytical and statistical tools to understand the complexities of brain changes associated with concurrent use of several licit and illicit substances.

CURRENT AND FUTURE ^{1}H MRS RESEARCH IN ALCOHOL USE DISORDERS

In this section, we touch on a few newly emerging lines of ^{1}H MRS research in AUD. As we have much neuroimaging data from individuals in early treatment and evidence that these relate to cognitive function, researchers have asked to what degree the observed neurobiologic and cognitive brain abnormalities are associated with relapse propensity after treatment or to what degree they may predict relapse. If significant correlates of treatment outcomes were identified, the critical brain phenotypes could become prime candidates for selective therapeutic targeting or cognitive and behavioral manipulations. In addition, as neuroimaging research has related brain structure and task-induced brain activity to specific genotypes, and as those who can maintain long-term abstinence appear to have some resilience to MRS-detectable brain abnormalities not seen in future relapsers (see below), efforts are also underway to evaluate if MRS measures in AUD are influenced by specific genes and common functional polymorphisms. Furthermore, the relative dearth of significant correlations between alcohol consumption and regional metabolite concentrations (assuming the consumption measures are good estimates of the true drinking behavior) is used to argue for the influence of common psychiatric and/or medical comorbidities on brain metabolite levels: these comorbidities are part and parcel of AUD so that excluding for them would eliminate the majority of our target population from study (e.g., depression, anxiety, stress, hypertension), rendering the research clinically irrelevant, or they are subclinical (e.g., liver dysfunction, metabolic syndrome) and therefore not detected at screening. In addition, certain personality traits commonly associated with substance abuse (e.g., impulsivity and anxiety) have been shown to affect brain metabolite levels directly (e.g., Glu in ACC of healthy controls (Hoerst et al., 2010; Grimm et al., 2012) and substance-dependent individuals (Schmaal et al., 2012)) and may also contribute to the large variance of metabolite concentrations in AUD samples. Thus, AUD is not a phenomenon only explained by chronic heavy drinking; therefore, for a better understanding of the metabolic brain abnormalities and associated brain functions in AUD, meaningful future study designs have to take salient confounds of MRS measures into account. The interested reader is referred to Chapter 33 on comorbidities in AUD, which details the literature on influences of psychiatric and medical comorbidities as well as personality traits on neuroimaging measures in AUD.

Here, we briefly summarize the applicability of proton MRS to the prediction of relapse and the relatively new field of neuroimaging genetics in AUD, and we outline future directions in MRS research.

^{1}H MRS correlates of relapse in AUD

Most individuals treated for AUD relapse within 6 months following treatment (e.g., Dawson et al., 2007; reviewed in Durazzo et al., 2010a). While the neuropsychologic, psychiatric, sociodemographic, and behavioral correlates of relapse in AUD have been studied extensively, potential neurobiologic factors as objective predictors of successful abstinence in AUD are not as well understood. Identifying ^{1}H MRS-detectable factors that distinguish abstainers from relapsers will help understand aspects of neurobiology that drive the often lifelong relapse/remit cycle so characteristic of alcohol dependence (Maisto and Connors, 2006; Zywiak et al., 2006; Dawson et al., 2007) and that are critical for maintaining long-term abstinence. Treatment-seeking alcohol-dependent individuals who relapsed within 3 weeks of study demonstrated lower than normal cerebellar NAA and Cho at 3–5 days of abstinence, whereas cerebellar and frontal metabolite levels were not reduced in individuals who relapsed after 3 weeks of abstinence (Parks et al., 2002). While these specific early findings were not confirmed by others, they demonstrate an attempt to link ^{1}H MRS measures to treatment outcome. We recently combined ^{1}H MR spectroscopic imaging with assessment of comorbid psychiatric disorders and comprehensive cognitive testing in individuals at 1 month of abstinence from alcohol (Durazzo et al., 2008). The study participants were classified into abstainers (no alcohol consumption) and relapsers (any alcohol consumption) based on their drinking behavior between 6 and 12 months after treatment and then contrasted on the regional metabolite concentrations and other outcome measures obtained at 1 month of abstinence. Independent classifiers were NAA in frontal white and temporal gray matter, Cho in frontal gray matter, processing speed, and comorbid unipolar mood disorder. Combined, these measures accurately classified 83% of abstainers and 90% of relapsers into their respective groups, accounting for 72% of the variance in drinking status at follow-up. In a largely similar population studied at 1 week of abstinence, ^{1}H MRSI metabolite levels in regions of the extended BREOS were related to relapse (Durazzo et al., 2010a): Future relapsers had lower NAA than future abstainers in DLPFC, ACC, insula, superior corona radiata, and cerebellar vermis and lower Cr in similar BREOS regions. Of note, baseline metabolite concentrations of future abstainers did not differ significantly from those in controls, although these individuals had consumed similar amounts of alcohol over their lifetimes as the metabolically abnormal relapsers. Moreover, regional baseline NAA levels and several measures of posttreatment alcohol consumption were strongly correlated in relapsers. Together these data suggest that metabolic abnormalities in anterior brain regions and specifically those in the BREOS early in treatment – in brain structures that subserve reward-related processes and behaviors – are associated with treatment outcome, even relapse severity. These MRS findings are consistent with other neurobiologic abnormalities in the ACC and DLPFC found to be associated with increased risk of relapse to substance use (Baler and Volkow, 2006; Goldstein et al., 2009; Durazzo et al., 2010b; Heatherton and Wagner, 2011; Volkow et al., 2011). Whether these MRS classifiers are useful in predicting relapse needs to be tested in independent samples. Clearly, more research is needed to understand how these brain abnormalities in treatment seekers contribute to treatment outcomes (Loeber et al., 2010), so that specific treatments can be developed that help mitigate these select abnormalities. Finally, the fact that regional MRS measures in abstainers did not differ significantly from those in non-alcoholic controls, whereas those in relapsers with similar long-term drinking histories did, suggests that other yet unexplored factors drive drinking behavior after treatment. A more comprehensive and integrated assessment is necessary to understand salient mechanisms contributing to relapse in AUD. One of those factors relating to vulnerability and resilience to relapse is neurogenetics.

Neuroimaging genetics

Neuroimaging genetics (or genomics) is an evolving field that aims to explain some of the variance observed in neuroimaging measures through genetic influences. Twin studies have shown that NAA, mIno, Cr, and Cho levels show substantial heritability, as high as 72% for NAA in the posterior cingulate cortex (Batouli et al., 2012), demonstrating the high degree of genetic determination of metabolite levels. Other recent reports have linked specific functional single nucleotide polymorphisms (SNPs) of certain genes to regional metabolite levels in healthy controls and other pathologies (e.g., brain-derived neurotrophic factor (BDNF) and regional NAA concentrations (Stern et al., 2008; Gallinat et al., 2010; Gruber et al., 2012)). Recent observations also support the possibility that genetic background contributes to the variance of ^{1}H MRS measures within alcohol-dependent groups: (1) the common finding that measures of alcohol drinking severity do not correlate well with ^{1}H MRS outcome measures, after correcting for the influences of age and common comorbidities, suggests that other factors must influence these measures; (2) the intriguing finding that brain metabolite levels in individuals in early treatment who can maintain long-term abstinence do not differ significantly from non-alcoholic control populations, while those in future relapsers with similar drinking severity do (Durazzo et al., 2010a) (see above); and (3) results of abnormal brain structure and cognitive processing in alcohol-naïve adolescents with a positive FHA, which are similar to abnormalities detected in adult alcoholics (Hanson et al., 2010; Herting et al., 2010; for review, see Tessner and Hill, 2010).

An early study suggested genetic influences on MRS metabolite measures by demonstrating that the low frontal white-matter NAA levels in treatment-naïve heavy drinkers vs controls were driven by those with a negative FHA (Meyerhoff et al., 2004). This suggested a protective effect of a positive FHA on frontal lobe neuronal integrity, consistent with the interpretation of higher frontal glucose metabolism in alcohol-naïve individuals with a positive vs a negative FHA (Volkow et al., 2006). Although this is in apparent contrast to the protective effect of a positive FHA on brain structure and cognition described above (Hanson et al., 2010; Herting et al., 2010), the studies together demonstrate a complex modulation of different phenotypes by genetic factors. Also among heavy drinkers, rare copy number deletions, a measure of overall genetic mutation load, correlated with NAA, Glx, and Cr levels in ACC, suggesting genetic influences on prefrontal cellular oxidative phosphorylation (Yeo et al., 2011). In another study of treatment-naïve heavy drinkers (Xia et al., 2012), low NAA/Cr in a prefrontal volume including ACC and adjacent white matter was driven by drinkers with the A/A genotype for metabotropic glutamate receptor 3 (GMR3) SNP rs6465084. Similar genotypic differences were observed in controls. Furthermore, alcoholic T-carriers for a different GMR3 SNP had lower NAA/Cr in prefrontal GM and poorer executive function than the A/A genotype. As the prefrontal NAA levels correlated with executive function, a common finding in AUD, cognitive performance in AUD may also be influenced by this genotype. Cognitive function in recently abstinent alcoholics has been shown to be affected by catechol-*O*-methyltransferase genotype (Durazzo et al., 2012). Furthermore, preliminary analyses indicate that the different alleles of the BDNF SNP rs6265 are not only associated with variable tissue volume in the frontal lobe (Mon et al., 2013) and hippocampi during abstinence from alcohol, but also with differential long-term metabolic recovery of frontal white-matter NAA and Cho (Meyerhoff et al., 2011).

While there is strong evidence for genetic factors influencing a person's risk for developing AUD (Bierut et al., 1998; Schuckit et al., 2001; Kendler et al., 2003; Mayfield et al., 2008), few studies have related genetics to relapse vulnerability (Bauer et al., 2007; Pinto et al., 2008; Wojnar et al., 2009; Hashimoto et al., 2011). The genes shown to be involved in risk to relapse also critically influence impulsivity, reward sensitivity, and inhibition (Boettiger et al., 2007; Mier et al., 2010). These complex relationships may be mediated through brain characteristics that can be assessed by MR methods, especially ^{1}H MRS with its ability to measure metabolites reflecting specific cell types and tissue metabolism (Wojnar et al., 2009). Establishing links among functional genetic SNPs, brain neuroimaging phenotypes, and behavior will help build a mechanistic foundation for individual differences in behavior and related psychiatric disease, as has been described (Hariri, 2009). In this context, ^{1}H MRS should be evaluated as a potential intermediate brain phenotype (endophenotype) of AUD, linking genotype to substance use behavior and psychopathology (Gottesman and Gould, 2003; Meyerhoff and Durazzo, 2008; Hutchison, 2010). With critical genetic effects on MRS-detectable metabolite measures forthcoming, research in MRS genetics will offer clues to the functional significance of genetic differences in AUD, and ^{1}H MRS can potentially influence the future of clinical management of AUD via monitoring the efficacy of pharmacologic and behavioral interventions as a function of genotype.

Knowledge gaps and outlook

Brain ^{1}H MRS has contributed significantly to our understanding of the pathophysiology and biologic basis of AUD at the tissue level and it has helped demonstrate and expand our understanding of the neuroplastic processes active during recovery from chronic drinking. ^{1}H MRS has contributed significantly to the realization that brain changes in AUD are not associated solely with chronic heavy drinking, but with numerous conditions that are highly comorbid with chronic alcohol consumption. Possibly because of these complexities, retrospective cross-sectional MRS studies have not been able to clarify drinking patterns that are most damaging to the brain in the long term. We also still do not know how to interpret some of the observed metabolic brain alterations in chronic heavy drinkers – this is an area where carefully designed mechanistic animal studies can help inform human research in a mutually cross-fertilizing way – and to what degree the observed metabolite abnormalities may be consistent with neuroinflammation (Sullivan and Zahr, 2008) or associated with chronic stress (Sinha, 2012).

Longitudinal studies during abstinence from alcohol have demonstrated some metabolic changes consistent with considerable brain plasticity even after several decades of chronic drinking. However, because of the difficulties associated with studying large numbers of abstinent alcohol-dependent individuals over extended periods of documented alcohol abstinence, it is unclear if brain metabolite levels recover completely into the "normal" control range or even if "recovered" alcohol-dependent individuals ought to be expected to display metabolite levels similar to those observed in age-matched non-alcoholic controls. Some of the metabolite abnormalities detected in alcohol dependence may be present before the development of AUD, may have little to do with heavy drinking *per se*, and should therefore not be expected to recover into the "normal" range anyway. Supporting this assertion are recent studies in alcohol-naïve adolescents with a positive FHA that show structure and function different from those of their FHA-negative counterparts, suggesting the possibility also of FHA-associated MRS abnormalities before significant alcohol exposure.

In addition, more longitudinal studies are needed to further understanding of risk factors versus consequences of excessive alcohol use and abuse. Such research may include studies during recovery from heavy drinking and during development of alcohol problems in adolescence and early adulthood as well as prospective studies of disease progression in non-treatment seekers. Any such studies need to be carefully controlled for comorbidities and personality traits that can affect common ^{1}H MRS outcome measures in a way both similar to that of chronic heavy drinking and distinctly different (Abé et al., 2013). Although alcohol dependence is more common in men (5.4%) than women (2.3%) (Grant et al., 2004), women have been found to be specifically sensitive to the neurobiologic and neurocognitive effects of chronic alcohol consumption at equivalent drinking severity levels (Mann et al., 1992; Parsons and Nixon, 1998; Sullivan et al., 2004). More data are needed to understand gender effects on ^{1}H MRS metabolite levels and their changes with heavy drinking and abstinence.

Understanding metabolic recovery processes, linking specific neurobiologic abnormalities and their cognitive correlates as well as their changes during abstinence to successful treatment outcome will help identify novel pharmacologic targets or pinpoint specific networks (e.g., frontal cortex exerting control over automatic behavior) that can be strengthened via cognitive behavioral therapy or cognitive remediation; such approaches are reminiscent of harnessing the brain's intrinsic regenerative powers seen with rehabilitation therapy after a stroke or traumatic brain injury. ^{1}H MRS can further help monitor treatment effects and clarify neuroplastic processes, monitor changes in levels of the neurotransmitters GABA and Glu that are common pharmacologic targets in AUD treatment research, and it can provide biomarkers of neurobiologic improvements as correlates of cognitive recovery. The clinical value of ^{1}H MRS will increase substantially if it can be demonstrated that a few select *in vivo* metabolite measures can increase our abilities to predict successful treatment outcome, as this would be one way of focusing treatment resources on those at greatest risk of relapse and repeated treatments. While this sounds expensive, we should recognize that early detection of Alzheimer's disease these days considers PET imaging a viable tool for screening those at greatest risk for the disease.

Most ^{1}H MRS studies have been unable to demonstrate strong correlations between high quantity/frequency of alcohol consumption and brain metabolic abnormalities. This perceived lack of specificity may have hindered ^{1}H MRS to become a more useful tool in understanding brain effects of AUD. However, the absence of such correlations should not be a reason for lack of concern about chronic heavy drinking nor for lack of a role of ^{1}H MRS in yielding useful biomarkers of AUD to help inform better treatment approaches. Substantial evidence suggests that the brain of an alcohol-dependent individual is metabolically abnormal, and that abstinence from chronic consumption can help reverse some of the brain injury. Nevertheless, stopping alcohol intake is not the panacea for optimal brain health. It is rather outside circumstances that "form" the alcohol-dependent individual, personality characteristics, common comorbidities, genetic background, lifestyle (exercise, diet, and nutrition), and living environment that all add up to an individual's neurometabolic profile. This has been woefully underestimated in the past, when brain alterations in alcohol dependence have often been understood as (or expected to be) straightforward consequences of chronic alcohol consumption. If we are to explicate these complex relationships in individuals struggling with AUD, we as researchers and program officials need to be better prepared to study complex relationships with appropriate scientific rigor that deal with the unique challenges of studying complex populations, including recruitment, diagnostic assessment, reproducibility, generalizability, and design of relevant clinical trials. Isolating chronic drinking experimentally has no doubt taught us a lot about brain abnormalities in alcohol dependence; however, studying our target population out of their social context does little to provide clinically useful biomarkers for informing and designing efficient and safe treatments that succeed in the real world. ^{1}H MRS has a unique role to play here as part of the non-invasive MR armamentarium that can be applied repeatedly without known harm and at comparatively low overall cost (when applied judiciously) to studying salient brain changes associated with injury and recovery from addiction for the purpose of informing better and more efficient treatments.

ACKNOWLEDGMENTS

This review was made possible through support from NIH R01 AA10788 (DJM).

References

Abé C, Mon A, Durazzo TC et al. (2013). Polysubstance and alcohol dependence: unique differences in magnetic resonance-derived brain metabolite concentrations. Drug Alcohol Depend 130: 30–37.

Agarwal N, Renshaw PF (2012). Proton MR spectroscopy-detectable major neurotransmitters of the brain: biology and possible clinical applications. AJNR Am J Neuroradiol 33: 595–602.

Baler RD, Volkow ND (2006). Drug addiction: the neurobiology of disrupted self-control. Trends Mol Med 12: 559–566.

Balestrino M, Lensman M, Parodi M et al. (2002). Role of creatine and phosphocreatine in neuronal protection from anoxic and ischemic damage. Amino Acids 23: 221–229.

Bartsch AJ, Homola G, Biller A et al. (2007). Manifestations of early brain recovery associated with abstinence from alcoholism. Brain 130: 36–47.

Baslow MH (2003a). Brain *N*-acetylaspartate as a molecular water pump and its role in the etiology of Canavan disease: a mechanistic explanation. J Mol Neurosci 21: 185–190.

Baslow MH (2003b). *N*-acetylaspartate in the vertebrate brain: metabolism and function. Neurochem Res 28: 941–953.

Baslow MH, Guilfoyle DN (2007). Using proton magnetic resonance imaging and spectroscopy to understand brain "activation". Brain Lang 102: 153–164.

Batouli SA, Sachdev PS, Wen W et al. (2012). The heritability of brain metabolites on proton magnetic resonance spectroscopy in older individuals. Neuroimage 62: 281–289.

Bauer LO, Covault J, Harel O et al. (2007). Variation in GABRA2 predicts drinking behavior in project MATCH subjects. Alcohol Clin Exp Res 31: 1780–1787.

Behar KL, Rothman DL, Petersen KF et al. (1999). Preliminary evidence of low cortical GABA levels in localized 1H-MR spectra of alcohol-dependent and hepatic encephalopathy patients. Am J Psychiatry 156: 952–954.

Benarroch EE (2008). *N*-acetylaspartate and *N*-acetylaspartylglutamate: neurobiology and clinical significance. Neurology 70: 1353–1357.

Bendszus M, Weijers HG, Wiesbeck G et al. (2001). Sequential MR imaging and proton MR spectroscopy in patients who underwent recent detoxification for chronic alcoholism: correlation with clinical and neuropsychological data. AJNR Am J Neuroradiol 22: 1926–1932.

Bertolino A, Callicott JH, Mattay VS et al. (2001). The effect of treatment with antipsychotic drugs on brain *N*-acetylaspartate measures in patients with schizophrenia. Biol Psychiatry 49: 39–46.

Bierut LJ, Dinwiddie SH, Begleiter H et al. (1998). Familial transmission of substance dependence: alcohol, marijuana, cocaine, and habitual smoking: a report from the Collaborative Study on the Genetics of Alcoholism. Arch Gen Psychiatry 55: 982–988.

Boettiger CA, Mitchell JM, Tavares VC et al. (2007). Immediate reward bias in humans: fronto-parietal networks and a role for the catechol-*O*-methyltransferase 158(Val/Val) genotype. J Neurosci 27: 14383–14391.

Brand A, Richter-Landsberg C, Leibfritz D (1993). Multinuclear NMR studies on the energy metabolism of glial and neuronal cells. Dev Neurosci 15: 289–298.

Brenner RE, Munro PM, Williams SC et al. (1993). The proton NMR spectrum in acute EAE: the significance of the change in the Cho:Cr ratio. Magn Reson Med 29: 737–745.

Choi C, Bhardwaj PP, Kalra S et al. (2007). Measurement of GABA and contaminants in gray and white matter in human brain in vivo. Magn Reson Med 58: 27–33.

Cosgrove KP, Esterlis I, Mason GF et al. (2011). Neuroimaging insights into the role of cortical GABA systems and the influence of nicotine on the recovery from alcohol dependence. Neuropharmacology 60: 1318–1325.

Daeppen JB, Smith TL, Danko GP et al. (2000). Clinical correlates of cigarette smoking and nicotine dependence in alcohol-dependent men and women. The Collaborative Study Group on the Genetics of Alcoholism. Alcohol Alcohol 35: 171–175.

Dawson DA, Goldstein RB, Grant BF (2007). Rates and correlates of relapse among individuals in remission from DSM-IV alcohol dependence: a 3-year follow-up. Alcohol Clin Exp Res 31: 2036–2045.

De Stefano N, Matthews PM, Arnold DL (1995). Reversible decreases in *N*-acetylaspartate after acute brain injury. Magn Reson Med 34: 721–727.

De Witte P (2004). Imbalance between neuroexcitatory and neuroinhibitory amino acids causes craving for ethanol. Addict Behav 29: 1325–1339.

Di Sclafani V, Finn P, Fein G (2008). Treatment-naive active alcoholics have greater psychiatric comorbidity than normal controls but less than treated abstinent alcoholics. Drug Alcohol Depend 98: 115–122.

Dlugos CA, Pentney RJ (1997). Morphometric evidence that the total number of synapses on Purkinje neurons of old F344 rats is reduced after long-term ethanol treatment and restored to control levels after recovery. Alcohol Alcohol 32: 161–172.

Duka T, Trick L, Nikolaou K et al. (2011). Unique brain areas associated with abstinence control are damaged in multiply detoxified alcoholics. Biol Psychiatry 70: 545–552.

Durazzo TC, Meyerhoff DJ (2007). Neurobiological and neurocognitive effects of chronic cigarette smoking and alcoholism. Front Biosci 12: 4079–4100.

Durazzo TC, Gazdzinski S, Banys P et al. (2004). Cigarette smoking exacerbates chronic alcohol-induced brain damage: a preliminary metabolite imaging study. Alcohol Clin Exp Res 28: 1849–1860.

Durazzo TC, Gazdzinski S, Banys P et al. (2006a). Brain metabolite concentrations and neurocognition during short-term recovery from alcohol dependence: preliminary evidence of the effects of concurrent chronic cigarette smoking. Alcohol Clin Exp Res 30: 539–551.

Durazzo TC, Rothlind JC, Gazdzinski S et al. (2006b). A comparison of neurocognitive function in nonsmoking and chronically smoking short-term abstinent alcoholics. Alcohol 39: 1–11.

Durazzo TC, Gazdzinski S, Meyerhoff DJ (2007). The neurobiological and neurocognitive consequences of chronic cigarette smoking in alcohol use disorders. Alcohol Alcohol 42: 174–185.

Durazzo TC, Gazdzinski S, Yeh PH et al. (2008). Combined neuroimaging, neurocognitive and psychiatric factors to predict alcohol consumption following treatment for alcohol dependence. Alcohol Alcohol 43: 683–691.

Durazzo TC, Pathak V, Gazdzinski S et al. (2010a). Metabolite levels in the brain reward pathway discriminate those who remain abstinent from those who resume hazardous alcohol consumption after treatment for alcohol dependence. J Stud Alcohol Drugs 71: 278–289.

Durazzo T, Gazdzinski S, Mon A et al. (2010b). Cortical perfusion in alcohol-dependent individuals during short-term abstinence: relationships to resumption of hazardous drinking after treatment. Alcohol 44: 201–210.

Durazzo T, Hutchison K, Fryer S et al. (2012). Associations of cigarette smoking and polymorphisms in brain-derived neurotrophic factor and catechol-O-methyltransferase with neurocognition in alcohol dependent individuals during early abstinence. Front Pharmacol 3 (178).

Durazzo TC, Mon A, Gazdzinski S et al. (2013). Chronic cigarette smoking in alcohol dependence: associations with cortical thickness and *N*-acetylaspartate levels in the extended brain reward system. Addict Biol 18: 379–391.

Ende G, Braus DF, Walter S et al. (2000). Effects of age, medication, and illness duration on the *N*-acetyl aspartate signal of the anterior cingulate region in schizophrenia. Schizophr Res 41: 389–395.

Ende G, Welzel H, Walter S et al. (2005). Monitoring the effects of chronic alcohol consumption and abstinence on brain metabolism: a longitudinal proton magnetic resonance spectroscopy study. Biol Psychiatry 58: 974–980.

Ende G, Walter S, Welzel H et al. (2006). Alcohol consumption significantly influences the MR signal of frontal choline-containing compounds. Neuroimage 32: 740–746.

Epperson CN, O'Malley S, Czarkowski KA et al. (2005). Sex, GABA, and nicotine: the impact of smoking on cortical GABA levels across the menstrual cycle as measured with proton magnetic resonance spectroscopy. Biol Psychiatry 57: 44–48.

Fabricius K, Pakkenberg H, Pakkenberg B (2007). No changes in neocortical cell volumes or glial cell numbers in chronic alcoholic subjects compared to control subjects. Alcohol Alcohol 42: 400–406.

Falkenberg LE, Westerhausen R, Specht K et al. (2012). Resting-state glutamate level in the anterior cingulate predicts blood-oxygen level-dependent response to cognitive control. Proc Natl Acad Sci U S A 109: 5069–5073.

Fein G, Landman B (2005). Treated and treatment-naive alcoholics come from different populations. Alcohol 35: 19–26.

Fein G, Meyerhoff DJ (2000). Ethanol in human brain by magnetic resonance spectroscopy: correlation with blood and breath levels, relaxation, and magnetization transfer. Alcohol Clin Exp Res 24: 1227–1235.

Fein G, Meyerhoff DJ, Di Sclafani V et al. (1994). 1H magnetic resonance spectroscopic imaging separates neuronal from glial changes in alcohol-related brain atrophy. Chapter in NIAAA Research Monograph No. 27/Alcohol and Glial Cells, pp. 227–241.

Fein G, Di Sclafani V, Cardenas VA et al. (2002). Cortical gray matter loss in treatment-naive alcohol dependent individuals. Alcohol Clin Exp Res 26: 558–564.

Ferguson KJ, MacLullich AM, Marshall I et al. (2002). Magnetic resonance spectroscopy and cognitive function in healthy elderly men. Brain 125: 2743–2749.

Fisher SK, Novak JE, Agranoff BW (2002). Inositol and higher inositol phosphates in neural tissues: homeostasis, metabolism and functional significance. J Neurochem 82: 736–754.

Friend KB, Malloy PF, Sindelar HA (2005). The effects of chronic nicotine and alcohol use on neurocognitive function. Addict Behav 30: 193–202.

Gallinat J, Kunz D, Lang UE et al. (2007a). Association between cerebral glutamate and human behaviour: the sensation seeking personality trait. Neuroimage 34: 671–678.

Gallinat J, Lang UE, Jacobsen LK et al. (2007b). Abnormal hippocampal neurochemistry in smokers: evidence from proton magnetic resonance spectroscopy at 3 T. J Clin Psychopharmacol 27: 80–84.

Gallinat J, Schubert F, Bruhl R et al. (2010). Met carriers of BDNF Val66Met genotype show increased *N*-acetylaspartate concentration in the anterior cingulate cortex. Neuroimage 49: 767–771.

Gazdzinski S, Durazzo TC, Weiner MW et al. (2008a). Are treated alcoholics representative of the entire population with alcohol use disorders? A magnetic resonance study of brain injury. Alcohol 42: 67–76.

Gazdzinski S, Durazzo TC, Yeh PH et al. (2008b). Chronic cigarette smoking modulates injury and short-term recovery of the medial temporal lobe in alcoholics. Psychiatry Res 162: 133–145.

Gazdzinski S, Durazzo TC, Mon A et al. (2010). Body mass index is associated with brain injury in alcohol dependence – a multimodal magnetic resonance study. Alcohol Clin Exp Res 34: 2089–2096.

Glass JM, Adams KM, Nigg JT et al. (2006). Smoking is associated with neurocognitive deficits in alcoholism. Drug Alcohol Depend 82: 119–126.

Goldstein RZ, Craig AD, Bechara A et al. (2009). The neurocircuitry of impaired insight in drug addiction. Trends Cogn Sci 13: 372–380.

Goto N, Yoshimura R, Moriya J et al. (2010). Critical examination of a correlation between brain gamma-aminobutyric acid (GABA) concentrations and a personality trait of extroversion in healthy volunteers as measured by a 3 Tesla proton magnetic resonance spectroscopy study. Psychiatry Res 182: 53–57.

Gottesman II, Gould TD (2003). The endophenotype concept in psychiatry: etymology and strategic intentions. Am J Psychiatry 160: 636–645.

Govindaraju V, Young K, Maudsley AA (2000). Proton NMR chemical shifts and coupling constants for brain metabolites. NMR Biomed 13: 129–153.

Grant BF, Dawson DA, Stinson FS et al. (2004). The 12-month prevalence and trends in DSM-IV alcohol abuse and dependence: United States, 1991–1992 and 2001–2002. Drug Alcohol Depend 74: 223–234.

Grimm S, Schubert F, Jaedke M et al. (2012). Prefrontal cortex glutamate and extraversion. Soc Cogn Affect Neurosci 7: 811–818.

Gruber O, Hasan A, Scherk H et al. (2012). Association of the brain-derived neurotrophic factor val66met polymorphism with magnetic resonance spectroscopic markers in the human hippocampus: in vivo evidence for effects on the glutamate system. Eur Arch Psychiatry Clin Neurosci 262: 23–31.

Haga KK, Khor YP, Farrall A et al. (2009). A systematic review of brain metabolite changes, measured with 1H magnetic resonance spectroscopy, in healthy aging. Neurobiol Aging 30: 353–363.

Hanson KL, Medina KL, Nagel BJ et al. (2010). Hippocampal volumes in adolescents with and without a family history of alcoholism. Am J Drug Alcohol Abuse 36: 161–167.

Harding AJ, Wong A, Svoboda M et al. (1997). Chronic alcohol consumption does not cause hippocampal neuron loss in humans. Hippocampus 7: 78–87.

Hariri AR (2009). The neurobiology of individual differences in complex behavioral traits. Annu Rev Neurosci 32: 225–247.

Harper C (1998). The neuropathology of alcohol-specific brain damage, or does alcohol damage the brain? J Neuropathol Exp Neurol 57: 101–110.

Harper C (2009). The neuropathology of alcohol-related brain damage. Alcohol Alcohol 44: 136–140.

Harper C, Kril J (1989). Patterns of neuronal loss in the cerebral cortex in chronic alcoholic patients. J Neurol Sci 92: 81–89.

Harris BR, Gibson DA, Prendergast MA et al. (2003). The neurotoxicity induced by ethanol withdrawal in mature organotypic hippocampal slices might involve cross-talk between metabotropic glutamate type 5 receptors and *N*-methyl-D-aspartate receptors. Alcohol Clin Exp Res 27: 1724–1735.

Hashimoto JG, Forquer MR, Tanchuck MA et al. (2011). Importance of genetic background for risk of relapse shown in altered prefrontal cortex gene expression during abstinence following chronic alcohol intoxication. Neuroscience 173: 57–75.

Hasin DS, Stinson FS, Ogburn E et al. (2007). Prevalence, correlates, disability, and comorbidity of DSM-IV alcohol abuse and dependence in the United States: results from the National Epidemiologic Survey on Alcohol and Related Conditions. Arch Gen Psychiatry 64: 830–842.

Heatherton TF, Wagner DD (2011). Cognitive neuroscience of self-regulation failure. Trends Cogn Sci 15: 132–139.

Hermann D, Weber-Fahr W, Sartorius A et al. (2012). Translational magnetic resonance spectroscopy reveals excessive central glutamate levels during alcohol withdrawal in humans and rats. Biol Psychiatry 71: 1015–1021.

Herting MM, Schwartz D, Mitchell SH et al. (2010). Delay discounting behavior and white matter microstructure abnormalities in youth with a family history of alcoholism. Alcohol Clin Exp Res 34: 1590–1602.

Hetherington HP, Newcomer BR, Pan JW (1998). Measurements of human cerebral GABA at 4.1 T using numerically optimized editing pulses. Magn Reson Med 39: 6–10.

Hoerst M, Weber-Fahr W, Tunc-Skarka N et al. (2010). Glutamate levels in the anterior cingulate cortex correlate with self-reported impulsivity in patients with borderline personality disorder and healthy controls. Arch Gen Psychiatry 67: 946–954.

Hugg JW, Kuzniecky RI, Gilliam FG et al. (1996). Normalization of contralateral metabolic function following temporal lobectomy demonstrated by h-1 magnetic resonance spectroscopic imaging. Ann Neurol V40: 236–239.

Hurd R, Sailasuta N, Srinivasan R et al. (2004). Measurement of brain glutamate using TE-averaged PRESS at 3T. Magn Reson Med 51: 435–440.

Hutchison KE (2010). Substance use disorders: realizing the promise of pharmacogenomics and personalized medicine. Annu Rev Clin Psychol 6: 577–589.

Jagannathan NR, Desai NG, Raghunathan P (1996). Brain metabolite changes in alcoholism: an in vivo proton magnetic resonance spectroscopy (MRS) study. Magn Reson Imaging 14: 553–557.

Jansen JF, Backes WH, Nicolay K et al. (2006). 1H MR spectroscopy of the brain: absolute quantification of metabolites. Radiology 240: 318–332.

Jensen GB, Pakkenberg B (1993). Do alcoholics drink their neurons away? Lancet 342: 1201–1204.

John U, Meyer C, Rumpf HJ et al. (2003). Strength of the relationship between tobacco smoking, nicotine dependence and the severity of alcohol dependence syndrome criteria in a population-based sample. Alcohol Alcohol 38: 606–612.

Johnson BA (2005). Recent advances in the development of treatments for alcohol and cocaine dependence: focus on topiramate and other modulators of GABA or glutamate function. CNS Drugs 19: 873–896.

Kaiser LG, Young K, Meyerhoff DJ et al. (2008). A detailed analysis of localized J-difference GABA editing: theoretical and experimental study at 4 T. NMR Biomed 21: 22–32.

Kalivas PW, O'Brien C (2008). Drug addiction as a pathology of staged neuroplasticity. Neuropsychopharmacology 33: 166–180.

Kam K, Nicoll R (2007). Excitatory synaptic transmission persists independently of the glutamate-glutamine cycle. J Neurosci 27: 9192–9200.

Ke Y, Cohen BM, Lowen S et al. (2002). Biexponential transverse relaxation (T(2)) of the proton MRS creatine resonance in human brain. Magn Reson Med 47: 232–238.

Ke Y, Streeter CC, Nassar LE et al. (2004). Frontal lobe GABA levels in cocaine dependence: a two-dimensional, J-resolved magnetic resonance spectroscopy study. Psychiatry Res 130: 283–293.

Kedia S, Sell MA, Relyea G (2007). Mono- versus polydrug abuse patterns among publicly funded clients. Subst Abuse Treat Prev Policy 2: 33.

Kendler KS, Prescott CA, Myers J et al. (2003). The structure of genetic and environmental risk factors for common psychiatric and substance use disorders in men and women. Arch Gen Psychiatry 60: 929–937.

Khiat A, Bard C, Lacroix A et al. (2000). Recovery of the brain choline level in treated Cushing's patients as monitored by proton magnetic resonance spectroscopy. Brain Res 862: 301–307.

Korbo L (1999). Glial cell loss in the hippocampus of alcoholics. Alcohol Clin Exp Res 23: 164–168.

Kril JJ, Halliday GM (1999). Brain shrinkage in alcoholics: a decade on and what have we learned? Prog Neurobiol 58: 381–387.

Kril JJ, Harding AJ, Wong A et al. (1996). Is the hippocampus damaged in alcoholics with memory loss? Alcohol Clin Exp Res 20: 76A.

Kroenke CD, Flory GS, Park B et al. (2013). Chronic ethanol (EtOH) consumption differentially alters gray and white matter EtOH methyl (1) H magnetic resonance intensity in the primate brain. Alcohol Clin Exp Res 37: 1325–1332.

Lee E, Jang DP, Kim JJ et al. (2007). Alteration of brain metabolites in young alcoholics without structural changes. Neuroreport 18: 1511–1514.

Licata SC, Renshaw PF (2010). Neurochemistry of drug action: insights from proton magnetic resonance spectroscopic imaging and their relevance to addiction. Ann N Y Acad Sci 1187: 148–171.

Loeber S, Duka T, Welzel Marquez H et al. (2010). Effects of repeated withdrawal from alcohol on recovery of cognitive impairment under abstinence and rate of relapse. Alcohol Alcohol 45: 541–547.

Maddock RJ, Buonocore MH (2012). MR spectroscopic studies of the brain in psychiatric disorders. Curr Top Behav Neurosci, (epub ahead of print) PubMed PMID: 22294088.

Maisto SA, Connors GJ (2006). Relapse in the addictive behaviors: integration and future directions. Clin Psychol Rev 26: 229–231.

Mann K, Hermann D (2010). Individualised treatment in alcohol-dependent patients. Eur Arch Psychiatry Clin Neurosci 260 (Suppl 2): S116–S120.

Mann K, Batra A, Gunthner A et al. (1992). Do women develop alcoholic brain damage more readily than men? Alcohol Clin Exp Res: 1052–1056.

Mansell D, Penk W, Hankin CS et al. (2006). The illness burden of alcohol-related disorders among VA patients: the veterans health study. J Ambul Care Manage 29: 61–70.

Mansfield P (1977). Multi-planar image formation using NMR spin echoes. J Phys C: Solid State Phys 10: L55–L58.

Mansfield P, Maudsley AA (1977). Medical imaging by NMR. Br J Radiol 50: 188–194.

Marjanska M, Auerbach EJ, Valabregue R et al. (2012). Localized 1H NMR spectroscopy in different regions of human brain in vivo at 7 T: T2 relaxation times and concentrations of cerebral metabolites. NMR Biomed 25: 332–339.

Martin PR, Gibbs SJ, Nimmerrichter AA et al. (1995). Brain proton magnetic resonance spectroscopy studies in recently abstinent alcoholics. Alcohol Clin Exp Res 19: 1078–1082.

Mason GF, Petrakis IL, de Graaf RA et al. (2006). Cortical gamma-aminobutyric acid levels and the recovery from ethanol dependence: preliminary evidence of modification by cigarette smoking. Biol Psychiatry 59: 85–93.

Maudsley AA, Govind V, Arheart KL (2012). Associations of age, gender and body mass with 1H MR-observed brain metabolites and tissue distributions. NMR Biomed 25: 580–593.

Mayfield RD, Harris RA, Schuckit MA (2008). Genetic factors influencing alcohol dependence. Br J Pharmacol 154: 275–287.

Medina KL, Shear PK, Schafer J et al. (2004). Cognitive functioning and length of abstinence in polysubstance dependent men. Arch Clin Neuropsychol 19: 245–258.

Mekle R, Mlynarik V, Gambarota G et al. (2009). MR spectroscopy of the human brain with enhanced signal intensity at ultrashort echo times on a clinical platform at 3T and 7T. Magn Reson Med 61: 1279–1285.

Mertens JR, Lu YW, Parthasarathy S et al. (2003). Medical and psychiatric conditions of alcohol and drug treatment patients in an HMO: comparison with matched controls. Arch Intern Med 163: 2511–2517.

Meyerhoff DJ, Durazzo TC (2008). Proton magnetic resonance spectroscopy in alcohol use disorders: a potential new endophenotype? Alcohol Clin Exp Res 32: 1146–1158.

Meyerhoff D, Blumenfeld R, Truran D et al. (2004). Effects of heavy drinking, binge drinking, and family history of alcoholism on regional brain metabolites. Alcohol Clin Exp Res 28: 650–661.

Meyerhoff DJ, Durazzo TC, Hutchison K et al. (2011). Associations between BDNF Val66Met genotype and brain metabolite concentrations in alcohol dependence. Alcohol Clin Exp Res 35 (134A).

Mier D, Kirsch P, Meyer-Lindenberg A (2010). Neural substrates of pleiotropic action of genetic variation in COMT: a meta-analysis. Mol Psychiatry 15: 918–927.

Miguel-Hidalgo JJ, Wei J, Andrew M et al. (2002). Glia pathology in the prefrontal cortex in alcohol dependence with and without depressive symptoms. Biol Psychiatry 52: 1121–1133.

Miguel-Hidalgo JJ, Overholser JC, Meltzer HY et al. (2006). Reduced glial and neuronal packing density in the orbitofrontal cortex in alcohol dependence and its relationship with suicide and duration of alcohol dependence. Alcohol Clin Exp Res 30: 1845–1855.

Miller BL (1991). A review of chemical issues in 1H NMR spectroscopy: *N*-Acetyl-L-aspartate, creatine and choline. NMR Biomed 4: 47–52.

Miller BL, Chang L, Booth R et al. (1996). In vivo 1H MRS choline: correlation with in vitro chemistry/histology. Life Sci 58: 1929–1935.

Modi S, Bhattacharya M, Kumar P et al. (2011). Brain metabolite changes in alcoholism: localized proton magnetic resonance spectroscopy study of the occipital lobe. Eur J Radiol 79: 96–100.

Moffett JR, Namboodiri MA, Cangro CB et al. (1991). Immunohistochemical localization of N-acetylaspartate in rat brain. Neuroreport 2: 131–134.

Mon A, Durazzo T, Meyerhoff DJ (2012). Glutamate, GABA, and other cortical metabolite concentrations during early abstinence from alcohol and their associations with neurocognitive changes. Drug Alcohol Depend 125: 27–36.

Mon A, Durazzo TC, Gazdzinski S et al. (2013). Brain-derived neurotrophic factor (BDNF) genotype is associated with lobar gray and white matter volume recovery in abstinent alcohol dependent individuals. Genes Brain Behav 12: 98–107.

Moss HB, Chen CM, Yi HY (2007). Subtypes of alcohol dependence in a nationally representative sample. Drug Alcohol Depend 91: 149–158.

O'Neill J, Cardenas VA, Meyerhoff DJ (2001). Effects of abstinence on the brain: quantitative magnetic resonance imaging and magnetic resonance spectroscopic imaging in chronic alcohol abuse. Alcohol Clin Exp Res 25: 1673–1682.

Öz G, Alger JR, Barker PB et al. (2014). Clinical proton MR spectroscopy in central nervous system disorders: The MRS Consensus Group. Radiology 279: 658–679, 2013.

Pan JW, Takahashi K (2005). Interdependence of *N*-acetyl aspartate and high-energy phosphates in healthy human brain. Ann Neurol 57: 92–97.

Parks MH, Dawant BM, Riddle WR et al. (2002). Longitudinal brain metabolic characterization of chronic alcoholics with proton magnetic resonance spectroscopy. Alcohol Clin Exp Res 26: 1368–1380.

Parsons OA, Nixon SJ (1998). Cognitive functioning in sober social drinkers: a review of the research since 1986. J Stud Alcohol 59: 180–190.

Pettegrew JW, Kopp SJ, Minshew NJ et al. (1987). 31P nuclear magnetic resonance studies of phosphoglyceride metabolism in developing and degenerating brain: preliminary observations. J Neuropathol Exp Neurol 46: 419–430.

Pettegrew JW, Panchalingam K, Withers G et al. (1990). Changes in brain energy and phospholipid metabolism during development and aging in the Fischer 344 rat. J Neuropathol Exp Neurol 49: 237–249.

Pfefferbaum A, Sullivan EV, Mathalon DH et al. (1995). Longitudinal changes in magnetic resonance imaging brain volumes in abstinent and relapsed alcoholics. Alcohol Clin Exp Res 19: 1177–1191.

Pinto E, Reggers J, Gorwood P et al. (2008). The short allele of the serotonin transporter promoter polymorphism influences relapse in alcohol dependence. Alcohol Alcohol 43: 398–400.

Pouwels PJ, Frahm J (1998). Regional metabolite concentrations in human brain as determined by quantitative localized proton MRS. Magn Reson Med 39: 53–60.

Prendergast MA, Harris BR, Mayer S et al. (2000). Chronic, but not acute, nicotine exposure attenuates ethanol withdrawal-induced hippocampal damage in vitro. Alcohol Clin Exp Res 24: 1583–1592.

Rigotti DJ, Inglese M, Gonen O (2007). Whole-brain *N*-acetylaspartate as a surrogate marker of neuronal damage in diffuse neurologic disorders. AJNR Am J Neuroradiol 28: 1843–1849.

Romberger DJ, Grant K (2004). Alcohol consumption and smoking status: the role of smoking cessation. Biomed Pharmacother 58: 77–83.

Room R (2004). Smoking and drinking as complementary behaviours. Biomed Pharmacother 58: 111–115.

Rosen Y, Lenkinski RE (2007). Recent advances in magnetic resonance neurospectroscopy. Neurotherapeutics 4: 330–345.

Ross B, Bluml S (2001). Magnetic resonance spectroscopy of the human brain. Anat Rec 265: 54–84.

Ross B, Michaelis T (1994). Clinical applications of magnetic resonance spectroscopy. Magn Reson Q 10: 191–247.

Rothman DL, Petroff OA, Behar KL et al. (1993). Localized 1H NMR measurements of gamma-aminobutyric acid in human brain in vivo. Proc Natl Acad Sci U S A 90: 5662–5666.

Sanacora G, Gueorguieva R, Epperson CN et al. (2004). Subtype-specific alterations of gamma-aminobutyric acid and glutamate in patients with major depression. Arch Gen Psychiatry 61: 705–713.

Sartorius A, Lugenbiel P, Mahlstedt MM et al. (2008). Proton magnetic resonance spectroscopic creatine correlates with creatine transporter protein density in rat brain. J Neurosci Methods 172: 215–219.

Schmaal L, Veltman DJ, Nederveen A et al. (2012). *N*-acetylcysteine normalizes glutamate levels in cocaine-dependent patients: a randomized crossover magnetic resonance spectroscopy study. Neuropsychopharmacology 37: 2143–2152.

Schubert F, Gallinat J, Seifert F et al. (2004). Glutamate concentrations in human brain using single voxel proton magnetic resonance spectroscopy at 3 Tesla. Neuroimage 21: 1762–1771.

Schuckit MA, Edenberg HJ, Kalmijn J et al. (2001). A genome-wide search for genes that relate to a low level of response to alcohol. Alcohol Clin Exp Res 25: 323–329.

Schuff N, Ezekiel F, Gamst AC et al. (2001). Region and tissue differences of metabolites in normally aged brain using multislice 1H magnetic resonance spectroscopic imaging. Magn Reson Med 45: 899–907.

Schwab ME, Bartholdi D (1996). Degeneration and regeneration of axons in the lesioned spinal cord. Physiol Rev 76: 319–370.

Schweinsburg BC, Taylor MJ, Videen JS et al. (2000). Elevated myo-inositol in gray matter of recently detoxified but not long-term alcoholics: a preliminary MR spectroscopy study. Alcohol Clin Exp Res 24: 699–770.

Schweinsburg BC, Taylor MJ, Alhassoon OM et al. (2001). Chemical pathology in brain white matter of recently detoxified alcoholics: a 1H magnetic resonance spectroscopy investigation of alcohol-associated frontal lobe injury. Alcohol Clin Exp Res 25: 924–934.

Schweinsburg BC, Alhassoon OM, Taylor MJ et al. (2003). Effects of alcoholism and gender on brain metabolism. Am J Psychiatry 160: 1180–1183.

Seitz D, Widmann U, Seeger U et al. (1999). Localized proton magnetic resonance spectroscopy of the cerebellum in detoxifing alcoholics. Alcohol Clin Exp Res 23: 158–163.

Simmons ML, Frondoza CG, Coyle JT (1991). Immunocytochemical localization of *N*-acetyl-aspartate with monoclonal antibodies. Neuroscience 45: 37–45.

Sinha R (2012). How does stress lead to risk of alcohol relapse? Alcohol Res Curr Rev 34: 432–440.

Stern AJ, Savostyanova AA, Goldman A et al. (2008). Impact of the brain-derived neurotrophic factor Val66Met polymorphism on levels of hippocampal *N*-acetyl-aspartate assessed by magnetic resonance spectroscopic imaging at 3 Tesla. Biol Psychiatry 64: 856–862.

Stinson FS, Grant BF, Dawson DA et al. (2005). Comorbidity between DSM-IV alcohol and specific drug use disorders in the United States: results from the National Epidemiologic Survey on Alcohol and Related Conditions. Drug Alcohol Depend 80: 105–116.

Sullivan EV (2000). NIAAA Research Monograph No. 34: Human brain vulnerability to alcoholism: Evidence from neuroimaging studies. In: A Noronha, M Eckardt, K Warren (Eds.), Review of NIAAA's neuroscience and behavioral research portfolio. National Institute on Alcohol Abuse and Alcoholism, Bethesda, MD, pp. 473–508.

Sullivan EV, Zahr NM (2008). Neuroinflammation as a neurotoxic mechanism in alcoholism: commentary on "Increased MCP-1 and microglia in various regions of human alcoholic brain". Exp Neurol 213: 10–17.

Sullivan EV, Rosenbloom M, Serventi KL et al. (2004). Effects of age and sex on volumes of the thalamus, pons, and cortex. Neurobiol Aging 25: 185–192.

Tessner KD, Hill SY (2010). Neural circuitry associated with risk for alcohol use disorders. Neuropsychol Rev 20: 1–20.

Thoma R, Mullins P, Ruhl D et al. (2011). Perturbation of the glutamate-glutamine system in alcohol dependence and remission. Neuropsychopharmacology 36: 1359–1365.

Tkác I, Andersen P, Adriany G et al. (2001). In vivo 1H NMR spectroscopy of the human brain at 7 T. Magn Reson Med 46: 451–456.

Umhau JC, Momenan R, Schwandt ML et al. (2010). Effect of acamprosate on magnetic resonance spectroscopy measures of central glutamate in detoxified alcohol-dependent individuals: a randomized controlled experimental medicine study. Arch Gen Psychiatry 67: 1069–1077.

Venkatesh SK, Gupta RK, Pal L et al. (2001). Spectroscopic increase in choline signal is a nonspecific marker for differentiation of infective/inflammatory from neoplastic lesions of the brain. J Magn Reson Imaging 14: 8–15.

Vion-Dury J, Meyerhoff DJ, Cozzone PJ et al. (1994). What might be the impact on neurology of the analysis of brain metabolism by in vivo magnetic resonance spectroscopy? J Neurol 241: 354–371.

Volkow ND, Wang GJ, Begleiter H et al. (2006). High levels of dopamine D2 receptors in unaffected members of alcoholic families: possible protective factors. Arch Gen Psychiatry 63: 999–1008.

Volkow ND, Wang G-J, Fowler JS et al. (2011). Addiction: Beyond dopamine reward circuitry. Proc Natl Acad Sci 108: 15037–15042.

Wang JJ, Durazzo TC, Gazdzinski S et al. (2009). MRSI and DTI: a multimodal approach for improved detection of white matter abnormalities in alcohol and nicotine dependence. NMR Biomed 22: 516–522.

Wojnar M, Brower KJ, Strobbe S et al. (2009). Association between Val66Met brain-derived neurotrophic factor (BDNF) gene polymorphism and post-treatment relapse in alcohol dependence. Alcohol Clin Exp Res 33: 693–702.

Xia Y, Ma D, Hu J et al. (2012). Effect of metabotropic glutamate receptor 3 genotype on *N*-acetylaspartate levels and neurocognition in non-smoking, active alcoholics. Behav Brain Funct 8: 42.

Xin L, Mlynarik V, Lei H et al. (2010). Influence of regional macromolecule baseline on the quantification of neurochemical profile in rat brain. Paper read at ISMRM, at Stockholm, Sweden.

Yeo RA, Gangestad SW, Gasparovic C et al. (2011). Rare copy number deletions predict individual variation in human brain metabolite concentrations in individuals with alcohol use disorders. Biol Psychiatry 70: 537–544.

Yeo RA, Thoma RJ, Gasparovic C et al. (2013). Neurometabolite concentration and clinical features of chronic alcohol use: a proton magnetic resonance spectroscopy study. Psychiatry Res 211: 141–147.

Zahr NM, Mayer D, Pfefferbaum A et al. (2008). Low striatal glutamate levels underlie cognitive decline in the elderly: evidence from in vivo molecular spectroscopy. Cereb Cortex 18: 2241–2250.

Zahr NM, Mayer D, Rohlfing T et al. (2010). Brain injury and recovery following binge ethanol: evidence from in vivo magnetic resonance spectroscopy. Biol Psychiatry 67: 846–854.

Zahr NM, Mayer D, Rohlfing T et al. (2013). In vivo glutamate measured with magnetic resonance spectroscopy: behavioral correlates in aging. Neurobiol Aging 34: 1265–1276.

Zhu PJ, Chiappinelli VA (1999). Nicotine modulates evoked GABAergic transmission in the brain. J Neurophysiol 82: 3041–3045.

Zywiak WH, Stout RL, Trefry WB et al. (2006). Alcohol relapse repetition, gender, and predictive validity. J Subst Abuse Treat 30: 349–353.

Section 8

Neuroimaging of brain function

Handbook of Clinical Neurology, Vol. 125 (3rd series)
Alcohol and the Nervous System
E.V. Sullivan and A. Pfefferbaum, Editors

Chapter 20

Cognition, emotion, and attention

EVA M. MÜLLER-OEHRING[1,2]* AND TILMAN SCHULTE[2]

[1]*Department of Psychiatry and Behavioral Sciences, Stanford University School of Medicine, Stanford, CA, USA*

[2]*Neuroscience Program, Center for Health Sciences, Biosciences, SRI International, Menlo Park, CA, USA*

COGNITION, ATTENTION, AND EMOTION IN ALCOHOL ABUSE AND DEPENDENCE

Heavy alcohol consumption can adversely affect a variety of cognitive abilities, including selective attention (Oscar-Berman and Bonner, 1985; Baribeau et al., 1987), processing speed (Glenn and Parsons, 1991), visuospatial abilities (Robertson et al., 1985; Stavro et al., 2012; Maurage et al., 2013), memory functions (Ryan, 1980), and decision making (Bechara et al., 2001). Emotional deficits have been also reported in alcoholics (Amenta et al., 2012; Kornreich et al., 2013), including deficits in decoding emotional facial expressions and in controlling impulsivity (Oscar-Berman et al., 1990; Townshend and Duka 2003; Montagne et al., 2006; Clark et al., 2007; Foisy et al., 2007; Salloum et al., 2007). Dysfunctional affective brain processing is considered a key factor of addiction and emotional imbalance (Makris et al., 2008; Gujar et al., 2011).

In recent years, investigators have identified frontal-lobe volume shrinkage (Pfefferbaum et al., 1997; Durazzo et al., 2011) and neuronal loss (Kril et al., 1997; Harper et al., 2003) in alcohol abuse and alcoholism as potential neural underpinnings for functional deficits in cognition, attention, and emotion regulation (Dao-Castellana et al., 1998; Evert and Oscar-Berman, 2001; Polo et al., 2003). Postmortem evidence in individuals with lifetime chronic alcohol abuse identified neuropathologic damage of white- and gray-matter structures, including synapse and axon degeneration (Harper, 2009). New developments in *in vivo* magnetic resonance imaging (MRI) have provided researchers with multiple neuroimaging methods, including conventional MRI, diffusion tensor imaging (DTI), and functional MRI (fMRI), to study the effects of alcohol on brain gray- and white-matter structure in conjunction with brain functions of cognition, attention, and emotion (Schulte et al., 2010, 2012). In addition, functional connectivity MRI (fcMRI) and DTI-based fiber tractography enable neural network analysis to explore which brain regions are functionally and structurally connected. White-matter fiber tractography studies revealed signs of neural disconnection in chronic alcohol abuse, such as corpus callosum fibers connecting the two cerebral hemispheres (Schulte et al., 2005b; Liu et al., 2010; Pfefferbaum et al., 2010) and degradation of other brain circuits (Schulte et al., 2012).

In this chapter, we focus on the effects of chronic alcohol consumption on attention, emotion, and cognition. Cognition describes non-motor, non-limbic aspects of behavior primarily related to planning and voluntary action (Nieoullon, 2002). It requires the integration of current environmental demands, recruitment of knowledge, including past events, the prediction of future events, and adequate timing of action also involving working memory processes. These processes interact with emotion processes for evaluation whether perceived information is harmless or dangerous. Also, the ability to regulate emotions is essential for controlling cognition and action. We will describe the effects of alcohol consumption on structural and functional properties of neural network connections subserving cognition, attention, and emotion, and discuss recovery of brain structure and functional networks with abstinence and how cognitive abilities are necessary for adequate judgment towards alcohol problems relevant for useful behavioral change (Le Berre et al., 2012).

CONCEPTS OF ATTENTION AND THEIR NEURAL CORRELATES

The executive attention system can be depicted as a supervisory attention and executive control mechanism.

*Correspondence to: Eva M. Müller-Oehring, Neuroscience Program, Center for Health Sciences, Biosciences, SRI International, Menlo Park, CA 94025, USA. Tel: +1-650-859-2767, Fax: +1-650-859-2743, E-mail: eva.moe@sri.com

Attentional control refers to the ability to choose voluntarily what to pay attention to and what to ignore and is closely linked to executive functions involving conflict resolution, reasoning, mental flexibility, task switching, working memory, planning, and the initiation and monitoring of actions. The ability to process conflict between stimulus attributes or between targets involves directing attention selectively and inhibiting competing responses. These processes are associated with midline frontal areas (anterior cingulate cortex: ACC) and dorsolateral prefrontal cortices (DLPFC) (Botvinick et al., 2001). In a computational model, Botvinick et al. (2001) proposed that the ACC detects the occurrence of conflict and signals other areas, such as the DLPFC, which has a key role in the dynamic tuning of executive control and, hence, in behavioral adaptation to changing environments. This type of control becomes important when automatic or previously learned behaviors are no longer effective to achieve a goal. Thus, the executive attention system plays an integral part for cognitive flexibility, the ability to switch easily between different mental operations, strategies, cognitive concepts, and to override habitual responses, processes associated with corticostriatal circuits that involve basal ganglia, ACC, lateral ventral and dorsal prefrontal cortices, and premotor areas (Alexander et al., 1986) (Fig. 20.1A).

Chronic alcoholism is associated with loss of control over one's actions, especially when it comes to alcohol, suggesting disturbance of brain systems involved in executive attention (Field et al., 2010) and decision making (Le Berre et al., 2012). Support for this assumption comes from animal and human studies on the effects of alcohol consumption on brain and behavior (Giancola and Moss, 1998; Noël et al., 2001; Pfefferbaum et al., 2001). For example, studying the effects of chronic ethanol exposure on the medial prefrontal cortex (mPFC) in a mouse model, Kroener and colleagues (2012) reported alterations in synaptic plasticity in the mPFC together with deficits in cognitive flexibility in an attentional set-shifting task (Holmes et al., 2012).

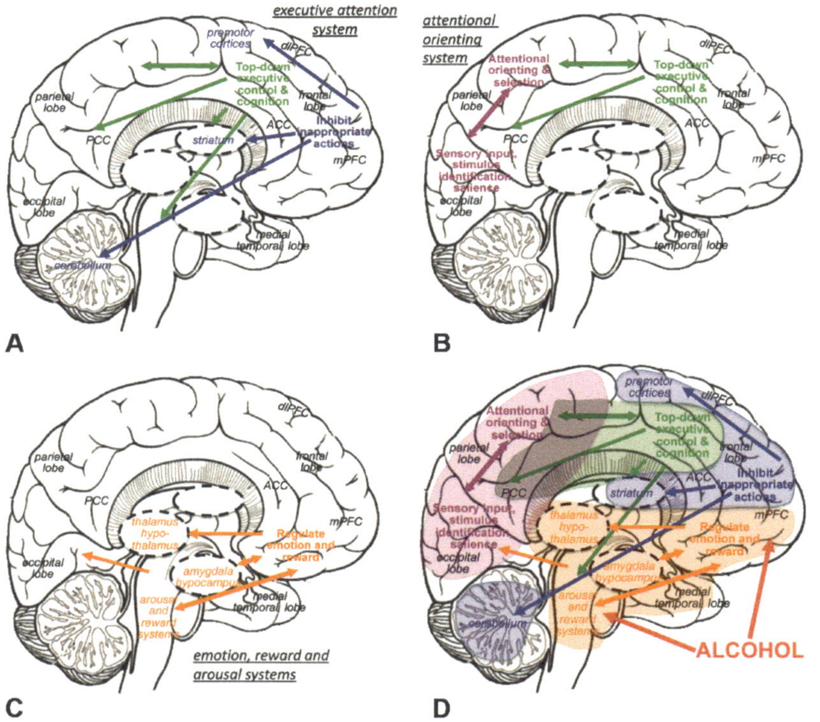

Fig. 20.1. Attention and cognition brain systems. (**A**) Frontally based top-down executive control and cognition systems (green) and behavioral control systems (blue) interact with (**B**) the parietally based attentional orienting and selection systems (pink). (**C**) Medial prefrontal emotion regulation (orange) and behavioral control systems (blue) interact with medial temporal limbic (amygdala, hippocampus), striatal, cerebellar, and midbrain-based reward and arousal systems (orange) to exhibit inhibitory control over both emotion- and reward-processing regions to prevent spontaneous and inappropriate responses. Emotion and saliency are closely linked, and attentional capture by saliency of a stimulus is influenced by the individual's current needs and motivations involving mesocorticolimbic brain regions that process emotion, aversion, and reward (orange). (**D**) Chronic alcohol consumption (red) compromises structural integrity and functional connectivity, and can affect the dynamic tuning between brain systems, e.g., the frontally based attention and executive control systems, the limbic emotion and midbrain-striatal reward systems. mPFC, medial prefrontal cortex; ACC, anterior cingulate cortex; PCC, posterior cingulate cortex; dlPFC, dorsolateral prefrontal cortex.

Dysfunctional impulsivity includes deficits in attention, lack of reflection, and insensitivity to consequences, all of which occur in alcohol addiction (Crews and Boettiger, 2009). In an acute alcohol experiment measuring event-related brain potentials (ERPs) during performance of an emotional Go/No-Go task (Euser and Franken, 2012), alcohol affected emotional modulation in the later stages of cognitive control. Here, larger No-Go N200 and smaller Go and No-Go P300 amplitudes were observed after alcohol consumption (0.65 g/kg alcohol) across all emotion conditions and especially during processing of angry facial expression, accompanied by decreased task performance. Thus, although early mechanism of prepotent response inhibition appeared intact in alcohol-intoxicated individuals, increased effort (Falkenstein et al., 1999) was needed to discriminate between Go and No-Go trials (enhanced N200), which then could have led to resource depletion for later inhibitory and execution processes (reduced P300). Accordingly, acute alcohol seemed to dampen emotional responsiveness and restrict the availability of attentional resources for cognitive control. Such effects would explain the lack of control in alcohol-intoxicated individuals in social-emotional situations. Moreover, chronic alcoholics appear to show emotion-related behavioral disinhibition regardless of being sober or intoxicated (Oscar-Berman et al., 1990; Townshend and Duka, 2003; Clark et al., 2007; Foisy et al., 2007).

In adolescence, heavy alcohol use was associated with poorer attention, memory, and processing speed, whereas deficits specific to visuospatial ability were observed even in non-alcohol-abusing adolescents with a positive familial history of alcoholism (FHP) suggesting, to some extent, premorbid vulnerability of parietal brain functions (Thoma et al., 2011). Poor attention and response inhibition in adolescence (Peeters et al., 2012) may also predict future problem behaviors, including alcohol use. A recent fMRI study on alcohol and other substance use in adolescents revealed significantly less activation in frontoparietal and striatal network nodes of the executive attention system during a Go/No-Go task of response inhibition and response selection in those youth who later transitioned into heavy use of alcohol (Norman et al., 2011). Thus, executive control dysfunction and compromised prefrontal-striatal brain network integrity can be both risk factors and consequences of alcohol abuse.

In chronic alcoholism, and even during abstinence, widespread volume shrinkage was observed in cortical gray matter (Jernigan et al., 1991; Fein et al., 2002) and white matter (Pfefferbaum et al., 1992), which was greatest in the prefrontal cortex, and prefrontal and frontal white matter (Kubota et al., 2001) and related to impairment of executive functions (Chanraud et al., 2007; Konrad et al., 2012). Impairment of cognitive and executive attention functions has been found during early abstinence (Zinn et al., 2004; Goudriaan et al., 2005; Loeber et al., 2009) with cognitive recovery over the first months to 1 year of abstinence (Mann et al., 1999; Munro et al., 2000; Sullivan et al., 2000). Despite recovery of most neurocognitive deficits with long-term abstinence (Fein et al., 2006; Loeber et al., 2010), impaired decision making persisted (mean of 6.6 years abstinence; Fein et al., 2004). What's more, compromise in executive attention functions likely affects the efficacy of alcoholism treatment programs and relapse risk even after years of abstinence (Jin et al., 1998; Noël et al., 2001, 2002; but see Loeber et al., 2010).

The attentional orienting system for visual events is engaged in the selection of information from sensory input and can be experimentally manipulated by cue presentations that direct attention to a location or feature (attentional allocation) (Posner and Petersen, 1990). This system has been associated with posterior brain areas, including the superior parietal lobe and temporal parietal junctions, the superior colliculus, and frontal areas such as frontal eye fields (Kastner and Ungerleider, 2000; Corbetta and Shulman, 2002; Szczepanski et al., 2010; Petersen and Posner, 2012). The ability to select stimuli out of many competing distractors and to suppress irrelevant stimuli is essential for making choices and guiding actions (Broadbent, 1958). Such top-down attention modulation occurs at the level of stimulus-driven processing with the posterior parietal cortex providing a link between automatic associative and controlled processing (King et al., 2012). Because distracting stimuli compete for access to visual short-term memory (Desimone and Duncan, 1995; Kastner and Ungerleider, 2001; Bundesen et al., 2005; Huang and Pashler, 2007; Beck and Kastner, 2009; Vandenberghe and Gillebert, 2009; Gazzaley, 2011; Peters et al., 2012), the capacity to screen them out selectively facilitates retention of information that is crucial for learning, habituation, and social interactions (Campanella and Belin, 2007). Thus, chronic alcohol consumption's effect of limiting the capacity to process goal-relevant information (Parsons, 1998; Yoder et al., 2009; Zoethout et al., 2009; Petit et al., 2012; Schulte et al., 2012) has deleterious effects on the ability to function flexibly and efficiently in everyday life (Fig. 20.1B).

Chronic alcoholism affects not only the frontally based executive attention functions but also the posterior parietal attentional selection and orienting functions (Fein et al., 2009; Müller-Oehring et al., 2009). Interaction of frontoparietal attentional orienting and executive control systems has been evidenced in neuroimaging studies, showing higher DLPFC activity and lower ACC activity when cognitive conflict occurred repeatedly relative to new conflict (Botvinick et al., 1999; Badre and Wagner, 2004; Kerns et al., 2004). Thus, conflict repetition likely enables attentional preparation, reduces the experience of conflict, and facilitates action selection. This is consistent with a study by Luks and colleagues (2007), who reported activation of left DLPFC and left intraparietal sulcus (IPS) regions when cues were informative for processing the upcoming conflict. This and other studies suggest involvement of the DLPFC in maintaining task-relevant information and deploying top-down attention to modulate incoming sensory information accordingly (Banich et al., 2000; Cohen et al., 2000; Milham et al., 2003). On the other hand, the IPS has been associated with deployment and sustaining of visuospatial attention.

A paradigm used widely to study the frontoparietal selective attention and cognitive control system is the Stroop task that describes the prolonged response when subjects are asked to name the ink color of a word printed in a color incongruent with the word's meaning (e.g., the word "blue" written in red ink) relative to when the word's meaning and ink color are the same (e.g., the word "blue" written in blue ink) (Carter et al., 1998; MacDonald et al., 2000; Salo et al., 2001; Bush et al., 2003). Alcoholics have shown enhanced Stroop word–color interference, with longer reaction times compared with control subjects, indicating compromise of executive functions of conflict monitoring and response inhibition between competing stimulus attributes (Duka et al., 2003; Schulte et al., 2005a, 2012; Pitel et al., 2007). Similar findings are seen in adolescents at risk for alcohol abuse (Silveri et al., 2011). In a recent fMRI study, abstinent alcoholics and controls showed similar frontoparietal activations during Stroop processing, but different posterior cingulate cortex (PCC) activations when executive control demands were manipulated. Here, greater deviations from the normal PCC activity in alcoholics correlated with higher amounts of lifetime alcohol consumption (Schulte et al., 2012).

The PCC is considered a hub involving both resting-state and task-related networks that function together to support complex behavior (Leech et al., 2011). The posterior and middle cingulate cortices appear to mediate functional connections between voluntary (top-down) attention and executive control network regions, and to interact with midbrain nodes of the reward network associated with automatic (bottom-up) attention (Schulte et al., 2012). The reward network is relevant for behavioral conditioning to optimize utilization of resources for response selection (Potts et al., 2006; Mohanty et al., 2008). Thus, the PCC may regulate activity and change the functional connectivity between executive control, automatic attention, and resource allocation sites and may mirror compromise in functional networking in alcoholism. Yet, recent fMRI studies in abstinent alcohols indicate that functional recovery can occur despite compromise of specific functional networks through compensatory recruitment of parallel networks enabling network adaptation or reorganization (Chanraud et al., 2013).

ATTENTIONAL CONTROL SYSTEMS INTERACT WITH EMOTION AND REWARD SYSTEMS

Evidence from an fMRI study (Yamasaki et al., 2002) suggests that attention and emotional functions are integrated in the ACC, located in the mPFC. Specifically, a ventral processing stream conveys emotional arousal information from the amygdala to ventral prefrontal cortex, and a dorsal stream relays selective attentional information from posterior parietal cortex to dorsal prefrontal cortex. Regional subdivisions of the ACC are specialized to process cognitive and affective information (Margulies et al., 2007; Etkin et al., 2011), possibly explaining why the structure as a whole was found to be the only brain region with equivalent responses to attentional selection and emotional arousal (Yamasaki et al., 2002). In particular, the cognitive (dorsal) division of the ACC influences processes such as response inhibition (Matthews et al., 2004) and error processing (Walton et al., 2007; Orr and Hester, 2012; Spunt et al., 2012). In comparison, the affective (ventral) subdivision is involved in the processing and integration of emotional information (Simmons et al., 2008; Etkin et al., 2011) and motivational significance of stimuli related to reward (Ortega et al., 2011), fear (Steenland et al., 2012), and social interaction (Mühlberger et al., 2011).

Mesocorticolimbic neurocircuitry underlies processing of emotion (Coccaro et al., 2007; LoPresti et al., 2008) and includes the hippocampus, parahippocampal and subcallosal gyri, and insular, orbitofrontal and cingulate cortices. The subcortical components of this circuit include the olfactory bulb, hypothalamus, amygdala, septal and some thalamic nuclei. Frontolimbic neural circuits involve a reciprocal prefrontal cortex–amygdala relationship that was associated with better regulation of negative emotion (Banks et al., 2007; Lee et al., 2012), a process that seems to be compromised in chronic alcoholics (Marinkovic et al., 2009; O'Daly et al., 2012). Further, deficits in decoding emotional facial expressions have been observed in alcoholics (Townshend and Duka, 2003; Foisy et al., 2007) and associated with blunted responses to emotional faces in rostral ACC (Salloum et al., 2007) and temporal limbic regions, i.e., the amygdala (Marinkovic et al., 2009). On the other hand, alcoholics were found to exhibit enhanced ventral striatal activation to alcohol-related stimuli (Wrase et al., 2007). The ventral striatum (nucleus accumbens, NAcc) is a core region of the frontostriatal reward network and is interconnected to dopaminergic midbrain structures, limbic regions (i.e., hippocampus), and prefrontal cortices. It has been functionally associated with incentive salience (Robinson and Berridge, 2000), associative learning (Young et al., 1998), and reward prediction (Yoder et al., 2009). A deficit in the recruitment of this dopaminergic reward circuitry by non-drug rewards has been assumed to play a role in addiction and to motivate drug intake to temporarily restore the ability to experience pleasure (Blum et al., 2000; Koob et al., 2004). Indeed, fMRI studies in alcoholics evidenced reduced ventral striatal brain activity for non-drug rewards (Wrase et al., 2007; Beck et al., 2009). However, another fMRI study showed similar ventral striatal responses to potential non-drug rewards in alcoholics and controls, but alcoholics showed increased NAcc activation to reward delivery and more severe deactivation of the NAcc when an expected reward was not given (Bjork et al., 2008). Also, poorer frontal cortex functional connectivity with the striatum was observed in alcoholics, which further suggests a weakened frontostriatal reward pathway of response inhibition (Courtney et al., 2013).

Thus, within the frontostriatal reward (Leh et al., 2007; Draganski et al., 2008) and frontolimbic emotion (Mark et al., 1993; Concha et al., 2005) circuits, the mPFC appears to exhibit inhibitory control over both emotion- and reward-processing regions to prevent spontaneous and inappropriate responses (Garcia et al., 1999; Kim et al., 2003; Quirk et al., 2003). Accordingly, not a single brain region but rather the interaction of various interconnected structures enables emotional and reward-related behavioral regulation (Fig. 20.1C).

The alerting system influences our capacity to maintain attention and generally enhances responsivity activation to stimuli; it is further engaged by emotionally or motivationally salient events and participates in the reward mechanisms, including aspects of drug abuse (Heimer et al., 1997). Rapid changes in the environment automatically elicit an orienting response to promptly evaluate the new situation. Alcoholics are impaired in locating visual events and slow on tasks requiring visually guided movement (Wilson et al., 1988), indicative of alcoholism-related attenuation of voluntary attentional control and automatic attentional capture mechanisms. Automatic attentional capture can be evoked by salient visual stimuli and processed without awareness (Kanai et al., 2006). This capacity is considered part of a bottom-up saliency map in the primary visual cortex (Zhaoping, 2008; Zhang et al., 2012) and also occurs with visually non-salient stimuli that convey threatening information (Lin et al., 2009). Accordingly, emotion and saliency are closely linked. Attentional capture by saliency of a stimulus is influenced by the individual's current needs and motivations involving mesocorticolimbic brain regions that process emotion, aversion, and reward (Krebs et al., 2011; Lammel et al., 2011; Pourtois et al., 2012). Specifically, alertness-related subcortical structures such as brainstem and thalamus are structurally linked to limbic areas such as hypothalamus, amygdala, and hippocampus (Nakano, 2000; Castle et al., 2005; Vandewalle et al., 2009). In alcoholism, bottom-up saliency maps (Melloni et al., 2012) may be selectively biased towards processing alcohol stimuli and are influenced by alterations in reward settings that automatically guide attention to meet individual needs and motivations (Vollstädt-Klein et al., 2012) (Fig. 20.1C).

ATTENTION AND MEMORY

Theories of attention propose that working memory representations play a key role in selection of task-relevant stimuli (Bundesen et al., 2005; Reinhart et al., 2012; Soto et al., 2012). For effective behavioral performance visual information needs to be stored in a short-term memory template to maintain (Xu and Chun, 2006) and to manipulate information (Joyce and Robbins, 1991; Ambrose et al., 2001) involved in functions of spatial processing (Cowan, 1999), phonologic store and rehearsal mechanisms (Baddeley, 1992; Cowan, 1999). In addition, emotion and memory systems are closely linked. Frontotemporal brain regions act conjointly to promote the retention of emotionally arousing events and retrieve them from long-term stores (Kensinger and Corkin, 2004). Memory-enhancing effects of emotional arousal involve interactions between subcortical and cortical structures that are coordinated by the amygdala (Cahill et al., 1996). Thus, the amygdala directly mediates aspects of emotional learning and facilitates memory operations in other regions, including the hippocampus and prefrontal cortex regions (orbitofrontal cortex, ACC) (LaBar and Cabeza, 2006). Binge alcohol drinking in young adults influences these operations, leading to increased impulsivity, impairments in spatial working memory, and impaired emotional learning (Stephens and Duka, 2008). These behavioral changes appear to be driven by facilitated excitatory neurotransmission and reduced plasticity (long-term potentiation) in amygdala and hippocampus, as evidenced in a rodent model in which rats were exposed to intermittent episodes of alcohol consumption and withdrawal (Stephens and Duka, 2008). In alcoholism, memory deficits have been indicated for visuospatial information (Silberstein and Parsons, 1980; Leber et al., 1981) by using the

Memory-for-Designs (Graham and Kendall, 1960), a measure of short-term reproductive memory (Mandes and Gessner, 1989). Working memory deficits in alcoholism have been associated with frontal, hippocampal, and cerebellar functional compromise (Joyce and Robbins, 1991; Oscar-Berman et al., 1992; Zhang et al., 1997; Pitel et al., 2012; Vollstädt-Klein et al., 2012). Prefrontal regions are more generally related to the maintenance of top-down cognitive control, and parietal, medial temporal, and cerebellar regions are more specifically related to working memory (Mackiewicz Seghete et al., 2012). Nonetheless, Desmond et al. (2003) suggested that brain activation in left frontal and right cerebellar regions can support the articulatory control system of verbal working memory activity and a compensatory increase in activation may help alcoholics to maintain the same level of performance as controls (Desmond et al., 2003). When compromised, however, working memory networks subserving learning and consolidation of information and the formation of new associations may contribute to the self-sustaining nature of alcoholism.

Associations between positive (hedonic) effects of alcohol and drinking events are automatically processed and implicitly stored in memory. As a result, alcohol-related stimuli, including context information (King et al., 2012), gain emotional valence by association with the rewarding effects of alcohol itself. With continued alcohol use, the regular association of these alcohol cues with the ingestion of alcohol leads to conditioned reactions (Carter and Tiffany, 1999) motivated by desire for reward or relief craving that activate drug intake (Robinson and Berridge, 2000). Cue-elicited alcohol seeking and consuming may be seen as an attentional orienting response (Wiers et al., 2006), i.e., a bias to attend selectively to alcohol-related stimuli at the expense of other stimuli (Sharma et al., 2001) that gates emotional responses and motivational states (Lindquist et al., 2012).

An attentional bias towards automatic processing of alcohol-related stimuli occurs in alcoholism (Beck et al., 2012) and may explain both the feeling that the illness is uncontrollable (Hyman and Malenka, 2001), and repeated hazardous drinking behavior despite knowledge of social, economic, and health consequences. Attentional bias toward alcohol-related stimuli has been studied using alcohol-related pictures or words in modified visual dot probe (Noël et al., 2006; Townshend and Duka, 2007), Go/No-Go tasks (Noël et al., 2007), modified Posner (Stormark et al., 2000), and dual-task paradigms (Waters and Green, 2003). In the modified Posner paradigm, alcohol cue and target appear consecutively, testing both allocation of attention towards alcohol-related stimuli and disengagement of that attention necessary for effective target processing. The visual orienting system selects information from sensory input and can be experimentally manipulated by cue presentations that direct attention to a location or feature (Schulte et al., 2005a). Thus, based on the assumption that abstinent alcoholics' attentional bias toward alcohol interferes with goal-directed behavior of avoiding alcohol, Stormark and colleagues (1997) modified the Posner paradigm to study this approach avoidance process. The subjects' task in the modified Posner paradigm was to respond as quickly as possible to an asterisk (target) that appeared on the left or right of a fixation point on a computer screen. The cue stimuli, both alcohol and neutral words presented prior to the target, were either located at the same visual field location (valid), or at the opposite visual field location (invalid) than the target. The semantic content of the cue had to be ignored. The findings from this approach avoidance Posner paradigm support the idea of an initial involuntary attentional bias towards alcohol cues (automatic approach) that was overcome when more processing time enabled alcoholics to withdraw attention from the alcohol cue (voluntarily avoidance).

In addition to semantic word cues, pairs of alcohol and neutral picture cues were used in modified visual probe tasks to study shifts in attentional allocation in alcoholism. These results also suggested that attention in alcoholics was preferentially allocated to the location of alcohol picture cues (Townshend and Duka, 2001; Field et al., 2004). Such attentional bias to alcohol cues was further reported in patients with a shorter duration of dependence (less than 9 years) but not in those with longer duration of dependence (Loeber et al., 2009). Yet, longer dependent patients showed greater deficits in functions of attention and working memory than shorter dependent patients. However, group differences in attentional bias were no longer significant after covarying for attention and working memory, indicating that the attentional bias towards alcohol is mediated by cognitive functions. In heavy social drinkers, alcohol cue reactivity has been proposed to be a significant predictor for vulnerability to develop alcohol misuse (Petit et al., 2012).

Disturbance of corticolimbic and striatal brain systems in chronic alcoholism may take the form of a maladaptive learning process in which enduring memories of alcohol experiences are formed (Bernier et al., 2011). Impairment in executive attention function impedes easy switching between different strategies and overriding habitual responses and alcohol-related memories. Such deficits consequently may be limiting factors for treatment efforts (Arnsten, 1997).

Thus, the impaired dynamic tuning between frontally based executive control and corticolimbic emotion and memory, and striatal reward systems seen in chronic alcoholism, impedes behavioral adaptation to changing environments (Fig. 20.1D). Dynamic tuning and efficiency of brain circuits rely on the integrity of white-matter fibers transmitting signals between network nodes (Greicius et al., 2009). In alcoholism, white-matter fiber compromise has been observed in selective white-matter fiber bundles involving limbic fiber tracts (fornix and cingulum) that connect corticolimbic and striatal nodes of emotion and reward circuits (Harris et al., 2008; Pfefferbaum et al., 2009). Recent evidence from a combined DTI–fMRI study supports the assumption that the robustness of functional network connectivity is related to fiber integrity in alcoholics (Schulte et al., 2012).

ALCOHOLISM – A NEURAL DISCONNECTION SYNDROME?

Chronic alcohol consumption is associated with subtle but significant compromise in the integrity of white-matter fibers, for example of the corpus callosum, a thick band of white-matter fiber connecting the two cerebral hemispheres (Pfefferbaum et al., 2010). Subtle interhemispheric fiber degradation in alcoholism can restrict interhemispheric information transfer and integration (Schulte et al., 2005b, 2010) and also affect attention, emotion, and cognition (Schulte et al., 2006; Müller-Oehring et al., 2013). To study how such fiber degradation in alcoholism may affect visual bottom-up processes, Schulte and colleagues (2010) used a behavioral paradigm that compared a visual display of paired with single targets presented to one or both visual hemifields. Typically, responses are faster to stimulus pairs than single stimuli, a phenomenon called the redundant targets effect (Miller, 1982) with performance advantages from bilateral but not unilateral redundant targets indicating hemispheric resource sharing to make this redundant targets process more efficient (Maertens and Pollmann, 2005). Alcoholics showed less bilateral processing advantage than controls when responding with the left hand and greater unilateral processing advantage when responding with the right hand. DTI-based tractography demonstrated degraded left posterior cingulate and posterior callosal fibers in these alcoholics. These microstructural findings were consistent with functional imaging results of less left posterior cingulate and extrastriate cortex activation when processing bilateral compared with unilateral visual field stimulation (Schulte et al., 2010). Thus, chronic alcoholics showed impaired efficiency of visual information processing and fast response execution on a very basic bottom-up processing level of integrating information from two visual hemifields.

However, when testing higher-order visual attention functions using ERP, Bijl and colleagues (2005) found no differences in light to heavy drinkers. Higher-order functions have preferential hemispheric processing: the left hemisphere for language and the right hemisphere for visuospatial attention. Also, the left hemisphere appears to preferentially process details and the right hemisphere the global gestalt (Han et al., 2002). Inhibitory projections between hemispheres through the corpus callosum ensure processing efficiency (Häberling et al., 2011). To study how callosal degradation in alcoholism may affect lateralized higher-order attention and cognitive control functions, Müller-Oehring and colleagues (2009) used a hierarchical letter paradigm, in which large global letters were composed of small local letters, to assess component global-local processes of visuospatial attention ability (Maunsell and Treue, 2006) in combination with DTI. Lateralized functions of global and local feature perception were generally spared in abstinent chronic alcoholics, but impairments arose when local and global information interfered and required interhemispheric communication (Müller-Oehring et al., 2009). Here, microstructural degradation of the corpus callosum correlated with degree of local-global interference, suggesting that compromise in callosal microstructural connectivity in alcoholics can alter the interhemispheric tuning of lateralized higher-order visuospatial functions.

FUNCTIONAL NETWORKS OF ATTENTION AND COGNITION

Functional connectivity MRI has enabled the investigation of neural network synchronous activation and how synchronous network dynamics are affected in alcoholics. Recent research highlights the selective impairment in chronic alcoholics of brain networks including the default mode network (DMN) in which distinct regions (i.e., PCC, mPFC, and lateral parietal cortex) are synchronously activated during a resting state and desynchronized during task performance (Chanraud et al., 2010; Anticevic et al., 2012). Although the PCC is typically deactivated in healthy subjects when focusing on tasks of higher attentional load, we recently found that alcoholics failed to exhibit task-induced deactivation within the DMN (Schulte et al., 2012). There is evidence that dopamine transporters are implicated in the modulation of neural activity in the DMN and the mPFC, including the ACC during visuospatial attention. In particular, children with an attention-deficit/hyperactivity disorder, exhibiting poor attention and inhibitory control, can profit from stimulants (methylphenidate, amphetamine) that increase dopamine levels in the brain (Tomasi et al., 2009). Adolescents with normal markers of dopamine neurotransmission exhibited the normal pattern of DMN deactivation when performing an attention task, whereas those with greater familial densities of alcohol use disorders were less successful at modulating activity of the DMN, suggesting that they were more engaged in task-independent thought, less focused on task-relevant thought, and less able to inhibit task-irrelevant cognition (Spadoni et al., 2008).

There is, however, limited understanding of the role of DMN on corticostriatal dysfunction and impaired attention and cognitive control in chronic alcoholism. The finding that FHP adolescents were less able to deactivate the ventral caudate during successful inhibition than controls with no familial risk may imply a pre-existing abnormality in ventral striatal function in FHP youth (Heitzeg et al., 2010). In addition, those FHP youth who had engaged in heavy alcohol use deactivated orbital and medial prefrontal regions, suggesting an efficiency loss of the prefrontal "control" mechanism with potential to further dysregulate the frontostriatal motivational circuitry (Heitzeg et al., 2010). An fMRI study in FHP and FHN (negative familial history of alcoholism) adolescents that tested reward network connectivity during a monetary-incentive delay task found deficits in NAcc reward-related functional connectivity in FHP adolescents as well as deficits in decoupling of the precuneus, a major DMN node (Weiland et al., 2012). Activity in DMN regions that are typically positively correlated in the resting state may have competing functions when task focus is necessary (Fox et al., 2005). Thus, failure to decouple DMN regions as attention shifts from an internal, self-referential focus during rest to an external focus during task can be a sign of inefficient internetwork communication (Weiland et al., 2012). Hence, altered patterns of functional connectivity in cognitive networks may serve as a pre-existing risk factor (Heitzeg et al., 2010; Herting et al., 2011; Wetherill et al., 2012) and as a marker for neurocognitive impairment (Sharp et al., 2011). In addition, recruitment of parallel networks may enable compensation for impairment and mark recovery in abstinent alcoholics (Schulte et al., 2012; Chanraud et al., 2013).

CONCLUSION

Taken together, chronic alcohol consumption affects brain systems of attention, emotion, and cognition, including automatic processes such as attentional bias to alcohol cues. Such implicit attention processes promote alcohol consumption and contribute to relapse through emotional reactions that influence behavior and decisions likely adding to the self-sustaining nature of the illness, and its resistance to treatment efforts. Thus, an understanding of the underlying neural mechanisms will likely provide the basis for better treatment strategies for developing skills that enhance alcoholism therapy adherence and quality of life, and reduce the propensity for relapse (Bates et al., 2002). Analysis of cognitive deficits in disturbances of top-down or bottom-up processing could be meaningful in designing therapeutic or rehabilitative strategies tailored to individual differences in locus of impairment.

References

Alexander GE, DeLong MR, Strick PL (1986). Parallel organization of functionally segregated circuits linking basal ganglia and cortex. Annu Rev Neurosci 9: 357–381.

Ambrose ML, Bowden SC, Whelan G (2001). Working memory impairments in alcohol-dependent participants without clinical amnesia. Alcohol Clin Exp Res 25 (2): 185–191.

Amenta S, Noel X, Verbanck P et al. (2012). Decoding of emotional components in complex communicative situations (irony) and its relation to empathic abilities in male chronic alcoholics: an issue for treatment. Alcohol Clin Exp Res 37: 339–347.

Anticevic A, Cole MW, Murray JD et al. (2012). The role of default network deactivation in cognition and disease. Trends Cogn Sci 16: 584–592.

Arnsten AF (1997). Catecholamine regulation of the prefrontal cortex. J Psychopharmacol 11: 151–162.

Baddeley A (1992). Working memory. Science 255: 556–559.

Badre D, Wagner AD (2004). Selection, integration, and conflict monitoring; assessing the nature and generality of prefrontal cognitive control mechanisms. Neuron 41 (3): 473–487.

Banich MT, Milham MP, Atchley R et al. (2000). fMRI studies of Stroop tasks reveal unique roles of anterior and posterior brain systems in attentional selection. J Cogn Neurosci 12: 988–1000.

Banks SJ, Eddy KT, Angstadt M et al. (2007). Amygdala-frontal connectivity during emotion regulation. Soc Cogn Affect Neurosci 2: 303–312.

Baribeau JC, Ethier M, Braun CM (1987). Neurophysiological assessment of selective attention in males at risk for alcoholism. Electroencephalogr Clin Neurophysiol 40 (suppl.): 651–656.

Bates ME, Bowden SC, Barry D (2002). Neurocognitive impairment associated with alcohol use disorders: implications for treatment. Exp Clin Psychopharmacol 10: 193–212.

Bechara A, Dolan S, Denburg N et al. (2001). Decision-making deficits, linked to a dysfunctional ventromedial prefrontal cortex, revealed in alcohol and stimulant abusers. Neuropsychologia 39: 376–389.

Beck DM, Kastner S (2009). Top-down and bottom-up mechanisms in biasing competition in the human brain. Vision Res 49: 1154–1165.

Beck A, Schlagenhauf F, Wustenberg T et al. (2009). Ventral striatal activation during reward anticipation correlates with impulsivity in alcoholics. Biol Psychiatry 66: 734–742.

Beck A, Wüstenberg T, Genauck A et al. (2012). Effect of brain structure, brain function, and brain connectivity on relapse in alcohol-dependent patients. Arch Gen Psychiatry 269: 842–852.

Bernier BE, Whitaker LR, Morikawa H (2011). Previous ethanol experience enhances synaptic plasticity of NMDA receptors in the ventral tegmental area. J Neurosci 31: 5205–5212.

Bijl S, de Bruin EA, Kenemans JL et al. (2005). Effects of chronic alcohol consumption in a visual attention task and an auditory oddball task: an event-related potential study. Alcohol Clin Exp Res 29: 2029–2038.

Bjork JM, Smith AR, Hommer DW (2008). Striatal sensitivity to reward deliveries and omissions in substance dependent patients. Neuroimage 42: 1609–1621.

Blum K, Braverman ER, Holder JM et al. (2000). Reward deficiency syndrome: a biogenetic model for the diagnosis and treatment of impulsive, addictive, and compulsive behaviors. J Psychoactive Drugs 32 (Suppl. i–iv): 1–112.

Botvinick M, Nystrom LE, Fissell K et al. (1999). Conflict monitoring versus selection-for-action in anterior cingulate cortex. Nature 402 (6758): 179–181.

Botvinick MM, Braver TS, Barch DM et al. (2001). Conflict monitoring and cognitive control. Psychol Rev 108: 624–652.

Broadbent DE (1958). Perception and communication, Pergamon Press, Oxford.

Bundesen C, Habekost T, Kyllingsbaek S (2005). A neural theory of visual attention: bridging cognition and neurophysiology. Psychol Rev 112: 291–328.

Bush G, Shin LM, Holmes J et al. (2003). The Multi-Source Interference Task: validation study with fMRI in individual subjects. Mol Psychiatry 8: 60–70.

Cahill L, Haier RJ, Fallon J et al. (1996). Amygdala activity at encoding correlated with long-term, free recall of emotional information. Proc Natl Acad Sci U S A 23: 8016–8021.

Campanella S, Belin P (2007). Integrating face and voice in person perception. Trends Cogn Sci 11: 535–543.

Carter BL, Tiffany ST (1999). Cue-reactivity and the future of addiction research. Addiction 94: 349–351.

Carter CS, Braver TS, Barch DM et al. (1998). Anterior cingulate cortex, error detection, and the online monitoring of performance. Science 280: 747–749.

Castle M, Comoli E, Loewy AD (2005). Autonomic brainstem nuclei are linked to the hippocampus. Neuroscience 134: 657–669.

Chanraud S, Martelli C, Delain F et al. (2007). Brain morphometry and cognitive performance in detoxified alcohol-dependents with preserved psychosocial functioning. Neuropsychopharmacology 32: 429–438.

Chanraud S, Pitel AL, Rohlfing T et al. (2010). Dual tasking and working memory in alcoholism: relation to frontocerebellar circuitry. Neuropsychopharmacology 35: 1868–1878.

Chanraud S, Pitel AL, Müller-Oehring EM et al. (2013). Remapping the brain to compensate for impairment in recovering alcoholics. Cereb Cortex 23: 97–104.

Clark US, Oscar-Berman M, Shagrin B et al. (2007). Alcoholism and judgments of affective stimuli. Neuropsychology 21: 346–362.

Coccaro EF, McCloskey MS, Fitzgerald DA et al. (2007). Amygdala and orbitofrontal reactivity to social threat in individuals with impulsive aggression. Biol Psychiatry 62: 168–178.

Cohen JD, Botvinick M, Carter CS (2000). Anterior cingulate and prefrontal cortex: who's in control? Nat Neurosci 3: 421–423.

Concha L, Gross DW, Beaulieu C (2005). Diffusion tensor tractography of the limbic system. AJNR Am J Neuroradiol 26: 2267–2274.

Corbetta M, Shulman GL (2002). Control of goal-directed and stimulus-driven attention in the brain. Nat Rev Neurosci 3: 201–215.

Courtney KE, Ghahremani DG, Ray LA (2013). Fronto-striatal functional connectivity during response inhibition in alcohol dependence. Addict Biol 18: 593–604.

Cowan N (1999). The differential maturation of two processing rates related to digit span. J Exp Child Psychol 72: 193–209.

Crews FT, Boettiger CA (2009). Impulsivity, frontal lobes and risk for addiction. Pharmacol Biochem Behav 93: 237–247.

Dao-Castellana MH, Samson Y, Legault F et al. (1998). Frontal dysfunction in neurologically normal chronic alcoholic subjects: metabolic and neuropsychological findings. Psychol Med 28: 1039–1048.

Desimone R, Duncan J (1995). Neural mechanisms of selective visual attention. Annu Rev Neurosci 18: 193–222.

Desmond JE, Chen SH, DeRosa E et al. (2003). Increased frontocerebellar activation in alcoholics during verbal working memory: an fMRI study. Neuroimage 19: 1510–1520.

Draganski B, Kherif F, Klöppel S et al. (2008). Evidence for segregated and integrative connectivity patterns in the human basal ganglia. J Neurosci 28: 7143–7152.

Duka T, Townshend JM, Collier K et al. (2003). Impairment in cognitive functions after multiple detoxifications in alcoholic inpatients. Alcohol Clin Exp Res 27: 1563–1572.

Durazzo TC, Tosun D, Buckley S et al. (2011). Cortical thickness, surface area, and volume of the brain reward system in alcohol dependence: relationships to relapse and extended abstinence. Alcohol Clin Exp Res 35: 1187–1200.

Etkin A, Egner T, Kalisch R (2011). Emotional processing in anterior cingulate and medial prefrontal cortex. Trends Cogn Sci 15: 85–93.

Euser AS, Franken IH (2012). Alcohol affects the emotional modulation of cognitive control: an event-related brain potential study. Psychopharmacology (Berl) 222: 459–476.

Evert DL, Oscar-Berman M (2001). Selective attentional processing and the right hemisphere: effects of aging and alcoholism. Neuropsychology 15: 452–461.

Falkenstein M, Hoormann J, Hohnsbein J (1999). ERP components in Go/Nogo tasks and their relation to inhibition. Acta Psychol (Amst) 101: 267–291.

Fein G, Di Sclafani V, Cardenas VA et al. (2002). Cortical gray matter loss in treatment-naïve alcohol dependent individuals. Alcohol Clin Exp Res 26: 558–564.

Fein G, Klein L, Finn P (2004). Impairment on a simulated gambling task in long-term abstinent alcoholics. Alcohol Clin Exp Res 28: 1487–1491.

Fein G, Torres J, Price LJ et al. (2006). Cognitive performance in long-term abstinent alcoholic individuals. Alcohol Clin Exp Res 30: 1538–1544.

Fein G, Shimotsu R, Chu R et al. (2009). Parietal gray matter volume loss is related to spatial processing deficits in long-term abstinent alcoholic men. Alcohol Clin Exp Res 33: 1806–1814.

Field M, Mogg K, Zetteler J et al. (2004). Attentional biases for alcohol cues in heavy and light social drinkers: the roles of initial orienting and maintained attention. Psychopharmacology (Berl) 176: 88–93.

Field M, Wiers RW, Christiansen P et al. (2010). Acute alcohol effects on inhibitory control and implicit cognition: implications for loss of control over drinking. Alcohol Clin Exp Res 34: 1346–1352.

Foisy ML, Kornreich C, Petiau C et al. (2007). Impaired emotional facial expression recognition in alcoholics: are th deficits specific to emotional cues? Psychiatry Res 150: 33–41.

Fox MD, Snyder AZ, Vincent JL et al. (2005). The human brain is intrinsically organized into dynamic, anticorrelated functional networks. Proc Natl Acad Sci U S A 5: 9673–9678.

Garcia R, Vouimba RM, Baudry M et al. (1999). The amygdala modulates prefrontal cortex activity relative to conditioned fear. Nature 402: 294–296.

Gazzaley A (2011). Influence of early attentional modulation on working memory. Neuropsychologia 49: 1410–1424.

Giancola PR, Moss HB (1998). Executive cognitive functioning in alcohol use disorders. Recent Dev Alcohol 14: 227–251.

Glenn SW, Parsons OA (1991). Effects of alcoholism and instructional conditions on speed/accuracy tradeoffs. Alcohol Clin Exp Res 15 (4): 612–619.

Goudriaan AE, Oosterlaan J, de Beurs E et al. (2005). Decision making in pathological gambling: a comparison between pathological gamblers, alcohol dependents, persons with Tourette syndrome, and normal controls. Brain Res Cogn Brain Res 23: 137–151.

Graham FK, Kendall BS (1960). Memory for Designs Test: Revised general manual. Percept Mot Skills 11: 147–188.

Greicius MD, Supekar K, Menon V et al. (2009). Resting-state functional connectivity reflects structural connectivity in the default mode network. Cereb Cortex 19: 72–78.

Gujar N, Yoo S-S, Hu P et al. (2011). Sleep deprivation amplifies reactivity of brain reward networks, biasing the appraisal of positive emotional experiences. J Neurosci 31: 4466–4474.

Häberling IS, Badzakova-Trajkov G, Corballis MC (2011). Callosal tracts and patterns of hemispheric dominance: a combined fMRI and DTI study. Neuroimage 54: 779–786.

Han S, Weaver JA, Murray SO et al. (2002). Hemispheric asymmetry in global/local processing: effects of stimulus position and spatial frequency. Neuroimage 17: 1290–1299.

Harper C (2009). The neuropathology of alcohol-related brain damage. Alcohol Alcohol 44: 136–140.

Harper C, Dixon G, Sheedy D et al. (2003). Neuropathological alterations in alcoholic brains. Studies arising from the New South Wales Tissue Resource Centre. Prog Neuropsychopharmacol Biol Psychiatry 27: 951–961.

Harris GJ, Jaffin SK, Hodge SM et al. (2008). Frontal white matter and cingulum diffusion tensor imaging deficits in alcoholism. Alcohol Clin Exp Res 32: 1001–1013.

Heimer L, Alheid GF, de Olmos JS et al. (1997). The accumbens: beyond the core-shell dichotomy. J Neuropsychiatry Clin Neurosci 9: 354–381.

Heitzeg MM, Nigg JT, Yau WY et al. (2010). Striatal dysfunction marks preexisting risk and medial prefrontal dysfunction is related to problem drinking in children of alcoholics. Biol Psychiatry 68: 287–295.

Herting MM, Fair D, Nagel BJ (2011). Altered fronto-cerebellar connectivity in alcohol-naïve youth with a family history of alcoholism. Neuroimage 14: 2582–2589.

Holmes A, Fitzgerald PJ, MacPherson KP et al. (2012). Chronic alcohol remodels prefrontal neurons and disrupts NMDAR-mediated fear extinction encoding. Nat Neurosci 15: 1359–1361.

Huang L, Pashler H (2007). Working memory and the guidance of visual attention: consonance-driven orienting. Psychon Bull Rev 14: 148–153.

Hyman SE, Malenka RC (2001). Addiction and the brain: the neurobiology of compulsion and its persistence. Nat Rev Neurosci 2: 695–703.

Jernigan TL, Butters N, DiTraglia G et al. (1991). Reduced cerebral grey matter observed in alcoholics using magnetic resonance imaging. Alcohol Clin Exp Res 15: 418–427.

Jin H, Rourke SB, Patterson TL et al. (1998). Predictors of relapse in long-term abstinent alcoholics. J Stud Alcohol 59: 640–646.

Joyce EM, Robbins TW (1991). Frontal lobe function in Korsakoff and non-Korsakoff alcoholics: planning and spatial working memory. Neuropsychologia 29: 709–723.

Kanai R, Tsuchiya N, Verstraten FA (2006). The scope and limits of top-down attention in unconscious visual processing. Curr Biol 16: 2332–2336.

Kastner S, Ungerleider LG (2000). Mechanisms of visual attention in the human cortex. Annu Rev Neurosci 23: 315–341.

Kastner S, Ungerleider LG (2001). The neural basis of biased competition in human visual cortex. Neuropsychologia 39: 1263–1276.

Kensinger EA, Corkin S (2004). Two routes to emotional memory: distinct neural processes for valence and arousal. Proc Natl Acad Sci U S A 101: 3310–3315.

Kerns JG, Cohen JD, MacDonald 3rd AW et al. (2004). Anterior cingulate conflict monitoring and adjustments in control. Science 303: 1023–1026.

Kim H, Somerville LH, Johnstone T et al. (2003). Inverse amygdala and medial prefrontal cortex responses to surprised faces. Neuroreport 14: 2317–2322.

King JA, Korb FM, Egner T (2012). Priming of control: implicit contextual cuing of top-down attentional set. J Neurosci 32: 8192–8200.

Konrad A, Vucurevic G, Lorscheider M et al. (2012). Broad disruption of brain white matter microstructure and relationship with neuropsychological performance in male patients with severe alcohol dependence. Alcohol Alcohol 47: 118–126.

Koob GF, Ahmed SH, Boutrel B et al. (2004). Neurobiological mechanisms in the transition from drug use to drug dependence. Neurosci Biobehav Rev 27: 739–749.

Kornreich C, Brevers D, Canivet D et al. (2013). Impaired processing of emotion in music, faces and voices supports a generalized emotional decoding deficit in alcoholism. Addiction 108: 80–88.

Krebs RM, Boehler CN, Egner T et al. (2011). The neural underpinnings of how reward associations can both guide and misguide attention. J Neurosci 29 (31): 9752–9759.

Kril JJ, Halliday GM, Svoboda MD et al. (1997). The cerebral cortex is damaged in chronic alcoholics. Neuroscience 79: 983–998.

Kroener S, Mulholland PJ, New NN et al. (2012). Chronic alcohol exposure alters behavioral and synaptic plasticity of the rodent prefrontal cortex. PLoS One 7: e37541.

Kubota M, Nakazaki S, Hirai S et al. (2001). Alcohol consumption and frontal lobe shrinkage: study of 1432 non-alcoholic subjects. J Neurol Neurosurg Psychiatry 71: 104–106.

LaBar KS, Cabeza R (2006). Cognitive neuroscience of emotional memory. Nat Rev Neurosci 7: 54–64.

Lammel S, Ion DI, Roeper J et al. (2011). Projection-specific modulation of dopamine neuron synapses by aversive and rewarding stimuli. Neuron 9: 855–862.

Le Berre AP, Rauchs G, La Joie R et al. (2012). Impaired decision-making and brain shrinkage in alcoholism. Eur Psychiatry S0924-9338: 00136–00138.

Leber WR, Jenkins RL, Parsons OA (1981). Recovery of visual-spatial learning and memory in chronic alcoholics. J Clin Psychol 37: 192–197.

Lee H, Heller AS, van Reekum CM et al. (2012). Amygdala-prefrontal coupling underlies individual differences in emotion regulation. Neuroimage 62: 1575–1581.

Leech R, Kamourieh S, Beckmann CF et al. (2011). Fractionating the default mode network: distinct contributions of the ventral and dorsal posterior cingulate cortex to cognitive control. J Neurosci 31: 3217–3224.

Leh SE, Ptito A, Chakravarty MM et al. (2007). Fronto-striatal connections in the human brain: a probabilistic diffusion tractography study. Neurosci Lett 419: 113–118.

Lin JY, Murray SO, Boynton GM (2009). Capture of attention to threatening stimuli without perceptual awareness. Curr Biol 14: 1118–1122.

Lindquist KA, Wager TD, Kober H et al. (2012). The brain basis of emotion: a meta-analytic review. Behav Brain Sci 35: 121–143.

Liu IC, Chiu CH, Chen CJ et al. (2010). The microstructural integrity of the corpus callosum and associated impulsivity in alcohol dependence: a tractography-based segmentation study using diffusionspectrum imaging. Psychiatry Res 184: 128–134.

Loeber S, Duka T, Welzel H et al. (2009). Impairment of cognitive abilities and decision making after chronic use of alcohol: the impact of multiple detoxifications. Alcohol Alcohol 44: 372–381.

Loeber S, Duka T, Welzel Márquez H et al. (2010). Effects of repeated withdrawal from alcohol on recovery of cognitive impairment under abstinence and rate of relapse. Alcohol Alcohol 45: 541–547.

LoPresti ML, Schon K, Tricarico MD et al. (2008). Working memory for social cues recruits orbitofrontal cortex and amygdala: a functional magnetic resonance imaging study of delayed matching to sample for emotional expressions. J Neurosci 28: 3718–3728.

Luks TL, Simpson GV, Dale CL et al. (2007). Preparatory allocation of attention and adjustments in conflict processing. Neuroimage 35: 949–958.

MacDonald 3rd AW, Cohen JD, Stenger VA et al. (2000). Dissociating the role of the dorsolateral prefrontal and anterior cingulate cortex in cognitive control. Science 288: 1835–1838.

Mackiewicz Seghete KL, Cservenka A, Herting MM et al. (2012). Atypical spatial working memory and task-general brain activity in adolescents with a family history of alcoholism. Alcohol Clin Exp Res 37: 390–398.

Maertens M, Pollmann S (2005). fMRI reveals a common neural substrate of illusory and real contours in V1 after perceptual learning. J Cogn Neurosci 17: 1553–1564.

Makris N, Oscar-Berman M, Jaffin SK et al. (2008). Decreased volume of the brain reward system in alcoholism. Biol Psychiatry 64: 192–202.

Mandes E, Gessner T (1989). The principle of additivity and its relation to clinical decision making. J Psychol 123: 485–490.

Mann K, Gunther A, Stetter F et al. (1999). Rapid recovery from cognitive deficits in abstinent alcoholics: a controlled test–retest study. Alcohol Alcohol 34: 567–574.

Margulies DS, Kelly AM, Uddin LQ et al. (2007). Mapping the functional connectivity of anterior cingulate cortex. Neuroimage 37: 579–588.

Marinkovic K, Oscar-Berman M, Urban T et al. (2009). Alcoholism and dampened temporal limbic activation to emotional faces. Alcohol Clin Exp Res 33: 1880–1892.

Mark LP, Daniels DL, Naidich TP et al. (1993). Limbic system anatomy: an overview. AJNR Am J Neuroradiol 14: 349–352.

Matthews SC, Paulus MP, Simmons AN et al. (2004). Functional subdivisions within anterior cingulate cortex and their relationship to autonomic nervous system function. Neuroimage 22: 1151–1156.

Maunsell JH, Treue S (2006). Feature-based attention in visual cortex. Trends Neurosci 29: 317–322.

Maurage P, Joassin F, Pesenti M et al. (2013). The neural network sustaining crossmodal integration is impaired in alcohol-dependence: an fMRI study. Cortex 49: 1610–1626.

Melloni L, van Leeuwen S, Alink A et al. (2012). Interaction between bottom-up saliency and top-down control: how saliency maps are created in the human brain. Cereb Cortex 22: 2943–2952.

Milham MP, Banich MT, Barad V (2003). Competition for priority in processing increases prefrontal cortex's involvement in top-down control: an event-related fMRI study of the stroop task. Brain Res Cogn Brain Res 17: 212–222.

Miller J (1982). Divided attention: evidence for coactivation with redundant signals. Cognit Psychol 14: 247–279.

Mohanty A, Gitelman DR, Small DM et al. (2008). The spatial attention network interacts with limbic and monoaminergic systems to modulate motivation-induced attention shifts. Cereb Cortex 18: 2604–2613.

Montagne B, Kessels RP, Wester AJ et al. (2006). Processing of emotional facial expressions in Korsakoff's syndrome. Cortex 42: 705–710.

Mühlberger A, Wieser MJ, Gerdes AB et al. (2011). Stop looking angry and smile, please: start and stop of the very same facial expression differentially activate threat- and reward-related brain networks. Soc Cogn Affect Neurosci 6: 321–329.

Müller-Oehring EM, Schulte T, Fama R et al. (2009). Global-local interference is related to callosal compromise in alcoholism: a behavior-DTI association study. Alcohol Clin Exp Res 33: 477–489.

Müller-Oehring EM, Schulte T, Rohlfing T et al. (2013). Visual search and the aging brain: discerning the effects of age-related brain volume shrinkage on alertness, feature binding, and attentional control. Neuropsychology 27: 48–59.

Munro CA, Saxton J, Butters MA (2000). The neuropsychological consequences of abstinence among older alcoholics: a cross-sectional study. Alcohol Clin Exp Res 24: 1510–1516.

Nakano K (2000). Neural circuits and topographic organization of the basal ganglia and related regions. Brain Dev 22 (Suppl. 1): S5–S16.

Nieoullon A (2002). Dopamine and the regulation of cognition and attention. Prog Neurobiol 67: 53–83.

Noël X, Van der Linden M, Schmidt N et al. (2001). Supervisory attentional system in nonamnesic alcoholic men. Arch Gen Psychiatry 58: 1152–1158.

Noël X, Sferrazza R, Van Der Linden M et al. (2002). Contribution of frontal cerebral blood flow measured by (99 m)Tc-Bicisate spect and executive function deficits to predicting treatment outcome in alcohol-dependent patients. Alcohol Alcohol 37: 347–354.

Noël X, Colmant M, Van Der Linden M et al. (2006). Time course of attention for alcohol cues in abstinent alcoholic patients: the role of initial orienting. Alcohol Clin Exp Res 30: 1871–1877.

Noël X, Van der Linden M, d'Acremont M et al. (2007). Alcohol cues increase cognitive impulsivity in individuals with alcoholism. Psychopharmacology (Berl) 192: 291–298.

Norman AL, Pulido C, Squeglia LM et al. (2011). Neural activation during inhibition predicts initiation of substance use in adolescence. Drug Alcohol Depend 119: 216–223.

O'Daly OG, Trick L, Scaife J et al. (2012). Withdrawal-associated increases and decreases in functional neural connectivity associated with altered emotional regulation in alcoholism. Neuropsychopharmacology 37: 2267–2276.

Orr C, Hester R (2012). Error-related anterior cingulate cortex activity and the prediction of conscious error awareness. Front Hum Neurosci 6: 177.

Ortega LA, Uhelski M, Fuchs PN et al. (2011). Impairment of recovery from incentive downshift after lesions of the anterior cingulate cortex: emotional or cognitive deficits? Behav Neurosci 125: 988–995.

Oscar-Berman M, Bonner RT (1985). Matching- and delayed matching-to-sample performance as measures of visual processing, selective attention, and memory in aging and alcoholic individuals. Neuropsychologia 23: 639–651.

Oscar-Berman M, Hancock M, Mildworf B et al. (1990). Emotional perception and memory in alcoholism and aging. Alcohol Clin Exp Res 14: 383–393.

Oscar-Berman M, Hutner N, Bonner RT (1992). Visual and auditory spatial and nonspatial delayed-response performance by Korsakoff and non-Korsakoff alcoholic and aging individuals. Behav Neurosci 106: 613–622.

Parsons OA (1998). Neurocognitive deficits in alcoholics and social drinkers: a continuum? Alcohol Clin Exp Res 22: 954–961.

Peeters M, Wiers RW, Monshouwer K et al. (2012). Automatic processes in at-risk adolescents: the role of alcohol-approach tendencies and response inhibition in drinking behavior. Addiction 107: 1939–1946.

Peters JC, Roelfsema PR, Goebel R (2012). Task-relevant and accessory items in working memory have opposite effects on activity in extrastriate cortex. J Neurosci 32: 17003–17011.

Petersen SE, Posner MI (2012). The attention system of the human brain: 20 years after. Annu Rev Neurosci 35: 73–89.

Petit G, Kornreich C, Maurage P et al. (2012). Early attentional modulation by alcohol-related cues in young binge drinkers: an event-related potentials study. Clin Neurophysiol 123: 925–936.

Pfefferbaum A, Lim KO, Zipursky RB et al. (1992). Brain gray and white matter volume loss accelerates with aging in chronic alcoholics: a quantitative MRI study. Alcohol Clin Exp Res 16: 1078–1089.

Pfefferbaum A, Sullivan EV, Mathalon DH et al. (1997). Frontal lobe volume loss observed with magnetic resonance imaging in older chronic alcoholics. Alcohol Clin Exp Res 21: 521–529.

Pfefferbaum A, Desmond JE, Galloway C et al. (2001). Reorganization of frontal systems used by alcoholics for spatial working memory: an fMRI study. Neuroimage 14: 7–20.

Pfefferbaum A, Rosenbloom M, Rohlfing T et al. (2009). Degradation of association and projection white matter systems in alcoholism detected with quantitative fiber tracking. Biol Psychiatry 65: 680–690.

Pfefferbaum A, Rosenbloom MJ, Fama R et al. (2010). Transcallosal white matter degradation detected with quantitative fiber tracking in alcoholic men and women: selective relations to dissociable functions. Alcohol Clin Exp Res 34: 1201–1211.

Pitel AL, Beaunieux H, Witkowski T et al. (2007). Genuine episodic memory deficits and executive dysfunctions in alcoholic subjects early in abstinence. Alcohol Clin Exp Res 31: 1169–1178.

Pitel AL, Chanraud S, Rohlfing T et al. (2012). Face-name association learning and brain structural substrates in alcoholism. Alcohol Clin Exp Res 36: 1171–1179.

Polo MD, Escera C, Yago E et al. (2003). Electrophysiological evidence of abnormal activation of the cerebral network of involuntary attention in alcoholism. Clin Neurophysiol 114: 134–146.

Posner MI, Petersen SE (1990). The attention system of the human brain. Annu Rev Neurosci 13: 25–42.

Potts GF, Martin LE, Burton P et al. (2006). When things are better or worse than expected: the medial frontal cortex and the allocation of processing resources. J Cogn Neurosci 18: 1112–1119.

Pourtois G, Notebaert W, Verguts T (2012). Cognitive and affective control. Front Psychol 3: 477.

Quirk GJ, Likhtik E, Pelletier JG et al. (2003). Stimulation of medial prefrontal cortex decreases the responsiveness of central amygdala output neurons. J Neurosci 23: 8800–8807.

Reinhart RM, Heitz RP, Purcell BA et al. (2012). Homologous mechanisms of visuospatial working memory maintenance in macaque and human: properties and sources. J Neurosci 32: 7711–7722.

Robertson LC, Stillman R, Delisn DC (1985). The effect of alcohol abuse on perceptual reference frames. Neuropsychologia 23: 69–76.

Robinson TE, Berridge KC (2000). The psychology and neurobiology of addiction: an incentive-sensitization view. Addiction 95 (Suppl. 2): S91–S117.

Ryan C (1980). Learning and memory deficits in alcoholics. J Stud Alcohol 41: 437–447.

Salloum JB, Ramchandani VA, Bodurka J et al. (2007). Blunted rostral anterior cingulate response during a simplified decoding task of negative emotional facial expressions in alcoholic patients. Alcohol Clin Exp Res 31: 1490–1504.

Salo R, Henik A, Robertson LC (2001). Interpreting Stroop interference: an analysis of differences between task versions. Neuropsychology 15: 462–471.

Schulte T, Mueller-Oehring EM, Rosenbloom MJ et al. (2005a). Differential effect of HIV infection and alcoholism on conflict processing, attentional allocation, and perceptual load: evidence from a Stroop Match-to-Sample task. Biol Psychiatry 57: 67–75.

Schulte T, Sullivan EV, Müller-Oehring EM et al. (2005b). Corpus callosal microstructural integrity influences interhemispheric processing: a diffusion tensor imaging study. Cereb Cortex 15: 1384–1392.

Schulte T, Müller-Oehring EM, Salo R et al. (2006). Callosal involvement in a lateralized stroop task in alcoholic and healthy subjects. Neuropsychology 20: 727–736.

Schulte T, Müller-Oehring EM, Rohlfing T et al. (2010). J Neurosci 30: 12168–12178.

Schulte T, Müller-Oehring EM, Sullivan EV et al. (2012). Synchrony of corticostriatal-midbrain activation enables normal inhibitory control and conflict processing in recovering alcoholic men. Biol Psychiatry 71: 269–278.

Sharma D, Albery IP, Cook C (2001). Selective attentional bias to alcohol related stimuli in problem drinkers and non-problem drinkers. Addiction 96: 285–295.

Sharp DJ, Beckmann CF, Greenwood R et al. (2011). Default mode network functional and structural connectivity after traumatic brain injury. Brain 134: 2233–2247.

Silberstein JA, Parsons OA (1980). Neuropsychological impairment in female alcoholics. In: M Galanter (Ed.), Currents in Alcoholism, vol. 7. Grune & Stratton, New York, pp. 481–496.

Silveri MM, Rogowska J, McCaffrey A et al. (2011). Adolescents at risk for alcohol abuse demonstrate altered frontal lobe activation during Stroop performance. Alcohol Clin Exp Res 35: 218–228.

Simmons A, Matthews SC, Feinstein JS et al. (2008). Anxiety vulnerability is associated with altered anterior cingulate response to an affective appraisal task. Neuroreport 19: 1033–1037.

Soto D, Llewelyn D, Silvanto J (2012). Distinct causal mechanisms of attentional guidance by working memory and repetition priming in early visual cortex. J Neurosci 7: 3447–3452.

Spadoni AD, Norman AL, Schweinsburg AD et al. (2008). Effects of family history of alcohol use disorders on spatial working memory BOLD response in adolescents. Alcohol Clin Exp Res 32: 1135–1145.

Spunt RP, Lieberman MD, Cohen JR et al. (2012). The phenomenology of error processing: the dorsal ACC response to stop-signal errors tracks reports of negative affect. J Cogn Neurosci 24: 1753–1765.

Stavro K, Pelletier J, Potvin S (2012). Widespread and sustained cognitive deficits in alcoholism: a meta-analysis. Addict Biol 18: 203–213.

Steenland HW, Li XY, Zhuo M (2012). Predicting aversive events and terminating fear in the mouse anterior cingulate cortex during trace fear conditioning. J Neurosci 32: 1082–1095.

Stephens DN, Duka T (2008). Cognitive and emotional consequences of binge drinking: role of amygdala and prefrontal cortex. Philos Trans R Soc Lond B Biol Sci 363: 3169–3179.

Stormark KM, Field NP, Hugdahl K et al. (1997). Selective processing of visual alcohol cues in abstinent alcoholics: an approach-avoidance conflict? Addict Behav 22: 509–519.

Stormark KM, Laberg JC, Nordby H et al. (2000). Alcoholics' selective attention to alcohol stimuli: automated processing? J Stud Alcohol 61: 18–23.

Sullivan EV, Rosenbloom MJ, Lim KO et al. (2000). Longitudinal changes in cognition, gait, and balance in abstinent and relapsed alcoholic men: relationships to changes in brain structure. Neuropsychology 14: 178–188.

Szczepanski SM, Konen CS, Kastner S (2010). Mechanisms of spatial attention control in frontal and parietal cortex. J Neurosci 30: 148–160.

Thoma RJ, Monnig MA, Lysne PA et al. (2011). Adolescent substance abuse: the effects of alcohol and marijuana on neuropsychological performance. Alcohol Clin Exp Res 35: 39–46.

Tomasi D, Volkow ND, Wang R et al. (2009). Dopamine transporters in striatum correlate with deactivation in the default mode network during visuospatial attention. PLoS One 4: e6102.

Townshend JM, Duka T (2001). Attentional bias associated with alcohol cues: differences between heavy and occasional social drinkers. Psychopharmacology (Berl) 157: 67–74.

Townshend JM, Duka T (2003). Mixed emotions: alcoholics' impairments in the recognition of specific emotional facial expressions. Neuropsychologia 41: 773–782.

Townshend JM, Duka T (2007). Avoidance of alcohol-related stimuli in alcohol-dependent inpatients'. Alcohol Clin Exp Res 31: 1349–1357.

Vandenberghe R, Gillebert CR (2009). Parcellation of parietal cortex: convergence between lesion-symptom mapping and mapping of the intact functioning brain. Behav Brain Res 199: 171–182.

Vandewalle G, Maquet P, Dijk DJ (2009). Light as a modulator of cognitive brain function. Trends Cogn Sci 13: 429–438.

Vollstädt-Klein S, Loeber S, Richter A et al. (2012). Validating incentive salience with functional magnetic resonance imaging: association between mesolimbic cue reactivity and attentional bias in alcohol-dependent patients. Addict Biol 17: 807–816.

Walton ME, Croxson PL, Behrens TE et al. (2007). Adaptive decision making and value in the anterior cingulate cortex. Neuroimage 36 (Suppl. 2): T142–T154.

Waters H, Green MW (2003). A demonstration of attentional bias, using a novel dual task paradigm, towards clinically salient material in recovering alcohol abuse patients? Psychol Med 33: 491–498.

Weiland BJ, Welsh RC, Yau WY et al. (2012). Accumbens functional connectivity during reward mediates sensation-seeking and alcohol use in high-risk youth. Drug Alcohol Depend 28: 130–139.

Wetherill RR, Bava S, Thompson WK et al. (2012). Frontoparietal connectivity in substance-naïve youth with and without a family history of alcoholism. Brain Res 13: 66–73.

Wiers RW, Cox WM, Field M et al. (2006). The search for new ways to change implicit alcohol-related cognitions in heavy drinkers. Alcohol Clin Exp Res 30: 320–331.

Wilson JT, Wiedmann KD, Phillips WA et al. (1988). Visual event perception in alcoholics. J Clin Exp Neuropsychol 10: 222–234.

Wrase J, Schlagenhauf F, Kienast T et al. (2007). Dysfunction of reward processing correlates with alcohol craving in detoxified alcoholics. Neuroimage 35: 787–794.

Xu Y, Chun MM (2006). Dissociable neural mechanisms supporting visual short-term memory for objects. Nature 440: 91–95.

Yamasaki H, LaBar KS, McCarthy G (2002). Dissociable prefrontal brain systems for attention and emotion. Proc Natl Acad Sci U S A 99: 11447–11451.

Yoder KK, Morris ED, Constantinescu CC et al. (2009). When what you see isn't what you get: alcohol cues, alcohol administration, prediction error, and human striatal dopamine. Alcohol Clin Exp Res 33: 139–149.

Young AM, Ahier RG, Upton RL et al. (1998). Increased extracellular dopamine in the nucleus accumbens of the rat during associative learning of neutral stimuli. Neuroscience 83: 1175–1183.

Zhang XL, Begleiter H, Porjesz B (1997). Do chronic alcoholics have intact implicit memory? An ERP study. Electroencephalogr Clin Neurophysiol 103 (4): 457–473.

Zhang X, Zhaoping L, Zhou T et al. (2012). Neural activities in v1 create a bottom-up saliency map. Neuron 12: 183–192.

Zhaoping L (2008). Attention capture by eye of origin singletons even without awareness–a hallmark of a bottom-up saliency map in the primary visual cortex. J Vis 7: 1.1–1.18.

Zinn S, Stein R, Swartzwelder HS (2004). Executive functioning early in abstinence from alcohol. Alcohol Clin Exp Res 28: 1338–1346.

Zoethout RW, Schoemaker RC, Zuurman L et al. (2009). Central nervous system effects of alcohol at a pseudo-steady-state concentration using alcohol clamping in healthy volunteers. Br J Clin Pharmacol 68: 524–534.

Handbook of Clinical Neurology, Vol. 125 (3rd series)
Alcohol and the Nervous System
E.V. Sullivan and A. Pfefferbaum, Editors

Chapter 21

The neurobiology of alcohol craving and relapse

DONGJU SEO AND RAJITA SINHA*
Department of Psychiatry, Yale University School of Medicine, New Haven, CT, USA

INTRODUCTION

A major problem jeopardizing recovery from alcoholism is high alcohol craving and the chronic, relapsing nature of the illness. Many individuals with alcohol use disorders (AUD) report significant difficulties with craving and struggle with relapse urges (Fox et al., 2007; Breese et al., 2011). Researchers have also identified that alcohol craving and relapse are critical factors perpetuating the malady of chronic alcoholism (Bottlender and Soyka, 2004; Sinha et al., 2011) and thus, mechanisms underlying alcohol craving and relapse have been a central focus of alcohol research, including in neuroimaging studies.

Neuronal processes underlying alcohol craving and relapse are complex due to an interplay between neuroadaptations from chronic alcohol abuse and environmental risk factors affecting the developmental course of alcoholism. After a prolonged period of alcohol abuse, neuroadaptations from repeated alcohol consumption increase allostatic load on key neural circuit of emotion and reward regulation (Breese et al., 2011; Seo and Sinha, 2011; Seo et al., 2013) and place AUD individuals in an emotionally susceptible state. In the face of emotional or alcohol-related challenges (e.g., stress and alcohol cues), vulnerable AUD individuals often lose control over alcohol craving and make poor decisions, thereby increasing relapse risk (Bechara et al., 2001; Goldstein et al., 2009; Sinha, 2012, 2013). Thus, to have a comprehensive understanding of neural mechanisms underlying craving and relapse, relevant brain functions should be considered in the context of neuroadaptations and interactions with environmental risk factors.

This chapter is intended to provide a neurobiologic understanding of brain structures and functions related to alcohol craving and relapse risk. It will review relevant neuroimaging studies and discuss neural mechanisms underlying craving and relapse as well as significant risk factors in the course of chronic alcoholism. The first section provides a review of the role of neuroadaptations in chronic alcoholism. In the second section, we review the basic concepts and role of alcohol craving and relapse in the course of alcoholism. In the third section, functional brain changes associated with alcohol craving and relapse will be reviewed in the context of environmental risk factors such as alcohol, alcohol-associated cues, and stress. In this section, the emphasis is on the functional mechanisms related to emotion regulation and decision making in the prefrontal-striatal-limbic circuit and their contributions to alcohol craving and relapse. The final section concludes the review and discusses future research directions. Elucidating the neural mechanisms underlying alcohol craving and relapse in the context of its risk factors will help understand complex vulnerability factors for alcohol-related pathology and could facilitate the development of effective treatments.

ALCOHOL-RELATED NEUROADAPTATIONS

One distinctive feature of chronic alcoholism is altered neurophysiologic functions and dysregulated emotional states resulting from continued alcohol use over an extensive period of time (Li and Sinha, 2008; Breese et al., 2011). Chronic alcohol abuse gradually produces neuroadaptive changes in neural circuits of emotional modulation, which makes AUD individuals vulnerable to affect dysregulation (Breese et al., 2011; Seo and Sinha, 2011; Seo et al., 2013). It is also known that repetitive alcohol use alters dopaminergic tone, hijacking the natural reward system and leading to compulsive

*Correspondence to: Rajita Sinha, Ph.D., Professor of Psychiatry, Neurobiology and Child Study, Departments of Psychiatry, Yale University School of Medicine, 2 Church Street South, Suite 209, New Haven, CT 06519, USA. Tel: +1-203-737-3398, E-mail: Rajita.sinha@yale.edu

alcohol-seeking behaviors (Volkow et al., 1996, 2002; Di Chiara, 1997). In addition, chronic alcohol use has an adverse influence on the peripheral and central nervous system (CNS), resulting in a shift in neuronal response and basal physiologic states. Such neuroadaptations have been associated with dysregulatory control over alcohol craving and urge to use alcohol after a period of abstinence in individuals with alcohol dependence (Breese et al., 2011).

Striatal dopamine transmission

Chronic alcoholism has been associated with disruption in reward-related neural system, especially dopamine (DA) transmission in the ventral striatum (VS) (Di Chiara, 1997; Volkow et al., 2002). The VS is involved in the processing of reward response (Everitt and Robbins, 2000; Knutson et al., 2001). Increased ventral striatal activity is elicited during the experience of pleasant feeling (Hare et al., 2005), and elevated DA levels in this region have been observed in response to rewarding stimuli, including food (Hernandez and Hoebel, 1988), sexual arousal (Hull et al., 1993), and drugs (Ranaldi et al., 1999). Preclinical studies have shown that acute alcohol administration stimulates DA release in the VS (for review, see Di Chiara, 1997). However, chronic alcohol abuse leads to deficient striatal DA activity, indicating DA alterations that may impact continued alcohol use (Martinez et al., 2005). Other preclinical studies further report that alcohol exposure for a prolonged period of time leads to altered dopaminergic neurotransmission in the VS and the ventral tegmental area (VTA), which, in turn, increases alcohol self-administration (Di Chiara, 1997; Koob et al., 2004; Cleck and Blendy, 2008). It follows that disrupted striatal DA system associated with chronic alcoholism contributes to the vulnerability of alcohol-dependent individuals to high craving and dyscontrolled reward and alcohol-seeking behaviors.

Neuronal hyperexcitability

Emerging evidence indicates that chronic alcohol-related neuronal adaptations may result in basal-state, neuronal hyperexcitability in AUD patients, which could contribute to high alcohol craving and relapse risk (Breese et al., 2011; Sinha et al., 2011; Seo et al., 2013). The neuronal hyperexcitability in chronic alcoholism has been proposed in multiple endocrine and physiologic studies and encompasses the well-known "kindling" phenomena (Adinoff et al., 1998; Porjesz and Begleiter, 2003; Breese et al., 2011; Sinha et al., 2011). Excessive alcohol use could potentiate neuroadaptations similar to a "kindling"-like process, first identified by Ballenger and Post (1978), and further supported by basic and clinical studies demonstrating that persistent alcohol administration leads to a hyperexcitable CNS activity, especially in frontal regions (Bauer, 2001; Porjesz and Begleiter, 2003).

Although alcohol has an inhibitory effect on the CNS, the nervous system counteracts this effect over the course of persistent alcohol exposures, facilitating the activity of excitatory (glutamatergic), and suppressing inhibitory neurotransmitter systems, leading to CNS hyperexcitability (for a review, see Becker, 1998). This kindling-like, hyperactive CNS response may contribute to high alcohol seeking, craving, and relapse risk in the context of specific environments such as stress or alcohol cues (Becker, 1998; Breese et al., 2011). This is also consistent with multiple physiologic and endocrine studies showing hyperactive autonomic and hypothalamic–pituitary–adrenal (HPA) axis systems at basal levels in early recovering AUD individuals, including elevated basal cortisol levels (Adinoff et al., 1998) and heart rate (Sinha, 2009), dysregulated heart rate variability (Thayer et al., 2006), and alcohol-related upregulation in corticotropin-releasing hormone and noradrenergic response (Koob et al., 2004; Richardson et al., 2008; Sinha, 2008). In addition, greater adrenal sensitivity during a relaxing, resting state has been associated with shorter time to relapse after treatment (Sinha et al., 2011), indicating upregulated HPA axis system associated with alcohol relapse.

Consistent with this, neuroimaging studies have also reported alteration in the brain system involved in emotional modulation. It is well documented that chronic alcoholism is associated with functional and structural alterations in the prefrontal-striatal-limbic circuit (for review, see Buhler and Mann, 2011). A recent functional magnetic resonance imaging (fMRI) study also reported that hyperactive brain response in the ventromedial prefrontal cortex (VmPFC)/anterior cingulate cortex (ACC) and VS during neutral-relaxing state is predictive of high alcohol craving and early relapse in abstinent alcohol-dependent patients (Seo et al., 2013), suggesting the important role of basal state hyperactivity in alcohol craving and relapse.

Taken together, neuroadaptive changes in the neural circuit modulating emotions and behavioral responses may significantly impact the emotion and behaviors of alcohol-dependent individuals throughout the course of alcohol addiction, thereby making it more difficult to overcome craving and desire for alcohol consumption after a period of abstinence. In the next section, we review the concept and roles of alcohol craving over the course of alcohol addiction and the neurobiology underlying early relapse behaviors in the context of neuroadaptative changes in the cortico-limbic-striatal circuit.

CLINICAL NEUROBIOLOGY OF CRAVING AND RELAPSE IN CHRONIC ALCOHOLISM

Alcohol craving

Craving or "wanting" for desired substance is a major clinical feature of addiction (including alcoholism) that makes it difficult for AUD individuals to overcome chronic, alcohol-seeking behaviors (Tiffany, 1990; Robinson and Berridge, 1993; Sinha, 2013). The concept of alcohol craving also involves a sense of dyscontrol: in other words, difficulties regulating a strong desire or impulse to drink alcohol (Verheul et al., 1999; Sinha, 2013). Alcohol craving is associated with continued alcohol consumption and contributes to alcohol relapse and severity of alcohol use after treatment or an extended period of abstinence (Sinha et al., 2011). High alcohol craving has also been associated with poor clinical outcome and repeated relapse behaviors by triggering maladaptive patterns of chronic alcohol intake (Sinha et al., 2011; Sinha, 2013). To understand the role of craving in addiction, a number of models have been proposed to account for craving's contributions to the progression and severity of addiction. Principal concepts motivating these models include incentive salience (Robinson and Berridge, 1993) and positive and negative reinforcement (Koob and Le Moal, 2008).

Incentive salience and craving

Craving has been described as a state of sensitized incentive salience, which progressively increases by repeated administration of addictive substances, including alcohol (Robinson and Berridge, 1993). The incentive sensitization theory (Robinson and Berridge, 1993) of drug wanting and addiction suggests that chronic drug (e.g., alcohol)-related neuroadaptation sensitizes the reward circuit as well as its interconnected cortico-striatal regions towards increased salience of the drug and drug-related stimuli. This process significantly promotes greater wanting, or the craving response, and increases risk of chronic alcoholism and relapse (Koob and Volkow, 2010). This idea is consistent with large epidemiologic studies showing the association among high levels of alcohol/drug use, craving, and increased risk of addiction and alcohol-related problems (Dawson et al., 2005; Room et al., 2005; Rehm et al., 2009). Considerable evidence also indicates that alcohol-dependent individuals have significantly higher alcohol craving than healthy social drinkers (Greeley et al., 1993; Myrick et al., 2004).

As described in the first section, allostatic changes in brain structure and function triggered by continued, excessive consumption of alcohol may underlie high craving and alcohol-seeking behaviors (Koob et al., 2004; Breese et al., 2011). Specifically, repetitive alcohol consumption results in adaptive changes in brain reward circuits; this alteration increases salience toward alcohol stimuli and amplifies desire for alcohol drinking (Sinha, 2001; Koob et al., 2004; Breese et al., 2005b). In addition, the nature of the craving response can be altered via this neuroadaptation during the course of alcohol illness, e.g., positively reinforced alcohol craving (pleasure, reward) vs negatively reinforced craving (anxiety, negative emotional arousal) (Koob et al., 2004; Koob and Le Moal, 2008).

Positively reinforced craving

Initially, substance of abuse, including alcohol, generates pleasant and rewarding feelings by promoting DA transmission in the VS. It has also been suggested that striatal DA transmission is associated with reward-related reinforcement learning (Schultz et al., 1997; Schultz, 1998; Suri, 2002); the neurotransmitter DA released from VS reacts to conditioned, reward stimuli (Schultz, 1998) and triggers reward-seeking behavior (Ikemoto and Panksepp, 1999; Everitt and Robbins, 2000; Arias-Carrion and Poppel, 2007).

The significant role of the striatum in rewarding aspect of alcohol craving has been reported in both animal and human studies. In animals, γ-aminobutyric acid and DA activity in the VS is known to mediate alcohol consumption (Di Chiara, 1997; Rewal et al., 2009). In healthy men, alcohol craving during alcohol cue exposure was predominantly associated with increased activity in the ventral and dorsal striatum (Seo et al., 2011) (Fig. 21.1). During reward learning process, the VS is involved in learning reward values, whereas the dorsal

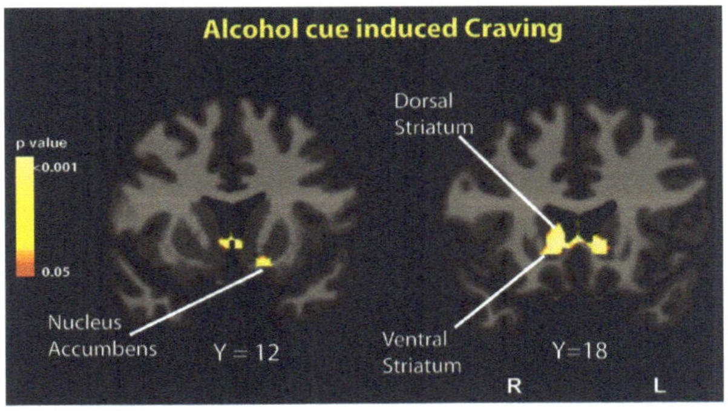

Fig. 21.1. Whole-brain voxel-based correlation between alcohol craving and neural responses to alcohol cue in healthy men. Alcohol craving was positively correlated with activity in a striatum cluster ($r = 0.74$) involving the ventral and dorsal striatum ($p < 0.05$, whole-brain family-wise error rate corrected). Coordinates are given in Montreal Neurological Institute space. R, right; L, left. (Reproduced from Seo et al., 2011, with permission of John Wiley.)

striatum modulates reward values for action (Kahnt et al., 2009). The dorsal striatum is also engaged in action initiation and in habit learning, including habits related to substance abuse (Porrino et al., 2004; Volkow et al., 2006; Kahnt et al., 2009). These results suggest that desire or craving for alcohol involves positive reinforcement mechanisms that promote pursuing rewarding stimuli in healthy individuals.

Negatively reinforced craving

With continued alcohol intake, the reward state is decreased, and other interconnected brain regions involved in stress and arousal (e.g., amygdala, insula, PFC) and habits (e.g., dorsal striatum) become sensitized and may play a role in increasing craving for alcohol along with negative emotional arousal and irritability (Weiss et al., 1992; Koob, 2009). More specifically, negatively reinforced craving can be illustrated with a state of compulsive alcohol/drug intake driven by high negative emotion and substantially decreased rewarding feeling (Koob et al., 2004; Koob, 2009). AUD patients frequently report negative affect, anxiety, and dysphoria during alcohol withdrawal (Koob and Le Moal, 1997). Especially in early abstinent AUD patients, high uncontrollable alcohol craving is often accompanied by anxiety and negative emotion and with other physiologic dysregulation such as high blood pressure, heart rate, and disruption in HPA axis system, as demonstrated by increased cortisol and adrenocorticotropic hormone (ACTH) (Cooney et al., 1997; Sinha and O'Malley, 1999; Breese et al., 2005b; Fox et al., 2007; Sinha et al., 2009). Importantly, negative emotion tied to alcohol craving strongly reinforces compulsive alcohol consumption in AUD patients (Austin and Smith, 2008). These results indicate that dyscontrol over high alcohol craving in early recovering AUD patients may be closely related to their inability to regulate emotional distress and altered physiologic states.

In terms of underlying neural regions, it has been suggested that negative reinforcement aspects of alcohol craving may involve altered response in the amygdala (Koob et al., 2004; Koob and Le Moal, 2008; Koob, 2009). Amygdala activity is elicited during emotional arousal (Fitzgerald et al., 2006) and especially reacts to emotion with aversive states such as fearful stimuli (Morris et al., 1996; Phan et al., 2002) or emotional distress (van Marle et al., 2009). Increased amygdala activity is often observed during the experience of negative emotion and hypervigilance (Etkin and Wager, 2007). The amygdala is also engaged in the modulation of stress-related bodily response via its abundant repository of corticotropin-releasing factor (CRF) receptors, cell bodies, and terminals (Merchenthaler et al., 1982; Koob, 2009). Sensitized CRF and norepinephrine systems, modulated by the amygdala, play a pivotal role in the negative reinforcement aspect of addiction (Koob, 1999, 2009). Further, the amygdala has anatomic connections with the VS (Pitkanen, 2000; Jackson and Moghaddam, 2001), suggesting that continued alcohol consumption could produce a neuroadaptive response in the limbic-striatal circuit involving these two regions.

In AUD patients, sensitized limbic-striatal circuit from repeated alcohol abuse could increase susceptibility to anxiety, negative emotion, and abnormal physiologic arousal in autonomic and HPA axis systems (Fox et al., 2008; Sinha et al., 2009). Furthermore, with prolonged alcohol abuse, overly stimulated limbic-striatal regions can overload the capacity of prefrontal cortical regulatory circuits, leading to PFC dysfunction. Neuroimaging studies on cognitive control indicate an inverse relationship between the PFC and the amygdala, and have demonstrated that increased PFC activity successfully reduces amygdala activity during the regulation of negative emotion (Ochsner et al., 2002, 2004). It follows that chronically disinhibited limbic-striatal activity could adversely impact prefrontal regulatory function, especially in the face of unexpected chronic life stressors or with repeated alcohol abuse (Kalivas, 2007; Sinha and Li, 2007; Volkow et al., 2007a). Consistent with these findings, PFC dysfunction, especially in the VmPFC and orbitofrontal cortex (OFC), was found in individuals with substance abuse as well as alcohol-dependent individuals (Volkow et al., 2007a, b), including the association of decreased striatal D_2 receptor availability with hypoactive OFC and ACC function in AUD patients (Volkow et al., 2007b). The prefrontal regulatory dysfunction could be detrimental to AUD patients, since it impairs general emotion regulation ability not only for craving but also for other emotionally challenging conditions (e.g., stress). In alcoholics, this can substantially weaken their ability for emotion regulation and coping, perpetuating compulsive alcohol seeking and poor treatment outcome, and increasing alcohol relapse risk (Li and Sinha, 2008; Seo and Sinha, 2011).

Alcohol relapse

Alcohol relapse, a return to alcohol drinking after a short or extended period of abstinence, is often preceded by uncontrollable craving, serving as a major impediment to recovery from alcoholism (Breese et al., 2011; Becker and Happel, 2012; Sinha, 2012). Over the course of withdrawal and protracted abstinence, alterations in neural circuits involved in emotion regulation and decision making (Bechara et al., 2001) play a key role in alcohol relapse behaviors, such as strong emotional urges and cravings that may bias prefrontal regulatory function, and dominate decision-making processes, leading to emotionally driven habitual behaviors. Further, this altered adaptation, especially disrupted emotion regulatory circuit, contributes to high alcohol craving, tolerance, and withdrawal and protracted abstinence symptoms (Li and Sinha, 2008; Becker and Happel, 2012). Various withdrawal symptoms followed by alcohol cessation include both psychologic (anxiety, negative emotion, emotional distress) and physiologic (autonomic and HPA axis system disruption, sleep disturbance, fatigue) symptoms (Breese et al., 2011; Becker and Happel, 2012). It has been suggested that these various symptoms result from neuroadaptations including both functional and structural changes in modulatory circuits of stress and reward response due to chronic and excessive alcohol intake (Breese et al., 2011; Buhler and Mann, 2011).

Functional alterations associated with relapse

Loci of chronic alcohol-related neuroadaptations include striatal-limbic regions as well as the PFC, which are involved in stress and reward modulation (Breese et al., 2011; Seo and Sinha, 2011). Neuroimaging studies with humans have found a relationship between altered striatal DA transmission and relapse vulnerability. In treatment-seeking AUD patients, reduced levels of DA D2 receptor in caudate did not recover after alcohol detoxification, and remained low both during early withdrawal and after protracted alcohol abstinence compared with controls (Volkow et al., 2002). Consistent with this, a [^{123}I]IBZM single-photon emission computed tomography (SPECT) study indicates that low DA levels, indicated by increased striatal DA D2 receptor density during detoxification, are found to be significantly associated with early relapse, highlighting the relationship between striatal DA activity and relapse vulnerability (Guardia et al., 2000). In another study, high alcohol craving and relapse risk were associated with alteration in the amygdala region; in 51 alcohol-dependent individuals, decreased brain volume in the amygdala was evident in relapsers compared with non-relapsers, and this was significantly associated with elevated alcohol craving and increased alcohol consumption during the follow-up (Wrase et al., 2008).

Chronic alcohol-induced, repeated sensitization in limbic-striatal regions places a significant burden on prefrontal regulatory functions and compromises its ability to control emotions, impulses, and alcohol craving and to maintain sobriety (Bechara, 2005; Everitt and Robbins, 2005; Goldstein and Volkow, 2011). In particular, alcohol-related PFC dysfunction could adversely affect higher-order executive functions, including decision making and inhibitory control, which substantially weaken self-control, the ability to overcome the urge to drink, and contribute to maladaptive patterns of multiple detoxifications and chronic relapse (Bechara and Damasio, 2002; Bechara, 2005; Goldstein and Volkow, 2011). In support of this, a recent fMRI study reported that hyperactivity in the VmPFC/ACC and VS during neutral-relaxing state is significantly associated with high alcohol craving responses to stress and alcohol exposure (Fig. 21.2). This relaxed, resting-state hyperactivity also predicted subsequent early return to both initial and heavy alcohol drinking after treatment as well as severity of alcohol abuse during the follow-up (Seo et al., 2013), highlighting the necessity for the intact VmPFC/ACC function in overcoming alcohol craving and relapse urges.

The VmPFC/ACC plays a vital role in emotional and reward regulation as well as emotional decision making (Hampton et al., 2006; Urry et al., 2006). VS activity is involved in craving and reward response in alcoholics (Grusser et al., 2004; Heinz et al., 2004). The VmPFC and VS are functionally connected through the glutamatergic pathways, which contributes to drug reinstatement (McFarland and Kalivas, 2001; Koob and Volkow, 2010). The glutamatergic pathways modulate dopaminergic activity in the VS, and upregulated glutamatergic systems have been associated with alcohol dependence (De Witte et al., 2005; Kalivas and Volkow, 2005). These results suggest that prolonged alcohol abuse sensitizes core regions of reward modulation (VmPFC and VS) during early abstinence, which disrupts regulatory control over high alcohol craving in the face of emotional challenge and results in poor decision making on relapse-related behaviors.

Structural alterations associated with relapse

Functional changes associated with alcohol craving and relapse may reflect underlying brain atrophy in AUD patients. In support of this, structural imaging studies have found significant neural substrates of alcohol relapse risk, especially in prefrontal-striatal-limbic regions. More specifically, several studies have reported an association between volume reduction in the medial PFC and OFC regions and relapse risk in AUD individuals. In a combined structural and functional MRI study with 46 detoxified alcohol-dependent patients, subsequent relapsers displayed greater brain atrophy in the bilateral OFC, the right medial PFC, and ACC relative to healthy controls as well as non-relapsers (Beck et al., 2012). In addition, the left medial PFC during alcohol cue was differentially activated in subsequently relapsed alcoholics (Beck et al., 2012).

In 75 treatment-seeking alcohol-dependent subjects, relapsers showed global reduction of cortical thickness

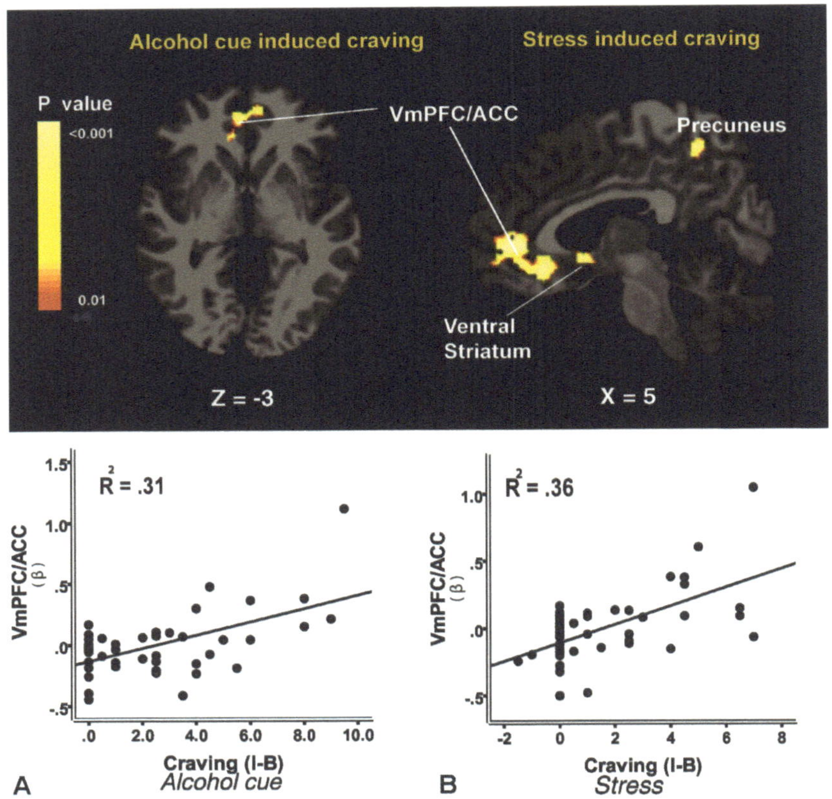

Fig. 21.2. Alcohol craving and hyperactivity during neutral-relaxing trials. During neutral-relaxing trials, increased activity in the VmPFC/ACC was associated with (**A**) alcohol cue-induced craving ($r = 0.56$) as well as (**B**) stress-induced craving ($r = 0.6$) in alcohol-dependent individuals ($p < 0.01$, whole-brain corrected). For stress-induced craving, enhanced activity in the ventral striatum ($r = 0.49$) and precuneus ($r = 0.52$) during neutral trials additionally predicted craving response. Hyperactivity in the VmPFC/ACC, ventral striatum, and precuneus was also found to be predictive of early relapse after treatment. VmPFC, ventromedial prefrontal cortex; ACC, anterior cingulate cortex. (Reproduced from Seo et al., 2013, with permission of American Medical Association.)

in the majority of brain regions compared with controls. However, compared to abstainers, relapsed alcoholics showed reduced volume in the bilateral OFC and right rostral middle and caudal middle frontal lobe (Durazzo et al., 2011). In 45 early recovering AUD patients, gray-matter volume reduction in medial PFC, and posterior parietal-occipital regions (including precuneus), was found in comparison with healthy controls. Furthermore this volume reduction was significantly associated with shorter time to both initial lapse and heavy drinking relapse after treatment (Rando et al., 2011) (Fig. 21.3).

Taken together, these structural imaging studies provide evidence of gray-matter atrophy in alcohol relapse risk, especially in the medial PFC, OFC, and ACC regions. This latent vulnerability may impact functional integrity of these regions that are involved in cognitive/behavioral control and emotion regulation that are critical in controlling intense urges to resume alcohol drinking in the face of significant risk factors (Sinha, 2008; Breese et al., 2011).

FACTORS INCREASING ALCOHOL CRAVING AND RELAPSE RISK

In chronic alcoholism, neuroadaptations in key regions of emotion regulation and decision making interact with environmental factors in modulating alcohol-seeking behaviors. These environmental risk factors (e.g., alcohol or stress) serve as triggers to initiate craving response and resumption of alcohol intake. Specifically, considerable evidence from preclinical and clinical studies on alcoholism converges on the conclusion that alcohol, alcohol-associated cues, and stress are major risk factors triggering alcohol craving and relapse (Cooney et al., 1997; Shaham et al., 2003; Sinha, 2008; Breese et al., 2011; Sinha et al., 2011). Preclinical studies have also demonstrated that these cues are driving forces to promote compulsive alcohol seeking and reinstatement of alcohol seeking in animals (Katner and Weiss, 1999; Ciccocioppo et al., 2001). For example, research showed that animals re-exposed to contextual cues previously

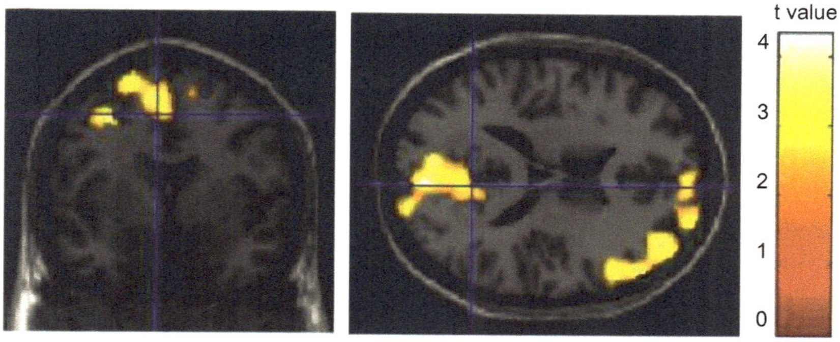

Fig. 21.3. Brain images showing gray-matter volume reduction in alcohol-dependent patients compared with healthy social drinkers (family-wise error, $p < 0.025$). Brain volume was significantly decreased in alcohol-dependent patients in areas encompassing the medial frontal cortex, anterior cingulate cortex, and the parietal-occipital sulcus, and this was also predictive of subsequent early- and heavy-drinking relapses after treatment. (Reproduced from Rando et al., 2011, with permission of the American Psychiatric Assocation.)

associated with alcohol self-administration significantly increased their responses to alcohol and modulated their responses to alcohol-conditioned cues, leading to subsequent reinstatement of alcohol seeking (Zironi et al., 2006; Chaudhri et al., 2008). Thus, neural correlates of alcohol craving and relapse can be best understood in the context of functional neuroadaptations and interaction with these environmental risk factors. Recently, significant advances in neuroimaging methods have been made in understanding functional changes associated with alcohol craving and relapse in the context of these risk factors, especially in the prefrontal-striatal-limbic circuit (Buhler and Mann, 2011).

Alcohol or alcohol cue

Alcohol itself or alcohol-associated cues (e.g., visual, olfactory, auditory) are risk factors for alcohol craving and relapse (Anton, 1999; Fox et al., 2007; Breese et al., 2011). Alcohol-related stimuli elicit craving response and increase desire to resume drinking. They also trigger conditioned responses and retrieval of memories pertaining to alcohol-related contexts (Ludwig and Stark, 1974; Ludwig et al., 1974). For this reason, AUD individuals are found to be more strongly responsive to alcohol-provoking stimuli compared with controls (Myrick et al., 2004). Neuroimaging studies report that alcohol cue compared to neutral cues elicited activation in the prefrontal-limbic-striatal circuit in AUD individuals, including the OFC, medial PFC, ACC, striatum and amygdala (for review, see Buhler and Mann, 2011). However, significant associations with alcohol craving during alcohol or alcohol cue exposure have been dominantly found in the striatum, with some additional associations found in the OFC and ACC.

For example, a SPECT study reported that increased blood flow in the right caudate nucleus was significantly associated with high alcohol craving in 9 alcohol-dependent subjects who received a drinking glass containing a low dose of alcoholic beverage (Modell and Mountz, 1995). During the viewing of alcohol cue, alcohol craving in AUD subjects has been positively correlated with brain activation in the left VS, ACC, and left OFC, but not in control social drinkers using fMRI (Myrick et al., 2004). More specifically, DA dysfunction in the VS has been associated with increased alcohol craving and salient response to alcohol-related stimuli. Using multimodal imaging, increased activity in the medial PFC and striatum was found during the viewing of alcohol-related stimuli, and alcohol craving was associated with reduced availability of DA D2-like receptors in the VS in detoxified, alcohol-dependent men (Heinz et al., 2004). The DA dysfunction is also consistent with previous findings showing reduced DA D2 receptors in striatal regions in alcoholics relative to controls (Volkow et al., 1996). Further, neuroimaging evidence also suggests that striatal activation in alcohol-dependent individuals may be more biased towards alcohol-related cue, rather than generic reward processing. In an fMRI study with 16 detoxified alcoholics and 16 controls, alcohol craving was associated with increased activity during alcohol cue, but decreased activity in the VS while expecting monetary reward only in alcoholics (Wrase et al., 2007).

Alcohol cue-induced functional changes have also been associated with early relapse in alcoholics. During the presentation of alcohol-related visual stimuli, activity in the ACC, medial PFC, and striatum was more pronounced in subsequently relapsed alcoholics. Especially the intensity of medial PFC activity was associated with amount of subsequent alcohol consumption in 10 recovering AUD individuals (Grusser et al., 2004). In addition, strong activity in the ventral putamen was elicited in 4 treatment-seeking AUD patients, who relapsed within

3 months during the viewing of alcohol-related visual cues (Braus et al., 2001).

In relation to relapse, some studies with varying emotion or memory paradigms also report associations of altered activity in the striatum and medial PFC with alcohol relapse. For example, during the viewing of emotionally positive vs neutral pictures, detoxified AUD patients displayed greater activation in the VS and thalamus, and this activation was associated with subsequent relapse (Heinz et al., 2007). During a working-memory task, a SPECT study reports that reduced cerebral blood flow in the medial frontal lobe was associated with impaired working-memory performance as well as early alcohol relapse (Noel et al., 2002).

Taken together, studies using alcohol-related stimuli have dominantly reported that altered activity in brain regions involving the VS, OFC, ACC, and medial PFC are significantly associated with alcohol craving and relapse risk, implicating a critical alcohol-related pathophysiology in this circuit that affects craving and relapse vulnerability in AUD individuals.

Stress

Stressful environments often trigger alcohol-seeking behaviors and lead to excessive consumption of alcohol use (Sillaber and Henniger, 2004; Breese et al., 2005b; Koob, 2009). A close link between stress and alcohol abuse has been found in both animal and human studies (Sinha, 2001; Breese et al., 2011). Preclinical studies showed that peer-reared monkeys consume substantially larger amounts of alcohol during adulthood compared with mother-reared monkeys (Higley et al., 1991). Following the September 11 terrorist attack, a substantial increase in substance abuse (alcohol, cigarettes, and marijuana) was reported among New York City residents (Vlahov et al., 2004). In alcohol-dependent individuals, stress-related emotional problems and negative emotional arousal have been associated with high levels of craving and desire to use alcohol (Cooney et al., 1997; Fox et al., 2007; Sinha et al., 2009).

As a neurobiologic explanation underlying this link, a shared neural mechanism between stress and reward-related behaviors has been proposed in the prefrontal-striatal-limbic circuit (Sinha and Li, 2007; Seo and Sinha, 2011). Repeated sensitization, either by chronic stress or alcohol/drug abuse, can lead to alterations in this circuit and increase susceptibility to craving and relapse (Sinha, 2001; Weiss and Porrino, 2002; Koob et al., 2004; Li and Sinha, 2008). Due to this shared loci of neuroadaptations, it is posited that a reciprocal relation between stress and addictive behaviors may exist; chronic stress can contribute to alcohol-seeking behaviors, and alcohol dependence can increase stress sensitivity and vulnerability to stress-related illnesses (Seo and Sinha, 2011). In order to understand the role of stress in alcohol craving and relapse risk, this section will review neural substrates underlying the interrelationship between stress and alcoholism.

Stress and the development of alcoholism

Stress triggers alcohol craving and increases susceptibility to early relapse (Sinha, 2001; Sinha et al., 2005). Preclinical studies showed that, in stressful situations, glucocorticoid hormones regulate DA release in the striatum (for review, see Marinelli and Piazza, 2002). This suggests that neuroadaptations in striatal reward pathways modulated by the stress system could be associated with vulnerability to addiction under chronic stress. Consistent with this, prenatally stressed male mice displayed excessive alcohol consumption along with altered patterns of DA and glutamate transmission in the limbic forebrain after reaching adulthood (Campbell et al., 2009). Human neuroimaging studies also provided evidence for adverse influence of acute and chronic stress in prefrontal regulatory function, including impaired attentional control (Liston et al., 2009). It has been suggested that, under stress, habitual and impulsive emotional response in the amygdala and striatum governs behaviors due to decreased PFC regulatory function (Arnsten, 1998; Sinha, 2001). Especially, emotional or physiologic life stressors (e.g., trauma, chronic pain) could oversensitize the reactivity in limbic-striatal regions, which could intensify negative emotional state and increase vulnerability to alcoholism for relief. For example, high rates of comorbid addiction, including alcoholism as well as altered prefrontal-striatal-limbic function, have been found in individuals experiencing high levels of stress, including posttraumatic stress disorder and chronic pain syndromes (for review, see Seo and Sinha, 2011). It follows that chronic and repeated stress exposure could overstimulate the prefrontal-limbic-striatal circuit, which may further exacerbate emotional distress and drive alcohol-seeking behaviors, likely due to weakened PFC control over both stress and reward-related behaviors.

Stress sensitivity, alcohol craving, and relapse

Chronic alcohol abuse generates a number of co-occurring symptoms, including emotional distress, negative emotions, sleep and appetite changes, and irritability (Sinha, 2001; Sinha et al., 2009). Furthermore, multiple studies show that AUD patients display significantly high levels of stress-induced alcohol craving and anxiety compared to controls (Fox et al., 2007, 2008; Sinha et al., 2009, 2011). In AUD patients, chronic alcohol-related neuroadaptations in peripheral and central stress systems significantly elevate stress-related anxiety and craving via disruption in autonomic and HPA axis functioning. For example, repeated alcohol use and acute alcohol withdrawal have been associated with elevated basal stress hormones, including CRF, norepinephrine, and cortisol (for review, see Breese et al., 2011). In terms of underlying brain system, preclinical studies showed alcohol-related alterations in neural circuits of stress regulation in the prefrontal-striatal-limbic-circuit (Di Chiara, 1997; Koob et al., 2004; Cleck and Blendy, 2008; Koob, 2009). In particular, it has been shown that a neural circuit connecting the amygdala, VS, and ACC plays a crucial role in cue and stress-related drug reinstatement in rats (for review, see Kalivas and McFarland, 2003).

In humans, laboratory and neuroimaging studies have found significant associations among high alcohol craving, early relapse risk, and altered response in stress-related neural pathways, including autonomic and HPA axis systems as well as brain regions of stress regulation. In a laboratory study, early recovering AUD patients showed high levels of basal heart rate and salivary cortisol compared to healthy controls. However, during stress and alcohol cue exposure, AUD patients displayed decreased stress-induced heart rate and cortisol response accompanied by high alcohol craving and emotional distress (Sinha et al., 2009). In a clinical outcome study with 93 early abstinent AUD patients, stress-induced alcohol craving was associated with poor clinical outcomes and early relapse. Additionally, greater adrenal sensitivity (cortisol-to-ACTH ratio) during the neutral condition was also associated with early relapse (Sinha et al., 2011). This is consistent with other studies reporting altered stress response and HPA axis function associated with alcoholism and relapse outcomes (Junghanns et al., 2003; Breese et al., 2005a).

In summary, these studies suggest that the disrupted stress system contributes to alcohol craving and relapse susceptibility in AUD patients. More specifically, altered autonomic and HPA axis systems in patients may reflect underlying dysfunction in key brain regions modulating autonomic and endocrine systems in early recovering alcoholics. For example, the VmPFC is an important regulatory region for endogenous modulation of autonomic and HPA axis systems during emotionally arousing or stressful situations (Figueiredo et al., 2003; Radley et al., 2006), suggesting potential VmPFC dysfunction in response to stress in AUD patients.

Consistent with this, an fMRI study demonstrated that early abstinent AUD patients displayed hypoactive VmPFC response to stress and alcohol cue compared with healthy controls. Furthermore, in this study, decreased VmPFC/ACC and insula responses to stress were associated with relapse severity, measured by the duration of alcohol use during follow-up (Seo et al., 2013) (Fig. 21.4).

The insula is involved in somatosensory integration during emotional and stress arousal (Damasio, 1996; Li and Sinha, 2008). In addition to autonomic and HPA axis modulation, the VmPFC is engaged in emotional decision making, which is crucial for relapse-related decision and behaviors (Bechara, 2005). In other neuroimaging studies, decreased VmPFC activity has been associated with difficulties with emotion regulation (van Reekum et al., 2007) as well as with decision-making impairments in addicted individuals, including alcoholics (for review, see Bechara, 2005). Collectively, these studies suggest a critical role of the VmPFC dysfunction in response to stress (emotion regulation and decision making) and in increasing alcohol relapse risk.

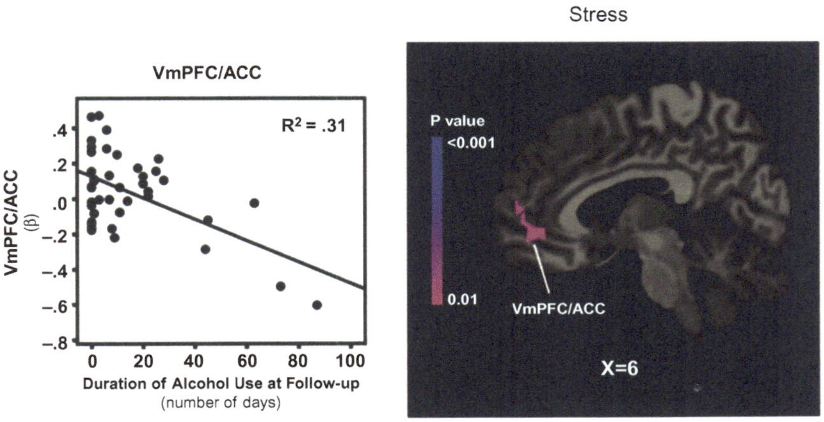

Fig. 21.4. Association of stress-induced brain activity with alcohol relapse severity. In 45 alcohol-dependent individuals, decreased VmPFC/ACC activity during stress exposure ($r = -0.56$, $R^2 = 0.31$) predicted the number of days of alcohol used during the subsequent 90-day follow-up period ($p < 0.01$, whole-brain corrected). VmPFC, ventromedial prefrontal cortex; ACC, anterior cingulate cortex; R, right; L, left. (Reproduced from Seo et al., 2013, with permission of American Medical Association.)

To summarize, endocrine and neuroimaging studies point to a role of altered peripheral system in stress and its interactions with the CNS reactivity in understanding alcohol craving and relapse. It is likely that altered function in core stress regulatory regions (e.g., VmPFC) may lead to disrupted autonomic and HPA axis response during stress, which could increase emotional distress, stress-induced alcohol craving, and relapse susceptibility in AUD patients.

CONCLUSIONS

This chapter reviews relevant structural and functional neuroimaging studies in order to understand neural substrates underlying alcohol craving and relapse. Because alcohol relapse is often preceded by high alcohol craving, common brain regions tend to be engaged in these emotional (craving) and behavioral (relapse) aspects of alcohol-related pathology. Specifically, studies report neuroadaptations in the prefrontal-striatal-limbic circuit play a role in both craving and relapse behaviors in individuals with chronic alcoholism. Review of structural imaging research also supports this idea and indicates that reduced brain volume in this circuit (i.e., the medial PFC, OFC, precuneus, and amydala) is related to craving and alcohol relapse.

In terms of functional changes, altered activity in regions of the medial PFC, ACC, and VS appears to play a significant role in explaining high alcohol craving and an early return to alcohol drinking after a period of abstinence. During alcohol or alcohol cue exposure, functional imaging studies showed significant association of the medial PFC, OFC, ACC, and VS with craving and relapse. During stress, the VmPFC, ACC, and insula regions were associated with these behaviors. In addition, hyperactivity in the VmPFC, ACC, and VS during the neural-relaxing condition was associated with alcohol craving during stress and alcohol cue as well as with subsequent relapse (Seo et al., 2013). Altered DA transmission, especially in the VS area, was also found to be an important correlate, such that low DA levels were associated with both alcohol craving (Heinz et al., 2004) and early relapse (Guardia et al., 2000), demonstrating the significant involvement of striatal DA transmission in craving, reward, and relapse behaviors in chronic alcoholism.

Across different paradigms, a number of studies report altered functional reponse in the medial PFC, OFC, ACC, and the VS in explaining alcohol craving and relapse-related behaviors (Fig. 21.5). The medial PFC and OFC are involved in emotional aspects of decision making, with their impairments often observed in addiction patients (Bechara et al., 2001). The OFC, especially the medial region, is closely connected with the

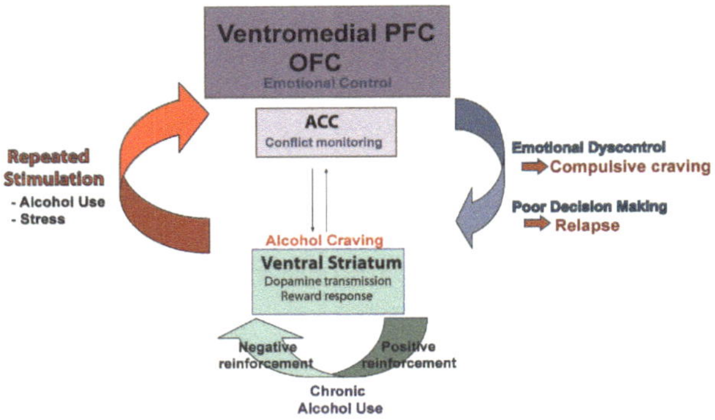

Fig. 21.5. A schematic diagram of a neural circuit underlying alcohol craving and relapse. Alcohol intake initially provides positive reinforcement, eliciting pleasant feelings via dopamine release in the ventral striatum. However, with continued alcohol use, altered ventral striatal function triggers negative reinforcement, characterized by compulsive alcohol craving accompanied by negative emotion and distress. Chronic alcohol abuse and stress also overstimulate the medial prefrontal cortex (PFC)–striatal circuit to progressively weaken prefrontal regulatory control, leading to craving, compulsive alcohol seeking, poor decision making, and early relapse. OFC, orbitofrontal cortex; ACC, anterior cingulate cortex.

amygdala (Carmichael and Price, 1995). It transfers cortical signals to the hypothalamus and also connects functionally with the VS, which permits regulation of reward response and behaviors (Nakano et al., 1999; Ongur and Price, 2000). The medial OFC also encodes and modulates emotional response from the amygdala and VS (Rolls, 2000; O'Doherty et al., 2001; Schoenbaum et al., 2007). The ACC has been associated with conflict monitoring for both cognitive (Botvinick et al., 1999; MacDonald et al., 2000) and emotional (Etkin et al., 2006) stimuli. Especially, the rostral ACC has close functional and anatomic connection with the medial OFC (Carmichael and Price, 1995), and its activity has been observed during emotional aspects of conflict monitoring and control (Mohanty et al., 2007).

Taken together, given the significant role of the medial frontal-limbic-striatal circuit in the emotional modulation, it can be inferred that the core circuit of reward and stress modulation (mPFC, OFC, ACC, amygdala, VS) is impaired in AUD individuals, and sensitization of this circuit from continued alcohol use could further contribute to high alcohol craving and early relapse in chronic alcoholism.

These results point to some clinical implications in that this core circuit could be explored further to develop effective treatments for reducing alcohol craving and to improve clinical outcomes. For example, a study with deep brain stimulation on the VS reported positive outcome in AUD individuals (Muller et al., 2009), and mindfulness training was also found to be beneficial for improving prefrontal function (Brewer et al., 2011). It suggests that recovering brain function in the prefrontal-striatal-limbic circuit can be a good treatment target for pathology related to alcohol craving and relapse. Thus, future research efforts should be made in effective utilization of neuroimaging techniques in examining treatment outcome for individuals with chronic alcoholism. In particular, a carefully designed clinical outcome study combined with pharmacologic imaging would be beneficial for early identification of individuals at risk for relapse as well as monitoring improvement of brain response in the treatment for AUD individuals suffering from craving and recurrent relapse behaviors.

References

Adinoff B, Iranmanesh A, Veldhuis J et al. (1998). Disturbances of the stress response: the role of the HPA axis during alcohol withdrawal and abstinence. Alcohol Health Res World 22: 67–72.

Anton RF (1999). What is craving? Models and implications for treatment. Alcohol Res Health 23: 165–173.

Arias-Carrion O, Poppel E (2007). Dopamine, learning, and reward-seeking behavior. Acta Neurobiol Exp (Wars) 67: 481–488.

Arnsten AFT (1998). The biology of being frazzled. Science 280: 1711–1712.

Austin JL, Smith JE (2008). Drinking for negative reinforcement: the semantic priming of alcohol concepts. Addict Behav 33: 1572–1580.

Ballenger JC, Post RM (1978). Kindling as a model for alcohol withdrawal syndromes. Br J Psychiatry 133: 1–14.

Bauer LO (2001). Predicting relapse to alcohol and drug abuse via quantitative electroencephalography. Neuropsychopharmacology 25: 332–340.

Bechara A (2005). Decision making, impulse control and loss of willpower to resist drugs: a neurocognitive perspective. Nat Neurosci 8: 1458–1463.

Bechara A, Damasio H (2002). Decision-making and addiction (part I): impaired activation of somatic states in substance dependent individuals when pondering decisions with negative future consequences. Neuropsychologia 40: 1675–1689.

Bechara A, Dolan S, Denburg N et al. (2001). Decision-making deficits, linked to a dysfunctional ventromedial prefrontal cortex, revealed in alcohol and stimulant abusers. Neuropsychologia 39: 376–389.

Beck A, Wustenberg T, Genauck A et al. (2012). Effect of brain structure, brain function, and brain connectivity on relapse in alcohol-dependent patients. Arch Gen Psychiatry 69: 842–852.

Becker HC (1998). Kindling in alcohol withdrawal. Alcohol Health Res World 22: 25–33.

Becker HC, Happel KI (2012). Effects of alcohol dependence and withdrawal on stress responsiveness and alcohol consumption. Alcohol Res Curr Rev 34: 448–458.

Bottlender M, Soyka M (2004). Impact of craving on alcohol relapse during, and 12 months following, outpatient treatment. Alcohol Alcohol 39: 357–361.

Botvinick M, Nystrom LE, Fissell K et al. (1999). Conflict monitoring versus selection-for-action in anterior cingulate cortex. Nature 402: 179–181.

Braus DF, Wrase J, Grusser S et al. (2001). Alcohol-associated stimuli activate the ventral striatum in abstinent alcoholics. J Neural Transm 108: 887–894.

Breese GR, Chu K, Dayas CV et al. (2005a). Stress enhancement of craving during sobriety: a risk for relapse. Alcohol Clin Exp Res 29: 185–195.

Breese GR, Overstreet DH, Knapp DJ (2005b). Conceptual framework for the etiology of alcoholism: a "kindling"/stress hypothesis. Psychopharmacology (Berl) 178: 367–380.

Breese GR, Sinha R, Heilig M (2011). Chronic alcohol neuroadaptation and stress contribute to susceptibility for alcohol craving and relapse. Pharmacol Ther 129: 149–171.

Brewer JA, Worhunsky PD, Gray JR et al. (2011). Meditation experience is associated with differences in default mode network activity and connectivity. Proc Natl Acad Sci U S A 108: 20254–20259.

Buhler M, Mann K (2011). Alcohol and the human brain: a systematic review of different neuroimaging methods. Alcohol Clin Exp Res 35: 1771–1793.

Campbell JC, Szumlinski KK, Kippin TE (2009). Contribution of early environmental stress to alcoholism vulnerability. Alcohol 43: 547–554.

Carmichael ST, Price JL (1995). Limbic connections of the orbital and medial prefrontal cortex in macaque monkeys. J Comp Neurol 363: 615–641.

Chaudhri N, Sahuque LL, Janak PH (2008). Context-induced relapse of conditioned behavioral responding to ethanol cues in rats. Biol Psychiatry 64: 203–210.

Ciccocioppo R, Angeletti S, Weiss F (2001). Long-lasting resistance to extinction of response reinstatement induced by ethanol-related stimuli: role of genetic ethanol preference. Alcohol Clin Exp Res 25: 1414–1419.

Cleck JN, Blendy JA (2008). Making a bad thing worse: adverse effects of stress on drug addiction. J Clin Invest 118: 454–461.

Cooney NL, Litt MD, Morse PA et al. (1997). Alcohol cue reactivity, negative-mood reactivity, and relapse in treated alcoholic men. J Abnorm Psychol 106: 243–250.

Damasio AR (1996). The somatic marker hypothesis and the possible functions of the prefrontal cortex. Philos Trans R Soc Lond B Biol Sci 351: 1413–1420.

Dawson DA, Grant BF, Stinson FS et al. (2005). Psychopathology associated with drinking and alcohol use disorders in the college and general adult populations. Drug Alcohol Depend 77: 139–150.

De Witte P, Littleton J, Parot P et al. (2005). Neuroprotective and abstinence-promoting effects of acamprosate: elucidating the mechanism of action. CNS Drugs 19: 517–537.

Di Chiara G (1997). Alcohol and dopamine. Alcohol Health Res World 21: 108–114.

Durazzo TC, Tosun D, Buckley S et al. (2011). Cortical thickness, surface area, and volume of the brain reward system in alcohol dependence: relationships to relapse and extended abstinence. Alcohol Clin Exp Res 35: 1187–1200.

Etkin A, Wager TD (2007). Functional neuroimaging of anxiety: a meta-analysis of emotional processing in PTSD, social anxiety disorder, and specific phobia. Am J Psychiatry 164: 1476–1488.

Etkin A, Egner T, Peraza DM et al. (2006). Resolving emotional conflict: a role for the rostral anterior cingulate cortex in modulating activity in the amygdala. Neuron 51: 871–882.

Everitt BJ, Robbins TW (2000). Second-order schedules of drug reinforcement in rats and monkeys: measurement of reinforcing efficacy and drug-seeking behaviour. Psychopharmacology (Berl) 153: 17–30.

Everitt BJ, Robbins TW (2005). Neural systems of reinforcement for drug addiction: from actions to habits to compulsion. Nat Neurosci 8: 1481–1489.

Figueiredo HF, Bruestle A, Bodie B et al. (2003). The medial prefrontal cortex differentially regulates stress-induced c-fos expression in the forebrain depending on type of stressor. Eur J Neurosci 18: 2357–2364.

Fitzgerald DA, Angstadt M, Jelsone LM et al. (2006). Beyond threat: amygdala reactivity across multiple expressions of facial affect. Neuroimage 30: 1441–1448.

Fox HC, Bergquist KL, Hong KI et al. (2007). Stress-induced and alcohol cue-induced craving in recently abstinent alcohol-dependent individuals. Alcohol Clin Exp Res 31: 395–403.

Fox HC, Hong KI, Siedlarz K et al. (2008). Enhanced sensitivity to stress and drug/alcohol craving in abstinent cocaine-dependent individuals compared to social drinkers. Neuropsychopharmacology 33: 796–805.

Goldstein RZ, Volkow ND (2011). Dysfunction of the prefrontal cortex in addiction: neuroimaging findings and clinical implications. Nat Rev Neurosci 12: 652–669.

Goldstein RZ, Craig AD, Bechara A et al. (2009). The neurocircuitry of impaired insight in drug addiction. Trends Cogn Sci 13: 372–380.

Greeley JD, Swift W, Prescott J et al. (1993). Reactivity to alcohol-related cues in heavy and light drinkers. J Stud Alcohol 54: 359–368.

Grusser SM, Wrase J, Klein S et al. (2004). Cue-induced activation of the striatum and medial prefrontal cortex is associated with subsequent relapse in abstinent alcoholics. Psychopharmacology (Berl) 175: 296–302.

Guardia J, Catafau AM, Batlle F et al. (2000). Striatal dopaminergic D(2) receptor density measured by [(123)I]iodobenzamide SPECT in the prediction of treatment outcome of alcohol-dependent patients. Am J Psychiatry 157: 127–129.

Hampton AN, Bossaerts P, O'Doherty JP (2006). The role of the ventromedial prefrontal cortex in abstract state-based inference during decision making in humans. J Neurosci 26: 8360–8367.

Hare TA, Tottenham N, Davidson MC et al. (2005). Contributions of amygdala and striatal activity in emotion regulation. Biol Psychiatry 57: 624–632.

Heinz A, Siessmeier T, Wrase J et al. (2004). Correlation between dopamine D(2) receptors in the ventral striatum and central processing of alcohol cues and craving. Am J Psychiatry 161: 1783–1789.

Heinz A, Wrase J, Kahnt T et al. (2007). Brain activation elicited by affectively positive stimuli is associated with a lower risk of relapse in detoxified alcoholic subjects. Alcohol Clin Exp Res 31: 1138–1147.

Hernandez L, Hoebel BG (1988). Food reward and cocaine increase extracellular dopamine in the nucleus accumbens as measured by microdialysis. Life Sci 42: 1705–1712.

Higley JD, Hasert MF, Suomi SJ et al. (1991). Nonhuman primate model of alcohol abuse: effects of early experience, personality, and stress on alcohol consumption. Proc Natl Acad Sci U S A 88: 7261–7265.

Hull EM, Eaton RC, Moses J et al. (1993). Copulation increases dopamine activity in the medial preoptic area of male rats. Life Sci 52: 935–940.

Ikemoto S, Panksepp J (1999). The role of nucleus accumbens dopamine in motivated behavior: a unifying interpretation with special reference to reward-seeking. Brain Res Brain Res Rev 31: 6–41.

Jackson ME, Moghaddam B (2001). Amygdala regulation of nucleus accumbens dopamine output is governed by the prefrontal cortex. J Neurosci 21: 676–681.

Junghanns K, Backhaus J, Tietz U et al. (2003). Impaired serum cortisol stress response is a predictor of early relapse. Alcohol Alcohol 38: 189–193.

Kahnt T, Park SQ, Cohen MX et al. (2009). Dorsal striatal-midbrain connectivity in humans predicts how reinforcements are used to guide decisions. J Cogn Neurosci 21: 1332–1345.

Kalivas PW (2007). Neurobiology of cocaine addiction: implications for new pharmacotherapy. Am J Addict 16: 71–78.

Kalivas PW, McFarland K (2003). Brain circuitry and the reinstatement of cocaine-seeking behavior. Psychopharmacology (Berl) 168: 44–56.

Kalivas PW, Volkow ND (2005). The neural basis of addiction: a pathology of motivation and choice. Am J Psychiatry 162: 1403–1413.

Katner SN, Weiss F (1999). Ethanol-associated olfactory stimuli reinstate ethanol-seeking behavior after extinction and modify extracellular dopamine levels in the nucleus accumbens. Alcohol Clin Exp Res 23: 1751–1760.

Knutson B, Adams CM, Fong GW et al. (2001). Anticipation of increasing monetary reward selectively recruits nucleus accumbens. J Neurosci 21: RC159.

Koob GF (1999). Corticotropin-releasing factor, norepinephrine, and stress. Biol Psychiatry 46: 1167–1180.

Koob GF (2009). Dynamics of neuronal circuits in addiction: reward, antireward, and emotional memory. Pharmacopsychiatry 42 (Suppl 1): S32–S41.

Koob GF, Le Moal M (1997). Drug abuse: hedonic homeostatic dysregulation. Science 278: 52–58.

Koob GF, Le Moal M (2008). Review. Neurobiological mechanisms for opponent motivational processes in addiction. Philos Trans R Soc Lond B Biol Sci 363: 3113–3123.

Koob GF, Volkow ND (2010). Neurocircuitry of addiction. Neuropsychopharmacology 35: 217–238.

Koob GF, Ahmed SH, Boutrel B et al. (2004). Neurobiological mechanisms in the transition from drug use to drug dependence. Neurosci Biobehav Rev 27: 739–749.

Li CS, Sinha R (2008). Inhibitory control and emotional stress regulation: neuroimaging evidence for frontal-limbic dysfunction in psycho-stimulant addiction. Neurosci Biobehav Rev 32: 581–597.

Liston C, McEwen BS, Casey BJ (2009). Psychosocial stress reversibly disrupts prefrontal processing and attentional control. Proc Natl Acad Sci U S A 106: 912–917.

Ludwig AM, Stark LH (1974). Alcohol craving. Subjective and situational aspects. Q J Stud Alcohol 35: 899–905.

Ludwig AM, Wikler A, Stark LH (1974). The first drink: psychobiological aspects of craving. Arch Gen Psychiatry 30: 539–547.

MacDonald 3rd AW, Cohen JD, Stenger VA et al. (2000). Dissociating the role of the dorsolateral prefrontal and anterior cingulate cortex in cognitive control. Science 288: 1835–1838.

Marinelli M, Piazza PV (2002). Interaction between glucocorticoid hormones, stress and psychostimulant drugs. Eur J Neurosci 16: 387–394.

Martinez D, Gil R, Slifstein M et al. (2005). Alcohol dependence is associated with blunted dopamine transmission in the ventral striatum. Biol Psychiatry 58: 779–786.

McFarland K, Kalivas PW (2001). The circuitry mediating cocaine-induced reinstatement of drug-seeking behavior. J Neurosci 21: 8655–8663.

Merchenthaler I, Vigh S, Petrusz P et al. (1982). Immunocytochemical localization of corticotropin-releasing factor (CRF) in the rat brain. Am J Anat 165: 385–396.

Modell JG, Mountz JM (1995). Focal cerebral blood flow change during craving for alcohol measured by SPECT. J Neuropsychiatry Clin Neurosci 7: 15–22.

Mohanty A, Engels AS, Herrington JD et al. (2007). Differential engagement of anterior cingulate cortex subdivisions for cognitive and emotional function. Psychophysiology 44: 343–351.

Morris JS, Frith CD, Perrett DI et al. (1996). A differential neural response in the human amygdala to fearful and happy facial expressions. Nature 383: 812–815.

Muller UJ, Sturm V, Voges J et al. (2009). Successful treatment of chronic resistant alcoholism by deep brain stimulation of nucleus accumbens: first experience with three cases. Pharmacopsychiatry 42: 288–291.

Myrick H, Anton RF, Li X et al. (2004). Differential brain activity in alcoholics and social drinkers to alcohol cues: relationship to craving. Neuropsychopharmacology 29: 393–402.

Nakano K, Kayahara T, Chiba T (1999). Afferent connections to the ventral striatum from the medial prefrontal cortex (area 25) and the thalamic nuclei in the macaque monkey. Ann N Y Acad Sci 877: 667–670.

Noel X, Sferrazza R, Van Der Linden M et al. (2002). Contribution of frontal cerebral blood flow measured by (99 m)Tc-Bicisate spect and executive function deficits to predicting treatment outcome in alcohol-dependent patients. Alcohol Alcohol 37: 347–354.

Ochsner KN, Bunge SA, Gross JJ et al. (2002). Rethinking feelings: an FMRI study of the cognitive regulation of emotion. J Cogn Neurosci 14: 1215–1229.

Ochsner KN, Ray RD, Cooper JC et al. (2004). For better or for worse: neural systems supporting the cognitive down- and up-regulation of negative emotion. Neuroimage 23: 483–499.

O'Doherty J, Kringelbach ML, Rolls ET et al. (2001). Abstract reward and punishment representations in the human orbitofrontal cortex. Nat Neurosci 4: 95–102.

Ongur D, Price JL (2000). The organization of networks within the orbital and medial prefrontal cortex of rats, monkeys and humans. Cereb Cortex 10: 206–219.

Phan KL, Wager T, Taylor SF et al. (2002). Functional neuroanatomy of emotion: a meta-analysis of emotion activation studies in PET and fMRI. Neuroimage 16: 331–348.

Pitkanen A (2000). Connectivity of the rat amygdaloid complex. In: J Aggleton (Ed.), The Amygdala: A Functional Analysis. Oxford University, New York.

Porjesz B, Begleiter H (2003). Alcoholism and human electrophysiology. Alcohol Res Health 27: 153–160.

Porrino LJ, Lyons D, Smith HR et al. (2004). Cocaine self-administration produces a progressive involvement of limbic, association, and sensorimotor striatal domains. J Neurosci 24: 3554–3562.

Radley J, Arias C, Sawchenkole PE (2006). Regional differentiation of the medial prefrontal cortex in regulating adaptive responses to acute emotional stress. J Neurosci 26: 12967–12976.

Ranaldi R, Pocock D, Zereik R et al. (1999). Dopamine fluctuations in the nucleus accumbens during maintenance, extinction, and reinstatement of intravenous D-amphetamine self-administration. J Neurosci 19: 4102–4109.

Rando K, Hong KI, Bhagwagar Z et al. (2011). Association of frontal and posterior cortical gray matter volume with time to alcohol relapse: a prospective study. Am J Psychiatry 168: 183–192.

Rehm J, Mathers C, Popova S et al. (2009). Global burden of disease and injury and economic cost attributable to alcohol use and alcohol-use disorders. Lancet 373: 2223–2233.

Rewal M, Jurd R, Gill TM et al. (2009). Alpha4-containing GABAA receptors in the nucleus accumbens mediate moderate intake of alcohol. J Neurosci 29: 543–549.

Richardson HN, Lee SY, O'Dell LE et al. (2008). Alcohol self-administration acutely stimulates the hypothalamic-pituitary-adrenal axis, but alcohol dependence leads to a dampened neuroendocrine state. Eur J Neurosci 28: 1641–1653.

Robinson TE, Berridge KC (1993). The neural basis of drug craving: an incentive-sensitization theory of addiction. Brain Res 18: 247–291.

Rolls ET (2000). The orbitofrontal cortex and reward. Cereb Cortex 10: 284–294.

Room R, Babor T, Rehm J (2005). Alcohol and public health. Lancet 365: 519–530.

Schoenbaum G, Saddoris MP, Stalnaker TA (2007). Reconciling the roles of orbitofrontal cortex in reversal learning and the encoding of outcome expectancies. Ann N Y Acad Sci 1121: 320–335.

Schultz W (1998). Predictive reward signal of dopamine neurons. J Neurophysiol 80: 1–27.

Schultz W, Dayan P, Montague PR (1997). A neural substrate of prediction and reward. Science 275: 1593–1599.

Seo D, Sinha R (2011). Neural mechanisms of stress and addiction. In: B Adinoff, EA Stein (Eds.), Neuroimaging in the Addictions. John Wiley, Chichester, pp. 211–233.

Seo D, Jia Z, Lacadie CM et al. (2011). Sex differences in neural responses to stress and alcohol context cues. Hum Brain Mapp 32: 1998–2013.

Seo D, Lacadie CM, Tuit K et al. (2013). Disrupted ventromedial prefrontal function, alcohol craving, and subsequent relapse risk. JAMA Psychiatry 70: 727–739.

Shaham Y, Shalev U, Lu L et al. (2003). The reinstatement model of drug relapse: history, methodology and major findings. Psychopharmacology (Berl) 168: 3–20.

Sillaber I, Henniger MS (2004). Stress and alcohol drinking. Ann Med 36: 596–605.

Sinha R (2001). How does stress increase risk of drug abuse and relapse? Psychopharmacology (Berl) 158: 343–359.

Sinha R (2008). Chronic stress, drug use, and vulnerability to addiction. Ann N Y Acad Sci 1141: 105–130.

Sinha R (2009). Modeling stress and drug craving in the laboratory: implications for addiction treatment development. Addict Biol 14: 84–98.

Sinha R (2012). How does stress lead to risk of alcohol relapse? Alcohol Res Curr Rev 34: 432–440.

Sinha R (2013). The clinical neurobiology of drug craving. Curr Opin Neurobiol 23: 649–654.

Sinha R, Li CS (2007). Imaging stress- and cue-induced drug and alcohol craving: association with relapse and clinical implications. Drug Alcohol Rev 26: 25–31.

Sinha R, O'Malley SS (1999). Craving for alcohol: findings from the clinic and the laboratory. Alcohol Alcohol 34: 223–230.

Sinha R, Lacadie C, Skudlarski P et al. (2005). Neural activity associated with stress-induced cocaine craving: a functional magnetic resonance imaging study. Psychopharmacology (Berl) 183: 171–180.

Sinha R, Fox HC, Hong KA et al. (2009). Enhanced negative emotion and alcohol craving, and altered physiological responses following stress and cue exposure in alcohol dependent individuals. Neuropsychopharmacology 34: 1198–1208.

Sinha R, Fox HC, Hong KI et al. (2011). Effects of adrenal sensitivity, stress- and cue-induced craving, and anxiety on subsequent alcohol relapse and treatment outcomes. Arch Gen Psychiatry 68: 942–952.

Suri RE (2002). TD models of reward predictive responses in dopamine neurons. Neural Netw 15: 523–533.

Thayer JF, Hall M, Sollers 3rd JJ et al. (2006). Alcohol use, urinary cortisol, and heart rate variability in apparently healthy men: evidence for impaired inhibitory control of the HPA axis in heavy drinkers. Int J Psychophysiol 59: 244–250.

Tiffany ST (1990). A cognitive model of drug urges and drug-use behavior: role of automatic and nonautomatic processes. Psychol Rev 97: 147–168.

Urry HL, van Reekum CM, Johnstone T et al. (2006). Amygdala and ventromedial prefrontal cortex are inversely coupled during regulation of negative affect and predict the diurnal pattern of cortisol secretion among older adults. J Neurosci 26: 4415–4425.

van Marle H, Hermans E, Qin S et al. (2009). From specificity to sensitivity: how acute stress affects amygdala processing of biologically salient stimuli. Biol Psychiatry 66: 649–655.

van Reekum CM, Urry HL, Johnstone T et al. (2007). Individual differences in amygdala and ventromedial prefrontal cortex activity are associated with evaluation speed and psychological well-being. J Cogn Neurosci 19: 237–248.

Verheul R, van den Brink W, Geerlings P (1999). A three-pathway psychobiological model of craving for alcohol. Alcohol Alcohol 34: 197–222.

Vlahov D, Galea S, Ahern J et al. (2004). Consumption of cigarettes, alcohol, and marijuana among New York City residents six months after the September 11 terrorist attacks. Am J Drug Alcohol Abuse 30: 385–407.

Volkow ND, Wang GJ, Fowler JS et al. (1996). Decreases in dopamine receptors but not in dopamine transporters in alcoholics. Alcohol Clin Exp Res 20: 1594–1598.

Volkow ND, Wang GJ, Maynard L et al. (2002). Effects of alcohol detoxification on dopamine D2 receptors in alcoholics: a preliminary study. Psychiatry Res 116: 163–172.

Volkow ND, Wang GJ, Telang F et al. (2006). Cocaine cues and dopamine in dorsal striatum: mechanism of craving in cocaine addiction. J Neurosci 26: 6583–6588.

Volkow N, Fowler J, Wang G et al. (2007a). Dopamine in drug abuse and addiction: results of imaging studies and treatment implications. Arch Neurol 64: 1575–1579.

Volkow ND, Wang GJ, Telang F et al. (2007b). Profound decreases in dopamine release in striatum in detoxified alcoholics: possible orbitofrontal involvement. J Neurosci 27: 12700–12706.

Weiss F, Porrino LJ (2002). Behavioral neurobiology of alcohol addiction: recent advances and challenges. J Neurosci 22: 3332–3337.

Weiss F, Markou A, Lorang MT et al. (1992). Basal extracellular dopamine levels in the nucleus accumbens are decreased during cocaine withdrawal after unlimited-access self-administration. Brain Res 593: 314–318.

Wrase J, Schlagenhauf F, Kienast T et al. (2007). Dysfunction of reward processing correlates with alcohol craving in detoxified alcoholics. Neuroimage 35: 787–794.

Wrase J, Makris N, Braus DF et al. (2008). Amygdala volume associated with alcohol abuse relapse and craving. Am J Psychiatry 165: 1179–1184.

Zironi I, Burattini C, Aicardi G et al. (2006). Context is a trigger for relapse to alcohol. Behav Brain Res 167: 150–155.

Handbook of Clinical Neurology, Vol. 125 (3rd series)
Alcohol and the Nervous System
E.V. Sullivan and A. Pfefferbaum, Editors

Chapter 22

Compensatory recruitment of neural resources in chronic alcoholism

SANDRA CHANRAUD[1]* AND EDITH V. SULLIVAN[2]

[1]*EPHE; INCIA, CNRS UMR 5287, Université Victor Segalen Bordeaux, Bordeaux, France*

[2]*Department of Psychiatry and Behavioral Sciences, Stanford University School of Medicine, Stanford, CA, USA*

Functional recovery occurs with sustained sobriety, but the neural mechanisms enabling recovery are only now emerging. Theories about promising mechanisms involve concepts of neuroadaptation, where excessive alcohol consumption results in untoward structural and functional brain changes which are subsequently candidates for reversal with sobriety. Neuropsychologic studies of chronic alcoholism have led to several hypotheses about neural substrates of functional change: (1) cognitive and motor deficits in chronic alcoholics are detected most consistently on tasks that tap executive, mnemonic, emotional, and postural stability processes mediated by frontal, limbic, striatal, and cerebellar systems; and (2) chronic alcoholism interacts with aging, resulting in accelerated aging, whereby the brains of older alcoholics either age prematurely or show accelerated decline; in either case, the age–alcoholism synergy curtails recovery. Additional views of functional adaptation in chronic alcoholism have emerged from neuroimaging studies, which have identified factors that influence compensation, such as brain integrity and reserve. Here, we set forth the argument that knowledge about the physiology of brain plasticity will reveal mechanisms of neural compensation that account for functional recovery with sobriety in chronic alcoholism.

Views on functional adaptation in chronic alcoholism have expanded with results from neuroimaging studies. Here, we first describe and define the concept of neuroadaptation derived from theories forthcoming from this growing literature on age-related cognitive functioning. Then we describe findings as they apply to chronic alcoholism and factors that could influence compensation, such as brain reserve (Stern, 2002) and the integrity of brain structure. Finally, we review brain plasticity based on physiologic mechanisms that could underlie mechanisms of neural compensation.

WHAT ARE COMPENSATORY MECHANISMS?

Observed symptoms of neurologic insult or disease, including alcoholism, are not necessarily the manifestation of focal brain injury traditionally considered to underlie the symptoms but can be the behavioral outcome of plastic changes occurring in multiple loci and affecting far-reaching brain regions through circuitry. "Plastic" changes are those reactive to experience or events. Once a brain region is damaged, parallel circuits can be recruited or activated to establish alternative functioning routes. An example comes from transcranial magnetic stimulation (TMS) studies, which have shown that, after a repetitive TMS-induced shift in activity in the primary motor cortex, parallel motor circuits can be activated to establish alternative, although limited, input to spinal motoneurons (for review, see Pascual-Leone et al., 2005). These parallel circuits may originate from the contralateral, undamaged cortex or from anatomically neighboring brain regions.

Considering lifespan activity of the brain, positron emotion tomography studies showed that aging commonly has two opposite effects: compared with young adults, older adults showed lower activity in some brain regions but higher activity in other brain regions (e.g., Grady et al., 1994; Cabeza et al., 1997). Similarly, neural activity increases in aging in frontal regions where

*Correspondence to: Sandra Chanraud, Laboratoire de Neurobiologie intégrative et adaptative, Equipe de Neurosciences Cognitives Humaines et Neuroimagerie, Université Segalen Bordeaux 2 Zone Nord, Case 22, 146, rue Léo Saignat, 33076 Bordeaux cedex, France. E-mail: sandra.chanraud@u-bordeaux.fr

functioning appears to be altered in older men and women, thereby presenting a paradox, given that these regions are highly vulnerable to the negative effects of aging. To accommodate these opposing actions, it was proposed that the brain is able to reorganize its functioning to counteract neural decline and to maintain performance. This post hoc interpretation and burgeoning novel concept indicated the need for objective and operational criteria defining functional compensation and recognition of guidelines to establish neuroadaptation.

Initially, criteria for functional change were related to performance because there was good reason to assume that compensatory mechanisms were naturally in place to preserve adequate functioning of the brain (Teuber and Weinstein, 1956; Gazzaniga, 2005). This possibility raised the question: for a process to be considered compensatory, is it required to result in an adequate to normal level of performance? Cabeza and Dennis (2013) note two concepts for defining brain changes following insult: (1) change in brain function may be adaptive and enable successful compensation; or (2) change in brain function may be poorly preserved, resulting in unsuccessful compensation. In progressive disorders, these concepts may proceed sequentially with degeneration. This distinction between successful compensation and attempted compensation also needed to be defined. To address these problems, Cabeza and Dennis proposed four criteria necessary for brain changes to be considered compensatory and distinguished attempted compensation from successful compensation. Below is a summary of their concept.

It is first reasonable to assume that attempts at compensation to overcome cognitive impairment evolve as a function of structural or functional decline and then as a result of cognitive demands (Fig. 22.1). Therefore, compensation, when initially attempted, should outweigh decline and be negatively related to brain decline. Thus compensation should permit adjustment to at least aspects of the deficits and allow attainment of a satisfactory level of functioning. With disease progression, compensatory effort should increase with task demands, which are defined by increases of cognitive demands exerted, the level of which is determined by unaffected individuals.

When attempting to compensate for brain decline, brain reserve mechanisms will be limited by neural resources and will reach asymptote before decreasing. In the case of high cognitive demands, it can be assumed that compensatory mechanisms could interfere at one point with simultaneous processing. For example, attentional resources might be recruited for compensatory activities and therefore would be less available for the primary task. In the case of successful compensation,

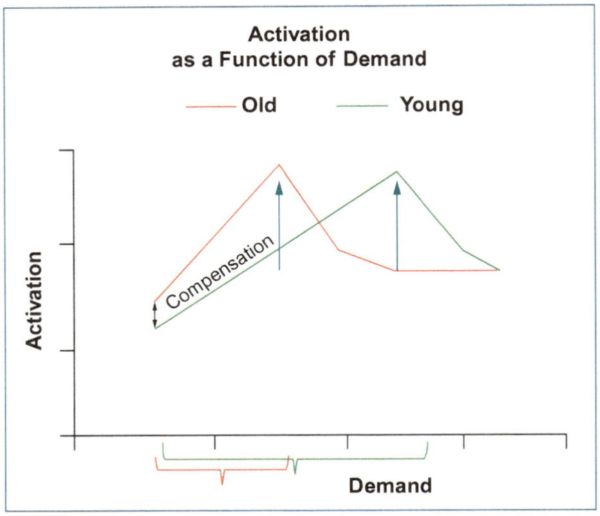

Fig. 22.1. Level of compensation as a function of cognitive demands. Compensation in elderly people (red), as represented here by higher activation for same demands as in younger people (green), allows a satisfactory level of functioning to be maintained but then reaches limits of resources (asymptote, represented by blue arrows) and starts to decline; therefore, before maximal capacity, as observed in younger individuals. Courtesy of Patricia A. Reuter-Lorenz. The model proposed by A. Reuter-Lorenz (Reuter-Lorenz and Park, 2010) describes a leftward shift in the demand–activity function, which occurs because older adults engage more neural resources at lower levels of task demands and, as a result, have fewer resources available to meet higher levels of task demands.

a direct positive relationship should be found between greater neural activity or functional connectivity and successful performance and also with decreases in physiologic markers observed in brain regions where alterations directly target cognitive performance.

A method used in chronic alcoholism to test successful compensation was to employ cognitive tasks known to engage regions to be altered and regions supposedly engaged in compensation. We (Chanraud et al., 2011) challenged the compensatory role of the regions by interfering with the function of the target regions using a secondary task. The assumed set-up system was that, during task X, healthy subjects needed to recruit region A, whereas alcoholics required regions A and B; thus, a simultaneously conducted task Y that depended on region B was used to interfere with the performance of chronic alcoholics to a greater extent than healthy subjects, resulting in a bottleneck effect on the resources. This system overload highlighted the possible cost of compensation: recruiting additional regions for one task makes them less available for other simultaneously engaged tasks, pointing out the probable impairment of chronic alcoholics for conducting two tasks simultaneously, described next.

ALCOHOLISM-RELATED INCREASES AND DIFFERENCES IN ACTIVITY AND FUNCTIONAL CONNECTIVITY

Here, we turn to specific examples revealed in functional magnetic resonance imaging (fMRI) studies of chronic alcoholism. We first review these findings and then discuss them in light of the criteria defining compensation used in aging studies. fMRI has permitted investigation of brain functioning in real time through exploration of changes of hemodynamic response in different brain regions between resting state and cognitive states. These studies have served to localize neural activity associated with the resting state, in contrast to the task-activated state induced by cognitive operations and task demands. Below we first discuss task-activated studies and then turn to resting-state fMRI studies to identify intrinsic neural circuitry that supports selective cognitive operations and is the target for disruption by disease, in this case chronic alcoholism.

Initial studies in alcoholism of brain activations recruited while engaged in a specific cognitive task focused on working memory processes. Neural substrates of observed neuropsychologic deficits were probed using a spatial working-memory task in a task-activated fMRI study of adult men with and without alcoholism (Pfefferbaum et al., 2001). Whereas controls activated a usual pattern of brain regions defining the "where" (dorsal) processing stream involving the occipital-parietal system typically invoked to accomplish visuospatial tasks, alcoholics exhibited a different pattern of activation. Instead, alcoholics activated the "what" (ventral) stream, involving the inferior occipital-temporal and limbic systems, while performing the spatial working-memory task. Given that performance was similar in both groups, the authors proposed the hypothesis that an atypical activation pattern indicates a functional reorganization of brain systems enabling alternative neural pathways to execute the task because of the restricted availability of the usual network. Similar results were found in young adult women with a history of alcoholism (Tapert et al., 2001). A potential negative outcome of displaced function would be to limit access to the compensatory region if required by simultaneous or parallel processing demands of both "what?" and "where?"-based information (Sullivan and Pfefferbaum, 2005; Oscar-Berman and Marinkovic, 2007).

Another study investigated the compensation hypothesis and observed abnormal activations in alcoholics while dual processing in performing a primary spatial working memory task (Chanraud et al., 2011). The results fulfilled criteria necessary for considering observed brain activations as compensatory (Davis et al., 2008). First, a relationship was revealed between a brain region that was less activated in alcoholics than controls (cerebellar region VI/Crus I) and a brain region that was more activated in alcoholics than controls (dorsolateral prefrontal cortex). This pair of relations satisfies the criterion requiring that abnormal increases of activity in the supposed compensatory region be related to abnormally low activation in other brain regions normally associated with the task, suggesting a recruitment of the compensatory region to boost performance. Second, even though activation within cerebellar region VI was related to accuracy in controls, this region was not activated at the control level in alcoholics; nonetheless, the same significant relation between level of activation in cerebellar region VI and accuracy occurred in the alcoholics. Positive correlations with accuracy in the two groups support involvement of this region as essential for performing the task. The dampened activation pattern associated with normal behavioral performance in the alcoholic group fulfills another criterion for a brain mechanism to be considered as compensatory (Chanraud et al., 2013; Fig. 22.2).

Following investigations of spatial working memory, verbal working memory was explored in adult alcoholics (Desmond et al., 2003). Once again, the two groups did not differ in performance levels but differed in patterns of brain activation during performance of the task. Here, the alcoholics activated the same frontal brain region as controls, i.e., the left frontal (BA44/45), but at greater levels of activation and bilaterally. Unlike the controls, the alcoholics also activated the right superior cerebellum. This excess in activation was hypothesized to underlie compensation enabling normal performance. These data were the first in alcoholism studies to highlight the potential role of cerebellum in compensating for functional impairment in frontal regions, consistent with anatomic connection between frontal and cerebellar regions (Schmahmann and Pandya, 1997; Zahr et al., 2010).

Rapid self-paced motor activity, such as finger tapping, is a motor cortex function. The motor cortex initiates a stimulus to the muscles of the hand, which is then coordinated by interplay between the cerebellum and the frontal lobes. Controls and chronic alcoholics were scanned during exercises of finger tapping with either dominant or non-dominant hand during both externally and self-paced timing to explore patterns of activation. The detoxified alcoholics performed the finger-tapping tasks significantly more slowly than the control group but, contrary to expectations, the slower tapping was not accompanied by proportionate decreases in brain activation in the cerebral cortex and cerebellum; rather, the alcoholics had a significant increase of activation in the cortical brain region on the same side as the active hand during dominant-hand tapping. In other words,

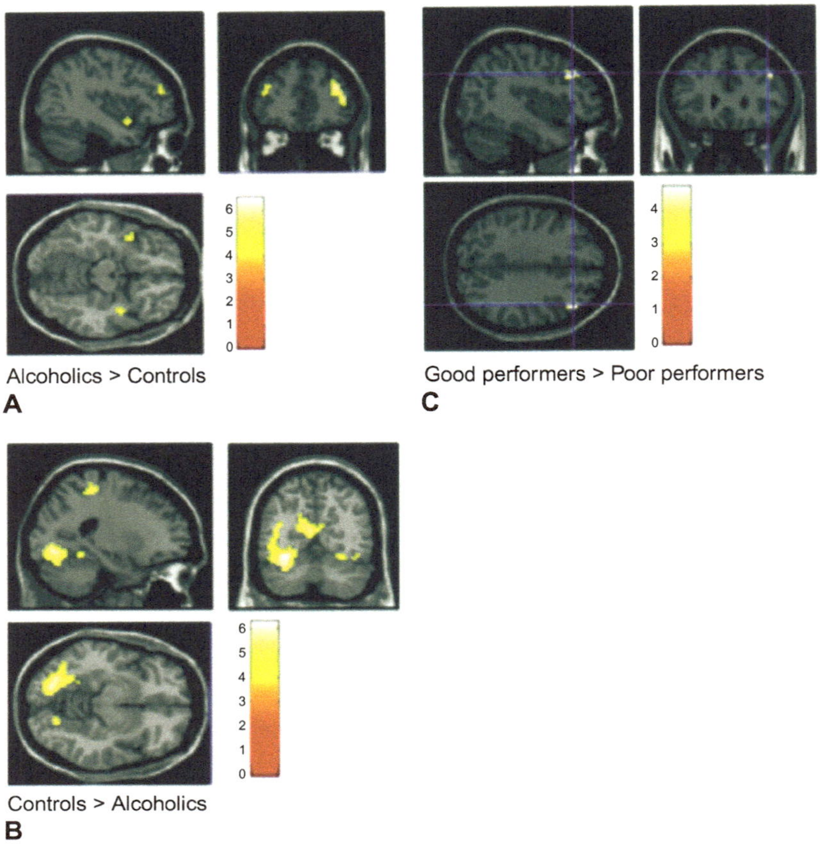

Fig. 22.2. Left: Brain regional activation in (**A**) alcoholics relative to controls and (**B**) controls relative to alcoholics, while performing on the task. (**C**) Brain regional activation in good alcoholic performers (8 subjects) relative to poor alcoholic performers (7 subjects). Cutoff for performance estimated with accuracy = 50% of good responses; $p < 0.05$ false discovery rate corrected and $k > 10$ voxels. (Reproduced from Chanraud et al., 2011b, with permission from the publishers.)

the alcoholics had to invoke more brain regions to do less. In the same study, self-paced finger tapping activated frontocerebellar networks in controls (e.g., anterior cingulate, anterior lobe, and vermis of the cerebellum) but only the parietal precuneus in alcoholics (Parks et al., 2010).

Another study revealed that the processing of redundant targets relative to a single target was associated with a significant hemodynamic response in bilateral extrastriate cortices in controls. By contrast, although alcoholics activated only the left extrastriate cortex, they, unlike controls, also showed significant activations in the thalamus, pallidum, and left cerebellum (Schulte et al., 2012). Compensatory mechanisms seem to be engaged also for performing tasks not specifically reported to be impaired in chronic alcoholism, such as language skills, which are relatively spared in alcoholics (for reviews, see Parsons, 1987; Oscar-Berman and Marinkovic, 2007) despite structural abnormalities in the language-related brain regions (Pfefferbaum et al., 1997; Chanraud et al., 2007). To expand these findings, an exploratory analysis of language skills revealed that, even though alcoholics showed comparable performance in terms of error rates and response time relative to controls, they exhibited greater fMRI response in the left middle frontal gyrus, right superior frontal gyrus, and cerebellar vermis relative to the controls (Chanraud-Guillermo et al., 2009). These findings suggest again that frontocerebellar neural activity could participate in compensatory mechanisms in chronic alcoholics.

In addition to cognitive compromise, alcoholism is associated with impaired emotional control. Consistent with findings in cognition, recent studies provide evidence for the presence of compensatory mechanisms during emotional processing by chronic alcoholics. Participants were presented with emotionally charged words and photographs of faces during deep (semantic) and shallow (perceptual) encoding tasks followed by a recognition task. Subjects were asked to perform four different encoding tasks that varied both type of material (faces or words) and depth of encoding (deep or shallow). The alcoholics were slower than controls in rating the emotion expressed on a face and in rating the words as abstract or concrete; however, the groups did not differ in accuracy or reaction time during recognition testing. Whereas the emotional network of alcoholics had lower responsivity to emotional cues, the alcoholics may have relied on prefrontal rather than temporal limbic areas when presented with emotionally or socially challenging situations (Marinkovic et al., 2009). Consistent with that possibility, another study reported that alcoholics activated more frontal, limbic, and temporal regions relative to controls when they had to make judgments about whether they liked or disliked an image. Increases were particularly robust in the frontal lobe and areas of the brain associated with language (Gilman and Hommer, 2008). Alcoholics were more likely than controls to recruit brain language areas for both language and emotional processing.

The functional or anatomic derangement of the corticocerebellar circuitry in alcoholics contributes to alcoholism-related cognitive impairment and also has the potential to contribute to compensation for compromised functions. Neuroimaging studies on functional connectivity (Habas et al., 2009; Krienen and Buckner, 2009) have confirmed the presence of segregated loops within corticocerebellar circuitry. Functional connectivity studies conducted in chronic alcoholism followed these findings and combined them with the idea that behavior could lead to changes in brain circuitry and ultimately to behavioral modification. Yet, useful compensatory changes in alcoholics might not always manifest as an increase in regional brain activity but might also be reflected in an increase in functional connectivity. Indeed, functional neuroimaging studies have now revealed that chronic alcoholism modulates activity and functional connectivity (Chanraud et al., 2011; Camchong et al., 2013c; Muller-Oehring et al., 2013; Pitel et al., 2013).

New focus on neural connectivity arose from the observation that a constellation of brain regions was often deactivated during traditional fMRI activation experiments entailing task performance when the expectation was to find greater activation related to engagement in the task (Raichle and Snyder, 2007). While at rest, i.e., not engaged in a specific externally imposed task, heightened functional connectivity between spontaneously active brain regions was consistently observed. Brain regions showing synchronization between their spontaneous fluctuations comprise intrinsic networks characterized by a complex activation–deactivation scheme launched from resting to performing a task. A major circuit specific to the rest condition is called the default mode network (DMN). Regions of this network present higher activation when the brain is resting and deactivate when engaged in a task (Raichle and Snyder, 2007).

Default mode network in alcoholism

The DMN was the first intrinsic functional network to be implicated in chronic alcoholism. Specifically, a deficit in DMN modulation was first revealed between the posterior cingulate and cerebellar regions in switching from rest to task conditions (Chanraud et al., 2013). This abnormal synchrony was interpreted as potentially reflecting processing inefficiency (Lawton-Craddock et al., 2003), where processing efficiency "refers to the effective use of accurate information while ignoring irrelevant, inaccurate or interfering information" (Nixon and Parsons, 1991).

Regions of the DMN are normally highly connected during rest and functionally decoupled during task performance, presumably for efficient processing by allocating resources to appropriate regions (Lewis et al., 2009). Compromised connectivity detected in alcoholics represents a decoupling of synchronization between regions that are functionally synchronized in controls. Corrupted synchronization could result in disturbed connectivity, precluding direct or efficient transmission between brain regions corroborating pathophysiologic models of alcoholism as a functional disconnection syndrome (Chanraud et al., 2010). The same study also noted, however, that brain networks other than the ones invoked by controls were involved when alcoholics engaged in a spatial working-memory task.

The addition of newly established, direct functional connections may have led to greater functional connectivity and better than expected performance in this sample of alcoholics who reached the same level of performance as controls. With further probing, these results might reveal the presence of anatomic connections or functional synchronization in alcoholics that are not present in controls. The presence of connections outside the DMN together with the observations of comparable performance of the controls and alcoholics led to the speculation that a compensatory system could operate through an alternative network, as also suggested by other studies (Pfefferbaum et al., 2001; Desmond et al., 2003; Chanraud-Guillermo, 2009), and therefore overcome compromise in the network skirted (Caplan et al., 2007).

Intrinsic functional networks

Other intrinsic networks include synchronized brain regions that activate together during specific cognitive processes. Alterations of one or several of these intrinsic networks may underlie cognitive performance and be directly related to cognitive impairments (Andrews-Hanna et al., 2007; Chanraud et al., 2013).

Neuroimaging studies have identified alcoholism-related changes in functional connectivity between distributed brain nodes in task-related networks for short-term abstinent (Schulte et al., 2012; Park, 2012; Pitel et al., 2013) and long-term abstinent alcoholics (Camchong et al., 2013b, d). During executive function tasks, Schulte et al. (2012) observed that connectivity of the middle cingulate cortex was modified in alcoholics compared with controls.

Consistent with previous investigations reporting the predictive value of functional connectivity during a task to performance level (Summerfield et al., 2006; Nagano-Saito et al., 2008; Hampson et al., 2010), functional connectivity in controls and alcoholics was investigated both at rest and during a face–name association learning task (Pitel et al., 2013). These novel explorations revealed information about functional connectivity during both conditions and also how functional connectivity is modulated during the switch from one to the other. Controls showed a significant correlation between cerebello-hippocampal-positive synchronization during the task and performance on associative recognition, whereas alcoholics with more negatively correlated activity during rest performed better on the associative recognition. These patterns of relations (Fig. 22.3) give credence to the possibility that compensatory mechanisms can be involved during associative memory to enable alcoholics to perform on par with controls despite different patterns of brain activations and, perhaps, because of these differences.

Regarding long-term abstinent alcoholics, it was observed that resting-state functional connectivity of the reward and executive control networks (investigated using separately the nucleus accumbens for reward and the subgenual cingulate cortex for executive control as a seed) differed in comparison with short-term abstinent alcoholics. This difference suggested that also during the abstinence period, functional changes are involved and probably set up in order to modify behaviors for sustaining abstinence (Camchong et al., 2013a). Furthermore, group differences in functional connectivity at rest within these networks were related to cognitive flexibility and length of abstinence. Long-term abstinent alcoholics exhibited lower functional connectivity than controls in the reward network (Camchong et al., 2013a), whereas short-term abstinent alcoholics exhibited lower functional connectivity than long-term abstinent alcoholics in executive network (Camchong et al., 2013c). Resting functional connectivity therefore might be adaptive for counteracting chronic alcoholism-related functional changes, including processes that underlie craving and relapse. Indeed, these findings may reflect adaptive mechanisms that support the long-term abstinent alcoholic who is ready to inhibit or successfully stop at-risk behaviors that may lead to further substance use.

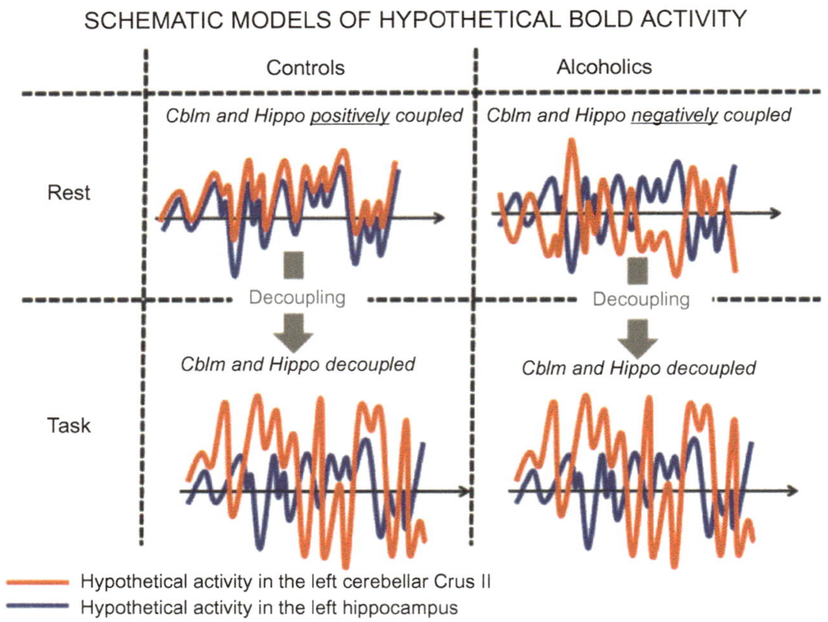

Fig. 22.3. Schematic hypothetic representation of the modulation of cerebello-hippocampal synchronization from positively (controls) and negatively (alcoholics) coupled activity at rest to decoupled activity when engaged in the associative learning task. Note that the waveforms represent hypothetic, idealized blood oxygen level-dependent (BOLD) responses to illustrate coupling and decoupling of cerebello-hippocampal time series. (Reproduced from Pitel et al., 2013, with permission from the publishers.)

Frontocerebellar structural and functional connectivity

The animal literature provides evidence for the existence of multiple structural connectivity loops, which have the potential of providing alternate pathways when damage occurs in one of the parallel loops. In non-human primates, two parallel closed loops within the frontocerebellar circuit have been identified structurally using viral tracing methods. Each loop supports different behaviors (Kelly and Strick, 2003). A motor loop, which connects cerebellar and premotor cortical systems, supports performance on motor tracking tests, whereas a cognitive loop, which connects cerebellar and dorsolateral prefrontal cortical systems, supports performance on a spatial working-memory task. In humans, imaging data analyzed using concepts of intrinsic connectivity networks identified four segregated frontocerebellar circuits. Dissociable functions of these loops are related to the sites of termination in the cerebral cortex of specific projections from the cerebellum. Examples of these divergent but parallel loops include the motor and executive loops (Fig. 22.4): (1) the motor network, involving motor lobules of the cerebellar vermis (e.g., IV, V, VI) and motor cerebral cortices (Biswal et al., 1995) affecting functions of gait and balance (Sullivan et al., 2005; Butler et al., 2006); and (2) the executive network, involving the cerebellar neocortex (e.g., lobule VII, lobule VIII, Crus I, and Crus II) and prefrontal cortical sites (e.g., BA9 and 46) contributing to cognitive functions, such as verbal (Desmond et al., 1997, 2003) and spatial (Pfefferbaum et al., 2001) working memory and set shifting (Seeley et al., 2007). These structural loops and functional circuitry set the stage for additional loci of compensatory escape of function.

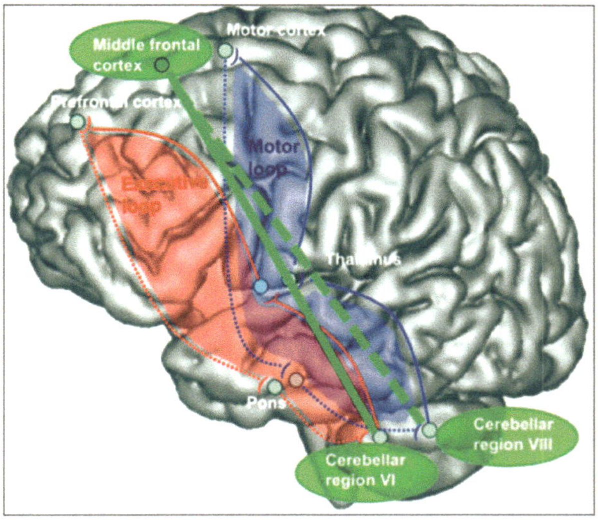

Fig. 22.4. The corticocerebellar circuit: two dissociable but associated loops, the motor loop (blue) connecting motor cortex, thalamus, and anterior cerebellum and the executive loop (red) connecting prefrontal-pontine-posterior cerebellar sites (adapted from Chanraud et al., 2011b). In green: hypothetic remapping of executive frontocerebellar loop in chronic alcoholism. (Reproduced from Chanraud et al., 2011b, with permission from the publishers.)

Evidence indicates that recruitment of functionally intact cerebellar loops can compensate for the alcoholic's otherwise impaired performance on tasks requiring, for example, visuospatial or working-memory skills (Sullivan et al., 2005). A biologic interpretation is that the recruitment of an unaffected loop parallel to the affected one leads to rapid changes in connectivity, which depend on already present but "silent" connections rather than the creation of new ones (Nakayama et al., 2005). This assumption is supported by findings which revealed that functional synchronization between prefrontal cortex and a cerebellar region (VIII) was observed in alcoholics adjacent to those measured in controls. Furthermore, results showing a relation between synchronization of this cerebellar region with the prefrontal cortex at rest and accuracy suggest that bottom-up resources recruited via cerebellocortical connections are likely to compensate for top-down corticocerebellar deficits, thereby enabling normal performance (Chanraud et al., 2011).

ARE ALCOHOLISM-RELATED INCREASES IN FUNCTIONAL ACTIVITY AND CONNECTIVITY COMPENSATORY?

Determination of whether differences in functional connectivity in alcoholism are compensatory required consideration of operational criteria. Among criteria established for defining compensatory mechanisms, two major criteria were proposed as mandatory for physiologic mechanisms to be considered as involved in compensation systems (Davis et al., 2008). One criterion requires that abnormal brain activity, that is, activity not seen in an unaffected comparison group, be associated with normal-level performance. A second criterion requires that abnormal increases of activity in the supposed compensatory region occur with lower activity in processing by the damaged regions. More detailed are criteria recently defined in aging research and described earlier in this review. In general, relationships between brain structural or functional compromise and decline in cognitive performance appear crucial for attributing a compensatory role to abnormal physiologic phenomena. Links between level of assumed compensation and behavioral performance have emerged from alcoholism studies (Chanraud et al., 2001, 2013; Camchong et al., 2013a, b; Pitel et al., 2013; Schulte et al., 2012). Neural damage in alcoholism has been depicted as "incomplete lesions," where neural nodes and white-matter connections are damaged but not severed or permanently destroyed (Sullivan and Pfefferbaum, 2005). This depiction of neural damage indicates the potential of structural and functional repair in alcoholism with sobriety or perhaps training.

Factors contributing to compensatory mechanisms

Total processing capacity of the brain for cognitive, motor, and sensory functions differs from one person to another and is likely a limiting factor to compensation. In addition to sources commonly recruited for a specific task, reserve resources may or may not be available depending on the extent, locus, and type of pathologic or non-pathologic compromise. These factors might be on different levels of analysis, for example, on a global level (such as connectivity of attentional system with other systems) or local level (activation of one specific brain region). Reuter-Lorenz and Park (2010) proposed the compensation-related utilization of neural circuits hypothesis, or CRUNCH, to account for functional compensation in normal aging. This model proposes a leftward shift in the demand–activity function, which occurs because older adults engage more neural resources at lower levels of task demands and, as a result, have fewer resources available to meet higher levels of task demands (Fig. 22.1). This theory can be adapted to chronic alcoholism but requires consideration of other contributing factors that can influence brain resources, such as age, education level, family history of alcoholism, and brain reserve. Consequently, a common question relative to why some alcoholics may not seem to establish compensatory mechanisms has to be considered in light of these different factors that could hamper successful compensation. Although this review does not discuss influences of these factors on compensation capacity, as few studies have explored the interaction between external factors and compensatory brain changes, we speculate on how inter- and intraindividual characteristics might interact with the compensatory mechanisms.

The concept of brain reserve refers to the ability to maintain an adequate level of brain functioning (brain resilience) despite age- or pathology-related changes; it can be considered as a reserve of function available in case of brain damage (Stern, 2002). However, brain reserve cannot be characterized as unidimensional and is usually parsed into two models, depending on whether passive or active processes are invoked. In the passive model, reserve is defined in terms of the amount of damage that can be sustained before reaching a threshold for clinical expression. On a physiologic level, it might include regional brain volume or synapse count. The model recognizes that level of passive reserve is individual-dependent and also assumes that, once the threshold of passive reserve is reached, functional deficits appear. In the active model, reserve depends on differences in how a task is processed and describes how the brain actively attempts to counteract or respond to consequences of damage. These two approaches are not mutually exclusive as exposure to an enriched environment (active processes), defined as a combination of more opportunities for physical activity, learning, and social interaction, may produce structural and functional changes in the brain and influence the rate of neural repair (passive processes). Relevant to alcoholism, brain reserve could explain, at least in part, compensatory mechanisms as it could take the form of using brain networks or cognitive processes that are not normally used by healthy individuals to perform the target task. More efficient use of brain networks or differential ability to recruit alternate networks might be highly related to the individual's passive and active reserves (Goldstein, 2005; Sullivan and Tapert, 2013).

Is there a continuum in compensatory mechanisms?

The contribution of passive brain reserve to compensatory mechanisms implies a continuum in recruiting resources that decreases with aging or other forms of brain compromise. However, compensatory mechanisms can be dependent on passive reserve and also on active reserve, which is plastic, that is, modified in response to experience. To the extent that physical and cognitive exercise enhances neuroplasticity (Valenzuela, 2008), such training could be considered in resolution of alcohol-related cognitive and motor impairments, as it is in militating aging-related deficits (for review, see Bates et al., 2013). Conceptually, it might be worth thinking of processes occurring after brain injury and leading to restoration of function as fitting different mechanisms that may proceed in parallel but not necessarily in lock step. Initial changes aim to minimize damage. Rapid functional improvement is likely to occur if brain components are dysfunctional but only partially damaged. Subsequent processes, once structural or functional damage is too advanced, would involve recovery in relation to relearning. These processes could be launched by recruiting latent functions, creating new functions through establishment of new neural pathways, or by strengthening existing circuitry. Next and finally, we speculate on how experiential, or plastic, brain changes might confer new brain function.

Neuroplastic changes possibly underlying compensatory mechanisms

Neuroplasticity appears to be an efficient way to recruit the brain's limited resources (Zatorre et al., 2012). Depending on the context, neural plasticity can: (1) confer no perceptible change in the behavioral output of the brain; (2) lead to changes demonstrated only under special testing conditions (high-demanding task); or (3) cause behavioral changes through takeover of other functional systems (Pascual-Leone et al., 2005). Plasticity is an intrinsic, permanent characteristic of the nervous system, and understanding of normal or abnormal psychologic function necessitates invoking the concept of brain plasticity (Goldstein, 2005; Pascual-Leone et al., 2005).

Formation of new pathways is possible only following initial reinforcement of pre-existing connections. In accordance with the brain reserve concept, plastic changes could be linked to passive reserve as defined by existing connections, and reinforcement could be linked to active reserve as defined by the consequence of environmental influences. With regard to compensatory mechanisms in chronic alcoholism, we reported regional or network functional changes, mostly attributable to long-term changes, but continuous and long-term changes most probably depend initially on short-term rapid changes. On a physiologic level, short-term changes would occur at cellular and molecular levels. Cellular changes can be broadly categorized into neuronal changes in gray matter, connectivity changes in white matter, and extraneuronal changes (Fig. 22.5). Neuronal changes in gray matter may include neurogenesis (the genesis of neurons in the adult human brain has been observed in postmortem human brain tissue using immunofluorescent labeling for neuronal markers. New neurons were generated from dividing progenitor cells in the dentate gyrus of adult humans (Eriksson et al., 1998), synaptogenesis, and changes in neuronal morphology. In white matter, changes in the number of axons, axon diameter, the packing density of fibers, axon branching, axon trajectories, and

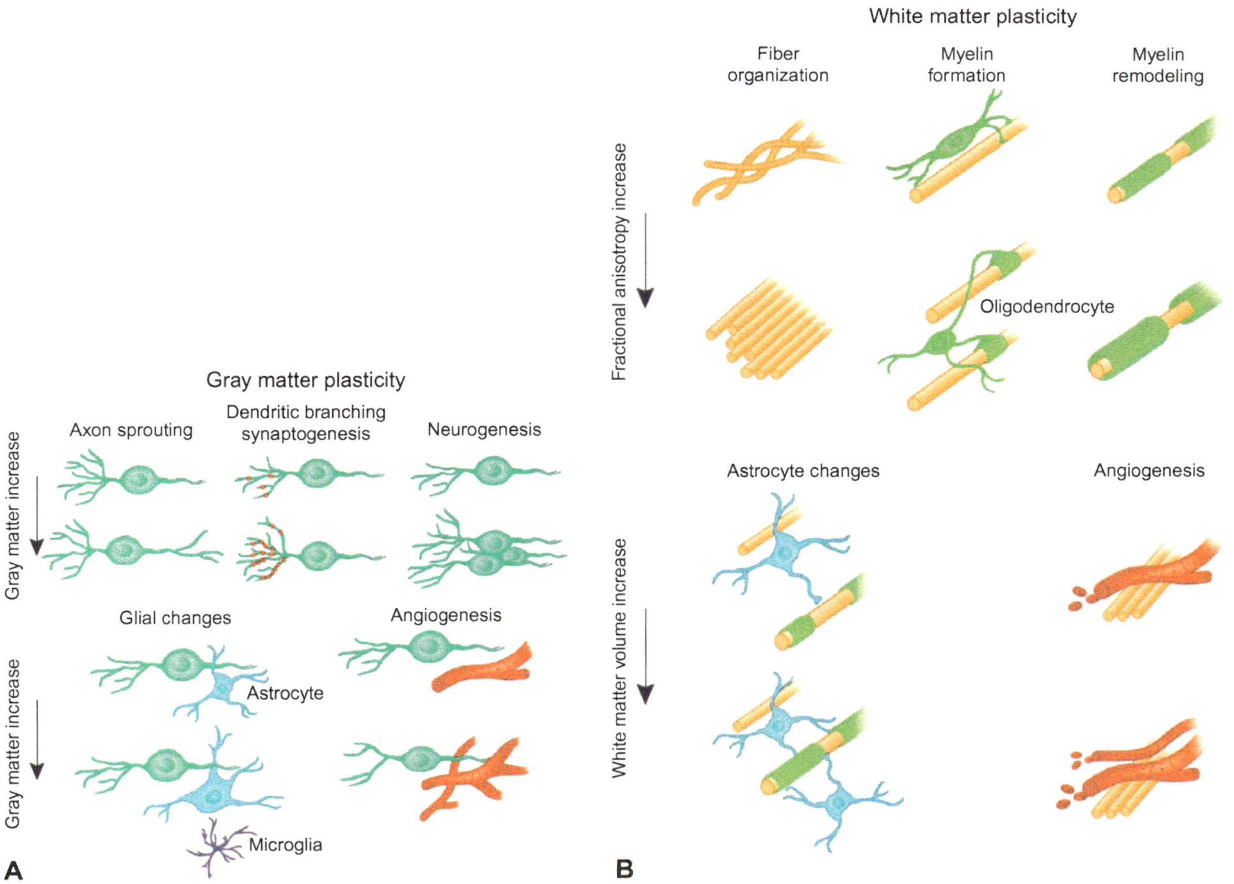

Fig. 22.5. Candidate cellular mechanisms. (**A**)/ Cellular events in gray-matter regions underlying changes detected by magnetic resonance imaging during learning include axon sprouting, dendritic branching and synaptogenesis, neurogenesis, changes in glial number and morphology, and angiogenesis. (**B**) Changes in white-matter regions include alterations in fiber organization, which could include axon branching, sprouting, packing density, axon diameter, fiber crossing, and the number of axons; myelination of unmyelinated axons; changes in myelin thickness and morphology; changes in astrocyte morphology or number; and angiogenesis. (Reproduced from Zatorre et al., 2012, with permission from the publishers.)

myelination can be found. Extraneuronal changes include increases in glial cell size and number, and angiogenesis (Sagi et al., 2012). Such neurostructural plasticity might be available to alcoholics and might be structural mechanisms of adaptive, plastic changes (Koob and Le Moal, 2005).

CONCLUSION

We reviewed findings in chronic alcoholism, highlighting the presence of physiologic changes that are likely involved in compensation. Even if criteria defining compensatory mechanisms are based on a relatively simple model, several factors are necessary for brain changes to be considered compensatory. Indeed, neural resources most certainly differ among individuals, and external factors unrelated to alcohol consumption have the potential to influence compensation level. In alcoholism, as in other conditions affecting brain integrity, task difficulty is a further consideration to discriminate at which point the availability of neural resources will no longer be sufficient to meet the increased demands of compensation. Evaluation of brain changes in terms of activation and connectivity in relation to performance has significant implications for concepts of neuroplasticity and reorganization in therapeutic settings. Self-directed control of compensatory responses brought under external or internal control may be useful in therapeutic efforts involving cognitive and motor remediation following brain insult (Le Berre et al., 2013). Specific patterns of functional activation or resting-state connectivity might also serve to distinguish alcoholics who maintain abstinence from those who do not (Camchong et al., 2013a; Seo et al., 2013). Confirmation of these speculations about the development and extent of functional compensatory mechanisms requires longitudinal study.

ACKNOWLEDGMENT

This work was supported in part by NIH/NIAAA grants U01 AA017923, R37 AA010723, K05 AA017168, R01 AA012388 and funding from the Ecole Pratique des Hautes Etudes and from the French Agence Nationale de la Recherche within the context of the Investments for the Future programme, referenced ANR-10-LABX-57 and named TRAIL.

References

Andrews-Hanna JR, Snyder AZ, Vincent JL et al. (2007). Disruption of large-scale brain systems in advanced aging. Neuron 56: 924–935.

Bates ME, Buckman JF, Nguyen TT (2013). A role for cognitive rehabilitation in increasing the effectiveness of treatment for alcohol use disorders. Neuropsychol Rev 23: 27–47.

Biswal B, Yetkin FZ, Haughton VM et al. (1995). Functional connectivity in the motor cortex of resting human brain using echo-planar MRI. Magn Reson Med 34: 537–541.

Butler EE, Druizin M, Sullivan EV (2006). Gait changes and adaptations in adulthood: Pregnancy, aging, and alcoholism. In: J Rose, JG Gamble (Eds.), Human Walking. 3rd edn. Lippincott, Williams and Wilkins, Philadelphia, MD.

Cabeza R, Dennis NA (2013). Frontal lobes and aging: deterioration and compensation. In: DT Stuss, RT Knight (Eds.), Principles of frontal lobe functioning. Oxford University Press, New York.

Cabeza R, Grady CL, Nyberg L et al. (1997). Age-related differences in neural activity during memory encoding and retrieval: a positron emission tomography study. J Neurosci 17: 391–400.

Camchong J, Stenger A, Fein G (2013a). Resting-state synchrony during early alcohol abstinence can predict subsequent relapse. Cereb Cortex 23: 2086–2099.

Camchong J, Stenger A, Fein G (2013b). Resting-state synchrony in long-term abstinent alcoholics. Alcohol Clin Exp Res 37: 75–85.

Camchong J, Stenger VA, Fein G (2013c). Resting-state synchrony in short-term versus long-term abstinent alcoholics. Alcohol Clin Exp Res 37: 794–803.

Camchong J, Stenger VA, Fein G (2013d). Resting state synchrony in long-term abstinent alcoholics with versus without comorbid drug dependence. Drug Alcohol Depend 131: 56–65.

Caplan JB, Mcintosh AR, De Rosa E (2007). Two distinct functional networks for successful resolution of proactive interference. Cereb Cortex 17: 1650–1663.

Chanraud S, Martelli C, Delain F et al. (2007). Brain morphometry and cognitive performance in detoxified alcohol-dependents with preserved psychosocial functioning. Neuropsychopharmacology 32: 429–438.

Chanraud S, Zahr NM, Sullivan EV et al. (2010). MR diffusion tensor imaging: A window into white matter integrity of the working brain. Neuropsychol Rev 20: 209–225.

Chanraud S, Pitel AL, Muller-Oehring EM et al. (2013). Remapping the brain to compensate for impairment in recovering alcoholics. Cereb Cortex 23: 97–104.

Chanraud S, Pitel AL, Pfefferbaum A et al. (2011). Disruption of functional connectivity of the default-mode network in alcoholism. Cereb Cortex 21: 2272–2281.

Chanraud-Guillermo S, Andoh J, Martelli C et al. (2009). Imaging of language-related brain regions in detoxified alcoholics. Alcohol Clin Exp Res 33: 977–984.

Davis SW, Dennis NA, Daselaar SM et al. (2008). Que PASA? The posterior-anterior shift in aging. Cereb Cortex 18: 1201–1209.

Desmond JE, Gabrieli JDE, Wagner AD et al. (1997). Lobular patterns of cerebellar activation in verbal working-memory and finger-tapping tasks as revealed by functional MRI. J Neurosci 17: 9675–9685.

Desmond JE, Chen SH, De Rosa E et al. (2003). Increased frontocerebellar activation in alcoholics during verbal working memory: an fMRI study. Neuroimage 19: 1510–1520.

Eriksson PS, Perfilieva E, Bjork-Eriksson T et al. (1998). Neurogenesis in the adult human hippocampus. Nat Med 4: 1313–1317.

Gazzaniga MS (2005). Forty-five years of split-brain research and still going strong. Nat Rev Neurosci 6: 653–659.

Gilman JM, Hommer DW (2008). Modulation of brain response to emotional images by alcohol cues in alcohol-dependent patients. Addict Biol 13: 423–434.

Goldstein K (Ed.), (2005). The Organism. Zone Books, New York.

Grady CL, Maisog JM, Horwitz B et al. (1994). Age-related changes in cortical blood flow activation during visual processing of faces and location. J Neurosci 14: 1450–1462.

Habas C, Kamdar N, Nguyen D et al. (2009). Distinct cerebellar contributions to intrinsic connectivity networks. J Neurosci 29: 8586–8594.

Hampson M, Driesen N, Roth JK et al. (2010). Functional connectivity between task-positive and task-negative brain areas and its relation to working memory performance. Magn Reson Imaging 28: 1051–1057.

Kelly RM, Strick PL (2003). Cerebellar loops with motor cortex and prefrontal cortex of a nonhuman primate. J Neurosci 23: 8432–8444.

Koob GF, Le Moal M (2005). Plasticity of reward neurocircuitry and the 'dark side' of drug addiction. Nat Neurosci 8: 1442–1444.

Krienen FM, Buckner RL (2009). Segregated fronto-cerebellar circuits revealed by intrinsic functional connectivity. Cereb Cortex 19: 2485–2497.

Lawton-Craddock A, Nixon SJ, Tivis R (2003). Cognitive efficiency in stimulant abusers with and without alcohol dependence. Alcohol Clin Exp Res 27: 457–464.

Le Berre AP, Vabret F, Cauvin C et al. (2013). Cognitive barriers to readiness to change in alcohol-dependent patients. Alcohol Clin Exp Res 36: 1542–1549.

Lewis CM, Baldassarre A, Committeri G et al. (2009). Learning sculpts the spontaneous activity of the resting human brain. Proc Natl Acad Sci U S A 106: 17558–17563.

Marinkovic K, Oscar-Berman M, Urban T et al. (2009). Alcoholism and dampened temporal limbic activation to emotional faces. Alcohol Clin Exp Res 33: 1880–1892.

Muller-Oehring EM, Jung YC, Sullivan EV et al. (2013). Midbrain-driven emotion and reward processing in alcoholism. Neuropsychopharmacology 38: 1844–1853.

Nagano-Saito A, Leyton M, Monchi O et al. (2008). Dopamine depletion impairs frontostriatal functional connectivity during a set-shifting task. J Neurosci 28: 3697–3706.

Nakayama K, Kiyosue K, Taguchi T (2005). Diminished neuronal activity increases neuron-neuron connectivity underlying silent synapse formation and the rapid conversion of silent to functional synapses. J Neurosci 25: 4040–4051.

Nixon SJ, Parsons O (1991). Alcohol-related efficiency deficits using an ecologically valid test. Alcohol Clin Exp Res 15: 601–606.

Oscar-Berman M, Marinkovic K (2007). Alcohol: effects on neurobehavioral functions and the brain. Neuropsychol Rev 17: 239–257.

Park DC (2012). Reduced fronto-cerebellar functional connectivity in chronic alcoholic patients. Alcohol Clin Exp Res 36: 294–301.

Parks MH, Greenberg DS, Nickel MK et al. (2010). Recruitment of additional brain regions to accomplish simple motor tasks in chronic alcohol-dependent patients. Alcohol Clin Exp Res 34: 1098–1099.

Parsons OA (1987). Neuropsychological consequences of alcohol abuse: Many questions – some answers. In: OA Parsons, N Butters, PE Nathan (Eds.), Neuropsychology of Alcoholism. Guilford Press, New York.

Pascual-Leone A, Amedi A, Fregni F et al. (2005). The plastic human brain cortex. Annu Rev Neurosci 28: 377–401.

Pfefferbaum A, Sullivan E, Rosenbloom MJ et al. (1997). Cortical gray matter and ventricular changes in alcoholic and control men over a 5 year interval (abs). Alcohol Clin Exp Res 21: 20A.

Pfefferbaum A, Desmond JE, Galloway C et al. (2001). Reorganization of frontal systems used by alcoholics for spatial working memory: an fMRI study. Neuroimage 14: 7–20.

Pitel AL, Chanraud S, Muller-Oehring EM et al. (2013). Modulation of limbic-cerebellar functional connectivity enables alcoholics to recognize who is who. Brain Struct Funct 218: 683–695.

Raichle ME, Snyder AZ (2007). A default mode of brain function: a brief history of an evolving idea. Neuroimage 37: 1083–1090, discussion 1097–1099.

Reuter-Lorenz PA, Park DC (2010). Human neuroscience and the aging mind: a new look at old problems. J Gerontol B Psychol Sci Soc Sci 65: 405–415.

Sagi Y, Tavor I, Hofstetter S et al. (2012). Learning in the fast lane: new insights into neuroplasticity. Neuron 73: 1195–1203.

Schmahmann JD, Pandya DN (1997). The cerebrocerebellar system. In: JD Schmahmann (Ed.), The Cerebellum and Cognition. Academic Press, San Diego.

Schulte T, Muller-Oehring EM, Sullivan EV et al. (2012). Synchrony of corticostriatal-midbrain activation enables normal inhibitory control and conflict processing in recovering alcoholic men. Biol Psychiatry 71: 269–278.

Seeley WW, Menon V, Schatzberg AF et al. (2007). Dissociable intrinsic connectivity networks for salience processing and executive control. J Neurosci 27: 2349–2356.

Seo D, Lacadie CM, Tuit K et al. (2013). Disrupted ventromedial prefrontal function, alcohol craving, and subsequent relapse risk. JAMA Psychiatry 70: 727–739.

Stern Y (2002). What is cognitive reserve? Theory and research application of the reserve concept. J Int Neuropsychol Soc 8: 448–460.

Sullivan EV, Pfefferbaum A (2005). Neurocircuitry in alcoholism: a substrate of disruption and repair. Psychopharmacology (Berl) 180: 583–594.

Sullivan EV, Tapert SF (2013). Introduction to the special issue of *Neuropsychology Review* on cognitive enhancement and rehabilitation. Neuropsychol Rev 23: 10–12.

Sullivan EV, Deshmukh A, De Rosa E et al. (2005). Striatal and forebrain nuclei volumes: contribution to motor function and working memory deficits in alcoholism. Biol Psychiatry 57: 768–776.

Summerfield C, Greene M, Wager T et al. (2006). Neocortical connectivity during episodic memory formation. PLoS Biol 4: e128.

Tapert SF, Brown GG, Kindermann SS et al. (2001). fMRI measurement of brain dysfunction in alcohol dependent young women. Alcohol Clin Exp Res 25: 236–245.

Teuber H-L, Weinstein S (1956). Ability to discover hidden figures after cerebral lesions. Arch Neurol Psychiatry 76: 369–379.

Valenzuela MJ (2008). Brain reserve and the prevention of dementia. Curr Opin Psychiatry 21: 296–302.

Zahr NM, Pitel AL, Chanraud S et al. (2010). Contributions of studies on alcohol use disorders to understanding cerebellar function. Neuropsychol Rev 20: 280–289.

Zatorre RJ, Fields RD, Johansen-Berg H (2012). Plasticity in gray and white: neuroimaging changes in brain structure during learning. Nat Neurosci 15: 528–536.

Section 9

Neuroelectrophysiology

Handbook of Clinical Neurology, Vol. 125 (3rd series)
Alcohol and the Nervous System
E.V. Sullivan and A. Pfefferbaum, Editors

Chapter 23

Understanding alcohol use disorders with neuroelectrophysiology

MADHAVI RANGASWAMY AND BERNICE PORJESZ*
Henri Begleiter Neurodynamics Laboratory, Department of Psychiatry and Behavioral Sciences, SUNY Downstate Medical Center, Brooklyn, NY, USA

INTRODUCTION

The literature is replete with compelling evidence of brain dysfunction in chronic alcoholics as well as their high-risk offspring from neuropsychologic, neuroimaging, neuropathologic, and neurophysiologic techniques; in particular frontal lobe changes have been highlighted (Moselhy et al., 2001; Zahr et al., 2010, 2011). Neuroimaging methods, such as structural/functional magnetic resonance imaging and positron emission tomography, have certain advantages that include excellent spatial resolution; yet they offer relatively poor temporal resolution compared to electrophysiologic methods that evaluate brain function in the millisecond range (Celesia and Brigell, 1992; Krieger et al., 1995). Neuroelectrophysiologic techniques have further advantages of being non-invasive and relatively inexpensive to implement. Electrophysiology has provided several excellent measures of acute and chronic effects of alcohol on the brain. While it was assumed that these aberrant characteristics in alcoholics were solely due to the neurotoxic effects of alcohol on the brain, the evidence indicates that some neuroelectrophysiologic characteristics reflect predispositions that antecede the development of alcoholism and related disorders.

Alcoholism is a complex neuropsychiatric condition with a multifactorial etiology that warrants the use of diverse neurobiologic methods. This disorder not only involves effects of alcohol on brain structures but also subsequent alteration in brain electrophysiology potentiated by the addiction cycle. Alcoholism or alcohol use disorder (AUD) is a common familial disorder with increased risk among biologic relatives of alcoholics (Goodwin et al., 1973; Cadoret et al., 1980; Bohman et al., 1987; Prescott, 2001). Family, twin, and adoption studies that highlight genetic contributions to AUDs suggest that both genders are equally vulnerable (Heath et al., 1997; Prescott et al., 1999). Yet AUDs may not be a specific disease but part of a spectrum of co-occurring disinhibitory disorders with overlapping genetic factors and shared underlying risk factors (Krueger et al., 2002; Kendler et al., 2003) and differential expression (Hicks et al., 2004). Thus, these behavioral phenomena – antisocial, impulsive traits, substance use disorders – are variable expressions of a disinhibitory complex (Gorenstein and Newman, 1980) with AUD as one possible outcome in this spectrum. Understanding addictive behavior is complex and involves interactions among behavioral, environmental, and genetic factors; neuroelectrophysiologic techniques allow dissection of some of these issues and provide hope for finding useful intervention loci.

Ongoing brain activity comprises action potentials and graded potentials like inhibitory and excitatory post synaptic potentials and is generated by a dynamically regulated collection of synapses on excitatory and inhibitory cells. This ensemble field of electrical activity can be recorded non-invasively using scalp electrodes. Since the first recorded human electroencephalogram (EEG), pioneered in 1924 by Hans Berger (1873–1941) (Haas, 2003), digitization methods have revolutionized EEG acquisition, generating new methods of analysis. Three general approaches can be used to record and analyze these neuroelectric phenomena: (1) continuous EEG; (2) event-related potentials (ERPs); and (3) event-related oscillations (EROs).

CONTINUOUS ELECTROENCEPHALOGRAM

Continuous EEG records brain activity when the subject is at rest or in relaxed wakefulness. Resting-state EEG is

*Correspondence to: Bernice Porjesz, Henri Begleiter Neurodynamics Laboratory, Department of Psychiatry and Behavioral Sciences, SUNY Downstate Medical Center, 450 Clarkson Avenue, Box 1203, Brooklyn, NY 11203, USA. Tel: +1-718-270-2024, Fax: +1-718-270-4081, E-mail: bp@hbnl.downstate.edu

primarily analyzed in the frequency domain, as no specific periodicity can be imposed using fast Fourier transform-based methods. Traditionally, EEG is decomposed into the following frequency bands: delta (0–3 Hz), theta (4–7 Hz), alpha (8–12 Hz), beta (13–28 Hz), and gamma (>29 Hz), and each band reflects different types of brain activity. Variations in relative power of these specific frequency bands can indicate level of consciousness, psychologic state, or presence of neurologic disorders (Nunez, 1995; Niedermayer and Lopes Da Silva, 1999). Resting theta rhythm has its maximum power in the posterior scalp region but is not prevalent in the normal adult waking EEG. Alpha band is a posterior dominant rhythm that emerges with closing of the eyes and relaxation and attenuates with eye opening or mental exertion. Hence, it is described as an index of relaxed wakefulness. Alpha is slower in young children (closer to theta frequency) and increases with age into high alpha frequencies and is a key feature of EEG maturation (Niedermayer and Lopes Da Silva, 1999); alpha power is stable throughout adult life. Beta rhythm is present all over the scalp but predominantly at frontocentral loci and is enhanced in response to certain barbiturates, sedatives, and tranquilizers.

EVENT-RELATED POTENTIALS

ERPs are averaged scalp EEG responses time-locked to specific events in a sensory, motor, or cognitive task. The averaged responses or waveforms are composed of characteristic negative and positive deflections (i.e., components). They reflect the summated activity of network ensembles active during the various processes involved in the task (Luck, 2005). Time domain analysis compares the amplitudes and latencies of these sequential peaks and troughs. Early components with a latency of less than 100 ms reflect sensory processes, followed by early components that are associated with attention processes, while later components reflect higher associative processes. See Figure 23.1 (top panel) for illustration of various ERP components. Most early studies used "oddball" paradigms and focused on the P3 or P300 component related to stimulus significance. Amplitude of P3 is taken to reflect central nervous system (CNS) inhibition (the larger the P3, the more the inhibition) (Birbaumer et al., 1990). While the P3 component, particularly in the oddball paradigm, is most widely used to study alcohol, more recent studies have focused on various other ERP components and cognitive tasks. This review is restricted to the examination of those ERP components that have proved most significant in the study of alcoholism and they are described in later sections.

EVENT-RELATED OSCILLATIONS

EROs are embedded in continuous scalp-recorded EEG activity acquired during cognitive tasks. A substantial literature indicates that some ERP features arise from changes in dynamics of ongoing EEG rhythms/oscillations of different frequency bands that reflect ongoing sensory and/or cognitive processes (see Figure 23.1 (bottom) for illustration of EROs during P3 response to targets in oddball task). While EROs may be partitioned into the same frequency bands as spontaneous resting EEG (e.g., delta, theta, alpha, beta, gamma), they are functionally different from spontaneous rhythms (Klimesch et al., 2007). Specific frequency oscillatory responses have been attributed to underlie various cognitive processes, as follows: delta: signal detection and decision making; theta: conscious awareness, recognition memory, and episodic retrieval; slow alpha: attribution of attentional resources; fast alpha: semantic memory and stimulus processing; beta and gamma: sensory integrative processes (Basar, 1999). Newer time by frequency transformations such as S-transform (Stockwell et al., 1996) or other wavelet-type analyses provide a time-based decomposition of the EEG signal associated with an event (van Vugt et al., 2007), and generate amplitude/power measures and phase information. EROs influence the timing of neural activity and coordinate synchronous activity in groups of active neurons (Fries, 2005). Synchronization of oscillations underlie self-organization of neural networks and are important indices of maturity and efficiency of these networks, providing an energy-efficient mechanism for coordination of distributed neural activity (Buzsaki and Draguhn, 2004). Phase relationships between signals from different brain regions provide a measure of temporal interactions between transient active cognitive networks; hence phase synchrony can be considered as an index of "crosstalk" or communication in the brain (Sauseng and Klimesch, 2008; Uhlhaas et al., 2010) and aids in the study of functional connectivity. High-frequency (i.e., beta and gamma) EROs are implicated in short-range communication, whereas low-frequency (i.e., delta, theta, and alpha) EROs are involved in longer-range communication between brain areas (von Stein and Sarnthein, 2000).

Advances in analytic techniques that adapt new mathematic methods have aided the study of complex cognitive processes and neural communication in normal (Hummel and Gerloff, 2005; Chen et al., 2008b; Chorlian et al., 2009) as well as pathologic conditions (Uhlhaas and Singer, 2006). Important state-of-the-art analysis tools based on mathematic approaches include: (1) phase synchrony in oscillations (Varela et al., 2001);

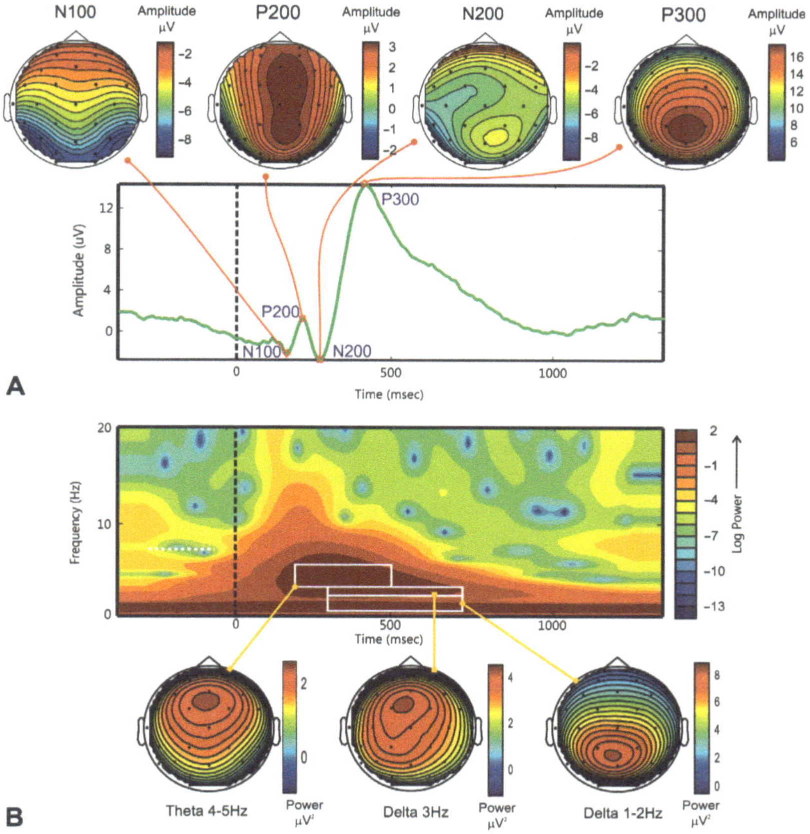

Fig. 23.1. Visual oddball event-related potential (ERP) and corresponding event-related oscillations (ERO) using time frequency transformation. (Data from Jones et al., 2006b.)

(**A**) Responses to rare targets in a visual oddball task at the vertex (Cz electrode): Typical grand mean waveform from 100 normal control subjects showing the major sequential ERP components and their topography, namely N100 (N1), P200 (P2), N200 (N2), and P300 (P3). These components are discussed in the text and described here:

The N1 (N100) component, a negative deflection occurring around 100 ms, is involved in an early attentional selection process; it is dampened to unattended stimuli and enhanced to attended ones (Hillyard et al., 1973; Näätänen et al., 1978). The vertex P200 component has been associated with visual feature discrimination (O'Donnell et al., 1997) and is sensitive to salient features of target stimuli (Luck and Hillyard, 1994a). The posterior N200 reflects the degree of attention required for processing stimuli and is sensitive to target probability (Folstein and van Petten, 2008). The P3 (P300) is a large positive deflection seen 300–700 ms after a rare stimulus embedded in a series of unattended standard stimuli. It is proposed to reflect attentional allocation and context updating (Polich and Herbst, 2000) and cognitive closure (Desmedt, 1980; Verleger, 1988), while its time of occurrence (latency) reflects mental processing speed; the earlier and larger the P3, the easier the processing. Evidence indicates that P3 has multiple sources, with contributions from frontal cortex (including anterior cingulate) and hippocampus (Halgren et al., 1980; Menon et al., 1997; Kiehl and Liddle, 2001; Ardekani et al., 2002).

(**B**) Time–frequency transformation using the S-transform (same data from top panel), illustrating the distribution of power over time during processing of the target. The white boxes indicate time–frequency regions of interest (TFROI) in specific band widths and their topography in the head plots at the bottom. It can be seen that during target detection theta (4–7 Hz) has a frontal distribution while delta (1–2 Hz) has a posterior distribution (Jones et al., 2006b).

and (2) methods of source localization (e.g., sLORETA (low-resolution brain electromagnetic tomography) which solves the "inverse problem" to generate anatomically constrained solutions of active sources within the brain that underlie the event-related scalp activity (Pascual-Marqui, 2002). The ERP and ERO analyses are supplemented by source localization techniques to infer anatomic substrates and have been successfully used to study several psychiatric disorders including AUDs (Coutin-Churchman and Moreno, 2008; Holmes and Pizzagalli, 2008; Kamarajan et al., 2010; Itoh et al., 2011; Pandey et al., 2012). Hence, these tools are instrumental in translating and comparing findings from electrophysiologic studies with those from imaging methodologies, or using them in conjunction with each other, to create a multimodal approach.

This review will focus on the effects of alcohol on brain function using non-invasive electrophysiologic techniques described above (EEG, ERP, and ERO), and will be restricted to measures that have proved most significant for the study of AUDs; it will be divided into four main sections: The first will examine the effects of acute doses of alcohol on brain function in social drinkers, and will include studies comparing those with a family history of alcoholism to those without a family history. The second section will address the newer studies dealing with the effects of binge drinking on brain function. The third section will review studies on chronic effects of alcohol on brain function, and will try and tease apart electrophysiologic indices of brain dysfunction that antecede the development of AUDs from those that are a consequence of AUDs. The fourth section will discuss the use of these electrophysiologic measures as endophenotypes for the development of AUDs and will review results with specific genes that have been identified with these methods.

ACUTE EFFECTS OF ALCOHOL ON THE BRAIN IN SOCIAL DRINKERS

The focus of alcohol challenge studies has been to investigate the effects of alcohol on normal brain function, as assessed by various electrophysiologic measures obtained at rest, and while engaged in sensory or cognitive tasks. These studies have also been very useful in examining whether naïve offspring of alcoholics who are at high risk (HR) respond differently to alcohol than offspring of non-alcoholics at low risk (LR), revealing an underlying neural liability with exposure to alcohol. No differences have been reported between HR and LR subjects in the uptake and clearance of alcohol in the blood on the blood alcohol curve (Newlin and Thomson, 1990). Past and current longitudinal studies have shown that a low level of response to alcohol predicted later heavier drinking and mediated a disposition for developing AUDs (Schuckit, 1994; Volavka et al., 1996; Trim et al., 2009). A meta-analysis suggests that a diminished response to alcohol is more frequently seen in family history-positive (FHP)/HR compared with family history-negative (FHN)/LR subjects (Pollock, 1992), but not all studies concur (see Newlin and Thomson, 1990, for review).

Acute effects of alcohol on EEG

The prominent effects of low doses of alcohol include increases in slow alpha activity or lowering of alpha peak frequency, while moderate doses show increases in slow alpha and theta bands (Ehlers et al., 2004). The effects of alcohol on beta band are more equivocal. Decreases in beta peak frequency (Ehlers et al., 1989) and increases in beta power have been reported (Ehlers and Schuckit, 1990; Stenberg et al., 1994); increased beta is also associated with moderate drinking (Ehlers and Schuckit, 1990). While changes in these frequency bands were marked at both posterior and frontal scalp loci, the alpha increase was very prominent in anterior regions (Ehlers et al., 1989). It has been proposed that acute ethanol administration disrupts the non-linear structure of EEG oscillations, thus increasing randomness (Ehlers et al., 1998b). The effects of alcohol on the EEG of subjects at risk for developing alcoholism as determined by spectral analysis methods are well known (Table 23.1). FHP individuals have shown greater increases than FHN in alpha (Pollock et al., 1983; Cohen et al., 1993a) and greater decreases of fast alpha after alcohol administration (Pollock et al., 1983). Alpha activity has also been positively associated with desire to drink in FHP before and after consumption (Kaplan et al., 1988), and beta power increases are also prominent (Ehlers and Schuckit, 1990).

Differences in EEG response to alcohol may have ethnic variations; Hispanic FHP young adults had decreased fast alpha while non-Hispanic adults showed an increase in the same band. Fast alpha (9–12 Hz) power at baseline was also found to be negatively associated with level of response to alcohol, with increased EEG alpha power at baseline being predictive of a less intense response to alcohol (Ehlers et al., 2004). An early prospective study (Volavka et al., 1996) showed that, in high-risk men, a diminished alpha-frequency EEG response to alcohol was related to the development of alcohol dependence 10 years later.

Newer studies using a multimodal approach, transcranial magnetic stimulation along with EEG, have shown a strong effect of ethanol on cortical connectivity especially over right frontal and also on left parietal areas (Kahkonen et al., 2001). In a recent interesting experiment that measured alcohol effects in real-world situations – a cocktail party and being in a driving simulator – an equation combining the beta and theta power was shown to be very successful in classifying alcohol and placebo, and the EEG score was significantly related to breath alcohol content (Gevins et al., 2012).

In summary, alcohol challenge studies in HR and LR subjects uncover a reactive alpha system. A tendency of slowing of peak frequency within reactive bands and reduced non-linearity of EEG postethanol is suggestive of a widespread synchronization of neuronal substrates, especially since low frequencies are proposed to be involved in long-range synchronizations in the brain (von Stein and Sarnthein, 2000). Both slow (theta and slow alpha) and fast (beta) frequencies appear to be affected postethanol and this is indicative of a modulation in thalamocortical networks. In their model, Llinas

Table 23.1

Alcohol challenge electroencephalogram (EEG) studies

Study	Sample characteristics	Alcohol dose	EEG variable	Summary of findings
Pollock et al., 1983	31 FHP males, 17 FHN control males; 19–21 years	0.5 g/kg consumed over 15 minutes	EEG from C3, C4, T3, T4, P3, P4, O1, O2 slow alpha (7.42–9.46 Hz), fast alpha (9.75–12.1 Hz), theta (3.51–7.03 Hz), mean alpha frequency	Decrease in fast alpha and mean alpha frequency; increase in slow alpha and theta bands in both groups (postingestion), greater increase seen in high-risk patients. High-risk patients had greater decrease of fast alpha and mean alpha frequency than low-risk patients at 120 minutes. Largest effects seen on posterior electrodes
Ehlers and Schuckit, 1990	24 FHP males and 24 FHN males; 21–25 years	0.75 mL/kg dose Mean BAC at 90 minutes – 0.0726 g/dL	EEG from four bipolar derivations: F3–C3, F4–C4, P3–O1, P4–O2 Power: beta (12–20 Hz) at P3–O1, P4–O2	Increased beta power at 90 minutes only in FHP. Moderate drinkers had higher beta power at baseline and 90 minutes postingestion
Lukas et al., 1989	6 right-handed females; FHN (5), FHP (1); 21–27 years	0.7 g/kg consumed over 18 minutes Plasma ethanol 70–80 mg/dL	EEG power: delta (0.5–3.5) Theta (4.0–7.5) Slow alpha (7.5–9.5) Fast alpha (9.5–12.5) Beta1 (12–15.5) Beta2 (16.0–19.5) Beta3 (20.0–23.5)	Increased slow alpha (7.5–9.5) and fast alpha (9.5–12.5) all over the scalp in FHN. FHP subject showed reduced alpha
Lukas et al., 1990	4 males; 25–39 years; light social drinkers; FHN (2), FHP (2)	0.7 g/kg consumed over 19 minutes	Power in alpha spectral band; auditory P300 amplitude and dipole	Increased alpha power all over the scalp in FHN. FHP showed no change. P3 amplitude lower in FHN at 30 minutes, no change in FHP. Dipole shows a shift to a more posterior distribution
Lukas and Mendelson, 1988	12 males; 21–35 years; light social drinkers	0.7 g/kg consumed over 15 minutes; plasma ethanol ~80 mg/dL	EEG power in four bands: 0.25–4 Hz; 4–8 Hz; 8–13 Hz, and 13–30 Hz	Increase in alpha power (8–13 Hz) 0–30 minutes; followed by a decrease in alpha power 35–60 minutes
Kaplan et al., 1988	25 high-risk + FHP males and 24 low-risk males; 21–28 years	Low dose; 2 12oz beer; ~18 g of alcohol	EEG alpha (9–12 Hz) power; subjective ratings of desire to drink and intoxication	Increase in alpha power after alcohol in both groups; the positive relationship of alpha activity with desire to drink was found only in FHP and not in FHN
Ehlers and Schuckit, 1988	30 males; 15 FHP and 15 FHN controls	0.75 mL/kg	EEG from two bipolar derivations: F4–C4, P4–O2 Power: theta (4–7.5 Hz), slow alpha (7.5–9 Hz), fast alpha (9–12 Hz), beta (12–20 Hz); EEG stability measure – coefficient of variation of mean power	Increase in alpha power at 90 minutes postethanol, especially in low alpha and a decrease in EEG stability for alpha band. This destabilizing effect was less intense in FHP

Continued

Table 23.1

Continued

Study	Sample characteristics	Alcohol dose	EEG variable	Summary of findings
Ehlers et al., 1989	24 males; 21–25 years; FHN	0.75 mL/kg; BAC at 90 minutes = 0.0726 g/dL; at 150 minutes = 0.0562 g/dL	EEG from four bipolar derivations: F3–C3, F4–C4, P3–O1, P4–O2 Power: theta (4–7 Hz), slow alpha (7.5–9 Hz), fast alpha (9–12 Hz), beta (12–20 Hz)	Increased power in theta (all loci) and slow alpha (right frontal locus) at 90 and 150 minutes postingestion. Peak frequency in fast alpha (all loci) theta and beta bands (frontocentral loci) also decreased after alcohol ingestion. Drinking quantity/frequency associated with beta band in EEG; subjective response to ethanol associated with fast alpha
Stenberg et al., 1994	37 males – 22 (ethanol group) and 15 (placebo group); 21–29 years	Dose 1 0.5g/kg; dose 2 1.0 g/kg; mean BAC 31 mg/dL and 78 mg/dL	EEG power: delta (0.5–3.5) theta (4.0–7.5), alpha (8–12), beta (12.5–30)Hz	Robust dose-dependent increase in theta (parietal) and beta (frontal and parietal) in alcohol group; increase in alpha seen in both groups; left hemisphere effects of alcohol; low beta at baseline was associated with larger alpha increase to alcohol
Ehlers et al., 2004	FHP subjects; 79 Hispanic – 20 males and 59 females; 206 non-Hispanic – 78 males and 128 females; mean (SD) age 22 (2.9)	0.75 mL/kg (females), 0.90 mL/kg (males) consumed over 7 minutes	Bipolar leads: P3–O1, P4–O2; slow alpha (7.5–9 Hz), fast alpha (9–12 Hz)	Increased slow alpha power after ethanol ingestion. Decrease in fast alpha power following alcohol in Hispanic subjects; increases in fast alpha power following alcohol in non-Hispanic subjects
Ehlers et al., 1998b		0.75 mL/kg consumed over 7 minutes	EEG from bipolar P4–O2; coherence time; relative alpha power; relative theta power; non-linear analysis measures – slope asymmetry, Kaplan's delta epsilon intercept and slope; 1% radius and redundancy	Increase in coherence time; decrease in alpha relative power and increase in theta relative power. Significant changes in non-linear measures suggesting a reduction in non-linearity in EEG
Ehlers and Schuckit, 1991	24 FHP males, 24 FHN males		Fast alpha (9–12 Hz)	Decreased fast alpha (9–12 Hz) energy at 90 minutes in FHN. FHP had higher fast alpha energy than FHN at baseline
Cohen et al., 1993b	21 males; 19–39 years (mean 22); light social drinkers; FHN	0.5 mL/kg (low dose); 0.8 mL/kg (high dose) consumed in 10 minutes	Slow alpha 7.5–10 Hz; fast alpha 10.5–13.0 Hz Slow beta 13.5–19.5 Hz Fast beta 20–26 Hz	Increase in slow alpha for both low- and high-dose ethanol at frontal regions F3, F4, C3, and C4 35–70 minutes postdrink
Cohen et al., 1993a	21 high-risk (FHP) males, light drinkers, 19–39 years (mean 22 years) vs 21 low-risk (FHN) males, 19–29 years (mean 22 years)	0.5 mL/kg (low dose) 0.8 mL/kg (high dose) consumed in 10 minutes	Slow alpha 7.5–10 Hz; fast alpha 10.5–13.0 Hz Slow beta 13.5–19.5 Hz Fast beta 20–26 Hz	Increase in slow alpha for both low- and high-dose ethanol at frontal regions F3, F4, and right parietal P4 for high-risk and low-risk patients during ascending limb. High-risk patients showed higher increase. The descending limb revealed steeper decrease in slow alpha in high-risk patients and no decrease in low-risk patients

BAC, blood alcohol curve; FHP, family history-positive; FHN, family history-negative.

and colleagues (1999) have proposed that the enhanced low-frequency (theta) oscillations in the thalamocortical module can affect the lateral inhibitory drive in the cortex and eventually result in high-frequency coherent activation of cortical modules. Stenberg et al. (1994) also observed that a possible harmonic relationship may exist between the theta and beta bands affected in their study and speculated the presence of related or even same oscillators underlying the ethanol effect on EEG. Hence, these findings provide a strong lead for examining the neurophysiologic and neurochemical bases of vulnerability to acute alcohol.

Acute effects of alcohol on ERPs

ERPs were the earliest tools used to study the effects of alcohol on the brain and more recently this has been supplemented by the ERO approach (see next subsection). The early studies were also interested in understanding the modulatory influence of family history and risk on ERPs (Porjesz and Begleiter, 1983). Primarily using auditory oddball tasks and male participants, changes in P3b characteristics after administration of alcohol have been observed in several studies (Table 23.2). Alcohol administration has the general effect of slowing latencies and reducing amplitudes of the P3b peak with both auditory and visual stimuli; a number of studies indicate that HR/FHP subjects manifest less P3b reduction and have a faster recovery of ERP features to baseline levels (Elmasian et al., 1982; Schuckit et al., 1988; Porjesz and Begleiter, 1990). Three-stimulus oddball tasks also generate more anterior P3a components to unattended rare non-target stimuli, which index orienting to novelty (Courchesne et al., 1975; Squires et al., 1975; Knight and Scabini, 1998). Decremental effects of alcohol on P3a amplitudes have also been significant (Campbell and Lowick, 1987; Ehlers et al., 1998a; Jaaskelainen et al., 1999; Marinkovic et al., 2001), suggesting an effect of alcohol on attention processes.

One fundamental question related to low-voltage P3 in HR offspring concerns the potential effect of alcohol in subjects with low P3 at baseline, the subjects who are presumably at risk. Ehlers et al. (1998a) noted that low P3 during a placebo condition was predictive of low level of change or an actual increase in P3 amplitude after alcohol challenge. However, the influence of variables such as ethnicity (Ehlers et al., 1998a) and gender has not been thoroughly explored. Taken together, these findings suggest that, while some electrophysiologic differences between HR and LR individuals are apparent without an alcohol challenge, others are seen only in response to ethanol challenges, possibly representing sensitization and tolerance in the HR subjects that may be innate.

The P3 results suggest a multifocal influence of alcohol on cognition, which is supported by later ERP studies examining the effect of alcohol on several cognitive domains, including attention and monitoring, leading to effective performance control. These domains are part of the rubric of executive control, and impairment in these processes has a widespread influence on cognition. Alcohol-induced changes have been reported in components related to visual processing (Colrain et al., 1993; Weschke and Niedeggen, 2012), covert attention (Jaaskelainen et al., 1995; Kenemans et al., 2010), sustained attention (Rohrbaugh et al., 1987), motor preparation (Marinkovic et al., 2000), response inhibition (Easdon et al., 2005), error/action monitoring (Ridderinkhof et al., 2002; Euser et al., 2011), and semantic memory (Marinkovic et al., 2004).

Among the early ERP components, the N100 amplitude is consistently reduced postalcohol, suggesting involvement of early sensory or attention-related process, particularly in auditory paradigms. Reporting on earlier components of the oddball task – N100 and P200 (Fig. 23.1) – greater sensitization of P2 on the ascending limb of the blood alcohol curve and faster recovery to baseline on the descending limb for both N1 and P2 in HR was seen (Cohen et al., 1998), in keeping with the differentiator model of Newlin and Thomson (1990). Some studies using visual stimuli did not report N1 amplitude reduction postalcohol (Rohrbaugh et al., 1987; Colrain et al., 1993). A preattentive component to spatial frequency (SFD80), which appears at 80 ms after the stimulus, was not affected by moderate alcohol (Kenemans et al., 2010). Alcohol was found to decrease the discrimination ability for visual contrast, resulting in reduced visual evoked potential amplitudes, but its effects on motion perception indicated impairment in visuospatial attention (Weschke and Niedeggen, 2012).

Similarly, covert attention is affected by alcohol, as indicated by the reduced mismatch negativity (MMN) amplitudes for both auditory and visual stimuli (Jaaskelainen et al., 1995, 1996; Kenemans et al., 2010). MMN is an automatic neuronal mismatch between a deviant auditory input and a sensory-memory trace representing the standard stimuli; it is a preattentive process engaging covert attention, and provides an objective measure of auditory discrimination and sensory memory (Näätänen, 1990). (See section on attention – N100 and mismatch negativity, below, for studies on MMN and chronic alcoholics.)

Two theoretic models have been proposed to evaluate the effects of alcohol on cognition: First, the attention allocation model, which suggests that alcohol enhances the focus and affects shifts of attention, such that only the most salient cues are attended to and other available cues are ignored (Steele and Josephs, 1988, 1990). This is

Table 23.2

Alcohol challenge studies using event-related potentials (ERP)

Study	Sample characteristics	Alcohol dose	Task and ERP components	Summary of findings
Elmasian et al., 1982	15 FHP males and 15 FHN males; 21–26 years. 1–84 drinks per month; three dose groups	Placebo, 0.56 g/kg and 0.94 g/kg consumed in 24 minutes. BAC peak 66 mg/100 mL and 97.6 mg/100 mL	Auditory oddball task; P3 peak latency and peak amplitude; non-target P2 peak	Reduction in P3 amplitude in FHN (most marked in high dose) and FHP (strong decrease at all three doses); alcohol increased latencies in both groups but FHP had more slowing
Teo and Ferguson, 1986	9 males and 4 females; 20–44 years	Placebo, 0.3 g/kg and 0.54 g/kg of total body weight	Auditory oddball task; Cz electrode; N100, P200, N200, and P300	High dose produced an increase in latency for N1, P2, N2 components for both stimuli. Target P3 amplitude reduced and latency increased (dose-dependent)
Rohrbaugh et al., 1987	Male social drinkers	Separate sessions of four doses ranging from 0 (placebo) to 1.05 g/kg lean body weight, with periodic maintenance dose of 0.12 g/kg	Visual sustained attention task requiring response to target digits in a stream of blurred non-target digits, presented one at a time. ERP – N1, P2, N2, and P3	Dose-related decrease of detection performance. Early ERP components (N1 and P2) not affected by alcohol. Ethanol caused dose-related delays in N2 and P3 latencies and reaction time increases. The amplitude of N2 also decreased over time on task, and P3 amplitude decreased both as a function of dose and time on task
Campbell and Lowick, 1987	10 subjects	Placebo; 1.0 mL/kg dosage of 94% ethanol	Auditory oddball task; ERP components – N1b, P2, and P3 to distractor and target waveforms	N1b and P2 amplitudes were significantly reduced. P3 to distractor stimuli was significantly reduced. No target P3 change
Schuckit et al., 1988	21 FHP males and 21 FHN males; 21–25 years	Placebo, 0.75 mL/kg, 1.1 mL/kg consumed in 10 minutes	Auditory oddball task; P3 peak latency	Increase in P3 latency postethanol in both groups for ascending BAC. For the high dose the FHP showed a faster recovery in P3 latency compared to FHN
Colrain et al., 1993	10 males; 18–25 years; social drinkers	Placebo; 0.28; 0.36; 0.54 and 0.72 g/kg consumed over 20 minutes	Reversing checkerboard in standard oddball paradigm. Peak and latency P1, P2, N2, P3. 11 Electrode sites (10–20 system)	Increase in reaction time, N2 and P3 latency with increasing dose. No effect on P1 and P2. Reduced RMS power of the P3 complex and N2–P3 amplitude difference at central and parietal sites at higher alcohol doses
Jaaskelainen et al., 1995	5 females; 5 males ($n = 10$); 20–30 years. Social drinkers	Placebo; 0.5 g/kg consumed in 40 minutes (BAC prior to test = 0.063%)	Two auditory oddball streams (target/non-target) presented to right and left ear. Subjects told to attend to one ear and ignore the other. ERP – N1(Cz), P2(Cz), MMN(Fz), N2b(Cz), P3b(Pz)	Attended and unattended N1 amplitudes were reduced after alcohol ingestion. Unattended P2 amplitudes also reduced after alcohol. MMN amplitude decreased significantly after alcohol, with no change on N2b and P3b amplitudes. Latency increased for MMN and N2b

Jaaskelainen et al., 1996	4 females, 5 males ($n = 9$); 23–29 years. Social drinkers	Placebo; 0.55 g/kg (BAC prior to test = 0.58); 0.85 g/kg (BAC prior to test = 1.04); consumed in 25 minutes	Auditory oddball with short and long interstimulus interval (ISI) while subjects read a book. Electrodes : F3, Fz, F4, C3, Cz, C4, Pz. ERP – MMN from Fz (rare-standard tone); N1 and P2 from Cz (standard tones)	N1, P2, and MMN amplitudes reduced for long ISI task. High alcohol dose reduced N1 amplitudes in both long and short ISI tasks. MMN decrease was primarily frontal
Porjesz and Begleiter, 1996	24 HR males and 24 LR males; 19–24 years	0.5 mL/kg (low dose) 0.8 mL/kg (high dose) consumed in 10 minutes	Visual oddball, with easy and difficult line discrimination. P3 amplitude, latency	Decrease in P3 amplitude in both; percent decrease from baseline to ascending limb was greater in HR. P3 latency was longer at peak BAC and descending limb for both; HR recovered faster to baseline level
Jaaskelainen et al., 1999	7 males and 4 females ($n = 11$) social drinkers; 24–30 years	Placebo; 0.3 g/kg; 0.6 g/kg; consumed in 25 minutes	Auditory forced choice task with two tones and rare deviations in tone frequency	Low dose: For standard tones N1, N2 amplitude decreased; N2 and P3 latency increased. P2 amplitude increased. For deviant tones P3a amplitude decreased. Higher dose did not produce any additive effects
Cohen et al., 1998	27 LR males, 23 HR males; 18–30 years; HR = light drinkers ($n = 14$), heavy drinkers ($n = 9$)	Placebo, 0.5 g/kg, 0.8 g/kg consumed in 10 minutes	AEP to binaural auditory clicks; measures: N100, P200	Within group – LR showed decreased N100 between placebo and high dose; HR showed decreased N100 and P200. Between group – HR N100 and P200 amplitude decrease is higher than LR during ascending BAC; HR – N100 and P200 amplitudes recover to baseline levels faster that LR in descending BAC
Ehlers et al., 1998a	28 FHP and 14 FHN males; 18–25 years; Native American heritage	Placebo, 0.75 mL/kg consumed over 7 minutes; BAC peak ~0.078 g/dL	Auditory oddball plus noise; ERP data from Fz, Cz, Pz; P3a (225–300 ms) and P3b (300–400 ms) amplitude	Alcohol significantly reduced P3a amplitude when compared to placebo
Marinkovic et al., 2000	12 males; 21–26 years; light social drinkers	Placebo, 0.4 g/kg consumed in 10–15 minutes	Word recognition task; 13 electrodes, 10–20 system. Lateralized readiness potential (LRP) (C3 subtracted from C4). Mu (12.5–17.5 Hz) band	Go (old words) condition shows no alcohol-related LRP changes, NoGo (novel words) condition shows enhanced LRP to alcohol. Mu band power reduced in NoGo trials
Marinkovic et al., 2001	12 males; 21–28 years; light social drinkers and non-smokers	Placebo, 0.4 g/kg consumed in 10–15 minutes	Three stimulus auditory oddball task; 13 electrodes, 10–20 system. ERP (average amplitude) – N1, N2, P3a, P3b, and slow wave	Increased frontal N2 and decreased frontal P3a average amplitude to ethanol. Decreased P3b to alcohol centroparietally
Ridderinkhof et al., 2002	14 right-handed males; social drinkers; age 20–40 years	Placebo, low-dose (0.45 g/kg), and high-dose (0.85 g/kg); BAC 0.4%, 0.97%	Subjects pretrained on Eriksen Flanker Task, then tested again after placebo and alcohol. ERP – N2 for congruent and incongruent correct responses; ERN for error responses	No effect of alcohol on N2; ERN was reduced significantly by both alcohol doses
Bartholow et al., 2003	42 adults (21 males and 21 females); age 21–30 years	Placebo; moderate dose (females 0.36 g/kg, males, 0.40 g/kg); high dose (females 0.72 g/kg, males 0.80 g/kg)	Version of Eriksen Flanker Task; EEG: 20 scalp electrodes (10–20 system). PCA was conducted on ERP data. Two poststimulus epochs, 200–350 ms and 400–500 ms, were evaluated. Reaction time and accuracy	Dose-related decrease in P3 amplitude. Moderate dose increased P3 latency; high susceptibility to alcohol moderated an increase in ERN amplitude to both alcohol doses

Continued

Table 23.2

Continued

Study	Sample characteristics	Alcohol dose	Task and ERP components	Summary of findings
Marinkovic et al., 2004	12 males; 21–26 years; light social drinkers	Placebo, 0.4 g/kg consumed in 10–15 minutes	Verbal/word recognition task; 13 electrodes, 10–20 system. ERP components – P/N180, N310, N450, P580	N180 at T5, T6 – reduced after alcohol ingestion. N450 and P580 latency was increased by alcohol
Easdon et al., 2005	Three groups with 4 males and 4 females: control (mean age = 27.75), low dose (mean age = 24.13) and high dose (mean age = 26.75)	Placebo, 0.56 g/kg (low dose) or 0.80 g/kg (moderate dose)	Sustained attention to response task (Go-NoGo) 64-channel EEG; stimulus-locked (P1,N170, N2, P3) and response-locked ERPs (correct: Nc, Pc; error: Ne, Pe)	Increase in error rate postethanol. Both doses of alcohol reduced N170 and P3 amplitudes during Go, No-Go, and error trials. Both doses of alcohol reduced the Ne amplitude whereas the Pe amplitude decreased only after moderate doses of alcohol
Euser et al., 2011	64 males; 30 males (placebo); 31 males (alcohol). Age 18–25 years	Placebo; 0.65 g/kg consumed over 8 minutes	Balloon analogue risk task (BART); Fz, FCz, and Cz; FRN (max. negativity 200–300 ms after feedback); P300	No alcohol effect on FRN, reduced amplitude of P300 to loss feedback. Alcohol affects later stages of outcome evaluation
Kenemans et al., 2010	8 females and 8 males; 18–25 years; social drinkers	Placebo, moderate alcohol (BAC level between 0.04% and 0.07%)	Auditory MMN task; visual task – 2.4 versus 0.6 c/d gratings in pseudorandom sequence (deviant (10%), standard (90%)); ERP component – rareness related negativity (RRN: difference between deviant and standard ERP between 120 and 170 ms at Oz), SFD80 – ERP response to spatial frequency; MMN	MMN – alcohol decreased amplitudes significantly. RRN – moderate-dose alcohol significantly reduced RRN. SFD80 – this preattentive sensory component was not affected by alcohol. Descending BAC analyzed
Weschke and Niedeggen, 2012	29 subjects in two groups: alcohol and placebo. Alcohol (n =15; 8 males, 7 females); Placebo (n = 14; 7 males, 7 females)	Placebo; alcohol: females 0.30 g/kg, males 0.35 g/kg consumed within 30 minutes. BAC level 0.047%	Contrast perception task (subjects identified high contrast stimulus); Motion perception task (subjects identify stimulus with dots moving up). EEG – O1, O2, P3, P4. ERP components – contrast evoked negativity (110–170 ms); motion evoked negativity (160–240 ms); late positive complex (P300)	Alcohol reduced contrast-evoked negativity and P300; Motion-evoked negativity was reduced contralaterally by alcohol

BAC, blood alcohol curve; ERN, error-related negativity; FHP, family history-positive; FHN, family history-negative; FRN, feedback-related negativity; HR, high risk; LR, low risk; MMN, mismatch negativity; AEP, auditory evoked potential; PCA, principal component analysis.

supported by results of studies on overt and covert attention, as described previously (Jaaskelainen et al., 1996). The second related model is the response inhibition model (Fillmore and Vogel-Sprott, 1999, 2000) based on the theory of cognitive control (Logan and Cowan, 1984), where behavioral activation and inhibition are two independent processes. Impaired behavioral inhibition processes underlie deficits of self-control observed postalcohol. The alcohol-related decline in performance may be related to difficulties in maintaining attention on the task at hand and/or deficits in inhibiting a prepotent response.

This has been studied with a "reverse oddball" Go/NoGo task, a paradigm widely used to estimate response inhibition (Jodo and Kayama, 1992; Eimer, 1993; Falkenstein et al., 1999) wherein a subject has to respond to a given stimulus (Go) and withhold it for another stimulus (NoGo). The ERPs to the two stimuli were examined to evaluate the neural correlates of response production and inhibition. Two significant ERP signatures of response inhibition have been found: an enlarged negative frontocentral N2 component (200–300 ms) on NoGo trials, and an augmented positive-going frontocentral "NoGo P3" (300–600 ms) (Pfefferbaum et al., 1985; Pfefferbaum and Ford, 1988; Eimer, 1993; Filipovic et al., 1999).

Low and moderate doses of alcohol increased the number of errors relative to alcohol-free performance (Easdon et al., 2005). Success in withholding a prepotent response was associated with an early-enhanced stimulus-locked negativity (N170) at inferior parietal sites, which was delayed when participants failed to inhibit the motor command. Moreover, both doses reduced N170 and P3 amplitudes during Go, NoGo, and error trials. In comparison with correct responses, errors generated large response-locked negative (Ne) and positive (Pe) waves at central sites to feedback, components reflecting motivation significance and recognition of error. Both doses of alcohol reduced Ne amplitude, whereas Pe amplitude decreased only after moderate doses. These results indicate that behavioral disinhibition following alcohol consumption involved alcohol-induced deficits in maintaining and allocating attention, thereby affecting the processing of incoming stimuli and the recognition that an errant response has been made.

A frontal negativity in the N2 range (variously named as ERN (error-related negativity), F-ERN, FMN (frontomedial negativity), ORN (outcome-related negativity)) has been identified as the ERP correlate of the feedback response (Nieuwenhuis et al., 2004; Hajcak et al., 2006). Moderate alcohol has been shown to increase the detrimental effects of incongruent flankers of a visual target in a response inhibition task (Bartholow et al., 2005), along with increased ERN and decreased P3 amplitudes, suggesting a restructuring of self-monitoring processes postalcohol use. To answer the question if alcohol causes a reduction in the efficiency of control processes in general, a study using the flanker task revealed impairments in both interference control and error detection (Ridderinkhof et al., 2002). The investigators found that the frontocentral ERN was attenuated significantly by alcohol consumption while N2 amplitude was not, indicating that alcohol in moderate doses produced a significant deterioration in detection of erroneous responses (ERN). Another study suggested that later stages of outcome processing were affected by alcohol, as indicated by reduced P300 to loss feedback (Euser et al., 2011). Motor preparation was evaluated using the lateralized readiness potential (LRP) in trials where the response had to be withheld (NoGo); the LRP was significantly higher postalcohol ingestion and the movement-related mu rhythm was also significantly attenuated, suggesting a premature activation of the motor system (Marinkovic et al., 2000). This also provides evidence for enhanced motor impulsivity that may be part of the conceptual framework of the response inhibition model.

Alcohol effects on cognitive domains such as semantic and mnemonic processes have not been studied extensively; however, one study showed that alcohol ingestion attenuated the temporoparietal negative potential (N180), revealing an effect of a moderately low alcohol dose on early prelexical stage of verbal processing (Marinkovic et al., 2004). Alcohol significantly increased the difficulty of semantic access and integration, as reflected in its effect on later potentials – larger N450 amplitude and longer P580 latency. This effect was particularly prominent in arousal-related trials, suggesting that alcohol impairs processes that modulate cognitive functioning related to semantic and integration systems rather than via memory processes.

These studies provide some evidence for both models, suggesting either a multifactorial pathway or a common upstream factor perhaps relating to structural–functional dynamics of the brain networks. Studies in the following section have attempted to explore the latter.

Acute effects of alcohol on EROs

More recently, with the use of EROs, a small number of studies have attempted to parse apart the effects of alcohol on various frequency bands and brain regions involved in aspects of cognitive processing.

Several studies have examined the effects of alcohol on cognitive processing in the auditory modality. The dose-related impact of alcohol on auditory transient evoked 40-Hz responses during a selective attention task was investigated; higher doses of alcohol significantly suppressed the early evoked gamma responses in both attended and non-attended conditions, suggesting cognitive impairment or lack of sensory binding (Jaaskelainen et al., 2000). Administration of alcohol was shown to decrease early synchronization during auditory encoding and increase later desynchronization in theta (4–6 Hz), low alpha (6–8 Hz), and high alpha (8–10 Hz) bands (Krause et al., 2002). This indicates that alcohol has disorganizing effects on the brain's electric oscillatory systems in theta and lower alpha frequency ranges during cognitive processing. Similarly, disorganizing effects of alcohol on phase synchronization of EROs during an auditory oddball task were recently reported in humans and rats (Ehlers et al., 2012). They demonstrated reduced synchrony within and between neuronal networks with ethanol, perhaps by increasing the level of noise in key interactions; reduced alpha-phase synchrony was also correlated with blood alcohol level.

There is also evidence to suggest acute effects of alcohol on theta oscillations during various cognitive tasks in the visual modality. Moderate alcohol intoxication modulated event-related theta activity during visual word processing, although alcohol was shown to attenuate theta power overall (Marinkovic et al., 2012). Moderate alcohol consumption was particularly deleterious to semantic retrieval since it reduced theta oscillations for real words but not pseudowords. Event-related theta power was also associated with sources in left-lateralized frontotemporal areas, reflecting lexical-semantic retrieval processes. This finding is in agreement with previous studies that suggest that executive functions are especially vulnerable to alcohol intoxication. Task-related EEG changes were studied during the performance of a mental arithmetic task for low "social" alcohol dosages with the presumption that even social drinking may have detrimental effects. While no detrimental alcohol effect was seen on behavioral indices of task performance, there was an ethanol-induced attenuation of the task-related frontally dominant theta increase; this effect was modest (Boha et al., 2009). Alcohol dose had an activating effect on autonomic measures of heart rate and electrodermal activity, and changes in non-linear measures – omega complexity and synchronization likelihood – indicate increased synchrony in the theta band corresponding to increased working memory effort (Molnar et al., 2009). Moderate alcohol consumption caused spatiotemporal changes to conflict in both early and late processing stages during a Stroop task, and attenuated total event-related theta power, suggesting that alcohol-induced deficits in cognitive control may result from theta suppression in the executive network (Kovacevic et al., 2012). Slower reaction time was associated with attenuated theta power that was attributed to anterior cingulate cortex, indicating an impairment in top-down control of motor preparation and execution. Alcohol significantly increased theta activity in subjects who were engaged in performing an attentional motor task of simulated automobile driving. Hence these results suggest that an alcohol-induced impairment of top-down strategic processing may underlie poor self-control, which in turn may affect drinking (Laukka et al., 1995).

Taken together, alcohol studies of EROs during various cognitive tasks indicate reduced-phase synchrony between and within neural networks in alpha and theta bands, and reduced theta oscillations, particularly implicating effects on frontal areas. The impact on the synchrony and generation of oscillations may influence the various cognitive operations leading to deficits in attention and response inhibition.

EFFECTS OF BINGE DRINKING ON ELECTROPHYSIOLOGY

Binge drinking is a relatively new term in the field of alcoholism and involves short periods of excessive drinking alternating with abstinence; it is generally defined as consumption of five or more drinks (four or more in females) during a 2-hour interval (Courtney and Polich, 2009). This behavior is widely prevalent among young adults and is associated with impairments in executive, visuospatial, and other domains (Hermens et al., 2013). Understanding the correlates of binge drinking in the developing brain is instrumental to designing prevention and management protocols. Animal studies have shown more brain damage from binge drinking in adolescent compared to adult animals. Regional damage to frontal association cortex and impaired hippocampal neurogenesis is greater in adolescent than in adult rats (Crews et al., 2000, 2006). Courtney and Polich (2009) have provided clear recommendations regarding coherent and precise definitions of binge drinking in order to pursue a proper evaluation of brain deficits.

Binge drinking and EEG

Investigations of resting EEG in binge drinkers have emerged in recent years, aided by a clear definition of binge drinking. In earlier studies, moderate drinkers showed greater spectral power and higher peak frequency in the beta (12–20 Hz) band when compared to low social drinkers, suggesting that beta activity might index quantity and frequency of alcohol consumption (Ehlers et al. 1989; Ehlers and Schuckit, 1990). More recently, Courtney and Polich (2010) examined male and female non-binge drinkers, low-binge drinkers, and high-binge drinkers who had been drinking alcohol at their respective levels for an average of 3 years. The non- and low-binge drinkers exhibited less spectral power than the high-binge drinkers in the delta (0–4 Hz) and fast beta (20–35 Hz) bands. Although the causal relationship between binge drinking and increased fast beta power is unclear, the authors suggest that the alteration of fast beta activity in high-binge drinkers is similar to the EEG spectral pattern seen in alcoholics (Rangaswamy et al., 2002), and may be a biomarker for potential future AUDs, even in the absence of familial alcoholism.

Functional connectivity during eyes-closed EEG was different between light and heavy drinkers in a study on EEG synchronization in heavy-drinking college students, defined as those who consume more than 30 units containing 12 grams of alcohol per week (de Bruin et al., 2004). Heavy drinkers exhibited abnormally increased synchronization in theta (4–8 Hz) and gamma (30–45 Hz) bands. These frequency bands have been associated with memory formation involving hippocampo-neocortical connections (Buzsaki, 1996). Altered synchronization could reflect structural changes in neural networks involving the hippocampus and cortex, as observed in pathologic studies (Harper, 2009). EEG synchronization in six frequency bands (delta, theta, alpha, slow beta, fast beta, and gamma) when compared between low, moderate, and heavy drinkers revealed a loss of lateralization in alpha and slow beta bands in male and female heavy drinkers (de Bruin et al., 2006). In addition, moderately and heavily drinking males had lower fast beta (20–30 Hz) synchronization than light-drinking males; synchronization in alpha and beta frequencies was impaired during rest and mental rehearsal in those drinking in excess of 21 alcoholic drinks per week.

Binge drinking and ERPs

Studies examining the neurophysiologic consequences of binge drinking are revealing both early and global effects on cognition as evaluated using event-related activity (Maurage et al., 2012). One of the earliest studies of binge drinking on cognitive changes investigated ERPs in young adult Southwest California Indians with a history of binge drinking during adolescence (Ehlers et al., 2007). Using a facial discrimination task, they found that adolescent binge drinking was associated with lower P450 (like P3b) amplitude and a longer P350 (like P3a) latency, in those with family histories of ethanol dependence.

Another study using a working-memory continuous performance task and a principal component analysis approach in analyzing ERPs examined attention and working-memory processes (Crego et al., 2009). They reported increased N2 components for matching stimuli in binge drinkers, interpreting these findings as more effortful processing in the performance of the task. They also reported no P3 amplitude differences between the matching and non-matching conditions in binge drinkers, indicating a deficit in differentiating relevant and irrelevant processes. Employing a similar task in a later study, they revealed reduced amplitudes of the late positive complex in binge drinkers when compared to controls (Crego et al., 2010). This was also associated with hypoactivation of right anterior prefrontal cortex, determined using source localization with eLORETA (exact low-resolution brain electromagnetic tomography).

A longitudinal study that compared young adult binge drinkers with age-matched non-drinking controls showed progressive cerebral dysfunction without marked behavioral deficits (Maurage et al., 2009). The baseline assessments showed no differences; however after 9 months, the binge drinkers had significantly delayed latencies for all ERP components (P1, N2, P3b) of emotional auditory processing, reflecting impairments in perceptive as well as decisional processes. However, a more recent visual oddball study reported an opposite result of increased P3b amplitudes in binge drinkers when compared to controls (Crego et al., 2012); participants included only those without a personal or family history of alcoholism, unlike the previously described Ehlers et al. study (2007), which may indicate a differential vulnerability in those with and without family history.

A study that investigated links between response inhibition in a visual Go/NoGo paradigm in social drinkers found that Go and NoGo N2 showed a strong trend of being smaller centrally for heavy drinkers compared to light drinkers, but the Go P3 showed no group differences (Oddy and Barry, 2009). Only the NoGo P3 reduction was correlated with alcohol consumption. A response inhibition longitudinal study from the same laboratory (Lopez-Caneda et al., 2012) also reported increased P3 amplitudes for both Go (at baseline and follow-up) and NoGo (only at follow-up), suggesting a progressing influence of binge drinking on response inhibition. The authors suggested that increased amplitudes of P3 may reflect compensatory mechanisms within the adolescent brain.

An alcohol cue reactivity study (Petit et al., 2012) on binge drinkers reported enhanced P100 amplitude for alcohol-related images versus neutral images. This finding is similar to that seen in adult alcoholics and was not observed in control subjects. The later components (N2 and P3) were not affected, suggesting an early attention bias for alcohol-related cues. Maurage et al. (2012), in a most detailed study on binge drinking and its ERP correlates, used a face detection oddball task to evaluate both early and late cognitive processes, effect of comorbid conditions, and alcohol consumption pattern and amount. The authors reported massive ERP impairments from the early P100/N100, N170/P2, N2b/P3a, and P3b. Alcohol intake amount and specific binge-drinking patterns were also associated with these impairments.

Taken together, these studies suggest an ongoing process of refinement in ascertaining binge drinking, and as this is still in development, a definitive picture of electrophysiologic deficits is yet to emerge. Binge drinking appears to be associated with a specific pattern of EEG activity (increased fast beta) in young adults that may reflect the future development of AUDs. The ERP measures are equivocal, particularly in the context of response inhibition; however a few studies do indicate deficits in perceptual as well as decisional processes, as reflected in decrements in early (P1, N1) and late (N2, P3b) ERP components, as well as changes in frontal sources of these activities.

CHRONIC ALCOHOLISM AND NEUROELECTROPHYSIOLOGY

Chronic alcoholism is associated with a broad spectrum of brain disturbances ranging from severe symptoms of Wernicke–Korsakoff syndrome to subtle but nonetheless significant cognitive disturbances characteristic of a majority of alcoholic patients. The etiology of alcohol-related brain damage/dysfunction is not entirely known, as there are brain changes during acute and chronic intoxication, as well as during withdrawal; some brain changes recover with prolonged abstinence and some brain anomalies antecede the development of AUDs and may be involved in the predisposition to develop AUDs. This section will focus on EEG, ERP, and ERO measures of brain dysfunction in AUDs in abstinent alcoholics, as well as in HR offspring of alcoholics, to help determine which are consequences of AUDs and which antecede its development. (For earlier reviews of other aspects of alcoholism and electrophysiology (e.g., sensory components), see Porjesz and Begleiter, 1983, 1985, 1993, 1996.)

Chronic alcoholism and resting EEG

Differences in both EEG power and coherence have been reported in alcoholics; some of these are the consequence of chronic alcoholism while others antecede its development. This section will review EEG power and coherence measures in alcoholics and those at risk.

THETA BAND

Increased resting theta power in alcoholics has been reported in a number of studies in the literature (Propping et al., 1981; Pollock et al., 1992; Rangaswamy et al., 2003); no relationship between the length of abstinence and theta power was found (Pollock et al., 1992). However another study reported a decrease in theta power in female alcoholics (Propping et al., 1981). Elevated resting theta activity observed in the EEG of alcohol-dependent individuals is indicative of a dysfunctional neurophysiologic status in these individuals. As indicated earlier in this chapter, we can speculate that the increases of theta produced by the acute administration of alcohol in healthy individuals may subsequently evolve into a more pervasive increase in theta in chronic alcoholics following prolonged exposure. No strong evidence of increased resting theta has been reported in offspring of alcoholics, suggesting that this measure may index a state-dependent condition.

Increases of theta rhythm have been seen in altered neurophysiologic states of the brain, involving altered cholinergic functioning, such as Alzheimer's disease, aging, and the transition from wakefulness to sleep (Niedermayer and Lopes Da Silva, 1999). Slow EEG activity (theta and delta) has been correlated with cholinergic activity and central cholinergic pathways (Steriade, 1990). Elevated tonic theta power in the EEG may reflect a deficiency in information-processing capacity of the CNS (Klimesch et al., 2001). The theta power increase may be an electrophysiologic index of the imbalance in the excitation-inhibition homeostasis in the cortex.

ALPHA BAND

There is extensive literature, dating back to the 1940s, indicating unstable or poor alpha rhythm in alcoholics; alcoholics manifest less prevalent and lower alpha than do non-alcoholics (for reviews, see Begleiter and Platz, 1972; Propping et al., 1981). However some more recent studies did not find the same results (Pollock et al., 1992; Enoch et al., 1999). A pronounced slow alpha decrease is associated with relapse (Saletu-Zyhlarz et al., 2004); there is an increase in slow alpha, a decrease in fast alpha, and a deceleration of the alpha centroid with 6 months of abstinence. On the other hand, participants who had a family history of alcoholism had significantly higher spectral power in the slow alpha frequencies (7.5–9 Hz) (Ehlers and Phillips, 2003); this was found for males with alcoholic fathers (Ehlers and Schuckit, 1991) and women at high risk for developing alcoholism (Ehlers et al., 1996). While reduced EEG alpha power in male and female offspring of alcoholics has been reported (Finn and Justus, 1999), this was not related to comorbid traits of anxiety or antisocial personality. A distinctive low-voltage alpha variant (LVA), has been reported to be associated with a subtype of alcoholism that is associated with anxiety disorder (Enoch et al.,

1999). LVA, which is characterized by an absence or very low-amplitude alpha rhythmicity, is found in 5–10% of individuals (Anokhin et al., 1992; Enoch et al., 1995). Ethnic variations may exist in the prevalence of LVA, EEG variants, and association with alcoholism or risk (Ehlers and Phillips, 2003).

Beta band

Beta frequency rhythms are also known as inhibition-based rhythms. Increased beta power in the EEG of alcoholics, particularly in the resting condition, has been well documented (Propping et al., 1981; Costa and Bauer, 1997; Winterer et al., 1998; Bauer, 2001; Rangaswamy et al., 2002). Increased beta power was observed at all scalp loci in the large Collaborative Study on the Genetics of Alcoholism (COGA) sample, but was most prominent in the central region for slow-medium-frequency beta (12–20 Hz) and over the frontal regions for fast beta (20–28 Hz) (Rangaswamy et al., 2002). Increased beta activity, particularly fast beta (19.5–39.8 Hz), has proved to be an excellent predictor of relapse (Bauer, 2001; Saletu-Zyhlarz et al., 2004). Desynchronized beta activity over frontal areas in relapsers has been suggested as a correlate of functional disturbance of prefrontal cortex (Winterer et al., 1998).

Increased beta power has also been described in the EEG of relatives of alcoholics (Gabrielli et al., 1982; Pollock et al., 1995; Finn and Justus, 1999; Rangaswamy et al., 2004b). However, some studies examining acute effects of alcohol on HR report an absence of pre-ethanol baseline differences in resting EEG between LR and HR subjects (Pollock et al., 1983; Kaplan et al., 1988; Cohen et al., 1991). A positive family history of alcoholism was related to increased beta power in HR (Gabrielli et al., 1982; Pollock et al., 1995; Finn and Justus, 1999), and when present along with a diagnosis of antisocial personality, it was associated with increased frontal beta power (Bauer and Hesselbrock, 1993). In the COGA study, increased beta power on the resting EEG was demonstrated in a large sample of offspring of alcoholics (Rangaswamy et al., 2004b). Taken together, as the increase in beta power in alcoholics was not related to length of abstinence (Rangaswamy et al., 2002) and was also present in individuals at risk (Rangaswamy et al., 2004b), this suggests that it may not be an effect of alcohol use, but perhaps antecede the development of AUDs.

Most studies reporting beta band differences in alcoholics and HR offspring also underscore the issue of gender in electrophysiologic research. In studies evaluating alcoholics and HR offspring of both genders, beta band changes were more robust in males while females showed either no elevation or only a modest increase (Gabrielli et al., 1982; Pollock et al., 1995; Finn and Justus, 1999; Rangaswamy et al., 2002, 2004b). Gender differences were highlighted in the COGA study, where male HR offspring had elevated slow beta (12–16 Hz), while female HR offspring showed significantly increased faster beta power (16–28 Hz), particularly those with two or more alcoholic first-degree relatives. Existing gender differences in the progression and pathology of alcoholism and spectral properties of EEG highlight the importance of studying risk indicators within the context of gender.

Interhemispheric coherence

EEG in alcoholics also reveals an increased interhemispheric coherence when compared to unaffected individuals (Kaplan et al., 1985; Michael et al., 1993). Bilateral intrahemispheric coherences in alpha and beta frequency bands were increased in both long-term abstinent and non-abstinent alcoholics compared to controls (Winterer et al., 2003a). These findings were strongest for the high alpha (10.5–12 Hz) frequency band, and were most pronounced at temporal, parietal, and occipital regions, particularly when depressiveness was included as a covariate; there was no effect of length of abstinence on these findings.

In summary, studies that have investigated the resting EEG composition indicate beta band increases as a primary characteristic feature in alcoholics and HR subjects, and, less significantly, theta band increases in the alcoholics. This reactivity of beta band has also been observed in studies assessing binge drinkers. The alpha band differences in chronic alcoholism are not conclusive and remain equivocal, although alcohol challenge studies in HR and LR subjects have uncovered a reactive alpha system that tends to shift to slower frequencies (see earlier in this chapter). The evidence of elevated beta power also provides strong support for the excitation-inhibition imbalance model proposed to underlie the predisposition to alcohol dependence (Begleiter and Porjesz, 1999), and future studies are required to clarify gender differences in EEG profiles in alcoholism.

Chronic alcoholism and event-related potentials

This section reviews ERP components reflecting various cognitive functions that are impaired in alcoholics.

Attention – N100 and mismatch negativity

Early attention selection processes are affected in both alcohol-dependent and unaffected HR individuals (Steinhauer et al., 1987), as indicated by diminished N100 component. Using a bimodal task (auditory and visual stimuli), a study on abstinent alcoholics, controls, and FHP offspring showed reduced visual N1 amplitude in alcoholics and reduced auditory N1 amplitude in the FHP individuals (Patterson et al., 1987). The dampening of N1 amplitudes to repetitive stimuli may be associated with the refractoriness (Cohen et al., 1996) or may be a reflection of lateral inhibition at the cortical level (Sable et al., 2004); however they were not very effective in differentiating HR from LR individuals.

Larger MMN amplitudes have been reported in recently detoxified alcoholics (Kathmann et al., 1995). The automatic stimulus change detector mechanism associated with MMN generation is impaired in chronic alcoholics over the age of 40, suggesting that the neurotoxic effects of chronic consumption of alcohol are more prone to appear after a critical age (Polo et al., 1999). One study showed no MMN differences between controls and alcoholics who were abstinent for an average of 6 years (Fein et al., 2004), while another (Pekkonen et al., 1998) observed that increasing durations of abstinence reduced the MMN amplitude, perhaps indicating improved efficiency of covert processes upon abstinence. Ahveninen et al. (2000) found significantly enhanced MMN amplitudes to deviant sounds that correlated with reaction time lag caused by deviants, indicating pronounced distractibility and impaired reorienting to the relevant task in alcoholics. The MMN enhancement predicted poorer hit rates in alcoholics and along with reaction time lag it also correlated with an early onset of AUDs. Impairment in neural inhibition of involuntary attention shifting may be more pronounced in early-onset alcoholics. Grau and colleagues (2001) found that, while the MMN component is abolished with more demanding tasks in chronic alcoholics, it is present in normal controls. It has been suggested that the mechanisms to detect auditory differences may be reorganized in the brains of alcoholics, as revealed by lower scalp current densities in left frontal and right temporal areas during MMN in alcoholics (Marco-Pallares et al., 2007). However, no group differences in MMN amplitude have been reported in young HR offspring (van der Stelt et al., 1997; Rodriguez Holguin et al., 1998).

In summary, attention effects seem more associated with state-related aspects and provide a better measure for recovery than a predisposition.

Target detection (oddball tasks)

It is well established that alcoholics manifest reduced amplitudes of P3b to task-relevant target stimuli, particularly over parietal regions; similarly, alcoholics manifest low frontally distributed P3a to rare non-target stimuli in both visual and auditory modalities (for reviews, see Porjesz and Begleiter, 1996; Porjesz et al., 2005). Koskinen et al. (2011) analyzed auditory P3 in a twin study, and lower P3 to novel stimuli (P3a) was consistently associated with alcohol use in adolescence. More recent studies have indicated that low P3b amplitudes are present not only in male alcoholics, but in female alcoholics as well, though not to the same extent as in males (Hill and Steinhauer, 1993; Prabhu et al., 2001; Suresh et al., 2003). The lower P3 amplitude is also significantly associated with increased impulsivity and decreased activity of frontal sources in alcoholics (Chen et al., 2007). However, in an unusual sample of treatment-naive actively drinking adolescents with alcohol dependence, no reduction of P3b amplitude was observed in comparison to matched controls (Cuzen et al., 2013), possibly due to absence of any family history, comorbidity, and short drinking history.

The results for N2 component, especially the amplitude, have been equivocal. Porjesz et al. (1987b) observed longer N2 latency but no changes in amplitude in a visual discrimination oddball task conducted on alcoholics. A multimodal study also reported increased N2 latency and P3 latency in an auditory paradigm (Cadaveira et al., 1991). N2 amplitudes in an auditory oddball task were significantly lower for alcohol-dependent individuals when compared to controls (Realmuto et al., 1993; Cristini et al., 2003) but contrary findings of increased N2 amplitude in alcoholics have also been reported (Olbrich et al., 2000).

Recovery with abstinence

The P3 amplitude in alcoholics did not completely recover with prolonged abstinence (Porjesz and Begleiter, 1985; Glenn et al., 1994) and remained lower when compared to controls. Also, relapse was associated with longer N2 latency and this was not modulated by family history (Glenn et al., 1993). Studies evaluating long-term abstinent alcoholics found that P3b amplitude was reduced even after 3–10 years of abstinence (Porjesz and Begleiter, 1985) and along with increased P3a and P3b latencies after an average abstinence of 6 years (Fein and Chang, 2006).

It has been hypothesized that an underlying CNS disinhibition (i.e., hyperexcitablity) is involved in a predisposition to develop alcoholism (Begleiter and Porjesz, 1999). As described earlier, low-amplitude P3 in alcoholics is also suggestive of reduced CNS inhibition. A collection of studies examining offspring of alcoholics who are at greater risk has helped to understand if the low P3 amplitudes are due to prolonged effects of alcohol on the brain, or if they are antecedent to its development, indicating an underlying predisposition. Young HR sons of alcoholics without prior alcohol exposure had significantly lower P3 voltages compared with matched LR boys from control families without first- or second-degree alcoholic relatives (Begleiter et al., 1984). These findings were replicated in many HR/LR samples (older/younger, male/female) under several experimental conditions (for review, see Porjesz et al., 2005). Furthermore, low P3 amplitude prior to puberty has been found to predict later substance abuse, including alcohol abuse in adolescence (Berman et al., 1993; Hill et al., 1995b; Iacono et al., 2002, 2003). In addition to P3b results, offspring of alcoholics also manifested low-amplitude P3a components in both visual (Rodriguez Holguin et al., 1999) and auditory paradigms (Hada et al., 2001).

There is evidence that the P3 amplitude is directly related to the number of first-degree alcoholic relatives and not the drinking history of an alcoholic (Pfefferbaum et al., 1991; Cohen et al., 1995) or FHP/HR individual (Benegal et al., 1995). In the COGA study, lower P3 was found in alcoholics and family members from densely affected families compared with those from control families (Porjesz et al., 1998). Family history was significantly and selectively associated with lower P3 amplitudes to alcohol-related stimuli in a group of young adult African American men and women; current use of marijuana and alcohol did not modify P3 amplitudes (Ehlers et al., 2003). In contrast, some studies have failed to find relationships between family history and P3b amplitude (Polich and Bloom, 1987; Ehlers et al., 2001; Houston et al., 2003). Most P3 studies on HR have focused on males; data for female offspring have been less consistent, suggesting a weaker effect (Hill and Steinhauer, 1993; Hill et al., 1995a). Data from the COGA study also showed lower P3 amplitude in female offspring from multiplex alcoholism families, yet to a lesser degree than in males (Porjesz and Begleiter, 1996; Porjesz et al., 1998). In a comprehensive meta-analysis of all published HR versus LR studies at the time, Polich et al. (1994) found that the strongest P3 group differences were obtained in young male offspring with relatively difficult visual tasks and concluded that low-voltage P3 may have predictive value as an index of vulnerability for alcoholism. Thus, the low-voltage P3 component is a robust finding that characterizes individuals at risk for alcoholism, and provides an excellent phenotypic "trait" marker.

Response inhibition (Go/NoGo tasks)

Understanding response inhibition lies at the core of behavioral control, which may be impaired across the spectrum of disinhibitory disorders (Zucker et al., 2011). Alcoholics not only manifest reduced P3 amplitudes to Go stimuli, but reduced P3 to NoGo stimuli as well (Pfefferbaum et al., 1991; Cohen et al., 1997b; Fallgatter et al., 1998; Rodriguez Holguin et al., 1999; Hada et al., 2000). Furthermore, chronic alcoholics manifest less differentiation between their responses to task-relevant target stimuli and task-irrelevant non-target stimuli, suggesting less effective inhibitory processes. Similarly, Cristini et al. (2003) reported reduced N2 in alcoholics in a Go/NoGo task. A recent study reported significantly reduced N2 amplitudes in alcohol-dependent subjects for Go and NoGo trials, particularly for NoGo trials in frontal regions where alcoholics did not show a more frontal distribution (Pandey et al., 2012) (Fig. 23.2); controls had significantly larger frontal amplitudes for NoGo, in line with a frontal generator for N2 (van Veen and Carter, 2002; Nieuwenhuis et al., 2004). The anteriorly distributed NoGo P3 potentials were also markedly reduced in amplitude in alcoholic subjects as well as in high-risk individuals, indicating impaired inhibitory control in these individuals (Cohen et al., 1997a, b; Kamarajan et al., 2005a, b; Colrain et al., 2011).

Error monitoring and response evaluation

Reward and feedback evaluation as a behavioral process has come under special scrutiny in the context of addictive and impulsive disorders. Correlates of valence attached to the negative and positive consequences of behavior have been studied using ERPs in alcoholics and those at risk for AUDs. While there have been very few ERP studies examining reward/feedback processing in alcoholics, they provide interesting insights. Probably the first study of this kind was done by Porjesz et al. (1987a), who reported decreased P3 amplitude in response to incentive stimuli in abstinent alcoholics. More recently, using the balloon analogue risk task, which measures risk-taking propensity, Fein and Chang (2008) reported smaller amplitude in feedback negativity in FHP treatment-naive alcoholics compared to controls. Although these findings support the notion that alcoholics have a specific deficiency in reward evaluation, the nature of these deficits is still not clear due to the paucity of such studies in alcoholics. Increased impulsivity and risk taking have been found in alcoholics with reduced components to outcome/feedback stimuli during a gambling task (Kamarajan et al., 2010).

Studies have shown that the feedback/outcome-related negativity (ORN) is localized to medial frontal areas (Gehring and Willoughby, 2002; Nieuwenhuis et al., 2004; Masaki et al., 2006). Source localization in healthy individuals reveals a medial frontal source for ORN for loss, and a medial posterior source for gain (Kamarajan et al., 2009) (Fig. 23.3). Alcoholics had significantly lower ORN amplitude than controls for loss trials (Kamarajan et al., 2010). However, the

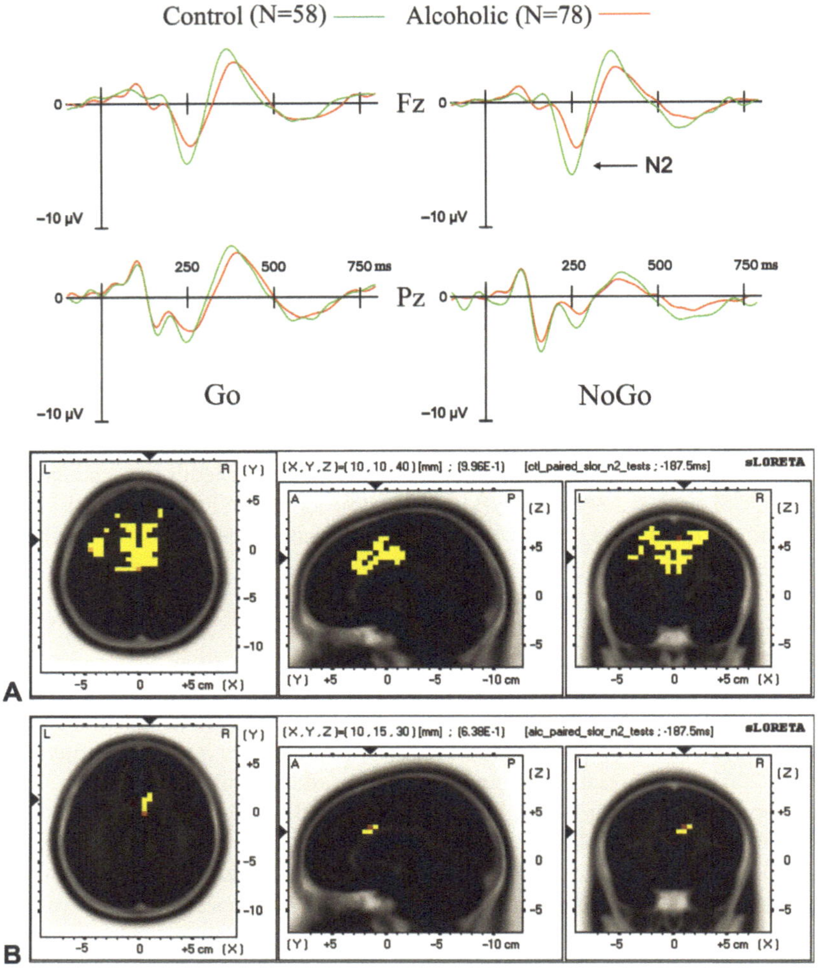

Fig. 23.2. Event-related potential waveforms and source localization of N2 component with low-resolution brain electromagnetic tomography (LORETA). (Data from Pandey et al., 2012.)

(Top panel) ERP waveforms from response inhibition task (Go/NoGo) comparing alcoholics and normal controls showing enhanced N2 wave and a frontal P3 during response inhibition (NoGo). Reduced N2 amplitudes are seen to both Go and NoGo condition, particularly for NoGo at frontal (Fz) lead in alcoholics.

(Bottom panel) sLORETA images for controls (**A**) and alcoholics (**B**) indicating areas in the brain which are significantly more activated for NoGo when compared to Go. Controls show a strong bilateral activation of anterior cingulate and also right medial frontal cortex. Alcoholics show a much smaller region of activation in anterior cingulate that does not involve bilateral areas.

outcome-related positivity (ORP) was lower for both gain and loss trials, suggesting that the negative and positive components subserve different aspects of outcome monitoring. The feedback/ORP in gambling paradigms is considered to index the subjective evaluation of the magnitude of outcome (Yeung and Sanfey, 2004; Overbeek et al., 2005; Toyomaki and Murohashi, 2005; Kamarajan et al., 2009). Contrary results have been noted in studies that tested alcoholics and matched controls (Schellekens et al., 2010; Padilla et al., 2011); however these studies were conducted on very small samples. Alcoholics generated larger ERN amplitudes than controls following incorrect and correct responses on the Eriksen flanker task (Padilla et al., 2011). Both groups showed evidence of posterror slowing. The amplitudes in the alcoholics were related to longer reaction times in correct trials, suggesting increased effort in alcoholics. Smaller negative amplitudes were associated with length of sobriety, suggesting a normalization of monitoring activity with extended abstinence.

Semantic processing

A negative ERP component, designated as N4 or N400 (300–650 ms) over centroparietal scalp, and initially elicited to semantic incongruency, has been the cornerstone of semantic ERP studies (Kutas and Hillyard, 1980; Bentin, 1989). The N400 varies systematically with the

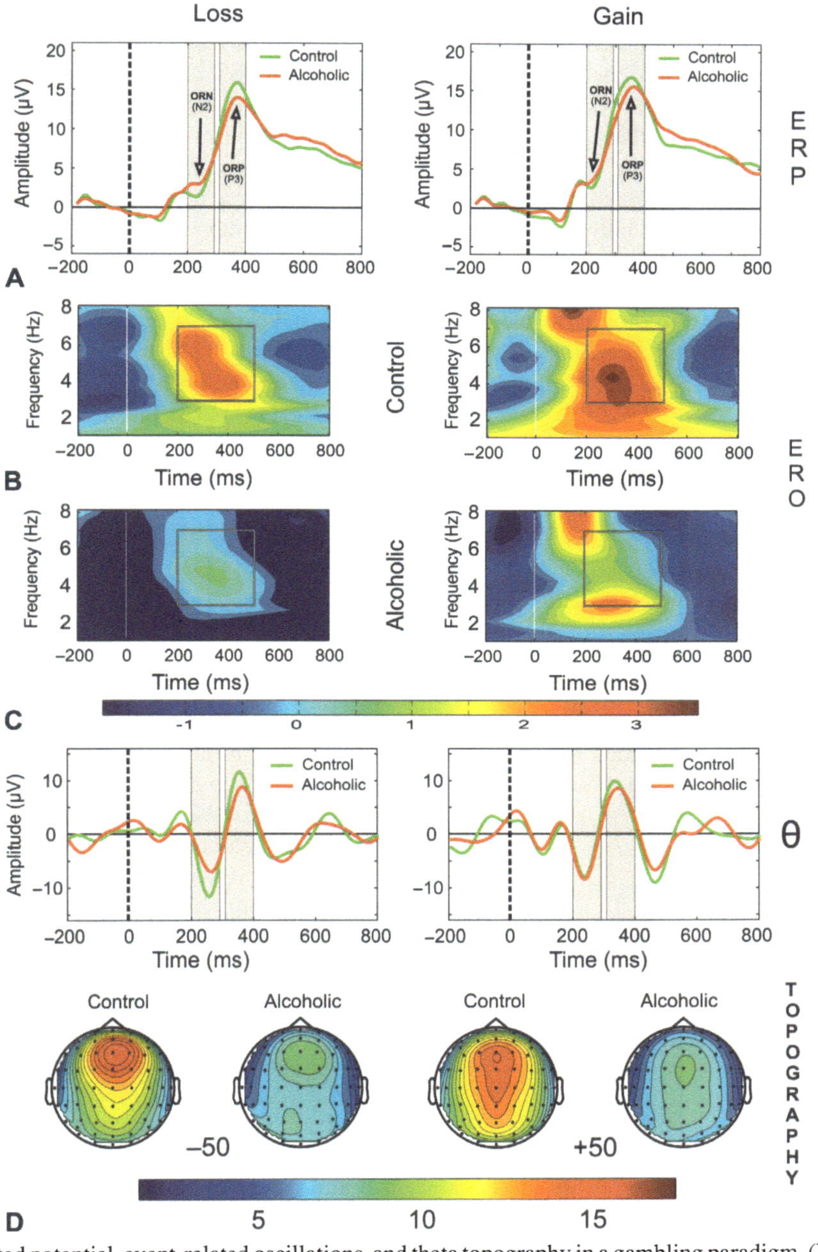

Fig. 23.3. Event-related potential, event-related oscillations, and theta topography in a gambling paradigm. (Data from Kamarajan et al., 2012.)

(**A**) Waveforms to loss and gain feedback stimuli in a gambling task showing error-related negativity (ERN)/outcome-related negativity (ORN) component and the outcome-related positivity (ORP) (P3) component in male controls ($n = 38$) and male alcoholics ($n = 38$).

(**B**) Time–frequency transformation of electroencephalogram epochs describing the total power in controls and alcoholics for loss and gain trials. Box indicates the region of interest – theta (3–7 Hz) band, 200–500 ms.

(**C**) Theta band response corresponding to the event-related potential traces in the above panel. Note differences between normal controls and alcoholics in the gray-shaded area.

(**D**) Theta power topography: note the sharply anterior peak for loss and a more diffuse slightly posterior spread for gain. Alcoholics have lower theta power, and the reduction is of greater magnitude for loss.

processing of potentially meaningful stimuli, where the amplitude is reduced by a number of factors (Kutas and Federmeier, 2000). Increased latency for N400 response to related/incongruent semantic information has been reported in alcoholics, especially in those with comorbid antisocial personality (Ceballos et al., 2003). Ceballos and colleagues (2005) found significantly less negative N4 amplitudes in alcohol-dependent individuals relative to non-dependent controls. Another widely used paradigm is the semantic priming paradigm, in which a word preceded by an unrelated word (unprimed) produces a larger N400 compared to a word preceded by a related word (primed) (Bentin, 1989; McCarthy and Nobre, 1993). Reduced N4 amplitude of the difference waveform between primed and unprimed words has been reported (Nixon et al., 2002). Similarly, in a recent semantic decision task there was less attenuation of N400 amplitudes to primed words when compared to unprimed words in alcoholics, a phenomenon that was intact in the controls (Porjesz et al., 2002b; Roopesh et al., 2010) (Fig. 23.4). Significant group differences were not seen for latency; however all subjects had slower reaction time for unprimed words compared to primed words, but significantly less reaction time savings between the unprimed and primed condition was noted for alcoholics. This lack of attenuation for the

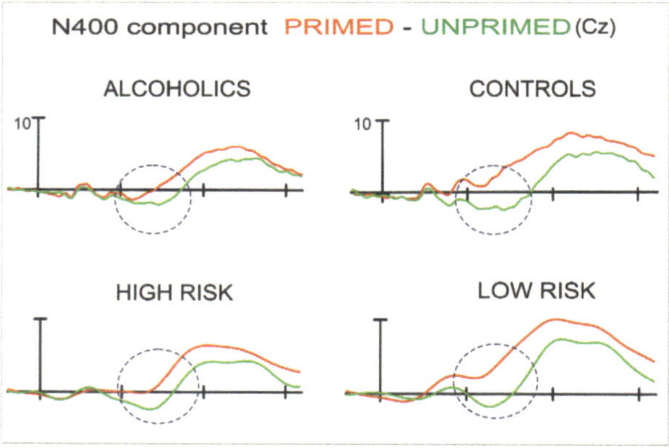

Fig. 23.4. Event-related potential waveforms from a semantic priming task. (Data from Roopesh et al., 2009, 2010.)

Waveforms from a semantic decision task for primed and unprimed words. The dotted circle marks the N400 component that is prominent in the case of unprimed words and highly attenuated for primed words, as seen in controls ($n = 57$) and low-risk ($n = 28$) individuals; alcoholics ($n = 87$) and high-risk ($n = 23$) individuals have less attenuation of this component.

N400 is a negative response seen between 200 and 600 ms in response to semantic incongruency (Kutas and Federmeier, 2011). The semantic priming task has been one of the most extensively used event-related potential paradigms to study the effect of priming on N400 (Bentin, 1989).

primed word suggests a deficiency of semantic priming process in the alcoholics where the expectancy for the second word of the antonym pair is not adequately generated. Similarly, young adult male HR offspring from alcoholic families manifested a lack of N400 attenuation, indicating deficits in semantic expectancy and post lexical semantic processing which may be present prior to alcohol dependence (Roopesh et al., 2009) (Fig. 23.4). These studies suggest that alcohol-dependent individuals and those at risk suffer from subtle impairments indicative of a reduced efficiency in resource optimization.

In summary, ERP studies reveal a pattern of deficits that affect primarily the domains of attention, response inhibition, and performance monitoring. More complex cognitive processes such as mnemonic, semantic, and lexical processes are also affected; however these may be associated with underlying attention and executive deficits.

Chronic alcoholism and event-related oscillations

Several studies have demonstrated that P3 responses are primarily the outcome of theta and delta oscillations elicited during cognitive processing of stimuli (Basar-Eroglu et al., 1992; Yordanova and Kolev, 1996; Basar, 1999; Karakas et al., 2000a, b), with delta oscillations more concentrated in the posterior region, while theta is more centered in the frontocentral region (Karakas et al., 2000b) (Fig. 23.1). ERO changes in chronic alcoholics reveal a neuronal state with altered excitability. This has also been suggested by some transcranial magnetic stimulation studies (Conte et al., 2008; Muralidharan et al., 2008).

In a visual oddball paradigm, alcoholics manifested significantly reduced theta and delta ERO amplitudes while processing the target stimuli (Jones et al., 2006b); theta differences have a frontal focus while delta tends to be more posterior (Fig. 23.1). Adolescent HR offspring also showed similar reductions in delta and theta power when compared to LR adolescents; however, the topography of theta is shifted more posteriorly to vertex and parietal regions, similar to the topography of delta oscillations (Rangaswamy et al., 2007). Interestingly, the EROs were superior to P3 amplitude in differentiating between HR and LR offspring. Similar to the observations from P3 studies, the results suggest that decreased theta and delta ERO to target stimuli may antecede the development of AUDs and represent an excellent trait marker.

These findings were replicated more recently (Andrew and Fein, 2010a) for evoked and total power in delta and theta ERO in long-term abstinent alcoholics;

the authors proposed that P300 and ERO measures provide comparable information. Phase locking enhances signal-to-noise ratios; however the increase in non-phase locked (induced) theta oscillations was suggested to be a marker of chronic alcohol abuse on the brain (Gilmore and Fein, 2012). This effect may recover, at least partially, with extended abstinence. The increased induced theta was also a strong predictor of alcoholism status (Andrew and Fein, 2010b).

Oscillatory responses associated with response inhibition were investigated using a Go/NoGo paradigm in abstinent alcoholics (Kamarajan et al., 2004). Decreased power in delta and theta oscillations was observed in alcoholics, particularly during NoGo processing, and prominent frontally. These changes were confirmed in another study that reported reduced delta oscillations for NoGo, which was correlated with white-matter degradation in the cingulate bundles (Colrain et al., 2011). Offspring of alcoholics showed significantly decreased activity in delta (1–3 Hz), theta (4–7 Hz), and alpha1 (8–9 Hz) bands during the NoGo condition, as well as reduced delta and theta activity during the Go condition (Kamarajan et al., 2006). Similar to alcoholics, differences were more prominent in the NoGo than in the Go condition. Thus it seems probable that these oscillatory responses may antecede the development of AUDs.

In a recent study on reward processing in alcoholics during a gambling task, event-related theta band (3.0–7.0 Hz) oscillations were evaluated during the loss/gain feedback (Kamarajan et al., 2012). The alcoholic group showed significantly decreased theta power during reward processing compared to controls, particularly during the evaluation of loss. Current source density maps of alcoholics revealed weaker and diffuse source activity for all conditions and weaker bilateral prefrontal sources during loss while the controls manifested stronger and more focused midline sources. Alcoholics also exhibited increased impulsivity, risk taking (as revealed by behavioral measures), and a strong association between reduced anterior theta power and impulsive task performance. Decreased power in theta oscillations and more diffuse current density may be due to reorganized and inefficient neural reward network in alcoholics.

Early attention selection impairments have been reported in alcoholics and HR (see section on attention – N100 and mismatch negativity), while some studies report no differences. The underlying reason for the variation in results may arise from an enhanced theta phase resetting in the absence of any N1 amplitude and power changes in alcoholics when compared to controls (Fuentemilla et al., 2009). The phase resetting defines excitability windows of phase-locked neurons, which in turn directs information flow, hence implying a hyperexcited neuronal state in alcoholics. Early phase-locked gamma is an important processing step for the selection/identification of target stimuli, indicative of a top-down mechanism involved in selective attention (Fell et al., 2003); it is larger to attended compared to unattended stimuli, particularly over frontal regions (Basar, 1999; Yordanova et al., 2001). Neuroimaging studies using attentional tasks have implicated the role of frontoparietal networks in this top-down control of selective attention (Corbetta et al., 2000; Giesbrecht et al., 2003). Early gamma (28–45 Hz) band response (1–150 ms) is significantly attenuated in the frontal region for target processing in abstinent alcoholics (Padmanabhapillai et al., 2006a) and children of alcoholics (HR) (7–17 years) (Padmanabhapillai et al., 2006b). Differences were only seen in the parietal region for the target condition in HR individuals. A dysfunctional frontoparietal attentional network may be associated with the impaired gamma band response and a deficient frontal top-down processing mechanism (Rangaswamy et al., 2004a).

ELECTROPHYSIOLOGIC MEASURES AS ENDOPHENOTYPES

Alcoholism is a common, complex (non-Mendelian) disorder with contributions from both genetic and environmental influences and their interactions. As seen in this review, neuroelectrophysiologic measures (e.g., P3, theta ERO, EEG beta) that differentiate between alcoholics and controls, and between HR offspring from densely affected alcoholic families and LR controls, serve as effective endophenotypes (intermediate phenotypes that correlate with diagnosis). These endophenotypes are under genetic control and are highly heritable, and have been successfully used in the search for genes associated with risk for AUDs and related disorders (Porjesz and Rangaswamy, 2007; Rangaswamy and Porjesz, 2008a, b). As the genomic technologies have evolved from linkage scans with microsatellites to candidate gene studies and genomewide association studies, these studies have highlighted targets that have proved to be relevant to understanding the pathophysiology of AUDs.

EEG phenotypes

Data on the heritability of EEG frequencies are quite compelling. The largest twin study estimates the heritability of theta and alpha power to be 0.89 and beta to be 0.86 (van Beijsterveldt and Boomsma, 1994; van Beijsterveldt et al., 1996); these heritability estimates are higher than those for AUDs and other psychiatric diagnoses. Heritability estimates for EEG coherence range between 0.5 and 0.7 (Stassen et al., 1988;

van Beijsterveldt and Boomsma, 1994; van Baal et al., 1998; van Beijsterveldt et al., 1998; Chorlian et al., 2007).

Using EEG power as an endophenotype, the COGA project reported genetic linkage and linkage disequilibrium between beta and a $GABA_A$ receptor gene (Porjesz et al., 2002a). Beta rhythm is generated in a network of excitatory pyramidal cells and inhibitory interneurons involving $GABA_A$ action as the pacemaker (Whittington et al., 2000). The same $GABA_A$ receptor gene (*GABRA2*) associated with beta EEG was also associated with alcohol dependence (Edenberg et al., 2004), a finding that has been replicated (Covault et al., 2004) and expanded to include other substance dependence in adults and conduct disorder in adolescents (Agrawal et al., 2006; Dick et al., 2006). The involvement of the GABAergic system in AUDs is supported by neuroimaging studies, which indicate deficient GABA benzodiazepine receptors in the brains of alcoholics (Abi-Dargham et al., 1998; Lingford-Hughes et al., 1998) and HR offspring (Volkow et al., 1995). Dysfunction in $GABA_A$ receptor genes may affect neural excitability, or the imbalance between excitation-inhibition (hyperexcitability) reflected in increased beta observed in alcoholics and HR offspring; this in turn may be involved in the predisposition to develop AUDs and related disinhibitory disorders.

A low-voltage alpha (LVA) phenotype, characterized by an absence or very low-amplitude alpha rhythmicity, is found in 5–10% of individuals (Anokhin et al., 1992; Enoch et al., 1995). The exon 7 variant of the $GABA_B$ receptor gene and EEG alpha voltage (LVA or normal) was significantly associated for control but not alcoholic subjects (Winterer et al., 2003b). LVA in females has been associated with a genetic variant resulting in low catechol-*O*-methyltransferase activity, which is involved in the dopaminergic system, yielding low levels of norepinephrine (Enoch et al., 2003). This may partly explain the association of LVA and anxiety disorders in alcoholic women.

Increased interhemispheric coherence has been a feature of EEG in alcoholism; when associated with increased depressiveness, it has been suggested to involve GABAergic and/or glutamatergic neurotransmission. Winterer et al. (2003c) revealed that three exonic variants of a $GABA_B$ receptor gene influence cortical synchronization (coherence). In the COGA study, significant linkage for theta (6–7 Hz) interhemispheric coherence at parieto-occipital regions led to significant association with several single nucleotide polymorphisms (SNPs) in *GABRA2*. Another significant linkage peak for theta (6–7 Hz) centroparietal coherence was significantly associated with SNPs in *CHRM2*, a cholinergic muscarinic receptor gene (Porjesz and Rangaswamy, 2007; Rangaswamy and Porjesz, 2008b).

EROs as endophenotypes

EROs have proved to be more useful than ERPs in the search for genes involved in AUDs and related disorders in COGA. Theta oscillations to targets have been instrumental in identifying two excellent candidate genes under a significant linkage peak – *CHRM2* and *GRM8*. Both encode subunits of neurotransmitter receptors: *CHRM2* encodes a cholinergic muscarinic receptor M2, whereas *GRM8* encodes the metabotropic glutamate receptor 8 in a family of G-protein-coupled receptors. Significant associations were observed between the frontal theta ERO and SNPs in *CHRM2*, and to parietal delta (Jones et al., 2004, 2006a). Several of the same SNPs were significantly associated with alcohol dependence along with depression, drug dependence and externalizing disorders (Wang et al., 2004; Dick et al., 2007), findings that were replicated by other groups (Comings et al., 2002; Luo et al., 2005). Theta oscillations were also significantly associated with several SNPs in *GRM8*, as well as alcohol dependence (Chen et al., 2008a).

Generation of theta and delta oscillations depends on level of activation of M2 muscarinic autoreceptors (Fellous and Sejnowski, 2000; Tiesinga et al., 2001) that inhibit further acetylcholine release by presynaptic cells, leading to inhibition of irrelevant networks. Acetylcholine plays a significant role in stimulus significance (Perry et al., 1999), selective attention (Mitrofanis and Guillery, 1993), P3 generation, and modified memory performance (Hammond et al., 1987; Dierks et al., 1994; Frodl-Bauch et al., 1999; Potter et al., 2000). The GRM8 gene encodes a presynaptic autoreceptor involved in modulating neuronal excitability by inhibiting glutamate release at the synapse (Schoepp, 2001). Hence, these findings implicate *CHRM2* and *GRM8* in the generation and modulation of these oscillations during the P3 response to target stimuli. GABAergic, cholinergic, and glutamatergic system interactions have been proposed to underlie these rhythms and P300 (Frodl-Bauch et al., 1999). Thus, the genetic underpinnings of these oscillations influence the excitability in neural networks.

This is further supported by a recent family genomewide association study in COGA with a genomewide significant finding in another neurotransmitter-related gene – *KCNJ6* (a potassium inward rectifier channel, GIRK2) and frontal theta EROs (Kang et al., 2012). These results suggest that *KCNJ6* or its product GIRK2 accounts for some of the variations in theta oscillations. GIRK2 receptor activation contributes to slow inhibitory postsynaptic potentials that modulate neuronal excitability, and therefore influence neuronal networks (Luscher and Slesinger, 2010). Animal models have shown that

GIRK channels are directly activated by ethanol and are important effectors in both opioid- and ethanol-induced analgesia and are considered a viable drug target; GIRK2 also modulates opioid effects on analgesia and addiction in humans. Thus these findings between theta EROs and *KCNJ6* have important implications for neural excitability and alcohol addiction.

Together, these results indicate that the neurophysiologic endophenotypes implicate some of the transmitter genes important for modulating and maintaining neural excitability; variations in this excitability may underlie the predisposition or susceptibility for AUDs and related disorders.

CONCLUSION

In conclusion, the vulnerability to alcohol effects and AUDs may be associated with a modulation of excitability of some neural circuits more than others. These changes affect networks associated with cognitive domains of attention and self-monitoring that are part of the rubric of the frontal executive function which is impacted with both acute and chronic alcohol use. Both ERP and ERO studies discussed here provide evidence for impaired attention, response inhibition, and monitoring functions. Source localization of these components highlights impaired loci in frontal lobes, suggesting the utility of a multimodal approach. Future studies that integrate neuroelectrophysiology and neuroimaging are essential to understanding these complex structure–function interactions.

Studies conducted so far suggest that there are several common substrates (e.g., theta oscillations, beta oscillations, P3 amplitude) that are influenced by alcohol in both acute and chronic use and this in turn may reflect the underlying vulnerability of the brain to alcohol. The potential to isolate genetic underpinnings of impaired neuroelectrophysiologic features associated with alcohol use is another exciting direction that may provide viable targets for intervention. Although no functional variant affecting the neuroelectrophysiologic characteristics has yet been identified at the molecular level, a large body of pharmacologic evidence attests to the relevance of these receptors for aspects of cognitive function. This approach has the unprecedented potential to unravel the complex interplay of various neural subsystems relevant to the generation of brain oscillations elicited under different cognitive conditions and in disease states.

Alcohol dependence results from a complex interaction of changing genetic and environmental liabilities across development. Genetic studies have successfully used these endophenotypes to reveal significantly associated SNPs from the same genes that are also associated with alcohol dependence and related disorders. Thus, genes underlying the variations in endophenotypes are also associated with the disease. Therefore, understanding genetic control of brain electrical activity can provide clues about cerebral function and also shed light on mechanisms involved in psychiatric disorders, such as AUDs, where impairment in brain electric activity is apparent. Prospective studies of young individuals with "risk genotypes" can lead to an improved understanding of how neural and cognitive changes contribute to susceptibility across development, which in turn can lead to the design of well-targeted prevention initiatives.

ACKNOWLEDGMENT

This study was supported by the National Institutes of Health (NIH) grants R01AA005524, R01AA002686 and U10AA008401 from the National Institute on Alcohol Abuse and Alcoholism (NIAAA).

We would like to thank Niklas Manz, Ph.D. and Carlene Haynes for editorial assistance.

References

Abi-Dargham A, Krystal JH, Anjilvel S et al. (1998). Alterations of benzodiazepine receptors in type II alcoholic subjects measured with SPECT and [123I]iomazenil. Am J Psychiatry 155: 1550–1555.

Agrawal A, Edenberg HJ, Foroud T et al. (2006). Association of GABRA2 with drug dependence in the collaborative study of the genetics of alcoholism sample. Behav Genet 36: 640–650.

Ahveninen J, Escera C, Polo MD et al. (2000). Acute and chronic effects of alcohol on preattentive auditory processing as reflected by mismatch negativity. Audiol Neurootol 5: 303–311.

Andrew C, Fein G (2010a). Event-related oscillations versus event-related potentials in a P300 task as biomarkers for alcoholism. Alcohol Clin Exp Res 34: 669–680.

Andrew C, Fein G (2010b). Induced theta oscillations as biomarkers for alcoholism. Clin Neurophysiol 121: 350–358.

Anokhin A, Steinlein O, Fischer C et al. (1992). A genetic study of the human low-voltage electroencephalogram. Hum Genet 90: 99–112.

Ardekani BA, Choi SJ, Hossein-Zadeh GA et al. (2002). Functional magnetic resonance imaging of brain activity in the visual oddball task. Brain Res Cogn Brain Res 14: 347–356.

Bartholow BD, Pearson M, Sher KJ et al. (2003). Effects of alcohol consumption and alcohol susceptibility on cognition: a psychophysiological examination. Biol Psychol 64: 167–190.

Bartholow BD, Pearson MA, Dickter CL et al. (2005). Strategic control and medial frontal negativity: beyond errors and response conflict. Psychophysiology 42: 33–42.

Basar E (1999). Brain function and oscillations. Integrative brain function, Neurophysiology and cognitive processes. Springer, Berlin, Vol.II.

Basar-Eroglu C, Basar E, Demiralp T et al. (1992). P300-response: possible psychophysiological correlates in delta and theta frequency channels. A review. Int J Psychophysiol 13: 161–179.

Bauer LO (2001). Predicting relapse to alcohol and drug abuse via quantitative electroencephalography. Neuropsychopharmacology 25: 332–340.

Bauer LO, Hesselbrock VM (1993). EEG, autonomic and subjective correlates of the risk for alcoholism. J Stud Alcohol 54: 577–589.

Begleiter H, Platz A (1972). The effects of alcohol on the central nervous system in humans. Physiology and Behavior, The Biology of Alcoholism In: B Kissin, H Begleiter (Eds.), Vol. 2. Plenum, New York.

Begleiter H, Porjesz B (1999). What is inherited in the predisposition toward alcoholism? A proposed model. Alcohol Clin Exp Res 23: 1125–1135.

Begleiter H, Porjesz B, Bihari B et al. (1984). Event-related brain potentials in boys at risk for alcoholism. Science 225: 1493–1496.

Benegal V, Jain S, Subbukrishna DK et al. (1995). P300 amplitudes vary inversely with continuum of risk in first degree male relatives of alcoholics. Psychiatr Genet 5: 149–156.

Bentin S (1989). Electrophysiological studies of visual word perception, lexical organization, and semantic processing: a tutorial review. Lang Speech 32: 205–220.

Berman SM, Whipple SC, Fitch RJ et al. (1993). P3 in young boys as a predictor of adolescent substance use. Alcohol 10: 69–76.

Birbaumer N, Elbert T, Canavan AG et al. (1990). Slow potentials of the cerebral cortex and behavior. Physiol Rev 70: 1–41.

Boha R, Molnar M, Gaal ZA et al. (2009). The acute effect of low-dose alcohol on working memory during mental arithmetic: I. Behavioral measures and EEG theta band spectral characteristics. Int J Psychophysiol 73: 133–137.

Bohman M, Cloninger R, Sigvardsson S et al. (1987). The genetics of alcoholisms and related disorders. J Psychiatr Res 21: 447–452.

Buzsaki G (1996). The hippocampo-neocortical dialogue. Cereb Cortex 6: 81–92.

Buzsaki G, Draguhn A (2004). Neuronal oscillations in cortical networks. Science 304: 1926–1929.

Cadaveira F, Grau C, Roso M et al. (1991). Multimodality exploration of event-related potentials in chronic alcoholics. Alcohol Clin Exp Res 15: 607–611.

Cadoret RJ, Cain CA, Grove WM (1980). Development of alcoholism in adoptees raised apart from alcoholic biologic relatives. Arch Gen Psychiatry 37: 561–563.

Campbell KB, Lowick BM (1987). Ethanol and event-related potentials: the influence of distractor stimuli. Alcohol 4: 257–263.

Ceballos NA, Nixon SJ, Phillips JA et al. (2003). Semantic processing in alcoholics with and without antisocial symptomatology. J Stud Alcohol 64: 286–291.

Ceballos NA, Houston RJ, Smith ND et al. (2005). N400 as an index of semantic expectancies: differential effects of alcohol and cocaine dependence. Prog Neuropsychopharmacol Biol Psychiatry 29: 936–943.

Celesia GG, Brigell M (1992). Event-related potentials. Curr Opin Neurol Neurosurg 5: 733–739.

Chen AC, Porjesz B, Rangaswamy M et al. (2007). Reduced frontal lobe activity in subjects with high impulsivity and alcoholism. Alcohol Clin Exp Res 31: 156–165.

Chen ACH, Tang Y, Rangaswamy M et al. (2008a). Association of single nucleotide polymorphisms in a glutamate receptor gene (GRM8) with theta power of event-related oscillations and alcohol dependence. Am J Med Genet 150B: 359–368.

Chen CC, Hsieh JC, Wu YZ et al. (2008b). Mutual-information-based approach for neural connectivity during self-paced finger lifting task. Hum Brain Mapp 29: 265–280.

Chorlian DB, Tang Y, Rangaswamy M et al. (2007). Heritability of EEG coherence in a large sib-pair population. Biol Psychol 75: 260–266.

Chorlian DB, Rangaswamy M, Porjesz B (2009). EEG coherence: topography and frequency structure. Exp Brain Res 198: 59–83.

Cohen HL, Porjesz B, Begleiter H (1991). EEG characteristics in males at risk for alcoholism. Alcohol Clin Exp Res 15: 858–861.

Cohen HL, Porjesz B, Begleiter H (1993a). The effects of ethanol on EEG activity in males at risk for alcoholism. Electroencephalogr Clin Neurophysiol 86: 368–376.

Cohen HL, Porjesz B, Begleiter H (1993b). Ethanol-induced alterations in electroencephalographic activity in adult males. Neuropsychopharmacology 8: 365–370.

Cohen HL, Wang W, Porjesz B et al. (1995). Auditory P300 in young alcoholics: regional response characteristics. Alcohol Clin Exp Res 19: 469–475.

Cohen HL, Projesz B, Begleiter H (1996). Temporal recovery of auditory evoked potentials in individuals at risk for alcoholism. Alcohol 13: 333–339.

Cohen HL, Porjesz B, Begleiter H et al. (1997a). Neuroelectric correlates of response production and inhibition in individuals at risk to develop alcoholism. Biol Psychiatry 42: 57–67.

Cohen HL, Porjesz B, Begleiter H et al. (1997b). Neurophysiological correlates of response production and inhibition in alcoholics. Alcohol Clin Exp Res 21: 1398–1406.

Cohen HL, Porjesz B, Stimus AT et al. (1998). Effects of ethanol on temporal recovery of auditory-evoked potentials in individuals at risk for alcoholism. Alcohol Clin Exp Res 22: 945–953.

Colrain IM, Taylor J, Mclean S et al. (1993). Dose dependent effects of alcohol on visual evoked potentials. Psychopharmacology (Berl) 112: 383–388.

Colrain IM, Sullivan EV, Ford JM et al. (2011). Frontally mediated inhibitory processing and white matter microstructure: age and alcoholism effects. Psychopharmacology (Berl) 213: 669–679.

Comings DE, Wu S, Rostamkhani M et al. (2002). Association of the muscarinic cholinergic 2 receptor (CHRM2) gene with major depression in women. Am J Med Genet 114: 527–529.

Conte A, Attilia ML, Gilio F et al. (2008). Acute and chronic effects of ethanol on cortical excitability. Clin Neurophysiol 119: 667–674.

Corbetta M, Kincade JM, Ollinger JM et al. (2000). Voluntary orienting is dissociated from target detection in human posterior parietal cortex. Nat Neurosci 3: 292–297.

Costa L, Bauer L (1997). Quantitative electroencephalographic differences associated with alcohol, cocaine, heroin and dual-substance dependence. Drug Alcohol Depend 46: 87–93.

Courchesne E, Hillyard SA, Galambos R (1975). Stimulus novelty, task relevance and the visual evoked potential in man. Electroencephalogr Clin Neurophysiol 39: 131–143.

Courtney KE, Polich J (2009). Binge drinking in young adults: Data, definitions, and determinants. Psychol Bull 135: 142–156.

Courtney KE, Polich J (2010). Binge drinking effects on EEG in young adult humans. Int J Environ Res Public Health 7: 2325–2336.

Coutin-Churchman P, Moreno R (2008). Intracranial current density (LORETA) differences in QEEG frequency bands between depressed and non-depressed alcoholic patients. Clin Neurophysiol 119: 948–958.

Covault J, Gelernter J, Hesselbrock V et al. (2004). Allelic and haplotypic association of GABRA2 with alcohol dependence. Am J Med Genet B Neuropsychiatr Genet 129: 104–109.

Crego A, Holguin SR, Parada M et al. (2009). Binge drinking affects attentional and visual working memory processing in young university students. Alcohol Clin Exp Res 33: 1870–1879.

Crego A, Rodriguez-Holguin S, Parada M et al. (2010). Reduced anterior prefrontal cortex activation in young binge drinkers during a visual working memory task. Drug Alcohol Depend 109: 45–56.

Crego A, Cadaveira F, Parada M et al. (2012). Increased amplitude of P3 event-related potential in young binge drinkers. Alcohol 46: 415–425.

Crews FT, Braun CJ, Hoplight B et al. (2000). Binge ethanol consumption causes differential brain damage in young adolescent rats compared with adult rats. Alcohol Clin Exp Res 24: 1712–1723.

Crews FT, Mdzinarishvili A, Kim D et al. (2006). Neurogenesis in adolescent brain is potently inhibited by ethanol. Neuroscience 137: 437–445.

Cristini P, Fournier C, Timsit-Berthier M et al. (2003). ERPs (N200, P300 and CNV) in alcoholics: relapse risk assessment. Neurophysiol Clin 33: 103–119.

Cuzen NL, Andrew C, Thomas KG et al. (2013). Absence of p300 reduction in South African treatment-naive adolescents with alcohol dependence. Alcohol Clin Exp Res 37: 40–48.

De Bruin EA, Bijl S, Stam CJ et al. (2004). Abnormal EEG synchronisation in heavily drinking students. Clin Neurophysiol 115: 2048–2055.

De Bruin EA, Stam CJ, Bijl S et al. (2006). Moderate-to-heavy alcohol intake is associated with differences in synchronization of brain activity during rest and mental rehearsal. Int J Psychophysiol 60: 304–314.

Desmedt JE (1980). P300 in serial tasks: An essential post–decision closure mechanism. In: HH Kornhuber, L Deecke (Eds.), Motivation, Motor and Sensory Processes of the Brain. Elsevier, Amsterdam.

Dick DM, Bierut L, Hinrichs A et al. (2006). The role of GABRA2 in risk for conduct disorder and alcohol and drug dependence across developmental stages. Behav Genet 36: 577–590.

Dick DM, Agrawal A, Wang JC et al. (2007). Alcohol dependence with comorbid drug dependence: genetic and phenotypic associations suggest a more severe form of the disorder with stronger genetic contribution to risk. Addiction 102: 1131–1139.

Dierks T, Frolich L, Ihl R et al. (1994). Event-related potentials and psychopharmacology. Cholinergic modulation of P300. Pharmacopsychiatry 27: 72–74.

Easdon C, Izenberg A, Armilio ML et al. (2005). Alcohol consumption impairs stimulus- and error-related processing during a Go/No-Go Task. Brain Res Cogn Brain Res 25: 873–883.

Edenberg HJ, Dick DM, Xuei X et al. (2004). Variations in GABRA2, encoding the alpha 2 subunit of the GABA(A) receptor, are associated with alcohol dependence and with brain oscillations. Am J Hum Genet 74: 705–714.

Ehlers CL, Phillips E (2003). EEG low-voltage alpha and alpha power in African American young adults: relation to family history of alcoholism. Alcohol Clin Exp Res 27: 765–772.

Ehlers CL, Schuckit MA (1988). EEG response to ethanol in sons of alcoholics. Psychopharmacol Bull 24: 434–437.

Ehlers CL, Schuckit MA (1990). EEG fast frequency activity in the sons of alcoholics. Biol Psychiatry 27: 631–641.

Ehlers CL, Schuckit MA (1991). Evaluation of EEG alpha activity in sons of alcoholics. Neuropsychopharmacology 4: 199–205.

Ehlers CL, Wall TL, Schuckit MA (1989). EEG spectral characteristics following ethanol administration in young men. Electroencephalogr Clin Neurophysiol 73: 179–187.

Ehlers CL, Phillips E, Parry BL (1996). Electrophysiological findings during the menstrual cycle in women with and without late luteal phase dysphoric disorder: relationship to risk for alcoholism? Biol Psychiatry 39: 720–732.

Ehlers CL, Garcia-Andrade C, Wall TL et al. (1998a). Determinants of P3 amplitude and response to alcohol in Native American Mission Indians. Neuropsychopharmacology 18: 282–292.

Ehlers CL, Havstad J, Prichard D et al. (1998b). Low doses of ethanol reduce evidence for nonlinear structure in brain activity. J Neurosci 18: 7474–7486.

Ehlers CL, Wall TL, Garcia-Andrade C et al. (2001). Auditory P3 findings in mission Indian youth. J Stud Alcohol 62: 562–570.

Ehlers CL, Phillips E, Sweeny A et al. (2003). Event-related potential responses to alcohol-related stimuli in African-American young adults: relation to family history of alcoholism and drug usage. Alcohol Alcohol 38: 332–338.

Ehlers CL, Phillips E, Wall TL et al. (2004). EEG alpha and level of response to alcohol in Hispanic- and non-Hispanic-American young adults with a family history of alcoholism. J Stud Alcohol 65: 301–308.

Ehlers CL, Phillips E, Finnerman G et al. (2007). P3 components and adolescent binge drinking in Southwest California Indians. Neurotoxicol Teratol 29: 153–163.

Ehlers CL, Wills DN, Havstad J (2012). Ethanol reduces the phase locking of neural activity in human and rodent brain. Brain Res 1450: 67–79.

Eimer M (1993). Effects of attention and stimulus probability on ERPs in a Go/Nogo task. Biol Psychol 35: 123–138.

Elmasian R, Neville H, Woods D et al. (1982). Event-related brain potentials are different in individuals at high and low risk for developing alcoholism. Proc Natl Acad Sci U S A 79: 7900–7903.

Enoch MA, Rohrbaugh JW, Davis EZ et al. (1995). Relationship of genetically transmitted alpha EEG traits to anxiety disorders and alcoholism. Am J Med Genet 60: 400–408.

Enoch MA, White KV, Harris CR et al. (1999). Association of low-voltage alpha EEG with a subtype of alcohol use disorders. Alcohol Clin Exp Res 23: 1312–1319.

Enoch MA, Xu K, Ferro E et al. (2003). Genetic origins of anxiety in women: a role for a functional catechol-*O*-methyltransferase polymorphism. Psychiatr Genet 13: 33–41.

Euser AS, Van Meel CS, Snelleman M et al. (2011). Acute effects of alcohol on feedback processing and outcome evaluation during risky decision-making: an ERP study. Psychopharmacology (Berl) 217: 111–125.

Falkenstein M, Hoormann J, Hohnsbein J (1999). ERP components in Go/Nogo tasks and their relation to inhibition. Acta Psychol (Amst) 101: 267–291.

Fallgatter AJ, Wiesbeck GA, Weijers HG et al. (1998). Event-related correlates of response suppression as indicators of novelty seeking in alcoholics. Alcohol Alcohol 33: 475–481.

Fein G, Chang M (2006). Visual P300s in long-term abstinent chronic alcoholics. Alcohol Clin Exp Res 30: 2000–2007.

Fein G, Chang M (2008). Smaller feedback ERN amplitudes during the BART are associated with a greater family history density of alcohol problems in treatment-naive alcoholics. Drug Alcohol Depend 92: 141–148.

Fein G, Whitlow B, Finn P (2004). Mismatch negativity: no difference between controls and abstinent alcoholics. Alcohol Clin Exp Res 28: 137–142.

Fell J, Fernandez G, Klaver P et al. (2003). Is synchronized neuronal gamma activity relevant for selective attention? Brain Res Brain Res Rev 42: 265–272.

Fellous JM, Sejnowski TJ (2000). Cholinergic induction of oscillations in the hippocampal slice in the slow (0.5–2 Hz), theta (5–12 Hz), and gamma (35–70 Hz) bands. Hippocampus 10: 187–197.

Filipovic SR, Jahanshahi M, Rothwell JC (1999). Cortical potentials related to decision-making: comparison of two types of go/no-go decision. Neuroreport 10: 3583–3587.

Fillmore MT, Vogel-Sprott M (1999). An alcohol model of impaired inhibitory control and its treatment in humans. Exp Clin Psychopharmacol 7: 49–55.

Fillmore MT, Vogel-Sprott M (2000). Response inhibition under alcohol: effects of cognitive and motivational conflict. J Stud Alcohol 61: 239–246.

Finn PR, Justus A (1999). Reduced EEG alpha power in the male and female offspring of alcoholics. Alcohol Clin Exp Res 23: 256–262.

Folstein JR, van Petten C (2008). Influence of cognitive control and mismatch on the N2 component of the ERP: a review. Psychophysiology 45: 152–170.

Fries P (2005). A mechanism for cognitive dynamics: neuronal communication through neuronal coherence. Trends Cogn Sci 9: 474–480.

Frodl-Bauch T, Bottlender R, Hegerl U (1999). Neurochemical substrates and neuroanatomical generators of the event-related P300. Neuropsychobiology 40: 86–94.

Fuentemilla L, Marco-Pallares J, Gual A et al. (2009). Impaired theta phase-resetting underlying auditory N1 suppression in chronic alcoholism. Neuroreport 20: 337–342.

Gabrielli Jr WF, Mednick SA, Volavka J et al. (1982). Electroencephalograms in children of alcoholic fathers. Psychophysiology 19: 404–407.

Gehring WJ, Willoughby AR (2002). The medial frontal cortex and the rapid processing of monetary gains and losses. Science 295: 2279–2282.

Gevins A, Chan CS, Sam-Vargas L (2012). Towards measuring brain function on groups of people in the real world. PLoS One 7: e44676.

Giesbrecht B, Woldorff MG, Song AW et al. (2003). Neural mechanisms of top-down control during spatial and feature attention. Neuroimage 19: 496–512.

Gilmore CS, Fein G (2012). Theta event-related synchronization is a biomarker for a morbid effect of alcoholism on the brain that may partially resolve with extended abstinence. Genes Brain Behav 2: 796–805.

Glenn SW, Sinha R, Parsons OA (1993). Electrophysiological indices predict resumption of drinking in sober alcoholics. Alcohol 10: 89–95.

Glenn S, Parsons OA, Sinha R (1994). Assessment of recovery of electrophysiological and neuropsychological functions in chronic alcoholics. Biol Psychiatry 36: 443–452.

Goodwin DW, Schulsinger F, Hermansen L et al. (1973). Alcohol problems in adoptees raised apart from alcoholic biological parents. Arch Gen Psychiatry 28: 238–243.

Gorenstein EE, Newman JP (1980). Disinhibitory psychopathology: a new perspective and a model for research. Psychol Rev 87: 301–315.

Grau C, Polo MD, Yago E et al. (2001). Auditory sensory memory as indicated by mismatch negativity in chronic alcoholism. Clin Neurophysiol 112: 728–731.

Haas LF (2003). Hans Berger (1873–1941), Richard Caton (1842–1926), and electroencephalography. J Neurol Neurosurg Psychiatry 74: 9.

Hada M, Porjesz B, Begleiter H et al. (2000). Auditory P3a assessment of male alcoholics. Biol Psychiatry 48: 276–286.

Hada M, Porjesz B, Chorlian DB et al. (2001). Auditory P3a deficits in male subjects at high risk for alcoholism. Biol Psychiatry 49: 726–738.

Hajcak G, Moser JS, Holroyd CB et al. (2006). The feedback-related negativity reflects the binary evaluation of good versus bad outcomes. Biol Psychol 71: 148–154.

Halgren E, Squires NK, Wilson CL et al. (1980). Endogenous potentials generated in the human hippocampal formation and amygdala by infrequent events. Science 210: 803–805.

Hammond EJ, Meador KJ, Aung-Din R et al. (1987). Cholinergic modulation of human P3 event-related potentials. Neurology 37: 346–350.

Harper C (2009). The neuropathology of alcohol-related brain damage. Alcohol Alcohol 44: 136–140.

Heath AC, Bucholz KK, Madden PA et al. (1997). Genetic and environmental contributions to alcohol dependence risk in a national twin sample: consistency of findings in women and men. Psychol Med 27: 1381–1396.

Hermens DF, Lagopoulos J, Tobias-Webb J et al. (2013). Pathways to alcohol-induced brain impairment in young people: A review. Cortex 49: 3–17.

Hicks BM, Krueger RF, Iacono WG et al. (2004). Family transmission and heritability of externalizing disorders: a twin-family study. Arch Gen Psychiatry 61: 922–928.

Hill SY, Steinhauer SR (1993). Event-related potentials in women at risk for alcoholism. Alcohol 10: 349–354.

Hill SY, Muka D, Steinhauer S et al. (1995a). P300 amplitude decrements in children from families of alcoholic female probands. Biol Psychiatry 38: 622–632.

Hill SY, Steinhauer S, Lowers L et al. (1995b). Eight-year longitudinal follow-up of P300 and clinical outcome in children from high-risk for alcoholism families. Biol Psychiatry 37: 823–827.

Hillyard SA, Hink RF, Schwent VL et al. (1973). Electrical signs of selective attention in the human brain. Science 182: 177–180.

Holmes AJ, Pizzagalli DA (2008). Spatiotemporal dynamics of error processing dysfunctions in major depressive disorder. Arch Gen Psychiatry 65: 179–188.

Houston RJ, Bauer LO, Hesselbrock VM (2003). Depression and familial risk for substance dependence: a P300 study of young women. Psychiatry Res 124: 49–62.

Hummel F, Gerloff C (2005). Larger interregional synchrony is associated with greater behavioral success in a complex sensory integration task in humans. Cereb Cortex 15: 670–678.

Iacono WG, Carlson SR, Malone SM et al. (2002). P3 event-related potential amplitude and the risk for disinhibitory disorders in adolescent boys. Arch Gen Psychiatry 59: 750–757.

Iacono WG, Malone SM, Mcgue M (2003). Substance use disorders, externalizing psychopathology, and P300 event-related potential amplitude. Int J Psychophysiol 48: 147–178.

Itoh T, Sumiyoshi T, Higuchi Y et al. (2011). LORETA analysis of three-dimensional distribution of delta band activity in schizophrenia: relation to negative symptoms. Neurosci Res 70: 442–448.

Jaaskelainen IP, Pekkonen E, Alho K et al. (1995). Dose-related effect of alcohol on mismatch negativity and reaction time performance. Alcohol 12: 491–495.

Jaaskelainen IP, Naatanen R, Sillanaukee P (1996). Effect of acute ethanol on auditory and visual event-related potentials: a review and reinterpretation. Biol Psychiatry 40: 284–291.

Jaaskelainen IP, Schroger E, Naatanen R (1999). Electrophysiological indices of acute effects of ethanol on involuntary attention shifting. Psychopharmacology (Berl) 141: 16–21.

Jaaskelainen IP, Hirvonen J, Saher M et al. (2000). Dose-dependent suppression by ethanol of transient auditory 40-Hz response. Psychopharmacology (Berl) 148: 132–135.

Jodo E, Kayama Y (1992). Relation of a negative ERP component to response inhibition in a Go/No-go task. Electroencephalogr Clin Neurophysiol 82: 477–482.

Jones KA, Porjesz B, Almasy L et al. (2004). Linkage and linkage disequilibrium of evoked EEG oscillations with CHRM2 receptor gene polymorphisms: implications for human brain dynamics and cognition. Int J Psychophysiol 53: 75–90.

Jones KA, Porjesz B, Almasy L et al. (2006a). A cholinergic receptor gene (CHRM2) affects event-related oscillations. Behav Genet 36: 627–639.

Jones KA, Porjesz B, Chorlian D et al. (2006b). *S*-transform time-frequency analysis of P300 reveals deficits in individuals diagnosed with alcoholism. Clin Neurophysiol 117: 2128–2143.

Kahkonen S, Kesaniemi M, Nikouline VV et al. (2001). Ethanol modulates cortical activity: direct evidence with combined TMS and EEG. Neuroimage 14: 322–328.

Kamarajan C, Porjesz B, Jones KA et al. (2004). The role of brain oscillations as functional correlates of cognitive systems: a study of frontal inhibitory control in alcoholism. Int J Psychophysiol 51: 155–180.

Kamarajan C, Porjesz B, Jones KA et al. (2005a). Alcoholism is a disinhibitory disorder: neurophysiological evidence from a Go/No-Go task. Biol Psychol 69: 353–373.

Kamarajan C, Porjesz B, Jones KA et al. (2005b). Spatial-anatomical mapping of NoGo-P3 in the offspring of alcoholics: evidence of cognitive and neural disinhibition as a risk for alcoholism. Clin Neurophysiol 116: 1049–1061.

Kamarajan C, Porjesz B, Jones K et al. (2006). Event-related oscillations in offspring of alcoholics: neurocognitive disinhibition as a risk for alcoholism. Biol Psychiatry 59: 625–634.

Kamarajan C, Porjesz B, Rangaswamy M et al. (2009). Brain signatures of monetary loss and gain: outcome-related potentials in a single outcome gambling task. Behav Brain Res 197: 62–76.

Kamarajan C, Rangaswamy M, Tang Y et al. (2010). Dysfunctional reward processing in male alcoholics: an ERP study during a gambling task. J Psychiatr Res 44: 576–590.

Kamarajan C, Rangaswamy M, Manz N et al. (2012). Topography, power, and current source density of theta oscillations during reward processing as markers for alcohol dependence. Hum Brain Mapp 33: 1019–1039.

Kang SJ, Rangaswamy M, Manz N et al. (2012). Family-based genome-wide association study of frontal theta oscillations identifies potassium channel gene KCNJ6. Genes Brain Behav 11: 712–719.

Kaplan RF, Glueck BC, Hesselbrock MN et al. (1985). Power and coherence analysis of the EEG in hospitalized alcoholics and nonalcoholic controls. J Stud Alcohol 46: 122–127.

Kaplan RF, Hesselbrock VM, O'connor S et al. (1988). Behavioral and EEG responses to alcohol in nonalcoholic men with a family history of alcoholism. Prog Neuropsychopharmacol Biol Psychiatry 12: 873–885.

Karakas S, Erzengin OU, Basar E (2000a). The genesis of human event-related responses explained through the theory of oscillatory neural assemblies. Neurosci Lett 285: 45–48.

Karakas S, Erzengin OU, Basar E (2000b). A new strategy involving multiple cognitive paradigms demonstrates that ERP components are determined by the superposition of oscillatory responses. Clin Neurophysiol 111: 1719–1732.

Kathmann N, Wagner M, Rendtorff N et al. (1995). Delayed peak latency of the mismatch negativity in schizophrenics and alcoholics. Biol Psychiatry 37: 754–757.

Kendler KS, Prescott CA, Myers J et al. (2003). The structure of genetic and environmental risk factors for common psychiatric and substance use disorders in men and women. Arch Gen Psychiatry 60: 929–937.

Kenemans JL, Hebly W, Van Den Heuvel EH et al. (2010). Moderate alcohol disrupts a mechanism for detection of rare events in human visual cortex. J Psychopharmacol 24: 839–845.

Kiehl KA, Liddle PF (2001). An event-related functional magnetic resonance imaging study of an auditory oddball task in schizophrenia. Schizophr Res 48: 159–171.

Klimesch W, Doppelmayr M, Wimmer H et al. (2001). Theta band power changes in normal and dyslexic children. Clin Neurophysiol 112: 1174–1185.

Klimesch W, Sauseng P, Hanslmayr S et al. (2007). Event-related phase reorganization may explain evoked neural dynamics. Neurosci Biobehav Rev 31: 1003–1016.

Knight RT, Scabini D (1998). Anatomic bases of event-related potentials and their relationship to novelty detection in humans. J Clin Neurophysiol 15: 3–13.

Koskinen SM, Ahveninen J, Kujala T et al. (2011). A longitudinal twin study of effects of adolescent alcohol abuse on the neurophysiology of attention and orienting. Alcohol Clin Exp Res 35: 1339–1350.

Kovacevic S, Azma S, Irimia A et al. (2012). Theta oscillations are sensitive to both early and late conflict processing stages: effects of alcohol intoxication. PLoS One 7: e43957.

Krause CM, Aromaki A, Sillanmaki L et al. (2002). Alcohol-induced alterations in ERD/ERS during an auditory memory task. Alcohol 26: 145–153.

Krieger S, Timmer J, Lis S et al. (1995). Some considerations on estimating event-related brain signals. J Neural Transm 99: 103–129.

Krueger RF, Hicks BM, Patrick CJ et al. (2002). Etiologic connections among substance dependence, antisocial behavior, and personality: modeling the externalizing spectrum. J Abnorm Psychol 111: 411–424.

Kutas M, Federmeier KD (2000). Electrophysiology reveals semantic memory use in language comprehension. Trends Cogn Sci 4: 463–470.

Kutas M, Federmeier KD (2011). Thirty years and counting: finding meaning in the N400 component of the event-related brain potential (ERP). Annu Rev Psychol 62: 621–647.

Kutas M, Hillyard SA (1980). Reading senseless sentences: brain potentials reflect semantic incongruity. Science 207: 203–205.

Laukka SJ, Jarvilehto T, Alexandrov Yu I et al. (1995). Frontal midline theta related to learning in a simulated driving task. Biol Psychol 40: 313–320.

Lingford-Hughes AR, Acton PD, Gacinovic S et al. (1998). Reduced levels of GABA-benzodiazepine receptor in alcohol dependency in the absence of grey matter atrophy. Br J Psychiatry 173: 116–122.

Llinas RR, Ribary U, Jeanmonod D et al. (1999). Thalamocortical dysrhythmia: a neurological and neuropsychiatric syndrome characterized by magnetoencephalography. Proc Natl Acad Sci U S A 96: 15222–15227.

Logan GD, Cowan WB (1984). On the ability to inhibit thought and action: A theory of an act of control. Psychol Rev 91: 295–327.

Lopez-Caneda E, Cadaveira F, Crego A et al. (2012). Hyperactivation of right inferior frontal cortex in young binge drinkers during response inhibition: a follow-up study. Addiction 107: 1796–1808.

Luck SJ (2005). An introduction to the event-related potential technique. MIT Press, Cambridge, MA.

Luck SJ, Hillyard SA (1994). Electrophysiological correlates of feature analysis during visual search. Psychophysiology 31: 291–308.

Lukas SE, Mendelson JH (1988). Electroencephalographic activity and plasma ACTH during ethanol-induced euphoria. Biol Psychiatry 23: 141–148.

Lukas SE, Mendelson JH, Woods BT et al. (1989). Topographic distribution of EEG alpha activity during ethanol-induced intoxication in women. J Stud Alcohol 50: 176–185.

Lukas SE, Mendelson JH, Kouri E et al. (1990). Ethanol-induced alterations in EEG alpha activity and apparent source of the auditory P300 evoked response potential. Alcohol 7: 471–477.

Luo X, Kranzler HR, Zuo L et al. (2005). CHRM2 gene predisposes to alcohol dependence, drug dependence and affective disorders: results from an extended case-control structured association study. Hum Mol Genet 14: 2421–2434.

Luscher C, Slesinger PA (2010). Emerging roles for G protein-gated inwardly rectifying potassium (GIRK) channels in health and disease. Nat Rev Neurosci 11: 301–315.

Marco-Pallares J, Ruffini G, Polo MD et al. (2007). Mismatch negativity impairment associated with alcohol consumption in chronic alcoholics: a scalp current density study. Int J Psychophysiol 65: 51–57.

Marinkovic K, Halgren E, Klopp J et al. (2000). Alcohol effects on movement-related potentials: a measure of impulsivity? J Stud Alcohol 61: 24–31.

Marinkovic K, Halgren E, Maltzman I (2001). Arousal-related P3a to novel auditory stimuli is abolished by a moderately low alcohol dose. Alcohol Alcohol 36: 529–539.

Marinkovic K, Halgren E, Maltzman I (2004). Effects of alcohol on verbal processing: an event-related potential study. Alcohol Clin Exp Res 28: 415–423.

Marinkovic K, Rosen BQ, Cox B et al. (2012). Event-related theta power during lexical-semantic retrieval and decision conflict is modulated by alcohol intoxication: anatomically constrained MEG. Front Psychol 3: 121.

Masaki H, Takeuchi S, Gehring WJ et al. (2006). Affective-motivational influences on feedback-related ERPs in a gambling task. Brain Res 1105: 110–121.

Maurage P, Pesenti M, Philippot P et al. (2009). Latent deleterious effects of binge drinking over a short period of time revealed only by electrophysiological measures. J Psychiatry Neurosci 34: 111–118.

Maurage P, Joassin F, Speth A et al. (2012). Cerebral effects of binge drinking: respective influences of global alcohol intake and consumption pattern. Clin Neurophysiol 123: 892–901.

McCarthy G, Nobre AC (1993). Modulation of semantic processing by spatial selective attention. Electroencephalogr Clin Neurophysiol 88: 210–219.

Menon V, Ford JM, Lim KO et al. (1997). Combined event-related fMRI and EEG evidence for temporal-parietal cortex activation during target detection. Neuroreport 8: 3029–3037.

Michael A, Mirza KA, Mukundan CR et al. (1993). Interhemispheric electroencephalographic coherence as a biological marker in alcoholism. Acta Psychiatr Scand 87: 213–217.

Mitrofanis J, Guillery RW (1993). New views of the thalamic reticular nucleus in the adult and the developing brain. Trends Neurosci 16: 240–245.

Molnar M, Boha R, Czigler B et al. (2009). The acute effect of low-dose alcohol on working memory during mental arithmetic: II. Changes of nonlinear and linear EEG-complexity in the theta band, heart rate and electrodermal activity. Int J Psychophysiol 73: 138–142.

Moselhy HF, Georgiou G, Kahn A (2001). Frontal lobe changes in alcoholism: a review of the literature. Alcohol Alcohol 36: 357–368.

Muralidharan K, Venkatasubramanian G, Pal PK et al. (2008). Abnormalities in cortical and transcallosal inhibitory mechanisms in subjects at high risk for alcohol dependence: a TMS study. Addict Biol 13: 373–379.

Näätänen R (1990). The role of attention in auditory information processing as revealed by event-related potentials and other brain measures of cognitive function. Behav Brain Sci 13: 201–233.

Näätänen R, Gaillard AW, Mäntysalo S (1978). Early selective-attention effect on evoked potential reinterpreted. Acta Psychol (Amst) 42: 313–329.

Newlin DB, Thomson JB (1990). Alcohol challenge with sons of alcoholics: a critical review and analysis. Psychol Bull 108: 383–402.

Niedermayer E, Lopes Da Silva FH (Eds.), (1999). Electroencephalography: Basic Principles, Clinical Applications and Related Fields. Williams & Wilkins, New York.

Nieuwenhuis S, Holroyd CB, Mol N et al. (2004). Reinforcement-related brain potentials from medial frontal cortex: origins and functional significance. Neurosci Biobehav Rev 28: 441–448.

Nixon SJ, Tivis R, Ceballos N et al. (2002). Neurophysiological efficiency in male and female alcoholics. Prog Neuropsychopharmacol Biol Psychiatry 26: 919–927.

Nunez PL (1995). Neocortical dynamics and Human EEG rhythms. Oxford University Press, New York.

Oddy BW, Barry RJ (2009). The relationship of N2 and P3 to inhibitory processing of social drinkers in a Go/NoGo task. Int J Psychophysiol 72: 323–330.

O'Donnell BF, Swearer JM, Smith LT et al. (1997). A topographic study of ERPs elicited by visual feature discrimination. Brain Topogr 10: 133–143.

Olbrich HM, Maes H, Gann H et al. (2000). Auditory and visual event-related potentials in alcoholics: abnormalities of components and brain electrical field. Eur Arch Psychiatry Clin Neurosci 250: 215–220.

Overbeek TJM, Nieuwenhuis S, Ridderinkhof KR (2005). Dissociable Components of Error Processing. Int J Psychophysiol 19: 319–329.

Padilla ML, Colrain IM, Sullivan EV et al. (2011). Electrophysiological evidence of enhanced performance monitoring in recently abstinent alcoholic men. Psychopharmacology (Berl) 213: 81–91.

Padmanabhapillai A, Porjesz B, Ranganathan M et al. (2006a). Suppression of early evoked gamma band response in male alcoholics during a visual oddball task. Int J Psychophysiol 60: 15–26.

Padmanabhapillai A, Tang Y, Ranganathan M et al. (2006b). Evoked gamma band response in male adolescent subjects at high risk for alcoholism during a visual oddball task. Int J Psychophysiol 62: 262–271.

Pandey AK, Kamarajan C, Tang Y et al. (2012). Neurocognitive deficits in male alcoholics: an ERP/sLORETA analysis of the N2 component in an equal probability Go/NoGo task. Biol Psychol 89: 170–182.

Pascual-Marqui RD (2002). Standardized low-resolution brain electromagnetic tomography (sLORETA): technical details. Methods Find Exp Clin Pharmacol 24 (Suppl. D): 5–12.

Patterson BW, Williams HL, Mclean GA et al. (1987). Alcoholism and family history of alcoholism: effects on visual and auditory event-related potentials. Alcohol 4: 265–274.

Pekkonen E, Ahveninen J, Jaaskelainen IP et al. (1998). Selective acceleration of auditory processing in chronic alcoholics during abstinence. Alcohol Clin Exp Res 22: 605–609.

Perry E, Walker M, Grace J et al. (1999). Acetylcholine in mind: a neurotransmitter correlate of consciousness? Trends Neurosci 22: 273–280.

Petit G, Kornreich C, Maurage P et al. (2012). Early attentional modulation by alcohol-related cues in young binge drinkers: an event-related potentials study. Clin Neurophysiol 123: 925–936.

Pfefferbaum A, Ford JM (1988). ERPs to stimuli requiring response production and inhibition: effects of age, probability and visual noise. Electroencephalogr Clin Neurophysiol 71: 55–63.

Pfefferbaum A, Ford JM, Weller BJ et al. (1985). ERPs to response production and inhibition. Electroencephalogr Clin Neurophysiol 60: 423–434.

Pfefferbaum A, Ford JM, White PM et al. (1991). Event-related potentials in alcoholic men: P3 amplitude reflects family history but not alcohol consumption. Alcohol Clin Exp Res 15: 839–850.

Polich J, Bloom FE (1987). P300 from normals and adult children of alcoholics. Alcohol 4: 301–305.

Polich J, Herbst KL (2000). P300 as a clinical assay: rationale, evaluation, and findings. Int J Psychophysiol 38: 3–19.

Polich J, Pollock VE, Bloom FE (1994). Meta-analysis of P300 amplitude from males at risk for alcoholism. Psychol Bull 115: 55–73.

Pollock VE (1992). Meta-analysis of subjective sensitivity to alcohol in sons of alcoholics. Am J Psychiatry 149: 1534–1538.

Pollock VE, Volavka J, Goodwin DW et al. (1983). The EEG after alcohol administration in men at risk for alcoholism. Arch Gen Psychiatry 40: 857–861.

Pollock VE, Schneider LS, Zemansky MF et al. (1992). Topographic quantitative EEG amplitude in recovered alcoholics. Psychiatry Res 45: 25–32.

Pollock VE, Earleywine M, Gabrielli WF (1995). Personality and EEG beta in older adults with alcoholic relatives. Alcohol Clin Exp Res 19: 37–43.

Polo MD, Escera C, Gual A et al. (1999). Mismatch negativity and auditory sensory memory in chronic alcoholics. Alcohol Clin Exp Res 23: 1744–1750.

Porjesz B, Begleiter H (1983). Brain dysfunction and alcohol. In: B Kissin, H Begleiter (Eds.), The pathogenesis of alcoholism: biological factors, Plenum Press, New York.

Porjesz B, Begleiter H (1985). The use of event-related potentials in the study of alcoholism: implications for the study of drugs of abuse. NIDA Res Monogr 62: 77–99.

Porjesz B, Begleiter H (1990). Individuals at risk for alcoholism: Neurophysiologic processes. In: CR Cloninger, H Begleiter (Eds.), Genetics and biology of alcoholism. Cold Spring Harbor Laboratory Press, New York.

Porjesz B, Begleiter H (1993). Neurophysiological factors associated with alcoholism. In: SJ Nixon, WA Hunt (Eds.), Alcohol-induced brain damage. NIH Publication, Rockville, MD.

Porjesz B, Begleiter H (1996). Effects of alcohol on electrophysiological activity of the brain. In: H Begleiter, B Kissin (Eds.), Alcohol and Alcoholism: The Pharmacology of Alcohol and Alcohol Dependence. Oxford University Press, New York.

Porjesz B, Rangaswamy M (2007). Neurophysiological endophenotypes, CNS disinhibition, and risk for alcohol dependence and related disorders. ScientificWorldJournal 7: 131–141.

Porjesz B, Begleiter H, Bihari B et al. (1987a). Event-related brain potentials to high incentive stimuli in abstinent alcoholics. Alcohol 4: 283–287.

Porjesz B, Begleiter H, Bihari B et al. (1987b). The N2 component of the event-related brain potential in abstinent alcoholics. Electroencephalogr Clin Neurophysiol 66: 121–131.

Porjesz B, Begleiter H, Reich T et al. (1998). Amplitude of visual P3 event-related potential as a phenotypic marker for a predisposition to alcoholism: preliminary results from the COGA Project. Collaborative Study on the Genetics of Alcoholism. Alcohol Clin Exp Res 22: 1317–1323.

Porjesz B, Begleiter H, Wang K et al. (2002a). Linkage and linkage disequilibrium mapping of ERP and EEG phenotypes. Biol Psychol 61: 229–248.

Porjesz B, Almasy L, Edenberg HJ et al. (2002b). Linkage disequilibrium between the beta frequency of the human EEG and a GABAA receptor gene locus. Proc Natl Acad Sci U S A 99: 3729–3733.

Porjesz B, Rangaswamy M, Kamarajan C et al. (2005). The utility of neurophysiological markers in the study of alcoholism. Clin Neurophysiol 116: 993–1018.

Potter DD, Pickles CD, Roberts RC et al. (2000). Scopolamine impairs memory performance and reduces frontal but not parietal visual P3 amplitude. Biol Psychol 52: 37–52.

Prabhu VR, Porjesz B, Chorlian DB et al. (2001). Visual p3 in female alcoholics. Alcohol Clin Exp Res 25: 531–539.

Prescott CA (2001). The genetic epidemiology of alcoholism – Sex differences and future directions. Alcohol in Health and Disease 125–149.

Prescott CA, Aggen SH, Kendler KS (1999). Sex differences in the sources of genetic liability to alcohol abuse and dependence in a population-based sample of U.S. twins. Alcohol Clin Exp Res 23: 1136–1144.

Propping P, Kruger J, Mark N (1981). Genetic disposition to alcoholism. An EEG study in alcoholics and their relatives. Hum Genet 59: 51–59.

Rangaswamy M, Porjesz B (2008a). From event-related potential to oscillations: genetic diathesis in brain (dys)function and alcohol dependence. Alcohol Res Health 31: 238–242.

Rangaswamy M, Porjesz B (2008b). Uncovering genes for cognitive (dys)function and predisposition for alcoholism spectrum disorders: a review of human brain oscillations as effective endophenotypes. Brain Res 1235: 153–171.

Rangaswamy M, Porjesz B, Chorlian DB et al. (2002). Beta power in the EEG of alcoholics. Biol Psychiatry 52: 831–842.

Rangaswamy M, Porjesz B, Chorlian DB et al. (2003). Theta power in the EEG of alcoholics. Alcohol Clin Exp Res 27: 607–615.

Rangaswamy M, Porjesz B, Ardekani BA et al. (2004a). A functional MRI study of visual oddball: evidence for frontoparietal dysfunction in subjects at risk for alcoholism. Neuroimage 21: 329–339.

Rangaswamy M, Porjesz B, Chorlian DB et al. (2004b). Resting EEG in offspring of male alcoholics: beta frequencies. Int J Psychophysiol 51: 239–251.

Rangaswamy M, Jones KA, Porjesz B et al. (2007). Delta and theta oscillations as risk markers in adolescent offspring of alcoholics. Int J Psychophysiol 63: 3–15.

Realmuto G, Begleiter H, Odencrantz J et al. (1993). Event-related potential evidence of dysfunction in automatic processing in abstinent alcoholics. Biol Psychiatry 33: 594–601.

Ridderinkhof KR, De Vlugt Y, Bramlage A et al. (2002). Alcohol consumption impairs detection of performance errors in mediofrontal cortex. Science 298: 2209–2211.

Rodriguez Holguin S, Corral M, Cadaveira F (1998). Mismatch negativity in young children of alcoholics from high-density families. Alcohol Clin Exp Res 22: 1363–1368.

Rodriguez Holguin S, Porjesz B, Chorlian DB et al. (1999). Visual P3a in male subjects at high risk for alcoholism. Biol Psychiatry 46: 281–291.

Rohrbaugh JW, Stapleton JM, Parasuraman R et al. (1987). Dose-related effects of ethanol on visual sustained attention and event-related potentials. Alcohol 4: 293–300.

Roopesh BN, Rangaswamy M, Kamarajan C et al. (2009). Priming deficiency in male subjects at risk for alcoholism: the N4 during a lexical decision task. Alcohol Clin Exp Res 33: 2027–2036.

Roopesh BN, Rangaswamy M, Kamarajan C et al. (2010). Reduced resource optimization in male alcoholics: N400 in a lexical decision paradigm. Alcohol Clin Exp Res 34: 1905–1914.

Sable JJ, Low KA, Maclin EL et al. (2004). Latent inhibition mediates N1 attenuation to repeating sounds. Psychophysiology 41: 636–642.

Saletu-Zyhlarz GM, Arnold O, Anderer P et al. (2004). Differences in brain function between relapsing and abstaining alcohol-dependent patients, evaluated by EEG mapping. Alcohol Alcohol 39: 233–240.

Sauseng P, Klimesch W (2008). What does phase information of oscillatory brain activity tell us about cognitive processes? Neurosci Biobehav Rev 32: 1001–1013.

Schellekens AF, De Bruijn ER, Van Lankveld CA et al. (2010). Alcohol dependence and anxiety increase error-related brain activity. Addiction 105: 1928–1934.

Schoepp DD (2001). Unveiling the functions of presynaptic metabotropic glutamate receptors in the central nervous system. J Pharmacol Exp Ther 299: 12–20.

Schuckit MA (1994). Low level of response to alcohol as a predictor of future alcoholism. Am J Psychiatry 151: 184–189.

Schuckit MA, Gold EO, Croot K et al. (1988). P300 latency after ethanol ingestion in sons of alcoholics and in controls. Biol Psychiatry 24: 310–315.

Squires NK, Squires KC, Hillyard SA (1975). Two varieties of long-latency positive waves evoked by unpredictable auditory stimuli in man. Electroencephalogr Clin Neurophysiol 38: 387–401.

Stassen HH, Lykken DT, Propping P et al. (1988). Genetic determination of the human EEG. Survey of recent results on twins reared together and apart. Hum Genet 80: 165–176.

Steele CM, Josephs RA (1988). Drinking your troubles away. II: An attention-allocation model of alcohol's effect on psychological stress. J Abnorm Psychol 97: 196–205.

Steele CM, Josephs RA (1990). Alcohol myopia. Its prized and dangerous effects. Am Psychol 45: 921–933.

Steinhauer SR, Hill SY, Zubin J (1987). Event-related potentials in alcoholics and their first-degree relatives. Alcohol 4: 307–314.

Stenberg G, Sano M, Rosen I et al. (1994). EEG topography of acute ethanol effects in resting and activated normals. J Stud Alcohol 55: 645–656.

Steriade M (1990). Cholinergic control of thalamic function. Arch Int Physiol Biochim 98: A11–A46.

Stockwell RG, Mansinha L, Lowe Rp (1996). Localization of the complex spectrum: The S-transform. IEEE Trans on Sig Proc 44: 998–1001.

Suresh S, Porjesz B, Chorlian DB et al. (2003). Auditory P3 in female alcoholics. Alcohol Clin Exp Res 27: 1064–1074.

Teo RK, Ferguson DA (1986). The acute effects of ethanol on auditory event-related potentials. Psychopharmacology (Berl) 90: 179–184.

Tiesinga PH, Fellous JM, Jose JV et al. (2001). Computational model of carbachol-induced delta, theta, and gamma oscillations in the hippocampus. Hippocampus 11: 251–274.

Toyomaki A, Murohashi H (2005). Discrepancy between feedback negativity and subjective evaluation in gambling. Neuroreport 16: 1865–1868.

Trim RS, Schuckit MA, Smith TL (2009). The relationships of the level of response to alcohol and additional characteristics to alcohol use disorders across adulthood: a discrete-time survival analysis. Alcohol Clin Exp Res 33: 1562–1570.

Uhlhaas PJ, Singer W (2006). Neural synchrony in brain disorders: relevance for cognitive dysfunctions and pathophysiology. Neuron 52: 155–168.

Uhlhaas PJ, Roux F, Rodriguez E et al. (2010). Neural synchrony and the development of cortical networks. Trends Cogn Sci 14: 72–80.

van Baal GC, De Geus EJ, Boomsma DI (1998). Genetic influences on EEG coherence in 5-year-old twins. Behav Genet 28: 9–19.

van Beijsterveldt CE, Boomsma DI (1994). Genetics of the human electroencephalogram (EEG) and event-related brain potentials (ERPs): a review. Hum Genet 94: 319–330.

van Beijsterveldt CE, Molenaar PC, De Geus EJ et al. (1996). Heritability of human brain functioning as assessed by electroencephalography. Am J Hum Genet 58: 562–573.

van Beijsterveldt CE, Molenaar PC, De Geus EJ et al. (1998). Genetic and environmental influences on EEG coherence. Behav Genet 28: 443–453.

van Der Stelt O, Gunning WB, Snel J et al. (1997). No electrocortical evidence of automatic mismatch dysfunction in children of alcoholics. Alcohol Clin Exp Res 21: 569–575.

van Veen V, Carter CS (2002). The anterior cingulate as a conflict monitor: fMRI and ERP studies. Physiol Behav 77: 477–482.

van Vugt MK, Sederberg PB, Kahana MJ (2007). Comparison of spectral analysis methods for characterizing brain oscillations. J Neurosci Methods 162: 49–63.

Varela F, Lachaux JP, Rodriguez E et al. (2001). The brainweb: phase synchronization and large-scale integration. Nat Rev Neurosci 2: 229–239.

Verleger R (1988). Event-related potentials and cognition: a critique of the context updating hypothesis and an alternative interpretation of P3. Behav Brain Sci 11: 343–427.

Volavka J, Czobor P, Goodwin DW et al. (1996). The electroencephalogram after alcohol administration in high-risk men and the development of alcohol use disorders 10 years later. Arch Gen Psychiatry 53: 258–263.

Volkow ND, Wang GJ, Begleiter H et al. (1995). Regional brain metabolic response to lorazepam in subjects at risk for alcoholism. Alcohol Clin Exp Res 19: 510–516.

Von Stein A, Sarnthein J (2000). Different frequencies for different scales of cortical integration: from local gamma to long range alpha/theta synchronization. Int J Psychophysiol 38: 301–313.

Wang JC, Hinrichs AL, Stock H et al. (2004). Evidence of common and specific genetic effects: association of the muscarinic acetylcholine receptor M2 (CHRM2) gene with alcohol dependence and major depressive syndrome. Hum Mol Genet 13: 1903–1911.

Weschke S, Niedeggen M (2012). Differential effects of moderate alcohol consumption on motion and contrast processing. Psychophysiology 49: 833–841.

Whittington MA, Traub RD, Kopell N et al. (2000). Inhibition-based rhythms: experimental and mathematical observations on network dynamics. Int J Psychophysiol 38: 315–336.

Winterer G, Kloppel B, Heinz A et al. (1998). Quantitative EEG (QEEG) predicts relapse in patients with chronic alcoholism and points to a frontally pronounced cerebral disturbance. Psychiatry Res 78: 101–113.

Winterer G, Enoch MA, White KV et al. (2003a). EEG phenotype in alcoholism: increased coherence in the depressive subtype. Acta Psychiatr Scand 108: 51–60.

Winterer G, Mahlberg R, Smolka MN et al. (2003b). Association analysis of exonic variants of the GABA(B)-receptor gene and alpha electroencephalogram voltage in normal subjects and alcohol-dependent patients. Behav Genet 33: 7–15.

Winterer G, Smolka M, Samochowiec J et al. (2003c). Association of EEG coherence and an exonic GABA(B) R1 gene polymorphism. Am J Med Genet 117B: 51–56.

Yeung N, Sanfey AG (2004). Independent coding of reward magnitude and valence in the human brain. J Neurosci 24: 6258–6264.

Yordanova J, Kolev V (1996). Brain theta response predicts P300 latency in children. Neuroreport 8: 277–280.

Yordanova J, Banaschewski T, Kolev V et al. (2001). Abnormal early stages of task stimulus processing in children with attention-deficit hyperactivity disorder – evidence from event-related gamma oscillations. Clin Neurophysiol 112: 1096–1108.

Zahr NM, Pitel AL, Chanraud S et al. (2010). Contributions of studies on alcohol use disorders to understanding cerebellar function. Neuropsychol Rev 20: 280–289.

Zahr NM, Kaufman KL, Harper CG (2011). Clinical and pathological features of alcohol-related brain damage. Nat Rev Neurol 7: 284–294.

Zucker RA, Heitzeg MM, Nigg JT (2011). Parsing the undercontrol/disinhibition pathway to substance use disorders: a multilevel developmental problem. Child Dev Perspect 5: 248–255.

Handbook of Clinical Neurology, Vol. 125 (3rd series)
Alcohol and the Nervous System
E.V. Sullivan and A. Pfefferbaum, Editors

Chapter 24

Alcohol and the sleeping brain

IAN M. COLRAIN[1,2]*, CHRISTIAN L. NICHOLAS[2], AND FIONA C. BAKER[1,3]

[1]*Center for Health Sciences, SRI International, Menlo Park, CA, USA*

[2]*Melbourne School of Psychological Sciences, The University of Melbourne, Parkville, Vic., Australia*

[3]*Brain Function Research Group, School of Physiology, University of the Witwatersrand, Johannesburg, South Africa*

INTRODUCTION

Sleep is a fundamental, necessary, and complex behavior in humans. The mammalian nervous system demonstrates two distinct states of sleep – rapid-eye-movement (REM) and non-rapid-eye-movement (NREM) sleep – that involve different patterns of neurologic activation and neurotransmitter release, and presumably different functions. Many of the neurotransmitters known to be involved in wake–sleep regulation are also affected by alcohol, and thus, not surprisingly, alcohol affects sleep in a number of ways. As with other drugs, alcohol demonstrates tolerance with repeated administration and dependence, and this tolerance is accompanied by adaptation of neurotransmitter systems, either by modulation of their release or by modifying the sensitivity of their response mechanisms. Withdrawal of alcohol in dependent individuals can be associated with neurologic manifestations of the consequent neurochemical imbalance. Over time, recovery can take place to restore a normal balance of inhibitory and excitatory systems; however, it is possible that some changes induced by alcohol may be resistant to restoration. In addition to the direct pharmacologic effect of alcohol on sleep-regulatory systems, the long-term consequences of alcohol abuse and dependence on brain macrostructure and microstructure may also themselves lead to changes in sleep regulation or changes in electroencephalographic (EEG) manifestations of sleep. The influence of alcohol on sleep thus needs to be evaluated in multiple contexts. This chapter reports the evidence for the acute effects of alcohol on sleep in humans and in animal models, the effects of short-term repeated administration of alcohol on sleep, and the observations of the long-term consequences of alcohol dependence on sleep.

Human sleep is defined on the basis of changes in EEG activity, augmented by measurement of eye movements and postural muscle tone to aid in the differentiation of REM sleep from wakefulness. The movement from wakefulness to deep sleep largely maps on to an EEG frequency continuum with very-high-frequency gamma oscillations (~40 Hz and higher) and desynchronized beta activity (~20–30 Hz) associated with active wakefulness, giving way to synchronized alpha (~10 Hz) in relaxed wakefulness, theta activity (4–7 Hz) in light stage 1 NREM sleep (now called N1 sleep: Iber et al., 2007). Stage 2 NREM sleep (N2) is characterized by phasic sleep spindles (synchronized sigma activity: 12–16 Hz) and K-complexes, high-amplitude delta waves, and constitutes 45–55% of the night. Stages 3 and 4 NREM sleep (N3 or slow-wave sleep, SWS) is characterized by slow delta activity (0.3–2 Hz), constituting about 18% of the night in young adults. REM sleep, constituting 20–25% of the night, demonstrates desynchronized activity in theta and beta ranges reflective of the partial reactivation of brainstem mechanisms that are fully active in wakefulness and deactivated in NREM sleep. NREM and REM sleep alternate across the night in approximately 90-minute cycles. As the sleep episode progresses, stage 2 sleep accounts for most of NREM sleep, and the amount of REM sleep increases.

According to the two-process model of sleep (Borbely, 1982), the timing, depth, and duration of sleep are controlled by the interaction between circadian (process C) and homeostatic (process S) processes. Process S reflects the buildup of sleep pressure during wake, which dissipates during sleep. EEG delta-wave activity in NREM sleep is considered a marker of process S and is highest in the first part of the sleep period when sleep pressure is highest. Process C reflects circadian

*Correspondence to: Ian M. Colrain, Ph.D., SRI International, 333 Ravenswood Ave., Menlo Park, CA 94025, USA. Tel: +1-650-859-3915, Fax: +1-650-859-2743, E-mail: ian.colrain@sri.com

regulation of sleep timing. While the overarching purpose of sleep remains unknown, evidence suggests that it serves several functions, including conservation of brain energy, facilitation of certain forms of learning and memory, and support of cognitive capacity, including pruning and maintenance of synaptic connectivity (Tononi and Cirelli, 2006). Insufficient sleep or insomnia (difficulty initiating or maintaining sleep or nonrestorative sleep) is associated with negative consequences to immune function, impaired cardiovascular and cerebrovascular health, cognitive impairment, and a change in emotional reactivity. Insomnia is a risk factor for the development of psychiatric disorders, including depression, anxiety, and alcohol abuse (Breslau et al., 1996). Insomnia complaints may also interfere with the recovery process and contribute to relapse among patients in recovery from alcohol dependence (Brower, 2001; Arnedt et al., 2007). Therefore, it has, been suggested that treating insomnia may aid recovery in alcoholics and prevent or reduce the likelihood of relapse (Brower, 2001; Arnedt et al., 2007).

ACUTE EFFECTS OF ALCOHOL ON SLEEP

Sleep occurs over a sustained period, typically lasting approximately 8 hours in humans. In the absence of continued dosing, alcohol consumed prior to the onset of sleep, therefore, will not be at a constant level throughout the sleep period. Depending on the timing of sleep onset relative to consumption, blood alcohol levels may continue to rise for some time during sleep, but inevitably they will start to fall as a function of metabolism, the time course of which is unaltered by sleep itself (Rundell et al., 1972). Sleep, therefore, could be expected to be affected differently during the initial period of high alcohol levels from the subsequent elimination phase. The presence of alcohol metabolites such as aldehyde need to be considered in terms of their own possible influence on sleep mechanisms, as do secondary effects of alcohol, such as diuresis.

Effects of an acute pre-bedtime dose of alcohol on sleep have been extensively studied, although methodology has varied greatly between studies in terms of dose and timing of alcohol administration, age and gender of subjects, and sample size. Alcohol initially acts as a sedative. Commonly reported phenomena include shortened sleep onset latency (Stone, 1980; MacLean and Cairns, 1982; Scrima et al., 1982; Williams et al., 1983; Roehrs et al., 1999) and increased slow-wave sleep (SWS) in the first half of the night (Rundell et al., 1972; Prinz et al., 1980; MacLean and Cairns, 1982; Williams et al., 1983; Feige et al., 2006; Van Reen et al., 2006; Arnedt et al., 2011b; Sagawa et al., 2011; Chan et al., 2013). REM sleep is suppressed, with a longer latency to REM sleep and decreased REM sleep in the first half of the night (Rundell et al., 1972; Williams et al., 1983; Van Reen et al., 2006; Sagawa et al., 2011; Chan et al., 2013) or across the whole night (Rundell et al., 1972; Williams et al., 1983; Roehrs et al., 1991, 1999; Van Reen et al., 2006; Arnedt et al., 2011b). In the second half of the night, sleep is disrupted, with increased wakefulness and/or stage 1 sleep. This pattern of initial sleep augmentation followed by a period of poor-quality sleep can lead to a downward spiral, with insomnia being self-treated with alcohol to produce a rapid sleep onset, subsequent poor sleep then leading to daytime sleepiness that is self-treated with caffeine, which exacerbates insomnia, requiring more alcohol to fall asleep, and so on. It is estimated that alcohol is used by more than one in 10 individuals as a hypnotic agent to self-medicate sleep problems (Arnedt et al., 2007).

Acute effects of alcohol on sleep: repeated administration

While there is reasonable consensus on the effects of acute alcohol administration on a single night's sleep in non-alcohol-dependent individuals, less is known regarding the effects of repeated administration over multiple nights, a situation that better represents the normal experience for non-dependent drinkers in the community. Studies of the effects of repeated alcohol administration over multiple nights are rare and suffer from small sample sizes. To our knowledge, only five such studies have been published, with a total of 19 men and 5 women evaluated in experiments that vary in the dose of alcohol administered, the timing of the alcohol relative to sleep, and the number of nights of consecutive usage.

Yules et al. (1966) studied three young, non-alcohol-dependent men over 5 nights of drinking, with 1 g/kg ethanol administered 15 minutes before bedtime. Yules et al. (1967) studied four young men over three or five nights of drinking, with 1 g/kg ethanol administered 4 hours before bedtime. Unfortunately, these studies predate the acceptance of a standardized sleep scoring system, and the data are difficult to interpret (e.g., stage 1 NREM and REM sleep are combined); however their data are consistent with REM suppression and enhancement of slow-wave sleep in the first part of the night.

Rundell et al. (1972) studied seven young men over three nights of drinking with alcohol administered over an hour, ending 30 minutes before bed, with blood alcohol concentrations at bedtime between 0.05 and 0.095 mg%. Data are presented from a baseline night, three drinking nights, and the mean of two recovery nights. Prinz et al. (1980) studied five young men over nine nights of drinking (seven of them at home) with a

0.8 g/kg dose (0.08 breath alcohol concentration (BAC) on the laboratory nights) consumed over the hour before bedtime. Data are reported from a baseline night; the first and ninth alcohol nights, and a recovery night. Feige et al. (2006) studied five young men and five young women over three nights of drinking. Alcohol was consumed before bed to obtain BAC of 0.03 or 1.0% in two different conditions. Data are presented from a baseline night, three drinking nights, and two recovery nights. The results for the first half of the night from these studies are summarized in Figure 24.1.

SWS was significantly increased over baseline on the first drinking night in the Prinz et al. (1980) and Feige et al. (2006) (0.10% BAC dose) studies, but not in the Feige et al. (2006) (0.03% BAC dose) or Rundell et al. (1972) studies. Prinz et al. (1980) show SWS as recovered to baseline on night nine; however, Feige et al. (2006) showed significant increases in SWS for the first half of the night on all three drinking nights, so if tolerance to the SWS induction effect occurs with repeated nightly use, it would appear to take some time to develop.

The percentage of REM sleep in the first half of the night was not decreased on the first drinking night at either the 0.03 or 0.10% BAC doses in the Feige et al. (2006) study. However, REM density (a measure of the number of eye movements per unit time) was significantly reduced on all three drinking nights for the 0.10% BAC relative to baseline, and recovered to baseline with no evidence of rebound on the first recovery night (Feige et al., 2006). Prinz et al. (1980) reported that REM sleep in the first half of the first drinking night ($7.0 \pm 3.1\%$) and of the ninth drinking night ($9.5 \pm 3.18\%$) was lower than baseline ($17.26 \pm 2.20\%$), although the difference was not statistically reliable in this small sample. Rundell et al. (1972) reported a decrease in REM sleep on the first drinking night in their study, but values on the second and third drinking nights were not different to baseline. While these studies support others showing a suppressing effect of REM sleep by a single dose of alcohol, more studies are needed to determine whether the effect persists after multiple drinking nights.

Given the clear importance of tolerance to sleep effects as a mechanism for the development of abusive drinking (Conroy and Brower, 2011), the lack of consistent results from the small number of studies and of subjects evaluated provides a clear argument for more work to be done to investigate the effects of repeated administration of alcohol on sleep.

Acute alcohol: sleep EEG data

Computerized analysis of the sleep EEG may reveal subtle effects of alcohol that may vary according to brain region and that are not evident from manual scoring of the polysomnogram. Few investigations have examined the acute effect of alcohol on EEG power spectra (Rundell et al., 1972; Dijk et al., 1992; Landolt et al., 1996; Van Reen et al., 2006). Rundell et al. (1972) conducted an automated period analysis of EEG in the first half of the night in young men following administration

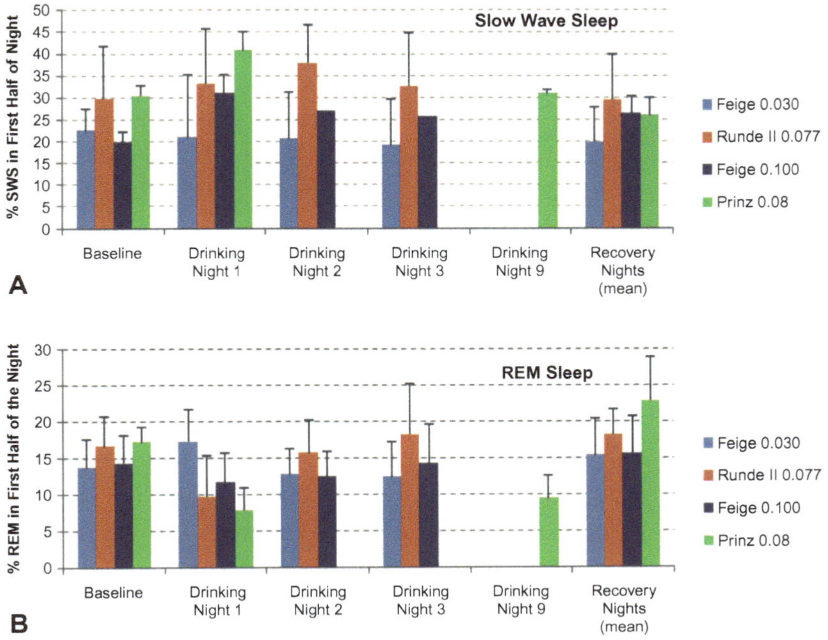

Fig. 24.1. The percentage of (**A**) slow-wave sleep (SWS) and (**B**) rapid eye movement (REM) sleep in the first half of the night across multiple nights of drinking. (Data from Rundell et al., 1972; Prinz et al., 1980; Feige et al., 2006.)

of 0.9 g/kg alcohol and reported decreases in beta along with increases in alphoid (first derivative of the zero crossing count). Dijk et al. (1992) observed increases in delta power in a single central derivation in young men following administration of 0.6 g/kg alcohol, consistent with increased SWS and decreases in sigma power only in the first 2 hours of the sleep period (Dijk et al., 1992). Landolt et al. (1996) reported elevated delta and theta power during SWS in frontal, central, parietal, and occipital derivations in older men, with the increase in delta appearing to be most prominent in the frontal and central derivations following moderate alcohol consumption (0.55 g/kg). Effects were only seen in the first sleep cycle, with significantly reduced delta power with alcohol in subsequent cycles. This cycle effect is complicated, however, by the fact that alcohol was administered 6 hours before sleep. Van Reen and colleagues (2006) reported that 0.49 g/kg of alcohol was associated with increased alpha activity in anterior derivations (Fz/Cz) with no difference in posterior derivations (Pz/Oz) during NREM sleep compared to baseline in young women, suggesting that the effect of alcohol on the sleep EEG differs according to region. Given that these studies differed in alcohol dose (0.49–0.90 g/kg body weight), timing (35 minutes to 6 hours prior to sleep), analysis methods, and results, it is difficult to determine if there is a consistent effect of alcohol on power spectra on the basis of these studies.

SLEEP IN ALCOHOLISM

Studies consistently show a high comorbidity of insomnia and alcoholism (reviewed in Arnedt et al., 2007a). Based on studies of clinical populations, 36–91% of patients report insomnia either while drinking or within several weeks of stopping (Brower and Perron, 2010). Many alcoholics develop poor sleep habits and irregular sleep–wake schedules when drinking, which may persist after quitting. In particular, napping and maintaining an irregular sleep–wake schedule are associated with greater wakefulness and poorer sleep quality at night (Currie et al., 2003).

Laboratory-based polysomnographic studies of abstinent alcoholics typically show a pattern of sleep disturbance with increased wakefulness consistent with self-reports of persistent sleep disturbance common in this population. Sleep efficiency is a simple index of the proportion of the time in bed spent asleep and thus a polysomnographic marker of general sleep quality. It is generally decreased in recently detoxified alcoholics (periods of sobriety, ranging from 16 to 46 days) (Snyder and Karacan, 1985; Gillin et al., 1990; Drummond et al., 1998; Irwin et al., 2000; Gann et al., 2001)' but seems to show recovery, with most studies examining alcoholics after longer periods of abstinence not showing differences between alcoholics and controls (Adamson and Burdick, 1973; Williams and Rundell, 1981; Drummond et al., 1998; Colrain et al., 2009c).

Laboratory studies also reveal decreased SWS (Allen and Wagman, 1975; Rundell et al., 1975; Wagman and Allen, 1975; Allen et al., 1977; Othmer et al., 1982; Gillin et al., 1990; Thompson et al., 1995; Le Bon et al., 1997; Drummond et al., 1998; Irwin et al., 2000, 2002; Gann et al., 2001; Feige et al., 2007; Colrain et al., 2009c) as a highly consistent finding (Benca et al., 1992; Ehlers, 2000) in alcoholics relative to controls. There is also evidence of increased REM sleep pressure (Gillin et al., 1990; Thompson et al., 1995; Drummond et al., 1998; Gann et al., 2001; Feige et al., 2007; Colrain et al., 2009c). It is reasonable to expect increased REM pressure in actively drinking or recently detoxified alcoholics, given that REM sleep is suppressed with high doses of alcohol (Aldrich, 1998). This form of REM rebound cannot explain the increased REM in those who have been abstinent for a long time, relative to controls. It is possible that increased REM sleep may represent a predisposition to altered sleep rather than a consequence of alcohol abuse, although REM is not elevated in adolescents with a positive family history of alcoholism (Tarokh et al., 2012a). Another possibility is that alcohol abuse leads to long-lasting neurochemical changes in the brainstem. Possible mechanisms will be discussed below. Figure 24.2 gives an example of the proportions of wakefulness (pre-sleep and throughout the night), and different sleep stages in alcoholic and control men and women.

Alcoholism: sleep EEG data

Spectral analysis of sleep EEG data has been conducted in a smaller number of studies. Consistent with the reports of reduced SWS, Irwin et al. (2000) reported significantly lower NREM delta power (0.75–4 Hz) across

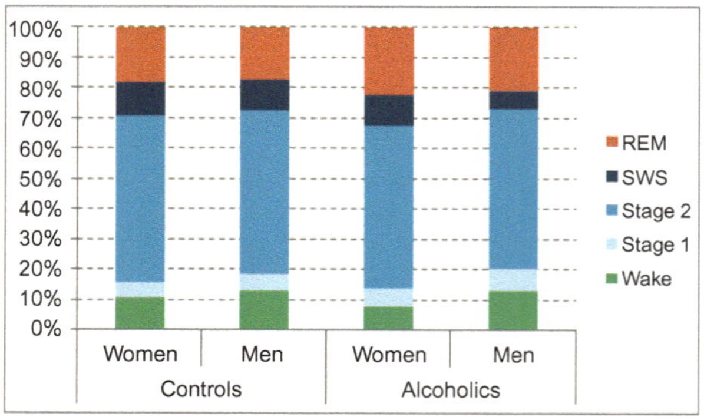

Fig. 24.2. The percentage of the night spent in different sleep stages (Rechtschaffen and Kales, 1968) in men and women with alcohol dependence and sex-matched controls. (Data from Colrain et al., 2009c.)

the entire night in alcoholics relative to controls, particularly in the first NREM period. Total power in the measured spectrum was also lower, with a trend for lower theta power in the first NREM period in alcoholics. The delta power result was confirmed in the baseline data in Irwin et al. (2002), although it appeared to be stronger in African American than in European American alcoholics. On the other hand, analysis of placebo condition sleep EEG data from a study investigating REM sleep induction tests using galanthamine hydrobromide did not show differences between alcoholics and controls in delta or theta frequencies during NREM or REM sleep (Feige et al., 2007). Our own data from multiple EEG sites (Colrain et al., 2009c) revealed that, relative to controls, alcoholics had significantly lower slow-wave activity in NREM sleep across the whole night and in the first NREM period in the slow (<1 Hz) band as well as in each of the 1 to <2, 2 to <3, and 3 to <4 Hz delta EEG bands. The slower end of theta activity (up to 6 Hz) displayed a similar pattern to delta frequencies. The effects were more prominent in frontal than posterior scalp derivations in alcoholic men and women. Interestingly, the delta effects were not seen in REM sleep but have been reported by others during wakefulness (Bauer, 2001; Saletu-Zyhlarz et al., 2004). These lower levels of delta power during NREM sleep could reflect underlying gray- and white-matter volume deficits and compromised connectivity in the brain that characterize chronic alcoholics, as described later.

Differences in activity in the fast frequency bands (beta and gamma) during sleep between alcoholics and controls are less consistent. Feige et al. (2007) reported elevated beta activity in REM and gamma activity in stage 2 NREM sleep, but only in data from the adaptation nights, with no differences for subsequent placebo nights from their drug study. Irwin et al. (2002) reported a trend for elevated beta activity in alcoholics across the entire night at baseline that became a significant difference during a recovery night following a night of partial sleep deprivation. Colrain et al. (2009c) did not see any differences between alcoholics and controls in high-frequency EEG activity during sleep. Because these analyses are performed on stable sleep epochs, results suggest that, once sleep is attained, it is not necessarily characterized by elevated fast frequency activity. By contrast, primary insomniacs have greater beta power during NREM sleep than normal sleepers, thought to reflect higher levels of cortical arousal (Riemann et al., 2010). Topographic differences in EEG spectral power during sleep evaluated in alcoholics compared with controls revealed that slow frequency activity was maximal over frontal scalp regions in both alcoholics and control subjects (Colrain et al., 2009c). This is consistent with previous studies of normal subjects (Finelli et al., 2001), positron emission tomography data (Dang-Vu et al., 2005), and high-density EEG array (Massimini et al., 2004) studies, which indicate that the frontal cortex is preferentially involved in the generation of slow-wave activity during sleep. Differences in slow frequency between alcoholics and controls were also more marked over the frontal scalp, with alcoholics showing lower delta EEG power (Fig. 24.3). This topographic pattern is consistent with the known frontal susceptibility to alcoholism-related alterations in brain structure and function (Oscar-Berman et al., 2013; Zahr et al., 2013).

Sex effects in the impact of alcohol and alcoholism on sleep

Few studies have investigated sex differences in the effect of acute alcohol intake or chronic alcoholism on sleep. There are sex differences in drinking patterns and pharmacokinetics of alcohol. Women tend to have less body water and more body fat than men and therefore tend to have higher blood alcohol levels than men after consuming the same amount of alcohol due to alcohol's hydrophilic properties (Ely et al., 1999). Alcohol

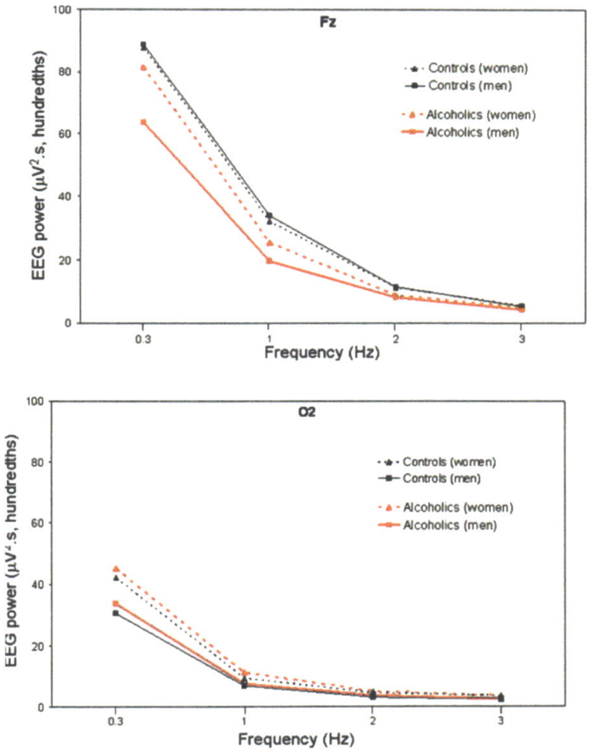

Fig. 24.3. All-night non-rapid-eye-movement sleep slow-wave activity (0.3–<4 Hz) at Fz (upper panel) and O2 (lower panel) for 34 alcoholics (21 men) and 41 controls (18 men). Values are averaged within the 0.3–1 Hz bin, and within each 1-Hz frequency bin thereafter. Bins are identified by their lower boundary. EEG, electroencephalogram. (Data from Colrain et al., 2009c.)

consumption and the response to alcohol in women may also be impacted by the hormonal and/or mood fluctuations associated with the menstrual cycle. There are reports of greater alcohol consumption in women with severe premenstrual syndrome, although not necessarily limited to the premenstrual phase when symptoms are prominent, and there are also reports of increased alcohol consumption premenstrually in a portion of alcoholic women (Kouri and Halbreich, 1998). The interaction between menstrual cycle phase, alcohol consumption, and premenstrual symptoms is further complicated by a family history of alcoholism (Evans and Levin, 2011). Historically, alcoholism has been more prevalent among men than women; however, the gender gap is decreasing with regard to both alcohol consumption and alcohol use disorders, at least in the United States (Keyes et al., 2008). Women are more vulnerable to the development of alcohol-related disease such as liver cirrhosis (Walter et al., 2003) and alcohol-dependent women have worse quality-of-life scores than alcohol-dependent men (Peters et al., 2003). With regard to sleep, there is limited evidence for sex differences in the impact of alcohol. Interestingly, despite the commonly reported sex difference in prevalence rates of insomnia in community samples, with the risk for developing insomnia being 1.41 times greater in women (Manber et al., 2006), no sex differences were identified in the frequency of insomnia in men and women undergoing treatment for alcohol dependence (Brower, 2001; Brower et al., 2001).

In a study of 93 healthy adults (59 women), Arnedt et al. (2011b) found that drinking to reach peak BAC of 0.1 g% before bedtime resulted in more disrupted sleep (decreased sleep efficiency and increased wakefulness) but similar impact on sleep architecture (% SWS and % REM sleep) in healthy women compared with men. Women also reported higher ratings of sleepiness after consuming alcohol than did men. No sex differences in the effects of alcohol on sleep were seen in the group of older adolescents studied by Chan et al. (2013). In a study of 42 recovering alcoholics (15 women) and 42 controls (23 women), we found that women had a better sleep efficiency and more delta activity during NREM sleep than men, regardless of diagnosis (Colrain et al., 2009b). Both male and female alcoholics had less delta activity during NREM sleep than sex-matched controls and, although the sex by diagnosis interaction term was not significant, alcoholic men appeared to show a more prominent reduction in delta activity during NREM sleep compared with male controls than female alcoholics did compared with female controls. Further, estimated lifetime alcohol consumption predicted percentage of SWS in alcoholic men but not alcoholic women (Colrain et al., 2009b). Estimated lifetime alcohol consumption was higher in alcoholic men than women, and the women had longer periods of sobriety prior to testing on average. Studies that include larger groups of male and female alcoholics are needed to further evaluate sex differences in the impact of alcohol dependence on sleep.

ALCOHOL DEPENDENCE AND SLEEP IN ADOLESCENCE

As described above, insomnia is a risk factor for the development of alcohol dependence in adults. Recent studies suggest that sleep problems may developmentally precede onset of alcohol use and alcohol problems in adolescents (reviewed in Shibley et al., 2008)). There are substantial changes in sleep behavior across adolescence, with a shift to later bedtimes, a shift in chronotype towards eveningness, with a preference for evening activities, and a shorter sleep duration (reviewed in Colrain and Baker, 2011). There are also changes in sleep architecture, the most notable of which is a dramatic decrease in the amount of SWS during the teenage years (Carskadon, 1982). Changes in adolescent sleep and sleep behavior may be driven by social and environmental factors as well as by physiologic changes in circadian and homeostatic sleep regulation. The loss of SWS and its associated sleep protective delta activity may relate to the high prevalence of sleep problems in adolescents; the lifetime prevalence of *Diagnostic and Statistical Manual of Mental Disorders* (Fourth Edition) (DSM-IV: American Psychiatric Association, 1994)-diagnosed insomnia in adolescents (aged 13–16 years) is estimated as 10.7% (Johnson et al., 2006). Adolescents reporting sleep problems also report more mood disturbances, inattention and memory problems, conduct disorders, and increased drug and alcohol use (Shibley et al., 2008). Based on survey data, adolescents who use alcohol are more likely to report sleep problems than those who abstain even after adjusting for internalizing and externalizing problems (Johnson and Breslau, 2001). Short (<6 h) sleep (Roberts et al., 2009) and poor sleep quality (Manni et al., 1997) have both significantly predicted the development of alcohol use disorders. Perceived tiredness and poor sleep habits accounted for up to 26% of the variance in psychoactive substance use in adolescents (Tynjala et al., 1997). A greater tendency for either circadian phase delay (Pieters et al., 2010) or evening chronotype (Negriff et al., 2011) is also predictive of alcohol use in adolescents. One of the few prospective studies found that chronic insomnia predicted alcohol use in adolescents aged 11–17 years, 12 months after an initial assessment (Roberts et al., 2008). Also, mothers' ratings of early childhood sleep disturbances significantly predicted an early onset of alcohol, marijuana, and illicit drug use in adolescence (Wong et al., 2004). These findings have led to the recommendation that clinicians should treat sleep disturbances in adolescents to prevent or delay the adverse effects of addiction (Shibley et al., 2008). Longitudinal studies that track the development of insomnia symptoms along with changes in alcohol use across adolescence are critical to establish definitively the impact of sleep behavior on subsequent alcohol abuse.

SLEEP HOMEOSTASIS AND CIRCADIAN PROBLEMS WITH ALCOHOL ABUSE

While several studies have evaluated sleep in alcoholics under baseline conditions, in the absence of a sleep challenge, few studies have evaluated sleep homeostasis in abstinent alcoholics. Irwin et al. (2002) studied 46 primary alcoholics in a partial sleep deprivation paradigm in which sleep was studied at baseline, on a night in which sleep was restricted to between 3:00 and 6:30 a.m. and on a recovery night. In both Caucasian and African American alcoholics there was no evidence of homeostatic recovery in either the amount of SWS or in delta EEG power. Two studies from the Michigan group, one in 10 alcoholics (Brower et al., 2011) and the other in 48 alcoholics (Armitage et al., 2012), used a mild sleep challenge of delaying sleep by 3 hours to assess homeostatic recovery. Both papers reported a blunted response in the alcoholics relative to controls, with the smaller study showing alcoholics to have a slower decay of slow-wave activity over the night than a group diagnosed with major depression (Brower et al., 2011).

Despite strong evidence for disruption of circadian rhythms in rodent chronic drinking models (Spanagel et al., 2005), few studies have evaluated circadian rhythms in chronic alcoholics. Most studies have been conducted in small groups and in the acute detoxification phase of withdrawal. Two studies have shown no melatonin rhythm in the first few days following drinking cessation (Fonzi et al., 1994; Schmitz et al., 1996), with restoration of normal rhythmicity after 2 weeks, and a third (Mukai et al., 1998) reported a delayed rhythm only in alcoholics who were suffering from delirium tremens. Kuhlwein et al. (2003) found a delay in the rise time of nocturnal melatonin as well as a delay in the peak, associated with a delay in sleep onset in African American alcoholics studied 2 weeks postwithdrawal. African American subjects were selected as they have previously shown more pronounced sleep disturbance (Irwin et al., 2002).

The apparently delayed melatonin rhythms are in contrast to the single study showing evidence of an advanced body temperature rhythm early in withdrawal (Kodama et al., 1988), although this was more pronounced in alcoholics with comorbid depression. The temperature rhythm had normalized by 3 weeks in most patients.

Cortisol rhythms show no evidence for disruption early in withdrawal or 2–4 weeks postdrinking in two studies (Fonzi et al., 1994; Mukai et al., 1998). However, those with delirium tremens did have altered rhythms (Fonzi et al., 1994; Mukai et al., 1998). Kuhlwein et al. (2003) reported lower cortisol early in the night and higher levels later in the night in their African American alcoholics after 2 weeks.

Two remaining studies focusing on REM sleep showed advanced timing of REM sleep which did not normalize for several months (Imatoh et al., 1986) and abnormal 5HIAA (serotonin metabolite) rhythms in those with delirium tremens or clouded sensorium but not alcoholics with only autonomic dysregulation symptoms (Sano et al., 1993).

The studies outlined above provide limited support for evidence of blunted sleep homeostasis that may persist long-term into periods of abstinence and possible circadian rhythm problems in the acute withdrawal period, especially in those suffering severe withdrawal symptom such as delirium tremens. Abnormalities in the timing of REM sleep would appear to last longer into the abstinence period. The role of circadian misalignment in disturbed brain reward function, and its role in the development of alcohol use disorders, is the subject of a recent review by Hasler and Clark (2013).

EVOKED POTENTIALS DURING SLEEP

Sleep event-related potentials (ERPs) provide an opportunity to evaluate the brain's ability to generate slow-frequency EEG activity during sleep as a function of acute changes in neurochemical state following consumption of alcohol or as a consequence of micro- and macrostructural changes in the brain caused by long-term alcohol dependence. The presentation of auditory or respiratory stimuli during sleep leads to a reliable series of modality-independent (Colrain et al., 1999) evoked potential responses: a P2 component at around 200 ms, and N350, N550, and P900 components. The major contributor to the sleep ERP response is the evoked K-complex (Colrain, 2005; Halasz, 2005), with the N350, N550, and P900 responses all substantially larger when K-complexes are included in the average (Fig. 24.4), although there is evidence that the N350 can be substantial in the absence of K-complexes if evoked vertex sharp waves are present in the average (Colrain et al., 2000).

The K-complex reflects a single instance of high-amplitude delta wave during sleep and probably reflects a sleep-protective process (De Gennaro et al., 2000;

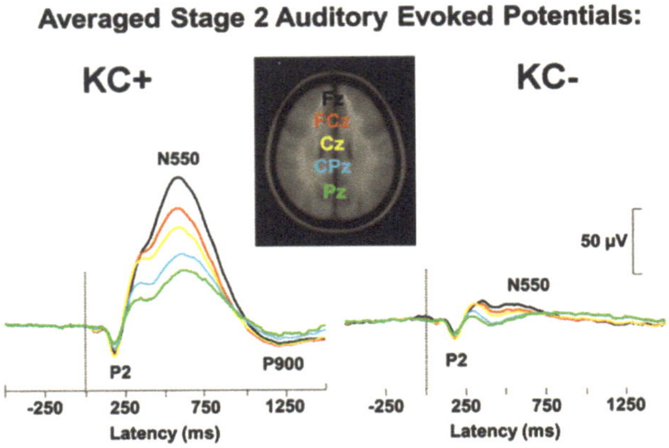

Fig. 24.4. Examples of averaged evoked responses from stage 2 non-rapid eye movement sleep. The left panel (KC+) shows the result of averaging responses that included K–complexes. The right panel (KC–) shows the result of averaging responses not including K–complexes. Waveforms are presented from Fz, FCz, Cz, CPz, and Pz.

Colrain, 2005; Czisch et al., 2009). Data linking the K-complex to delta generation are supported by both animal (Amzica and Steriade, 1997, 1998b, 2002) and human (De Gennaro et al., 2000; Nicholas et al., 2006) experiments. Further, the scalp distribution of the prominent negative (N550) component of the averaged evoked K-complex (Colrain, 2005) shows the same bilaterally symmetrical, frontal predominant distribution seen in spontaneous delta activity in SWS (Finelli et al., 2001). Converging evidence suggests that the K-complex is a cortically generated EEG phenomenon because it is associated with a voltage polarity inversion between superficial (negative voltage) and deeper cortical layers (positive voltage) observed in animal studies (Amzica and Steriade, 1998a) and intracranial recordings in human epileptic patients (Wennberg and Lozano, 2003). This polarity inversion would appear not to support hypotheses of a subcortical generator (Sassin and Johnson, 1968; Dang-Vu et al., 2008; Czisch et al., 2009). The large amplitude of the K-complex (and the N550) is typically an order of magnitude greater than even relatively large event-related potential components such as the P300 seen during wakefulness. This implies a need for large numbers of healthy neurons to be involved in its generation. These neurons need to be simultaneously active and the synchronization of their firing is contingent on healthy white-matter tracts. It is therefore not surprising that the amplitude of the prominent negative component of the averaged evoked K-complex (N550) declines with normal aging in healthy men and women (Crowley et al., 2002, 2004; Colrain et al., 2010). This decrease parallels age-related decreases in gray-matter volume (Pfefferbaum et al., 1994). We have thus hypothesized that the K-complex may also be a sensitive marker of the brain degradation seen in chronic alcoholism (Colrain et al., 2010).

Sleep-evoked responses in alcoholism

Two studies have evaluated sleep-evoked responses in abstinent long-term alcoholics. Nicholas et al. (2002) studied 7 abstinent long-term alcoholic men meeting DSM-IV criteria for alcohol dependence and 8 normal control men. Alcoholics were less likely to generate a K-complex in response to a tone than matched controls. The alcoholic group also showed a significantly smaller-amplitude N550 component at a frontal site compared with controls; however, the latency of the component did not differ between the groups. The P2, N350, and P900 components measured at Cz showed no group differences for amplitude or latency.

In a larger study, Colrain et al. (2009a) studied 42 abstinent long-term alcoholics (27 men) and 42 controls (19 men). As in the previous study (Nicholas et al., 2002), alcoholics were significantly less likely to produce K-complexes than controls. P2 amplitude was, however, smaller in alcoholics than controls, with the difference being largest at Cz, where the component was maximal, but smaller at other sites (Fig. 24.5). P2 latency was longer in alcoholics. There were no sex differences or interactions between diagnosis and sex for K-complex incidence, P2 amplitude or P2 latency. Frontal (but not posterior) N550 and P900 amplitudes were smaller in alcoholics than controls and smaller in men than women, but the sex difference was not related to diagnosis. Latencies of N550 and P900 did not differ as a function of diagnosis or sex.

Chronic alcoholism is thus associated with both an impaired ability to produce evoked delta frequency responses (lower K-complex incidence) and smaller magnitudes of the responses when evoked (smaller N550 and P900 amplitudes).

Links between sleep EEG effects and altered brain structure in alcoholism

In vivo magnetic resonance imaging (MRI) data show that alterations in both cortical gray- and white-matter volumes are associated with chronic, excessive alcohol consumption, with the prefrontal cortex being the supratentorial area most affected (Pfefferbaum and Sullivan, 2013). These data are supported by evidence for reduced postmortem neuronal count in frontal cortex in alcoholics (Harper et al., 1987; Harper and Kril, 1989; Sutherland et al., 2013) (see Chapter 35 in this volume for review). A reduced number of neurons in alcoholics could result in reduced delta EEG amplitude, as fewer neural columns are available for synchronized burst firing. Chronic heavy alcohol use also disrupts

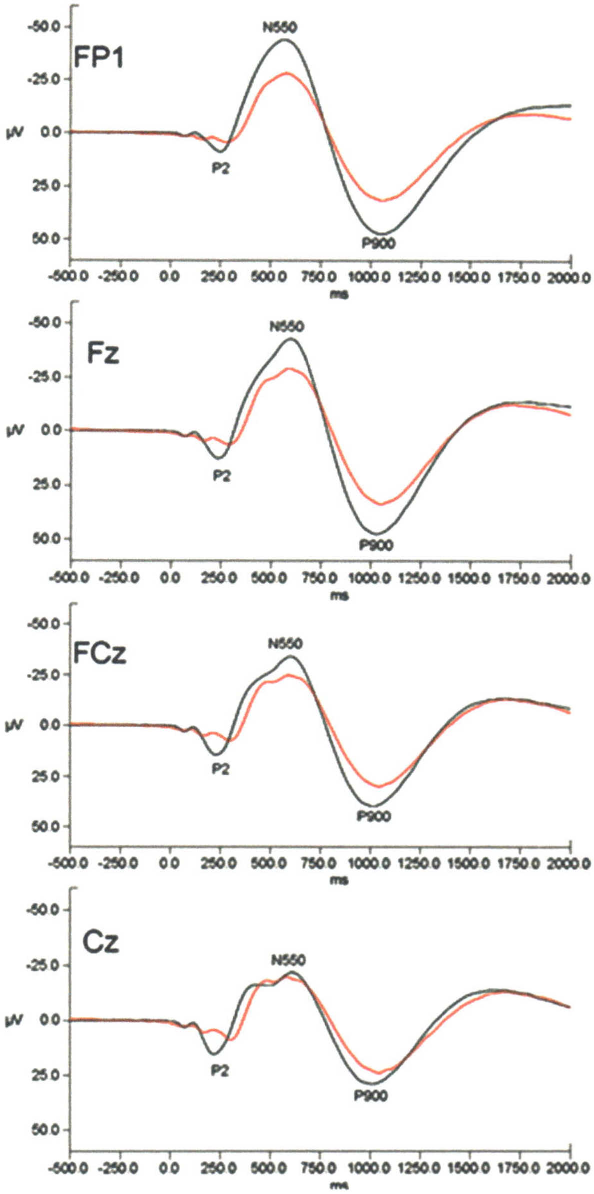

Fig. 24.5. Grand mean evoked potential waveforms for alcoholics (red lines) and control subjects (black lines) for the FP1, Fz, FCz, and Cz electrode sites. Data are presented with negative voltages up the *y*-axis. (Data from Colrain et al., 2009a.)

white-matter microstructure, as manifest by abnormally low fractional anisotropy, both in localized regions of white matter, such as the corpus callosum and centrum semiovale, and more broadly throughout the cortex (Pfefferbaum et al., 2006, 2009, 2010; Pfefferbaum and Sullivan, 2013). Compromised white-matter integrity could conceivably reduce the ability of the brain to coordinate synchronized burst firing across large cortical areas.

An indirect test of the neuronal loss hypothesis of K-complex amplitude deficit in chronic alcoholism was conducted using gray-matter volumes from structural MRI data acquired from the subjects in Colrain et al. (2009a). Statistical models were constructed to determine the extent to which cortical and subcortical volumes could predict evoked potential component amplitudes in sleeping alcoholics and controls. Stepwise multiple regression entering age, intracranial volume, diagnosis, lobar gray-matter volumes, and subcortical tissue volumes to predict N550 amplitude at Fz produced different models in men and women (Colrain et al., 2011). For men, sensorimotor gray-matter volume made a significant independent contribution to N550 amplitude, with the amount of variance explained significantly improving with the addition of diagnostic group. These data support the hypothesis that diminished gray-matter volume in chronic alcoholism contributes to an impaired ability to generate large-amplitude slow waves, although not all the variance could be explained by loss of volume. Poor connectivity (i.e., deficits in white-matter integrity) likely also contributes, although relations between evoked potential amplitude and diffusion tensor imaging measures of white-matter integrity are yet to be tested. Interestingly, in women, while age and temporal gray-matter volume provided the best model, the addition of diagnosis did not improve the model.

POSSIBLE NEUROCHEMICAL MECHANISMS OF THE ACUTE AND CHRONIC ALCOHOL EFFECTS ON SLEEP EEG

Sleep is associated with a complex set of interactions between different neurotransmitter systems (see Fuller et al., 2006; Espana and Scammell, 2011, for reviews). NREM sleep is a period in which cholinergic and noradrenergic brainstem arousal mechanisms are dramatically reduced. The acute effects of alcohol on sleep EEG can arguably be explained by alcohol's γ-aminobutyric acid (GABA) agonist properties. The subsequent withdrawal of tonic input to the reticular nucleus of thalamus allows the release of GABA and inhibition of thalamocortical circuits (Steriade, 1999). The hyperpolarizing effect of GABA causes the opening of low-threshold calcium ion channels and a pattern of synchronized burst firing that manifests as sleep spindles in the sleep EEG. Further release of GABA causes greater levels of hyperpolarization and the production of delta EEG waveforms (Steriade et al., 1993).

The movement between NREM and REM sleep involves a complex interaction between REM-on and REM-off neuronal groups in the brainstem. The

REM-on groups largely consist of cholinergic cells in the lateral dorsal tegmentum and the pedunculopontine tegemental nuclei. REM-off cells involve the serotonergic dorsal raphe nucleus and noradrenergic locus coeruleus. The model originally developed by McCarley and Hobson (1975) proposed a set of reciprocal interactions between the two groups of neurons, whereby REM-on neurons are influenced by a self-excitatory loop but also have an excitatory link to REM-off neurons. Once a threshold level of activation is reached in the REM-off cells, they become dominant. These have an inhibitory action on REM-on cells but also a self-inhibitory feedback loop that progressively decreases their activity. Eventually, activity drops below a threshold point and REM-on cells regain dominance. Recent work has identified an important role for GABAergic interneurons that act to facilitate the REM-off process (McCarley, 2011). It is, therefore, plausible, that alcohol could influence this REM-off process through its effects on GABA, leading to the suppression of REM sleep in the short term.

Neurochemistry of acute alcohol effects

Animal data indicate that administration of GABAergic antagonists leads to increased REM (Xi et al., 1999, 2001; Sanford et al., 2003). Alcohol leads to presynaptic release of GABA in the brainstem and spinal cord (Kelm et al., 2011) and thus, it is reasonable to hypothesize that this sequence plays a role in alcohol's suppression of REM sleep in the context of high doses of alcohol.

The increase in delta activity is also consistent with alcohol's GABA agonist properties. GABA-mediated hyperpolarization of cortical and thalamocortical neurons is thought to underlie the calcium channel-mediated burst firing that results in EEG delta activity (Steriade, 1999). While alcohol does not lead to presynaptic GABA release in the thalamus or cortex the way it does in some other brain regions (Kelm et al., 2011), it does enhance the function of $GABA_A$ receptors. Further, there is evidence for acute ethanol modulation of metobatropic glutamate receptor (mGluR)-mediated slow currents (Su et al., 2010) that are thought to underlie the slow oscillation in thalamocortical cells underlying delta generation (Hughes et al., 2002).

Sleep spindles (associated with sigma frequency power) are produced when thalamocortical cells become hyperpolarized, resulting in low-threshold spikes and the generation of spindle frequency activity in thalamic reticular cells (Steriade, 1993). Further hyperpolarization leads to cessation of spindle activity and the development of delta activity (Steriade, 1993). Alcohol preferentially facilitates tonic GABAergic neurotransmission (Wei et al., 2004), which is more likely to lead to delta than sigma production. The observed decrease in sigma power, following alcohol consumption in Dijk et al. (1992), is in contrast to the spindle and sigma facilitation seen following administration of benozdiazepines (Johnson et al., 1976). In addition, no studies have found altered sigma power in alcoholics. These findings are consistent with data showing that alcohol facilitates benzodiazepine-insensitive $GABA_A$ receptor subtypes (Krystal et al., 2006) and older data showing inverse effects of benzodiazepines on sigma and delta activity (Johnson et al., 1979).

Neurochemistry of alcoholism effects

The sleep EEG effects in those with long-term alcohol dependence are the opposite to those following acute alcohol administration. One possible mechanism is long-term alteration in responsiveness of GABA mechanisms. There is evidence of allosteric modification of GABA receptors (Kang et al., 1998; Follesa et al., 2006) and reduced $GABA_A$ receptor function (Mihic and Harris, 1995; Valenzuela and Harris, 1997) in rodent models of alcohol dependence. Thus downregulation of brainstem GABAergic systems following development of alcohol dependence would lead to diminished activity in REM-off systems (Fig. 24.6). leading to an increased propensity for REM. This hypothesis has not been directly tested, and it should be noted that other factors may play a role in the increased REM seen in long-term abstinent alcoholics. For example, administration of the tumor necrosis factor α antagonist etanercept led to normalization of REM sleep in 18 abstinent alcoholics (Irwin et al., 2009).

In addition to loss of gray-matter volume and reduced connectivity, downregulation of GABA systems could also partially explain the decrease in both delta power and the amplitude of evoked delta responses in abstinent alcoholics. However, again, there are other possible mechanisms that may also contribute to these effects. For example, there is evidence of altered amplitude of T-type calcium current following chronic alcohol consumption in non-human primates (Carden et al., 2006; Welsh et al., 2011), evidence of blocked T-type calcium currents in reticular thalamocortical neurons *in vitro* (Joksovic et al., 2005), and evidence for chronic ethanol modulation of mRNA of mGluR1 (Simonyi et al., 1996), all of which could lead to attenuated delta generation.

FAMILIAL PREDISPOSITION FOR ALCOHOLISM EFFECTS ON SLEEP?

As has been discovered in awake EEG and evoked potentials, it is also possible that sleep EEG effects in

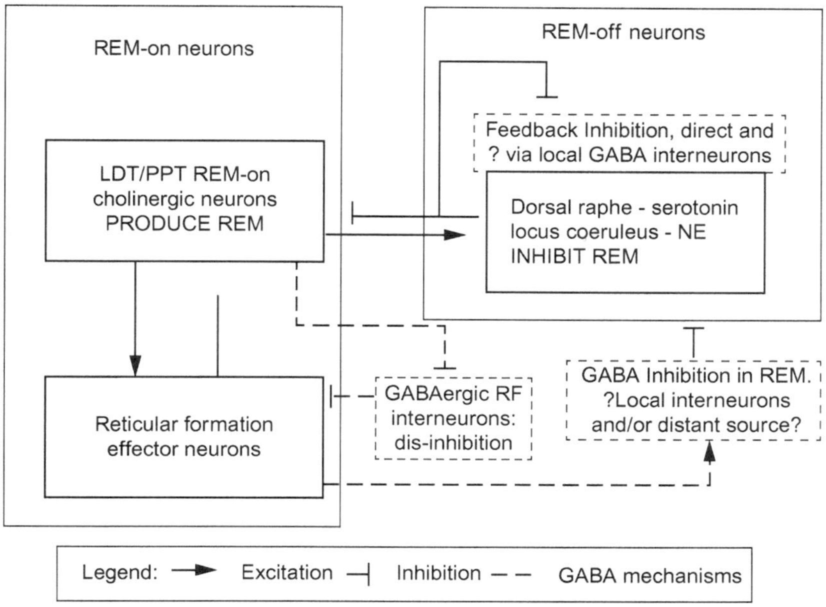

Fig. 24.6. A structural model of rapid eye movement (REM) sleep control highlighting the role of GABAergic interneurons (McCarley, 2011). LDT, laterodorsal tegmental nucleus; PPT, pedunculopontine tegmental nucleus; RF, reticular formation; GABA, γ-aminobutyric acid; NE, norepinephrine.

alcoholism may partially reflect a genetic predisposition to alcoholism. The studies addressing this possibility in sleep have used a comparison of those with positive and negative alcoholism family histories rather than genetic analysis. Arendt et al. (2011b) did not find any differences in the acute alcohol effects on sleep in family history-positive or -negative men and women. Tarokh et al.'s (2012b) study of sleep EEG in family history-positive and -negative youth provides evidence that a familial predisposition is not associated with greater REM sleep, suggesting that increased REM sleep is more likely to be a consequence of alcohol abuse *per se*. Comparison of the child and adolescent cohorts in Tarokh et al. (2012b) reveals a tendency for family history-positive children to have lower levels of NREM delta, with more consistent differences emerging in the adolescent cohort. However, Tarokh et al. (2012b) did not show family history differences in sleep homeostasis using NREM delta power as the dependent measure; nor did Colrain et al. (2011) find a family history effect for NREM evoked delta amplitude.

While there have been no studies evaluating genotype influences on sleep EEG, the data from wake EEG studies are clearly relevant, especially the known association between the GABA gene *GABRA2* and beta oscillations (Porjesz et al., 2002) and between the cholinergic gene *CHRM2* and delta oscillations (Porjesz and Rangaswamy, 2007; Rangaswamy and Porjesz, 2013). It should be noted, however, that EEG during wakefulness that is measured in association with cognitive or behavioral tasks reflects the activity in task-specific pathways and networks, whereas sleep EEG is measuring a baseline state. Family history linkages with wake but not sleep EEG phenomena may thus reflect specificity in the predisposition to altered circuitry associated with the awake behaviors, rather than with EEG generation *per se*. There are data showing relations between genes associated with circadian rhythm regulation and alcoholism-related insomnia. Brower et al. (2012) reported that a specific PER3 genotype (PER3 (4/4)) conferred a small but significantly increased risk of insomnia over and above drinking frequency, mental and physical health status, and childhood abuse in a large sample of alcohol-dependent patients. PER3 (4/4) is associated with evening circadian preference and lower homeostatic sleep drive (Brower et al., 2012).

A ROLE FOR SLEEP IN TREATMENT, RECOVERY, AND RELAPSE

Given that sleep problems persist during alcohol recovery, it raises the possibility that treatment of sleep problems in alcoholic patients could aid recovery and decrease relapse rates (Brower, 2001). Self-reported sleep problems as well as sleep continuity and REM sleep measures derived from the polysomnogram are significant predictors of relapse (reviewed in Arnedt et al., 2007). Periodic leg movements, seen in recently detoxified alcoholics are also more common in those who go on to relapse (Gann et al., 2002). A recent pilot study showed that cognitive behavioral treatment for insomnia (CBTI) in alcohol-dependent patients was more effective than placebo in improving subjective sleep and in reducing daytime fatigue (Arnedt et al., 2011a). However, there were no differences in relapse rates between those treated with placebo and those treated with CBTI in this small sample ($n = 17$). In a larger controlled trial of CBTI in recovering alcoholic men and women, Currie and colleagues (2004) reported that patients treated with CBTI had improved subjective sleep quality, reduced sleep onset latency, and fewer awakenings compared to pretreatment, whereas wait-listed controls showed no significant changes in sleep. Improvements in the CBTI group were maintained 3 and 6 months after treatment. There was no difference in alcoholism relapse rates between those treated with CBTI and wait list controls.

We have found intriguing evidence that the amplitude of the N550 component of the K-complex may act as a marker of brain recovery with abstinence. After correction for age and estimated lifetime alcohol consumption, time sober demonstrated significant correlations with increased N550 and P900 amplitudes in recovering alcoholics (Colrain et al. 2009a). Additional data were collected from 15 alcohol-dependent subjects (12 men) who had been studied previously (Colrain et al. 2009a). N550 and P900 amplitude were significantly larger at frontal sites with the additional 12 months of abstinence (Colrain et al., 2012) (Fig. 24.7). These data are suggestive of the N550 and P900 reflecting functional improvement associated with recovery in gray-matter volume and white-matter integrity (Cardenas et al., 2007). However, this apparent marker of brain recovery should not be interpreted as a recovery of sleep. Sleep problems persist even after several months of abstinence (Arnedt et al., 2007).

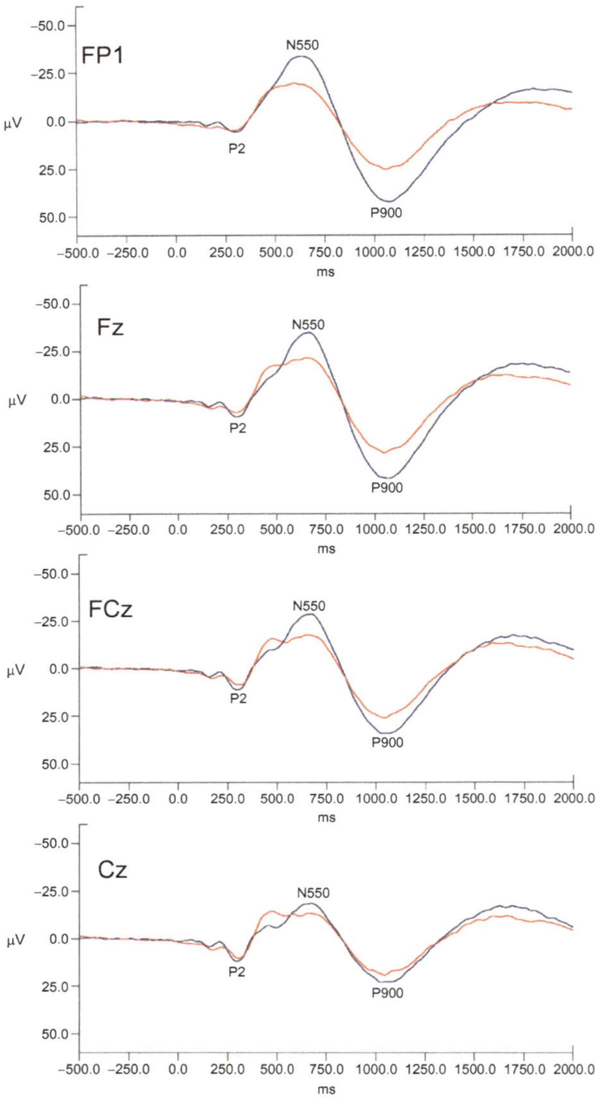

Fig. 24.7. Grand mean evoked potential waveforms for alcoholics at initial assessment (red lines) and at 12-month follow-up (blue lines) Fz, FCz, Cz, CPz and Pz. Data are presented with negative voltages up the *y*-axis. (Data from Colrain et al., 2012.)

CONCLUSION

Alcohol has a profound impact on sleep, with effects dependent on acute versus chronic use and dependence. While alcohol is initially sedating, this effect disappears after a few hours, resulting in a fragmented and disturbed sleep in the second half of the night. Sustained use of alcohol in chronic alcoholism is associated with major sleep problems. Ongoing sleep and circadian disruption are features of alcohol-drinking binges. For those who are homeless, there are almost insurmountable obstacles to obtaining appropriately restorative sleep. When abstinent, sleep issues persist, with insomnia and vivid dreams being common complaints, which can be a factor leading to relapse.

Animal (typically rodent) models of acute or chronic alcohol administration on sleep have provided valuable insights into the neurochemical effects of alcohol on brain structures and systems that play a role in sleep regulation; however, many questions remain unanswered. Further studies are needed in humans and animal models to establish whether there are genetic predispositions to sleep differences or susceptibilities to alcohol, whether the REM increase seen in alcoholics is a reflection of an irreversible change in the brainstem, and whether there are periods of vulnerability to the onslaught of alcohol on sleep regulatory systems, such as during adolescence. It will also be valuable for future studies of adolescents and family history-positive individuals to explore further what aspects of altered brain structure and sleep EEG predate the onset of alcohol abuse compared to changes that occur as a result of the impact of alcohol on the brain. It is critically important from a clinical perspective to determine whether preventing sleep disorders may help prevent the development of alcohol use disorders in high-risk individuals and which treatments are most effective in alcoholics, both in terms of improving sleep quality and supporting continued abstinence. At this time when polysubstance dependence is common, it also is becoming increasingly relevant to investigate the interactive effects of substances of abuse on sleep behavior and regulation.

ACKNOWLEDGMENTS

The grants are National Institute on Alcohol Abuse and Alcoholism (AA020565 to IMC; AA021696 to FB and IMC) and National Health & Medical Research Council of Australia (1012195 to CLN).

References

Adamson J, Burdick JA (1973). Sleep of dry alcoholics. Arch Gen Psychiatry 28: 146–149.

Aldrich M (1998). Effects of alcohol on sleep. National Institute on Alcohol Abuse and Alcoholism Research Monograph 33 Alcohol Problems and Aging. U.S. Department of Health and Human Services, Bethesda, MD, pp. 281–300.

Allen RP, Wagman AM (1975). Do sleep patterns relate to the desire for alcohol? In: MM Gross (Ed.), Alcohol Intoxication and Withdrawal. Plenum Press, New York, pp. 495–508.

Allen RP, Wagman AM, Funderburk FR (1977). Slow wave sleep changes: alcohol tolerance and treatment implications. In: MM Gross (Ed.), Alcohol Intoxication and Withdrawal IIIA. Plenum Press, New York, pp. 629–640.

American Psychiatric Association (1994). Diagnostic and Statistical Manual of Mental Disorders. Fourth Edition, APA, Washington DC.

Amzica F, Steriade M (1997). The K-complex: its slow (<1-Hz) rhythmicity and relation to delta waves. Neurology 49: 952–959.

Amzica F, Steriade M (1998a). Cellular substrates and laminar profile of sleep K-complex. Neuroscience 82: 671–686.

Amzica F, Steriade M (1998b). Electrophysiological correlates of sleep delta waves. Electroencephalogr Clin Neurophysiol 107: 69–83.

Amzica F, Steriade M (2002). The functional significance of K-complexes. Sleep Med Rev 6: 139–149.

Armitage R, Hoffmann R, Conroy DA et al. (2012). Effects of a 3-hour sleep delay on sleep homeostasis in alcohol dependent adults. Sleep 35: 273–278.

Arnedt JT, Conroy DA, Brower KJ (2007). Treatment options for sleep disturbances during alcohol recovery. J Addict Dis 26: 41–54.

Arnedt JT, Conroy DA, Armitage R et al. (2011a). Cognitive-behavioral therapy for insomnia in alcohol dependent patients: a randomized controlled pilot trial. Behav Res Ther 49: 227–233.

Arnedt JT, Rohsenow DJ, Almeida AB et al. (2011b). Sleep following alcohol intoxication in healthy, young adults: effects of sex and family history of alcoholism. Alcohol Clin Exp Res 35: 870–878.

Bauer LO (2001). Predicting relapse to alcohol and drug abuse via quantitative electroencephalography. Neuropsychopharmacology 25: 332–340.

Benca RM, Obermeyer WH, Thisted RA et al. (1992). Sleep and psychiatric disorders. A meta-analysis. Arch Gen Psychiatry 49: 651–668.

Borbely AA (1982). A two process model of sleep regulation. Hum Neurobiol 1: 195–204.

Breslau N, Roth T, Rosenthal L et al. (1996). Sleep disturbance and psychiatric disorders: a longitudinal epidemiological study of young adults. Biol Psychiatry 39: 411–418.

Brower KJ (2001). Alcohol's effects on sleep in alcoholics. Alcohol Res Health 25: 110–125.

Brower KJ, Perron BE (2010). Prevalence and correlates of withdrawal-related insomnia among adults with alcohol dependence: results from a national survey. Am J Addict 19: 238–244.

Brower KJ, Aldrich MS, Robinson EA et al. (2001). Insomnia, self-medication, and relapse to alcoholism. Am J Psychiatry 158: 399–404.

Brower KJ, Hoffmann R, Conroy DA et al. (2011). Sleep homeostasis in alcohol-dependent, depressed and healthy control men. Eur Arch Psychiatry Clin Neurosci 261: 559–566.

Brower KJ, Wojnar M, Sliwerska E et al. (2012). PER3 polymorphism and insomnia severity in alcohol dependence. Sleep 35: 571–577.

Carden WB, Alexander GM, Friedman DP et al. (2006). Chronic ethanol drinking reduces native T-type calcium current in the thalamus of nonhuman primates. Brain Res 1089: 92–100.

Cardenas VA, Studholme C, Gazdzinski S et al. (2007). Deformation-based morphometry of brain changes in alcohol dependence and abstinence. Neuroimage 34: 879–887.

Carskadon MA (1982). The second decade. In: C Guilleminault (Ed.), Sleeping and waking disorders: indications and techniques. Addison Wesley, Menlo Park, CA, pp. 99–125.

Chan JK, Trinder J, Andrewes HE et al. (2013). The acute effects of alcohol on sleep architecture in late adolescence. Alcohol Clin Exp Res 37: 1720–1728.

Colrain IM (2005). The K-complex: a seven-decade history. Sleep 28: 255–273.

Colrain IM, Baker FC (2011). Changes in sleep as a function of adolescent development. Neuropsychol Rev 21: 5–21.

Colrain IM, Webster KE, Hirst G (1999). The N550 component of the evoked K-complex: a modality non-specific response? J Sleep Res 8: 273–280.

Colrain IM, Webster KE, Hirst G et al. (2000). The roles of vertex sharp waves and K-complexes in the generation of N300 in auditory and respiratory-related evoked potentials during early stage 2 NREM sleep. Sleep 23: 97–106.

Colrain IM, Crowley KE, Nicholas CL et al. (2009a). The impact of alcoholism on sleep evoked delta frequency responses. Biol Psychiatry 66: 177–184.

Colrain IM, Turlington S, Baker FC (2009b). Impact of alcoholism on sleep architecture and EEG power spectra in men and women. Sleep 32: 1341–1352.

Colrain IM, Turlington SR, Baker FC (2009c). The impact of alcoholism on sleep architecture and EEG power spectra in men and women. Sleep 32: 1341–1352.

Colrain IM, Crowley KE, Nicholas CL et al. (2010). Sleep evoked delta frequency responses show a linear decline in amplitude across the adult lifespan. Neurobiol Aging 31: 874–883.

Colrain IM, Sullivan EV, Rohlfing T et al. (2011). Independent contributions of cortical gray matter, aging, sex and alcoholism to K-complex amplitude evoked during sleep. Sleep 34: 787–795.

Colrain IM, Padilla ML, Baker FC (2012). Partial recovery of alcohol dependence-related deficits in sleep evoked potentials following twelve months of abstinence. Frontiers Neurol 3: 13.

Conroy DA, Brower KJ (2011). Alcohol, toxins, and medications as a cause of sleep dysfunction. Handb Clin Neurol 98: 587–612.

Crowley KE, Trinder J, Colrain IM (2002). An examination of evoked K-complex amplitude and frequency of occurrence in the elderly. J Sleep Res 11: 129–140.

Crowley KE, Trinder J, Colrain IM (2004). Evoked K-complex generation: the impact of sleep spindles and age. Clin Neurophysiol 115: 471–476.

Currie SR, Clark S, Rimac S et al. (2003). Comprehensive assessment of insomnia in recovering alcoholics using daily sleep diaries and ambulatory monitoring. Alcohol Clin Exp Res 27: 1262–1269.

Currie SR, Clark S, Hodgins DC et al. (2004). Randomized controlled trial of brief cognitive-behavioural interventions for insomnia in recovering alcoholics. Addiction 99: 1121–1132.

Czisch M, Wehrle R, Stiegler A et al. (2009). Acoustic oddball during NREM sleep: a combined EEG/fMRI study. PLoS One 4: e6749.

Dang-Vu TT, Desseilles M, Laureys S et al. (2005). Cerebral correlates of delta waves during non-REM sleep revisited. Neuroimage 28: 14–21.

Dang-Vu TT, Schabus M, Desseilles M et al. (2008). Spontaneous neural activity during human slow wave sleep. Proc Natl Acad Sci U S A 105: 15160–15165.

De Gennaro L, Ferrara M, Bertini M (2000). The spontaneous K-complex during stage 2 sleep: is it the 'forerunner' of delta waves? Neurosci Lett 291: 41–43.

Dijk DJ, Brunner DP, Aeschbach D et al. (1992). The effects of ethanol on human sleep EEG power spectra differ from those of benzodiazepine receptor agonists. Neuropsychopharmacology 7: 225–232.

Drummond SP, Gillin JC, Smith TL et al. (1998). The sleep of abstinent pure primary alcoholic patients: natural course and relationship to relapse. Alcohol Clin Exp Res 22: 1796–1802.

Ehlers CL (2000). Alcohol and sleep. In: A Noronha, A Eckardt, K Warren (Eds.), Review of NIAA's Neuroscience and Behavioral Research Portfolio. U.S. Department of Health and Human Services, Bethesda, MD, pp. 417–436.

Ely M, Hardy R, Longford NT et al. (1999). Gender differences in the relationship between alcohol consumption and drink problems are largely accounted for by body water. Alcohol Alcohol 34: 894–902.

Espana RA, Scammell TE (2011). Sleep neurobiology from a clinical perspective. Sleep 34: 845–858.

Evans SM, Levin FR (2011). Response to alcohol in women: role of the menstrual cycle and a family history of alcoholism. Drug Alcohol Depend 114: 18–30.

Feige B, Gann H, Brueck R et al. (2006). Effects of alcohol on polysomnographically recorded sleep in healthy subjects. Alcohol Clin Exp Res 30: 1527–1537.

Feige B, Scaal S, Hornyak M et al. (2007). Sleep electroencephalographic spectral power after withdrawal from alcohol in alcohol-dependent patients. Alcohol Clin Exp Res 31: 19–27.

Finelli LA, Borbely AA, Achermann P (2001). Functional topography of the human nonREM sleep electroencephalogram. Eur J Neurosci 13: 2282–2290.

Follesa P, Biggio F, Talani G et al. (2006). Neurosteroids, GABAA receptors, and ethanol dependence. Psychopharmacology (Berl) 186: 267–280.

Fonzi S, Solinas GP, Costelli P et al. (1994). Melatonin and cortisol circadian secretion during ethanol withdrawal in chronic alcoholics. Chronobiologia 21: 109–112.

Fuller PM, Gooley JJ, Saper CB (2006). Neurobiology of the sleep-wake cycle: sleep architecture, circadian regulation, and regulatory feedback. J Biol Rhythms 21: 482–493.

Gann H, Feige B, Hohagen F et al. (2001). Sleep and the cholinergic rapid eye movement sleep induction test in patients with primary alcohol dependence. Biol Psychiatry 50: 383–390.

Gann H, Feige B, Fasihi S et al. (2002). Periodic limb movements during sleep in alcohol dependent patients. Eur Arch Psychiatry Clin Neurosci 252: 124–129.

Gillin JC, Smith TL, Irwin M et al. (1990). EEG sleep studies in "pure" primary alcoholism during subacute withdrawal: relationships to normal controls, age, and other clinical variables. Biol Psychiatry 27: 477–488.

Halasz P (2005). K-complex, a reactive EEG graphoelement of NREM sleep: an old chap in a new garment. Sleep Med Rev 9: 391–412.

Harper C, Kril JJ (1989). Patterns of neuronal loss in the cerebral cortex in chronic alcoholic patients. J Neurol Sci 92: 81–89.

Harper CG, Kril JJ, Daly JM (1987). Are we drinking our neurones away? Br Med J 294: 534–536.

Hasler BP, Clark DB (2013). Circadian misalignment, reward-related brain function, and adolescent alcohol involvement. Alcohol Clin Exp Res 37: 558–565.

Hughes SW, Cope DW, Blethyn KL et al. (2002). Cellular mechanisms of the slow (<1 Hz) oscillation in thalamocortical neurons in vitro. Neuron 33: 947–958.

Iber C, Ancoli-Israel S, Chesson A et al. (2007). The AASM manual for the scoring of sleep and associated events: rules, terminology and technical specifciation. American Academy of Sleep Medicine, Westchester, IL.

Imatoh N, Nakazawa Y, Ohshima H et al. (1986). Circadian rhythm of REM sleep of chronic alcoholics during alcohol withdrawal. Drug Alcohol Depend 18: 77–85.

Irwin M, Miller C, Gillin JC et al. (2000). Polysomnographic and spectral sleep EEG in primary alcoholics: an interaction between alcohol dependence and African-American ethnicity. Alcohol Clin Exp Res 24: 1376–1384.

Irwin M, Gillin JC, Dang J et al. (2002). Sleep deprivation as a probe of homeostatic sleep regulation in primary alcoholics. Biol Psychiatry 51: 632–641.

Irwin MR, Olmstead R, Valladares EM et al. (2009). Tumor necrosis factor antagonism normalizes rapid eye movement sleep in alcohol dependence. Biol Psychiatry 66: 191–195.

Johnson EO, Breslau N (2001). Sleep problems and substance use in adolescence. Drug Alcohol Depend 64: 1–7.

Johnson LC, Hanson K, Bickford RG (1976). Effect of flurazepam on sleep spindles and K-complexes. Electroencephalogr Clin Neurophysiol 40: 67–77.

Johnson LC, Seales DM, Naitoh P et al. (1979). The effects of flurazepam hydrochloride on brain electrical activity during sleep. Electroencephalogr Clin Neurophysiol 47: 309–321.

Johnson EO, Roth T, Schultz L et al. (2006). Epidemiology of DSM-IV insomnia in adolescence: lifetime prevalence, chronicity, and an emergent gender difference. Pediatrics 117: e247–e256.

Joksovic PM, Brimelow BC, Murbartian J et al. (2005). Contrasting anesthetic sensitivities of T-type Ca^{2+} channels of reticular thalamic neurons and recombinant Ca(v)3.3 channels. Br J Pharmacol 144: 59–70.

Kang MH, Spigelman I, Olsen RW (1998). Alteration in the sensitivity of GABA(A) receptors to allosteric modulatory drugs in rat hippocampus after chronic intermittent ethanol treatment. Alcohol Clin Exp Res 22: 2165–2173.

Kelm MK, Criswell HE, Breese GR (2011). Ethanol-enhanced GABA release: a focus on G protein-coupled receptors. Brain Res Rev 65: 113–123.

Keyes KM, Grant BF, Hasin DS (2008). Evidence for a closing gender gap in alcohol use, abuse, and dependence in the United States population. Drug Alcohol Depend 93: 21–29.

Kodama H, Nakazawa Y, Kotorii T et al. (1988). Biorhythm of core temperature in depressive and non-depressive alcoholics. Drug Alcohol Depend 21: 1–6.

Kouri EM, Halbreich U (1998). Effects of alcohol and other drugs in women of reproductive age: hormonal interactions. Drugs Today (Barc) 34: 837–843.

Krystal JH, Staley J, Mason G et al. (2006). Gamma-aminobutyric acid type A receptors and alcoholism: intoxication, dependence, vulnerability, and treatment. Arch Gen Psychiatry 63: 957–968.

Kuhlwein E, Hauger RL, Irwin MR (2003). Abnormal nocturnal melatonin secretion and disordered sleep in abstinent alcoholics. Biol Psychiatry 54: 1437–1443.

Landolt HP, Roth C, Dijk DJ et al. (1996). Late-afternoon ethanol intake affects nocturnal sleep and the sleep EEG in middle-aged men. J Clin Psychopharmacol 16: 428–436.

Le Bon O, Verbanck P, Hoffmann G et al. (1997). Sleep in detoxified alcoholics: impairment of most standard sleep parameters and increased risk for sleep apnea, but not for myoclonias—A controlled study. J Stud Alcohol 58: 30–36.

MacLean AW, Cairns J (1982). Dose-response effects of ethanol on the sleep of young men. J Stud Alcohol 43: 434–444.

Manber R, Baker FC, Gress JL (2006). Sex differences in sleep and sleep disorders: a focus on women's sleep. Int J Sleep Disorders 1: 7–15.

Manni R, Ratti MT, Marchioni E et al. (1997). Poor sleep in adolescents: a study of 869 17-year-old Italian secondary school students. J Sleep Res 6: 44–49.

Massimini M, Huber R, Ferrarelli F et al. (2004). The sleep slow oscillation as a traveling wave. J Neurosci 24: 6862–6870.

McCarley RW (2011). Neurobiology of REM sleep. Handb Clin Neurol 98: 151–171.

McCarley RW, Hobson JA (1975). Neuronal excitability modulation over the sleep cycle: a structural and mathematical model. Science 189: 58–60.

Mihic SJ, Harris RA (1995). Alcohol actions at the GABAA receptor/chloride channels complex. In: RA Deitrich, G Erwin (Eds.), Pharmacological Effects of Ethanol on the Nervous System. CRC Press, Boca Raton, FL, pp. 51–71.

Mukai M, Uchimura N, Hirano T et al. (1998). Circadian rhythms of hormone concentrations in alcohol withdrawal. Psychiatry Clin Neurosci 52: 238–240.

Negriff S, Dorn LD, Pabst SR et al. (2011). Morningness/eveningness, pubertal timing, and substance use in adolescent girls. Psychiatry Res 185: 408–413.

Nicholas CL, Sullivan EV, Pfefferbaum A et al. (2002). The effects of alcoholism on auditory evoked potentials during sleep. J Sleep Res 11: 247–253.

Nicholas CL, Trinder J, Crowley KE et al. (2006). The impact of slow wave sleep proximity on evoked K-complex generation. Neurosci Lett 404: 127–131.

Oscar-Berman M, Valmas MM, Sawyer KS et al. (2013). Profile of spared, impaired, and recovered neuropsychological processes. In: A Pfefferbaum, EV Sullivan (Eds.), Handbook of Clinical Neurology: Alcohol and the Nervous System. Elsevier, Amsterdam.

Othmer O, Dauhaday WH, Goodin DW et al. (1982). Sleep and growth hormone secretion in adults. J Clin Psychiatry 43: 411–414.

Peters TJ, Millward LM, Foster J (2003). Quality of life in alcohol misuse: comparison of men and women. Arch Womens Ment Health 6: 239–243.

Pfefferbaum A, Sullivan EV (2013). DTI and MRI. In: A Pfefferbaum, EV Sullivan (Eds.), Handbook of Clinical Neurology: Alcohol and the Nervous System. Elsevier, Amsterdam.

Pfefferbaum A, Mathalon DH, Sullivan EV et al. (1994). A quantitative magnetic resonance imaging study of changes in brain morphology from infancy to late adulthood. Arch Neurol 51: 874–887.

Pfefferbaum A, Adalsteinsson E, Sullivan EV (2006). Supratentorial profile of white matter microstructural integrity in recovering alcoholic men and women. Biol Psychiatry 59: 364–372.

Pfefferbaum A, Rosenbloom MJ, Rohlfing T et al. (2009). Degradation of association and projection white matter systems in alcoholism detected with quantitative fiber tracking. Biol Psychiatry 65: 680–690.

Pfefferbaum A, Rosenbloom MJ, Fama R et al. (2010). Transcallosal white matter degradation detected with quantitative fiber tracking in alcoholics men and women: selective relations to dissociable functions. Alcohol Clin Exp Res 34: 1201–1211.

Pieters S, Van Der Vorst H, Burk WJ et al. (2010). Puberty-dependent sleep regulation and alcohol use in early adolescents. Alcohol Clin Exp Res 34: 1512–1518.

Porjesz B, Rangaswamy M (2007). Neurophysiological endophenotypes, CNS disinhibition, and risk for alcohol dependence and related disorders. ScientificWorldJournal 7: 131–141.

Porjesz B, Almasy L, Edenberg HJ et al. (2002). Linkage disequilibrium between the beta frequency of the human EEG and a GABAA receptor gene locus. Proc Natl Acad Sci U S A 99: 3729–3733.

Prinz PN, Roehrs TA, Vitaliano PP et al. (1980). Effect of alcohol on sleep and nighttime plasma growth hormone and cortisol concentrations. J Clin Endocrinol Metab 51: 759–764.

Rangaswamy M, Porjesz B (2013). Understanding alcohol use disorders with neuroelectrophysiology. In: A Pfefferbaum, EV Sullivan (Eds.), Handbook of Clinical Neurology: Alcohol and the Nervous System. Elsevier, Amsterdam.

Rechtschaffen A, Kales A (1968). A Manual of Standardised Terminology, Techniques and Scoring Systems for Sleep Stages of Human Subjects. U.S. Government Printing Office, Washington D.C.

Riemann D, Spiegelhalder K, Feige B et al. (2010). The hyperarousal model of insomnia: a review of the concept and its evidence. Sleep Med Rev 14: 19–31.

Roberts RE, Roberts CR, Duong HT (2008). Chronic insomnia and its negative consequences for health and functioning of adolescents: a 12-month prospective study. J Adolesc Health 42: 294–302.

Roberts RE, Roberts CR, Duong HT (2009). Sleepless in adolescence: prospective data on sleep deprivation, health and functioning. J Adolesc 32: 1045–1057.

Roehrs T, Yoon J, Roth T (1991). Nocturnal and next-day effects of ethanol and basal level of sleepiness. Hum Psychopharmacol 6: 307–312.

Roehrs T, Papineau K, Rosenthal L et al. (1999). Ethanol as a hypnotic in insomniacs: self administration and effects on sleep and mood. Neuropsychopharmacology 20: 279–286.

Rundell OH, Lester BK, Griffiths WJ et al. (1972). Alcohol and sleep in young adults. Psychopharmacologia 26: 201–218.

Rundell OH, Williams HL, Lester BK (1975). Sleep in alcoholic patients: Longitudinal findings. In: MM Gross (Ed.), Alcohol Intoxication and Withdrawal. Plenum Press, New York, pp. 389–402.

Sagawa Y, Kondo H, Matsubuchi N et al. (2011). Alcohol has a dose-related effect on parasympathetic nerve activity during sleep. Alcohol Clin Exp Res 35: 2093–2100.

Saletu-Zyhlarz GM, Arnold O, Anderer P et al. (2004). Differences in brain function between relapsing and abstaining alcohol-dependent patients, evaluated by EEG mapping. Alcohol Alcohol 39: 233–240.

Sanford LD, Tang X, Xiao J et al. (2003). GABAergic regulation of REM sleep in reticularis pontis oralis and caudalis in rats. J Neurophysiol 90: 938–945.

Sano H, Suzuki Y, Yazaki R et al. (1993). Circadian variation in plasma 5-hydroxyindoleacetic acid level during and after alcohol withdrawal: phase advances in alcoholic patients compared with normal subjects. Acta Psychiatr Scand 87: 291–296.

Sassin JF, Johnson LC (1968). Body motility during sleep and its relation to the K-complex. Exp Neurol 22: 133–144.

Schmitz MM, Sepandj A, Pichler PM et al. (1996). Disrupted melatonin-secretion during alcohol withdrawal. Prog Neuropsychopharmacol Biol Psychiatry 20: 983–995.

Scrima L, Broudy M, Nay KN et al. (1982). Increased severity of obstructive sleep apnea after bedtime alcohol ingestion: diagnostic potential and proposed mechanism of action. Sleep 5: 318–328.

Shibley HL, Malcolm RJ, Veatch LM (2008). Adolescents with insomnia and substance abuse: consequences and comorbidities. J Psychiatr Pract 14: 146–153.

Simonyi A, Zhang JP, Sun AY et al. (1996). Chronic ethanol on mRNA levels of IP3R1, IP3 3-kinase and mGluR1 in mouse Purkinje neurons. Neuroreport 7: 2115–2118.

Snyder S, Karacan I (1985). Sleep patterns of sober chronic alcoholics. Neuropsychobiology 13: 97–100.

Spanagel R, Rosenwasser AM, Schumann G et al. (2005). Alcohol consumption and the body's biological clock. Alcohol Clin Exp Res 29: 1550–1557.

Steriade M (1993). Cellular substrates of brain rhythms. In: E Niedermeyer, F Lopes da Silva (Eds.), Electroencephalography: Basic Principles, Clinical Applications and Related Fields. Williams and Wilkins, Baltimore, MD, pp. 27–62.

Steriade M (1999). Cellular substrates of brain rhythms. In: E Niedermeyer, F Lopes da Silva (Eds.), Electroencephalography: Basic Principles, Clinical Applications and Related Fields. Williams and Wilkins, Baltimore, MD, pp. 28–75.

Steriade M, McCormick DA, Sejnowski TJ (1993). Thalamocortical oscillations in the sleeping and aroused brain. Science 262: 679–685.

Stone BM (1980). Sleep and low doses of alcohol. Electroencephalogr Clin Neurophysiol 48: 706–709.

Su LD, Sun CL, Shen Y (2010). Ethanol acutely modulates mGluR1-dependent long-term depression in cerebellum. Alcohol Clin Exp Res 34: 1140–1145.

Sutherland GT, Sheedy D, Kril JJ (2013). Neuropathology of alcoholism. In: A Pfefferbaum, EV Sullivan (Eds.), Handbook of Clinical Neurology: Alcohol and the Nervous System. Elsevier, Amsterdam.

Tarokh L, Carskadon MA, Achermann P (2012a). Dissipation of sleep pressure is stable across adolescence. Neuroscience 216: 167–177.

Tarokh L, Van Reen E, Acebo C et al. (2012b). Adolescence and parental history of alcoholism: insights from the sleep EEG. Alcohol Clin Exp Res 36: 1530–1541.

Thompson PM, Gillin JC, Golshan S et al. (1995). Polygraphic sleep measures differentiate alcoholics and stimulant abusers during short-term abstinence. Biol Psychiatry 38: 831–836.

Tononi G, Cirelli C (2006). Sleep function and synaptic homeostasis. Sleep Med Rev 10: 49–62.

Tynjala J, Kannas L, Levalahti E (1997). Perceived tiredness among adolescents and its association with sleep habits and use of psychoactive substances. J Sleep Res 6: 189–198.

Valenzuela CF, Harris RA (1997). Alcohol neurobiology. In: JH Lowinson, P Ruiz, RB Millman et al. (Eds.), Substance Abuse: A comprehensive Textbook. Williams & wilkins, Baltimore, MD, pp. 119–142.

Van Reen E, Jenni OG, Carskadon MA (2006). Effects of alcohol on sleep and the sleep electroencephalogram in healthy young women. Alcohol Clin Exp Res 30: 974–981.

Wagman AMI, Allen RP (1975). Effects of alcohol ingestion and abstinence on slow wave sleep of alcoholics. In: MM Gross (Ed.), Alcohol Intoxication and Withdrawal, Plenum Press, New York, pp. 453–466.

Walter H, Gutierrez K, Ramskogler K et al. (2003). Gender-specific differences in alcoholism: implications for treatment. Arch Womens Ment Health 6: 253–258.

Wei W, Faria LC, Mody I (2004). Low ethanol concentrations selectively augment the tonic inhibition mediated by delta subunit-containing GABAA receptors in hippocampal neurons. J Neurosci 24: 8379–8382.

Welsh JP, Han VZ, Rossi DJ et al. (2011). Bidirectional plasticity in the primate inferior olive induced by chronic ethanol intoxication and sustained abstinence. Proc Natl Acad Sci U S A 108: 10314–10319.

Wennberg RA, Lozano AM (2003). Intracranial volume conduction of cortical spikes and sleep potentials recorded with deep brain stimulating electrodes. Clin Neurophysiol 114: 1403–1418.

Williams HL, Rundell Jr OH (1981). Altered sleep physiology in chronic alcoholics: reversal with abstinence. Alcohol Clin Exp Res 5: 318–325.

Williams DL, MacLean AW, Cairns J (1983). Dose-response effects of ethanol on the sleep of young women. J Stud Alcohol 44: 515–523.

Wong MM, Brower KJ, Fitzgerald HE et al. (2004). Sleep problems in early childhood and early onset of alcohol and other drug use in adolescence. Alcohol Clin Exp Res 28: 578–587.

Xi MC, Morales FR, Chase MH (1999). A GABAergic pontine reticular system is involved in the control of wakefulness and sleep. Sleep Res Online 2: 43–48.

Xi MC, Morales FR, Chase MH (2001). Induction of wakefulness and inhibition of active (REM) sleep by GABAergic processes in the nucleus pontis oralis. Arch Ital Biol 139: 125–145.

Yules RB, Freedman DX, Chandler KA (1966). The effect of ethyl alcohol on man's electroencephalographic sleep cycle. Electroencephalogr Clin Neurophysiol 20: 109–111.

Yules RB, Lippman ME, Freedman DX (1967). Alcohol administration prior to sleep. The effect on EEG sleep stages. Arch Gen Psychiatry 16: 94–97.

Zahr NM, Pitel AL, Chanraud S et al. (2013). Contributions of studies on alcohol use disorders to understanding cerebellar function. In: A Pfefferbaum, EV Sullivan (Eds.), Handbook of Clinical Neurology: Alcohol and the Nervous System. Elsevier, Amsterdam.

Section 10

Fetal alcohol spectrum disorder

Handbook of Clinical Neurology, Vol. 125 (3rd series)
Alcohol and the Nervous System
E.V. Sullivan and A. Pfefferbaum, Editors

Chapter 25

Neurobehavioral, neurologic, and neuroimaging characteristics of fetal alcohol spectrum disorders

LEILA GLASS, ASHLEY L. WARE, AND SARAH N. MATTSON*
Center for Behavioral Teratology, San Diego State University, San Diego, CA, USA

INTRODUCTION

The use of alcohol during pregnancy can have deleterious consequences for the fetus, including changes in central nervous system (CNS) development leading to permanent neurologic alterations and both cognitive and behavioral deficits. Alcohol's teratogenic effects were highlighted initially by a series of case studies of children born to alcoholic women (Lemoine et al., 1968, Jones et al., 1973). These studies identified common features present in these children, leading to the definition of the fetal alcohol syndrome (FAS: Jones and Smith, 1973). Subsequent diagnostic criteria have retained the core features defined by Jones and Smith, which are listed in Table 25.1 and illustrated in Figure 25.1. In all diagnostic schemata, the diagnosis of FAS relies primarily on physical features; however, the majority of children exposed prenatally to alcohol do not have sufficient physical features required for this diagnosis. Individuals affected by prenatal alcohol exposure, including those with and without FAS are identified under the umbrella of FAS disorders (FASD: Stratton et al., 1996, Bertrand et al., 2004, Hoyme et al., 2005). Prevalence estimates suggest rates of at least 0.2–0.7% for FAS in typical, mixed racial, and mixed socioeconomic populations in the United States and 2–5% for FASD in younger school children in the United States and Western European countries (May et al., 2009). While clinical and preclinical (animal models) studies confirm that even low to moderate levels of prenatal alcohol exposure can have deleterious effects, critical doses of such exposure have yet to be specified and the most clinically significant and consistent consequences occur following heavy exposure. This chapter summarizes the neurobehavioral, neurologic, and neuroimaging characteristics of FASD, focusing primarily on clinical research of individuals with a history of heavy prenatal alcohol exposure, although studies of lower levels of exposure, particularly prospective, longitudinal studies, will be discussed where relevant.

NEUROPSYCHOLOGIC CHARACTERISTICS

Cognitive abilities

While external physical features are indicative of prenatal alcohol exposure and are the basis for an FAS diagnosis, the effects on CNS are clinically significant given their relation to behavioral, social, and cognitive functioning. Numerous neuropsychologic studies have documented deficits that are widespread, specific, and potentially severe, supporting significant functional consequences in individuals with a history of prenatal alcohol exposure.

General intellectual function

One of the most robust neuropsychologic findings in children with heavy prenatal exposure to alcohol is diminished general intellectual performance (Mattson and Riley, 1998). FASDs are a leading identifiable cause of intellectual disability (Pulsifer, 1996), although not all children with heavy prenatal alcohol exposure exhibit this disability. IQ scores of alcohol-exposed children vary significantly from severe intellectual disability to above average (e.g., 40–112), with the average IQ score for a child with FAS being approximately 70. Furthermore, verbal and non-verbal IQ scores are both affected in this population and IQ deficits occur in children with

*Correspondence to: Sarah N. Mattson, Ph.D., Center for Behavioral Teratology, San Diego State University, 6330 Alvarado Court, Suite 100, San Diego, CA 92120, USA. Tel: +1-619-594-7228, E-mail: sarah.mattson@sdsu.edu

Table 25.1

Summary and comparison of the various diagnostic schemas for prenatal alcohol-related disorders

	4-Digit Code[1]	Revised IOM[2]	Canadian[3]	National Task Force on FAS/FAE[4]
Fetal alcohol syndrome (FAS)				
Facial characteristics	Simultaneous presentation of short palpebral fissures (≤2 SD), thin vermillion border, smooth philtrum	Two of the following: short palpebral fissures (≤10th percentile), thin vermillion border, smooth philtrum	Simultaneous presentation of short palpebral fissures (≤2 SD), thin vermillion border, smooth philtrum	Simultaneous presentation of short palpebral fissures (≤10th percentile), thin vermillion border, smooth philtrum
Growth retardation	Height or weight ≤10th percentile	Height or weight ≤10th percentile	Height or weight or disproportionately low weight-to-height ratio (≤10th percentile)	Height or weight ≤10th percentile
Central nervous system (CNS) involvement	Head circumference (occipital-frontal circumference: OFC) ≥2 SD below norm or significant abnormalities in brain structure or evidence of hard neurologic findings or significant impairment in three or more domains of brain function (≥2 SD below the mean), as assessed by validated and standardized tools	Head circumference (OFC) ≤10th percentile or structural brain abnormality	Evidence of three or more impairments in the following CNS domains: hard and soft neurologic signs; brain structure; cognition; communication; academic achievement; memory; executive functioning and abstract reasoning; attention deficit/ hyperactivity; adaptive behavior, social skills, social communication	Head circumference (OFC) ≤10th percentile or structural brain abnormality or neurologic problems or other soft neurologic signs outside normal limits or functional impairment as evidenced by global cognitive or intellectual deficits, below the third percentile (2 SD) below the mean or functional deficits below the 16th percentile (1 SD) below the mean in at least three domains: cognitive or developmental markers, executive functioning, motor, social skills, attention/hyperactivity, and other (i.e., sensory, memory, language)
Alcohol exposure	Confirmed or not confirmed	Confirmed or not confirmed	Confirmed or not confirmed	Confirmed or not confirmed
Partial FAS (pFAS)				Not proposed
Facial characteristics	Short palpebral fissures (≤2 SD) and either a smooth philtrum or thin vermillion border, with the other being normal or palpebral fissure (≤1 SD) and both a smooth philtrum and thin vermillion	Two or more of the following: short palpebral fissures (≤10th percentile), thin vermillion border, smooth philtrum	Two or more of the following: short palpebral fissures, thin vermillion border, smooth philtrum	Not applicable
Growth retardation	Not required	Either height or weight ≤10th percentile or see CNS involvement	Not required	Not applicable

CNS involvement	Same as for FAS	Head circumference ≤10th percentile or structural brain abnormality or behavioral and cognitive abnormalities inconsistent with developmental level	Same as for FAS	Not applicable
Alcohol exposure	Confirmed	Confirmed or not confirmed	Confirmed	Felt that there were insufficient data to provide guidance for this diagnosis. Formed group to discuss
Alcohol-related neurodevelopmental disorder (ARND)	Does not propose this diagnostic category, but rather has several categories assessing functional deficits			Not applicable
CNS involvement	Same as for FAS	Either structural brain anomaly or OFC ≤10th percentile or evidence of a complex pattern of behavioral or cognitive abnormalities inconsistent with developmental level that cannot be explained by genetics, family background, or environment alone	Same as for FAS	Not applicable
Alcohol exposure	Confirmed	Confirmed	Confirmed	Not applicable
Notes	The 4-Digit Code provides an assessment of effects in four areas (growth, face, CNS, and alcohol exposure) that results in 256 different codes and 22 diagnostic categories. A specific pattern or level of alcohol exposure is not required, just that alcohol exposure is confirmed or not	Alcohol exposure is defined as a pattern of excessive intake or heavy episodic drinking	Alcohol exposure is defined as a pattern of excessive intake or heavy episodic drinking. A domain is considered "impaired" when on a standardized measure: Scores are ≥2 SD below the mean, or there is a discrepancy of at least 1 SD between subdomains or there is a discrepancy of at least 1.5–2 SD among subtests on a measure.	Alcohol exposure levels are not defined, but the authors cite evidence of alcohol exposure based upon clinical observation; self-report; reports of heavy alcohol use during pregnancy by a reliable informant; medical records documenting positive blood alcohol levels, or alcohol treatment; or other social, legal, or medical problems related to drinking during pregnancy

All of the diagnostic schemes assume that genetic or medical causes have been ruled out and that appropriate norms are used when available.

All of the diagnostic schemes use the University of Washington Lip-Philtrum Guide (http://depts.washington.edu/fasdpn/htmls/lip-philtrum-guides.htm).

For palpebral fissure norms, the 4-Digit Code uses Hall et al. (1989), Hoyme utilizes Thomas et al. (1987), and Chudley provides both the Thomas and Hall charts; the National Task Force guidelines do not mention which chart to use. Hall recently wrote that her charts underrepresented normal palpebral fissure length (Hall, 2010) and should be replaced by those from Clarren et al. (2010).

Note that < 2 sd = 2.3rd percentile in a normal distribution.

[1]Astley and Clarren (2000).

[2]Hoyme et al. (2005).

[3]Chudley et al. (2005).

[4]Bertrand et al. (2004).

Reproduced from Warren et al. (2011).

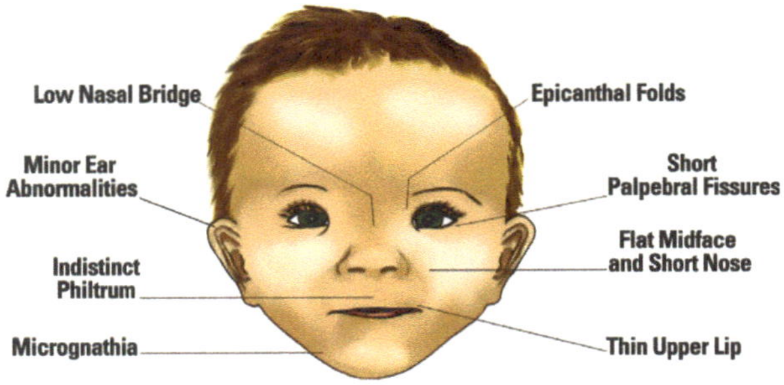

Fig. 25.1. Facial characteristics associated with fetal alcohol exposure. (Reproduced from Warren et al., 2011.)

and without alcohol-related dysmorphia (Mattson et al., 1997, 1998). These IQ deficits are relatively stable and lifelong (Streissguth et al., 1999) and relate to poor educational, psychiatric, and overall life outcomes (Steinhausen and Spohr, 1998).

Even relatively lower maternal alcohol consumption levels are associated with decreased IQ scores after demographic variables are considered; drinking in excess of 1 oz of alcohol per day on average shifts the IQ distribution approximately seven points lower, resulting in a nearly 3.5-fold increase in the number of children with subnormal IQ scores (i.e., below 85; Streissguth et al., 1990).

Academic achievement

Academic achievement deficits exist in all three domains: math, reading, and spelling; exist beyond the effects of low IQ; and rates of learning disabilities in these areas are elevated in children with FASD (Goldschmidt et al., 1996; Mattson et al., 1998; Chudley et al., 2005; Jirikowic et al., 2008a; Rasmussen and Bisanz, 2011). Alcohol-exposed children consistently demonstrate impairments in mathematic functioning (Coles et al., 2009), including basic numerical processing and cognitive estimation (Kopera-Frye et al., 1996; Meintjes et al., 2010; Jacobson et al., 2011). Children with FASD also demonstrate dysfunction in reading and spelling (Streissguth et al., 1994a; Adnams et al., 2007; Kodituwakku, 2007). These deficits occur in pre-reading and pre-spelling abilities, such as phonologic processing, in addition to standardized reading and spelling measures, including comprehension and reading speed (Streissguth et al., 1994c; Carmichael Olson et al., 1998; Mattson et al., 1998; Adnams et al., 2007).

The relationship between dose and timing of prenatal alcohol exposure and academic achievement was analyzed in a longitudinal prospectively identified sample (Goldschmidt et al., 1996). At age 6, there was a linear dose–response relationship between prenatal alcohol exposure and performance on standardized arithmetic tasks; however, the relationship between exposure level and verbal academic domains (reading and spelling) was better modeled by a threshold model of approximately one drink per day in the second trimester. Targeted interventions (Adnams et al., 2007; Coles et al., 2009; Peadon et al., 2009) and patient advocacy (Duquette et al., 2006) can facilitate the remediation of academic achievement deficits across domains in FASD.

Executive function

Executive function encompasses a wide range of higher-order cognitive processes that facilitate goal-directed behavior. Parent reports and standardized tests of higher-order cognition using a wide range of methods provide strong convergent support for executive dysfunction as a hallmark feature of FASD (Kodituwakku et al., 1995; Mattson et al., 1999; Ware et al., 2012) that occurs in alcohol-exposed children with and without the FAS diagnosis and may exist above the influence of IQ (Schonfeld et al., 2001). Deficits exist in multiple executive function domains, including planning, set shifting, cognitive flexibility, concept formation, working memory, and response inhibition (Kodituwakku et al., 1995; Kerns et al., 1997; Mattson et al., 1999; Connor et al., 2000) and relate to clinically significant impairments in adaptive behavior, social skills, social problem solving, and impulsivity (Schonfeld et al., 2006; McGee et al., 2008; Ware et al., 2012). However, executive function deficits are not universally supported by all studies. For example, in one study children with FASD were unimpaired on tasks requiring response inhibition and rule learning (Kodituwakku et al., 1995).

Language

Basic language functioning and the ability to communicate are important in various aspects of life, including verbal expression, comprehension, and social interactions. The effects of prenatal alcohol exposure on both language and communication are unclear given relatively limited study. Children with FASD have demonstrated deficits in expressive language, the ability to produce verbal or written expression, and receptive language, the ability to understand language, in addition to delays in language acquisition (Adnams et al., 2007; Aragón et al., 2008a; McGee et al., 2009; Wyper and Rasmussen, 2011). The underlying mechanisms of these deficits have yet to be fully elucidated, although they may relate to physical changes related to prenatal alcohol exposure or to impaired auditory and verbal processing. Deficits in the physical aspects of language may contribute to poor expressive language, as children with FASD demonstrate dysarthria, lisping, and articulation difficulties (Becker et al., 1990).

An additional underlying component of language that has been evaluated is auditory and phonologic processing; however the results are inconclusive. One prospective, longitudinal study of children with lower levels of exposure (averaging 0.34 oz of absolute alcohol/day) suggested dose-dependent deficits in phonologic processing (Streissguth et al., 1994b, c), while a later study comparing children with FAS and average IQ to controls found no group differences (Carmichael Olson et al., 1998). Other studies involving lower levels of exposure found language deficits in basic abilities (Gusella and Fried, 1984; Fried and Watkinson, 1988), while others did not (Fried and Watkinson, 1990; Greene et al., 1990; Fried et al., 1992; O'Leary et al., 2009). It is unclear whether inconsistent results are due to differences in exposure levels, developmental trajectories, within-group variability, or subjectivity of the measures assessing the underlying language abilities.

These basic language difficulties contribute to poor communication abilities in children with FASD, which may explain social and behavioral issues (Crocker et al., 2009; Ware et al., 2012) as well as contribute to poor academic performance, social communication, and general abilities within the verbal domain (Church and Kaltenbach, 1997; McGee et al., 2009).

Learning and memory

Learning and memory have been extensively studied in children with FASD and there is robust empiric support for deficits in these areas, particularly in the verbal domain (Richardson et al., 2002; Pei et al., 2008; Willoughby et al., 2008; Manji et al., 2009; Wheeler et al., 2012). The California Verbal Learning Test-Children's Version (CVLT-C) allows for the study of encoding, retrieval, and retention abilities. Research using this measure found both encoding and retrieval deficits in children with FASD. However, retention – the amount of material recalled after adjusting for initial learning – is spared (Mattson et al., 1996a; Mattson and Roebuck, 2002), suggesting that, while alcohol-exposed children struggle with the initial learning of verbal information, they are able to retain the information they do learn. These results have been replicated in different cohorts of heavily exposed children and adults, in children with lower levels of exposure, and using different tasks, further providing evidence for spared retention (Kaemingk et al., 2003; Willoughby et al., 2008). Further examination of this finding indicated that spared retention may be related to use of an implicit strategy and, when no strategy is available, retention is impaired (Mattson and Roebuck, 2002; Roebuck-Spencer and Mattson, 2004). These deficits may present as latent effects of alcohol exposure.

A longitudinal study of low (<3 drinks/week) to moderate (up to 1 drink/day) prenatal exposure levels did not detect verbal learning and memory deficits at younger (through ages 5 and 6 years) ages (Fried and Watkinson, 1990; Fried et al., 1992); however, these impairments became apparent later (ages 10 and 14 years) with FASD (Richardson et al., 2002; Willford et al., 2004). In typically developing children, there is often improvement on executive functioning across adolescent development, possibly related to maturation of the frontal-lobe structures and accompanying organizational strategies that occur during this time period (Anderson, 2002; Blakemore and Choudhury, 2006). However, this trajectory has not been explicitly reviewed in the FASD population.

Non-verbal memory deficits are less clear in this population, as the existing literature is sparse and sometimes inconsistent (Mattson and Roebuck, 2002; Kaemingk et al., 2003; Aragón et al., 2008b). For instance, studies both support (Wheeler et al., 2012) and refute (Autti-Rämö et al., 2002) facial memory impairments in children with FASD. Similar controversy exists in the area of spatial and object memory as some studies found deficits in spatial, but not object, memory (e.g., Uecker and Nadel, 1998; Hamilton et al., 2003), whereas another study found no significant differences between object and spatial memory performance (Pei et al., 2008). However, the latter study lacked a control comparison group against which to gauge overall performance of the FASD group.

Inconsistencies may result from discrepancies in sample sizes, tasks, definitions of impairment, and statistic approaches. For example, variables related to the retention of learned information, which are important in verbal memory, are often not included in non-verbal memory studies (Uecker and Nadel, 1996, 1998). Discrepant findings may also result from inconsistent attention to lower-order cognitive processes. When addressed, lower-order processes can help account for higher-order deficits. For example, spatial memory impairments in FASD were explained by deficits in lower-order processes, including visual processing and verbal memories (Kaemingk and Halverson, 2000).

Cognitive processing speed and attention

Cognitive processing speed deficits are considered a core impairment in children with FASD (Jacobson et al., 1993, 1994; Kable and Coles, 2004; Burden et al., 2005a; Kodituwakku, 2009) and are exhibited throughout development, in infants, children, and adolescents (Nanson and Hiscock, 1990; Streissguth et al., 1994b; Jacobson, 1998). These deficits are most apparent within the context of complex cognition and are less affected when engaged in automatic processing (Burden et al., 2005a).

Children with FASD also exhibit deficits in vigilance, reaction time, sustained attention, and information processing (Jacobson et al., 1994;, Streissguth et al., 1994c; Burden et al., 2005b, 2009). Parent and teacher reports corroborate laboratory findings of attention deficits (Brown et al., 1991; Janzen et al., 1995; Mattson and Riley, 2000; Nash et al., 2006). Attention deficits may be domain-specific: two studies found greater impairments in sustained visual attention compared to auditory attention (Coles et al., 2002; Mattson et al., 2006), while another study found the opposite pattern (Connor et al., 1999). Recent studies have attempted to further clarify the specificity of visual attention deficits in children with prenatal alcohol exposure and found impaired alerting (ability to achieve and maintain attention, e.g., a readiness to respond) and executive control (the ability to control behavior, e.g., to monitor, detect, and resolve conflicts to achieve goals) (Kooistra et al., 2011; Crocker et al., 2012).

Behavioral characteristics

In addition to the wide range of cognitive deficits, behavioral problems are highly prevalent in children with FASD and may be more salient to parents and other caregivers.

Problem behaviors

High rates of externalizing problems are consistently reported in children with FASD (Sood et al., 2001; Jirikowic et al., 2008b). Externalizing behaviors in FASD include defiance, aggressiveness, and delinquency. Children with FASD demonstrate more delinquent behavior (Roebuck et al., 1999), exhibit impaired moral decision-making abilities (Schonfeld et al., 2005), and are also more likely to lie about their behavior at a young age (Rasmussen et al., 2008) compared to non-exposed peers. Delinquent behavior persists into adulthood (Famy et al., 1998; Spohr et al., 2007). Elevations of internalizing behaviors, such as anxiety, depression, and being withdrawn, are also reported in this population, and exist beyond the presence of facial dysmorphology, intelligence, or demographic factors (Mattson and Riley, 2000; O'Connor and Kasari, 2000; Sood et al., 2001; O'Connor and Paley, 2006).

Both internalizing and externalizing behavioral problems relate to long-term negative life outcomes, including increased legal issues (Fast et al., 1999; Streissguth et al., 2004), substance abuse issues (Baer et al., 2003; Alati et al., 2008), and suicide attempts (Streissguth et al., 1996; O'Malley and Huggins, 2005; Baldwin, 2007). A dose–response effect has also been found between prenatal alcohol exposure and childhood behavioral problems, even after controlling for contributing factors such as education, family environment, exposure to violence, socioeconomic status, and psychopathology (Sood et al., 2001).

Psychopathology

In addition to high rates of problematic behaviors, FASD are associated with increased risk for psychiatric diagnoses, such as oppositional defiant disorder and conduct disorder (Streissguth et al., 1996; O'Connor et al., 2002; Burd et al., 2003; Lynch et al., 2003; Fryer et al., 2007a; Disney et al., 2008). Additionally, they have higher rates of negative affect and an increased risk for major depressive disorder compared to typically developing controls (O'Connor and Paley, 2006; Fryer et al., 2007a).

The most common psychiatric diagnosis in individuals with FASD is attention-deficit hyperactivity disorder (ADHD), with estimates suggesting that that more than 60% of children with FASD receive a diagnosis of ADHD (LaDue et al., 1992; Fryer et al., 2007a). Alcohol-exposed children with and without an ADHD diagnosis have been compared to disentangle the behavioral and neuropsychologic effects of prenatal alcohol exposure and concomitant ADHD (Glass et al., 2013; Ware et al., 2013a). Behaviorally, the presence of an ADHD diagnosis exacerbated alcohol-related deficits in communication, attention problems, and externalizing behavior (Ware et al., 2013a, b); however, there were no increased neuropsychologic deficits related to having both disorders (Glass et al., 2013).

Social skills and social communication

There is strong evidence of impaired interpersonal and social skills in children with FASD, such as reduced social competence, clinginess, not getting along with others, being teased or bullied, poor social judgment and responding to social cues (Whaley et al., 2001; Coggins et al., 2007; Fagerlund et al., 2012; Kully-Martens et al., 2012; Malone and Koren, 2012), which fail to improve with age (Thomas et al., 1998; Crocker et al., 2009). These deficits may relate to deficient social problem solving, as adolescents with FASD in one study approached problems with a pessimistic orientation, had a low frustration threshold when encountering a difficult task, and self-rated themselves as less effective at identifying problems, generating alternative solutions, and making decisions (McGee et al., 2008). Children with prenatal alcohol exposure struggle to balance linguistic and social-cognitive task demands (Coggins et al., 2003), often fail to provide sufficient organization and information for communication partners during narratives (Coggins et al., 2007), use ambiguous references and inappropriately distinguish concepts in their stories (Thorne et al., 2007), and fail to consider the perspective of the listener during interactions (Timler et al., 2005). Theory of mind, the ability to attribute mental states of another to oneself (Greenbaum et al., 2009), is also impaired in children with FASD (Rasmussen et al., 2009; Graham et al., 2013), although this impairment is not supported by all literature (Rasmussen et al., 2012). These deficits may affect school performance and academic achievement, as well as contribute to psychopathology in this population. Fortunately, these deficits have been able to be effectively remediated with intervention (O'Connor et al., 2006) (Table 25.2).

Neurologic characteristics

Aside from neurobehavioral deficits detailed above, additional neurologic findings and CNS dysfunction may bring individuals with FASD to medical attention.

Sensory integration and processing

Sensory processing is an area of CNS functioning that includes processing cognitive responses to sensory input. Children with FASD have sensory-processing irregularities across multiple modalities, particularly in the auditory and visual domains (Church and Gerkin, 1988; Scher et al., 1998; Kable et al., 2009). Auditory filtering and processing are impaired in younger preschool-aged children with FASD (Jirikowic et al., 2008a; Carr et al., 2010; Stephen et al., 2012). FASD have also been associated with developmental delays in auditory maturity, sensorineural hearing loss, conductive hearing loss due to frequent episodes of otitis media, and central hearing loss that may warrant early intervention (Stratton et al., 1996; Church and Kaltenbach, 1997; Cohen-Kerem et al., 2007).

In the visual domain, children with prenatal alcohol exposure have impaired oculomotor control and increased saccadic reaction times (Green et al., 2009). Further, children with FAS have a high frequency of ocular pathology and visual impairments, including refractive errors, strabismus, and optic nerve hypoplasia that may result in reduced visual acuity (Strömland, 2004). This population exhibits deficits on simple spatial construction tests (Conry, 1990; Janzen et al., 1995; Korkman et al., 1998; Jirikowic et al., 2008a), in addition to complex visuospatial functioning (Janzen et al., 1995; Uecker and Nadel, 1996; Aronson and Hagberg, 1998; Pei et al., 2011; Coffman et al., 2013). When tested on a complex visuospatial functioning task, the Rey-Osterrieth Complex Figure, children with FASD have difficulty with initial encoding, recalling structural components over time, demonstrate increased misplacement of design structures, and have lower organizational levels compared to controls. These results support the patterns seen in the verbal domain of poor recall of information being impacted primarily by limitations in initial encoding (Pei et al., 2011).

Neurophysiologic techniques have allowed for further support of somatosensory disruptions resulting from prenatal alcohol exposure. In both auditory and visual domains, alcohol-exposed children exhibit prolonged wave latencies consistent with developmental delay (N2 waves), slower processing, and attention problems (N1 and P1 waves; Scher et al., 1998). Older children with FASD also exhibit atypical wave latencies on evoked-potential attention tasks (Kaneko et al., 1996). However, unlike infant findings, where N1 and P3 wave latencies (related to decision making) discriminated between alcohol-exposed and non-exposed children, only P3 wave latencies differentiated these groups in older children (Kaneko et al., 1996). These findings, along with others from longitudinal studies, suggest that some pathology observed in infants with FASD may resolve with development (Spohr and Steinhausen, 1987). Additionally, when matched to controls on an executive control task, children with FASD exhibit slower P2 and P3 latencies (Burden et al., 2009, 2011), which may indicate impairments in early visual processing and increased cognitive effort. In these studies, event-related potential (ERP) abnormalities were positively correlated with prenatal alcohol exposure levels, suggesting dose-dependent abnormalities.

Another electrophysiologic technique, magnetoencephalography, has been used to examine auditory and visual response latency and processing in children with prenatal exposure to alcohol. When responding to auditory stimuli, young children (3–6 years of age) with FASD display significant M100 and M200 latencies compared to controls (Stephen et al., 2012). In terms of visual response, alcohol-exposed adolescents exhibit visual M100 response latencies (Coffman et al., 2013) that are indicative of slower primary visual cortex activity and delayed processing of peripheral, as opposed to central, visual stimuli compared to age-matched controls. Additionally, saccadic reaction times correlate with M100 latency, signifying an association between sensory deficits and motor impairments in this population (Coffman et al., 2013). Deficient somatosensory processing may contribute to poorer outcomes in adaptive function and academic achievement in FASD (Franklin et al., 2008; Jirikowic et al., 2008a; Carr et al., 2010).

Motor skills

Motor impairment is another area of CNS dysfunction associated with prenatal alcohol exposure. Robust

Table 25.2

Neurobehavioral characteristics of fetal alcohol spectrum disorders (FASD)

Cognitive domain		FASD vs controls
General intellectual function		Decreased general intellectual functioning (Mattson et al., 1998). IQ deficits are similar with and without dysmorphology (Mattson et al., 1997, 1998), and persist into adulthood (Streissguth et al., 1999)
Academic achievement		Academic achievement impaired beyond diminished IQ (Mattson et al., 1998; Jirikowic et al., 2008a; Rasmussen and Bisanz, 2011). Impaired math (Coles et al., 2009), reading, and spelling (Streissguth et al., 1994a; Adnams et al., 2007; Kodituwakku, 2007). High rate of learning disabilities (Chudley et al., 2005)
Executive function		Executive dysfunction (Kodituwakku et al., 1995; Ware et al., 2012), including planning, set shifting, cognitive flexibility, mental manipulation, working memory, and reasoning deficits. Deficits exist after controlling for IQ and regardless of facial dysmorphology (Schonfeld et al., 2001). Response inhibition deficits inconsistently reported (Kodituwakku et al., 1995; Connor et al., 2000; Burden et al., 2010)
Language		Dysarthria, lisping, and articulation difficulties (Becker et al., 1990). Expressive and receptive language deficits (Adnams et al., 2007; Aragón et al., 2008a; McGee et al., 2009; Wyper and Rasmussen, 2011). Impaired phonologic processing (Streissguth et al., 1994c; Carmichael Olson et al., 1998)
Learning and memory		Verbal learning and memory deficits characterized by encoding and retrieval deficits with spared retention (Mattson and Roebuck, 2002; Pei et al., 2008; Willoughby et al., 2008; Crocker et al., 2011; Mattson et al., 2011; Wheeler et al., 2012). Deficits in non-verbal (Mattson and Roebuck, 2002; Kaemingk et al., 2003; Aragón et al., 2008b), facial (Autti-Rämö et al., 2002; Wheeler et al., 2012), and object and spatial memory (Uecker and Nadel, 1998; Kaemingk and Halverson, 2000; Hamilton et al., 2003; Pei et al., 2008) are inconsistently reported
Cognitive processing speed and attention		Impaired processing speed (Kable and Coles, 2004; Burden et al., 2005a; Kodituwakku, 2009), particularly within the context of complex cognition and less apparent in automatic processing (Burden et al., 2005a). Impaired vigilance, reaction time, organizing and maintaining attention, and information processing (Jacobson et al., 1994; Streissguth et al., 1994c; Burden et al., 2005b; Burden et al., 2009). Parent and teacher reports corroborate cognitive findings (Janzen et al., 1995; Mattson and Riley, 2000; Nash et al., 2006). Some studies report poorer visual compared to auditory attention (Connor et al., 1999; Coles et al., 2002; Mattson et al., 2006)
Behavioral characteristics	Problem behaviors	Elevated externalizing (Jirikowic et al., 2008b; Ware et al., 2013b) and internalizing behaviors (Mattson and Riley, 2000; Sood et al., 2001; O'Connor and Paley, 2006). Higher rates of delinquent behavior (Roebuck et al., 1999). Impaired moral decision making (Schonfeld et al., 2005) and more likely to lie about behavior (Rasmussen et al., 2008). Higher rates of legal issues (Fast et al., 1999; Streissguth et al., 2004), substance abuse (Baer et al., 2003; Alati et al., 2008), and suicide attempts (O'Malley and Huggins, 2005; Baldwin, 2007)
	Psychopathology	High rate of concomitant attention-deficit hyperactivity disorder (LaDue et al., 1992; Fryer et al., 2007a), psychiatric diagnoses, and secondary disabilities (Streissguth et al., 1996; Burd et al., 2003; Lynch et al., 2003; O'Connor and Paley, 2006; Fryer et al., 2007a; Disney et al., 2008)
	Social skills and communication	Diminished interpersonal and social skills (Whaley et al., 2001; Coggins et al., 2007; Crocker et al., 2009; Fagerlund et al., 2012; Kully-Martens et al., 2012; Malone and Koren, 2012). Poor ability to balance linguistic and social-cognitive demands (Coggins et al., 2003). Failure to provide information in narrative communication (Coggins et al., 2007) and consider perspective taking (Timler et al., 2005). Theory of mind deficits are inconsistently reported (Rasmussen et al., 2012; Graham et al., 2013)

convergent evidence supports deficits in fine motor speed and coordination (Mattson et al., 1998; Roebuck-Spencer et al., 2004; Jirikowic et al., 2008a; Chiodo et al., 2009). Motor reaction timing is slowed (Simmons et al., 2002, 2010; Wass et al., 2002; Green et al., 2009) and more variable (Simmons et al., 2009). Gross motor deficits are also reported in children with FASD, including dysfunctional force regulation (Simmons et al., 2012), atypical trajectories in goal-directed arm movements (Dömellof et al., 2010), poor balance (Barr et al., 1990), atypical gait (Marcus, 1987), and weak grasp (Conry, 1990). Gross motor control results are nuanced, as one study found deficits in static postural stability (Kooistra et al., 2009), whereas a previous study found that, when sensorimotor input was reliable, there were no differences between children with FASD and controls (Roebuck et al., 1998). Further, there are some studies that suggest that fine motor skills may be more affected than gross motor skills (Adnams et al., 2001; Kalberg et al., 2006) and that motor skill deficits may only occur in the most severely affected (Connor et al., 2006; Kalberg et al., 2006) or youngest individuals (Simmons et al., 2006).

Seizure activity

While the prevalence of epilepsy in the general population is less than 1%, Bell and colleagues (2010) found that 5.9% of children with FASD had a confirmed diagnosis of epilepsy and 11.8% have had at least one documented seizure episode during their lifespan. Other earlier studies indicate that individuals with FAS develop epilepsy at three to 20 times the rate of the general population (Spohr and Steinhausen, 1987; O'Malley and Barr, 1998). There are similar demographic variables, such as maternal malnutrition, prematurity, birth complications, and low prenatal care, which are associated with both FASD and seizure disorders (Bell et al., 2010). Further, there are hypothesized similarities in neural structures affected by the teratogenicity of alcohol and those related to epileptiform activity. For example, the hippocampus is associated with the genesis and spread of epileptiform activity in the human brain (Badawy and Jackson, 2012) and hippocampal size and function are affected in individuals with FASD (Willoughby et al., 2008). More direct evidence for this relation comes from one study using an animal model in which daily neonatal exposure to alcohol resulted in a dose-dependent decrease in seizure threshold and a selective loss of CA1 pyramidal cells in the hippocampus. The reduced threshold for seizures was related to the hippocampal pathology (Bonthius et al., 2001).

Sleep pathology

Children with FASD often present with sleep abnormalities (Steinhausen and Spohr, 1998; Jan et al., 2010; Wengel et al., 2011). The animal literature has suggested that prenatal exposure to alcohol alters the expression of genes that underlie circadian functioning (Chen et al., 2006; Sakata-Haga et al., 2006). Novel neuroimaging methods, including neurophysiologic techniques, have allowed for further investigation of prenatal alcohol exposure's disruptive effects on neural functioning. Early electroencephalography (EEG) studies elucidated specific neuronal mechanisms that contribute to aberrant behavioral outcomes in infants and children with FASD. Newborns with FASD exhibit hypersynchrony during all sleep stages (Havlicek et al., 1977; Chernick et al., 1983; Ioffe and Chernick, 1988), which predict developmental problems (Ioffe and Chernick, 1990). Compared to non-exposed infants, those with prenatal alcohol exposure have more disturbed sleep and frequent arousals (Scher et al., 1998). Clinical implications of sleep disturbances include increased hyperactivity, aggression, inattention, depression, executive dysfunction, motor impairment, diabetes, and obesity (Jan et al., 2010). While these symptoms are often seen in children with FASD without sleep loss, poor sleep may exacerbate these areas of dysfunction. Due to the increased risk of poor sleep in children with FASD, proper sleep hygiene and a comprehensive evaluation of sleep environment are recommended (Jan et al., 2010) (Table 25.3).

Using neurobehavioral data to identify individuals affected by prenatal alcohol exposure

As discussed, there have been well-documented, robust findings of deficits across a variety of neuropsychologic and behavioral domains. However, multiple factors including overlap with other clinical conditions make accurate identification of alcohol-affected children difficult, particularly in the absence of facial dysmorphology. Accurate identification is paramount to the development and testing of clinical interventions. As a result, there has been a strong effort to determine a neuropsychologic or neurobehavioral profile to improve identification of children with FASD (Mattson and Riley, 2011). There has been a recent successful attempt to distinguish children with heavy prenatal alcohol exposure from typically developing controls using neurobehavioral tasks and latent profile analysis (Mattson et al., 2010). While the classification rates were high and specific profile measures were identified, it is also possible that more than one phenotype associated with prenatal alcohol exposure exists (Gray et al., 2009). To further determine the

Table 25.3

Neurologic characteristics of fetal alcohol spectrum disorders (FASD)

Domain		FASD vs controls
Seizure activity		Increased rate of epilepsy (Spohr and Steinhausen, 1987; O'Malley and Barr, 1998) and seizures (Bell et al., 2010)
Sensory processing	Auditory	Impaired auditory filtering and processing (Jirikowic et al., 2008a; Carr et al., 2010; Stephen et al., 2012). Developmental delays in auditory maturity, sensorineural hearing loss, central hearing loss, and conductive hearing loss due to frequent otitis media (Stratton et al., 1996; Church and Kaltenbach, 1997; Cohen-Kerem et al., 2007)
	Visual	High rate of ocular pathology and visual impairments (i.e., refractive errors, strabismus, and optic nerve hypoplasia (Strömland, 2004). Impaired oculomotor control and increased saccadic reaction times (Green et al., 2009). Visuospatial dysfunction for both simple (Conry, 1990; Janzen et al., 1995; Korkman et al., 1998; Jirikowic et al., 2008a) and complex (Uecker and Nadel, 1996; Aronson and Hagberg, 1998; Coffman et al., 2013) tasks
	Neurophysiologic studies	Prolonged wave latencies, in auditory and visual domains (Scher et al., 1998). Atypical evoked-potential wave latencies during an attention task (Kaneko et al., 1996). Abnormal auditory response latencies (Coffman et al., 2013; Stephen et al., 2012)
Motor skills	Overall	Delayed motor reaction timing (Simmons et al., 2002; Wass et al., 2002; Green et al., 2009; Simmons et al., 2010) and increased motor timing variability (Simmons et al., 2009). Parent-reported delayed motor abilities, especially for fine motor compared to gross motor control (Adnams et al., 2001; Kalberg et al., 2006)
	Gross	Dysfunctional force regulation (Simmons et al., 2012), atypical trajectories in goal-directed arm movements (Dömellof et al., 2010), poor balance (Barr et al., 1990), atypical gait (Marcus, 1987), and weak grasp (Conry, 1990). Impaired static postural control (Kooistra et al., 2009), unless reliable sensorimotor input is present (Roebuck et al., 1998)
	Fine	Fine motor speed and coordination deficits (Mattson et al., 1998; Roebuck-Spencer et al., 2004; Jirikowic et al., 2008a; Chiodo et al., 2009)
Sleep pathology		High rate of sleep abnormalities (Steinhausen and Spohr, 1998; Jan et al., 2010; Wengel et al., 2011). More disturbed sleep and more frequent arousals (Scher et al., 1998). Hypersynchrony during all sleep stages in newborns (Havlicek et al., 1977; Chernick et al., 1983; Ioffe and Chernick, 1988)

specificity of deficits in children with FASD, additional research has focused on discriminating alcohol-exposed children from contrast groups with similar clinical presentations such as ADHD (Kooistra et al., 2011; Mattson et al., 2011), autism (Strickland et al., 2007; Mukherjee et al., 2011), specific learning disability (Fagerlund et al., 2012), or groups matched on IQ (Mattson and Riley, 2000; Vaurio et al., 2011). A second study using latent profile analysis of neurobehavioral measures in alcohol-exposed children and non-exposed children with ADHD resulted in a 70% accuracy rate for classification (Mattson et al., 2013), indicating that children with FASD can be accurately distinguished from children with ADHD, although room for improvement still exists.

NEUROIMAGING CHARACTERISTICS

Global structural abnormalities

The neurologic and neuropsychologic abnormalities described likely relate to underlying CNS anomalies, especially structural and functional brain abnormalities. Early autopsy studies of FAS described diffuse brain abnormalities, including microcephaly and microencephaly, callosal dysgenesis and malformations, reduced cerebellar volume, and neuronal abnormalities (Jones and Smith, 1973; Clarren et al., 1978; Clarren and Smith, 1978). Advanced *in vivo* neuroimaging techniques, particularly magnetic resonance imaging (MRI), have allowed for in-depth investigation of neural

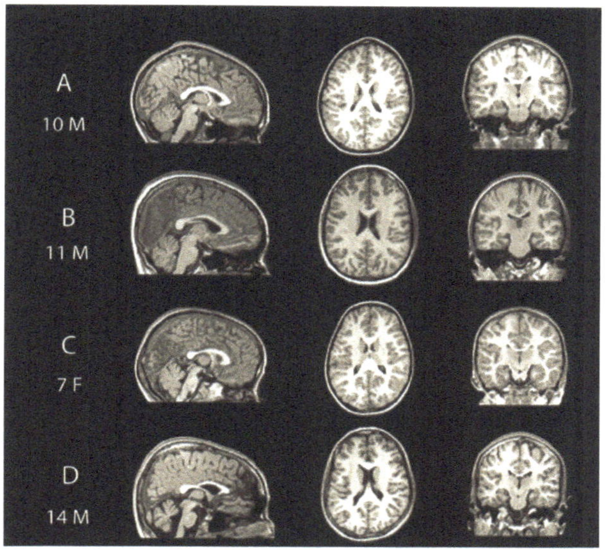

Fig. 25.2. Significant variability in brain structure is demonstrated among these T1-weighted structural magnetic resonance imaging scans of three children with fetal alcohol spectrum disorders relative to an unexposed child: (**A**) unexposed child (10-year-old male); (**B**) alcohol-exposed child without fetal alcohol syndrome (11-year-old male); (**C**) child with fetal alcohol syndrome (7-year-old female); and (**D**) child with fetal alcohol syndrome (14-year-old male). (Reproduced from Nuñez et al., 2011.)

structural deficits associated with prenatal exposure to alcohol (Fig. 25.2).

Cerebral volume and shape

As reported in early autopsy studies, microcephaly is the most common consequence of prenatal exposure to alcohol (e.g., Roussotte et al., 2012b). Case and group studies have also revealed cortical atrophy (Riikonen et al., 1999), schizencephaly (Spalice et al., 2011), and polymicrogyria (Reinhardt et al., 2010) in heavily exposed individuals.

In addition to total brain volume reduction, the frontal (Sowell et al., 2002a; Astley et al., 2009b), temporal, and parietal (Sowell et al., 2002a) lobes show regional volume deficits; however, only the parietal lobes remain disproportionately reduced beyond total brain volume reduction in FASD (Sowell et al., 2002a) and FAS (Archibald et al., 2001). Increased parietal-lobe displacement is also associated with prenatal alcohol exposure (Sowell et al., 2002a). Occipital-lobe alterations have less support in the literature, with the majority of findings suggesting that lobule volume, shape, and placement (Archibald et al., 2001; Sowell et al., 2002a, b) are similar in alcohol-exposed and non-exposed youth, with the exception of the occipital/temporal region (Li et al., 2008). One recent study, however, found disproportionate occipital reduction in young adults with FASD (Chen et al., 2012).

Youth with FASD show cortical surface area reductions in anterior and orbital frontal regions (Sowell et al., 2008b), and disproportionately reduced volumes are present in dysmorphic adults with prenatal alcohol exposure in the left cuneus and bilateral temporal-lobe regions (Chen et al., 2012). Structural abnormalities in frontal-lobe regions are associated with facial dysmorphology in children with FASD; children with facial dysmorphology have greater reductions in frontal-lobe regions compared to non-dysmorphic children with alcohol exposure (Astley et al., 2009b). This is not true in adults (Chen et al., 2012), perhaps due to age-related changes in the degree of facial dysmorphology.

Cognitive correlates of structural brain abnormalities following prenatal alcohol exposure have been examined. While some studies found associations between volumetric or structural abnormalities and cognitive impairment in this population, it is important to note that not all studies have found significant associations (Autti-Rämö et al., 2002; Nardelli et al., 2011). As shown in Table 25.4, diminished brain volume is associated with prenatal exposure level and decreased intellectual functioning in children (Roussotte et al., 2012b) and young adults (Chen et al., 2012) with FASD. Cortical surface area reductions in anterior and orbital frontal regions and cortical thickness in dorsal frontal regions are associated with verbal learning and memory deficits in FASD (Coles et al., 2011; Fryer et al., 2012).

Cerebrospinal fluid

Findings of cerebrospinal fluid (CSF) volume abnormalities following prenatal alcohol exposure are

Table 25.4

Neuroimaging characteristics of fetal alcohol spectrum disorders (FASD)

Brain area	Volumetric findings (FASD vs controls)	Structural abnormalities	Brain–behavior relations	Functional abnormalities (FASD vs behaviorally matched controls)
Total brain volume	Reduced overall (Mattson et al., 1996b; Willoughby et al., 2008; Nardelli et al., 2011; Roussotte et al., 2011). Reduced white, deep and cortical gray matter (Bjorkquist et al., 2010; Nardelli et al., 2011; Roussotte et al., 2012b)	Schizencephaly (Spalice et al., 2011) and polymicrogyria (Reinhardt et al., 2010)	Volume negatively associated with IQ in children (Roussotte et al., 2011) and young adults (Chen et al., 2012)	**MRI and PET** Abnormal regional cerebral metabolic rates (Clark et al., 2000) **SPECT** Abnormal cerebral blood perfusion (Bhatara et al., 2002) and blood flow (Riikonen et al., 1999; Clark et al., 2000) **EEG** Abnormalities at rest (Mattson et al., 1992)
Frontal lobe	Reduced overall (Sowell et al., 2002a). Reduced white (Sowell et al., 2002a; Astley et al., 2009b) and gray matter (Astley et al., 2009b) Reduced in caudal middle, medial orbital, superior, rostral medial, pars triangularis, pars orbital regions (Coles et al., 2011)	Anterior and orbital surface area reductions (Astley et al., 2009b). Cortical atrophy (Riikonen et al., 1999). Increased cortical thickness (Sowell et al., 2008b; Yang et al., 2012b)	Surface area, volume, and cortical thickness associated with verbal learning and memory (Fryer et al., 2012)	**MRI and PET** Prefrontal blood perfusion abnormalities (Riikonen et al., 1999). Increased medial-frontal serotonin transporter binding (Riikonen et al., 2005) **fMRI** Greater BOLD in dorsal frontal during verbal memory (Sowell et al., 2007) and verbal (O'Hare et al., 2009) and non-verbal (Roussotte et al., 2011; Diwadkar et al., 2013) WM, insula during spatial WM (Norman et al., 2013), and medial and middle frontal during RI (Fryer et al., 2007b; O'Brien et al., 2013). Lesser BOLD in prefrontal with increasing non-verbal WM task difficulty (Astley et al., 2009a), in middle, medial, and superior frontal with non-verbal WM efficiency and reaction time (Spadoni et al., 2009), in prefrontal math-related regions during number processing (Santhanam et al., 2009). Reduced task-related DMN deactivation in medial prefrontal cortex (Santhanam et al., 2011) **fcMRI** Lower connectivity in paracentral region of FASD children with static encephalopathy (Wozniak et al., 2011)
Temporal lobe	Reduced (Sowell et al., 2002a). Decreased occipital/temporal region (Li et al., 2008). Decreased fusiform gyrus (Coles et al., 2011)	Increased gray-matter density (Sowell et al., 2001b, 2002a). Increased cortical thickness (Sowell et al., 2008b; Yang et al., 2012b)		**fMRI** Blood perfusion differences (Bhatara et al., 2002). Greater BOLD in posterior regions during verbal WM (O'Hare et al., 2009). Less BOLD in medial and posterior regions during verbal WM (Sowell et al., 2007)

Parietal lobe	Reduced (Archibald et al., 2001; Sowell et al., 2002a; Chen et al., 2012). Reduced cortical, white (Archibald et al., 2001) and gray matter (Sowell et al., 2002a)	Increased gray-matter density (Sowell et al., 2002a). Increased cortical thickness (Sowell et al., 2008b; Yang et al., 2012b)		**fMRI** Greater BOLD in inferior parietal during verbal (O'Hare et al., 2009) and non-verbal (Roussotte et al., 2011; Diwadkar et al., 2013) WM. Lower BOLD in parietal regions with increasing non-verbal WM difficulty (Astley et al., 2009a) and in math-related regions during number processing (Santhanam et al., 2009)
Occipital lobe	No reductions (Sowell et al., 2002b; Yang et al., 2012b). Decreased occipital/temporal region (Li et al., 2008)	Increased cortical thickness in lateral regions (Yang et al., 2012b)		**MRI and SPECT** Blood perfusion abnormalities in parieto-occipital regions (Riikonen et al., 1999)
Corpus callosum	Reduced (Riley et al., 1995; Yang et al., 2012a). Partial and full agenesis (Mattson et al., 1992; Riley et al., 1995; Astley et al., 2009b) DTI Lower FA: genu, isthmus, splenium (Li et al., 2009; Wozniak et al., 2009), splenium tract (Lebel et al., 2008), and callosal body (Fryer et al., 2009). Higher MD: isthmus (Wozniak et al., 2006; Li et al., 2009), genu, and splenium regions (Ma et al., 2005)	Decreased length and area (Riley et al., 1995; Astley et al., 2009b; Yang et al., 2012a), increased displacement (Sowell et al., 2001a) and shape variability (Bookstein et al., 2002a, 2002b; Yang et al., 2012a). Decreased area in midsagittal (Astley et al., 2009b) and splenium (Sowell et al., 2001a)	Posterior–anterior displacement is associated with verbal learning and recall (Sowell et al., 2001a). Thickness associated with executive function (Bookstein et al., 2002b). Motor abilities correlate with surface area (Roebuck et al., 2002) and thinning (Bookstein et al., 2002b). IQ associated with surface area (Roebuck et al., 2002). Splenium MD values associated with WM and perceptual organization (Wozniak et al., 2009); Splenium FA values positively associated with visuomotor performance (Sowell et al., 2008a)	**fcMRI** Low interhemispheric connectivity associated with higher isthmus tract MD values and associated with the posterior mid-body tract (Wozniak et al., 2011). Low connectivity during arithmetic (Santhanam et al., 2011). Perceptual reasoning associated with interhemispheric connectivity and isthmus and midbody MD values (Wozniak et al., 2011)
Cerebellum	Reduced (Mattson et al., 1996b; Sowell et al., 1996; Archibald et al., 2001). Reduced anterior regions (Sowell et al., 1996; O'Hare et al., 2005) and anterior vermis (O'Hare et al., 2005)	Atrophy (Riikonen et al., 1999), hypoplasia (Autti-Rämö et al., 2002), and sulci distinction (Codreanu et al., 2012). Displaced anterior vermis (O'Hare et al., 2005)	Volume associated with verbal learning and memory (O'Hare et al., 2005). Abnormalities correlate with motor impairments (Roebuck et al., 1998; Roebuck-Spencer et al., 2004)	
Hippocampus	Reduced overall (Willoughby et al., 2008; Nardelli et al., 2011) across development (Willoughby et al., 2008). Reduced hippocampal commissure (Swayze et al., 1997) and parahippocampal and fusiform gyri (Coles et al., 2011)		Right volume associated with verbal and non-verbal recall (Coles et al., 2011). Left volume associated with verbal recall (Willoughby et al., 2008)	

Continued

Table 25.4

Continued

Brain area	Volumetric findings (FASD vs controls)	Structural abnormalities	Brain–behavior relations	Functional abnormalities (FASD vs behaviorally matched controls)
Basal ganglia	Reduced caudate (Riikonen et al., 2005), putamen (Mattson et al., 1996b; Astley et al., 2009b; Roussotte et al., 2011), globus pallidus (Nardelli et al., 2011), and lenticular nucleus (Mattson et al., 1996b)	Moderate exposure associated with caudate asymmetry (Willford et al., 2010)	Putamen volume positively correlated with IQ (Roussotte et al., 2011); caudate volume associated with cognitive control and verbal learning and recall (Fryer et al., 2012)	**MRI and PET** Increased striatal dopamine transporter binding (Riikonen et al., 2005) **fMRI** Greater BOLD: in caudate and putamen during non-verbal WM (Spadoni et al., 2009; Diwadkar et al., 2013; Norman et al., 2013) and RI (O'Brien et al., 2013). Lower BOLD: in basal ganglia during RI (Fryer et al., 2007b) and non-verbal WM (Spadoni et al., 2009; Roussotte et al., 2011; Diwadkar et al., 2013; Norman et al., 2013) **fcMRI** Weaker dorsal caudate and stronger posterior putamen and prefrontal connections and connectivity during spatial WM (Roussotte et al., 2012a)
Thalamus	Reduced (Mattson et al., 1992; Nardelli et al., 2011; Roussotte et al., 2012b)			**MRI and PET** Abnormal regional cerebral metabolic rates in thalamus (Clark et al., 2000) **fMRI** Greater BOLD on cue-dependent and cued trials (O'Brien et al., 2013)

DTI, diffusion tensor imaging; RI, response inhibition; WM, working memory; DMN, default mode network; MRI, magnetic resonance imaging; fMRI, functional MRI; fcMRI, functional connectivity MRI; BOLD, blood oxygen level dependent; MD, mean diffusivity; FA, fractional anisotropy; EEG, electroencephalogram; SPECT, single-photon emission computed tomography.

inconsistent. An initial case study reported increased ventricular volume (Mattson et al., 1992), while subsequent larger studies indicated that CSF reductions did not differ from controls (Archibald et al., 2001) or did not persist when supratentorial intracranial volume was controlled (Sowell et al., 2001b).

White matter

In relation to total brain volume reductions, white-matter volumes are disproportionately reduced in children with FASD (Bjorkquist et al., 2010). Alcohol-exposed children have less white matter in frontal, parietal, and temporal lobes (Archibald et al., 2001; Sowell et al., 2002a; Astley et al., 2009b) and lower white-matter density in left posterior temporoparietal cortex (Sowell et al., 2001b). Although group findings indicate that occipital regions are not especially sensitive to alcohol's teratogenicity, a case study reported decreased white matter in occipital horns and parieto-occipital regions as well (Riikonen et al., 1999).

Gray matter

Children with FASD have aberrant cortical (Nardelli et al., 2011; Roussotte et al., 2012b) and total gray-matter volume (Sowell et al., 2001b, 2002a) compared to controls. Cross-sectional studies indicate reduced total and deep gray-matter volumes (Sowell, 2001b; Nardelli et al., 2011). Despite total gray-matter reductions, children with FASD have disproportionately greater gray-matter density in specific brain regions, including temporal and parietal cortices, relative to total brain volume reductions (Sowell et al., 2001b, 2002a). In a cross-sectional study, children with FASD displayed significant volumetric reductions in deep gray-matter regions (hippocampus, thalamus, and globus pallidus) relative to controls across three age ranges. While both groups demonstrated age-related volume increases, the FASD group remained consistently smaller than the control group and no group–age interactions were noted (Nardelli et al., 2011).

A limited number of investigations have examined cortical thickness in children with FASD. Findings suggest that alcohol-exposed individuals have thicker cortices in the frontal, parietal, and temporal lobes compared to controls (Sowell et al., 2008b; Fernandez-Jaen et al., 2011). Similar findings occur in dysmorphic and non-dysmorphic children, particularly in the left inferior frontal and right superior and middle temporal cortices (Yang et al., 2012b), and results correlated with facial dysmorphology (Yang et al., 2012b) and cognitive outcomes (Sowell et al., 2008b). Increased cortical thickness may indicate that developmental synaptic pruning and myelination associated with efficient cognitive performance are disrupted in FASD (Hensch, 2004). However, not all findings support increased cortical thickness in this population, as Zhou et al. (2011) found cortical thinning in a separate cross-sectional study of children and adults with FASD and no relation to cognition.

Regional brain abnormalities

Corpus callosum

As described in early autopsy studies (e.g., Clarren et al., 1978), structural neuroimaging has indicated complete and partial agenesis of the corpus callosum in FASD (Mattson et al., 1992; Riley et al., 1995). As shown in Figure 25.3, children and adults with FASD exhibit structural abnormalities of the corpus callosum, including decreased volume, length, and area, displacement and increased shape variability (Swayze et al., 1997; Autti-Rämö et al., 2002; Bookstein et al., 2002a, b; Yang et al., 2012a). The splenium and midline callosal regions (Swayze et al., 1997; Yang et al., 2012a) may be especially sensitive to alcohol's teratogenicity as these regions exhibit the greatest abnormalities following prenatal alcohol exposure. Callosal thickness is associated with facial dysmorphology in children with FAS after accounting for differences in white-matter volume (Yang et al., 2012a). Callosal abnormalities are also associated with behavioral outcomes, mainly between motor abilities and callosal area reduction (Roebuck et al., 2002) and thinning (Bookstein et al., 2002b), verbal learning and displacement (Sowell et al., 2001a), and executive functioning and thickness (Bookstein et al., 2002b).

In addition to volumetric white-matter abnormalities in FASD, recent MRI advances, including diffusion tensor imaging (DTI), have furthered the understanding of how alcohol's teratogenicity disrupts white-matter integrity. Generally, lower functional anisotropy (FA) and higher mean diffusivity (MD) represent abnormal brain development, as these measures suggest less efficient and more disorganized white matter, respectively. DTI findings in FASD support callosal abnormalities, particularly in the splenium in adults (Li et al., 2009) and children (Sowell et al., 2008a), that may negatively impact cognitive functioning (Sowell et al., 2008a). Children with FASD have lower FA in the genu, isthmus, splenium (Wozniak et al., 2009), splenium tract (Lebel et al., 2008), and body of the corpus callosum (Fryer et al., 2009) compared to controls. Higher MD is also reported in the isthmus (Li et al., 2009), genu, and splenium (Ma et al., 2005) regions of the corpus callosum in FASD. Adults with FASD have higher radial diffusivity (RD) in the isthmus (Li et al., 2009).

In children with FASD, lower FA values are associated with lower visuomotor scores (Sowell et al., 2008a) and higher splenium MD values are associated with lower

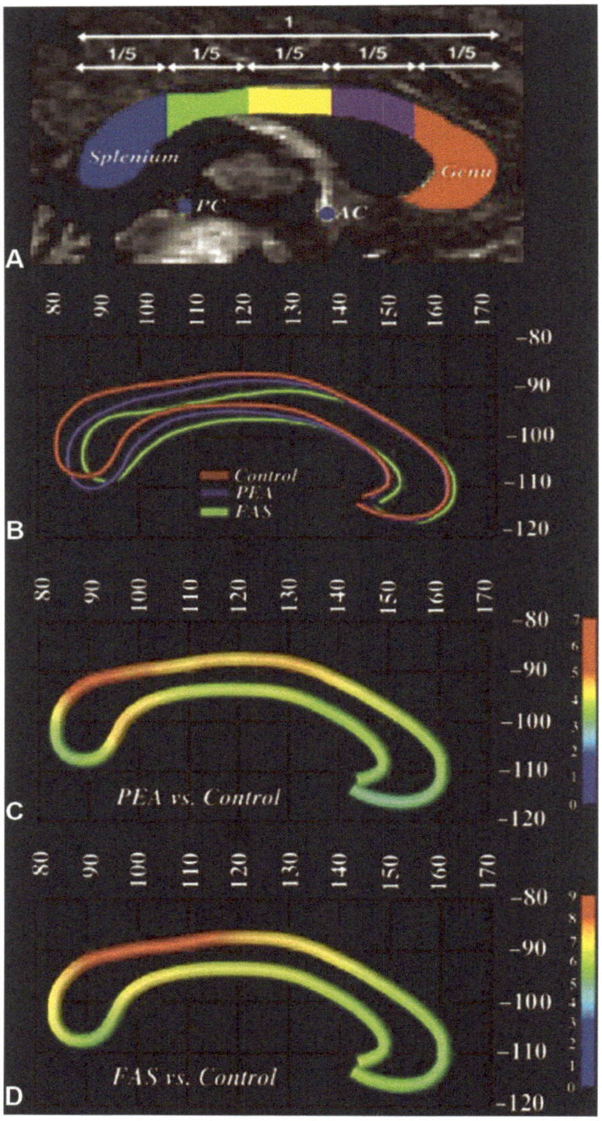

Fig. 25.3. Displacement of the corpus callosum seen in alcohol-exposed subjects. (**A**) The corpus callosum divided into five equal lengths along the horizontal axis. Splenium = blue, isthmus = green, posterior midbody = yellow, anterior midbody = purple, and the genu = red, superimposed over the gray scale midsagittal slice. (**B**) Callosal distinctions, yet similar patterns of displacement are seen between fetal alcohol syndrome (FAS), non-dysmorphic individuals with prenatal exposure to alcohol (PEA), and control groups using average callosal lines in ICBM-305 standard space. Map (in ICBM-305 space) shows average displacement vectors (mm) for PEA vs controls (**C**) and FAS vs controls (**D**). Displacement of the corpus callosum for the PEA group falls between the control and FAS groups. (Reproduced from Nuñez et al., 2011.)

working memory and perceptual organization scores (Wozniak et al., 2009). Not all studies, however, have reported significant relations between white matter and behavior (Lebel et al., 2008; Li et al., 2009). White-matter anomalies (lower FA) in the corpus callosum (Fryer et al., 2009) correlate with craniofacial dysmorphology associated with FAS in children (Wozniak et al., 2006), but not in adults (Li et al., 2009).

Although the literature generally reports lower FA and higher MD in this population, some findings have indicated the opposite pattern (higher FA and lower MD values). Examples include lower MD in the genu-related tract (Lebel et al., 2008) and higher FA in several regions, including the right cingulum and right posterior limb of the internal capsule (Fryer et al., 2009). Though typically interpreted to reflect myelination, DTI may also be sensitive to other properties of axonal integrity, including the organization and alignment of axon and fiber tracts within white matter, which could explain the conflicting reports of FA and MD values within the FASD literature. Higher than average FA and lower than average MD may also indicate damaged white-matter fibers or cortical reorganization (Jacobsen et al., 2007). Additional discrepancies in findings may result from differences in DTI methodology. Specifically, studies have utilized voxel-based morphometric (e.g., Sowell et al., 2008a), semiautomated tractography (e.g., Wozniak et al., 2009), and tensor-based spatial statistics (TBSS; Fryer et al., 2009). While some of these methods examine voxel-wise group differences, other techniques, such as TBSS, examines tract-based group differences. Taken together, the DTI results support dysfunction of synchrony as well as myelin-related insult in children with FASD. The presence of aberrant white-matter microstructure may result in clinically significant altered neural functioning, connectivity, and inefficient patterning during cognitive activity and may contribute to neuropsychologic deficits in this population.

Cerebellum

The cerebellar vault is reduced in alcohol-exposed individuals with FAS (Mattson et al., 1996b). While some studies have found disproportionate cerebellar reductions (O'Hare et al., 2005), others (Archibald et al., 2001; Astley et al., 2009b) did not find differences after total brain volume was taken into account. Within the cerebellum, the anterior regions, particularly the anterior vermis, may be most affected, whereas the later-developing posterior regions are relatively spared in FASD (Sowell et al., 1996). Other studies have reported cerebellar atrophy (Riikonen et al., 1999), hypoplasia (Autti-Rämö et al., 2002), and increased cerebellar sulci distinction (Codreanu et al., 2012) in children with FASD. While prenatal alcohol exposure levels correlate with cerebellar reduction in this population (Chen et al., 2012), there is also a combined effect of alcohol and nicotine exposure on cerebellar volume in children (de Zeeuw et al., 2012), which confounds this finding. Cerebellar volumetric reductions in FASD correlate with verbal learning and memory (O'Hare et al., 2005), and attention (Mattson et al., 2006), and may contribute to impaired balance and bimanual coordination observed in this population (Roebuck-Spencer et al., 2004).

Heavy prenatal alcohol exposure is associated with disruption of white matter in the left cerebellum (Lebel et al., 2010) and left middle cerebellar peduncle and increases perpendicular diffusivity in left middle cerebellar peduncle compared to controls, which may represent poorer axon-packing density (Spottiswoode et al., 2011). Cerebellar white-matter microstructure may contribute to cognitive abilities in children with FASD. FA values in cerebellar white matter are associated with mathematic scores (Lebel et al., 2010) and are qualitatively associated with an eye-blink conditioning task in alcohol-exposed children (Spottiswoode et al., 2011). White-matter anomalies (lower FA) in the cerebellum (Spottiswoode et al., 2011) also correlate with craniofacial dysmorphology associated with FAS.

Hippocampus

Hippocampal volume is disproportionately reduced in individuals with prenatal alcohol exposure relative to total brain volume (Willoughby et al., 2008; Nardelli et al., 2011), although this has not been replicated in all studies (e.g., Archibald et al., 2001; Roussotte et al., 2012b). In addition, a cross-sectional study indicated that children with FASD do not show normal age-related increases in the volume of this structure (Willoughby et al., 2008). Hippocampal reductions may be individually lateralized in this population, as some children exhibit unilateral hemispheric reductions (Autti-Rämö et al., 2002) and a group-level study reported disproportionate reduction in left hippocampal volume only (Willoughby et al., 2008). In addition, some hippocampal subregions are relatively more affected by alcohol's teratogenicity, including the hippocampal commissure (Swayze et al., 1997) and parahippocampal and fusiform gyri (Coles et al., 2011). Hippocampal volumetric reductions are associated with verbal functioning, verbal learning, verbal recall, and non-verbal memory in some (O'Hare et al., 2005; Willoughby et al., 2008; Coles et al., 2011), but not all, studies (Autti-Rämö et al., 2002).

Basal ganglia

The basal ganglia are especially sensitive to alcohol's teratogenicity. Specific basal ganglia nuclei, particularly the caudate (Mattson et al., 1996b), putamen (Roussotte et al., 2012b), and globus pallidus (Nardelli et al., 2011), are disproportionately reduced in children with FASD compared to controls. The lenticular nucleus, comprising the putamen and globus pallidus, is smaller in children with FASD, although not in all studies after accounting for total brain volume (Mattson et al., 1996b; Archibald et al., 2001).

In general, studies report percent reductions between 12% and 17% for the caudate and 11–13% for the putamen relative to total brain volume in alcohol-exposed children compared to non-exposed controls (Mattson et al., 1996b; Roussotte et al., 2012b), with children with FAS showing the greatest reduction in the caudate (e.g., Mattson et al., 1996b). Abnormalities in this region relate to alcohol exposure levels (Cortese et al., 2006), although not in all studies (Chen et al., 2012). Basal ganglia alterations also correlate with cognition, learning, memory, and cognitive control, even after variance associated with IQ, age, intracranial volume, and FAS diagnosis are controlled (Fryer et al., 2012; Roussotte et al., 2012b).

Additional subcortical structures

Heavy prenatal alcohol exposure also affects other subcortical structures, although fewer studies exist. Structures of the diencephalon, including the thalamus, are proportionately reduced (Mattson et al., 1996b; Roussotte et al., 2012b). Similarly, the amygdala may not be particularly sensitive to prenatal alcohol exposure, as volumetric reductions are proportional to total brain volume reductions (Roussotte et al., 2012b).

Functional brain abnormalities

Electrophysiologic studies

Advances in functional studies of the brain, including those involving EEG, have allowed researchers to investigate the effects of prenatal alcohol exposure on brain function. One study using EEG and MRI found moderate abnormalities at rest in two adolescents with FAS (Mattson et al., 1992). Task-related findings suggest that youth with FASD exhibit aberrant ERP latencies during visual and auditory response inhibition (Burden et al., 2009, 2011; Steinmann et al., 2011). When matched to controls on performance, children with FASD exhibit slower P2 latency and smaller N2 amplitudes (Burden et al., 2009), which suggests impaired discrimination between Go and inhibitory (No-Go) conditions and less inhibitory control in both domains (Steinmann et al., 2011). These studies found positive associations between ERP abnormalities and prenatal alcohol exposure levels; greater exposure levels were correlated with more aberrant neural functioning. Additionally, P3 latencies distinguished non-exposed adolescents with ADHD from adolescents with FASD (with and without FAS), indicating distinct etiologies of cognitive impairment and attention deficits in these clinical populations (Burden et al., 2010). Overall, EEG findings support increased cognitive effort, developmental delay, and attention impairment following prenatal alcohol exposure.

Functional magnetic resonance imaging

Functional MRI (fMRI) studies have examined cognitive domains, including attention, memory, and executive control, in children with histories of heavy prenatal exposure using blood oxygen level-dependent (BOLD) response.

Attention

Although attention has been relatively well studied neuropsychologically, only one study has examined attention during fMRI. During a sustained visual attention task, adults with prenatal alcohol exposure showed reduced volume and activation in ventral occipital-temporal regions (Li et al., 2008). These results indicate that regions known to be associated with sustained visual attention are sensitive to prenatal alcohol exposure and that damage to these regions may explain alcohol-related attention deficits.

Verbal learning

During verbal learning tasks, youth with FASD exhibit less BOLD response in left superior temporal regions and greater left middle frontal gyrus response relative to controls (Sowell et al., 2007). As shown in Figure 25.4, when matched on IQ, alcohol-exposed children show greater BOLD response in middle, inferior, and superior frontal regions, but lower BOLD response in left superior temporal gyrus relative to controls. This pattern of results supports frontal compensation for inefficient verbal learning systems.

During verbal working memory performance, youth with FASD show increased BOLD response in left inferior parietal, bilateral posterior temporal, and left dorsal frontal regions (O'Hare et al., 2009) (Fig. 25.4). As in the verbal learning study, these results suggest altered neural recruitment in FASD. In this case, a larger neural network may be compensating for inefficient frontal-parietal functioning during verbal working memory. Results in this domain may relate to facial dysmorphology; in one study of verbal working memory, different patterns of activation were observed in controls, heavily exposed children without facial dysmorphology, and children with facial dysmorphology consistent with FAS. In this study, control subjects activated Broca's area, non-dysmorphic subjects activated frontostriatal regions, and dysmorphic subjects activated parietal and cereballar regions (Diwadkar et al., 2013).

Spatial working memory

Spatial working memory is perhaps the most commonly studied domain in fMRI studies of FASD. During spatial working memory tasks, alcohol-exposed youth exhibit abnormal BOLD response in prefrontal, parietal, and subcortical regions compared to controls (Malisza et al., 2005; Spadoni et al., 2009; Roussotte et al., 2011; Diwadkar et al., 2013). However, results are mixed. Hyperactivation in some frontal and striatal regions has been reported in three studies (Malisza et al., 2005, 2012; Spadoni et al., 2009), while one study reported hypoactivation (Roussotte et al., 2011) of frontal and striatial regions, although differences from controls were not statistically significant. Reductions in bilateral temporal activation during spatial working memory performance have also been reported (Malisza et al., 2012). Although

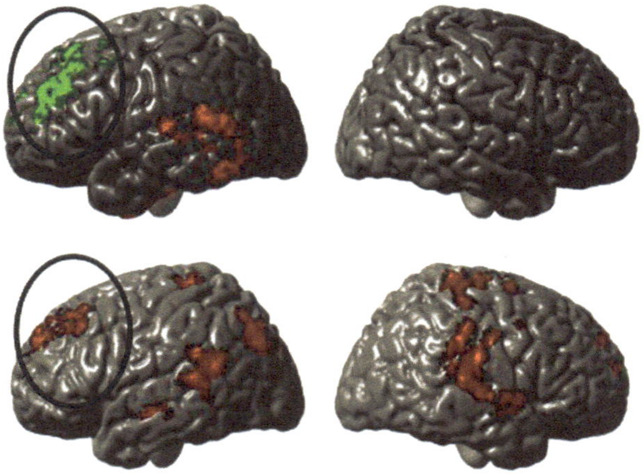

Fig. 25.4. Brain activation maps showing differences between alcohol-exposed subjects and controls. (**A**) Brain activation differences by group (alcohol-exposed vs entire control group) during a verbal learning task. Colored Z values (green indicates regions showing significant differences, with alcohol-exposed > control; red indicates regions showing significant differences, with control > alcohol-exposed) are displayed for voxel thresholds of Z > 1.7 and a (corrected) cluster significance threshold of $p = 0.05$. A significant increase in activation is seen in the left frontal lobes (in green inside the black oval) in the alcohol-exposed group relative to controls. (**B**) Brain activation differences by group (alcohol-exposed and entire control group) during a verbal working memory task with IQ differences statistically controlled. Regions of significant (Z > 1.7 and a corrected cluster significance threshold of $p = 0.05$) activation for verbal working memory vs rest contrast are in color. Red indicates regions showing significant differences with alcohol-exposed > control. There is anatomic similarity of increased activation in dorsal frontal lobes (green in **A** and red in **B**) for both functional magnetic resonance imaging tasks. (Reproduced from Nuñez et al., 2011.)

correlations between task performance and activation patterns can be informative, in the two spatial working memory studies that examined these correlations, none were significant in the alcohol-exposed groups (Spadoni et al., 2009; Roussotte et al., 2011). Discrepancies in activation findings in these studies may result from task differences or differences in subject characteristics. Differences between dysmorphic and nondysmorphic subjects were not evaluated, likely due to small sample sizes or absence of subjects with FAS (Malisza et al., 2012).

Activation patterns during spatial working memory also distinguish children with FASD from other clinical populations, including those with prenatal alcohol and methamphetamine exposure, positive family history of substance use disorders, and ADHD. Relative to children with prenatal alcohol and methamphetamine exposure, children with prenatal alcohol exposure show decreased activation in the anterior cingulate, right frontal regions, right basal ganglia, right insula, and left precuneus (Roussotte et al., 2011).

The specificity of the effects of prenatal alcohol exposure was also tested in a comparison with non-exposed children with a positive familial history of substance use disorders and controls. Activation in the left middle and superior frontal region during a spatial working memory task was specific to the effects of prenatal alcohol exposure while other differences in the insula and lentiform were more likely related to family history of alcoholism (Norman et al., 2013). Finally, a recent study compared children with prenatal alcohol exposure to non-exposed children with ADHD and controls during spatial working memory (Malisza et al., 2012) and supported the specificity of alcohol-related findings. Relative to baseline, the alcohol-exposed group displayed greater activation than the ADHD group in the middle and posterior cingulate, inferior parietal, and precuneus, although activation in the alcohol-exposed group was similar to controls.

Memory for faces likely represents a specific aspect of memory, as described above. In a 2-back working memory task using faces as stimuli, activation relative to baseline differed between youth with FAS or partial FAS and controls in right posterior parietal, right dorsolateral prefrontal, and right middle frontal regions (Astley et al., 2009a). Activation in these regions was moderately but significantly negatively correlated with errors on the 2-back task.

Response inhibition

The neural basis of response inhibition has been assessed using fMRI in youth with prenatal alcohol exposure. In two studies examining response inhibition using a visual Go/No-Go paradigm, children with FASD had greater BOLD response in frontal and parietal regions when matched to controls on behavioral performance (Fryer et al., 2007b; O'Brien et al., 2013), suggesting possible neural compensation for ineffiecient processing in these regions or, alternatively, immature inhibitory processing. Results of subcortical activation during response inhibition differed between studies, however, with one study reporting less activation in the right caudate nucleus (Fryer et al., 2007b) and the other finding greater activation in the caudate and thalamus (O'Brien et al., 2013). Further examination of response inhibition focused on cue-dependent responding during inhibitory control as activation patterns differ when response inhibition trials are preceded by predictive cues (Arce et al., 2006; Leland et al., 2008). In this study, alcohol-exposed children were less accurate and more conservative in their responses and showed less activation in the pre- and postcentral gyrus than controls, suggesting that they did not benefit as much from the implicit cues (O'Brien et al., 2013).

Number processing

Two studies have examined number processing using fMRI and both found activation differences between alcohol-exposed subjects and controls. Santhanam et al. (2009) reported reduced activation in alcohol-exposed adults in the left superior and right inferior parietal regions and medial frontal gyrus, which are associated with arithmetic (Dehaene et al., 2003, 2004). Reductions in relation to controls were observed in both dysmorphic and non-dysmorphic subjects, although the differences were significant only for the dysmorphic group. Task performance was also reduced in the alcohol-exposed subjects, although math ability was not associated with BOLD activation (Santhanam et al., 2009). In children with FAS and partial FAS, a more diffuse pattern of activation was noted during two different numerical processing tasks (Meintjes et al., 2010). While controls activated frontal and parietal regions, as in the previous study, the children with FAS and partial FAS activated more diffuse areas, including the cerebellar vermis, again suggesting a more immature pattern of response.

Default mode network

The default mode network (DMN), which represents areas that are more active during times of rest compared to times of cognitive activity (e.g., the medial prefrontal cortex, posterior cingulate cortex, precuneus, inferior parietal lobules, and medial temporal regions), is aberrant in FASD. Santhanam and colleagues (2011) found that, in comparison to controls, dysmorphic alcohol-exposed adults had significantly less deactivation of the DMN in the medial prefrontal cortex during arithmetic tasks relative to a baseline task. Interestingly, the degree of DMN deactivation during the task was correlated with task accuracy in the posterior cingulate and medial prefrontal cortices in dysmorphic and non-dysmorphic adults with prenatal alcohol exposure, suggesting that attention modulation is related to cognitive performance in this population.

Functional connectivity

Three studies have used functional connectivity MRI (fcMRI) techniques in subjects with histories of prenatal alcohol exposure. Wozniak et al. (2011) examined interhemispheric functional connectivity in seven white-matter tracts in children with FASD and found reduced connectivity in medial parietal regions connected by posterior callosal fiber projections. The degree of functional disruption was correlated with DTI mean diffusivity in controls but not FASD. Although formal comparisons could not be conducted due to small sample size, connectivity abnormalities did not appear to be related to degree of facial dysmorphology. In a separate study, both reduced and increased connectivity were found in children with prenatal alcohol exposure during a working memory task (Roussotte et al., 2012a). Connectivity with frontal regions was increased in the putamen and decreased in the caudate and both regions showed weaker connections and with fewer brain regions in alcohol-exposed children relative to controls. In adults, functional connectivity is reduced in regions associated with the DMN, as described above (Santhanam et al., 2011). Overall findings suggest neural reorganization that could relate to impaired cognition in individuals with FASD.

Metabolic neuroimaging studies

Aberrant neural functioning in children with FASD is also supported by metabolic studies. Single-photon emission computed tomography (SPECT) studies indicate cerebral blood perfusion and metabolic rate abnormalities. Blood perfusion differences exist in the temporal lobe (Bhatara et al., 2002) and parieto-occipital and prefrontal regions (Riikonen et al., 1999). Positron emission tomography and SPECT studies in alcohol-exposed children found group-level abnormalities in regional cerebral metabolic rates in frontal-striatal regions, including the thalamus, basal ganglia, and medial frontal cortex (Clark et al., 2000; Riikonen et al., 2005). Group-level findings are supported by a pediatric case study, which found decreased activity in the basal ganglia, thalamus, and temporal lobes using SPECT (Codreanu et al., 2012).

CONCLUSION

The effects of prenatal alcohol exposure are significant, pervasive, and lifelong. Cognitive and behavioral deficits have been documented in most domains studied.

Additional research is needed to further refine the FASD phenotype, determine the specificity and underlying mechanisms of observed behavioral and brain changes, examine developmental trajectories, and identify effective and targeted treatments for affected individuals.

ACKNOWLEDGMENTS

Preparation of this chapter was supported in part by National Institute on Alcohol Abuse and Alcoholism grants U01 AA014834, R01 AA019605, and T32 AA013525. We gratefully acknowledge the assistance and support of the Center for Behavioral Teratology, San Diego State University.

REFERENCES

Adnams CM, Kodituwakku PW, Hay A et al. (2001). Patterns of cognitive-motor development in children with fetal alcohol syndrome from a community in South Africa. Alcohol Clin Exp Res 25: 557–562.

Adnams CM, Sorour P, Kalberg WO et al. (2007). Language and literacy outcomes from a pilot intervention study for children with fetal alcohol spectrum disorders in South Africa. Alcohol 41: 403–414.

Alati R, Clavarino A, Najman JM et al. (2008). The developmental origin of adolescent alcohol use: findings from the Mater University Study of Pregnancy and its outcomes. Drug Alcohol Depend 98: 136–143.

Anderson P (2002). Assessment and development of executive function (EF) during childhood. Child Neuropsychol 8: 71–82.

Aragón AS, Coriale G, Fiorentino D et al. (2008a). Neuropsychological characteristics of Italian children with fetal alcohol spectrum disorders. Alcohol Clin Exp Res 32: 1909–1919.

Aragón AS, Kalberg WO, Buckley D et al. (2008b). Neuropsychological study of FASD in a sample of American Indian children: processing simple versus complex information. Alcohol Clin Exp Res 32: 2136–2148.

Arce E, Leland DS, Miller DA et al. (2006). Individuals with schizophrenia present hypo- and hyperactivation during implicit cueing in an inhibitory task. Neuroimage 32: 704–713.

Archibald SL, Fennema-Notestine C, Gamst A et al. (2001). Brain dysmorphology in individuals with severe prenatal alcohol exposure. Dev Med Child Neurol 43: 148–154.

Aronson M, Hagberg B (1998). Neuropsychological disorders in children exposed to alcohol during pregnancy: a follow-up study of 24 children to alcoholic mothers in Goteborg, Sweden. Alcohol Clin Exp Res 22: 321–324.

Astley SJ, Clarren SK (2000). Diagnosing the full spectrum of fetal alcohol-exposed individuals: introducing the 4-digit diagnostic code. Alcohol Alcohol 35: 400–410.

Astley SJ, Aylward EH, Carmichael Olson H et al. (2009a). Functional magnetic resonance imaging outcomes from a comprehensive magnetic resonance study of children with fetal alcohol spectrum disorders. J Neurodev Disord 1: 61–80.

Astley SJ, Aylward EH, Carmichael Olson H et al. (2009b). Magnetic resonance imaging outcomes from a comprehensive magnetic resonance study of children with fetal alcohol spectrum disorders. Alcohol Clin Exp Res 33: 1671–1689.

Autti-Rämö I, Autti T, Korkman M et al. (2002). MRI findings in children with school problems who had been exposed prenatally to alcohol. Dev Med Child Neurol 44: 98–106.

Badawy RA, Jackson GD (2012). Temporal lobe epilepsy and the hippocampus. In: T Bartsch (Ed.), The Clinical Neurobiology of the Hippocampus: an Integrative View, Oxford University Press, Oxford, pp. 245–256.

Baer JS, Sampson PD, Barr HM et al. (2003). A 21-year longitudinal analysis of the effects of prenatal alcohol exposure on young adult drinking. Arch Gen Psychiatry 60: 377–385.

Baldwin MR (2007). Fetal alcohol spectrum disorders and suicidality in a healthcare setting. Int J Circumpolar Health 66: 54–60.

Barr HM, Streissguth AP, Darby BL et al. (1990). Prenatal exposure to alcohol, caffeine, tobacco, and aspirin: effects on fine and gross motor performance in 4-year-old children. Dev Psychol 26: 339–348.

Becker M, Warr-Leeper GA, Leeper HA (1990). Fetal alcohol syndrome: a description of oral motor, articulatory, short-term memory, grammatical, and semantic abilities. J Commun Disord 23: 97–124.

Bell SH, Stade B, Reynolds JN et al. (2010). The remarkably high prevalence of epilepsy and seizure history in fetal alcohol spectrum disorders. Alcohol Clin Exp Res 34: 1084–1089.

Bertrand J, Floyd RL, Weber MK et al. (2004). National Task Force on FAS/FAE: guidelines for Referral and Diagnosis. Centers for Disease Control and Prevention, Atlanta, GA.

Bhatara VS, Lovrein F, Kirkeby J et al. (2002). Brain function in fetal alcohol syndrome assessed by single photon emission computed tomography. S D J Med 55: 59–62.

Bjorkquist OA, Fryer SL, Reiss AL et al. (2010). Cingulate gyrus morphology in children and adolescents with fetal alcohol spectrum disorders. Psychiatry Res Neuroimaging 181: 101–107.

Blakemore S-J, Choudhury S (2006). Development of the adolescent brain: implications for executive function and social cognition. J Child Psychol Psychiatry 47: 296–312.

Bonthius DJ, Woodhouse J, Bonthius NE et al. (2001). Reduced seizure threshold and hippocampal cell loss in rats exposed to alcohol during the brain growth spurt. Alcohol Clin Exp Res 25: 70–82.

Bookstein FL, Sampson PD, Connor PD et al. (2002a). Midline corpus callosum is a neuroanatomical focus of fetal alcohol damage. Anat Rec 269: 162–174.

Bookstein FL, Streissguth AP, Sampson PD et al. (2002b). Corpus callosum shape and neuropsychological deficits in adult males with heavy fetal alcohol exposure. Neuroimage 15: 233–251.

Brown RT, Coles CD, Smith IE et al. (1991). Effects of prenatal alcohol exposure at school age. II. Attention and behavior. Neurotoxicol Teratol 13: 369–376.

Burd L, Klug MG, Martsolf JT et al. (2003). Fetal alcohol syndrome: neuropsychiatric phenomics. Neurotoxicol Teratol 25: 697–705.

Burden MJ, Jacobson SW, Jacobson JL (2005a). Relation of prenatal alcohol exposure to cognitive processing speed and efficiency in childhood. Alcohol Clin Exp Res 29: 1473–1483.

Burden MJ, Jacobson SW, Sokol RJ et al. (2005b). Effects of prenatal alcohol exposure on attention and working memory at 7.5 years of age. Alcohol Clin Exp Res 29: 443–452.

Burden MJ, Andrew C, Saint-Amour D et al. (2009). The effects of fetal alcohol syndrome on response execution and inhibition: an event-related potential study. Alcohol Clin Exp Res 33: 1994–2004.

Burden MJ, Jacobson JL, Westerlund A et al. (2010). An event-related potential study of response inhibition in ADHD with and without prenatal alcohol exposure. Alcohol Clin Exp Res 34: 617–627.

Burden MJ, Westerlund A, Muckle G et al. (2011). The effects of maternal binge drinking during pregnancy on neural correlates of response inhibition and memory in childhood. Alcohol Clin Exp Res 35: 69–82.

Carmichael Olson H, Feldman JJ, Streissguth AP et al. (1998). Neuropsychological deficits in adolescents with fetal alcohol syndrome: clinical findings. Alcohol Clin Exp Res 22: 1998–2012.

Carr JL, Agnihotri S, Keightley M (2010). Sensory processing and adaptive behavior deficits of children across the fetal alcohol spectrum disorder continuum. Alcohol Clin Exp Res 34: 1022–1032.

Chen CP, Kuhn P, Advis JP et al. (2006). Prenatal ethanol exposure alters the expression of period genes governing the circadian function of beta-endorphin neurons in the hypothalamus. J Neurochem 97: 1026–1033.

Chen X, Coles CD, Lynch ME et al. (2012). Understanding specific effects of prenatal alcohol exposure on brain structure in young adults. Hum Brain Mapp 33: 1663–1676.

Chernick V, Childiaeva R, Ioffe S (1983). Effects of maternal alcohol intake and smoking on neonatal electroencephalogram and anthropometric measurements. Am J Obstet Gynecol 146: 41–47.

Chiodo LM, Janisse J, Delaney-Black V et al. (2009). A metric of maternal prenatal risk drinking predicts neurobehavioral outcomes in preschool children. Alcohol Clin Exp Res 33: 634–644.

Chudley AE, Conry J, Cook JL et al. (2005). Fetal alcohol spectrum disorder: Canadian guidelines for diagnosis. Can Med Assoc J 172: S1–S21.

Church MW, Gerkin KP (1988). Hearing disorders in children with fetal alcohol syndrome: findings from case reports. Pediatrics 82: 147–154.

Church MW, Kaltenbach JA (1997). Hearing, speech, language, and vestibular disorders in the fetal alcohol syndrome: a literature review. Alcohol Clin Exp Res 21: 495–512.

Clark CM, Li D, Conry J et al. (2000). Structural and functional brain integrity of fetal alcohol syndrome in nonretarded cases. Pediatrics 105: 1096–1099.

Clarren SK, Smith DW (1978). The fetal alcohol syndrome. N Engl J Med 298: 1063–1067.

Clarren SK, Alvord Jr EC, Sumi SM et al. (1978). Brain malformations related to prenatal exposure to ethanol. J Pediatr 92: 64–67.

Clarren SK, Chudley AE, Wong L et al. (2010). Normal distribution of palpebral fissure lengths in Canadian school age children. Can J Clin Pharmacol 17: e67–e78.

Codreanu I, Yang J, Zhuang H (2012). Brain single-photon emission computed tomography in fetal alcohol syndrome: a case report and study implications. J Child Neurol 27: 1580–1584.

Coffman BA, Kodituwakku P, Kodituwakku EL et al. (2013). Primary visual response (M100) delays in adolescents with FASD as measured with MEG. Hum Brain mapping 34 (11): 2852–2862.

Coggins TE, Olswang LB, Carmichael Olson H et al. (2003). On becoming socially competent communicators: the challenge for children with fetal alcohol exposure. Int J Lang Commun Disord 27: 121–150.

Coggins TE, Timler GR, Olswang LB (2007). A state of double jeopardy: impact of prenatal alcohol exposure and adverse environments on the social communicative abilities of school-age children with fetal alcohol spectrum disorder. Lang Speech Hear Serv Sch 38: 117–127.

Cohen-Kerem R, Bar-Oz B, Nulman I et al. (2007). Hearing in children with fetal alcohol spectrum disorder (FASD). Can J Clin Pharmacol 14: e307–e312.

Coles CD, Platzman KA, Lynch ME et al. (2002). Auditory and visual sustained attention in adolescents prenatally exposed to alcohol. Alcohol Clin Exp Res 26: 263–271.

Coles CD, Kable JA, Taddeo E (2009). Math performance and behavior problems in children affected by prenatal alcohol exposure: intervention and follow-up. J Dev Behav Pediatr 30: 7–15.

Coles CD, Goldstein FC, Lynch ME et al. (2011). Memory and brain volume in adults prenatally exposed to alcohol. Brain Cogn 75: 67–77.

Connor PD, Streissguth AP, Sampson PD et al. (1999). Individual differences in auditory and visual attention among fetal alcohol-affected adults. Alcohol Clin Exp Res 23: 1395–1402.

Connor PD, Sampson PD, Bookstein FL et al. (2000). Direct and indirect effects of prenatal alcohol damage on executive function. Dev Neuropsychol 18: 331–354.

Connor PD, Sampson PD, Streissguth AP et al. (2006). Effects of prenatal alcohol exposure on fine motor coordination and balance: a study of two adult samples. Neuropsychologia 44: 744–751.

Conry J (1990). Neuropsychological deficits in fetal alcohol syndrome and fetal alcohol effects. Alcohol Clin Exp Res 14: 650–655.

Cortese BM, Moore GJ, Bailey BA et al. (2006). Magnetic resonance and spectroscopic imaging in prenatal alcohol-exposed children: preliminary findings in the caudate nucleus. Neurotoxicol Teratol 28: 597–606.

Crocker N, Vaurio L, Riley EP et al. (2009). Comparison of adaptive behavior in children with heavy prenatal alcohol exposure or attention-deficit/hyperactivity disorder. Alcohol Clin Exp Res 33: 2015–2023.

Crocker N, Vaurio L, Riley EP et al. (2011). Comparison of verbal learning and memory in children with heavy prenatal alcohol exposure or attention deficit/hyperactivity disorder. Alcohol Clin Exp Res 35: 1114–1121.

Crocker N, Deweese BN, Dudley JD et al. (2012). Attention networks in children with heavy prenatal alcohol exposure: neurobehavioral performance. Alcohol Clin Exp Res 36 (Suppl. S1): 206A.

de Zeeuw P, Zwart F, Schrama R et al. (2012). Prenatal exposure to cigarette smoke or alcohol and cerebellum volume in attention-deficit/hyperactivity disorder and typical development. Translat Psychiatry 2: e84.

Dehaene S, Piazza M, Pinel P, Cohen L (2003). Three parietal circuits for number processing. Cognit Neuropsychol 20: 487–506.

Dehaene S, Molko N, Cohen L et al. (2004). Arithmetic and the brain. Curr Opin Neurobiol 14: 218–224.

Disney ER, Iacono W, McGue M et al. (2008). Strengthening the case: prenatal alcohol exposure is associated with increased risk for conduct disorder. Pediatrics 122: e1225–e1230.

Diwadkar VA, Meintjes EM, Goradia D et al. (2013). Differences in cortico-striatal-cerebellar activation during working memory in syndromal and nonsyndromal children with prenatal alcohol exposure. Hum Brain Mapping 34 (8): 1931–1945.

Dömellof E, Fagard J, Jacquet AY et al. (2010). Goal-directed arm movements in children with fetal alcohol syndrome: a kinematic approach. Eur J Neurol 18: 312–320.

Duquette C, Stodel E, Fullarton S et al. (2006). Persistence in high school: experiences of adolescents and young adults with fetal alcohol spectrum disorder. J Intellect Dev Disabil 31: 219–231.

Fagerlund Å, Autti-Rämö I, Kalland M et al. (2012). Adaptive behaviour in children and adolescents with foetal alcohol spectrum disorders: a comparison with specific learning disability and typical development. Eur Child Adolesc Psychiatry 21: 221–231.

Famy C, Streissguth AP, Unis AS (1998). Mental illness in adults with fetal alcohol syndrome or fetal alcohol effects. Am J Psychiatry 155: 552–554.

Fast DK, Conry J, Loock CA (1999). Identifying fetal alcohol syndrome among youth in the criminal justice system. J Dev Behav Pediatr 20: 370–372.

Fernandez-Jaen A, Fernandez-Mayoralas DM, Quinones Tapia D et al. (2011). Cortical thickness in fetal alcohol syndrome and attention deficit disorder. Pediatr Neurol 45: 387–391.

Franklin L, Deitz J, Jirikowic T et al. (2008). Children with fetal alcohol spectrum disorders: problem behaviors and sensory processing. Am J Occup Ther 62: 265–273.

Fried PA, Watkinson B (1988). 12- and 24-month neurobehavioural follow-up of children prenatally exposed to marihuana, cigarettes and alcohol. Neurotoxicol Teratol 10: 305–313.

Fried PA, Watkinson B (1990). 36- and 48-month neurobehavioral follow-up of children prenatally exposed to marijuana, cigarettes, and alcohol. J Dev Behav Pediatr 11: 49–58.

Fried PA, O'Connell CM, Watkinson B (1992). 60- and 72-month follow-up of children prenatally exposed to marijuana, cigarettes, and alcohol: cognitive and language assessment. J Dev Behav Pediatr 13: 383–391.

Fryer SL, McGee CL, Matt GE et al. (2007a). Evaluation of psychopathological conditions in children with heavy prenatal alcohol exposure. Pediatrics 119: e733–e741.

Fryer SL, Tapert SF, Mattson SN et al. (2007b). Prenatal alcohol exposure affects frontal-striatal BOLD response during inhibitory control. Alcohol Clin Exp Res 31: 1415–1424.

Fryer SL, Schweinsburg BC, Bjorkquist OA et al. (2009). Characterization of white matter microstructure in fetal alcohol spectrum disorders. Alcohol Clin Exp Res 33: 514–521.

Fryer SL, Mattson SN, Jernigan TL et al. (2012). Caudate volume predicts neurocognitive performance in youth with heavy prenatal alcohol exposure. Alcohol Clin Exp Res 36: 1932–1941.

Glass L, Ware AL, Crocker N et al. (2013). Neuropsychological deficits associated with heavy prenatal alcohol exposure are not exacerbated by ADHD. Neuropsychology 27: 713–724.

Goldschmidt L, Richardson GA, Stoffer DS et al. (1996). Prenatal alcohol exposure and academic achievement at age six: a nonlinear fit. Alcohol Clin Exp Res 20: 763–770.

Graham DG, Brusati GM, Glass L et al. (2013). Theory of mind ability in children with heavy prenatal alcohol exposure: visuospatial processing mediates influence of affect recognition. Manuscript submitted for publication.

Gray R, Mukherjee RA, Rutter M (2009). Alcohol consumption during pregnancy and its effects on neurodevelopment: what is known and what remains uncertain. Addiction 104: 1270–1273.

Green CR, Mihic AM, Brien DC et al. (2009). Oculomotor control in children with fetal alcohol spectrum disorders assessed using a mobile eye-tracking laboratory. Eur J Neurosci 29: 1302–1309.

Greenbaum RL, Stevens SA, Nash K et al. (2009). Social cognitive and emotion processing abilities of children with fetal alcohol spectrum disorders: a comparison with attention deficit hyperactivity disorder. Alcohol Clin Exp Res 33: 1656–1670.

Greene T, Ernhart CB, Martier SS et al. (1990). Prenatal alcohol exposure and language development. Alcohol Clin Exp Res 14: 937–945.

Gusella JL, Fried PA (1984). Effects of maternal social drinking and smoking on offspring at 13 months. Neurobehav Toxicol Teratol 6: 13–17.

Hall JG (2010). New palpebral fissure measurements. Am J Med Genet 152A: 1870.

Hall JG, Froster-Iskenius UG, Allanson JE (1989). Handbook of Normal Physical Measurements. Oxford University Press, New York.

Hamilton DA, Kodituwakku P, Sutherland RJ et al. (2003). Children with fetal alcohol syndrome are impaired at place learning but not cued-navigation in a virtual Morris water task. Behav Brain Res 143: 85–94.

Havlicek V, Childiaeva R, Chernick V (1977). EEG frequency spectrum characteristics of sleep states in infants of alcoholic mothers. Neuropadiatrie 8: 360–373.

Hensch TK (2004). Critical period regulation. Annu Rev Neurosci 27: 549–579.

Hoyme HE, May PA, Kalberg WO et al. (2005). A practical clinical approach to diagnosis of fetal alcohol spectrum disorders: clarification of the 1996 Institute of Medicine criteria. Pediatrics 115: 39–47.

Ioffe S, Chernick V (1988). Development of the EEG between 30 and 40 weeks gestation in normal and alcohol-exposed infants. Dev Med Child Neurol 30: 797–807.

Ioffe S, Chernick V (1990). Prediction of subsequent motor and mental retardation in newborn infants exposed to alcohol in utero by computerized EEG analysis. Neuropediatrics 21: 11–17.

Jacobsen LK, Picciotto MR, Heath CJ et al. (2007). Prenatal and adolescent exposure to tobacco smoke modulates the development of white matter microstructure. J Neurosci 27: 13491–13498.

Jacobson SW (1998). Specificity of neurobehavioral outcomes associated with prenatal alcohol exposure. Alcohol Clin Exp Res 22: 313–320.

Jacobson SW, Jacobson JL, Sokol RJ et al. (1993). Prenatal alcohol exposure and infant information processing ability. Child Dev 64: 1706–1721.

Jacobson SW, Jacobson JL, Sokol RJ (1994). Effects of fetal alcohol exposure on infant reaction time. Alcohol Clin Exp Res 18: 1125–1132.

Jacobson JL, Dodge NC, Burden MJ et al. (2011). Number processing in adolescents with prenatal alcohol exposure and ADHD: differences in the neurobehavioral phenotype. Alcohol Clin Exp Res 35: 431–442.

Jan JE, Asante KO, Conry JL et al. (2010). Sleep health issues for children with FASD: clinical considerations. Intl J Pediatrics. Published online 2010 July 14. doi: 10.1155/2010/639048 Available online at: http://downloads.hindawi.com/journals/ijped/2010/639048.pdf.

Janzen LA, Nanson JL, Block GW (1995). Neuropsychological evaluation of preschoolers with fetal alcohol syndrome. Neurotoxicol Teratol 17: 273–279.

Jirikowic T, Carmichael Olson H, Kartin D (2008a). Sensory processing, school performance, and adaptive behavior of young school-age children with fetal alcohol spectrum disorders. Phys Occup Ther Pediatr 28: 117–136.

Jirikowic T, Kartin D, Carmichael Olson H (2008b). Children with fetal alcohol spectrum disorders: a descriptive profile of adaptive function. Can J Occup Ther 75: 238–248.

Jones KL, Smith DW (1973). Recognition of the fetal alcohol syndrome in early infancy. Lancet 2: 999–1001.

Jones KL, Smith DW, Ulleland CN et al. (1973). Pattern of malformation in offspring of chronic alcoholic mothers. Lancet 1: 1267–1271.

Kable JA, Coles CD (2004). The impact of prenatal alcohol exposure on neurophysiological encoding of environmental events at six months. Alcohol Clin Exp Res 28: 489–496.

Kable JA, Coles CD, Lynch ME et al. (2009). The impact of maternal smoking on fast auditory brainstem responses. Neurotoxicol Teratol 31: 216–224.

Kaemingk KL, Halverson PT (2000). Spatial memory following prenatal alcohol exposure: more than a material specific memory deficit. Child Neuropsychol 6: 115–128.

Kaemingk KL, Mulvaney S, Halverson PT (2003). Learning following prenatal alcohol exposure: performance on verbal and visual multitrial tasks. Arch Clin Neuropsychol 18: 33–47.

Kalberg WO, Provost B, Tollison SJ et al. (2006). Comparison of motor delays in young children with fetal alcohol syndrome to those with prenatal alcohol exposure and with no prenatal alcohol exposure. Alcohol Clin Exp Res 30: 2037–2045.

Kaneko WM, Phillips EL, Riley EP et al. (1996). EEG findings in fetal alcohol syndrome and Down syndrome children. Electroencephalogr Clin Neurophysiol 98: 20–28.

Kerns KA, Don A, Mateer CA et al. (1997). Cognitive deficits in nonretarded adults with fetal alcohol syndrome. J Learn Disabil 30: 685–693.

Kodituwakku PW (2007). Defining the behavioral phenotype in children with fetal alcohol spectrum disorders: a review. Neurosci Biobehav Rev 31: 192–201.

Kodituwakku PW (2009). Neurocognitive profile in children with fetal alcohol spectrum disorders. Ment Retard Dev Disabil Res Rev 15: 218–224.

Kodituwakku PW, Handmaker NS, Cutler SK et al. (1995). Specific impairments in self-regulation in children exposed to alcohol prenatally. Alcohol Clin Exp Res 19: 1558–1564.

Kooistra L, Ramage B, Crawford S et al. (2009). Can attention deficit hyperactivity disorder and fetal alcohol spectrum disorder be differentiated by motor and balance deficits? Hum Mov Sci 28: 529–542.

Kooistra L, Crawford S, Gibbard B et al. (2011). Comparing attentional networks in fetal alcohol spectrum disorder and the inattentive and combined subtypes of attention deficit hyperactivity disorder. Dev Neuropsychol 36: 566–577.

Kopera-Frye K, Dehaene S, Streissguth AP (1996). Impairments of number processing induced by prenatal alcohol exposure. Neuropsychologia 34: 1187–1196.

Korkman M, Autti-Rämö I, Koivulehto H et al. (1998). Neuropsychological effects at early school age of fetal alcohol exposure of varying duration. Child Neuropsychol 4: 199–212.

Kully-Martens K, Denys K, Treit S et al. (2012). A review of social skills deficits in individuals with fetal alcohol spectrum disorders and prenatal alcohol exposure: profiles, mechanisms, and interventions. Alcohol Clin Exp Res 36: 568–576.

LaDue RA, Stressiguth AP, Randels SP (1992). Clinical considerations pertaining to adolescents and adults with fetal alcohol syndrome. In: TB Sonderegger (Ed.), Perinatal Substance Abuse: Research Findings and Clinical Implications. The Johns Hopkins University Press, Baltimore, pp. 104–131.

Lebel C, Rasmussen C, Wyper K et al. (2008). Brain diffusion abnormalities in children with fetal alcohol spectrum disorder. Alcohol Clin Exp Res 32: 1732–1740.

Lebel C, Rasmussen C, Wyper K et al. (2010). Brain microstructure is related to math ability in children with fetal alcohol spectrum disorder. Alcohol Clin Exp Res 34: 354–363.

Leland DS, Arce E, Miller DA et al. (2008). Anterior cingulate cortex and benefit of predictive cueing on response inhibition in stimulant dependent individuals. Biol Psychiatry 63: 184–190.

Lemoine P, Harousseau H, Borteyru J-P et al. (1968). Les enfants de parents alcooliques. Anomalies observées. A propos de 127 cas [Children of alcoholic parents. Abnormalities observed in 127 cases.]. Ouest Med 21: 476–482.

Li Z, Ma X, Peltier S et al. (2008). Occipital-temporal reduction and sustained visual attention deficit in prenatal alcohol exposed adults. Brain Imag Behav 2: 39–48.

Li L, Coles CD, Lynch ME et al. (2009). Voxelwise and skeleton-based region of interest analysis of fetal alcohol syndrome and fetal alcohol spectrum disorders in young adults. Hum Brain Mapp 30: 3265–3274.

Lynch ME, Coles CD, Corley T et al. (2003). Examining delinquency in adolescents differentially prenatally exposed to alcohol: the role of proximal and distal risk factors. J Stud Alcohol 64: 678–686.

Ma X, Coles CD, Lynch ME et al. (2005). Evaluation of corpus callosum anisotropy in young adults with fetal alcohol syndrome according to diffusion tensor imaging. Alcohol Clin Exp Res 29: 1214–1222.

Malisza KL, Allman AA, Shiloff D et al. (2005). Evaluation of spatial working memory function in children and adults with fetal alcohol spectrum disorders: a functional magnetic resonance imaging study. Pediatr Res 58: 1150–1157.

Malisza KL, Buss JL, Bolster BR et al. (2012). Comparison of spatial working memory in children with prenatal alcohol exposure and those diagnosed with ADHD; a functional magnetic resonance imaging study. J Neurodev Disord 4: 12.

Malone M, Koren G (2012). Alcohol-induced behavioural problems in fetal alcohol spectrum disorders versus confounding behavioural problems. J Population Ther Clin Pharmacol 19: e32–e40.

Manji S, Pei J, Loomes C et al. (2009). A review of the verbal and visual memory impairments in children with foetal alcohol spectrum disorders. Dev Neurorehabil 12: 239–247.

Marcus JC (1987). Neurological findings in the fetal alcohol syndrome. Neuropediatrics 18: 158–160.

Mattson SN, Riley EP (1998). A review of the neurobehavioral deficits in children with fetal alcohol syndrome or prenatal exposure to alcohol. Alcohol Clin Exp Res 22: 279–294.

Mattson SN, Riley EP (2000). Parent ratings of behavior in children with heavy prenatal alcohol exposure and IQ-matched controls. Alcohol Clin Exp Res 24: 226–231.

Mattson SN, Riley EP (2011). The quest for a neurobehavioral profile of heavy prenatal alcohol exposure. Alcohol Res Health 34: 51–55.

Mattson SN, Roebuck TM (2002). Acquisition and retention of verbal and nonverbal information in children with heavy prenatal alcohol exposure. Alcohol Clin Exp Res 26: 875–882.

Mattson SN, Riley EP, Jernigan TL et al. (1992). Fetal alcohol syndrome: a case report of neuropsychological, MRI and EEG assessment of two children. Alcohol Clin Exp Res 16: 1001–1003.

Mattson SN, Riley EP, Delis DC et al. (1996a). Verbal learning and memory in children with fetal alcohol syndrome. Alcohol Clin Exp Res 20: 810–816.

Mattson SN, Riley EP, Sowell ER et al. (1996b). A decrease in the size of the basal ganglia in children with fetal alcohol syndrome. Alcohol Clin Exp Res 20: 1088–1093.

Mattson SN, Riley EP, Gramling LJ et al. (1997). Heavy prenatal alcohol exposure with or without physical features of fetal alcohol syndrome leads to IQ deficits. J Pediatr 131: 718–721.

Mattson SN, Riley EP, Gramling LJ et al. (1998). Neuropsychological comparison of alcohol-exposed children with or without physical features of fetal alcohol syndrome. Neuropsychology 12: 146–153.

Mattson SN, Goodman AM, Caine C et al. (1999). Executive functioning in children with heavy prenatal alcohol exposure. Alcohol Clin Exp Res 23: 1808–1815.

Mattson SN, Calarco KE, Lang AR (2006). Focused and shifting attention in children with heavy prenatal alcohol exposure. Neuropsychology 20: 361–369.

Mattson SN, Roesch SC, Fagerlund Å et al. (2010). Toward a neurobehavioral profile of fetal alcohol spectrum disorders. Alcohol Clin Exp Res 34: 1640–1650.

Mattson SN, Crocker N, Nguyen TT (2011). Fetal alcohol spectrum disorders: neuropsychological and behavioral features. Neuropsychol Rev 21: 81–101.

Mattson SN, Roesch SC, Glass L et al. (2013). Further development of a neurobehavioral profile of fetal alcohol spectrum disorders. Alcohol Clin Exp Res 37: 517–528.

May PA, Gossage JP, Kalberg WO et al. (2009). Prevalence and epidemiologic characteristics of FASD from various research methods with an emphasis on recent in-school studies. Ment Retard Dev Disabil Res Rev 15: 176–192.

McGee CL, Fryer SL, Bjorkquist OA et al. (2008). Deficits in social problem solving in adolescents with prenatal exposure to alcohol. Am J Drug Alcohol Abuse 34: 423–431.

McGee CL, Bjorkquist OA, Riley EP et al. (2009). Impaired language performance in young children with heavy prenatal alcohol exposure. Neurotoxicol Teratol 31: 71–75.

Meintjes EM, Jacobson JL, Molteno CD et al. (2010). An FMRI study of number processing in children with fetal alcohol syndrome. Alcohol Clin Exp Res 34: 1450–1464.

Mukherjee R, Layton M, Yacoub E et al. (2011). Autism and autistic traits in people exposed to heavy prenatal alcohol: data from a clinical series of 21 individuals and nested case control study. Adv Ment Hth Intellect Disabil 5: 42–49.

Nanson JL, Hiscock M (1990). Attention deficits in children exposed to alcohol prenatally. Alcohol Clin Exp Res 14: 656–661.

Nardelli A, Lebel C, Rasmussen C et al. (2011). Extensive deep gray matter volume reductions in children and adolescents with fetal alcohol spectrum disorders. Alcohol Clin Exp Res 35: 1404–1417.

Nash K, Rovet J, Greenbaum R et al. (2006). Identifying the behavioural phenotype in fetal alcohol spectrum disorder: sensitivity, specificity and screening potential. Arch Womens Ment Health 9: 181–186.

Norman AL, O'Brien JW, Spadoni AD et al. (2013). A functional magnetic resonance imaging study of spatial working memory in children with prenatal alcohol exposure: Contribution of familial history of alcohol use disorders. Alcohol Clin Exp Res 37: 132–140.

Nuñez SC, Roussotte F, Sowell ER (2011). Focus on: structural and functional brain abnormalities in fetal alcohol spectrum disorders. Alcohol Res Health 34: 121–131.

O'Brien JW, Norman AL, Fryer SL et al. (2013). Effect of predictive cuing on response inhibition in children with heavy prenatal alcohol exposure. Alcohol Clin Exp Res 37: 644–654.

O'Connor MJ, Kasari C (2000). Prenatal alcohol exposure and depressive features in children. Alcohol Clin Exp Res 24: 1084–1092.

O'Connor MJ, Paley B (2006). The relationship of prenatal alcohol exposure and the postnatal environment to child depressive symptoms. J Pediatr Psychol 31: 50–64.

O'Connor MJ, Shah B, Whaley S et al. (2002). Psychiatric illness in a clinical sample of children with prenatal alcohol exposure. Am J Drug Alcohol Abuse 28: 743–754.

O'Connor MJ, Frankel F, Paley B et al. (2006). A controlled social skills training for children with fetal alcohol spectrum disorders. J Consult Clin Psychol 74: 639–648.

O'Hare ED, Kan E, Yoshii J et al. (2005). Mapping cerebellar vermal morphology and cognitive correlates in prenatal alcohol exposure. NeuroReport 16: 1285–1290.

O'Hare ED, Lu LH, Houston SM et al. (2009). Altered frontal-parietal functioning during verbal working memory in children and adolescents with heavy prenatal alcohol exposure. Hum Brain Mapp 30: 3200–3208.

O'Leary C, Zubrick SR, Taylor CL et al. (2009). Prenatal alcohol exposure and language delay in 2-year-old children: the importance of dose and timing on risk. Pediatrics 123: 547–554.

O'Malley KD, Barr H (1998). Fetal alcohol syndrome and seizure disorder. Can J Psychiatry 43: 1051 [letter].

O'Malley K, Huggins J (2005). Suicidality in adolescents and adults with fetal alcohol spectrum disorders. Can J Psychiatry 50: 125 [letter].

Peadon E, Rhys-Jones B, Bower C et al. (2009). Systematic review of interventions for children with fetal alcohol spectrum disorders. BMC Pediatr 9: 35.

Pei JR, Rinaldi CM, Rasmussen C et al. (2008). Memory patterns of acquisition and retention of verbal and nonverbal information in children with fetal alcohol spectrum disorders. Can J Clin Pharmacol 15: e44–e56.

Pei J, Job J, Kully-Martens K et al. (2011). Executive function and memory in children with fetal alcohol spectrum disorder. Child Neuropsychol 17: 290–309.

Pulsifer MB (1996). The neuropsychology of mental retardation. J Intl Neuropsychol Soc 2: 159–176.

Rasmussen C, Bisanz J (2011). The relation between mathematics and working memory in young children with fetal alcohol spectrum disorders. J Special Educ 45: 184–191.

Rasmussen C, Talwar V, Loomes C et al. (2008). Brief report: Lie-telling in children with fetal alcohol spectrum disorder. J Pediatr Psychol 33: 220–225.

Rasmussen C, Wyper K, Talwar V (2009). The relation between theory of mind and executive functions in children with fetal alcohol spectrum disorders. Can J Clin Pharmacol 16: e370–e380.

Rasmussen C, Tamana S, Baugh L et al. (2012). Neuropsychological impairments on the NEPSY-II among children with FASD. Child Neuropsychol 19: 337–349.

Reinhardt K, Mohr A, Gärtner J et al. (2010). Polymicrogyria in fetal alcohol syndrome. Teratology 88: 128–131.

Richardson GA, Ryan C, Willford J et al. (2002). Prenatal alcohol and marijuana exposure: effects on neuropsychological outcomes at 10 years. Neurotoxicol Teratol 24: 309–320.

Riikonen R, Salonen I, Partanen K et al. (1999). Brain perfusion SPECT and MRI in foetal alcohol syndrome. Dev Med Child Neurol 41: 652–659.

Riikonen RS, Nokelainen P, Valkonen K et al. (2005). Deep serotonergic and dopaminergic structures in fetal alcoholic syndrome: a study with nor-β-CIT-single-photon emission computed tomography and magnetic resonance imaging volumetry. Biol Psychiatry 57: 1565–1572.

Riley EP, Mattson SN, Sowell ER et al. (1995). Abnormalities of the corpus callosum in children prenatally exposed to alcohol. Alcohol Clin Exp Res 19: 1198–1202.

Roebuck TM, Simmons RW, Mattson SN et al. (1998). Prenatal exposure to alcohol affects the ability to maintain postural balance. Alcohol Clin Exp Res 22: 252–258.

Roebuck TM, Mattson SN, Riley EP (1999). Behavioral and psychosocial profiles of alcohol-exposed children. Alcohol Clin Exp Res 23: 1070–1076.

Roebuck TM, Mattson SN, Riley EP (2002). Interhemispheric transfer in children with heavy prenatal alcohol exposure. Alcohol Clin Exp Res 26: 1863–1871.

Roebuck-Spencer TM, Mattson SN (2004). Implicit strategy affects learning in children with heavy prenatal alcohol exposure. Alcohol Clin Exp Res 28: 1424–1431.

Roebuck-Spencer TM, Mattson SN, Marion SD et al. (2004). Bimanual coordination in alcohol-exposed children: role of the corpus callosum. J Intl Neuropsychol Soc 10: 536–548.

Roussotte FF, Bramen JE, Nunez SC et al. (2011). Abnormal brain activation during working memory in children with prenatal exposure to drugs of abuse: the effects of methamphetamine, alcohol, and polydrug exposure. Neuroimage 54: 3067–3075.

Roussotte FF, Rudie JD, Smith L et al. (2012a). Frontostriatal connectivity in children during working memory and the effects of prenatal methamphetamine, alcohol, and polydrug exposure. Dev Neurosci 34: 43–57.

Roussotte FF, Sulik KK, Mattson SN et al. (2012b). Regional brain volume reductions relate to facial dysmorphology and neurocognitive function in fetal alcohol spectrum disorders. Hum Brain Mapp 33: 920–937.

Sakata-Haga H, Dominguez HD, Sei H et al. (2006). Alterations in circadian rhythm phase shifting ability in rats

following ethanol exposure during the third trimester brain growth spurt. Alcohol Clin Exp Res 30: 899–907.
Santhanam P, Li Z, Hu X et al. (2009). Effects of prenatal alcohol exposure on brain activation during an arithmetic task: an fMRI study. Alcohol Clin Exp Res 33: 1901–1908.
Santhanam P, Coles CD, Li Z et al. (2011). Default mode network dysfunction in adults with prenatal alcohol exposure. Psychiatry Res 194: 354–362.
Scher MS, Richardson GA, Robles N et al. (1998). Effects of prenatal substance exposure: altered maturation of visual evoked potentials. Pediatr Neurol 18: 236–243.
Schonfeld AM, Mattson SN, Lang AR et al. (2001). Verbal and nonverbal fluency in children with heavy prenatal alcohol exposure. J Stud Alcohol 62: 239–246.
Schonfeld AM, Mattson SN, Riley EP (2005). Moral maturity and delinquency after prenatal alcohol exposure. J Stud Alcohol 66: 545–554.
Schonfeld AM, Paley B, Frankel F et al. (2006). Executive functioning predicts social skills following prenatal alcohol exposure. Child Neuropsychol 12: 439–452.
Simmons RW, Wass T, Thomas JD et al. (2002). Fractionated simple and choice reaction time in children with prenatal exposure to alcohol. Alcohol Clin Exp Res 26: 1412–1419.
Simmons RW, Thomas JD, Levy SS et al. (2006). Motor response selection in children with fetal alcohol spectrum disorders. Neurotoxicol Teratol 28: 278–285.
Simmons RW, Levy SS, Riley EP et al. (2009). Central and peripheral timing variability in children with heavy prenatal alcohol exposure. Alcohol Clin Exp Res 33: 400–407.
Simmons RW, Thomas JD, Levy SS et al. (2010). Motor response programming and movement time in children with heavy prenatal alcohol exposure. Alcohol 44: 371–378.
Simmons RW, Nguyen TT, Levy SS et al. (2012). Children with heavy prenatal alcohol exposure exhibit deficits when regulating isometric force. Alcohol Clin Exp Res 36: 302–309.
Sood B, Delaney-Black V, Covington C et al. (2001). Prenatal alcohol exposure and childhood behavior at age 6 to 7 years: I. Dose-response effect Pediatrics 108: e34–e42.
Sowell ER, Jernigan TL, Mattson SN et al. (1996). Abnormal development of the cerebellar vermis in children prenatally exposed to alcohol: size reduction in lobules I-V. Alcohol Clin Exp Res 20: 31–34.
Sowell ER, Mattson SN, Thompson PM et al. (2001a). Mapping callosal morphology and cognitive correlates: effects of heavy prenatal alcohol exposure. Neurology 57: 235–244.
Sowell ER, Thompson PM, Mattson SN et al. (2001b). Voxel-based morphometric analyses of the brain in children and adolescents prenatally exposed to alcohol. NeuroReport 12: 515–523.
Sowell ER, Thompson PM, Mattson SN et al. (2002a). Regional brain shape abnormalities persist into adolescence after heavy prenatal alcohol exposure. Cereb Cortex 12: 856–865.
Sowell ER, Thompson PM, Peterson BS et al. (2002b). Mapping cortical gray matter asymmetry patterns in adolescents with heavy prenatal alcohol exposure. Neuroimage 17: 1807–1819.
Sowell ER, Lu LH, O'Hare ED et al. (2007). Functional magnetic resonance imaging of verbal learning in children with heavy prenatal alcohol exposure. NeuroReport 18: 635–639.
Sowell ER, Johnson A, Kan E et al. (2008a). Mapping white matter integrity and neurobehavioral correlates in children with fetal alcohol spectrum disorders. J Neurosci 28: 1313–1319.
Sowell ER, Mattson SN, Kan E et al. (2008b). Abnormal cortical thickness and brain-behavior correlation patterns in individuals with heavy prenatal alcohol exposure. Cereb Cortex 18: 136–144.
Spadoni AD, Bazinet AD, Fryer SL et al. (2009). BOLD response during spatial working memory in youth with heavy prenatal alcohol exposure. Alcohol Clin Exp Res 33: 2067–2076.
Spalice A, Del Balzo F, Nicita F et al. (2011). Teaching neuroimages: schizencephaly in fetal alcohol syndrome. Neurology 77: e96.
Spohr H-L, Steinhausen H-C (1987). Follow-up studies of children with fetal alcohol syndrome. Neuropediatrics 18: 13–17.
Spohr H-L, Willms J, Steinhausen H-C (2007). Fetal alcohol spectrum disorders in young adulthood. J Pediatr 150: 175–179.
Spottiswoode BS, Meintjes EM, Anderson AW et al. (2011). Diffusion tensor imaging of the cerebellum and eyeblink conditioning in fetal alcohol spectrum disorder. Alcohol Clin Exp Res 35: 2174–2183.
Steinhausen H-C, Spohr H-L (1998). Long-term outcome of children with fetal alcohol syndrome: psychopathology, behavior and intelligence. Alcohol Clin Exp Res 22: 334–338.
Steinmann TP, Andrew CM, Thomsen CE et al. (2011). An auditory Go/No-Go study of event-related potentials in children with fetal alcohol spectrum disorders. Conference proceedings: annual International Conference of the IEEE Engineering in Medicine and Biology Society. IEEE Engineering in Medicine and Biology Society. Conference 2011, pp. 789–792.
Stephen JM, Kodituwakku PW, Kodituwakku EL et al. (2012). Delays in auditory processing identified in preschool children with FASD. Alcohol Clin Exp Res 36: 1720–1727.
Stratton K, Howe C, Battaglia F (Eds.), (1996). Fetal alcohol syndrome: diagnosis, epidemiology, prevention, and treatment. Institute of Medicine, National Academy Press, Washington, DC.
Streissguth AP, Barr HM, Sampson PD (1990). Moderate prenatal alcohol exposure: effects on child IQ and learning problems at age 7 1/2 years. Alcohol Clin Exp Res 14: 662–669.
Streissguth AP, Barr HM, Carmichael Olson H et al. (1994a). Drinking during pregnancy decreases word attack and arithmetic scores on standardized tests: adolescent data from a population-based prospective study. Alcohol Clin Exp Res 18: 248–254.
Streissguth AP, Barr HM, Sampson PD et al. (1994b). Prenatal alcohol and offspring development: the first fourteen years. Drug Alcohol Depend 36: 89–99.

Streissguth AP, Sampson PD, Carmichael Olson H et al. (1994c). Maternal drinking during pregnancy: attention and short-term memory in 14-year-old offspring – a longitudinal prospective study. Alcohol Clin Exp Res 18: 202–218.

Streissguth AP, Barr HM, Kogan J et al. (1996). Understanding the occurrence of secondary disabilities in clients with fetal alcohol syndrome (FAS) and fetal alcohol effects (FAE). Final Report to the Centers for Disease Control and Prevention (CDC). University of Washington Publication Services, Seattle, WA.

Streissguth AP, Barr HM, Bookstein FL et al. (1999). The long-term neurocognitive consequences of prenatal alcohol exposure: A 14-year study. Psychol Sci 10: 186–190.

Streissguth AP, Bookstein FL, Barr HM et al. (2004). Risk factors for adverse life outcomes in fetal alcohol syndrome and fetal alcohol effects. J Dev Behav Pediatr 25: 228–238.

Strickland DC, McAllister D, Coles CD et al. (2007). An evolution of virtual reality training designs for children with autism and fetal alcohol spectrum disorders. Topics Lang Disord 27: 226–241.

Strömland K (2004). Visual impairment and ocular abnormalities in children with fetal alcohol syndrome. Addict Biol 9: 153–157.

Swayze VW, Johnson VP, Hanson JW et al. (1997). Magnetic resonance imaging of brain anomalies in fetal alcohol syndrome. Pediatrics 99: 232–240.

Thomas LT, Gaitantzis YA, Frias JL (1987). Palpebral fissure length from 29 weeks to 14 years. J Pediatr 111: 267–268.

Thomas SE, Kelly SJ, Mattson SN et al. (1998). Comparison of social abilities of children with fetal alcohol syndrome to those of children with similar IQ scores and normal controls. Alcohol Clin Exp Res 22: 528–533.

Thorne JC, Coggins TE, Carmichael Olson H et al. (2007). Exploring the utility of narrative analysis in diagnostic decision making: picture-bound reference, elaboration, and fetal alcohol spectrum disorders. J Speech Hear Res 50: 459–474.

Timler GR, Olswang LB, Coggins TE (2005). "Do I know what I need to do?" A social communication intervention for children with complex clinical profiles. Lang Speech Hear Serv Sch 36: 73–85.

Uecker A, Nadel L (1996). Spatial locations gone awry: object and spatial memory deficits in children with fetal alcohol syndrome. Neuropsychologia 34: 209–223.

Uecker A, Nadel L (1998). Spatial but not object memory impairments in children with fetal alcohol syndrome. Am J Ment Retard 103: 12–18.

Vaurio L, Riley EP, Mattson SN (2011). Neuropsychological comparison of children with heavy prenatal alcohol exposure and an IQ-matched comparison group. J Intl Neuropsychol Soc 17: 463–473.

Ware AL, Crocker N, O'Brien JW et al. (2012). Executive function predicts adaptive behavior in children with histories of heavy prenatal alcohol exposure and attention-deficit/hyperactivity disorder. Alcohol Clin Exp Res 36: 1431–1441.

Ware AL, Glass L, Crocker N et al. (2013a). The interaction between prenatal alcohol exposure and ADHD on adaptive functioning in children. Alcohol Clin Exp Res 38: 1439–1447.

Ware AL, O'Brien JW, Crocker N et al. (2013b). The effects of prenatal alcohol exposure and attention-deficit/hyperactivity disorder on psychopathology and behavior. Alcohol Clin Exp Res 37: 507–516.

Warren KR, Hewitt BG, Thomas JD (2011). Fetal alcohol spectrum disorders: research challenges and opportunities. Alcohol Res Health 34: 4–14.

Wass TS, Simmons RW, Thomas JD et al. (2002). Timing accuracy and variability in children with prenatal exposure to alcohol. Alcohol Clin Exp Res 26: 1887–1896.

Wengel T, Hanlon-Dearman AC, Fjeldsted B (2011). Sleep and sensory characteristics in young children with fetal alcohol spectrum disorder. J Dev Behav Pediatr 32: 384–392.

Whaley SE, O'Connor MJ, Gunderson B (2001). Comparison of the adaptive functioning of children prenatally exposed to alcohol to a nonexposed clinical sample. Alcohol Clin Exp Res 25: 1018–1024.

Wheeler SM, Stevens SA, Sheard ED et al. (2012). Facial memory deficits in children with fetal alcohol spectrum disorders. Child Neuropsychol 18: 339–346.

Willford JA, Richardson GA, Leech SL et al. (2004). Verbal and visuospatial learning and memory function in children with moderate prenatal alcohol exposure. Alcohol Clin Exp Res 28: 497–507.

Willford JA, Chandler LS, Goldschmidt L et al. (2010). Effects of prenatal tobacco, alcohol and marijuana exposure on processing speed, visual-motor coordination, and interhemispheric transfer. Neurotoxicol Teratol 32: 580–588.

Willoughby KA, Sheard ED, Nash K et al. (2008). Effects of prenatal alcohol exposure on hippocampal volume, verbal learning, and verbal and spatial recall in late childhood. J Intl Neuropsychol Soc 14: 1022–1033.

Wozniak JR, Mueller BA, Chang P-N et al. (2006). Diffusion tensor imaging in children with fetal alcohol spectrum disorders. Alcohol Clin Exp Res 30: 1799–1806.

Wozniak JR, Muetzel RL, Mueller BA et al. (2009). Microstructural corpus callosum anomalies in children with prenatal alcohol exposure: an extension of previous diffusion tensor imaging findings. Alcohol Clin Exp Res 33: 1825–1835.

Wozniak JR, Mueller BA, Muetzel RL et al. (2011). Interhemispheric functional connectivity disruption in children with prenatal alcohol exposure. Alcohol Clin Exp Res 35: 849–861.

Wyper KR, Rasmussen CR (2011). Language impairments in children with fetal alcohol spectrum disorders. J Population Ther Clin Pharmacol 18: e364–e376.

Yang Y, Phillips OR, Kan E et al. (2012a). Callosal thickness reductions relate to facial dysmorphology in fetal alcohol spectrum disorders. Alcohol Clin Exp Res 36: 798–806.

Yang Y, Roussotte F, Kan E et al. (2012b). Abnormal cortical thickness alterations in fetal alcohol spectrum disorders and their relationships with facial dysmorphology. Cereb Cortex 22: 1170–1179.

Zhou D, Lebel C, Lepage C et al. (2011). Developmental cortical thinning in fetal alcohol spectrum disorders. Neuroimage 58: 16–25.

Handbook of Clinical Neurology, Vol. 125 (3rd series)
Alcohol and the Nervous System
E.V. Sullivan and A. Pfefferbaum, Editors

Chapter 26

Fetal alcohol spectrum disorder: pathogenesis and mechanisms

KATHLEEN K. SULIK*
Department of Cell Biology and Physiology and Bowles Center for Alcohol Studies, University of North Carolina, Chapel Hill, NC, USA

INTRODUCTION

With over four decades of clinical investigation and 100 years of basic research, much has been learned about the birth defects that result from prenatal alcohol (ethanol) exposure in people and animal models. This is especially true for congenital abnormalities involving the nervous system. Our current understanding of the cascades of altered developmental processes that occur subsequent to an initial interaction of alcohol and/or its metabolites with an endogenous molecule and that contribute to structural and functional birth defects (i.e., alcohol's teratopathogenesis) is substantial and growing. In this chapter, no distinction is made between the impact of alcohol and its major metabolite, acetaldehyde, which is also teratogenic. Among the known major components of the cascades of alcohol-induced developmental damage are excessive cell death, changes in the cell cycle and proliferation, cell migration, cell morphogenesis, and gene expression, as well as free radical damage and interference with cell signaling. For these, upstream and downstream events are described as they are considered pertinent and informative. While the molecular events that initiate alcohol's teratopathogenesis (i.e., its primary mechanisms) remain for the most part unidentified, a synopsis of emerging mechanistic data is presented.

As reflected in the term, fetal alcohol spectrum disorder (FASD), prenatal alcohol exposure can result in myriad abnormal endpoints. While the emphasis in this chapter is on alcohol-induced neuroteratogenesis, non-neural tissues and organs, including facial and ocular structures, the heart, and limbs, are also frequently involved as part of FASD. Alcohol-induced birth defects involving one or more structures or systems may result from one predominant primary mechanism and/or pathogenic cascade or from simultaneous insult to more than one distinct pathway. This, along with the fact that an initial insult to a selected cell population is expected to subsequently adversely affect other cell populations and/or developmental events, commonly makes interpretation of experimental data complex.

As for other teratogens, alcohol's adverse prenatal effects are developmental stage-dependent, with many critical periods having been identified. Study of animal models has shown that a variety of different structural abnormalities, including those of the brain, can result from alcohol insult limited to specific times during the embryonic period. Exemplary are studies of a mouse FASD model in which differing patterns of brain and facial dysmorphology result from acute maternal alcohol exposure on gestational day (GD) 7 versus 8.5 (Lipinski et al., 2012), and in which heart, kidney, and limb defects follow alcohol insult on GDs 8.5, 9, and 9.25, respectively (Daft et al., 1986; Gage and Sulik, 1991; Kotch et al., 1992). Developmental stages present in mice on GDs 7 through 9 correspond to those in humans during the third through fourth weeks after fertilization. Additionally, alcohol exposure limited to even narrow windows of time during early embryogenesis or involving any or all of the three trimester equivalents can yield functional (including behavioral) defects.

In addition to developmental stage dependency, the type and severity of alcohol-induced abnormality are dependent upon exposure pattern and dosage, genetic background, and other environmental influences, including maternal nutrition, stress, and other coadministered drugs. Research employing model systems has facilitated definition of these modulating factors and has provided a wealth of information regarding alcohol's effects on both the mother and the conceptus (reviewed by Goodlett and Horn, 2001; Chen et al., 2003; Cudd,

*Correspondence to: Kathleen K. Sulik, Ph.D., Department of Cell Biology and Physiology and Bowles Center for Alcohol Studies, University of North Carolina, Chapel Hill, NC 27599-7178, USA. Tel: +1-919-966-3208, E-mail: mouse@med.unc.edu

2005; Miller, 2006; Schneider et al., 2011). The focus of this chapter is on alcohol-induced damage to the developing baby, itself, providing an overview of exemplary data rather than an exhaustive literature review.

CELL DEATH

Cell death is an event common to the teratogenic process initiated by many agents. Early reports of prenatal alcohol exposure and teratogenesis in model systems describe cell degeneration (Sandor et al., 1980; Bannigan and Burke, 1982; Bannigan and Cottell, 1984), as do more recent basic FASD studies, the latter of which include distinction between two types of alcohol-induced cell death – necrosis and apoptosis (Smith, 1997). Necrosis is characterized by swelling of cytoplasmic organelles, rupture of the plasma membrane with uncontrolled release of cellular components, and subsequent local inflammation. Typically, alcohol-induced necrosis involving cells of the nervous system is produced by a relatively fulminant insult or following alcohol withdrawal. While necrosis is traditionally viewed as a non-regulated form of cell death, it is now recognized that some necrotic cell death may be executed through a signaling pathway-dependent mechanism termed necroptosis or programmed necrosis, a process that has been implicated in mediating neuronal excitotoxicity (Christofferson and Yuan, 2010).

In the developing brain, apoptosis, a controlled form of cellular self-destruction, appears to be the most typical form of alcohol-induced cell death. Following alcohol-mediated modulation of extracellular or intracellular signals the apoptotic cascade is triggered, with subsequent release of mitochondrial proteins and cytochrome c and activation of cysteine proteases called caspases. Apoptosis involves nuclear condensation (pyknosis), DNA fragmentation, and the formation of membrane-bound cellular fragments termed apoptotic bodies. Excessive apoptotic cell death has been identified in neuronal as well as non-neuronal cell types in FASD models (Cartwright et al., 1998; Dunty et al., 2001; Olney et al., 2002a, b; Johnson et al., 2007). Demonstrating the significant role of apoptosis in alcohol teratogenesis are studies showing the antiteratogenic effect of preventing alcohol-induced reduction in the antiapoptotic Bcl2 protein (Sari et al., 2012), bolstering Bcl2 levels genetically (Heaton et al., 1999) or pharmacologically (Ullah et al., 2012), and treatment with the caspase inhibitor, CED-3 protease (Cartwright et al., 1998).

FASD research conducted primarily in rodents has highlighted the first- and third-trimester equivalents as times when excessive cell death plays a major role in alcohol-induced neuroteratogenesis. Findings from both exposure periods illustrate that the excessive cell death is not uniformly distributed, but is temporally, regionally, and cell type-specific. With an eye toward defining the involved pathogenetic cascade(s), studies employing acute alcohol exposure paradigms and identification of cell death patterns present within hours following treatment have been particularly informative.

With respect to acute exposures occurring during the first trimester, vital staining of mouse embryos within 12 hours of maternal alcohol treatment has allowed a relatively comprehensive temporospatial determination of cell populations that are killed both within and outside the CNS (Sulik et al., 1988; Kotch et al., 1992; Kotch and Sulik, 1992; Dunty et al., 2001, 2002). As illustrated in Figure 26.1, relatively large proportions of the cells in selected regions of the developing forebrain and hindbrain are vulnerable at early stages of development to alcohol-induced cell death. Included are the cells at the rim of the forebrain neural folds; cells that upon closure of the anterior neuropore normally form midline tissues. In the hindbrain the alar plate as well as premigratory neural crest cells are vulnerable. In addition to Nile blue sulfate uptake, TUNEL assays confirm the cell death to be apoptotic. Comparable findings in studies employing a chick embryo model (Cartwright and Smith, 1995a, b; Cartwright et al., 1998) have been followed by experiments showing that teratogenic alcohol levels rapidly elicit a G protein-stimulated rise in phospholipase C activity and an intracellular Ca^{2+} transit (Garic-Stankovic et al., 2005). Recent studies have shown that downstream of the alcohol-induced calcium transient is repression of the transcriptional effector, β-catenin. Blockade of the calcium transient normalizes β-catenin activity and prevents cell death (Flentke et al., 2011).

The third trimester is a time when differentiating neurons are sending out axons and making synaptic connections and when approximately one-third of the postmitotic neuronal population undergoes naturally occurring apoptotic cell death, thus eliminating surplus cells. Cell death induced by alcohol during this time period has been most studied in neonatal (third-trimester equivalent) rodents (Olney et al., 2002a; Young and Olney, 2006; Idrus and Napper, 2012). The results of this work suggest that the alcohol-induced cell death represents an abnormal augmentation of the normal cell death process, a premise upon which hypotheses regarding a role for alcohol-mediated alterations in the function (e.g., via changes in receptor activity) or abundance of neuronal survival factors are largely based. Implicated are alterations in nerve growth factor, brain-derived neurotrophic factor (BDNF), glia-derived neurotrophic factor, insulin-like growth factor-1 (IGF-1), heparin-binding epidermal growth factor *N*-methyl-D-aspartate (NMDA), and pituitary adenylyl cyclase activating polypeptide (Kilburn et al., 2006; Miller, 2006).

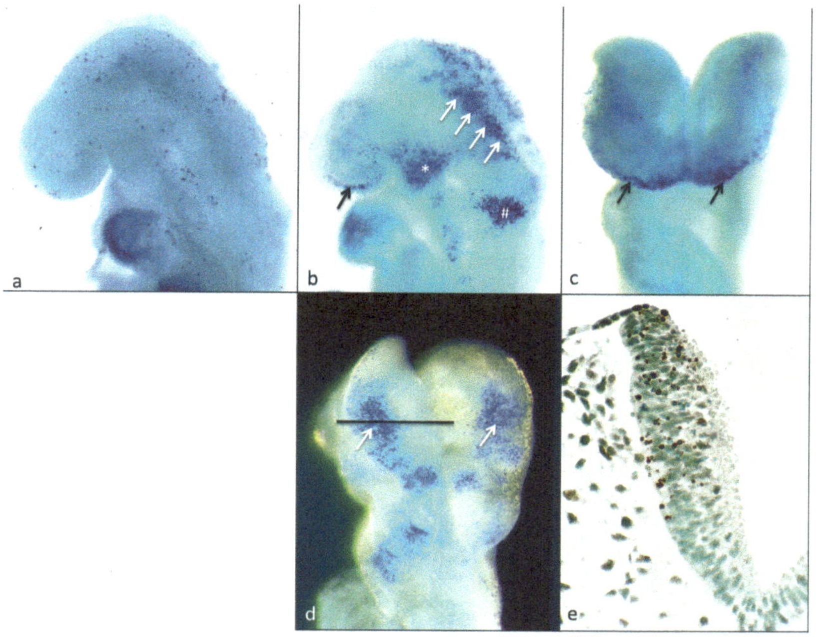

Fig. 26.1. Vital staining with Nile blue sulfate (NBS) in control (**a**) and alcohol-exposed mouse embryos (lateral view in **b**, frontal view in **c**, and dorsal view in **d**) illustrates excessive stain uptake (dark blue stipple), as is indicative of apoptotic cell death, in the latter. The NBS staining pattern illustrates excessive cell death in selected cell populations, including those in the rim of the forebrain neural folds (black arrows), the first branchial arch (*), the otic placode (#), and the alar plate of the hindbrain (white arrows). TUNEL labeling of a histologic section through the hindbrain (**E**) confirms apoptotic cell death in the alar versus the basal plate. (Modified from Dunty et al., 2001, and Kotch and Sulik, 1992.)

As reviewed by Goodlett et al. (2005), studies employing animal models have shown that, during the third-trimester equivalent period of synaptogenesis, postmitotic neurons in the cerebral cortex, hypothalamus, cerebellum, and brainstem are killed by alcohol. For the neurons in these various brain regions, temporal windows of vulnerability vary, as exemplified by results of acute (binge) exposure-based experiments showing that maximal numbers of Purkinje neurons are killed approximately 3 days earlier than are cerebrocortical neurons. While large alcohol doses have been employed in most of these studies, Young and Olney (2006) have reported that, in infant mice, a rise in blood alcohol concentration to a moderate level of approximately 50 mg/dL for a duration of 30–45 minutes is sufficient to cause widespread neuroapoptosis. Further research (Farber et al., 2010) has also included examination of alcohol-induced neuroapoptosis in fetal macaque brains following high-dose, 8-hour maternal alcohol treatment at various times between days 105 and 155 of their 160–165-day gestation period. A 60-fold increase in apoptosis above control levels was found in numerous fetal brain regions, with patterns of effect being consistent with those in rodents.

Study of alcohol-induced apoptotic cell death as occurs during the period of synaptogenesis has shown that, while patterns of caspase 3 activation and cell death are corresponding (Olney et al., 2002a), caspase 3 knockout mice remain vulnerable to alcohol-induced neuronal death (Young et al., 2005). Thus, caspase 3 does not appear to be an essential factor in determining whether cell death will occur. Indeed, a caspase-3-independent cell death pathway that is related to Bax translocation into mitochondria followed by the mitochondrial permeability transition and release of death-promoting factors including Smac/DIABLO, apoptosis-inducing factor and EndoG, has been reported (Miller, 1997; Stefanis et al., 1999; Chang et al., 2003). Alcohol-induced developmental neuroapoptosis does not occur in Bax knockout mice, indicating that translocation of Bax protein to mitochondria is an essential step (Young et al., 2003). It is speculated that, upstream of this event, alcohol-mediated suppression of neuronal activity via a combination of NMDA receptor blockade and hyperactivation of GABA-A receptors triggers the cell death process (Ikonomidou et al., 2000). Another upstream event that may be involved in alcohol-induced neuroapoptosis is suppression of the phosphorylation of ERK, an extracellular signal-regulated protein kinase that is believed to play a role in cell survival (Young et al., 2008). This is supported by the finding that lithium, which stimulates ERK phosphorylation, prevents the alcohol-induced phosphorylation suppression.

Studies of neuronal cell death resulting from alcohol exposure during the third-trimester equivalent have also focused on excitotoxic cell death as occurs during withdrawal and results from overactivity of NMDA receptors. Supporting a role for this effect are studies of neonatal rats in which NMDA receptor antagonists have been shown to effectively attenuate alcohol-induced abnormalities, including motor and learning deficits and reductions in Purkinje cell numbers (Thomas et al., 2002; Idrus et al., 2011).

CELL CYCLE AND PROLIFERATION

Animal model-based studies of alcohol-induced cell cycle and proliferation alterations have focused, to a large extent, on insult occurring during the second-trimester equivalent, when most neurogenesis occurs. In rats, a species in which cortical neurogenesis begins on GD 11 and ends on GD 21 (comparable to weeks 6 through 17 of human development), prenatal alcohol exposure delays and extends the period of neuronal generation by 1 day (Miller, 1986, 2006). Notably, increases in the time spent by neurons in the G1 phase of the cell cycle following prenatal alcohol exposure appears to result from interference with growth factors. Supporting studies show that IGF receptor function/signaling is reduced in the presence of alcohol (Resnicoff et al., 1993) and that alcohol prevents tyrosine autophosphorylation of type 1 IGF receptors, and inhibits cyclins and cyclin-dependent kinases (Li et al., 2002).

Of interest, a computational model indicates that cell cycle perturbations occurring in rodent models during the second-trimester equivalent have a more significant effect than synaptogenesis-related insult occurring during the third-trimester equivalent (Gohlke et al., 2005). Additionally, these investigators suggest that the developing human neocortex may be more sensitive to the effects of alcohol than the rodent as a result of the relative increase in the length of neurogenesis and subsequent increased neocortical size in humans. They also predict that neuronal death during synaptogenesis may play a greater role in shaping the brain in primates than in rodents.

CELL MIGRATION

In animal models, alcohol has been shown to reduce the rate of neuronal migration during the inside-out development of the cortex and to cause migrating neurons to finish their migration in inappropriate locations. This is evidenced by studies showing that, following prenatal alcohol exposure from GD 6 to term in rats (comparable to weeks 3 through 18 of human development), many of the neurons that are generated late in gestation and that should populate cerebral cortical layer II/III instead either terminate their migrations in layers IV, V, or VI or overmigrate into layer I (Miller, 1986, 1993). Speculation regarding likely mechanisms involved in alcohol-induced neuronal migration errors includes disruption of cell adhesion, specifically involving L1 cell adhesion molecule, and interference with glial-guided migration or detachment of neurons from glial fibers. Notably, alcohol causes premature transformation of radial glia into astrocytes, with associated migrating neurons being unable to complete their migration (Miller and Robertson, 1993). Also implicated in neuronal migrational errors are effects on cytoskeletal elements, integrins, extracellular matrix components, and growth factors (Miller, 2006).

In addition to migration error-based intracortical defects, supracortical abnormalities presenting as suprapial/leptomeningeal heterotopias result from prenatal alcohol exposure. Results of *in vivo* and *in vitro* studies of fetal rat cortices suggest that these defects result from alcohol-induced alterations in the pial membrane and glia limitans (Komatsu et al., 2001; Mooney et al., 2004). In turn, these alterations are hypothesized to be secondary to an initial interference with growth factor regulation, leading to reduced expression of reelin and subsequent aberrant movement of neurons into the cortical marginal zone and beyond (Mooney et al., 2004). A recent study of mice in which prenatal alcohol exposure was limited to just GD 7 (comparable to the mid third week of human development), a time when the neural plate is just beginning to form, has shown that even at this very early stage of development alcohol can cause severe cortical migration errors, including leptomeningeal heterotopias (Fig. 26.2; Godin et al., 2010). These results suggest that alcohol damages or kills the progenitors of cells that are key to cerebrocortical cell migration. Recognizing that, on GD 7 in mice, the developing septal region is particularly vulnerable and that this brain region is an important source of reelin-producing Cajal-Retzius cells, it is hypothesized that adverse effects on this cell population underlie the cerebrocortical migration errors that follow this early insult.

CELL MORPHOGENESIS

Both *in vivo* and *in vitro* animal studies have shown that prenatal alcohol exposure affects dendritic morphology. Exemplary are the results of examination of the apical dendrite of hippocampal CA1 pyramidal cells from postnatal rats whose mothers had been chronically alcohol-exposed prior to and throughout pregnancy and nursing (Tarelo-Acuna et al., 2000). In the alcohol-exposed dendrites a relatively high proportion of stubby or wide spines were found, a change that is expected to affect electrical excitability and firing pattern. Cui

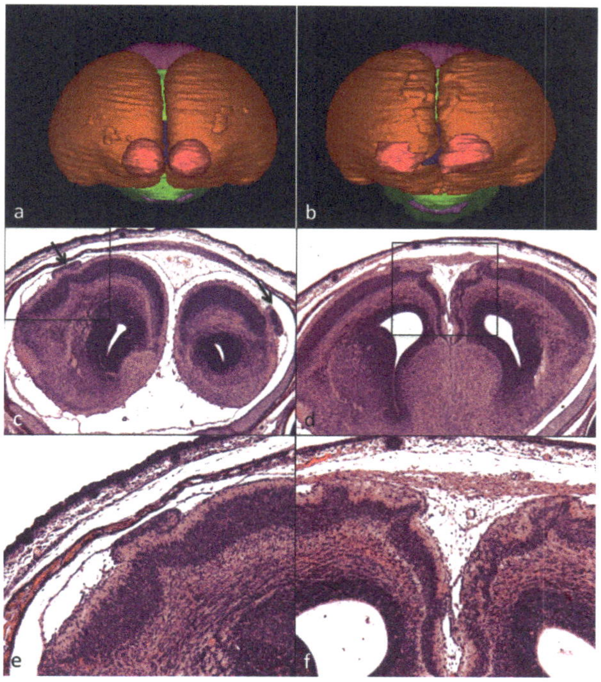

Fig. 26.2. Magnetic resonance imaging-based brain reconstructions (**a, b**) and histologic sections (**c–f**) of alcohol-affected fetal mice that had been acutely treated with alcohol on their 7th gestational day (comparable to mid third week of human development) illustrate cerebrocortical migration errors. In the specimen shown in **a, c,** and **e** the defects are leptomeningeal heterotopias (arrows in **c**) while cortical dysplasia is present in the specimen in **b, d,** and **f**. Boxed areas in **c** and **d** are shown at higher magnification in **e** and **f**, respectively. Color codes in **a** and **b** are as follows: cerebral cortex, red; olfactory bulbs, pink; mesencephalon, magenta; diencephalon, green; septal region, blue; pituitary, purple. (Modified from Godin et al., 2010.)

et al. (2010) also report dendritic spine changes in their study of pyramidal cells in the visual cortex of postnatal mice following chronic prenatal (first- and second-trimester equivalent) alcohol exposure. While this study showed decreased numbers of spines, their mean length was increased. Additionally noted were changes in synaptic ultrastructure. Histologic analyses also reveal alcohol-induced changes in overall dendritic form. For example, in the shell of the nucleus accumbens of adult rats that had been chronically alcohol-exposed during their gestation, the dendrites of medium spiny neurons were found to be reduced in length and to be less branched than in controls; results that are suggestive of effects on reward processing (Rice et al., 2012). In keeping with the *in vivo* studies, cultured rat hippocampal pyramidal neurons exhibit alcohol-induced correlative decreases in total dendritic length and synapse number (Yanni and Lindsley, 2000). Other *in vitro* studies of this cell type have shown that alcohol induces alterations in axonal growth dynamics and have provided evidence suggesting that this is a result of disruption of neurotrophic factor regulation (Lindsley et al., 2011).

In addition to *in vitro* studies of hippocampal pyramidal neurons, cerebellar granule cells cultured in the presence of alcohol have been examined. In these neurons, L1 cell adhesion molecule (L1) has a neurite outgrowth-promoting activity that appears to result from its activation of Src family kinases, the scaffolding protein Cas, and extracellular signal-regulated kinases 1 and 2 (ERK 1/2) (Chen and Charness, 2012). L1-mediated axon outgrowth can be inhibited by pharmacologic concentrations of alcohol, an effect that appears at least in part to involve alcohol-induced retardation of L1 trafficking through lipid rafts (Tang et al., 2011). Lending support to this premise are the results of an acute postnatal day 6 rat (comparable to 31–32 weeks of human prenatal development) *in vivo* alcohol exposure study showing a significant proportional increase of L1 in cerebellar lipid rafts (Littner et al., 2013). Importantly, the peptide, NAP (an active motif of activity dependent neuroprotective protein) which, like L1, can stimulate axon outgrown in cultured cerebellar granule cells, reduces alcohol's inhibition of L1 adhesion (Wilkemeyer et al., 2002). *In vivo* and in whole embryo culture systems this peptide, along with SAL, a short peptide from activity-dependent neurotrophic factor protein, are antiteratogenic (Spong et al., 2001; Poggi et al., 2003; Wilkemeyer et al., 2003, 2004; Chen et al., 2005; Sari and Gozes, 2006; Parnell et al., 2007; Zhou et al., 2008). Protection afforded by these peptides has been linked to prevention of an alcohol-induced rise in BDNF (Incerti et al., 2010).

Of major importance in our increasing understanding of alcohol's mechanism of action is recognition of its molecular interactions. At the forefront of this work is that related to defining the interaction between alcohol and L1. L1 homophilic binding, as occurs in L1 mediated cell-cell adhesion, is potentiated by folding of the extracellular domain of the molecule into a horseshoe structure in which the immunoglobulin 1 and 4 domains lie opposed to each other. Photolabeling experiments have illustrated that alcohol interacts with a binding pocket at the Ig1–Ig4 interface (Dou et al., 2011). Phosphorylation of the L1 cytoplasmic domain regulates the conformation and function of the extracellular domain, a phenomenon known as inside-out signaling. Recently, sensitivity of L1 to alcohol has been shown to be regulated by phosphorylation of an ERK2 substrate (S1248) on L1's intracellular domain. The relevance of this finding to FASD is supported by the fact that ERK2 is a component of the MAPK signaling pathway and that *in vitro* as well as *in vivo* studies show that

MAPK activity differences correspond to sensitivity to alcohol-induced cell adhesion inhibition and teratogenesis (Dou et al., 2013).

GENE EXPRESSION CHANGES

Genetic

New genomic methodologies, especially microarray analyses, have provided a way to screen the expression (conversion to a gene product) of large numbers of genes in order to identify alcohol-induced changes. Green et al. (2007) used this technique in comparing alcohol-induced gene expression changes in two closely related mouse substrains (C57Bl/6N and C57Bl/6J) that show differing teratogenic responses to an acute early (GD 8; comparable to early in the fourth week of human development) insult. Gene expression and pathway analyses revealed downregulation of ribosomal proteins and upregulation of glycolysis and the pentose phosphate pathway just in the C57Bl/6 N embryos and upregulation of tight junction, focal adhesion, adherens junction and actin cytoskeleton regulation pathways in both substrains. Based on gene expression changes, entry points for alcohol were identified at several hubs, including MAPK1, ALDH3A2, IGF-1, and EGFR. Overall the findings from this study support the view that developmental exposure to alcohol alters signaling pathways linking receptor activation to cytoskeletal reorganization.

A similar and more recent comparison of DBA/2J mice, which are relatively resistant to alcohol teratogenesis, and the susceptible C57Bl/6J strain showed that alcohol exposure decreased expression of many genes and increased relatively few (Downing et al., 2012). Genes found to be preferentially downregulated in the C57Bl/6J mouse strain are those involved in transcription, translation, and mRNA splicing. Additionally, genes involved in protein synthesis, the ubiquitin pathway, and epigenetic genes involved in chromatin remodeling and methylation were found to be important alcohol targets. Supporting these findings are results of whole embryo and neuronal cell culture-based microarray analyses (Gutala et al., 2004; Liu et al., 2009; Zhou et al., 2011b). Notably, in the Downing et al. study, no interstrain expression differences in genes coding for alcohol or aldehyde dehydrogenase, or the p450 enzyme Cyp2e1 that is involved in alcohol metabolism, were found.

Epigenetic

Epigenetic modifications entail changes in gene expression via mechanisms other than changes in DNA sequences (Kobor and Weinberg, 2011). The gene-silencing or enhancing activity can result from methylation, histone modification, and the effects of micro RNAs. The latter are single-stranded, non-protein-coding RNA molecules of about 17–25 nucleotides in length that control gene networks by translation repression. On one hand, the expression of certain micro RNAs is controlled by DNA methylation and chromatin modifications, and on the other, micro RNAs can affect the methylation machinery and the expression of proteins involved in histone modifications. Epigenetic changes may remain for the duration of a cell's life through many mitotic cell divisions, for multiple cell generations, and may also be passed on transgenerationally following meiotic cell division. It is notable that, as early as 1913, Dr. Charles Stockard concluded from studies of rats that alcohol had a paternally mediated transgenerational effect on perinatal mortality (Stockard, 1913).

In recent teratogenesis studies employing mice that have an epigentically sensitive allele as a biomarker, alcohol has been shown to cause hypermethylation and gene silencing, resulting in phenotypic changes (Kaminen-Ahola et al., 2010a, b). More specifically, mice carrying the Agouti viable yellow (A^{vy}) allele as a biomarker normally show variable coat colors, ranging from yellow to dark brown. In yellow mice the Agouti gene regulation region is hypomethylated and the gene is active, while in brown mice, the gene is silent as a result of a highly methylated regulation region. In mice that had been exposed to alcohol prenatally from the time of fertilization to GD 8 (comparable in humans to fertilization until early in the fourth week), increased methylation occurred, as indicated by darker coat colors.

Additionally, studies examining alcohol's effect on developing neurons and neuronal stem cells have shown epigenetic changes. More specifically, in cerebellar neurons from third-trimester equivalent alcohol-exposed rats, reductions were identified in histone acetylation and CREB-binding protein (CBP) (Guo et al., 2011). CBP is a histone acetyltransferase and a component of the epigenetic mechanism that controls neuronal gene expression. In cultured neural stem cells, a binge-like alcohol exposure paradigm prevented normal DNA methylation programming of key neural stem cell genes and retarded cell differentiation (Zhou et al., 2011a). In second-trimester human fetal brain neurospheres alcohol has also been shown to alter the fate of neuronal stem and progenitor cells (Vangipuram and Lyman, 2010).

Regarding micro RNAs, a relatively small subset of those expressed at various developmental stages show alcohol sensitivity (Sathyan et al., 2007; Miranda, 2012). Among the sensitive molecules are those in the miR-9 family, which plays a significant role in early neural tube patterning and telencephalic neurogenesis. During early prenatal development, miR-9 expression is diminished by alcohol, while at later developmental stages and in adults it is induced. Alcohol is hypothesized to affect micro RNA expression by regulating the epigenetic landscape of the genome via alteration of methylation patterns and also by influencing the activity of ligand-gated ion channels.

Of interest, transgenerational alcohol-induced epigenetic modifications have been reported in a rat model. Offspring of rats fed an alcohol-containing liquid diet from GD 7 to GD 21 (corresponding to weeks 3–17 of human prenatal development) showed a deficit in proopiomelanocortin (Pomc)-related neural functions along with hypermethylationin of the *Pomc* promoter and altered levels of histone-modifying proteins and DNA methyltransferase levels in POMC neurons. These Pomc gene methylation and expression effects and functional deficits persisted in the male germline. Speculation has been advanced regarding a role for these prenatal alcohol exposure-induced changes in genetic and sex-related risks for alcoholism (Govorko et al., 2012).

Additional support for an epigenetic role in alcohol-induced neuroteratogenesis stems from rodent studies in which choline supplementation has been shown to be ameliorative. This effect is presumably mediated through the effects of choline on DNA methylation (Thomas et al., 2004, 2007, 2009, 2010; Ryan et al., 2008; Monk et al., 2012; Thomas and Tran, 2012). Indeed, a recent study entailing administration of this nutrient to alcohol-exposed neonatal rats and subsequent assessment of the brains of these animals on postnatal day 21 found that alcohol treatment-induced hypermethylation in the hippocampus and prefrontal cortex was reduced as a result of choline supplementation (Otero et al., 2012). Deficiency of another nutrient, folic acid, which is important in the production of methyl groups as are required for DNA methylation, has also been suggested as playing a role in alcohol's teratogenesis via an epigenetic mechanism (Ballard et al., 2012).

REACTIVE OXYGEN SPECIES-MEDIATED DAMAGE

Free radicals are highly reactive molecules, many of which contain oxygen. In cells they can cause oxidative stress, subsequent damage to cellular components, including membranes, DNA, and proteins, and can induce apoptosis.

There is considerable evidence supporting a role for oxidative stress in alcohol's teratogenesis, including a number of studies illustrating the protective capacity of both exogenous and endogenous antioxidants (Davis et al., 1990; Kotch et al., 1995; Chen and Sulik, 1996, 2000; Henderson et al., 1999; Ramachandran et al., 2003; Chen et al., 2004; Heaton et al., 2006; Wentzel and Eriksson, 2006; Parnell et al., 2010; Antonio et al., 2011). Regarding the latter, studies of cultured fetal rat cortical neurons suggest that alcohol causes glutathione depletion and that the highly differing sensitivity to alcohol-induced cell death within specific neuronal populations is related to their glutathione content (Maffi et al., 2008; Narasimhan et al., 2011). Importantly, chemically or virally induced transcriptional activation of Nrf2, a transcription factor that regulates the induction of detoxifying and antioxidant genes, has been shown to prevent alcohol-induced glutathione depletion and the related apoptotic death of cultured cortical neurons (Narasimhan et al., 2011); to protect against alcohol-induced oxidative stress and apoptosis in early mouse embryos (Dong et al., 2008, 2010; Yan et al., 2010); and to increase the survival of cerebellar granule neurons in third-trimester equivalent alcohol-exposed rats (Kumar et al., 2011). The effectiveness of antioxidants affords a practical, nutrition-based approach to reducing alcohol's teratogenicity in human populations.

Many pathways have been suggested as contributing to the ability of alcohol to induce a state of oxidative stress, with potential sources of radical oxygen species (ROS) including mitochondrial respiratory chain enzymes, xanthine oxidase, cytochrome P450 enzymes and NADPH oxidase (NOX). Recent studies of early mouse embryos have identified the latter as playing an important role in alcohol's teratogenicity, with alcohol exposure inducing NOX enzyme activity and increasing mRNA expression of NOX regulatory subunits. The NOX inhibitor, diphenyleneiodonium, was shown to be protective, diminishing alcohol-induced ROS generation, caspase-3 activity, and apoptosis (Dong et al., 2010).

RETINOID AND SONIC HEDGEHOG SIGNALING

Retinoic acid (RA) signaling is critical for normal development (Rhinn and Dolle, 2012), with both diminished and excessive RA concentrations resulting in birth defects. Recognizing that RA synthesis depends on oxidation of retinol and retinal by enzymes that also metabolize alcohol (alcohol and aldehyde dehydrogenase), Pullarkat (1991) and Duester (1991) hypothesized that alcohol's teratogenicity results from its competitive inhibition of these enzymes. Supporting this as a potential mechanism is work by Deltour et al. (1996) demonstrating in cultured mouse embryos that alcohol exposure reduces RA concentrations. Additionally, in several developmental systems RA receptors (RARs) and cellular retinol-binding protein are disregulated by alcohol (Zachman and Grummer, 1998). Illustrating RA involvement in alcohol teratogenicity, Johnson et al. (2007) employed a pan-RAR antagonist and disulfiram, an ALDH inhibitor, showing that temporary abrogation of RA signaling results in limb defects comparable to those resulting from alcohol exposure. In these experiments RA was shown to diminish excessive cell death as underlies alcohol-induced limb defects; and alcohol was shown to reduce the expression of sonic hedgehog (Shh), a key developmental control gene. Previous work had shown that Shh signaling is stimulated by, and dependent upon, RA (Helms et al., 1994; Stratford et al., 1996).

Mounting evidence implicates interference with Shh signaling in alcohol-induced birth defects, especially those resulting from early gestational exposure. Studies conducted using an acute (high-binge) exposure mouse model have shown that the brain and face are concurrently affected following alcohol insult occurring at early gastrulation stages of development (corresponding to the third week postfertilization in humans) and that the widely ranging degrees of effect include all of the salient craniofacial features of FAS and corpus callosum abnormalities as also characterize the severe end of FASD (Sulik et al., 1988; Sulik, 2005; O'Leary-Moore et al., 2011) (Figs. 26.3 and 26.4). Severely affected animals present with median facial and forebrain deficiencies that fall within the spectrum termed holoprosencephaly. In humans holoprosencephaly is recognized as resulting from environmental exposures and/or genetic abnormalities involving the Shh gene itself, as well as genes that modulate the Shh pathway (Roessler and Muenke, 2010; Solomon et al., 2012). Studies employing fish and chick models have shown that provision of exogenous shh protein (Ahlgren et al., 2002) or Shh mRNA overexpression (Loucks and Ahlgren, 2009) in alcohol-exposed embryos prevents many of the gross physical phenotypes, suggesting that the suppression of Shh signaling, as is required for normal ventral forebrain induction and development, is one of the major effects of alcohol exposure. Supporting this premise is recent research showing that the severity and incidence of FAS-like facial and brain abnormalities, including those falling within the holoprosencephaly spectrum, are exacerbated in prenatal alcohol-exposed mice by haploinsufficiency of Shh as well as Gli2, which is a transcriptional activator in the Shh pathway (Kietzman et al, 2014). Additionally, mutation of the Shh coreceptor, Cdon, synergizes with alcohol to yield holoprosencephaly in mice (Fig. 26.5) (Hong and Krauss, 2012). While the molecular mechanism of alcohol's synergistic action remains unknown, findings from these gene–environment interaction studies highlight the promise of continuing to seek genes whose modification/mutation may exacerbate or attenuate alcohol's impact on development.

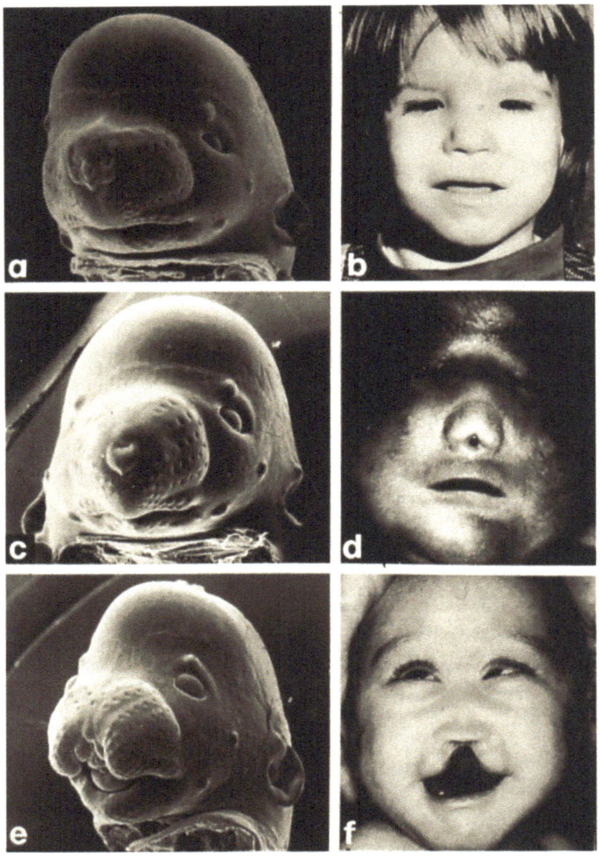

Fig. 26.3. The facial features of mouse fetuses acutely exposed to alcohol on their 7th day of gestation (comparable to the mid third week of human development), showing varying degrees of effect (**a**, **c**, and **e**), can be compared with those of an individual with fetal alcohol syndrome (**b**) and those whose defects are recognized as falling within the holoprosencephaly spectrum (**d** and **f**). (Modified from Sulik, 2005.)

CONCLUSION

Animal model-based studies have provided a wealth of information regarding the genesis of prenatal alcohol exposure-induced structural and functional abnormalities and are expected to continue to play a prominent role in this regard. As evidenced by the selected studies considered herein, a number of differing teratogenic cascades account for a wide variety of alcohol-induced developmental endpoints, with the developmental stage and genetic background of the exposed individual as well as alcohol exposure patterns and dosages playing critical modulating roles. Overall, it is clear that the developing brain is vulnerable to alcohol insult at virtually every prenatal stage. Technologic advances and rapidly increasing knowledge in the fields of genetics, cell, developmental, and neurobiology are essential to accurately piecing together experimental evidence in refining

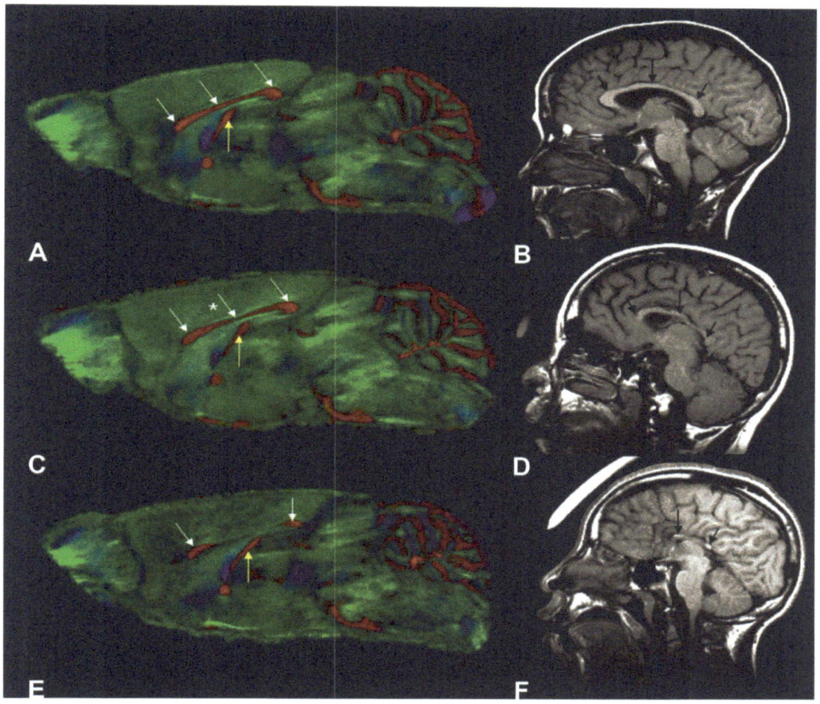

Fig. 26.4. Alcohol exposure on gestational day (GD) 7 (comparable to the mid third week of human development) in mice yields permanent central nervous system dysmorphology comparable to that in individuals with fetal alcohol syndrome (FAS). Diffusion tensor images are from a control adolescent mouse (**A**) and two that had been alcohol-exposed on GD 7 (**C**, **E**). The alcohol-exposed mice show varying degrees of brain dysmorphology, including thinning (star in **C**) to median absence of the corpus callosum (**E**). The hippocampal commissure (yellow arrows) is also reduced in the more severely affected mouse (**E**). Dysmorphology of the corpus callosum in this mouse model is remarkably similar to that seen in individuals with FAS. As compared to a control individual (**B**), those with FAS (**D**, **F**) also have considerable dysmorphology of the corpus callosum. (Human images courtesy of Dr. S. Mattson. Reproduced from O'Leary-Moore et al., 2011, by permission of Springer Science and Business Media.)

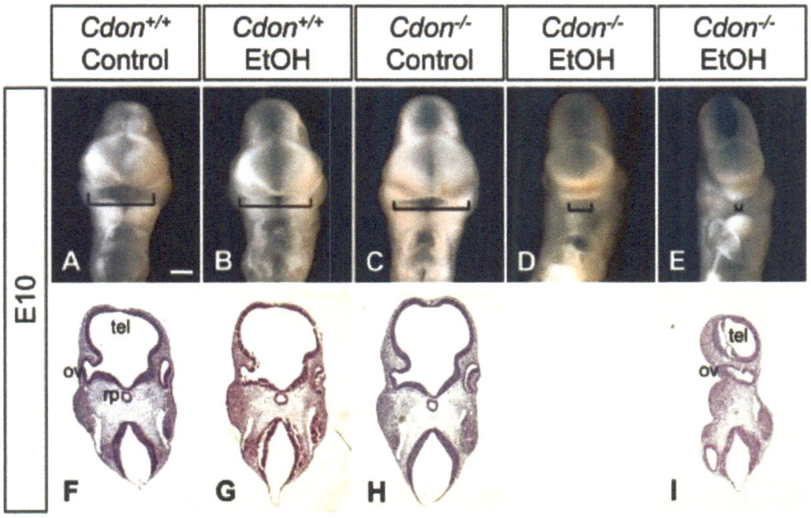

Fig. 26.5. Synergy between loss of Cdon, which encodes a Shh coreceptor, and prenatal alcohol exposure produces holoprosencephaly. (**A–E**) Frontal views of gestational day (GD) 10.0 mouse embryos (comparable to week 5 of human development) of the indicated genotype that had been treated *in utero* on GD 7 (comparable to the mid third week of human development) with alcohol (EtOH) or saline (control). Brackets indicate distance between the left and right nasal pits, a measure of midline formation. Scale bar, 250 mm. (**F–I**) Coronal sections stained with hematoxylin and eosin. Some alcohol-treated Cdon2/2 embryos (**E**, **I**) displayed severe holoprosencephaly with loss of telencephalic structure (tel), a single optic vesicle (ov), and absence of the anterior pituitary precursor, Rathke's pouch (rp). (Reproduced from Hong and Krauss, 2012.)

our understanding of the genesis of alcohol-induced birth defects, to the planning and execution of future studies, and to applying the knowledge gained to diminish the severity or occurrence of FASD.

References

Ahlgren SC, Thakur V, Bronner-Fraser M (2002). Sonic hedgehog rescues cranial neural crest from cell death induced by ethanol exposure. Proc Natl Acad Sci U S A 99: 10476–10481.

Antonio AM, Gillespie RA, Druse-Manteuffel MJ (2011). Effects of lipoic acid on antiapoptotic genes in control and ethanol-treated fetal rhombencephalic neurons. Brain Res 1383: 13–21.

Ballard MS, Sun M, Ko J (2012). Vitamin A, folate, and choline as a possible preventive intervention to fetal alcohol syndrome. Med Hypotheses 78: 489–493.

Bannigan J, Burke P (1982). Ethanol teratogenicity in mice: a light microscopic study. Teratology 26: 247–254.

Bannigan J, Cottell D (1984). Ethanol teratogenicity in mice: an electron microscopic study. Teratology 30: 281–290.

Cartwright MM, Smith SM (1995a). Increased cell death and reduced neural crest cell numbers in ethanol-exposed embryos: partial basis for the fetal alcohol syndrome phenotype. Alcohol Clin Exp Res 19: 378–386.

Cartwright MM, Smith SM (1995b). Stage-dependent effects of ethanol on cranial neural crest cell development: partial basis for the phenotypic variations observed in fetal alcohol syndrome. Alcohol Clin Exp Res 19: 1454–1462.

Cartwright MM, Tessmer LL, Smith SM (1998). Ethanol-induced neural crest apoptosis is coincident with their endogenous death, but is mechanistically distinct. Alcohol Clin Exp Res 22: 142–149.

Chang LK, Schmidt RE, Johnson Jr EM (2003). Alternating metabolic pathways in NGF-deprived sympathetic neurons affect caspase-independent death. J Cell Biol 162: 245–256.

Chen S, Charness ME (2012). Ethanol disrupts axon outgrowth stimulated by netrin-1, GDNF, and L1 by blocking their convergent activation of Src family kinase signaling. J Neurochem 123: 602–612.

Chen SY, Sulik KK (1996). Free radicals and ethanol-induced cytotoxicity in neural crest cells. Alcohol Clin Exp Res 20: 1071–1076.

Chen SY, Sulik KK (2000). Iron-mediated free radical injury in ethanol-exposed mouse neural crest cells. J Pharmacol Exp Ther 294: 134–140.

Chen WJ, Maier SE, Parnell SE et al. (2003). Alcohol and the developing brain: neuroanatomical studies. Alcohol Res Health 27: 174–180.

Chen SY, Dehart DB, Sulik KK (2004). Protection from ethanol-induced limb malformations by the superoxide dismutase/catalase mimetic, EUK-134. Faseb J 18: 1234–1236.

Chen SY, Charness ME, Wilkemeyer MF et al. (2005). Peptide-mediated protection from ethanol-induced neural tube defects. Dev Neurosci 27: 13–19.

Christofferson DE, Yuan J (2010). Necroptosis as an alternative form of programmed cell death. Curr Opin Cell Biol 22: 263–268.

Cudd TA (2005). Animal model systems for the study of alcohol teratology. Exp Biol Med (Maywood) 230: 389–393.

Cui ZJ, Zhao KB, Zhao HJ et al. (2010). Prenatal alcohol exposure induces long-term changes in dendritic spines and synapses in the mouse visual cortex. Alcohol Alcohol 45: 312–319.

Daft PA, Johnston MC, Sulik KK (1986). Abnormal heart and great vessel development following acute ethanol exposure in mice. Teratology 33: 93–104.

Davis WL, Crawford LA, Cooper OJ et al. (1990). Ethanol induces the generation of reactive free radicals by neural crest cells in vitro. J Craniofac Genet Dev Biol 10: 277–293.

Deltour L, Ang HL, Duester G (1996). Ethanol inhibition of retinoic acid synthesis as a potential mechanism for fetal alcohol syndrome. Faseb J 10: 1050–1057.

Dong J, Sulik KK, Chen SY (2008). Nrf2-mediated transcriptional induction of antioxidant response in mouse embryos exposed to ethanol in vivo: implications for the prevention of fetal alcohol spectrum disorders. Antioxid Redox Signal 10: 2023–2033.

Dong J, Sulik KK, Chen SY (2010). The role of NOX enzymes in ethanol-induced oxidative stress and apoptosis in mouse embryos. Toxicol Lett 193: 94–100.

Dou X, Menkari CE, Shanmugasundararaj S et al. (2011). Two alcohol binding residues interact across a domain interface of the L1 neural cell adhesion molecule and regulate cell adhesion. J Biol Chem 286: 16131–16139.

Dou X, Wilkemeyer MF, Menkari CE et al. (2013). Mitogen-activated protein kinase modulates ethanol inhibition of cell adhesion mediated by the L1 neural cell adhesion molecule. Proc Natl Acad Sci U S A 110: 5683–5688.

Downing C, Flink S, Florez-McClure ML et al. (2012). Gene expression changes in C57BL/6 J and DBA/2 J mice following prenatal alcohol exposure. Alcohol Clin Exp Res 36: 1519–1529.

Duester G (1991). A hypothetical mechanism for fetal alcohol syndrome involving ethanol inhibition of retinoic acid synthesis at the alcohol dehydrogenase step. Alcohol Clin Exp Res 15: 568–572.

Dunty Jr WC, Chen SY, Zucker RM et al. (2001). Selective vulnerability of embryonic cell populations to ethanol-induced apoptosis: implications for alcohol-related birth defects and neurodevelopmental disorder. Alcohol Clin Exp Res 25: 1523–1535.

Dunty Jr WC, Zucker RM, Sulik KK (2002). Hindbrain and cranial nerve dysmorphogenesis result from acute maternal ethanol administration. Dev Neurosci 24: 328–342.

Farber NB, Creeley CE, Olney JW (2010). Alcohol-induced neuroapoptosis in the fetal macaque brain. Neurobiol Dis 40: 200–206.

Flentke GR, Garic A, Amberger E et al. (2011). Calcium-mediated repression of beta-catenin and its transcriptional signaling mediates neural crest cell death in an avian model of fetal alcohol syndrome. Birth Defects Res A Clin Mol Teratol 91: 591–602.

Gage JC, Sulik KK (1991). Pathogenesis of ethanol-induced hydronephrosis and hydroureter as demonstrated following in vivo exposure of mouse embryos. Teratology 44: 299–312.

Garic-Stankovic A, Hernandez MR, Chiang PJ et al. (2005). Ethanol triggers neural crest apoptosis through the selective activation of a pertussis toxin-sensitive G protein and a phospholipase Cbeta-dependent Ca2+ transient. Alcohol Clin Exp Res 29: 1237–1246.

Godin EA, O'Leary-Moore SK, Khan AA et al. (2010). Magnetic resonance microscopy defines ethanol-induced brain abnormalities in prenatal mice: effects of acute insult on gestational day 7. Alcohol Clin Exp Res 34: 98–111.

Gohlke JM, Griffith WC, Faustman EM (2005). A systems-based computational model for dose-response comparisons of two mode of action hypotheses for ethanol-induced neurodevelopmental toxicity. Toxicol Sci 86: 470–484.

Goodlett CR, Horn KH (2001). Mechanisms of alcohol-induced damage to the developing nervous system. Alcohol Res Health 25: 175–184.

Goodlett CR, Horn KH, Zhou FC (2005). Alcohol teratogenesis: mechanisms of damage and strategies for intervention. Exp Biol Med (Maywood) 230: 394–406.

Govorko D, Bekdash RA, Zhang C et al. (2012). Male germline transmits fetal alcohol adverse effect on hypothalamic proopiomelanocortin gene across generations. Biol Psychiatry 72: 378–388.

Green ML, Singh AV, Zhang Y et al. (2007). Reprogramming of genetic networks during initiation of the Fetal Alcohol Syndrome. Dev Dyn 236: 613–631.

Guo W, Crossey EL, Zhang L et al. (2011). Alcohol exposure decreases CREB binding protein expression and histone acetylation in the developing cerebellum. PLoS One 6: e19351.

Gutala R, Wang J, Kadapakkam S et al. (2004). Microarray analysis of ethanol-treated cortical neurons reveals disruption of genes related to the ubiquitin-proteasome pathway and protein synthesis. Alcohol Clin Exp Res 28: 1779–1788.

Heaton MB, Moore DB, Paiva M et al. (1999). Bcl-2 overexpression protects the neonatal cerebellum from ethanol neurotoxicity. Brain Res 817: 13–18.

Heaton MB, Paiva M, Madorsky I et al. (2006). Effect of bax deletion on ethanol sensitivity in the neonatal rat cerebellum. J Neurobiol 66: 95–101.

Helms J, Thaller C, Eichele G (1994). Relationship between retinoic acid and sonic hedgehog, two polarizing signals in the chick wing bud. Development 120: 3267–3274.

Henderson GI, Chen JJ, Schenker S (1999). Ethanol, oxidative stress, reactive aldehydes, and the fetus. Front Biosci 4: D541–D550.

Hong M, Krauss RS (2012). Cdon mutation and fetal ethanol exposure synergize to produce midline signaling defects and holoprosencephaly spectrum disorders in mice. PLoS Genet 8: e1002999.

Idrus NM, Napper RM (2012). Acute and long-term Purkinje cell loss following a single ethanol binge during the early third trimester equivalent in the rat. Alcohol Clin Exp Res 36: 1365–1373.

Idrus NM, McGough NN, Riley EP et al. (2011). Administration of memantine during ethanol withdrawal in neonatal rats: effects on long-term ethanol-induced motor incoordination and cerebellar Purkinje cell loss. Alcohol Clin Exp Res 35: 355–364.

Ikonomidou C, Bittigau P, Ishimaru MJ et al. (2000). Ethanol-induced apoptotic neurodegeneration and fetal alcohol syndrome. Science 287: 1056–1060.

Incerti M, Vink J, Roberson R et al. (2010). Prevention of the alcohol-induced changes in brain-derived neurotrophic factor expression using neuroprotective peptides in a model of fetal alcohol syndrome. Am J Obstet Gynecol 202 (457): e451–e454.

Johnson CS, Zucker RM, Hunter 3rd ES et al. (2007). Perturbation of retinoic acid (RA)-mediated limb development suggests a role for diminished RA signaling in the teratogenesis of ethanol. Birth Defects Res A Clin Mol Teratol 79: 631–641.

Kaminen-Ahola N, Ahola A, Flatscher-Bader T et al. (2010a). Postnatal growth restriction and gene expression changes in a mouse model of fetal alcohol syndrome. Birth Defects Res A Clin Mol Teratol 88: 818–826.

Kaminen-Ahola N, Ahola A, Maga M et al. (2010b). Maternal ethanol consumption alters the epigenotype and the phenotype of offspring in a mouse model. PLoS Genet 6: e1000811.

Kietzman HW, Everson JL, Sulik KK et al. (2014). The teratogenic effects of prenatal ethanol exposure are exacerbated by Sonic hedgehog or Gli2 haploinsufficiency in the mouse. PLoS One 9: e89448.

Kilburn BA, Chiang PJ, Wang J et al. (2006). Rapid induction of apoptosis in gastrulating mouse embryos by ethanol and its prevention by HB-EGF. Alcohol Clin Exp Res 30: 127–134.

Kobor M, Weinberg J (2011). Focus on: epigenetics and fetal alcohol spectrum disorders. Alcohol Res Health 34: 29–37.

Komatsu S, Sakata-Haga H, Sawada K et al. (2001). Prenatal exposure to ethanol induces leptomeningeal heterotopia in the cerebral cortex of the rat fetus. Acta Neuropathol 101: 22–26.

Kotch LE, Sulik KK (1992). Patterns of ethanol-induced cell death in the developing nervous system of mice; neural fold states through the time of anterior neural tube closure. Int J Dev Neurosci 10: 273–279.

Kotch LE, Dehart DB, Alles AJ et al. (1992). Pathogenesis of ethanol-induced limb reduction defects in mice. Teratology 46: 323–332.

Kotch LE, Chen SY, Sulik KK (1995). Ethanol-induced teratogenesis: free radical damage as a possible mechanism. Teratology 52: 128–136.

Kumar A, Singh CK, Lavoie HA et al. (2011). Resveratrol restores Nrf2 level and prevents ethanol-induced toxic effects in the cerebellum of a rodent model of fetal alcohol spectrum disorders. Mol Pharmacol 80: 446–457.

Li Z, Miller MW, Luo J (2002). Effects of prenatal exposure to ethanol on the cyclin-dependent kinase system in the developing rat cerebellum. Brain Res Dev Brain Res 139: 237–245.

Lindsley TA, Shah SN, Ruggiero EA (2011). Ethanol alters BDNF-induced Rho GTPase activation in axonal growth cones. Alcohol Clin Exp Res 35: 1321–1330.

Lipinski RJ, Hammond P, O'Leary-Moore SK et al. (2012). Ethanol-induced face-brain dysmorphology patterns are correlative and exposure-stage dependent. PLoS One 7: e43067.

Littner Y, Tang N, He M et al. (2013). L1 cell adhesion molecule signaling is inhibited by ethanol in vivo. Alcohol Clin Exp Res 37: 383–389.

Liu Y, Balaraman Y, Wang G et al. (2009). Alcohol exposure alters DNA methylation profiles in mouse embryos at early neurulation. Epigenetics 4: 500–511.

Loucks EJ, Ahlgren SC (2009). Deciphering the role of Shh signaling in axial defects produced by ethanol exposure. Birth Defects Res A Clin Mol Teratol 85: 556–567.

Maffi SK, Rathinam ML, Cherian PP et al. (2008). Glutathione content as a potential mediator of the vulnerability of cultured fetal cortical neurons to ethanol-induced apoptosis. J Neurosci Res 86: 1064–1076.

Miller MW (1986). Effects of alcohol on the generation and migration of cerebral cortical neurons. Science 233: 1308–1311.

Miller MW (1993). Migration of cortical neurons is altered by gestational exposure to ethanol. Alcohol Clin Exp Res 17: 304–314.

Miller MW (1997). Effects of prenatal exposure to ethanol on callosal projection neurons in rat somatosensory cortex. Brain Res 766: 121–128.

Miller MW (2006). Brain Development: Normal Processes and the Effects of Alcohol and Nicotine. Oxford University Press, Oxford.

Miller MW, Robertson S (1993). Prenatal exposure to ethanol alters the postnatal development and transformation of radial glia to astrocytes in the cortex. J Comp Neurol 337: 253–266.

Miranda RC (2012). MicroRNAs and fetal brain development: implications for ethanol teratology during the second trimester period of neurogenesis. Front Genet 3: 77.

Monk BR, Leslie FM, Thomas JD (2012). The effects of perinatal choline supplementation on hippocampal cholinergic development in rats exposed to alcohol during the brain growth spurt. Hippocampus 22: 1750–1757.

Mooney SM, Siegenthaler JA, Miller MW (2004). Ethanol induces heterotopias in organotypic cultures of rat cerebral cortex. Cereb Cortex 14: 1071–1080.

Narasimhan M, Mahimainathan L, Rathinam ML et al. (2011). Overexpression of Nrf2 protects cerebral cortical neurons from ethanol-induced apoptotic death. Mol Pharmacol 80: 988–999.

O'Leary-Moore SK, Parnell SE, Lipinski RJ et al. (2011). Magnetic resonance-based imaging in animal models of fetal alcohol spectrum disorder. Neuropsychol Rev 21: 167–185.

Olney JW, Tenkova T, Dikranian K et al. (2002a). Ethanol-induced apoptotic neurodegeneration in the developing C57BL/6 mouse brain. Brain Res Dev Brain Res 133: 115–126.

Olney JW, Wozniak DF, Farber NB et al. (2002b). The enigma of fetal alcohol neurotoxicity. Ann Med 34: 109–119.

Otero NK, Thomas JD, Saski CA et al. (2012). Choline supplementation and DNA methylation in the hippocampus and prefrontal cortex of rats exposed to alcohol during development. Alcohol Clin Exp Res 36: 1701–1709.

Parnell SE, Chen SY, Charness ME et al. (2007). Concurrent dietary administration of D-SAL and ethanol diminishes ethanol's teratogenesis. Alcohol Clin Exp Res 31: 2059–2064.

Parnell SE, Sulik KK, Dehart DB et al. (2010). Reduction of ethanol-induced ocular abnormalities in mice through dietary administration of *N*-acetylcysteine. Alcohol 44: 699–705.

Poggi SH, Goodwin K, Hill JM et al. (2003). The role of activity-dependent neuroprotective protein in a mouse model of fetal alcohol syndrome. Am J Obstet Gynecol 189: 790–793.

Pullarkat RK (1991). Hypothesis: prenatal ethanol-induced birth defects and retinoic acid. Alcohol Clin Exp Res 15: 565–567.

Ramachandran V, Watts LT, Maffi SK et al. (2003). Ethanol-induced oxidative stress precedes mitochondrially mediated apoptotic death of cultured fetal cortical neurons. J Neurosci Res 74: 577–588.

Resnicoff M, Sell C, Ambrose D et al. (1993). Ethanol inhibits the autophosphorylation of the insulin-like growth factor 1 (IGF-1) receptor and IGF-1-mediated proliferation of 3 T3 cells. J Biol Chem 268: 21777–21782.

Rhinn M, Dolle P (2012). Retinoic acid signalling during development. Development 139: 843–858.

Rice JP, Suggs LE, Lusk AV et al. (2012). Effects of exposure to moderate levels of ethanol during prenatal brain development on dendritic length, branching, and spine density in the nucleus accumbens and dorsal striatum of adult rats. Alcohol 46: 577–584.

Roessler E, Muenke M (2010). The molecular genetics of holoprosencephaly. Am J Med Genet C Semin Med Genet 154C: 52–61.

Ryan SH, Williams JK, Thomas JD (2008). Choline supplementation attenuates learning deficits associated with neonatal alcohol exposure in the rat: effects of varying the timing of choline administration. Brain Res 1237: 91–100.

Sandor S, Checiu M, Fazakas-Todea I et al. (1980). The effect of ethanol upon early development in mice and rats. I. In vivo effect upon preimplantation and early postimplantation stages. Morphol Embryol (Bucur) 26: 265–274.

Sari Y, Gozes I (2006). Brain deficits associated with fetal alcohol exposure may be protected, in part, by peptides derived from activity-dependent neurotrophic factor and activity-dependent neuroprotective protein. Brain Res Rev 52: 107–118.

Sari Y, Weedman JM, Ge S (2012). Activity-dependent neurotrophic factor-derived peptide prevents alcohol-induced apoptosis, in part, through Bcl2 and c-Jun N-terminal kinase signaling pathways in fetal brain of C57BL/6 mouse. Neuroscience 202: 465–473.

Sathyan P, Golden HB, Miranda RC (2007). Competing interactions between micro-RNAs determine neural progenitor survival and proliferation after ethanol exposure: evidence from an ex vivo model of the fetal cerebral cortical neuroepithelium. J Neurosci 27: 8546–8557.

Schneider ML, Moore CF, Adkins MM (2011). The effects of prenatal alcohol exposure on behavior: rodent and primate studies. Neuropsychol Rev 21: 186–203.

Smith SM (1997). Alcohol-induced cell death in the embryo. Alcohol Health Res World 21: 287–297.

Solomon BD, Bear KA, Wyllie A et al. (2012). Genotypic and phenotypic analysis of 396 individuals with mutations in Sonic Hedgehog. J Med Genet 49: 473–479.

Spong CY, Abebe DT, Gozes I et al. (2001). Prevention of fetal demise and growth restriction in a mouse model of fetal alcohol syndrome. J Pharmacol Exp Ther 297: 774–779.

Stefanis L, Park DS, Friedman WJ et al. (1999). Caspase-dependent and -independent death of camptothecin-treated embryonic cortical neurons. J Neurosci 19: 6235–6247.

Stockard C (1913). The effect on the offspring of intoxicating the male parent and the transmission of the defects to subsequent generations. Am Nat 47: 641–682.

Stratford T, Horton C, Maden M (1996). Retinoic acid is required for the initiation of outgrowth in the chick limb bud. Curr Biol 6: 1124–1133.

Sulik KK (2005). Genesis of alcohol-induced craniofacial dysmorphism. Exp Biol Med (Maywood) 230: 366–375.

Sulik KK, Cook CS, Webster WS (1988). Teratogens and craniofacial malformations: relationships to cell death. Development 103 (Suppl.): 213–231.

Tang N, Farah B, He M et al. (2011). Ethanol causes the redistribution of L1 cell adhesion molecule in lipid rafts. J Neurochem 119: 859–867.

Tarelo-Acuna L, Olvera-Cortes E, Gonzalez-Burgos I (2000). Prenatal and postnatal exposure to ethanol induces changes in the shape of the dendritic spines from hippocampal CA1 pyramidal neurons of the rat. Neurosci Lett 286: 13–16.

Thomas JD, Tran TD (2012). Choline supplementation mitigates trace, but not delay, eyeblink conditioning deficits in rats exposed to alcohol during development. Hippocampus 22: 619–630.

Thomas JD, Fleming SL, Riley EP (2002). Administration of low doses of MK-801 during ethanol withdrawal in the developing rat pup attenuates alcohol's teratogenic effects. Alcohol Clin Exp Res 26: 1307–1313.

Thomas JD, Garrison M, O'Neill TM (2004). Perinatal choline supplementation attenuates behavioral alterations associated with neonatal alcohol exposure in rats. Neurotoxicol Teratol 26: 35–45.

Thomas JD, Biane JS, O'Bryan KA et al. (2007). Choline supplementation following third-trimester-equivalent alcohol exposure attenuates behavioral alterations in rats. Behav Neurosci 121: 120–130.

Thomas JD, Abou EJ, Dominguez HD (2009). Prenatal choline supplementation mitigates the adverse effects of prenatal alcohol exposure on development in rats. Neurotoxicol Teratol 31: 303–311.

Thomas JD, Idrus NM, Monk BR et al. (2010). Prenatal choline supplementation mitigates behavioral alterations associated with prenatal alcohol exposure in rats. Birth Defects Res A Clin Mol Teratol 88: 827–837.

Ullah I, Ullah N, Naseer MI et al. (2012). Neuroprotection with metformin and thymoquinone against ethanol-induced apoptotic neurodegeneration in prenatal rat cortical neurons. BMC Neurosci 13: 11.

Vangipuram SD, Lyman WD (2010). Ethanol alters cell fate of fetal human brain-derived stem and progenitor cells. Alcohol Clin Exp Res 34: 1574–1583.

Wentzel P, Eriksson UJ (2006). Ethanol-induced fetal dysmorphogenesis in the mouse is diminished by high antioxidative capacity of the mother. Toxicol Sci 92: 416–422.

Wilkemeyer MF, Menkari CE, Spong CY et al. (2002). Peptide antagonists of ethanol inhibition of l1-mediated cell-cell adhesion. J Pharmacol Exp Ther 303: 110–116.

Wilkemeyer MF, Chen SY, Menkari CE et al. (2003). Differential effects of ethanol antagonism and neuroprotection in peptide fragment NAPVSIPQ prevention of ethanol-induced developmental toxicity. Proc Natl Acad Sci U S A 100: 8543–8548.

Wilkemeyer MF, Chen SY, Menkari CE et al. (2004). Ethanol antagonist peptides: structural specificity without stereospecificity. J Pharmacol Exp Ther 309: 1183–1189.

Yan D, Dong J, Sulik KK et al. (2010). Induction of the Nrf2-driven antioxidant response by tert-butylhydroquinone prevents ethanol-induced apoptosis in cranial neural crest cells. Biochem Pharmacol 80: 144–149.

Yanni PA, Lindsley TA (2000). Ethanol inhibits development of dendrites and synapses in rat hippocampal pyramidal neuron cultures. Brain Res Dev Brain Res 120: 233–243.

Young C, Olney JW (2006). Neuroapoptosis in the infant mouse brain triggered by a transient small increase in blood alcohol concentration. Neurobiol Dis 22: 548–554.

Young C, Klocke BJ, Tenkova T et al. (2003). Ethanol-induced neuronal apoptosis in vivo requires BAX in the developing mouse brain. Cell Death Differ 10: 1148–1155.

Young C, Roth KA, Klocke BJ et al. (2005). Role of caspase-3 in ethanol-induced developmental neurodegeneration. Neurobiol Dis 20: 608–614.

Young C, Straiko MM, Johnson SA et al. (2008). Ethanol causes and lithium prevents neuroapoptosis and suppression of pERK in the infant mouse brain. Neurobiol Dis 31: 355–360.

Zachman RD, Grummer MA (1998). The interaction of ethanol and vitamin A as a potential mechanism for the pathogenesis of Fetal Alcohol syndrome. Alcohol Clin Exp Res 22: 1544–1556.

Zhou FC, Fang Y, Goodlett C (2008). Peptidergic agonists of activity-dependent neurotrophic factor protect against prenatal alcohol-induced neural tube defects and serotonin neuron loss. Alcohol Clin Exp Res 32: 1361–1371.

Zhou FC, Balaraman Y, Teng M et al. (2011a). Alcohol alters DNA methylation patterns and inhibits neural stem cell differentiation. Alcohol Clin Exp Res 35: 735–746.

Zhou FC, Zhao Q, Liu Y et al. (2011b). Alteration of gene expression by alcohol exposure at early neurulation. BMC Genomics 12: 124.

Handbook of Clinical Neurology, Vol. 125 (3rd series)
Alcohol and the Nervous System
E.V. Sullivan and A. Pfefferbaum, Editors

Chapter 27

Current hypotheses on the mechanisms of alcoholism

R.P. VETRENO* AND F.T. CREWS
Bowles Center for Alcohol Studies, Department of Pharmacology and Psychiatry, University of North Carolina, Chapel Hill, NC, USA

ALCOHOLISM IS A DISEASE CHARACTERIZED BY CONTINUED USE DESPITE NEGATIVE CONSEQUENCES

Alcoholism is a major societal problem due to its negative impact on health and socioeconomics. Excessive alcohol use and abuse are among the leading causes of premature death and disability in the United States (Michaud et al., 2006), and are ranked third behind tobacco use and poor diet/obesity as a leading cause of preventable mortality (Mokdad et al., 2004). According to a large, population-based study by the National Institute on Alcohol Abuse and Alcoholism on alcohol drinking habits in the United States (the National Epidemiological Survey on Alcohol and Related Conditions: NESARC), approximately 65% of American adults report consuming alcohol (i.e., one or more drinks in the last 12-month period). Within this population, 72% of individuals reported consuming low, safe levels of alcohol (men: fewer than 5 drinks daily, 14 drinks weekly; women: fewer than 4 drinks daily, 7 drinks weekly) while the remaining 28% of individuals reported heavy drinking (frequent consumption of 5 or more drinks per day: Grant et al., 2004, 2007). Approximately three-quarters of all alcohol ingested in the United States is consumed by 10% of the population (Li, 2008). In addition to the increased risk of developing a variety of health problems, heavy-drinking patterns also increase the likelihood of developing an addiction to alcohol.

According to the *Diagnostic and Statistical Manual of Mental Disorders-5* (DSM-5), a psychiatric diagnosis of substance abuse involves repeated use of alcohol or other drugs despite problems related to use of the substance (American Psychiatric Association, 2013). This is exemplified by the requirement that the individual's drinking habit lead to clinically significant impairment or distress. Central to a medical diagnosis of alcohol dependence is loss of control over use and preoccupation with alcohol. Diminished control over alcohol use refers to escalating use, inability to reduce or control use, and continued, persistent use despite harm to the individual, family, and community due to impaired ability to alter behavior. Preoccupation refers to increased time spent obtaining, using, and recovering from alcohol as well as abandoning important social and family obligations in favor of alcohol use. The acute effects of alcohol on behavior include disinhibition and increased impulsivity, which are indicative of reduced prefrontal cortex (PFC) executive function (Weafer and Fillmore, 2008; Dick et al., 2010).

Although the acute effects of alcohol ingestion on brain and behavior have received considerable attention (Volkow et al., 1990; Varlinskaya and Spear, 2002), we are only beginning to unravel the long-term consequences of prolonged alcohol on neurobiology. In support of persistent alcohol-induced changes to the brain, alcoholics return to their heavy-drinking habits during relapse and do not initiate their drinking behaviors from baseline, earlier levels. A key component of the behavioral pathology of alcohol dependence and addiction centers on loss of cortical executive function and cognitive flexibility, and mounting limbic anxiety and impulsivity. In the chapter that follows, mechanisms that may underlie the shift from moderate-drinking practices to heavy-drinking patterns and alcoholism will be discussed in relation to alcohol-induced changes to brain structure and function.

*Correspondence to: Ryan P. Vetreno, Bowles Center for Alcohol Studies, University of North Carolina, The School of Medicine, CB #7178, 1021 Thurston-Bowles Building, Chapel Hill, NC 27599-71780, USA. E-mail: rvetreno@email.unc.edu

DIMINISHED EXECUTIVE FUNCTION IN THE ALCOHOLIC IS CONSISTENT WITH A COMPROMISE OF PREFRONTAL CORTEX FUNCTION

Chronic repetitive use of alcohol culminates in changes to brain and behavior that are characterized by diminished behavioral control, difficulty avoiding negative consequences, and increased preoccupation that is associated with mounting craving and limbic negative affect. Alcohol dependence and addiction involve a disruption of the normal balance between self-control mechanisms and emotional drive. The PFC and limbic system structures are particularly vulnerable to the neurotoxic effects of alcohol, and are critically involved in the aforementioned processes. The PFC modulates decision making and other executive functions, such as motivation, planning and goal setting, and impulse inhibition. The limbic system, which consists of the amygdala, hippocampus, and diencephalon, contributes to regulation of memory, emotion, and mood. Continued use of alcohol and other drugs of abuse alters cortical and limbic system neurobiology, culminating in a loss of attention, impaired decision making, and increased impulsivity and anxiety that drive repeated cycles of binge drinking, further disrupting cognitive-limbic circuits through progressive neuroimmune activation and loss of control over alcohol use (Fig. 27.1).

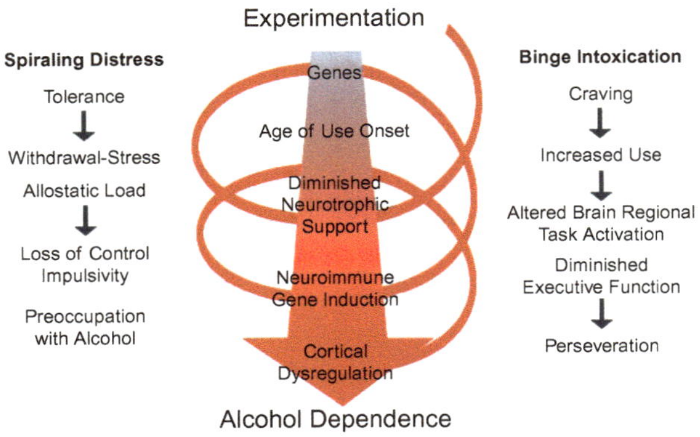

Fig. 27.1. Spiral of distress modeling the progression from alcohol experimentation to alcohol dependence. Left: spiraling distress. Cycles of alcohol intoxication and withdrawal stress increase tolerance and increased allostatic load that promote craving and increased use, leading to further cycles of binge intoxication that progress to loss of control over alcohol use (Koob and Le Moal, 1997; Koob et al., 2004). Middle: Genetics and age of drinking onset are factors that influence binge drinking which, over cycles, alters brain gene expression, e.g., decreased trophic factor expression and increased neuroimmune gene expression that ultimately lead to cortical dysfunction. Right: Craving increases use and binge drinking that disrupts frontal cortical function, reduces behavioral flexibility and executive functions, increasing perseverative-compulsive actions and decreasing goal orientation. The spiral of distress links binge intoxication cycles to alcohol dependence, characterized by preoccupation with alcohol and impulsive drug taking that is likely due in part to diminished frontal cortical executive function. These mechanisms promote alcohol dependence and addiction through increased craving and other negative feelings as well as perseveration and loss of control over drinking. (Adapted from Koob and Le Moal, 1997.)

Human prefrontal cortex case studies

The PFC regulates behavior through decision-making processes and other executive functions, such as motivation, planning and goal setting, impulse inhibition, and regulation of limbic drive. Executive function encompasses abstract thinking, decision making, motivation, planning, attention to tasks, and inhibition of impulsive responses. Impulsivity, which is regulated by the PFC, refers to actions that are poorly conceived, prematurely expressed, and unduly risky or inappropriate to the situation and that often result in undesirable consequences (de Wit, 2009) that increase during intoxication and alcoholism. In addition, the PFC is involved in the retention of long-term emotional memory and modification of emotions (Bechara et al., 2000) to fit societal norms necessary for functional integration into society.

Some of the behavioral symptoms of alcoholism (e.g., loss of attention, poor decision making, and increased impulsivity) are similar to deficits observed in individuals with lesions of the PFC (Bechara et al., 1994; Bechara, 2005). The PFC consists of several structures, including the dorsolateral PFC (dlPFC), orbitofrontal cortex (OFC), ventromedial PFC (vmPFC), and anterior cingulate cortex (ACC). Although precise functional contributions of each of these intersecting structures are difficult to delineate due to considerable interconnections and overlapping functions, human case studies involving lesions and insults to these cortical regions have offered some insight into their respective contributions to executive function and addiction-like behavior. The dlPFC (Brodmann areas 9 and 46) contributes to executive function through modulation of attention as well as behavioral organization (Cummings, 1998; Stuss and Alexander, 2000). In humans, lesions of the dlPFC produce a "frontal dysexecutive syndrome" characterized by impaired planning, strategy development, and working memory as well as diminished cognitive flexibility (Clark and Manes, 2004). Similar working-memory deficits have been observed in non-human primates with lesions of this structure (Baddeley, 1992; Goldman-Rakic, 1992). The dlPFC is also associated with behavioral regulation due to its ability to maintain and integrate sensory and affective information (Carmichael and Price, 1995a, b), allowing for the representation of expected outcomes used to guide behavior (Schoenbaum et al., 2006). The OFC (Brodmann areas 11, 12, and 13) is also critically involved in the generation and use of outcome expectancies (Schoenbaum et al., 2006) as damage to this region results in profound deficits in self-regulation and control. These deficits were exemplified in the famous case of Phineas Gage, a railroad worker who survived the passage of a tamping rod through his OFC. He experienced several cognitive changes as a consequence of this insult that included behavioral disinhibition and loss of behavioral flexibility (Harlow, 1848, 1868). Other studies of OFC lesions have identified behavioral changes involving increased impulsivity and perseverative behaviors as well as deficits in emotional recognition and impaired judgment in social contexts (Rolls et al., 1994; Berlin et al., 2004; Bechara, 2005; Coutlee and Huettel, 2012; Rosenbloom et al., 2012). Studies of humans with bilateral damage of the vmPFC (Brodmann area 10) reveal severe impairments in decision making (Eslinger et al., 1984; Damasio et al., 1991; Bechara et al., 1998), defined as selecting the most advantageous response from an array of available options. Finally, the ACC (Brodmann areas 24, 32, and 33), especially the dorsal subregion, has been implicated in inhibitory control (Yucel and Lubman, 2007), whereas the anterior component plays a role in linking decision making to emotion (Cella et al., 2010). These human case studies suggest that damage to the PFC might contribute mechanistically to the development and maintenance of alcoholism. Thus, subregions of the PFC each contribute to executive functioning, and damage to these regions results in behavioral impairments that are commonly observed in the alcoholic individual (e.g., diminished cognitive flexibility and decision making, perseveration, behavioral dysregulation, and emotive dysfunction).

Alcohol-induced neurodegeneration

Human lesion studies suggest that loss of PFC function contributes to cognitive inflexibility, impulsivity, and loss of behavioral control, which are hallmark features common in alcoholism. Alcoholics are capable of ingesting large quantities of alcohol, which can result in PFC neurodegeneration (Crews and Boettiger, 2009) and might contribute to the loss of executive function associated with alcoholism. In the alcoholic brain, neuroimaging studies find reduced gray- and white-matter volumes across multiple regions, which are greatest in the PFC as well as white-matter tracts of the frontal cortex (Pfefferbaum et al., 1995; Kubota et al., 2001; Sullivan and Pfefferbaum, 2005). Although neuronal loss does not appear to account for all of the volumetric loss observed in the alcoholic brain, both the superior frontal cortex (Harper and Kril, 1989) and OFC (Miguel-Hidalgo et al., 2006; Qin and Crews, 2012b) show neuronal loss, while the temporal lobes do not show evidence of similar loss (Kril and Harper, 1989). This occurs in the absence of major nutritional deficiencies (e.g., thiamine (vitamin B_1)), although they can also cause neurodegeneration and potentiate alcoholic neurodegeneration (Bowden et al., 2001; Qin and Crews, 2014). It has been proposed that progressive increases in alcohol consumption, a common feature of alcoholism, lead to PFC alterations that reduce executive functioning that promote further alcohol abuse and neurodegeneration (Crews et al., 2004). Alcohol-induced PFC degeneration leads to loss of executive function and diminished limbic system regulation, leading to increased impulsive behavior that contributes to the persistence of addiction (Bechara, 2005; Crews et al., 2005).

Alcoholic individuals display several components of impulsivity, including deficits in delay discounting and behavioral inhibition (de Wit, 2009). Connections between the vmPFC, ACC, and amygdala are thought to be critical for impulse control (Bechara, 2005). Human neuroimaging studies have also found hyperactivation in the amygdala in response to stimuli that elicit craving, which is indicative of impulsivity (Breiter et al., 1997; Childress et al., 1999; Kilts et al., 2001). In contrast, there is evidence for hypoactivation of the PFC in the addicted brain. Employing positron emission tomography in conjunction with 2-deoxy-2[^{18}F] fluoro-D-glucose, Volkow and colleagues (1997) found reductions in PFC metabolism that were evident 2–3 months following detoxification, suggesting that cortical hypoactivity represents a persistent consequence of alcoholism. It is possible that PFC hypoactivity could lead to compulsive drug use in response to drug-related stimuli (Volkow et al., 2004). Thus, hyperactivity in the amygdala, coupled with a hypofunctional PFC, might be a mechanism underlying the development of alcoholism.

Frontal cortex dysfunction is common in the human alcoholic brain (Crews and Boettiger, 2009; Crews and Nixon, 2009). Compromised frontal cortical executive function could lead to alcohol abuse, or alcohol use might insult PFC function. Using animal models with high blood levels of ethanol induces limbic and cortical brain damage (Crews et al., 2004; Crews and Nixon, 2009). Studies investigating changes in human cortical thickness find thinning in many cortical brain regions, including entorhinal cortex, piriform cortex, and perirhinal cortex (Fortier et al., 2011) that are also found to be sensitive to binge-drinking rat models of brain damage (Fig. 27.2) (Obernier et al., 2002a, b). The brain damage in the rat binge ethanol study did not alter learning ability assessed using the Morris water maze special

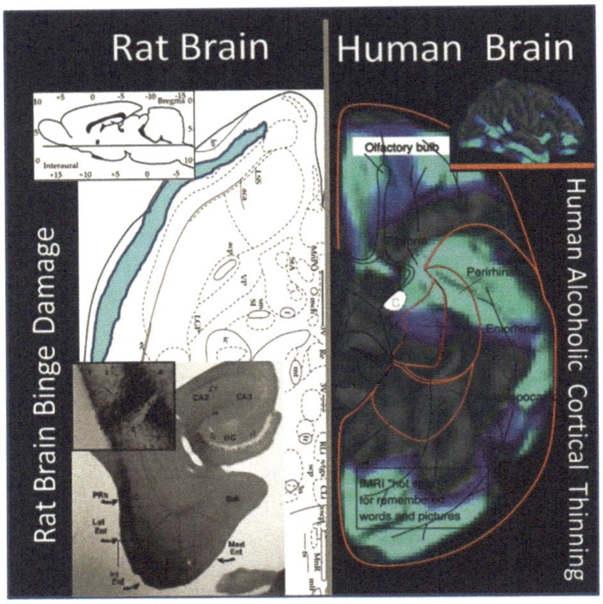

Fig. 27.2. Comparison of human alcoholic cortical thinning with rat binge alcohol treatment-induced neurodegeneration. Shown are a horizontal section through the ventral rat brain and the ventral surface of the human brain. (Left side) Rat brain anatomic section in the upper left indicates location of ventral horizontal section. The bottom left shows a brain section with silver cell death stain (black areas) showing neurodegeneration from a binge-drinking model that includes association cortical areas, piriform and perirhinal cortex, as well as entorhinal cortex. Blue shading indicates that rat piriform neurodegeneration extends to frontal regions. (Adapted from Obernier et al., 2002a; Crews et al., 2004.) On the right side is a ventral view of the cortex from the human brain. Human alcoholic ventral cortical thinning is shown as a significance map of group differences in cortical thickness in abstinent alcoholics as compared to non-alcoholic controls with areas showing brain regions with significant cortical thinning in alcoholics highlighted in blue (light blue, $p < 0.01$; dark blue, $p < 0.05$). (Adapted from Fortier et al., 2011.) Note perirhinal, entorhinal, and piriform cortex are areas with cortical thinning in humans and areas of binge-drinking neurodegeneration in rats, suggesting binge-drinking human alcoholics are damaging association cortical areas. (Adapted from Fortier et al., 2011.)

learning task; however, it did alter reversal learning. Reversal learning tasks assess behavioral or cognitive flexibility. In these tasks, which are applicable across species, subjects first learn a stimulus–reward association, and then the reward association is switched. This ostensibly simple process taxes multiple executive functions, including attention, working memory, and response inhibition. Alcoholic humans and binge-treated animals show a loss of flexibility. Cognitive or behavioral flexibility is a critical executive function that can be broadly defined as the ability to adapt behaviors in response to changes in the environment. Binge ethanol treatment for 4 days did not alter learning ability, but 3 weeks after the binge treatment ethanol-treated rats took more time and a greater number of trials to learn the reversal location (Fig. 27.3) (Obernier et al., 2002b). Reversal learning provides a measure of rule shifting, an executive function involving the ability to revise a previously learned behavioral response with a new response when outcomes do not match those predicted by the preceding cues (Stalnaker et al., 2009). Loss of control of drinking alcohol as a key pathology of alcoholism likely includes a loss of behavioral flexibility related to binge drinking-induced changes in brain. More recent studies have found that adolescent binge ethanol exposure of mice (Coleman et al., 2009, 2011) or adolescent exposure of rats (Vetreno et al., 2013) shows similar deficits in behavioral flexibility as adults assessed using reversal learning tasks. This behavioral paradigm models the perseveration and inability of human alcoholics to learn new, adaptive behaviors. Taken together, these human and animal studies implicate dysfunctional PFC and limbic system structures in the behavioral symptoms of addiction.

Alcohol alters the neurotrophin/ neuroimmune balance

Although the cellular and molecular mechanisms underlying addiction remain to be fully elucidated, recent evidence suggests the potential involvement of a shift in neurotrophic/neuroimmune signaling (Fig. 27.4). Transcription factors, such as cyclic AMP-responsive element-binding protein (CREB) and nuclear factor kappa-light-chain enhancer of activated B cells (NF-κB) regulate expression of a diversity of central nervous system (CNS) genes, and have recently been implicated as a mechanism underlying alcoholism. CREB and its many target genes, including neuropeptide Y (NPY) and brain-derived neurotrophic factor (BDNF), are involved in promoting neuronal survival and protecting neurons from excitotoxicity and apoptosis (Lonze and Ginty, 2002). Glutamatergic *N*-methyl-D-aspartate (NMDA) receptors regulate synaptic plasticity and can be coupled to CREB. However, high levels of extrasynaptic NMDA receptor activation and NMDA receptor-mediated Ca^{2+} flux trigger excitotoxicity, leading to either rapid or delayed neuronal death, which is associated with diminished CREB activation (Hardingham and Bading, 2010). Levels of CREB-DNA binding and phosphorylated CREB, as well as the target gene BDNF, are decreased in the rat frontal cortex following a 24-hour withdrawal from chronic ethanol exposure (Pandey et al., 1999, 2001). In addition, NPY levels are reduced in the cortex following ethanol treatment, an effect that was complemented by reduced phosphorylated CREB

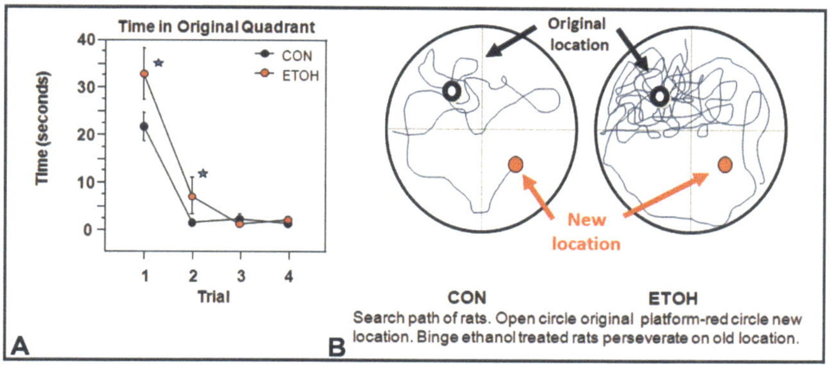

Fig. 27.3. Binge drinking reduces behavioral flexibility as assessed by reversal learning. This study treated rats with a 4-day binge-drinking model and then assessed learning and reversal learning after weeks of abstinence to assess long-term changes induced by binge drinking. All animals (both control (CON) and alcohol binge-treated (EtOH)) learned the initial location without any apparent learning differences. Both CON and EtOH animals learned the location of the platform in the Morris water maze task, suggesting binge drinking did not alter learning. For reversal learning the submerged platform was moved to the opposite quadrant and animals were given four trials (21 days postbinge treatment). Binge-treated animals had difficulty. They took more time to locate the new location, spent more time in the previous location consistent with perseveration on the old location, and took longer to reach criteria in the reversal task. (**A**) Shown is the mean ± SEM of the time required to find the reversal location during the reversal trials. Reversal criterion was set at 2 sd from the distance to platform on the last day of the reference memory task. EtOH animals required a significantly greater number of trials to reach criterion than CON animals [$t(14) = 2.376$; $*p < 0.05$]. (**B**) A vertical view of the track taken by a CON and an EtOH rat during the first reversal trial. The open circle represents the location of the submerged platform the animals initially learned. The red circle represents the location of the platform during the reversal learning task. The EtOH animal has numerous reentries into the original goal quadrant due to perseverative behavior and poor search strategy and fails to reach the new platform location within the 90-second time allowed. The binge-treated animals have not received ethanol for several weeks. The reversal learning deficits are consistent with a loss of behavioral flexibility due to disruption of frontal cortical circuits. (Data summarized from Obernier et al., 2002b.)

(Bison and Crews, 2003). Using *in vitro* slice culture, our laboratory found that ethanol dose-dependently reduces CREB-DNA binding while simultaneously increasing NF-κB-DNA binding (Zou and Crews, 2006).

In the CNS, NF-κB is well known for its ubiquitous roll in inflammatory and neuroimmune responses (Mattson and Camandola, 2001). This transcription factor is expressed in neurons and glia (Kaltschmidt et al., 1993; Mattson and Camandola, 2001) where it regulates both physiologic processes and neurodegeneration (Grilli and Memo, 1999; Schneider et al., 1999; Pizzi et al., 2002). Nuclear activation of NF-κB is associated with induction of several downstream proinflammatory cytokines (e.g., tumor necrosis factor-alpha (TNF-α), interleukin (IL)-1β, and IL-6), chemokines (e.g., monocyte chemoattractant protein-1 (MCP-1, CCL-2)), and oxidases (inducible nitric oxide synthase (iNOS), cyclooxygenase-2 (COX-2), and NADPH oxidase (NOX2)) that lead to positive loops of amplification (Fig. 27.5). Activation of the proinflammatory NF-κB cascade with simultaneous reductions of neurotropic CREB signaling results in hyperexcitability, neuronal injury, and network reorganization that might also represent a mechanism that contributes to the development of alcoholism.

ALCOHOL AND INNATE IMMUNE SYSTEM ACTIVATION

Over the last decade, the innate immune system has received considerable attention for its involvement in alcoholism. Innate immunity, together with other immune-regulating nervous system components, regulates the neuroimmune response in the CNS. Alcohol exerts its deleterious effects first on the periphery through proinflammatory cascades (Crews et al., 2006b) that activate both the adaptive and innate immune system (Szabo and Mandrekar, 2009; McClain et al., 2011; Vetreno et al., 2013). The adaptive immune system consists of a class of highly specialized lymphocytic cells that recognize, remember, and target specific pathogens through the production of antibodies. It is through the formation of an immunologic memory of the pathogen that the adaptive immune system protects the organism from future insults. In contrast to adaptive immunity, the innate immune system mounts non-specific immune responses to pathogen invasion via secretion of proinflammatory cytokines and chemokines that activate neuroimmune cells (e.g., microglia: Dantzer et al., 2008).

Microglia, the resident innate immune macrophage-like cells of the CNS, along with astrocytes, modulate

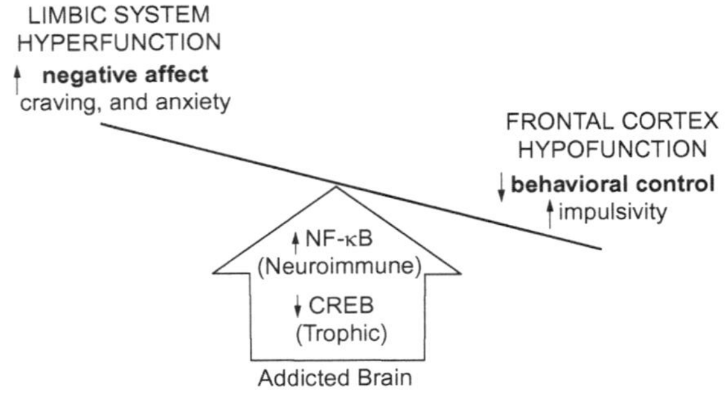

Fig. 27.4. Alcohol-induced shift in neurotrophic/neuroimmune axis. A simplified schematic distinguishing the frontal cortical and limbic system behavioral changes that characterize drug-induced activation of the innate immune system. Both alcohol addiction and stress upregulate nuclear factor kappa-light-chain enhancer of activated B cells (NF-κB) signaling while reducing cyclic AMP-responsive element-binding protein (CREB) expression, which reduces glutamate transporters, leading to prefrontal cortex hyperexcitability (Crews et al., 2006a; Zou and Crews, 2006; Reissner and Kalivas, 2010) and diminished cortical behavioral control and behavioral inflexibility (Stalnaker et al., 2009; Gruber et al., 2010). Concurrently, induction of innate immune genes in limbic brain regions increases negative affect, craving, and anxiety-like behaviors, prompting further drug abuse and self-medication (Kelley and Dantzer, 2011). The long-term deleterious effects of prolonged alcohol abuse involve diminished activation of frontal cortical behavioral control circuits that lead to a progressive loss of attention and poor decision making that combine with increased negative affect and anxiety that motivate further drug-taking behaviors. Taken together, these innate immune gene-induced behavioral changes characterize the drug-addicted brain (Crews and Vetreno, 2011).

important metabolic, trophic, and synaptic functions in addition to responding to and amplifying neuroimmune responses in the CNS (Streit et al., 2004; Farina et al., 2007). In the healthy brain, ramified or "resting" microglia contribute to the integration of sensory systems and overall survey of the brain milieu (Raivich, 2005). In response to an endogenous or exogenous insult, microglia undergo distinct morphologic and gene expression changes associated with cellular activation (Graeber, 2010). During the initial stages of activation, microglia become enlarged and increase expression of cell matrix and cell adhesion proteins. They also begin to secrete neuroimmune signaling molecules that upregulate expression of major histocompatibility complex (MHC) proteins and Toll-like receptors (TLRs), key signaling molecules that respond to bacterial endotoxins (e.g., lipopolysaccharide (LPS)). In general, the TLR family of pattern recognition receptors consists of an extracellular domain of leucine-rich repeat motifs, a transmembrane domain that determines receptor location, and an intracellular domain (Akira et al., 2001; Takeda et al., 2003). To date, 13 TLRs have been identified (TLRs 1–13) in mammals (Medzhitov, 2001; Takeda et al., 2003) that signal through similar cascades to activate NF-κB transcription of additional proinflammatory cytokines through paracrine and autocrine amplification. Activated microglia secrete many proinflammatory cytokines (e.g., TNF-α, IL-1β, IL-6). Cytokines are signaling molecules that regulate many processes and were first discovered for their roles in orchestrating complex multicellular processes involved in defending against pathogen infection and clearing damaged cells and tissue, including signals that attract specific cells toward sites of tissue damage-initiating proinflammatory-oxidative cascades that resolve and are followed by trophic, wound-healing, cell growth cascades (Merrill and Benveniste, 1996; Pollmacher et al., 2002). Cytokine signals interact in proinflammatory signaling cascades that spread and amplify in an autocrine and paracrine fashion. As neuroimmune signaling cascades continue to intensify, microglia progress toward mitosis, proliferation, and phagocytic oxidative surges that dissolve and engulf cellular debris (Fig. 27.6). Highly activated microglia are termed "ameboid" and are characterized by a phagocytic, rounded macrophage-like morphology associated with maximal proinflammatory activation and microglial proliferation (Graeber, 2010). Microglia are further characterized as either proinflammatory (M1), or neurotrophic (M2), depending on whether they are functioning to "kill" or "repair," respectively, although the morphologic characterization of these microglial subtypes is not entirely understood (Colton, 2009; Michelucci et al., 2009). Astrocytes are also pivotal components of the neuroimmune system as they secrete cytokines and inflammatory modulators following insult-induced activation (Aschner, 1998; Aschner et al., 2002). Indeed, S100B, which is predominantly expressed in astrocytes, is an agonist at the receptor for advanced glycation end products (RAGE), another neuroimmune receptor (Sparvero et al., 2009; Sims et al., 2010; Han et al., 2011). Together, microglia and astrocytes are the major constituents of the innate immune system in the CNS.

TLRs and RAGE recognize and respond to a variety of endogenous neuroimmune signaling molecules. Research has begun to identity several endogenous neuroimmune signaling molecules, including high-mobility group box 1 (HMGB1: Garg et al., 2010), heat shock proteins (Vabulas et al., 2002), extracellular matrix breakdown proteins (Scheibner et al., 2006), and S100 proteins (Sparvero et al., 2009). HMGB1 has received the most attention because it is an endogenous neuroimmune agonist at both TLRs and RAGE. It is a ubiquitous evolutionarily conserved nuclear protein present in most eukaryotic cells. It consists of three distinct domains:

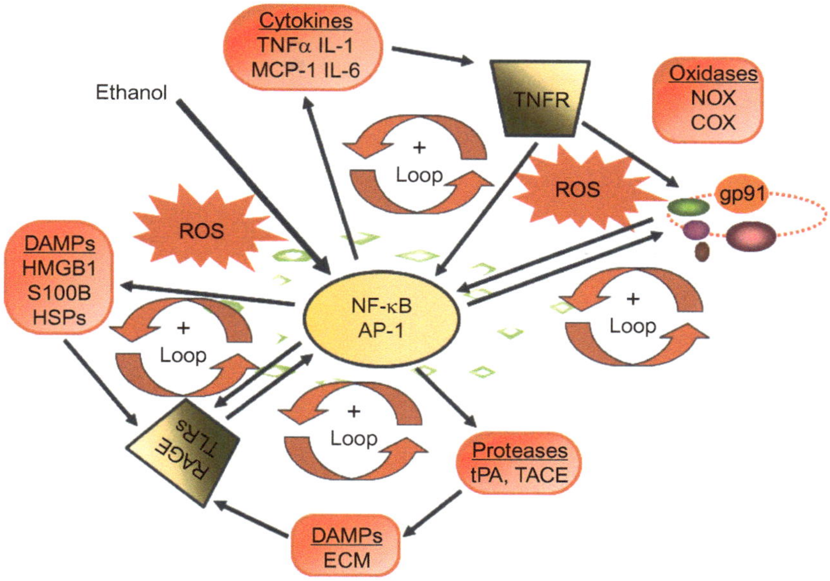

Fig. 27.5. Loops of signaling converge on NF-κB, increasing the expression of chemokines, cytokines, oxidases, and proteases. The proinflammatory transcription factor NF-κB is involved in the induction of neuroimmune genes (Knapp and Crews, 1999; He and Crews, 2008; Qin et al., 2008; Alfonso-Loeches et al., 2010; Zou and Crews, 2010). Alcohol (ethanol) increases NF-κB-DNA binding and transcription. Positive loops of activation occur through induction of genes that stimulate further NF-κB activation, leading to autocrine and paracrine amplification and persistent signals. Cytokines and chemokines, such as TNF-α, IL-1β, IL-6, and MCP-1, as well as their receptors (TNFR in figure), are induced, resulting in amplification loops. Reactive oxygen species (ROS) resulting from oxidases such as NADPH oxidase or ethanol metabolism increase NF-κB transcription of NOX2^{phox} (gp91), a key NOX catalytic subunit (Cao et al., 2005) that produces ROS (Qin et al., 2008). Toll-like receptors (TLRs) and the receptor for advanced glycation end product (RAGE) are increased by ethanol (Alfonso-Loeches et al., 2010; Crews et al., 2013; Vetreno et al., 2013), as are other neuroimmune signaling molecules, resulting in the formation of positive activation loops (Crews et al., 2013; Vetreno et al., 2013). These loops spread neuroimmune signaling across the brain, causing altered neurocircuitry and neurobiology. AP-1, activator protein-1; COX, cyclooxygenase; DAMPs, damage- (or danger-) associated molecular patterns; ECM, extracellular matrix; HMGB1, high-mobility group box 1; HSPs, heat shock proteins; IL-1, interleukin-1; MCP-1, monocyte chemoattractant protein-1; NF-κB, nuclear factor kappa-light-chain-enhancer of activated B cells; NOX, nicotinamide adenine dinucleotide phosphate (NADPH) oxidases; TACE,TNF-α converting enzyme; TNF-α, tissue necrosis factor-alpha; tPA, tissue plasminogen activator.

two homologous HMG boxes (box A and box B), and a C-terminal tail (Huang et al., 2010). Several studies have identified box B as the domain responsible for the proinflammatory properties of HMGB1 (Li et al., 2003; Messmer et al., 2004), whereas box A is anti-inflammatory and capable of attenuating box B-induced inflammation (Yang et al., 2004). Physiologically, HMGB1is involved in DNA bending and transcription stability (Paull et al., 1993; Thomas, 2001), but following secretion from cells, by either active release from neuroimmune cells or passive release by leaky necrotic cells, HMGB1exerts cytokine-like activity (Sims et al., 2010). Although defense against invading pathogens is key to the survival of all mammals, timely resolution of the neuroimmune response is critical to prevent inflammation-induced tissue damage. Indeed, persistent activation of this system leads to chronic inflammation, which may contribute to the development and progression of multiple diseases, including Parkinson's disease, Alzheimer's disease, and multiple sclerosis (Glass et al., 2010), and possibly alcoholism and addiction.

Alcohol activates the neuroimmune signaling system

Following ingestion, alcohol causes the stomach to become permeable or "leaky" (Ferrier et al., 2006). Recent studies report that intragastric administration of ethanol at doses >2 g/kg (Ferrier et al., 2006) potentiates gut innate immune signaling, disrupting gut tight junctions and opening sites that allow the gut biome bacteria and their fragments to enter portal circulation (Sims et al., 2010). Biome leakage contains endotoxins including LPS, which potentiates alcohol-induced liver inflammation and secretion of proinflammatory cytokines, including the proinflammatory cytokine TNF-α, which is released into the blood (Qin et al., 2007).

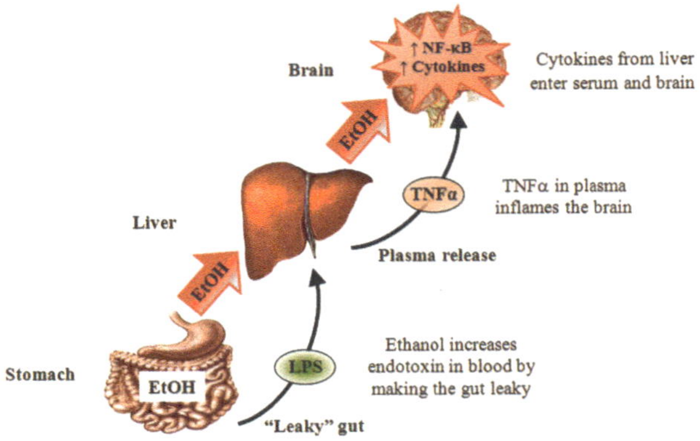

Fig. 27.6. Systemic ethanol (EtOH) activation of cytokine signaling in the brain. Consumed EtOH enters the stomach and makes it "leaky," allowing lipopolysaccharide (LPS) to enter the plasma. The circulating EtOH and LPS lead to liver proinflammatory cytokine induction, which results in the production and secretion of tumor necrosis factor-alpha (TNF-α) and other proinflammatory cytokines. TNF-α is transported into the brain from the blood and increases neuroimmune expression through increased activation and the synthesis of nuclear factor kappa-light-chain enhancer of activated B cells (NF-κB) and other cytokines. See Qin et al. (2007) for details.

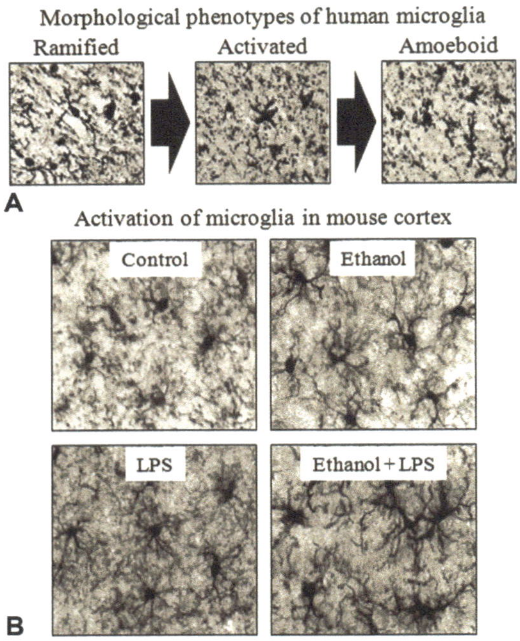

Fig. 27.7. Microglial morphology, lipopolysaccharide (LPS) and alcohol in the brain. (**A**) Representative figures depicting characteristic stages of microglial activation. Ramified or "resting" microglia are characterized by long, highly ramified processes with comparatively small cell bodies. Activated microglia are characterized by swollen, truncated processes and enlarged cell bodies. Ameboid or "phagocytic" microglia are characterized by large, ameba-like cell body with no or few small processes (Kreutzberg, 1996; Raivich et al., 1999). Photomicrographs depict stages of microglial activation in postmortem human brain tissue (He and Crews, 2008). (**B**) Representative figures depicting microglial activation in the mouse cortex following ethanol and/or LPS treatment. Male C57BL/6 mice were treated with saline or ethanol (5 g/kg, i.g.) for 10 days. Half of the subjects received a dose of LPS (3 mg/kg, i.p.) 24 hours after ethanol/saline treatment, and were sacrificed 1 hour later. Ten daily doses of ethanol exposure potentiated LPS-induced microglial activation. (Adapted from Qin et al., 2008.)

Proinflammatory cytokines in the blood are then either transported across the blood–brain barrier (e.g., TNF receptor (TNFR): Qin et al., 2007; Banks and Erickson, 2010) or activate endothelial cells to release cytokines (Watkins et al., 1995) that, combined with ethanol, enter the brain where they induce neuroimmune activation (Fig. 27.7). The precise mechanism for the movement of proinflammatory cytokines from the periphery to the CNS remains to be fully elucidated, but does not appear to involve a compromise of the blood–brain barrier (BBB). Employing a 4-day binge rat model of alcoholic brain damage, Marshall and colleagues (2013) assessed BBB integrity using immunohistochemistry for albumin, a 140-kDa protein that does not cross an intact BBB, and found no evidence of BBB dysfunction. The resolution of liver and blood cytokine increases in the periphery occurs over time, whereas marked cytokine induction within the brain, once activated, remains active and primed for additional activation that persists for months. The enduring and progressive induction of the neuroimmune response in brain is consistent with the cumulative progression and persistence of addiction, and might represent a mechanism contributing to the development of alcoholism.

Although there is an increasing body of literature supporting the hypothesis that alcohol is a potent immunomodulatory drug, it has opposing effects depending on whether administration is acute or chronic (Szabo and Mandrekar, 2009). Indeed, acute ethanol was found to suppress the innate immune response in both *in vivo* and *in vitro* models. For instance, LPS-induced TNF-α and IL-1β production was blunted in blood monocytes obtained from healthy human volunteers 16 hours after acute alcohol exposure (2 mL vodka/kg body weight: Crews et al., 2006b). In animal models, acute ethanol exposure attenuated the TNF-α, IL-1β, and IL-6 immune response to a pathogen challenge (Pruett et al., 2004). Similarly, Szabo's group reported that acute ethanol (25 mM) inhibited LPS-induced TNF-α expression, but not TLR2-induced upregulation of the TNF-α in the presence of ethanol (Crews et al., 2006b). In contrast, chronic ethanol exposure is associated with increased neuroimmune activation. Through a series of elegant animal studies, Guerri's laboratory demonstrated both *in vivo* and *in vitro* that chronic ethanol treatment induces astrocytic and microglial activation (Valles et al., 2004; Blanco et al., 2005; Fernandez-Lizarbe et al., 2009; Alfonso-Loeches et al., 2010; Pascual et al., 2011). For instance, *in vitro* administration of ethanol at physiologically relevant levels activates Toll-IL-1 receptor (TIR) signaling pathways in cultured astrocytes in a time-dependent fashion similar to what occurs following LPS and IL-1β administration, although at a much smaller response. Similarly, our laboratory found increased TLR2, TLR3, and TLR4 expression in the OFC of adult mice 24 hours after exposure to 10 days of 5 g/kg ethanol (Crews et al., 2013). Although ethanol and stress activate microglia, they do not produce the phagocytic oxidative bursts associated with maximal microglial activation by LPS or severe neurodegeneration. Thus, ethanol generally suppresses neuroimmune responses when it is present, but causes adaptations, particularly with chronic treatment, that sensitize responses during abstinence.

Involvement of the neuroimmune signaling system in alcoholism is supported by findings in the postmortem human alcoholic brain. Our laboratory recently discovered microglial activation as well as upregulated levels of MCP-1 (CCR-2) in the ventral tegmental area, substantia nigra, hippocampus, and amygdala of postmortem human alcoholic brain tissue, relative to moderate drinking controls (He and Crews, 2008). Our laboratory recently found upregulated expression of TLR2, TLR3, TLR4, and RAGE in the postmortem OFC of human alcoholics, which correlated with lifetime alcohol consumption (Crews et al., 2013). Although the mechanisms underlying ethanol-induced upregulation of neuroimmune signaling remains unknown, a potential signaling molecule that likely plays a role is HMGB1. Upon upregulation of HMGB1, it is released from the nucleus to the extracellular space where it exerts its cytokine-like effects (Muller et al., 2004; Hock et al., 2007; Maroso et al., 2010). Following its release, HMGB1 activates TLRs and RAGE, which facilitate proinflammatory responses in a cyclic fashion (Fig. 27.8) (Park et al., 2004; Yang et al., 2010; Rauvala and Rouhiainen, 2010; Volz et al., 2010). We discovered that alcohol causes increased expression and extracellular release of HMGB1 in brain slice cultures and in animal models of chronic ethanol exposure (Fig. 27.9) (Crews et al., 2013). Activation of this pathway leads to induction of NF-κB and activator protein-1, which is accompanied

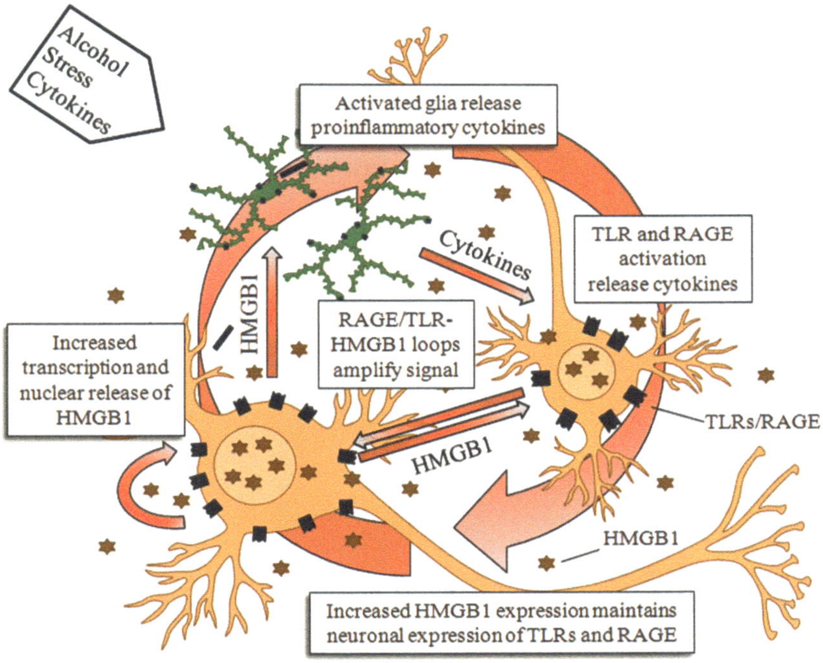

Fig. 27.8. Neuroimmune activation through HMGB1-TLR signaling loops. A single injection of lipopolysaccharide (LPS: 5 mg/kg, i.p.) in C57BL/6 mice caused a long-lasting increase in protein expression of tumor necrosis factor-alpha (TNF-α). (See Qin et al., 2007.)

The persistent priming of neuroimmune activation lasts for long periods due to positive loops of signaling between neurons and glia. HMGB1 released from neurons stimulates glial activation and glial synthesis of proinflammatory cytokines that further induce additional neuroimmune genes consistent with the long-lasting impact of alcoholism and the persistence, e.g., the folklore that once an alcoholic always an alcoholic (e.g., the neuroimmune priming persists and does not go away, like the risk of relapse to alcoholism). (Adapted from Vetreno et al., 2013.)

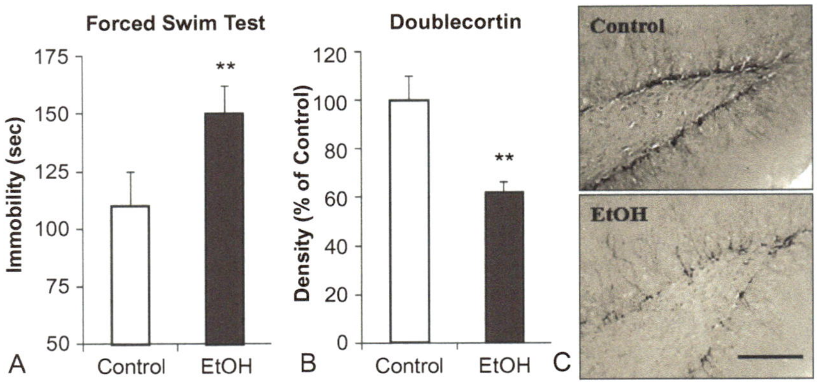

Fig. 27.9. Chronic ethanol (EtOH) self-administration induces depression-like behavior and inhibits hippocampal neurogenesis. C57BL/6J mice self-administered either ethanol (10% v/v) or water for 28 days. (**A**) Increased immobility (seconds) on the forced-swim test provides an index of depression-like behavior. Following a period of abstinence from chronic ethanol consumption, mice evidenced increased immobility time relative to controls. (**B**) Doublecortin (DCX) expression, a marker of neurogenesis, was reduced in the dentate gyrus of mice exposed to chronic ethanol self-administration. (**C**) Representative photomicrographs depicting the reduced DCX expression. Alcohol-induced reduced neurogenesis and neuroprogenitor cell proliferation are associated with increased depression-like behavior. Antidepressants reverse both (Stevenson et al., 2009). Scale bar = 200 μm. $^{**}p < 0.01$, relative to control. (Adapted from Stevenson et al., 2009.)

by an upregulation of iNOS and COX-2 expression in astrocytes that, in turn, leads to further downstream activation of the NF-κB signaling (Blanco et al., 2005). In support of the role of neuroimmunity in alcoholism, treatment with indomethacin, an anti-inflammatory drug, reduces chronic intermittent ethanol induction of brain innate immune genes (iNOS and COX-2) in astrocytes and reduces markers of cell death and behavioral dysfunction (Pascual et al., 2007). In addition, administration of minocycline, an antibiotic that blocks microglial activation (Plane et al., 2010; Qin and Crews, 2012a), reduces ethanol-induced neuroimmune activation and neural cell death (Qin and Crews, 2012a). These data provide evidence that ethanol-induced induction of neuroimmune signaling contributes to the development and maintenance of alcoholism.

Induction of neuroimmune cascades contributes to addiction-like behaviors

Executive functioning is mediated by neural circuitry, including the PFC and interconnected subcortical regions, which regulate mood and cognition through reciprocal glutamatergic connections with multiple limbic system brain regions. In the frontal cortex, positive loops of neuroimmune activation likely lead to hyperexcitability (Crews et al., 2006a). Ethanol-induced activation of microglial NF-κB cascades is relatively small compared to the bursts of innate immune genes responding to cell death in brain associated with stroke, traumatic brain injury, and other diseases involving major neurodegeneration (Qin et al., 2008). In astrocytes, ethanol exposure induces NF-κB transcription leading to increased expression of proinflammatory innate immune genes (Zou and Crews, 2006, 2010; Pascual et al., 2007) and reduced astrocyte glutamate transport (Zou and Crews, 2005). This increase in extracellular glutamate leads to increased neuronal excitation, microglial activation, and excitotoxicity (Zou and Crews, 2006; (Ward et al., 2009). The increase in glutamate excitotoxicity might also contribute to ethanol-induced increases of caspase-3 and COX-2 in the frontal cortex (Knapp and Crews, 1999; Alfonso-Loeches et al., 2010).

Innate immune genes also play a critical role in modulating neural plasticity by affecting long-term potentiation (LTP) and long-term depression (LTD), both of which are neural correlates of learning and memory (Stanton, 1996; Dantzer et al., 2008; Boulanger, 2009). During periods of health, the delicate neuronal–glial communication balance maintains neural plasticity through secretion of several neuroimmune mediators, including IL-1β, IL-6, and IL-18, which are induced during hippocampal LTP and related to synaptic strengthening (Schneider et al., 1998; Balschun et al., 2004; del Rey et al., 2013). However, innate immune system activation by endogenous or exogenous stimuli results in a neuroimmune imbalance, resulting in a shift in the neuronal–glial communication balance (Besedovsky and del Rey, 2011), and might impair LTP and LTD, leading to deficits in plasticity. Indeed, Maggio et al. (2013) found that a brief exposure to systemic LPS (1 mg/kg, i.p. twice a week for 1 week) transiently impaired hippocampal LTP, which was recovered to control levels 7 days following the conclusion of inflammagen exposure. In contrast, longer exposure to LPS (1 mg/kg, i.p. twice a week for 1 month) persistently altered LTP, which was evident 2 months following the conclusion of inflammagen exposure. Thus, ethanol-induced disruption of neural plasticity, particularly following binge drinking, might contribute to the long-term deficits in learning and memory associated with alcoholism.

Disruption of frontal cortex and limbic system circuits due to ethanol-induced induction of neuroimmune signaling might contribute to the loss of cognitive flexibility and other addiction-like behaviors. Neurogenesis, a process whereby newborn neurons are generated and functionally integrated into the existing hippocampal circuitry (Altman and Das, 1965; Zhao et al., 2006), is critically involved in learning and emotive function. This process is also particularly vulnerable to ethanol-induced neuroimmune activation. Using an *ex vivo* model of organotypic hippocampal-entorhinal cortex brain slice culture, (Zou and Crews, 2012) found that 4 days of 100 mM ethanol reduced cellular proliferation and neurogenesis. This effect was accompanied by a shift in the neuroimmune/neurotrophin balance, wherein mRNA levels of the proinflammatory cytokine IL-1β were increased while expression of CREB and BDNF was reduced. Interestingly, ethanol-induced reductions of neurogenesis were reversed by administration of a neutralizing anti-IL-1β antibody as well as blockade of IL-1 surface receptors. Further, administration of rolipram, a phosphodiesterase 4 inhibitor that increases cAMP leading to increased PKA activation and inhibition of NF-κB induction (Takahashi et al., 2002), and BHT, an antioxidant that blocks NF-κB induction of neuroimmune genes (Crews et al., 2006a; Zou and Crews, 2010), reversed ethanol-induced inhibition of neurogenesis. Similarly, using an *in vivo* 4-day chronic binge model of ethanol exposure, Nixon and Crews (2002) found diminished hippocampal neural stem cell proliferation and neurogenesis. Diminished neurogenesis in this structure likely contributes to deficits in learning and alterations in mood. A study following chronic alcohol administration to mice found reduced neurogenesis and increased immobility in the forced-swim test, a finding consistent with negative affect – depression and hopelessness behavior (Stevenson et al., 2009) (Fig. 27.9). Interestingly, antidepressant treatment reversed both the lost neurogenesis and the depression-like behavior. Thus, components of the negative affect and bad feelings in alcoholism may be related to alcohol-induced changes in neurogenesis.

The involvement of innate immune genes in the reduction of glutamate transporters and subsequent hyperexcitability inactivates the frontal cortex and might also contribute to the neurobiology of addiction (Crews et al., 2006a), as evidenced by decreased ability to adapt behavior to successful outcomes when there are changes in the situation. Alcohol activates microglia and astrocytes (Lee et al., 2004; Qin et al., 2007; Alfonso-Loeches et al., 2010), which might lead to glial adaptive conditioning (Shpargel et al., 2008). Adaptive conditioning occurs when the brain is exposed to persistent low levels of noxious stimuli that culminate in neuroprotection by priming of the brain for future insults (i.e., increased expression of anti-inflammatory molecules). However, repeated bouts of exposure to these stimuli (e.g., intermittent) leads to persistent activation of microglia and astrocytes (McClain et al., 2011) that secrete proinflammatory cytokines that disrupt neurocircuits, and promote neuronal damage and cell death (Valles et al., 2004). In support of a glial contribution to addiction, administration of prednisone, a synthetic glucocorticoid, increases anxiety-like behavior on the elevated plus-maze as well as microglial populations in the frontal cortex and hippocampus (Gonzalez-Perez et al., 2001). In human postmortem frontal cortical tissue of depressed individuals that committed suicide, astrocytes were found to exhibit enlarger cell bodies and more ramified processes, which is indicative of activated astrocytes (Torres-Platas et al., 2011). Taken together, there is mounting evidence to support a role for neuroimmune signaling in the development and maintenance of alcoholism (Fig. 27.10). Further, the neuroimmune signaling system might offer a novel target in the treatment of alcoholism and other addictions.

RISK FACTORS THAT CONTRIBUTE TO THE PROGRESSION OF ALCOHOL DEPENDENCE

Genetic contribution to alcohol dependence

Similar to other psychopathologies (e.g., depression and schizophrenia), alcohol dependence stems from a combination of precipitating environmental factors and inherent biologic/genetic predispositions (Sander et al., 1999). Specific variants of multiple neuroimmune genes are associated with the genetic risk for alcohol dependence in humans (Crews, 2012) which further supports the connection between the innate immune system of the CNS and alcoholism. Approximately 50% of the risk for alcohol dependence stems from genetic factors (Schuckit, 2009). Variations on chromosomes at specific locations on DNA (i.e., polymorphisms) result in gene variants (i.e., alleles) that may diminish or increase the risk for alcohol dependence. The enzyme CYP2E1 is involved in alcohol metabolism, and polymorphisms of the gene that encodes for this enzyme are associated with increased risk for alcoholism (Webb et al., 2011). In the periphery, CYP2E1 is localized to monocytes (immune cells), and ethanol metabolism by this enzyme leads to

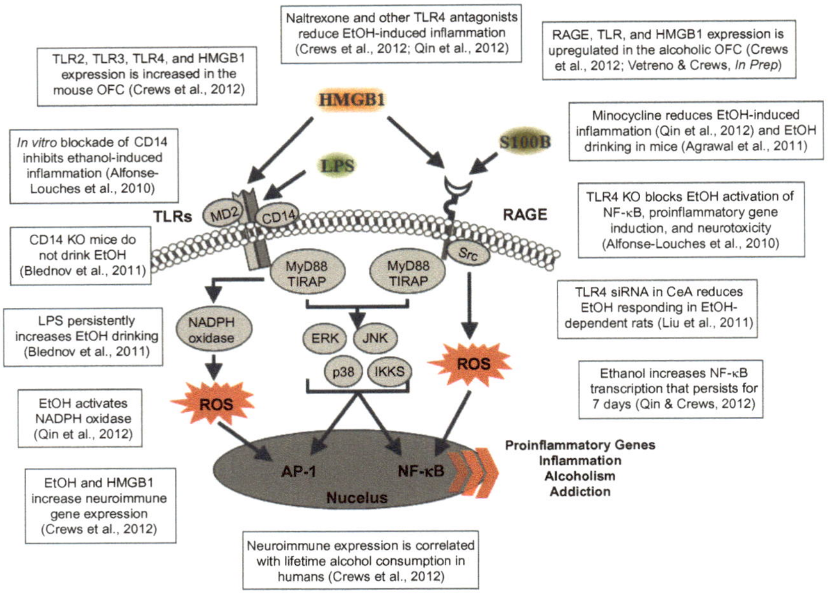

Fig. 27.10. Neuroimmune signaling cascade activation and evidence for involvement in alcohol-induced neurodegeneration. A simplified schematic of the TLR and RAGE receptor signaling cascades. Stimulation of TLRs leads to the generation of ROS and downstream activation of NF-κB. Similarly, activation of the RAGE receptor leads to caspase-3 induction and downstream activation of NF- κB. The production of NF-κB leads to the secretion of proinflammatory gene expression, neuroimmune induction, and cell death. AP-1, activator protein-1; CD14, cluster of differentiation 14; ERK, extracellular signal-regulated kinase; HMGB1, high-mobility group box-1; IKK, inhibitor of nuclear factor kappa-B; JNK, c-Jun N-terminal kinases; LPS, lipopolysaccharide; MyD88, myeloid differentiation primary response gene 88; NADPH oxidase, nicotinamide adenine dinucleotide phosphate-oxidase; NF-κB, nuclear factor kappa-light-chain enhancer of activated B cells; RAGE, receptor for advanced glycation end products; ROS, reactive oxygen species; S100B, S100 calcium-binding protein B; Src, proto-oncogene tyrosine-protein kinase; TIRAP, Toll/interleukin-1 receptor domain-containing adaptor protein; TLRs, Toll-like receptors.

activation and consequent production of reactive oxygen species and activation of NF-κB cascades in response to LPS (Cao et al., 2005). Within the CNS, ethanol upregulates CYP2E1 expression in astrocytes (Montoliu et al., 1994, 1995), which likely contribute to astrocytic activation and induction of NF-κB cascades. Polymorphisms of the gene that encodes NF-κB have also been implicated in alcohol dependence (Flatscher-Bader et al., 2005; Okvist et al., 2007). Alleles of many other genes linked to innate immune factors are associated with alcoholism (Crews, 2012). The interactions are complex due to combinations of alleles that promote proinflammatory signals with alleles that blunt anti-inflammatory signals (Crews, 2012). Thus, human genetic factors related to alcoholism might include innate immune genes.

Gene expression studies of genetically paired rats and mice that differ primarily in their preference for ethanol consumption reveal that NF-κB, its regulatory proteins, and many innate immune genes are central to high-ethanol-drinking behaviors (Mulligan et al., 2006). Beta-2 microglobulin, an NF-κB target gene involved in MHC immune signaling (Pahl, 1999), and other transcription regulation transcriptomes (i.e., RNA molecules), were significantly overexpressed in high-ethanol-preferring mice which might induce a predisposition for excessive ethanol consumption (Mulligan et al., 2006). In addition, work from Blednov and colleagues provided interesting data supporting the hypothesis that innate immune genes regulate ethanol-drinking behavior (Blednov et al., 2005, 2011, 2012). In null mutant mice for specific genes associated with TLR and NF-κB signaling, including β_2-macroglobulin, cathepsin S, cathepsin F, IL-1 receptor antagonist (*Il1rn*), CD14, and IL-6, these animals universally evidenced altered ethanol preference and intake relative to wild-type controls across multiple ethanol-drinking paradigms. However, across measures, the male and female *Il1rn* mutant mice evidenced reduced ethanol preference and intake, supporting a role for neuroimmune involvement in ethanol-drinking behavior (Blednov et al., 2012). Taken together, these findings are consistent with a modulatory role of innate immune genes in alcohol preference and consumption.

Polymorphisms of several genes related to the dopaminergic system, which is involved in frontal cortical and limbic system regulation of behavior, have been implicated in the development of impulsive behaviors (Kreek et al., 2005). For instance, genetic variation at the Val158Met locus of the catechol-*O*-methyltransferase (COMT) gene influences decision-making behavior and underlying activity in brain activity associated with impulsive choice (Boettiger et al., 2007). The COMT enzyme plays a significant role in regulating frontal cortex dopaminergic function (Chen et al., 2004), and individuals homozygous for the enzymatically more active 158Val allele show an increased tendency to choose immediate over delayed rewards. Such genetic variation in COMT function may also contribute to other forms of impulsive behavior (Congdon and Canli, 2005; Kreek et al., 2005), and might contribute mechanistically to alcoholism (Ishiguro et al., 1999; Samochowiec et al., 2006). Thus, genes that alter impulsivity likely contribute to risk for alcoholism and other mental diseases that have overlapping psychopathology.

Adolescent onset of alcohol consumption

Adolescence is a highly conserved critical developmental period that encompasses the transition from childhood to adulthood. It is defined by a constellation of behaviors that encompass high levels of risk taking, increased exploration, novelty and sensation seeking, and social interaction that promote the acquisition of skills necessary for maturation and independence. In addition, it is a period when the adolescent brain, particularly the frontal cortex and limbic system, undergoes both progressive and regressive changes involving synaptogenesis, finetuning of projections and inputs, and maturation of neurotransmitter systems (Spear, 2000; Ernst et al., 2009). Indeed, absolute PFC gray-matter volumes decline in humans (Sowell et al., 1999, 2001) as well as in rats (van Eden et al., 1990) during adolescence. In addition, there is considerable synaptic pruning during adolescence, especially of excitatory glutamatergic inputs to the PFC in humans and non-human primates (Huttenlocher, 1984; Zecevic et al., 1989). By contrast, dopaminergic, serotonergic, and cholinergic inputs to the PFC increase, ultimately reaching mature levels (Kalsbeek et al., 1988; Kostovic, 1990; Gould et al., 1991; Rosenberg and Lewis, 1994). Greater incidence of novelty/sensation seeking is also a strong predictor of alcohol and drug use (Baumrind, 1987; Wills et al., 1994) that might alter cortical maturation, leading to addiction due to the vulnerability of the developing PFC (Crews et al., 2007).

In humans, initiation of alcohol use typically occurs during adolescence and is highly prevalent as 8% of eighth-grade, 16% of 10th-grade, and 25% of 12th-grade adolescents report heavy episodic drinking (i.e., >5 consecutive alcohol drinks per episode) over the past 2 weeks (Johnston et al., 2009). This heavy-drinking pattern persists into college, as 44% of students report binge drinking every 2 weeks and 19% report >3 binge-drinking episodes per week (Wechsler et al., 1995; O'Malley et al., 1998). The effects of alcohol on the adolescent brain are different from those observed in adulthood. For instance, adolescents are less sensitive to the sedative effects of alcohol (Silveri and Spear, 1998), which allows them to binge drink. In contrast, they are more vulnerable to alcohol-induced neurotoxicity (Monti et al., 2005; Crews et al., 2007). Heavy drinking among human adolescent males was found to increase impulsivity the following year in those individuals predisposed to adolescent-typical impulsivity (White et al., 2011). Furthermore, early onset of alcohol use (<13 years of age) is associated with increased drinking frequency and physical violence (Gruber et al., 1996), which is consistent with diminished impulse inhibition. In addition, adolescents and adults with alcohol use disorders evidence deficits in executive functioning (Brown et al., 2000; Tapert and Brown, 2000; Hanson et al., 2011). Using animal models of adolescent binge drinking, we found impaired reversal learning (Coleman et al., 2011; Vetreno and Crews, 2012), The increased sensitivity of the adolescent brain to alcohol-induced toxicity, coupled with the dynamic synaptic remodeling that characterizes this stage of development, might strengthen the learning components of heavy-drinking behaviors and perpetuate the loss of important self-control and goal-setting components of the maturing brain's executive centers. Indeed, an early age of drinking onset is highly predictive of later development of alcoholism, regardless of family history (Grant, 1998).

Neuroimmune gene activation in the brain remains elevated for long periods (Qin et al., 2007, 2008), consistent with the enduring nature of addiction. The persistence of neuroimmune activation is likely amplified in the adolescent brain (Spear, 2000; Vetreno and Crews, 2012; Vetreno et al., 2013) because of their increased ethanol consumption (Silveri and Spear, 1998). Similar to our adult findings, our laboratory recently discovered that adolescent binge ethanol exposure increases TLRs, RAGE, and HMGB1 neuroimmune expression in the adolescent brain, an effect that persists into adulthood. These increases were accompanied by persistent upregulation of proinflammatory cytokines (TNF-α and MCP-1), oxidases (COX-2 and NOX2), and neuroimmune signaling molecules (HMGB1 and S100β: Vetreno and Crews, 2012; Vetreno et al., 2013). The induction of neuroimmune signaling includes neuronal-glial signaling that leads to hyperexcitability in neuronal networks that contribute to addiction (Fig. 27.11). Consistent with the increased excitability and excitotoxic vulnerability of the PFC during adolescence, an earlier age of drinking onset correlated with increased expression of TLRs, RAGE, and HMGB1 in the postmortem human alcoholic

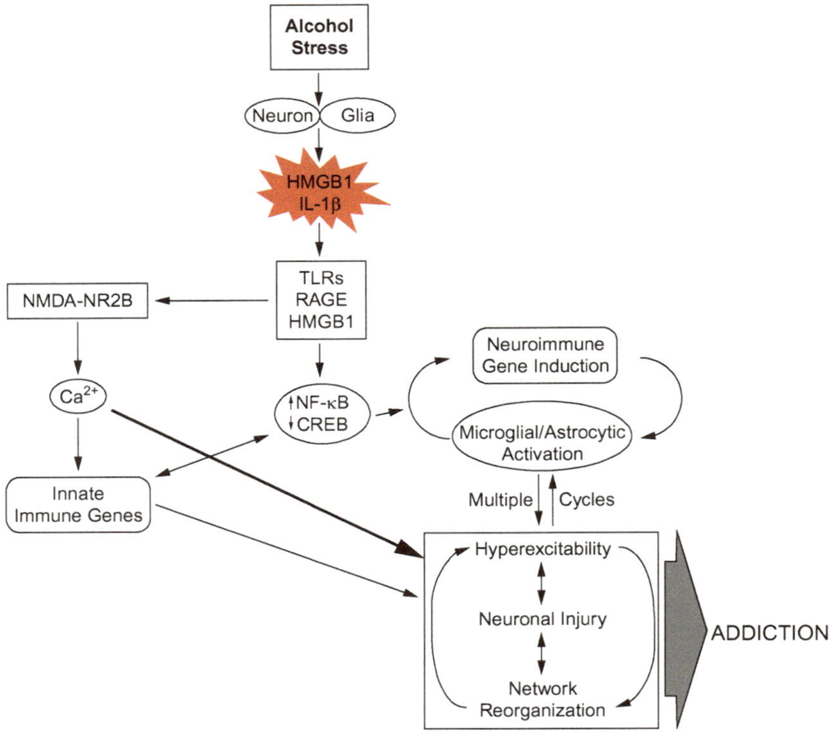

Fig. 27.11. Hyperexcitability contributes to the neurobiology of addiction. A simplified schematic depicting how neuroimmune signaling leads to hyperexcitability and the neurobiology of addiction. Alcohol and stress activate neurons and glia in the central nervous system, resulting in the release of various neuroimmune signals (e.g., high-mobility group box 1 (HMGB1) and interleukin-1β (IL-1β)) that activate neuroimmune receptors (i.e., Toll-like receptors (TLRs) and receptor for advanced glycation end products (RAGE)). Neuroimmune receptor stimulation leads to activation of glutamatergic *N*-methyl-D-aspartate (NMDA) receptors, e.g., NR2B (Maroso et al., 2010; Iori et al., 2013), which increases Ca^{2+} flux, triggering induction of neuroimmune genes. In addition, TLR/RAGE activation leads to downstream transcription of nuclear factor kappa-light-chain enhancer of activated B cells (NF-κB) signaling that might be accompanied by diminished cyclic AMP-responsive element-binding protein (CREB) expression, which contributes to neuroimmune gene induction. These two pathways converge, leading to cycles of neuroimmune gene induction that lead to hyperexcitability, neuronal cell death, and network reorganization that culminates in addiction.

OFC (Fig. 27.12). Although the mechanism underlying the persistence of neuroimmune activation remains to be elucidated, it does appear to involve positive loops of activation (Crews et al., 2011). This hypothesis would predict a relationship between TLR/RAGE and HMGB1 expression. Indeed, expression of HMGB1 is positively correlated with RAGE and TLR immunoreactivity in the postmortem human alcoholic OFC (Vetreno et al., 2013) (Fig. 27.13) and in the rat PFC following adolescent binge ethanol exposure (Vetreno and Crews, 2012; Vetreno et al., 2013).

Disorders of PFC function

Innate PFC dysfunction represents another risk factor that could contribute to the development of alcohol and other addictions. Loss of PFC control over impulsivity is a cardinal feature of both attention-deficit hyperactivity disorder (ADHD) and addiction. ADHD is a highly prevalent psychiatric disorder (NIH, 2000) characterized by symptoms of inattention, hyperactivity, and impulsivity, as well as diminished executive function (Biederman, 1998; APA, 2000; Vaurio et al., 2008). Neuroimaging studies have revealed abnormal prefrontal cortical and striatal activation in ADHD individuals (Herrmann et al., 2010) as well as significant gray-matter reductions in the frontal gyrus and cingulate gyrus (Overmeyer et al., 2001). Animal models of ADHD implicate serotonergic and dopaminergic function in impulsivity and attention (Puumala and Sirvio, 1998), suggesting that dysfunctional catecholaminergic signaling might underlie some of the impulsivity symptoms. A number of studies have been conducted to assess the potential relationship between ADHD and the development of alcoholism and other drug addictions (Wilson and Marcotte, 1996; Biederman et al., 1997; Murphy

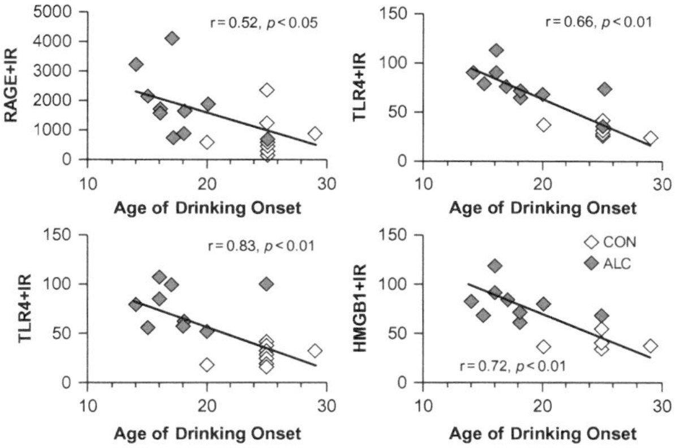

Fig. 27.12. Receptor for advanced glycation end product (RAGE), Toll-like receptor (TLR) 3 and 4, and high-mobility group box 1 (HMGB1) expression in the human alcoholic postmortem orbitofrontal cortex correlates with age of drinking onset. The younger the age of drinking onset, the greater the risks of developing alcohol dependence (Grant, 1998). Neuroimmune signaling shows a similar correlation consistent with alcohol-induced increased neuroimmune signaling, increasing the risk for alcoholism. Depicted are individual self and family reports of age of drinking onset vs RAGE ($\times$10 000 pixels/mm^2), TLR3 (cells/mm^3), TLR4 (cells/mm^3), and HMGB1 (cells/mm^3) immunoreactivity. Across subjects, age of drinking onset negatively correlated with neuroimmune signal immunoreactivity ($r=-0.42$, $p<0.001$). Moderate alcohol drinking controls (CON) tended to self-report a later age of drinking onset (25 ± 1 years of age) in comparison to individuals that met criteria for alcoholism (18 ± 1 years of age). (Adapted from Vetreno et al., 2013).

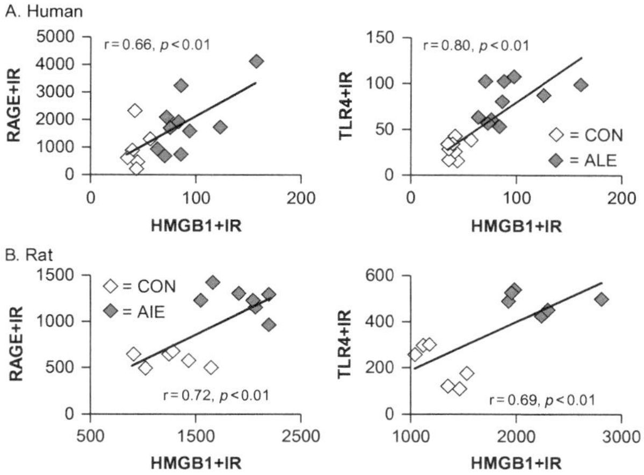

Fig. 27.13. Receptor for advanced glycation end product (RAGE) and Toll-like receptor 4 (TLR4) expression correlates with high-mobility group box 1 (HMGB1) expression in the prefrontal cortex. The persistent nature of neuroimmune gene induction is likely due to upregulation of both the agonist, HMGB1, and receptors, e.g., TLR receptors that progressively increase in concert with chronic alcohol abuse. (**A**) Depicted are correlations of individual human RAGE ($\times$10 000 pixels/mm^2) and TLR4 (cells/mm^3) immunoreactivity versus HMGB1 (cells/mm^3) expression in the orbitofrontal cortex (OFC) of alcoholics and moderate-drinking controls. Across subjects, RAGE and TLR4 expression correlated with HMGB1. The correlation of HMGB1 with RAGE/TLR4 is consistent with neuroimmune loops of amplification. (**B**) Depicted are correlations of RAGE ($\times$10 000 pixels/mm^2) and TLR4 (cells/mm^2) immunoreactivity versus HMGB1 (cells/mm^2) expression in the OFC of adult rats exposed to adolescent binge ethanol. Across subjects, RAGE and TLR4 expression correlated with HMGB1. The correlation of RAGE/TLR4 with HMGB1 indicates the persistence of neuroimmune loops of activation. (Adapted from Vetreno and Crews, 2012; Crews et al., 2013; Vetreno et al., 2013).

et al., 2002). Although there is a lack of agreement regarding the association with ADHD and addiction, diminished PFC function is likely a contributing factor to heavy drinking (Span and Earleywine, 1999). This seems especially true given the involvement of the PFC in executive functioning and its dysregulation in the alcoholic brain. Indeed, perhaps due to increased impulsivity, college students with ADHD are more likely than others to abuse alcohol and other drugs (Murphy et al., 2002).

CONCLUSIONS

Over time, continued alcohol use leads to the development of alcoholism-promoting behaviors. Diminished executive functioning due to hypoactive PFC executive control and hyperactive limbic system anxiety and negative emotion might contribute mechanistically to the shift from experimental use to alcoholism and dependence. Human neuroimaging studies and animal models find that chronic alcohol exposure causes neurodegeneration that is most prominent in the PFC, which might contribute to cortical hypofunction and diminished executive function. Further, alcohol diminishes CREB/neurotrophin support while increasing transcription of NF-κB, which in turn leads to downstream induction of proinflammatory chemokines, cytokines, oxidases, proteases, and other neuroimmune genes in a cyclic fashion that spreads across the CNS. These findings, coupled with the neuroimmune genetic finds of Blednov and colleagues (2012), support the role for the involvement of innate immune gene expression in the development of alcoholism. Furthermore, hyperexcitability and behavioral inflexibility are associated with innate immune gene expression in the frontal cortex (Vetreno et al., 2013). The emerging view that innate immunity underlies addiction has important ramifications for the interaction of early onset of drug and alcohol abuse with genetics. Furthermore, it suggests that therapies aimed at suppressing innate immune activation in the CNS might represent a novel therapeutic approach to prevent the development, and perhaps treat addiction and/or anxiety disorders.

References

Akira S, Takeda K, Kaisho T (2001). Toll-like receptors: critical proteins linking innate and acquired immunity. Nat Immunol 2: 675–680.

Alfonso-Loeches S, Pascual-Lucas M, Blanco AM et al. (2010). Pivotal role of TLR4 receptors in alcohol-induced neuroinflammation and brain damage. J Neurosci 30: 8285–8295.

Altman J, DAS GD (1965). Autoradiographic and histological evidence of postnatal hippocampal neurogenesis in rats. J Comp Neurol 124: 319–335.

APA (2000). Diagnostic and statistical manual of mental disorders: DSM-IV-TR. Washington, D.C., American Psychiatric Association.

American Psychiatric Association (2013). Diagnostic and statistical manual of mental disorders, 5th edn. American Psychiatric Publishing, Arlington, VA.

Aschner M (1998). Astrocytes as mediators of immune and inflammatory responses in the CNS. Neurotoxicology 19: 269–281.

Aschner M, Sonnewald U, Tan KH (2002). Astrocyte modulation of neurotoxic injury. Brain Pathol 12: 475–481.

Baddeley A (1992). Working memory. Science 255: 556–559.

Balschun D, Wetzel W, Del Rey A et al. (2004). Interleukin-6: a cytokine to forget. FASEB J 18: 1788–1790.

Banks WA, Erickson MA (2010). The blood–brain barrier and immune function and dysfunction. Neurobiol Dis 37: 26–32.

Baumrind D (1987). A developmental perspective on adolescent risk taking in contemporary America. New Dir Child Adolesc Dev 93–125.

Bechara A (2005). Decision making, impulse control and loss of willpower to resist drugs: a neurocognitive perspective. Nat Neurosci 8: 1458–1463.

Bechara A, Damasio AR, Damasio H et al. (1994). Insensitivity to future consequences following damage to human prefrontal cortex. Cognition 50: 7–15.

Bechara A, Damasio H, Tranel D et al. (1998). Dissociation of working memory from decision making within the human prefrontal cortex. J Neurosci 18: 428–437.

Bechara A, Damasio H, Damasio AR (2000). Emotion, decision making and the orbitofrontal cortex. Cereb Cortex 10: 295–307.

Berlin HA, Rolls ET, Kischka U (2004). Impulsivity, time perception, emotion and reinforcement sensitivity in patients with orbitofrontal cortex lesions. Brain 127: 1108–1126.

Besedovsky HO, Del Rey A (2011). Central and peripheral cytokines mediate immune-brain connectivity. Neurochem Res 36: 1–6.

Biederman J (1998). Attention-deficit/hyperactivity disorder: a life-span perspective. J Clin Psychiatry 59 (Suppl 7): 4–16.

Biederman J, Wilens T, Mick E et al. (1997). Is ADHD a risk factor for psychoactive substance use disorders? Findings from a four-year prospective follow-up study. J Am Acad Child Adolesc Psychiatry 36: 21–29.

Bison S, Crews F (2003). Alcohol withdrawal increases neuropeptide Y immunoreactivity in rat brain. Alcohol Clin Exp Res 27: 1173–1183.

Blanco AM, Valles SL, Pascual M et al. (2005). Involvement of TLR4/type I IL-1 receptor signaling in the induction of inflammatory mediators and cell death induced by ethanol in cultured astrocytes. J Immunol 175: 6893–6899.

Blednov YA, Bergeson SE, Walker D et al. (2005). Perturbation of chemokine networks by gene deletion alters the reinforcing actions of ethanol. Behav Brain Res 165: 110–125.

Blednov YA, Benavidez JM, Geil C et al. (2011). Activation of inflammatory signaling by lipopolysaccharide produces a prolonged increase of voluntary alcohol intake in mice. Brain Behav Immun 25 (Suppl 1): S92–S105.

Blednov YA, Ponomarev I, Geil C et al. (2012). Neuroimmune regulation of alcohol consumption: behavioral validation of genes obtained from genomic studies. Addict Biol 17: 108–120.

Boettiger CA, Mitchell JM, Tavares VC et al. (2007). Immediate reward bias in humans: fronto-parietal networks and a role for the catechol-*O*-methyltransferase 158(Val/Val) genotype. J Neurosci 27: 14383–14391.

Boulanger LM (2009). Immune proteins in brain development and synaptic plasticity. Neuron 64: 93–109.

Bowden SC, Crews FT, Bates ME et al. (2001). Neurotoxicity and neurocognitive impairments with alcohol and drug-use disorders: potential roles in addiction and recovery. Alcohol Clin Exp Res 25: 317–321.

Breiter HC, Gollub RL, Weisskoff RM et al. (1997). Acute effects of cocaine on human brain activity and emotion. Neuron 19: 591–611.

Brown SA, Tapert SF, Granholm E et al. (2000). Neurocognitive functioning of adolescents: effects of protracted alcohol use. Alcohol Clin Exp Res 24: 164–171.

Cao Q, Mak KM, Lieber CS (2005). Cytochrome P4502E1 primes macrophages to increase TNF-alpha production in response to lipopolysaccharide. Am J Physiol 289: G95–G107.

Carmichael ST, Price JL (1995a). Limbic connections of the orbital and medial prefrontal cortex in macaque monkeys. J Comp Neurol 363: 615–641.

Carmichael ST, Price JL (1995b). Sensory and premotor connections of the orbital and medial prefrontal cortex of macaque monkeys. J Comp Neurol 363: 642–664.

Cella M, Dymond S, Cooper A (2010). Impaired flexible decision-making in major depressive disorder. J Affect Disord 124: 207–210.

Chen J, Lipska BK, Halim N et al. (2004). Functional analysis of genetic variation in catechol-*O*-methyltransferase (COMT): effects on mRNA, protein, and enzyme activity in postmortem human brain. Am J Hum Genet 75: 807–821.

Childress AR, Mozley PD, Mcelgin W et al. (1999). Limbic activation during cue-induced cocaine craving. Am J Psychiatry 156: 11–18.

Clark L, Manes F (2004). Social and emotional decision-making following frontal lobe injury. Neurocase 10: 398–403.

Coleman Jr LG, Jarskog LF, Moy SS et al. (2009). Deficits in adult prefrontal cortex neurons and behavior following early post-natal NMDA antagonist treatment. Pharmacol Biochem Behav 93: 322–330.

Coleman Jr LG, He J, Lee J et al. (2011). Adolescent binge drinking alters adult brain neurotransmitter gene expression, behavior, brain regional volumes, and neurochemistry in mice. Alcohol Clin Exp Res 35: 671–688.

Colton CA (2009). Heterogeneity of microglial activation in the innate immune response in the brain. J Neuroimmune Pharmacol 4: 399–418.

Congdon E, Canli T (2005). The endophenotype of impulsivity: reaching consilience through behavioral, genetic, and neuroimaging approaches. Behav Cogn Neurosci Rev 4: 262–281.

Coutlee CG, Huettel SA (2012). The functional neuroanatomy of decision making: prefrontal control of thought and action. Brain Res 1428: 3–12.

Crews FT (2012). Immune function genes, genetics, and the neurobiology of addiction. Alcohol Res Curr Rev 34: 355–361.

Crews FT, Boettiger CA (2009). Impulsivity, frontal lobes and risk for addiction. Pharmacol Biochem Behav 93: 237–247.

Crews FT, Nixon K (2009). Mechanisms of neurodegeneration and regeneration in alcoholism. Alcohol Alcohol 44: 115–127.

Crews FT, Vetreno RP (2011). Addiction, adolescence, and innate immune gene induction. Frontiers Psychiatry 2: 19.

Crews FT, Collins MA, Dlugos C et al. (2004). Alcohol-induced neurodegeneration: when, where and why? Alcohol Clin Exp Res 28: 350–364.

Crews FT, Buckley T, Dodd PR et al. (2005). Alcoholic neurobiology: changes in dependence and recovery. Alcohol Clin Exp Res 29: 1504–1513.

Crews F, Nixon K, Kim D et al. (2006a). BHT blocks NF-kappaB activation and ethanol-induced brain damage. Alcohol Clin Exp Res 30: 1938–1949.

Crews FT, Bechara R, Brown LA et al. (2006b). Cytokines and alcohol. Alcohol Clin Exp Res 30: 720–730.

Crews F, He J, Hodge C (2007). Adolescent cortical development: a critical period of vulnerability for addiction. Pharmacol Biochem Behav 86: 189–199.

Crews FT, Zou J, Qin L (2011). Induction of innate immune genes in brain create the neurobiology of addiction. Brain Behav Immun 25 (Suppl 1): S4–S12.

Crews FT, Qin L, Sheedy D et al. (2013). High mobility group box 1/Toll-like receptor danger signaling increases brain neuroimmune activation in alcohol dependence. Biol Psychiatry 73: 602–612.

Cummings JL (1998). Frontal-subcortical circuits and human behavior. J Psychosom Res 44: 627–628.

Damasio H, Kuljis RO, Yuh W et al. (1991). Magnetic resonance imaging of human intracortical structure in vivo. Cereb Cortex 1: 374–379.

Dantzer R, O'Connor JC, Freund GG et al. (2008). From inflammation to sickness and depression: when the immune system subjugates the brain. Nat Rev Neurosci 9: 46–56.

De Wit H (2009). Impulsivity as a determinant and consequence of drug use: a review of underlying processes. Addict Biol 14: 22–31.

Del Rey A, Balschun D, Wetzel W et al. (2013). A cytokine network involving brain-borne IL-1beta, IL-1ra, IL-18, IL-6, and TNFalpha operates during long-term potentiation and learning. Brain Behav Immun 33: 15–23.

Dick DM, Smith G, Olausson P et al. (2010). Understanding the construct of impulsivity and its relationship to alcohol use disorders. Addict Biol 15: 217–226.

Ernst M, Romeo RD, Andersen SL (2009). Neurobiology of the development of motivated behaviors in adolescence: a window into a neural systems model. Pharmacol Biochem Behav 93: 199–211.

Eslinger PJ, Damasio H, Graff-Radford N et al. (1984). Examining the relationship between computed tomography and neuropsychological measures in normal and demented elderly. J Neurol Neurosurg Psychiatry 47: 1319–1325.

Farina C, Aloisi F, Meinl E (2007). Astrocytes are active players in cerebral innate immunity. Trends Immunol 28: 138–145.

Fernandez-Lizarbe S, Pascual M, Guerri C (2009). Critical role of TLR4 response in the activation of microglia induced by ethanol. J Immunol 183: 4733–4744.

Ferrier L, Berard F, Debrauwer L et al. (2006). Impairment of the intestinal barrier by ethanol involves enteric microflora and mast cell activation in rodents. Am J Pathol 168: 1148–1154.

Flatscher-Bader T, Van Der Brug M, Hwang JW et al. (2005). Alcohol-responsive genes in the frontal cortex and nucleus accumbens of human alcoholics. J Neurochem 93: 359–370.

Fortier CB, Leritz EC, Salat DH et al. (2011). Reduced cortical thickness in abstinent alcoholics and association with alcoholic behavior. Alcohol Clin Exp Res 35: 2193–2201.

Garg AD, Nowis D, Golab J et al. (2010). Immunogenic cell death, DAMPs and anticancer therapeutics: an emerging amalgamation. Biochim Biophys Acta 1805: 53–71.

Glass CK, Saijo K, Winner B et al. (2010). Mechanisms underlying inflammation in neurodegeneration. Cell 140: 918–934.

Goldman-Rakic PS (1992). Working memory and the mind. Sci Am 267: 110–117.

Gonzalez-Perez O, Ramos-Remus C, Garcia-Estrada J et al. (2001). Prednisone induces anxiety and glial cerebral changes in rats. J Rheumatol 28: 2529–2534.

Gould E, Woolf NJ, Butcher LL (1991). Postnatal development of cholinergic neurons in the rat: I. Forebrain. Brain Res Bull 27: 767–789.

Graeber MB (2010). Changing face of microglia. Science 330: 783–788.

Grant BF (1998). The impact of a family history of alcoholism on the relationship between age at onset of alcohol use and DSM-IV alcohol dependence: results from the National Longitudinal Alcohol Epidemiologic Survey. Alcohol Health Res World 22: 144–147.

Grant B, Moore TC, Shepard J et al. (2004). Source and Accuracy Statement: Wave 1 National Epidemiologic Survey on Alcohol and Related Conditions NESARC. National Institute on Alcohol Abuse and Alcoholism. Bethesda, MD. Available online at www.niaaa.nih.gov.

Grant BF, Harford TC, Muthen BO et al. (2007). DSM-IV alcohol dependence and abuse: further evidence of validity in the general population. Drug Alcohol Depend 86: 154–166.

Grilli M, Memo M (1999). Possible role of NF-kappaB and p53 in the glutamate-induced pro-apoptotic neuronal pathway. Cell Death Differ 6: 22–27.

Gruber E, Diclemente RJ, Anderson MM et al. (1996). Early drinking onset and its association with alcohol use and problem behavior in late adolescence. Prev Med 25: 293–300.

Gruber AJ, Calhoon GG, Shusterman I et al. (2010). More is less: a disinhibited prefrontal cortex impairs cognitive flexibility. J Neurosci 30: 17102–17110.

Han SH, Kim YH, Mook-Jung I (2011). RAGE: the beneficial and deleterious effects by diverse mechanisms of actions. Mol Cells 31: 91–97.

Hanson KL, Medina KL, Padula CB et al. (2011). Impact of adolescent alcohol and drug use on neuropsychological functioning in young adulthood: 10-year outcomes. J Child Adolesc Substance Abuse 20: 135–154.

Hardingham GE, Bading H (2010). Synaptic versus extrasynaptic NMDA receptor signalling: implications for neurodegenerative disorders. Nat Rev Neurosci 11: 682–696.

Harlow JM (1848). Passage of an iron rod through the head. Boston Med Surg J 39: 389–393.

Harlow JM (1868). Recovery from the passage of an iron rod through the head. Massachus Med Soc 2: 327–346.

Harper C, Kril J (1989). Patterns of neuronal loss in the cerebral cortex in chronic alcoholic patients. J Neurol Sci 92: 81–89.

He J, Crews FT (2008). Increased MCP-1 and microglia in various regions of the human alcoholic brain. Exp Neurol 210: 349–358.

Herrmann MJ, Biehl SC, Jacob C et al. (2010). Neurobiological and psychophysiological correlates of emotional dysregulation in ADHD patients. Attention Deficit Hyperactivity Disord 2: 233–239.

Hock R, Furusawa T, Ueda T et al. (2007). HMG chromosomal proteins in development and disease. Trends Cell Biol 17: 72–79.

Huang W, Tang Y, Li L (2010). HMGB1, a potent proinflammatory cytokine in sepsis. Cytokine 51: 119–126.

Huttenlocher PR (1984). Synapse elimination and plasticity in developing human cerebral cortex. Am J Ment Defic 88: 488–496.

Iori V, Maroso M, Rizzi M et al. (2013). Receptor for advanced glycation endproducts is upregulated in temporal lobe epilepsy and contributes to experimental seizures. Neurobiol Dis http:/dx.doi.org/10.1016/j.nbd.2013.03.006.

Ishiguro H, Haruo Shibuya T, Toru M et al. (1999). Association study between high and low activity polymorphism of catechol-*O*-methyltransferase gene and alcoholism. Psychiatr Genet 9: 135–138.

Johnston L, O'Malley P, Bachman J et al. (2009). Monitoring the future: National survey results on drug use, 1975–2008. Vol.1: Seconday school students, 2008. National Institute on Drug Abuse, Bethesda, MD.

Kalsbeek A, Voorn P, Buijs RM et al. (1988). Development of the dopaminergic innervation in the prefrontal cortex of the rat. J Comp Neurol 269: 58–72.

Kaltschmidt B, Baeuerle PA, Kaltschmidt C (1993). Potential involvement of the transcription factor NF-kappa B in neurological disorders. Mol Aspects Med 14: 171–190.

Kelley KW, Dantzer R (2011). Alcoholism and inflammation: neuroimmunology of behavioral and mood disorders. Brain Behav Immun 25 (Suppl 1): S13–S20.

Kilts CD, Schweitzer JB, Quinn CK et al. (2001). Neural activity related to drug craving in cocaine addiction. Arch Gen Psychiatry 58: 334–341.

Knapp DJ, Crews FT (1999). Induction of cyclooxygenase-2 in brain during acute and chronic ethanol treatment and ethanol withdrawal. Alcohol Clin Exp Res 23: 633–643.

Koob GF, Le Moal M (1997). Drug abuse: hedonic homeostatic dysregulation. Science 278: 52–58.

Koob GF, Ahmed SH, Boutrel B et al. (2004). Neurobiological mechanisms in the transition from drug use to drug dependence. Neurosci Biobehav Rev 27: 739–749.

Kostovic I (1990). Structural and histochemical reorganization of the human prefrontal cortex during perinatal and postnatal life. Prog Brain Res 85: 223–239, discussion 239–240.

Kreek MJ, Nielsen DA, Butelman ER et al. (2005). Genetic influences on impulsivity, risk taking, stress responsivity and vulnerability to drug abuse and addiction. Nat Neurosci 8: 1450–1457.

Kreutzberg GW (1996). Microglia: a sensor for pathological events in the CNS. Trends Neurosci 19: 312–318.

Kril JJ, Harper CG (1989). Neuronal counts from four cortical regions of alcoholic brains. Acta Neuropathol 79: 200–204.

Kubota M, Nakazaki S, Hirai S et al. (2001). Alcohol consumption and frontal lobe shrinkage: study of 1432 non-alcoholic subjects. J Neurol Neurosurg Psychiatry 71: 104–106.

Lee H, Jeong J, Son E et al. (2004). Ethanol selectively modulates inflammatory activation signaling of brain microglia. J Neuroimmunol 156: 88–95.

Li TK (2008). Quantifying the risk for alcohol-use and alcohol-attributable health disorders: present findings and future research needs. J Gastroenterol Hepatol 23 (Suppl 1): S2–S8.

Li J, Kokkola R, Tabibzadeh S et al. (2003). Structural basis for the proinflammatory cytokine activity of high mobility group box 1. Mol Med 9: 37–45.

Lonze BE, Ginty DD (2002). Function and regulation of CREB family transcription factors in the nervous system. Neuron 35: 605–623.

Maggio N, Shavit-Stein E, Dori A et al. (2013). Prolonged systemic inflammation persistently modifies synaptic plasticity in the hippocampus: modulation by the stress hormones. Frontiers Mol Neurosci 6: 46.

Maroso M, Balosso S, Ravizza T et al. (2010). Toll-like receptor 4 and high-mobility group box-1 are involved in ictogenesis and can be targeted to reduce seizures. Nat Med 16: 413–419.

Marshall SA, Mcclain JA, Kelso ML et al. (2013). Microglial activation is not equivalent to neuroinflammation in alcohol-induced neurodegeneration: The importance of microglia phenotype. Neurobiol Dis 54: 239–251.

Mattson MP, Camandola S (2001). NF-kappaB in neuronal plasticity and neurodegenerative disorders. J Clin Invest 107: 247–254.

Mcclain JA, Morris SA, Deeny MA et al. (2011). Adolescent binge alcohol exposure induces long-lasting partial activation of microglia. Brain Behav Immun 25 (Suppl 1): S120–S128.

Medzhitov R (2001). Toll-like receptors and innate immunity. Nat Rev Immunol 1: 135–145.

Merrill JE, Benveniste EN (1996). Cytokines in inflammatory brain lesions: helpful and harmful. Trends Neurosci 19: 331–338.

Messmer D, Yang H, Telusma G et al. (2004). High mobility group box protein 1: an endogenous signal for dendritic cell maturation and Th1 polarization. J Immunol 173: 307–313.

Michaud CM, Mckenna MT, Begg S et al. (2006). The burden of disease and injury in the United States 1996. Population Hlth Metrics 4: 11.

Michelucci A, Heurtaux T, Grandbarbe L et al. (2009). Characterization of the microglial phenotype under specific pro-inflammatory and anti-inflammatory conditions: Effects of oligomeric and fibrillar amyloid-beta. J Neuroimmunol 210: 3–12.

Miguel-Hidalgo JJ, Overholser JC, Meltzer HY et al. (2006). Reduced glial and neuronal packing density in the orbitofrontal cortex in alcohol dependence and its relationship with suicide and duration of alcohol dependence. Alcohol Clin Exp Res 30: 1845–1855.

Mokdad AH, Marks JS, Stroup DF et al. (2004). Actual causes of death in the United States, 2000. JAMA 291: 1238–1245.

Monti PM, Miranda Jr R, Nixon K et al. (2005). Adolescence: booze, brains, and behavior. Alcohol Clin Exp Res 29: 207–220.

Montoliu C, Valles S, Renau-Piqueras J et al. (1994). Ethanol-induced oxygen radical formation and lipid peroxidation in rat brain: effect of chronic alcohol consumption. J Neurochem 63: 1855–1862.

Montoliu C, Sancho-Tello M, Azorin I et al. (1995). Ethanol increases cytochrome P4502E1 and induces oxidative stress in astrocytes. J Neurochem 65: 2561–2570.

Muller S, Ronfani L, Bianchi ME (2004). Regulated expression and subcellular localization of HMGB1, a chromatin protein with a cytokine function. J Intern Med 255: 332–343.

Mulligan MK, Ponomarev I, Hitzemann RJ et al. (2006). Toward understanding the genetics of alcohol drinking through transcriptome meta-analysis. Proc Natl Acad Sci U S A 103: 6368–6373.

Murphy KR, Barkley RA, Bush T (2002). Young adults with attention deficit hyperactivity disorder: subtype differences in comorbidity, educational, and clinical history. J Nerv Ment Dis 190: 147–157.

NIH (2000). National Institutes of Health Consensus Development Conference Statement: diagnosis and treatment of attention-deficit/hyperactivity disorder (ADHD). J Am Acad Child Adolesc Psychiatry 39: 182–193.

Nixon K, Crews FT (2002). Binge ethanol exposure decreases neurogenesis in adult rat hippocampus. J Neurochem 83: 1087–1093.

O'Malley PM, Johnston LD, Bachman JG (1998). Alcohol use among adolescents. Alcohol Health Res World 22: 85–93.

Obernier JA, Bouldin TW, Crews FT (2002a). Binge ethanol exposure in adult rats causes necrotic cell death. Alcohol Clin Exp Res 26: 547–557.

Obernier JA, White AM, Swartzwelder HS et al. (2002b). Cognitive deficits and CNS damage after a 4-day binge ethanol exposure in rats. Pharmacol Biochem Behav 72: 521–532.

Okvist A, Johansson S, Kuzmin A et al. (2007). Neuroadaptations in human chronic alcoholics: dysregulation of the NF-kappaB system. PLoS One 2: e930.

Overmeyer S, Bullmore ET, Suckling J et al. (2001). Distributed grey and white matter deficits in hyperkinetic disorder: MRI evidence for anatomical abnormality in an attentional network. Psychol Med 31: 1425–1435.

Pahl HL (1999). Activators and target genes of Rel/NF-kappaB transcription factors. Oncogene 18: 6853–6866.

Pandey SC, Zhang D, Mittal N et al. (1999). Potential role of the gene transcription factor cyclic AMP-responsive element binding protein in ethanol withdrawal-related anxiety. J Pharmacol Exp Ther 288: 866–878.

Pandey SC, Roy A, Mittal N (2001). Effects of chronic ethanol intake and its withdrawal on the expression and phosphorylation of the creb gene transcription factor in rat cortex. J Pharmacol Exp Ther 296: 857–868.

Park JS, Svetkauskaite D, He Q et al. (2004). Involvement of toll-like receptors 2 and 4 in cellular activation by high mobility group box 1 protein. J Biol Chem 279: 7370–7377.

Pascual M, Blanco AM, Cauli O et al. (2007). Intermittent ethanol exposure induces inflammatory brain damage and causes long-term behavioural alterations in adolescent rats. Eur J Neurosci 25: 541–550.

Pascual M, Balino P, Alfonso-Loeches S et al. (2011). Impact of TLR4 on behavioral and cognitive dysfunctions associated with alcohol-induced neuroinflammatory damage. Brain Behav Immun 25 (Suppl 1): S80–S91.

Paull TT, Haykinson MJ, Johnson RC (1993). The nonspecific DNA-binding and -bending proteins HMG1 and HMG2 promote the assembly of complex nucleoprotein structures. Genes Dev 7: 1521–1534.

Pfefferbaum A, Sullivan EV, Mathalon DH et al. (1995). Longitudinal changes in magnetic resonance imaging brain volumes in abstinent and relapsed alcoholics. Alcohol Clin Exp Res 19: 1177–1191.

Pizzi M, Goffi F, Boroni F et al. (2002). Opposing roles for NF-kappa B/Rel factors p65 and c-Rel in the modulation of neuron survival elicited by glutamate and interleukin-1beta. J Biol Chem 277: 20717–20723.

Plane JM, Shen Y, Pleasure DE et al. (2010). Prospects for minocycline neuroprotection. Arch Neurol 67: 1442–1448.

Pollmacher T, Haack M, Schuld A et al. (2002). Low levels of circulating inflammatory cytokines–do they affect human brain functions? Brain Behav Immun 16: 525–532.

Pruett SB, Zheng Q, Fan R et al. (2004). Ethanol suppresses cytokine responses induced through Toll-like receptors as well as innate resistance to *Escherichia coli* in a mouse model for binge drinking. Alcohol 33: 147–155.

Puumala T, Sirvio J (1998). Changes in activities of dopamine and serotonin systems in the frontal cortex underlie poor choice accuracy and impulsivity of rats in an attention task. Neuroscience 83: 489–499.

Qin L, Crews FT (2012a). Chronic ethanol increases systemic TLR3 agonist-induced neuroinflammation and neurodegeneration. J Neuroinflamm 9: 130.

Qin L, Crews FT (2012b). NADPH oxidase and reactive oxygen species contribute to alcohol-induced microglial activation and neurodegeneration. J Neuroinflamm 9: 5.

Qin L, Crews FT (2014). Focal thalamic degeneration from ethanol and thiamine deficiency is associated with neuroimmune gene induction, microglial activation, and lack of monocarboxylic acid transporters. Alcohol Clin Exp Res 38: 657–671.

Qin L, Wu X, Block ML et al. (2007). Systemic LPS causes chronic neuroinflammation and progressive neurodegeneration. Glia 55: 453–462.

Qin L, He J, Hanes RN et al. (2008). Increased systemic and brain cytokine production and neuroinflammation by endotoxin following ethanol treatment. J Neuroinflamm 5: 10.

Raivich G (2005). Like cops on the beat: the active role of resting microglia. Trends Neurosci 28: 571–573.

Raivich G, Bohatschek M, Kloss CU et al. (1999). Neuroglial activation repertoire in the injured brain: graded response, molecular mechanisms and cues to physiological function. Brain Res Brain Res Rev 30: 77–105.

Rauvala H, Rouhiainen A (2010). Physiological and pathophysiological outcomes of the interactions of HMGB1 with cell surface receptors. Biochim Biophys Acta 1799: 164–170.

Reissner KJ, Kalivas PW (2010). Using glutamate homeostasis as a target for treating addictive disorders. Behav Pharmacol 21: 514–522.

Rolls ET, Hornak J, Wade D et al. (1994). Emotion-related learning in patients with social and emotional changes associated with frontal lobe damage. J Neurol Neurosurg Psychiatry 57: 1518–1524.

Rosenberg DR, Lewis DA (1994). Changes in the dopaminergic innervation of monkey prefrontal cortex during late postnatal development: a tyrosine hydroxylase immunohistochemical study. Biol Psychiatry 36: 272–277.

Rosenbloom MH, Schmahmann JD, PRICE BH (2012). The functional neuroanatomy of decision-making. J Neuropsychiatry Clin Neurosci 24: 266–277.

Samochowiec J, Kucharska-Mazur J, Grzywacz A et al. (2006). Family-based and case-control study of DRD2, DAT, 5HTT, COMT genes polymorphisms in alcohol dependence. Neurosci Lett 410: 1–5.

Sander T, Ball D, Murray R et al. (1999). Association analysis of sequence variants of GABA(A) alpha6, beta2, and gamma2 gene cluster and alcohol dependence. Alcohol Clin Exp Res 23: 427–431.

Scheibner KA, Lutz MA, Boodoo S et al. (2006). Hyaluronan fragments act as an endogenous danger signal by engaging TLR2. J Immunol 177: 1272–1281.

Schneider H, Pitossi F, Balschun D et al. (1998). A neuromodulatory role of interleukin-1beta in the hippocampus. Proc Natl Acad Sci U S A 95: 7778–7783.

Schneider A, Martin-Villalba A, Weih F et al. (1999). NF-kappaB is activated and promotes cell death in focal cerebral ischemia. Nat Med 5: 554–559.

Schoenbaum G, Roesch MR, Stalnaker TA (2006). Orbitofrontal cortex, decision-making and drug addiction. Trends Neurosci 29: 116–124.

Schuckit MA (2009). An overview of genetic influences in alcoholism. J Subst Abuse Treat 36: S5–S14.

Shpargel KB, Jalabi W, Jin Y et al. (2008). Preconditioning paradigms and pathways in the brain. Cleve Clin J Med 75 (Suppl 2): S77–S82.

Silveri MM, Spear LP (1998). Decreased sensitivity to the hypnotic effects of ethanol early in ontogeny. Alcohol Clin Exp Res 22: 670–676.

Sims GP, Rowe DC, Rietdijk ST et al. (2010). HMGB1 and RAGE in inflammation and cancer. Annu Rev Immunol 28: 367–388.

Sowell ER, Thompson PM, Holmes CJ et al. (1999). Localizing age-related changes in brain structure between childhood and adolescence using statistical parametric mapping. Neuroimage 9: 587–597.

Sowell ER, Delis D, Stiles J et al. (2001). Improved memory functioning and frontal lobe maturation between childhood and adolescence: a structural MRI study. J Int Neuropsychol Soc 7: 312–322.

Span SA, Earleywine M (1999). Cognitive functioning moderates the relation between hyperactivity and drinking habits. Alcohol Clin Exp Res 23: 224–229.

Sparvero LJ, Asafu-Adjei D, Kang R et al. (2009). RAGE (Receptor for Advanced Glycation Endproducts), RAGE ligands, and their role in cancer and inflammation. J Translat Med 7: 17.

Spear LP (2000). The adolescent brain and age-related behavioral manifestations. Neurosci Biobehav Rev 24: 417–463.

Stalnaker TA, Takahashi Y, Roesch MR et al. (2009). Neural substrates of cognitive inflexibility after chronic cocaine exposure. Neuropharmacology 56 (Suppl 1): 63–72.

Stanton PK (1996). LTD, LTP, and the sliding threshold for long-term synaptic plasticity. Hippocampus 6: 35–42.

Stevenson JR, Schroeder JP, Nixon K et al. (2009). Abstinence following alcohol drinking produces depression-like behavior and reduced hippocampal neurogenesis in mice. Neuropsychopharmacology 34: 1209–1222.

Streit WJ, Mrak RE, Griffin WS (2004). Microglia and neuroinflammation: a pathological perspective. J Neuroinflamm 1: 14.

Stuss DT, Alexander MP (2000). Executive functions and the frontal lobes: a conceptual view. Psychol Res 63: 289–298.

Sullivan EV, Pfefferbaum A (2005). Neurocircuitry in alcoholism: a substrate of disruption and repair. Psychopharmacology (Berl) 180: 583–594.

Szabo G, Mandrekar P (2009). A recent perspective on alcohol, immunity, and host defense. Alcohol Clin Exp Res 33: 220–232.

Takahashi N, Tetsuka T, Uranishi H et al. (2002). Inhibition of the NF-kappaB transcriptional activity by protein kinase A. Eur J Biochem 269: 4559–4565.

Takeda K, Kaisho T, Akira S (2003). Toll-like receptors. Annu Rev Immunol 21: 335–376.

Tapert SF, Brown SA (2000). Substance dependence, family history of alcohol dependence and neuropsychological functioning in adolescence. Addiction 95: 1043–1053.

Thomas JO (2001). HMG1 and 2: architectural DNA-binding proteins. Biochem Soc Trans 29: 395–401.

Torres-Platas SG, Hercher C, Davoli MA et al. (2011). Astrocytic hypertrophy in anterior cingulate white matter of depressed suicides. Neuropsychopharmacology 36: 2650–2658.

Vabulas RM, Braedel S, Hilf N et al. (2002). The endoplasmic reticulum-resident heat shock protein Gp96 activates dendritic cells via the Toll-like receptor 2/4 pathway. J Biol Chem 277: 20847–20853.

Valles SL, Blanco AM, Pascual M et al. (2004). Chronic ethanol treatment enhances inflammatory mediators and cell death in the brain and in astrocytes. Brain Pathol 14: 365–371.

Van Eden CG, Kros JM, Uylings HB (1990). The development of the rat prefrontal cortex. Its size and development of connections with thalamus, spinal cord and other cortical areas. Prog Brain Res 85: 169–183.

Varlinskaya EI, Spear LP (2002). Acute effects of ethanol on social behavior of adolescent and adult rats: role of familiarity of the test situation. Alcohol Clin Exp Res 26: 1502–1511.

Vaurio L, Riley EP, Mattson SN (2008). Differences in executive functioning in children with heavy prenatal alcohol exposure or attention-deficit/hyperactivity disorder. J Int Neuropsychol Soc 14: 119–129.

Vetreno RP, Crews FT (2012). Adolescent binge drinking increases expression of the danger signal receptor agonist HMGB1 and Toll-like receptors in the adult prefrontal cortex. Neuroscience 226: 475–488.

Vetreno RP, Qin L, Crews FT (2013). Increased receptor for advanced glycation end product expression in the human alcoholic prefrontal cortex is linked to adolescent drinking. Neurobiol Dis 59: 52–62.

Volkow ND, Hitzemann R, Wolf AP et al. (1990). Acute effects of ethanol on regional brain glucose metabolism and transport. Psychiatry Res 35: 39–48.

Volkow ND, Wang GJ, Overall JE et al. (1997). Regional brain metabolic response to lorazepam in alcoholics during early and late alcohol detoxification. Alcohol Clin Exp Res 21: 1278–1284.

Volkow ND, Fowler JS, Wang GJ et al. (2004). Dopamine in drug abuse and addiction: results from imaging studies and treatment implications. Mol Psychiatry 9: 557–569.

Volz HC, Kaya Z, Katus HA et al. (2010). The role of HMGB1/RAGE in inflammatory cardiomyopathy. Semin Thromb Hemost 36: 185–194.

Ward RJ, Colivicchi MA, Allen R et al. (2009). Neuroinflammation induced in the hippocampus of 'binge drinking' rats may be mediated by elevated extracellular glutamate content. J Neurochem 111: 1119–1128.

Watkins LR, Maier SF, Goehler LE (1995). Cytokine-to-brain communication: a review & analysis of alternative mechanisms. Life Sci 57: 1011–1026.

Weafer J, Fillmore MT (2008). Individual differences in acute alcohol impairment of inhibitory control predict ad libitum alcohol consumption. Psychopharmacology (Berl) 201: 315–324.

Webb A, Lind PA, Kalmijn J et al. (2011). The investigation into CYP2E1 in relation to the level of response to alcohol through a combination of linkage and association analysis. Alcohol Clin Exp Res 35: 10–18.

Wechsler H, Dowdall GW, Davenport A et al. (1995). Correlates of college student binge drinking. Am J Public Health 85: 921–926.

White HR, Marmorstein NR, Crews FT et al. (2011). Associations between heavy drinking and changes in impulsive behavior among adolescent boys. Alcohol Clin Exp Res 35: 295–303.

Wills TA, Vaccaro D, Mcnamara G (1994). Novelty seeking, risk taking, and related constructs as predictors of adolescent substance use: an application of Cloninger's theory. J Subst Abuse 6: 1–20.

Wilson JM, Marcotte AC (1996). Psychosocial adjustment and educational outcome in adolescents with a childhood diagnosis of attention deficit disorder. J Am Acad Child Adolesc Psychiatry 35: 579–587.

Yang H, Ochani M, Li J et al. (2004). Reversing established sepsis with antagonists of endogenous high-mobility group box 1. Proc Natl Acad Sci U S A 101: 296–301.

Yang L, Seifert A, Wu D et al. (2010). Role of phospholipase D2/phosphatidic acid signal transduction in micro- and delta-opioid receptor endocytosis. Mol Pharmacol 78: 105–113.

Yucel M, Lubman DI (2007). Neurocognitive and neuroimaging evidence of behavioural dysregulation in human drug addiction: implications for diagnosis, treatment and prevention. Drug Alcohol Rev 26: 33–39.

Zecevic N, Bourgeois JP, Rakic P (1989). Changes in synaptic density in motor cortex of rhesus monkey during fetal and postnatal life. Brain Res Dev Brain Res 50: 11–32.

Zhao C, Teng EM, Summers Jr RG et al. (2006). Distinct morphological stages of dentate granule neuron maturation in the adult mouse hippocampus. J Neurosci 26: 3–11.

Zou JY, Crews FT (2005). TNF alpha potentiates glutamate neurotoxicity by inhibiting glutamate uptake in organotypic brain slice cultures: neuroprotection by NF kappa B inhibition. Brain Res 1034: 11–24.

Zou J, Crews F (2006). CREB and NF-kappaB transcription factors regulate sensitivity to excitotoxic and oxidative stress induced neuronal cell death. Cell Mol Neurobiol 26: 385–405.

Zou J, Crews F (2010). Induction of innate immune gene expression cascades in brain slice cultures by ethanol: key role of NF-kappaB and proinflammatory cytokines. Alcohol Clin Exp Res 34: 777–789.

Zou J, Crews FT (2012). Inflammasome-IL-1beta signaling mediates ethanol inhibition of hippocampal neurogenesis. FrontiersNeurosci 6: 77.

Section 11

Adolescent drinking

Handbook of Clinical Neurology, Vol. 125 (3rd series)
Alcohol and the Nervous System
E.V. Sullivan and A. Pfefferbaum, Editors

Chapter 28

The effect of alcohol use on human adolescent brain structures and systems

LINDSAY M. SQUEGLIA[1], JOANNA JACOBUS[1,2], AND SUSAN F. TAPERT[1,2]*
[1]*Department of Psychiatry, University of California, San Diego, CA, USA*
[2]*Psychology Service, VA San Diego Healthcare System, San Diego, CA, USA*

INTRODUCTION

Adolescent brain development

The brain undergoes significant structural and functional changes between childhood and adolescence, with maturation commencing in early adulthood (Pfefferbaum et al., 1994). Gray-matter volume generally decreases after puberty, which has been linked to elimination of weak synaptic connections. At the same time, white-matter fiber tracts appear to continue developing, which allows for more efficient and rapid communication between brain regions (Giedd, 2004; Lebel et al., 2012). Maturation typically begins in lower-order sensorimotor regions, while brain regions associated with higher-order cognitive functioning (e.g., frontal-subcortical brain regions) develop later in adolescence (Sowell et al., 1999, 2004; Shaw et al., 2008; Giedd and Rapoport, 2010; Stiles and Jernigan, 2010). Healthy brain development throughout childhood and adolescence is important for optimal neurocognitive performance, with even subtle changes in neurodevelopmental trajectories (e.g., changes in brain structural volume, cortical thickness, demyelination) affecting cognitive, emotional, and social functioning (Nagy et al., 2004; Casey et al., 2008). Altered brain development due to exposure of neurotoxins during adolescence, particularly alcohol, could set the stage for cognitive problems into adulthood, conferring functional consequences throughout life.

Prevalence of adolescent alcohol use and drinking patterns

Alcohol is the most commonly used substance during adolescence. According to the 2013 Monitoring the Future Study, 30% of youth in the United States have used alcohol by eighth grade. These rates more than double during adolescence, with 69% of adolescents reporting alcohol use by the time they graduate high school. Heavy episodic drinking is also common among youth, with 24% of 12th-graders endorsing binge drinking (i.e., four or more drinks for females and five or more drinks for males) in the past 2 weeks and 28% admitting being drunk in the past 30 days (Johnston et al., 2013). Given the substantial neural development occurring during this time period, along with findings on neurotoxic associates of alcoholism in many adult studies (for review, see Sullivan and Pfefferbaum, 2005; Oscar-Berman and Marinkovic, 2007), further exploration of the impact of alcohol use on brain maturational changes is critical.

This chapter will review the most recent literature regarding the effects of adolescent alcohol use on neurocognition. We will present data from neuropsychologic and neuroimaging studies examining the relationship between alcohol use during adolescence and brain structure and function.

BRAIN STRUCTURAL CHANGES IN ADOLESCENT ALCOHOL USE

Gray-matter volume

Brain regions that undergo significant neurodevelopment during adolescence (e.g., cerebral cortex, particularly the prefrontal region; limbic system; and cerebellum) have been identified as being vulnerable to the effects of alcohol use during adolescence (Squeglia et al., 2009a).To elucidate the effect of alcohol use on adolescent gray-matter development, several

*Correspondence to: Susan F. Tapert, Ph.D., VA San Diego Healthcare System, Psychology Service (116B), 3350 La Jolla Village Drive, San Diego, CA 92161, USA. Tel: +1-858-552-8585, Fax: +1-858-552-7414, E-mail: stapert@ucsd.edu

cross-sectional magnetic resonance imaging (MRI) studies have been completed focusing on these regions.

The hippocampus has been of interest because of its involvement in memory consolidation. The first investigations of the effect of alcohol use on human adolescent hippocampi reported that adolescents and young adults (ages 13–21) diagnosed with alcohol use disorders had smaller bilateral hippocampal volumes than demographically matched controls; those initiating drinking at a younger age and who had a longer duration of alcohol use disorder had even smaller hippocampal volume (De Bellis et al., 2000). Another study (Nagel et al., 2005) reported similar results, with smaller left hippocampal volumes observed in alcohol-using adolescents compared to controls, even after excluding teens with co-occurring conduct disorder. However, hippocampal volume did not correlate with alcohol use (i.e., age of onset of regular drinking, years of regular drinking, drinks consumed per month, alcohol withdrawal symptoms, estimated typical peak blood alcohol concentration, or lifetime number of abuse/dependence criteria), suggesting possible premorbid differences between groups in hippocampal volume. Taken together, these studies suggest that the hippocampus plays a role in alcohol use disorders among adolescents; however, it is unclear if aberrations in hippocampal volume precede, or are a consequence of, alcohol use.

The frontal lobe is associated with higher-order executive functioning, emotional regulation, integration of novel stimuli, and cognitive flexibility. In adults, heavy alcohol use is associated with smaller prefrontal regions (Pfefferbaum et al., 1997). Initial findings in adolescent populations have found that the prefrontal cortex may be more vulnerable to the effects of alcohol than other brain regions, because this area of the brain continues to develop throughout adolescence and young adulthood. Significantly smaller gray- and white-matter volume has been reported in the prefrontal cortex in adolescents with comorbid alcohol use and psychiatric disorders when compared with demographically matched controls (De Bellis et al., 2005). These findings were replicated in a study examining adolescents who did not meet criteria for comorbid mood or attention disorders (Medina et al., 2008). A gender by alcohol interaction was found, with alcohol-dependent girls having smaller prefrontal cortex and white-matter volumes than controls, and alcohol-dependent boys having larger prefrontal and white-matter volumes than controls. These findings suggest that gender may moderate the relationship between adolescent alcohol use on prefrontal neuromaturation, with females having more pronounced negative sequelae from continued drinking.

The cerebellum, which plays a crucial role in both motor and cognitive functions, has also been associated with alcohol-related damage (Sullivan and Pfefferbaum, 2005); however, the effect of alcohol use on adolescent cerebellar structure is less clear. In a group of healthy, non-alcohol use-disordered teens, greater binge drinking during adolescence was linked to smaller cerebellar volumes, above and beyond variability attributable to risk factors for binge drinking (Lisdahl et al., 2013). These findings replicated previous work showing boys with an alcohol use disorder had smaller cerebellar volumes than control boys.

Although these studies are not prospective in nature, they suggest smaller gray-matter volume is related to alcohol use, whether it is as a predisposing factor or a consequence of heavy drinking. The underlying biologic mechanism of alcohol-related changes in brain development remains unclear and speculative without longitudinal study.

Cortical thickness

Normal adolescent brain development involves substantial cortical thinning, which has been linked to elimination of weak synaptic connections (Giedd, 2004; Stiles and Jernigan, 2010). Yet, limited work has examined how alcohol affects normal cortical thinning during this developmental period. Squeglia and colleagues (2011b) examined how heavy episodic drinking in adolescents aged 16–19 related to indices of cortical thickness. Binge-by-gender interactions were seen for cortical thickness in four left frontal brain regions, where female bingers had thicker cortices than female controls and male bingers had thinner cortices than male controls. Thicker left frontal cortices (i.e., less mature cortices) corresponded with poorer visuospatial, inhibition, and attention performances for female bingers and worse attention for male bingers. The finding that female drinkers had thicker cortices than male drinkers is particularly notable, considering female cortices begin thinning 1–2 years earlier than males on average (Giedd, 2004; Lenroot and Giedd, 2006), and could suggest an even greater deleterious effect of drinking on female brain maturation than statistical comparisons indicated. Longitudinal investigations are needed to elucidate the temporal sequence of events, as well as the underlying mechanism driving cortical thinning during adolescence.

White-matter integrity

White matter allows for efficient information processing and quick communication between cortical regions. Unlike gray-matter maturation, which follows an inverted U-shaped curve with reductions in cortical volume beginning in preadolescence (Pfefferbaum et al., 1994) and continuing throughout early adolescence (Giedd et al., 1999), white matter continues to increase linearly in volume until late adolescence and early adulthood (Lebel and Beaulieu, 2011). Structural MRI studies have found smaller overall white-matter volumes in adolescents with alcohol use disorders compared to matched controls, particularly prefrontal cortex white matter (De Bellis et al., 2005), although there is some evidence that the pattern varies by gender, with female drinkers showing smaller prefrontal volumes than males (Medina et al., 2008).

Diffusion tensor imaging (DTI) has provided a way to look at more subtle changes in white-matter fiber structure, which may be more appropriate for this developmental time period. Two common indices of the integrity of a fiber tract, fractional anisotropy (FA) and mean diffusivity (MD), provide an inference of white-matter coherence, compactness, and directional dependence of fiber structure (Le Bihan, 2003). Typically, higher FA and lower MD are seen with healthy white-matter development (Lebel et al., 2012). Given the more subtle differences (e.g., myelination, intracellular or extracellular fluid differences) that may be occurring during this window of development, microstructural changes may not be observable with traditional MRI, and therefore studies have focused on FA and MD.

Differences in white-matter integrity have been found in adolescent binge drinkers when compared to healthy controls. Specifically, binge drinkers (approximately age 18) showed less white-matter integrity (i.e., lower FA) than controls in several white-matter regions, including cortical and subcortical projection fibers. Lower FA in six of these regions was linked to severity of substance use variables, specifically greater lifetime hangover symptoms and higher estimated peak blood alcohol concentrations (McQueeny et al., 2009). Worse white-matter integrity has also been found in adolescent marijuana users (ages 16–19) with co-occurring alcohol use in widespread fiber tracts across the brain, including cortical and subcortical tracts, along with corpus callosum (Bava et al., 2009, 2010). To disentangle the relationship between binge drinking and marijuana use during adolescence, Jacobus et al. (2009) examined white-matter integrity in 42 adolescents (ages 16–19) who were classified as binge drinkers, marijuana users who also were binge drinkers, and matched controls with minimal substance use. Both substance-using groups showed poorer white-matter health compared to non-using controls; however, in several regions (i.e., superior corona radiata, inferior fronto-occipital fasciculus, middle cerebellar peduncle, and the superior longitudinal fasciculus), binge drinkers demonstrated poorer white-matter health compared with the binge + marijuana users and controls (Jacobus et al., 2009), suggesting alcohol may have a more deleterious effect on white-matter maturation than marijuana.

Poorer white-matter integrity may be both a precursor and a consequence of heavy drinking. In 11–15-year olds who had never engaged in alcohol use, youth with positive family history of alcohol use disorders had poorer white-matter integrity in cortical and subcortical areas, and poorer performance on a task of delay discounting and reaction time when compared with youth who had no family history of alcohol use disorders (Herting et al., 2010). These pre-existing differences in white-matter integrity may contribute to future substance use and risk-taking behaviors. For example, lower white-matter integrity measured in 16–19-year old adolescents was related to more self-reported substance use and delinquency or aggression at an 18-month follow-up, highlighting the importance and potential clinical utility of the observed cross-sectional differences discussed above (Jacobus et al., 2013). In a prospective study (Bava et al., 2013), alcohol use from ages 18 to 20 predicted poorer white-matter integrity at follow-up 1.5 years later in young adults, suggesting that regular heavy alcohol use over development likely has deleterious effects on brain tissue above and beyond baseline differences that may relate to genetic predispositions.

BRAIN FUNCTION DIFFERENCES IN ADOLESCENT ALCOHOL USERS

Functional MRI (fMRI) is a widely used, non-invasive, safe technique that estimates brain activation by measuring differences in blood oxygen level-dependent (BOLD) signals between pairs of experimental conditions. Higher BOLD response contrast signal indicates regions of the brain with greater neural activity to one condition compared with another.

Tapert and colleagues (2004c) found that adolescents who drank heavily for 1–2 years showed greater brain response than controls during a spatial working-memory task in bilateral parietal cortices, with lower response than controls in other regions, including the left precentral gyrus and bilateral cerebellar areas, despite no behavioral differences on task performance. This finding was more pronounced in adolescents who reported more severe withdrawal and hangover symptoms and had greater lifetime alcohol consumption. These abnormalities, in spite of similar task performance, indicate that subtle neural reorganization might occur in heavy-drinking adolescents early in their drinking trajectory. In contrast to these findings, alcohol-dependent female adolescents with 4–5 years of drinking history exhibited significantly less BOLD response than controls during the same spatial working-memory task in parietal and prefrontal regions of the brain (Tapert et al., 2001). The alcohol-dependent women also performed worse than controls on spatial working-memory trials, which was associated with greater alcohol withdrawal symptomatology, lower spatial working-memory response in the right parietal cortex, and poorer scores on neuropsychologic tests of memory and cognitive flexibility (Tapert et al., 2001). Although adolescents with shorter drinking histories were able to perform similarly to controls on spatial working-memory tasks despite differential neural activation (Tapert et al., 2004c), adolescents who had more long-term alcohol exposure showed greater behavioral deficits (Tapert et al., 2001). This suggests that the brain may initially be resilient to neural insults or compensate for them, whereas continued alcohol use may more severely interfere with normal neural functioning.

Adolescent drinkers differ from controls on tasks of verbal encoding, specifically, learning of novel word pairs. Relative to controls, 16–18-year old binge drinkers showed greater response in right superior frontal and bilateral inferior parietal cortices, but less response in occipital cortex during verbal encoding, suggesting greater engagement of working-memory systems during this task (Schweinsburg et al., 2010). In addition, controls showed activation in the left hippocampus during novel encoding, whereas binge drinkers did not, suggestive of disadvantaged processing of novel verbal information for bingers. Binge drinkers did not show performance differences during imaging, but had poorer recall of the word pairs they had seen during imaging, suggesting subtle differences in learning tactics. A follow-up investigation in 16–18-year-olds with or without histories of binge drinking and heavy marijuana use similarly found greater right superior frontal and inferior parietal BOLD response among binge drinkers, but no main effects for marijuana use (Schweinsburg et al., 2011). Taken together, these findings suggest binge drinking is associated with aberrant neural activity during verbal working memory.

Binge drinkers also show differences in processing of affective information. In a study examining more complex cognitive processing with a decision-making task (the Iowa Gambling Task) in binge drinkers ages 16–18, binge drinkers showed greater BOLD activity in the limbic brain regions (i.e., amygdala and insula) than matched controls. Greater drinking severity was positively related to BOLD activity in the insula and negatively related to BOLD activity in the orbitofrontal cortex (Xiao et al., 2013). This pattern of activation differences suggests that hyperreactivity in brain regions responsible for emotional and incentive-related behaviors is associated with risky behaviors like binge drinking.

Several recent prospective studies have shown pre-existing differences in neural activation prior to adolescent engagement in substance use. Specifically, brain activation patterns have been able to predict future alcohol use, alcohol-related impairments (e.g., "blackouts"), and substance use disorder symptomatology. In 12–14-year-old adolescents who were scanned before initiation of alcohol use, those adolescents who transitioned into to heavy alcohol use by age 18 had lower BOLD activation at baseline in frontal, temporal, and parietal cortices in response to a Go/No-Go task of response inhibition than matched controls (Norman et al., 2011). On the same Go/No-Go task in a different sample of 12–14-year-old substance-naïve adolescents, greater baseline activation during inhibitory processing was related to more alcohol-induced blackouts at a 5-year follow-up, suggesting a neurobiologic vulnerability to alcohol-induced memory impairment (Wetherill et al., 2012). In a slightly older group of 16–19-year-old teens, less ventromedial prefrontal activation and greater left angular gyrus activation predicted greater substance use (both alcohol and marijuana) at an 18-month follow-up (Mahmood et al., 2013). An interesting study by Nees and colleagues (2012) found that reward-related brain activation (e.g., ventral striatum), along with personality and behavioral traits, aided in the prediction of early-onset drinking in adolescents, further supporting the notion that pre-existing traits may leave certain youth vulnerable to addictive behaviors (Nees et al., 2012).

To our knowledge, only one study exists that has longitudinal fMRI data points examining adolescent alcohol use. In this study, 12–16-year-old adolescents were scanned before they ever used any alcohol or drugs and then were rescanned approximately 3 years later on the same scanner (Squeglia et al., 2012; Fig. 28.1). Those who transitioned into heavy use ($n = 20$) were matched to continuous non-drinkers ($n = 20$) over the follow-up on several alcohol risk and developmental factors. Adolescents who transitioned into heavy drinking over the follow-up showed less BOLD response during a visual working-memory task at baseline in frontal and parietal regions than matched controls; this local BOLD response increased after the initiation of heavy alcohol use. By contrast, continuous non-drinkers showed decreasing activation in these regions, as would be expected with maturation. Lower baseline activation predicted future alcohol use above and beyond common predictors of substance use (e.g., family history, age, gender), consistent with previous findings suggesting neural abnormalities predate initiation of heavy substance use (Mahmood et al., 2013; Wetherill et al., 2012). Taken together, these findings suggest that neural response patterns may be risk factors for future initiation of substance use, and that heavy drinking during adolescence is associated with subtle alterations in brain functioning.

Gender and comorbid substance use are important considerations when examining the effect of alcohol

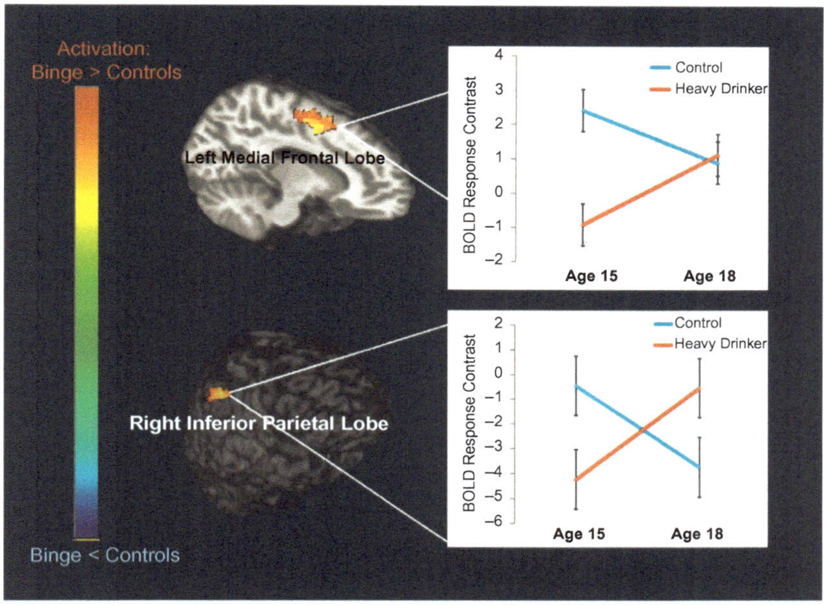

Fig. 28.1. Adolescents aged 12–16 years old were scanned before they ever used any alcohol or drugs and then were rescanned approximately 3 years later on the same scanner. Those who transitioned into heavy use ($n = 20$) were matched to continuous non-drinkers ($n = 20$) over the follow-up on several alcohol risk and developmental factors. Adolescents who transitioned into heavy drinking over the follow-up showed less blood oxygen level-dependent (BOLD) response during a visual working-memory task at baseline in frontal and parietal regions than matched controls; this local BOLD response increased after the initiation of heavy alcohol use. By contrast, continuous non-drinkers showed decreasing activation in these regions, as would be expected with maturation.

use on brain functioning. Squeglia and colleagues (2011a) found binge drinking by gender interactions in BOLD activation to a spatial working-memory task, with female binge drinkers showing less activation than female controls, and male binge drinkers exhibiting greater BOLD response than male controls. For female binge drinkers, less activation was associated with poorer neurocognitive performance; however, the opposite was found for male binge drinkers (Squeglia et al., 2011a). These findings are consistent with previous work that suggests differences in brain activation between males and females, and increased vulnerability of females to alcohol-related brain changes (Caldwell et al., 2005). In teens with co-occurring alcohol and marijuana use disorder, Schweinsburg et al. (2005) found less inferior frontal and temporal activation and increased medial frontal activation compared to controls in response to the spatial working-memory task, suggesting altered neural processing in teens with comorbid alcohol and marijuana use (Schweinsburg et al., 2005).

Genetics, vulnerability, and brain function

Several studies have found pre-existing differences in neural structure and function in adolescents with a positive family history of substance use even before they initiated heavy drinking. These findings suggest neural structure and functioning may be a phenotypic risk factor for initiation of substance use. In a study of adolescents ages 12–14, youth who had a positive family history of alcoholism showed decreased inhibitory response in frontal brain regions during a task of inhibition when compared to youth with negative family history (Schweinsburg et al., 2004). Pre-existing differences in BOLD response were also found in youth who were family history-positive for substance use, such that family history-positive youth had greater BOLD activation in frontal regions, cingulate, and insula on a Stroop color-word task compared to controls (Silveri et al., 2011). These findings suggest family history-positive adolescents may have reduced inhibitory capacity prior to the onset of alcohol use, which could increase the likelihood of their engaging in future substance use.

Family history of alcoholism is also a factor affecting working-memory systems. Teens with a positive family history have lower BOLD activation during a verbal working-memory task relative to teens with a negative family history (Cservenka et al., 2012). Family history-positive teens also showed aberrations on a spatial working-memory task in prefrontal, parietal, and cerebellar regions, suggesting these alterations may be a general risk factor for family history-positive youth (Mackiewicz et al., 2013). Futhermore, youth with a denser family history of alcohol use disorders had alterations in spatial working memory, exhibiting less BOLD activation in response to a simple vigilance condition in frontal and cingulate regions and greater activation during the resting condition (Spadoni et al., 2008).

Functional connectivity analysis of brain activation enables questions to be addressed regarding relations between activation sites and forms the basis for making inferences about functional brain circuitry. In a multimodal imaging study using fMRI and DTI indices, lower functional and structural connectivity between frontal and cerebellar regions was found in family history-positive teens when compared to matched controls (Herting et al., 2011). This study supported preliminary evidence that adolescents with a family history may also have differences in regulating their default-mode network. A similar investigation examining connectivity between frontoparietal areas in teens ages 12–14 found poorer connectivity between posterior parietal and dorsolateral prefrontal brain regions in family history-positive youth compared to controls without a family history of alcoholism; however, differences in corresponding white-matter tracts were not observed (Wetherill et al., 2012).

Several more studies have looked at BOLD response to more complex reward anticipation and decision-making tasks in order to understand whether altered reward-related circuitry in mesolimbic brain regions contributes to alcohol use disorder vulnerabilities. During a decision-making task, family history-positive youth showed atypical neural response in the dorsolateral prefrontal cortex and cerebellar regions when compared to demographically matched youth (Cservenka and Nagel, 2012). During a task of monetary incentive, brain activation differences were not found between children of alcoholics and matched controls (Bjork et al., 2008). However, in a similar study examining children (ages 18–22) of alcoholics using a monetary incentive delay task, differences were found in the mesolimbic substrates in response to this task in children of alcoholics (Yau et al., 2012). Decreased nucleus accumbens response was found in response to reward anticipation in children of alcoholics with a positive family history and low alcohol use. Positive relationships between externalizing, nucleus accumbens activity, and current and lifetime alcohol consumption in children of alcoholics were also found. Discrepancies in these two studies are likely related to developmental trajectories of the reward system, where risk taking varies by age and interacts with environment (Bjork et al., 2008; Yau et al., 2012).

In a prospective investigation, teens who were higher in resiliency (flexible and adaptive behavioral responses) during early adolescence (ages 12–15) had later onset of drinking (ages 18–22) (Weiland et al., 2012). Negative relationships were identified between resiliency and BOLD activity in basal ganglia and thalamus during a working-memory task. The authors suggest that resiliency may be related to pre-existing neuroprotective factors, as resiliency was predictive of brain activation differences (Weiland et al., 2012). In sum, genetic and personality factors may predispose adolescents to initiate heavy drinking, and should be targeted in prevention efforts.

Cue reactivity, level of response, and brain function

Cue reactivity is a classically conditioned response to drug stimuli. There have been few studies of cue reactivity in the literature with adolescents, despite reports of craving in the presence of alcohol and other drugs. Understanding brain regions subserving reactivity and craving is important because it may elucidate brain functioning that is pre-existing and genetically linked (e.g., family history status). Cue reactivity may also result in shifted attention to alcohol cues in the environment after heavy use and compromised coping skills in the presence of alcohol. One study looked at relationships between drinking models (i.e., family, friends), family history of an alcohol use disorder, and personal alcohol use on valence to alcohol cues (Pulido et al., 2009b). The sample included social-drinking young adults ages 18–23 and found that positive response to cues was dependent on more personal experience (greater current drinking) compared to family history or environment (Pulido et al., 2009b). In a follow-up study, family history was not associated with brain activation to positive or negative alcohol expectancies per se in non-drinking youth; however, differences in brain response were related to more positive alcohol use expectancies, in that less neural differentiation was found between tasks (Pulido et al., 2009a).

In two of the earlier cue reactivity studies, alcohol-dependent young women (ages 18–24) showed greater BOLD response compared to non-abusers during alcohol word presentation trials in prefrontal, insular, subcallosal, and anterior cingulate regions. Increased craving was related to increased BOLD response among the alcohol-dependent women (Tapert et al., 2004a). Another investigation examined differences in activation in response to alcohol pictures. Teens with alcohol use disorders (ages 14–17) showed increased activation to alcoholic beverage pictures in anterior cingulate and prefrontal and limbic brain regions compared to demographically matched controls. Increased brain response was related to more drinks per month (Tapert et al., 2003), highlighting differences in brain activation in heavy users in response to cues, which then likely enhances the vulnerability of teens for future use. More work prior to transition into alcohol abuse and dependence is needed in this area.

Level of response to alcohol is linked to pre-existing genetic characteristics. Research has shown that individuals requiring higher doses of alcohol to feel desired effects are more likely to develop an alcohol use disorder (Schuckit and Smith, 1996), and there are likely underlying neural substrates that impact cognitive and affective processes related to level of response. In a sample of adolescents ages 15–17 with varying degrees of alcohol use (over half with a family history of alcohol use disorder), BOLD activation to a visual working-memory task predicted level of response to alcohol. In several cases (e.g., frontal gyrus, temporal gyrus, insula), increased BOLD activation was related to needing more drinks to achieve an effect similar to that of early-drinking episodes (Tapert et al., 2004b).

NEUROCOGNITIVE PERFORMANCE IN ADOLESCENT ALCOHOL USERS

The current literature suggests that heavy drinking during adolescence has a subtle but significant deleterious effect on adolescent neurocognitive functioning. Several cross-sectional investigations have identified poorer performance in many neurocognitive domains, including attention and information processing (Tarter et al., 1995; Ferrett et al., 2010; Thoma et al., 2011), memory (Brown et al., 2000; Ferrett et al., 2010), visuospatial functioning (Sher et al., 1997; Giancola et al., 1998), language abilities (Moss et al., 1994), and executive functioning (Moss et al., 1994; Thoma et al., 2011), in heavy adolescent drinkers when compared with non-drinking controls. Taken together, these findings suggest a wide-reaching pattern of deleterious effects of alcohol use on neurocognitive abilities spanning childhood to late adolescence (Ferrett et al., 2010; Thoma et al., 2011; Parada et al., 2012).

In addition to cross-sectional studies examining neurocognitive functioning, several longitudinal investigations have been conducted, which have helped disentangle premorbid factors from consequences of heavy alcohol use. In 1999, Tapert and colleagues examined adolescents ages 13–19 recruited from inpatient treatment programs. Over 4 years of follow-up neurocognitive assessments, teens with greater alcohol and drug use following treatment had poorer neurocognitive functioning, particularly on measures of attention and visuospatial functioning (Tapert and Brown, 1999). In a follow-up study tracking participants for 8 years, from approximately age 16 to 24, greater alcohol and drug use predicted worsened attention, learning, and visuospatial functioning over the follow-up period (Tapert et al., 2002).

These findings have been replicated in a prospective study that characterized at-risk adolescents prior to ever initiating alcohol use. Alcohol-naïve adolescents, 12–14 years old, performed neuropsychological testing in follow-up study over 3 years. At follow-up (around age 16), approximately half of the sample transitioned to heavy or moderate drinking. For girls, more drinking days over the follow-up interval predicted worse visuospatial performance at follow-up; for boys, more hangover symptoms were associated with worse performance on a test of sustained attention at follow-up (Squeglia et al., 2009b). Assessment of long-term (10-year) patterns of alcohol and other drug use in teenagers from ages 16 to 27 revealed that heavier substance use patterns were related to poorer learning, memory, attention, and visuospatial processing, similar to findings in adult alcoholics (Sullivan and Pfefferbaum, 2005; Oscar-Berman and Marinkovic, 2007). For alcohol specifically, more alcohol use (with little other substance use) was related to worse short-term verbal memory. Interestingly, the authors also found that withdrawal symptoms at each follow-up visit were associated with poorer learning and memory performance (Hanson et al., 2011).

The idea that postdrinking withdrawal and hangover symptoms play a deleterious role in brain functioning is increasingly being explored in the literature, as these symptoms may be more detrimental than the quantity of alcohol consumed and may present a more accurate marker of drinking quantity. Greater hangover symptoms have been found to be predictive of worsened attention and visuospatial functioning (Tapert et al., 2002; Squeglia et al., 2009b). Negative relationships between alcohol withdrawal symptoms and learning and memory performance have also been found. For instance, among adolescent alcohol users ages 15–19, greater hangover symptoms were associated with poorer performance on measures of immediate and delayed recall (Mahmood et al., 2010).

Adolescent alcohol use affects higher-order, more complex cognitive functioning. Tenth-grade adolescent binge drinkers in China showed poorer performance on a task of risky decision making, particularly in their hypersensitivity to reward. Youth with a family history of alcoholism may be genetically linked to pre-existing differences in cognitive development (Johnson et al., 2008), mirroring findings from the neuroimaging literature. Weaknesses in executive functioning abilities have been found in youth between the ages of 3 and 14 years with a family history of alcoholism, suggesting that pre-existing differences contribute to risk for alcoholism (Nigg et al., 2004). Even with pre-existing differences, studies have clearly shown a link between substance use and poorer neurocognitive functioning with varying degrees of substance misuse (e.g., binge drinking, dependence).

SUMMARY AND CONCLUSIONS

Alcohol use is common during adolescence, with almost a quarter of all American high school seniors endorsing recent binge drinking. This trend is concerning, as adolescents undergo significant neural development, particularly in frontal regions of the brain that are involved in higher-order cognition. Adolescents who engage in heavy alcohol use show alterations in both gray- and white-matter brain structures, and aberrations in brain activity (Table 28.1). These differences in structure and function are related to poorer neurocognitive performance on tests of attention, working memory, spatial functioning, verbal and visual memory, and executive functioning. Adolescents who engage in heavy episodic drinking (binge drinking), as well as those who have greater withdrawal and hangover symptoms, tend to show the greatest deviations in functioning. Family history of alcohol is related to cognitive disadvantages that may leave youth more vulnerable to the neurotoxic effects of alcohol use.

Large-scale, prospective studies should help identify pre-existing brain differences in youth prior to initiating alcohol use, as well as identification of imaging indices and biomarkers that are able to predict future problematic drinking and co-occurring psychopathology. Gender is also an important variable to consider in both structural and functional neuroimaging because girls and boys peak in brain development at different ages and girls may be slightly more vulnerable to the effects of alcohol than boys. As alcohol is commonly used with other drugs, future studies should look at the co-occurring effect of alcohol with marijuana and other drug use. Use of these findings should be implemented in intervention and prevention measures. Remediation programs must be developed to help enhance cognition and decrease the likelihood of engagement in future substance use behaviors. Although these programs are employed with adults (Rupp et al., 2012) and are beginning to be explored prior to initiation of problematic alcohol use (Hollen, 1998), limited work has been done with adolescents with alcohol use histories. Prospective neuroimaging investigations to help disentangle pre-existing differences from toxicity continue to be important for future investigations. The investigation of how cognitive intervention may prevent future substance use involvement by targeting neural substrates underlying alcohol-related neurotoxicity is an essential next step in translating imaging and cognitive findings to clinical application.

Table 28.1

Summary of findings on the effect of alcohol on adolescent brain development

Cognition	↓ Visuospatial functioning, learning and memory, attention
Brain volume	↓ Hippocampus, prefrontal cortex, cerebellum volume
Cortical thickness	↑ Frontal cortices (females), ↓ frontal cortices (males)
White-matter integrity	↓ White-matter integrity
Brain activation	↓ Activation before initiating alcohol use; ↑ activation post heavy drinking

ACKNOWLEDGMENTS

This review was supported in part by NIAAA (R01 AA013419, 1U01 AA021692, and 1U01 AA021695) and NIDA (F32 DA032188).

DISCLOSURE STATEMENT

The authors are not aware of any affiliations, memberships, funding, or financial holdings that might be perceived as affecting the objectivity of this review.

References

Bava S, Frank LR, McQueeny T et al. (2009). Altered white matter microstructure in adolescent substance users. Psychiatry Res 173: 228–237.

Bava S, Jacobus J, Mahmood O et al. (2010). Neurocognitive correlates of white matter quality in adolescent substance users. Brain Cogn 72: 347–354.

Bava S, Jacobus J, Thayer R et al. (2013). Longitudinal changes in white matter integrity among adolescent substance users. Alcohol Clin Exp Res 37: E181–E189.

Bjork JM, Knutson B, Hommer DW (2008). Incentive-elicited striatal activation in adolescent children of alcoholics. Addiction 103: 1308–1319.

Brown SA, Tapert SF, Granholm E et al. (2000). Neurocognitive functioning of adolescents: effects of protracted alcohol use. Alcohol Clin Exp Res 24: 164–171.

Caldwell LC, Schweinsburg AD, Nagel BJ et al. (2005). Gender and adolescent alcohol use disorders on BOLD (blood oxygen level dependent) response to spatial working memory. Alcohol Alcohol 40: 194–200.

Casey BJ, Jones RM, Hare TA (2008). The adolescent brain. Ann N Y Acad Sci 1124: 111–126.

Cservenka A, Nagel BJ (2012). Risky decision-making: an FMRI study of youth at high risk for alcoholism. Alcohol Clin Exp Res 36: 604–615.

Cservenka A, Herting MM, Nagel BJ (2012). Atypical frontal lobe activity during verbal working memory in youth with a family history of alcoholism. Drug Alcohol Depend 123: 98–104.

De Bellis MD, Clark DB, Beers SR et al. (2000). Hippocampal volume in adolescent-onset alcohol use disorders. Am J Psychiatry 157: 737–744.

De Bellis MD, Narasimhan A, Thatcher DL et al. (2005). Prefrontal cortex, thalamus, and cerebellar volumes in adolescents and young adults with adolescent-onset alcohol use disorders and comorbid mental disorders. Alcohol Clin Exp Res 29: 1590–1600.

Ferrett HL, Carey PD, Thomas KG et al. (2010). Neuropsychological performance of South African treatment-naive adolescents with alcohol dependence. Drug Alcohol Depend 110: 8–14.

Giancola PR, Mezzich AC, Tarter RE (1998). Executive cognitive functioning, temperament, and antisocial behavior in conduct disordered adolescent females. J Abnorm Psychol 107: 629–641.

Giedd JN (2004). Structural magnetic resonance imaging of the adolescent brain. Ann N Y Acad Sci 1021: 77–85.

Giedd JN, Rapoport JL (2010). Structural MRI of pediatric brain development: what have we learned and where are we going? Neuron 67: 728–734.

Giedd JN, Blumenthal J, Jeffries NO et al. (1999). Brain development during childhood and adolescence: a longitudinal MRI study. Nat Neurosci 2: 861–863.

Hanson KL, Cummins K, Tapert SF et al. (2011). Changes in neuropsychological functioning over 10 years following adolescent substance abuse treatment. Psychol Addict Behav 25: 127–142.

Herting MM, Schwartz D, Mitchell SH et al. (2010). Delay discounting behavior and white matter microstructure abnormalities in youth with a family history of alcoholism. Alcohol Clin Exp Res 34: 1590–1602.

Herting MM, Fair D, Nagel BJ (2011). Altered fronto-cerebellar connectivity in alcohol-naive youth with a family history of alcoholism. Neuroimage 54: 2582–2589.

Hollen PJ (1998). Intervention booster: adding a decision-making module to risk reduction and other health care programs for adolescents. J Pediatr Health Care 12: 247–255.

Jacobus J, McQueeny T, Bava S et al. (2009). White matter integrity in adolescents with histories of marijuana use and binge drinking. Neurotoxicol Teratol 31: 349–355.

Jacobus J, Thayer RE, Trim RS et al. (2013). White matter integrity, substance use, and risk taking in adolescence. Psychol Addict Behav 27: 431–442.

Johnson CA, Xiao L, Palmer P et al. (2008). Affective decision-making deficits, linked to a dysfunctional ventromedial prefrontal cortex, revealed in 10th grade Chinese adolescent binge drinkers. Neuropsychologia 46: 714–726.

Johnston LD, O'Malley PM, Bachman JG et al. (2013). Monitoring the future national results on adolescent drug use: Overview of key findings, 2012. Institute for Social Research, The University of Michigan, Ann Arbor.

Le Bihan D (2003). Looking into the functional architecture of the brain with diffusion MRI. Nat Rev Neurosci 4: 469–480.

Lebel C, Beaulieu C (2011). Longitudinal development of human brain wiring continues from childhood into adulthood. J Neurosci 31: 10937–10947.

Lebel C, Gee M, Camicioli R et al. (2012). Diffusion tensor imaging of white matter tract evolution over the lifespan. Neuroimage 60: 340–352.

Lenroot RK, Giedd JN (2006). Brain development in children and adolescents: insights from anatomical magnetic resonance imaging. Neurosci Biobehav Rev 30: 718–729.

Lisdahl KM, Thayer R, Squeglia LM et al. (2013). Recent binge drinking predicts smaller cerebellar volumes in adolescents. Psychiatry Res 211: 17–23.

Mackiewicz Seghete KL, Cservenka A, Herting MM et al. (2013). Atypical spatial working memory and task-general brain activity in adolescents with a family history of alcoholism. Alcohol Clin Exp Res 37: 390–398.

Mahmood OM, Jacobus J, Bava S et al. (2010). Learning and memory performances in adolescent users of alcohol and marijuana: interactive effects. J Stud Alcohol Drugs 71: 885–894.

Mahmood OM, Goldenberg D, Thayer R et al. (2013). Adolescents' fMRI activation to a response inhibition task predicts future substance use. Addict Behav 38: 1435–1441.

McQueeny T, Schweinsburg BC, Schweinsburg AD et al. (2009). Altered white matter integrity in adolescent binge drinkers. Alcohol Clin Exp Res 33: 1278–1285.

Medina KL, McQueeny T, Nagel BJ et al. (2008). Prefrontal cortex volumes in adolescents with alcohol use disorders: unique gender effects. Alcohol Clin Exp Res 32: 386–394.

Moss HB, Kirisci L, Gordon HW et al. (1994). A neuropsychologic profile of adolescent alcoholics. Alcohol Clin Exp Res 18: 159–163.

Nagel BJ, Schweinsburg AD, Phan V et al. (2005). Reduced hippocampal volume among adolescents with alcohol use disorders without psychiatric comorbidity. Psychiatry Res 139: 181–190.

Nagy Z, Westerberg H, Klingberg T (2004). Maturation of white matter is associated with the development of cognitive functions during childhood. J Cogn Neurosci 16: 1227–1233.

Nees F, Tzschoppe J, Patrick CJ et al. (2012). Determinants of early alcohol use in healthy adolescents: the differential contribution of neuroimaging and psychological factors. Neuropsychopharmacology 37: 986–995.

Nigg JT, Glass JM, Wong MM et al. (2004). Neuropsychological executive functioning in children at elevated risk for alcoholism: findings in early adolescence. J Abnorm Psychol 113: 302–314.

Norman AL, Pulido C, Squeglia LM et al. (2011). Neural activation during inhibition predicts initiation of substance use in adolescence. Drug Alcohol Depend 119: 216–223.

Oscar-Berman M, Marinkovic K (2007). Alcohol: effects on neurobehavioral functions and the brain. Neuropsychol Rev 17: 239–257.

Parada M, Corral M, Mota N et al. (2012). Executive functioning and alcohol binge drinking in university students. Addict Behav 37: 167–172.

Pfefferbaum A, Mathalon DH, Sullivan EV et al. (1994). A quantitative magnetic resonance imaging study of changes in brain morphology from infancy to late adulthood. Arch Neurol 51: 874–887.

Pfefferbaum A, Sullivan EV, Mathalon DH et al. (1997). Frontal lobe volume loss observed with magnetic resonance imaging in older chronic alcoholics. Alcohol Clin Exp Res 21: 521–529.

Pulido C, Anderson KG, Armstead AG et al. (2009a). Family history of alcohol-use disorders and spatial working memory: effects on adolescent alcohol expectancies. J Stud Alcohol Drugs 70: 87–91.

Pulido C, Mok A, Brown SA et al. (2009b). Heavy drinking relates to positive valence ratings of alcohol cues. Addict Biol 14: 65–72.

Rupp CI, Kemmler G, Kurz M et al. (2012). Cognitive remediation therapy during treatment for alcohol dependence. J Stud Alcohol Drugs 73: 625–634.

Schuckit MA, Smith TL (1996). An 8-year follow-up of 450 sons of alcoholic and control subjects. Arch Gen Psychiatry 53: 202–210.

Schweinsburg AD, Paulus MP, Barlett VC et al. (2004). An fMRI study of response inhibition in youths with a family history of alcoholism. Ann N Y Acad Sci 1021: 391–394.

Schweinsburg AD, Schweinsburg BC, Cheung EH et al. (2005). fMRI response to spatial working memory in adolescents with comorbid marijuana and alcohol use disorders. Drug Alcohol Depend 79: 201–210.

Schweinsburg AD, McQueeny T, Nage BJ et al. (2010). A preliminary study of functional magnetic resonance imaging response during verbal encoding among adolescent binge drinkers. Alcohol 44: 111–117.

Schweinsburg AD, Schweinsburg BC, Nagel BJ et al. (2011). Neural correlates of verbal learning in adolescent alcohol and marijuana users. Addiction 106: 564–573.

Shaw P, Kabani NJ, Lerch JP et al. (2008). Neurodevelopmental trajectories of the human cerebral cortex. J Neurosci 28: 3586–3594.

Sher KJ, Martin ED, Wood PK et al. (1997). Alcohol use disorders and neuropsychological functioning in first-year undergraduates. Exp Clin Psychopharmacol 5: 304–315.

Silveri MM, Rogowska J, McCaffrey A et al. (2011). Adolescents at risk for alcohol abuse demonstrate altered frontal lobe activation during Stroop performance. Alcohol Clin Exp Res 35: 218–228.

Sowell ER, Thompson PM, Holmes CJ et al. (1999). In vivo evidence for post adolescent brain maturation in frontal and striatal regions. Nat Neurosci 2: 859–861.

Sowell ER, Thompson PM, Leonard CM et al. (2004). Longitudinal mapping of cortical thickness and brain growth in normal children. J Neurosci 24: 8223–8231.

Spadoni AD, Norman AL, Schweinsburg AD et al. (2008). Effects of family history of alcohol use disorders on spatial working memory BOLD response in adolescents. Alcohol Clin Exp Res 32: 1135–1145.

Squeglia LM, Jacobus J, Tapert SF (2009a). The influence of substance use on adolescent brain development. Clin EEG Neurosci 40: 31–38.

Squeglia LM, Spadoni AD, Infante MA et al. (2009b). Initiating moderate to heavy alcohol use predicts changes in neuropsychological functioning for adolescent girls and boys. Psychol Addict Behav 23: 715–722.

Squeglia LM, Schweinsburg AD, Pulido C et al. (2011a). Adolescent binge drinking linked to abnormal spatial working memory brain activation: differential gender effects. Alcohol Clin Exp Res 35: 1831–1841.

Squeglia LM, Sorg SF, Schweinsburg AD et al. (2011b). Binge drinking differentially affects adolescent male and female brain morphometry. Psychopharmacology (Berl) 220: 529–539.

Squeglia LM, Pulido C, Wetherill RR et al. (2012). Brain response to working memory over three years of adolescence: Influence of initiating heavy drinking. J Stud Alcohol Drugs 73: 749–760.

Stiles J, Jernigan TL (2010). The basics of brain development. Neuropsychol Rev 20: 327–348.

Sullivan EV, Pfefferbaum A (2005). Neurocircuitry in alcoholism: A substrate of disruption and repair. Psychopharmacology (Berl) 180: 583–594.

Tapert SF, Brown SA (1999). Neuropsychological correlates of adolescent substance abuse: four-year outcomes. J Int Neuropsychol Soc 5: 481–493.

Tapert SF, Brown GG, Kindermann SS et al. (2001). fMRI measurement of brain dysfunction in alcohol-dependent young women. Alcohol Clin Exp Res 25: 236–245.

Tapert SF, Granholm E, Leedy NG et al. (2002). Substance use and withdrawal: neuropsychological functioning over 8 years in youth. J Int Neuropsychol Soc 8: 873–883.

Tapert SF, Cheung EH, Brown GG et al. (2003). Neural response to alcohol stimuli in adolescents with alcohol use disorder. Arch Gen Psychiatry 60: 727–735.

Tapert SF, Brown GG, Baratta MV et al. (2004a). fMRI BOLD response to alcohol stimuli in alcohol dependent young women. Addict Behav 29: 33–50.

Tapert SF, Pulido C, Paulus MP et al. (2004b). Level of response to alcohol and brain response during visual working memory. J Stud Alcohol 65: 692–700.

Tapert SF, Schweinsburg AD, Barlett VC et al. (2004c). Blood oxygen level dependent response and spatial working memory in adolescents with alcohol use disorders. Alcohol Clin Exp Res 28: 1577–1586.

Tarter RE, Mezzich AC, Hsieh YC et al. (1995). Cognitive capacity in female adolescent substance abusers. Drug Alcohol Depend 39: 15–21.

Thoma RJ, Monnig MA, Lysne PA et al. (2011). Adolescent substance abuse: the effects of alcohol and marijuana on neuropsychological performance. Alcohol Clin Exp Res 35: 39–46.

Weiland BJ, Nigg JT, Welsh RC et al. (2012). Resiliency in adolescents at high risk for substance abuse: flexible adaptation via subthalamic nucleus and linkage to drinking and drug use in early adulthood. Alcohol Clin Exp Res 36: 1355–1364.

Wetherill RR, Bava S, Thompson WK et al. (2012). Frontoparietal connectivity in substance-naive youth with and without a family history of alcoholism. Brain Res 1432: 66–73.

Xiao L, Bechara A, Gong Q et al. (2013). Abnormal affective decision making revealed in adolescent binge drinkers using a functional magnetic resonance imaging study. Psychol Addict Behav 27: 443–454.

Yau WY, Zubieta JK, Weiland BJ et al. (2012). Nucleus accumbens response to incentive stimuli anticipation in children of alcoholics: relationships with precursive behavioral risk and lifetime alcohol use. J Neurosci 32: 2544–2551.

Section 12

Other topics

Handbook of Clinical Neurology, Vol. 125 (3rd series)
Alcohol and the Nervous System
E.V. Sullivan and A. Pfefferbaum, Editors

Chapter 29

Peripheral systems: neuropathy

RAJANI P. MAIYA AND ROBERT O. MESSING*
College of Pharmacy, University of Texas, Austin, TX, USA

HISTORY AND PREVALENCE

Despite evidence of alcoholic beverage production as early as 7000 BC in China (McGovern et al., 2004) and Iran (This et al., 2006), it was not until 1787 that John Coakley Lettsome first described alcohol-related polyneuropathy with key symptoms of pain, hyperalgesia, and weakness affecting mainly the lower extremities, and associated with gastrointestinal dysfunction (Minot et al., 1933). In 1822, James Jackson reported a similar syndrome, which he termed "athrodynia a potu," a progressive disease arising from the use of "ardent spirits." He noted the prominence of severe pain in the feet early in the course, "distressing" numbness in the feet and later in the hands, progressive weakness and wasting of the limbs, gastrointestinal symptoms, and improvement with abstinence from alcohol.

Alcoholic neuropathy is now known to be common among individuals with alcohol use disorders (Victor and Adams, 1953; Juntunen, 1985; Scholz et al., 1986; Monforte et al., 1995; Wetterling et al., 1999; Herson et al., 2003; Koike et al., 2003; Zambelis et al., 2005). The prevalence is difficult to ascertain, with rates varying widely, from 9% to 76% among clinical series, in part because of non-uniform criteria for diagnosis. The condition can cause significant disability, resulting in repeated limb injury, infections, falls, and severe, chronic pain. Assuming that 3.5% of people suffer from alcohol use disorders (World Health Organization, 2011), 45% have neuropathy, and the 2013 world population is 7 billion (http:/www.census.gov/population/popclockworld.html), then approximately 110 million people currently suffer from alcoholic neuropathy worldwide.

CLINICAL FEATURES

The symptoms of alcoholic neuropathy develop slowly over months in a pattern typical of a "dying back" axonal sensorimotor polyneuropathy (Juntunen, 1985; D'Amour and Butterworth, 1994; Monforte et al., 1995; Herson et al., 2003; Koike et al., 2003). Some investigators have found symptoms more common among alcoholic men than women (Vittadini et al., 2001; Zambelis et al., 2005) whereas others find no sex differences (Wetterling et al., 1999; Ammendola et al., 2001) or that they are more severe in women (Ammendola et al., 2000). The total lifetime dose and duration of alcohol intake correlate with the presence and severity of alcoholic neuropathy (Monforte et al., 1995; Agelink et al., 1998; Wetterling et al., 1999; Vittadini et al., 2001). Consumption of 100 mL of ethanol per day for at least 3 years may be the minimum required for onset of symptoms (Behse and Buchthal, 1977). The pattern of drinking also appears to be an important factor. Wetterling and colleagues (1999) examined drinking patterns in 241 patients admitted for detoxification and found that 30% of continuous (almost daily) drinkers and frequent heavy drinkers (more than three times per week with frequent intoxication) had polyneuropathy, compared with 11% of episodic drinkers (irregular alcohol consumption with sober intervals >5 days and less than one binge per week). Therefore, instruments such as the Alcohol Use Disorders Identification Test (AUDIT), which provide data on drinking patterns, may be helpful for screening patients at risk (Bohn et al., 1995).

Symptoms typically begin in the feet, with paresthesias and pain described as mainly dull and aching or as burning (Victor, 1992; Koike et al., 2003). The sensation of burning can be particularly distressing. A subjective sense of coldness in the feet is common. These unpleasant sensations can be worsened by touch and light pressure, making standing and walking difficult, and disturbing sleep due to bedclothes touching the feet. As sensory symptoms worsen and progress proximally in the legs, the hands may become affected, making it difficult to hold utensils.

*Correspondence to: Robert O. Messing, M.D., University of Texas at Austin, College of Pharmacy C0875, 107 W. Dean Keeton, BME 3.510E, Austin, TX 78712-0125, USA. Tel: +1-512-471-1735, E-mail: romessing@austin.utexas.edu

On examination there is a symmetric impairment of sensation in the lower extremities, particularly for thermal sensation and pain. Vibratory sensation and proprioception are commonly impaired as well. Deep tendon reflexes are often depressed or absent at the ankles. Later, distal weakness and atrophy may appear. Compression neuropathies may complicate the clinical course, affecting typically the radial, but also the ulnar and peroneal nerves, or the brachial plexus; these occur sometimes as the presenting symptom (Kemppainen et al., 1982).

Autonomic symptoms are less common than sensory and may be present in approximately 25% of alcohol-dependent patients (Monforte et al., 1995; Agelink et al., 1998). They include excessive sweating, dizziness, poor exercise tolerance, dysphonia, dysphagia, sleep apnea, impotence, and chronic (especially nocturnal) diarrhea (Tan et al., 1985; Barter and Tanner, 1987; Victor, 1992; Ravaglia et al., 2004). Erectile dysfunction may be particularly common as an early symptom in men (Ravaglia et al., 2004). There is a significant correlation between peripheral and autonomic neuropathy; approximately 41% of patients with alcoholic neuropathy also have signs of autonomic dysfunction (Monforte et al., 1995; Agelink et al., 1998). In addition, there is a positive correlation between autonomic neuropathy and total lifetime dose of alcohol or duration of alcohol dependence (Monforte et al., 1995; Agelink et al., 1998). By contrast, nutritional status, as determined by nutritional history, lean body mass, serum protein levels, blood folate and B_{12} levels, and erythrocyte transketolase activity, was not an important risk factor in at least one study (Monforte et al., 1995). Sympathetic and parasympathetic findings can be divergent, with only some patients showing deficits in both systems (Agelink et al., 1998). Alcoholic liver disease is associated with autonomic neuropathy and its presence may favor the development of parasympathetic dysfunction (Agelink et al., 1998; Milovanovic et al., 2009).

Many patients with alcoholic neuropathy also have postural instability and gait ataxia, and show deficits on tests of postural control. However, the severity of these deficits does not correlate with electrodiagnostic findings of neuropathy, suggesting that cerebellar degeneration rather than neuropathy is usually the major cause of ataxia in these patients (Scholz et al., 1986).

ELECTRODIAGNOSTIC STUDIES

The most common finding on nerve conduction studies is a striking reduction in the amplitude of sensory potentials in the lower extremities with moderate slowing of conduction (Behse and Buchthal, 1977; Ammendola et al., 2000; Koike et al., 2003; Zambelis et al., 2005). Compound muscle action potentials and motor conduction velocity may be moderately reduced in association with increased motor latencies. These abnormalities are less severe in the upper limbs.

Electromyographic (EMG) findings support the diagnosis of an axonal neuropathy. Behse and Buchthal (1977), in their study of 37 patients with symptoms of alcoholic neuropathy, found signs of denervation (diminished recruitment pattern, fibrillations, and positive sharp waves) in muscles from all 19 with severe leg weakness and in about half of muscles from patients with mild or no weakness. There was also evidence of reinnervation (increased motor unit duration and number of polyphasic potentials) in 80–90% of muscles from patients with severe and 67% of patients with minimal or no weakness. Scholz and colleagues (1986) examined 78 inpatients undergoing alcohol detoxification, 45% of whom had signs of mild neuropathy, and found increased polyphasic potentials in tibialis anterior in 67% of patients, but signs of denervation in only 5%. Thiele and Stalberg (1975) reported single-fiber EMG findings in 18 patients with mild to moderate alcoholic neuropathy that included weakness and muscle atrophy and found increased fiber density, increased jitter and impulse blocking, and prolonged action potential duration compared with normal control subjects and consistent with reinnervation following denervation. Therefore, EMG evidence of reinnervation is common, even in patients without weakness, in support of the diagnosis of axonal neuropathy.

Among subjects with electrodiagnostic evidence of neuropathy, up to 37% do not have neuropathic symptoms (Scholz et al., 1986; Ammendola et al., 2001; Zambelis et al., 2005). Asymptomatic alcoholic neuropathy may correspond to an early phase of the disease process (Niewiadomska et al., 1986; Scholz et al., 1986), although not all agree with this hypothesis (Zambelis et al., 2005).

OTHER LABORATORY STUDIES

Several studies of autonomic function can be helpful in documenting signs of autonomic neuropathy. Tests of parasympathetic function include measurements of the heart rate response to Valsalva's maneuver, deep breathing, or systemic administration of atropine, and the ratio of R-R intervals measured at the 30th and 15th beats after standing (Tan et al., 1984; Barter and Tanner, 1987; Monforte et al., 1995). Pupillary dilatation and supersensitivity to the pupillary constrictive action of methacholine suggest parasympathetic dysfunction (Tan et al., 1984). Screens of sympathetic nervous system function include measurements of the fall in systolic blood pressure after standing and of the increase in diastolic blood pressure in response to sustained hand grip (Barter and Tanner, 1987; Monforte et al., 1995). Computer-assisted analysis of variation in heart rate can provide information about both parasympathetic and sympathetic function and may be more accurate than bedside tests since it depends less on patient participation (Agelink et al., 1998; Milovanovic et al., 2009).

Although there is no definitive test for alcoholic neuropathy, laboratory studies indicating high levels of chronic ethanol consumption and associated organ damage can be helpful for diagnosis. Since there is significant association of alcoholic neuropathy with liver dysfunction (Vittadini et al., 2001; Zambelis et al., 2005), routine measurement of liver enzymes such as γ-glutamyltransferase (GGT), alanine aminotransferase, and aspartate aminotransferase can be helpful (Wetterling et al., 1999). Other indirect indicators of heavy alcohol consumption include increases in serum cardohydrate-deficient transferrin (CDT) and in mean corpuscular volume (Ingall, 2012; Torrente et al., 2012). The combination of GGT and CDT appears to provide higher sensitivity and specificity than any of these biomarkers alone (Hietala et al., 2006; Hock et al., 2005). More recently, several direct biomarkers derived from metabolism of ethanol have become available, including ethyl glucuronide and its metabolite ethyl sulfate, both of which can be measured in urine, and phosphatidylethanol, which can be measured in blood (Ingall, 2012; Torrente et al., 2012). A high incidence of false positives limits the use of ethyl glucuronide in screening for heavy drinking. Phosphatidylethanol is of particular interest because it has very high sensitivity and specificity, and it is not influenced by the presence of liver disease (Jo and Schlichter, 1999; Hartmann et al., 2007; Stewart et al., 2009; Wurst et al., 2010).

Since dietary deficiency of niacin, riboflavin, pyridoxine, vitamin B_{12}, or folate may cause neuropathy, serum levels of these vitamins can be useful for documenting nutritional causes, particularly in malnourished patients (Koike et al., 2003, 2012). Macrocytosis is a common finding and correlates with the presence of alcoholic neuropathy in some series (Wetterling et al., 1994; Vittadini et al., 2001; Zambelis et al., 2005). Although ethanol alone may cause macrocytosis, the presence of macrocytosis should raise suspicion of folate or B_{12} deficiency as contributing factors (Behse and Buchthal, 1977; Zambelis et al., 2005; Koike et al., 2003, 2012).

NEUROPATHOLOGY

Sural nerve biopsies from patients with alcoholic neuropathy commonly show axonal degeneration with reduced density of small and large fibers, clumping of myelin, short internodes, and axonal sprouting (Walsh and McLeod, 1970; Behse and Buchthal, 1977; Koike et al., 2001, 2003). Koike and colleagues (2001, 2003) found abundant regeneration only in patients with symptoms for more than 5 years (Fig. 29.1). Short internodes indicating degeneration and partial regeneration can also be found in myelinated fibers within the paravertebral sympathetic chain (Koike et al., 2001, 2003) (Fig. 29.2). Segmental demyelination is rare (Walsh and McLeod, 1970; Behse and Buchthal, 1977). Electron microscopy shows axonal degeneration with disorganization or complete breakdown of organelles in the axoplasm, secondary changes in the myelin sheath similar to those seen in early Wallerian degeneration, and small regenerating fibers, some in clusters, indicating regeneration (Tredici and Minazzi, 1975; Behse and Buchthal, 1977) (Fig. 29.3).

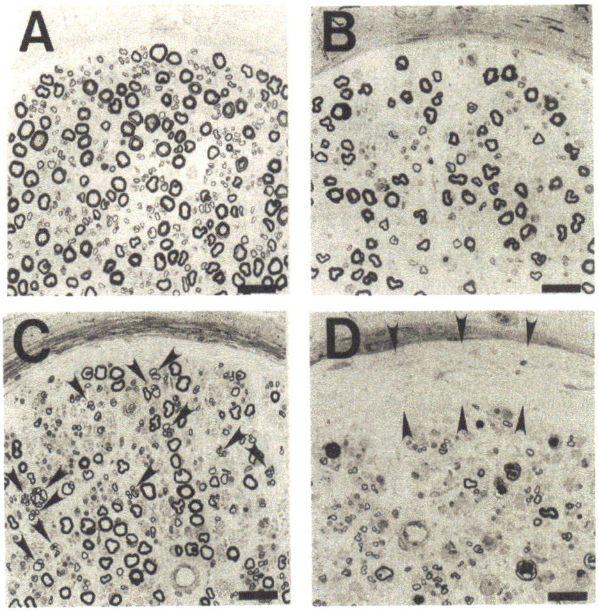

Fig. 29.1. Sural nerve pathology of alcoholic neuropathy. Light microscopic images of sural nerve specimens from control, alcoholic neuropathy, and thiamine deficiency-neuropathy patients are shown. (**A**) Transverse section of a sural nerve from a control patient is shown. (**B**) Transverse section of the sural nerve from a patient with a 4-month history of alcoholic neuropathy without thiamine deficiency is shown. Small myelinated fibers display more loss than large myelinated fibers. (**C**) Sural nerve specimen from a patient with a 5-year history of alcoholic neuropathy without thiamine deficiency is depicted. Extensive axonal sprouting (arrowheads) indicative of regenerating axons is seen, with no indication of small fiber-predominant axonal loss, as seen in **B**. (**D**) A representative specimen from a patient with non-alcoholic thiamine-deficiency neuropathy is shown. Unlike patients with alcoholic neuropathy without thiamine deficiency, both small and large myelinated fibers are significantly reduced, with large myelinated fibers being more severely affected than small myelinated fibers. Images reproduced from (Koike et al., 2003).

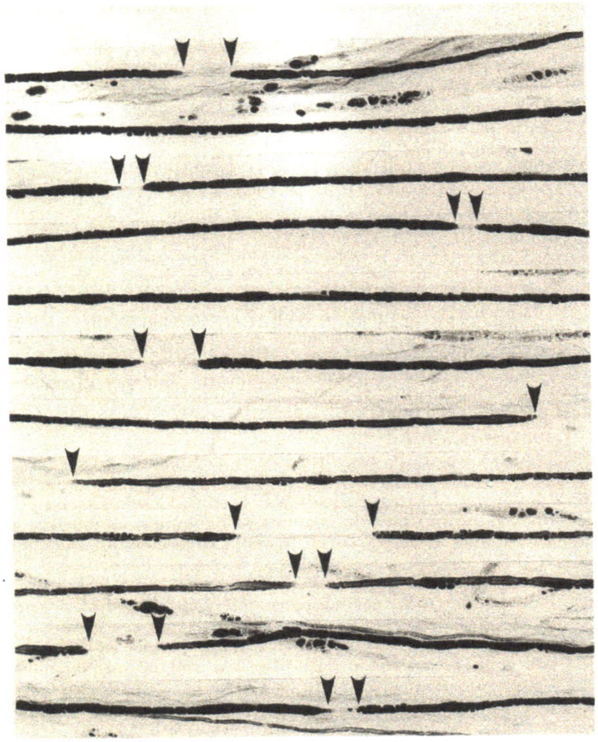

Fig. 29.2. Myelin abnormalities in alcoholic neuropathy. A single teased fiber from a patient with alcoholic neuropathy is depicted. Irregularities in the myelin sheath and segmental demyelination (arrowheads) due to widening of the nodes of Ranvier can be observed. (Reproduced from Koike et al., 2003.)

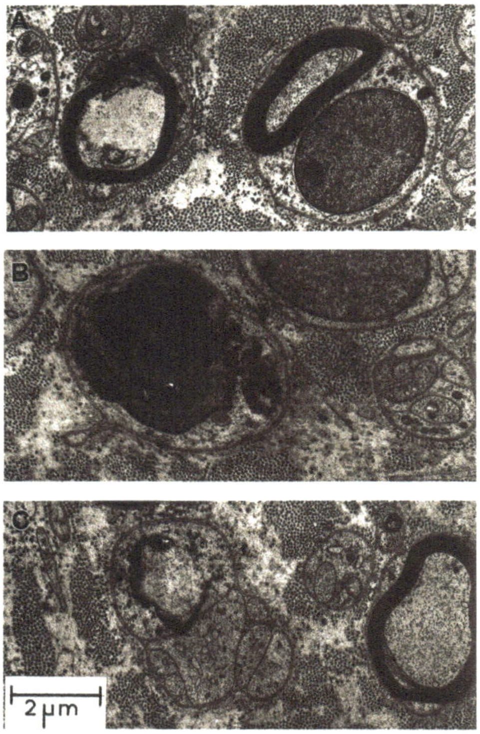

Fig. 29.3. Electron micrographs illustrating axonal degeneration of myelinated fibers of the sural nerve in alcoholic neuropathy. (A) A myelinated fiber (left) with diminished number of axonal organelles is depicted. (B) A Schwann cell with disorganized myelin and no axon is depicted. (C) A band of Bügner with 10 Schwann cell processes is depicted on the left. Remnants of myelin can be seen in one of the Schwann cell processes. In addition, the section also depicts two small normal myelinated fibers as well as unmyelinated axons surrounded by Schwann cells. (Adapted from Behse and Buchthal, 1977.)

DIFFERENTIAL DIAGNOSIS

In considering the diagnosis of alcoholic neuropathy, one must investigate other types of slowly progressive axonal sensorimotor polyneuropathy, such as those due to nutritional, metabolic, hereditary, infectious, or toxic factors (Juntunen, 1985). The diagnosis is based on finding symptoms and signs of polyneuropathy in a patient with a documented alcohol use disorder of at least 3 years' duration, without evidence of other causes. Therefore, diagnosis requires a thorough history of drinking habits along with a detailed medical history, a general physical examination with a focused examination of the peripheral nervous system, and a tailored laboratory assessment to exclude other causes. As noted above, the diagnosis is more likely if other medical complications of alcoholism are present, such as alcoholic liver disease or pancreatitis. However, since there is no definitive diagnostic test for alcoholic polyneuropathy, it is very important to search for other treatable causes of polyneuropathy in all patients with an alcohol use disorder and neuropathic symptoms.

PATHOGENESIS

Thiamine deficiency in humans

There has been a historic controversy about the etiology of alcoholic neuropathy (Mellion et al., 2011). Given the common occurrence of poor nutrition among persons with alcohol use disorders, and the apparent similarity of clinical features and pathology to beriberi, alcoholic neuropathy was thought for much of the 20th century to be due to nutritional deficiency, particularly of thiamine (Victor, 1992). Although this idea was supported by some early reports of clinical improvement following administration of B vitamins, more recent evidence supports an important role for neurotoxicity of ethanol. Koike and colleagues (2003) have provided some of the strongest clinical evidence for the contribution of ethanol neurotoxicity to alcoholic neuropathy. They examined 64 alcohol-dependent patients with neuropathy and divided them into two groups based on thiamine status, 28 being thiamine-deficient and 36 with normal thiamine levels. Thiamine levels in blood samples were measured directly by high-performance liquid chromatography, as well as indirectly by monitoring erythrocyte transketolase activity. In addition, they examined 32 additional patients with thiamine deficiency but no history of an alcohol use disorder. The group of alcoholics without thiamine deficiency showed slowly progressive symptoms that were mainly sensory, with pain as the primary complaint in most. By contrast, patients with thiamine deficiency alone showed symptoms that were more acutely progressive and included weakness with both superficial and deep sensory symptoms. Complaints in patients with both an alcohol use disorder and thiamine deficiency were highly variable and included the spectrum of symptoms experienced by the other two groups. Additional findings by Zambelis and collegaues (2005) support the existence of alcoholic neuropathy in the absence of nutritional deficiency. They examined 98 alcohol-dependent subjects without signs of malnutrition; 46 reported symptoms of a predominantly sensory polyneuropathy, 19 had autonomic symptoms, and there was evidence for subclinical neuropathy on electrophysiologic examination in another 11; none had motor complaints. These findings support the conclusion that the neurotoxicity from ethanol or its metabolites causes a mainly sensory polyneuropathy that can be distinguished clinically from neuropathy due to thiamine deficiency. The involvement of motor fibers in thiamine deficiency neuropathy is likely not a feature of more advanced neuropathy. In the study conducted by Koike and colleagues (2003), patients in the alcoholic neuropathy group displayed neuropathic symptoms for a longer period of time than those in the non-alcoholic thiamine deficiency neuropathy group.

Nevertheless, it is clear that poor nutrition is common among alcohol-dependent individuals and can result in deficiency of multiple vitamins, which can complicate the course of alcoholic neuropathy. Thiamine is a cofactor for several enzymes that function in glucose metabolism, and deficiency of thiamine is a well-recognized complication of alcoholism. The requirement for thiamine depends on metabolic rate and the minimum is 0.33 mg/1000 cal, a level which is often not achieved by alcoholics, particularly during a drinking binge (Hoyumpa, 1983). Alcohol also inhibits absorption of thiamine from the intestine, which accentuates the impact of dietary insufficiency (Hoyumpa, 1983). Moreover, thiamine is mainly stored in the liver, and in patients with severe alcoholic liver disease, hepatic thiamine levels may be reduced as much as 73%. Finally, thiamine must be phosphorylated by thiamine phosphokinase to be active as a cofactor, and this process appears to be impaired in chronic heavy drinkers (Poupon et al., 1990; Tallaksen et al., 1992). Therefore reduced thiamine intake, absorption, hepatic storage, and phosphorylation are mechanisms that underlie the strong association of thiamine deficiency with chronic alcoholism.

Deficiency of B vitamins other than thiamine

Other vitamin deficiencies may contribute to neuropathy in alcoholism. A recent case report has highlighted the potential role of folate deficiency (Koike et al., 2012), and deficiency in folate may also exacerbate thiamine deficiency by impairing intestinal uptake (Hoyumpa, 1983). Folate and vitamin B_{12} deficiency can also increase mean corpuscular volume, which is very common among chronic alcoholics (Ingall, 2012). Although macrocytosis in alcoholics may be due to a direct toxic effect of acetaldehyde on red cell maturation (Peterson et al., 1988; Niemela and Parkkila, 2004; Koivisto et al., 2006), nevertheless, given the association of deficiency in either vitamin with neuropathy and macrocytosis, it is prudent to check blood levels of B_{12} and folate and provide supplements if indicated to patients with alcoholic neuropathy. Plasma folate levels however may be normal in the setting of functional folate deficiency, as measured by tests such as formimino-l-glutamic acid excretion into the urine and deoxyuridine suppression of thymidine incorporation into lymphocytes, and in one study (Gimsing et al., 1989) both functional tests correlated better with abnormal neurophysiological tests than did plasma folate levels. Deficiency of niacin, riboflavin, and pyridoxine can also cause neuropathy and intake of these vitamins is commonly deficient in individuals with severe alcoholism (Koike et al., 2003), which provides the rationale for supplementing their diet with multiple B vitamins.

Lead in wine

Vittadini and colleagues (2001) noted in their study of 296 alcohol-dependent patients that neuropathic symptoms and abnormalities on electrophysiologic studies were more common and more severe in those who drank wine. They proposed that impurities such as lead may account for this finding. Lead toxicity causes a neuropathy in adults with prominent motor features, and among those more severely affected in the study by Vittadini and colleagues (2001), weakness and muscle cramps were common. High lead concentrations in wine can result from environmental pollution and contamination during wine production (Tariba, 2011). The concentration of lead in wine varies widely, ranging from 0.01 to 1.253 mg/L (Tariba, 2011) with an average of 130 ± 256 μg/L (Vittadini et al., 2001). Lead concentrations are higher in red than white wines and are highest in wines from central and eastern Europe (Tariba, 2011). The maximum acceptable limit has been set at 0.15 mg/L (Tariba, 2011). In adults, a blood lead level >10 μg/dL should be considered elevated, although clinical symptoms from lead alone are rarely seen below 60 μg/dL (Needleman, 2004). An average blood lead level of 26.5 ± 14.7 μg/dL was documented in one Italian study of lead levels in alcoholics (Vittadini et al., 2001).

Animal studies of alcoholic neuropathy and nutritional deficiency

Experiments in animals have identified differences and synergies between alcohol and nutritional deprivation in the development of neuropathy. In several studies, alcohol alone was neurotoxic. For example, axonal degeneration was observed in the distal axons of rats fed ethanol chronically with a nutritionally balanced diet (Bosch et al., 1979). In another study, distal nerve branches of rats fed an ethanol-containing diet showed more remyelinating axons and bands of regeneration than nerves from malnourished rats (Ortiz-Plata et al., 1998). Interestingly, however, the greatest pathology was in a group of ethanol-fed rats that were also malnourished. When switched to a normal diet, there was a marked reduction in nerve injury, and this effect was more striking when alcohol was withdrawn. Some studies have examined the effects specifically of thiamine on the development of alcoholic neuropathy. For example, in rats exposed to 10–25% ethanol for 36 weeks, neuropathic histologic and ultrastructural abnormalities were seen more frequently in the sciatic nerves and sciatic innervated muscles of animals that were thiamine-deficient compared with rats fed a normal or surplus thiamine diet, whether they were provided ethanol or not (Juntunen et al., 1979). It is important to mention that, in a few studies, no evidence of neuropathy could be found in animals fed ethanol with a nutritious diet. For example, Hallett and colleagues (1987) provided macaque male monkeys with a nutritionally balanced diet in which they received 50% of their calories from ethanol, but after 5 years found no electrophysiologic or histologic evidence of peripheral neuropathy in these animals. They observed similar results in a group of female monkeys maintained for 3 years on a diet in which they received 30% of their calories from ethanol. Taken together, the results of all of these studies indicate an important synergistic, and sometimes dominant, role for nutritional deficiency in the pathogenesis of alcoholic neuropathy. These findings parallel those in clinical studies where features of alcoholic neuropathy are worsened by inadequate nutrition (Koike et al., 2003).

MOLECULAR MECHANISMS

Acetaldehyde and oxidative stress

After consumption, ethanol is rapidly distributed throughout the body and metabolized to acetaldehyde in the liver by the enzyme alcohol dehydrogenase. Acetaldehyde interacts with proteins in the liver to form acetaldehyde-protein adducts and exerts deleterious effects on liver function by impairing the mitochondrial electron transport chain, inhibiting DNA repair, impairing microtubule function, and causing oxidative stress by depleting glutathione levels (Lieber, 1998; Masaki et al., 2004). Acetaldehyde-derived advanced glycation end products have been shown to induce cell death in cultured neuronal cells (Takeuchi and Saito, 2005). Some of the acetaldehyde produced in the liver can reach the peripheral nerves where it could exert toxic effects. While the direct effects of acetaldehyde on peripheral nerve physiology and function have not yet been explored, given evidence for acetaldehyde toxicity in the liver, it is possible that acetaldehyde plays a role in the pathogenesis of alcoholic neuropathy.

The conversion of ethanol to acetaldehyde also results in the generation of free radicals and reactive oxygen species. Oxidative stress is thought to play a causative role in several animal models of neuropathic pain (Lee et al., 2007). There is evidence that chronic ethanol consumption can lead to oxidative stress in the rat peripheral nerve. Increased accumulation of lipid peroxidation end products and reduced levels of glutathione were reported in the sciatic nerve of rats maintained on a chronic ethanol diet (Bosch-Morell et al., 1998). Administration of tocotrienol, an isoform of vitamin E, was not only found to reduce levels of these markers of oxidative stress in ethanol-fed rats, but also to ameliorate symptoms of painful neuropathy (Tiwari et al., 2012). In another study, co-administration of epigallocatechin-3-gallate, a polyphenolic antioxidant isolated from green tea, with ethanol significantly attenuated hyperalgesia induced by chronic alcohol consumption (Tiwari et al., 2011). Antioxidants such as quercetin, curcumin, and α-tocopherol were also effective in reducing painful symptoms and restoring nerve function in animal models of alcoholic neuropathy (Kandhare et al., 2012; Raygude et al., 2012). These results suggest that treatments that restore redox balance may be of therapeutic benefit in alcoholic neuropathy.

Glial dysfunction

Chronic consumption of ethanol can also affect microglial cells in the spinal cord. In a rat model of alcoholic neuropathy, hypertrophy of microglial cells was observed upon chronic ethanol exposure and withdrawal from ethanol led to marked microglial activation, suggesting that activation and proliferation of microglia could contribute to the development of alcoholic neuropathy (Narita et al., 2007). A recent study found patchy demyelination of motor and mixed nerves, suggesting impaired Schwann cell function, in rats chronically fed ethanol (Nguyen et al., 2012). Consistent with these *in vivo* findings, *in vitro* studies indicate that ethanol can inhibit the proliferation of Schwann cells and the formation of myelin (Tordoff et al., 2002).

The role of insulin and insulin-like growth factor (IGF) in alcoholic neuropathy

Alcohol impairs insulin and IGF signaling in the brain and the liver (Cohen et al., 2007; de la Monte et al., 2008a, b). Impaired insulin/IGF signaling leads to increased oxidative stress, which contributes to ethanol-induced neurodegeneration in the hippocampus and temporal cortex (Cohen et al., 2007). Insulin and IGF signaling are also important for myelin sheath maintenance in both central and peripheral nerves (Ogata et al., 2006; Chesik et al., 2008; Joseph and Ye, 2008). A recent study found reduced levels of mRNA for several components of insulin/IGF signaling, including insulin, IGF-1, and IGF-2, in the sciatic nerves of alcohol-fed rats (Nguyen et al., 2012). In addition, there was reduced immunoreactivity to insulin, IGF-1, insulin receptor substrates 1 and 4, and myelin-associated proteins in sciatic nerves from alcohol-fed rats compared with controls. These results led the authors to hypothesize that reduced insulin/insulin growth factor signaling underlies myelin defects associated with alcoholic neuropathy. A caveat with these findings is that in humans the primary pathology is axonal degeneration with secondary demyelination. Nevertheless, given these initial preclinical findings, it might be interesting to investigate in animals whether stimulating insulin/IGF signaling *in vivo* attenuates alcoholic neuropathy.

Nociceptor signaling via protein kinase C (PKC)

PKC belongs to a family of serine threonine kinases that transduce intracellular signals carried by lipid second messengers (Hardcastle et al., 1999). PKC has been widely studied as a regulator of nociceptor function. Specifically, the ε isofom of PKC has been identified as a key mediator in inflammatory (Khasar et al., 1999) and neuropathic pain, including pain associated with alcoholic polyneuropathy (Dina et al., 2000, 2006, 2007). Chronic alcohol consumption increases PKCε levels in dorsal root ganglia in a rat model of painful alcoholic neuropathy, and ethanol-induced mechanical hyperalgesia in this model is attenuated by acute administration of PKCε inhibitors locally at the site of nociceptive testing (Dina et al., 2000). Rats fed an ethanol-containing diet intermittently with periods of ethanol withdrawal developed hyperalgesia faster than rats exposed to a continuous diet of 6.5% ethanol, suggesting that processes active during repeated episodes of ethanol withdrawal are critical for the development of alcohol-induced hyperalgesia (Dina et al., 2006). Administration of antisense oligonucleotides against PKCε attenuated ethanol withdrawal-induced hyperalgesia, demonstrating the key role of PKCε in this process (Dina et al., 2006).

In one human study, alcoholic neuropathy was more severe in women than in men (Ammendola et al., 2000), and the same appears to be true for experimental alcoholic neuropathy in rats. Dina and colleagues (2007) reported that female rats maintained on an ethanol-containing diet develop more severe mechanical hyperalgesia than males maintained on a similar diet. Strikingly, females subjected to ovariectomy showed a great delay in the development of alcohol-induced hyperalgesia, and this effect could be reversed upon estrogen replacement. Sex differences were also observed in signaling pathways involved in alcohol-induced hyperalgesia. PKCε inhibition decreased hyperalgesia in both sexes but the effect was greater in female rats. Inhibition of the extracellular signal-activated kinases ERK1/2 also reduced hyperalgesia in both sexes, but administration of a protein kinase A (PKA) inhibitor was effective only in females. These results indicate that PKA, ERK 1/2, and PKCε contribute to painful alcoholic neuropathy in female rats, but only ERK1/2 and PKCε are important in males.

What are molecular mechanisms by which ethanol activates PKCε? Chronic ethanol exposure and withdrawal result in increased circulating levels of corticosterone and epinephrine (Adinoff et al., 1998; Errico et al., 2002; Thayer et al., 2006) and glucocorticoids exacerbate pain in some animal models of peripheral neuropathy (Takasaki et al., 2005). Primary afferent nociceptors express receptors for these stress hormones. In a rat model of binge drinking and painful neuropathy, adrenal medullectomy as well as pharmacologic or antisense-mediated antagonism of glucocorticoid or β_2-adrenergic receptors, which activate PKCε in nociceptors (Khasar et al., 1999), reduced mechanical hyperalgesia (Dina et al., 2008). Experiments also revealed that alcohol and stress act independently on the primary nociceptor to produce hyperalgesia (Ferrari et al., 2013).

PKCε interacts with the cytoplasmic polyadenylation element-binding protein (CPEB), which is a master regulator of protein translation and is highly expressed in subsets of nociceptors. The interaction of PKCε with CPEB is necessary for hyperalgesic priming, which is a long-lasting hyperresponsiveness of nociceptors to inflammatory mediators of pain such as cytokines (Gunthorpe et al., 2002). Hence, an intriguing possibility is that PKCε signaling mediates its effects by altering gene expression in nociceptors via interactions with CPEB. PKCε also promotes hyperalgesia by phosphorylating the polymodal receptor TRPV1 (Numazaki et al., 2002; Bhave et al., 2003) and the sodium channel $Na_V1.8$ in nociceptors (Wu et al., 2012). PKCε modulation of these channels could contribute to hyperalgesia in alcoholic neuropathy.

Axonal transport and the cytoskeleton

The axonal transport system, which is critical for the delivery of essential proteins and other cellular components to distal nerve terminals, is vulnerable to the neurotoxic effects of ethanol. Hooper and colleagues (1997) found that ethanol impairs anterograde axonal transport by as much as 70% in an *in vitro* rat dorsal root ganglia-sciatic nerve preparation incubated with a high dose of ethanol. These results were recapitulated in sciatic nerves isolated from rats chronically exposed to an ethanol-containing liquid diet. Chronic ethanol exposure also can impair retrograde transport in axons. Sensory nerves from rats maintained chronically on an ethanol-containing diet displayed a 45% decrease in retrograde transport in comparison to controls, whereas thiamine-deficient rats interestingly failed to show abnormalities in axonal transport in sensory or motor nerves (McLane, 1990). Likewise, rats maintained on a nutritionally deficient, ethanol-free diet do not show deficits in axonal transport, indicating that the deficits in axonal transport observed in rats fed an alcohol-containing diet are due to the direct toxic effects of ethanol (Malatova and Cizkova, 2002).

Ethanol's effects on transport may be selective for the type of transport. For example, differences have been observed in fast axonal transport of acetylcholinesterase (AChE) and slow transport of choline acetyltransferase (ChAT). In one study, consumption of ethanol significantly reduced retrograde transport of AChE in the sciatic nerves of rats but had no effect on anterograde transport (Malatova and Cizkova, 2002). By contrast, another study reported an increase in anterograde transport of AChE but no change in the transport of ChAT in the sciatic nerves of rats exposed chronically to ethanol (Bosch et al., 1979). The discrepancies between the two studies may reflect the different paradigms of alcohol exposure used. Nevertheless, these results suggest that chronic exposure of peripheral nerves to high concentrations of ethanol can impair both anterograde and retrograde transport.

Chronic alcohol consumption increases levels of phosphorylated neurofilament proteins in the spinal cord in association with a decrease in neurofilament-associated phosphatase (Guru et al., 1991). *In vitro* studies using hepatocytes and cultured neurons suggest that chronic ethanol consumption can increase levels of other cytoskeletal proteins such as neuronal protein 22 and phosphorylated microtubule-associated proteins, MAP1 and MAP2 (Ahluwalia et al., 2000; Depaz et al., 2005). Hence, the axonal transport deficits observed after chronic ethanol consumption could be due to alcohol-induced increases in the levels or phosphorylation of cytoskeletal proteins in peripheral neurons.

PKCε could also regulate alcohol's effects on axonal transport through its interactions with the cytoskeleton. PKCε can bind actin through protein interactions mediated by its C1 domain (Zeidman et al., 2002; Ling et al., 2005). There is increased association of PKCε with the actin cytoskeleton during neuronal differentiation and this interaction is required for PKCε to induce neurite outgrowth (Zeidman et al., 2002). PKCε's interactions with the cytoskeleton may also be important for its role in regulating neuropathic pain as pharmacologic agents that disrupt cytoskeletal integrity can block hyperalgesia induced by activating PKCε (Khasar et al., 1999).

PROGNOSIS AND TREATMENT

Alcoholic neuropathy is a reversible condition that improves with abstinence from alcohol (Hillbom and Wennberg, 1984; Tan et al., 1984; Weise et al., 1986; Villalta et al., 1989). The presence of autonomic neuropathy is associated with increased mortality (Johnson and Robinson, 1988; Milovanovic et al., 2009), which may reflect a causal role of autonomic dysfunction. Given the common occurrence of nutritional deficiency in severe alcoholism, practice recommendations generally include treatment with multivitamins and thiamine. Some have argued that the neurotoxicity of alcohol dominates in most cases (Koike and Sobue, 2006), making B vitamin supplementation unlikely to be of much benefit unless there is frank evidence of vitamin deficiency (Juntunen, 1985). However, two randomized trials of vitamin supplementation have lent support to treatment with B vitamins. Because alcohol impairs thiamine absorption and previous studies of thiamine replacement had given conflicting results, Woelk and colleagues (1998) examined the effectiveness of benfotiamine, an S-acyl derivative of thiamine that is more readily absorbed, in 84 alcoholic patients with marked symptoms of neuropathy. Treatment with 320 mg benfotiamine daily for 4 weeks followed by 120 mg daily for another 4 weeks improved strength, coordination, and overall neuropathy score compared with placebo, but did not have a significant effect on pain or vibratory thresholds. Also, subjects treated with a preparation containing benfotiamine plus vitamins B_6 and B_{12} failed to improve for unclear reasons. In a more recent randomized multicenter double-blind study, Peters and colleagues (2006) examined 204 patients with early symptoms and electrophysiologic signs of alcoholic neuropathy. Subjects were administered multivitamin tablets containing 250 mg vitamin B_1, 10 mg vitamin B_2, and 250 mg vitamin B_6 with or without 1 mg folic acid, three times daily. The B vitamin combination improved vibratory sensation and two-point touch discrimination, reduced pain, and improved coordination compared with placebo. Given the strong safety profile of these preparations, it would seem appropriate to administer combination B vitamins to patients with alcoholic neuropathy.

Supportive treatment of alcoholic neuropathy includes physical and occupational therapy, including, when necessary, gait training, training in activities of daily living, use of assistive devices, and range-of-motion and strengthening exercises. Nutritional consulting should be provided as well as referral for substance dependence treatment, including evaluation for treatment with naltrexone or acamprosate.

In patients with pain, symptomatic relief may be obtained by using drugs that have proven efficacy in other forms of peripheral neuropathy, most of which have been tested in postherpetic neuralgia or diabetic neuropathy (Dworkin et al., 2007). First-line drugs include tricyclic antidepressants and selective serotonin/norepinephrine reuptake inhibitors (SSNRIs) such as venlafaxine. However, duloxetine, another SSNR that is used to treat neuropathy, is specifically contraindicated in patients with alcohol abuse. Anticonvulsants acting at $\alpha 2/\delta$ subunits of voltage-gated calcium channels, gabapentin and pregabalin, have a better safety profile and may also be useful. Although opioids can be of benefit in other forms of neuropathy, their chronic use is

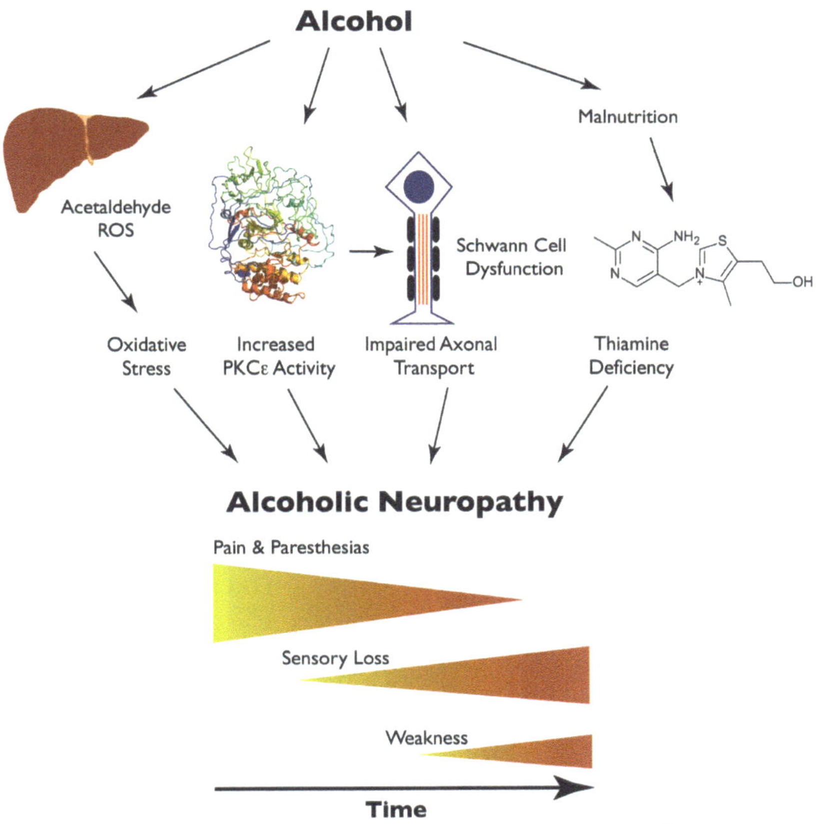

Fig. 29.4. Pathophysiology of alcoholic neuropathy. Alcohol can exert direct and indirect effects on nerve cell function. The direct effects of alcohol result from its metabolism to its toxic byproduct acetaldehyde. Acetaldehyde produced in the liver is transported to the periphery where it can form protein adducts and cause nerve damage. The conversion of alcohol to acetaldehyde also results in the formation of reactive oxygen species, which can exert deleterious effects on peripheral nerve physiology and function. Alcohol enhances protein kinase C ε (PKCε) levels and activity in dorsal root ganglia. Administration of PKCε inhibitors reduces ethanol-induced mechanical hyperalgesia. Chronic ethanol exposure can impair both anterograde and retrograde axonal transport. PKCε could also regulate alcohol's effects on axonal transport through its interactions with the cytoskeleton. Ethanol impairs Schwann cell proliferation and myelin sheath formation. Alcohol can also mediate indirect effects on nerve cell function by reducing absorption, hepatic storage, and phosphorylation of thiamine. ROS, reactive oxygen species. PKCε structure was created using the Phyre server (Kelley and Sternberg, 2009).

contraindicated in patients with substance dependence. Topical lidocaine can be helpful for localized foot pain, but its use is contraindicated if there is significant hepatic dysfunction. Of note, alcohol is analgesic, and in alcohol-dependent animals and humans, alcohol withdrawal increases pain sensitivity (Gatch, 2009; Jochum et al., 2010). The possibility that neuropathic pain promotes self-medication with alcohol is suggested by evidence that several young adults with chronic pain disorders report use of alcohol for pain relief (Brennan et al., 2005; Riley and King, 2009), although such behavior may be less common among older adults (Brennan et al., 2011). Therefore, strategies designed to relieve pain may be beneficial as adjuncts to treating the underlying addiction, particularly in younger patients with painful alcoholic neuropathy.

SUMMARY AND FUTURE DIRECTIONS

Evidence from both clinical studies and animal models support a primary role for neurotoxic effects of ethanol on peripheral nerve structure and function (Fig. 29.4). Nutritional deficits commonly associated with chronic alcoholism serve to worsen the neuropathy. Ultrastructural studies reveal features typical of an axonal "dying back" neuropathy. Concomitant with the ultrastructural findings, electrodiagnostic studies reveal a reduction in the amplitude of sensory potentials and more moderate decrease in nerve conduction velocities of distal nerves.

Animal models recapitulate several of the ultrastructural and electrophysiologic features of alcoholic neuropathy and have facilitated the identification of molecular mechanisms that underlie the deleterious effects of alcohol on peripheral nerves. Alcohol can exert direct neurotoxic effects through its metabolite acetaldehyde and by enhancing production of reactive oxygen species. Alcohol-induced activation of PKCε signaling plays an important role in reducing nociceptive threshold. Chronic alcohol consumption can also cause microglial activation and deficits in Schwann cell proliferation leading to myelin sheath abnormalities. Further, chronic alcohol consumption can cause cytoskeletal dysfunction and impair axonal transport.

Symptoms of alcoholic neuropathy are reversible and generally improve upon cessation of drinking. Nutritional deficiencies are frequently observed in patients of alcoholic neuropathy, and administration of B vitamins is beneficial to some patients. Treatment of painful symptoms of alcoholic neuropathy may provide symptomatic relief and reduce self-medication that sustains the cycle of alcohol addiction.

All of the existing treatment options suffer from variable efficacy. Hence there is a need for identification of novel therapeutic targets. Findings from animal studies provide impetus for exploring treatment strategies geared towards restoring redox balance in peripheral nerves. Antioxidants such as tocotrienol, which have been shown to reduce hyperalgesia associated with alcoholic neuropathy in animals, may be of therapeutic benefit in a clinical setting. Current evidence from animal models also suggests that PKCε is an attractive candidate target for therapeutic intervention.

References

Adinoff B, Iranmanesh A, Veldhuis J et al. (1998). Disturbances of the stress response: the role of the HPA axis during alcohol withdrawal and abstinence. Alcohol Health Res World 22: 67–72.

Agelink MW, Iranmanesh A, Veldhuis J et al. (1998). Alcoholism, peripheral neuropathy (PNP) and cardiovascular autonomic neuropathy (CAN). J Neurol Sci 161: 135–142.

Ahluwalia B, Ahmad S, Adeyiga O et al. (2000). Low levels of ethanol stimulate and high levels decrease phosphorylation in microtubule-associated proteins in rat brain: an in vitro study. Alcohol Alcohol 35: 452–457.

Ammendola A, Gemini D, Iannaccone S et al. (2000). Gender and peripheral neuropathy in chronic alcoholism: a clinical-electroneurographic study. Alcohol Alcohol 35: 368–371.

Ammendola A, Tata MR, Aurilio C et al. (2001). Peripheral neuropathy in chronic alcoholism: a retrospective cross-sectional study in 76 subjects. Alcohol Alcohol 36: 271–275.

Barter F, Tanner AR (1987). Autonomic neuropathy in an alcoholic population. Postgrad Med J 63: 1033–1036.

Behse F, Buchthal F (1977). Alcoholic neuropathy: Clinical, electrophysiological, and biopsy findings. Ann Neurol 2: 95–110.

Bhave G, Hu H, Glauner KS et al. (2003). Protein kinase C phosphorylation sensitizes but does not activate the capsaicin receptor transient receptor potential vanilloid 1 (TRPV1). Proc Natl Acad Sci U S A 100: 12480–12485.

Bohn MJ, Babor TF, Kranzler HR (1995). The Alcohol Use Disorders Identification Test (AUDIT): validation of a screening instrument for use in medical settings. J Stud Alcohol 56: 423–432.

Bosch EP, Pelham RW, Rasool CG et al. (1979). Animal models of alcoholic neuropathy: morphologic, electrophysiologic, and biochemical findings. Muscle Nerve 2: 133–144.

Bosch-Morell F, Martínez-Soriano F, Colell A et al. (1998). Chronic ethanol feeding induces cellular antioxidants decrease and oxidative stress in rat peripheral nerves. Effect of S-adenosyl-L-methionine and N-acetyl-L-cysteine. Free Radic Biol Med 25: 365–368.

Brennan PL, Schutte KK, Moos RH (2005). Pain and use of alcohol to manage pain: prevalence and 3-year outcomes among older problem and non-problem drinkers. Addiction 100: 777–786.

Brennan PL, Schutte KK, SooHoo S et al. (2011). Painful medical conditions and alcohol use: a prospective study among older adults. Pain Med 12: 1049–1059.
Chesik D, De Keyser J, Wilczak N (2008). Insulin-like growth factor system regulates oligodendroglial cell behavior: therapeutic potential in CNS. J Mol Neurosci 35: 81–90.
Cohen AC, Tong M, Wands JR et al. (2007). Insulin and insulin-like growth factor resistance with neurodegeneration in an adult chronic ethanol exposure model. Alcohol Clin Exp Res 31: 1558–1573.
D'Amour ML, Butterworth RF (1994). Pathogenesis of alcoholic peripheral neuropathy: direct effect of ethanol or nutritional deficit? Metab Brain Dis 9: 133–142.
de la Monte SM, Tong M, Cohen AC et al. (2008a). Insulin and insulin-like growth factor resistance in alcoholic neurodegeneration. Alcohol Clin Exp Res 32: 1630–1644.
de la Monte SM, Yeon J, Tong M et al. (2008b). Insulin resistance in experimental alcohol-induced liver disease. J Gastroenterol Hepatol 23: e477–e486.
Depaz IM, de las Heras R, Kroon P et al. (2005). Changes in neuronal protein 22 expression and cytoskeletal association in the alcohol-dependent and withdrawn rat brain. J Neurosci Res 81: 253–260.
Dina OA, Barletta J, Chen X et al. (2000). Key role for the epsilon isoform of protein kinase C in painful alcoholic neuropathy in the rat. J Neurosci 20: 8614–8619.
Dina OA, Messing RO, Levine JD (2006). Ethanol withdrawal induces hyperalgesia mediated by PKCepsilon. Eur J Neurosci 24: 197–204.
Dina OA, Gear RW, Messing RO et al. (2007). Severity of alcohol-induced painful peripheral neuropathy in female rats: role of estrogen and protein kinase (A and Cepsilon). Neuroscience 145: 350–356.
Dina OA, Khasar SG, Alessandri-Haber N et al. (2008). Alcohol-induced stress in painful alcoholic neuropathy. Eur J Neurosci 27: 83–92.
Dworkin RH, O'Connor AB, Backonja M et al. (2007). Pharmacologic management of neuropathic pain: evidence-based recommendations. Pain 132: 237–251.
Errico AL, King AC, Lovallo WR et al. (2002). Cortisol dysregulation and cognitive impairment in abstinent male alcoholics. Alcohol Clin Exp Res 26: 1198–1204.
Ferrari LF, Levine E, Levine JD (2013). Independent contributions of alcohol and stress axis hormones to painful peripheral neuropathy. Neuroscience 228: 409–417.
Gatch MB (2009). Ethanol withdrawal and hyperalgesia. Curr Drug Abuse Rev 2: 41–50.
Gimsing P, Melgaard B, Andersen K et al. (1989). Vitamin B-12 and folate function in chronic alcoholic men with peripheral neuropathy and encephalopathy. J Nutr 119: 416–424.
Gunthorpe MJ, Benham CD, Randall A et al. (2002). The diversity in the vanilloid (TRPV) receptor family of ion channels. Trends Pharmacol Sci 23: 183–191.
Guru SC, Shetty KT, Shankar SK (1991). Effect of chronic ethanol ingestion on phosphate content of neurofilament proteins and neurofilament associated protein phosphatase in rat spinal cord. Neurochem Res 16: 1193–1197.
Hallett M, Fox JG, Rogers AE et al. (1987). Controlled studies on the effects of alcohol ingestion on peripheral nerves of macaque monkeys. J Neurol Sci 80: 65–71.
Hardcastle IR, Rowlands MG, Barber AM et al. (1999). Inhibition of protein prenylation by metabolites of limonene. Biochem Pharmacol 57: 801–809.
Hartmann S, Aradottir S, Graf M et al. (2007). Phosphatidylethanol as a sensitive and specific biomarker: comparison with gamma-glutamyl transpeptidase, mean corpuscular volume and carbohydrate-deficient transferrin. Addict Biol 12: 81–84.
Herson PS, Virk M, Rustay N et al. (2003). A mouse model of episodic ataxia type-1. Nat Neurosci 6: 378–383.
Hietala J, Koivisto H, Latvala J et al. (2006). IgAs against acetaldehyde-modified red cell protein as a marker of ethanol consumption in male alcoholic subjects, moderate drinkers, and abstainers. Alcohol Clin Exp Res 30: 1693–1698.
Hillbom M, Wennberg A (1984). Prognosis of alcoholic peripheral neuropathy. J Neurol Neurosurg Psychiatry 47: 699–703.
Hock B, Soyka M, Schwarz M et al. (2005). Validity of carbohydrate-deficient transferrin (%CDT), gamma-glutamyltransferase (gamma-GT) and mean corpuscular erythrocyte volume (MCV) as biomarkers for chronic alcohol abuse: a study in patients with alcohol dependence and liver disorders of non-alcoholic and alcoholic origin. Addiction 100: 1477–1486.
Hooper NM, Karran EH, Turner AJ (1997). Membrane protein secretases. Biochem J 321: 265–279.
Hoyumpa AMJ (1983). Alcohol and thiamine metabolism. Alcohol Clin Exp Res 7: 11–14.
Ingall GB (2012). Alcohol biomarkers. Clin Lab Med 32: 391–406.
Jackson J (1822). On a peculiar disease resulting from the use of ardent spirits. N Engl J Med 11: 351–353.
Jo Y-H, Schlichter R (1999). Synaptic corelease of ATP and GABA in cultured spinal neurons. Nat Neurosci 2: 241–245.
Jochum T, Boettger MK, Burkhardt C et al. (2010). Increased pain sensitivity in alcohol withdrawal syndrome. Eur J Pain 14: 713–718.
Johnson RH, Robinson BJ (1988). Mortality in alcoholics with autonomic neuropathy. J Neurol Neurosurg Psychiatry 51: 476–480.
Joseph DA, Ye P (2008). Expanding the mind: insulin-like growth factor I and brain development. Endocrinology 149: 5958–5962.
Juntunen J (1985). Alcoholic polyneuropathy. Acta Med Scand Suppl 703: 265–272.
Juntunen J, Teräväinen H, Eriksson K et al. (1979). Peripheral neuropathy and myopathy. An experimental study of rats on alcohol and variable dietary thiamine. Virchows Arch A Pathol Anat Histol 383: 241–252.
Kandhare AD, Raygude KS, Ghosh P et al. (2012). Therapeutic role of curcumin in prevention of biochemical and behavioral aberration induced by alcoholic neuropathy in laboratory animals. Neurosci Lett 511: 18–22.

Kelley LA, Sternberg MJ (2009). Protein structure prediction on the Web: a case study using the Phyre server. Nat Protoc 4: 363–371.

Kemppainen R, Juntunen J, Hillbom M (1982). Drinking habits and peripheral alcoholic neuropathy. Acta Neurol Scand 65: 11–18.

Khasar SG, McCarter G, Green PG et al. (1999). A novel nociceptor signaling pathway revealed in protein kinase C epsilon mutant mice. Neuron 24: 253–260.

Koike H, Sobue G (2006). Alcoholic neuropathy. Curr Opin Neurol 19: 481–486.

Koike H, Mori K, Misu K et al. (2001). Painful alcoholic polyneuropathy with predominant small-fiber loss and normal thiamine status. Neurology 56: 1727–1732.

Koike H, Iijima M, Sugiura M et al. (2003). Alcoholic neuropathy is clinicopathologically distinct from thiamine-deficiency neuropathy. Ann Neurol 54: 19–29.

Koike H, Hama T, Kawagashira Y et al. (2012). The significance of folate deficiency in alcoholic and nutritional neuropathies: analysis of a case. Nutrition 28: 821–824.

Koivisto H, Hietala J, Anttila P et al. (2006). Long-term ethanol consumption and macrocytosis: diagnostic and pathogenic implications. J Lab Clin Med 147: 191–196.

Lee I, Kim HK, Kim JH et al. (2007). The role of reactive oxygen species in capsaicin-induced mechanical hyperalgesia and in the activities of dorsal horn neurons. Pain 133: 9–17.

Lieber CS (1998). Hepatic and other medical disorders of alcoholism: from pathogenesis to treatment. J Stud Alcohol 59: 9–25.

Ling M, Trollér U, Zeidman R et al. (2005). Identification of conserved amino acids N-terminal of the PKC epsilon C1b domain crucial for protein kinase C epsilon-mediated induction of neurite outgrowth. J Biol Chem 280: 17910–17919.

Malatova Z, Cizkova D (2002). Effect of ethanol on axonal transport of cholinergic enzymes in rat sciatic nerve. Alcohol 26: 115–120.

Masaki T, Mochizuki H, Matsushita S et al. (2004). Association of aldehyde dehydrogenase-2 polymorphism with alcoholic polyneuropathy in humans. Neurosci Lett 363: 288–290.

McGovern PE, Wang C, Wang C et al. (2004). Fermented beverages of pre- and proto-historic China. Proc Natl Acad Sci U S A 101: 17593–17598.

McLane JA (1990). Retrograde axonal transport in chronic ethanol-fed and thiamine-deficient rats. Alcohol 7: 103–106.

Mellion M, Gilchrist JM, de la Monte S (2011). Alcohol-related peripheral neuropathy: nutritional, toxic, or both? Muscle Nerve 43: 309–316.

Milovanovic B, Milinic N, Trifunovic D et al. (2009). Autonomic dysfunction in alcoholic cirrhosis and its relation to sudden cardiac death risk predictors. Gen Physiol Biophys 28: 251–261.

Minot GR, Strauss MB, Cobb S (1933). "Alcoholic" polyneuritis; dietary deficiency as a factor in its production. N Engl J Med 208: 1244–1249.

Monforte R, Estruch R, Valls-Sole J et al. (1995). Autonomic and peripheral neuropathies in patients with chronic alcoholism. A dose-related toxic effect of alcohol. Arch Neurol 52: 45–51.

Narita M, Narita M, Miyoshi K et al. (2007). Involvement of microglia in the ethanol-induced neuropathic pain-like state in the rat. Neurosci Lett 414: 21–25.

Needleman H (2004). Lead poisoning. Annu Rev Med 55: 209–222.

Nguyen VA, Le T, Tong M et al. (2012). Impaired insulin/IGF signaling in experimental alcohol-related myopathy. Nutrients 4: 1058–1075.

Niemela O, Parkkila S (2004). Alcoholic macrocytosis - is there a role for acetaldehyde and adducts? Addict Biol 9: 3–10.

Niewiadomska M, Wochnik-Dyjas D, Woronowicz B et al. (1986). Examination of the peripheral sensory neurone in subjects with alcohol dependence syndrome. Electromyogr Clin Neurophysiol 26: 579–592.

Numazaki M, Tominaga T, Toyooka H et al. (2002). Direct phosphorylation of capsaicin receptor VR1 by protein kinase Cepsilon and identification of two target serine residues. J Biol Chem 277: 13375–13378.

Ogata T, Yamamoto S, Nakamura K et al. (2006). Signaling axis in schwann cell proliferation and differentiation. Mol Neurobiol 33: 51–62.

Ortiz-Plata A, Palencia G, Garcia E et al. (1998). Ultrastructural changes in limb distal nerves of rats with alcoholism and/or malnutrition before and after dietary correction. J Appl Toxicol 18: 89–92.

Peters TJ, Kotowicz J, Nyka W et al. (2006). Treatment of alcoholic polyneuropathy with vitamin B complex: a randomised controlled trial. Alcohol Alcohol 41: 636–642.

Peterson CM, Jovanovic-Peterson L, Schmid-Formby F (1988). Rapid association of acetaldehyde with hemoglobin in human volunteers after low dose ethanol. Alcohol 5: 371–374.

Poupon RE, Gervaise G, Riant P et al. (1990). Blood thiamine and thiamine phosphate concentrations in excessive drinkers with or without peripheral neuropathy. Alcohol Alcohol 25: 605–611.

Ravaglia S, Marchioni E, Costa A et al. (2004). Erectile dysfunction as a sentinel symptom of cardiovascular autonomic neuropathy in heavy drinkers. J Peripher Nerv Syst 9: 209–214.

Raygude KS, Kandhare AD, Ghosh P et al. (2012). Evaluation of ameliorative effect of quercetin in experimental model of alcoholic neuropathy in rats. Inflammopharmacology 20: 331–341.

Riley JL, King C (2009). Self-report of alcohol use for pain in a multi-ethnic community sample. J Pain 10: 944–952.

Scholz E, Diener HC, Diener HC et al. (1986). Incidence of peripheral neuropathy and cerebellar ataxia in chronic alcoholics. J Neurol 233: 212–217.

Stewart SH, Reuben A, Brzezinski WA et al. (2009). Preliminary evaluation of phosphatidylethanol and alcohol consumption in patients with liver disease and hypertension. Alcohol Alcohol 44: 464–467.

Takasaki I, Kurihara T, Saegusa H et al. (2005). Effects of glucocorticoid receptor antagonists on allodynia and hyperalgesia in mouse model of neuropathic pain. Eur J Pharmacol 524: 80–83.

Takeuchi M, Saito T (2005). Cytotoxicity of acetaldehyde-derived advanced glycation end-products (AA-AGE) in alcoholic-induced neuronal degeneration. Alcohol Clin Exp Res 29: 220S–224S.

Tallaksen CM, Bohmer T, Bell H (1992). Blood and serum thiamin and thiamin phosphate esters concentrations in patients with alcohol dependence syndrome before and after thiamin treatment. Alcohol Clin Exp Res 16: 320–325.

Tan ET, Johnson RH, Lambie DG et al. (1984). Alcoholic vagal neuropathy: recovery following prolonged abstinence. J Neurol Neurosurg Psychiatry 47: 1335–1337.

Tan ET, Lambie DG, Johnson RH et al. (1985). Sleep apnoea in alcoholic patients after withdrawal. Clin Sci (Lond) 69: 655–661.

Tariba B (2011). Metals in wine - impact on wine quality and health outcomes. Biol Trace Elem Res 144: 143–156.

Thayer JF, Hall M, Sollers 3rd J et al. (2006). Alcohol use, urinary cortisol, and heart rate variability in apparently healthy men: Evidence for impaired inhibitory control of the HPA axis in heavy drinkers. Int J Psychophysiol 59: 244–250.

Thiele B, Stalberg E (1975). Single fibre EMG findings in polyneuropathies of different aetiology. J Neurol Neurosurg Psychiatry 38: 881–887.

This P, Lacombe T, Thomas MR (2006). Historical origins and genetic diversity of wine grapes. Trends Genet 22: 511–519.

Tiwari V, Kuhad A, Chopra K (2011). Amelioration of functional, biochemical and molecular deficits by epigallocatechin gallate in experimental model of alcoholic neuropathy. Eur J Pain 15: 286–292.

Tiwari V, Kuhad A, Chopra K (2012). Neuroprotective effect of vitamin E isoforms against chronic alcohol-induced peripheral neurotoxicity: possible involvement of oxidative-nitrodative stress. Phytother Res 26: 1738–1745.

Tordoff MG, Pilchak DM, Williams JA et al. (2002). The maintenance diets of C57BL/6 J and 129X1/SvJ mice influence their taste solution preferences: implications for large-scale phenotyping projects. J Nutr 132: 2288–2297.

Torrente MP, Freeman WM, Vrana KE (2012). Protein biomarkers of alcohol abuse. Expert Rev Proteomics 9: 425–436.

Tredici G, Minazzi M (1975). Alcoholic neuropathy. An electron-microscopic study. J Neurol Sci 25: 333–346.

Victor M (1992). The effects of alcohol on the nervous system. In: JH Mendelson, NK Mello (Eds.), Medical Diagnosis and Treatment of Alcoholism. McGraw-Hill, Inc., New York, pp. 201–262.

Victor M, Adams RD (1953). The effect of alcohol on the nervous system. Res Publ Assoc Res Nerv Ment Dis 32: 526–573.

Villalta J, Estruch R, Antunez E et al. (1989). Vagal neuropathy in chronic alcoholics: relation to ethanol consumption. Alcohol Alcohol 24: 421–428.

Vittadini G, Buonocore M, Colli G et al. (2001). Alcoholic polyneuropathy: a clinical and epidemiological study. Alcohol Alcohol 36: 393–400.

Walsh JC, McLeod JG (1970). Alcoholic neuropathy. Proc Aust Assoc Neurol 7: 77–83.

Weise F, Müller D, Krell D et al. (1986). Heart rate variability in chronic alcoholics: a follow-up study. Drug Alcohol Depend 17: 365–368.

Wetterling T, Kanitz RD, Veltrup C et al. (1994). Clinical predictors of alcohol withdrawal delirium. Alcohol Clin Exp Res 18: 1100–1102.

Wetterling T, Veltrup C, Driessen M et al. (1999). Drinking pattern and alcohol-related medical disorders. Alcohol Alcohol 34: 330–336.

Woelk H, Lehrl S, Bitsch R et al. (1998). Benfotiamine in treatment of alcoholic polyneuropathy: an 8-week randomized controlled study (BAP I Study). Alcohol Alcohol 33: 631–638.

World Health Organization (2011). Global status report on alcohol and health. http://nla.gov.au/nla.cat-vn5155205, Available online at:.

Wu DF, Messing RO, Chandra D et al. (2012). PKCepsilon phosphorylation of the sodium channel NaV1.8 increases channel function and produces mechanical hyperalgesia in mice. J Clin Invest 122: 1306–1315.

Wurst FM, Alling C, Thon N et al. (2010). Phosphatidylethanol: normalization during detoxification, gender aspects and correlation with other biomarkers and self-reports. Addict Biol 15: 88–95.

Zambelis T, Karandreas N, Tzavellas E et al. (2005). Large and small fiber neuropathy in chronic alcohol-dependent subjects. J Peripher Nerv Syst 10: 375–381.

Zeidman R, Trollér U, Raghunath A et al. (2002). Protein kinase Cepsilon actin-binding site is important for neurite outgrowth during neuronal differentiation. Mol Biol Cell 13: 12–24.

Handbook of Clinical Neurology, Vol. 125 (3rd series)
Alcohol and the Nervous System
E.V. Sullivan and A. Pfefferbaum, Editors

Chapter 30

Pharmacologic treatment of alcoholism

RAYMOND F. ANTON*, JOSEPH P. SCHACHT, AND SARAH W. BOOK
Center for Drug and Alcohol Programs, Alcohol Research Center, Department of Psychiatry and Behavioral Sciences, Medical University of South Carolina, Charleston, SC, USA

INTRODUCTION

While the societal burden of alcohol use disorders (AUD) is huge (Substance Abuse and Mental Health Services Administration, 2004; Rehm et al., 2009), the number of individuals with alcohol dependence who actually seek treatment is considerably lower (Hasin et al., 2007). Of those, only a relatively small number actually receive treatment with a medication for their drinking problem (Mark et al., 2009). One of the barriers to pharmacotherapy is the relatively small treatment effect that current medications can provide compared to placebo. Although treatment effects are similar to medications for other conditions, expectations of success for medications for alcoholism are often higher. Further, alcohol dependence is likely to be a heterogeneous disease, both genetically and phenotypically. This suggests that a wide variation in response to a specific medication is likely, both in clinical trials and during routine clinical care. This chapter will review mechanisms of actions of medications demonstrated to be effective for the treatment of alcohol dependence and will also suggest new targets for pharmacologic development to treat this condition. Where possible, emphasis will be placed on subtypes (endophenotypes) of alcoholics who might respond better to certain drugs or classes of drugs. Given the above perspective, this "personalized medicine" approach to alcoholism treatment is in its infancy, but it is likely to have a growing impact and importance in coming years.

NEUROCHEMICAL TARGETS FOR MEDICATION TREATMENT

Other chapters in this volume address alcohol's action on the brain; here we address how alcohol's myriad effects might be altered by medication treatment. Table 30.1 provides a brief overview of the neurochemical systems affected by medications and the effects these systems might have on different aspects of the alcohol dependence syndrome. Since alcohol dependence develops over many years and with increasing amount and frequency of consumption, it is also likely that the neurochemical effects of alcohol have varying importance during different stages of the development of dependence. This section will focus on the primary, well-studied neurochemical systems affected by alcohol.

It is now well established that alcohol, like most drugs of abuse, elevates dopamine in the nucleus accumbens (Nac), the seat of a brain network that underlies salient attention, reinforcement, and reward (Weiss and Porrino, 2002; Volkow et al., 2007). The Nac has neuronal connections to prefrontal cortical areas involved with decision making and judgment. Importantly, alcohol-induced opioid release and subsequent transmission either directly in the Nac or through connections from the ventral tegmental area or arcuate nucleus, or perhaps other brain areas, can lead to dopamine release in the Nac (for review, see Mendez and Morales-Mulia, 2008). Several additional neurochemical systems that modulate dopamine release in the Nac are also sensitive to alcohol effects, environmental stimuli (cues) associated with alcohol consumption, or both (Fig. 30.1). For instance, it is clear that alcohol's effects on gamma-aminobutyric acid (GABA) and glutamate receptors and transmission mediate alcohol behaviors and cognitive effects and likely play an important role in the development of dependence (Rassnick et al., 1992; Roberto et al., 2003). Serotonin, which has been variably associated with alcohol effects, also appears to be a potential

*Correspondence to: Raymond F. Anton, M.D., Center for Drug and Alcohol Programs, Alcohol Research Center, Department of Psychiatry and Behavioral Sciences, Medical University of South Carolina, 67 President Street, Charleston, SC 29425, USA. Tel: + 1-843-792-1226, E-mail: antonr@musc.edu

Table 30.1

Medications tested for efficacy in alcohol dependence by potential mode of action and neurochemical target (see text for detail)

Phenomenology	Neurochemistry	Pharmacology
Reward	Dopamine	Aripiprazole
	Opioid	Naltrexone*/nalmefene
	5-HT_3	Ondansetron
	Glutamate	Acamprosate*/topiramate
	Nicotine	Varenicline
Protracted withdrawal	Glutamate	Acamprosate*/topiramate
	GABA	Valproate/baclofen
		Gabapentin
Affective	5-HT_2	SSRIs
	5-HT_{1b}	Buspirone
	GABA	Valproate
Impulsive	Dopamine	Aripiprazole

*Food and Drug Administration-approved for treatment of alcoholism.
GABA, gamma-aminobutyric acid; SSRIs, selective serotonin reuptake inhibitors.

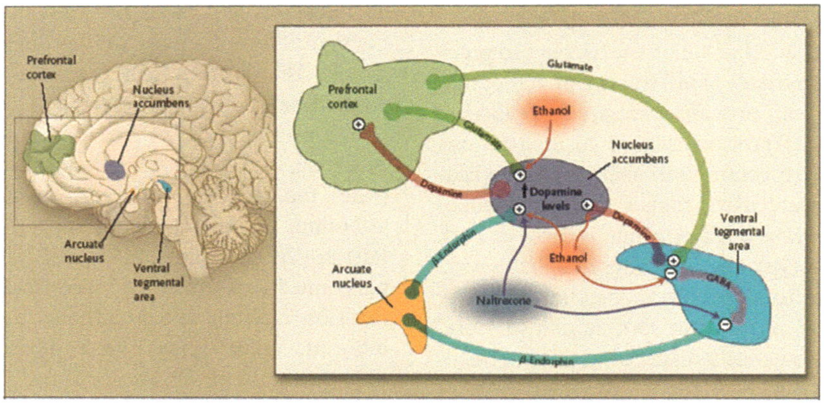

Fig. 30.1. Neurochemical circuits involved in alcohol dependence and craving. The brain opioid system is impacted by alcohol. It works in conjunction with other brain neurotransmitter systems to modulate alcohol reward/reinforcement as well as the salience of alcohol cues. Naltrexone, and other opioid receptor agonists, can moderate these effects, leading to their clinical utility. GABA, gamma-aminobutyric acid. (Reproduced from Anton, 2008.)

medication target. Specifically, the serotonin-3 (5-HT_3) receptor on Nac neurons might modulate alcohol-induced dopamine release (Campbell and McBride, 1995) and provide a rationale for the use of medications that block this receptor.

Work in animals has elucidated many of alcohol's effects in these neurochemical systems across certain brain areas. The current challenge is to improve understanding of the time course of these effects and how long-term adaptations in these systems, and the brain connections influenced by them, might lead to the compulsive heavy drinking, craving, and relapse that are the hallmarks of alcohol dependence. Pharmacologic modification of the acute or chronic effects of alcohol on these systems could lead to reduction in drinking, craving, and relapse rates. Also, more advanced animal drinking models (see further in this chapter) can provide proof of concept prior to clinical experiments and clinical trials.

OPIOID ANTAGONISTS

In the early 1990s, there was great excitement, mixed with some skepticism, regarding the first reports that naltrexone, an opioid receptor antagonist approved for the treatment of opiate addiction, had also been found effective in preventing relapse in alcoholics (O'Malley et al., 1992; Volpicelli et al., 1992). In many ways, this discovery ushered in the new era of drug discovery that has continued to the present. Alcohol engenders the release of beta-endorphin (Olive et al., 2001) which subsequently binds to mu opioid receptors at various loci in the brain reward system and either directly or indirectly leads to an increase in dopamine (Fig. 30.1). Blocking the mu opioid receptor with naltrexone reverses this pathophysiology and diminishes the reinforcing effects of alcohol, as well as craving stimulated by its related environmental cues. Extinction of conditioned alcohol-seeking behavior subsequently leads to less drinking, craving, and reduced relapse to heavy drinking.

Since the original reports, several meta-analyses of the many randomized clinical trials conducted worldwide have concluded that naltrexone is effective in reducing relapse drinking (Kranzler and Van Kirk, 2001; Bouza et al., 2004; Srisurapanont and Jarusuraisin, 2005; Rosner et al., 2010). Interestingly, the US Food and Drug Administration (FDA) approved naltrexone for the treatment of alcohol dependence without data from the customary two multisite randomized controlled trials normally required. Subsequently, the results of some multisite trials began to cast doubt on its efficacy (Chick et al., 2000; Krystal et al., 2001). While several factors could have accounted for these negative findings (e.g., subject severity, ancillary treatments), the results could not be summarily dismissed. However, around this time, several important results emerged that further supported naltrexone's efficacy. First, a large federally (National Institute on Alcohol Abuse and Alcoholism (NIAAA))-funded multisite randomized clinical trial, the COMBINE study (Anton et al., 2006), firmly established the efficacy of naltrexone; second, initial (Oslin et al., 2003) and subsequent (Anton et al., 2008b) reports of a genetic variant in the mu opioid receptor gene predicting naltrexone response appeared; and, third, neuroimaging data emerged that supported the effects of naltrexone (Myrick et al., 2008) on brain reward systems (see later in this chapter for further discussion of the latter two findings). These results, along with substantial support from animal (Gonzales and Weiss, 1998; Job et al., 2007; Mendez and Morales-Mulia, 2008) and clinical lab studies (O'Malley et al., 2002; Drobes et al., 2003; Anton et al., 2004a; Ray and Hutchison, 2007), have bolstered the scientific case for naltrexone's clinical efficacy.

Other opioid antagonists have been evaluated in clinical trials, but only nalmefene appears to have shown efficacy in some (though not all) randomized clinical trials (Anton et al., 2004b; Karhuvaara et al., 2007). Two recent large multisite studies supported its approval for marketing in Europe (Mann et al., 2012; Gual et al. 2013). Interestingly, nalmefene appears to have greater binding potential for delta and kappa opioid receptors as compared with naltrexone, which could lead to different, and possibly improved, efficacy. There have been no direct comparison trials of these medications, but at least one clinical lab study suggested similar effects (Drobes et al., 2003). While naltrexone has a "black box" cautionary warning about liver toxicity in its FDA-approved labeling, nalmefene is not noted to cause liver toxicity.

GABA/GLUTAMATERGIC MEDICATIONS

The GABA and glutamate systems, the primary inhibitory and excitatory systems in the brain, are sensitive to the pharmacologic effects of alcohol. They have been studied extensively in regard to the pharmacology of alcohol, its effects, and to some degree their alcohol-mediated effects on other neurotransmitter systems, including dopamine and endogenous opioids (for reviews, see Criswell and Breese, 2005; Gass and Olive, 2008). GABA and glutamate each have a number of pre- and postsynaptic receptors with a host of different actions on cells differentiated by both type and anatomic region; in this regard, these neurotransmitters are similar to alcohol itself, which has widespread effects on a variety of receptors. In general, excess GABA tone (increased release or greater receptor effects) is related to sedation, antianxiety, antiseizure, as well as coordination and cognitive impairment effects, whereas excess glutamate tone is related to stimulation, anxiety, cognitive disorganization, and seizures. This constellation of effects led scientists to suggest that these systems are neuromodulators whose counterbalancing mechanisms maintain overall brain homeostasis. Given this conceptualization and the types of behavior that emerge when these systems are dysregulated, perturbation of GABA and glutamate systems could account for many alcohol effects, including those observed under acute intoxication (increased GABA tone and decreased glutamate tone leading to sedation, incoordination, cognitive impairment) or in alcohol withdrawal (reduced GABA tone and increased glutamate tone underlying tremor, anxiety, seizures).

While these gross effects of GABA and glutamate on alcohol-related phenotypes are easily recognized, more subtle effects on alcohol reward and reinforcement also exist (Kalivas and Volkow, 2005; Gass and Olive, 2008). Therefore, there are two possible mechanisms by which GABA and glutamatergic medications can be useful for alcoholism treatment. One mechanism is the reduction of craving and relapse related to alcohol withdrawal (both acute and protracted), whereas the other is the reduction of alcohol-induced reward and habitual or compulsive drinking (more similar to naltrexone, as discussed above). With respect to the first mechanism, a number of studies and reviews have highlighted the effectiveness of anticonvulsants for the treatment of alcohol withdrawal (Malcolm et al., 2001; Book and

Myrick, 2005; Leggio et al., 2008). Medications that enhance GABA tone (e.g., barbiturates, benzodiazepines) effectively reduce alcohol withdrawal. However, another class of medications, anticonvulsants, putatively have more antiseizure efficacy and less abuse potential, do not themselves cause withdrawal, might cause less cognitive impairment, and might be useful as antirelapse agents. Accordingly, a number of studies have indicated that anticonvulsants, including gabapentin (Furieri and Nakamura-Palacios, 2007; Brower et al., 2008; Anton et al., 2009, 2011; Mason et al., 2014) and baclofen (Addolorato et al., 2007; Garbutt et al., 2010), might be useful for prolonged treatment after detoxification in reducing relapse risk and promoting abstinence. Whether this is through normalization of GABA/glutamate tone, leading to reduction of protracted withdrawal symptoms or craving, or due to another mechanism, is still unclear. In several studies, the anticonvulsant topiramate was found effective at reducing drinking and promoting abstinence in actively drinking outpatient alcoholics (Johnson et al., 2003, 2007), with an effect size equal to or larger than naltrexone. While not FDA-approved for alcohol dependence, with the proper dose adjustment and in the right patient, efficacy could be promising. Anticonvulsant use for the treatment of alcohol dependence is a promising area in need of more study.

Acamprosate, a unique glutamatergic agent without anticonvulsant properties, is thought to stabilize the glutamate system. The FDA approved acamprosate for the treatment of alcohol dependence based mainly on studies performed in Europe on hospitalized and currently abstinent patients (for review, see Mason and Ownby, 2000; Mann et al., 2004). In the United States, several trials did not support efficacy in outpatients (Anton et al., 2006; Mason et al., 2006). It is unclear why different results were obtained. One hypothesis is that acamprosate may only work if taken for a protracted period of time without any alcohol consumption in order to normalize the brain glutamate system, thereby maintaining abstinence, and that any "slip drinking" might neutralize this effect, leading to lack of efficacy. Indeed, in one US trial that showed no overall effect of acamprosate, those with a strong commitment to abstinence and who were compliant with medication seemed to show benefit (Mason et al., 2006). Also a study in recently detoxified inpatient alcoholics showed acamprosate could decrease brain glutamate, suggesting it might be uniquely useful for relapse prevention in those with putative alcohol withdrawal symptoms (Umhau et al., 2010).

SEROTONERGIC MEDICATIONS

Serotonin and its receptors are also ubiquitous in the human brain. Serotonin is involved in mood regulation, anxiety, stress response, appetite, and satiety; thus, to the extent that these processes contribute to the pathophysiology of AUD, serotonergic medications that mitigate them might be effective for patients with AUD, especially those with comorbid psychiatric diagnoses. Clinical trials evaluating selective serotonin reuptake inhibitors (SSRIs), one of the more widely studied classes of medications for the treatment of AUD, have been extensively reviewed by others (Pettinati, 2001; Miller et al., 2011). Although SSRIs (e.g., sertraline), when used alone, may effectively ameliorate some psychiatric symptoms in alcoholics, effects on drinking are less evident (see the section on treatment of co-occurring disorders, below, for further discussion of this issue). On the other hand, serotonin might play a role in reward and reinforcement independently of its role in anxiety, mood, and stress. Because 5-HT$_3$ receptors are thought to influence dopamine output from the Nac (Campbell and McBride, 1995), 5-HT$_3$ antagonists might play a role in the treatment of alcohol dependence. Several studies have suggested efficacy for the 5-HT$_3$ antagonist ondansetron (Sellers et al., 1994; Johnson et al., 2000; Kranzler et al., 2003), but the strength of this effect, the most appropriate dose, and the moderating effects of a certain endophenotype (early-onset alcoholism) or genotype (the serotonin transporter polymorphism) remain unresolved (see the section on pharmacogenetics, below, for greater detail).

DOPAMINERGIC MEDICATIONS

It has been long recognized that dopamine plays a crucial role in addiction and dependence. Accordingly, medications that act on various aspects of the dopamine system have been explored for efficacy in treating alcohol dependence. Historically, these have been small, single-site studies, but recently, several multisite studies have explored this issue more thoroughly using modern clinical trial methods in primary (non-comorbid) alcoholics. Aripiprazole, a partial dopamine agonist, is considered a "dopamine stabilizer" because it works at both pre- and postsynaptic receptors to moderate both deficient and excess dopamine transmission. A multisite study in primary alcoholics found that, while aripiprazole did not increase abstinence, it did have positive effects on a number of other drinking and biomarker variables (Anton et al., 2008a). However, side effects at higher doses were limiting, suggesting the need to study lower doses in alcoholics. In a recent multisite study of primary non-comorbid alcoholics, quetiapine, an atypical antipsychotic that acts on dopamine and serotonin receptors, was found to be well tolerated but did not reduce alcohol consumption (Litten et al., 2012a).

CHOLINERGIC/NICOTINE ACTING DRUGS

While there are not many studies in this area, a few have been reported. Galantamine, a cholinesterase inhibitor, was reported to reduce drinks per day (Mann et al., 2006). Varenicline, a partial alpha4beta2 nicotinic acetylcholine agonist approved by the FDA for smoking cessation, was shown to reduce heavy-drinking days, independently of smoking status (Litten et al., 2013).

COMBINATION PHARMACOTHERAPY

Given the complexity of alcohol actions on the brain, a combination of medications might work better than single agents alone by (1) inducing additive effects through actions on multiple neurochemical targets and (2) addressing phenotypic heterogeneity within a given population of individuals with AUD. To that end, several recent studies have evaluated this approach. The COMBINE study found that the combination of naltrexone and acamprosate was not superior to naltrexone alone (Anton et al., 2006). In addition, a study done with depressed alcoholics found that sertraline plus naltrexone was superior to either medication alone in both improving drinking outcomes and reducing depression (Pettinati et al., 2010). Further, a recent study completed by our group showed that gabapentin plus naltrexone was superior to naltrexone alone in reducing alcohol consumption but worked best in alcoholics with a history of alcohol withdrawal (Anton et al., 2011). Clearly, this area needs further exploration and might prove a fertile area of study, especially in alcoholics with clinically more complicated presentations.

TREATMENT OF AUD WITH CO-OCCURRING MOOD OR ANXIETY DISORDERS

Having a mood or anxiety disorder increases the likelihood of developing alcohol dependence (Martins and Gorelick, 2011). Indeed, many individuals seeking treatment for AUD also suffer from a co-occurring psychiatric disorder (Ross et al., 1988; Verheul et al., 2000). Because studies have shown that alcohol outcomes are often worse for patients with psychiatric comorbidity (Bradizza et al., 2006), efforts have been made to optimize alcohol treatments for this population. Unfortunately, the results of these efforts, which are reviewed below, have been largely mixed.

Depression

In the past 15 years, the main focus of treatment studies of alcohol dependence with comorbid depression has been on SSRIs, such as fluoxetine and sertraline. In an early study of fluoxetine among depressed alcoholic inpatients (Cornelius et al., 1997), fluoxetine, relative to placebo, resulted in significantly greater improvement on one measure of depression severity (Hamilton Rating Scale for Depression) but not another (Beck Depression Inventory), and improved alcohol outcomes over 12 weeks of treatment. Another study (Moak et al., 2003) randomized depressed alcohol-dependent outpatients to receive either sertraline or placebo over 12 weeks while also receiving cognitive behavioral therapy. Relative to placebo, sertraline was associated with less depression at the end of treatment only among female subjects and significantly reduced drinks per drinking day among all subjects, but did not affect other drinking measures. However, in a large multisite study of depressed alcoholic outpatients (Kranzler et al., 2006), sertraline, relative to placebo, did not improve any measure of depression or alcohol use severity. While SSRIs appear to be safe and well tolerated in alcohol-dependent individuals, on balance, clinical trials do not provide strong support for the use of SSRI monotherapy for the treatment of depressed alcoholics.

More recently, however, it has been shown that combining sertraline with naltrexone may be more efficacious for both depression and alcohol outcomes than either treatment alone (Pettinati et al., 2010). In that study, sertraline and naltrexone, as compared to placebo, or either medication alone, increased abstinence and time to heavy drinking as well as the percentage of patients not depressed at end of treatment. The idea that a combination of a medication for depression and a medication for alcohol may work better when combined for depressed alcoholics has heuristic appeal and needs further study.

Bipolar disorder

Conducting clinical trials on the seriously mentally ill is difficult given recruitment and retention challenges. Nevertheless, in the past 10 years, several research groups have investigated whether mood stabilizers and medications for alcohol dependence might reduce drinking in bipolar alcoholics. Among bipolar alcohol-dependent subjects randomized to either valproate or placebo (while also on lithium and a psychosocial intervention), those on valproate, although showing no differences in mood symptoms, had a significantly lower proportion of heavy-drinking days as compared to placebo (Salloum et al., 2005). In another study, bipolar alcoholics were randomized to either naltrexone or placebo while also taking a mood stabilizer and receiving cognitive behavioral therapy (Brown et al., 2009). Although not having a significant impact, those on naltrexone tended to have fewer drinking days and less alcohol craving. Most recently, a study of bipolar alcoholics reported that acamprosate added to mood stabilizers did not show any better drinking or mood outcomes compared to placebo (Tolliver et al., 2012). While limited in sample size, these trials show that these complicated patients could be studied. In at least two of them (despite limited statistical power) there was some indication that medications can play an important role in reducing alcohol consumption in this hard-to-treat population. Multisite trials with larger sample sizes are necessary to more fully understand the utility of pharmacotherapy in bipolar alcoholics.

Anxiety disorders

Anxiety in alcoholics is difficult to diagnose and treat. It is often unclear whether the anxiety is related to alcohol use or withdrawal and whether it should be treated separately from the AUD. It is clear that benzodiazepine drugs, a mainstay of anxiety pharmacotherapy, are not indicated for alcoholics due to both their negative interaction with alcohol and risk of precipitating dependence. Therefore, several research groups have evaluated the efficacy of safer anxiolytic medications (i.e., sertraline or buspirone) for treating anxious alcoholics. Buspirone, approved for the treatment of generalized anxiety, was hypothesized to be useful for comorbid anxiety and alcohol dependence because it does not interact with alcohol or have abuse liability. However, in a study of anxious alcoholic veterans, buspirone was not superior to placebo on either anxiety or alcohol outcomes (Malcolm et al., 1992). By contrast, another study found that community-recruited anxious alcoholics who received buspirone had better anxiety outcomes, were less likely to drop out of treatment, and had fewer drinking days over a 6-month follow-up as compared to those who received placebo (Kranzler et al., 1994).

Given their safety profile and approval for treatment of various anxiety disorders, SSRIs have also been studied in alcoholics with anxiety disorders. For example, in subjects with posttraumatic stress disorder (PTSD) with comorbid alcohol dependence treated with sertraline or placebo, there were no significant differences in alcohol consumption or PTSD symptoms in the intent-to-treat sample (Brady et al., 2005). However, a post hoc analysis found an interaction between medication effect and age of onset of PTSD as well as alcohol severity. Relative to placebo, sertraline reduced drinks per day in those with early-onset PTSD and less severe alcohol dependence, but increased drinking in those with later-onset PTSD and more severe alcohol dependence.

Among subjects with social anxiety disorder and comorbid alcohol abuse or dependence, paroxetine, another SSRI, ameliorated anxiety relative to placebo (Book et al., 2008), but had no significant effect on alcohol outcomes (Thomas et al., 2008). A follow-up study evaluating the addition of a brief alcohol psychotherapeutic intervention to the paroxetine treatment similarly showed no decreases in alcohol quantity or frequency (Book et al., 2012). The paucity of studies in this important clinical area speaks to the obvious need for more data.

Summary

Results to date suggest that the treatment of comorbid alcohol dependence with medications targeted to the psychiatric condition has only limited effect on alcohol consumption. Although some initial data suggested the addition of an alcohol consumption medication to specific pharmacotherapy targeted at the comorbid psychiatric condition might be efficacious, this position has received little support. What is clear is that comorbid psychiatric conditions should be considered in evaluating new medications for alcohol dependence and in determining the best clinical approach. The presence of comorbid disorders facilitates a "personalized medication" approach. As such, as more information becomes available regarding the underlying genetic basis of psychiatric conditions, this might inform specific pharmacologic strategies for both the psychiatric condition and the AUD. Recent findings on serotonin genes (see below) suggest such an approach is possible.

PHARMACOGENETICS AND ENDOPHENOTYPE PREDICTORS

Endophenotypes, which are measurable characteristics intermediate between a disease and a gene associated with it, and genetic predictors of alcohol treatment response have gained increasing attention. The goal of both is to "personalize" treatment. In the endophenotype area, it has been suggested that serotonergic medications might work differentially based on "age of onset," loosely defined as those meeting criteria for alcohol dependence before (early-onset) versus after (late-onset) age 25. Both the 5-HT_3 antagonist ondansetron (Johnson et al., 2000; Kranzler et al., 2003) and the 5-HT_2 antagonist quetiapine (Kampman et al., 2007) have been reported to work best in early-onset alcoholics. On the other hand, SSRIs that were reported to increase alcohol consumption in early-onset alcoholics (Kranzler et al., 1996) might actually be useful in the treatment of late-onset alcoholics (Pettinati et al., 2000). Variance in the serotonin transporter gene may further moderate this interaction between age of onset and SSRI response

(Kranzler et al., 2011), as well as the efficacy of ondansetron in reducing drinking (Johnson et al., 2011).

A recent promising finding, with growing literature support, is the discovery that a single-nucleotide polymorphism (SNP) at the 118 locus of the mu opioid receptor (*OPRM1*) gene (the A118G SNP encodes for an asparagine to aspartate amino acid substitution in the receptor protein) might be associated with altered alcohol reactivity (see Ray et al., 2012, for review) and naltrexone treatment response (Oslin et al., 2003; Anton et al., 2008b). While some studies do not support this finding (McGeary et al., 2006; Gelernter et al., 2007; Coller et al., 2011), evidence in animals and humans suggests the overall importance of this pharmacogenetic interaction (both for alcohol effects themselves and for the impact on naltrexone treatment response). For instance, data suggest that the Asp40 allele of the SNP is associated with more alcohol-induced craving, reinforcement, activation of brain reward centers to alcohol cues, stress reactivity, and differential naltrexone response, both in the laboratory and in the clinic (see Ray et al., 2012 and Setiawan et al., 2012, for reviews). Challenges remain, however, in determining: (1) potential epistatic gene interactions; (2) *OPRM1* genetic effects at different stages of alcohol dependence (e.g., early/mild vs later/severe); and (3) how race, culture, and developmental experience influence the effects of *OPRM1* genetic differences on alcohol/naltrexone interactions.

As an example of these challenges, our group recently reported that naltrexone and *OPRM1* A118G genotype interacted in their effects on alcohol consumption and brain activation to alcohol cues in early-stage alcoholics only when variation in the dopamine transporter (*DAT1/SLC6A3*) gene was also taken into account (Anton et al., 2012a; Schacht et al., 2013b). Since these individuals were not severe alcoholics and were not seeking treatment, this suggests that epistatic gene effects might also be more salient to pharmacologic treatment at different stages of alcohol dependence development.

USE OF NEUROIMAGING TO IDENTIFY NEW DRUGS AND TARGETS

Most medications for AUD are intended to alter the function of brain systems that underlie reward processing and behavioral control (Koob et al., 2009). Rapid methodologic advances in animal research (e.g., microdialysis, patch clamp recording, optogenetics) have allowed in vivo examination of these systems, but these techniques cannot be used in human subjects. Instead, functional neuroimaging has emerged as a powerful, non-invasive method of studying the effects of medications on human brain function. Most studies have used functional magnetic resonance imaging (fMRI), which measures differences in the magnitude of the blood oxygenation level-dependent signal (henceforth, "activation"). Others have used positron emission tomography (PET) to identify differences in neuroreceptor tone in AUD. Neuroimaging can be used to dissect the mechanism of action of a medication known to be effective for AUD, to probe the effects of a novel medication, and to identify new neural targets for pharmacotherapy. In each case, imaging can allow definition and analysis of brain-based endophenotypes for AUD.

fMRI studies typically require subjects to engage in a task intended to elicit activation of the brain system of interest. Since many approved and investigational medications for AUD are hypothesized to act through reduction of alcohol craving, many studies have employed alcohol cue reactivity paradigms, in which subjects are exposed to alcohol-related and "neutral" stimuli (e.g., pictures of alcoholic and non-alcoholic beverages), and brain activation elicited by these types of stimuli is compared, usually after active or placebo medication ingestion (Table 30.2). In these studies, the general hypothesis is that, if a medication can reduce alcohol cue-elicited activation in reward-related brain regions, then it may have the potential to reduce craving, drinking, and relapse in a randomized controlled trial.

Alcohol-related cues elicit craving, a phenomenon thought to be at least partially mediated by dopamine release within the mesolimbic and mesocortical pathways (Robinson and Berridge, 1993) and modulated by the actions of other neurotransmitter systems on these pathways (e.g., Gonzales and Weiss, 1998), as depicted in Figure 30.1 for alcohol itself. Accordingly, a recent meta-analysis of 28 functional imaging studies of alcohol cue reactivity found that alcohol cues most reliably elicited activation of a variety of areas along the mesolimbic and mesocortical pathways, including the ventral and dorsal striatum (VS and DS), medial prefrontal cortex (mPFC), anterior cingulate cortex (ACC), insula, and thalamus (Schacht et al., 2013a).

Interestingly, several medications that affect dopamine and opioid signaling have been reported to reduce alcohol cue-elicited activation in these areas, relative to placebo treatment. The first study to use functional neuroimaging to analyze medication reported that an acute dose of amisulpride, an atypical dopamine D_2 and D_3 receptor antagonist, reduced thalamic activation among detoxified, abstinent alcoholics (Hermann et al., 2006). Among non-treatment-seeking, actively drinking alcoholics with less severe dependence, both naltrexone and aripiprazole were reported to reduce VS activation after 1-2 weeks of medication (Myrick et al., 2008, 2010). However, as noted previously, a follow-up study

Table 30.2

Medication effects on alcohol cue-reactivity neuroimaging paradigms

First author, year	*n* (active/ placebo)	Dose and duration of medication	Scan timing*	Results
Hermann et al., 2006	10/0	400 mg amisulpride (one dose)	Pre/post	Amisulpride reduced activation of right thalamus
Myrick et al., 2008	23/24	50 mg naltrexone (NTX) × 7 days	Post	NTX reduced activation of right VS, right mPFC, right supramarginal gyrus, and bilateral OFC
	23/24	0.5 mg ondansetron (OND) × 7 days		OND reduced activation of right superior frontal gyrus, cingulate gyrus, cerebellar vermis, and primary visual cortex
	20/24	50 mg NTX, 0.5 mg OND × 7 days		NTX/OND combination reduced activation of right VS, right DS, and right middle frontal cortex gyrus
Myrick et al., 2010	14/16	15 mg aripiprazole (APZ) × 14 days	Post	APZ reduced activation of right VS Greater VS activation was related to drinking during time on medication in APZ group but not placebo
Langosch et al., 2012	12/10	Acamprosate (1332/1998 mg) × 14 days	Pre/post	No activation differences between acamprosate and placebo or between pre- and posttreatment scans
Schacht et al., 2013b	35/39	50 mg NTX × 6 days	Post	Among OPRM1 A118G G allele carriers, NTX reduced VS activation among DAT1 VNTR 10-repeat (10R) allele carriers relative to 9-repeat (9R) allele carriers NTX reduced mPFC activation in 10R carriers relative to 9R carriers
Schacht et al., 2013c	28/20	1200 mg gabapentin (GBP) × 14-21 days + 2 mg flumazenil (FMZ) infusions on each of first 2 days of treatment	Post	GBP/FMZ combination increased dorsal ACC activation among subjects with higher pretreatment alcohol withdrawal Dorsal ACC activation was associated with greater resistance to craving Greater dorsolateral PFC activation predicted subsequent heavy drinking across all subjects

*Indicates whether scan was conducted pre- and/or postmedication.

VS, ventral striatum; mPFC, medial prefrontal cortex; OFC, orbitofrontal cortex; DS, dorsal striatum; ACC, anterior cingulate cortex.

replicated the naltrexone finding only among individuals who carried a specific combination of *OPRM1* and *DAT1/ SLC6A3* variants (Schacht et al., 2013b). In the original naltrexone study, naltrexone, either alone or in combination with ondansetron, also reduced DS, mPFC, and orbitofrontal cortex (OFC) activation relative to placebo, while ondansetron monotherapy reduced activation of the cerebellar vermis, primary visual cortex, and superior frontal and cingulate gyri (Myrick et al., 2008).

In contrast to the work with dopaminergic and opioidergic medications, two studies of medications believed to affect glutamate and GABA balance (acamprosate, and a combination of gabapentin and flumazenil) reported that treatment-seeking alcoholics treated for 2-3 weeks with these medications did not display reduced cue-elicited activation relative to those treated with placebos (Langosch et al., 2012; Schacht et al., 2013c). However, the gabapentin-flumazenil combination relatively increased activation of dorsal ACC among subjects with high pretreatment alcohol withdrawal (a group previously shown to preferentially benefit from these medications; Anton et al., 2009), and greater activation of this area was associated with greater "cognitive resistance" to alcohol craving (Schacht et al., 2013c). Differences between these studies and those evaluating dopaminergic and opioidergic medications may represent differences in study populations, methodology (e.g., whether scanning was performed both before and after treatment or only afterwards), or duration and/or dose of medication. Further work among recently drinking, treatment-seeking alcoholics is sorely needed.

Relative to the number of fMRI studies of medication effects on alcohol cue reactivity, the effects of medications on other tasks relevant to AUD, such as affective processing and impulsivity, have not been well studied. With respect to affective processing, a study of recently detoxified, high trait-anxiety alcoholic inpatients reported that treatment with 4 weeks of a neurokinin-1 receptor antagonist, LY686107, reduced activation elicited by negative affective images in the insula, inferior frontal gyrus, middle temporal gyrus, and visual processing areas and increased activation elicited by positive images in the thalamus, VS, and inferior and middle temporal gyri, suggesting a possible shift from negative to positive emotionality (George et al., 2008). With respect to impulsivity, two studies have reported medication effects on brain activation elicited by a delay discounting task, in which subjects are asked to choose between a smaller amount of money available immediately or a larger amount available in the future (i.e., to evaluate impulsive vs restrained choices) (Boettiger et al., 2009), and a response inhibition task, in which subjects are asked to attempt to inhibit an already initiated response when presented with a "stop" signal (Schmaal et al., 2013). Among abstinent, non-treatment-seeking alcoholics, an acute dose of naltrexone increased restrained choice-elicited activation of lateral OFC, an area previously shown to activate more to restrained than impulsive choices, but did not suppress impulsive choice-related activation of areas believed to mediate impulsive choice (i.e., posterior parietal cortex, superior frontal gyrus, and parahippocampal gyrus) (Boettiger et al., 2009). Among abstinent, treatment-seeking alcoholics, an acute dose of the stimulant modafinil improved response inhibition and increased activation of the thalamus and supplementary motor area during successful inhibition, but only among subjects who initially had poor response inhibition (Schmaal et al., 2013). Given the disparate medications, experimental paradigms, and subject populations used in these studies, further research on similar tasks and medications, especially agents that may enhance cognition control and/or reduce impulsive responding, is clearly warranted.

Finally, functional imaging can also indicate new neural targets for medication development. For example, several recent PET studies reported that alcoholics displayed elevated cannabinoid receptor 1 (CB_1) binding throughout the brain (Hirvonen et al., 2012, Neumeister et al., 2012). In concert with preclinical data suggesting that CB_1 regulates the reinforcing properties of alcohol and that chronic alcohol disrupts endocannaboid signaling (Pava and Woodward, 2012), these studies suggest that agents that target the endocannabinoid system may represent a target of opportunity for medication development. Similarly, PET studies suggest reduced striatal availability of D_2 and D_3 receptors among recently abstinent alcoholics (Heinz et al., 2004, 2005; Rominger et al., 2012), suggesting that dopaminergic agents specific to these receptor subtypes may also have utility for AUD treatment.

While neuroimaging studies hold great promise for the evaluation of medication efficacy, research to date in this area has been limited by small sample sizes and disparate methods. Replication of extant findings is critical for future progress. Further, relatively little work has been done to demonstrate whether imaging endophenotypes can predict AUD treatment response. Such prediction is the ultimate goal of the "personalized medicine" approach, which aims to use individual patient characteristics to match patients to the medications from which they are most likely to benefit.

NEWER TARGETS FOR MEDICATION DEVELOPMENT

While the medications now available provide some efficacy to reduce drinking and promote abstinence, they are far from "silver bullets" in that they do not work for everyone. Since alcohol is known to affect so many brain biologic targets and pathways, and knowledge of their location in the specific brain regions involved with addiction is steadily advancing, it makes sense to evaluate how medications might impact alcohol use through these varied, newly elucidated mechanisms (Edwards et al., 2011; Davies et al., 2013). The search for new medications has been greatly assisted by the development of advanced animal models of alcohol consumption, reinforcement, and cue conditioning (Crabbe et al., 2011). This review is not meant to be exhaustive or to provide the detail necessary to evaluate these newer targets in great depth. Instead this section is meant to provide an awareness of the promise that these and other targets, yet to be discovered, could bring for improved treatments of AUD. Indeed, for many of them there is limited information and the reader is directed to the references provided and should search for the most current information as this area is quickly evolving. Table 30.3 provides a brief overview of the brain/neuronal targets of interest, the type of compound/biologic component tested, its hypothetical mode of action, and in what animal models it was found effective.

In general, the targets tested and the compounds being used to alter these targets fit into two broad categories of receptor targets: those affected by natural compounds (e.g. proteins) and those affected by small synthesized molecules. Examples of the former are the nociceptin and neuropeptide Y (NPY) receptors, where the natural protein products have been reported to alter alcohol-conditioned place preference (Ciccocioppo

Table 30.3

Novel medication targets

Target	Action	Drug type	CPP	Operant responding	Non-dependent drinking	Dependent drinking
Nociceptin	Agonist	Natural protein	+		+	
CRF1	Antagonist	e.g. Antalarmin		+	+	+
Orexin-1	Antagonist	SB-334867		+ (relapse)	+	
NPY	Agonist	Natural protein		+		
Vasopressin-1b	Antagonist	SSR149415			–	+
GDNF	Activation	Cabergoline		+	+	
P2X4R	Modulation	Ivermectin			+	
Sigma-1R	Antagonist	BD-1063		+	–	+
Canabinoid 1	Antagonist	SR 141716,		+	+	+
$GABA_AR$	Agonist	Ganaxolone		+/–		
$GABA_AR$	Agonist	DHM			+	
mGluR5	Antagonist	MTEP/MPEP	+	+	+	
mGluR7	Agonist	AMN082	+		+	
Kappa opioid	Agonist	U50,488H	+			
Kappa opioid	Antagonist	nor-BNI		+		
Alpha 1	Antagonist	Prazosin		+	+	

CPP, conditioned place preference; CRF, corticotropin-releasing factor; NPY, neuropeptide Y; GDNF, glial cell line-derived neurotropic factor; GABA, gamma-aminobutyric acid; DHM, dihydromyricetin. +, tested with positive result; –, tested with negative result; blank space, not tested.

et al., 1999) and non-dependent drinking (Ciccocioppo et al., 2004) or operant responding for alcohol (Gilpin et al., 2011). Because these natural proteins are not likely to easily cross the blood–brain barrier, might not survive peripheral metabolism, and might have off-target effects, it is not likely they will be useful in and of themselves. However, as a model to evaluate small-molecule drugs they provide important leads. Ongoing work in both the nociception and NPY systems is continuing.

Corticotropin-releasing factor (CRF) is similar to nociception and NPY in that it is a protein that binds to specific receptors. It differs, however, in that several small molecules have been developed that block the CRF1 receptor. It has been hypothesized that the main role of CRF1 receptors antagonists is to reduce: (1) the stress-induced activation of the hypothalamic-pituitary-adrenocortical hormone axis; and (2) the CRF-induced emotional and dysphoric effects in other areas of the brain, such as the amygdala (Heilig and Koob, 2007). As such, CRF1 antagonists might be most useful in blocking stress-induced drinking or drinking to prevent alcohol withdrawal dysphoria and craving seen in more severe dependence and associated with the stress response. A number of animal studies have shown that CRF1 antagonism can block alcohol consumption in dependent (Funk et al., 2007; Roberto et al., 2010) and non-dependent (Roberto et al., 2010) rats. CRF1 antagonists also blocked stress-induced alcohol-reinforced responding (Marinelli et al., 2007).

Vasopressin is a peptide that has multiple functions in the body, the most prominent being control of kidney blood flow and function. In the brain, it has actions similar to CRF in the regulation of stress and emotional state. A small molecule, SSR149415, which blocks the vasopressin type 1b receptor, has been shown to reduce drinking in dependent rats, but not in non-dependent rats (Edwards et al., 2012), consistent with the theory that stress system dysregulation is important in maintaining alcohol consumption in dependent animals and that blockade of targets fostering stress might have a role in the treatment of severe alcohol dependence.

Orexin is a small peptide that has been shown to be involved with sleep and appetite regulations and may play a role in satiety. Like CRF, there have been small molecules developed that bind to the type 1 orexin receptor. For example, SB-334867 has shown the ability to block alcohol operant responding and a relapse model (Dhaher et al., 2010) and to reduce drinking in a non-dependent rodent drinking model (Moorman and Aston-Jones, 2009).

Glial cell line-derived neurotropic factor (GDNF) is a protein that is known to support the survival and function of dopamine cells in the brain. Recent work suggests that changes in GDNF might change alcohol consumption and response. It has been reported that cabergoline, a D_2 agonist drug that is approved for use of hyperprolactinemia, activates the GDNF pathway and also reduces operant responding for alcohol and drinking in non-dependent rats (Carnicella et al., 2009).

Perhaps one of the more novel targets is a P2X4 receptor modulating an adenosine triphosphate-gated membrane ion channel that is sensitive to alcohol. Ivermectin, a drug that has been used for many years in the treatment of parasitic disorders, has been reported to activate this channel and reduce alcohol consumption in non-dependent mice (Yardley et al., 2012).

Sigma receptors are ubiquitous in the brain and seem to be the target of many substances and drugs. Despite having a less-understood role, some work has suggested that an antagonist of the sigma 1 receptor, BD-10163 can reduce operant responding for alcohol and consumption in dependent rats but not in non-dependent rats (Sabino et al., 2009).

One of the more extensively studied newer targets is CB_1, for which antagonist compounds have been developed and studied in humans for other conditions, such as obesity and nicotine dependence (Cahill and Ussher, 2007; Christopoulou and Kiortsis, 2011). On balance, there is considerable evidence that the endogenous cannabinoid system interacts with alcohol to modulate its effects (see Pava and Woodward, 2012, for extensive review). Studies in animals have shown that the CB1 antagonist SR 141716 can reduce alcohol consumption in non-dependent mice and rats, in dependent rats, and also reduce operant (reward) responding for alcohol. These encouraging results were leading to planned randomized controlled trials in individuals with AUD when reports of CB1 antagonists causing neuropsychiatric effects, including suicidal thinking and behavior, led to the abandonment of clinical development.

The next group of targets are the more traditional ones, often thought of in relationship to alcohol effects, i.e., specific GABA, glutamate, opioid, adrenergic receptors, and/or modulators. Since $GABA_A$ receptors are known to be sensitive to alcohol and drugs that affect them directly, and medications that indirectly affect them have shown some promise in the treatment of AUD, novel compounds working at this target might also hold promise. In this regard, since it is known that naturally occurring neurosteroids modulate $GABA_A$ receptors, small-molecule drugs such as ganaxalone that have similar action have been shown to reduce alcohol operant responding at certain doses (Besheer et al., 2010). Also, a natural product dihydromyricetin, which also modulates $GABA_A$ receptors, was found to reduce alcohol consumption in non-dependent rats (Shen et al., 2012).

Like $GABA_A$ receptors, glutamate receptors are also sensitive to alcohol effects and are widely distributed in the brain throughout cortical and subcortical regions, where a number of different receptor subtypes are known to exist. Small molecules targeting two glutamate subtypes have been studied in animal drinking models. The mGluR5 receptor antagonists MTEP and MPEP have been shown to block alcohol-conditioned place preference, operant responding, and non-dependent drinking (Lominac et al., 2006; Schroeder et al., 2008; Sinclair et al., 2012) whereas the mGluR7 agonist AMN082 inhibited alcohol-conditioned place preference and consumption in non-dependent mice and rats (Bahi, 2012; Bahi et al., 2012).

As mentioned in an earlier section, the brain opioid system has received considerable attention given the proven efficacy of naltrexone and, to some degree, nalmefene. Those drugs predominantly block mu opioid receptors and secondarily delta opioid receptors; other attention has been paid to novel drugs that might work predominantly on kappa opioid receptors. Although the exact role of the kappa opioid system in the acquisition and maintenance of alcohol dependence is controversial, it known that when kappa opioid receptors located on Nac dopamine cells are pharmacologically activated there is a reduction in dopamine output (the opposite of mu and delta stimulation). An example of a kappa agonist is U50,488H, which blocked the acquisition of alcohol-conditioned place preference at a dose which itself was not aversive (Logrip et al., 2009) (one of the historical limitations to the use of opioid kappa receptor agonists). On the other hand, possibly to counterbalance excessive mu stimulation during alcohol consumption, kappa receptors become upregulated and might provide the rationale for why kappa receptor antagonists might block alcohol-reinforced drinking during alcohol withdrawal (Walker and Koob, 2008).

There is interest in potentially "repurposing" already-approved FDA pharmaceuticals for the treatment of AUD. Another potential FDA-approved medication for another condition is the adrenergic alpha 1 antagonist prazosin that has been reported to interfere with alcohol operant responding, as well as reduce drinking in non-dependent alcohol-preferring rats. There has also been some preliminary evidence that it might reduce alcohol intake in alcohol-dependent individuals. A small study in 17 men showed a reduction of drinking days and the total alcohol consumed by the end of the 6-week trial (Simpson et al., 2009).

THE FUTURE

The development of medications for AUD faces a number of challenges. Ultimately the focus is on the assessment of risk by pharmaceutical companies for developmental success, which includes costs of development and return on investment (i.e., reimbursement policies). Here is a short list of issues that should be considered:

1. The need for better scientific-based predictive models of success in the clinic. This includes advances in neurochemical, animal behavioral, and preclinical laboratory (including brain imaging) studies in humans.
2. Improved and standardized clinical trial methodologies, including a consensus on appropriate and acceptable clinical outcomes.
3. A more complete and applied understanding of "subtypes" of alcoholics who might respond to different medications. Examples are genetic diversity, comorbid pathologies, alcohol symptoms (such as withdrawal symptoms, craving, tolerance, and sleep disturbance), and more subtle traits such as impulsivity and cognitive impairment.
4. A greater recognition of alcohol harm (both medical and psychosocial) that occurs in individuals presenting in various healthcare and social (e.g., judicial) settings, with appropriate referral for intervention and treatment.
5. A greater recognition by our payer systems that AUD should be treated and reimbursed in a similar fashion (if not better, given the potential cost savings) to other medical conditions.
6. The awareness of the public (including healthcare providers) that alcohol dependence is a brain disease with behavioral components that could benefit from pharmacologic intervention.
7. An eventual transition to the notion that alcohol dependence is a chronic illness needing extended treatment rather than one in which short-term solutions are applied.

Initiatives at NIAAA (Litten et al., 2012b) and in public-private partnerships (Anton et al., 2012b) are underway to improve and standardize clinical trial methodology while also attempting to reduce development and regulatory risk. Recent US healthcare reforms, such as the Mental Health and Substance Abuse Treatment parity legislation and the Affordable Health Care Act, might lead to better recognition of AUD and support appropriate treatments, including current pharmacotherapies and new ones that might evolve.

References

Addolorato G, Leggio L, Ferrulli A et al. (2007). Effectiveness and safety of baclofen for maintenance of alcohol abstinence in alcohol-dependent patients with liver cirrhosis: randomised, double-blind controlled study. Lancet 370: 1915–1922.

Anton RF (2008). Naltrexone for the management of alcohol dependence. N Engl J Med 359: 715–721.

Anton RF, Drobes DJ, Voronin K et al. (2004a). Naltrexone effects on alcohol consumption in a clinical laboratory paradigm: temporal effects of drinking. Psychopharmacology (Berl) 173: 32–40.

Anton RF, Pettinati H, Zweben A et al. (2004b). A multisite dose ranging study of nalmefene in the treatment of alcohol dependence. J Clin Psychopharmacol 24: 421–428.

Anton RF, O'Malley SS, Ciraulo DA et al. (2006). Combined pharmacotherapies and behavioral interventions for alcohol dependence: the COMBINE study: a randomized controlled trial. JAMA 295: 2003–2017.

Anton RF, Kranzler H, Breder C et al. (2008a). A randomized, multicenter, double-blind, placebo-controlled study of the efficacy and safety of aripiprazole for the treatment of alcohol dependence. J Clin Psychopharmacol 28: 5–12.

Anton RF, Oroszi G, O'Malley SS et al. (2008b). An evaluation of μ-opioid receptor (OPRM1) as a predictor of naltrexone response in the treatment of alcohol dependence: results from the combined pharmacotherapies and behavioral interventions for alcohol dependence (COMBINE) study. Arch Gen Psychiatry 65: 135–144.

Anton RF, Myrick H, Baros AM et al. (2009). Efficacy of a combination of flumazenil and gabapentin in the treatment of alcohol dependence: relationship to alcohol withdrawal symptoms. J Clin Psychopharmacol 29: 334–342.

Anton RF, Myrick H, Wright TM et al. (2011). Gabapentin combined with naltrexone for the treatment of alcohol dependence. Am J Psychiatry 168: 709–717.

Anton RF, Voronin KK, Randall PK et al. (2012a). Naltrexone modification of drinking effects in a subacute treatment and bar-lab paradigm: influence of OPRM1 and dopamine transporter (SLC6A3) genes. Alcohol Clin Exp Res 36: 2000–2007.

Anton RF, Litten RZ, Falk DE et al. (2012b). The Alcohol Clinical Trials Initiative (ACTIVE): purpose and goals for assessing important and salient issues for medications development in alcohol use disorders. Neuropsychopharmacology 37: 402–411.

Bahi A (2012). The selective metabotropic glutamate receptor 7 allosteric agonist AMN082 prevents reinstatement of extinguished ethanol-induced conditioned place preference in mice. Pharmacol Biochem Behav 101: 193–200.

Bahi A, Fizia K, Dietz M et al. (2012). Pharmacological modulation of mGluR7 with AMN082 and MMPIP exerts specific influences on alcohol consumption and preference in rats. Addict Biol 17: 235–247.

Besheer J, Lindsay TG, O'Buckley TK et al. (2010). Pregnenolone and ganaxolone reduce operant ethanol self-administration in alcohol-preferring p rats. Alcohol Clin Exp Res 34: 2044–2052.

Boettiger CA, Kelley EA, Mitchell JM et al. (2009). Now or later? An fMRI study of the effects of endogenous opioid blockade on a decision-making network. Pharmacol Biochem Behav 93: 291–299.

Book SW, Myrick H (2005). Novel anticonvulsants in the treatment of alcoholism. Expert Opin Investig Drugs 14: 371–376.

Book SW, Thomas SE, Randall PK et al. (2008). Paroxetine reduces social anxiety in individuals with a co-occurring alcohol use disorder. J Anxiety Disord 22: 310–318.

Book SW, Thomas SE, Smith JP et al. (2012). Severity of anxiety in mental health versus addiction treatment settings when social anxiety and substance abuse are comorbid. Addict Behav 37: 1158–1161.

Bouza C, Angeles M, Munoz A et al. (2004). Efficacy and safety of naltrexone and acamprosate in the treatment of alcohol dependence: a systematic review [see comment] [erratum appears in Addiction. 2005;100:573 Note: Magro, Angeles corrected to Angeles, Magro]. Addiction 99: 811–828.

Bradizza CM, Stasiewicz PR, Paas ND (2006). Relapse to alcohol and drug use among individuals diagnosed with co-occurring mental health and substance use disorders: a review. Clin Psychol Rev 26: 162–178.

Brady KT, Sonne S, Anton R et al. (2005). Sertraline in the treatment of co-occurring alcohol dependence and post-traumatic stress disorder. Alcohol Clin Exp Res 29: 395–401.

Brower KJ, Kim HM, Strobbe S et al. (2008). A randomized double-blind pilot trial of gabapentin versus placebo to treat alcohol dependence and comorbid insomnia. Alcohol Clin Exp Res 32: 1429–1438.

Brown ES, Carmody TJ, Schmitz JM et al. (2009). A randomized, double-blind, placebo-controlled pilot study of naltrexone in outpatients with bipolar disorder and alcohol dependence. Alcohol Clin Exp Res 33: 1863–1869.

Cahill K, Ussher M (2007). Cannabinoid type 1 receptor antagonists (rimonabant) for smoking cessation. Cochrane Database Syst Rev, CD005353.

Campbell AD, McBride WJ (1995). Serotonin-3 receptor and ethanol-stimulated dopamine release in the nucleus accumbens. Pharmacol Biochem Behav 51: 835–842.

Carnicella S, Ahmadiantehrani S, He DY et al. (2009). Cabergoline decreases alcohol drinking and seeking behaviors via glial cell line-derived neurotrophic factor. Biol Psychiatry 66: 146–153.

Chick J, Anton R, Checinski K et al. (2000). A multicentre, randomized, double-blind, placebo-controlled trial of naltrexone in the treatment of alcohol dependence or abuse. Alcohol Alcohol 35: 587–593.

Christopoulou FD, Kiortsis DN (2011). An overview of the metabolic effects of rimonabant in randomized controlled trials: potential for other cannabinoid 1 receptor blockers in obesity. J Clin Pharm Ther 36: 10–18.

Ciccocioppo R, Panocka I, Polidori C et al. (1999). Effect of nociceptin on alcohol intake in alcohol-preferring rats. Psychopharmacology (Berl) 141: 220–224.

Ciccocioppo R, Economidou D, Fedeli A et al. (2004). Attenuation of ethanol self-administration and of conditioned reinstatement of alcohol-seeking behaviour by the antiopioid peptide nociceptin/orphanin FQ in alcohol-preferring rats. Psychopharmacology (Berl) 172: 170–178.

Coller JK, Cahill S, Edmonds C et al. (2011). OPRM1 A118G genotype fails to predict the effectiveness of naltrexone treatment for alcohol dependence. Pharmacogenet Genomics 21: 902–905.

Cornelius JR, Salloum IM, Ehler JG et al. (1997). Fluoxetine in depressed alcoholics. A double-blind, placebo-controlled trial. Arch Gen Psychiatry 54: 700–705.

Crabbe JC, Harris RA, Koob GF (2011). Preclinical studies of alcohol binge drinking. Ann N Y Acad Sci 1216: 24–40.

Criswell HE, Breese GR (2005). A conceptualization of integrated actions of ethanol contributing to its GABAmimetic profile: a commentary. Neuropsychopharmacology 30: 1407–1425.

Davies DL, Bortolato M, Finn DA et al. (2013). Recent advances in the discovery and preclinical testing of novel compounds for the prevention and/or treatment of alcohol use disorders. Alcohol Clin Exp Res 37: 8–15.

Dhaher R, Hauser SR, Getachew B et al. (2010). The orexin-1 receptor antagonist SB-334867 reduces alcohol relapse drinking, but not alcohol-seeking, in alcohol-preferring (P) rats. J Addict Med 4: 153–159.

Drobes DJ, Anton RF, Thomas SE et al. (2003). A clinical laboratory paradigm for evaluating medication effects on alcohol consumption: naltrexone and nalmefene. Neuropsychopharmacology 28: 755–764.

Edwards S, Kenna GA, Swift RM et al. (2011). Current and promising pharmacotherapies, and novel research target areas in the treatment of alcohol dependence: a review. Curr Pharm Des 17: 1323–1332.

Edwards S, Guerrero M, Ghoneim OM et al. (2012). Evidence that vasopressin V1b receptors mediate the transition to excessive drinking in ethanol-dependent rats. Addict Biol 17: 76–85.

Funk CK, Zorrilla EP, Lee MJ et al. (2007). Corticotropin-releasing factor 1 antagonists selectively reduce ethanol self-administration in ethanol-dependent rats. Biol Psychiatry 61: 78–86.

Furieri FA, Nakamura-Palacios EM (2007). Gabapentin reduces alcohol consumption and craving: a randomized, double-blind, placebo-controlled trial. J Clin Psychiatry 68: 1691–1700.

Garbutt JC, Kampov-Polevoy AB, Gallop R et al. (2010). Efficacy and safety of baclofen for alcohol dependence: a randomized, double-blind, placebo-controlled trial. Alcohol Clin Exp Res 34: 1849–1857.

Gass JT, Olive MF (2008). Glutamatergic substrates of drug addiction and alcoholism. Biochem Pharmacol 75: 218–265.

Gelernter J, Gueorguieva R, Kranzler HR et al. (2007). Opioid receptor gene (OPRM1, OPRK1, and OPRD1) variants and response to naltrexone treatment for alcohol dependence: results from the VA Cooperative Study. Alcohol Clin Exp Res 31: 555–563.

George DT, Gilman J, Hersh J et al. (2008). Neurokinin 1 receptor antagonism as a possible therapy for alcoholism. Science 319: 1536–1539.

Gilpin NW, Misra K, Herman MA et al. (2011). Neuropeptide Y opposes alcohol effects on gamma-aminobutyric acid release in amygdala and blocks the transition to alcohol dependence. Biol Psychiatry 69: 1091–1099.

Gonzales RA, Weiss F (1998). Suppression of ethanol-reinforced behavior by naltrexone is associated with attenuation of the ethanol-induced increase in dialysate dopamine levels in the nucleus accumbens. J Neurosci 18: 10663–10671.

Gual A, He Y, Torup L et al. (2013). A randomised, double-blind, placebo-controlled efficacy study of nalmefene, as-needed use, in patients with alcohol dependence. Eur Neuropsychopharmacol 23 (11): 1432–1442.

Hasin DS, Stinson FS, Ogburn E et al. (2007). Prevalence, correlates, disability, and comorbidity of DSM-IV alcohol abuse and dependence in the United States: results from the National Epidemiologic Survey on Alcohol and Related Conditions. Arch Gen Psychiatry 64: 830–842.

Heilig M, Koob GF (2007). A key role for corticotropin-releasing factor in alcohol dependence. Trends Neurosci 30: 399–406.

Heinz A, Siessmeier T, Wrase J et al. (2004). Correlation between dopamine D(2) receptors in the ventral striatum and central processing of alcohol cues and craving. Am J Psychiatry 161: 1783–1789.

Heinz A, Siessmeier T, Wrase J et al. (2005). Correlation of alcohol craving with striatal dopamine synthesis capacity and D2/3 receptor availability: a combined [18 F]DOPA and [18 F]DMFP PET study in detoxified alcoholic patients. Am J Psychiatry 162: 1515–1520.

Hermann D, Smolka MN, Wrase J et al. (2006). Blockade of cue-induced brain activation of abstinent alcoholics by a single administration of amisulpride as measured with fMRI. Alcohol Clin Exp Res 30: 1349–1354.

Hirvonen J, Zanotti-Fregonara P, Umhau JC et al. (2012). Reduced cannabinoid CB(1) receptor binding in alcohol dependence measured with positron emission tomography. Mol Psychiatry, PMCID: 3594469.

Job MO, Tang A, Hall FS et al. (2007). Mu (mu) opioid receptor regulation of ethanol-induced dopamine response in the ventral striatum: evidence of genotype specific sexual dimorphic epistasis. Biol Psychiatry 62: 627–634.

Johnson BA, Roache JD, Javors MA et al. (2000). Ondansetron for reduction of drinking among biologically predisposed alcoholic patients. JAMA 284: 963–971.

Johnson BA, Ait-Daoud N, Bowden CL et al. (2003). Oral topiramate for treatment of alcohol dependence: a randomized controlled trial. Lancet 361: 1677–1685.

Johnson BA, Rosenthal N, Capece JA et al. (2007). Topiramate for treating alcohol dependence: a randomized controlled trial. JAMA 298: 1641–1651.

Johnson BA, Ait-Daoud N, Seneviratne C et al. (2011). Pharmacogenetic approach at the serotonin transporter gene as a method of reducing the severity of alcohol drinking. Am J Psychiatry 168: 265–275.

Kalivas PW, Volkow ND (2005). The neural basis of addiction: a pathology of motivation and choice. Am J Psychiatry 162: 1403–1413.

Kampman KM, Pettinati HM, Lynch KG et al. (2007). A double-blind, placebo-controlled pilot trial of quetiapine for the treatment of Type A and Type B alcoholism. J Clin Psychopharmacol 27: 344–351.

Karhuvaara S, Simojoki K, Virta A et al. (2007). Targeted nalmefene with simple medical management in the treatment of heavy drinkers: a randomized double-blind placebo-controlled multicenter study. Alcohol Clin Exp Res 31: 1179–1187.

Koob GF, Kenneth Lloyd G, Mason BJ (2009). Development of pharmacotherapies for drug addiction: a Rosetta stone approach. Nat Rev Drug Discov 8: 500–515.

Kranzler HR, Van Kirk J (2001). Efficacy of naltrexone and acamprosate for alcoholism treatment: a meta-analysis. Alcohol Clin Exp Res 25: 1335–1341.

Kranzler HR, Burleson JA, Del Boca FK et al. (1994). Buspirone treatment of anxious-alcoholics: a placebo-controlled trial. Arch Gen Psychiatry 51: 720–731.

Kranzler HR, Burleson JA, Brown J et al. (1996). Fluoxetine treatment seems to reduce the beneficial effects of cognitive-behavioral therapy in type B alcoholics. Alcohol Clin Exp Res 20: 1534–1541.

Kranzler HR, Pierucci-Lagha A, Feinn R et al. (2003). Effects of ondansetron in early-versus late-onset alcoholics: a prospective, open-label study. Alcohol Clin Exp Res 27: 1150–1155.

Kranzler HR, Mueller T, Cornelius J et al. (2006). Sertraline treatment of co-occurring alcohol dependence and major depression. J Clin Psychopharmacol 26: 13–20.

Kranzler HR, Armeli S, Tennen H et al. (2011). A double-blind, randomized trial of sertraline for alcohol dependence: moderation by age and 5-hydroxytryptamine transporter-linked promoter region genotype. J Clin Psychopharmacol 31: 22–30.

Krystal JH, Cramer JA, Krol WF et al. (2001). Naltrexone in the treatment of alcohol dependence. N Engl J Med 345: 1734–1739.

Langosch JM, Spiegelhalder K, Jahnke K et al. (2012). The impact of acamprosate on cue reactivity in alcohol dependent individuals: a functional magnetic resonance imaging study. J Clin Psychopharmacol 32: 661–665.

Leggio L, Kenna GA, Swift RM (2008). New developments for the pharmacological treatment of alcohol withdrawal syndrome. A focus on non-benzodiazepine GABAergic medications. Prog Neuropsychopharmacol Biol Psychiatry 32: 1106–1117.

Litten RZ, Fertig JB, Falk DE et al. (2012a). A double-blind, placebo-controlled trial to assess the efficacy of quetiapine fumarate XR in very heavy-drinking alcohol-dependent patients. Alcohol Clin Exp Res 36: 406–416.

Litten RZ, Egli M, Heilig M et al. (2012b). Medications development to treat alcohol dependence: a vision for the next decade. Addict Biol 17: 513–527.

Litten RZ, Ryan ML, Fertig JB et al. (2013). A double-blind, placebo-controlled trial assessing the efficacy of varenicline tartrate for alcohol dependence. J Addict Med (in press).

Logrip ML, Janak PH, Ron D (2009). Blockade of ethanol reward by the kappa opioid receptor agonist U50,488H. Alcohol 43: 359–365.

Lominac KD, Kapasova Z, Hannun RA et al. (2006). Behavioral and neurochemical interactions between Group 1 mGluR antagonists and ethanol: potential insight into their anti-addictive properties. Drug Alcohol Depend 85: 142–156.

Malcolm R, Anton RF, Randall CL et al. (1992). A placebo-controlled trial of buspirone in anxious inpatient alcoholics. Alcohol Clin Exp Res 16: 1007–1013.

Malcolm R, Myrick H, Brady KT et al. (2001). Update on anticonvulsants for the treatment of alcohol withdrawal. Am J Addict 10: 16–23.

Mann K, Lehert P, Morgan MY (2004). The efficacy of acamprosate in the maintenance of abstinence in alcohol-dependent individuals: results of a meta-analysis. Alcohol Clin Exp Res 28: 51–63.

Mann K, Ackermann K, Diehl A et al. (2006). Galantamine: a cholinergic patch in the treatment of alcoholism: a randomized, placebo-controlled trial. Psychopharmacology (Berl) 184 (1): 115–121.

Mann K, Bladstrom A, Torup L et al. (2012). Extending the treatment options in alcohol dependence: a randomized controlled study of as-needed nalmefene. Biol Psychiatry 73: 706–713.

Marinelli PW, Funk D, Juzytsch W et al. (2007). The CRF1 receptor antagonist antalarmin attenuates yohimbine-induced increases in operant alcohol self-administration and reinstatement of alcohol seeking in rats. Psychopharmacology (Berl) 195: 345–355.

Mark TL, Kassed CA, Vandivort-Warren R et al. (2009). Alcohol and opioid dependence medications: prescription trends, overall and by physician specialty. Drug Alcohol Depend 99: 345–349.

Martins SS, Gorelick DA (2011). Conditional substance abuse and dependence by diagnosis of mood or anxiety disorder or schizophrenia in the U.S. population. Drug Alcohol Depend 119: 28–36.

Mason BJ, Ownby RL (2000). Acamprosate for the treatment of alcohol dependence: a review of double-blind, placebo-controlled trials. CNS Spectr 5: 58–69.

Mason BJ, Goodman AM, Chabac S et al. (2006). Effect of oral acamprosate on abstinence in patients with alcohol dependence in a double-blind, placebo-controlled trial: The role of patient motivation. J Psychiatr Res 40: 383–393.

Mason BJ, Quello S, Goodell V et al. (2014). Gabapentin treatment for alcohol dependence: a randomized clinical trial. JAMA Intern Med 174: 70–77.

McGeary JE, Monti PM, Rohsenow DJ et al. (2006). Genetic moderators of naltrexone's effects on alcohol cue reactivity. Alcohol Clin Exp Res 30: 1288–1296.

Mendez M, Morales-Mulia M (2008). Role of mu and delta opioid receptors in alcohol drinking behaviour. Curr Drug Abuse Rev 1: 239–252.

Miller PM, Book SW, Stewart SH (2011). Medical treatment of alcohol dependence: a systematic review. Int J Psychiatry Med 42: 227–266.

Moak DH, Anton RF, Latham PK et al. (2003). Sertraline and cognitive behavioral therapy for depressed alcoholics: results of a placebo-controlled trial. J Clin Psychopharmacol 23: 553–562.

Moorman DE, Aston-Jones G (2009). Orexin-1 receptor antagonism decreases ethanol consumption and preference selectively in high-ethanol-preferring Sprague-Dawley rats. Alcohol 43: 379–386.

Myrick H, Anton RF, Li X et al. (2008). Effect of naltrexone and ondansetron on alcohol cue-induced activation of the ventral striatum in alcohol-dependent people. Arch Gen Psychiatry 65: 466–475.

Myrick H, Li X, Randall PK et al. (2010). The effect of aripiprazole on cue-induced brain activation and drinking parameters in alcoholics. J Clin Psychopharmacol 30: 365–372.

Neumeister A, Normandin MD, Murrough JW et al. (2012). Positron emission tomography shows elevated cannabinoid cb (1) receptor binding in men with alcohol dependence. Alcohol Clin Exp Res 36: 2104–2109.

Olive MF, Koenig HN, Nannini MA et al. (2001). Stimulation of endorphin neurotransmission in the nucleus accumbens by ethanol, cocaine, and amphetamine. J Neurosci 21: 184RC.

O'Malley SS, Jaffe AJ, Chang G et al. (1992). Naltrexone and coping skills therapy for alcohol dependence. A controlled study. Arch Gen Psychiatry 49: 881–887.

O'Malley SS, Krishnan-Sarin S, Farren C et al. (2002). Naltrexone decreases craving and alcohol self-administration in alcohol-dependent subjects and activates the hypothalamo-pituitary-adrenocortical axis. [see comment]. Psychopharmacology (Berl) 160: 19–29.

Oslin DW, Berrettini W, Kranzler HR et al. (2003). A functional polymorphism on the m-opioid receptor gene is associated with naltrexone response in alcohol-dependent patients. Neuropsychopharmacology 28: 1546–1552.

Pava MJ, Woodward JJ (2012). A review of the interactions between alcohol and the endocannabinoid system: implications for alcohol dependence and future directions for research. Alcohol 46: 185–204.

Pettinati HM (2001). The use of selective serotonin reuptake inhibitors in treating alcoholic subjects. J Clin Psychiatry 62: 26–31.

Pettinati HM, Volpicelli JR, Kranzler HR et al. (2000). Sertraline treatment for alcohol dependence: interactive effects of medication and alcoholic subtype. Alcohol Clin Exp Res 24: 1041–1049.

Pettinati HM, Oslin DW, Kampman KM et al. (2010). A double-blind, placebo-controlled trial combining sertraline and naltrexone for treating co-occurring depression and alcohol dependence. Am J Psychiatry 167: 668–675.

Rassnick S, D'Amico E, Riley E et al. (1992). GABA and nucleus accumbens glutamate neurotransmission modulate ethanol self-administration in rats. Ann N Y Acad Sci 654: 502–505.

Ray LA, Hutchison KE (2007). Effects of naltrexone on alcohol sensitivity and genetic moderators of medication response: a double-blind placebo-controlled study. Arch Gen Psychiatry 64: 1069–1077.

Ray LA, Barr CS, Blendy JA et al. (2012). The role of the Asn40Asp polymorphism of the mu opioid receptor gene (OPRM1) on alcoholism etiology and treatment: a critical review. Alcohol Clin Exp Res 36: 385–394.

Rehm J, Mathers C, Popova S et al. (2009). Global burden of disease and injury and economic cost attributable to alcohol use and alcohol-use disorders. Lancet 373: 2223–2233.

Roberto M, Madamba SG, Moore SD et al. (2003). Ethanol increases GABAergic transmission at both pre- and post-synaptic sites in rat central amygdala neurons. Proc Natl Acad Sci U S A 100: 2053–2058.

Roberto M, Cruz MT, Gilpin NW et al. (2010). Corticotropin releasing factor-induced amygdala gamma-aminobutyric acid release plays a key role in alcohol dependence. Biol Psychiatry 67: 831–839.

Robinson TE, Berridge KC (1993). The neural basis of drug craving: an incentive-sensitization theory of addiction. Brain Res Brain Res Rev 18: 247–291.

Rominger A, Cumming P, Xiong G et al. (2012). Fallypride PET measurement of striatal and extrastriatal dopamine D 2/3 receptor availability in recently abstinent alcoholics. Addict Biol 17: 490–503.

Rosner S, Hackl-Herrwerth A, Leucht S et al. (2010). Opioid antagonists for alcohol dependence. Cochrane Database Syst Rev, CD001867.

Ross HE, Glaser FB, Germanson T (1988). The prevalence of psychiatric disorders in patients with alcohol and other drug problems. Arch Gen Psychiatry 45: 1023–1031.

Sabino V, Cottone P, Zhao Y et al. (2009). The sigma-receptor antagonist BD-1063 decreases ethanol intake and reinforcement in animal models of excessive drinking. Neuropsychopharmacology 34: 1482–1493.

Salloum IM, Cornelius JR, Daley DC et al. (2005). Efficacy of valproate maintenance in patients with bipolar disorder and alcoholism. Arch Gen Psychiatry 62: 37–45.

Schacht JP, Anton RF, Myrick H (2013a). Functional neuroimaging studies of alcohol cue reactivity: a quantitative meta-analysis and systematic review. Addict Biol 18: 121–133.

Schacht JP, Anton RF, Voronin KE et al. (2013b). Interacting effects of naltrexone and OPRM1 and DAT1 variation on the neural response to alcohol cues. Neuropsychopharmacology 38: 414–422.

Schacht JP, Anton RF, Randall PK et al. (2013c). Effects of a GABA-ergic medication combination and initial alcohol withdrawal severity on cue-elicited brain activation among treatment-seeking alcoholics. Psychopharmacology (Berl) 227: 627–637.

Schmaal L, Joos L, Koeleman M et al. (2013). Effects of modafinil on neural correlates of response inhibition in alcohol-dependent patients. Biol Psychiatry 73: 211–218.

Schroeder JP, Spanos M, Stevenson JR et al. (2008). Cue-induced reinstatement of alcohol-seeking behavior is associated with increased ERK1/2 phosphorylation in specific limbic brain regions: blockade by the mGluR5 antagonist MPEP. Neuropharmacology 55: 546–554.

Sellers EM, Toneatto T, Romach MK et al. (1994). Clinical efficacy of the 5-HT3 antagonist ondansetron in alcohol abuse and dependence. Alcohol Clin Exp Res 18: 879–885.

Setiawan E, Pihl RO, Benkelfat C et al. (2012). Influence of the OPRM1 A118G polymorphism on alcohol-induced euphoria, risk for alcoholism and the clinical efficacy of naltrexone. Pharmacogenomics 13: 1161–1172.

Shen Y, Lindemeyer AK, Gonzalez C et al. (2012). Dihydromyricetin as a novel anti-alcohol intoxication medication. J Neurosci 32: 390–401.

Simpson TL, Saxon AJ, Meredith CW et al. (2009). A pilot trial of the alpha-1 adrenergic antagonist, prazosin, for alcohol dependence. Alcohol Clin Exp Res 33: 255–263.

Sinclair CM, Cleva RM, Hood LE et al. (2012). mGluR5 receptors in the basolateral amygdala and nucleus accumbens regulate cue-induced reinstatement of ethanol-seeking behavior. Pharmacol Biochem Behav 101: 329–335.

Srisurapanont M, Jarusuraisin N (2005). Naltrexone for the treatment of alcoholism: a meta-analysis of randomized controlled trials. Int J Neuropsychopharmacol 8: 267–280.

Substance Abuse and Mental Health Services Administration (2004). Results from the 2003 National Survey on Drug Use and Health: National Findings. NSDUH Series H-25, DHHS Publication No. SMA 04-3694, Office of Applied Studies, Rockville, MD.

Thomas SE, Randall PK, Book SW et al. (2008 Jan). A complex relationship between co-occurring social anxiety and alcohol use disorders: what effect does treating social anxiety have on drinking? Alcohol Clin Exp Res 32 (1): 77–84.

Tolliver BK, Desantis SM, Brown DG et al. (2012). A randomized, double-blind, placebo-controlled clinical trial of acamprosate in alcohol-dependent individuals with bipolar disorder: a preliminary report. Bipolar Disord 14: 54–63.

Umhau JC, Momenan R, Schwandt ML et al. (2010). Effect of acamprosate on magnetic resonance spectroscopy measures of central glutamate in detoxified alcohol-dependent individuals: a randomized controlled experimental medicine study. Arch Gen Psychiatry 67: 1069–1077.

Verheul R, Kranzler HR, Poling J et al. (2000). Axis I and Axis II disorders in alcoholics and drug addicts: fact or artifact? J Stud Alcohol 61: 101–110.

Volkow ND, Fowler JS, Wang G-J et al. (2007). Dopamine in drug abuse and addiction: results of imaging studies and treatment implications. Arch Neurol 64: 1575–1579.

Volpicelli JR, Alterman AI, Hayashida M et al. (1992). Naltrexone in the treatment of alcohol dependence. Arch Gen Psychiatry 49: 876–880.

Walker BM, Koob GF (2008). Pharmacological evidence for a motivational role of kappa-opioid systems in ethanol dependence. Neuropsychopharmacology 33: 643–652.

Weiss F, Porrino LJ (2002). Behavioral neurobiology of alcohol addiction: recent advances and challenges. J Neurosci 22: 3332–3337.

Yardley MM, Wyatt L, Khoja S et al. (2012). Ivermectin reduces alcohol intake and preference in mice. Neuropharmacology 63: 190–201.

Handbook of Clinical Neurology, Vol. 125 (3rd series)
Alcohol and the Nervous System
E.V. Sullivan and A. Pfefferbaum, Editors

Chapter 31

Alcohol–medical drug interactions

BANKOLE A. JOHNSON[1]* AND CHAMINDI SENEVIRATNE[2]

[1]*Department of Psychiatry and Brain Science Research Consortium at the University of Maryland School of Medicine, Baltimore, MD, USA*

[2]*Department of Psychiatry, Institute for Genome Sciences, University of Maryland School of Medicine, Baltimore, MD, USA*

INTRODUCTION

Many over-the-counter and prescription medications interact with alcohol when drinking is moderate to heavy. As a result, alcohol can reduce the desired effects, or enhance the adverse effects, of the interacting medications. Risk of developing adverse effects is greatest when medications are used in combination with alcohol, as often occurs in people with pre-existing alcoholism or multiple illicit drug addictions. The data from several US national surveys indicate a growing trend of concomitant use of alcohol and medications, resulting in potentially serious medical conditions. For example, the 2012 Drug Abuse Warning Network report showed a 115% increase in emergency department (ED) visits related to medical drug misuse between 2004 and 2010 (http:/www.samhsa.gov/data/2k12/DAWN096/SR096EDHighlights2010.htm). Among adults, alcohol accounted for 21% of benzodiazepine-, tranquilizer-, and opioid-related ED visits (White et al., 2011). Fifty-five percent of all alcohol-related ED visits in underaged drinkers (12–20 years) involved a medical drug (http:/www.samhsa.gov/data/2k10/WEBSR025AlcDrugCombo/alcdrugscombo.htm). In young adults (aged 18–24 years), ED visits due to combined alcohol and medication overdose increased by 76% from 1999 to 2008 (White et al., 2011). Another demographic group that is particularly vulnerable to alcohol-medication interactions is the elderly (aged ≥ 65 years). The risk of adverse reactions arising from alcohol-medication interactions is much greater in elderly individuals due to physiologic and social reasons. Physiologic processes in aging affect distribution and metabolism of alcohol and various medications. Brain tissue also becomes sensitive to chemical compounds in older adults. The greater use of medications in this age group and the presence of chronic disease conditions that are often worsened by alcohol increase the potential for alcohol-medication adverse events (Moore et al., 2007a).

Alcohol is a small polarized molecule with both lipophilic and hydrophilic properties (Ferreira and Willoughby, 2008). Amphipathic properties facilitate absorption across biologic membranes and a similar volume of distribution to that of total body water. These properties allow alcohol to reach the central nervous system (CNS), exerting its physical and behavioral effects in a dose-dependent manner. Pathophysiologic effects of alcohol are determined by the concentration of alcohol present in a person's blood (blood alcohol concentration: BAC) and parameters that influence its interaction at the effector sites. Likewise, the blood concentration of some medications and their interaction with their effector sites may also be influenced by alcohol. Based on these mechanisms, alcohol-medication interactions are classified into two main categories: (1) pharmacokinetic interactions and (2) pharmacodynamic interactions. This chapter focuses on mechanisms underlying pharmacokinetic and pharmacodynamic interactions and their implications. All examples for alcohol-medication interactions discussed here are limited to the ones that have been verified by independent studies.

PHARMACOKINETIC INTERACTIONS

Alcohol-medication pharmacokinetic interactions refer to the influence of alcohol on the absorption,

*Correspondence to: Bankole A. Johnson, DSc, MD, MBChB, MPhil, FRCPsych, DFAPA, The Dr. Irving J. Taylor Professor and Chair in the Department of Psychiatry, Professor of Pharmacology and Professor of Anatomy & Neurobiology, Director of the Brain Science Research Consortium at the University of Maryland School of Medicine, Baltimore, MD 21201, USA. Tel: +1-410-328-6735, Fax: +1-410-328-3693, E-mail: bjohnson@psych.umaryland.edu

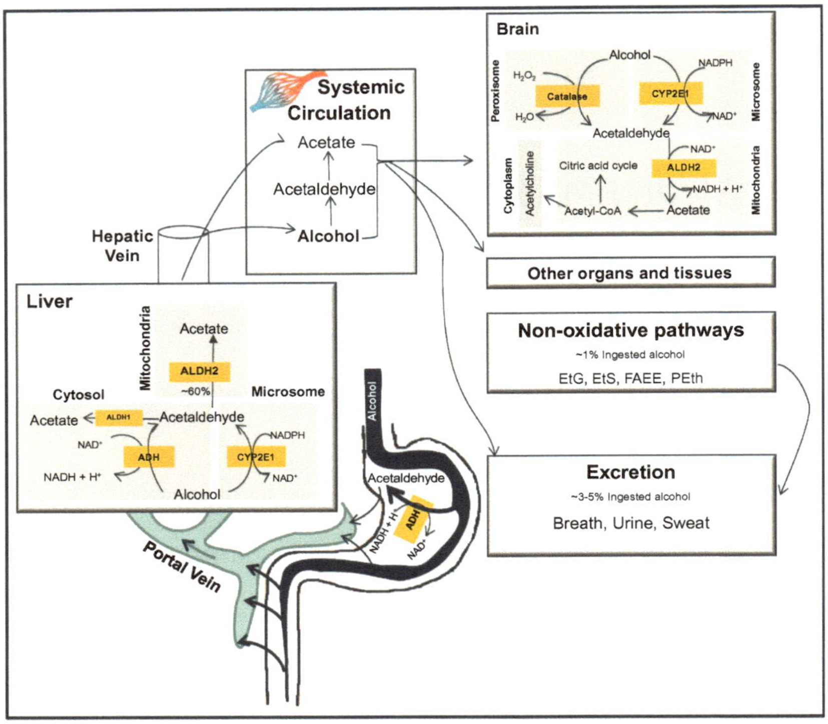

Fig. 31.1. Pharmacokinetic pathways of alcohol.

distribution, metabolism, and excretion of a medication or vice versa. All of these factors determine the BAC and, consequently, the concentrations at their effector sites. Pathways involved in pharmacokinetic mechanisms are depicted in Figure 31.1, and common factors affecting BAC are presented in Table 31.1.

Absorption and distribution

As alcohol passes along the digestive tract, a small amount is metabolized in the oral and gastric mucosa by the class IV isoform of the enzyme alcohol dehydrogenase (ADH), and about another 10–20% is absorbed into the gastric veins. The rest of the ingested alcohol is rapidly absorbed in the small intestines. Because the rate of absorption is much faster in the small intestine compared with the stomach, any delay in gastric emptying (GE) (Table 31.1) will reduce the speed of alcohol absorption. On the other hand, factors that increase intestinal transit time will enhance absorption. GE is the most important contributing factor determining the kinetic parameters of time to reach peak concentration (t_{max}) as well as the peak concentration of alcohol in blood (C_{max}). Alcohol absorbed in the stomach and small intestines enters the portal vein and is carried to the liver. In the liver, a portion of the alcohol is metabolized, as shown in Figure 31.1. Alcohol that escapes first-pass metabolism (FPM) in the liver enters the systemic circulation and is distributed throughout the water compartment of the body unbound to plasma proteins. Under normal conditions, tissue distribution occurs until an equilibrium is reached between plasma BAC and the tissue concentrations of alcohol (Ferreira and Willoughby, 2008). Time to reach equilibrium generally ranges from 1 to 2 hours after the ingestion of alcohol but is influenced by various factors (Table 31.1). In the absorptive phase, BAC rises more slowly than alcohol concentration in tissues, and during the elimination phase, BAC falls more slowly than alcohol concentrations in tissues. So, BAC may not reflect the actual impact of alcohol on tissues such as the brain.

Elimination

The main method of elimination of alcohol is through oxidation in the liver. The non-oxidative metabolism of alcohol accounts only for about 1% of total ingested alcohol, and it is then excreted in the urine. Only about 2–5% of absorbed alcohol is excreted unchanged in breath, sweat, or urine (Vonghia et al., 2008). The amount of ethanol eliminated per unit of time depends on the initial amount that was ingested. This pattern of elimination is called zero-order kinetics, which follows the Michaelis-Menten model of enzyme kinetics. Thus, the half-life of ethanol will be lower with small quantities of drinking and higher with severe drinking. Studies

Table 31.1

Factors affecting blood alcohol concentration (BAC)

Variable	Effect
Factors affecting absorption	
Stomach contents	Gastrectomy increases GE: ↑ BAC Fed state, amount of sugar in an alcohol beverage, cigarette smoking prolongs GE: ↓ BAC
Alcohol concentration	Absorption rates are greatest at concentrations between 20% and 30%; alcohol concentrations >30% irritate gastric mucosa, resulting in lower absorption rates
Autonomic nervous activity	SNS – decrease by slowing GE and portal blood flow PNS – increase GE
Age	Prolonged GE with aging
Osmolarity	Increases absorption (e.g., carbonated drinks)
Medications	See Table 31.2
Portal blood flow	A good blood flow in the portal system enhances absorption
Factors affecting distribution	
BMI	Increased body water facilitates bioavailability
Gender	Lower volume of distribution in females per unit of body weight due to smaller blood volume and lean body mass: ↑ BAC
Age	Older people tend to have more body fat
Time to drink	Affects time to peak BAC
Alterations to peripheral circulation	Peripheral vasoconstriction (e.g., cold temperatures, peripheral vascular disease, and cardiac disease): ↑ BAC Increased peripheral blood flow (e.g., high temperatures, exercise, and antihypertensive medications): ↓ BAC
Factors affecting metabolism	
Gender	Gastric ADH levels are lower in women than in men of similar weight: ↑ BAC
Age	Aging reduces gastric ADH levels
Race	Variations in enzyme levels (see Table 31.2 for genetic variations)
Gastric morphology	Atrophic gastritis and *Helicobactor pylori* infection reduce gastric ADH activity
Alcoholism	Gastric ADH is lower in alcoholics compared with moderate drinkers

GE, gastric emptying; BMI, body mass index; SNS, sympathetic nervous system; PNS, peripheral nervous system; ADH, alcohol dehydrogenase.

comparing different controlled alcohol doses have shown that the average rate of elimination of alcohol is generally around 0.01–0.035 g/100 mL/hour (Jones, 2010). Factors that lower or accelerate these rates are listed in Table 31.1. Although the zero-order kinetic model is widely used in forensic sciences, it is an oversimplified model that does not reflect complexities of pathways involved in alcohol metabolism, such as individual differences. Several models have been introduced to improve the understanding of alcohol metabolism following both oral and intravenous administration, including first-order and combined zero-and-first-order kinetic models. Improving on these models, Breton et al. proposed the Minimal Model of Ethanol Dynamics that includes metabolic mechanisms occurring in the stomach and gut (Breton, 2010; Kovatchev et al., 2012). Applying this model, Kovatchev et al. (2012) demonstrated the pattern of blood alcohol kinetics based on different levels of intake. As shown in Figure 31.2, during daytime, with 4 standard drinks/day or less, BAC levels were zero implicating complete metabolism of the ingested alcohol; at five or more drinks per day, there was a linear increase of the minimum BAC indicating a certain residual amount. According to this model, when a person consumed 11 or more standard drinks/day, BAC reached a level that never went down to zero within 24 hours. Although an earlier study has shown a lack of linear relationship between self-repots of drinking and BAC after 7 standard drinks (Bond et al., 2010), the simulated results in the study by Kovatchev et al. (2012) need experimental replication for establishing critical points of transition from stable to unstable dynamics.

The alcohol that is absorbed by gastric and intestinal veins enters the liver through the portal system. The metabolism of alcohol within both the stomach and the liver is known as FPM as it occurs prior to entering the systemic circulation (Levitt et al., 1997). With the gastric FPM accounting for only 2% or less, liver is undisputedly the main site of FPM. FPM reduces the availability of alcohol in the blood. Slow GE accelerates gastric FPM by prolonging exposure to gastric ADH. Several other main factors that influence alcohol metabolism are listed in

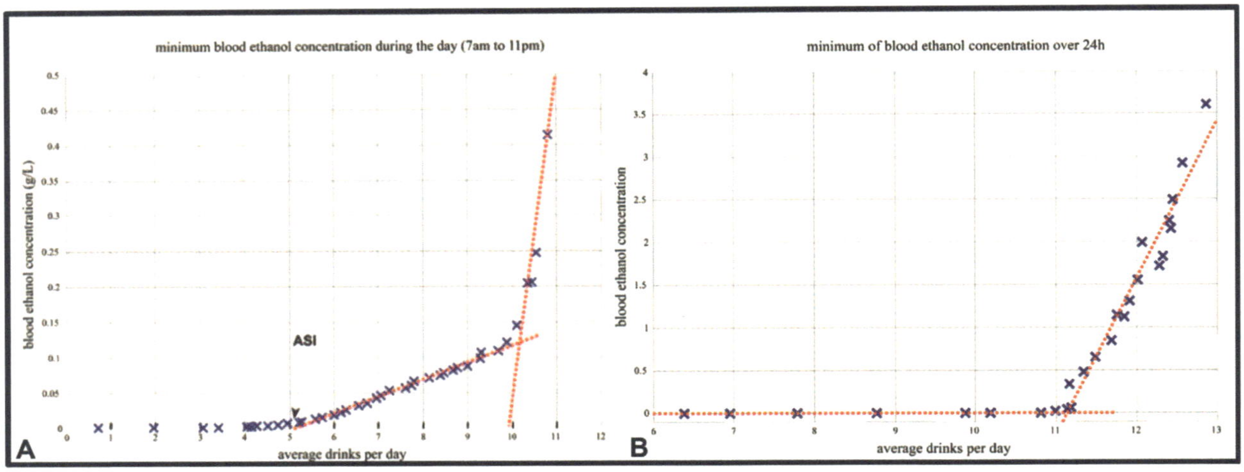

Fig. 31.2. Minimum blood alcohol levels during (A) daytime (7 AM–11 PM) as a function of average number of drinks per day and, (B) night (11 PM–7 AM) as a function of average number of drinks consumed during the day (7 AM–11 PM). (Figure as originally published in Kovatchev B, Breton M, Johnson B. (2012). *In silico* Models of Alcohol Dependence and Treatment Front. Psychiatry 3:4. http://dx.doi.org/10.3389/fpsyt.2012.00004).

Table 31.1. However, it should be noted that, despite the findings of many early studies that demonstrated a correlation of gastric ADH and FPM, the role of gastric ADH in oxidizing significant amounts of alcohol is still debatable, mainly because of its relatively low affinity (high K_m) for alcohol compared with the hepatic ADH.

Both FPM and the systemic portion of alcohol metabolism occur through oxidative and non-oxidative pathways.

Oxidative pathways of alcohol metabolism

Oxidative pathways consist of the enzymes ADH, cytochrome P450 (CYPs), aldehyde dehydrogenase (ALDH), and catalase. Medications that compete with alcohol for binding to these enzymes are discussed in Table 31.2.

ADH family of enzymes

ADH oxidizes alcohol into its primary metabolite, acetaldehyde (Fig. 31.1). There are five main classes of ADH enzyme isoforms (isozymes I–V) in humans. The class I isozyme (ADH1) catalyzes the rate-limiting step of the oxidative pathway of ethanol metabolism in the liver (Osier et al., 2002; Edenberg, 2007). ADH1 is formed by heteromers of subunits α, ß, and γ. Class II and III ADH isozymes are also expressed in the liver in smaller amounts (Edenberg, 1998). Class II ADH4 and, to a lesser degree, class I ADH1C contribute to gastric alcohol oxidation (Lee et al., 2006). ADH4 that is present in the liver and gastric mucosa may oxidize about 40% of total alcohol at intoxicating levels (Li et al., 1977; Lewis et al., 2012). Class III consists of ADH5, which oxidizes long-chain primary alcohols, but no activity has been reported in ethanol oxidation (Adinolfi et al., 1984). Medications that inhibit ADH are shown to increase BAC. When ADH is deficient, as seen in Asian populations, the effect of ADH inhibitors on alcohol oxidation will be negligent (Gugler, 1994).

CYP 450 system

The enzymes of the CYP 450 system are also involved in oxidizing alcohol to acetaldehyde. The main CYP 450 enzyme involved in alcohol oxidation is CYP2E1 (Heit et al., 2013; Osna and Donohue, 2013). Some recent findings have indicated the involvement of other CYP enzymes such as CYP1A2 (Salmela et al., 1998) and CYP3A4 (Salmela et al., 1998; Lytton et al., 1999; Feierman et al., 2003) that warrant further investigation. CYP2E1 activity depends on the pattern of drinking. When alcohol is consumed occasionally in small amounts, only an insignificant amount of alcohol is oxidized by CYP2E1. With chronic heavy drinking, CYP2E1 activity is induced by about 10-fold. This ability to oxidize more alcohol accelerates alcohol clearance from blood, leading to development of chronic tolerance (Pizon et al., 2007). Hence, the medications metabolized by CYP2E1 are oxidized faster in chronic heavy drinkers, requiring higher dosages. In contrast, if a medication metabolized by CYP2E1 is administered following a bout of acute heavy drinking, the therapeutic effects of the medication may become prolonged as a consequence

Table 31.2

Medications that commonly interact with alcohol

Drug class	Substance	Mechanism	Outcome	Comment
Analgesics	NSAIDs (e.g., aspirin, ibuprofen, phenylbutazone)	Low doses of NSAIDs delay GE (Kechagias et al., 1997); aspirin inhibits gastric ADH (Weathermon and Crabb, 1999); competes for binding sites on CYP2E1	↓ BAC intensifies adverse effects of the drug such as gastric bleeding (Kaufman et al., 1999; Neutel and Appel, 2000)	Adverse effects are worse when NSAIDs are taken on an empty stomach; avoid NSAIDs within 8–10 hours of drinking
	Acetaminophen	Binge drinking increases metabolism of acetaminophen to NAPQ1 in chronic alcoholics by induction of CYP2E1 and depletion of glutathione that inactivates NAPQ1 (Zimmerman and Maddrey, 1995; Hersh et al., 2007; Gomez-Moreno et al., 2008) and inhibits gastric ADH (Weathermon and Crabb, 1999)	NAPQ1 impairs calcium homeostasis and causes oxidative stress and cell death, leading to liver failure	Increased risk for hepatotoxicity with regular or high doses (>4 g/24 hours) of acetaminophen (Prescott, 2000)
	Opioids (e.g., meperidine, methadone, propoxyphene)	Agonist effects of opioids on μ opioid receptors are facilitated by alcohol through GABA and NMDA receptors (McCance-Katz et al., 2010)	Opioids combined with alcohol can cause excessive central nervous system (CNS) depression and impaired psychomotor function, leading to respiratory depression; overdoses of propoxyphene combined with alcohol have been linked to fatal reactions	Avoid alcohol when taking opioids
Antibacterial	Erythromycin	Accelerates GE through agonist effects on smooth-muscle motilin receptors Urbain et al., 1990; (Larson et al., 2010)	↑ BAC	
	Chloramphenicol	Inhibits low-K_m ALDH2 in the brain and liver (Vasiliou et al., 1986)	↑ BAC	
	Isoniazid	Inhibits low-K_m ALDH2 in the brain and liver (Vasiliou et al., 1986)	↑ BAC	↑ Hepatotoxicity in chronic alcoholics; monitor liver function; limit or avoid alcohol use
	Cephalosporin drugs with an *N*-methylthio-tetrazole moiety (NMTT) (e.g., cefoperazone and cefotetan, moxalactam, cefamandole)	Ceftriaxone shown to reduce ethanol drinking in rodent models via activation of glutamate transporter 1 (GLT1) (Sari et al., 2011; Qrunfleh et al., 2013)	Disulfiram-like reaction to alcohol (McMahon, 1980)	

Continued

Table 31.2

Continued

Drug class	Substance	Mechanism	Outcome	Comment
Anticoagulants	Warfarin	Acute alcohol ↑ warfarin availability; chronic alcohol use ↓ warfarin metabolism	↑ Risk of life-threatening hemorrhages with heavy alcohol use (Baker et al., 2004)	Monitor international normalized ratios if the patient consumes >3 alcoholic drinks/day or if there has been a significant change in the amount of recent alcohol consumption
Anticonvulsants	Phenytoin	Alcohol induces hepatic metabolism of phenytoin	Chronic alcohol consumption increases phenytoin metabolism (Tanaka, 2002)	
Antidepressants	Tricyclics (e.g., amitriptyline and doxepin)	Alcohol interferes with the first-pass metabolism of amitriptyline in the liver	↑ Amitriptyline levels in blood; excessive CNS depression and impaired psychomotor function	
	Monoamine oxidase inhibitors (MAOIs) (e.g., phenelzine)	MAOIs interact with alcoholic beverages containing tyrosine	Tyrosine increases blood pressure, causing severe hypertension	Avoid alcohol when taking MOAIs
Antidiabetic agents	Chlorpropamide	Inhibits low-K_m ALDH2 in the brain and liver (Vasiliou et al., 1986)	↑ Acetaldehyde in blood (Jerntorp, 1983; Jerntorp et al., 1986)	
	Glipizide	Alcohol delays glipizide absorption and elimination	Hypoglycemia	Limit alcohol to occasional use; completely avoid alcohol if signs of hypoglycemia occur
	Metformin	Alcohol potentiates metformin's effects on lactate metabolism	Alcohol plus metformin may increase risk for lactic acidosis (Shenoy, 2006)	Limit alcohol to occasional use and monitor for signs of lactic acidosis
	Insulin	Alcohol enhances the glucose-lowering action of insulin	Hypoglycemia	Should not drink alcohol in excess or on an empty stomach
Antihistamine	Ranitidine	Accelerates GE by antagonizing H_2 receptors (Oneta et al., 1998)	Concurrent antihistamine and alcohol use ↑ BAC (Amir et al., 1996; Arora et al., 2000)	
	Cimetidine	Inhibition of ADH2 in small intestines and liver; inhibition of class 1 ADH and ALDH in the liver under therapeutic doses of cimetidine and physiologic levels of ethanol (Lai et al., 2013)	Concurrent medication and alcohol use ↑ BAC (Caballeria et al., 1991; Weinberg et al., 1998; Monroe and Doering, 2001) and can cause hepatotoxicity (Devuyst et al., 1993)	
Antiprotozoal	Metronidazole	Acts not by inhibiting ALDH (Karamanakos et al., 2007) but by increasing ADH activity in intestines	Disulfiram-like reaction to alcohol (McMahon, 1980)	Should not drink ethanol within 2–3 days of taking metronidazole

Antipsoriatic	Acitretin	Alcohol converts acitretin to etretinate (Larsen et al., 1993; Gronhoj Larsen et al., 2000; Vena and Cassano, 2012)	Etretinate is teratogenic	
Antipsychotics	Phenothiazines (e.g., chlorpromazine); atypical antipsychotics (e.g., risperidone)		Excessive CNS depression and impaired psychomotor performance	Avoid alcohol
Gastroprokinetic agents	Cisapride	Accelerates gastric emptying by stimulation of presynaptic 5-HT_4 receptors in the myenteric plexus (Kechagias et al., 1999; Quigley, 2011)	↑ BAC (Dziekan et al., 1997; Oneta et al., 1998; Kechagias et al., 1999)	
	Metoclopramide	Accelerates gastric emptying by inhibition of pre- and postsynaptic D_2 receptors and stimulation of presynaptic 5-HT_4 receptors in the myenteric plexus (Lee and Kuo, 2010)	↑ BAC	
Nitrates	Nitroglycerine	ALDH2 bioactivates nitroglycerine to produce nitric oxide (Chen et al., 2005, 2008)	Acute alcohol enhances the adverse effects of nitroglycerin, such as dizziness or syncope	
Sedative agents	Barbiturates	Competes for binding to CYP2E1	↑ Barbiturate levels in blood, causing CNS depression	Avoid alcohol use as death has been reported with concomitant use
	Benzodiazepines, e.g.,diazepam lorazepam	Alcohol can increase absorption of diazepam and reduce hepatic metabolism; can reduce anxiolytic effects of some benzodiazepines	Excessive CNS depression and impaired psychomotor performance	Avoid moderate to high amounts of alcohol consumption

NSAIDs, non-steroidal anti-inflammatory drugs; GE, gastric emptying; ADH, alcohol dehydrogenase; BAC, blood alcohol concentration; GABA, gamma-aminobutyric acid; NMDA, *N*-methyl-d-aspartate; ALDH, aldehyde dehydrogenase.

of alcohol and the medication competing for the binding sites on CYP2E1 (Jones, 2010).

ALDH system

Acetaldehyde produced in the initial step of alcohol oxidation is classified by the International Agency for Research on Cancer as a known carcinogen (Secretan et al., 2009). The toxic acetaldehyde is further metabolized by ALDH into the benign secondary metabolite, acetate. The human ALDH superfamily consists of 19 isozymes with differing kinetic parameters (Song et al., 2011). The two major human liver isozymes – the cytosolic ALDH1 and the mitochondrial ALDH2 (Impraim et al., 1982) – are differentiated by their subcellular localization, kinetic parameters, and electrophoretic mobility. There are six enzyme subtypes identified in the ALDH1 family. Of these, ALDH1A1 contributes to alcohol oxidation, but with a lower catalytic efficacy (K_m for oxidizing acetaldehyde = 50–100 μM) than the mitochondrial ALDH2 (Nuutinen et al, 1983). Medications that block the hepatic ALDH can cause accumulation of acetaldehyde, which increases the risk of developing cancer. The most common examples are presented in Table 31.2. Among these medications, disulfiram is designed specifically to target the acetaldehyde metabolism pathway. Disulfiram acts by inhibiting low-K_m ALDH2 in the brain and liver (Vasiliou et al., 1986), resulting in about 5-10-fold higher levels of acetaldehyde following alcohol use.

Non-oxidative pathways of alcohol metabolism

Metabolism via non-oxidative pathways in humans accounts for about 1% of the total ingested alcohol. Non-oxidative pathways include:

- Conjugation of alcohol with activated glucuronic acid to form ethyl glucuronide (EtG) and,conjugation with sulfate to produce ethyl sulfate (EtS)
- Formation of fatty acid ethyl esters (FAEE)
- Formation of phosphatidylethanol (PEth).

EtG/EtS

EtG formation is catalyzed by the endoplasmic reticulum membrane-bound enzyme UDP-glucuronosyltransferase. Both EtG and EtS are being utilized as biomarkers of alcohol abstinence and moderate/heavy drinking. Studies have shown that alcohol concentrations as low as 5 g are detectable with human EtG and that liver disease does not significantly modify urine EtG/EtS readings (Stewart et al., 2013).

FAEE

FAEE is produced by esterification of ethanol with endogenous fatty acids or fatty coenzyme A (Soderberg et al., 2003). The esterification reaction is catalyzed by many enzymes. FAEEs are detectable in the serum, liver, adipose tissue, hair, and meconium of neonates (Pragst and Yegles, 2008) as direct markers for screening of alcohol use or abstinence.

PEth

PEth comprises a group of phospholipids carrying two variable fatty acid chains connected to a common nonpolar phosphoethanol head (Gunnarsson et al., 1998). They are formed in cell membranes from phosphatidylcholine only in the presence of alcohol (Maenhout et al., 2013). The transphosphatidylation reaction is catalyzed by the enzyme phospholipase D, which, in the absence of alcohol, hydrolyzes membrane phospholipids. Hence, PEth is used as a direct biomarker for the screening of recent alcohol consumption. PEth is measurable in blood for up to 2 weeks after cessation of drinking (Litten et al., 2010). A single bout of alcohol drinking will not produce detectable amounts of PEth; this would require continuous drinking of approximately 1000 g of alcohol over about 2 weeks. The sensitivity of PEth for the screening of alcohol use is greater than for traditional biomarkers.

Whether the medications interfering with alcohol pharmacokinetics cause a discrepancy between the actual amounts of alcohol consumed and the amounts metabolized into non-oxidative products is unclear. Furthermore, the variability of these markers among individuals is also not known.

Alcohol metabolism in the CNS

Alcohol metabolism in the extrahepatic sites is much smaller. Early research studies alleged that the oxidation of alcohol in the brain into its first metabolite, acetaldehyde, was non-enzymatic (Deitrich et al., 2006). This assertion was based primarily on the fact that most studies have failed to detect the main alcohol-metabolizing enzyme found in the liver, the class I ADH, in healthy human brain tissue (Estonius et al., 1996; Galter et al., 2003). In subsequent studies, it was found that the ADH class III isoform is predominant in the healthy human brain (Galter et al., 2003; Laniewska-Dunaj et al., 2013) but that it possesses low activity due to its high affinity to ethanol ($K_m > 2.5$ M). The other two known enzymes that oxidize alcohol into acetaldehyde, catalase and CYP2E1, only play a minor role in eliminating ethanol in the periphery at moderate doses of alcohol (Correa et al., 2012). However, acetaldehyde produced in the periphery does not enter the brain because of the presence of ALDH in the blood-brain barrier. Hence, the acetaldehyde present in the brain appears to be locally synthesized. Supporting this hypothesis, Aragon et al. (1991) and Gill et al. (1992) demonstrated the oxidation of acetaldehyde in the rat brain by peroxisomal enzyme system catalase-H_2O_2. At high concentrations or in chronic use of alcohol, CYP2E1 may oxidize alcohol into acetaldehyde. CYP2E1 expression in the brain is region-specific and found in peroxisomes and mitochondria. Acetaldehyde is a potent, biologically active compound responsible for many behavioral effects of alcohol. Several studies have implied a multitude of roles for acetaldehyde in alcohol preference, ranging from being the sole compound in the brain responsible for reinforcing effects of alcohol to having a modulatory effect on alcohol or endogenous dopamine. Contrary to these findings in the brain, in the periphery, acetaldehyde functions as a reinforcer only at low concentrations; at normal to high blood acetaldehyde concentrations, voluntary ethanol consumption is reduced in animals, indicating an aversive effect. Targeting the mechanism of elevated blood acetaldehyde via inhibition of hepatic low-K_m ALDH2, disulfiram was first introduced by Hald and Jacobsen in 1948 as an adjunct for treatment of chronic alcoholism (Hald and Jacobsen, 1948; Hald et al., 1949). The disulfiram-ethanol reaction, commonly known as "disulfiram-like-reaction" is characterized by nausea, vomiting, headaches, and a flushed face and neck; rare signs include vertigo, blurred vision, hypotension, and syncope (McMahon, 1980). In addition to inhibition of hepatic ALDH2, disulfiram reduces norepinephrine (NE) synthesis in the brain by blocking dopamine β-hydroxylase (DBH). As alcohol induces sedation via the same NE–DBH pathway, disulfiram given after a bout of drinking potentiates alcohol's sedative effects. Medications that affect acetaldehyde metabolism in the brain and periphery are presented in Table 31.2.

Unlike acetaldehyde, acetate, the secondary metabolite of alcohol, is readily transported across the blood-brain barrier into the brain. Acetate is then converted into acetyl-CoA, which is utilized in either the citric acid cycle or the production of the neurotransmitter acetylcholine (Fig. 31.1). Women may have higher sensitivity to alcohol in the CNS (Ammon et al., 1996).

Regulators of alcohol metabolism

The rate of alcohol metabolism is regulated by factors that influence BAC (Table 31.1). At a BAC of <0.02%, the rate of alcohol metabolism is exponential and involves ADH because of its low K_m for alcohol (2–10 mg/100 mL) (Jones, 2010); between BACs of 0.02% and 0.065%, the metabolic rate is constant as ADH becomes saturated. When the BAC is above 0.065%, CYP2E1 with relatively higher K_m (60–80 mg/100 mL) becomes active (Jones, 2010), resulting in greater metabolic rates. Functional genetic variations in the genes of FPM pathways may lead to variations in the volume of distribution of alcohol (Zuo et al., 2010). Some of the most tested genetic variants with an impact on alcohol-metabolizing enzymes are discussed below.

Influence of genetic variability in genes encoding alcohol-metabolizing enzymes

ADH GENES

Quantitative or qualitative changes in ADH enzymes can result in altered levels of acetaldehyde in tissues. Such inherent differences in enzyme activity might have a varied impact on alcohol-medication interactions in different individuals. As mentioned above, ADH1 and ADH4 are the main ADH isozymes involved in alcohol oxidation. ADH1 is formed by heteromers of subunits α, ß, and γ that are encoded by genes *ADH1A, ADH1B*, and *ADH1C*, respectively. The three genes are tightly clustered to an approximately 77-kb region on chromosome 4q21-23 in the 5′-*ADH1A-ADH1B-ADH1C*-3′ direction (Osier et al., 2002). Among these genes, *ADH1B* and *ADH1C* genetic variations, compared with those in *ADH1A*, are widely studied in different ethnic populations for associations with alcoholism-related traits and carcinogenesis. The A allele compared with the G allele of single-nucleotide polymorphism (SNP) rs1229984 in *ADH1B* and the A allele of rs698 in *ADH1C* were found to increase the capacity of ADH to oxidize alcohol into acetaldehyde by severalfold (Wu et al., 2013). In fact, rs1229984 is particularly important as both population-based candidate gene and genomewide association studies have associated rs1229984 with an increased risk for alcoholism-related traits and gastrointestinal tract cancers in Asian and European populations (McKay et al., 2011; Park et al., 2013).

ALDH-CODING GENES

The most consistently replicated genetic association with alcohol metabolism in the ALDH family of enzymes is the exonic non-synonymous SNP rs671 in the *ALDH2* gene. The *ALDH2* gene encodes the mitochondrial ALDH isozyme ALDH2 and is mapped to chromosome 12q24.2. The SNP rs671, also referred to as ALDH2*1/2, is now designated as Glu504Lys from the earlier nomenclature of Glu487Lys (Li et al., 2006). The substitution of glutamine residue with a lysine residue arises from a G to A allele transition. The protein subunit with the Lys504 residue encoded by the rs671:A allele, compared with the Glu504 residue encoded by the rs671:G allele (the wild type), has a shorter half-life (Xiao et al., 1996), causing an ALDH2 enzyme deficiency (Xiao et al., 1995) and leading to a restrained ability to metabolize acetaldehyde. The rs671:A allele is present in about 20% of Asians and rarely in African and European populations (http:/www.ncbi.nlm.nih.gov/projects/SNP/snp_ref.cgi?rs=671). Population-based candidate and genomewide association analyses have consistently reported a protective effect of the rs671:A allele for developing alcoholism in Asians (Kim et al., 2008; Li et al., 2012; Tan et al., 2012) as a result of the slower metabolism of acetaldehyde, which in turn leads to an aversive disulfiram-like-reaction (Goedde et al., 1983; Liu et al., 2005).

Genetic associations of the *ALDH1A1* gene, which encodes the cytoplasmic isozyme ALDH1A1, are relatively less explored. The *ALDH1A1* gene is located on chromosome 9q21.13. Spence et al. (2003) detected two *ALDH1A1* promoter region variants regulating its expression levels. They were the 17-bp deletion polymorphism ALDH1A1*2, found in Asian, European, and African populations at a frequency of <5%, and the 3-bp insertion polymorphism ALDH1A1*3, found only in individuals with African ancestry. In a later study, Ehlers et al. (2004) showed that Native American individuals carrying the ALDH1A1*2 allele drank less and their rate of alcoholism was lower. In individuals of African descent, ALDH1A1*2 was associated with a greater risk of developing alcoholism (Spence et al., 2003; Moore et al., 2007b). Several other studies reported a lack of an association with alcoholism in Asian populations (Otto et al., 2013). The inconclusive associations indicate the need for larger replication studies. Furthermore, to establish the link between drinking levels and individual variations in alcohol metabolism, the effects of ALDH1A1*2 or ALDH1A1*3 alleles on cytosolic ALDH1A1 isozyme levels need to be characterized.

CYP2E1-coding genes

The CYP2E1 gene, which was first sequenced by Umeno et al. (1988), is mapped to chromosome 10q24.3 (McBride et al., 1987; Kolble, 1993). Several functional polymorphisms in the CYP2E1 gene have been reported in the literature, but their effects remain inconclusive. The 5′- flanking alleles rs3813867:C and rs2031920:T, which are in linkage disequilibrium with each other, were shown to have greater expression levels of CYP2E1 enzymes *in vitro* (Hayashi et al., 1991; Wang et al., 2009). These alleles are collectively referred to as the CYP2E1*5B polymorphism. Both rs3813867:C and rs2031920:T alleles are common in Asian populations but rare in Caucasians and Africans (Zavras et al., 2002). Another functional allele in CYP2E1, located in intron 6, rs6413432:A, which is more common in Asians, was also shown to increase the expression levels of CYP2E1 (Vodicka et al., 2001). Population-based genetic association analyses, however, have indicated both a greater susceptibility to alcohol dependence associated with these functional alleles and the absence of an association (Iwahashi et al., 1998; Guo et al., 2011; Celorrio et al., 2012).

Catalase-coding genes

The catalase gene (CAT) is located on chromosome 11p13 (Junien et al., 1980). *In vitro* studies have shown an allele-based transcription level difference in the promoter region SNP rs1001179(C/T) of the CAT gene, leading to a genotype-dependent dose response in catalase activity (Forsberg et al., 2001). In Asian and European populations, the homozygous minor allele carriers (TT genotype) had the lowest levels of catalase in red blood cells, causing a hypo/acatalasemic state (Forsberg et al., 2001). Interestingly, the rs1001179 genotype effects were stronger in women than in men, were influenced by dietary and environmental factors, and are not reported in African populations (Nadif et al., 2005; Ahn et al., 2006; Bastaki et al., 2006).

PHARMACODYNAMIC INTERACTIONS

There are two main groups of alcohol-medication pharmacodynamic interactions:

1. Pharmacodynamic effects of alcohol on a medication: the influence of acute or chronic alcohol use on a medication at its effector site(s), altering the desired effects of the medication.
2. Pharmacodynamic effects of the medication on alcohol: the effects of a single or multiple doses of a medication on alcohol at its effector site(s), altering the motor and behavioral effects of alcohol (Jones, 2012).

Enzyme reactions are not involved in mechanisms underlying pharmacodynamic interactions. Some of the common pharmacodynamic mechanisms include: opening or closing of ion channels; binding to drug transport proteins, receptors, or transcription factors, and release of neurotransmitters. Most of the alcohol-medication pharmacodynamic interactions, especially those occurring in the CNS, are synergistic; that is, the combined effect of alcohol and the interacting medication is greater than either of their single effects. Some medications, such as phenobarbital, interact with alcohol both pharmacodynamically and pharmacokinetically. For such medications, the type of interaction that potentially occurs with alcohol is believed to vary on the severity of drinking (Weathermon and Crabb, 1999).

INTERACTION OF ALCOHOL WITH CURRENTLY TESTED MEDICATIONS FOR THE TREATMENT OF ALCOHOLISM

A list of medications tested for the treatment of alcohol addiction and their mechanisms of action are given in Table 31.3. The oldest medication for the treatment of alcohol addiction, disulfiram, was discovered in Denmark in 1948. Disulfiram is marketed as Antabuse or Antabus in various countries and remains one of the most widely prescribed medications for alcoholism. As discussed earlier in this chapter, disulfiram acts through developing an aversive reaction to alcohol, with the goal of scaring the patient from drinking. Since the discovery of disulfiram, several medications targeting the corticomesolimbic dopaminergic system have been tested for reducing drinking severity and craving for alcohol in alcoholics. Among these medications, oral naltrexone, ondansetron, sertraline, finasteride and olanzapine may have indirect effects on alcohol metabolism as they undergo significant first pass metabolism in the liver. Apart from the potential liver toxicity reported with prolonged and/or high dose naltrexone, naltrexone is metabolized in the liver by dihydrodiol dehydrogenase (DD4) into 6b-naltrexol (Porter et al., 2000), which belongs to a subgroup of the short-chain alcohol dehydrogenase/reductase family (Jouanneau and Meyer, 2006). In the liver, CYP450 enzymes that metabolize alcohol are involved in metabolizing ondansetron (CYP1A2 and CYP3A4), sertraline (CYP3A4), Finasteride (CYP3A4) (Figg et al., 1996; Obach et al., 2005; Huskey et al., 1995). Olanzapine is non-oxidatively metabolized into N-10-glucuronide in the liver (Murphy et al., 1998). Topiramate and Acamprosate are not known to have direct or indirect effects on alcohol metabolism, presumably because they are excreted as unchanged by the kidneys.

Alcoholism can be conceptualized as several different diseases based on pathophysiology, course, and treatment outcome. Therefore, a single treatment intervention does not work for all subtypes of alcoholism. Examples of currently tested medications to reduce drinking severity include ondansetron, baclofen, naltrexone, and topiramate. Selection of optimal therapy should be based on the subtype of alcoholism, such as early- vs late-onset alcoholism - the former characterized by chronic, more severe symptoms with a greater biologic predisposition. Other considerations are the age of the patient, the presence of other psychiatric disorders or conditions, and pharmacogenetic effects. A concise overview of currently available subtype-based alcoholism treatment is presented in the format of several clinical vignettes in a review by Johnson (2010). For example, for a currently drinking, chronic alcoholic patient presenting with medical complications including liver impairment, the drug of choice would be topiramate. Baclofen can be prescribed to an alcoholic patient with a similar history but without liver impairment. For young adults with problem drinking, either ondansetron or naltrexone can be prescribed if they carry the genetic variants mentioned in Table 31.3. Additionally, naltrexone can be prescribed to elderly alcoholic patients with no major depression.

Another population requiring specific attention is pregnant women who are dependent on alcohol. Alcohol use during pregnancy can cause fetal alcohol syndrome (FAS), intrauterine growth retardation, spontaneous abortions, stillbirths, and preterm births (Heberlein et al., 2012). As there is no safe limit to drinking during pregnancy, the goal of treatment should be to achieve total abstinence or to reduce drinking to a minimum if achieving abstinence is not realistic. In fact, the Centers for Disease Control advise women who are planning to become pregnant to avoid drinking, given the severity of consequences during the first trimester. Drinking reductions during pregnancy should be gradual; withdrawal is best treated with benzodiazepine, taking into account the risk factors during the first and third trimesters. Benzodiazepines may cause permanent functional damage on the developing brain, and therefore, the dose should be limited to minimally adequate amount to prevent major complications of withdrawal (McElhatton, 1994; Bhuvaneswar et al., 2007; Heberlein et al., 2012). Anticonvulsant use for withdrawal is associated with greater teratogenicity and thus should be avoided during pregnancy for treatment of withdrawal (Heberlein et al., 2012; Bromley et al., 2013).

CONCLUDING REMARKS

Knowledge of the metabolic pathways, absorption, and elimination of a drug is important to predict its interactions with alcohol. The past few decades have provided a wealth of information on the molecular mechanisms underlying these pathways. However, several gaps still remain to be addressed on certain aspects of alcohol-medication interactions. These include further characterization of the contribution of non-oxidative pathways of metabolism, alcohol-medication interactions under various pathologic conditions such as diabetes, obesity, or liver disease, and further elucidation of pharmacodynamic interactions, which requires a deeper understanding of the mechanisms of action of medications and alcohol.

Genetic studies have identified variants that alter the activity of enzymes or receptors in alcohol-medication interaction pathways. These findings can be incorporated to provide a tool for identifying high-risk groups to prevent the occurrence of adverse events arising from

Table 31.3

Pharmacogenetics of medications for alcohol treatment

Medication	Mechanism	Pharmacogenetic findings: Gene	Pharmacogenetic findings: Effect
Acamprosate	$GABA_A$ receptor agonist, mGluR5 receptor antagonist	GATA4	Efficacious for relapse prevention in rs13273672:A allele carriers (Kiefer et al., 2011)
Naltrexone	Active metabolite 6-β-naltrexol is a competitive antagonist at μ- and κ-opioid receptors; low-affinity antagonist at δ-opioid receptors	OPRM1$^{\pm}$	Greater efficacy in rs1799971:G (Asp40) allele subjects (Oslin et al., 2003; Anton et al., 2008; Ray et al., 2012); reduces craving in rs4654327:A allele carriers (Ashenhurst et al., 2012)
Ondansetron	5-HT_{3A} receptor antagonist	SLC6A4$^{\pm}$	Greater efficacy for reducing drinking severity in 5′-HTTLPR:LL (Kenna et al., 2009; Johnson et al., 2011)* and rs1042173:TT (Johnson et al., 2011) genotype carriers
		HTR3A	Greater efficacy for reducing drinking severity in rs1150226-AG, rs1176713-GG carriers (Johnson et al., 2013)
		HTR3B	Greater efficacy for reducing drinking severity in rs17614942-AC carriers (Johnson et al., 2013).
Sertraline	Primarily binds to the serotonin transporter and selectively inhibits serotonin reuptake (SSRI); shows lower affinity for dopamine transporter inhibition, sigma-1 receptor (σ_1R) antagonist, α1-adrenergic antagonists	SLC6A4	Sertraline showed a significant "5′-HTTLPR:LL and rs25531:AA genotype × age of alcoholism onset" effect on drinking reductions (Kranzler et al., 2012)
Finasteride	Type II 5α-reductase inhibitor	GABRA2	Reduced subjective effects of alcohol in rs279858:AA genotype individuals (Pierucci-Lagha et al., 2005)
Olanzapine	High-affinity antagonist/inverse agonist at 5-HT_2 serotonin receptors; lower affinity for M_3 receptor, serotonin 5-HT_1, $GABA_A$, beta-adrenergic receptors, and benzodiazepine binding sites, H_1 histamine receptors, serotonin 5-HT_{2C} and dopamine D_2 receptors	DRD4	VNTR long (L)-allele individuals showed greater reductions of craving (Hutchison et al., 2003)
Topiramate	Facilitate $GABA_A$ receptors and antagonize AMPA and Kainate glutamate receptors	GRIK1	Preliminary evidence of greater efficacy of topiramate in rs2832407:CC compared to non-carriers (Kranzler et al., 2013)

OPRM1 = μ opioid receptor gene; SLC6A4 = serotonin transporter gene; HTR3A = serotonin receptor subunit A gene; HTR3B = serotonin receptor subunit B gene; GABRA2 = gamma-aminobutyric acid (GABA) A receptor, alpha 2 gene; GATA4 = transcription factor GATA-4 gene; DRD4 = dopamine receptor subtype 4 gene; GRIK1 = glutamate receptor, ionotropic, kainate 1; VNTR = variable number of tandem repeats; 5-HT = 5-hydroxytryptamine (serotonin); *study participants randomized by genotype; $^{\pm}$Results replicated in independent trials.

alcohol-medication reactions. For instance, medications that lower alcohol-metabolizing enzymes such as ADH or ALDH (see Table 31.2 for examples of such medications) should be prescribed with caution to individuals who are already genetically predisposed to have low levels of enzymes, as they might become vulnerable to the toxic effects of alcohol. In a pharmacodynamic context, prescribing medications to treat alcohol addiction based on genetic variations that alter the function or expression of the target receptor molecules may enhance the efficacy of medications to reduce drinking.

References

Adinolfi A, Adinolfi M, Hopkinson DA (1984). Immunological and biochemical characterization of the human alcohol dehydrogenase chi-ADH isozyme. Ann Hum Genet 48 (Pt 1): 1–10.

Ahn J, Nowell S, McCann SE et al. (2006). Associations between catalase phenotype and genotype: modification by epidemiologic factors. Cancer Epidemiol Biomarkers Prev 15 (6): 1217–1222.

Amir I, Anwar N, Baraona E et al. (1996). Ranitidine increases the bioavailability of imbibed alcohol by accelerating gastric emptying. Life Sci 58 (6): 511–518.

Ammon E, Schafer C, Hofmann U et al. (1996). Disposition and first-pass metabolism of ethanol in humans: is it gastric or hepatic and does it depend on gender? Clin Pharmacol Ther 59 (5): 503–513.

Anton RF, Oroszi G, O'Malley S et al. (2008). An evaluation of mu-opioid receptor (OPRM1) as a predictor of naltrexone response in the treatment of alcohol dependence: results from the Combined Pharmacotherapies and Behavioral Interventions for Alcohol Dependence (COMBINE) study. Arch Gen Psychiatry 65 (2): 135–144.

Aragon CM, Abitbol M, Amit Z (1991). Ethanol-induced CTA mediated by acetaldehyde through central catecholamine activity. Psychopharmacology (Berl) 103 (1): 74–77.

Arora S, Baraona E, Lieber CS (2000). Alcohol levels are increased in social drinkers receiving ranitidine. Am J Gastroenterol 95 (1): 208–213.

Ashenhurst JR, Bujarski S, Ray LA (2012). Delta and kappa opioid receptor polymorphisms influence the effects of naltrexone on subjective responses to alcohol. Pharmacol Biochem Behav 103 (2): 253–259.

Baker RI, Coughlin PB, Gallus AS et al. (2004). Warfarin reversal: consensus guidelines, on behalf of the Australasian Society of Thrombosis and Haemostasis. Med J Aust 181 (9): 492–497.

Bastaki M, Huen K, Manzanillo P et al. (2006). Genotype-activity relationship for Mn-superoxide dismutase, glutathione peroxidase 1 and catalase in humans. Pharmacogenet Genomics 16 (4): 279–286.

Bhuvaneswar CG, Chang G, Epstein LA et al. (2007). Alcohol use during pregnancy: prevalence and impact. Prim Care Companion J Clin Psychiatry 9 (6): 455–460.

Bond J, Ye Y, Cherpitel CJ et al. (2010). The relationship between self- reported drinking and BAC level in emergency room injury cases: is it a straightline? Alcohol Clin Exp Res 34 (6): 1118–1125.

Breton MD (2010). In silico models of alcohol kinetics: deterministic approach. In: B Johnson (Ed.), Addiction Medicine: Science and Practice. Springer, New York, pp. 1265–1276.

Bromley RL, Mawer GE, Briggs M et al. (2013). The prevalence of neurodevelopmental disorders in children prenatally exposed to antiepileptic drugs. J Neurol Neurosurg Psychiatry 84 (6): 637–643.

Caballeria J, Baraona E, Deulofeu R et al. (1991). Effects of H2-receptor antagonists on gastric alcohol dehydrogenase activity. Dig Dis Sci 36 (12): 1673–1679.

Celorrio D, Bujanda L, Caso C et al. (2012). A comparison of Val81Met and other polymorphisms of alcohol metabolising genes in patients and controls in Northern Spain. Alcohol 46 (5): 427–431.

Chen Z, Foster MW, Zhang J et al. (2005). An essential role for mitochondrial aldehyde dehydrogenase in nitroglycerin bioactivation. Proc Natl Acad Sci U S A 102 (34): 12159–12164.

Chen CH, Budas GR, Churchill EN et al. (2008). Activation of aldehyde dehydrogenase-2 reduces ischemic damage to the heart. Science 321 (5895): 1493–1495.

Correa M, Salamone JD, Segovia KN et al. (2012). Piecing together the puzzle of acetaldehyde as a neuroactive agent. Neurosci Biobehav Rev 36 (1): 404–430.

Deitrich R, Zimatkin S, Pronko S (2006). Oxidation of ethanol in the brain and its consequences. Alcohol Res Health 29 (4): 266–273.

Devuyst O, Lefebvre C, Geubel A et al. (1993). Acute cholestatic hepatitis with rash and hypereosinophilia associated with ranitidine treatment. Acta Clin Belg 48 (2): 109–114.

Dziekan G, Contesse J, Werth B et al. (1997). Cisapride increases peak plasma and saliva ethanol levels under fasting conditions. J Intern Med 242 (6): 479–482.

Edenberg HJ (1998). Genetics of alcoholism. Science 282 (5392): 1269.

Edenberg HJ (2007). The genetics of alcohol metabolism: role of alcohol dehydrogenase and aldehyde dehydrogenase variants. Alcohol Res Health 30 (1): 5–13.

Ehlers CL, Spence JP, Wall TL et al. (2004). Association of ALDH1 promoter polymorphisms with alcohol-related phenotypes in southwest California Indians. Alcohol Clin Exp Res 28 (10): 1481–1486.

Estonius M, Svensson S, Hoog JO (1996). Alcohol dehydrogenase in human tissues: localisation of transcripts coding for five classes of the enzyme. FEBS Lett 397 (2–3): 338–342.

Feierman DE, Melinkov Z, Nanji AA (2003). Induction of CYP3A by ethanol in multiple in vitro and in vivo models. Alcohol Clin Exp Res 27 (6): 981–988.

Ferreira MP, Willoughby D (2008). Alcohol consumption: the good, the bad, and the indifferent. Appl Physiol Nutr Metab 33 (1): 12–20.

Figg WD, Dukes GE, Pritchard JF et al. (1996). Pharmacokinetics of ondansetron in patients with hepatic insufficiency. J Clin Pharmacol 36 (3): 206–215.

Forsberg L, Lyrenas L, de Faire U et al. (2001). A common functional C-T substitution polymorphism in the promoter region of the human catalase gene influences transcription factor binding, reporter gene transcription and is correlated to blood catalase levels. Free Radic Biol Med 30 (5): 500–505.

Galter D, Carmine A, Buervenich S et al. (2003). Distribution of class I, III and IV alcohol dehydrogenase mRNAs in the adult rat, mouse and human brain. Eur J Biochem 270 (6): 1316–1326.

Gill K, Menez JF, Lucas D et al. (1992). Enzymatic production of acetaldehyde from ethanol in rat brain tissue. Alcohol Clin Exp Res 16 (5): 910–915.

Goedde HW, Agarwal DP, Harada S et al. (1983). Population genetic studies on aldehyde dehydrogenase isozyme deficiency and alcohol sensitivity. Am J Hum Genet 35 (4): 769–772.

Gomez-Moreno G, Guardia J, Cutando A (2008). Interaction of paracetamol in chronic alcoholic patients. Importance for odontologists. Med Oral Patol Oral Cir Bucal 13 (4): E235–E238.

Gronhoj Larsen F, Steinkjer B, Jakobsen P et al. (2000). Acitretin is converted to etretinate only during concomitant alcohol intake. Br J Dermatol 143 (6): 1164–1169.

Gugler R (1994). H2-antagonists and alcohol. Do they interact? Drug Saf 10 (4): 271–280.

Gunnarsson T, Karlsson A, Hansson P et al. (1998). Determination of phosphatidylethanol in blood from alcoholic males using high-performance liquid chromatography and evaporative light scattering or electrospray mass spectrometric detection. J Chromatogr B Biomed Sci Appl 705 (2): 243–249.

Guo WJ, Wang Q, Lanzi G et al. (2011). Gender-specific interactions between alcohol metabolism genes and severity of quantitative alcohol-related-traits in a Tibetan population. Neurosci Lett 495 (1): 22–25.

Hald J, Jacobsen E (1948). A drug sensitizing the organism to ethyl alcohol. Lancet 2 (6539): 1001–1004.

Hald J, Jacobsen E, Larsen V (1949). Formation of acetaldehyde in the organism in relation to dosage of antabuse (tetraethylthiuramdisulphide) and to alcohol-concentration in blood. Acta Pharmacol Toxicol (Copenh) 5 (2): 179–188.

Hayashi S, Watanabe J, Kawajiri K (1991). Genetic polymorphisms in the 5'-flanking region change transcriptional regulation of the human cytochrome P450IIE1 gene. J Biochem 110 (4): 559–565.

Heberlein A, Leggio L, Stichtenoth D et al. (2012). The treatment of alcohol and opioid dependence in pregnant women. Curr Opin Psychiatry 25 (6): 559–564.

Heit C, Dong H, Chen Y et al. (2013). The Role of CYP2E1 in Alcohol Metabolism and Sensitivity in the Central Nervous System. Subcell Biochem 67: 235–247.

Hersh EV, Pinto A, Moore PA (2007). Adverse drug interactions involving common prescription and over-the-counter analgesic agents. Clin Ther 29 (Suppl.): 2477–2497.

Huskey SW, Dean DC, Miller RR et al. (1995). Identification of human cytochrome P450 isozymes responsible for the in vitro oxidative metabolism of finasteride. Drug Metab Dispos. 23 (10): 1126–1135. Erratum in: Drug Metab Dispos 1996;24(6):695.

Hutchison KE, Wooden A, Swift RM et al. (2003). Olanzapine reduces craving for alcohol: a DRD4 VNTR polymorphism by pharmacotherapy interaction. Neuropsychopharmacology 28 (10): 1882–1888.

Impraim C, Wang G, Yoshida A (1982). Structural mutation in a major human aldehyde dehydrogenase gene results in loss of enzyme activity. Am J Hum Genet 34 (6): 837–841.

Iwahashi K, Ameno S, Ameno K et al. (1998). Relationship between alcoholism and CYP2E1 C/D polymorphism. Neuropsychobiology 38 (4): 218–221.

Jerntorp P (1983). The chlorpropamide alcohol flush test in diabetes mellitus: methods for objective evaluation. Scand J Clin Lab Invest 43 (3): 249–254.

Jerntorp P, Ohlin H, Sundkvist G et al. (1986). Effects of chlorpropamide and alcohol on aldehyde dehydrogenase activity and blood acetaldehyde concentration. Diabetes Res 3 (7): 369–372.

Johnson BA (2010). Medication treatment of different types of alcoholism. Am J Psychiatry 167 (6): 630–639.

Johnson BA, Ait-Daoud N, Seneviratne C et al. (2011). Pharmacogenetic approach at the serotonin transporter gene as a method of reducing the severity of alcohol drinking. Am J Psychiatry 168 (3): 265–275.

Johnson BA, Seneviratne C, Wang XQ et al. (2013). Determination of genotype combinations that can predict the outcome of the treatment of alcohol dependence using the 5-HT(3) antagonist ondansetron. Am J Psychiatry 170 (9): 1020–1031.

Jones AW (2010). Evidence-based survey of the elimination rates of ethanol from blood with applications in forensic casework. Forensic Sci Int 200 (1–3): 1–20.

Jones AW (2012). Pharmacokinetic and Pharmacodynamic interactions between alcohol and other drugs. Handbook of drug and interactions. Springer Science+Business media, New York, p. 87.

Jouanneau Y, Meyer C (2006). Purification and characterization of an arene is-dihydrodiol dehydrogenase endowed with broad substrate specificity toward polycyclic aromatic hydrocarbon dihydrodiols. Appl Environ Microbiol 72 (7): 4726–4734.

Junien C, Turleau C, de Grouchy J et al. (1980). Regional assignment of catalase (CAT) gene to band 11p13. Association with the aniridia-Wilms' tumor-Gonadoblastoma (WAGR) complex. Ann Genet 23 (3): 165–168.

Karamanakos PN, Pappas P, Boumba VA et al. (2007). Pharmaceutical agents known to produce disulfiram-like reaction: effects on hepatic ethanol metabolism and brain monoamines. Int J Toxicol 26 (5): 423–432.

Kaufman DW, Kelly JP, Wiholm BE et al. (1999). The risk of acute major upper gastrointestinal bleeding among users of aspirin and ibuprofen at various levels of alcohol consumption. Am J Gastroenterol 94 (11): 3189–3196.

Kechagias S, Jonsson KA, Norlander B et al. (1997). Low-dose aspirin decreases blood alcohol concentrations by delaying gastric emptying. Eur J Clin Pharmacol 53 (3–4): 241–246.

Kechagias S, Jonsson KA, Jones AW (1999). Impact of gastric emptying on the pharmacokinetics of ethanol as influenced by cisapride. Br J Clin Pharmacol 48 (5): 728–732.

Kenna GA, Zywiak WH, McGeary JE et al. (2009). A within-group design of nontreatment seeking 5-HTTLPR genotyped alcohol-dependent subjects receiving ondansetron and sertraline. Alcohol Clin Exp Res 33 (2): 315–323.

Kiefer F, Witt SH, Frank J et al. (2011). Involvement of the atrial natriuretic peptide transcription factor GATA4 in alcohol dependence, relapse risk and treatment response to acamprosate. Pharmacogenomics J 11 (5): 368–374.

Kim DJ, Choi IG, Park BL et al. (2008). Major genetic components underlying alcoholism in Korean population. Hum Mol Genet 17 (6): 854–858.

Kolble K (1993). Regional mapping of short tandem repeats on human chromosome 10: cytochrome P450 gene CYP2E, D10S196, D10S220, and D10S225. Genomics 18 (3): 702–704.

Kovatchev B, Breton M, Johnson B (2012). In silico models of alcohol dependence and treatment. Front Psychiatry3: 4. http://dx.doi.org/10.3389/fpsyt.2012.00004. eCollection 2012. PMID: 22347195.

Kranzler HR, Armeli S, Tennen H (2012). Post-treatment outcomes in a double-blind, randomized trial of sertraline for alcohol dependence. Alcohol Clin Exp Res 36 (4): 739–744.

Kranzler HR, Armeli S, Tennen H et al. (2013). 5-HTTLPR genotype and daily negative mood moderate the effects of sertraline on drinking intensity. Addict Biol 18: 1024–1031.

Lai CL, Li YP, Liu CM et al. (2013). Inhibition of human alcohol and aldehyde dehydrogenases by cimetidine and assessment of its effects on ethanol metabolism. Chem Biol Interact 202: 275–282.

Laniewska-Dunaj M, Jelski W, Orywal K et al. (2013). The activity of class I, II, III and IV of alcohol dehydrogenase (ADH) isoenzymes and aldehyde dehydrogenase (ALDH) in brain cancer. Neurochem Res 38 (7): 1517–1521.

Larsen FG, Jakobsen P, Knudsen J et al. (1993). Conversion of acitretin to etretinate in psoriatic patients is influenced by ethanol. J Invest Dermatol 100 (5): 623–627.

Larson JM, Tavakkoli A, Drane WE et al. (2010). Advantages of azithromycin over erythromycin in improving the gastric emptying half-time in adult patients with gastroparesis. Neurogastroenterol Motil 16 (4): 407–413.

Lee A, Kuo B (2010). Metoclopramide in the treatment of diabetic gastroparesis. Expert Rev Endocrinol Metab 5 (5): 653–662.

Lee SL, Chau GY, Yao CT et al. (2006). Functional assessment of human alcohol dehydrogenase family in ethanol metabolism: significance of first-pass metabolism. Alcohol Clin Exp Res 30 (7): 1132–1142.

Levitt MD, Li R, DeMaster EG et al. (1997). Use of measurements of ethanol absorption from stomach and intestine to assess human ethanol metabolism. Am J Physiol 273 (4 Pt 1): G951–G957.

Lewis SJ, Zuccolo L, Davey Smith G et al. (2012). Fetal alcohol exposure and IQ at age 8: evidence from a population-based birth-cohort study. PLoS One 7 (11): e49407.

Li TK, Bosron WF, Dafeldecker WP et al. (1977). Isolation of pi-alcohol dehydrogenase of human liver: is it a determinant of alcoholism? Proc Natl Acad Sci U S A 74 (10): 4378–4381.

Li Y, Zhang D, Jin W et al. (2006). Mitochondrial aldehyde dehydrogenase-2 (ALDH2) Glu504Lys polymorphism contributes to the variation in efficacy of sublingual nitroglycerin. J Clin Invest 116 (2): 506–511.

Li D, Zhao H, Gelernter J (2012). Strong protective effect of the aldehyde dehydrogenase gene (ALDH2) 504lys (*2) allele against alcoholism and alcohol-induced medical diseases in Asians. Hum Genet 131 (5): 725–737.

Litten RZ, Bradley AM, Moss HB (2010). Alcohol biomarkers in applied settings: recent advances and future research opportunities. Alcohol Clin Exp Res 34 (6): 955–967.

Liu Y, Yoshimura K, Hanaoka T et al. (2005). Association of habitual smoking and drinking with single nucleotide polymorphism (SNP) in 40 candidate genes: data from random population-based Japanese samples. J Hum Genet 50 (2): 62–68.

Lytton SD, Helander A, Zhang-Gouillon ZQ et al. (1999). Autoantibodies against cytochromes P-4502E1 and P-4503A in alcoholics. Mol Pharmacol 55 (2): 223–233.

Maenhout TM, De Buyzere ML, Delanghe JR (2013). Non-oxidative ethanol metabolites as a measure of alcohol intake. Clin Chim Acta 415: 322–329.

McBride OW, Umeno M, Gelboin HV et al. (1987). A Taq I polymorphism in the human P450IIE1 gene on chromosome 10 (CYP2E). Nucleic Acids Res 15 (23): 10071.

McCance-Katz EF, Sullivan LE, Nallani S (2010). Drug interactions of clinical importance among the opioids, methadone and buprenorphine, and other frequently prescribed medications: a review. Am J Addict 19 (1): 4–16.

McElhatton PR (1994). The effects of benzodiazepine use during pregnancy and lactation. Reprod Toxicol 8 (6): 461–475.

McKay JD, Truong T, Gaborieau V et al. (2011). A genome-wide association study of upper aerodigestive tract cancers conducted within the INHANCE consortium. PLoS Genet 7 (3): e1001333.

McMahon FG (1980). Disulfiram-like reaction to a cephalosporin. JAMA 243 (23): 2397.

Monroe ML, Doering PL (2001). Effect of common over-the-counter medications on blood alcohol levels. Ann Pharmacother 35 (7–8): 918–924.

Moore AA, Whiteman EJ, Ward KT (2007a). Risks of combined alcohol/medication use in older adults. Am J Geriatr Pharmacother 5 (1): 64–74.

Moore S, Montane-Jaime K, Shafe S et al. (2007b). Association of ALDH1 promoter polymorphisms with alcohol-related phenotypes in Trinidad and Tobago. J Stud Alcohol Drugs 68 (2): 192–196.

Murphy AT, Lake BG, Bernstein JR et al. (1998). Characterization of olanzapine (LY170053) in human liver slices by liquid chromatography/tandem mass spectrometry. J Mass Spectrom 33 (12): 1237–1245.

Nadif R, Mintz M, Jedlicka A et al. (2005). Association of CAT polymorphisms with catalase activity and exposure to environmental oxidative stimuli. Free Radic Res 39 (12): 1345–1350.

Neutel CI, Appel WC (2000). The effect of alcohol abuse on the risk of NSAID-related gastrointestinal events. Ann Epidemiol 10 (4): 246–250.

Nuutinen H, Lindros KO, Salaspuro M (1983). Determinants of blood acetaldehyde level during ethanol oxidation in chronic alcoholics. Alcohol Clin Exp Res 7 (2): 163–168.

Obach RS, Cox LM, Tremaine LM (2005). Sertraline is metabolized by multiple cytochrome P450 enzymes, monoamine oxidases, and glucuronyl transferases in human: an in vitro study. Drug Metab Dispos 33 (2): 262–270. Epub 2004 Nov 16.

Oneta CM, Simanowski UA, Martinez M et al. (1998). First pass metabolism of ethanol is strikingly influenced by the speed of gastric emptying. Gut 43 (5): 612–619.

Osier MV, Pakstis AJ, Soodyall H et al. (2002). A global perspective on genetic variation at the ADH genes reveals unusual patterns of linkage disequilibrium and diversity. Am J Hum Genet 71 (1): 84–99.

Oslin DW, Berrettini W, Kranzler HR et al. (2003). A functional polymorphism of the mu-opioid receptor gene is associated with naltrexone response in alcohol-dependent patients. Neuropsychopharmacology 28 (8): 1546–1552.

Osna NA, Donohue Jr TM (2013). CYP2E1-Catalyzed Alcohol Metabolism: Role of Oxidant Generation in Interferon Signaling, Antigen Presentation and Autophagy. Subcell Biochem 67: 177–197.

Otto JM, Hendershot CS, Collins SE et al. (2013). Association of the ALDH1A1*2 promoter polymorphism with alcohol phenotypes in young adults with or without ALDH2*2. Alcohol Clin Exp Res 37 (1): 164–169.

Park BL, Kim JW, Cheong HS et al. (2013). Extended genetic effects of ADH cluster genes on the risk of alcohol dependence: from GWAS to replication. Hum Genet 132 (6): 657–668.

Pierucci-Lagha A, Covault J, Feinn R et al. (2005). GABRA2 alleles moderate the subjective effects of alcohol, which are attenuated by finasteride. Neuropsychopharmacology 30 (6): 1193–1203.

Pizon AF, Becker CE, Bikin D (2007). The clinical significance of variations in ethanol toxicokinetics. J Med Toxicol 3 (2): 63–72.

Porter SJ, Somogyi AA, White JM (2000). Kinetics and inhibition of the formation of 6beta-naltrexol from naltrexone in human liver cytosol. Br J Clin Pharmacol 50 (5): 465–471.

Pragst F, Yegles M (2008). Determination of fatty acid ethyl esters (FAEE) and ethyl glucuronide (EtG) in hair: a promising way for retrospective detection of alcohol abuse during pregnancy? Ther Drug Monit 30 (2): 255–263.

Prescott LF (2000). Paracetamol, alcohol and the liver. Br J Clin Pharmacol 49 (4): 291–301.

Qrunfleh AM, Alazizi A, Sari Y (2013). Ceftriaxone, a beta-lactam antibiotic, attenuates relapse-like ethanol-drinking behavior in alcohol-preferring rats. J Psychopharmacol 27: 541–549.

Quigley EM (2011). Cisapride: what can we learn from the rise and fall of a prokinetic? J Dig Dis 12 (3): 147–156.

Ray LA, Bujarski S, Chin PF et al. (2012). Pharmaco-genetics of naltrexone in asian americans: a randomized placebo-controlled laboratory study. Neuropsychopharmacology 37 (2): 445–455.

Salmela KS, Kessova IG, Tsyrlov IB et al. (1998). Respective roles of human cytochrome P-4502E1, 1A2, and 3A4 in the hepatic microsomal ethanol oxidizing system. Alcohol Clin Exp Res 22 (9): 2125–2132.

Sari Y, Sakai M, Weedman JM et al. (2011). Ceftriaxone, a beta-lactam antibiotic, reduces ethanol consumption in alcohol-preferring rats. Alcohol Alcohol 46 (3): 239–246.

Secretan B, Straif K, Baan R et al. (2009). A review of human carcinogens–Part E: tobacco, areca nut, alcohol, coal smoke, and salted fish. Lancet Oncol 10 (11): 1033–1034.

Shenoy C (2006). Metformin-associated lactic acidosis precipitated by acute renal failure. Am J Med Sci 331 (1): 55–57.

Soderberg BL, Salem RO, Best CA et al. (2003). Fatty acid ethyl esters. Ethanol metabolites that reflect ethanol intake. Am J Clin Pathol 119 (Suppl.): S94–S99.

Song BJ, Abdelmegeed MA, Yoo SH et al. (2011). Post-translational modifications of mitochondrial aldehyde dehydrogenase and biomedical implications. J Proteomics 74 (12): 2691–2702.

Spence JP, Liang T, Eriksson CJ et al. (2003). Evaluation of aldehyde dehydrogenase 1 promoter polymorphisms identified in human populations. Alcohol Clin Exp Res 27 (9): 1389–1394.

Stewart SH, Koch DG, Burgess DM et al. (2013). Sensitivity and specificity of urinary ethyl glucuronide and ethyl sulfate in liver disease patients. Alcohol Clin Exp Res 37 (1): 150–155.

Tan A, Sun J, Xia N et al. (2012). A genome-wide association and gene-environment interaction study for serum triglycerides levels in a healthy Chinese male population. Hum Mol Genet 21 (7): 1658–1664.

Tanaka E (2002). Toxicological interactions between alcohol and benzodiazepines. J Toxicol Clin Toxicol 40 (1): 69–75.

Umeno M, McBride OW, Yang CS et al. (1988). Human ethanol-inducible P450IIE1: complete gene sequence, promoter characterization, chromosome mapping, and cDNA-directed expression. Biochemistry 27 (25): 9006–9013.

Urbain JL, Vantrappen G, Janssens J et al. (1990). Intravenous erythromycin dramatically accelerates gastric emptying in gastroparesis diabeticorum and normals and abolishes the emptying discrimination between solids and liquids. J Nucl Med 31 (9): 1490–1493.

Vasiliou V, Malamas M, Marselos M (1986). The mechanism of alcohol intolerance produced by various therapeutic agents. Acta Pharmacol Toxicol (Copenh) 58 (5): 305–310.

Vena GA, Cassano N (2012). The effects of alcohol on the metabolism and toxicology of anti-psoriasis drugs. Expert Opin Drug Metab Toxicol 8 (8): 959–972.

Vodicka P, Soucek P, Tates AD et al. (2001). Association between genetic polymorphisms and biomarkers in styrene-exposed workers. Mutat Res 482 (1–2): 89–103.

Vonghia L, Leggio L, Ferrulli A et al. (2008). Acute alcohol intoxication. Eur J Intern Med 19 (8): 561–567.

Wang SM, Zhu AP, Li D et al. (2009). Frequencies of genotypes and alleles of the functional SNPs in CYP2C19 and CYP2E1 in mainland Chinese Kazakh, Uygur and Han populations. J Hum Genet 54 (6): 372–375.

Weathermon R, Crabb DW (1999). Alcohol and medication interactions. Alcohol Res Health 23 (1): 40–54.

Weinberg DS, Burnham D, Berlin JA (1998). Effect of histamine-2 receptor antagonists on blood alcohol levels: a meta-analysis. J Gen Intern Med 13 (9): 594–599.

White AM, Hingson RW, Pan IJ et al. (2011). Hospitalizations for alcohol and drug overdoses in young adults ages 18-24 in the United States, 1999-2008: results from the Nationwide Inpatient Sample. J Stud Alcohol Drugs 72 (5): 774–786.

Wu M, Chang SC, Kampman E et al. (2013). Single nucleotide polymorphisms of ADH1B, ADH1C and ALDH2 genes and esophageal cancer: a population-based case-control study in China. Int J Cancer 132 (8): 1868–1877.

Xiao Q, Weiner H, Johnston T et al. (1995). The aldehyde dehydrogenase ALDH2*2 allele exhibits dominance over ALDH2*1 in transduced HeLa cells. J Clin Invest 96 (5): 2180–2186.

Xiao Q, Weiner H, Crabb DW (1996). The mutation in the mitochondrial aldehyde dehydrogenase (ALDH2) gene responsible for alcohol-induced flushing increases turnover of the enzyme tetramers in a dominant fashion. J Clin Invest 98 (9): 2027–2032.

Zavras AI, Wu T, Laskaris G et al. (2002). Interaction between a single nucleotide polymorphism in the alcohol dehydrogenase 3 gene, alcohol consumption and oral cancer risk. Int J Cancer 97 (4): 526–530.

Zimmerman HJ, Maddrey WC (1995). Acetaminophen (paracetamol) hepatotoxicity with regular intake of alcohol: analysis of instances of therapeutic misadventure. Hepatology 22 (3): 767–773.

Zuo L, Gelernter J, Kranzler HR et al. (2010). ADH1A variation predisposes to personality traits and substance dependence. Am J Med Genet B Neuropsychiatr Genet 153B (2): 376–386.

Handbook of Clinical Neurology, Vol. 125 (3rd series)
Alcohol and the Nervous System
E.V. Sullivan and A. Pfefferbaum, Editors

Chapter 32

Genetics of alcoholism

HOWARD J. EDENBERG[1]* AND TATIANA FOROUD[2]

[1]*Department of Biochemistry and Molecular Biology, Indiana University School of Medicine, Indianapolis, IN, USA*

[2]*Department of Medical and Molecular Genetics, Indiana University School of Medicine, Indianapolis, IN, USA*

GENETIC VARIATIONS CONTRIBUTE TO RISK OF ALCOHOLISM

Alcohol use disorders (AUD), including alcohol dependence (alcoholism) and alcohol abuse, are defined on the basis of behavioral characteristics as patterns of drinking that lead to significant impairment or distress. Many factors combine to result in an AUD. While some risk factors are environmental (including availability of alcohol, peer use of alcohol, and stressors), it is clear that a substantial portion of the difference in risk between individuals has a genetic basis. We know this from several converging lines of evidence. Alcoholism has a tendency to run in families, although it is important to note that not only genes but also substantial environmental influences are shared within families. Adoption studies can partially disentangle familial environment from genetics, and they show that the risk for alcoholism in adoptees is better correlated with their biologic parents (the source of genes) than with their adoptive parents (the environmental influences) (Heath, 1995; Sigvardsson et al., 1996; Bohman et al., 1981; Cloninger et al., 1981). Twin studies show that monozygotic twins are more alike in risk for alcoholism than fraternal twins, also pointing to a substantial role for genes. Twin and family studies suggest that approximately 45-65% of the liability is genetic (Pickens et al., 1991; Kendler et al., 1994; Heath et al., 1997, Prescott and Kendler, 1999). The fact that a monozygotic twin of an alcoholic is often not alcoholic demonstrates that, although genes affect risk, other non-genetic factors also contribute to the susceptibility toward alcohol dependence.

Animal models also provide evidence to support a genetic contribution to alcohol dependence and related phenotypes. Many lines of mice and rats have been selectively bred for traits associated with alcoholism, including alcohol preference, consumption, sensitivity to withdrawal, and sensitivity to sedating effects. That mouse and rat lines can be bred for these traits directly demonstrates that there is a genetic basis for them (McBride and Li, 1998; Foroud et al., 2010). The animal studies also demonstrate that different alcoholism-related traits are due to different collections of genetic variants, because lines bred for one trait may not show an effect on the other traits (Blednov et al., 2012; Crabbe et al., 2012).

Although no single line of evidence is itself conclusive, given the multifaceted nature of the disease, the convergence of evidence overwhelmingly indicates that the risk for alcoholism and related traits is influenced in part by genetic differences. This is emphatically not to say that any one genetic variation or collection of variations leads inevitably to alcoholism. There is no "gene for alcoholism." Rather, alcoholism is a complex genetic trait in which variations in many genes plus variations in the environment (including social factors) affect risk for the disease. This is one key reason that the search for genetic differences that affect risk for alcoholism is difficult. Additional complexity arises from the possibility that genetic variations may affect risk in a particular environment, which can differ among groups being studied.

DEFINING THE PHENOTYPE

A key consideration when designing a study to identify risk factors for AUD is evaluating what is the best phenotype to analyze. A phenotype could be a diagnosis, a combination of diagnoses (e.g., alcohol dependence comorbid with depression), a quantitative trait related to AUD, or an endophenotype such as a neurobiologic

*Correspondence to: Howard J. Edenberg, Ph.D., Department of Biochemistry and Molecular Biology, Indiana University School of Medicine, 635 Barnhill Drive, MS4063, Indianapolis, IN 46202-5122, USA. Tel: +1-317-274-2353, E-mail: edenberg@iu.edu

trait more common in alcoholics. There are advantages and disadvantages to each when designing a study to identify genetic factors contributing to AUD.

The *Diagnostic and Statistical Manual* of the American Psychiatric Association (DSM-IV-TR, henceforth DSM-IV) (American Psychiatric Association, 2000) established criteria for both alcohol dependence (alcoholism) and alcohol abuse. Studies based on diagnosis compare those who meet or do not meet criteria for alcohol dependence. It is important to recognize that AUDs, and substance use disorders (SUDs) in general, are heterogeneous disorders. Alcohol (and substance) dependence is defined by the presence of three or more of a set of seven criteria. Alcohol abuse is defined by the presence of two out of four criteria in the absence of meeting criteria for dependence. One can theoretically meet DSM-IV criteria for alcohol dependence 99 ways: there are 35 different ways to reach three criteria out of seven, 35 more ways to pick four of seven, etc. DSM-5 will no longer separate alcohol dependence from abuse but rather will define AUD on the basis of meeting at least two of 11 criteria, 10 drawn from the DSM-IV definitions of abuse and dependence, with the addition of craving (American Psychiatric Association, 2013). DSM-5 AUDs are classified into moderate (two or three criteria) or severe (four or more criteria), leaving even more possible combinations. The heterogeneity of the diagnoses, shared with other SUDs, is a challenge for both the clinician and the geneticist. The different combinations of criteria may reflect underlying heterogeneity in both the genetic and environmental contributions to the disease as well as in its course (Kendler et al., 2012). Some studies have been based on meeting diagnostic criteria for any SUD; the hypothesis underlying such studies is that there are common genetic factors that underlie addiction to different substances. In a study based upon diagnosis, there is some concern about how to classify young individuals who have not passed through the age of risk, or who meet some criteria for alcohol dependence, but not enough to be diagnosed with AUD.

Another strategy that has been used to explore the genetics of AUD is to explore quantitative phenotypes that are related to the diagnosis of AUD, but which provide greater variability among individuals. This strategy classifies individuals along a continuum rather than classifying them into affected and unaffected groups that might be quite heterogeneous, and allows incorporation of more subjects into the analysis. Among the phenotypes that have been explored is the maximum number of drinks consumed within a 24-hour period, and quantity/frequency measures of alcohol consumption in a typical week or other period. This strategy can increase the power to detect variants.

A third approach also uses a quantitative phenotype but focuses on those which are defined as an endophenotype: a measurable, heritable trait related to the disease, generally found at higher frequency even in family members who do not have the disease than in the general population. Neurophysiologic endophenotypes used in the study of alcoholism include event-related potentials and event-related oscillations (Porjesz et al., 2005). This approach has the advantage of potentially placing individuals who may not even drink alcohol along a quantitative continuum that can further improve the power of genetic studies.

GENETIC APPROACHES FOR IDENTIFYING VARIANTS THAT AFFECT RISK FOR ALCOHOLISM

Given the strong evidence that part of the risk for alcoholism is genetic, there have been significant efforts to find specific genes in which variations affect the risk. Knowing genes and pathways that affect risk should aid in better diagnosis, the search for drugs to treat the disease, and better-matching treatments to individuals by examining subgroups of patients who share specific pathways of vulnerability. But the heterogeneity of the disease, the likelihood that many genes (perhaps hundreds) can contribute to risk, and the heterogeneity of the environment make identification of relevant genes a very difficult task.

Several different approaches have been taken. One, candidate gene studies, relies upon *a priori* hypotheses that a particular gene or pathway is involved. Linkage studies, following the inheritance of regions of the chromosome to determine which are segregating with the disease in a set of families, can point to regions of the genome in which risk factors lie; however, extensive additional studies are needed to narrow the search to only a few genes within the broad interval. Another approach, genomewide association studies (GWAS), tests hundreds of thousands to millions of markers across the genome to determine which are associated with the disease. All three have led to interesting, yet incomplete, results.

CANDIDATE GENE STUDIES

Genes involved in alcohol metabolism

Early candidate gene studies were focused on genes that encode enzymes involved in ethanol metabolism. This was a logical strategy, because the effects of any drug are in part determined by its metabolism and clearance. Ethanol metabolism follows a short pathway (Fig. 32.1): oxidation to acetaldehyde, followed by further oxidation of the acetaldehyde to acetate (the same compound to

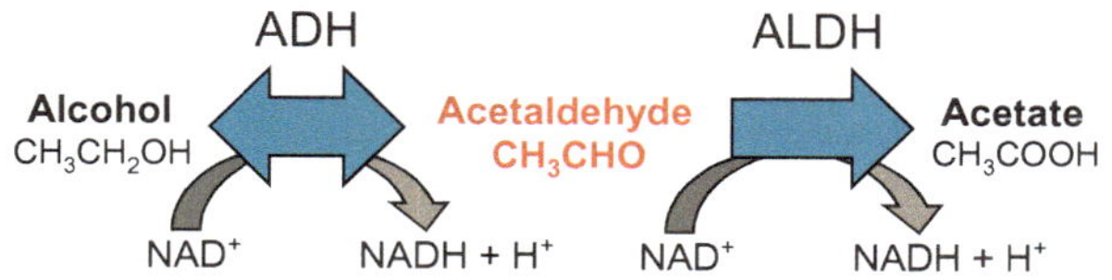

Fig. 32.1. Alcohol metabolism. Alcohol (ethanol) is oxidized to acetaldehyde, a toxic intermediate, in a reversible reaction catalyzed primarily by one of several alcohol dehydrogenases (ADH), with the reduction of NAD^+ to NADH. The acetaldehyde is irreversibly oxidized to acetate by an aldehyde dehydrogenase (ALDH), reducing another NAD^+; the primary aldehyde dehydrogenase in mammals that acts on acetaldehyde is the mitochondrial ALDH2.

Table 32.1

Kinetic constants of human alcohol dehydrogenases*

Gene	Other name†	Enzyme‡	K_m (mM)	V_{max} (/min)
ADH1A	*ADH1*	αα	4	20
*ADH1B*1*	*ADH2*1*	β1β1 (Arg48,Arg370)	0.05	4
*ADH1B*2*	*ADH2*2*	β2β2 (His48,Arg370)	0.9	350
*ADH1B*3*	*ADH2*3*	β3β3 (Arg48,Cys370)	40	300
*ADH1C*1*	*ADH3*1*	γ1γ1 (Arg272,Ile350)	1	90
*ADH1C*2*	*ADH3*2*	γ2γ2 (Gln272,Val350)	0.6	40
ADH4	*ADH2*†	ππ	30	20
ADH5	*ADH3*†	ξξ	>1000	100
ADH7	*ADH4*†	σσ	30	1800

*For oxidation of ethanol. Data from Hurley et al. (2002).
†In the older literature, *ADH1A*, *ADH1B*, and *ADH1C* were known as *ADH1*, *ADH2*, and *ADH3*, respectively. Some groups currently use non-standard names for *ADH4*, *ADH5*, and *ADH7* that can cause confusion, and are shown here to alert readers.
‡Alcohol dehydrogenase enzymes function as homo- and heterodimers; data are for homodimers.

which other foods are converted) (Hurley and Edenberg, 2012). The key enzymes involved in ethanol metabolism are alcohol dehydrogenases, which catalyze the bulk of the oxidation to acetaldehyde, and aldehyde dehydrogenases, which oxidize acetaldehyde to acetate. Cytochrome P450 2E1 (CYP2E1) and catalase can also oxidize ethanol to acetaldehyde, but normally make only a small contribution to ethanol elimination, although CYP2E1 may contribute to some toxic effects of ethanol by generating reactive oxygen species, and is a site at which ethanol and drug metabolism interact. Most (perhaps 90%) of the oxidation of ethanol occurs in the liver, although some occurs in the stomach (Hurley et al., 2002).

There is a family of seven closely related alcohol dehydrogenases, encoded by genes that are clustered in a small region of chromosome 4q. ADH1B is the major cytosolic alcohol dehydrogenase in liver and carries out most of the ethanol oxidation; enzymes encoded by *ADH1A* and *ADH1C* contribute at low concentrations of ethanol, and ADH4 contributes an increasing proportion as the concentration rises (reviewed by Hurley et al., 2002; Edenberg and Bosron, 2010; Hurley and Edenberg, 2012). Early work on purified enzymes demonstrated that there are common variants of ADH1B and ADH1C that differed in activity. ADH1B*2 (ADH1B*48His), relatively common in East Asians but rare elsewhere, metabolizes ethanol at a much faster rate than the ADH1B*1 found in most other people. ADH1B*3 (ADH1B*370Cys) is relatively common in many African populations, but rare elsewhere; it also metabolized ethanol at a much faster rate than ADH1B*1 (Table 32.1). A variant form of ADH1C, ADH1C*2 (ADH1C*Gln272Val350) oxidizes ethanol at a somewhat slower rate than ADH1C*1 (Arg272Ile350).

Early studies of populations from Taiwan demonstrated that *ADH1B*2* is strongly protective against alcohol dependence (Thomasson et al., 1991, 1994) (Fig. 32.2). The odds ratio for someone carrying the *ADH1B*2* allele (rs1229984) in heterozygous form was in the range of 0.2-0.4, and it was even lower for someone homozygous for that allele (Chen et al., 1999; Whitfield, 2002; Luczak et al., 2006; Li et al., 2012). The low frequency of *ADH1B*2* in most populations from Europe or Africa made it difficult to detect a significant protective effect. However, a recent large study on Americans of European descent demonstrated that the odds ratio was very similar, 0.34 (Bierut et al., 2012). The effect of *ADH1C*2* is smaller and has been harder to detect, but there is evidence that *ADH1C* variants are protective (Thomasson et al., 1991; Treutlein

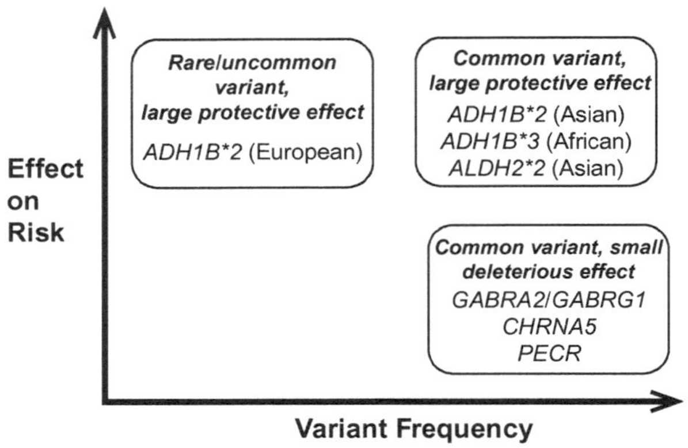

Fig. 32.2. Allelic spectrum of risk factors contributing to alcohol dependence. Variants contributing to the risk of alcohol dependence fall along an allelic spectrum from rare to common. The effect of these variants varies from small effects typically found through genomewide association studies to large effects of variants such as ADH1B*2 and ALDH2*2; these protective variants differ greatly in allele frequency in different populations.

et al., 2009; Frank et al., 2012; Biernacka et al., 2013). There have been fewer studies of *ADH1B*3*, but a protective effect has been found (Edenberg et al., 2006). Other *ADH* and *ALDH* genes have also been shown to affect risk for alcoholism, particularly *ADH4* (below), but the effects are not as strong.

Variations in ALDH2, the mitochondrial acetaldehyde dehydrogenase that carries out most of the oxidation of acetaldehyde to acetate, were also found. ALDH2 functions as a homotetramer, and a variant form of the enzyme, *ALDH2*2* (ALDH2*504 K; rs671), found nearly exclusively in people from East Asia, leads to a nearly inactive enzyme even when present along with *ALDH2*1*, the active form (Crabb et al., 1989; Xiao et al., 1996). Thus even heterozygotes for *ALDH2*2* have very low levels of mitochondrial aldehyde dehydrogenase activity and build up acetaldehyde when they consume ethanol. This leads to the "Asian flush" reaction, which includes prominent flushing, tachycardia, and nausea. Needless to say, this reaction is highly aversive, and tends to limit the amount of alcohol consumed. It is essentially equivalent to having disulfiram on board all of the time. The earliest studies showed a strong protective effect of the *ALDH2*2* allele against alcoholism (Thomasson et al., 1991), and this finding is highly reproducible (Chen et al., 1999; Whitfield, 2002; Luczak et al., 2006, Eng et al., 2007; Hurley and Edenberg, 2012; Li et al., 2012) (Fig. 32.2). Homozygotes usually get severely sick with one drink or less, and are essentially completely protected against alcoholism. As strong as the reaction among heterozygotes is, the degree of protection is still susceptible to environmental and social factors. This was dramatically demonstrated by the reduction of the protective effect of a single copy of *ALDH2*2* between 1970 and 1992 in Japan (Higuchi, 1994), a period during which social pressure for drinking as part of the business culture increased. *ADH1B* and *ALDH2* have by far the strongest known effects on risk for alcoholism of any genes yet identified.

Nicotinic acetylcholine receptors

Nicotine and alcohol dependence are frequently comorbid disorders, suggesting that there may be common pathways contributing to the risk of both disorders. There is consistent evidence that variations in the cluster of nicotinic receptors on chromosome 15 (*CHRNA5, CHRNA3, CHRNB4*) are important risk factors for nicotine addiction as well as other smoking-related phenotypes such as cigarettes per day (Saccone et al., 2010; Munafo et al., 2012). Analyses, primarily in populations of European ancestry, have detected association between phenotypes related to nicotine addiction and a missense single-nucleotide polymorphism (SNP) in the *CHRNA5* gene that changes a conserved amino acid from aspartic acid to asparagine (D398N) (Wang et al., 2009a).

There is evidence for additional risk factors within the 3'UTR of *CHRNA3* that also contribute to nicotine dependence and cigarettes per day (Bierut et al., 2008). Because of the frequent comorbidity of smoking and alcohol dependence, this gene cluster has been studied for possible contribution to risk of alcohol dependence. Wang et al. (2009b) reported in the Collaborative Study of the Genetics of Alcoholism (COGA) sample that eight variants spanning 52 kb in the *CHRNA5–CHRNA3* gene cluster were significantly associated with alcohol dependence, although the D398N mutation in *CHRNA5* was not. Subsequent analyses suggested that the association with these SNPs was independent of smoking status. In another study, two SNPs in *CHRNA5* were associated with alcohol dependence, but only in African Americans (Sherva et al., 2010); the same SNPs were not associated in the European Americans. An SNP in high-linkage disequilibrium with the missense mutation *CHRNA5*D398N* is associated with the level of response to alcohol, an alcoholism-related endophenotype (Joslyn et al., 2008). Thus, there is consistent evidence that variation within the nicotinic receptors contributes to alcohol-related phenotypes (Fig. 32.2), with some evidence that different variations contribute to different phenotypes, and perhaps in different ethnic groups, but these observations must be replicated in additional samples.

Candidate gene studies have implicated several genes in the opioid system, including *OPRK1* and *PDYN*,

OPRM1, *OPRD1*, and *POMC*, in risk for alcohol dependence or other substance dependence (Luo et al., 2003; Zhang et al., 2006, 2008, 2009; Edenberg et al., 2008a). There has been much interest and controversy over the years about the potential role of variants in the *DRD2* gene. The most recent evidence suggests variations in the nearby *ANKK1* gene and *TTC12* are associated with alcohol dependence (Dick et al., 2007; Munafo et al., 2007; Yang et al., 2007, 2008). These candidate genes have not all been consistently replicated, possibly in part because their effects are small and may differ in different populations. There has been nominal support, however, for some of the candidate genes in GWAS studies.

LINKAGE STUDIES FOLLOWED BY POSITIONAL CANDIDATE GENE ANALYSES

Linkage studies published in 1998 identified several chromosomal regions that appeared to contain genes affecting risk for alcoholism. COGA analyzed 987 individuals selected from 105 families that had many members diagnosed with alcohol dependence (Reich et al., 1998). COGA found evidence for linkage in regions on chromosomes 1, 2, 4, and 7. A sib-pair study of 152 subjects from a Southwest American Indian tribe provided evidence for linkage on chromosomes 4 and 11 (Long et al., 1998). The regions identified by linkage studies are very large, containing hundreds to thousands of genes. In a complex disease such as alcoholism, recombination-based approaches could not narrow the region, so researchers chose to analyze the association of candidate genes within these large linkage regions. A comprehensive review of linkage studies can be found in Treutlein et al. (2009); we will illustrate some highlights.

GABA-A receptor genes

A cluster of four genes encoding subunits of the type A receptor for gamma-amino butyric acid (GABA), the major inhibitory neurotransmitter in mammalian brains, was within the linkage peak on chromosome 4p identified by both the COGA group (Reich et al., 1998) and studies in a Southwest American Indian tribe (Long et al., 1998). These GABA-A receptors were strong candidate genes; GABA-A receptors are sensitive to ethanol. SNPs across this set of four genes were tested in the COGA sample, and many SNPs within and adjacent to *GABRA2* significantly affected risk for alcoholism (Edenberg et al., 2004). *GABRA2* encodes the $\alpha 2$ subunit of the GABA receptor, a heteromeric ligand-gated chloride channel. None of the associated SNPs, however, was obviously functional. This association has been replicated in samples of European (Covault et al., 2004; Lappalainen et al., 2005; Fehr et al., 2006; Soyka et al., 2008; Villafuerte et al., 2012) and African ancestry (Ittiwut et al., 2012). The association is strongest in alcoholics with early-onset, high-severity, or comorbid drug dependence (Agrawal et al., 2006; Fehr et al., 2006). The association may extend beyond *GABRA2* and include the adjacent *GABRG1* gene (Covault et al., 2008; Enoch et al., 2009); there is evidence for distinct effects in each gene (Covault et al., 2008; Ittiwut et al., 2012) and also for long-range haplotypes that contribute to the risk of alcohol dependence (Enoch et al., 2009; Ittiwut et al., 2012). *Gabra2* deletion in mice resulted in hypersensitivity to sedative and ataxic effects of ethanol but did not affect self-administration in an operant setting (Dixon et al., 2012). A point mutation introduced into the mouse *Gabra2* gene that makes it insensitive to ethanol potentiation led to many changes in the effects of ethanol, including loss of motor stimulant effects (Blednov et al., 2011). *GABRA2* is also associated with two endophenotypes, the power in the β frequency range (12-30 Hz) of the electroencephalogram (EEG) (Edenberg et al., 2004), which reflects an activated state of neural networks, and excess EEG fast activity (Lydall et al., 2011).

Given the association of *GABRA2* with alcoholism, other GABA-receptor genes have been studied as candidate genes. There is evidence for the involvement of *GABRG3, GABRA1, GABRA5, GABRR1,* and *GABRR2* in risk for alcoholism or traits closely related to it (Song et al., 2003; Dick et al., 2004, 2006; Xuei et al., 2010). There is evidence that a GABRA5 inverse agonist inhibits ethanol drinking in rhesus monkeys (Ruedi-Bettschen et al., 2013). A study comparing mice lacking six different GABA-A receptors showed that only *Gabra2* deletion reduced conditioned taste aversion to ethanol, whereas both *Gabra2* and *Gabra3* deletions slowed recovery from ethanol incoordination (Blednov et al., 2013).

Genes on chromosome 4

The broad linkage peak on chromosome 4q, identified in several studies (Long et al., 1998; Reich et al., 1998; Williams et al., 1999; Saccone et al., 2000; Prescott et al., 2006), includes the region containing the cluster of *ADH* genes. In the European American subjects from COGA, *ADH4* showed the strongest association with alcoholism (Edenberg et al., 2006). This association has been replicated in other samples (Whitfield, 2002; Guindalini et al., 2005; Luo et al., 2005, 2006, 2007; Kuo et al., 2008; Hurley and Edenberg, 2012). Among the African American subset of the COGA sample, the *ADH1B*3* allele was significantly associated with alcohol dependence (Edenberg et al., 2006); this was also found for Native American Mission Indians (Wall et al., 2003). Associations between alcohol dependence, drinking, and other alcoholism-related traits have been found for the other *ADH* genes as well (Whitfield, 2002; Edenberg et al., 2006; Luo et al., 2007; Birley et al., 2008, 2009; Kuo et al., 2008; Macgregor et al., 2009; van Beek et al., 2010; Frank et al., 2012; Hurley and Edenberg, 2012).

Pursuing the hypothesis that a broad linkage peak reflects the combined contributions of several genes, additional genes within the broad linkage peak on chromosome 4 were studied in COGA. There was evidence that *NFKB1*, encoding a subunit of the transcription factor NF-κB, was associated with alcoholism (Edenberg et al., 2008b). There was also evidence that *SNCA*, which encodes α-synuclein, was associated with alcohol craving (Foroud et al., 2007), and that *TACR3*, encoding a tachykinin (neurokinin B) receptor, was associated with severe alcohol dependence and codependence with cocaine (Foroud et al., 2008). *DKK2* (dickkopf homolog 2) was found to be associated with alcohol dependence in the Irish Affected Sib Pair Study of Alcohol Dependence and confirmed in the COGA sample (Kalsi et al., 2010).

GENOMEWIDE ASSOCIATION STUDIES TO IDENTIFY COMMON VARIANTS

The analysis of hundreds of thousands to millions of SNPs has led to the identification of many genes and SNPs associated with the risk for common disorders such as diabetes, cardiovascular disease, and many types of cancer. Initial studies that included up to a few thousand samples were typically underpowered. As meta-analysis has been used to combine results from multiple studies, sample sizes attained for many common disorders now may reach tens of thousands of individuals, and results have become robust. Such sample sizes, however, are still a challenge to obtain in studies of alcohol dependence. Why is it so difficult to accrue large numbers of subjects with alcohol dependence when this disorder is so common? The answer appears in part to lie in the time needed for careful and consistent assessment of individuals to determine whether they meet diagnostic criteria for alcohol dependence, thereby reducing the rate of subject accrual, limiting the potential sample size of a study.

Despite these challenges, there have been several studies that assessed the role of common variation in the risk of alcohol dependence using a genomewide approach (Rietschel and Treutlein, 2013). Because several GWAS datasets are publicly available through dbGaP, some of the studies have used overlapping samples, so the results are not independent. A summary review published by Rietschel and Treutlein (2013) provides comprehensive details of each study and its findings. Therefore, rather than review all GWAS reports here, we summarize the results from several of the larger independent studies that illustrate some key points.

Treutlein et al. (2009) focused on males with early-onset alcohol dependence who had been hospitalized for the condition, a combination that indicates a high level of severity of AUD. Using 476 patients and 1358 controls, they detected association to a region on chromosome 2q35. A joint analysis of this sample and a replication sample detected genomewide evidence of association with SNPs in an intergenic (between two genes) region close to the peroxisomal *trans*-2-enoyl-CoA reductase (*PECR*) gene, which is involved in fatty-acid metabolism. Later analysis of an expanded sample identified genomewide significance in rs1789891, an SNP in complete linkage disequilibrium with the functional SNP in *ADH1C*, described above (Frank et al., 2012).

A study by COGA, based on cases drawn from families with multiple alcoholic members and on community controls, identified several promising genes, including bobby sox homolog (*BBX*) and a cluster of genes on chromosome 11, but none reached genomewide significance levels (Edenberg et al., 2010). Another study that included COGA subjects along with alcohol-dependent subjects from studies that recruited based upon cocaine or nicotine also found interesting genes, including PBX/knotted 1 homeobox 2 (*PKNOX2*) and a metabotropic glutamate receptor (*GRM5*). These genes did not reach genomewide significance (Bierut et al., 2010). Efforts are currently ongoing to combine all of these studies and several others into a larger meta-analysis to identify genes, SNPs, or both associated with the risk of alcohol dependence.

Some studies used quantitative phenotypes related to alcohol dependence. A GWAS of several quantitative phenotypes in an Australian sample revealed genes relevant to a phenotype based on heavy drinking, although again without reaching genomewide significance (Heath et al., 2011). Analysis of a factor score related to alcohol dependence symptoms in a sample originally ascertained for schizophrenia indicated that the calcium channel gene *KCNMA1* gave suggestive evidence for involvement (Kendler et al., 2011).

It has proven relatively easier to accrue larger samples sizes for AUD-related phenotypes other than diagnosis, that do not require so much effort to collect, for example, the quantity and frequency of alcohol consumed. Even such a seemingly simple question can be challenging, given that these values typically change across an individual's lifetime and different studies gather data on different periods (Agrawal et al., 2011).

A large meta-analysis of alcohol consumption (g/day/kg body weight) in 12 European population-based samples detected genomewide significant evidence ($P = 4 \times 10^{-8}$) of association with SNPs in the autism susceptibility candidate 2 gene (*AUTS2*) (Schumann et al., 2011). The molecular function of *AUTS2* is not known.

GWAS studies have also explored other alcohol-related phenotypes. In the COGA sample, a family-based GWAS detected genomewide evidence of association with SNPs in *KCNJ6* and frontal theta-band event-related oscillations (theta ERO) (Kang et al., 2012). *KCNJ6* encodes a potassium inward rectifier channel, GIRK2, whose activation contributes to slow inhibitory postsynaptic potentials that modulate neuronal excitability, and therefore influences neuronal networks (Luscher et al., 1997). Unfortunately, few studies collect electrophysiology, so finding large enough independent studies in which to replicate this finding may prove difficult.

Most of these studies were underpowered and did not detect genes that reached current standards for genomewide significance. However, many of the studies provided at least nominal significance for genes that were significant in candidate gene studies and for genes found interesting in other GWAS studies. This suggests that some of those genes represent real findings, and with a large increase in sample size might reach genomewide significance. Therefore, the next critical direction for studies of alcohol dependence and alcohol-related phenotypes is to develop ways to combine data across studies in more powerful meta-analyses. Several have begun, although the critical sample sizes, likely to be over 20 000 cases and at least that number of controls, will require either the inclusion of more studies or the use of non-diagnostic criteria for alcohol dependence.

RARE VARIANTS

Most studies to date have focused on common genetic variations, generally defined as having a minor allele frequency of at least 5%. Recently, more attention has been focused upon the potential contributions of uncommon and rare variants to risk of alcohol dependence or alcohol-related phenotypes. One approach to identify rare variants of greater effect on disease risk is to resequence the exome (the sum of the coding and nearby flanking regions of known genes) to discover more variants. Whole-exome sequencing has revealed a very large number of new variations, and because they are targeted to the transcribed regions of the genome, it is easier to predict the effects of such variants. For example, they could change the amino acid at particular sites in the encoded protein so as to result in major (or only minor) changes in protein function. Some of this variation is being cataloged across populations, but it is also important to look at cases affected by a particular disease, such as AUD. Another related approach is to resequence candidate genes or regions in large numbers of cases and controls to catalog all the variants found in the two groups and test whether the burden or number of rare variants in cases is greater than in controls.

Different strategies can be used to increase the likelihood that a rare variant of larger effect on disease risk may be identified. Focusing on alcohol-dependent individuals from families densely affected with alcoholism may be advantageous, because these families are more likely to segregate such a risk variant. An additional benefit of examining families is the ability to test for segregation of the rare variant among family members with alcohol dependence. Another approach to increase the likelihood of a rare variant is to explore individuals and families who have an extreme phenotype. This might be defined as dependence on multiple substances or might include extreme values for a critical endophenotype.

When rare variants in particular genes that may affect disease risk are identified, additional studies are required to further delineate the role of this gene. Once exonic variants are identified and thought to contribute to disease risk, it is important to examine other regions around the gene, including regulatory regions that are not transcribed, which may harbor additional variants contributing to risk. Then, custom microarrays can be developed that allow much larger samples of cases and controls to be genotyped for the identified variants.

Few, if any, published studies have explored the role of rare variation in the risk of alcohol dependence. However, sequencing studies to identify rare variants has been fruitful in particular candidate genes. Thus, it is likely that a balanced approach exploring the full allelic spectrum of potential variants will provide the greatest likelihood of identifying key genes and variants that contribute to the risk of alcohol dependence and/or other alcohol-related phenotypes. Tests of the overall burden of variants in a gene or pathway can be carried out, although there are complexities in addressing the full range of variants.

CONCLUSIONS

The genes identified to date account for only a small fraction of the genetic risk for alcoholism. It is clear that much larger numbers of subjects, carefully characterized as to symptomology and drinking behavior, are needed to find additional genes that affect risk. Most studies to date have been underpowered and have focused on common variants. Studies of rare variants are beginning. Looking forward, meta-analyses and large collaborative studies will be needed to identify and replicate the presumably large number of genes that contribute to risk for this common, complex disease.

ACKNOWLEDGMENTS

Related research in the authors' laboratories is supported by grants from the National Institute on Alcohol Abuse and Alcoholism AA006460, AA008401, AA02892, and AA07611. The funding agency had no role in the drafting, editing, or publishing of this manuscript.

REFERENCES

Agrawal A, Edenberg HJ, Foroud T et al. (2006). Association of GABRA2 with drug dependence in the collaborative study of the genetics of alcoholism sample. Behav Genet 36 (5): 640–650.

Agrawal A, Freedman ND, Bierut LJ (2011). Genome-wide association studies of alcohol intake–a promising cocktail? Am J Clin Nutr 93 (4): 681–683.

American Psychiatric Association (2000). Diagnostic and Statistical Manual of Mental Disorders, Fourth Edition (Text Revision), American Psychiatric Association, Washington, DC.

American Psychiatric Association (2013). Diagnostic and Statistical Manual of Mental Disorders: DSM-5, American Psychiatric Association, Washington, DC.

Biernacka JM, Geske JR, Schneekloth TD et al. (2013). Replication of Genome Wide Association Studies of Alcohol Dependence: Support for Association with Variation in ADH1C. PLoS One 8 (3): e58798.

Bierut LJ, Stitzel JA, Wang JC et al. (2008). Variants in nicotinic receptors and risk for nicotine dependence. Am J Psychiatry 165 (9): 1163–1171.

Bierut LJ, Agrawal A, Bucholz KK et al. (2010). A genome-wide association study of alcohol dependence. Proc Natl Acad Sci U S A 107 (11): 5082–5087.

Bierut LJ, Goate AM, Breslau N et al. (2012). ADH1B is associated with alcohol dependence and alcohol consumption in populations of European and African ancestry. Mol Psychiatry 17 (4): 445–450.

Birley AJ, James MR, Dickson PA et al. (2008). Association of the gastric alcohol dehydrogenase gene ADH7 with variation in alcohol metabolism. Hum Mol Genet 17 (2): 179–189.

Birley AJ, James MR, Dickson PA et al. (2009). ADH single nucleotide polymorphism associations with alcohol metabolism in vivo. Hum Mol Genet 18 (8): 1533–1542.

Blednov YA, Borghese CM, McCracken ML et al. (2011). Loss of ethanol conditioned taste aversion and motor stimulation in knockin mice with ethanol-insensitive alpha2-containing GABA (A) receptors. J Pharmacol Exp Ther 336 (1): 145–154.

Blednov YA, Mayfield RD, Belknap J et al. (2012). Behavioral actions of alcohol: phenotypic relations from multivariate analysis of mutant mouse data. Genes Brain Behav 11 (4): 424–435.

Blednov YA, Benavidez JM, Black M et al. (2013). Linking GABA (A) receptor subunits to alcohol-induced conditioned taste aversion and recovery from acute alcohol intoxication. Neuropharmacology 67: 46–56.

Bohman M, Sigvardsson S, Cloninger CR (1981). Maternal inheritance of alcohol abuse. Cross-fostering analysis of adopted women. Arch Gen Psychiatry 38 (9): 965–969.

Chen CC, Lu RB, Chen YC et al. (1999). Interaction between the functional polymorphisms of the alcohol-metabolism genes in protection against alcoholism. Am J Hum Genet 65 (3): 795–807.

Cloninger CR, Bohman M, Sigvardsson S (1981). Inheritance of alcohol abuse: cross-fostering analysis of adopted men. Arch Gen Psychiatry 38: 861–868.

Covault J, Gelernter J, Hesselbrock V et al. (2004). Allelic and haplotypic association of GABRA2 with alcohol dependence. Am J Med Genet B Neuropsychiatr Genet 129 (1): 104–109.

Covault J, Gelernter J, Jensen K et al. (2008). Markers in the 5'-region of GABRG1 associate to alcohol dependence and are in linkage disequilibrium with markers in the adjacent GABRA2 gene. Neuropsychopharmacology 33 (4): 837–848.

Crabb DW, Edenberg HJ, Bosron WF et al. (1989). Genotypes for aldehyde dehydrogenase deficiency and alcohol sensitivity. The inactive ALDH2 (2) allele is dominant. J Clin Invest 83 (1): 314–316.

Crabbe JC, Colville AM, Kruse LC et al. (2012). Ethanol tolerance and withdrawal severity in high drinking in the dark selectively bred mice. Alcohol Clin Exp Res 36 (7): 1152–1161.

Dick DM, Edenberg HJ, Xuei X et al. (2004). Association of GABRG3 with alcohol dependence. Alcohol Clin Exp Res 28 (1): 4–9.

Dick DM, Plunkett J, Wetherill LF et al. (2006). Association between GABRA1 and drinking behaviors in the collaborative study on the genetics of alcoholism sample. Alcohol Clin Exp Res 30 (7): 1101–1110.

Dick DM, Wang JC, Plunkett J et al. (2007). Family-Based Association Analyses of Alcohol Dependence Phenotypes Across DRD2 and Neighboring Gene ANKK1. Alcohol Clin Exp Res 31 (10): 1645–1653.

Dixon CI, Walker SE, King SL et al. (2012). Deletion of the gabra2 gene results in hypersensitivity to the acute effects of ethanol but does not alter ethanol self administration. PLoS One 7 (10): e47135.

Edenberg HJ, Bosron WF (2010). Alcohol Dehydrogenases. In: CA McQueen (Ed.), Comprehensive Toxicology. Academic Press, Oxford, pp. 111–130.

Edenberg HJ, Dick DM, Xuei X et al. (2004). Variations in GABRA2, encoding the alpha 2 subunit of the GABA (A) receptor, are associated with alcohol dependence and with brain oscillations. Am J Hum Genet 74 (4): 705–714.

Edenberg HJ, Xuei X, Chen HJ et al. (2006). Association of alcohol dehydrogenase genes with alcohol dependence: a comprehensive analysis. Hum Mol Genet 15 (9): 1539–1549.

Edenberg HJ, Wang J, Tian H et al. (2008a). A regulatory variation in OPRK1, the gene encoding the kappa-opioid receptor, is associated with alcohol dependence. Hum Mol Genet 17 (12): 1783–1789.

Edenberg HJ, Xuei X, Wetherill LF et al. (2008b). Association of NFKB1, which encodes a subunit of the transcription factor NF-kappaB, with alcohol dependence. Hum Mol Genet 17 (7): 963–970.

Edenberg HJ, Koller DL, Xuei X et al. (2010). Genome-wide association study of alcohol dependence implicates a region on chromosome 11. Alcohol Clin Exp Res 34 (5): 840–852.

Eng MY, Luczak SE, Wall TL (2007). ALDH2, ADH1B, and ADH1C genotypes in Asians: a literature review. Alcohol Res Health 30 (1): 22–27.

Enoch MA, Hodgkinson CA, Yuan Q et al. (2009). GABRG1 and GABRA2 as independent predictors for alcoholism in two populations. Neuropsychopharmacology 34 (5): 1245–1254.

Fehr C, Sander T, Tadic A et al. (2006). Confirmation of association of the GABRA2 gene with alcohol dependence by subtype-specific analysis. Psychiatr Genet 16 (1): 9–17.

Foroud T, Wetherill LF, Liang T et al. (2007). Association of alcohol craving with alpha-synuclein (SNCA). Alcohol Clin Exp Res 31 (4): 537–545.

Foroud T, Wetherill LF, Kramer J et al. (2008). The tachykinin receptor 3 is associated with alcohol and cocaine dependence. Alcohol Clin Exp Res 32 (6): 1023–1030.

Foroud T, Edenberg HJ, Crabbe JC (2010). Genetic research: who is at risk for alcoholism? Alcohol Res Health 33: 64–75.

Frank J, Cichon S, Treutlein J et al. (2012). Genome-wide significant association between alcohol dependence and a variant in the ADH gene cluster. Addict Biol 17 (1): 171–180.

Guindalini C, Scivoletto S, Ferreira RG et al. (2005). Association of genetic variants in alcohol dehydrogenase 4 with alcohol dependence in Brazilian patients. Am J Psychiatry 162 (5): 1005–1007.

Heath AC (1995). Genetic influences on alcoholism risk: a review of adoption and twin studies. Alcohol Health Res World 19: 166–171.

Heath AC, Bucholz KK, Madden PA et al. (1997). Genetic and environmental contributions to alcohol dependence risk in a national twin sample: consistency of findings in women and men. Psychol Med 27 (6): 1381–1396.

Heath AC, Whitfield JB, Martin NG et al. (2011). A quantitative-trait genome-wide association study of alcoholism risk in the community: findings and implications. Biol Psychiatry 70 (6): 513–518.

Higuchi S (1994). Polymorphisms of ethanol metabolizing enzyme genes and alcoholism. Alcohol Alcohol Suppl 2: 29–34.

Hurley TD, Edenberg HJ (2012). Genes encoding enzymes involved in ethanol metabolism. Alcohol Res Health 34 (3): 339–344.

Hurley TD, Edenberg HJ, Li T-K (2002). The Pharmacogenomics of alcoholism. In: Pharmacogenomics: The Search for Individualized Therapies, Germany Wiley-VCH, Weinheim, pp. 417–441.

Ittiwut C, Yang BZ, Kranzler HR et al. (2012). GABRG1 and GABRA2 Variation Associated with Alcohol Dependence in African Americans. Alcohol Clin Exp Res 36 (4): 588–593.

Joslyn G, Brush G, Robertson M et al. (2008). Chromosome 15q25.1 genetic markers associated with level of response to alcohol in humans. Proc Natl Acad Sci U S A 105 (51): 20368–20373.

Kalsi G, Kuo PH, Aliev F et al. (2010). A systematic gene-based screen of chr4q22-q32 identifies association of a novel susceptibility gene, DKK2, with the quantitative trait of alcohol dependence symptom counts. Hum Mol Genet 19 (12): 2497–2506.

Kang SJ, Rangaswamy M, Manz N et al. (2012). Family-based genome-wide association study of frontal theta oscillations identifies potassium channel gene KCNJ6. Genes Brain Behav 11 (6): 712–719.

Kendler KS, Neale MC, Heath AC et al. (1994). A twin-family study of alcoholism in women. Am J Psychiatry 151: 707–715.

Kendler KS, Kalsi G, Holmans PA et al. (2011). Genomewide Association Analysis of Symptoms of Alcohol Dependence in the Molecular Genetics of Schizophrenia (MGS2) Control Sample. Alcohol Clin Exp Res 35 (5): 963–975.

Kendler KS, Aggen SH, Prescott CA et al. (2012). Evidence for multiple genetic factors underlying the DSM-IV criteria for alcohol dependence. Mol Psychiatry 17 (12): 1306–1315.

Kuo PH, Kalsi G, Prescott CA et al. (2008). Association of ADH and ALDH genes with alcohol dependence in the Irish Affected Sib Pair Study of alcohol dependence (IASPSAD) sample. Alcohol Clin Exp Res 32 (5): 785–795.

Lappalainen J, Krupitsky E, Remizov M et al. (2005). Association between alcoholism and gamma-amino butyric acid alpha2 receptor subtype in a Russian population. Alcohol Clin Exp Res 29 (4): 493–498.

Li D, Zhao H, Gelernter J (2012). Strong protective effect of the aldehyde dehydrogenase gene (ALDH2) 504lys (*2) allele against alcoholism and alcohol-induced medical diseases in Asians. Hum Genet 131 (5): 725–737.

Long JC, Knowler WC, Hanson RL et al. (1998). Evidence for genetic linkage to alcohol dependence on chromosomes 4 and 11 from and autosome-wide scan in an American Indian population. Am J Med Genet 81: 216–221 (Neuropsych. Genet.).

Luczak SE, Glatt SJ, Wall TL (2006). Meta-analyses of ALDH2 and ADH1B with alcohol dependence in Asians. Psychol Bull 132 (4): 607–621.

Luo X, Kranzler HR, Zhao H et al. (2003). Haplotypes at the OPRM1 locus are associated with susceptibility to substance dependence in European-Americans. Am J Med Genet B Neuropsychiatr Genet 120 (1): 97–108.

Luo X, Kranzler HR, Zuo L et al. (2005). ADH4 gene variation is associated with alcohol and drug dependence: results from family controlled and population-structured association studies. Pharmacogenet Genomics 15 (11): 755–768.

Luo X, Kranzler HR, Zuo L et al. (2006). ADH4 gene variation is associated with alcohol dependence and drug dependence in European Americans: results from HWD tests and case-control association studies. Neuropsychopharmacology 31 (5): 1085–1095.

Luo X, Kranzler HR, Zuo L et al. (2007). Multiple ADH genes modulate risk for drug dependence in both African- and European-Americans. Hum Mol Genet 16 (4): 380–390.

Luscher C, Jan LY, Stoffel M et al. (1997). G protein-coupled inwardly rectifying K+channels (GIRKs) mediate post-synaptic but not presynaptic transmitter actions in hippocampal neurons. Neuron 19 (3): 687–695.

Lydall GJ, Saini J, Ruparelia K et al. (2011). Genetic association study of GABRA2 single nucleotide polymorphisms and electroencephalography in alcohol dependence. Neurosci Lett 500 (3): 162–166.

Macgregor S, Lind PA, Bucholz KK et al. (2009). Associations of ADH and ALDH2 gene variation with self report alcohol reactions, consumption and dependence: an integrated analysis. Hum Mol Genet 18 (3): 580–593.

McBride WJ, Li TK (1998). Animal models of alcoholism: neurobiology of high alcohol-drinking behavior in rodents. Crit Rev Neurobiol 12 (4): 339–369.

Munafo MR, Matheson IJ, Flint J (2007). Association of the DRD2 gene Taq1A polymorphism and alcoholism: a meta-analysis of case-control studies and evidence of publication bias. Mol Psychiatry 12 (5): 454–461.

Munafo MR, Timofeeva MN, Morris RW et al. (2012). Association between genetic variants on chromosome 15q25 locus and objective measures of tobacco exposure. J Natl Cancer Inst 104 (10): 740–748.

Pickens RW, Svikis DS, McGue M et al. (1991). Heterogeneity in the inheritance of alcoholism: a study of male and female twins. Arch Gen Psychiatry 48: 19–28.

Porjesz B, Rangaswamy M, Kamarajan C et al. (2005). The utility of neurophysiological markers in the study of alcoholism. Clin Neurophysiol 116 (5): 993–1018.

Prescott CA, Kendler KS (1999). Genetic and environmental contributions to alcohol abuse and dependence in a population-based sample of male twins. Am J Psychiatry 156 (1): 34–40.

Prescott CA, Sullivan PF, Kuo PH et al. (2006). Genomewide linkage study in the Irish affected sib pair study of alcohol dependence: evidence for a susceptibility region for symptoms of alcohol dependence on chromosome 4. Mol Psychiatry 11 (6): 603–611.

Reich T, Edenberg HJ, Goate A et al. (1998). Genome-wide search for genes affecting the risk for alcohol dependence. Am J Med Genet 81 (3): 207–215.

Rietschel M, Treutlein J (2013). The genetics of alcohol dependence. Ann N Y Acad Sci 1282: 39–70.

Ruedi-Bettschen D, Rowlett JK, Rallapalli S et al. (2013). Modulation of alpha5 Subunit-Containing GABA (A) Receptors Alters Alcohol Drinking by Rhesus Monkeys. Alcohol Clin Exp Res 37: 624–634.

Saccone NL, Kwon JM, Corbett J et al. (2000). A genome screen of maximum number of drinks as an alcoholism phenotype. Am J Med Genet 96 (5): 632–637.

Schumann G, Coin LJ, Lourdusamy A et al. (2011). Genome-wide association and genetic functional studies identify autism susceptibility candidate 2 gene (AUTS2) in the regulation of alcohol consumption. Proc Natl Acad Sci U S A 108 (17): 7119–7124.

Sherva R, Kranzler HR, Yu Y et al. (2010). Variation in nicotinic acetylcholine receptor genes is associated with multiple substance dependence phenotypes. Neuropsychopharmacology 35 (9): 1921–1931.

Sigvardsson S, Bohman M, Cloninger CR (1996). Replication of the Stockholm Adoption Study of alcoholism. Confirmatory cross-fostering analysis. Arch Gen Psychiatry 53 (8): 681–687.

Song J, Koller DL, Foroud T et al. (2003). Association of GABA (A) receptors and alcohol dependence and the effects of genetic imprinting. Am J Med Genet B Neuropsychiatr Genet 117B (1): 39–45.

Soyka M, Preuss UW, Hesselbrock V et al. (2008). GABA-A2 receptor subunit gene (GABRA2) polymorphisms and risk for alcohol dependence. J Psychiatr Res 42 (3): 184–191.

Thomasson HR, Edenberg HJ, Crabb DW et al. (1991). Alcohol and aldehyde dehydrogenase genotypes and alcoholism in Chinese men. Am J Hum Genet 48 (4): 677–681.

Thomasson HR, Crabb DW, Edenberg HJ et al. (1994). Low frequency of the ADH2*2 allele among Atayal natives of Taiwan with alcohol use disorders. Alcohol Clin Exp Res 18: 640–643.

Treutlein J, Cichon S, Ridinger M et al. (2009). Genome-wide association study of alcohol dependence. Arch Gen Psychiatry 66 (7): 773–784.

van Beek JH, Willemsen G, de Moor MH et al. (2010). Associations between ADH gene variants and alcohol phenotypes in Dutch adults. Twin Res Hum Genet 13 (1): 30–42.

Villafuerte S, Heitzeg MM, Foley S et al. (2012). Impulsiveness and insula activation during reward anticipation are associated with genetic variants in GABRA2 in a family sample enriched for alcoholism. Mol Psychiatry 17 (5): 511–519.

Wall TL, Carr LG, Ehlers CL (2003). Protective association of genetic variation in alcohol dehydrogenase with alcohol dependence in Native American Mission Indians. Am J Psychiatry 160 (1): 41–46.

Wang JC, Cruchaga C, Saccone NL et al. (2009a). Risk for nicotine dependence and lung cancer is conferred by mRNA expression levels and amino acid change in CHRNA5. Hum Mol Genet 18 (16): 3125–3135.

Wang JC, Grucza R, Cruchaga C et al. (2009b). Genetic variation in the CHRNA5 gene affects mRNA levels and is associated with risk for alcohol dependence. Mol Psychiatry 14 (5): 501–510.

Whitfield JB (2002). Alcohol dehydrogenase and alcohol dependence: variation in genotype-associated risk between populations. Am J Hum Genet 71 (5): 1247–1250.

Williams JT, Begleiter H, Porjesz B et al. (1999). Joint multipoint linkage analysis of multivariate qualitative and quantitative traits. II. Alcoholism and event-related potentials. Am J Hum Genet 65 (4): 1148–1160.

Xiao Q, Weiner H, Crabb DW (1996). The mutation in the mitochondrial aldehyde dehydrogenase (ALDH2) gene responsible for alcohol-induced flushing increases turnover of the enzyme tetramers in a dominant fashion. J Clin Invest 98 (9): 2027–2032.

Xuei X, Flury-Wetherill L, Dick D et al. (2010). GABRR1 and GABRR2, encoding the GABA-A receptor subunits rho1 and rho2, are associated with alcohol dependence. Am J Med Genet B Neuropsychiatr Genet 153B (2): 418–427.

Yang BZ, Kranzler HR, Zhao H et al. (2007). Association of haplotypic variants in DRD2, ANKK1, TTC12 and NCAM1 to alcohol dependence in independent case control and family samples. Hum Mol Genet 16 (23): 2844–2853.

Yang BZ, Kranzler HR, Zhao H et al. (2008). Haplotypic variants in DRD2, ANKK1, TTC12, and NCAM1 are associated with comorbid alcohol and drug dependence. Alcohol Clin Exp Res 32 (12): 2117–2127.

Zhang H, Luo X, Kranzler HR et al. (2006). Association between two {micro}-opioid receptor gene (OPRM1) haplotype blocks and drug or alcohol dependence. Hum Mol Genet 15 (6): 807–819.

Zhang H, Kranzler HR, Yang BZ et al. (2008). The OPRD1 and OPRK1 loci in alcohol or drug dependence: OPRD1 variation modulates substance dependence risk. Mol Psychiatry 13 (5): 531–543.

Zhang H, Kranzler HR, Weiss RD et al. (2009). Pro-opiomelanocortin gene variation related to alcohol or drug dependence: evidence and replications across family- and population-based studies. Biol Psychiatry 66 (2): 128–136.

Handbook of Clinical Neurology, Vol. 125 (3rd series)
Alcohol and the Nervous System
E.V. Sullivan and A. Pfefferbaum, Editors

Chapter 33

Co-occurring psychiatric disorders and alcoholism

J. STEPHEN RICH AND PETER R. MARTIN*
Department of Psychiatry, Vanderbilt University Medical Center, Nashville, TN, USA

INTRODUCTION

Alcohol use disorder (AUD), a term that comprises both alcohol abuse and alcohol dependence, is a highly prevalent psychiatric disorder. Over 50% of treated AUD patients were found to also suffer from other psychiatric disorder(s) (Grant et al., 2004a). While any psychiatric disorder can co-occur with AUD because it is so highly prevalent in the general population, detailed study has revealed disorders across multiple psychiatric domains, with rates of co-occurrence far greater than chance. This suggests a synergistic relationship between AUD and these co-occurring disorders. The basis of this synergy has been found to take multiple forms, including behavioral and neurobiologic. These factors can play a role in all phases of AUD, including the predisposition to both AUD and co-occurring psychopathology, the vulnerability to environmental risk factors that exacerbate these predispositions, the nature of reinforcement in acute intoxication, progressive alcohol use leading to dependence, neuroadaptation underpinning chronic dependence, the manifestations of acute and protracted withdrawal, emergence of medical and psychiatric complications, and ultimately the potential for relapse once abstinence has been achieved (Fig. 33.1).

In this chapter, we will explore the individual psychiatric disorders that most commonly co-occur with AUD and how to properly diagnose them and to conceptualize their bidirectional relationship with AUD. We will also review clinical course in patients who present with co-occurring disorders, as well as how comorbidity should influence treatment. Most importantly, however, throughout each of these sections we will emphasize the significance of recognizing co-occurrence and discuss how, both neurobiologically and clinically, the synergies between co-occurring disorders yield a result far more complex than a mere sum of the component disorders.

THE RELATIONSHIP BETWEEN ALCOHOL USE DISORDERS AND OTHER PSYCHOPATHOLOGY

Of the psychiatric disorders that co-occur with AUD, mood and anxiety disorders are statistically among the most highly associated (Table 33.1). The strong causal relationship of bipolar (affective) disorder (BPAD) with AUD is supported by an odds ratio of 3.5 (Grant et al., 2004a), whereas the associations of AUD with major depressive disorder (MDD) and anxiety disorders are comparatively weaker, as suggested by their odds ratios of 1.8 and 1.7, respectively.

A stronger association than that found in the general population of any two psychiatric disorders may indicate a causal or mechanistically meaningful linkage between the disorders in question. The high rate of AUD in patients with BPAD strongly suggests that AUD may play a significant role in the pathogenesis of BPAD or vice versa, or that both are influenced by a common factor such as genetic polymorphisms. The smaller odds ratios found for MDD and anxiety disorders indicate that, although these disorders have a higher prevalence among those suffering from AUD than in the general population, the association may be of less relevance in explaining causality. Moreover, this lesser association may even be inflated, at least for MDD, by the frequent misdiagnosis of BPAD as MDD in the earliest stage of the illness (Bschor et al., 2012) and other reasons for not recognizing BPAD in mood disorder patients (Judd and Akiskal, 2003b).

Several other conditions that can manifest with psychiatric symptoms have also been found to have a greater than chance association with AUD. Personality

*Correspondence to: Peter R. Martin, M.D., Professor of Psychiatry and Pharmacology, Director, Division of Addiction Psychiatry and Vanderbilt Addiction Center, Vanderbilt Psychiatric Hospital, Suite 3068, 1601 23rd Avenue South, Nashville, TN 37212-8645, USA. Tel: +1-615-343-4527, Fax: +1-615-322-0175, E-mail: peter.martin@vanderbilt.edu

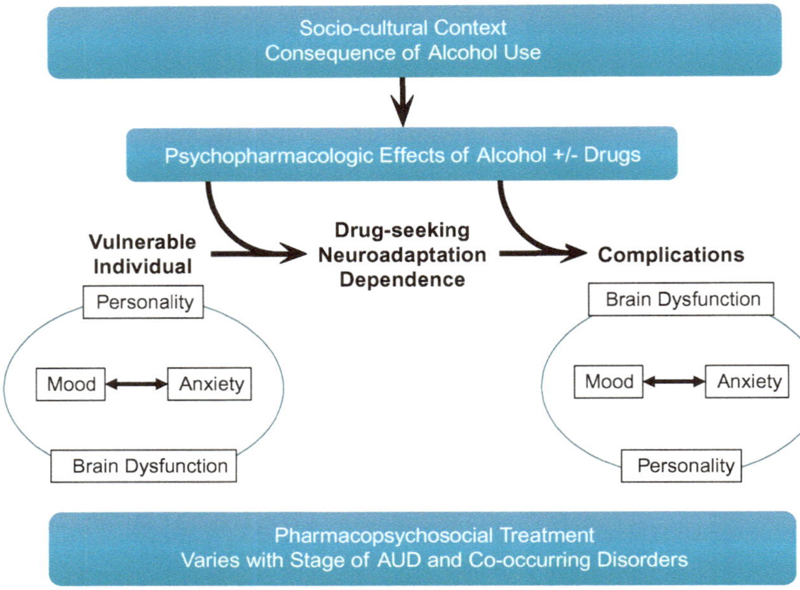

Fig. 33.1. The interaction between alcohol use disorder (AUD) and psychopathology is complex and involves elements from multiple domains of life. When patients consume alcohol, each of these elements plays a role in determining whether their use will progress to dependence. Their co-occurring psychopathology can both confer vulnerability and modify the complications of use and dependence. The stage of progression as well as the complications and co-occurring psychopathology largely guide choices of treatment and their efficacy.

Table 33.1

Epidemiology of psychiatric disorders significantly associated with alcohol use disorder (AUD)

Disorder	12-month prevalence	12-month prevalence in AUD	Odds ratio for AUD
Bipolar (affective) disorder	2.82%	30.97–43.6%	3.5
Major depressive disorder	7.06%	13.70%	1.8
Anxiety disorders	11.08%	17.05%	1.7
Personality disorders (Grant et al., 2004b)	14.8%	28.6%	2.6
Drug use disorders (Compton et al., 2007)	2.0%	33.05%	9.0

Source: (Grant et al., 2004a), except where otherwise noted.

disorders and drug use disorders also have high odds ratios of 2.6 and 9.0, respectively (Grant et al., 2004b; Compton et al., 2007), suggesting significant behavioral and neurobiologic synergy. Finally, organic brain dysfunction likely plays a significant role in AUD, both as a predisposition to alcohol use as well as a consequence of brain injury associated with alcohol dependence, although comparable epidemiologic data on this issue are not readily available.

IDENTIFYING PSYCHOPATHOLOGY IN AUD PATIENTS

For a variety of reasons, including stigma associated with mental disorders or inadequate insurance coverage for mental healthcare, patients often present to a primary care provider or physician of another specialty before they seek out a psychiatrist (Grembowski et al.. 2002; Abrahams et al., 2006). One of the most important considerations in care for the AUD patient with a co-occurring other psychiatric disorder may be recognizing the presence of the comorbid condition and understanding how it should modify management. In order to facilitate this, the core elements of the psychiatric disorders that most commonly co-occur with AUD are discussed below.

Mood disorders

The most common presentation for the patient with a mood disorder involves an episode of depression (Judd

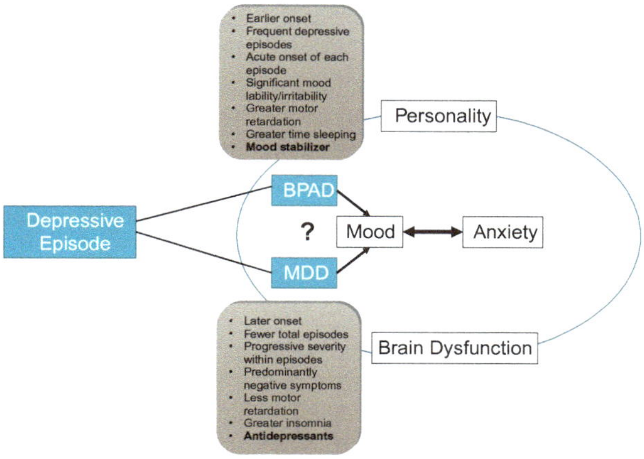

Fig. 33.2. Depressive episodes co-occurring in alcohol use disorder often represent underlying bipolar (affective) disorder (BPAD), though they can appear superficially similar to major depressive disorder (MDD). Given the important implications on outcome and treatment, proper differentiation is crucial. Closer inspection of the time, frequency, acuity, and nature of these episodes can aid diagnosis and guide appropriate management.

and Akiskal, 2003a) (Fig. 33.2). As defined by the criteria in the *Diagnostic and Statistical Manual* (DSM), a major depressive episode consists of either depressed mood or decreased level of interest in previously enjoyed activities combined with four other symptoms involving thoughts of guilt, hopelessness, or worthlessness, as well as neurovegetative changes, including disturbed sleep, concentration, appetite, or energy (American Psychiatric Association, 2000). Depressive episodes occur in both BPAD and MDD. To warrant the diagnosis of MDD, these episodes must occur in the absence of any history of mania or hypomania, as well as any causative medical condition (American Psychiatric Association, 2000). However, if the first episode of mood disturbance in a BPAD patient is a depressive episode, or if the patient presents for medical care only when experiencing depressive episodes, there is a distinct danger of misdiagnosis. Indeed, recent research has found that clinicians often miss the BPAD diagnosis if they do not explicitly attempt to elicit BPAD symptoms (Bschor et al., 2012). Because of this many patients wait 10 years or more from the onset of mood symptoms before an accurate BPAD diagnosis is ultimately made (Hirschfeld, 2001). This error in diagnosis is even more likely to occur if the patient also suffers from drug use disorders because both the number and severity of depressive episodes in patients with bipolar symptoms have been associated with severity of substance abuse (Hoertel et al., 2012) and it is exceedingly difficult to make a mood diagnosis in the midst of active substance use (Franken and Hendriks, 2001), especially in the adolescent population, when both mood and drug disorders tend to begin almost at once (Judd and Akiskal, 2003a; Duffy et al., 2012). Most importantly, since patients rarely visit the physician when their mood is normal or elevated as they feel "well and in control," the probability of seeking care for both AUD and co-occurring BPAD tends to be postponed to the depressive phase, further enhancing the likelihood of misdiagnosis.

Experiencing episodes of elevated or dysphoric mood outside the context of substance effects indicates the diagnosis of BPAD. As defined by the DSM, a hypomanic or manic (here termed hyperthymic) episode consists of three or more symptoms lasting 4 or more days (or prompting hospitalization), generally in the domains of decreased need for sleep, difficulty concentrating due to racing thoughts, pressured (increased rate of) speech, increased (often delusional) sense of self-worth, and substantial impulsivity (American Psychiatric Association, 2000). The changes cannot be due to a medical condition or directly resulting from substance intoxication/withdrawal. The episode is termed mania if these changes cause significant impairment in one or more domains of functioning; lesser degrees of such mood changes are termed hypomania (American Psychiatric Association, 2000). Of note, some patients may experience symptoms of mania or hypomania that do not fully meet the DSM criteria for the BPAD diagnosis, largely due to a failure to meet the 4-day duration criterion. Recent research has found such cases eventually progress to full BPAD (Swann et al., 2000; Homish et al., 2012); indeed, since these cases have not qualified for BPAD under current diagnostic criteria, many have been erroneously diagnosed as MDD or dysfunctional personality traits, but more importantly, are difficult to distinguish from the effects of active drug abuse. Nonetheless, there is evidence of significantly poorer outcome if such patients are misdiagnosed as having MDD and are not treated appropriately as a hyperthymic disorder, including worse AUD (Merikangas et al., 2008), injudicious pharmacotherapy with antidepressants, and shortened periodicity of BPAD cycles (Ghaemi et al., 2003). The prevalence of BPAD doubles when the duration criterion is not required for diagnosis (Judd and Akiskal, 2003b). Therefore, such subthreshold cases warrant careful monitoring and ongoing evaluation, and, especially in the context of co-occurring AUD, antidepressant treatment may be ill advised.

BPAD patients spend approximately half of their time in a symptomatic phase (Judd et al., 2003). The typical course of BPAD involves more time in the depressive phase than in the hypomanic or manic (Judd et al., 2002), ranging from 3:1 in BPAD I (defined by the presence of full manic states) to as much as 37:1 in BPAD II (defined by depressive and hypomanic, but no full manic, states) (Judd et al., 2003). As a general rule, a manic episode is often followed by a depressive episode that lasts longer than the preceding mania (Judd et al., 2002). Often, patients may initially experience low-grade mania as a positive, indeed welcome (American Psychiatric Association, 2000), experience even as the associated disorganization, irritability, and impulsivity progressively lead them to actions which they ultimately regret. Consistent with this, a systematic review has found that BPAD patients experience a lower quality of life than the general population even after an acute mood episode has resolved (Dean et al., 2004), with some uncommon exceptions (Akiskal and Akiskal, 2005). Importantly, the impulsivity and poor judgment engendered in episodes of hyperthymia typically lead patients to engage in increased substance abuse (Swann et al., 2004). Moreover, attendant denial surrounding alcohol use associated with mood instability may impair insight concerning the disorder(s), including poor understanding and a failure to disclose these hyperthymic (and heavily substance-abusing) episodes to treating physicians. Indeed, patients may only realize the detrimental impact of the manic episode once it is complete, when their euphoria has faded, their thinking is no longer disorganized, and they fully appreciate and feel guilty about what they have done. At that point, patients commonly progress to a full depressive episode in which previous hyperthymia may be forgotten or idealized and the role of alcohol consumed is not fully appreciated (Bowden, 2001). In fact, patients present to a physician primarily or even exclusively during depressive episodes, and rarely during hyperthymic states (Judd and Akiskal, 2003a). This can result in misdiagnosis of these patients as having a depressive, rather than bipolar, illness; an unfortunate consequence may be inappropriate treatment which may adversely affect clinical outcomes.

Given that alcohol intake may worsen the depressed mood that defines such states (Davis et al., 2010), patients may not even recognize the extent of their past elevated mood so it is not surprising that physicians fail to elicit BPAD symptoms. A significant family history of BPAD is strongly suggestive of the diagnosis in such patients (Swann, 1997). Additionally, BPAD is suggested if patients manifest early (especially prepubertal or pubertal: Akiskal, 1995) and/or there is acute onset of mood symptoms, rapid resolution of symptoms after an episode, and greater overall disease burden (in terms of both frequency and duration of episodes, including depressive ones) (Dunner, 2002). Proper differentiation is crucially important, however, because the treatments required for BPAD and MDD are significantly different; indeed, unopposed antidepressants are well known to actually induce manic episodes in BPAD patients (Ghaemi et al., 2000) and shorten the periodicity of these episodes, thus reducing the overall time spent in a euthymic state over the life course of the disorder (Ghaemi et al., 2003). The risk of misdiagnosis appears to be especially high in the adolescent population, creating particular concern since in this population a depressive episode is often the first indication of illness (Merikangas et al., 2012). Given the propensity towards increased substance abuse in the context of a manic episode (Swann et al., 2004), inducing such an episode would be especially dangerous in the comorbid patient. Furthermore, multiple trials of antidepressants in the BPAD patient may ultimately reduce the efficacy of mood stabilizers once the diagnosis is corrected (Swann et al., 2000). As a result, carefully collecting a complete and accurate history when patients are first seen is paramount in those presenting with a depressive episode, even after multiple similar presentations previously.

In some instances, however, even a detailed history may not prove conclusive. Many patients, ultimately diagnosed with BPAD, first experience a depressive episode at disease onset (Lish et al., 1994). At this point, there may be little to differentiate such patients from those with MDD. BPAD often has its onset in mid-adolescence (Merikangas et al., 2012), and by this point, more than 20% of BPAD adolescents have already attempted suicide (Aivar et al., 2012). Because of this often rapid progression to a potentially lethal outcome, patients recently diagnosed with MDD must still be carefully observed for possible emergence of BPAD. This is especially important given the evidence that subthreshold symptoms (often due to failure to meet the 4-day duration criterion) frequently progress to full BPAD (Judd and Akiskal, 2003b; Bschor et al., 2012; Hoertel et al., 2012; Homish et al., 2012), although the course of symptoms in adolescents may be less predictable (Aivar et al., 2012).

A final possible presentation is that of the mixed-mood episode (also known as “dysphoric mania”). If such episodes are present outside the context of substance use, the BPAD diagnosis should be made. According to DSM criteria, an episode can only be considered a mixed episode if it meets full criteria for both a manic and a depressive episode simultaneously (American Psychiatric Association, 2000). Even so, given that subsyndromal manic symptoms are being increasingly recognized as having clinical significance in isolation (Swann et al., 2000; Homish et al., 2012), the significance of a subsyndromal mixed episode has not yet been well studied but may follow the same pattern. This episode type is particularly significant because the combination of depressive thoughts with manic agitation, disorganization, and impulsivity represents a significantly increased risk of suicidal behavior (Jamison, 2000) and substance use (Strakowski et al., 2000) compared to any of the other mood states. Furthermore, mixed-mood episodes appear to respond even more robustly to most mood stabilizers (McElroy et al., 1992), making misdiagnosis as MDD all the more detrimental. Finally, mixed states are particularly difficult to distinguish from the effects of active substance abuse. Although any form of BPAD is exacerbated by the added disinhibition associated with alcohol use, the volatility and potential danger inherent in a mixed episode are uniquely synergistic with judgment-impairing psychopharmacologic effects of alcohol, making this mood state especially important to recognize and aggressively treat in the comorbid patient. Furthermore, some patients may present with significant impulsivity and irritability but without strong indication of substantial mood shifts; such patients also require careful evaluation to differentiate mood from personality disorders or disinhibition due to organic brain dysfunction, discussed in further detail below.

Differentiating primary vs secondary mood symptoms in co-occurring conditions

Reliable diagnosis is extremely challenging when patients are actively drinking. Often one must rely on probabilistic considerations derived from the history in selecting the most likely diagnosis and an appropriate management plan. Both alcohol intoxication and the withdrawal syndrome that emerges when drinking is discontinued can result in a range of signs and symptoms that may closely resemble other psychiatric disorders. For example, intoxication with alcohol can often cause significant alterations of mood, including depression and social disinhibition, whereas withdrawal is characterized by anxiety and disturbed sensorium (Koch-Weser et al., 1976). Accordingly, in the DSM, substance-induced mood disorders are considered in the differential diagnosis of a primary mood disorder, and vice versa. When AUD patients present with symptoms suggestive of a mood disorder, the critical element of the history is often the timeline of symptoms in relation to alcohol use. Generally, the diagnosis of a primary mood disorder can only be made with confidence if present during an extended period of sobriety (6 months or more). The closer to drinking one identifies mood symptoms, the greater the possibility that one is dealing with a substance-induced mood disorder rather than a primary disorder. Furthermore, depressive symptoms should be considered in a lifelong context for each AUD patient because an isolated depressive episode likely belies other prior mood changes and thus might represent the downward pole of BPAD or an externalizing personality disorder rather than simply MDD.

Physicians must also remain aware of enduring brain adaptive or allostatic changes associated with chronic use of alcohol or other drugs which can present as a protracted abstinence syndrome after signs of acute withdrawal have abated (Koob, 2009). Patients with protracted abstinence syndrome may continue to exhibit symptoms of altered brain functioning, including mood and anxiety symptoms, driven by neurobiologic perturbations that may not normalize for a year or more after stopping drinking (Soto et al., 1985). These changes are highly reactive to stressful environmental cues and may lead to dysphoric responses that ultimately enhance the likelihood of relapse (Fig. 33.3). The etiology of these disturbed responses to the environment is not fully clear, but may represent progressive, long-term allostatic changes in the extended amygdala, especially corticotropin-releasing factor (CRF)-induced responses to stress and changes in gamma-aminobutyric acid (GABA) and *N*-methyl-D-aspartate (NMDA) neurotransmission associated with exposure to alcohol (Koob, 2009).

Due to the close similarities of alcohol-induced symptoms to primary mood disorders, these disorders may not always be differentiable, especially if the symptoms have only occurred in the context of substance abuse, e.g., if patients have chronically used alcohol since an early age. Ultimately, for a definitive primary mood disorder diagnosis to be possible, patients must have exhibited mood (depressive or hyperthymic) disturbances during a period when they were completely free of effects (intoxication or withdrawal) of all substances for a significant period of time on the order of months to years, with diagnostic accuracy improving as more time passes (Franken and Hendriks, 2001). In addition, although initially presenting with depressed mood, detailed history may reveal that the stimulus for relapse may actually have been the onset of irritability, or impulsiveness, or other hyperthymic behavior. Due to the high prevalence of AUD in patients with BPAD or externalizing personality disorders, the importance of considering hyperthymia in the AUD population is evident even when characteristic depressive symptoms appear compelling.

Close observation longitudinally once treatment has begun is also important. Given the episodic nature of mood disorders, improvement with abstinence without specific treatment may merely represent the conclusion of the most recent episode of a primary mood disorder rather than conclusive evidence that the patient has a substance-induced mood disorder (Franken and Hendriks, 2001). Furthermore, appropriate management of significant mood symptoms increasingly appears to be warranted even before a definitive diagnosis can be made; recent research has indicated that failing to identify and manage such mood symptoms (and, thus, leaving patients potentially exposed to a worsening mood

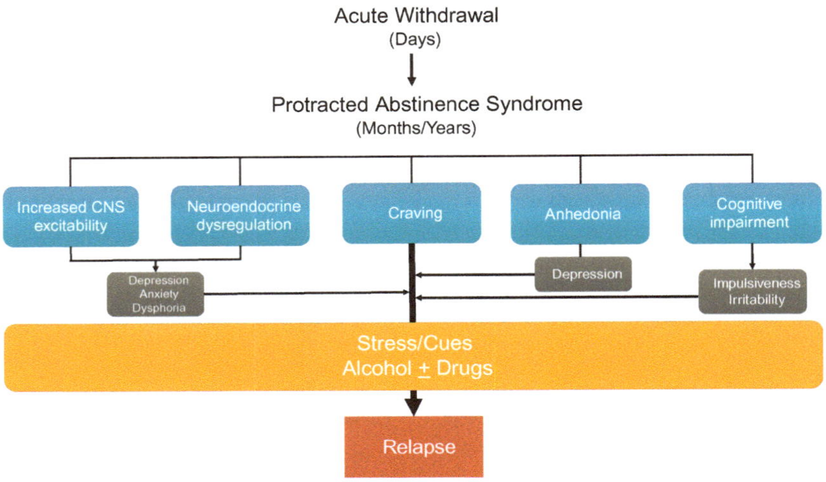

Fig. 33.3. The protracted abstinence syndrome follows acute withdrawal from alcohol, and can last for months or years. It represents a constellation of multiple neurophysiologic manifestations that can mimic or exacerbate primary mood symptoms. In the context of environmental stress or alcohol-related cues, heightened by co-occurring primary psychopathology, neuroadaptation caused by chronic exposure to substances can quickly drive the individual with alcohol use disorder to relapse. CNS, central nervous system.

disorder) can strongly promote relapse to alcohol as well as suicide (Hasin et al., 2002). Of note, the risk of suicide is particularly acute in the adolescent population, so careful evaluation and management of this population are especially important (Merikangas et al., 2012). Given recent research showing both BPAD and substance abuse family history to be among the most predictive risk factors for adolescent BPAD (Scott et al., 2012), early intervention provides a unique opportunity to both reduce suicide risk as well as to identify and begin early management of, or possibly even prevent, comorbid AUD.

Anxiety disorders

Anxiety is a non-specific cluster of symptoms associated with the experience of psychologic distress and/or physical tension closely akin to fear, associated with activation of autonomic stress responses. Anxiety is part of the presentation of many physical and psychiatric disorders, including mood disorders and AUD. Accordingly, the approach to treatment must be determined by the underlying diagnosis rather than simply a goal of symptom relief. The anxiety disorders, though characterized by the presence of anxiety, actually consist of multiple distinct syndromes, many of which have notable comorbidity with AUD. Posttraumatic stress disorder (PTSD) is likely the most commonly associated with AUD, as over 20% of those with PTSD used substances to self-medicate, making it the anxiety disorder most associated with self-medication (Leeies et al., 2010). Patients can develop PTSD when they have been subjected to an acutely traumatic experience, which is then persistently re-experienced through intrusive and involuntary flashbacks, recollection of the event through nightmares, or intense avoidance of any reminder of the event. The anxiety cannot be better accounted for by another condition such as alcohol withdrawal, making diagnosis of this condition challenging if comorbid AUD is present (American Psychiatric Association, 2000).

Anxiety disorders other than PTSD co-occur much less frequently with AUD, possibly because they tend to be found in those with internalizing, rather than externalizing, personality traits (see below). Generalized anxiety disorder consists of significant anxiety concerning multiple unrelated domains. Phobias are much more focused and involve a very specific fear, usually of a type of creature or situation. One common (and distinct) variant of this is social phobia, most commonly in the form of a fear of public speaking or, in some forms, directly engaging with strangers in any way. Obsessive-compulsive disorder occurs when patients have overriding concerns (obsessions) that are very specific and may repeatedly engage in similarly specific behaviors (compulsions). In some cases, patients may have anxiety that is so severe that they develop a panic attack, involving multiple elements of autonomic nervous system activation. Panic disorder occurs when patients have such attacks regularly, ultimately leading to a cyclic fear of another attack as a key source of ongoing anxiety and avoidance of the outside world (American Psychiatric Association, 2000).

Alcohol may be consumed to alleviate any form of anxiety and, when alcohol is acutely discontinued, much more severe anxiety ensues in the context of withdrawal. Patients without panic attacks may acquire them for the first time during alcohol withdrawal due to the severity of their anxiety, and then panic disorder may persist as patients develop a self-propagating fear of anxiety and avoidance behaviors.

Differentiating anxiety from mood disorders

Patients with anxiety disorders, especially PTSD or social phobia, may attempt to structure their lives so as to avoid distressing anxiety symptoms. In addition, they may increase alcohol use in order to be able to cope with activities that previously were within their behavioral repertoire (American Psychiatric Association, 2000). This avoidance can be challenging to differentiate from loss of interest, a core element of a depressive episode, especially if AUD is also present, because avoidance, anhedonia, and dysphoric mood are common symptoms of alcohol withdrawal (Heinz et al., 1994) and anxiety and mood disorders often co-occur (Brown et al., 2001). Indeed, except for disorders in which a specific stimulus plays a pathogenic role (e.g., PTSD and specific phobias), the frequent co-occurrence and often identical first-line pharmacologic treatments suggest that general anxiety disorder and depression may share sufficient elements to be considered part of the same overall syndrome, although this concept remains a subject of debate (Myrick and Brady, 2003).

Another relevant distinction is that between anxiety and manic behavior. Patients often use whatever vocabulary they have to convey their emotions. Someone who has a new and foreign experience such as racing thoughts with significant agitation and resulting inability to think clearly may describe this as "anxiety." Some patients may also experience significant difficulty sleeping or may even experience their emotional pain somatically and mistake it for physical pain. Despite their own summary, however, these symptoms are often most consistent with a diagnosis of hyperthymia (American Psychiatric Association, 2000). This distinction is especially important in the comorbid patient, because these experiences often serve as stimuli for further alcohol use, which both exacerbates the AUD and also could dampen elevated energy and thus make the recognition of BPAD even more difficult without further information. As a result, such patient reports should be clearly delineated in terms of specific symptoms, rather than accepting the informant's understanding of anxiety at face value.

Personality disorders (Axis II in DSM)

It has become increasingly apparent that the risk for, development of, and life course of AUD are powerfully influenced by personality, which is a fundamental, likely biologically shaped characteristic of each person (a trait or style of behaving) which influences inner experience of and coping with one's environment. It is only when a personality trait deviates substantially from cultural expectations and leads to distress or impairment due to pervasive effects on coping and enduring nature that the term personality disorder is appropriate. In the DSM, personality disorders are grouped into clusters based on similar overall characteristics and possibly pathogenesis. Of specific relevance to AUD are cluster B (externalizing) and cluster C (internalizing) personality disorders.

Externalizing personalities are those who direct their thoughts and stress outwards. They may often project their emotions on to others or blame others for their problems, perceiving their own negative self-judgment as coming from someone else even if that person never expressed a negative thought at all. Furthermore, they typically try to manage stress by modifying their external environment, often through manipulating others or even violence, in an attempt to tightly control perceptions (both others' and their own) of themselves. They may even try to "externalize" emotional pain by cutting themselves, since self-inflicted pain is under their control, in contrast to the emotional pain that is not.

The cluster B personality disorders include borderline, narcissistic, histrionic, and antisocial. Borderline personalities are those who often have a pattern of highly unstable and tumultuous social interactions wherein they become highly attached to individuals then rapidly reject them with little overt cause, seeing individuals and situations as either perfectly good or entirely bad. Narcissistic personalities are those who create the impression of self-absorption and an inability to recognize the needs of others unless they primarily serve their own needs. Histrionic personalities are those who are typically dramatic and sexually provocative and often thrive from creating social tension in those around them. Antisocial personalities show a general disregard for the rights and interests of others and indeed often seem to take pleasure in violating them (American Psychiatric Association, 2000).

Individuals with cluster B personality disorders seek to modify the environment to cope with internal stress and accompanying anxiety and often do so with self-medication, using alcohol (Colder, 2001). Alcohol initially diminishes experience of stress and anxiety by attenuating activation of the hypothalamic-pituitary-adrenal axis via inhibitory GABAergic neurotransmission in relevant limbic circuits (Herman, 2012). Careful assessment of a patient's reasons for alcohol use can therefore yield important clues to properly recognizing and treating AUD. Furthermore, cluster B personality disorders often include significant elements of impulsivity and irritability, which are challenging to distinguish from the mood instability noted in BPAD II or

AUD-associated organic brain injury. Especially in the context of acute stress, impulsivity and irritability may promote both initiation as well as progression of AUD.

Individuals with internalizing personalities tend to direct their thoughts and actions inward. They absorb responsibility for anything that goes wrong in the world around them, generating progressively greater internal stress rather than attempting to effect any change in the environment around them. As external factors in their lives become increasingly stressful, internalizing patients will often attempt to identify how their own faults might have contributed to these factors even when such factors may have been completely outside their control. As stress mounts, these patients progressively demonstrate rumination and withdrawal from the outside world.

The cluster C personality disorders include avoidant, dependent, and obsessive-compulsive. Avoidant personalities are those who desire social interaction but fear rejection and hence do not engage. Dependent personalities are those who require the near-constant presence of another person from whom they often demand an extremely high level of commitment and approval. Obsessive-compulsive personalities are highly perfectionistic, require their surroundings to match well-defined criteria, and become distressed when these criteria are violated (American Psychiatric Association, 2000). Although less commonly than for externalizing disorders, alcohol can be employed by individuals suffering from internalizing personality disorders as self-medication leading to AUD.

Personality characteristics have been fundamental in typologic classifications of forms of alcohol dependence in an attempt to conceptualize pathogenesis and rationalize treatment of AUD. In the Cloninger classification, alcohol use in internalizing individuals (type 1) is characterized by an overall loss of control, driven by high reward dependence, harm avoidance, and low novelty seeking, with onset typically after age 25 and often involving a combination of environmental and genetic factors (Cloninger, 1987). By contrast, externalizing individuals (type 2) demonstrate an opposite pattern, characterized by impulsive behavior with inability to abstain, driven by high novelty seeking, low harm avoidance, and low reward dependence, with onset typically before age 25 and with a greater genetic heritability (Cloninger, 1987). The personality dimensions of each subtype of AUD are compared, along with the neurotransmitter dysfunctions hypothesized in pathogenesis, in Table 33.2.

Table 33.2

Personality traits and neurotransmitters in the Cloninger typology

Personality trait	Type 1	Type 2	Proposed neurotransmitter
Novelty seeking	Low	High	Dopamine
Harm avoidance	High	Low	Serotonin
Reward dependence	High	Low	Norepinephrine

Source: Cloninger et al. (1993).

The Lesch typology expands on this approach by recognizing additional categories. Of particular significance are separate subtypes of alcohol dependence related to underlying organic brain damage or presence of mood symptoms (Lesch and Walter, 1996). The key contribution of both of these classificatory systems has been their goal of predicting effective pharmacotherapeutic strategies, based on clinical presentation, that presumably reflect neurobiologic underpinnings. For example, alcohol use in those with significant externalizing personality elements, mood symptoms, or organic brain changes driving their alcohol use appears to respond well to naltrexone, whereas those with more internalizing elements appear to respond preferentially to acamprosate (Hillemacher and Bleich, 2008). Therefore, personality characteristics, while not considered by DSM as major psychiatric disorders, play a significant role in understanding and management of AUD patients, especially if they have additional comorbid psychiatric diagnoses.

Differentiating mood from personality disorders

The personality disorders have many similarities to, and in some ways blend with and modify expression of, primary Axis I disorders. Similar to BPAD, cluster B personality disorders present with significant mood instability, impulsivity, and irritability, and may even have significant anxiety that can sometimes be misinterpreted by patients as "racing thoughts" (a cardinal symptom of BPAD) (Clark et al., 1994). However, in the context of personality disorders, racing thoughts (although upsetting) are not as disorganized, nor are they associated with significant pressured speech, decreased need for sleep, or episodic euphoria seen in BPAD. Personality disorders can also be characterized by an enduring sense of emptiness and fractured sense of identity (American Psychiatric Association, 2000). The stress these "existential" symptoms generate may be greatly exacerbated by the presence of AUD, characterized by futile "self-medication" of painful and seemingly immutable affects (Colder, 2001). Indeed, manifestations of mood instability in externalizing personality disorders and BPAD, combined with shared phenomenologic, developmental, and genetic features, may all reflect common neurobiologic underpinnings, i.e., these conditions may represent different degrees of expression of a unified underlying pathogenesis (Akiskal et al., 1985).

Comparable to the close relationship of BPAD to cluster B personality disorders, elements of MDD are shared with cluster C personality disorders (Clark et al., 1994). Cluster C individuals often have high levels of anxiety that can progress to worsening mood (Clark et al., 1994). A "depressed" mood in this context, however, does not represent true MDD unless that depression clearly extends beyond an acutely stressful event and is expressed with some consistency throughout the entire depressive episode (American Psychiatric Association, 2000). Even so, as with patients with other forms of anxiety/depression, the presence of stressful life events requires close attention in the AUD patient, since such acute stresses represent a high risk for self-medication (Colder, 2001). Furthermore, as with cluster B and BPAD, the nature of cluster C and depression are pathophysiologically closely related and have been found to be mediated by shared neurobiologic (impaired monoaminergic neurotransmission) and psychosocial mechanisms (Cloninger et al., 1993).

Therefore, given the significant neurobiologic relationship between them, the presence of personality traits or a disorder should not exclude the possibility of a related mood disorder, as mood and personality are strongly intertwined (Skodol et al., 1999). Even so, the clinician must clarify that patients with a personality disorder truly warrant an additional mood diagnosis, especially if AUD is also present. This is because the appropriate treatment for personality disorders is often different, with psychopharmacology playing much less of a role than psychotherapy (Evershed, 2011).

Organic brain disease

Organic brain changes are important due to both their high prevalence among AUD patients and their potential for contribution to the development and clinical progression of the disorder. Up to 50–70% of patients with AUD demonstrate cognitive deficits on formal neuropsychologic testing (Eckardt and Martin, 1986; Bates et al., 2002), with 10% showing severe deficits (Eckardt and Martin, 1986; Parsons and Nixon, 1993). Of particular importance, these changes often predate the onset of the AUD, suggesting that the changes themselves might predispose patients to developing AUD (Begleiter et al., 1984; Tapert et al., 2002; Tarter et al., 2003). Recent research, both *in vivo* and at autopsy, has implicated brain volume, functional, and metabolic changes, especially those in the frontal lobe and cerebellum, as the most common underlying neuropathologic changes associated with these deficits (Nicolas et al., 2000; Kubota et al., 2001).

In addition to these highly prevalent cognitive changes, some related clinical pathologic syndromes have also been well characterized. For example, Wernicke -Korsakoff syndrome is an important consideration in patients who present with acute changes from their normal state in mental status that may initially appear to be a neuropsychiatric condition (Martin et al., 1986). This represents the most severe form of alcoholism-associated brain injury with damage to the same regions that are often less compromised in the vast majority of alcoholics, due to both the direct effects of chronic alcohol exposure as well as resulting nutritional deficiencies, particularly of thiamine (Martin et al., 2003). Furthermore, the clinical presentation of AUD can be modified by a lifestyle of multiple traumatic head injuries (Weinstein and Martin, 1995) and compromised liver function through hepatotoxicity of alcohol and associated hepatitis C (Maddrey, 2000), as well as other conditions that tend to occur in patients with AUD (Eckardt and Martin, 1986). In such patients, the classic presentations of mood and personality disturbances discussed above are often difficult to differentiate from subtle neurocognitive deficits; additionally, signs and symptoms in the affective, personality, and organic domains may be highly synergistic in their influence on clinical course of AUD. In particular, common manifestations of organic brain disease can be disinhibition and impulsive behavior. As described above, impulsiveness and disinhibition can explain how mood and personality disorders can lead to relapses and progressively increased alcohol consumption; thus, by further exacerbating impulse regulation and poor decision making, organic brain dysfunction would accelerate the progression of each of these closely linked disorders (Bowden et al., 2001; Goudriaan et al., 2006; Kamarajan et al., 2006).

Drug use disorders

Drug use disorders are also present in 33.05% of AUD patients according to one estimate (Compton et al., 2007). As other drugs of abuse serve much the same purpose in the behavioral repertoire as does drinking alcohol, modify brain functioning in related and synergistic ways, and also can impair the sensorium acutely and cause long-term decrements in brain functioning, it is not surprising that most individuals with AUD have tried or are actively using other drugs. If the scope of drugs used by an individual with AUD also includes cigarette smoking, almost all qualify as having a co-occurring drug use disorder (Reich et al., 2008). The identification of co-occurring drug use is important as treatment over the long term must include them as well if a successful recovery is to be achieved (Crowe and Reeves, 1994). Attempts to attain abstinence with one substance, while continuing to use another that is readily substituted in dysfunctional addictive behaviors, is inconsistent with a cognitive model of recovery (Crowe and Reeves, 1994). For example, substances, such as benzodiazepines and barbiturates, similarly target GABA receptors and are cross-tolerant with alcohol (indeed, this shared characteristic is the basis for using these agents in alcohol detoxification) (Asplund et al., 2004). The pattern of substituting another substance for alcohol includes not only illicit and prescription drug use - cigarette smoking also enhances impulsive behaviors that can lead to relapse due to closely related pharmacologic actions within brain reward circuits (Heffner et al., 2012). As a result, any and all substance use disorders must be identified and appropriately addressed before recovery can successfully be attained and clinical progression of AUD avoided (Jaffe, 1992).

NEUROBIOLOGIC MECHANISMS INVOLVED IN COMORBIDITIES

The epidemiologic associations between AUD and many of the comorbid psychiatric disorders discussed in this chapter suggest potentially causal linkages. In this section we summarize certain mechanistic elements that are consistent with a certain shared causality among clinical expressions of co-occurring psychiatric disorders and AUD. However, since the neurobiologic underpinnings of each disorder remain to be fully elucidated, potential causality can only be inferred.

Some of the well-characterized effects of alcohol involve enhancement of GABAergic (Lingford-Hughes et al., 1998; McBride and Li, 1998; Lê and Shaham, 2002; Koob, 2004) and inhibition of glutamatergic (Hoffman et al., 1990; Tsai et al., 1995; Tsai and Coyle, 1998; Vengeliene et al., 2005) neurotransmission. Variations in these neurotransmitter systems have been implicated in both vulnerability to and progression from casual drinking to addiction (Tsai et al., 1995; Tsai and Coyle, 1998; Koob, 2004), as well as in maintaining dependence by promoting relapse in patients who attempt abstinence (Lê and Shaham, 2002; Vengeliene et al., 2005). Alcohol generally inhibits neuronal firing in the brain by allosteric modulation of $GABA_A$ receptors, causing enhanced opening and intracellular flux through chloride ion channels to hyperpolarize neuronal membranes. As a consequence, there is reduced activation of downstream glutamatergic channels that then translates to reduced excitatory sodium ion channel opening and flux (Santhakumar et al., 2007). Neuroadaptation to this level of inhibition occurs with chronic use of alcohol, such that, once alcohol drinking is discontinued, GABAergic neurotransmission is inhibited and glutamatergic neurotransmission is enhanced. These changes in neuronal firing result from reduced chloride influx through downregulated and altered GABA-activated chloride channels and increases in sodium channel activation via uninhibited and upregulated glutamatergic channels. The net result is a state of central nervous system hyperexcitability, namely the acute alcohol withdrawal syndrome, which is characterized by anxiety, mood disturbances, altered sensorium, and potential seizures, which typically last for 3-7 days (Koch-Weser et al., 1976), until GABAergic and glutamatergic systems can re-equilibrate to a significant degree (Faingold et al., 2000).

There is accumulating evidence for much longer-term neuronal alterations akin to learning (McCool, 2011) in specific regions of the limbic system that may explain the clinical characteristics of the protracted abstinence syndrome (Fig. 33.3). After repeated cycles of intoxication and withdrawal, as the AUD patient transitions from occasional use to alcohol dependence, a shift is thought to occur in the motivational drive for drinking alcohol (Wills and Winder, 2013). Alcohol ceases to have primarily positive rewarding effects, but rather, its negative reinforcing properties, specifically postwithdrawal negative affective states (anxiety and depression), become the major motivational force behind compulsive drinking. This transition is mediated by progressive and enduring changes of excitatory neuronal signaling in the extended amygdala as a result of long-term potentiation of glutamatergic neurons (NMDA receptors). The result is dysfunctional, responding to stressors and enhanced cueing of the brain to alcohol environmental triggers. Stress- and cue-induced cravings ultimately can precipitate relapse to drinking in recovering AUD patients. Therefore, the major challenge in treatment of AUD patients is assistance with avoidance or improved coping with these internal and external stimuli.

The molecular mechanisms whereby chronic alcohol consumption modifies hedonic state, stress responses, and facilitates reinitiation of drinking are thought to be via CRF modulation of NMDA tone in the extended amygdala, specifically the basal nucleus of the stria terminalis and its projections to the ventral tegmental area (Fig. 33.4). Upstream CRF release can also be mediated by dopamine, released by use of other substances of abuse and possibly also alcohol drinking. Dopamine neurotransmission has been strongly implicated in co-occurring mood disorders, specifically euphoria in

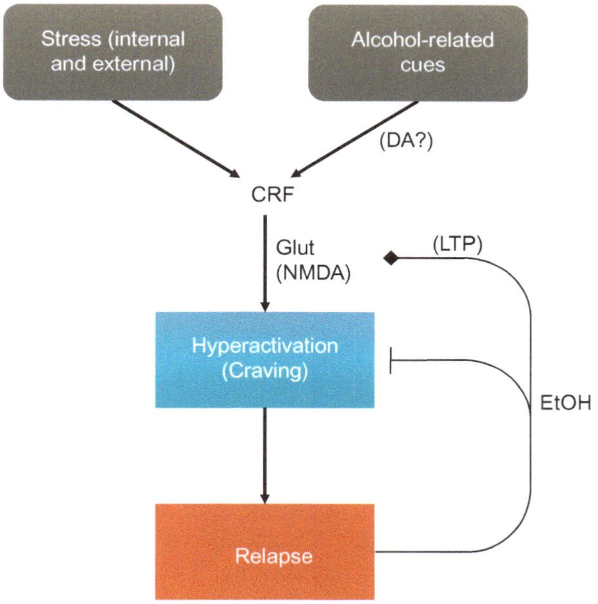

Fig. 33.4. Projections in the extended amygdala, especially from basal nucleus of the stria terminalis to ventral tegmental area, are thought to play a significant role in relapse and dependence. Through corticotropin-releasing factor (CRF) modulation, chronic alcohol use potentiates these glutamatergic projections via upregulation of *N*-methyl-D-aspartate (NMDA) receptors. Stress (such as that caused by psychopathology) and alcohol-related cues cause CRF release, engaging the potentiated glutamatergic projections, resulting in craving caused by hyperactivation. The individual may then relapse, using alcohol to mitigate this hyperactivation. Ongoing use, however, only serves to further strengthen the glutamatergic projections through long-term potentiation (LTP), thus generating a cycle of dependence. DA, dopamine; Glut, glutamate; EtOH, ethyl alcohol.

mania (a hyperdopaminergic state) and dysphoria associated with depression and anxiety (a hypodopaminergic state). Of particular importance for patients with primary mood disorders with co-occurring substance abuse is that dopamine mediates the perceived reward and subsequent reinforcement of both alcohol and other substances, thus exacerbating both drug and mood disorders simultaneously (Salloum and Thase, 2002). There is also evidence for alterations in brain opioid, serotonin, as well as GABAergic neurotransmission as underpinning alterations in hedonic tone of recovering AUD patients during the protracted abstinence period which dovetails with primary mood disorders (Koob, 2009).

Disorders co-occurring with AUD can serve as an independent source of the relapse-inducing stress described above, especially BPAD (Petty et al., 1990; Bowden, 1998; Clinton and Meador-Woodruff, 2004). This is not surprising considering that 60–80% of patients with BPAD also have a drug use disorder and the very high odds ratio for co-occurrence of BPAD with AUD. The significant mechanistic relationship between BPAD and alcohol/drug dependence is further highlighted by the fact that many of the mood stabilizers effective in treatment of BPAD have been found to directly affect alcohol use (Johnson et al., 2008; De Sousa, 2010) primarily via effects on NMDA receptors (Dixon and Hokin, 1997, 1998; Brambilla et al., 2003). As expected, measures of GABAergic tone may predict the efficacy of mood stabilizers in BPAD (Petty et al., 1996) and may similarly be relevant in response of AUD to treatment (Cosgrove et al., 2011). Recent reports of the efficacy of a neurokinin 1 receptor antagonist in blunting alcohol cravings in abstinent AUD subjects support the putative role of behavioral stress responses in relapse (George et al., 2008) and complement analogous findings with CRF antagonists (Heilig and Koob, 2007).

TREATMENT AND OUTCOMES IN COMORBIDITY

As discussed in the preceding sections, each co-occurring disorder requires careful assessment for proper diagnosis and subsequent management. Recognition of the presence and contributions of each disorder to the total clinical picture is vital because, without considering these complexities, treatment will be ineffective at best, often countertherapeutic at worst. In this section, we will discuss the appropriate first-line treatment of the mood and anxiety disorders in a patient who has AUD, as well as potential outcomes that make appropriate management so crucial.

BPAD is typically managed with a mood stabilizer as first-line treatment. With the exception of lithium, many agents in this class were initially used as anticonvulsant drugs but were later found to be effective in BPAD patients as well. Some of the most common treatment choices for BPAD include lithium, valproic acid, carbamazepine, and lamotrigine (Hirschfeld et al., 2003); the specific choice is often driven by side-effect profile and tolerability. Given the activated state of the brain during both acute and protracted withdrawal, as described above, it is predicted that anticonvulsant mood stabilizers might benefit AUD patients with respect to both alcohol craving and relapse. Recent evidence suggests that oxcarbazepine may be a particularly attractive choice in AUD patients due to its mood-stabilizing effects, with less toxicity (Hirschfeld et al., 2003), and the added benefit of reducing cravings for alcohol (Martinotti et al., 2007) and other drugs of abuse (Salloum and Thase, 2002; Weiss et al., 2005; Llopis and Castillo, 2008). The atypical antipsychotic drugs are used for stabilization of acute mania, but are typically not used if possible during the maintenance phase (Hirschfeld et al., 2003). There is little evidence to support the use of antipsychotics as the primary treatment of co-occurring AUD and hyperthymia, although they are used empirically in patients with intense anxiety or serious cluster B personality disorders (McIntyre and Katzman, 2003).

First-line treatment of MDD and most anxiety disorders typically relies on selective serotonin reuptake inhibitors (SSRIs) (Koenig and Thase, 2009). These medications have been found to be much safer than some of their earlier antidepressant counterparts, such as the tricyclic antidepressants and monoamine oxidase inhibitors, many of which can be especially dangerous in overdose – an important consideration in depressed or severely anxious patients with AUD. However, SSRIs have been found to be only minimally effective in drug-using populations (Nunes and Levin, 2004), despite suggestions that they may have a role in the comorbid population (Davis et al., 2010). Of note, however, most studies in adults (Kranzler et al., 2006) or adolescents (Cornelius et al., 2009) indicate that AUD patients with MDD did not experience significant benefit from an SSRI. This modest efficacy of SSRIs should be weighed against the potential for activation and a precipitated manic episode when given to AUD patients who are very likely to have hyperthymia despite presenting with a depressive syndrome (Ghaemi et al., 2000). Indeed, if patients have experienced such activation after starting an SSRI, hyperthymia is strongly suggested (Peet, 1994), and SSRIs should be avoided.

Psychotherapy has been demonstrated to be the treatment of choice for the personality disorders, but also is a highly effective adjunct in BPAD (Miklowitz, 2008), MDD (De Jonghe et al., 2001), and anxiety disorders (Hunot et al., 2007). Multiple forms of psychotherapy exist, but the superiority of one psychotherapeutic approach over another often remains unclear and a subject of significant debate (Leichsenring, 2001; Llopis and Castillo, 2008). Among the comorbid population, motivational interviewing, a technique in which the provider explores the patient's ambivalence towards consequences of drinking and benefits of recovery, has been found to be particularly effective in enhancing motivation for abstinence and reducing alcohol use (Burke et al., 2003; Hettema et al., 2005; Rubak et al., 2005). Alcoholics Anonymous (AA) remains an effective, popular, economic, and often well-received approach for treatment of AUD (Morgenstern et al., 1997; Moos and Moos, 2006). Comorbid patients may face challenges compared to uncomplicated alcoholics, as there is tension between needed pharmacologic treatments for co-occurring disorders and an imperative to live "totally drug-free." Fortunately, AA now actively collaborates with the medical community and recent evidence indicates that efficacy of AA and similar community programs can be augmented by, and indeed is most effective when initiated in, the context of formal addiction treatment (Tonigan et al., 2003).

Failure to properly manage comorbid conditions associated with AUD can result in significant disability, lost productivity, and healthcare costs. Likely the most concerning outcome in AUD patients with co-occurring other psychiatric disorders is suicide. Of BPAD patients, 29.2% have attempted suicide (Chen and Dilsaver, 1996). However, when BPAD co-occurs with AUD, prevalence only rises to 31.4% (Dalton et al., 2003), suggesting comorbidity is the rule rather than the exception for BPAD. Among MDD patients, 15.9% have attempted suicide. However, in contrast to the situation with BPAD, the rate of suicide rises to 27.8% when MDD co-occurs with AUD, suggesting an additive contribution of each disorder (Preuss et al., 2002). Similarly, for the anxiety disorders, 23.7% attempt suicide in the general population, compared to 47.0% of patients in which anxiety disorders are comorbid with substance use disorders (Nepon et al., 2010). In addition, comorbidity has been shown to be additive with respect to the magnitude of the decline on measures of disability, life role functioning, and social functioning compared to when each disorder occurs by itself (Grant et al., 2004a). As a result, co-occurring disorders generate a significant added burden that can only be alleviated with accurate diagnosis and proper treatment of both disorders.

CONCLUSION

Psychiatric comorbidity is a significant issue among most patients with AUD. There is a significantly increased risk of relapse and prognosis is worse with respect to suicidal behavior and various other outcomes among the comorbid population. Nevertheless, such patients can do well with appropriate management. Without the aid of laboratory tests or radiologic studies afforded many conditions in medicine, psychiatric disorders require careful attention to the history as well as to the longitudinal pattern of expression of the disorder(s) over a lifetime. Under the best scenario, appropriate choice of medication for the co-occurring psychiatric disorder can significantly improve long-term prognosis for both disorders. Indeed, properly recognizing and treating comorbid disorders, especially in the case of BPAD, has been shown to result in improved functioning in multiple domains (Weiss et al., 2005), as well as reduced relapse (Salloum and Thase, 2002).

Psychotherapy and social interventions should be combined judiciously with psychopharmacology in an integrated pharmacopsychosocial approach to treatment

(Martin et al., 2007). Nevertheless, a major research challenge is to investigate how best to combine these tripartite treatment modalities to optimize outcomes because few investigators have the requisite expertise in more than one of these modalities. Additionally, it will be important to elucidate further how these disorders interact with each other at multiple levels, from pathophysiologic to clinical expression, and ultimately clinical outcome. It will be necessary to determine how best to differentiate primary ("driving") from secondary ("following") psychiatric conditions, and how the management of the comorbid patient might differ from the management of patients who have each of the disorders individually. Current consensus is that these disorders are synergistic and by virtue of their co-occurrence in the patient they create a range of difficulties and challenges that are not simply a sum of its parts.

References

Abrahams S, Da A, Sj B (2006). Better practices in collaborative mental health care: an analysis of the evidence base. Can J Psychiatry 51 (6 Suppl 1): 7S–72S.

Aivar P, Anne-Liis K, Gunilla O et al. (2012). Hypomania spectrum disorders from adolescence to adulthood: a 15-year follow-up of a community sample. J Affect Disord 145 (2): 190–199.

Akiskal HS (1995). Depressive onset in pre-pubertal, pubertal, and possibly teenage and early adult years (age 21 and earlier) has been shown to presage eventual bipolarity. J Am Acad Child Adolesc Psychiatry 34: 754–763.

Akiskal KK, Akiskal HS (2005). The theoretical underpinnings of affective temperaments: implications for evolutionary foundations of bipolar disorder and human nature. J Affect Disord 85 (1): 231–239.

Akiskal HS, Yerevanian BI, Davis GC et al. (1985). The nosologic status of borderline personality: clinical and polysomnographic study. Am J Psychiatry 142 (2): 192–198.

American Psychiatric Association (2000). Diagnostic and Statistical Manual of Mental Disorders, 4th ed., text rev. ed. APA Press, Washington, DC.

Asplund CA, Aaronson JW, Aaronson HE (2004). 3 Regimens for alcohol withdrawal and detoxification. J Fam Pract 53 (7): 545–554.

Bates ME, Bowden SC, Barry D (2002). Neurocognitive impairment associated with alcohol use disorders: implications for treatment. Exp Clin Psychopharmacol 10 (3): 193.

Begleiter H, Porjesz B, Bihari B et al. (1984). Event-related brain potentials in boys at risk for alcoholism. Science 225 (4669): 1493–1496.

Bowden CL (1998). New concepts in mood stabilization: evidence for the effectiveness of valproate and lamotrigine. Neuropsychopharmacology 19 (3): 194–199.

Bowden C (2001). Strategies to reduce misdiagnosis of bipolar depression. Psychiatr Serv 52 (1): 51–55.

Bowden SC, Crews FT, Bates ME et al. (2001). Neurotoxicity and neurocognitive impairments with alcohol and drug-use disorders: potential roles in addiction and recovery. Alcohol Clin Exp Res 25 (2): 317–321.

Brambilla P, Perez J, Barale F et al. (2003). GABAergic dysfunction in mood disorders. Mol Psychiatry 8 (8): 721–737.

Brown TA, Campbell LA, Lehman CL et al. (2001). Current and lifetime comorbidity of the DSM-IV anxiety and mood disorders in a large clinical sample. J Abnorm Psychol 110 (4): 585.

Bschor T, Angst J, Azorin JM et al. (2012). Are bipolar disorders underdiagnosed in patients with depressive episodes? Results of the multicenter BRIDGE screening study in Germany. J Affect Disord 142 (1–3): 45–52.

Burke BL, Arkowitz H, Menchola M (2003). The efficacy of motivational interviewing: a meta-analysis of controlled clinical trials. J Consult Clin Psychol 71 (5): 843.

Chen YW, Dilsaver SC (1996). Lifetime rates of suicide attempts among subjects with bipolar and unipolar disorders relative to subjects with other Axis I disorders. Biol Psychiatry 39 (10): 896–899.

Clark LA, Watson D, Mineka S (1994). Temperament, personality, and the mood and anxiety disorders. J Abnorm Psychol 103 (1): 103.

Clinton SM, Meador-Woodruff JH (2004). Abnormalities of the NMDA receptor and associated intracellular molecules in the thalamus in schizophrenia and bipolar disorder. Neuropsychopharmacology 29 (7): 1353.

Cloninger CR (1987). Neurogenetic adaptive mechanisms. Science 236: 410–416.

Cloninger CR, Svrakic DM, Przybeck TR (1993). A psychobiological model of temperament and character. Arch Gen Psychiatry 50 (12): 975.

Colder CR (2001). Life stress, physiological and subjective indexes of negative emotionality, and coping reasons for drinking: Is there evidence for a self-medication model of alcohol use? Psychol Addict Behav 15 (3): 237.

Compton WM, Thomas YF, Stinson FS et al. (2007). Prevalence, correlates, disability, and comorbidity of DSM-IV drug abuse and dependence in the United States: results from the national epidemiologic survey on alcohol and related conditions. Arch Gen Psychiatry 64 (5): 566.

Cornelius JR, Bukstein OG, Wood DS et al. (2009). Double-blind placebo-controlled trial of fluoxetine in adolescents with comorbid major depression and an alcohol use disorder. Addict Behav 34 (10): 905–909.

Cosgrove KP, Esterlis I, Mason GF et al. (2011). Neuroimaging insights into the role of cortical GABA systems and the influence of nicotine on the recovery from alcohol dependence. Neuropharmacology 60 (7): 1318–1325.

Crowe AH, Reeves R (1994). Treatment for alcohol and other drug abuse: Opportunities for coordination. US Department of Health and Human Services, Public Health Service, Substance Abuse and Mental Health Services Administration, Center for Substance Abuse Treatment, Rockville, MD.

Dalton EJ, Cate-Carter TD, Mundo E et al. (2003). Suicide risk in bipolar patients: the role of co-morbid substance use disorders. Bipolar Disord 5 (1): 58–61.

Davis LL, Wisniewski SR, Howland RH et al. (2010). Does comorbid substance use disorder impair recovery from major depression with SSRI treatment? An analysis of the STAR* D level one treatment outcomes. Drug Alcohol Depend 107 (2): 161.

De Jonghe F, Kool S, Van Aalst G et al. (2001). Combining psychotherapy and antidepressants in the treatment of depression. J Affect Disord 64 (2–3): 217–229.

De Sousa A (2010). The role of topiramate and other anticonvulsants in the treatment of alcohol dependence: a clinical review. CNS Neurol Disord Drug Targets 9 (1): 45–49.

Dean BB, Gerner D, Gerner RH (2004). A systematic review evaluating health-related quality of life, work impairment, and healthcare costs and utilization in bipolar disorder. Curr Med Res Opin 20 (2): 139–154.

Dixon JF, Hokin LE (1997). The antibipolar drug valproate mimics lithium in stimulating glutamate release and inositol 1, 4, 5-trisphosphate accumulation in brain cortex slices but not accumulation of inositol monophosphates and bisphosphates. Proc Natl Acad Sci U S A 94 (9): 4757–4760.

Dixon JF, Hokin LE (1998). Lithium acutely inhibits and chronically up-regulates and stabilizes glutamate uptake by presynaptic nerve endings in mouse cerebral cortex. Proc Natl Acad Sci U S A 95 (14): 8363–8368.

Duffy A, Horrocks J, Milin R et al. (2012). Adolescent substance use disorder during the early stages of bipolar disorder: a prospective high-risk study. J Affect Disord 142 (1–3): 57–64.

Dunner DL (2002). "A two-illness model of bipolar disorder"– by RT Joffe, LT Young, and GM MacQueen: a commentary. Bipolar Disord 1 (1): 36–37.

Eckardt MJ, Martin PR (1986). Clinical assessment of cognition in alcoholism. Alcohol Clin Exp Res 10 (2): 123–127.

Evershed S (2011). Treatment of personality disorder: skills-based therapies. Adv Psychiatr Treat 17 (3): 206–213.

Faingold C, Li Y, Evans MS (2000). Decreased GABA and increased glutamate receptor-mediated activity on inferior colliculus neurons in vitro are associated with susceptibility to ethanol withdrawal seizures. Brain Res 868 (2): 287–295.

Franken IHA, Hendriks VM (2001). Screening and diagnosis of anxiety and mood disorders in substance abuse patients. Am J Addict 10 (1): 30–39.

George DT, Gilman J, Hersh J et al. (2008). Neurokinin 1 receptor antagonism as a possible therapy for alcoholism. Science Signalling 319 (5869): 1536.

Ghaemi SN, Boiman EE, Goodwin FK (2000). Diagnosing bipolar disorder and the effect of antidepressants: a naturalistic study. J Clin Psychiatry 61 (10): .

Ghaemi SN, Hsu DJ, Soldani F et al. (2003). Antidepressants in bipolar disorder: the case for caution. Bipolar Disord 5 (6): 421–433.

Goudriaan AE, Oosterlaan J, De Beurs E et al. (2006). Neurocognitive functions in pathological gambling: a comparison with alcohol dependence, Tourette syndrome and normal controls. Addiction 101 (4): 534–547.

Grant BF, Stinson FS, Dawson DA et al. (2004a). Prevalence and co-occurrence of substance use disorders and independent mood and anxiety disorders: results from the National Epidemiologic Survey on Alcohol and Related Conditions. Arch Gen Psychiatry 61 (8): 807.

Grant BF, Stinson FS, Dawson DA et al. (2004b). Co-occurrence of 12-month alcohol and drug use disorders and personality disorders in the United States: results from the National Epidemiologic Survey on Alcohol and Related Conditions. Arch Gen Psychiatry 61 (4): 361.

Grembowski DE, Martin D, Patrick DL et al. (2002). Managed care, access to mental health specialists, and outcomes among primary care patients with depressive symptoms. J Gen Intern Med 17 (4): 258–269.

Hasin D, Liu X, Nunes E et al. (2002). Effects of major depression on remission and relapse of substance dependence. Arch Gen Psychiatry 59 (4): 375.

Heffner JL, Fleck DE, DelBello MP et al. (2012). Cigarette smoking and impulsivity in bipolar disorder. Bipolar Disord 14 (7): 735–742.

Heilig M, Koob GF (2007). A key role for corticotropin-releasing factor in alcohol dependence. Trends Neurosci 30 (8): 399–406.

Heinz A, Schmidt LG, Reischies FM (1994). Anhedonia in schizophrenic, depressed, or alcohol-dependent patients–neurobiological correlates. Pharmacopsychiatry 27: 7.

Herman JP (2012). Neural pathways of stress integration: relevance to alcohol abuse. Alcohol Res Curr Rev 34 (4).

Hettema J, Steele J, Miller WR (2005). Motivational interviewing. Annu Rev Clin Psychol 1: 91–111.

Hillemacher T, Bleich S (2008). Neurobiology and treatment in alcoholism—recent findings regarding Lesch's typology of alcohol dependence. Alcohol Alcohol 43 (3): 341–346.

Hirschfeld RMA (2001). Bipolar spectrum disorder: improving its recognition and diagnosis. J Clin Psychiatr 62: 5–9.

Hirschfeld R, Bowden CL, Gitlin MJ et al. (2003). Practice guideline for the treatment of patients with bipolar disorder (revision). Focus 1 (1): 64.

Hoertel N, Le Strat Y, Angst J et al. (2012). Subthreshold bipolar disorder in a US national representative sample: prevalence, correlates and perspectives for psychiatric nosography. J Affect Disord 146 (3): 338–347.

Hoffman PL, Rabe CS, Grant KA et al. (1990). Ethanol and the NMDA receptor. Alcohol 7 (3): 229–231.

Homish GG, Marshall D, Dubovsky SL et al. (2012). Predictors of later bipolar disorder in patients with subthreshold symptoms. J Affect Disord 144 (1–2): 129–133.

Hunot V, Churchill R, Silva de Lima M et al. (2007). Psychological therapies for generalised anxiety disorder. Cochrane Database Syst Rev, CD001848.

Jaffe JH (1992). Current concepts of addiction. Addictive States 70: 1–21.

Jamison KR (2000). Suicide and bipolar disorder. J Clin Psychiatry 61 (Suppl 9): 47–51.

Johnson BA, Rosenthal N, Capece JA et al. (2008). Improvement of physical health and quality of life of alcohol-dependent individuals with topiramate treatment: US multisite randomized controlled trial. Arch Intern Med 168 (11): 1188.

Judd LL, Akiskal HS (2003a). Depressive episodes and symptoms dominate the longitudinal course of bipolar disorder. Curr Psychiatry Rep 5 (6): 417–418.

Judd LL, Akiskal HS (2003b). The prevalence and disability of bipolar spectrum disorders in the US population: re-analysis of the ECA database taking into account subthreshold cases. J Affect Disord 73 (1–2): 123–131.

Judd LL, Akiskal HS, Schettler PJ et al. (2002). The long-term natural history of the weekly symptomatic status of bipolar I disorder. Arch Gen Psychiatry 59 (6): 530.

Judd LL, Schettler PJ, Akiskal HS et al. (2003). Long-term symptomatic status of bipolar I vs. bipolar II disorders. Int J Neuropsychopharmacol 6 (2): 127–137.

Kamarajan C, Porjesz B, Jones K et al. (2006). Event-related oscillations in offspring of alcoholics: neurocognitive disinhibition as a risk for alcoholism. Biol Psychiatry 59 (7): 625–634.

Koch-Weser J, Sellers EM, Kalant H (1976). Alcohol intoxication and withdrawal. N Engl J Med 294 (14): 757–762.

Koenig AM, Thase ME (2009). First-line pharmacotherapies for depression – what is the best choice? Pol Arch Med Wewn 119 (7–8): 478–486.

Koob GF (2004). A role for GABA mechanisms in the motivational effects of alcohol. Biochem Pharmacol 68 (8): 1515–1525.

Koob GF (2009). Dynamics of neuronal circuits in addiction: reward, antireward, and emotional memory. Pharmacopsychiatry 42 (1): 32.

Kranzler HR, Mueller T, Cornelius J et al. (2006). Sertraline treatment of co-occurring alcohol dependence and major depression. J Clin Psychopharmacol 26 (1): 13–20.

Kubota M, Nakazaki S, Hirai S et al. (2001). Alcohol consumption and frontal lobe shrinkage: study of 1432 nonalcoholic subjects. J Neurol Neurosurg Psychiatry 71 (1): 104–106.

Lê AD, Shaham Y (2002). Neurobiology of relapse to alcohol in rats. Pharmacol Ther 94 (1–2): 137–156.

Leeies M, Pagura J, Sareen J et al. (2010). The use of alcohol and drugs to self-medicate symptoms of posttraumatic stress disorder. Depress Anxiety 27 (8): 731–736.

Leichsenring F (2001). Comparative effects of short-term psychodynamic psychotherapy and cognitive-behavioral therapy in depression: a meta-analytic approach. Clin Psychol Rev 73 (1–2): 123–131.

Lesch OM, Walter H (1996). Subtypes of alcoholism and their role in therapy. Alcohol Alcohol 31 (supp 1): 63–67.

Lingford-Hughes AR, Acton PD, Gacinovic S et al. (1998). Reduced levels of GABA-benzodiazepine receptor in alcohol dependency in the absence of grey matter atrophy. Br J Psychiatry 173 (2): 116–122.

Lish JD, Dime-Meenan S, Whybrow PC, Price RA et al. (1994). The National Depressive and Manic-depressive Association (DMDA) survey of bipolar members. J Affect Disord 31 (4): 281–294.

Llopis LJJ, Castillo AA (2008). Efficacy of oxcarbazepine treatment in patients diagnosed with cocaine abuse/dependence. Adicciones 20 (3): 263.

Maddrey WC (2000). Alcohol-induced liver disease. Clin Liver Dis 4 (1): 115–131.

Martin PR, Adinoff B, Weingartner H et al. (1986). Alcoholic organic brain disease: nosology and pathophysiologic mechanisms. Prog Neuropsychopharmacol Biol Psychiatry 10 (2): 147–164.

Martin PR, Singleton CK, Hiller-Sturmhofel S (2003). The role of thiamine deficiency in alcoholic brain disease. Alcohol Res Health 27 (2): 134–142.

Martin P, Weinberg BA, Bealer BK (2007). Healing addiction: An integrated pharmacopsychosocial approach to treatment. Wiley, Hoboken, NJ.

Martinotti G, Di Nicola M, Romanelli R et al. (2007). High and low dosage oxcarbazepine versus naltrexone for the prevention of relapse in alcohol-dependent patients. Hum Psychopharmacol Clin Exp 22 (3): 149–156.

McBride WJ, Li TK (1998). Animal models of alcoholism: neurobiology of high alcohol-drinking behavior in rodents. Crit Rev Neurobiol 12 (4): 339.

McCool BA (2011). Ethanol modulation of synaptic plasticity. Neuropharmacology 61 (7): 1097–1108.

McElroy SL, Keck PE, Pope HG et al. (1992). Clinical and research implications of the diagnosis of dysphoric or mixed mania or hypomania. Am J Psychiatry 149: 1633–1644.

McIntyre R, Katzman M (2003). The role of atypical antipsychotics in bipolar depression and anxiety disorders. Bipolar Disord 5 (s2): 20–35.

Merikangas KR, Herrell R, Swendsen J et al. (2008). Specificity of bipolar spectrum conditions in the comorbidity of mood and substance use disorders: results from the Zurich cohort study. Arch Gen Psychiatry 65 (1): 47.

Merikangas KR, Cui L, Kattan G et al. (2012). Mania with and without depression in a community sample of US adolescents. Arch Gen Psychiatry, 2012.38 v1.

Miklowitz DJ (2008). Adjunctive psychotherapy for bipolar disorder: state of the evidence. Am J Psychiatry 165 (11): 1408.

Moos RH, Moos BS (2006). Participation in treatment and Alcoholics Anonymous: a 16-year follow-up of initially untreated individuals. J Clin Psychol 62 (6): 735–750.

Morgenstern J, Labouvie E, McCrady BS et al. (1997). Affiliation with Alcoholics Anonymous after treatment: a study of its therapeutic effects and mechanisms of action. J Consult Clin Psychol 65 (5): 768.

Myrick H, Brady K (2003). Current review of the comorbidity of affective, anxiety, and substance use disorders. Curr Opin Psychiatry 16 (3): 261–270.

Nepon J, Belik SL, Bolton J et al. (2010). The relationship between anxiety disorders and suicide attempts: findings from the National Epidemiologic Survey on Alcohol and Related Conditions. Depress Anxiety 27 (9): 791–798.

Nicolas JM, Fernández-Solà J, Robert J et al. (2000). High ethanol intake and malnutrition in alcoholic cerebellar shrinkage. QJM 93 (7): 449–456.

Nunes EV, Levin FR (2004). Treatment of depression in patients with alcohol or other drug dependence. JAMA 291 (15): 1887–1896.

Parsons OA, Nixon SJ (1993). Neurobehavioral sequelae of alcoholism. Neurol Clin 11 (1): 205–218.

Peet M (1994). Induction of mania with selective serotonin re-uptake inhibitors and tricyclic antidepressants. Br J Psychiatry 164 (4): 549–550.

Petty F, Kramer GL, Dunnam D et al. (1990). Plasma GABA in mood disorders. Psychopharmacol Bull 26 (2): 157–161.

Petty F, Rush AJ, Davis JM et al. (1996). Plasma GABA predicts acute response to divalproex in mania. Biol Psychiatry 39 (4): 278–284.

Preuss UW, Schuckit MA, Smith TL et al. (2002). Comparison of 3190 alcohol-dependent individuals with and without suicide attempts. Alcohol Clin Exp Res 26 (4): 471–477.

Reich MS, Dietrich MS, Finlayson R et al. (2008). Coffee and cigarette consumption and perceived effects in recovering alcoholics participating in Alcoholics Anonymous in Nashville, Tennessee. Alcohol Clin Exp Res 32 (10): 1799–1806.

Rubak S, Sandbæk A, Lauritzen T et al. (2005). Motivational interviewing: a systematic review and meta-analysis. Br J Gen Pract 55 (513): 305.

Salloum IM, Thase ME (2002). Impact of substance abuse on the course and treatment of bipolar disorder. Bipolar Disord 2 (3p2): 269–280.

Santhakumar V, Wallner M, Otis TS (2007). Ethanol acts directly on extrasynaptic subtypes of $GABA_A$ receptors to increase tonic inhibition. Alcohol 41 (3): 211–221.

Scott EM, Hermens DF, Naismith SL et al. (2012). Distinguishing young people with emerging bipolar disorders from those with unipolar depression. J Affect Disord 144 (3): 208–215.

Skodol AE, Stout RL, McGlashan TH et al. (1999). Co-occurrence of mood and personality disorders: a report from the collaborative longitudinal personality disorders study (CLPS). Depress Anxiety 10 (4): 175–182.

Soto CBD, O'Donnell WE, Allred LJ et al. (1985). Symptomatology in alcoholics at various stages of abstinence. Alcohol Clin Exp Res 9 (6): 505–512.

Strakowski SM, DelBello MP, Fleck DE et al. (2000). The impact of substance abuse on the course of bipolar disorder. Biol Psychiatry 48 (6): 477–485.

Swann AC (1997). Is bipolar depression a specific biological entity? In: Bipolar Disorder: Biological Models and Their Clinical Applications, 255–285.

Swann AC, Bowden CL, Calabrese JR et al. (2000). Mania: differential effects of previous depressive and manic episodes on response to treatment. Acta Psychiatr Scand 101 (6): 444–451.

Swann AC, Dougherty DM, Pazzaglia PJ et al. (2004). Impulsivity: a link between bipolar disorder and substance abuse. Bipolar Disord 6 (3): 204–212.

Tapert SF, Baratta MV, Abrantes AM et al. (2002). Attention dysfunction predicts substance involvement in community youths. J Am Acad Child Adolesc Psychiatry 41 (6): 680–686.

Tarter RE, Kirisci L, Mezzich A et al. (2003). Neurobehavioral disinhibition in childhood predicts early age at onset of substance use disorder. Am J Psychiatry 160 (6): 1078–1085.

Tonigan JS, Connors GJ, Miller WR (2003). Participation and involvement in Alcoholics Anonymous: treatment matching in alcoholism. In: TF Babor, FK Del Boca (Eds.), International research monographs in the addictions. Cambridge University Press, New York, NY, pp. 184–204.

Tsai G, Coyle JT (1998). The role of glutamatergic neurotransmission in the pathophysiology of alcoholism. Annu Rev Med 49 (1): 173–184.

Tsai G, Gastfriend DR, Coyle JT (1995). The glutamatergic basis of human alcoholism. Am J Psychiatry 152 (3): 332–340.

Vengeliene V, Bachteler D, Danysz W et al. (2005). The role of the NMDA receptor in alcohol relapse: a pharmacological mapping study using the alcohol deprivation effect. Neuropharmacology 48 (6): 822–829.

Weinstein DD, Martin PR (1995). Psychiatric implications of alcoholism and traumatic brain injury. Am J Addict 4 (4): 285–296.

Weiss RD, Ostacher MJ, Otto MW et al. (2005). Does recovery from substance use disorder matter in patients with bipolar disorder? J Clin Psychiatr 66 (6): 730–735.

Wills T, Winder D (2013). Ethanol effects on N-methyl-D-aspartate receptors in the bed nucleus of the stria terminalis. Cold Spring Harb Perspect Med 3: a012161.

Handbook of Clinical Neurology, Vol. 125 (3rd series)
Alcohol and the Nervous System
E.V. Sullivan and A. Pfefferbaum, Editors

Chapter 34

Hepatic encephalopathy in alcoholic cirrhosis

ROGER F. BUTTERWORTH*
Neuroscience Research Unit, Hôpital St-Luc (CHUM), University of Montreal, Montreal, Qc., Canada

INTRODUCTION

Hepatic encephalopathy (HE) is a serious neuropsychiatric complication of liver diseases that is characterized by a wide range of neurologic and psychiatric symptoms, including alterations in psychomotor, cognitive, and fine motor functions. Deficits in attention, visual perception, visuospatial construction, as well as motor speed and accuracy are symptoms of early HE in cirrhotic patients (Weissenborn, 2012). Symptoms may be episodic or persistent and are, in most cases, precipitated by factors such as a gastrointestinal bleed, sedatives, dietary protein overload, infection, and/or electrolyte imbalance.

Transjugular intrahepatic portosystemic shunt (TIPS) is an effective procedure used for the prevention and treatment of complications of cirrhosis, including symptomatic portal hypertension, gastrointestinal (variceal) bleeding, and refractory ascites (Pomier-Layrargues, 1996). However, HE following the TIPS procedure may occur in up to 50% of cases. The incidence of post-TIPS encephalopathy is influenced by patient age, occurrence of pre-TIPS HE, as well as serum sodium and the design (particularly the diameter) of the stent used (Rossle and Euringer, 2012).

Differential diagnosis of HE in alcoholic cirrhosis is essential since the neurologic symptoms associated with HE in cirrhosis are non-specific. It is therefore important to confirm the presence of liver disease and to exclude other causes and comorbidities. Lumbar puncture may be performed in some cases in order to exclude altered consciousness due to encephalitis. Neuroimaging studies in general and magnetic resonance imaging (MRI), in particular, are increasingly being applied for the differential diagnosis of HE in alcoholic patients. Multiple episodes of overt HE in patients with cirrhosis may result in persistent and cumulative deficits of working memory and learning (Sotil et al., 2009; Bajaj et al., 2010; Riggio et al., 2011). Multiple episodes of HE may also result in the neurodegenerative disorder known as acquired hepatocerebral degeneration that is characterized by neuronal necrosis in basal ganglia, cerebellum, and cerebral cortex in addition to Alzheimer type II astrocytosis. Other neuropathologic findings that are particularly prevalent in alcoholic cirrhosis include Wernicke' encephalopathy and cerebellar degeneration (Fig. 34.1) and their increased incidence is attributed, in part, to a loss of liver stores of thiamine (Kril and Butterworth, 1997). Other cirrhosis-related neuropathology in basal ganglia and spinal cord has been attributed to "parkinsonism in cirrhosis" and postshunt myelopathy respectively (Butterworth, 2007, 2013).

Clinical grading of HE remains problematic in patients with cirrhosis. Although it is well established that ammonia plays a key role in the pathogenesis of HE in cirrhosis (described in detail in the section on pathophysiology of HE in alcoholic cirrhosis), measurement of venous ammonia concentrations is of limited value for HE grading and electrophysiologic techniques such as visual/auditory evoked potentials or electroencephalography (EEG) are not commonly used, due to their time-consuming and costly nature that often require highly trained personnel.

The West Haven scale (Table 34.1) continues to be used to grade the severity of altered consciousness in HE; briefly, grade 1 is characterized by lethargy and psychomotor slowing, and grade 2 principally involves disorientation. Somnolence and stupor are classified as grade 3 and appearance of the comatose state defines grade 4. Motor symptoms that accompany these changes in level of consciousness include asterixis, ataxia, dysarthria, tremor, rigidity, and hypokinesia, as well as spasticity and extensor plantar responses, all of which increase with increasing HE grade, as assessed by West Haven (Weissenborn, 2012). However, it should be borne

*Correspondence to: Dr. Roger F. Butterworth, Neuroscience Research Unit, Hôpital St-Luc (CHUM), 1058 St-Denis, Montreal, Qc., Canada, H2X 3 J4. E-mail: rb@enceph.com

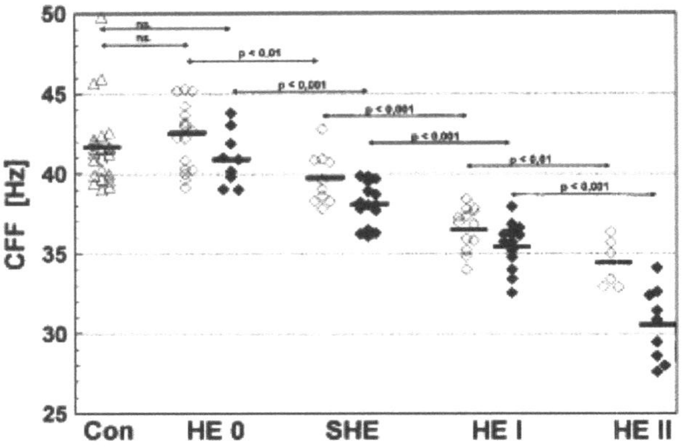

Fig. 34.1. Critical flicker frequency (CFF) values in Hz in control subjects (CON) and in cirrhotic patients without hepatic encephalopathy (HE 0), with minimal or subclinical HE (SHE), or with overt HE (HE I, HE II) defined by Westhaven criteria (Table 34.1). Open circles, non-alcoholics; filled circles, alcoholics. (Reproduced from Kircheis et al., 2002, with permission.)

Table 34.1

West Haven scale for grading the severity of altered consciousness in hepatic encephalopathy

Grade	Criteria
1	Trivial lack of awareness
	Shortened attention span
	Impaired performance of addition
	Euphoria or anxiety
2	Lethargy or apathy
	Minimal disorientation for time or place
	Subtle personality change
	Inappropriate behavior
	Impaired performance of subtraction
3	Somnolence to semistupor, responsive to verbal stimuli
	Confusion
	Gross disorientation, bizarre behavior
4	Coma (unresponsive to verbal or noxious stimuli)

in mind that the use of West Haven criteria has been criticized on account of its inherent subjective nature, with the potential to result in high interexaminer variability. Consequently, alternative clinical grading scales are starting to replace West Haven (Bajaj et al., 2009; Cordoba, 2011).

An alternative technique to evaluate neurologic dysfunction in patients with cirrhosis makes use of the measurement of critical flicker frequency (CFF), a quantitative measure that is based on the correlation between cerebral processing of oscillatory visual stimuli and its impairment due to worsening encephalopathy. The CFF procedure has been extensively used for the evaluation and grading of neurocognitive changes in a wide range of neurologic disorders, including low-grade HE in cirrhosis (Kircheis et al., 2002), where a CFF cutoff of 39 Hz led to a 100% differentiation between cirrhotic patients with clinically overt HE compared to patients without clinical evidence of HE. This held true for both alcoholic and non-alcoholic cirrhotics (Fig. 34.1), was only minimally affected by age or time of day, and did not require lengthy training of personnel. On the other hand, a wide range of neuroactive and psychoactive drugs may affect CFF performance (Smith and Misiak, 1976), including benzodiazepines that may have been prescribed as sedatives or used in association with endoscopy in cirrhotic patients.

Minimal HE (MHE) is now a well-recognized phenomenon with major impact on quality of life in patients with cirrhosis (Dhiman et al., 2010). Patients with MHE have no detectable cognitive or psychomotor abnormalities upon routine clinical assessment but manifest clear deficits upon psychometric testing. Loss of ability to perform complex tasks, operate machinery, or to drive an automobile (Bajaj et al., 2007) have also been reported in these patients. Psychometric testing reveals abnormalities in attention such as impairments of vigilance and orientation, executive functions, visuospatial coordination, and reaction times. There are currently several available psychometric tests for the diagnosis of MHE. The Psychometric Hepatic Encephalopathy Score (PHES) is a multitest paradigm made up of number connection tests (NCT-A and NCT-B) together with a line-drawing test and a serial dotting test (Weissenborn et al., 2001). This test is simple and inexpensive to apply, relying on non-verbal tasks, and has wide international applicability, requiring minimal translation (Randolph et al., 2009). The definition of MHE using the PHES is a score of 2 SDs below normal on at least two of the component tests. More recently, a computerized test known as the Inhibitory Control Test (ICT) has been gaining in popularity. This easily administered test assesses sustained attention, vigilance, working memory, and response inhibition by measurement of responses to targets but not to so-called "lures" (Bajaj et al., 2008). The CFF technique (see above) has also been employed for the diagnosis of MHE (Romero-Gomez et al., 2007).

Positron emission tomography (PET) using ligands that bind to dopamine receptors or transporters in brain has been employed to accurately diagnose and assess the degree of transmitter changes in cases of parkinsonism in cirrhosis (Weissenborn et al., 2000). However, costs, together with lack of availability of the necessary equipment, have so far precluded its wide diagnostic application in liver failure.

NEUROPATHOLOGY OF HE

Given that the most conspicuous neuropathologic changes are glial rather than neuronal, HE has been referred to as a "primary gliopathy"; in cirrhosis, morphologic changes to astrocytes and microglia predominate. Changes to astrocytes, known as "Alzheimer type II astrocytosis" are consistently observed in brain sections from cirrhotic patients who died in grade 4 HE (Butterworth et al., 1987); such changes (Fig. 34.2 and Table 34.2) are characterized by a large pale nucleus, prominent nucleolus with margination of chromatin, and glycogen accumulation. Cells manifesting Alzheimer type II changes vary in size and shape with spherical forms in cerebral cortex to irregular forms in basal ganglia; they may also appear as doublets or multiplets, suggestive of hyperplasia (Norenberg, 1987). The Alzheimer type II phenotype has been described in a wide range of hyperammonemic syndromes (Butterworth et al., 1987) associated with congenital urea cycle enzyme defects, as well as in experimental animals with urease-induced hyperammonemia and in primary cultures of astrocytes exposed to ammonia. Since activation of microglial cells occurs in relation to the presence of HE in patients with cirrhosis (Zemtsova et al., 2011), a neuroinflammatory component in the pathogenesis of HE is suggested.

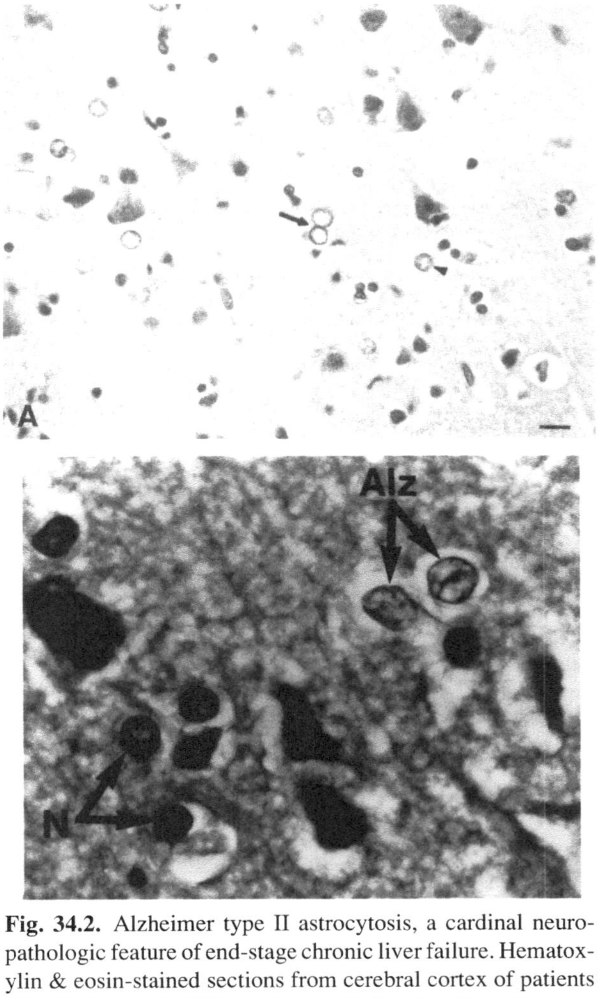

Fig. 34.2. Alzheimer type II astrocytosis, a cardinal neuropathologic feature of end-stage chronic liver failure. Hematoxylin & eosin-stained sections from cerebral cortex of patients who died in grade IV hepatic encephalopathy. Note the presence of normal astrocyte nuclei (N) compared to Alzheimer type II changes (Alz) in astrocyte nuclei characterized by swollen, watery appearance, a prominent nucleolus, and margination of chromatin. Note the presence of a doublet (arrow in lower magnification image), suggestive of hyperplasia. (Composite figure from Norenberg, 1987 and Butterworth et al., 1987, with permission.)

PATHOPHYSIOLOGY OF HE IN ALCOHOLIC CIRRHOSIS

Ammonia

Liver is the major organ responsible for the removal of excess ammonia in the form of urea or glutamine. Hepatic ammonia removal is highly compartmentalized, involving two distinct cell types; urea is synthesized from ammonia via the urea cycle in periportal hepatocytes whereas perivenous hepatocytes transform ammonia into glutamine via glutamine synthetase (GS). Cirrhotic patients commonly have intra- and extrahepatic portal-systemic shunts and this portal-systemic shunting, in addition to the impaired residual hepatocyte function, results in decreased ammonia removal. Consequently, hyperammonemia and increased brain ammonia content are characteristic of cirrhosis.

Evidence for an association between HE and ammonia dates back over a century to the studies of Eck, who fed meat to portacaval-shunted dogs leading to severe neurologic impairment progressing to coma. Then, in the 1950s, the treatment of ascites in cirrhotic patients using ammonium ion exchange resins led to reduction of ascitic volume but also led to neurologic symptoms that were indistinguishable from HE (Gabuzda et al., 1952). Arterial blood and brain ammonia concentrations are frequently increased in patients with HE and PET studies using $^{13}NH_3$ reveal increases in cerebral ammonia uptake and metabolism in cirrhotic patients with mild HE independently of the etiology of cirrhosis (Lockwood et al., 1991; Keiding et al., 2006).

Concentrations of ammonia equivalent to those reported in brain in HE have deleterious effects on central nervous system (CNS) excitability (Felipo and Butterworth, 2002) and direct effects of the NH_4^+ ion on both inhibitory and excitatory neurotransmission have been reported. Concentrations of ammonia as low as 0.5 mM impair postsynaptic inhibition in brain by inactivation of the extrusion of chloride ion from

Table 34.2

Occurrence of Alzheimer type II astrocytosis and unsuspected thalamic and cerebellar pathology in alcoholic cirrhotic patients

	Patient		Neuropathology			
	Age	Sex	Acute WE	Chronic WE	Cerebellar Degeneration	Alzheimer II
1	62	M	–	–	+	+
2	42	F	–	–	–	++
3	61	M	–	–	–	+
4	47	M	–	–	–	+
5	44	M	–	–	–	+++
6	65	M	–	–	++	++
7	42	M	–	–	–	++
8	46	F	–	–	+++	++
9	59	F	present	present	+	+
10	45	M	–	–	–	++
11	77	M	–	–	++	++
12	52	M	–	present	+++	+
13	61	F	–	present	+++	+++
14	28	M	–	–	–	++
15	48	M	–	–	+	++
16	57	M	–	–	+	++
17	37	F	–	present	++	++
18	59	M	–	–	–	+
19	54	M	–	–	+	++
20	62	F	–	–	–	+
21	44	M	–	–	++	++
22	54	M	–	–	–	+++
23	51	M	–	–	+	++
24	58	M	present	–	+	++
25	56	M	–	present	–	+
26	51	F	–	–	+	++
27	45	F	present	–	–	+
28	47	M	present	–	–	++
29	57	M	–	present	++	+
30	60	M	–	–	++	++

Rating of pathologic changes: – absent; + mild; ++ moderate; +++ severe.
Adapted from Kril and Butterworth (1997), with permission.

neurons (Raabe, 1987). Ammonia also has deleterious effects on excitatory neurotransmission, where both presynaptic and postsynaptic effects have been reported (Felipo and Butterworth, 2002).

Submillimolar concentrations of ammonia inhibit the tricarboxylic acid cycle enzyme alpha-ketoglutarate dehydrogenase (Lai and Cooper, 1986) and concomitantly stimulate the glycolytic enzyme phosphofructokinase, resulting in increased lactate production in brain (Tofteng et al., 2002). Cerebrospinal fluid (CSF) lactate concentrations correlate well with the deterioration and subsequent recovery of neurologic status in experimental HE (Therrien et al., 1991) and in cirrhotic patients with mild to moderate HE (Yao et al., 1987).

Manganese

Manganese is normally eliminated via the hepatobiliary route and blood manganese concentrations are increased in cirrhotic patients. A highly consistent finding on MRI of cirrhotic patients is bilateral symmetric hyperintensities in globus pallidus on T_1-weighted imaging (Spahr et al., 1996), a representative example of which is shown in Figure 34.3.

There is evidence to suggest that these hyperintense images are the result of pallidal manganese deposition. Similar MR pallidal signals have been reported in other conditions, including Alagille's syndrome (Devenyi et al., 1994) (a hepatic disorder characterized by

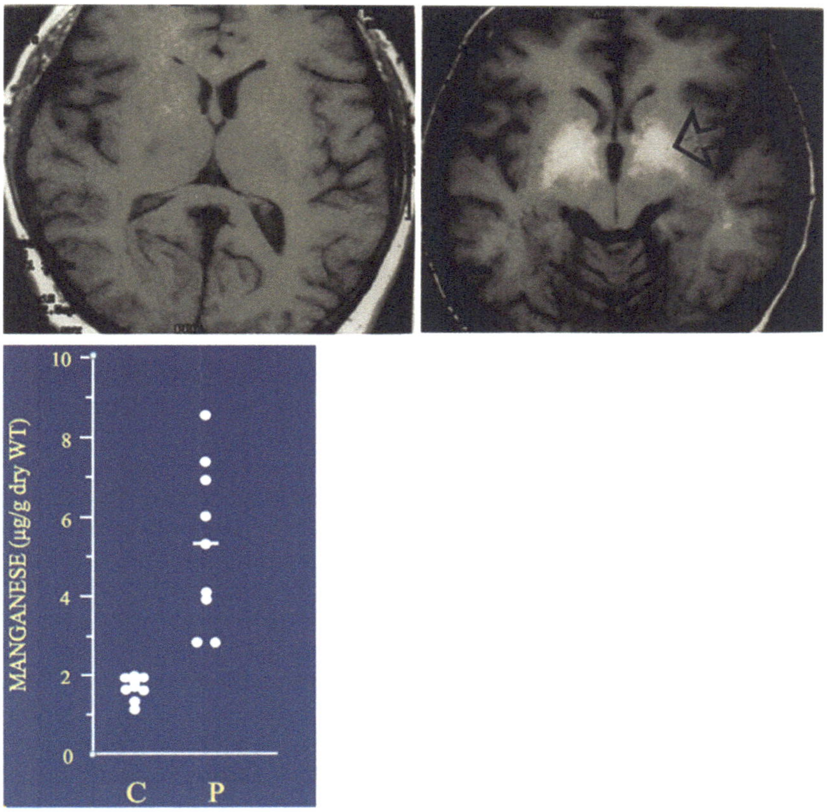

Fig. 34.3. Magnetic resonance imaging in cirrhosis. Upper panel: T1-weighted signal hyperintensities in globus pallidus of an alcoholic cirrhotic patient with mild hepatic encephalopathy compared to an age-matched control. Lower panel: Increased manganese content of globus pallidus samples obtained at autopsy from cirrhotic patients (P) compared to age-matched controls (C).

cholestasis and intrahepatic bile duct paucity) as well as in patients during total parenteral nutrition (Mirowitz et al., 1991). Both of these conditions are associated with increased blood manganese. Direct measurements reveal up to sevenfold increases in manganese content of pallidum obtained from cirrhotic patients who died in hepatic coma (Pomier-Layrargues et al., 1995; Fig. 34.3) and from animals with surgical portacaval shunts where accumulation of manganese was observed not only in pallidum but, to a lesser extent, in other basal ganglia structures (Rose et al., 1999). More recently, MR signal hyperintensities were observed in substantia nigra in 22% of patients with cirrhosis awaiting liver transplantation (LT) (Burkhard et al., 2003) and, again, these signal hyperintensities were attributed to manganese deposition. It was proposed that these changes were responsible for the extrapyramidal signs attributed to the disorder known as "parkinsonism in cirrhosis" (Burkhard et al., 2003; Butterworth, 2013).

Inflammation

Patients with cirrhosis are susceptible to infection (due to impaired host defense mechanisms) that may lead to a

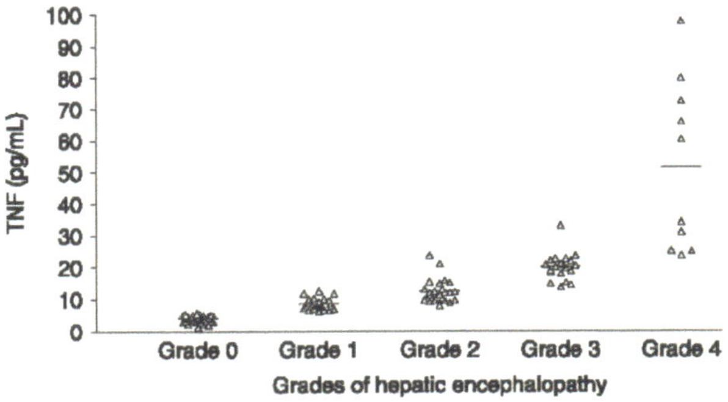

Fig. 34.4. Circulating levels of tumor necrosis factor-α (TNF-α) in cirrhotic patients as a function of encephalopathy grade. (Reproduced from Odeh, 2007, with permission.)

systemic inflammatory response, the presence of which may precipitate HE. Hepatic tissue damage due to toxins, particularly alcohol, may exacerbate this inflammatory response. Circulating levels of the proinflammatory cytokine tumor necrosis factor-alpha are invariably increased in patients with cirrhosis and the extent of the increase correlates significantly with severity of HE in these patients (Odeh, 2007; Fig. 34.4). Additive (and possibly synergistic) effects of inflammation and ammonia in the pathogenesis of HE in cirrhosis have been suggested (Odeh et al., 2005; Shawcross et al., 2007). Inflammation of the brain *per se* (neuroinflammation), characterized by microglial activation and increased brain cytokine production, has been reported in brain tissue from alcoholic patients with HE (Zemtsova et al., 2011).

Brain glucose and energy metabolism

There is no convincing evidence that HE in cirrhosis is primarily the consequence of reduced brain levels of high-energy phosphates, whether assessed biochemically (Hindfelt et al., 1977) or spectroscopically (Bates et al., 1989). On the other hand, alterations of brain glucose utilization, of cerebral blood flow, and of brain glycolytic and oxidative pathways have consistently been reported in patients with cirrhosis and HE. These alterations of brain metabolism are dependent on the brain region under investigation. For example, PET studies using ^{18}F-deoxyglucose reveal significant decreases in glucose utilization in the anterior cingulate cortex in cirrhotic patients with mild HE (Lockwood et al., 2002). This brain structure is implicated in the control of the anterior attention system responsible for the monitoring of, for example, response selection to visual stimuli. A significant correlation was observed in this study between the extent of decreased glucose utilization in anterior cingulate and the impaired performance in attention-demanding tasks such as NCT-B and digit symbol psychometric tests. Measurement of cerebral perfusion using ^{15}O-H_2O-PET revealed a redistribution of blood flow from regions such as the anterior cingulate cortex to subcortical structures, including thalamus (Lockwood et al., 1997).

Precipitation of severe encephalopathy and coma in experimental animals with chronic liver failure is accompanied by impaired brain glucose oxidation, leading to accumulation of lactate and alanine (Therrien et al., 1991), together with reduced concentrations of glutamate and aspartate sufficiently severe to impair the malate-aspartate shuttle (Hindfelt et al., 1977). Increased brain lactate and alanine concentrations in HE are consistent with impaired oxidation of pyruvate, a finding that has been attributed to ammonia-induced inhibition of the tricarboxylic acid cycle enzyme α-ketoglutarate dehydrogenase (α-KGDH) (Lai and Cooper, 1986). Neurologic disorders such as Wernicke's encephalopathy due to thiamine deficiency, as well as some cases of congenital lactic acidosis, manifest a similar spectrum of α-KGDH inhibition, lactate accumulation, and encephalopathy (Butterworth and Heroux, 1989; Navarro et al., 2008), suggesting a common mechanism with that of HE in cirrhosis. Increased CSF and brain lactate is a well-established predictor of severity of HE in cirrhosis (Hindfelt et al., 1977; Yao et al., 1987; Therrien et al., 1991).

The GABA system and neurosteroids

Throughout the 1980s a great deal of attention was devoted to the theory that HE was the result of "increased GABAergic tone" in the brain. Abnormalities of the brain gamma-aminobutyric acid (GABA) system were initially reported in experimental animal models of acute liver failure (Schafer and Jones, 1982). However, in animal models of cirrhosis and portal-systemic shunting, no alterations of the GABA system were observed (Butterworth and Giguère, 1986; Roy et al., 1988). Furthermore, autopsied brain tissue from cirrhotic patients who died in hepatic coma was found to contain normal activities of GABA-related enzymes and unchanged densities and affinities of GABA receptors (Lavoie et al., 1987; Butterworth et al., 1988). This lack of evidence of intrinsic alterations of the GABA system *per se* led to interest in modulatory components of the GABA-A receptor complex (Fig. 34.5).

Benzodiazepine-binding sites form part of the GABA-A receptor complex and stimulation of the benzodiazepine-binding site "facilitates" the action of GABA on the functionally linked GABA-A site of the complex. In this way, chloride channel opening is increased, leading to hyperpolarization and inhibition. Improvement of neurologic status in patients with

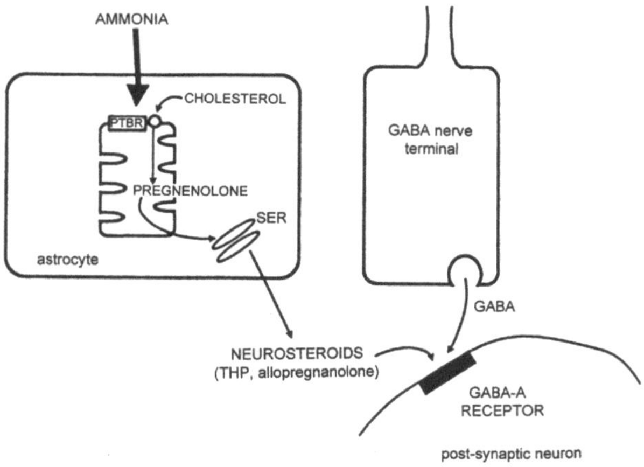

Fig. 34.5. GABA-A receptor activation by the neurosteroid allopregnanolone: astrocyte neuron signaling in the pathogenesis of hepatic encephalopathy in cirrhosis. Mitochondrial translocator protein activated by ammonia or manganese leads to uptake of cholesterol and its conversion to allopregnanolone that stimulates the steroid modulatory site on the postsynaptic GABA-A receptor, leading to increased neuroinhibition. GABA, gamma-aminobutyric acid; PTBR, peripheral benzodiazepine receptor; SER, smooth endoplasmic reticulum; THP, tetrahydroprogesterone.

cirrhosis, many of alcoholic etiology, with HE was reported following administration of the central benzodiazepine receptor antagonist flumazenil (Pomier-Layrargues et al., 1994; Gyr et al., 1996). It was suggested that the beneficial action of flumazenil in these patients was due to inhibition of the action of an "endogenous ligand." An earlier investigation had failed to show any significant alterations of densities or affinities of benzodiazepine modulatory sites in material from cirrhotic patients with HE (Butterworth et al., 1988), leaving open the possibility that the beneficial effect of flumazenil was the result of the blocking of the action of "endogenous benzodiazepines" (Mullen et al., 1990).

Initial studies demonstrated that CSF from patients with advanced HE did contain significant amounts of "benzodiazepine-like" substances (Olasmaa et al., 1990) and a known pharmaceutical benzodiazepine, diazepam, as well as its *N,N*-desmethyl metabolite, both of which are active positive allosteric modulators of GABA neurotransmission, were isolated from serum and CSF of cirrhotic patients with HE. The interest generated by these reports was, unfortunately, tempered by the fact that the concentrations of benzodiazepines reported in blood, CSF, and brain extracts of these HE patients were very low (an order of magnitude below levels associated with their sedative actions).

An alternative explanation for the phenomenon of "increased GABAergic tone" in HE has recently emerged, an explanation that involves translocator protein (TLP), a mitochondrial protein expressed by neuroglia. Endogenous agonist ligands for TLP include the neuropeptide diazepam-binding inhibitor and its processing peptide octadecaneuropeptide and increases of these peptides have been reported in HE (Rothstein et al., 1989; Butterworth et al., 1991). Activation of the TLP by these ligands in brain results in increased synthesis of a class of compounds known as neurosteroids that are synthesized independently of peripheral steroidal sources (adrenals and gonads). Moreover, the neurosteroid allopregnanolone, a potent positive allosteric modulator of the GABA-A receptor complex, is increased severalfold in autopsied brain tissue from cirrhotic patients (mostly alcoholic etiology), who died in grade 4 HE (Ahboucha et al., 2005; Fig. 34.6).

Meanwhile, further studies of the precise relationship between TLP activation and HE were ongoing in material from patients with HE and from animal models. Increased densities of TLP sites were reported in autopsied frontal cortex and caudate nuclei of cirrhotic patients who died in grade 4 HE (Lavoie et al., 1990). More recently, these findings were confirmed in cirrhotic patients with PET and the TLP ligand ^{11}C-PK11195 (Cagnin et al., 2006), where the magnitude

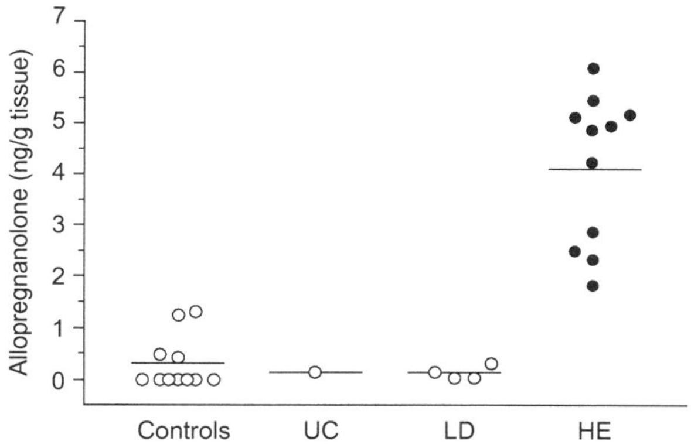

Fig. 34.6. Allopregnanolone content of cortical brain tissue from cirrhotic patients who died in grade IV encephalopathy. Panel A: Structure of allopregnanolone; panel B: brain allopregnanolone concentrations in hepatic encephalopathy patients (HE), in cirrhotic patients without HE (LD), in a patient who died in uremic coma (UC), and in age-matched controls (Controls). (Adapted from Ahboucha et al., 2005, with permission).

of the increased signals was positively correlated with the severity of cognitive impairment. Since TLP is localized predominantly on glial (microglial and astrocytic) mitochondria in mammalian brain, it is not surprising that increased TLP expression in HE is associated with mitochondrial changes. Indeed, portacaval anastomosis in the rat leads to astrocytic mitochondrial proliferation, which also results when cultured cortical astrocytes (Norenberg and Lapham, 1974) or C6 glioma cells (Shiraishi et al., 1990) are exposed to TLP agonist ligands. Moreover, expression of TLP is significantly correlated with the presence of Alzheimer type II changes in autopsied brain tissue from cirrhotic patients who died in grade 4 HE (Bélanger et al., 2004).

Serotonin

Evidence of increased serotonin (5-HT) synthesis and turnover in brain in HE includes the report of increased concentrations of the 5-HT metabolite 5-hydroxyindoleacetic acid (5-HIAA) in CSF of cirrhotic patients in grade 4 HE (Young et al., 1975), in autopsied brain tissue from similar patients (Bergeron et al., 1989) and in the brains of portacaval-shunted rats (Bergeron et al., 1990). Moreover, these studies showed that 5-HIAA/5-HT concentration ratios (considered to be a measure of 5-HT turnover) are also increased in brain tissue from cirrhotic patients who died in grade 4 HE (Bergeron et al., 1989). Direct measurement using an *in vivo* decarboxylase inhibition assay was used to confirm increased 5-HT turnover in the brains of portacaval-shunted rats (Bengtsson et al., 1991) and increased brain 5-HT turnover was found to correlate with the degree of portal-systemic shunting and blood ammonia in rats following variable portal vein stenosis (Lozeva et al., 2004), an animal model of minimal HE.

Other evidence consistent with increased 5-HT turnover in human HE includes the report of increased activities of monoamine oxidase MAO_A and of MAO_A mRNA in autopsied brain tissue from cirrhotic patients who died in grade 4 HE (Mousseau et al., 1997) with concomitant increases of the 5-HT metabolite, 5-HIAA (Bergeron et al., 1989). Increased densities of postsynaptic serotonin ($5\text{-}HT_2$) receptors were also reported in this material (Raghavendra Rao and Butterworth, 1994). Taken together, these findings suggest a 5-HT synaptic deficit in HE. Serotonin plays a key role in the regulation of sleep mechanisms and altered 5-HT neurotransmission has been implicated in a range of psychiatric disorders, leading to the possibility that increased 5-HT turnover in brain could relate to the early signs and symptoms of HE in cirrhosis, including sleep disorders and depression.

Histamine

Histamine is involved in a wide range of physiologic functions, including sleep, arousal, and circadian rhythmicity. Alterations of the brain histamine system in experimental HE include increases in brain histamine and its metabolite tele-methylhistamine, as well as increased histamine release from the hypothalamus (Lozeva-Thomas et al., 2004), suggesting increased histaminergic activity. Increased brain histamine could result from increased brain availability of the precursor amino acid L-histidine. Activation of postsynaptic histamine H_1 receptors is known to suppress deep slow-wave sleep (Tasaka, 1994) and these receptors are implicated in the entrainment of circadian rhythms to the light-dark cycle in mammals (Jacobs et al., 2000). HE in chronic liver failure is characterized by decreased total duration of slow-wave sleep and disorganization of the normal sleep cycle (Cordoba et al., 1998). Reports of upregulation of histamine H_1 receptors in cerebral cortex of HE patients together with changes in serotonin receptors in these patients could account for the sleep disturbances. Histamine H_1 receptor blockers improve sleep quality in humans and significant improvements in circadian rhythms have been reported in portacaval-shunted rats administered the H_1 receptor blocker mepyramine (Lozeva et al., 2000).

Dopamine

The presence of extrapyramidal signs and symptoms in patients with end-stage chronic liver failure has prompted, by analogy with the well-established dopamine (DA) deficit in Parkinson's disease, evaluation of the DA system in HE. Studies in autopsied brain tissue from cirrhotic patients who died in hepatic coma reveal severalfold increases of the DA metabolite homovanillic acid (Bergeron et al., 1989). Similar findings, consistent with increased DA turnover, were reported in the brains of rats following portacaval anastomosis (Bergeron et al., 1990). A possible explanation for these findings could relate to the report of increased activities of the monoamine-metabolizing enzyme MAO-A (Mousseau et al., 1997). Densities of the postsynaptic DA (D_2) receptor are reduced in autopsied pallidum of cirrhotic patients (Mousseau et al., 1993) and this finding was confirmed in HE patients using single-photon emission computed tomography (Weissenborn et al., 2000). A disorder known as "parkinsonism in cirrhosis" (a form of acquired hepatolenticular degeneration with many clinical and biochemical similarities to Parkinson's disease) has been described (Butterworth, 2013) and estimates of its incidence in patients with cirrhosis range from 2% to 22% (Burkhard et al., 2003). MRI evaluation reveals T1-weighted signal hyperintensities in substantia nigra in addition to globus pallidus and its cause is suspected to be the deposition of manganese in these brain structures, leading to dopaminergic neuronal dysfunction and cell death (Burkhard et al., 2003).

THERAPEUTIC ADVANCES

Ammonia-lowering strategies

Hyperammonemia in cirrhosis results from impaired interorgan trafficking of ammonia (Chatauret and Butterworth, 2004), as shown in Figure 34.7. Unlike the liver, skeletal muscle and brain do not have an effective urea cycle and must rely on glutamine synthesis for ammonia removal so that, in severe liver failure, muscle becomes the major route for ammonia detoxification (Ganda and Ruderman, 1976). A molecular basis for this has been proposed involving a posttranslational increase of GS (Desjardins et al., 1999). Brain does not adapt to liver failure by induction of GS activity, a finding that has been attributed to oxidative/nitrosative stress leading to protein tyrosine nitration of the enzyme (Schleiss et al., 2002). In spite of this lack of induction of GS, brain glutamine concentrations are invariably increased in cirrhosis and increased CSF glutamine concentrations correlate well with HE grade in cirrhotic patients (Laubenberger et al., 1997).

Restriction of dietary protein is no longer recommended as a means of prevention of HE in cirrhotic patients (Cordoba et al., 2004). Long-term nitrogen restriction is potentially harmful and a positive nitrogen balance is essential in order to promote liver regeneration and to increase the capacity for ammonia removal by skeletal muscle. Protein intake in the

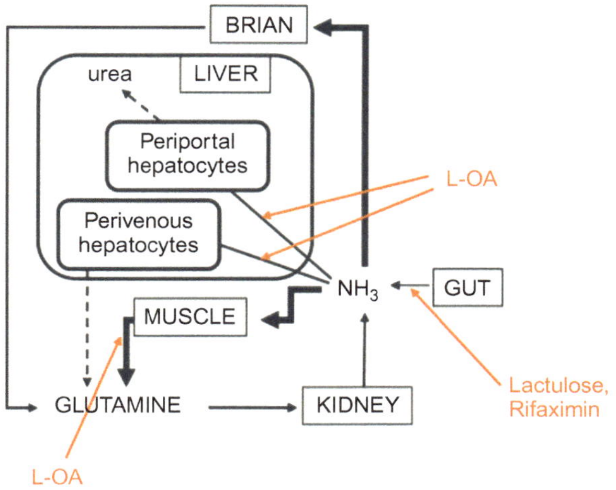

Fig. 34.7. Interorgan trafficking of ammonia in end-stage chronic liver failure. Ammonia removal by the liver as urea or glutamine is reduced by 80% in cirrhosis; ammonia uptake by brain and skeletal muscle is concomitantly increased. Non-absorbable disaccharides (e.g., lactulose) and antibiotics (e.g., rifaximin) act by reducing gut ammonia production. L-ornithine L-aspartate (L-OA) acts by optimizing liver ammonia removal by residual hepatocytes and stimulates ammonia removal by muscle.

1–2 g/kg/day range is now generally recommended in order to maintain an adequate nitrogen balance (Plauth et al., 2006).

The non-absorbable disaccharide, lactulose, is routinely used for lowering of blood ammonia in patients with cirrhosis. The ammonia-lowering effect involves increased fecal nitrogen excretion by facilitation of the incorporation of ammonia into bacteria as well as a cathartic effect. A controlled trial demonstrated efficacy of lactulose in cirrhotic patients with minimal HE (Prasad et al., 2007) as well as for secondary prophylaxis (Sharma et al., 2009). Antibiotics continue to be used for reduction of circulating ammonia in cirrhosis. Mechanisms of action include reduction of gut ammonia production but inhibition of intestinal glutaminase has been suggested in some cases (Mullen and Prakash, 2012). Aminoglycosides (neomycin, paromomycin) are effective but used for short periods of time due to adverse events such as ototoxicity. More recently, the antibiotic rifaximin has been shown to be effective for secondary HE prophylaxis in patients with cirrhosis (Bass et al., 2010). Probiotics increase the intestinal production of urease-negative bacteria, thereby decreasing gut ammonia production. Lactic acid-producing probiotics have the advantage of reducing gut ammonia absorption. A controlled clinical trial with these agents led to significant neurologic improvement and lowering of venous ammonia in cirrhotic patients (Agrawal et al., 2012).

An alternative means of lowering blood ammonia in cirrhosis targets the stimulation of ammonia fixation. Ammonia is normally removed by urea formation (periportal hepatocytes) as well as by glutamine formation (perivenous hepatocytes, muscle, and brain). L-ornithine L-aspartate is effective in lowering blood ammonia and concomitantly improving neuropsychiatric status in cirrhotic patients (Kircheis et al., 1997). Studies in experimental animals reveal that the beneficial effect of L-ornithine L-aspartate is primarily due to the stimulation of urea formation by residual hepatocytes in the liver, in addition to stimulation of glutamine formation by skeletal muscle (Rose et al., 1998). Successful lowering of blood ammonia concentrations in cirrhotic patients has also been accomplished using sodium benzoate (Sushma et al., 1992) and, following up on studies in experimental animals (Therrien et al., 1997), a protective effect of L-carnitine agonist ammonia-precipated encephalopathy in cirrhotic patients has been described (Malaguarnera et al., 2005).

Neuropharmacology

Clinical studies continue to focus on the reduction of circulating ammonia, with regrettably little attention paid to therapeutic targeting of the brain. However, as the precise pathophysiologic mechanisms and alterations in neurotransmitter systems responsible for the pathogenesis of HE become more clearly defined, more innovative pharmacologic approaches are expected to emerge.

A number of controlled clinical trials have evaluated the efficacy of the benzodiazepine receptor antagonist flumazenil in cirrhotic patients with HE. In a subset of HE patients, improvements following flumazenil were spectacular (Pomier-Layrargues et al., 1994). However, as discussed in the previous section on the GABA system, enthusiasm for this approach was tempered by the possible confounding effects of prior exposure of patients to pharmaceutic benzodiazepines (frequently used as sedatives or as part of an endoscopic workup in cirrhotic patients) and by the poor correlation between the clinical response and blood levels of benzodiazepines in these patients (Butterworth et al., 1995). Adding to these difficulties is the short half-life and lack of an oral formulation for flumazenil. Both L-DOPA (Michel et al., 1980) and the dopamine receptor agonist bromocriptine (Uribe et al., 1979; Morgan et al., 1980) have been used in clinical trials in cirrhotic patients with HE. While results were not encouraging in terms of overall cognitive improvement, it is possible that they have a beneficial effect on motor performance (Burkhard et al., 2003). Following up on the demonstration of a beneficial effect of histamine H_1 antagonist on neurologic function in an experimental animal model of HE (Lozeva et al., 2000), results of a controlled clinical trial revealed improvements in sleep quality in cirrhotic patients with minimal HE following administration of the H_1 blocker hydroxyzine (Spahr et al., 2007).

Liver support systems

Artificial liver support systems aimed at the normalization of plasma composition of essential factors together with removal of toxins have been evaluated for the treatment of HE in cirrhosis. Such systems are generally based upon bioreactors loaded with hepatocytes. However, controlled clinical trials using such systems are limited. Albumin dialysis systems such as molecular absorbant recirculating system (MARS) remove manganese and benzodiazepines as well as certain cytokines and have been the subject of controlled clinical trials (Hassanein et al., 2007).

Liver transplantation

LT is the only definitive treatment for end-stage liver disease and its complications, including HE, are either reversed or significantly improved following LT (Tarter et al., 1992), although complete improvements generally take some time. In a study of the effects of LT in cirrhotic patients with MHE, visuospatial function improved 3 months post-LT, with improved spatial short- and long-term memory at 6 months, selective attention at 18 months, followed some time later by improvement in verbal short-term memory (Mattarozzi et al., 2004). Moreover, responses to LT in MHE are not uniform. On the other hand, parameters such as MR signal hyperintensities (Pujol et al., 1996), EEG (Ciancio et al., 2002), and brain glucose utilization (Senzolo et al., 2009) generally resolve promptly following LT in patients with cirrhosis.

The presence and severity of HE in cirrhosis affect LT outcomes, including neurologic recovery, quality of life, and survival. In a study of the effects of recurrent episodes of overt HE on transplant outcome in cirrhotic patients, Bajaj et al. (2010) reported persistent and cumulative deficits of cognitive function as a function of the number of HE episodes. An effect of previous episodes of overt HE on posttransplant survival has also been reported (Bajaj and Saeian, 2005). Some patients with parkinsonism in cirrhosis respond to LT (Burkhard et al., 2003). However, posttransplant residual extrapyramidal symptoms have been reported in some cases, suggesting an element of irreversibility (Butterworth, 2013). Some post-LT neurologic "sequelae" may, in fact, be due to existing, unrecognized pre-LT degenerative disorders associated with chronic alcoholism in cirrhosis (Kril and Butterworth, 1997).

SUMMARY

There can be little doubt, given the overwhelming body of evidence from both clinical and experimental studies, that ammonia plays a major role in the pathogenesis of HE in cirrhosis. Levels of the ammonia detoxification product, glutamine, are increased in brain and correlate with the degree of neurologic impairment and Alzheimer type II astrocytosis, the neuropathologic feature characteristic of HE in cirrhosis, is also encountered in chronic hyperammonemic syndromes of non-hepatic origin. Ammonia has direct toxic effects on both CNS inhibition and excitation. It is unlikely, however, that ammonia's effects on brain function are the sole cause of HE in cirrhosis. Other neurotoxic substances generated in liver failure, such as manganese and proinflammatory cytokines, may gain entry to brain and act independently or synergistically with ammonia.

Studies of cerebral energy metabolism provide little convincing evidence that HE is the consequence of primary brain energy failure. However, brain glucose utilization is selectively decreased in anterior cingulate cortex, a brain structure associated with visual processing and attention in HE. Brain glucose oxidation is decreased in HE in cirrhosis, resulting in lactate accumulation. Liver failure leads to significant changes in synthesis, degradation, or function of a wide range of neurotransmitters, including GABA, and monoamines (serotonin, dopamine, histamine). The discovery of neurosteroids with potent agonist modulatory action at the GABA-A receptor may explain the notion of "increased GABAergic tone" in HE. Alterations of serotonin and histamine systems have been implicated in the pathogenesis of the sleep disturbances in early HE in cirrhosis.

Management and treatment of HE continue to rely heavily on reduction of circulating ammonia using a range of agents, including those aimed at reducing gut ammonia production (lactulose, antibiotics, probiotics) and those aimed at increased ammonia fixation (L-ornithine L-aspartate). Protein restriction is no longer recommended, except for short periods in cases of severe protein intolerance. Extrapyramidal symptoms suggestive of parkinsonism in cirrhosis may respond to L-DOPA treatment, and thiamine supplementation for the prevention of Wernicke's encephalopathy is recommended particularly in alcoholic cirrhotic patients. Flumazenil may be useful, particularly if ingestion of benzodiazepines is suspected. As the central mechanisms responsible for HE in cirrhosis become better elucidated, further therapies targeting the brain will undoubtedly appear. The majority of cases of HE in cirrhosis respond very well to LT. However, restoration of cognitive dysfunction may take several months and, in a small number of cases, symptoms may be unresponsive, particularly in patients having had multiple episodes of HE or other neurodegenerative conditions associated with chronic alcoholism.

ACKNOWLEDGMENTS

Funds from the author's research unit were provided by grants from The Canadian Institutes of Health Research (CIHR). The author is grateful to Sarah Henzi for her assistance with preparation of the manuscript and figures.

References

Agrawal A, Sharma BC, Sharma P et al. (2012). Secondary prophylaxis of hepatic encephalopathy in cirrhosis: an open-label, randomised controlled trial of lactulose, probiotics and no therapy. Am J Gastroenterol 107 (7): 1043–1050.

Ahboucha S, Pomier-Layrargues G, Mamer O et al. (2005). Increased brain concentrations of a neuroinhibitory steroid in human hepatic encephalopathy. Ann Neurol 58: 169–170.

Bajaj JS, Saeian K (2005). MELD score does not discriminate against patients with hepatic encephalopathy. Dig Dis Sci 50: 753–756.

Bajaj JS, Hafeezullah M, Hoffman RG et al. (2007). Minimal hepatic encephalopathy: a vehicle for accidents and traffic violations. Am J Gastroenterol 102: 1903–1909.

Bajaj JS, Hafeezullah M, Franco J et al. (2008). Inhibitory control test for the diagnosis of minimal hepatic encephalopathy. Gastroenterology 135: 1591–1600.

Bajaj JS, Wade JB, Sanyal AJ (2009). Spectrum of neurocognitive impairment in cirrhosis: implications for the assessment of hepatic encephalopathy. Hepatology 50: 2014–2021.

Bajaj JS, Schubert CM, Heuman DM et al. (2010). Persistence of cognitive impairment after resolution of overt hepatic encephalopathy. Gastroenterology 138: 2332–2340.

Bass NM, Mullen KD, Sanyal A et al. (2010). Rifaximin treatment in hepatic encephalopathy. N Engl J Med 362: 1071–1081.

Bates TE, Williams SR, Kauppinen RA et al. (1989). Observation of cerebral metabolites in an animal model of ALF *in vivo*: ^{1}H and ^{31}P nuclear magnetic resonance study. J Neurochem 53: 102–110.

Bélanger M, Ahboucha S, Desjardins P et al. (2004). Upregulation of peripheral-type (mitochondrial) benzodiazepine receptors in hyperammonemic syndromes: consequences for neuronal excitability. Advances in Molecular and Cell Biology 31 (3): 983–997.

Bengtsson F, Bugge M, Johansen KH et al. (1991). Brain tryptophan hydroxylation in the portacaval shunted rat: a hypothesis for the regulation of serotonin turnover in vivo. J Neurochem 56: 1069–1074.

Bergeron M, Reader TA, Pomier Layrargues G et al. (1989). Monoamines and metabolites in autopsied brain tissue from cirrhotic patients with hepatic encephalopathy. Neurochem Res 14: 853–859.

Bergeron M, Swain MS, Reader TA et al. (1990). Effect of ammonia on brain serotonin metabolism in relation to function in the portacaval shunted rat. J Neurochem 55: 222–229.

Burkhard PR, Delavelle J, Du Pasquier R et al. (2003). Chronic parkinsonism associated with cirrhosis: a distinct subset of acquired hepatocerebral degeneration. Arch Neurol 60: 521–528.

Butterworth RF (2007). Neuronal cell death in hepatic encephalopathy. Metab Brain Dis 22: 309–320.

Butterworth RF (2013). Parkinsonism in cirrhosis: pathogenesis and current therapeutic options. Metab Brain Dis 28: 261–267.

Butterworth RF, Giguère JF (1986). Cerebral amino acids in portal-systemic encephalopathy: lack of evidence for altered γ-aminobutyric acid (GABA) function. Metab Brain Dis 1: 221–228.

Butterworth RF, Heroux M (1989). Effect of pyrithiamine treatment and subsequent thiamine rehabilitation on regional cerebral amino acids and thiamine-dependent enzymes. J Neurochem 52: 1079–1084.

Butterworth RF, Giguère JF, Michaud J et al. (1987). Ammonia: key factor in the pathogenesis of hepatic encephalopathy. Neurochem Pathol 6: 1–12.

Butterworth RF, Lavoie J, Giguère JF et al. (1988). Affinities and densities of high affinity ^{3}H-muscimol (GABA-A) binding sites and central benzodiazepine receptors are unchanged in autopsied brain tissue from cirrhotic patients with hepatic encephalopathy. Hepatology 8: 1084–1088.

Butterworth RF, Tonon MC, Désy L et al. (1991). Increased brain content of the endogenous benzodiazepine receptor ligand octadecaneuropeptide (ODN) following portacaval anastomosis in the rat. Peptides 12: 119–125.

Butterworth RF, Wells J, Pomier Layrargues G (1995). Detection of benzodiazepines in hepatic encephalopathy. Reply. Hepatology 21: 605.

Cagnin A, Taylor-Robinson SD, Forton DM et al. (2006). In vivo imaging of cerebral peripheral benzodiazepine binding sites in patients with hepatic encephalopathy. Gut 55: 547–553.

Chatauret N, Butterworth RF (2004). Effects of liver failure on inter-organ trafficking of ammonia: implications for the treatment of encephalopathy. J Gastroenterol Hepatol 19: S219–S223.

Ciancio A, Marchet A, Saracco G et al. (2002). Spectral electroencephalogram analysis in hepatic encephalopathy and liver transplantation. Liver Transpl 8: 630–635.

Cordoba J (2011). New assessment of hepatic encephalopathy. J Hepatol 54: 1030–1040.

Cordoba J, Cabrera J, Lataif L et al. (1998). High prevalence of sleep disturbance in cirrhosis. Hepatology 27 (2): 339–345.

Cordoba J, López-Hellín J, Planas M et al. (2004). Normal protein diet for episodic hepatic encephalopathy: results of a randomized study. J Hepatol 41: 38–43.

Desjardins P, Rama Rao VK et al. (1999). Effect of portacaval anastomosis on glutamine synthetase protein and gene expression in brain, liver and skeletal muscle. Metab Brain Dis 14: 273–282.

Devenyi AG, Barron TF, Mamourian AC (1994). Dystonia, hyperintense basal ganglia, and high whole blood manganese levels in Alagilles's syndrome. Gastroenterology 106: 1068–1071.

Dhiman RK, Kurmi R, Thumburu KK et al. (2010). Diagnosis and prognostic significance of minimal hepatic encephalopathy in patients with cirrhosis of liver. Dig Dis Sci 55: 2381–2390.

Felipo V, Butterworth RF (2002). Neurobiology of Ammonia. Prog Neurobiol 67: 259–279.

Gabuzda Jr G, Philips GB, Davidson CS (1952). Reversible toxic manifestations in patients with cirrhosis of the liver given cation-exchange resins. N Engl J Med 246: 124–130.

Ganda OP, Ruderman NB (1976). Muscle nitrogen metabolism in chronic hepatic insufficiency. Metabolism 25 (4): 427–435.

Gyr K, Meier R, Haussler J et al. (1996). Evaluation of the efficacy and safety of flumazenil in the treatment of portal systemic encephalopathy: a double blind, randomised, placebo controlled multicentre study. Gut 39: 319–324.

Hindfelt B, Plum F, Duffy TE (1977). Effects of acute ammonia intoxication on cerebral metabolism in rats with portacaval shunts. J Clin Invest 59: 386–396.

Jacobs EH, Yamatodani A, Timmerman H (2000). Is histamine the final neurotransmitter in the entrainment of circadian rhythms in mammals? Trends Pharmacol Sci 21: 293–298.

Keiding S, Sørensen M, Bender D et al. (2006). Brain metabolism of ^{13}N-ammonia during acute hepatic encephalopathy in cirrhosis measured by positron emission tomography. Hepatology 43: 42–50.

Kircheis G, Nilius R, Held C et al. (1997). Therapeutic efficacy of L-ornithine-L-aspartate infusions in patients with cirrhosis and hepatic encephalopathy: results of a placebo-controlled, double-blind study. Hepatology 25 (6): 1351–1360.

Kircheis G, Wettstein M, Timmermann L et al. (2002). Critical flicker frequency for quantification of low-grade hepatic encephalopathy. Hepatology 35: 357–366.

Kril JJ, Butterworth RF (1997). Diencephalic and cerebellar pathology in alcoholic and non-alcoholic patients with end-stage liver disease. Hepatology 26: 837–841.

Lai JCK, Cooper AJL (1986). Brain α-ketoglutarate dehydrogenase: kinetic properties, regional distribution and effects of inhibitors. J Neurochem 47: 1376–1386.

Laubenberger J, Haussinger D, Boyer S et al. (1997). Proton magnetic resonance spectroscopy of brain in symptomatic and asymptomatic patients with liver cirrhosis. Gastroenterology 112: 1610–1616.

Lavoie J, Giguère JF, Pomier Layrargues G et al. (1987). Amino acid changes in autopsied brain tissue from cirrhotic patients with hepatic encephalopathy. J Neurochem 49: 692–697.

Lavoie J, Pomier Layrargues G, Butterworth RF (1990). Increased densities of "peripheral-type" benzodiazepine receptors in autopsied brain tissue from cirrhotic patients with hepatic encephalopathy. Hepatology 11: 874–878.

Lockwood AH, Yap EWH, Wong W-H (1991). Cerebral ammonia metabolism in patients with severe liver disease and minimal hepatic encephalopathy. J Cereb Blood Flow Metab 11: 337–341.

Lockwood AH, Weissenborn K, Butterworth RF (1997). An image of the brain in patients with liver disease. Curr Opin Neurol 10: 525–533.

Lockwood AH, Weissenborn K, Bokemeyer M et al. (2002). Correlations between cerebral glucose metabolism and neuropsychological test performance in non-alcoholic cirrhotics. Metab Brain Dis 17: 29–40.

Lozeva V, Valjakka A, Lecklin A et al. (2000). Effects of the histamine H(1) receptor blocker, pyrilamine, on spontaneous locomotor activity of rats with long-term portacaval anastomosis. Hepatology 31 (2): 336–344.

Lozeva V, Montgomery JA, Tuomisto L et al. (2004). Increased brain serotonin turnover correlates with the degree of shunting and hyperammonemia in rats following variable portal vein stenosis. J Hepatol 40: 742–748.

Lozeva-Thomas V, Ahonen P, Chatauret N et al. (2004). Brain histamine in experimental acute liver failure: effects of L-histidine loading. Inflamm Res 53 (Suppl. 1): S55–S56.

Malaguarnera M, Pistone G, Elvira R et al. (2005). Effects of L-carnitine in patients with hepatic encephalopathy. World J Gastroenterol 11 (45): 7197–7202.

Mattarozzi K, Stracciari A, Vignatelli L et al. (2004). Minimal hepatic encephalopathy: longitudinal effects of liver transplantation. Arch Neurol 61: 242–247.

Michel H, Solere M, Granier P et al. (1980). Treatment of cirrhotic hepatic encephalopathy with L-DOPA. A controlled trial. Gastroenterology 29: 555–561.

Mirowitz SA, Westrich TJ, Hirsch JD (1991). Hyperintense basal ganglia on T_1-weighted MR images in patients receiving parenteral nutrition. Radiology 191: 117–120.

Morgan MY, Jakobovits AW, James IM et al. (1980). Successful use of bromocriptine in the treatment of chronic hepatic encephalopathy. Gastroenterology 78: 663–670.

Mousseau DD, Perney P, Pomier Layrargues G et al. (1993). Selective loss of pallidal dopamine D2 receptor density in hepatic encephalopathy. Neurosci Lett 162: 192–196.

Mousseau DD, Baker GB, Butterworth RF (1997). Increased density of catalytic sites and expression of brain monoamine oxidase A in humans with hepatic encephalopathy. J Neurochem 68: 1200–1208.

Mullen KD, Prakash RK (2012). Antibiotic treatment for hepatic encephalopathy. In: KV Mullen, RV Prakash (Eds.), Hepatic Encephalopathy. Humana Press/Springer, New York, pp. 159–164.

Mullen KD, Szauter KM, Kaminsky-Russ K (1990). "Endogenous" benzodiazepine activity in body fluids of patients with hepatic encephalopathy. Lancet 336: 81–83.

Navarro D, Zwingmann C, Chatauret N et al. (2008). Glucose loading precipitates focal lactic acidosis in vulnerable medial thalamus of thiamine-deficient rats. Metab Brain Dis 23: 115–122.

Norenberg MD (1987). The role of astrocytes in hepatic encephalopathy. Neurochem Pathol 6: 13–33.

Norenberg MD, Lapham LW (1974). The astrocyte response in experimental portal-systemic encephalopathy: an electron microscope study. J Neuropathol Exp Neurol 33: 422–435.

Odeh M (2007). Pathogenesis of hepatic encephalopathy: the tumour necrosis factor-alpha theory. Eur. J Clin Invest 37: 291–304.

Odeh M, Sabo E, Srugo I et al. (2005). Relationship between tumour necrosis factor-alpha and ammonia in patients with hepatic encephalopathy due to chronic liver failure. Ann Med 37: 603–612.

Olasmaa M, Rothstein JD, Guidotti A et al. (1990). Endogenous benzodiazepine receptor ligands in human and animal hepatic encephalopathy. J Neurochem 55: 2015–2023.

Plauth M, Cabré E, Riggio O et al. (2006). ESPEN Guidelines on enteral nutrition in liver disease. Clin Nutr 25: 285–294.

Pomier-Layrargues G (1996). TIPS and hepatic encephalopathy. Semin Liver Dis 16: 315–320.

Pomier-Layrargues G, Giguère JF, Lavoie J et al. (1994). Clinical efficacy of benzodiazepine antagonist RO 15-1788 (flumazenil) in cirrhotic patients with hepatic coma: results of a randomized double-blind placebo-controlled cross-over trial. Hepatology 19: 32–37.

Pomier-Layrargues G, Spahr L, Butterworth RF (1995). Increased manganese concentrations magnetic resonance hyperintensity? Lancet 345: 735.

Prasad S, Dhiman RK, Duseja A et al. (2007). Lactulose improves cognitive functions and health-related quality of life in patients with cirrhosis who have minimal hepatic encephalopathy. Hepatology 45: 549–559.

Pujol J, Kulisevsky J, Moreno A et al. (1996). Neurospectroscopic alterations and globus pallidus hyperintensity as related magnetic resonance markers of reversible hepatic encephalopathy. Neurology 47: 1526–1530.

Raabe W (1987). Synaptic transmission in ammonia intoxication. Neurochem Pathol 6: 145–166.

Raghavendra Rao VL, Butterworth RF (1994). Alterations of [^{3}H]8-OH-DPAT and [^{3}H]-ketanserin binding sites in autopsied brain tissue from cirrhotic patients with hepatic encephalopathy. Neurosci Lett 182: 69–72.

Randolph C, Hilsabeck R, Kato A et al. (2009). Neuropsychological assessment of hepatic encephalopathy: ISHEN practice guidelines. Liver Int 29: 629–635.

Riggio O, Ridola L, Pasquale C et al. (2011). Evidence of persistent cognitive impairment after resolution of overt hepatic encephalopathy. Clin Gastroenterol Hepatol 9: 181–183.

Romero-Gomez M, Córdoba J, Jover R et al. (2007). Value of the critical flicker frequency in patients with minimal hepatic encephalopathy. Hepatology 45: 879–885.

Rose C, Michalak A, Pannunzio P et al. (1998). L-ornithine-L-aspartate in experimental portal-systemic encephalopathy: therapeutic efficacy and mechanism of action. Metab Brain Dis 13 (2): 24–32.

Rose C, Butterworth RF, Zayed J et al. (1999). Manganese deposition in basal ganglia structures results from both portal-systemic shunting and liver dysfunction. Gastroenterology 117: 640–644.

Rossle M, Euringer W (2012). Hepatic encephalopathy in patients with transjugular intrahepatic portosystemic shunt (TIPS). In: KD Mullen, RK Prakash (Eds.), Hepatic encephalopathy: Clinical Gastroenterology. Humana Press/Springer, New York, pp. 211–220.

Rothstein JD, McKhann G, Guarneri P et al. (1989). Hepatic encephalopathy and cerebrospinal fluid content of diazepam binding inhibitor (DBI). Ann Neurol 26: 57–62.

Roy S, Pomier-Layrargues G, Butterworth RF et al. (1988). Hepatic encephalopathy in cirrhotic and portacaval shunted dogs: lack of changes in brain GABA uptake, brain GABA levels, brain glutamic acid decarboxylase and brain post-synaptic GABA receptors. Hepatology 8: 845–849.

Schafer DF, Jones EA (1982). Hepatic encephalopathy and the γ-aminobutyric acid system. Lancet 1: 18–20.

Schleiss F, Gorg B, Fischer R et al. (2002). Ammonia induces MK-801 sensitive nitration and phosphopylation of protein tyrosine residues in rat astrocytes. FASEB J 16: 739–741.

Senzolo M, Pizzolato G, Ferronato C et al. (2009). Long term evaluation of cognitive function and cerebral metabolism in liver transplanted patients. Transplant Proc 41: 1295–1296.

Sharma BC, Sharma P, Agrawal A et al. (2009). Secondary prophylaxis of hepatic encephalopathy: an open-label randomized controlled trial of lactulose versus placebo. Gastroenterology 137: 885–891.

Shawcross DL, Wright G, Olde Daminck SWM et al. (2007). Role of ammonia and inflammation in minimal hepatic encephalopathy. Metab Brain Dis 22: 125–138.

Shiraishi T, Black KL, Ikesaki K (1990). Peripheral benzodiazepine receptor ligands induce morphological changes in mitochondria of cultured glioma cells. Soc Neurosci Abs 16: 214–216.

Smith JM, Misiak H (1976). Critical flicker frequency (CFF) and psychotropic drugs in normal human subjects – a review. Psychopharmacologia 47 (2): 175–182.

Sotil EU, Gottstein J, Ayala E et al. (2009). Impact of preoperative overt hepatic encephalopathy on neurocognitive function after liver transplantation. Liver Transpl 15 (2): 184–192.

Spahr L, Butterworth RF, Fontaine S et al. (1996). Increased blood manganese in cirrhotic patients: relationship to pallidal magnetic resonance signal hyperintensity and neurological symptoms. Hepatology 24: 1116–1120.

Spahr L, Coeytaux A, Giostra E et al. (2007). Histamine H_1 blocker Hydroxyzine improves sleep in patients with cirrhosis and minimal hepatic encephalopathy: a randomized controlled pilot trial. Am J Gastroenterol 102: 744–753.

Sushma S, Dasarathy S, Tandon RK et al. (1992). Sodium benzoate in the treatment of acute hepatic encephalopathy: a double-blind randomized trial. Hepatology 16 (1): 138–144.

Tarter RE, Switala J, Plail J et al. (1992). Severity of hepatic encephalopathy before liver transplantation is associated with quality of life after transplantation. Arch Intern Med 152: 2097–2101.

Tasaka K (1994). New advances in histamine research. Spinger Verlag, Tokyo.

Therrien G, Giguère JF, Butterworth RF (1991). Increased cerebrospinal fluid lactate reflects deterioration of neurological status in experimental portal-systemic encephalopathy. Metab Brain Dis 6: 225–231.

Therrien G, Butterworth J, Rose C et al. (1997). Protective effect of L-carnitine in ammonia-precipitated encephalopathy in portacaval shunted rats: evidence for a central mechanism of action. Hepatology 25: 551–556.

Tofteng F, Jorgensen L, Hansen BA et al. (2002). Cerebral microdialysis in patients with fulminant hepatic failure. Hepatology 36: 1333–1340.

Uribe M, Farca A, Marquez MA et al. (1979). Treatment of chronic portal-systemic encephalopathy with bromocriptine: a double-blind controlled trial. Gastroenterology 76: 1347–1351.

Weissenborn K (2012). Diagnosis of overt hepatic encephalopathy. In: KD Mullen, RK Prakash (Eds.), Hepatic Encephalopathy: Clinical Gastroenterology. Humana Press/Springer, New York, pp. 97–102.

Weissenborn K, Berding G, Kostler H (2000). Altered striatal dopamine D2 receptor density and dopamine transport in a patient with hepatic encephalopathy. Metab Brain Dis 15: 173–178.

Weissenborn K, Ennen JC, Schomerus H et al. (2001). Neuropsychological characterization of hepatic encephalopathy. J Hepatol 34: 768–773.

Yao H, Sadoshima S, Fujii K et al. (1987). Cerebrospinal fluid lactate in patients with hepatic encephalopathy. Eur Neurol 27: 182–187.

Young SN, Lal S, Feldmuller F et al. (1975). Relationships between tryptophan in serum and CSF and 5-hydroxyindoleacetic acid in CSF of man: effects of cirrhosis of the liver and probenecid administration. J Neurol Neurosurg Psychiatry 38: 322–330.

Zemtsova I, Gorg B, Keitel V et al. (2011). Microglial activation in hepatic encephalopathy in rats and humans. Hepatology 54: 204–215.

Handbook of Clinical Neurology, Vol. 125 (3rd series)
Alcohol and the Nervous System
E.V. Sullivan and A. Pfefferbaum, Editors

Chapter 35

Neuropathology of alcoholism

GREG T. SUTHERLAND[1], DONNA SHEEDY[1], AND JILLIAN J. KRIL[1,2]*
[1]*Department of Pathology, Sydney Medical School, University of Sydney, Sydney, Australia*
[2]*Department of Medicine, Sydney Medical School, University of Sydney, Sydney, Australia*

INTRODUCTION

One of the more serious consequences of alcohol abuse is alcohol-related brain damage (ARBD). An extensive body of literature points to the brain as an organ particularly sensitive to damage through both direct and indirect mechanisms. Over the past three decades our understanding of the factors that contribute to, and are the consequences of, ARBD has been advanced through studies of the neuroimaging, neuropsychology, neuropathology, genetics, and molecular pathology of the brain. Such studies have been conducted in rodents and, to a lesser extent, primates, and cell culture and other model systems as well as in living human subjects and postmortem human brain tissue samples. The use of postmortem human brain tissue offers both advantages and challenges. Not only is the human brain anatomically and functionally complex but a wide array of factors must be taken into consideration when designing and interpreting the results of studies of human brain. Humans are genetically diverse, each has a different exposure to disease risk factors, including alcohol, and a different history of nutrition, coexisting diseases, and other factors that have the potential to affect the brain and its functions. Careful selection and characterization of study cohorts are therefore essential if our knowledge in this area is to be advanced.

ALCOHOL NEUROTOXICITY

Functional brain impairment following acute alcohol ingestion is well recognized. Additionally, a significant body of literature points to neuroradiologically identified microstructural changes in the brain following acute alcohol administration in humans (Rickenbacher et al., 2011; Kong et al., 2012) and in rodent models of binge drinking (Zahr et al., 2013). However, the neuropathology of acute alcohol intoxication is often unremarkable, although congestion and edema may occur in some cases. Understanding the effects of chronic alcohol and distinguishing the direct toxic effects of alcohol from the secondary effects of additional complicating pathologies poses a greater challenge. One way in which the direct effects of alcohol toxicity can be studied is through selection of study subjects that consume alcohol beyond a defined threshold, but whose brains are normal on routine neuropathologic examination. Such thresholds are applied independently of other risks such as family history. The most commonly employed threshold is 80 g of ethanol (eight standard drinks) per day (Harper and Kril, 1985; Dodd et al., 1992; Kril et al., 1997; Harding et al., 2000; Lewohl et al., 2000; Matsumoto, 2009), although many alcoholics drink far more than this amount. It is also recognized that many alcoholics consume alcohol in variable patterns over their life. For these alcoholics, calculation of average or lifetime consumption (Harding et al., 2011) can be a better correlate of the severity of neuropathology (Harding et al., 1996; Kril et al., 1997). The calculation of lifetime alcohol consumption better accommodates periods of abstinence and the compound effects of heavy alcohol consumption over a longer period of time.

COMPLICATING PATHOLOGIES

As noted, many alcoholics also have coexisting medical conditions or lifestyle choices that can damage the brain independently of alcohol or exacerbate the damaging effects of alcohol on the brain. Nutritional deficiencies, both undernutrition, where the total calorie intake is reduced, resulting in an inadequate supply of macro- and micronutrients, and malnutrition, where particular

*Correspondence to: Professor Jillian J. Kril, Department of Pathology DO6, Sydney Medical School, University of Sydney, Sydney NSW 2006, Australia. Tel: 612-9036 7118, Fax: 612-9351 3429, E-mail: jillian.kril@sydney.edu.au

elements of the diet, such as vitamins, are inadequate, are common in alcoholics (Harris et al., 2008). Alcoholics have a lower body weight and lower fat mass than social drinkers (Addolorato et al., 1998; Ross et al., 2012) due, in part, to reduced dietary intake and, in part, to the metabolic effects of excessive alcohol consumption (Lieber, 2003). Furthermore, specific deficiencies of micronutrients are commonly identified in alcoholics (Harris et al., 2008; Ross et al., 2012). Pellagra (vitamin B_3, or niacin, deficiency) is almost exclusively seen in alcoholics. Although its prevalence has not been assessed for many years and would be difficult to accurately determine due to declining autopsy rates, clinical and pathologic case reports and small series continue to appear in the literature, suggesting ongoing relevance to the alcoholic population. Pellagra is characterized by chromatolysis of neurons in the pontine nuclei and dentate of the cerebellum (Hauw et al., 1988). The Wernicke-Korsakoff syndrome (WKS: vitamin B_1 or thiamine, deficiency) is characterized acutely by hemorrhages in the mammillary bodies, medial thalamus, and periventricular regions of the brainstem (Kril, 1996; Kril and Harper, 2012) and chronically by atrophy, discoloration, and neuronal loss in these regions (Fig. 35.1). The prevalence of WKS is greatest in the alcoholic population (Harper, 1979, 1983). In the most recent prevalence study, Harper and colleagues (1998) found 25 cases of WKS among 2212 forensic autopsies (population prevalence of 1.1%), or 19% of the alcoholic population. Of these 25 cases, acute WKS was rare (8%) compared with acute-on-chronic (24%) and chronic cases (68%). Other nutritional abnormalities such as an increased prevalence of diabetes and dyslipidemia have been described in alcoholics, although the contribution of these to ARBD is yet to be fully evaluated. There is, however, evidence of degeneration of the white matter in alcoholics linked to hepatosteatosis and insulin resistance (de la Monte et al., 2009).

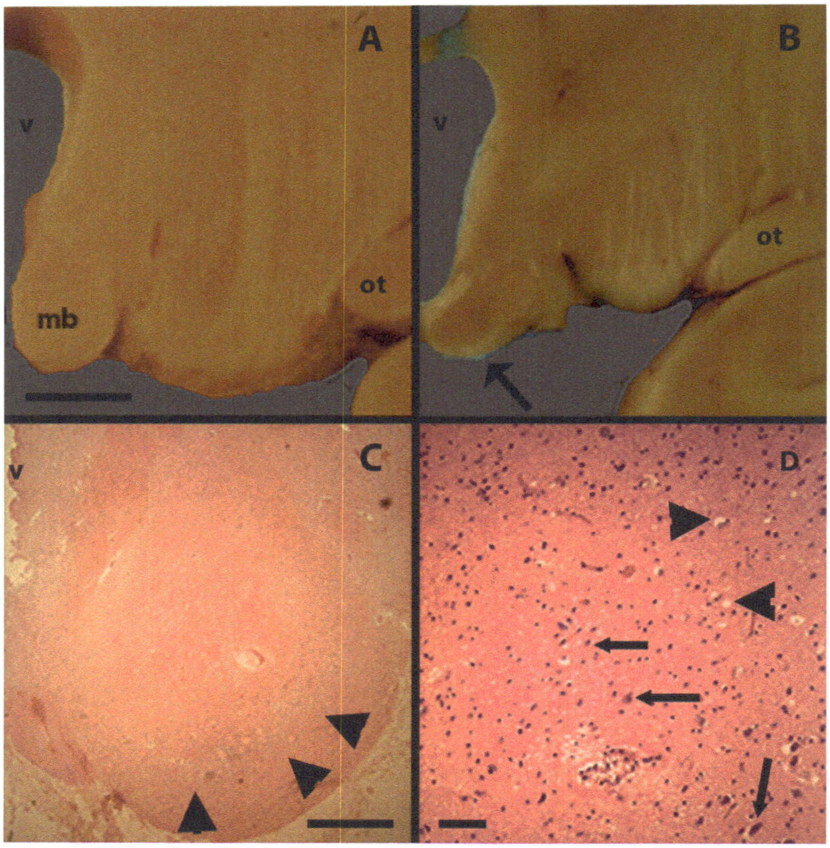

Fig. 35.1. (**A**) A photograph of the right mammillary body (mb) and adjacent hypothalamic and thalamic nuclei surrounding the third ventricle (v) in a non-alcoholic individual. Scale bar = 5 mm; ot, optic tract. (**B**) A section at a similar anatomic level from a patient with Wernicke-Korsakoff syndrome (WKS) showing shrinkage and brown discoloration of the mammillary body (arrow). Scale equivalent to (**A**). (**C**) A low-power micrograph of the mammillary body in WKS showing vacuolation and pallor of the neuropil (arrowheads). Scale bar = 1 mm. (**D**) A higher-power micrograph of the periventricular mediodorsal nucleus of the thalamus in WKS showing sparse neurons (long arrows), areas of vacuolation (arrowheads), and numerous small glial nuclei (gliosis). Scale bar = 50 μm.

Another major complication of alcoholism is the development of hepatic encephalopathy (HE), a metabolic encephalopathy secondary to liver disease. Some 20% of alcoholics have steatosis that may progress to hepatitis, cirrhosis, or both, and between 10 and 15% of alcoholics develop cirrhosis of the liver. The pathogenesis of HE is unclear, although elevated circulating ammonia levels are believed to be a key factor (Butterworth, 1995). HE can result in coma and death due to cerebral edema. Interestingly, HE is largely reversible on correction of the underlying liver disease, a feature that is supported by the neuropathology. Neuronal loss is not seen in HE; rather, astrocytic pathology called Alzheimer type II change is the prominent histopathology. Alzheimer type II astrocytes show swollen nuclei with marginated chromatin and are commonly seen in deep gray-matter structures and in the lower layers of the cerebral cortex. Although most commonly seen in HE, Alzheimer type II astrocytes are seen in other metabolic encephalopathies, such as uremic encephalopathy (Harris et al., 2008).

Other neuropathologies are also frequently observed in alcoholics. Head injury is common and contusions, particularly of the basal surfaces of the brain, may be seen. Similarly, excessive alcohol consumption, and comorbid cigarette smoking, are risk factors for stroke and several mechanisms linking alcohol use and stroke, including cardiac arrhythmias and impaired coagulation, have been suggested (Gorelick, 1987; Allen and Bayraktutan, 2008). At autopsy the prevalence of stroke among alcoholics is higher than in non-alcoholic individuals (Walbran et al., 1981). Alcohol abuse also alters susceptibility to infection (Saieva et al., 2012) and this may affect the brain. The range and prevalence of complicating pathologies highlight the need for a thorough postmortem examination of the brain, although this poses a challenge with the declining number of autopsies being performed in most countries. Improvements in imaging techniques and the availability of neuroimaging have meant that *in vivo* studies of brain structure and function of increasing sophistication are now possible (Gazdzinski et al., 2005; Sullivan and Pfefferbaum, 2009; Chanraud et al., 2011; Colrain et al., 2011; Abe et al., 2013; Camchong et al., 2013). Nonetheless, postmortem examination of the brain is essential to exclude coexisting pathologies and confirm diagnoses. The provision of tissue for research studies is now almost exclusively from brain banks, such as the New South Wales Tissue Resource Centre (Sheedy et al., 2008).

STRUCTURAL CHANGES IN ARBD

A consistent finding in chronic alcoholics is smaller brain size than in non-alcoholic controls. During development and maturation the human brain and skull grow in unison, meaning there is a tight relationship between intracranial cavity volume and brain volume. Brain atrophy, or indeed any other change in brain size, is therefore clearly demonstrated by comparing these two volumes.

Brain atrophy in alcoholics was initially identified in a postmortem study of brain weight by Harper and Blumbergs (1982). Alcoholic men had a 75 g (5%) smaller brain weight than non-alcoholic men (Harper and Blumbergs, 1982), although Torvik and colleagues (1982) showed that this varied with age and was less obvious in the oldest age group (70–79 years). The presence of atrophy was substantiated in a study of brain and intracranial cavity volumes where the pericerebral space, the difference between the two measures, was $8.3 \pm 3.3\%$ in controls and significantly greater at $11.3 \pm 5.2\%$ in alcoholics (Harper and Kril, 1985). Subsequent computed tomography and magnetic resonance imaging (MRI) studies have confirmed the presence of smaller brain volume in alcoholics than controls. Postmortem studies have found larger ventricular size (Harper et al., 1985) and smaller white-matter volume (Harper et al., 1985; Kril et al., 1997), particularly of the frontal lobes (Kril et al., 1997). The magnitude of white-matter loss correlates with mean daily alcohol consumption. Early postmortem volumetric studies where the entire cortex was measured revealed no (Harper et al., 1985), or a modest (de la Monte, 1988), atrophy of the cerebral cortex in alcoholics; however, a subsequent analysis of anatomically discrete cortical regions revealed shrinkage of the dorsolateral prefrontal cortex, albeit somewhat variably between cases (Kril et al., 1997). In addition, there is atrophy of the hippocampus that is restricted to shrinkage of white matter and most marked in those alcoholics still actively drinking at the time of their death (Harding et al., 1997). By contrast, quantitative MRI studies reveal shrinkage of volume restricted to the cortex in younger alcoholics but shrinkage of both white matter and cortex from older alcoholics (Pfefferbaum et al., 1997). As in the postmortem studies, smaller volume is most marked in the frontal lobes (Rosenbloom et al., 2003; Cardenas et al., 2005). More recently, compromise of the structural and functional integrity of white matter has been demonstrated using diffusion tensor imaging (Chanraud et al., 2011), suggesting that white-matter abnormalities resulting in a regional disconnection of the cortex may underlie the cognitive and neurologic deficits seen in chronic alcoholics.

In alcoholics with coexisting complicating pathologies, brain pathology is more severe and more widespread than is observed in alcoholics without these complicating pathologies. For example, when alcoholics with WKS are compared to those without WKS, atrophy of the white matter is more marked and extends beyond the frontal lobes (Kril et al., 1997). In addition, there is significant atrophy of amygdala and a trend toward atrophy of the thalamus and hippocampus in WKS alcoholics which is not seen in non-WKS alcoholics (Kril et al., 1997). These studies support the findings of quantitative MRI studies (reviewed by Jung et al., 2012). Few studies have examined the effect of additional liver disease or HE on brain atrophy in alcoholics. Our earlier study of pericerebral space showed, in a secondary analysis, that alcoholics with either steatosis or cirrhosis had a larger mean pericerebral space value (16.2%) than alcoholics with normal livers (10.2%) (Harper et al., 1985). Although, to some degree, this may reflect a more severe drinking history, there is also the possibility of additional brain damage as a result of liver disease. However, this is not supported by a quantitative MRI study of nine patients with HE, six of whom had a history of alcoholism (Patel et al., 2004).

In addition to the recognized central nervous system (CNS) pathologies complicating ARBD, evidence has emerged that cigarette smoking also contributes to brain atrophy in alcoholics. The prevalence of smoking among alcoholics is higher, with ~50% being smokers (Sussman et al., 2011) compared with ~20% of the general community. MRI studies have shown greater atrophy (Gazdzinski et al., 2005) and evidence of greater neuronal injury (Durazzo et al., 2004) in alcoholic smokers than non-smokers. The neurotoxic effects of cigarette smoking have also been demonstrated in the non-alcoholic population. Heavy smokers have cortical (Yu et al., 2011) and subcortical (Das et al., 2012) atrophy and disruption of white-matter microstructure (Hudkins et al., 2012) compared with controls. Confirmatory postmortem studies have yet to be performed in either population. Similar compound effects on brain atrophy have been found in studies of those who abuse alcohol and illicit drugs such as cocaine (Liu et al., 1998; Bjork et al., 2003) and in those with mood disorders (Bielau et al., 2005), suggesting that careful screening of study participants is required to clearly demonstrate the effect of alcohol abuse on brain structure.

Cerebellar atrophy is also frequently seen in alcoholics. In a prevalence study, 26% of alcoholics examined neuropathologically were found to have atrophy of the cerebellar vermis (Torvik et al., 1982). This is predominantly of the anterior, superior lobes (lobes I-V) of the vermis, anterior to the primary fissure, and is clearly visible on routine neuropathologic examination and using neuroimaging in the sagittal plane. A postmortem study of the area of the vermis in alcoholics showed a deficit in total area of the lobes of 8–26% relative to non-alcoholic controls (Phillips et al., 1987). In a subsequent study a similar degree of atrophy of the whole vermis (~5–10%) was identified in alcoholics with and without WKS (Phillips et al., 1990). Interestingly, Torvik (1987) also found cerebellar atrophy in a third of controls >70 years, suggesting contributors other than alcohol in this age group. Smaller cerebellar volumes in aged individuals and those with alcoholism have also been shown *in vivo* using MRI volumetrics (Sullivan et al., 2000).

NEURONAL LOSS IN ARBD

Despite clearly demonstrable cognitive and motor functional deficits in alcoholics, the degree of neuronal loss in alcoholics without additional nutritional or metabolic pathologies is surprisingly subtle. Routine neuropathologic examination fails to identify neuronal loss from the cortex of alcoholics. Quantitative studies have revealed a reduced neuronal density in the superior frontal gyrus of the dorsolateral prefrontal cortex (Brodmann area (BA) 8), but not from the anterior cingulate (BA32), inferior temporal (BA20), or primary motor (BA4) cortices (Kril and Harper, 1989). This finding was confirmed using unbiased quantitative techniques (Kril et al., 1997). The magnitude of the loss using both neuronal density and unbiased estimation of total neuron number methodologies is ~20% (Fig. 35.2). Greater glial density and smaller mean neuronal size in alcoholics relative to controls were also demonstrated in the superior frontal cortex. Interestingly, significantly smaller neuronal size, without neuronal loss, was seen in the primary motor and anterior cingulate cortices (Kril and Harper, 1989). Neither the presence of cirrhosis of the liver nor WKS resulted in a further reduction in neuron number (Kril and Harper, 1989; Kril et al., 1997).

Neuronal counts of all subregions of the hippocampal formation have not identified any neuronal loss in alcoholics with or without WKS (Harding et al., 1997). This is despite a reduction in white-matter volume (see above) and is in contrast to studies of alcohol administration in rodents. Quantitation of neuron number in a variety of subcortical and brainstem nuclei also reveals absence of neuronal loss in alcoholics without complicating pathologies (Fig. 35.2). These structures include the anterior principal and mediodorsal (Fig. 35.1) nuclei of the thalamus (Harding et al., 2000), the medial mammillary nucleus of the hypothalamus (Harding et al., 2000), the serotonergic median raphe (Baker et al., 1996) and noradrenergic locus ceruleus (Halliday et al., 1992; Baker et al., 1994) in the brainstem, and the cholinergic basal forebrain (Cullen et al., 1997). One interesting exception is the vasopressin-containing neurons of the hypothalamus. These neurons occur predominantly in the supraoptic and paraventricular nuclei of the hypothalamus and were found to be fewer in number in alcoholics who

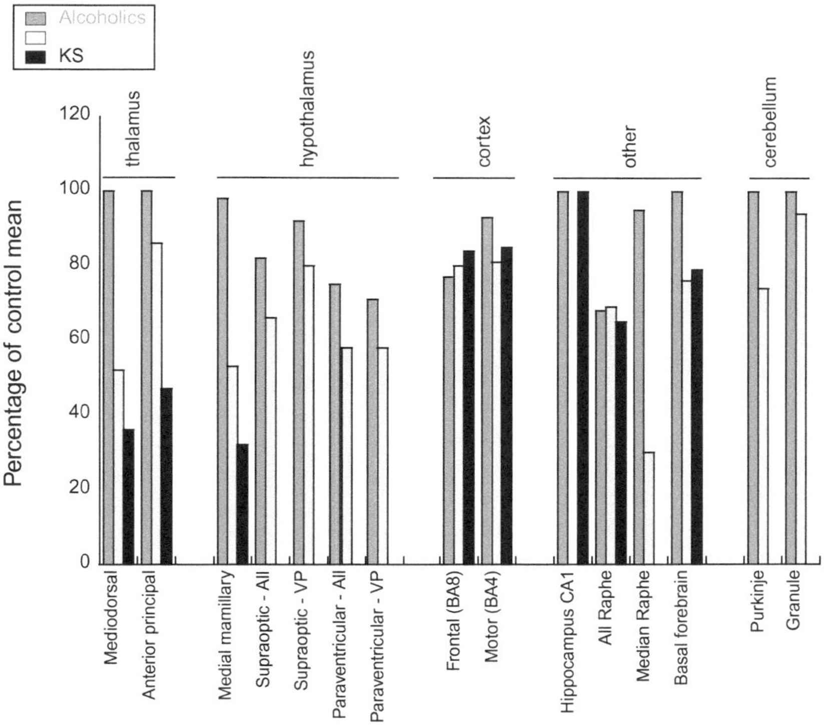

Fig. 35.2. Summary of neuronal counts for cortical and subcortical brain regions in alcoholics as a percentage of control values (gray bars). Significant neuronal loss is seen from the superior frontal cortex (Brodmann area 8) and the supraoptic and paraventricular nuclei of the hypothalamus. Other brain regions examined do not differ significantly from controls. Alcoholics with the additional thiamine deficiency of Wernicke-Korsakoff syndrome (WKS; pale bars) and those with the memory impairment of Korsakoff's syndrome (KS; black bars) show more marked and widespread neuronal loss.

consumed >100 g ethanol per day compared with those consuming less alcohol (Harding et al., 1996). Fewer neurons and smaller volumes of the nuclei correlated with higher daily alcohol intake.

The majority of these studies also examined alcoholics with pathologically confirmed thiamine deficiency (i.e., WKS). In some studies alcoholics with the profound amnesia of Korsakoff's syndrome (KS) were considered separately from those with WKS without amnesia (Kril and Harper, 2012). Neuronal loss was not detected in the hippocampal formation (Harding et al., 1997) or locus ceruleus (Halliday et al., 1992) in WKS. By contrast, alcoholics with WKS with and without amnesia had neuronal loss of similar magnitude from the serotonergic raphe nucleus (30%: Halliday et al., 1993) and the cholinergic basal forebrain (20%: Cullen et al., 1997). In the diencephalon, while all alcoholics with WKS have fewer neurons in the medial mammillary nucleus of the hypothalamus and the mediodorsal nucleus of the thalamus, only those with amnesia (KS) had additional loss from the anterior principal nucleus of the thalamus (Fig. 35.2) (Harding et al., 2000). Taken together these studies show that thiamine deficiency in alcoholics results in significant and widespread neuronal loss and that the amnesia of KS results from damage to a number of nuclei subserving memory circuits in the brain (Kril and Harper, 2012).

Studies of the cerebellum reveal conflicting data. Early studies of neuronal density revealed impoverished Purkinje cell density in the vermis of alcoholics without WKS (Phillips et al., 1987) and those with WKS (Phillips et al., 1990). The magnitude of loss (~20%) was similar in both groups. A secondary analysis of cases with cirrhosis of the liver showed a more marked reduction in Purkinje cell density with cirrhosis in the alcoholic and WKS groups (Phillips et al., 1990). These findings contrast with those from a second quantitative study using unbiased stereologic methods where no volume or neuronal loss was found from any of the function zones of the cerebellum of alcoholics (Baker et al., 1999). Alcoholics with WKS showed reduced Purkinje cell number in most zones; however, this only reached significance in the flocculus. No volume loss or change in granule cell number was seen (Baker et al., 1999). The reason for these differences is unclear, although the broad variance among alcoholics suggests vulnerability to cerebellar damage in some, but not all, cases.

MOLECULAR DIFFERENCES

Despite the well-known clinical effects of acute alcohol intoxication the exact mechanism of action remains obscure. Ethanol is an amphipathic molecule, soluble in both lipids and water, resulting in almost universal distribution throughout the body. In combination with its distribution alcohol can interact with many other biologic molecules, giving it an almost infinite range of potential effects. In the CNS alcohol is known to affect the makeup of neuronal membranes (Melgaard, 1983), the integrity of the blood-brain barrier, and the transport of substrates across it (Haorah et al., 2008), glucose uptake into astrocytes, the main energy supply chain for neurons (Abdul Muneer et al., 2011), and to enhance the production of radical nitrogen and oxygen species while depressing endogenous antioxidant activity (Nordmann et al., 1990; Bondy, 1992). The latter can potentially induce proinflammatory cytokine signaling and neuroinflammation (Alfonso-Loeches et al., 2010). With such wide-ranging effects, the study of alcohol toxicity and the adaptation that occurs following chronic exposure is complex and requires a variety of approaches.

NEUROTRANSMITTERS AND THEIR RECEPTORS

Alcohol has wide-ranging effects on neurotransmission, including on the serotonin, dopamine (Heinz and Goldman, 2000), and endogenous opioid systems (Sirohi et al., 2012). However, it is alcohol's ability to simultaneously suppress excitatory (glutaminergic) and stimulate inhibitory (GABAergic) neurotransmission that is believed to be responsible for many of the acute effects observed clinically.

Alcohol is both a non-competitive antagonist of excitatory *N*-methyl-D-aspartate receptors (Dingledine et al., 1999) and a gamma-aminobutyric acid (GABA) receptor agonist. Alcohol also acts indirectly to increase GABAergic activity receptor subunit expression by upregulating membrane trafficking and modulating receptor sensitivity through both posttranslational modifications such as dephosphorylation and stimulating neurosteroid synthesis (Kumar et al., 2009). The latter in turn can bind to the GABA receptor and prolong activity. The chronic effects of alcohol are also thought to revolve around a region-specific alteration in $GABA_A$ receptor subunit expression and receptor activity (Kumar et al., 2009).

The $GABA_A$ receptor was the major target of early molecular studies that utilized human brain autopsy tissue. In particular, receptor density was shown to correlate with regional neuronal loss in the alcoholic brain, being substantially higher in the superior frontal cortex of chronic alcoholics compared with the motor cortex (Tran et al., 1981; Dodd et al., 1992). However, subsequent studies have failed to reproduce these differences consistently for either the alpha (Lewohl et al., 2001) or beta $GABA_A$ (Buckley et al., 2006) subunits.

GENETICS AND GENOMICS

It has been known for some time that offspring of alcoholics are at greater risk of developing alcoholism, with twin studies suggesting that the overall heritability of alcoholism is approximately 50% (Goldman et al., 2005). The $GABA_A$ receptor gene cluster and the alpha 2 subunit *GABRA2* gene in particular have been shown, by both genetic linkage and association studies, to affect the susceptibility to alcohol dependence (Rietschel and Treutlein, 2013). However, significant associations with *GABRA2* have not been replicated in the latest genome-wide association studies (GWAS) (Rietschel and Treutlein, 2013). Rietschel and Treutlein suggest that, apart from the alcohol metabolism genes, alcohol dehydrogenase and aldehyde dehydrogenase, the genetic architecture of alcohol dependence is likely to be made up of multiple genes with small effect sizes, referred to as genetic heterogeneity.

In contrast to alcohol dependence there have been few studies looking at the genetic susceptibility to alcohol neurotoxicity, but a GABRA2 single-nucleotide polymorphism (SNP) was shown to modify sensitivity to the acute effects of alcohol (Haughey et al., 2008). *GABRA2* itself has not been specifically investigated in ARBD but there were no associations between three other $GABA_A$ receptor subunit genes and WKS. In their review of all WKS association studies, Guerrini and colleagues (2009) noted that significant associations have only been described for the gut thiamine transporter-encoding genes *SLC19A2* and *SLC19A3*. In terms of alcohol-related dementia, gene association studies have been rare. One study found an association between the APOE $\epsilon4$ genotype and hippocampal volume that was seen only in female alcoholics (Bleich et al., 2003). This common variant confers a large risk for sporadic Alzheimer's disease and accelerated hippocampal atrophy (Ashford, 2004), suggesting it may represent a non-specific determinant of susceptibility to brain damage (Mahley and Huang, 2012). Other studies have so far failed to find associations with genes encoding presenilin 1 (Isoe et al., 1996), the estrogen receptor (Ji et al., 2000), and mitochondrial aldehyde dehydrogenase (Shin et al., 2005). The latter was in a Korean population and part of a wider dementia study.

A recent paper in a mouse model suggested that alcohol exerts its neurotoxic effect by impairing one-carbon metabolism, with roll-on effects on DNA repair and DNA methylation (Fowler et al., 2012). The effects of alcohol on DNA repair were more severe in mice deficient in methylenetetrahydrofolate reductase (MTHFR), an enzyme that catalyzes the rate-determining step in methionine production, and noted that *MTHFR* mutations were known to increase the risk of alcohol-induced cancers. DNA methylation deficits have also been seen in transcriptomic studies and these are discussed further below.

GWAS have been made possible due to the technologic advances associated with the human genome project and subsequent studies to characterize the common genetic variation in the human population (Alfonso-Loeches et al., 2010). Genomewide arrays can now simultaneously interrogate 2.5 million SNPs in a single experiment, although large cohort sizes are usually required because of the inherent genetic heterogeneity of most complex diseases. Nevertheless, GWAS are still likely to miss rare functional variants that are postulated to cause or modify disease risk in certain affected individuals. A newer technology, referred to as next-generation sequencing, is currently superseding array platforms and, although prohibitively expensive today, this technology will ultimately allow the entire genome of all individuals in case-control cohorts to be sequenced.

GENE EXPRESSION

The other major spin-off from the knowledge and technology developed around the human genome project is genomewide expression or transcriptomic studies. The ability to investigate the total gene expression of a cell or tissue and then compare individuals with and without a disease gives us an unbiased platform both to test existing hypotheses and discover hitherto unknown pathobiology. Again, the original probe hybridization-based array technology is currently being replaced by next-generation sequencing, called RNA-Seq, that promises a more accurate and comprehensive analysis of neurodegenerative diseases (Sutherland et al., 2011). However, the success of both techniques is highly dependent on the quality of the starting template, ribonucleic acid (RNA).

RNA quality or integrity is commonly measured using an algorithm that combines multiple features from an RNA electrophoresis gel scan and represents them as a total RNA integrity number (RIN) (Schroeder et al., 2006). It was originally thought that RNA degradation was largely due to postmortem factors, but it is now known, for postmortem brain tissue in particular, that it is the premortem or agonal period that is critical for the quality of the RNA that is extracted (Monoranu et al., 2009; Durrenberger et al., 2010). RIN is highly correlated with brain pH (Monoranu et al., 2009) and brain pH and RIN decrease with the length of the agonal period and the number of adverse, often hypoxic, events within that period, such as hospitalization, coma, artificial respiration, or respiratory illness (Durrenberger et al., 2010). In addition, the agonal period is associated with novel gene expression patterns that are not directly related to the disease (Li et al., 2004). A major issue for case-control transcriptomic studies is that patients with neurologic disorders generally have a more protracted and averse agonal period and subsequently a lower pH and RIN compared with controls who have often died suddenly (Yates et al., 1990; Monoranu et al., 2009). The removal of low-RIN samples is a standard practice in most studies using postmortem brain tissue but it remains prudent to evaluate the potential effects of RIN and other confounders in transcriptomic data before concluding that there are disease-specific differences.

Notwithstanding these technical issues, genomewide expression studies of brain tissue from alcoholics without comorbidities such as liver disease or WKS have revealed very subtle and largely divergent data (Flatscher-Bader et al., 2006). An exception has been the dysregulation of the myelin-associated genes, although the individual genes themselves and their direction of change have varied between studies (Lewohl et al., 2000; Iwamoto et al., 2004; Li et al., 2004). Other major themes have been DNA repair (Mayfield et al., 2002; Iwamoto et al., 2004), the ubiquitin-proteasome system, and immune function (Sokolov et al., 2003; Liu et al., 2004).

The proinflammatory or neuroimmune-activating effects of alcohol may initially involve cytokine production by resident microglia and astrocytes via the innate immune system, Toll-like receptors (TLR) (Blanco et al., 2005; Blanco and Guerri, 2007). This in turn induces the proinflammatory transcription factor NF-κB (nuclear factor of kappa light polypeptide gene enhancer in B cells) transcription, with increased levels of the cytokines interleukin 1beta, tumor necrosis factor-alpha, and interleukin-6 (Alfonso-Loeches et al., 2010). Research from experimental animals has shown that the inhibition of TLR4-mediated *NF-κB* signaling attenuates both the toxic (Alfonso-Loeches et al., 2010) and acute behavioral (Wu et al., 2012) actions of alcohol. Gene expression studies suggest that the activation of *NF-κB* transcription with acute alcohol exposure eventually gives way to downregulation, with subsequent cycles of intoxication and withdrawal (Okvist et al., 2007). In keeping with the known regional loss of neurons in ARBD, the genes with NF-κβ elements were generally downregulated in the prefrontal cortex of chronic alcoholics but not the motor cortex (Okvist et al., 2007).

The common comorbidity HE is another important potential confounder for transcriptomic analysis of ARBD. Like hypoxic episodes in the agonal period HE is associated with a metabolic acidosis, and alcoholics with HE have a significantly lower brain pH and RIN than those without HE (Sheedy et al., 2012). Our own work comparing the white-matter transcriptome of alcoholics with HE and non-HE alcoholics with controls also found that transcriptomic changes were largely confined to the HE group (Sutherland et al., 2014). Furthermore, like hypoxia, HE was associated with a downregulation of energy metabolism pathways, and the additive effect of HE and agonal hypoxic events on lowering brain pH, RIN, and the magnitude of gene downregulation could be seen. HE was also characterized by an upregulation of proinflammatory pathways consistent with the increasing realization that both systemic inflammation and elevated brain ammonia levels probably act in combination to cause this condition (Shawcross et al., 2005, 2011; Rama Rao et al., 2010). The term systemic is important here as the resident microglia, although activated, do not appear to be a prominent source of inflammatory cytokines in HE (Zemtsova et al., 2011).

Two relatively recent paradigm shifts in our knowledge of gene expression regulation have been recognition of the roles played by epigenetics and non-coding RNA species, along with interactions between the two. Ponomarev et al. (2012) showed that the alcoholic prefrontal cortex was characterized by global DNA hypomethylation and postulated that this was due to alcohol-induced deficiencies in folate and its derivative, *S*-adenosyl methionine, the major substrate for DNA methylation. Interestingly, the GC content of genes appeared to be the major factor that determined whether alcohol affected their expression rather than their membership of a specific cellular pathway.

The interest in non-coding RNA species has evolved with the realization that transcription is pervasive and not limited to just the protein-coding region of the genome (Clark et al., 2011). Non-coding RNAs come in many forms and sizes, including the relatively well-known microRNAs, small oligonucleotides, 22 bases in length, that are able to control protein expression by restricting access to translational machinery or augmenting transcript degradation (Fabian et al., 2010). Lewohl et al. (2011) found that 35 microRNAs were upregulated in chronic alcoholic brains and their putative targets coincided with the downregulated mRNAs from their previous microarray data (Liu et al., 2006). MicroRNAs can modulate TLR (and NF-κB) signaling and they display regulatory reciprocity with epigenetic factors (Nunez and Mayfield, 2012). A special property of microRNAs is their promiscuity, which means they are able to bind to and downregulate multiple gene targets simultaneously. This is not only ideal for coordinating complex gene expression networks in organisms but it also raises the possibility that microRNAs may be highly effective therapeutic targets for complex disorders such as ARBD (Nunez and Mayfield, 2012).

PROTEOMICS

The term proteome was coined from genome and means all the proteins, including their posttranslational modifications, that are produced in a particular cell or tissue at any one time (Wilkins et al., 1996). In reality the technology available generally limits findings to hundreds of molecules as opposed to the thousands simultaneously analyzed in transcriptomic analyses. Consistent with histologic studies (Kril and Harper, 1989; Harper and Kril, 1991) and particularly transcriptomic studies, noticeable changes in protein levels have generally been restricted to alcoholics with coexisting pathologies (Matsumoto, 2009). For example, Lewohl and colleagues (2005) demonstrated consistent decreases in myelin-related proteins only in cirrhotic alcoholics, but not those without coexisting disease. It has been suggested that these changes may reflect the impact of HE (Butterworth, 2003), although we were unable to detect such myelin-related changes in the transcriptomic analysis of white matter of HE alcoholics (Sutherland et al., 2014). Another possibility, given that hepatic cirrhosis alters thiamine homeostasis (Butterworth, 1995), is that a proportion of alcoholics with cirrhosis also have subclinical WKS. This is supported by a series of proteomic studies carried out by the Matsumoto group across various gray- and white-matter regions that showed a consistent downregulation of the thiamine-dependent enzymes (Matsumoto et al., 2007). In particular, the enzyme transketolase, that requires the cofactor thiamine pyrophosphate, was reduced in many regions of the alcoholic brain (Matsumoto et al., 2007). One specific impact of a reduction in transketolase activity is a decrease in the reducing agent nicotinamide adenine dinucleotide phosphate-oxidase (NADPH). NAPDH maintains the cellular levels of the antioxidant, glutathione, and its paucity renders the brain more susceptible to pro-oxidative effects of alcohol (Alexander-Kaufman and Harper, 2009).

Along with cirrhosis and subclinical thiamine deficiency, cigarette smoking is another potential confounder that can complicate the molecular picture of ARBD. Given that the co-abuse of alcohol and nicotine is common, their differential effects can be difficult to tease apart, but at least one study suggests that nicotine exposure, rather than alcohol, might be the major driver of increased glutaminergic signaling (Flatscher-Bader et al., 2008).

Outside of the degenerative aspects of ARBD, there are also important changes associated with tolerance and dependence. Transcriptomic studies of the regions of the mesolimbocortical (reward) pathway have shown that, unlike the prefrontal cortex, where genes associated with energy production, oxidative stress, and DNA damage are differentially expressed, the gene expression patterns of the ventral tegmentum area (VTA) and nucleus accumbens (NA) reflect changes in brain plasticity (Flatscher-Bader et al., 2010). Again, both smoking and alcohol affected the genes involved in cell structure and neuron remodeling, but here smoking co-abuse had the opposite effect on gene expression in the NA versus the VTA, suggesting that nicotine modulates alcohol-induced changes in a region-specific manner.

In conclusion, molecular studies have made us more aware of the comorbidities and potential confounders associated with ARBD than providing hard and fast evidence of the direct effects of long-term alcohol consumption. Of course, the severity of comorbidities such as hepatic cirrhosis are correlated with the lifetime alcohol dose, but it does mean that ARBD should not be viewed in isolation.

SUMMARY

The damaging effects of long-term alcohol abuse are incontrovertible and marked functional and cognitive abnormalities are commonly seen in many alcoholics. However, structural studies of the postmortem brain reveal relatively modest pathology, compared, for example, with the degree of pathology seen in patients with cognitive decline due to early Alzheimer-type dementia. This suggests that deficits in alcoholics are predominantly due to functional or subtle morphologic changes in neurons rather than neurodegeneration *per se*. This suggestion is supported by neuroimaging studies that show marked abnormalities in white-matter connectivity and subtle, anatomically discrete, cortical volume changes. They are also supported by the relative paucity of change identified in genomic and proteomic studies of alcoholic postmortem brain. The absence of marked neuronal loss offers the potential for recovery of function with abstinence, as has also been shown in neuroimaging studies, at least for younger alcoholics.

In contrast to those alcoholics without coexisting pathologies, those with nutritional deficiencies or liver disease resulting in HE show more marked and widespread brain pathology. In alcoholics with WKS, neuronal loss is identified from diencephalic nuclei known to be vulnerable to thiamine deficiency and subcortical and brainstem nuclei (Fig. 35.2). Although similar quantitative studies have not been performed on alcoholics with HE, those studies that have included cases with HE report more marked damage. It should be acknowledged that alcoholics who develop these secondary complications may have a more severe drinking history; yet, other factors such as genetic susceptibility to disease should not be excluded. Furthermore, as both WKS and HE result in impairment of vital energy metabolism pathways, it is not surprising that these diseases, even in the absence of alcohol dependence, have detrimental effects on the brain.

Distinguishing the damaging effects of alcohol abuse from other lifestyle and medical factors is difficult but important if effective treatment and prevention strategies are to be developed. While the damaging effects of WKS and HE are well established, more recent studies have shown that other factors, such as cigarette smoking, illicit drug use, and mood disorders, may also play a role in the neuropathology of alcoholism. Further studies on well-characterized subject cohorts are required for ful elucidation of all the contributors to ARBD.

ACKNOWLEDGMENTS

The authors are members of the New South Wales Tissue Resource Centre, which is supported by the University of Sydney, National Health and Medical Research Council of Australia (605210), Schizophrenia Research Institute, and the National Institute of Alcohol Abuse and Alcoholism (R24AA012725 and R28AA012725).

References

Abdul Muneer PM, Alikunju S, Szlachetka AM et al. (2011). Ethanol impairs glucose uptake by human astrocytes and neurons: protective effects of acetyl-L-carnitine. Int J Physiol Pathophysiol Pharmacol 3: 48–56.

Abe C, Mon A, Durazzo TC et al. (2013). Polysubstance and alcohol dependence: unique abnormalities of magnetic resonance-derived brain metabolite levels. Drug Alcohol Depend 130: 30–37.

Addolorato G, Capristo E, Greco AV et al. (1998). Influence of chronic alcohol abuse on body weight and energy metabolism: is excess ethanol consumption a risk factor for obesity or malnutrition? J Intern Med 244: 387–395.

Alexander-Kaufman K, Harper C (2009). Transketolase: observations in alcohol-related brain damage research. Int J Biochem Cell Biol 41: 717–720.

Alfonso-Loeches S, Pascual-Lucas M, Blanco AM et al. (2010). Pivotal role of TLR4 receptors in alcohol-induced neuroinflammation and brain damage. J Neurosci 30: 8285–8295.

Allen CL, Bayraktutan U (2008). Risk factors for ischaemic stroke. Int J Stroke 3: 105–116.

Ashford JW (2004). APOE genotype effects on Alzheimer's disease onset and epidemiology. J Mol Neurosci 23: 157–165.

Baker KG, Halliday GM, Harper CG (1994). Effect of chronic alcohol consumption on the human locus coeruleus. Alcohol Clin Exp Res 18: 1491–1496.

Baker KG, Halliday GM, Kril JJ et al. (1996). Chronic alcoholics without Wernicke-Korsakoff syndrome or cirrhosis do not lose serotonergic neurons in the dorsal raphe nucleus. Alcohol Clin Exp Res 20: 61–66.

Baker KG, Harding AJ, Halliday GM et al. (1999). Neuronal loss in functional zones of the cerebellum of chronic alcoholics with and without Wernicke's encephalopathy. Neuroscience 91: 429–438.

Bielau H, Trubner K, Krell D et al. (2005). Volume deficits of subcortical nuclei in mood disorders: a postmortem study. Eur Arch Psychiatry Clin Neurosci 255: 401–412.

Bjork JM, Grant SJ, Hommer DW (2003). Cross-sectional volumetric analysis of brain atrophy in alcohol dependence: effects of drinking history and comorbid substance use disorder. Am J Psychiatry 160: 2038–2045.

Blanco AM, Guerri C (2007). Ethanol intake enhances inflammatory mediators in brain: role of glial cells and TLR4/IL-1RI receptors. Front Biosci 12: 2616–2630.

Blanco AM, Valles SL, Pascual M et al. (2005). Involvement of TLR4/type I IL-1 receptor signaling in the induction of inflammatory mediators and cell death induced by ethanol in cultured astrocytes. J Immunol 175: 6893–6899.

Bleich S, Wilhelm J, Graesel E et al. (2003). Apolipoprotein E epsilon 4 is associated with hippocampal volume reduction in females with alcoholism. J Neural Transm 110: 401–411.

Bondy SC (1992). Ethanol toxicity and oxidative stress. Toxicol Lett 63: 231–241.

Buckley ST, Foley PF, Innes DJ et al. (2006). GABA(A) receptor beta isoform protein expression in human alcoholic brain: interaction with genotype. Neurochem Int 49: 557–567.

Butterworth RF (1995). Hepatic encephalopathy. Neurologist 1: 95–104.

Butterworth RF (2003). Hepatic encephalopathy–a serious complication of alcoholic liver disease. Alcohol Res Health 27: 143–145.

Camchong J, Stenger A, Fein G (2013). Resting-state synchrony during early alcohol abstinence can predict subsequent relapse. Cereb Cortex 23 (9): 2086–2099.

Cardenas VA, Studholme C, Meyerhoff DJ et al. (2005). Chronic active heavy drinking and family history of problem drinking modulate regional brain tissue volumes. Psychiatry Res 138: 115–130.

Chanraud S, Pitel AL, Pfefferbaum A et al. (2011). Disruption of functional connectivity of the default-mode network in alcoholism. Cereb Cortex 21: 2272–2281.

Clark MB, Amaral PP, Schlesinger FJ et al. (2011). The reality of pervasive transcription. PLoS Biol 9: e1000625.

Colrain IM, Sullivan EV, Ford JM et al. (2011). Frontally mediated inhibitory processing and white matter microstructure: age and alcoholism effects. Psychopharmacology (Berl) 213: 669–679.

Cullen KM, Halliday GM, Caine D et al. (1997). The nucleus basalis (Ch4) in the alcoholic Wernicke-Korsakoff syndrome: reduced cell number in both amnesic and non-amnesic patients. J Neurol Neurosurg Psychiatry 63: 315–320.

Das D, Cherbuin N, Anstey KJ et al. (2012). Lifetime cigarette smoking is associated with striatal volume measures. Addict Biol 17: 817–825.

de la Monte SM (1988). Disproportionate atrophy of cerebral white matter in chronic alcoholics. Arch Neurol 45: 990–992.

de la Monte SM, Longato L, Tong M et al. (2009). The liver-brain axis of alcohol-mediated neurodegeneration: role of toxic lipids. Int J Environ Res Public Health 6: 2055–2075.

Dingledine R, Borges K, Bowie D et al. (1999). The glutamate receptor ion channels. Pharmacol Rev 51: 7–61.

Dodd PR, Thomas GJ, Harper CG et al. (1992). Amino acid neurotransmitter receptor changes in cerebral cortex in alcoholism: effect of cirrhosis of the liver. J Neurochem 59: 1506–1515.

Durazzo TC, Gazdzinski S, Banys P et al. (2004). Cigarette smoking exacerbates chronic alcohol-induced brain damage: a preliminary metabolite imaging study. Alcohol Clin Exp Res 28: 1849–1860.

Durrenberger PF, Fernando S, Kashefi SN et al. (2010). Effects of antemortem and postmortem variables on human brain mRNA quality: a BrainNet Europe study. J Neuropathol Exp Neurol 69: 70–81.

Fabian MR, Sonenberg N, Filipowicz W (2010). Regulation of mRNA translation and stability by microRNAs. Annu Rev Biochem 79: 351–379.

Flatscher-Bader T, van der Brug MP, Landis N et al. (2006). Comparative gene expression in brain regions of human alcoholics. Genes Brain Behav 5 (Suppl. 1): 78–84.

Flatscher-Bader T, Zuvela N, Landis N et al. (2008). Smoking and alcoholism target genes associated with plasticity and glutamate transmission in the human ventral tegmental area. Hum Mol Genet 17: 38–51.

Flatscher-Bader T, Harrison E, Matsumoto I et al. (2010). Genes associated with alcohol abuse and tobacco smoking in the human nucleus accumbens and ventral tegmental area. Alcohol Clin Exp Res 34: 1291–1302.

Fowler AK, Hewetson A, Agrawal RG et al. (2012). Alcohol-induced one-carbon metabolism impairment promotes dysfunction of DNA base excision repair in adult brain. J Biol Chem 287: 43533–43542.

Gazdzinski S, Durazzo TC, Studholme C et al. (2005). Quantitative brain MRI in alcohol dependence: preliminary evidence for effects of concurrent chronic cigarette smoking on regional brain volumes. Alcohol Clin Exp Res 29: 1484–1495.

Goldman D, Oroszi G, Ducci F (2005). The genetics of addictions: uncovering the genes. Nature reviews. Genetics 6: 521–532.

Gorelick PB (1987). Alcohol and stroke. Stroke 18: 268–271.

Guerrini I, Thomson AD, Gurling HM (2009). Molecular genetics of alcohol-related brain damage. Alcohol Alcohol 44: 166–170.

Halliday G, Ellis J, Harper C (1992). The locus coeruleus and memory: a study of chronic alcoholics with and without the memory impairment of Korsakoff's psychosis. Brain Res 598: 33–37.

Halliday G, Ellis J, Heard R et al. (1993). Brainstem serotonergic neurons in chronic alcoholics with and without the memory impairment of Korsakoff's psychosis. J Neuropathol Exp Neurol 52: 567–579.

Haorah J, Schall K, Ramirez SH et al. (2008). Activation of protein tyrosine kinases and matrix metalloproteinases causes blood-brain barrier injury: novel mechanism for neurodegeneration associated with alcohol abuse. Glia 56: 78–88.

Harding AJ, Halliday GM, Ng JLF et al. (1996). Loss of vasopressin-immunoreactive neurons in alcoholics is dose-related and time-dependent. Neuroscience 72: 699–708.

Harding AJ, Wong A, Svoboda M et al. (1997). Chronic alcohol consumption does not cause hippocampal neuron loss in humans. Hippocampus 7: 78–87.

Harding A, Halliday G, Caine D et al. (2000). Degeneration of anterior thalamic nuclei differentiates alcoholics with amnesia. Brain 123 (Pt 1): 141–154.

Harding AJ, McCrossin T, Sheedy DL et al. (2011). Quantifying alcohol consumption for research cohorts. Alcohol Clin Exp Res 35: 164A–164A.

Harper C (1979). Wernicke's encephalopathy: a more common disease than realised. J Neurol Neurosurg Psychiatry 42: 226–231.

Harper CG (1983). The incidence of Wernicke's encephalopathy in Australia – a neuropathological study of 131 cases. J Neurol Neurosurg Psychiatry 46: 593–598.

Harper CG, Blumbergs PC (1982). Brain weights in alcoholics. J Neurol Neurosurg Psychiatry 45: 838–840.

Harper C, Kril J (1985). Brain atrophy in chronic alcoholic patients: a quantitative pathological study. J Neurol Neurosurg Psychiatry 48: 211–217.

Harper C, Kril J (1991). If you drink your brain will shrink. Neuropathological considerations. Alcohol Alcohol Suppl 1: 375–380.

Harper CG, Kril JJ, Holloway RL (1985). Brain shrinkage in chronic alcoholics – a pathological study. Br Med J 290: 501–504.

Harper CG, Sheedy DL, Lara AI et al. (1998). Prevalence of Wernicke-Korsakoff syndrome in Australia: has thiamine fortification made a difference? Med J Aust 168: 542–545.

Harris J, Chimelli L, Kril JJ et al. (2008). Nutritional deficiencies, metabolic disorders and toxins affecting the nervous system. In: S Love, DN Louis, DW Ellison (Eds.), Greenfield's Neuropathology, 8th edn. Edward Arnold, London, pp. 675–731.

Haughey HM, Ray LA, Finan P et al. (2008). Human gamma-aminobutyric acid A receptor alpha2 gene moderates the acute effects of alcohol and brain mRNA expression. Genes Brain Behav 7: 447–454.

Hauw JJ, De Baecque C, Hausser-Hauw C et al. (1988). Chromatolysis in alcoholic encephalopathies. Pellagra-like changes in 22 cases. Brain 111: 843–857.

Heinz A, Goldman D (2000). Genotype effects on neurodegeneration and neuroadaptation in monoaminergic neurotransmitter systems. Neurochem Int 37: 425–432.

Hudkins M, O'Neill J, Tobias MC et al. (2012). Cigarette smoking and white matter microstructure. Psychopharmacology (Berl) 221: 285–295.

Isoe K, Urakami K, Ji Y et al. (1996). Presenilin-1 polymorphism in patients with Alzheimer's disease, vascular dementia and alcohol-associated dementia in Japanese population. Acta Neurol Scand 94: 326–328.

Iwamoto K, Bundo M, Yamamoto M et al. (2004). Decreased expression of NEFH and PCP4/PEP19 in the prefrontal cortex of alcoholics. Neurosci Res 49: 379–385.

Ji Y, Urakami K, Wada-Isoe K et al. (2000). Estrogen receptor gene polymorphisms in patients with Alzheimer's disease, vascular dementia and alcohol-associated dementia. Dement Geriatr Cogn Disord 11: 119–122.

Jung YC, Chanraud S, Sullivan EV (2012). Neuroimaging of Wernicke's encephalopathy and Korsakoff's syndrome. Neuropsychol Rev 22: 170–180.

Kong LM, Zheng WB, Lian GP et al. (2012). Acute effects of alcohol on the human brain: diffusion tensor imaging study. AJNR Am J Neuroradiol 33: 928–934.

Kril JJ (1996). Neuropathology of thiamine deficiency disorders. Metab Brain Dis 11: 11–19.

Kril JJ, Harper CG (1989). Neuronal counts from four regions of alcoholic brains. Acta Neuropathol 79: 200–204.

Kril JJ, Harper CG (2012). Neuroanatomy and neuropathology associated with Korsakoff's syndrome. Neuropsychol Rev 22: 72–80.

Kril JJ, Halliday GM, Svoboda MD et al. (1997). The cerebral cortex is damaged in chronic alcoholics. Neuroscience 79: 983–998.

Kumar S, Porcu P, Werner DF et al. (2009). The role of GABA(A) receptors in the acute and chronic effects of ethanol: a decade of progress. Psychopharmacology (Berl) 205: 529–564.

Lewohl JM, Wang L, Miles MF et al. (2000). Gene expression in human alcoholism: microarray analysis of frontal cortex. Alcohol Clin Exp Res 24: 1873–1882.

Lewohl JM, Huygens F, Crane DI et al. (2001). GABA(A) receptor alpha-subunit proteins in human chronic alcoholics. J Neurochem 78: 424–434.

Lewohl JM, Wixey J, Harper CG et al. (2005). Expression of MBP, PLP, MAG, CNP, and GFAP in the Human Alcoholic Brain. Alcohol Clin Exp Res 29: 1698–1705.

Lewohl JM, Nunez YO, Dodd PR et al. (2011). Up-regulation of microRNAs in brain of human alcoholics. Alcohol Clin Exp Res 35 (11): 1928–1937.

Li JZ, Vawter MP, Walsh DM et al. (2004). Systematic changes in gene expression in postmortem human brains associated with tissue pH and terminal medical conditions. Hum Mol Genet 13: 609–616.

Lieber CS (2003). Relationships between nutrition, alcohol use, and liver disease. Alcohol Res Health 27: 220–231.

Liu X, Matochik JA, Cadet JL et al. (1998). Smaller volume of prefrontal lobe in polysubstance abusers: a magnetic

resonance imaging study. Neuropsychopharmacology 18: 243–252.
Liu J, Lewohl JM, Dodd PR et al. (2004). Gene expression profiling of individual cases reveals consistent transcriptional changes in alcoholic human brain. J Neurochem 90: 1050–1058.
Liu J, Lewohl JM, Harris RA et al. (2006). Patterns of gene expression in the frontal cortex discriminate alcoholic from nonalcoholic individuals. Neuropsychopharmacology 31: 1574–1582.
Mahley RW, Huang Y (2012). Apolipoprotein e sets the stage: response to injury triggers neuropathology. Neuron 76: 871–885.
Matsumoto I (2009). Proteomics approach in the study of the pathophysiology of alcohol-related brain damage. Alcohol Alcohol 44: 171–176.
Matsumoto I, Alexander-Kaufman K, Iwazaki T et al. (2007). CNS proteomes in alcohol and drug abuse and dependence. Expert Rev Proteomics 4: 539–552.
Mayfield RD, Lewohl JM, Dodd PR et al. (2002). Patterns of gene expression are altered in the frontal and motor cortices of human alcoholics. J Neurochem 81: 802–813.
Melgaard B (1983). The neurotoxicity of ethanol. Acta Neurol Scand 67: 131–142.
Monoranu CM, Apfelbacher M, Grunblatt E et al. (2009). pH measurement as quality control on human post mortem brain tissue: a study of the BrainNet Europe consortium. Neuropathol Appl Neurobiol 35: 329–337.
Nordmann R, Ribiere C, Rouach H (1990). Ethanol-induced lipid peroxidation and oxidative stress in extrahepatic tissues. Alcohol Alcohol 25: 231–237.
Nunez YO, Mayfield RD (2012). Understanding alcoholism through microRNA signatures in brains of human alcoholics. Frontiers in genetics 3: 43.
Okvist A, Johansson S, Kuzmin A et al. (2007). Neuroadaptations in human chronic alcoholics: dysregulation of the NF-kappaB system. PLoS One 2: e930.
Patel N, White S, Dhanjal NS et al. (2004). Changes in brain size in hepatic encephalopathy: a coregistered MRI study. Metab Brain Dis 19: 431–445.
Pfefferbaum A, Sullivan EV, Mathalon DH et al. (1997). Frontal lobe volume loss observed with magnetic resonance imaging in older chronic alcoholics. Alcohol Clin Exp Res 21: 521–529.
Phillips SC, Harper CG, Kril JJ (1987). A quantitative histological study of the cerebellar vermis in alcoholic patients. Brain 110: 301–314.
Phillips SC, Harper CG, Kril JJ (1990). The contribution of Wernicke's encephalopathy to alcohol-related cerebellar damage. Drug Alcohol Rev 9: 53–60.
Ponomarev I, Wang S, Zhang L et al. (2012). Gene coexpression networks in human brain identify epigenetic modifications in alcohol dependence. J Neurosci 32: 1884–1897.
Rama Rao KV, Jayakumar AR, Tong X et al. (2010). Marked potentiation of cell swelling by cytokines in ammonia-sensitized cultured astrocytes. Journal of neuroinflammation 7: 66.
Rickenbacher E, Greve DN, Azma S et al. (2011). Effects of alcohol intoxication and gender on cerebral perfusion: an arterial spin labeling study. Alcohol 45: 725–737.
Rietschel M, Treutlein J (2013). The genetics of alcohol dependence. Ann N Y Acad Sci 1282: 39–70.
Rosenbloom M, Sullivan EV, Pfefferbaum A (2003). Using magnetic resonance imaging and diffusion tensor imaging to assess brain damage in alcoholics. Alcohol Res Health 27: 146–152.
Ross LJ, Wilson M, Banks M et al. (2012). Prevalence of malnutrition and nutritional risk factors in patients undergoing alcohol and drug treatment. Nutrition 28: 738–743.
Saieva C, Bardazzi G, Masala G et al. (2012). General and cancer mortality in a large cohort of Italian alcoholics. Alcohol Clin Exp Res 36: 342–350.
Schroeder A, Mueller O, Stocker S et al. (2006). The RIN: an RNA integrity number for assigning integrity values to RNA measurements. BMC Mol Biol 7: 3.
Shawcross DL, Olde Damink SW, Butterworth RF et al. (2005). Ammonia and hepatic encephalopathy: the more things change, the more they remain the same. Metab Brain Dis 20: 169–179.
Shawcross DL, Sharifi Y, Canavan JB et al. (2011). Infection and systemic inflammation, not ammonia, are associated with Grade 3/4 hepatic encephalopathy, but not mortality in cirrhosis. J Hepatol 54: 640–649.
Sheedy D, Garrick T, Dedova I et al. (2008). An Australian Brain Bank: a critical investment with a high return! Cell Tissue Bank 9: 205–216.
Sheedy D, Say M, Stevens J et al. (2012). Influence of liver pathology on markers of postmortem brain tissue quality. Alcohol Clin Exp Res 36: 55–60.
Shin IS, Stewart R, Kim JM et al. (2005). Mitochondrial aldehyde dehydrogenase polymorphism is not associated with incidence of Alzheimer's disease. Int J Geriatr Psychiatry 20: 1075–1080.
Sirohi S, Bakalkin G, Walker BM (2012). Alcohol-induced plasticity in the dynorphin/kappa-opioid receptor system. Frontiers in molecular neuroscience 5: 95.
Sokolov BP, Jiang L, Trivedi NS et al. (2003). Transcription profiling reveals mitochondrial, ubiquitin and signaling systems abnormalities in postmortem brains from subjects with a history of alcohol abuse or dependence. J Neurosci Res 72: 756–767.
Sullivan EV, Pfefferbaum A (2009). Neuroimaging of the Wernicke-Korsakoff syndrome. Alcohol Alcohol 44: 155–165.
Sullivan EV, Deshmukh A, Desmond JE et al. (2000). Cerebellar volume decline in normal aging, alcoholism, and Korsakoff's syndrome: relation to ataxia. Neuropsychology 14: 341–352.
Sussman S, Lisha N, Griffiths M (2011). Prevalence of the addictions: a problem of the majority or the minority? Eval Health Prof 34: 3–56.
Sutherland GT, Janitz M, Kril JJ (2011). Understanding the pathogenesis of Alzheimer's disease: will RNA-Seq realize the promise of transcriptomics? J Neurochem 116: 937–946.

Sutherland GT, Sheedy D, Kril JJ (2014). Using autopsy tissue to study alcohol-related brain damage in the genomic age. Alcohol Clin Exp Res 38: 994–1001.

Torvik A (1987). Brain lesions in alcoholics: neuropathological observations. Acta Med Scand Suppl 717: 47–54.

Torvik A, Lindboe CF, Rogde S (1982). Brain lesions in alcoholics. A neuropathological study with clinical correlations. J Neurol Sci 56: 233–248.

Tran VT, Snyder SH, Major LF et al. (1981). GABA receptors are increased in brains of alcoholics. Ann Neurol 9: 289–292.

Walbran BB, Nelson JS, Taylor JR (1981). Association of cerebral infarction and chronic alcoholism: an autopsy study. Alcohol Clin Exp Res 5: 531–535.

Wilkins MR, Sanchez JC, Gooley AA et al. (1996). Progress with proteome projects: why all proteins expressed by a genome should be identified and how to do it. Biotechnol Genet Eng Rev 13: 19–50.

Wu Y, Lousberg EL, Moldenhauer LM et al. (2012). Inhibiting the TLR4-MyD88 signalling cascade by genetic or pharmacological strategies reduces acute alcohol-induced sedation and motor impairment in mice. Br J Pharmacol 165: 1319–1329.

Yates CM, Butterworth J, Tennant MC et al. (1990). Enzyme activities in relation to pH and lactate in postmortem brain in Alzheimer-type and other dementias. J Neurochem 55: 1624–1630.

Yu R, Zhao L, Lu L (2011). Regional grey and white matter changes in heavy male smokers. PLoS One 6: e27440.

Zahr NM, Mayer D, Rohlfing T et al. (2013). A mechanism of rapidly reversible cerebral ventricular enlargement independent of tissue atrophy. Neuropsychopharmacology 38 (6): 1121–1129.

Zemtsova I, Gorg B, Keitel V et al. (2011). Microglia activation in hepatic encephalopathy in rats and humans. Hepatology 54: 204–215.

Handbook of Clinical Neurology, Vol. 125 (3rd series)
Alcohol and the Nervous System
E.V. Sullivan and A. Pfefferbaum, Editors

Chapter 36

Genetic differences in response to alcohol

SACHIO MATSUSHITA* AND SUSUMU HIGUCHI
National Hospital Organization, Kurihama Medical and Addiction Center, Kanagawa, Japan

RESPONSE TO ALCOHOL AS AN ENDOPHENOTYPE OF ALCOHOL USE DISORDER

Alcohol use disorder (AUD) is a typical complex, genetically influenced disorder. Similar to many medical conditions, multiple genes are thought to influence a range of intermediate characteristics that subsequently interact with the environment to produce the disorder (Kendler et al., 1995). Twin and adoption studies have clearly demonstrated that 50–60% of phenotypic variations are accounted for by a genetic component (Dick and Foroud, 2003; Tyndale, 2003), and two twin studies, which controlled for the effects of sociodemographic and psychiatric predictors, found that nearly 50% of the variance in the risk for AUD and the age at the time of diagnosis can be attributed to genetic factors (Knopik et al., 2004; Liu et al., 2004). Despite strong evidence of the contribution of genetic effects to the risk of AUD, the identification of specific genes that alter the risk of AUD has been difficult. First, in addition to the genetic component, environmental factors play a significant role in the susceptibility to AUD. Second, multiple genes with modest effects are involved in AUD, and different sets of genes may contribute to the disorder in different individuals. Therefore, any gene that affects a specific intermediate characteristic or phenotype is likely to contribute to the vulnerability toward AUD but explain only a small proportion of the vulnerability.

To be considered an endophenotype, an intermediate phenotype must meet the following established criteria: specificity, state independence, heritability, familial association, cosegregation (Gottesman and Gould, 2003), and biologic/clinical plausibility (Tsuang et al., 1993). These rigorously defined, somewhat less complicated, and more homogeneous constructs might be useful to elucidate the genetic risk of AUD.

Potential endophenotypes that have received attention in the research literature thus far include, but are not limited to, the following: comorbid psychiatric disorders, enzymes responsible for metabolizing alcohol (Edenberg, 2007; Schuckit, 2009), and personality factors such as antisociality (Cadoret et al., 1995) impulsivity, sensation seeking (Magid et al., 2007), and response to alcohol.

The subjective response (SR) to alcohol, which reflects individual differences in sensitivity to its pharmacologic effects, may be an important endophenotype. SR satisfies at least four of the six endophenotype criteria: it is specific, state-independent, heritable, and biologically/clinically plausible (Morean and Corbin, 2010).

HERITABILITY OF LEVEL OF RESPONSE TO ALCOHOL

Since SR was first identified as a potentially relevant factor in determining vulnerability to AUD, especially in younger individuals (Schuckit, 1980), the number of studies examining individual differences in SR has increased. Empirical estimates suggesting that SR accounts for 40–60% of the variance in the AUD risk conferred by genetics have further increased interest in this potential mechanism of genetic risk (Schuckit, 1999).

The level of response (LR) to alcohol is genetically influenced, with heritability accounting for about 60% (as measured by comparing alcohol challenge results in monozygotic and dizygotic twins) (Heath et al., 1999; Viken et al., 2003). Similar heritabilities for the LR to alcohol have been estimated using the retrospective Self-Report of the Effects of Alcohol (SRE) measure and comparing correlations among first-degree relatives versus unrelated individuals (Schuckit et al., 2005a). A more recent sibling study in a predominantly African

*Correspondence to: Sachio Matsushita, M.D., Ph.D., National Hospital Organization, Kurihama Medical and Addiction Center, 5-3-1 Nobi, Yokosuka, Kanagawa, 239-0841, Japan. Tel: +81-46-848-1550, Fax: +81-46-849-7743, E-mail: sachio-m@wa2.so-net.ne.jp

American, non-alcohol-dependent population concluded that the heritability of the SRE was 0.67 (Kalu et al., 2012).

GENETIC INFLUENCES ON LEVEL OF RESPONSE TO ALCOHOL

With a moderate heritability, genetic influences on the LR to alcohol should be substantial, although this may well result from the involvement of numerous genes, each with a small effect (Enoch, 2013). Several genes have been implicated as potentially contributing to heavier drinking, alcohol-related problems, and AUD through a low LR to alcohol.

Gamma-aminobutyric acid A (GABA) receptor genes

$GABA_A$ receptors represent the major inhibitory class of neurotransmission receptors in the mammalian brain. The receptors mediate fast inhibitory neurotransmission by increasing chloride influx, which results in the hyperpolarization of neurons (Klausberger et al., 2000). Among alcohol's wide range of central nervous system (CNS) effects, it potently influences GABA transmission in multiple ways, including the modulation of presynaptic GABA release and postsynaptic chloride flux. These actions are thought to contribute to some of the subjective effects of alcohol, such as behavioral disinhibition at lower doses and sedation and ataxia at higher doses. One of the few robust genetic predictors for AUD is GABRA2, which encodes the $GABA_A$ α_2 subunit that forms the pentameric $GABA_A$ receptor $\alpha2\beta1\gamma1$, found in the mesolimbic dopamine (DA) reward pathway. Several studies have associated tightly linked single nucleotide polymorphisms (SNPs) in a distal GABRA2 haplotype block with AUD (Enoch, 2008). A few studies have associated GABRA2 variations within this haplotype block with the SR to alcohol. To explore the association between the GABRA2 gene polymorphism and the SR to alcohol, we examined 110 Japanese non-alcoholic college students, stratified according to ALDH2*1/*1 and ALDH2*1/*2 genotype, using the alcohol clamp method (O'Connor et al., 1998). This is an intravenous alcohol-loading procedure which is designed to maintain target blood alcohol levels. SRs to alcohol were measured using a sensation scale and the Biphasic Alcohol Effects Scale (BAES). As a result, minor (less common) homozygotes of SNPs within the GABRA2 haplotype block were found to be associated with a lower initial response to alcohol (Roh et al., 2011). Similar results were found in a much smaller study in social drinkers (Pierucci-Lagha et al., 2005). Recently, a study of 69 healthy volunteers who participated in a cumulative oral dosing procedure with continuous subjective assessments using a visual analog drug effect scale, the BAES, and the Subjective High Assessment Scale (SHAS) showed that carriers of the minor alleles of SNPs within this haplotype block had lower negative alcohol effect scores than carriers of the major homozygotes (Uhart et al., 2013). Moreover, a recent animal study examining the consequences of deleting the GABRA2 gene on the ataxic properties of alcohol reported that GABRA2 knockout mice showed a hypersensitivity to the acute effects of alcohol compared with wild-type mice (Dixon et al., 2012). These results are congruent, and an association between GABRA2 and AUD is plausible. However, it should be noted that no functional variant has yet been found, either in GABRA2 or in the tightly linked and closely adjacent GABRG1 gene.

Apart from GABRA2, the GABA gene cluster on chromosome 5, especially GABRA6 and GABRA1, is also reportedly associated with AUD. A secondary analysis of the multisite Collaborative Study on the Genetics of Alcoholism (COGA) study showed the association of GABRA1 with measures of drinking severity, such as history of blackouts, age at first drunkenness, and LR to alcohol (Dick et al., 2006). The Pro385Ser substitution in GABRA6 has been associated with a lower LR to the sedating effects of alcohol (Schuckit et al., 1999; Hu et al., 2005).

Serotonin transporter gene

A number of association studies examining the functional promoter polymorphism 5-HTTLPR in SLC6A4, the gene encoding the serotonin transporter (5-HTT), and the LR have been reported. 5-HTTLPR is a common polymorphism that includes a variable number of tandem repeats (44-basepair insertion/deletion) in the promoter region. This polymorphism alters transcription such that the less common, short S allele is associated with an approximately 50% reduction in 5-HTT availability and a concomitant increase in synaptic serotonin, a reduction in the neuronal uptake of serotonin, and a decrease in 5-HIAA (Heils et al., 1996). Triallelic genotyping (incorporating the functional SNP rs25531) of 5-HTTLPR has identified a low transcriptional activity S allele but has also separated a high-activity L_A allele from a third allele, L_G, which displays a low transcriptional activity similar to that of the S allele (Hu et al., 2006). One study examined the relationship between 5-HTTLPR and the LR and AUD in 85 men followed up for 20 years. The LR was evaluated using the oral consumption of alcohol (0.75 mL/kg); as a result, the HTTLPR L allele was found to be significantly related to both the lowest LR and alcohol dependence (Hu et al., 2005). In another study consisting of 243 German adolescents who had completed the SRE questionnaire, the LR to alcohol was significantly lower in carriers of the 5-HTTLPR high-activity variant (Hinckers et al., 2006).

Opioid receptor gene

OPRM1 (opioid receptor, mu 1) is a key candidate gene contributing to AUD, since β-endorphin and the μ opioid receptor have been shown to play important roles in the rewarding or reinforcing effects of alcohol (Thorsell, 2013). A polymorphism in OPRM1, A118G (Asn40Asp) rs1799971, has been reported (Bergen et al., 1997), and this polymorphism specifically influences β-endorphin binding, resulting in a threefold affinity for β-endorphin (Bond et al., 1998). Moreover, positron emission tomography (PET) studies have shown that this polymorphism alters μ opioid receptor availability in healthy controls, and abstinent alcoholics and carriers of the G allele had a lower global binding potential at the basal scan than subjects who were homozygous for the A allele (Weerts et al., 2013). A recent report combined a placebo-controlled intravenous alcohol administration with a PET methodology to examine the ventral striatum (VS) DA response to alcohol in male social drinkers divided according to those with and those without the G allele of the A118G polymorphism. Using 11C-raclopride, this study showed that the carriers of the minor G allele displayed greater VS DA release than carriers of the major AA homozygote (Ramchandani et al., 2011). In this report, the authors generated two humanized mouse lines carrying the respective human sequence variant. Brain microdialysis showed a fourfold greater peak DA response to an alcohol challenge in the GG mice, compared with the AA mice. Taken together, these genetic studies consistently implicate the G allele with greater DA release in response to alcohol, and these findings suggest increased subjective effects in response to alcohol. In fact, in accordance with these imaging results, an earlier study consisting of 38 students showed that G allele carriers experienced increased levels of subjective intoxication and higher levels of alcohol-induced sedation relative to AA homozygotes (Ray and Hutchison, 2004). Recently, the same author examined the effect of the A118G polymorphism in non-treatment-seeking problem drinkers and found that G allele carriers reported greater stimulation and a positive mood in response to the intravenous administration of alcohol (Ray et al., 2013). Moreover, in another study, men with the G allele who were heavy drinkers reported a significantly greater craving for alcohol in response to alcohol cues (van den Wildenberg et al., 2007).

These findings suggest that the OPRM1 A118G variation appears to modulate the LR to alcohol. In addition to these effects, a recent meta-analysis of studies examining how the effect of naltrexone, a μ opioid receptor antagonist, in patients with alcohol dependence might be moderated by the A118G polymorphism showed that naltrexone-treated patients carrying the G allele had lower relapse rates than those who were homozygous for the A allele (Chamorro et al., 2012). The results of this meta-analysis correspond with those of other studies suggesting that the OPRM1 A118G variation might modulate the LR to alcohol.

Nicotinic acetylcholine receptors gene

Nicotine and alcohol dependence are highly comorbid disorders that are likely to share overlapping genetic components (Swan et al., 1996). Nicotinic acetylcholine receptors (nAChRs) are pentameric cholinergic receptors that form ligand-gated ion channels. They are key mediators for the effect of nicotine on the CNS. Neuronal subtypes of nAChRs include various homomeric or heteromeric combinations of 12 different nicotinic subunits: α2 through α10 and β2 through β4. The CHRNA5-CHRNA3-CHRNB4 gene cluster on chromosome 15 encodes the α5, α3, and β4 subunits. In addition to numerous studies showing an association between genetic variations within this region with nicotine dependence, the LR to alcohol is also reportedly associated with SNPs within this region. In a study of 367 siblings, two sentinel SNPs within a region of high linkage disequilibrium were associated with the LR to alcohol (Joslyn et al., 2008). An association between this region and alcohol dependence was also reported in families included in the COGA data set (Wang et al., 2009). In addition, anotherstudy reported an association between the CHRNB2 gene, which encodes the β2 subunit, and the SR to alcohol and nicotine (Ehringer et al., 2007). Although these studies have suggested an association between nAChRs and the LR to alcohol, the number of studies examining this association is still small, and further studies examining such an association are essential to better understand the role of these genes in the LR to alcohol.

Interaction of gene–environment and other factors

Since all the above-mentioned genes together explain only about 50% of the variance in the risk for AUDs and for LR itself, evaluating how the LR and the associated genes relate to a range of environmental and cultural factors in producing heavy drinking and associated problems is also important (Schuckit, 2009). These factors include drinking patterns among peers, levels of stress, attitudes toward alcohol, and ways of coping with stress (Schuckit et al., 2004, 2005b, 2006, 2008). Structural equation models can be used to evaluate how endophenotypes and related genes might relate to these additional factors, with results supporting the conclusion that a low LR partially explains, or mediates, the relationship between a family history of AUDs and alcohol-related outcomes (Schuckit and Smith, 2000; Schuckit et al., 2004). In these modeling studies, the LR appears to have a direct impact on alcohol-related outcomes, as peer pressure, expectations of the effects of alcohol, and ways of coping with stress influence.

ALCOHOL METABOLISM AND GENETIC VARIATIONS OF ALCOHOL-METABOLIZING ENZYMES

Alcohol dehydrogenase

After alcohol is swallowed, it is absorbed primarily from the small intestine into the veins that collect blood from the stomach and bowels and from the portal vein, which leads to the liver. From there it is carried to the liver, where it is exposed to enzymes and metabolized. The liver is the main organ responsible for metabolizing ingested alcohol, and the primary enzymes that are involved are alcohol dehydrogenase (ADH), aldehyde dehydrogenase (ALDH), cytochrome P450 (CYP2E1), and catalase. However, alcohol is also metabolized in non-liver tissues that do not contain ADH, such as the brain, by the enzymes cytochrome P450 and catalase. In general, alcohol metabolism is achieved through oxidative pathways, which either add oxygen or remove hydrogen, and through non-oxidative pathways. ADH, cytochrome P450, and catalase all contribute to the oxidative metabolism of alcohol.

The major pathway for the oxidative metabolism of alcohol in the liver involves ADH. The metabolism of alcohol by ADH produces acetaldehyde, a highly reactive and toxic byproduct that may contribute to tissue damage and, possibly, the addictive process. As shown in Table 36.1, ADH constitutes a complex enzyme family; in humans, five classes have been categorized based on their kinetic and structural properties.

Humans have seven ADHs, and the genes encoding these enzymes are all localized on chromosome 4 in a head-to-tail array that is about 370 kb long (Edenberg, 2007). As shown in Table 36.1, the enzymes produced from these genes all differ slightly in their activities. These enzymes are classified as class I through V based on their enzymatic properties and the extent of sequence similarities (Crabb et al., 2004). Classes I (ADH1A, ADH1B, and ADH1C), II (ADH2), and possibly IV (ADH4) are thought to be involved in alcohol oxidation in vivo (Crabb et al., 2004). Among the class I genes, two are known to be polymorphic, and ADH1B and ADH1C each have three alleles. The ADH1A, ADH1B, and ADH1C genes produce closely related proteins that function as homo- and heterodimers and account for most of the alcohol-oxidizing capacity in the liver (Edenberg, 2007). The kinetic properties and population distribution of the variant alleles of ADH1 and ADH2 are presented in Table 36.1. ADH1B*1 encodes the β1 form of the enzyme and carries the amino acid arginine (Arg) at positions 48 and 370 in the amino acid chain, while ADH1B*2 encodes β1 and carries histidine (His) at position 48 (rs1229984) and ADH1B*3 encodes β3 and carries cysteine (Cys) at position 370 (rs2066702). As shown in Table 36.1, larger differences in the kinetic properties across three ADH1B variants are observed relative to the difference between the γ1 and γ2 subunits. The newly identified ADH1C*3 allele among Native Americans (Osier et al., 2002) has not been enzymatically characterized (Edenberg, 2007). Functional polymorphisms have been identified in the coding region of the ADH2 gene and in the promoter region of the ADH4 gene (Edenberg et al., 1999; Stromberg et al., 2002). The former polymorphism has been reported to show a slightly altered K_m (the Michaelis-Menten constant) and stability, while the latter exhibits an approximately twofold change in expression.

Despite substantial kinetic differences in vitro, in vivo experiments based on an oral administration paradigm detected few interindividual differences across ADH1B genotypes (Mizoi et al., 1994). Recently, however, use of the alcohol-clamping technique demonstrated a significant difference in alcohol elimination rates (AER) among Jewish men with different ADH1B genotypes. The results demonstrated that AER significantly increased with the number of ADH1B*2 copies, whereas minimal AER differences were observed across ADH1C genotypes (Neumark et al., 2004). Moreover, a recent study of Japanese alcoholics who voluntarily admitted to a hospital one day after heavy drinking showed that those who carried two copies of the ADH1B*1 allele still had significant blood alcohol concentrations (BAC), whereas those who carried at least one ADH1B*2 allele had very low BACs, consistent with a more rapid alcohol metabolism by ADH1B*2 (Yokoyama et al., 2007).

The ADH1B*2 allele is highly prevalent in Asian ethnic groups, particularly in northeast Asians (i.e., Chinese, Japanese and Koreans). Among Han Chinese, Taiwanese and Chinese Americans, 84–92% possessed at least one ADH1B*2 allele, including 40–60% who were homozygous for ADH1B*2 (Eng et al., 2007). The ADH1B*2 allele was also commonly found in Japanese people, and 81–100% of participants in 10 Japanese studies possessed at least one ADH1B*2 allele, including 34–71% who were homozygous for the allele (Eng et al., 2007). This allele is also relatively common in the Middle East (with a frequency of 20%), but it is uncommon elsewhere in Europe and Africa. In general, its allele frequency is less than 4% in populations of

Table 36.1

Properties of polymorphic forms of human alcohol dehydrogenase

Class	Gene nomenclature New	Former	Polymorphic site	Subunit type	K_m (mM)	V_{max} (/min)	Populations*	Tissue
I	ADH1A	ADH1		A	4.0	30	Caucasians	Liver
	ADH1B*1	ADH2*1	Arg48, Arg370	β_1	0.05	4	African Americans	
	ADH1B*2	ADH2*2	His48, Arg370	β2	0.9	350	Asians	Liver, lung
	ADH1B*3	ADH2*3	Arg48, Cys370	β3	40.0	300	African Americans	
	ADH1C*1	ADH3*1	Arg272, Ile350	γ1	1.0	90	All groups	Liver, stomach
	ADH1C*2	ADH3*2	Gln272, Val350	γ2	0.6	40	Caucasians	
	ADH1C*3	ADH3*3	Thr352§	γ3	–†	–†	Native Americans	
II	ADH2*1	ADH4		Π	9	10‡	–†	Liver, cornea
	ADH2*2		Ile308Val	Not shown	10.6	10.5‡	Swedes	
III	ADH3	ADH5		X	> 1000	100	–†	Most tissues
IV	ADH4	ADH7		σ (μ)	30	1800	–†	Stomach
V	ADH5	ADH6		Not applicable	30	Not applicable	Not applicable	Liver, stomach

Data are from Kendler et al. (1995), Crabb et al. (2004), Higuchi et al. (2006) and Edenberg (2007).
*Populations that have high allele frequencies for these variants. The alleles are not limited to those populations.
†Data not available.
‡*k*cat (dissociation rate, unit of values is the same as that of V_{max}).
§ADH1C*352Thr has been found in Native Americans as an additional variation on chromosomes with the Val350 characteristic of ADH1C*2.

European descent, and it was not found in any of the 90 individuals of European descent who were studied in a large genotyping project (iHapMap). Likewise, the allele is absent in most African populations (Hurley and Edenberg, 2012).

In Asian populations where ADH1B*2 is common, sufficient evidence indicates that this allele is protective against alcohol dependence. In most studies of diverse ethnic groups, the frequencies of the ADH1B*2 allele were significantly lower among the alcoholic probands than among the controls (Whitfield, 2002; Wall et al., 2003; Higuchi et al., 2004; Konishi et al., 2004; Li et al., 2007). An analysis of Han Chinese individuals living in Taiwan showed that the relative risk of alcohol dependence was reduced to 0.2 if a person carried a single ADH1B*2 allele and to 0.12 for homozygotes (Chen et al., 1999a). A more recent meta-analysis similarly showed a very strong protective effect for the ADH1B*2 allele in Asians (odds ratio, ~0.44; $P < 10^{-36}$) (Li et al., 2007). Moreover, the involvement of the ADH1B gene in susceptibility to alcohol dependence has been suggested by a population-based longitudinal study in Taiwan, and the less active ADH1B*1 allele was identified as a risk factor for alcoholism with an odds ratio of 5.87 over other ADH1B alleles (Cheng et al., 2004). Other than Asian populations, the Jewish population also has a higher frequency of the ADH1B*2 allele. A study of 152 Jewish-American subjects showed more unpleasant reactions following alcohol consumption and less frequent drinking in men with the ADH1B*2 allele than in those without the allele (Carr et al., 2002).

Although it has been difficult to detect the effect of ADH1B*2 in Europeans, a meta-analysis by Li and colleagues (2007) showed an odds ratio of 0.65 and $P = 0.04$ in alcohol-dependent people without secondary disease. A study in European Americans found that each ADH1B*2 allele lowered the number of symptoms of alcoholism, as specified in the *Diagnostic and Statistical Manual of Mental Disorders,* 4th edition (DSM-IV), and also reduced the maximum number of drinks consumed in one sitting (Sherva et al., 2009). Another analysis across three large, well-characterized samples of European Americans demonstrated a strong effect of ADH1B*2, close to what is seen in Asian populations (odds ratio < 0.4; $P < 10^{-8}$) (Bierut et al., 2012).

A study of 4597 Australians reported a significant association of ADH1B with flushing and alcohol intake and a borderline association with alcohol dependence (Macgregor et al., 2009). A study of 9080 males and females from Copenhagen found that ADH1B*1 homozygotes drank more alcohol and had a higher prevalence of daily drinking, heavy drinking, and excessive drinking in addition to a higher risk of developing alcoholism (Tolstrup et al., 2008). A study in a large Eastern European Slavic population consisting of 3016 males and 3481 females also found homozygotes for the ADH1B*1 allele to have a significantly higher drinking frequency and annual and weekly intakes of alcohol than ADH1B*2 carriers. The odds ratio of heavy drinking in ADH1B*1 homozygotes versus ADH1B*2 carriers was 2.1 (95% confidence intervals (CI) 1.1–3.2) in men and 2.2 (1.0–4.7) in women. However, no consistent association with problem drinking and the negative consequences of drinking was seen (Hubacek et al., 2012). A study examining 267 white college students in the United States also showed that individuals with an ADH1B*2 allele consumed a lower maximum number of drinks within a 24-hour period, reported a greater LR to alcohol, and were more likely to have experienced alcohol-induced headaches and more severe hangovers than those without an ADH1B*2 allele (Wall et al., 2005).

The mechanism of this protective effect in the ADH1B gene is not well understood. As the ADH1B*2 allele encodes the highly active β2 ADH isozyme, it has been postulated that the more rapid oxidation of alcohol at least transiently elevates acetaldehyde levels and that this accumulation of acetaldehyde could be protective against alcohol dependence (Crabb et al., 2004). However, an alcohol clamp study in healthy Japanese detected no difference in blood acetaldehyde concentrations among individuals with ADH1B genotypes and either active or inactive ALDH2 carriers (Hesselbrock et al., 2005), suggesting that the ADH1B*2 allele might exert its protective effect against alcohol dependence not as a result of acetaldehyde accumulation, but in a high LR to alcohol. Although the number of studies is limited, some reports have suggested an effect of ADH1B*2 on the LR to alcohol. Using objective and subjective measures before and after the ingestion of alcohol or a placebo, one study observed an effect of ADH1B*2 on the LR to alcohol among Asian Americans. In this study, participants with the ALDH2*1/*2 genotype and those with the ADH1B*2/*2 genotype were more likely to experience alcohol-induced vomiting and to report feeling less “great overall” 30 minutes after the ingestion of alcohol than those with the ADH1B*1/*2 genotype (Cook et al., 2005). The ADH1B*2 allele also was associated with hypersensitivity to alcohol in a Scandinavian population (Linneberg et al., 2010). In this study, consisting of 8000 adults living in Copenhagen, 3.78% of the study participants had the ADH1B*2 allele, and drinkers with the allele had an increased risk of alcohol-induced hypersensitivity reactions, defined as asthma and flushing induced by alcohol (odds ratios for ADH1B*1/*1 and ADH1B*1/*2 vs ADH1B*2/*2, 1.82; 95% CI, 1.04–3.17) (Linneberg et al., 2010). The SNP encoding ALDH1B1 (rs2228093) was also found to be associated with alcohol-induced hypersensitivity, but the exact function of the SNP was unknown.

The ADH1B*3 allele is relatively common in Eastern African populations, with frequencies of 27% among the Yoruba, 14% among the Luhya, and 24% among people of African ancestry living in the southwestern United States, but only 4% among the Maasai in Kenya (Hurley and Edenberg, 2012). By contrast, the allele is rare elsewhere and was not found in European or Asian HapMap samples (Hurley and Edenberg, 2012). Fewer studies of alcohol dependence have been conducted among people of African ancestry, but ADH1B*3 has been shown to be protective against alcohol dependence (Edenberg, 2007; Edenberg et al., 2010). The ADH1B*3 polymorphism has been shown to be associated with a 10% increase over ADH1B*1 in AER (Thomasson et al., 1995). This is consistent with the findings of a recent study showing that differences in ethanol-oxidizing activities among the recombinant variants of each gene under different conditions confirmed a role of the ADH1B*3 gene variant, but not the ADH2 gene variants, in protection against alcohol dependence (Lee et al., 2004).

Acetaldehyde dehydrogenase

Of the more than 10 ALDH subtypes known to exist in humans, a significant role in acetaldehyde oxidation has been identified only for ALDH1A1, ALDH1B1, and ALDH2, with respective K_m values for acetaldehyde of 180 μM, 55 μM, and less than 2 μM (Table 36.2). Because of its low K_m for acetaldehyde, mitochondrial ALDH2 plays a central role in human acetaldehyde metabolism. The variant allele ALDH2*2, resulting from a single nucleotide exchange, causes the substitution of lysine (i.e., ALDH2K in variant type) for glutamic acid (ALDH2E in normal type) at position 487 of the mature subunit polypeptide (Hempel et al., 1984; Yoshida et al., 1984). ALDH2*2 is a common variant occurring in approximately 40% of East Asians, including Han Chinese, Japanese, and Koreans, but is rare in other ethnic groups (Lai et al., 2014). An estimated 540 million people carry ALDH2*2, representing approximately 8% of the world's population (Brooks et al., 2009).

Table 36.2

Human aldehyde dehydrogenase involved in alcohol metabolism

Class	Gene	Allelic variant	Structure	K_m (μM)	V_{max} (/min)	V_{max} (/min/μM)	Cellular localization
I	ALDH1A1		Tetramer	180	380	2.1	Cytosol
	ALDH1B1		Tetramer	55	40	0.7	Mitochondria
II	ALDH2	ALDH2*1	Tetramer	0.2	280	1400	Mitochondria
		ALDH2*2	Tetramer	1.4	20	14	Mitochondria

Data are from Farres et al. (1994), Klyosov et al. (1996), Peng and Yin (2009), Stagos et al. (2010) and Hurley and Edenberg (2012).

ALDH2*2 homozygotes exhibit essentially no ALDH2 activity, while heterozygotes exhibit only a markedly reduced activity; hence, ALDH2 deficiency is considered to be a dominant negative trait (Yoshida and Dave, 1985; Crabb et al., 1989). East Asians deficient for ALDH2 exhibit a striking accumulation of acetaldehyde in the blood circulation following alcohol ingestion (Enomoto et al., 1991; Mizoi et al., 1994), and individuals with this allele exhibit facial flushing, tachycardia, headache, nausea, and other discomforting reactions as a result of the high acetaldehyde concentration in their blood (Harada et al., 1981).

Genetic epidemiologic studies indicate that a homozygous status for the ALDH2*2 allele almost fully protects against the development of alcoholism, whereas a heterozygous allele status partially protects against alcoholism (approximately 60%) (Higuchi et al., 1994; Chen et al., 1999a; Kim et al., 2008). The suppressive effects of the ALDH2*2 allele on alcohol consumption and the occurrence of alcohol dependence appear to become stronger with an increasing gene dose of the ALDH2*2 allele. Among 3171 Japanese alcoholic probands, 322 (10.5%) had the ALDH2*2 allele. All were heterozygous for ALDH2*1/*2, suggesting that homozygosity for ALDH2*2 almost completely inhibits the development of alcohol dependence because of the severe flushing response it produces, whereas the suppressive effect of heterozygosity for ALDH2*1/*2 is incomplete (Chen et al., 1999b; Higuchi et al., 2004).

The tissue ALDH2 activity in heterozygotes was 17% of that in the ALDH2*1/*1 genotype, whereas the activity of ALDH2*2/*2 was too low to be precisely determined. The protein amounts of tissue ALDH2 in variant homozygotes and heterozygotes were similar but only 30–40% of the value in normal homozygotes. Linear regression analyses show that the ALDH2 activities were significantly correlated with the protein contents in normal homozygotes and heterozygotes, respectively. The specific activity of ALDH2 per enzyme protein in ALDH2*1/*2 was 38% of that in ALDH2*1/*1 (Lai et al., 2014).

The physiological basis for overcoming the protection in heterozygotic alcoholics remains unclear. After chronic alcohol consumption, ALDH2*1/*2 alcoholics may metabolically adapt to a lower accumulation of blood acetaldehyde, may physiologically adapt to a lower intensity in cardiovascular responses, or may display both metabolic and physiologic adaptations to alleviate the discomfort sensitivity reaction. To test this hypothesis, an alcohol loading test was performed on alcoholics with ALDH2*1/*1 and ALDH2*1/*2 genotypes, and the blood alcohol, acetaldehyde, and acetate concentrations as well as the cardiovascular hemodynamic responses, and subjective sensations were measured (Chen et al., 2009). This study found comparable blood acetaldehyde profiles between alcoholic subjects with ALDH2*1/*2 and non-alcoholic subjects with the same ALDH2 genotype and revealed significantly lower intensities in both physiologic and psychologic responses in alcoholics with ALDH2*1/*2 than in non-alcoholics with the same ALDH2 genotype, suggesting that physiologic tolerance and/or an innately low sensitivity may play crucial roles in overcoming the deterring response of acetaldehyde.

Since the blood-brain barrier limits the diffusion of peripherally produced acetaldehyde to the brain and the brain has long been believed to lack any physiologically relevant alcohol-metabolizing system, earlier studies suggested that acetaldehyde was unlikely to affect the CNS (Lindros and Hillbom, 1979). However, recent studies have shown that the brain produces acetaldehyde from alcohol metabolism involving catalase, cytochrome P4502E1 (CYP2E1), and other pathways that require further clarification (Dick and Foroud, 2003). Moreover, intraperitoneal injections of acetaldehyde result in a range of behavioral effects, such as rewarding effects, sedation, and amnesia, that are difficult to ascribe to peripheral effects in animal studies (Gottesman and Gould, 2003; Quertemont et al., 2004; McLaughlin et al., 2008). While studies on the effects of acetaldehyde in humans are limited, a decreased amplitude in P300 was reported in subjects with ALDH2*1/*2 (Wall and Ehlers, 1995). Recently, the acute effects of acetaldehyde were reported in 12 men with ALDH2*1/*1 and 12 men with ALDH2*1/*2 in a double-blind placebo-controlled crossover alcohol-loading study. The study found that the performances of the subjects with ALDH2*1/*2 were poorer than those of the subjects with ALDH2*1/*1 and identified a significant correlation between psychomotor performance and blood acetaldehyde concentration (Kim et al., 2010). These studies were conducted on subjects with ALDH2*1/*2 who had very high blood acetaldehyde levels (more than 50 μM in the subjects in the reference (Kim et al., 2010)), and the ALDH2*2 allele exists almost exclusively in the Asian population. However, the activity of ALDH was significantly reduced by chronic alcohol consumption even in a non-Asian population (Nuttinen et al., 1983), and a recent study examining Japanese alcoholics who had experienced heavy drinking one day before the blood acetaldehyde measurements were performed showed no significant differences in blood acetaldehyde levels between alcoholics with ALDH2*1/*1 and those with ALDH2*1/*2 (Yokoyama et al., 2010).

Acetaldehyde has long been suggested to be involved in a number of alcohol's pharmacologic and behavioral effects, such as its reinforcing, aversive, sedative, amnestic, and stimulant properties. However, the role of acetaldehyde in alcohol's effects has been an extremely controversial topic during the past two decades. Opinions have ranged from those who virtually deny any role of acetaldehyde in alcohol's effects to those who claim that alcoholism is, in fact, "acetaldehydism" (Dick and Foroud, 2003). A number of recent studies, mainly on animals, have demonstrated that acetaldehyde administration induces a range of behavioral effects, and taken together with the high blood acetaldehyde levels in alcoholic patients, the role of acetaldehyde in alcohol's effects should attract more attention.

CONCLUSIONS

The LR to alcohol is considered to be an important endophenotype of AUD and a low LR to alcohol is associated with a high risk of AUD.

Several genes have been implicated as potentially contributing to heavier drinking, alcohol-related problems, and AUD through a low LR to alcohol. Identified are genes for GABA receptors, serotonin transporter, opioid receptor, and nicotinic acetylcholine receptors, but the most well-characterized genes that have a strong impact on the LR to alcohol are those for alcohol-metabolizing enzymes.

The major pathway for the oxidative metabolism of alcohol in the liver involves ADH and it produces acetaldehyde. Acetaldehyde is oxidized by ALDH and mitochondrial ALDH2 plays a central role in human acetaldehyde metabolism.

It is widely accepted that genetic polymorphisms of ADH1B and ALDH2 affect susceptibility to AUD. The less active form of ADH1B encoded by ADH1B*1 increases the risk of AUD in comparison with the superactive forms of ADH1B encoded by the ADH1B*2 allele, and the inactive forms of ALDH2 encoded by ALDH2*2 have been found to have an inhibitory effect on drinking due to the severe acetaldehyde accumulation in comparison with the active ALDH2 encoded by ALDH2*1.

Although the mechanism of protective effects of the ADH1B*2 allele is not well understood, the ADH1B*2 allele is associated with hypersensitivity to alcohol. The ALDH2*2 allele is associated with high LR to alcohol as a result of acetaldehyde accumulation. A number of animal studies have approved the role of acetaldehyde in a wide range of alcohol's pharmacologic and behavioral effects and the role of acetaldehyde in alcohol addiction should attract more attention.

References

Bergen AW, Kokoszka J, Peterson R et al. (1997). Mu opioid receptor gene variants: lack of association with alcohol dependence. Mol Psychiatry 2: 490–494.

Bierut LJ, Goate AM, Breslau N et al. (2012). ADH1B is associated with alcohol dependence and alcohol consumption in populations of European and African ancestry. Mol Psychiatry 17: 445–450.

Bond C, LaForge KS, Tian M et al. (1998). Single-nucleotide polymorphism in the human mu opioid receptor gene alters beta-endorphin binding and activity: possible implications for opiate addiction. Proc Natl Acad Sci U S A 95: 9608–9613.

Brooks PJ, Enoch MA, Goldman D et al. (2009). The alcohol flushing response: an unrecognized risk factor for esophageal cancer from alcohol consumption. PLoS Med 6: e50.

Cadoret RJ, Yates WR, Troughton E et al. (1995). Adoption study demonstrating two genetic pathways to drug abuse. Arch Gen Psychiatry 52: 42–52.

Carr LG, Foroud T, Stewart T et al. (2002). Influence of ADH1B polymorphism on alcohol use and its subjective effects in a Jewish population. Am J Med Genet 112: 138–143.

Chamorro AJ, Marcos M, Miron-Canelo JA et al. (2012). Association of micro-opioid receptor (OPRM1) gene polymorphism with response to naltrexone in alcohol dependence: a systematic review and meta-analysis. Addict Biol 17: 505–512.

Chen CC, Lu RB, Chen YC et al. (1999a). Interaction between the functional polymorphisms of the alcohol-metabolism genes in protection against alcoholism. Am J Hum Genet 65: 795–807.

Chen YC, Lu RB, Peng GS et al. (1999b). Alcohol metabolism and cardiovascular response in an alcoholic patient homozygous for the ALDH2*2 variant gene allele. Alcohol Clin Exp Res 23: 1853–1860.

Chen YC, Peng GS, Tsao TP et al. (2009). Pharmacokinetic and pharmacodynamic basis for overcoming acetaldehyde-induced adverse reaction in Asian alcoholics, heterozygous for the variant ALDH2*2 gene allele. Pharmacogenet Genomics 19: 588–599.

Cheng AT, Gau SF, Chen TH et al. (2004). A 4-year longitudinal study on risk factors for alcoholism. Arch Gen Psychiatry 61: 184–191.

Cook TA, Luczak SE, Shea SH et al. (2005). Associations of ALDH2 and ADH1B genotypes with response to alcohol in Asian Americans. J Stud Alcohol 66: 196–204.

Crabb DW, Edenberg HJ, Bosron WF et al. (1989). Genotypes for aldehyde dehydrogenase deficiency and alcohol sensitivity. The inactive ALDH2(2) allele is dominant. J Clin Invest 83: 314–316.

Crabb DW, Matsumoto M, Chang D et al. (2004). Overview of the role of alcohol dehydrogenase and aldehyde dehydrogenase and their variants in the genesis of alcohol-related pathology. Proc Nutr Soc 63: 49–63.

Dick DM, Foroud T (2003). Candidate genes for alcohol dependence: a review of genetic evidence from human studies. Alcohol Clin Exp Res 27: 868–879.

Dick DM, Plunkett J, Wetherill LF et al. (2006). Association between GABRA1 and drinking behaviors in the collaborative study on the genetics of alcoholism sample. Alcohol Clin Exp Res 30: 1101–1110.

Dixon CI, Walker SE, King SL et al. (2012). Deletion of the gabra2 gene results in hypersensitivity to the acute effects of ethanol but does not alter ethanol self administration. PLoS One 7: e47135.

Edenberg HJ (2007). The genetics of alcohol metabolism: role of alcohol dehydrogenase and aldehyde dehydrogenase variants. Alcohol Res Health 30: 5–13.

Edenberg HJ, Jerome RE, Li M (1999). Polymorphism of the human alcohol dehydrogenase 4 (ADH4) promoter affects gene expression. Pharmacogenetics 9: 25–30.

Edenberg HJ, Koller DL, Xuei X et al. (2010). Genome-wide association study of alcohol dependence implicates a region on chromosome 11. Alcohol Clin Exp Res 34: 840–852.

Ehringer MA, Clegg HV, Collins AC et al. (2007). Association of the neuronal nicotinic receptor beta2 subunit gene (CHRNB2) with subjective responses to alcohol and nicotine. Am J Med Genet B Neuropsychiatr Genet 144B: 596–604.

Eng MY, Luczak SE, Wall TL (2007). ALDH2, ADH1B, and ADH1C genotypes in Asians: a literature review. Alcohol Res Health 30: 22–27.

Enoch MA (2008). The role of GABA(A) receptors in the development of alcoholism. Pharmacol Biochem Behav 90: 95–104.

Enoch MA (2013). Genetic influences on response to alcohol and response to pharmacotherapies for alcoholism. Pharmacol Biochem Behav, pii: S0091-3057(13)00284.

Enomoto N, Takase S, Yasuhara M et al. (1991). Acetaldehyde metabolism in different aldehyde dehydrogenase-2 genotypes. Alcohol Clin Exp Res 15: 141–144.

Farres J, Wang X, Takahashi K et al. (1994). Effects of changing glutamate 487 to lysine in rat and human liver mitochondrial aldehyde dehydrogenase. A model to study human (Oriental type) class 2 aldehyde dehydrogenase. J Biol Chem 269: 13854–13860.

Gottesman II, Gould TD (2003). The endophenotype concept in psychiatry: etymology and strategic intentions. Am J Psychiatry 160: 636–645.

Harada S, Agarwal DP, Goedde HW (1981). Aldehyde dehydrogenase deficiency as cause of facial flushing reaction to alcohol in Japanese. Lancet 2: 982.

Heath AC, Madden PA, Bucholz KK et al. (1999). Genetic differences in alcohol sensitivity and the inheritance of alcoholism risk. Psychol Med 29: 1069–1081.

Heils A, Teufel A, Petri S et al. (1996). Allelic variation of human serotonin transporter gene expression. J Neurochem 66: 2621–2624.

Hempel J, Kaiser R, Jornvall H (1984). Human liver mitochondrial aldehyde dehydrogenase: a C-terminal segment positions and defines the structure corresponding to the one reported to differ in the Oriental enzyme variant. FEBS Lett 173: 367–373.

Hesselbrock V, Higuchi S, Soyka M (2005). Recent developments in the genetics of alcohol-related phenotypes. Alcohol Clin Exp Res 29: 1321–1324.

Higuchi S, Matsushita S, Imazeki H et al. (1994). Aldehyde dehydrogenase genotypes in Japanese alcoholics. Lancet 343: 741–742.

Higuchi S, Matsushita S, Masaki T et al. (2004). Influence of genetic variations of ethanol-metabolizing enzymes on phenotypes of alcohol-related disorders. Ann N Y Acad Sci 1025: 472–480.

Higuchi S, Matsushita S, Kashima H (2006). New findings on the genetic influences on alcohol use and dependence. Curr Opin Psychiatry 19: 253–265.

Hinckers AS, Laucht M, Schmidt MH et al. (2006). Low level of response to alcohol as associated with serotonin transporter genotype and high alcohol intake in adolescents. Biol Psychiatry 60: 282–287.

Hu X, Oroszi G, Chun J et al. (2005). An expanded evaluation of the relationship of four alleles to the level of response to alcohol and the alcoholism risk. Alcohol Clin Exp Res 29: 8–16.

Hu XZ, Lipsky RH, Zhu G et al. (2006). Serotonin transporter promoter gain-of-function genotypes are linked to obsessive-compulsive disorder. Am J Hum Genet 78: 815–826.

Hubacek JA, Pikhart H, Peasey A et al. (2012). ADH1B polymorphism, alcohol consumption, and binge drinking in Slavic Caucasians: results from the Czech HAPIEE study. Alcohol Clin Exp Res 36: 900–905.

Hurley TD, Edenberg HJ (2012). Genes encoding enzymes involved in ethanol metabolism. Alcohol Res 34: 339–344.

Joslyn G, Brush G, Robertson M et al. (2008). Chromosome 15q25.1 genetic markers associated with level of response to alcohol in humans. Proc Natl Acad Sci U S A 105: 20368–20373.

Kalu N, Ramchandani VA, Marshall V et al. (2012). Heritability of level of response and association with recent drinking history in nonalcohol-dependent drinkers. Alcohol Clin Exp Res 36: 1034–1041.

Kendler KS, Walters EE, Neale MC et al. (1995). The structure of the genetic and environmental risk factors for six major psychiatric disorders in women. Phobia, generalized anxiety disorder, panic disorder, bulimia, major depression, and alcoholism. Arch Gen Psychiatry 52: 374–383.

Kim DJ, Choi IG, Park BL et al. (2008). Major genetic components underlying alcoholism in Korean population. Hum Mol Genet 17: 854–858.

Kim SW, Bae KY, Shin HY et al. (2010). The role of acetaldehyde in human psychomotor function: a double-blind placebo-controlled crossover study. Biol Psychiatry 67: 840–845.

Klausberger T, Fuchs K, Mayer B et al. (2000). GABA (A) receptor assembly. Identification and structure of gamma(2) sequences forming the intersubunit contacts with alpha(1) and beta(3) subunits. J Biol Chem 275: 8921–8928.

Klyosov AA, Rashkovetsky LG, Tahir MK et al. (1996). Possible role of liver cytosolic and mitochondrial aldehyde dehydrogenases in acetaldehyde metabolism. Biochemistry 35: 4445–4456.

Knopik VS, Heath AC, Madden PA et al. (2004). Genetic effects on alcohol dependence risk: re-evaluating the importance of psychiatric and other heritable risk factors. Psychol Med 34: 1519–1530.

Konishi T, Luo HR, Calvillo M et al. (2004). ADH1B*1, ADH1C*2, DRD2 (-141C Ins), and 5-HTTLPR are associated with alcoholism in Mexican American men living in Los Angeles. Alcohol Clin Exp Res 28: 1145–1152.

Lai CL, Yao CT, Chau GY et al. (2014). Dominance of the inactive Asian variant over activity and protein contents of mitochondrial aldehyde dehydrogenase 2 in human liver. Alcohol Clin Exp Res 38: 44–50.

Lee SL, Hoog JO, Yin SJ (2004). Functionality of allelic variations in human alcohol dehydrogenase gene family: assessment of a functional window for protection against alcoholism. Pharmacogenetics 14: 725–732.

Li H, Mukherjee N, Soundararajan U et al. (2007). Geographically separate increases in the frequency of the derived ADH1B*47His allele in eastern and western Asia. Am J Hum Genet 81: 842–846.

Lindros KO, Hillbom ME (1979). Acetaldehyde in cerebrospinal fluid: its near-absence in ethanol-intoxicated alcoholics. Med Biol 57: 246–247.

Linneberg A, Gonzalez-Quintela A, Vidal C et al. (2010). Genetic determinants of both ethanol and acetaldehyde metabolism influence alcohol hypersensitivity and drinking behaviour among Scandinavians. Clin Exp Allergy 40: 123–130.

Liu IC, Blacker DL, Xu R et al. (2004). Genetic and environmental contributions to the development of alcohol dependence in male twins. Arch Gen Psychiatry 61: 897–903.

Macgregor S, Lind PA, Bucholz KK et al. (2009). Associations of ADH and ALDH2 gene variation with self report alcohol reactions, consumption and dependence: an integrated analysis. Hum Mol Genet 18: 580–593.

Magid V, Maclean MG, Colder CR (2007). Differentiating between sensation seeking and impulsivity through their mediated relations with alcohol use and problems. Addict Behav 32: 2046–2061.

McLaughlin PJ, Chuck TL, Arizzi-LaFrance MN et al. (2008). Central vs. peripheral administration of ethanol, acetaldehyde and acetate in rats: effects on lever pressing and response initiation. Pharmacol Biochem Behav 89: 304–313.

Mizoi Y, Yamamoto K, Ueno Y et al. (1994). Involvement of genetic polymorphism of alcohol and aldehyde dehydrogenases in individual variation of alcohol metabolism. Alcohol Alcohol 29: 707–710.

Morean ME, Corbin WR (2010). Subjective response to alcohol: a critical review of the literature. Alcohol Clin Exp Res 34: 385–395.

Neumark YD, Friedlander Y, Durst R et al. (2004). Alcohol dehydrogenase polymorphisms influence alcohol-elimination rates in a male Jewish population. Alcohol Clin Exp Res 28: 10–14.

Nuutinen H, Lindros KO, Salaspuro M (1983). Determinants of blood acetaldehyde level during ethanol oxidation in chronic alcoholics. Alcohol Clin Exp Res 7: 163–168.

O'Connor S, Morzorati S, Christian J et al. (1998). Clamping breath alcohol concentration reduces experimental variance: application to the study of acute tolerance to alcohol and alcohol elimination rate. Alcohol Clin Exp Res 22: 202–210.

Osier MV, Pakstis AJ, Goldman D et al. (2002). A proline-threonine substitution in codon 351 of ADH1C is common in Native Americans. Alcohol Clin Exp Res 26: 1759–1763.

Peng GS, Yin SJ (2009). Effect of the allelic variants of aldehyde dehydrogenase ALDH2*2 and alcohol dehydrogenase ADH1B*2 on blood acetaldehyde concentrations. Hum Genomics 3: 121–127.

Pierucci-Lagha A, Covault J, Feinn R et al. (2005). GABRA2 alleles moderate the subjective effects of alcohol, which are attenuated by finasteride. Neuropsychopharmacology 30: 1193–1203.

Quertemont E, Tambour S, Bernaerts P et al. (2004). Behavioral characterization of acetaldehyde in C57BL/6 J mice: locomotor, hypnotic, anxiolytic and amnesic effects. Psychopharmacology (Berl) 177: 84–92.

Ramchandani VA, Umhau J, Pavon FJ et al. (2011). A genetic determinant of the striatal dopamine response to alcohol in men. Mol Psychiatry 16: 809–817.

Ray LA, Hutchison KE (2004). A polymorphism of the mu-opioid receptor gene (OPRM1) and sensitivity to the effects of alcohol in humans. Alcohol Clin Exp Res 28: 1789–1795.

Ray LA, Bujarski S, MacKillop J et al. (2013). Subjective response to alcohol among alcohol-dependent individuals: effects of the mu-opioid receptor (OPRM1) gene and alcoholism severity. Alcohol Clin Exp Res 37 (Suppl. 1): E116–E124.

Roh S, Matsushita S, Hara S et al. (2011). Role of GABRA2 in moderating subjective responses to alcohol. Alcohol Clin Exp Res 35: 400–407.

Schuckit MA (1980). Self-rating of alcohol intoxication by young men with and without family histories of alcoholism. J Stud Alcohol 41: 242–249.

Schuckit MA (1999). New findings in the genetics of alcoholism. JAMA 281: 1875–1876.

Schuckit MA (2009). An overview of genetic influences in alcoholism. J Subst Abuse Treat 36: S5–S14.

Schuckit MA, Smith TL (2000). The relationships of a family history of alcohol dependence, a low level of response to alcohol and six domains of life functioning to the development of alcohol use disorders. J Stud Alcohol 61: 827–835.

Schuckit MA, Mazzanti C, Smith TL et al. (1999). Selective genotyping for the role of 5-HT2A, 5-HT2C, and GABA alpha 6 receptors and the serotonin transporter in the level of response to alcohol: a pilot study. Biol Psychiatry 45: 647–651.

Schuckit MA, Smith TL, Anderson KG et al. (2004). Testing the level of response to alcohol: social information processing model of alcoholism risk–a 20-year prospective study. Alcohol Clin Exp Res 28: 1881–1889.

Schuckit MA, Smith TL, Danko G et al. (2005a). Correlations among first-degree relatives for responses on the Self-Rating of the Effects of Alcohol Questionnaire in teenagers. J Stud Alcohol 66: 62–65.

Schuckit MA, Smith TL, Danko GP et al. (2005b). Evaluation of a level of response to alcohol-based structural equation model in adolescents. J Stud Alcohol 66: 174–184.

Schuckit M, Smith T, Pierson J et al. (2006). Relationships among the level of response to alcohol and the number of alcoholic relatives in predicting alcohol-related outcomes. Alcohol Clin Exp Res 30: 1308–1314.

Schuckit MA, Smith TL, Trim R et al. (2008). Testing the level of response to alcohol-based model of heavy drinking and alcohol problems in offspring from the San Diego Prospective Study. J Stud Alcohol Drugs 69: 571–579.

Sherva R, Rice JP, Neuman RJ et al. (2009). Associations and interactions between SNPs in the alcohol metabolizing genes and alcoholism phenotypes in European Americans. Alcohol Clin Exp Res 33: 848–857.

Stagos D, Chen Y, Brocker C et al. (2010). Aldehyde dehydrogenase 1B1: molecular cloning and characterization of a novel mitochondrial acetaldehyde-metabolizing enzyme. Drug Metab Dispos 38: 1679–1687.

Stromberg P, Svensson S, Hedberg JJ et al. (2002). Identification and characterisation of two allelic forms of human alcohol dehydrogenase 2. Cell Mol Life Sci 59: 552–559.

Swan GE, Carmelli D, Cardon LR (1996). The consumption of tobacco, alcohol, and coffee in Caucasian male twins: a multivariate genetic analysis. J Subst Abuse 8: 19–31.

Thomasson HR, Beard JD, Li TK (1995). ADH2 gene polymorphisms are determinants of alcohol pharmacokinetics. Alcohol Clin Exp Res 19: 1494–1499.

Thorsell A (2013). The mu-opioid receptor and treatment response to naltrexone. Alcohol Alcohol 48: 402–408.

Tolstrup JS, Nordestgaard BG, Rasmussen S et al. (2008). Alcoholism and alcohol drinking habits predicted from alcohol dehydrogenase genes. Pharmacogenomics J 8: 220–227.

Tsuang MT, Faraone SV, Lyons MJ (1993). Identification of the phenotype in psychiatric genetics. Eur Arch Psychiatry Clin Neurosci 243: 131–142.

Tyndale RF (2003). Genetics of alcohol and tobacco use in humans. Ann Med 35: 94–121.

Uhart M, Weerts EM, McCaul ME et al. (2013). GABRA2 markers moderate the subjective effects of alcohol. Addict Biol 18: 357–369.

van den Wildenberg E, Wiers RW, Dessers J et al. (2007). A functional polymorphism of the mu-opioid receptor gene (OPRM1) influences cue-induced craving for alcohol in male heavy drinkers. Alcohol Clin Exp Res 31: 1–10.

Viken RJ, Rose RJ, Morzorati SL et al. (2003). Subjective intoxication in response to alcohol challenge: heritability and covariation with personality, breath alcohol level, and drinking history. Alcohol Clin Exp Res 27: 795–803.

Wall TL, Ehlers CL (1995). Acute effects of alcohol on P300 in Asians with different ALDH2 genotypes. Alcohol Clin Exp Res 19: 617–622.

Wall TL, Carr LG, Ehlers CL (2003). Protective association of genetic variation in alcohol dehydrogenase with alcohol dependence in Native American Mission Indians. Am J Psychiatry 160: 41–46.

Wall TL, Shea SH, Luczak SE et al. (2005). Genetic associations of alcohol dehydrogenase with alcohol use disorders and endophenotypes in white college students. J Abnorm Psychol 114: 456–465.

Wang JC, Grucza R, Cruchaga C et al. (2009). Genetic variation in the CHRNA5 gene affects mRNA levels and is associated with risk for alcohol dependence. Mol Psychiatry 14: 501–510.

Weerts EM, McCaul ME, Kuwabara H et al. (2013). Influence of OPRM1 Asn40Asp variant (A118G) on [11C]carfentanil binding potential: preliminary findings in human subjects. Int J Neuropsychopharmacol 16: 47–53.

Whitfield JB (2002). Alcohol dehydrogenase and alcohol dependence: variation in genotype-associated risk between populations. Am J Hum Genet 71: 1247–1250, author reply 50-1.

Yokoyama A, Tsutsumi E, Imazeki H et al. (2007). Contribution of the alcohol dehydrogenase-1B genotype and oral microorganisms to high salivary acetaldehyde concentrations in Japanese alcoholic men. Int J Cancer 121: 1047–1054.

Yokoyama A, Tsutsumi E, Imazeki H et al. (2010). Polymorphisms of alcohol dehydrogenase-1B and aldehyde dehydrogenase-2 and the blood and salivary ethanol and acetaldehyde concentrations of Japanese alcoholic men. Alcohol Clin Exp Res 34: 1246–1256.

Yoshida A, Dave V (1985). Enzymatic activity of atypical Oriental types of aldehyde dehydrogenases. Biochem Genet 23: 585–590.

Yoshida A, Huang IY, Ikawa M (1984). Molecular abnormality of an inactive aldehyde dehydrogenase variant commonly found in Orientals. Proc Natl Acad Sci U S A 81: 258–261.

Handbook of Clinical Neurology, Vol. 125 (3rd series)
Alcohol and the Nervous System
E.V. Sullivan and A. Pfefferbaum, Editors

Chapter 37

Epidemiology of drinking, alcohol use disorders, and related problems in US ethnic minority groups

RAUL CAETANO*, PATRICE A.C. VAETH, KAREN G. CHARTIER, AND BRITAIN A. MILLS
School of Public Health, Dallas Regional Campus, University of Texas Health Science Center at Houston, Dallas, TX, USA

INTRODUCTION

This chapter describes the epidemiology of drinking, alcohol use disorders, including abuse and dependence, and other alcohol-related problems (e.g., legal, interpersonal, job-related) among US ethnic minorities. The groups that have been traditionally included in these studies are Whites, Blacks, Hispanics, Asian Americans, American Indians, and Alaska Natives. To some extent, this grouping follows US census classifications of race and ethnicity, but there are also relevant differences. For instance, the US Census Bureau has treated Hispanics (or Latinos) as an ethnic group and not as a racial group, such as Whites or Blacks. The Bureau correctly argues that Hispanics can be of any race, and indeed they are. Yet, all studies in alcohol epidemiology treat Hispanics as a racially uniform group, without differentiating between, for instance, White Hispanics, Black Hispanics, and Asian Hispanics. Also, studies often treat ethnic minority groups as if they were culturally homogeneous, as if they all had come from the same (for example) European, Asian, or Latin American country. In reality, immigrant groups come to the United States from a variety of countries with different drinking traditions. Life in the United States can lead to some uniformity in drinking through acculturation, a process by which immigrant groups adopt drinking norms and drinking patterns dominant in US society. But this process never totally erases national or ethnic-specific drinking patterns and norms. Consequently, there is considerable heterogeneity in drinking across the three larger Hispanic national groups in the United States: Mexican Americans, Puerto Ricans, and Cuban Americans (Caetano et al., 2008; Mills and Caetano, 2010; Ramisetty-Mikler et al., 2010; Vaeth et al., 2012).

Most of the studies reviewed in this chapter analyze data collected from random samples of the adult household population (18 years of age and older) in particular ethnic groups. Because of this random selection, these studies provide results that are representative of drinking and its associated problems in the general population. An exception to the focus on adult drinking is the National Survey on Drug Use and Health, which is conducted by the Research Triangle Institute under contract with the Office of Applied Studies of the Substance Abuse and Mental Health Administration (SAMHSA, 2011). This yearly survey interviews over 70 000 people 12 years of age and older to capture drinking and drug use among the adolescent and adult household population. These alcohol epidemiologic studies show a number of health disparities across US ethnic groups. Besides well-known genetic differences in response to drinking, such as the flushing reaction to alcohol among Asian groups (Crabb et al., 1989; Kimura and Higuchi, 2011), many researchers have explained these health disparities in terms of the cumulative social disadvantages that characteristically affect ethnic minority groups in the United States.

BRIEF HISTORIC OVERVIEW

Alcohol research with ethnic minorities in the United States began approximately 50 years ago in the mid 20th century with studies of drinking by Italian, Jewish, and Irish Americans (e.g., Bales, 1946; Lolli et al., 1958; Snyder, 1958). In general, these studies did not analyze representative samples of these groups but described indepth analysis of drinking patterns, norms regulating the use of alcohol, and cultural traditions associated with drinking in these groups. During the 1960s, epidemiologic

*Correspondence to: Raul Caetano, University of Texas School of Public Health, Dallas Regional Campus, 5323 Harry Hines Blvd, V8.112, Dallas, TX 75390-9128, USA. Tel: +1-214-648-1080, E-mail: Raul.Caetano@utsouthwestern.edu

surveys of drinking patterns with local or national representative samples of the household population were developed and analyzed (Knupfer and Room, 1967; Cahalan et al., 1969). These surveys had broad descriptive goals and their attention to ethnic differences in drinking was limited. However, in the late 1970s a new era in alcohol research with ethnic groups began. The National Institute on Mental Health funded the Epidemiology Catchment Area (ECA) study, which examined the prevalence and correlates of psychiatric disorders in five metropolitan areas of the country: Los Angeles, Saint Louis, Baltimore, Durham, and New Haven. Considerable attention was given to the epidemiology of alcohol abuse and dependence in these surveys, with special attention paid to Blacks and Hispanics (see, for instance, Helzer and Canino, 1992). The ECA sample interviewed in Los Angeles was designed to examine psychiatric disorders among Mexican Americans and the sample in Saint Louis had a strong focus on Blacks (Burnam et al., 1987a, b; Burnam, 1989). In 1984, the then Berkeley-based Alcohol Research Group conducted the first national survey of drinking with oversamples of Blacks and Hispanics (Caetano, 1989; Herd, 1990). Respondents were a representative sample of the US household population of adults, and several papers analyzing this survey appeared in the literature with indepth analysis of drinking among Blacks and Hispanics. Also in the 1980s, the 1985 National Household Survey on Drug Abuse (NHSDA), which interviewed a nationally representative sample of the US household population, oversampled Blacks and Hispanics, providing detailed national-level data on drinking and alcohol use disorders in these two groups (Carson-DeWitt, 2001). In 2003, the NHSDA became the National Survey on Drug Use and Health (NSDUH), which does not oversample ethnic minorities, but because it interviews a large number of respondents annually (around 70 000), it can provide detailed information on drinking (past 30 days) and alcohol use disorders (past year) among ethnic minorities.

Other national surveys of relevance to the study of the association between drinking and ethnicity are the 1995 National Alcohol Survey (NAS), which oversampled Blacks and Hispanics nationally, the 1992 National Longitudinal Alcohol Epidemiologic Survey (NLAES), which oversampled Blacks, and the 2002 National Epidemiologic Survey of Alcohol Related Conditions (NESARC), which oversampled Blacks and Hispanics. The NAS was conducted by the Alcohol Research Group, then at Berkeley, now in Emeryville, California. The second and third surveys, NLAES and NESARC, were conducted by the National Institute on Alcohol Abuse and Alcoholism (NIAAA). In addition to the 1995 NAS, the Alcohol Research Group has conducted national telephone surveys in 2000, 2005, and 2010. All these telephone surveys are large enough and have oversamples of Blacks and Hispanics that enable comparative analyses of drinking among Whites, Blacks, and Hispanics in the United States.

WHAT THE CHAPTER WILL COVER

This chapter is organized as follows: first, drinking among Whites, Blacks, and Hispanics is described. The focus is on these three groups because these are the ethnic groups most frequently studied in alcohol epidemiology and because these are the three largest ethnic groups represented in the US population. This is followed by a discussion of alcohol use disorders (alcohol abuse and dependence), as defined in the fourth edition of the *Diagnostic and Statistical Manual* of the American Psychiatric Association (DSM-IV: American Psychiatric Association, 1994), among Whites, Blacks, and Hispanics. Next, the chapter addresses drinking and alcohol use disorders among subgroups of Hispanics, defined by national origin: Mexican Americans, Puerto Ricans, Cuban Americans, and South/Central Americans. The same structure is applied to the examination of drinking and alcohol use disorders among Asian American subgroups. Presently, there is more information at the national-group level for Hispanics and Asian Americans than other groups. This is followed by a review of alcohol studies with Native Americans (i.e., American Indians and Alaska Natives), focusing on both drinking and associated problems in this group. Finally, theories proposed to explain health disparities in the alcohol field are reviewed and application to the neurology clinic considered.

DRINKING AMONG WHITES, BLACKS, AND HISPANICS

Most of the recent cross-sectional analyses that describe drinking among Whites, Blacks, and Hispanics have been based on data from the NESARC (e.g., Grant et al., 2007; Hasin et al., 2008; Keyes et al., 2009) or the NSDUH. There have also been trend analyses based on NLAES and NESARC data (e.g., Caetano et al., 2010, 2011; Chartier and Caetano, 2011). Caetano et al. (2010) described trends in the mean number of drinks consumed per month among Whites, Blacks, and Hispanics using NLAES and NESARC data. Table 37.1 shows these data, indicating that, among both men and women, there were no statistically significant differences for any of the three ethnic groups between 1992 and 2002. Within-year analyses showed that, for 2002, the mean number of drinks for White men was significantly higher than the mean for Black and Hispanic men. Among women, analysis of data for 1992 and 2002 showed statistically significant differences in the mean number of

Table 37.1

Mean number of drinks consumed per month by Whites, Blacks, and Hispanics: Trends between survey years (NLAES = 1992; NESARC = 2002)

	White		Black		Hispanic	
	1992	2002	1992	2002	1992	2002
Males*	21.3	22.3	19.8	18.9	18.5	17.8
Females†‡	6.2	6.2	4.9	5.2	3.3	3.9

NESARC, National Epidemiologic Study on Alcohol and Related Conditions; NLAES, National Longitudinal Alcohol Epidemiology Study.
**t*-tests significant between White and Black men ($P<0.01$), and White and Hispanic men ($P<0.001$) in 2002.
†*t*-tests significant between Hispanic and Black women ($P<0.01$), and Hispanic and White women ($P<0.001$) in 1992.
‡*t*-tests significant between Hispanic and White women ($P<0.001$), and Black and White women ($P<0.05$) in 2002.

drinks between Hispanic and Black women and between Hispanic and White women.

Caetano et al. (2010) also described trends in the proportion of drinkers, showing a statistically significant increase in the proportion of current drinkers for all three ethnic groups between 1992 and 2002. Table 37.2 from that work shows that the increase in percentage points was similar in the three ethnic groups: five points among Whites, six among Blacks, and seven among Hispanics. As a result, White and Hispanic men have similar rates of current drinking in 2002 that are higher than that of Blacks. Among men, there are statistically significant differences in the frequency of drinking five or more drinks in a day for all three ethnic groups between 1992 and 2002. This is best seen in Table 37.2 by examining the proportion of men reporting this behavior once a week or more in 2002 compared to 1992. At the same time, there was also an increase in the proportion of drinkers who did not report drinking five or more in a day from 1992 to 2002. This increase was larger for Whites and Blacks (10 and 13 percentage points, respectively) than for Hispanics (nine percentage points). There also were significant trends between 1992 and 2002 in drinking to intoxication less than once per month or more frequently. However, in contrast to data on drinking five or more drinks, the proportion in each frequency category of intoxication was more stable. The overall proportion of men reporting intoxication was the same in 1992 and 2002 for Whites (29% and 30%), Blacks (21% and 22%), and Hispanics (26% and 25%).

Data for women in Table 37.2 show that there also was a statistically significant increase in the proportion of current drinkers in all three ethnic groups. About a third of Black and Hispanic women were drinkers in 2002, compared to almost half of White women. Variations in the frequency of drinking five or more drinks per day between 1992 and 2002 were also significant for all three ethnic groups. There also were differences in the proportion of White, Black, and Hispanic women who drank but did not report consuming five or more drinks in a day, which increased in all three groups. The overall proportion of drinkers reporting drinking five or more drinks among women declined among Whites (16% to 12%) and Blacks (8% to 6%), but was relatively stable among Hispanics (10% to 9%). In all three ethnic groups, variations in the frequency of drinking to intoxication in each specific frequency category from 1992 to 2002 were relatively small.

Data from the NLAES and NESARC cover alcohol consumption in the 12 months prior to the survey interview. Data for the NSDUH cover the past 30 days. This narrower window may miss drinkers with an infrequent pattern of drinking or binge drinking but is less susceptible to recall biases than 12-month data (SAMHSA, 2012b). Data from the 2011 NSDUH show that any alcohol use in adults (i.e., ages 18 or older) is most prevalent for Whites (60.5%) and similar for Hispanics (47.2%) and Blacks (46.2%). Binge drinking and heavy drinking are high-risk consumption patterns that contribute to a variety of alcohol-related social and health problems (Naimi et al., 2003; Rehm et al., 2003). In 2011, Whites reported the highest prevalence of heavy drinking (7.6%), followed by Hispanics (5.7%), and Blacks (4.4%). Rates of binge drinking were 25.3% for Whites and 26.0% for Hispanics, with a lower percentage for Blacks (4.4%; NIAAA, 2004; Chen et al., 2010).

The data discussed so far clearly show that men drink more than women, a trend common among all ethnic groups in the US population. Age is also an significant predictor of drinking and binge drinking. Epidemiologic research in the United States has repeatedly shown a traditional pattern of drinking by age, indicating that those who are younger drink more than those who are older (Cahalan et al., 1969; Clark and Midanik, 1982). These cross-sectional data for the US general population show an increase in drinking during the 20s followed by an

Table 37.2

Current drinkers, frequency of drinking five or more drinks in a day, and frequency of intoxication among Whites, Blacks, and Hispanics by gender and survey year (NLAES = 1992; NESARC = 2002)

	White		Black		Hispanic	
Males	1992	2002	1992	2002	1992	2002
Drinking status*						
Current drinkers (*n*)	58.31 (7937)	63.52 (6885)	46.56 (967)	52.50 (1554)	52.95 (635)	60.28 (2281)
5+ drinks in a day (*n*)	13 538	10 782†	2091	3025†	1191	3700†
Once a week or more	9.2	13.6	9.6	11.4	9.8	14.1
1–3 times a month	9.2	7.0	6.8	3.6	11.0	9.3
< Once a month	18.7	11.5	11.5	5.8	16.8	11.9
Drinker/No 5+ drinks	30.0	31.3	18.3	31.5	15.3	24.6
Ex-drinkers/abstainers	41.9	36.7	53.7	47.7	47.1	40.2
Drinking to intoxication (*n*)	13 555	10 766†	2094	3023‡	1192	3698‡
Once a week or more	2.7	5.2	3.2	4.4	2.8	4.3
1–3 times a month	5.4	5.1	4.2	4.9	5.2	4.5
< Once a month	21.0	20.3	13.7	12.7	18.6	16.4
Drinker/ no intoxication	29.1	32.7	25.2	30.3	26.3	34.5
Ex-drinkers/abstainers	41.8	36.7	53.7	47.7	47.1	40.2
	White		**Black**		**Hispanic**	
Females	1992	2002	1992	2002	1992	2002
Drinking status*						
Current drinkers (*n*)	37.69 (6848)	46.98 (6321)	21.26 (871)	30.31 (1507)	23.78 (411)	31.67 (1473)
5+ drinks in a day (*n*)	18 295	13 631†	3850	5194†	1617	4576†
Once a week or more	2.2	3.2	2.4	2.3	1.2	2.6
1–3 times a month	3.4	2.9	2.4	1.3	2.1	2.2
< Once a month	10.5	6.2	3.5	2.3	7.0	4.2
Drinker/ not 5+ drinks	21.5	34.5	12.8	24.3	13.5	22.6
Ex-drinkers/abstainers	62.4	53.2	78.9	69.8	76.3	68.5
Drinking to intoxication (*n*)	18 303	13 618†	3851	5192†	1618	4573†
Once a week or more	0.8	1.7	0.8	1.6	0.4	1.3
1–3 times a month	1.9	2.7	1.4	2.1	1.3	1.9
< Once a month	12.9	15.2	4.6	6.7	8.1	8.7
Drinker/ no intoxication	22.0	27.2	14.4	19.9	14.0	19.6
Ex-drinkers/abstainers	62.4	53.2	78.8	69.8	76.3	68.5

Proportions and means are weighted; numbers in parentheses reflect current drinker sample size.
NESARC, National Epidemiologic Study on Alcohol and Related Conditions; NLAES, National Longitudinal Alcohol Epidemiology Study,
*Significant between ethnic groups within survey years.
†$P < 0.001$, ‡$P < 0.01$: indicates significant difference between survey years within the given ethnic/gender group.
Reproduced from Caetano et al. (2010).

abrupt decline in the 30s that is sustained in older age groups. However, data on drinking by age for Blacks and Hispanics depart from this traditional pattern (Caetano, 1991; Hilton, 1991; Ramisetty-Mikler et al., 2010), slowing with age in both ethnic groups (e.g., Caetano, 1991). Such a pattern can lead to a higher prevalence of alcohol problems during middle age among Black and Hispanic men than Whites. Other predictors of drinking in US general population data, and among ethnic minorities, include marital status, level of education, and employment status. Specifically, those who are single or divorced, those who are less educated, and those who are unemployed reported more drinking and more drinking of five or more drinks in a single day or at a sitting (Hilton, 1987; Dawson et al., 1995; Caetano et al., 2010). For Hispanics, birthplace was also an important predictor of drinking, with US-born Hispanics drinking more than those Hispanics born abroad.

These same variables also predicted higher levels of binge drinking (Caetano et al., 2010).

Different trajectories of drinking are also evident across Whites, Blacks, and Hispanics. For example, NESARC data show that slightly more Hispanics (7.9%) and Whites (7.1%) than Blacks (5.5%) reported an early onset of drinking (i.e., before age 15: NIAAA, 2004; Chen et al., 2006). The rate of drinking onset before age 15 was lower for women compared to men in all ethnic groups, and lowest for Black women, at 3.9%. Longitudinal analyses of changes in drinking also show disparities between Whites, Blacks, and Hispanics. For example, Caetano and Kaskutas' (1995, 1996) analyses of 8-year changes in drinking and binge drinking between 1984 and 1992 showed that heavy drinking (drinking five or more drinks in a day at least once a week) decreased only among White men (from 19% to 12%). Among men, the incidence of heavy drinking was 7% among Whites, 10% among Blacks, and 17% among Hispanics. The persistence of heavy drinking (proportion of heavy drinkers at time 1 who remained heavy drinkers at time 2) was greater among Black (51%) and Hispanic men (43%) than among White men (32%). Blacks and Hispanics were 2 times and 1.5 times, respectively, more likely than Whites to remain heavy drinkers between time 1 and time 2 in the analysis. The mean number of drinks consumed per month by heavy drinkers was also higher for Blacks and Hispanics than for Whites. Unfortunately, however, these results are now 20 years old, and it is not clear whether findings from 1984 to 1992 would be applicable today given changes in consumption and rates of alcohol use since 1992 (Grant et al., 2004; Schmidt et al., 2007; Caetano et al., 2010, 2011).

Longitudinal analyses of NESARC data (Chen et al., 2010), comparing 2002 wave 1 to the 2005 follow-up wave 2, provide crude results for change in drinking status, and for the proportion of various groups of wave 1 respondents who exceed low-risk drinking limits at wave 2 by ethnicity. Exceeding low-risk drinking limits for men include drinking five or more drinks per day and more than 14 drinks per week, and for women drinking four or more drinks per day and more than seven drinks per week. For instance, Hispanics were more likely than Whites and Blacks to return to drinking or become drinkers at wave 2. Hispanic and Black non-current drinkers at wave 1 were more likely than Whites to exceed daily or weekly low-risk guidelines at wave 2, and did so at older ages (45 and more). Black and Hispanic current drinkers at wave 1 were more likely than Whites to exceed daily and/or weekly low-risk guidelines at wave 2. Black and Hispanic drinkers who exceeded low-drinking guidelines at wave 1 were more likely than Whites to remain drinking above low-risk guidelines.

ALCOHOL USE DISORDERS AND OTHER PROBLEMS AMONG WHITES, BLACKS, AND HISPANICS

Alcohol use disorders reviewed in this section include alcohol abuse and dependence, as defined by DSM-IV. Grant et al. (2004) examined unadjusted prevalence rates of alcohol use disorders for ethnic groups in NLAES and NESARC. This analysis showed higher rates for DSM-IV alcohol abuse among Whites (NLAES, 3.0%; NEARC, 4.6%) than among Blacks (NLAES, 1.4%; NESARC, 3.2%) and Hispanics (NLAES, 2.5%; NESARC, 3.9%). These results were independent of gender. Overall rates of dependence were similar across Whites (NLAES, 4.3%; NESARC, 3.8%), Blacks (NLAES, 3.8%; NESARC, 3.6%), and Hispanics (NLAES, 5.8%; NESARC, 3.9%). The reduction in rates between 1992 and 2002 for Whites and Hispanics was statistically significant. Other national data, such as those in SAMHSA's 2011 NSDUH, show similar 12-month rates for abuse plus dependence in Whites (7.0%) and Hispanics (7.5%) and lower rates for Blacks (5.6%: SAMHSA, 2012b).

Other trend analyses of NLAES and NESARC data on DSM-IV alcohol abuse and dependence have been reported by Caetano et al. (2011). The data in Table 37.3 are from those analyses and show results for the overall prevalence of abuse among men and women as well as prevalence for each of the DSM-IV abuse indicators. First, the prevalence of alcohol abuse is lower than expected from the indicator-level data in the table because some respondents reporting a particular abuse indicator are also alcohol-dependent and, as such, are not counted in estimation of the overall diagnosis of abuse. In NLAES, 63% of the respondents with a dependence diagnosis were also alcohol abusers. In NESARC, the proportion was 65%. Given DSM-IV hierarchic rules for diagnosis identification, dependence takes precedence over abuse. Turning to the results in Table 37.3, from 1992 to 2002, the prevalence of alcohol abuse increased significantly among both men and women in all three ethnic groups, with the exception of Hispanic women. Also, abuse was more prevalent among men than women both in 1992 and 2002. Trends in alcohol abuse indicators show a significant increase in the proportion of hazardous use among White and Black men. Among White men, there also was a significant decrease in the proportion of legal problems. The prevalence of "failure to fulfill role obligations" decreased in all three groups for men. Among women, hazardous use prevalence increased from 1992 to 2002 in all three ethnic groups. The prevalence of the indicator "failure to fulfill major role obligations" decreased among White women only. A comparison of the overall prevalence rate of

Table 37.3

Prevalence of alcohol abuse and abuse indicators among Whites, Blacks, and Hispanics in the US population by gender and survey year (NLAES = 1992; NESARC = 2002)

	White		Black		Hispanic	
Males	1992 (n = 13 600)	2002 (n = 10 845)	1992 (n = 2100)	2002 (n = 3041)	1992 (n = 1194)	2002 (n = 3722)
Alcohol abuse	5.5	7.4‡	2.7	5.7‡	4.4	6.2*
Abuse indicators						
Interpersonal problems	2.3	2.4	2.5	2.6	2.9	3.9
Hazardous use	8.1	10.4‡	4.0	6.9‡	6.8	7.5
Legal problems	1.3	0.9†	1.2	0.9	1.4	1.8
Major role obligations	3.2	1.0‡	2.1	1.0*	3.0	1.5*
	White		Black		Hispanic	
Females	1992 (n = 18 338)	2002 (n = 13 662)	1992 (n = 3855)	2002 (n = 5204)	1992 (n = 1620)	2002 (n = 4586)
Alcohol abuse	1.8	2.9‡	0.7	1.4†	1.0	1.7
Abuse indicators						
Interpersonal problems	0.6	0.8	0.7	0.8	0.3	0.6
Hazardous use	2.5	3.7‡	1.0	2.0†	1.0	2.1†
Legal problems	0.2	0.3	0.2	0.3	0.2	0.1
Major role obligations	1.2	0.5‡	0.7	0.5	0.4	0.3

Cells include weighted proportions.
NESARC, National Epidemiologic Study on Alcohol and Related Conditions; NLAES, National Longitudinal Alcohol Epidemiology Study.
$^{*}P < 0.05$, $^{†}P < 0.01$, $^{‡}P < 0.001$: indicates significant difference between survey years within the given ethnic/gender group.
Reproduced from Caetano et al. (2011).

abuse with the prevalence rate for each abuse criterion shows that most abuse diagnoses were probably due to the hazardous drinking criterion in all three ethnic groups, independently of gender, because a positive diagnosis of abuse requires the presence of only one criterion.

In addition to examining prevalence and trend data on alcohol abuse and dependence, other recent epidemiologic studies have used longitudinal data from waves 1 and 2 of the NESARC to examine the incidence of these two alcohol use disorders. Grant et al. (2007) examined predictors of first incidence of alcohol abuse and dependence, while Grant et al. (2009) examined the onset, persistence, and recurrence of alcohol use disorders. Grant et al. (2007) reported a slightly higher incidence of alcohol dependence for Blacks (2.7%) and Hispanics (2.4%) compared to Whites (1.5%), as well as a higher incidence among those who were younger, never married, and those who had lower income. Grant et al. (2009) reported a higher risk of persistence of alcohol dependence (but not abuse) among US-born Hispanic men under 40 years of age compared to Whites. This was not true for Black men who were also compared to Whites. Among women, persistence of alcohol dependence was elevated among older (40 and more) Blacks and Hispanics relative to Whites.

Other recent epidemiologic studies of alcohol dependence have also focused on recovery from this condition in the US population, showing that Whites hold some advantage over ethnic minorities in this area. For instance, in Dawson et al.'s (2005) analysis of longitudinal data from the NESARC, Whites had a decreased risk of continued/recurrent dependence compared to other ethnic groups. Persistence of prior to past year dependence was highest among Blacks (35%), followed by Hispanics (33%), then Whites (23%). Cross-sectional analyses of the symptom profiles among Whites, Blacks, and Hispanics can also provide clues about persistence of alcohol use disorders, especially alcohol dependence. Profiles characterized by physiologic signs and symptoms, such as withdrawal, tolerance, relief drinking, and unsuccessful attempts to cut down, may be more stable than others because of their association with physiologic dependence. Grant (2000) showed that profiles of dependence among Blacks were more frequently characterized by signs of physiologic dependence (withdrawal, tolerance) and impairment of control over drinking than those among Whites. Caetano et al. (2008) reported similar findings, showing that the dependence symptom profile for Blacks and Hispanics was characterized by a higher prevalence of tolerance and unsuccessful attempts to cut down.

Like many other health problems, alcohol abuse and dependence do not affect all individuals equally. The prevalence data reviewed above show that men had higher rates of abuse and dependence than women. In fact, men were about 2.5 times more likely than women to have alcohol abuse or dependence. Compared to those 65 years of age and older, the odds of developing an alcohol use disorder (12-month data) were 13.2 (99% confidence interval (CI): 9.44–18.5) in the 18–29-year age group, 8.1 (99% CI: 5.8–11.2) in the 30–44-year age group, and 4.1 (99% CI: 2.9–5.7) in the 45–64-year age group (Hasin et al., 2007). Others who were also more at risk for alcohol use disorders were those who never married or who were widowed/separated/divorced, those with lower education, and those with an annual family income below $35 000 (Grant and Dawson, 1997; Hasin et al., 2007).

Hispanic and Black ethnicity have also been associated with other alcohol-related problems. Hispanics are at a high risk for alcohol-related motor vehicle fatalities (Hilton, 2006), drinking and driving (Caetano and McGrath, 2005; SAMHSA, 2005), intimate-partner violence (Caetano et al., 2000, 2005), and alcoholic liver disease (Stinson et al., 2001; Flores et al., 2008; Yoon and Yi, 2008). Hispanics do not seem to profit from the protective effect that has been associated with moderate drinking on all-cause mortality (Kerr et al., 2011). Blacks are at high risk for alcohol-related intimate-partner violence (Caetano et al., 2000, 2005), fetal alcohol syndrome (FAS) (CDC, 2002; Russo et al., 2004), alcoholic liver disease (Flores et al., 2008), and mortality from oral cavity and pharynx (National Cancer Institute, 2011b). They also have a high incidence and a high mortality from larynx cancer (National Cancer Institute, 2011a) and esophageal cancer (Day et al., 1993; Brown et al., 1994). Like Hispanics, Blacks do not seem to benefit from a protective effect of moderate drinking on cardiac diseases and all-cause mortality, hypertension, and coronary heart diseases (Fuchs et al., 2001, 2004; Sempos et al., 2003; Kerr et al., 2011).

DRINKING, ALCOHOL USE DISORDERS, AND OTHER PROBLEMS AMONG HISPANIC NATIONAL GROUPS

As stated above, much of the alcohol research with Hispanics has described drinking and associated problems in this group as a whole, without consideration for potential differences across the various Latin American national groups represented in the United States. Yet there are considerable differences in drinking and alcohol use disorders across Hispanic national groups. A random sample household survey systematically examined these differences in drinking across US Hispanic national groups. The 2006 Hispanic Americans Baseline Alcohol Survey (HABLAS) interviewed 5224 Hispanic adults in five metropolitan areas of the United States: Miami, New York, Philadelphia, Houston, and Los Angeles. Trained bilingual interviewers conducted computer-assisted personal interviews at the respondents' home that lasted about 1 hour. As can be seen in Table 37.4, Puerto Rican and Mexican American men generally reported higher drinking rates, weekly consumption, and binge drinking than South/Central and Cuban Americans (Ramisetty-Mikler et al., 2010). Women drank significantly less than men (Table 37.5). Mexican American women reported the highest abstention rate (61%), while Puerto Rican women drank more and binged more frequently compared to their counterparts in other national groups. Logistic analyses showed that being younger, Puerto Rican or South/Central American, single, initiating drinking at < 14 years, US- or foreign-born men, and US-born women were significant predictors of binge drinking. Puerto Rican origin, initiating drinking at < 17 years, and male gender (US- or foreign-born) were significant predictors of volume of drinking.

Besides differences in drinking patterns and binge drinking, there also were substantial differences in 12-month rates of alcohol use disorders across Hispanic national groups. Mexican American (5.6%) and Puerto Rican men (5.2%) had higher rates than Cuban Americans (1.8%) and South/Central Americans (4.2%: Caetano et al., 2008). Among women, rates of alcohol abuse were lower, ranging from 1.1% to 0.2% across the different groups. Among men, rates of alcohol dependence were also higher among Puerto Ricans (15.3%) and Mexican Americans (15.1%) compared to Cuban Americans (5.3%) and South/Central Americans (9%). Puerto Rican women had the highest rates of alcohol dependence (6.4%), followed by Mexican Americans (2.1%), Cuban Americans (1.6%), and South/Central Americans (0.8%). Further, while the highest rates of abuse and dependence were among those in their 20s, the rate decline with age was not as strong as in the US population. Thus, Hispanics at older ages (i.e., 40–49 and 50–59 years) were at considerably more risk of dependence and its health consequences than the US general population. This was particularly true of Puerto Rican and Mexican American men.

In addition to alcohol abuse and dependence, there were also significant national group differences in

Table 37.4

Current drinking and binge drinking among males by Hispanic national group and age: Proportions and means

Hispanic national group	Drinking variable	Total	18–29 years	30–39 years	40–49 years	50+ years
Puerto Rican	Current drinker*	68.3 (687)	72.2 (202)	79.2 (148)	58.1 (134)	59.4 (203)
	Mean number of drinks per week ‡§¶¶	16.9 ± 1.6 (448)	18.1 ± 2.8 (150)	16.1 ± 2.8 (112)	24.3 ± 6.1 (80)	10.8 ± 2.1 (106)
	Frequency of binge†‡	(447)	(148)	(111)	(80)	(108)
	One or more times per month	6.3	7.9	4.1	10.2	4.6
	<Once a month	42.3	60.7	40.0	29.7	29.3
	Not in the past 12 months	51.4	31.5	55.9	60.1	66.1
Cuban American	Current drinker†	66.1 (662)	70.7 (84)	77.1 (97)	75.2 (134)	57.4 (347)
	Mean number of drinks per week‡¶¶**	8.4 ± 0.9 (427)	11.1 ± 3.3 (61)	7.9 ± 1.6 (76)	8.1 ± 1.3 (100)	7.7 ± 1.4 (190)
	Frequency of binge‡	(425)	(61)	(75)	(100)	(189)
	One or more times per month	5.6	11.4	9.1	2.8	3.2
	<Once a month	21.7	27.0	28.8	29.9	11.9
	Not in the past 12 months	72.8	61.7	62.2	67.4	84.9
Mexican American	Current drinker†	67.3 (635)	74.6 (226)	67.3 (179)	68.1 (130)	51.3 (100)
	Mean number of drinks per week‡**††	15.9 ± 1.7 (435)	15.3 ± 2.6 (172)	17.2 ± 3.8 (119)	16.8 ± 3.4 (97)	13.1 ± 3.6 (47)
	Frequency of binge‡	(434)	(174)	(120)	(95)	(45)
	One or more times per month	6.7	5.2	10.3	6.9	2.1
	<Once a month	39.5	42.8	45.0	35.2	23.9
	Not in the past 12 months	53.8	52.1	44.7	58.0	74.0
Dominican/South Central (D/SC) American	Current drinker†	68.2 (637)	76.9 (212)	74.3 (150)	69.8 (117)	52.5 (158)
	Mean number of drinks per week‡¶†† ‡‡§§	8.9 ± 0.8 (442)	10.7 ± 1.1 (163)	9.7 ± 1.6 (116)	8.9 ± 2.3 (79)	5.1 ± 1.2 (84)
	Frequency of binge†‡	(441)	(163)	(116)	(78)	(84)
	One or more times per month	6.5	11.5	4.1	6.7	0.8
	<Once a month	36.4	37.2	44.4	36.6	26.0
	Not in the past 12 months	57.1	51.3	51.5	56.8	73.2

Numbers in parentheses are denominators used to compute proportions/means for a given subgroup; the proportions and means are weighted.

*$P < 0.001$, †$P < 0.05$, indicates significant difference between age groups.

‡among current drinkers.

§*t*-test significant between 40–49 and 50+ years.

¶¶*t*-test significant between Puerto Rican and Cuban American.

¶*t*-test significant between Puerto Rican and D/SC.

***t*-test significant between Mexican American and Cuban American.

††*t*-test significant between Mexican American and D/SC American.

‡‡*t*-test significant between 18–29 and 50+ years.

§§*t*-test significant between 30–39 and 50+ years.

Table 37.5

Current drinking and binge drinking among females by Hispanic national group and age: Proportions and means

Hispanic national group	Drinking variable	Total	18–29 years	30–39 years	40–49 years	50+ years
Puerto Rican	Current drinker*	43.1 (640)	53.1 (159)	46.0 (130)	44.8 (122)	30.7 (229)
	Mean number of drinks per week‡§¶**	9.5 ± 2.3 (267)	9.2 ± 2.0 (94)	4.3 ± 1.6 (58)	13.2 ± 7.2 (51)	11.9 ± 7.0 (64)
	Frequency of binge‡	(266)	(92)	(60)	(51)	(63)
	One or more times per month	5.3	5.3	1.3	16.5	0.0
	<Once a month	45.8	44.4	58.2	44.5	35.7
	Not in the past 12 months	48.9	50.3	40.5	39.0	64.3
Cuban American	Current drinker†	43.5 (661)	62.9 (66)	65.9 (108)	47.8 (119)	27.8 (368)
	Mean number of drinks per week‡¶¶	3.4 ± 1.1 (230)	2.9 ± 1.0 (38)	3.4 ± 1.4 (57)	1.2 ± 0.4 (48)	5.0 ± 2.9 (87)
	Frequency of binge*‡	(232)	(38)	(59)	(48)	(87)
	One or more times per month	0.6	2.3	0.0	0.0	0.0
	<Once a month	21.8	28.7	42.2	8.7	7.7
	Not in the past 12 months	77.6	69.0	57.8	91.3	92.3
Mexican American	Current drinker	39.3 (644)	45.5 (190)	38.5 (221)	30.8 (126)	40.0 (107)
	Mean number of drinks per week‡ ¶	3.0 ± 1.0 (214)	1.9 ± 0.4 (87)	1.9 ± 0.6 (70)	2.6 ± 1.2 (30)	7.1 ± 4.7 (27)
	Frequency of binge‡	(215)	(86)	(71)	(31)	(27)
	One or more times per month	1.0	1.9	0.9	0.9	0.0
	<Once a month	25.1	28.4	29.8	22.7	13.9
	Not in the past 12 months	73.9	69.8	69.3	76.4	86.1
Dominican/South Central (D/SC) American	Current drinker†	46.7 (635)	50.1 (122)	48.4 (152)	55.6 (171)	32.7 (190)
	Mean number of drinks per week‡**	3.8 ± 0.6 (286)	4.2 ± 0.9 (69)	5.4 ± 1.9 (74)	2.6 ± 0.8 (93)	3.3 ± 1.0 (50)
	Frequency of binge‡	(286)	(69)	(74)	(93)	(50)
	One or more times per month	0.6	0.9	1.4	0.0	0.0
	<Once a month	26.7	32.9	33.7	17.5	26.4
	Not in the past 12 months	72.8	66.2	65.0	82.5	73.6

Numbers in parentheses are denominators used to compute proportions/means for a given subgroup; the proportions and means are weighted.

*$P < 0.01$, †$P < 0.001$: indicates significant difference between age groups.

‡among current drinkers.

§*t*-test significant between 18–29 and 30–39 years.

¶¶*t*-test significant between Puerto Rican and Cuban American.

¶*t*-test significant between Puerto Rican and Mexican American.

***t*-test significant between Puerto Rican and D/SC American.

alcohol-related problems. Alcohol problems can be dependence-related, as discussed above (i.e., increased tolerance, impaired control, prolonged intoxication), or social in nature (i.e., legal, interpersonal, or job-related). Although US Hispanics as a whole were less likely to report problems than non-Hispanics (Hilton, 1991; Caetano, 1997), Mexican Americans and Puerto Ricans appeared to be at greater risk than Cuban Americans (Caetano, 1998; Vaeth et al., 2009). Particularly, data from HABLAS show that, among men, 28% of Mexican Americans and 23% of Puerto Ricans reported two or more problems compared to 18% of South/Central Americans and 9% of Cuban Americans (Vaeth et al., 2009). Among women, Puerto Ricans were most likely to report two or more problems (i.e., 9%) versus 4% of Mexican Americans and South/Central Americans and 2% of Cuban Americans. Regarding the prevalence of specific types of problems, men, in general, regardless of national group, appeared to experience more dependence-related problems than social consequences. These included increased tolerance, impaired control, and salience of drinking. The most commonly reported social consequences included financial problems, belligerence, and problems with one's spouse.

Two additional points regarding Hispanic national group differences are worth noting. First, acculturation to life in the United States is a known positive predictor of drinking and alcohol problems (Caetano, 1987; Caetano et al., 2009). Moreover, its effect is also known to vary by gender, with stronger effects typically observed for women (Zemore, 2007). Results from the HABLAS sample have revealed that, although acculturation effects on drinking are comparable across Hispanic national groups (Mills and Caetano, 2010), effects on alcohol problems (abuse in particular) were slightly stronger among Puerto Ricans (Caetano et al., 2009).

Future alcohol research with Hispanics must take this heterogeneity across national Hispanic group into consideration by conducting national-group specific analyses as often as possible. Because US Hispanics are still geographically divided, with different national groups concentrated in different areas of the country (e.g., Mexican Americans mostly in the Southwest, Puerto Ricans mostly in the Northeast, Cuban American mostly in Florida), local prevention and treatment efforts must effectively rely on research results specific to the dominant national group in that local area. Public health actions must also be guided by findings indicating that, in spite of many similarities, Puerto Ricans and Mexican Americans are at higher risk for abuse, dependence, and the associated consequences than the other two groups of US Hispanics.

DRINKING, ALCOHOL USE DISORDERS, AND OTHER PROBLEMS AMONG ASIAN AMERICANS

Studies examining differences in drinking and alcohol use disorders across Asian American and Pacific Islander groups are less extensive than those for Hispanic national groups. For instance, the first two waves of the NESARC oversampled Blacks and Hispanics but not Asian Americans, limiting the utility of this national dataset for subgroup analyses by Asian American national groups. This apparent neglect results in part from the fact that, in spite of continuous population growth, Asian Americans are still about 5% of the US population. They are also seen as a group with low risk for alcohol problems because of their overall lower use of alcohol than other US ethnic groups. This perceived lower risk for excessive alcohol use and problems in Asian Americans is partially linked to research on alcohol-metabolizing gene variants, including the ALDH2*2 allele, which is observed at higher frequencies in Asians and reported to be protective for alcohol dependence in Japanese, Koreans, and Chinese (Kimura and Higuchi, 2011).

Based on the NESARC, Asian Americans had a lower rate of past-year drinking (48.4%) when compared to other ethnicities (53.2–69.5%: Chen et al., 2006). However, there was considerable variation in drinking and binge drinking across Asian American national groups. Figure 37.1 shows data from the 2004-2008 NSDUH for adults 18 years and older (SAMHSA, 2010). Korean Americans had the highest rates of both past 30-day drinking and binge drinking. Japanese Americans also had higher rates of drinking than other groups, but similar rates of binge drinking to Filipino Americans and Vietnamese Americans. Asian Indians reported lower rates of both drinking measures, while Chinese Americans had the lowest rate of binge drinking but

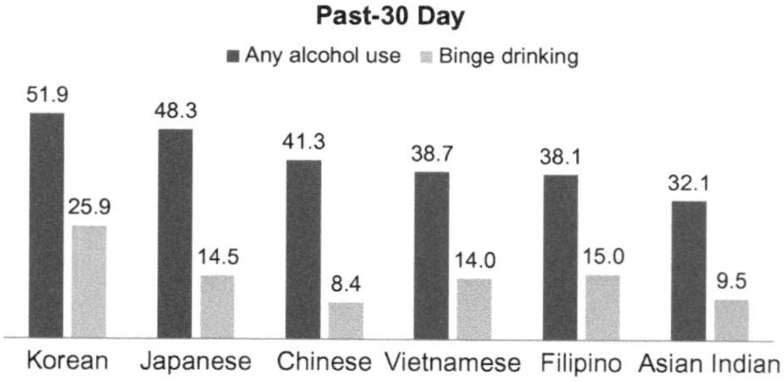

Fig. 37.1. Drinking by Asian American subgroups, 2004–2008. Data from the National Survey on Drug Use and Health; numbers shown are percentages; respondents are adults (18 years or older); data for Pacific Islanders not available. (Reproduced from Substance Abuse and Mental Health Services Administration [SAMHSA], 2010.)

similar drinking rates to Vietnamese and Filipino Americans. This report further showed that Asian Americans born in the United States had higher prevalence of drinking and binge drinking compared to those born outside the United States.

Sakai et al. (2010), using 1999–2002 NSDUH data for respondents 12 years and older, compared rates of drinking between multiracial and single-race Asian Americans. Most multiracial Native Hawaiians, other Pacific Islanders, and Asian Americans reported a higher prevalence of lifetime alcohol use when compared to their respective single-race subgroups, but a lower prevalence compared to single-race Whites. The exception was for Japanese Americans, with single-race Japanese reporting higher drinking rates than multiracial Japanese. Data from the 2006 NSDUH described results for Native Hawaiians and Pacific Islanders separate from other Asian groups. Prevalence rates of 30-day binge (26.8%) and heavy alcohol use (12.6%) for Pacific Islander groups ages 18 or older were more than double that of other Asian groups (12.5% and 2.6%: SAMHSA, 2007).

Regarding alcohol use disorders, trend data on alcohol abuse and dependence comparing NLAES and NESARC show that Asian Americans had the lowest rate of 12-month abuse (1% in NLAES, 2% in NESARC) and dependence (2.3% in NLAES, 2.4% in NESARC). The 2% rate of abuse in NESARC, the most recent survey, was 2.5 times lower than the rate for Whites and Native Americans, two times lower than the rate for Hispanics, and 1.6 times lower than the rate for Blacks. The 2.4% rate for dependence was 2.6–1.6 times lower than the rate in other ethnic groups. Hasin et al. (2007), reported similar findings when comparing Asian Americans and Whites, indicating that the former were less likely than Whites to meet criteria for 12-month and lifetime alcohol abuse and dependence. Takeuchi et al. (2007), using data from the 2002–2003 National Latino and Asian American Study, reported lifetime prevalence rates for substance use disorders among Asian American national groups. Filipinos were 2.3 times more likely to report a lifetime substance use disorder than Chinese Americans, with no statistical differences found between Chinese, Vietnamese, and other Asian Americans. Chae et al. (2008), based on data from the same survey, reported a 3.6% lifetime estimate of alcohol disorders in Asian Americans. Comparatively, Filipino Americans (20.2%) had a lower prevalence of lifetime alcohol disorders than other Asian Americans (39.3%), but a higher prevalence than Chinese Americans (10.3%) and Vietnamese Americans (2.5%). Price et al. (2002), also using US national data, reported a 12.8% lifetime rate of alcohol dependence for Japanese Americans, followed by 10.1% for Filipino Americans, and 9.7% for Korean Americans, and with lower rates for Chinese (4.5%) and Vietnamese Americans (3.4%). For past-year alcohol dependence, Sakai et al. (2010) showed that multiracial Native Hawaiians and Filipinos had higher prevalence than both their respective single-race Asian subgroups (Native Hawaiians and Filipinos) and Whites (Filipinos only).

Regarding other problems associated with alcohol use, data for liver cancers and motor vehicle deaths are relevant to Asian Americans. The majority (approximately 90%) of all primary liver cancers are hepatocellular carcinomas (HCC: Altekruse et al., 2009). Alcohol-related and non-alcohol-related liver cirrhosis usually precede HCC and are the two most common risk factors (Pelucchi et al., 2006; Altekruse et al., 2009; El-Serag, 2011). The relative risk for developing this cancer increases with increased levels of alcohol consumption (Pelucchi et al., 2006). By ethnic group, 2003–2005 age-adjusted incidence rates for HCC per 100 000 persons were highest among Asians (11.7), followed by Hispanics (8.0), Blacks (7.0), Native Americans (6.6), and Whites (3.9: Altekruse et al., 2009). Death rates for HCC per 100 000 people also were higher among minority groups, particularly Asian Americans (i.e., 8.9, 6.7, 5.8, 4.9, and 3.5 for Asians, Hispanics, Blacks, Native Americans, and Whites, respectively). Finally, the Centers for Disease Control and Prevention (CDC, 2009) statistics on alcohol-related motor vehicle crash deaths point to an important subgroup difference for Asian Americans. In 2006, the overall death rate among Asians (1.8 per 100 000 population) obscured the death rate among Native Hawaiians and other Pacific Islanders (5.9), which was less than the rate for Native Americans but similar to that for Hispanics (14.5 and 5.2, respectively).

DRINKING, ALCOHOL USE DISORDERS, AND OTHER PROBLEMS AMONG AMERICAN INDIANS AND ALASKA NATIVES

American Indians/Alaskan Natives are highlighted here separately as a high-risk ethnic minority group for heavier drinking and adverse alcohol consequences. Unlike some other ethnic groups, in which men are primarily at greater risk for alcohol problems, both Native American men and women are high-risk groups. It should be acknowledged that differences in drinking and alcohol problems for Native Americans by tribe and US geographic region have been identified (May and Gossage, 2001a, b; Beals et al., 2003, 2005; Koss et al., 2003).

In the US adult population, American Indians/Alaskan Natives have lower rates of abstinence than other ethnic minority groups. Seventeen percent (17.1%) of

Native Americans reported lifetime abstinence from alcohol, according to the NESARC (Chen et al., 2006), a rate that is higher than Whites (13.4%), but lower than Hispanics (25.7%), Blacks (24.7%), and Asian Americans (39.1%). Among US adult drinkers (Fig. 37.2), American Indians/Alaskan Natives reported lower rates of daily heavy drinking (28.4%, exceeding daily drinking guidelines) than Hispanics (33.9%), similar to Whites (27.3%), and higher than Blacks (22.5%) and Asian Americans (19.2%). However, weekly heavy drinking (exceeding weekly drinking guidelines) was highest among American Indians/Alaskan Natives (21.9%) compared to all other ethnic groups (Whites, 16.3%; Blacks, 16.4%; Hispanics, 11.8%; Asians, 9.8%). The 2011 NSDUH (SAMHSA, 2012a) reported rates of binge drinking in the past 30 days of 24.3% for American Indians/Alaskan Natives. This rate was similar to that for Whites (23.9%) and Hispanics (23.4%), and higher than that for Blacks (19.4%) and Asian Americans (11.6%). Heavy drinking (binge drinking on at least 5 days in the past 30 days) was at least two times higher among American Indians/Alaskan Natives (over 10%) than among Blacks, Hispanics, and Asian Americans.

Drinking among American Indians only (excluding Alaskan Natives) has been examined in the American Indian-Service Utilization, Psychiatric Epidemiology, Risk and Protective Factors Project (AI-SUPERPFP). This epidemiologic study interviewed a sample of randomly selected (from tribal rolls) American Indians 18-54 years of age living in the Southwest and Northern Plains (O'Connell et al., 2005). The Southwest sample drank less than the Northern Plains sample. For instance, in the heaviest-drinking category (most drinks consumed in a day were 18 or more and the number of drinking days during the past month was higher than four), 16% of the male group was from the Southwest, compared to 84% from the Northern Plains. Among women the percentages in this heaviest-drinking category were 12% for Southwest Indians and 88% for Northern Plains. Other analyses of AI-SUPERPFP focusing on binge drinking showed that, among men, Northern Plains and Southwest Indians were six and two times more likely, respectively, to report binge drinking than a US reference population from the NLAES sample (all races excluding American Indians: O'Connell et al., 2005). Among women, Northern Plains and Southwest Indians were six times and 1.7 times more likely than the US NLAES reference group to engage in binge drinking.

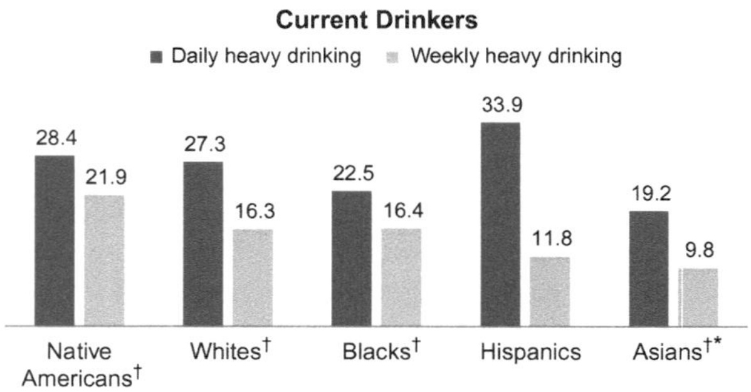

Fig. 37.2. Heavy drinking for US ethnic groups by gender. Data are from the 2002 National Epidemiologic Survey on Alcohol and Related Conditions; numbers shown are percentages; †Non-Hispanic; *includes Pacific Islanders. (Reproduced from Chen CM, Yi H-y, Falk DE, et al., 2006.)

These high rates of heavy drinking increase risk in Native Americans for experiencing adverse alcohol consequences, including alcohol abuse and dependence. Hasin et al. (2007) reported for American Indians/Alaskan Natives a higher prevalence of 12-month (12.1%) and lifetime (43.0%) alcohol use disorder compared to other ethnic groups (12-month, 4.5–8.9%; lifetime, 11.6–34.1%). Grant et al. (2004) reported rates of 12-month abuse and dependence in the 2002 NESARC by ethnicity. American Indians/Alaskan Natives had the highest rates of abuse (5.7%), followed by Whites (5.1%), Hispanics (3.9%), Blacks (3.3%), and Asian Americans (2.1%). A similar distribution of rates was reported for dependence. American Indians/Alaskan Natives had the highest rate (6.3%), followed by Hispanics (3.9%), Whites (3.8%), Blacks (3.6%), and Asian Americans (2.4%). Twelve-month prevalence rates for DSM-III-R abuse and dependence (American Psychiatric Association, 1987) for the AI-SUPERPFP dataset compared to the National Comorbidity Study (NCS: Kessler et al., 1994) were reported by Beals et al. (2005). These results show that, among men, the rates of abuse and dependence were 3.2% and 12.2% for Southwest Indians, 4.5% and 13% for Northern Plain Indians, and 3.5% and 10.7% for the NCS US national sample. Thus, rates of abuse were similar across the three groups, but rates of dependence were slightly higher (one or two percentage points) among the American Indians. Rates for women for abuse and dependence were 0.6% and 1.3% for Southwest Indians, 2.5% and 7.6% for Northern Plain Indians, and 1.6% and 3.7% for the NCS US national sample. Thus, the NCS national sample fell in the middle of the prevalence range, between the two Indian samples, with Northern Plain Indians reporting the highest prevalence rates.

American Indians/Alaskan Natives are also at high risk for a series of other problems associated with excessive drinking. Driving under the influence (DUI) is an important alcohol-related problem. Caetano and McGrath (2005) reported higher 12-month self-report rates for DUI in White (22.0%) and Native American (20.8%) men and men of mixed race (22.5%) compared to other ethnicities (16.8–12.0%). Rates of self-reported DUI for women were highest for Native Americans and women of mixed race (15.2% and 14.2%, respectively; other ethnicities 11.8–6.8%). The DUI arrest rate for Native Americans in 2001 was 479 arrestees per 100 000 residents compared with 332 for all other US ethnic groups (Perry, 2004). Rates of intoxication (i.e., blood alcohol concentration (BAC) $\geq$0.08%) for drivers who were fatally injured in a motor vehicle crash were highest for Native Americans (57%) and lowest for Asians (approximately 20%), with Hispanics, Whites, and Blacks falling in between (Hilton, 2006).

Native Americans are also overrepresented in national estimates of alcohol-involved suicides and crime. Recent alcohol use, according to the CDC (2009), was reported among suicide victims in 46% of Native Americans compared to 30% of Hispanics, 26% of Whites, 16% of Blacks, and 15% of Asians. A high proportion (47%) of Native American suicide victims tested positive for intoxication (BAC$\geq$0.08%: Caetano et al., 2013). These proportions were lower in Hispanics (38%) and Whites (33%), and particularly Blacks (26%) and Asians (23%). Native American suicide victims testing positive for alcohol had the highest mean BAC among these ethnic groups. Alcohol use also contributes to violence victimization among Native Americans. According to the US Department of Justice (Perry, 2004), Native American violent crime victims are more likely (62%) than other violent crime victims, including Whites (43%), Blacks (35%), and Asians (33%), to report alcohol use by their offender.

FAS and liver disease are often-cited examples of the negative health effects for Native Americans. From 1995 to 1997, FAS rates averaged 0.4 per 1000 live births according to the Fetal Alcohol Syndrome Surveillance Network, with the highest rates reported by Native Americans (3.2 per 1000: CDC, 2002). Native Americans have higher mortality rates for alcoholic liver disease than other US ethnic groups. The National Vital Statistical Reports (Miniño et al., 2011) on 2008 US deaths presented age-adjusted rates for alcoholic liver disease. Death rates for alcoholic liver disease in Native American men and women were 20.4 and 15.3 per 100 000 people, respectively, compared with 6.9 and 2.4 for men and women in the general population. From 2001 to 2005, alcohol-attributed deaths accounted for 11.7% of all Native American deaths, greater than twice the rate in the general population (CDC, 2008).

A THEORY ABOUT HEALTH DISPARITIES: CUMULATIVE ADVERSITIES

Health disparities across ethnic groups in US society have been recognized for over 30 years. In 1990, the National Institutes of Health (NIH) created the Office of Research on Minority Health (NIMHD, 2013). In 1993, Public Law 103-43, the Health Revitalization Act of 1993, established the Office of Research on Minority Health in the Office of the Director, NIH. In 2000, the National Center on Minority Health and Health Disparities was established by the passage of the Minority Health and Health Disparities Research and Education Act of 2000. The Institute of Medicine (IOM) of the National Academy of Sciences has released at least three reports examining health disparities in the United States (IOM, 2001, 2003, 2012). The 2003 IOM report defined health disparities as "racial or ethnic differences in the quality of healthcare that are not due to access-related factors, or clinical needs, preferences, and appropriateness of interventions." In spite of this recognition and a number of studies identifying specific areas of disparity, differences in access to health services and quality treatment persist (IOM, 2012).

An important theoretic framework from which health disparities across US ethnic groups can be understood is that of cumulative adversity (Turner and Lloyd, 1995; Hatch, 2005). The cumulative adversity theory proposes that certain health problems, such as alcohol use disorders, are the consequence of long-term or simultaneous cumulative hardships that frequently affect ethnic minorities. According to this model, ethnic minority status shapes the opportunities to which one is exposed over the life course and influences health status. Hardships to which minority groups are exposed include, but are not limited to, income disparities (Lynch et al., 2000; Gallo and Matthews, 2003; Meyer et al., 2008), unemployment, residential segregation, substandard housing, discrimination (Krieger and Sidney, 1996; Brondolo et al., 2003; Mays et al., 2007), and decreased access to healthcare (Smedley et al., 2003). Within this framework, an accumulation of adversities acting over time leads to an increased allostatic load and results in negative coping behaviors such as excessive alcohol use and greater vulnerability to ill health (McEwen, 2004).

Much of the research on health disparities in the United States is based on this model. For example, studies have demonstrated how the cumulative adversities of low socioeconomic status, ethnic stigma/discrimination, and other indicators of disadvantage impact drinking and substance use outcomes. Zemore et al. (2011) showed that perceived ethnic stigma was associated with alcohol dependence problems among those living below the poverty line. Lo and Cheng (2012) also showed that perceived discrimination had a positive association with substance use disorders among Hispanics, but that this association was strongest among those with low income. Mulia et al. (2008) similarly found that extreme disadvantage, a measure that included ethnic stigma, perceived unfair treatment, and poverty, was associated with problem drinking among Hispanics. Cerdá et al. (2010) found that, among White and Black men and women, increased neighborhood poverty was associated with an increased likelihood of binge drinking and an increased number of drinks consumed per week. Neighborhood disadvantage has also been associated with heavy drinking (Karriker-Jaffe et al., 2012) and alcohol-related problems (Jones-Webb et al., 1997; Karriker-Jaffe et al., 2012).

The research evidence on individual-level correlates (Hilton, 1991; Caetano and Kaskutas, 1995; Grant and Dawson, 1997; Grant et al., 2001; Hingson et al., 2006; Mulia et al., 2008) and neighborhood-level studies supports that socioeconomic adversities such as unemployment, lower education, lower family income, and negative life events are associated with a higher likelihood of heavier drinking and alcohol-related problems. Grant and Dawson (1997) and Hasin et al. (2007) reported that those with lower education and those with an annual family income below $35 000 have a higher risk of alcohol use disorders compared to those with higher education and higher income. Grant et al.'s (2007) results also showed a higher incidence of alcohol dependence among Blacks and Hispanics and those with a lower income. Although exceptions exist (Ennett et al., 1997; Pollack et al., 2005; Galea et al., 2007), studies of the effect of socioeconomic disadvantage at the neighborhood level have also reported positive associations between neighborhood poverty, drinking, and intimate-partner violence (Cunradi et al., 2000), neighborhood poverty, and binge drinking for men but not for women (McKinney et al., 2009), neighborhood poverty and alcohol-related problems among Black men (Jones-Webb et al., 1997), and alcohol outlet density (number of outlets per capita) and an increased risk of alcohol-related harms such as motor vehicle accidents, unintentional injuries, and sexually transmitted diseases (Scribner et al., 1999; Treno et al., 2001, 2007; Escobedo and Ortiz, 2002; Zhu et al., 2004; Cohen et al., 2006).

There are a number of other factors that characterize ethnicity and alcohol use and alcohol use disorder. These include attitudes towards alcohol consumption, biologic vulnerabilities, and norms associated with drinking. These factors may provide other explanations for the disparities that are reported across ethnic groups. Some ethnic groups are more likely to consume high-alcohol-content beverages (e.g., malt liquor), which could have greater social and health consequences (Vilamovska et al., 2009). Preference for such beverages seems to be more common in lower-income ethnic minority communities (Bluthenthal et al., 2005). Chartier and Caetano (2011) showed that Blacks and Hispanics underutilized some alcohol services compared to Whites, and Blacks and Hispanics with higher alcohol severity were less likely to use alcohol services. Studies also show a strong positive association between acculturation to US society and drinking among Hispanic women (Caetano, 1987; Caetano and Medina-Mora, 1988). Discrimination experiences have also received increasing attention in health disparities research as a means of explaining ethnic group differences in health outcomes that remain after accounting for socioeconomic inequalities between Whites and ethnic minorities (Brondolo et al., 2009; Williams et al., 2010). Based on the 2005 NESARC, 25% of Blacks and 15% of Hispanics reported experiencing discrimination in the past year (McLaughlin et al., 2010). Experiences of ethnic discrimination are associated with any drinking, greater drinking frequency, and problem drinking among Blacks (Martin et al., 2003; Borrell et al., 2007, 2010; Tran et al., 2010) and greater drinking frequency, binge, and heavy drinking among Hispanics (Borrell et al., 2010; Tran et al., 2010).

Mulia et al. (2008) examined the related variables of perceived ethnic group stigma and unfair treatment (not specifically based on ethnicity), and found positive relationships for these variables in predicting problem drinking among Blacks and Hispanics. Experiencing multiple forms of social adversities (stigma, unfair treatment, and poverty) also increased risk for problem drinking across ethnic groups. Zemore et al. (2011) showed that adversities accumulate in their association with alcohol problems. Unfair treatment for Blacks and perceived ethnic stigma for Hispanics was associated with alcohol dependence problems among those living below the poverty level but not above. Ethnic discrimination for Asians was associated with a greater likelihood of alcohol use disorders among individuals with low levels of ethnic identification (Chae et al., 2008).

CONCLUSIONS

The existing epidemiologic evidence indicates that there are serious alcohol-related health disparities across racial and ethnic groups in the United States. Some of the areas in which disparities are evident are, for example, rates of alcohol abuse and dependence, liver cirrhosis, FAS, drinking and driving, alcohol-related suicides, and certain types of cancer. There also are disparities in access to treatment and recovery from alcohol use disorders. Among ethnic minority groups, American Indians and Alaska Natives seem to be disproportionately affected by alcohol problems, having some of the highest rates for alcohol abuse, dependence, and drinking and driving. Asian Americans, on the other hand, are less affected than other ethnic groups and even the White majority. Blacks and Hispanics are sometimes at a disadvantage, and sometimes have problem rates similar to Whites.

The diversity in drinking and problems rates that is seen across these various groups also exists within groups, especially immigrant groups such as Hispanics and Asian Americans. This is because of the variety of drinking cultures that exist in the various countries from which these immigrants originate, as well as because of the variety of experiences that they have after their arrival in the United States. However, American Indians and Alaskan Natives are also quite diverse, and different tribes have very different patterns of alcohol use and problems. Finally, the White majority is not uniform either. Many Whites are immigrants, who came to the United States from a variety of countries in Europe and Scandinavia. These countries too, like those in Asia and Latin America, have different drinking cultures and thus the potential for difference in rates of alcohol-related problems.

The reasons for these health disparities are complex. Some are rooted in patterns of racial discrimination which themselves are linked to broader historic and persistent socioeconomic disadvantage characterizing minorities. This socioeconomic disadvantage is present at the individual level, being translated then as lack of opportunities for personal development through education, professional training, having secure and rewarding jobs, access to good healthcare, and also opportunities for creative and stimulating leisure activities. Socioeconomic disadvantage also affects the living environment, the neighborhoods where minorities live. These neighborhoods are frequently in inner-city areas, which are racially and economically segregated and characterized by substandard housing, poor schools, high crime, an excessive number of alcohol outlets, gang-related violence, and general social disorganization. Excessive drinking can be used as a coping mechanism for life in these conditions.

Research has an important role to play in society's attempts to ameliorate the problems associated with excessive alcohol use. First, research can identify the problems and provide accurate unbiased descriptions of their magnitude and nature. Alcohol epidemiology studies in the United States have provided such descriptions for the past 30 years. Much has been learned about drinking among ethnic groups during these years. Yet, as seen in the text above, much of the information collected applies only to ethnic groups combined in such a broad way that limits considerably the use of this knowledge. As ethnic minority group representation in US society grows, research must become more specific, describing alcohol-related problems at national-group level for immigrant groups and tribes or groups of tribes for American Indians and Alaska Natives. Research must also go beyond a purely descriptive level to include testing of specific hypotheses about the origin of health disparities. Theories about why such disparities exist have been developed but not tested in a systematic manner. A number of unique epidemiologic studies that can clarify important questions related to health disparities are yet to be conducted. For instance, important questions related to the role of immigration, acculturation, acculturation stress, and birthplace can only be appropriately answered with longitudinal studies of recently arrived immigrants, perhaps with comparison groups formed by members of the same immigrant group born in the United States. These studies are difficult to conduct and costly, which makes them especially challenging in an era of diminishing funding for health research. As they become more specific in description or hypothesis testing, these studies must meet the challenge of remaining relevant to a broader context in US society.

References

Altekruse SF, McGlynn KA, Reichman ME (2009). Hepatocellular carcinoma incidence, mortality, and survival trends in the United States From 1975 to 2005. J Clin Oncol 27: 1485–1491.

American Psychiatric Association (1987). Diagnostic and Statistical Manual of Mental Disorders, third edition. American Psychiatric Association, Washington, DC.

American Psychiatric Association (1994). Diagnostic and Statistical Manual of Mental Disorders, fourth edition. American Psychiatric Association, Washington, DC.

Bales RF (1946). Cultural differences in rates of alcoholism. Q J Stud Alcohol 6: 480–499.

Beals J, Spicer P, Mitchell CM et al. (2003). Racial disparities in alcohol use: comparison of 2 American Indian reservation populations with national data. Am J Public Health 93: 1683–1685.

Beals J, Manson SM, Whitesell NR et al. (2005). Prevalence of DSM-IV disorders and attendant help-seeking in 2 American Indian reservation populations. Arch Gen Psychiatry 62: 99–108.

Bluthenthal RN, Browntaylor D, Guzman-Becerra N et al. (2005). Characteristics of malt liquor beer drinkers in a low-income, racial minority community sample. Alcohol Clin Exp Res 29: 402–409.

Borrell LN, Jacobs Jr DR, Williams DR et al. (2007). Self-reported racial discrimination and substance use in the Coronary Artery Risk Development in Adults Study. Am J Epidemiol 166: 1068–1079.

Borrell LN, Roux AV, Jacobs Jr DR et al. (2010). Perceived racial/ethnic discrimination, smoking and alcohol consumption in the Multi-Ethnic Study of Atherosclerosis (MESA). Prev Med 51: 307–312.

Brondolo E, Rieppi R, Kelly KP et al. (2003). Perceived racism and blood pressure: a review of the literature and conceptual and methodological critique. Ann Behav Med 25: 55–65.

Brondolo E, Gallo LC, Myers HF (2009). Race, racism and health: disparities, mechanisms, and interventions. J Behav Med 32: 1–8.

Brown LM, Hoover RN, Greenberg RS et al. (1994). Are racial differences in squamous cell esophageal cancer explained by alcohol and tobacco use? J Natl Cancer Inst 86: 1340–1345.

Burnam MA (1989). Prevalence of alcohol abuse and dependence among Mexican Americans and Non-Hispanic Whites in the community. In: DL Spiegler, DA Tate, SS Aitken, CM Christian (Eds.), Alcohol Use Among US Ethnic Minorities. Research Monograph No 18, National Institute on Alcohol Abuse and Alcoholism, Washington, D.C., pp. 163–177.

Burnam AR, Hough RL, Karno M et al. (1987a). Acculturation and lifetime prevalence of psychiatric disorders among Mexican Americans in Los Angeles. J Health Soc Behav 28: 89–102.

Burnam MA, Hough RL, Escobar J (1987b). Six-month prevalence of specific psychiatric disorders among Mexican Americans and non-Hispanic whites in Los Angeles. Arch Gen Psychiatry 44: 687–694.

Caetano R (1987). Acculturation and drinking patterns among U.S. Hispanics. Br J Addict 82: 789–799.

Caetano R (1989). Concepts of alcoholism among whites, blacks and Hispanics in the United States. J Stud Alcohol 50: 580–582.

Caetano R (1991). Findings from the 1984 National Survey of Alcohol Use among U.S. Hispanics. In: WB Clark, ME Hilton (Eds.), Alcohol in America: drinking practices and problems, State University of New York Press, Albany, pp. 293–308.

Caetano R (1997). Prevalence, incidence and stability of drinking problems among White, Black and Hispanics: 1984-1992. J Stud Alcohol 58: 565–572.

Caetano R (1998). Alcohol use among Mexican Americans and in the U.S. population. In: MJ Gilbert (Ed.), Alcohol Consumption Among Mexicans and Mexican Americans: A Binational Perspective, Spanish Speaking Mental Health Research Center, University of California Los Angeles, Los Angeles, CA, pp. 53–84.

Caetano R, Kaskutas LA (1995). Changes in drinking patterns among Whites, Blacks and Hispanics: 1984-1992. J Stud Alcohol 56: 558–565.

Caetano R, Kaskutas LA (1996). Changes in drinking problems among Whites, Blacks and Hispanics: 1984-1992. Subst Use Misuse 31: 1547–1571.

Caetano R, McGrath C (2005). Driving under the influence (DUI) among U.S. ethnic groups. Accid Anal Prev 37: 217–224.

Caetano R, Medina-Mora ME (1988). Acculturation and drinking among people of Mexican descent in Mexico and the United States. J Stud Alcohol 49: 462–471.

Caetano R, Cunradi CB, Clark CL et al. (2000). Intimate partner violence and drinking patterns among White, Black, and Hispanic couples in the U.S. J Subst Abuse 11: 123–138.

Caetano R, Field CA, Ramisetty-Mikler S et al. (2005). The five-year course of intimate partner violence among White, Black and Hispanic couples in the United States. J Interpers Violence 20: 1039–1057.

Caetano R, Ramisetty-Mikler S, Rodriguez LA (2008). The Hispanic Americans Baseline Alcohol Survey (HABLAS): rates and predictors of alcohol abuse and dependence across Hispanic national groups. J Stud Alcohol Drugs 69: 441–448.

Caetano R, Ramisetty-Mikler S, Rodriguez LA (2009). The Hispanic Americans Baseline Alcohol Survey (HABLAS): the association between birthplace, acculturation and alcohol abuse and dependence across Hispanic national groups. Drug Alcohol Depend 99: 215–221.

Caetano R, Baruah J, Ramisetty-Mikler S et al. (2010). Sociodemographic predictors of pattern and volume of alcohol consumption across Hispanics, Blacks, and Whites: 10-year trend (1992-2002). Alcohol Clin Exp Res 34: 1782–1792.

Caetano R, Baruah J, Chartier KG (2011). Ten-year trends (1992 to 2002) in sociodemographic predictors and indicators of alcohol abuse and dependence among Whites, Blacks, and Hispanics in the United States. Alcohol Clin Exp Res 35: 1458–1466.

Caetano R, Kaplan MS, Huguet N et al. (2013). Acute alcohol intoxication and suicide among U.S. ethnic minorities: Findings from the National Violent Death Reporting System. Alcohol Clin Exp Res 37: 839–846.

Cahalan D, Cisin IH, Crossley HM (1969). American drinking practices: A national study of drinking behavior and attitudes, vol 6. Rutgers Center of Alcohol Studies, New Brunswick, NJ.

Carson-DeWitt R (2001). National Household Survey On Drug Abuse (NHSDA). In: R Carson-DeWitt (Ed.), Encyclopedia of Drugs, Alcohol, and Addictive Behavior, 2nd edn Macmillan Reference USA, New York.

Centers for Disease Control and Prevention [CDC] (2002). Fetal alcohol syndrome–Alaska, Arizona, Colorado, and New York, 1995–1997. Morb Mortal Wkly Rep 51: 433–435.

Centers for Disease Control and Prevention [CDC] (2008). Alcohol-attributable deaths and years of potential life lost among American Indians and Alaska Natives–United States, 2001–2005. Morb Mortal Wkly Rep Surveill Summ 57: 938–941.

Centers for Disease Control, Prevention [CDC] (2009). Compressed Mortality File (CMF) compiled from 1999-2006, Series 20, No. 2 L. Natl Vital Stat Rep. Available at, http://wonder.cdc.gov/cmf-icd10.html, Accessed January 9, 2012.

Cerdá M, Diez-Roux AV, Tchetgen Tchetgen E et al. (2010). The relationship between neighborhood poverty and alcohol Use: Estimation by marginal structural models. Epidemiology 21: 482–489.

Chae DH, Takeuchi DT, Barbeau EM et al. (2008). Alcohol disorders among Asian Americans: associations with unfair treatment, racial/ethnic discrimination, and ethnic identification (the national Latino and Asian Americans study, 2002-2003). J Epidemiol Community Health 62: 973–979.

Chartier KG, Caetano R (2011). Trends in alcohol services utilization from 1991-1992 to 2001-2002: Ethnic group differences in the U.S. population. Alcohol Clin Exp Res 35: 1485–1497.

Chen CM, H-y Yi, Falk DE et al. (2006). Alcohol use and alcohol use disorders in the United States: Main findings from the 2001-2002 National Epidemiologic Survey on Alcohol and Related Conditions (NESARC). National Institutes of Health, Bethesda, MD. Available at: http://pubs.niaaa.nih.gov/publications/manual.htm.

Chen CM, Yi H-y, Dawson DA et al. (2010). Alcohol Use and Alcohol Use Disorders in the United States, A 3-Year Follow-Up: Main Findings From the 2004–2005 Wave 2 National Epidemiologic Survey On Alcohol And Related Conditions (NESARC). National Institute on Alcohol Abuse and Alcoholism. US Alcohol Epidemiologic Data Reference Manual. Available at: http://pubs.niaaa.nih.gov/publications/NESARC_DRM2/NESARC2DRM.htm, Accessed January 24, 2012.

Clark WB, Midanik L (1982). Alcohol use and alcohol problems among U.S. adults: Results of the 1979 national survey. NIAAA. Alcohol consumption and related problems, Document (ADM) 82-1190, Washington, DC, pp. 3–52.

Cohen DA, Ghosh-Dastidar B, Scribner R et al. (2006). Alcohol outlets, gonorrhea, and the Los Angeles civil unrest: a longitudinal analysis. Soc Sci Med 62: 3062–3071.

Crabb DW, Edenberg HJ, Bosron WF et al. (1989). Genotypes for aldehyde dehydrogenase deficiency and alcohol sensitivity. The inactive ALDH2(2) allele is dominant. J Clin Invest 83: 314–316.

Cunradi CB, Caetano R, Clark CL et al. (2000). Neighborhood poverty as a predictor of intimate partner violence among White, Black, and Hispanic couples in the United States: A multilevel analysis. Ann Epidemiol 10: 297–308.

Dawson DA, Grant BF, Chou SP et al. (1995). Subgroup variation in U.S. drinking patterns: Results of the 1992 national longitudinal alcohol epidemiologic study. J Subst Abuse 7: 331–344.

Dawson DA, Grant BF, Stinson FS et al. (2005). Recovery from DSM-IV alcohol dependence: United States, 2001-2002. Addiction 100: 281–292.

Day GL, Blot WJ, Austin DF et al. (1993). Racial differences in risk of oral and pharyngeal cancer: alcohol, tobacco, and other determinants. J Natl Cancer Inst 85: 465–473.

El-Serag HB (2011). Hepatocellular carcinoma. N Engl J Med 365: 1118–1127.

Ennett ST, Flewelling RL, Lindrooth RC et al. (1997). School and neighborhood characteristics associated with school rates of alcohol, cigarette, and marijuana use. J Health Soc Behav 38: 55–71.

Escobedo LG, Ortiz M (2002). The relationship between liquor outlet density and injury and violence in New Mexico. Accid Anal Prev 34: 689–694.

Flores YN, Yee Jr HF, Leng M et al. (2008). Risk factors for chronic liver disease in Blacks, Mexican Americans, and Whites in the United States: results from NHANES IV, 1999-2004. Am J Gastroenterol 103: 2231–2238.

Fuchs FD, Chambless LE, Whelton PK et al. (2001). Alcohol consumption and the incidence of hypertension: the Atherosclerosis Risk in Communities Study. Hypertension 37: 1242–1250.

Fuchs FD, Chambless LE, Folsom AR et al. (2004). Association between alcoholic beverage consumption and incidence of coronary heart disease in Whites and Blacks. Am J Epidemiol 160: 466–474.

Galea S, Ahern J, Tracy M et al. (2007). Neighborhood income and income distribution and the use of cigarettes, alcohol, and marijuana. Am J Prev Med 32: S195–S202.

Gallo LC, Matthews KA (2003). Understanding the association between socioeconomic status and physical health: do negative emotions play a role? Psychol Bull 129: 10–51.

Grant BF (2000). Theoretical and observed subtypes of DSM-IV alcohol abuse and dependence in a general population sample. Drug Alcohol Depend 60: 287–293.

Grant BF, Dawson DA (1997). Age at onset of alcohol use and its association with DSM-IV alcohol abuse and dependence: results from the National Longitudinal Alcohol Epidemiologic Survey. J Subst Abuse 9: 103–110.

Grant BF, Stinson FS, Harford TC (2001). Age at onset of alcohol use and DSM-IV alcohol abuse and dependence: a 12-year follow-up. J Subst Abuse 13: 493–504.

Grant BF, Dawson DA, Stinson FS et al. (2004). The 12-month prevalence and trends in DSM-IV alcohol abuse and dependence: United States, 1991-1992 and 2001-2002. Drug Alcohol Depend 74: 223–234.

Grant BF, Harford TC, Muthén B et al. (2007). DSM-IV alcohol dependence and abuse: further evidence of validity in the general population. Drug Alcohol Depend 86: 154–166.

Grant BF, Goldstein RB, Chou SP et al. (2009). Sociodemographic and psychopathologic predictors of first incidence of DSM-IV substance use, mood and anxiety disorders: results from the Wave 2 National Epidemiologic Survey on Alcohol and Related Conditions. Mol Psychiatry 14: 1051–1066.

Hasin DS, Stinson FS, Ogburn E et al. (2007). Prevalence, correlates, disability, and comorbidity of DSM-IV alcohol abuse and dependence in the United States. Arch Gen Psychiatry 64: 830–842.

Hasin DS, Keyes KM, Alderson D et al. (2008). Cannabis withdrawal in the United States: results from NESARC. J Clin Psychiatry 69: 1354–1363.

Hatch SL (2005). Conceptualizing and identifying cumulative adversity and protective resources: implications for understanding health inequalities. J Gerontol B Psychol Sci Soc Sci 60: S130–S134.

Helzer JE, Canino GJ (Eds.), (1992). Alcoholism in North America, Europe, and Asia. Oxford University Press, Oxford, UK.

Herd D (1990). Subgroup differences in drinking patterns among Black and White men: results from a national survey. J Stud Alcohol 51: 221–232.

Hilton ME (1987). Demographic characteristics and the frequency of heavy drinking as predictors of self-reported drinking problems. Br J Addict 82: 913–925.

Hilton ME (1991). Demographic characteristics and the frequency of heavy drinking as predictors of self-reported drinking problems. In: WB Clark, M Hilton (Eds.), Alcohol in America: drinking practices and problems. State University of New York Press, Albany, NY, pp. 194–212.

Hilton J (2006). Race and ethnicity in fatal motor vehicle traffic crashes 1999-2004. US National Highway Traffic Safety Administration, Washington, DC.

Hingson RW, Heeren T, Winter MR (2006). Age at drinking onset and alcohol dependence – age at onset, duration, and severity. Arch Pediatr Adolesc Med 160: 739–746.

Institute of Medicine (2003). Unequal Treatment: Confronting Racial and Ethnic Disparities in Healthcare, Washington, DC. Available at: http://www.iom.edu/Reports/2002/Unequal-Treatment-Confronting-Racial-and-Ethnic-Disparities-in-Health-Care.aspx.

Institute of Medicine [IOM] (2001). Crossing the Quality Chasm: A New Health System for the 21st Century, Washington, DC. Available at: http://www.iom.edu/Reports/2001/Crossing-the-Quality-Chasm-A-New-Health-System-for-the-21st-Century.aspx.

Institute of Medicine [IOM] (2012). How far have we come in reducing health disparities?: Progress Since 2000. Washington, DC. Available at: http://www.healthequityks.org/download/IOM_Hlth_Disparities_Report_2012.pdf.

Jones-Webb RJ, Snowden L, Herd D et al. (1997). Alcohol-related problems among Black, Hispanic, and White men: The contribution of neighborhood poverty. J Stud Alcohol 58: 539–545.

Karriker-Jaffe KJ, Zemore SE, Mulia N et al. (2012). Neighborhood disadvantage and adult alcohol outcomes: differential risk by race and gender. J Stud Alcohol Drugs 73: 865–873.

Kerr WC, Greenfield TK, Bond J et al. (2011). Racial and ethnic differences in all-cause mortality risk according to alcohol consumption patterns in the national alcohol surveys. Am J Epidemiol 174: 769–778.

Kessler R, McGonagle K, Zhao S et al. (1994). Lifetime and 12-month prevalence of DSM-III-R psychiatric disorders in the United States: results from the National Comorbidity Survey. Arch Gen Psychiatry 51: 8–9.

Keyes KM, Geier T, Grant BF et al. (2009). Influence of a drinking quantity and frequency measure on the prevalence and demographic correlates of DSM-IV alcohol dependence. Alcohol Clin Exp Res 33: 761–771.

Kimura M, Higuchi S (2011). Genetics of alcohol dependence. Psychiatry Clin Neurosci 65: 213–225.

Knupfer G, Room R (1967). Drinking patterns and attitudes of Irish, Jewish and white Protestant American men. Q J Stud Alcohol 28: 676–699.

Koss MP, Yuan NP, Dightman D et al. (2003). Adverse childhood exposures and alcohol dependence among seven Native American tribes. Am J Prev Med 25: 238–244.

Krieger N, Sidney S (1996). Racial discrimination and blood pressure: the CARDIA study of young Black and White adults. Am J Public Health 86: 1370–1378.

Lo CC, Cheng TC (2012). Discrimination's role in minority groups' rates of substance-use disorder. Am J Addict 21: 150–156.

Lolli G, Serianni E, Golder GM et al. (1958). Alcohol in Italian culture: food and wine in relation to sobriety among Italians and Italian Americans. Rutgers Center of Alcohol Studies, New Brunswick, NJ.

Lynch JW, Smith GD, Kaplan GA et al. (2000). Income inequality and mortality: importance to health of individual income, psychosocial environment, or material conditions. Br Med J 320: 1200–1204.

Martin JK, Tuch SA, Roman PM (2003). Problem drinking patterns among African Americans: the impacts of reports of discrimination, perceptions of prejudice, and "risky" coping strategies. J Health Soc Behav 44: 408–425.

May PA, Gossage JP (2001a). Estimating the prevalence of fetal alcohol syndrome. A summary. Alcohol Res Health 25: 159–167.

May PA, Gossage P (2001b). New data on the epidemiology of adult drinking and substance use among American Indians of the northern states: male and female data on prevalence, patterns, and consequences. Am Indian Alsk Native Ment Health Res 10: 1–26.

Mays VM, Cochran SD, Barnes NW (2007). Race, race-based discrimination, and health outcomes among African Americans. Annu Rev Psychol 58: 201–225.

McEwen BS (2004). Protection and damage from acute and chronic stress: allostasis and allostatic overload and relevance to the pathophysiology of psychiatric disorders. Ann N Y Acad Sci 1032: 1–7.

McKinney CM, Caetano R, Harris TR et al. (2009). Alcohol availability and intimate partner violence among US couples. Alcohol Clin Exp Res 33: 169–176.

McLaughlin KA, Hatzenbuehler ML, Keyes KM (2010). Responses to discrimination and psychiatric disorders among black, Hispanic, female, and lesbian, gay, and bisexual individuals. Am J Public Health 100: 1477–1484.

Meyer IH, Schwartz S, Frost DM (2008). Social patterning of stress and coping: does disadvantaged social statuses confer more stress and fewer coping resources? Soc Sci Med 67: 368–379.

Mills BA, Caetano R (2010). The Hispanic Americans Baseline Alcohol Survey (HABLAS): predictors of alcohol attitudes and expectancies in Hispanic national groups. Alcohol Clin Exp Res 34: 790–799.

Miniño AM, Xu J, Kochanek KD (2011). Deaths: final data for 2008. Natl Vital Stat Rep 59: 1–52.

Mulia N, Ye Y, Zemore SE et al. (2008). Social disadvantage, stress, and alcohol use among Black, Hispanic, and White Americans: findings from the 2005 U.S. National Alcohol Survey. J Stud Alcohol Drugs 69: 824–833.

Naimi TS, Brewer RD, Mokdad A et al. (2003). Binge drinking among US adults. JAMA 289: 70–75.

National Cancer Institute (2011a). SEER Stat Fact Sheets: Larynx. Surveillance epidemiology and end results. Available at, http://seer.cancer.gov/statfacts/html/laryn.html, Accessed January 3, 2012.

National Cancer Institute (2011b). SEER Stat Fact Sheets: Oral cavity and pharynx. Surveillance epidemiology and end results. Available at, http://seer.cancer.gov/statfacts/html/oralcav.html, Accessed January 3, 2012.

National Institute on Alcohol Abuse and Alcoholism [NIAAA] (2004). NIAAA council approves definition of binge drinking. National Institute of Alcohol Abuse and Alcoholism, NIAAA Newsletter Bethesda, MD.

O'Connell JM, Novins DK, Beals J et al. (2005). Disparities in patterns of alcohol use among reservation-based and geographically dispersed American Indian populations. Alcohol Clin Exp Res 29: 107–116.

Pelucchi C, Gallus S, Garavello W et al. (2006). Cancer risk associated with alcohol and tobacco use: Focus on upper aero-digestive tract and liver. Alcohol Res Health 29: 193–198.

Perry SW (2004). American Indians and Crime: A BJS Statistical Profile, 1992-2002. US Department of Justice, Bureau of Justice Statistics. Available at: http://www.bjs.gov/content/pub/pdf/aic.pdf.

Pollack CE, Cubbin C, Ahn D et al. (2005). Neighbourhood deprivation and alcohol consumption: does the availability of alcohol play a role? Int J Epidemiol 34: 772–780.

Price RK, Risk NK, Wong MM et al. (2002). Substance use and abuse by Asian Americans and Pacific Islanders: preliminary results from four national epidemiologic studies. Public Health Rep 117 (Suppl. 1): S39–S50.

Ramisetty-Mikler S, Caetano R, Rodriguez LA (2010). The Hispanic Americans Baseline Alcohol Survey (HABLAS): alcohol consumption and sociodemographic predictors across Hispanic national groups. J Subst Use 15: 402–416.

Rehm J, Room R, Graham K et al. (2003). The relationship of average volume of alcohol consumption and patterns of drinking to burden of disease: an overview. Addiction 98: 1209–1228.

Russo D, Purohit V, Foudin L et al. (2004). Workshop on alcohol use and health disparities 2002: a call to arms. Alcohol 32: 37–43.

Sakai JT, Wang C, Price RK (2010). Substance use and dependence among Native Hawaiians, other Pacific Islanders, and Asian ethnic groups in the United States: contrasting multiple-race and single-race prevalence rates from a national survey. J Ethn Subst Abuse 9: 173–185.

Schmidt L, Ye Y, Greenfield TK et al. (2007). Ethnic disparities in clinical severity and services for alcohol problems: results from the National Alcohol Survey. Alcohol Clin Exp Res 31: 48–56.

Scribner RA, Cohen D, Kaplan S et al. (1999). Alcohol availability and homicide in New Orleans: conceptual considerations for small area analysis of the effect of alcohol outlet density. J Stud Alcohol 60: 310–316.

Sempos CT, Rehm J, Wu T et al. (2003). Average volume of alcohol consumption and all-cause mortality in African Americans: the NHEFS cohort. Alcohol Clin Exp Res 27: 88–92.

Smedley BD, Stith AY, Nelson AR (2003). Unequal treatment: Confronting racial and ethnic disparities in health care, National Academies Press, Washington, DC.

Snyder CR (1958). Alcohol and the Jews: A cultural study of drinking and sobriety. Free Press, New York, NY, US.

Stinson FS, Grant BF, Dufour MC (2001). The critical dimension of ethnicity in liver cirrhosis mortality statistics. Alcohol Clin Exp Res 25: 1181–1187.

Substance Abuse and Mental Health Services Administration [SAMHSA] (2005). Arrests for Driving Under the Influence Among Adult Drivers. Office of Applied Studies, Rockville, MD. Available at: http://oas.samhsa.gov/2k5/DUIarrests/DUIarrests.pdf.

Substance Abuse and Mental Health Services Administration [SAMHSA] (2007). Results from the 2006 National Survey on Drug Use and Health: Detailed Tables, USDoHaH Services, Rockville, MD.

Substance Abuse and Mental Health Services Administration [SAMHSA] (2010). The NSDUH Report: Substance Use among Asian Adults, Rockville, MD. Available at, http://www.samhsa.gov/data/2k10/179/SUAsianAdults.htm, Accessed January 15, 2013.

Substance Abuse and Mental Health Services Administration [SAMHSA] (2011). Results from the 2010 National Survey on Drug Use and Health: Mental Health Findings. NSDUH Series H-42, HHS Publication No. (SMA) 11-4667, U. S. Department of Health and Human Services, Rockville, MD.

Substance Abuse and Mental Health Services Administration [SAMHSA] (2012a). Results from the 2011 National Survey on Drug Use and Health: Detailed Tables, Table 2.46B. U. S. Department of Health and Human Services, Rockville, MD.

Substance Abuse and Mental Health Services Administration [SAMHSA] (2012b). Results from the 2011 National Survey on Drug Use and Health: Summary of National Findings. NSDUH Series H-44, HHS Publication No. (SMA) 12-4713. U. S. Department of Health and Human Services, Rockville, MD.

Takeuchi DT, Zane N, Hong S et al. (2007). Immigration-related factors and mental disorders among Asian Americans. Am J Public Health 97: 84–90.

The National Institute on Minority Health and Health Disparities [NIMHD] (2013). Important Events in NCMHD History. Available at, http://www.nimhd.nih.gov/about_ncmhd/history.asp, Accessed Februrary 18, 2013.

Tran AG, Lee RM, Burgess DJ (2010). Perceived discrimination and substance use in Hispanic/Latino, African-born Black, and Southeast Asian immigrants. Cultur Divers Ethnic Minor Psychol 16: 226–236.

Treno AJ, Gruenewald PJ, Johnson FW (2001). Alcohol availability and injury: the role of local outlet densities. Alcohol Clin Exp Res 25: 1467–1471.

Treno AJ, Johnson FW, Remer LG et al. (2007). The impact of outlet densities on alcohol-related crashes: a spatial panel approach. Accid Anal Prev 39: 894–901.

Turner RJ, Lloyd DA (1995). Lifetime traumas and mental health: the significance of cumulative adversity. J Health Soc Behav 36: 360–376.

Vaeth PAC, Caetano R, Ramisetty-Mikler S et al. (2009). Hispanic Americans Baseline Alcohol Survey (HABLAS): alcohol-related problems across Hispanic national groups. J Stud Alcohol Drugs 70: 991–999.

Vaeth PAC, Caetano R, Rodriguez LA (2012). The Hispanic Americans Baseline Alcohol Survey (HABLAS): the association between acculturation, birthplace and alcohol consumption across Hispanic national groups. Addict Behav 37: 1029–1037.

Vilamovska A-M, Brown Taylor D, Bluthenthal RN (2009). Adverse drinking-related consequences among lower income, racial, and ethnic minority drinkers: cross-sectional results. Alcohol Clin Exp Res 33: 645–653.

Williams DR, Mohammed SA, Leavell J et al. (2010). Race, socioeconomic status, and health: complexities, ongoing challenges, and research opportunities. Ann N Y Acad Sci 1186: 69–101.

Yoon Y, Yi H (2008). Surveillance report #83: Liver cirrhosis mortality in the United States, 1970-2005. NIAAA, Division of Epidemiology and Prevention Research, Bethesda, MD. Available at: http://pubs.niaaa.nih.gov/publications/surveillance83/Cirr05.htm, Accessed March 2, 2009.

Zemore SE (2007). Acculturation and alcohol among Latino adults in the United States: a comprehensive review. Alcohol Clin Exp Res 31: 1968–1990.

Zemore SE, Karriker-Jaffe KJ, Keithly S et al. (2011). Racial prejudice and unfair treatment: interactive effects with poverty and foreign nativity on problem drinking. J Stud Alcohol Drugs 72: 361–370.

Zhu L, Gorman DM, Horel S (2004). Alcohol outlet density and violence: a geospatial analysis. Alcohol Alcohol 39: 369–375.

Handbook of Clinical Neurology, Vol. 125 (3rd series)
Alcohol and the Nervous System
E.V. Sullivan and A. Pfefferbaum, Editors

Chapter 38

Alcohol and the law

ARIELA O. KARASOV[1] AND MICHAEL J. OSTACHER[1,2]*

[1]*Department of Psychiatry and Behavioral Sciences, Stanford University School of Medicine, Palo Alto, CA, USA*

[2]*Department of Psychiatry, Veterans Affairs Palo Alto Health Care System, Palo Alto, CA, USA*

Society has had an interest in controlling the production, distribution, and use of alcohol for millennia. The use of alcohol has always had consequences, be they positive or negative, and the role of government in the regulation of alcohol is now universal. This is accomplished at several levels, first through controls on production, importation, distribution, and use of alcoholic beverages, and second, through criminal laws, the aim of which is to address the behavior of users themselves. A number of interventions and policies reduce alcohol-related consequences to society by regulating alcohol pricing, targeting alcohol-impaired driving, and limiting alcohol availability. The legal system defines criminal responsibility in the context of alcohol use. In recent years, recovery-oriented policies have aimed to expand social supports for recovery and improve access to treatment for substance use disorders within the criminal justice system. Although this chapter focuses on alcohol and the law within the United States (rather than in other countries), it should be noted that a rich international literature has shaped policy in the United States and continues to do so.

Most people drink without causing significant problems for themselves or their communities, and alcohol plays an important social role in many societies. Alcohol intoxication, however, plays a key role in a high proportion of violent crimes, motor vehicle accidents, and trauma in general. Due to the cost to not only the individual consumer but also to society as a whole, the legal system and government policies are tailored to minimize the negative effects of alcohol intoxication. The legal system defines how crimes committed while intoxicated are prosecuted. Federal and state policies attempt to regulate the sale of and access to alcohol, enact countermeasures to alcohol-impaired driving, and develop interventions with at-risk drinkers and treatment of drinkers with alcohol dependence.

Perhaps the most significant public health issue related to alcohol use is alcohol-impaired driving. Alcohol-impaired driving fatalities account for approximately a third of total motor vehicle traffic fatalities in the United States(National Highway Traffic Safety Administration, 2008) with a total of approximately 10 000 deaths per year (National Highway Traffic Safety Admininstration, 2012). A wealth of research illustrates the relationship between blood alcohol content (BAC) and impaired driving, a relationship which is based on alcohol-induced deficits likely modulated through orbitofrontal/anterior cingulate, motor, and cerebellar regions (Calhoun et al., 2004). BAC is the most commonly used measure of alcohol intoxication. The level at which a person is considered legally impaired and unable to operate a vehicle varies by country, but the current 0.08% cutoff as defined in the United States is among the more lenient legal limits. Epidemiologic research suggests that drivers are substantially impaired at 0.08% in basic driving tasks and judgment and that these impairments begin at 0.02 g/dL. As drivers' BACs approach 0.08%, their risk of being involved in a crash increases significantly. It has been estimated that each 0.02 percentage increase in the BAC of a driver almost doubles the risk of being in a fatal crash (Zador, 1991).

INTERVENTIONS TO REDUCE ALCOHOL-RELATED CONSEQUENCES

The complex legal treatment of alcohol use has been evolving since the end of the temperance era. Federal, state, and local governments in the United States establish policies governing alcohol consumption and defining responses to alcohol-related problems. During the years of prohibition, between 1919 and 1933, alcohol was banned under the 18th amendment. Congress

*Correspondence to: Michael J. Ostacher, M.D., M.P.H., M.M.Sc., Department of Psychiatry, VA Palo Alto Health Care System, 3801 Miranda Avenue, Palo Alto, CA 94304-1290, USA. Tel: +1-650-493-5000, E-mail: ostacher@stanford.edu

ratified the 21st amendment in 1933, repealing national prohibition and endowing states with broad power to regulate alcoholic beverages, including regulation of various aspects of alcohol sales and possession (Heath, 2009; Okrent, 2010). Individual States also control alcohol use and its consequences through criminal law and tort liability. However, the federal government retains the ability to "regulate commerce" and to influence state-based alcohol policies through financial inducement.

A number of strategies have been implemented over the years to minimize the negative effects of alcohol on the individual and society. Cost-effective strategies include taxation that increases prices, alcohol-impaired driving countermeasures, restrictions on the availability of alcohol, brief interventions with at-risk drinkers, and treatment of drinkers with alcohol dependence (Table 38.1).

ALCOHOL TAXES AND PRICE CONTROLS

One approach to reduce the demand for alcohol is to increase the price relative to similar commodities.

Table 38.1

Effective policy interventions to reduce alcohol-related consequences

Alcohol taxes and price controls	Excise taxes Minimum unit pricing
Policies targeting alcohol-impaired driving	Blood alcohol content (BAC) limits Lower BAC laws for young or inexperienced drivers Administrative license suspension Sobriety checkpoints, breathalyzer tests Ignition interlock devices Social host liability/dram shop liability
Restrictions on alcohol availability	Limited hours of sale Outlet density controls, frequently through outlet licensing and government monopolies Minimum legal drinking age (minimum purchase age) laws
Policies to improve access to substance use treatment	The Affordable Care Act – expanding insurance coverage for substance use treatment Support for community-based organizations

Excise and sales taxes are the most common methods of affecting retail prices of alcohol, but the US government has historically been exceedingly slow to enact such taxes. Title XI of the Omnibus Budget Reconciliation Act of 1990 increased federal excise tax rates on beer and wine for the first time since November 1951 and the tax on distilled spirits for the second time during this period. It is quite clear that increases in taxes on alcohol have not kept up with inflation over the years, likely due to both perceived unpopularity of such a change among voters and pressures from industry through lobbying and acts such as the Excise and Economic Relief Act of 2013 (Latham, 2013). In addition to taxes, minimum unit pricing has been proposed as a means of decreasing hazardous drinking. Setting a minimum price per unit of alcohol appears to be particularly effective for decreasing alcohol consumption among heavy drinkers (Meier et al., 2010; Sharma et al., 2014).

Increases in pricing and taxes on alcohol are clearly effective as numerous studies have demonstrated that beverage alcohol prices and taxes are related inversely to drinking (Stout et al., 2000; Wagenaar et al., 2009). The positive effects of such taxes are far-reaching, including measurable impacts on violent crime (Markowitz, 2000a; Cook and Durrance, 2013) and lower incidence of sexually transmitted infections among certain age groups (Harrison et al., 2000).

With regard to the effects of alcohol pricing on criminal behavior, studies show that increasing the price of alcohol is an effective method of decreasing physical violence, including child abuse (Markowitz and Grossman, 2000) and spousal abuse, particularly the abuse of female partners (Markowitz, 2000b). Reporting biases limit the ability of researchers to parse out the relationship between alcohol pricing and rates of intimate-partner violence (Durrance et al., 2011), but domestic violence is likely a silent giant in the research on alcohol and crime. Overall, violence is responsive to changes in alcohol pricing. There appears to be a linear relationship such that a 1% increase in the beer tax decreases the general incidence of assaults by 0.45% (Markowitz, 2000a).

Alcohol pricing also appears to affect rates of sexually transmitted infection. Given alcohol's effects on impulsivity and decision making, this finding is not surprising. In a study in 2000, it was found that a $1 increase in the state liquor tax per gallon correlated with a reduction in gonorrhea rates of 2.1%, with slightly higher estimated reductions in syphilis rates (Harrison et al., 2000). A 20-cent increase in the tax on a six-pack of beer was associated with an 8.9% reduction in gonorrhea rates and a 32.7% reduction in syphilis rates (Harrison et al., 2000).

POLICIES TARGETING ALCOHOL-IMPAIRED DRIVING

A number of strategies have proven effective in reducing alcohol-impaired driving and its costs to society, including 0.08% BAC limits for adults, sobriety checkpoints (Shults et al., 2001), administrative license suspension (Wagenaar and Maldonado-Molina, 2007), and lower BAC laws for young or inexperienced drivers (Shults et al., 2001). The Alcohol Traffic Safety Act of 1983 eased the standard for arresting and convicting alcohol-impaired drivers, imposed more severe and certain penalties for alcohol-impaired driving, and increased resources to apprehend alcohol-impaired drivers. However, the United States has been more lenient than much of the world and slow to change BAC cutoffs, possibly due to the comparatively heavy dependence on driving in the United States.

All 50 states have legal statutes addressing impaired driving. Ever since the 1950s, there has been a gradual trend for states to raise the legal drinking age limit and increase sanctions for driving while intoxicated (DWI). BACs were first used as evidence in impaired-driving prosecutions in the 1950s, with an initial legal limit of 0.15%. Eventually, states enacted laws such that operating a vehicle over the limit was an offense in itself, and the limit was lowered from 0.15 to 0.10%. In 1992, the National Highway Traffic Safety Administration (NHTSA) proposed that all states should adopt 0.08 BAC laws, but five years later only 15 states had 0.08 laws. As late as 2000, 31 states had 0.10 BAC laws. The 1998 transportation reauthorization legislation (The Transportation Equity Act for the 21st Century) provided incentives of over $500 million dollars in grants for states to adopt a 0.08 BAC cutoff.

More stringent BAC laws have a significant impact on alcohol-related motor vehicle accidents. In a meta-analysis of 19 jurisdictions across the United States, there was a decrease of 14.8% in the rate of drinking drivers in fatal crashes after the 0.08 laws were introduced. Furthermore, the reduction was greater in states that had an administrative license suspension/revocation law and implemented frequent sobriety checkpoints (Tippetts et al., 2005). An amendment of the National Minimum Drinking Age Act in 1995 mandated withholding federal highway funds from states that failed to adopt a "zero-tolerance" legislation setting BAC limits of 0–0.02 for underage drinkers. This resulted in an approximately 20% decline in single-vehicle and night-time fatal crashes (Hingson et al., 1994).

Ideally, policies geared towards reducing impaired driving would focus on a BAC level likely to affect the majority of drivers. It may be reasonable to consider further lowering the BAC limit in the United States. The relative risk of being involved in a fatal crash as a driver is 4-10 times greater for drivers with BACs between 0.05 and 0.07, compared to drivers with 0.00 BACs (Zador et al., 2000). In addition, based on studies in the Netherlands, France, and Australia, lowering the legal BAC limit to 0.05 contributes to significant decreases in alcohol-related motor vehicle accidents.(Fell and Voas, 2006). Although one might assume that the 0.05 BAC limits target moderate drinkers, research indicates that the lower BAC limit also has a substantial effect on drivers, with BACs higher than 0.15 who are at the greatest risk of being involved in a fatal crash (Fell and Voas, 2006).

In addition to enforcing a BAC cutoff of 0.08 for DWI offenses, states are required to enforce sanctions for repeat DWI offenders. Some jurisdictions have additional ordinances to deter alcohol-impaired driving. These sanctions include a minimum of 5 days of imprisonment or 30 days of community service, license suspension for a year or longer, installation of an ignition interlock on or immobilization of the offender's car, and evaluation and treatment as deemed appropriate (National Highway Traffic Safety Administration, 2013). A number of states also have enhanced sanctions for high-BAC offenders. What constitutes "high-BAC" is defined by the individual state, but levels range from BACs of 0.10 or greater to 0.20 or greater. Enhanced sanctions appear to have a positive effect on recidivism rates among first-time offenders (McCartt and Northrup, 2004).

The effectiveness of license suspension policies depends on the timing of the license suspension relative to the DWI offense. License suspension laws in a number of states include statutes for preconviction license revocation. Under these "administrative license suspension" laws, law enforcement may confiscate licenses automatically at the time of a 0.08% or greater BAC test result or if the offender refuses testing. Studies have demonstrated limited efficacy for postconviction license suspension (Wagenaar and Maldonado-Molina, 2007; Wagenaar et al., 2007), whereas administrative license suspension laws appear to be effective deterrents (Wagenaar and Maldonado-Molina, 2007).

A 2004 pilot project in South Dakota also highlights the importance of immediate consequences in reducing problem drinking. The "24/7 Sobriety Project" implemented twice-daily breathalyzer tests for individuals with a repeated DWI arrest. Individuals who tested positive for alcohol or failed to comply with testing were immediately placed in jail for a brief period of time. The pilot appears to have been effective in that there was a 12% decrease in repeat DWI arrests and a 9% decrease in domestic violence arrests (Kilmer et al., 2013).

A growing body of evidence supports the use of ignition interlock devices in decreasing recidivism rates. In order to start a vehicle, the driver must pass an alcohol breath test through the device installed in the car. A 2004 Cochrane review found that interlock devices decreased recidivism rates by about 36% in prior DWI offenders relative to controls (Willis et al., 2004). Depending on the method of implementation of the interlock program, studies have demonstrated reductions in repeat DWI offenses of up to 65% (Beck et al. 1999; Roth et al., 2007). Unfortunately, recidivism rates appear to return to baseline once interlock devices are removed (Elder et al., 2011), so the device must be maintained continuously in order to have a lasting effect.

Social host tort liability holds an individual legally responsible for providing alcohol to minors or to adults who are obviously intoxicated. By providing alcohol to someone who is obviously intoxicated, the individual or business that serves as the "social host" may face legal consequences or civil penalties for indirectly endangering the lives of others (National Highway Traffic Safety Administration, 2013). Social host laws vary by state. Based on data from the Behavioral Risk Factor Surveys, a national survey of alcohol use and behaviors, the presence of tort liability decreased the probability of alcohol-impaired driving among all drinkers (Stout et al., 2000).

RESTRICTIONS ON ALCOHOL AVAILABILITY

Restricting the availability of alcohol is an effective means of reducing consumption as well as alcohol-related problems (Anderson et al., 2009a). Availability of alcohol may be regulated by limiting hours of sale or decreasing the density of outlets selling alcohol. All states, except for Nevada, either license private businesses to sell alcohol or have a monopoly system in which alcohol must be sold in state-owned stores, thereby exerting some control over the number and density of alcohol establishments. More hours of access to purchasing translates to increased consumption. After laws restricting the sale of alcohol on Sundays were repealed in five states in the United States, three states had significant increases in alcohol consumption (Yörük, 2013). Following the lift of the ban in New Mexico, there was a 29% increase in alcohol-related crashes and a 42% increase in crash fatalities on Sundays (McMillan and Lapham, 2006). Furthermore, in a cross-sectional study of communities in New York, longer business hours for alcohol-serving establishments were associated with higher rates of misdemeanor-level DWI charges (Schofield and Denson, 2013b).

MINIMUM LEGAL DRINKING AGE LAWS

With the repeal of prohibition in 1933, most states adopted policies with a minimum legal drinking age (MLDA) of 21. Then, around the time of the passing of the 26th amendment, which lowered the required voting age to 18, many states lowered the MLDA. In 1984, the federal uniform Drinking Age Act pressured states into raising the MLDA to 21 years – the government would withhold part of their federal highway funding if they failed to comply – and, by 1988, all states were in compliance.

In recent years, discussion has reemerged around the value of an MLDA of 21. Concern has been raised – particularly by some college administrators (Carpenter and Dobkin, 2011) – that legal sanctions not only fail to prevent drinking by those aged 18–20, but also may actually exacerbate problem drinking by increasing the allure of alcohol as "forbidden fruit" and pushing drinking behind closed doors (Fell, 2008).

However, the data indicate that a higher MLDA has had a positive and lasting impact (Carpenter and Dobkin, 2011), particularly on alcohol consumption and motor vehicle accidents (Wagenaar and Toomey, 2002). NHTSA has estimated that imposing an MLDA of 21 has prevented approximately 700–1000 deaths each year (National Highway Traffic Safety Admininstration, 2008, 2012) and reduced drinking among people aged 21 through 25 who grew up in states with an MLDA of 21 (NIAAA, 2000). Growing up in an environment in which alcohol consumption is illegal before the age of 21 appears to decrease problem drinking later in life. When adult alcohol consumption was compared between groups in which alcohol was legal during the ages of 18 through 20 versus illegal during those years, legal consumption before age 21 was associated with a 20–33% increase in alcohol consumption and a 10% increase in fatal accidents for men (Kaestner and Yarnoff, 2011). For women, exposure to a lower legal drinking age was associated with higher rates of suicide and homicide (Grucza et al., 2012). Youth suicide rates are also responsive to age-related alcohol policies (Markowitz et al., 2003).

INFORMATION AND EDUCATION ON ALCOHOL USE

Theoretically, health information that increases knowledge and changes attitudes will reduce alcohol-related problems. In reality, many efforts aimed at improving public knowledge and awareness, from classroom-based education to public information campaigns, have a limited impact (Anderson et al., 2009a). In 1989, it was mandated that alcoholic beverages in the United States have a label warning of the dangers of alcohol-impaired driving, alcohol consumption during pregnancy, and risks of health problems. Unfortunately, warning labels do not appear to have had an impact on alcohol-related problems (Wilkinson and Room, 2009). Studies do suggest, though, that bans on alcohol advertising are likely effective policy approaches in minimizing societal harm from alcohol use (Anderson et al., 2009a, b; Engels et al., 2009).

RECOVERY-ORIENTED POLICY

Studies demonstrate repeatedly that the legal deterrents to alcohol-impaired driving, restrictions on availability of alcohol, and increases in price that are effective overall have less of an impact on drinking-related problems among heavy drinkers struggling with alcohol use disorders (Stout et al., 2000; Schofield and Denson, 2013b). Given our understanding of the nature of addiction and chronic alcohol dependence, the limited effect of legal deterrents on alcohol-related behaviors among heavy users is not surprising. There has been a growing trend toward the development of recovery-oriented policies that may be particularly beneficial for this group.

Although the birth of recovery-oriented policy and the growing trend in governmental support are recent, a number of studies suggest that recovery-oriented policy interventions are a cost-effective means of impacting alcohol use and related problems (Humphreys and Lembke, 2014). Based out of the Substance Abuse Mental Health Services Administration, the Recovering Community Support Program of 1998 focused on social supports for recovery through grants to community organizations (Kaplan et al., 2010). The Access to Recovery Program of 2005 and subsequent Recovery-Oriented Systems of Care expanded support for access to community-based recovery organizations through vouchers for substance use treatment and development of networks of support.

Over the past 5 years, government policies have taken further steps to address alcohol and drug-related problems through a combination of public health programs and law enforcement strategies that are based upon an understanding of addiction as a disease. The Affordable Care Act (ACA), also known as "ObamaCare," made substantial changes to access to substance abuse treatment by mandating that health insurance include services for substance use disorders (Humphreys and McLellan, 2010). Previously, the 2008 Mental Health Parity and Addiction Equity Act only required substance use disorder coverage for existing employment-related and self-funded plans (Busch, 2012). The ACA expanded substance use treatment equity to Medicaid and state health insurance exchange plans (Barry and Huskamp, 2011) in addition to requiring that substance use disorder treatment coverage be comparable to coverage for medical treatments. Based on the positive effects of state parity legislation on substance use treatment rates (Wen et al., 2013), the ACA will likely have a significant effect on access nationwide. Finally, the 2013 National Drug Control Strategy outlines a goal of developing evidence-based policies for dealing with alcohol and substance use issues (Executive Office of the President of the United States and Kerlikowske, 2013). There is an emphasis on building support for community organizations, fostering collaboration between criminal justice agencies and substance use prevention organizations, and focusing on treatment for individuals involved in the criminal justice system.

ALCOHOL AND VIOLENCE

Alcohol intoxication is among the most common problems associated with violence, abuse, neglect, or other criminal activities. According to the Bureau of Justice review of alcohol and crime in 1996, 36% of convicted offenders were estimated to have been drinking at the time of the offense (Greenfeld, 1998). This appears to have held steady, since in the 2004 Survey of Inmates in State and Federal Correctional Facilities, an estimated 37% of state prisoners serving time for a violent offense said they were under the influence of alcohol at the time of the offense. A study by Haggård-Grann et al. (2006) found a 13.2-fold increase in risk of violence after alcohol consumption. Underreporting of intimate-partner abuse and reporting biases likely minimize the monumental role of alcohol in domestic violence.

In addition to affecting consumption and alcohol-impaired driving, increased alcohol availability impacts criminal behavior. In New York state (excluding New York City), extension of on-premises alcohol business hours was associated with increases in aggravated assaults per police reports, resulting in a predicted annual cost to those communities of close to 200 million dollars (Schofield and Denson, 2013a). Based on data from the National Crime Victimization Survey, a 1% decrease in the number of outlets that sell alcohol decreases the probability of rape by 1.75% (Markowitz, 2000a).

ALCOHOL AND CRIMINAL RESPONSIBILITY

An individual who chooses to drink is generally held legally responsible for criminal acts committed while intoxicated (Marlowe et al., 1999; Ogundipe and Weiss, 2009). The effects of alcohol on decision-making capacity are recognized, but "voluntary intoxication" is considered a willful acceptance of the risks of engaging in criminal behaviors in this altered mental state and any related legal ramifications (Anderson and Bokor, 2012). As recovery-oriented policy develops further over the next few decades, there will likely be changes to the criminal justice system's execution of the law in cases involving alcohol and substance use.

There are nuances to criminal responsibility in the setting of voluntary intoxication and the degree of criminal responsibility is largely dependent on the determination of "specific intent" versus "general intent" of the crime considered (Watterson, 1991). Specific-intent crimes require a mental state ("mens rea") to break the law and are planned and deliberate. By contrast, general-intent crimes do not require the intent to break the law, just an unlawful or guilty act ("actus reus"). Specific-intent requirements must be proven beyond a reasonable doubt; otherwise the crime charged is normally reduced to its general-intent equivalent, which carries a lesser sentence. In cases of intoxication, sentences may therefore be reduced if the defense can challenge the "specific-intent" nature of the crime due to the offender's diminished capacity (Watterson, 1991; Marlowe et al., 1999; Anderson and Bokor, 2012).

Due to alcohol's ability to decrease activity in the frontal lobe and thereby affect decision making and impulsivity, the question has been raised as to whether intoxication may in some cases be considered a state of legal insanity. Most states define insanity by the M'Naghten standard of 1843 (M'Naghten, 1843), which dictates that a person could be found not guilty by reason of insanity if the individual suffered from "a defect of reason, from disease of the mind" that inhibited the ability to understand the nature and quality of his or her actions (8 Eng. Rep. 718, 1843). Based on our knowledge of the effects of alcohol intoxication on the brain and the nature of alcohol dependence, it would certainly appear to rob the individual of reason. Despite this, the legal system generally considers alcohol use disorders separate from other forms of mental illness, with some exceptions.

There are cases involving intoxication in which a legal insanity defense may apply, such as when mental illness contributes to the effects of the substance. Notably, "mental illness" in this case excludes substance use disorders and the mental illness must be able to independently account for the insanity defense (Feix and Wolber, 2007). In the case of involuntary intoxication, an individual coerced or tricked into a state of intoxication may not be considered legally responsible for crimes committed while temporarily insane. It has been suggested that chronic alcoholism robs an individual of the choice to drink and that intoxication may be considered involuntary. However, the theory of chronic alcoholism as a mental disease sufficient to render alcohol consumption involuntary has never commanded significant legal support (Watterson, 1991). Finally, criminal acts committed in a state of delirium tremens are considered distinct from those committed while intoxicated and an insanity defense can be used (Watterson, 1991).

Enforcement of criminal responsibility in cases of alcohol intoxication is clearly a more complex issue than one might imagine. The medical system frequently interfaces with the legal system in the management of patients who face legal repercussions related to alcohol use. For individuals who are in a motor vehicle accident, hospitalization may be protective against the potential legal ramifications of driving intoxicated. In a retrospective study of trauma patients who had a BAC over the legal limit (average BAC of 0.218 g/dL) at the time of the motor vehicle accident, only 18% received a charge of DWI (Cheek et al., 2013).

In addition, a common strategy used by criminal suspects to obtain more lenient sentencing is to claim having experienced some form of amnesia. While alcohol blackouts are reported in cases of misconduct not under the scrutiny of the legal system, studies suggest that claims of alcoholic blackouts frequently serve a strategic legal function. In one Dutch study, individuals who were stopped for suspicion of driving under the influence were retrospectively evaluated for their experience of alcoholic blackout. Although BACs of offenders claiming blackout did not differ from those not claiming blackout, 85% of offenders who claimed blackout had caused an accident, compared to only 35% of offenders who did not claim to have experienced a blackout (van Oorsouw et al., 2004). Individuals with BACs below 250 mg/100 mL infrequently claim alcoholic blackout outside of a legal context. By contrast, in a legal setting, the majority of offenders claiming alcoholic blackout had BACs below 250 mg/100 mL (van Oorsouw et al., 2004).

ALCOHOL AND DECISION MAKING

Capacity and competency are not synonymous in that competence is a legal determination. In the healthcare setting, the capacity of an individual who is intoxicated to make decisions frequently comes into question and any licensed physician may evaluate a patient's capacity to make a specific decision. A court must determine overall decision-making competency and a determination of incompetency leads to the loss of legal rights (Tunzi, 2001; Ganzini et al., 2005). Because alcohol intoxication is a transient state, concerns about decision making are generally related to a specific question during the time period when the individual is significantly impaired.

Capacity is assessed for a specific question and capacity will vary over time based on the individual's mental status (e.g., a delirious or intoxicated patient may have periods of clarity). In evaluating capacity, the clinician assesses: (1) the individual's ability to understand information about the proposed treatment or decision; (2) the individual's ability to appreciate how that information applies to his or her situation; (3) the ability to reason with that information; and (4) the ability to make a choice and express it. If the individual lacks capacity to make a decision, the decision should be discussed with a surrogate decision maker (Tunzi, 2001).

CONCLUSION

The important role government plays an in moderating the costs of alcohol-related problems to society and a variety of policies has shaped alcohol use in the United States. The law affects alcohol use in meaningful ways and multiple combined approaches have been more effective than individual measures alone in reducing the negative consequences of alcohol use (Shults et al., 2009). Excise taxes that increase alcohol prices lead to decreases in drinking and violent crime. Alcohol-impaired driving countermeasures such as enforcement of more stringent BAC laws, license suspension for repeat DWI offenders, and repeated breathalyzer tests also appear to be effective. Restrictions on the availability of alcohol and MLDA laws have a positive impact on alcohol consumption, rates of assault, and alcohol-impaired driving and motor vehicle accidents.

The relatively recent and continued evolution of recovery-oriented policies within the criminal justice system may prove to be an effective avenue for the reduction of the consequences of alcohol misuse. Rather than a simplified "war on drugs" approach, there appears to be an increasing emphasis on evidence-based policy development that approaches alcohol use disorders with hope for treatment and prevention. Individuals are held responsible for crimes committed while intoxicated, but the legal system is slowly adapting to the realization that offering substance use treatment and connecting individuals with community support systems may prove to be more effective in reducing recidivism than simply imprisoning people for the consequences of their behavior.

Alcohol is not going away as a part of American society, but there remains a societal interest in limiting the negative consequences of its use. Laws will continue to evolve to achieve this end, and the future may hold a place for the continued incorporation of rational public health policy integrating treatment into laws governing alcohol.

ACKNOWLEDGMENT

This material is the result of work supported with resources and the use of facilities at the Veterans Affairs Palo Alto Health Care System and by NIH grant K23AA016340 (MJO).

The views expressed in this article are those of the authors and do not necessarily reflect the position or policy of the Department of Veterans Affairs or the United States government.

Websites

Affordable Care Act: http://www.whitehouse.gov/ondcp/healthcare.
Alcohol Policy Information System: http://alcoholpolicy.niaaa.nih.gov/.
Center for Alcohol Policy: http://www.centerforalcoholpolicy.org/.
National Highway Traffic Safety Administration: http://www.nhtsa.gov/Impaired.
Wikipedia: http://en.wikipedia.org/wiki/Alcohol_laws_of_the_United_States_by_state.

References

8 Eng. Rep. 718. (1843). M'Naghten's Case.
Anderson PD, Bokor G (2012). Forensic aspects of drug-induced violence. J Pharm Pract 25: 41–49.
Anderson P, Chisholm D, Fuhr DC (2009a). Effectiveness and cost-effectiveness of policies and programmes to reduce the harm caused by alcohol. Lancet 373 (9682): 2234–2246.
Anderson P, de Bruijn A, Angus K et al. (2009b). Impact of alcohol advertising and media exposure on adolescent alcohol use: a systematic review of longitudinal studies. Alcohol Alcohol 44 (3): 229–243.
Barry CL, Huskamp HA (2011). Moving beyond parity - mental health and addiction care under ACA. N Engl J Med 365: 973–975.
Beck KH, Rach WJ, Baker EA, Williams AF (1999). Effects of ignition interlock license restrictions on drivers with multiple alcohol offenses: a randomized trial in Maryland. Am J Public Health 89 (11): 1696–1700.
Busch S (2012). Implications of the Mental Health Parity and Addiction Equity Act. Am J Psychiatry 169: 1–3.
Calhoun VD, Pekar JJ, Pearlson GD (2004). Alcohol intoxication effects on simulated driving: exploring alcohol-dose effects on brain activation using functional MRI. Neuropsychopharmacology 29: 2097–2107.
Carpenter C, Dobkin C (2011). The minimum legal drinking age and public health. J Econom Perspect 25: 133–156.
Cheek SM, Murry JS, Truitt MS et al. (2013). Driving intoxicated: is hospital admission protective against legal ramifications? J Trauma Acute Care Surg 75: 1081–1084.
Cook PJ, Durrance CP (2013). The virtuous tax: lifesaving and crime-prevention effects of the 1991 federal alcohol-tax increase. J Health Econ 32: 261–267.
Durrance CP, Golden S, Perreira K et al. (2011). Taxing sin and saving lives: can alcohol taxation reduce female homicides? (1982). Soc Sci Med 73 (1): 169–176.
Elder RW, Voas R, Beirness D et al. (2011). Effectiveness of ignition interlocks for preventing alcohol-impaired driving

and alcohol-related crashes: a Community Guide systematic review. Am J Prev Med 40: 362–376.

Engels RCME, Hermans R, van Baaren RB et al. (2009). Alcohol portrayal on television affects actual drinking behaviour. Alcohol Alcohol 44 (3): 244–249.

Executive Office of the President of the United States, Kerlikowske RG (2013). National drug control stategy. National drug control stategy 1–95. http://www.whitehouse.gov/sites/default/files/ndcs_2013.pdf.

Feix J, Wolber G (2007). Intoxication and settled insanity: a finding of not guilty. J Am Acad Psychiatry Law 35: 172–182.

Fell JC (2008). Examination of the Criticisms of the Minimum Legal Drinking Age 21 Laws in the United States from a Traffic-Safety Perspective. National Highway Traffic Safety Administration, Washington, DC.

Fell JC, Voas RB (2006). The effectiveness of reducing illegal blood alcohol concentration (BAC) limits for driving: evidence for lowering the limit to.05 BAC. J Safety Res 37: 233–243.

Ganzini L, Volicer L, Nelson W et al. (2005). Ten myths about decision-making capacity. J Am Med Dir Assoc 6 (3 Suppl.): S100–S104.

Greenfeld LA (1998). Alcohol and Crime. U.S. Department of Justice, Bureau of Justice Statistics, NCJ 168632, Washington, DC.

Grucza RA, Hipp PR, Norberg KE et al. (2012). The legacy of minimum legal drinking age law changes: long-term effects on suicide and homicide deaths among women. Alcohol Clin Exp Res 36: 377–384.

Haggård-Grann U, Hallqvist J, Långström N et al. (2006). The role of alcohol and drugs in triggering criminal violence: a case-crossover study. Addiction 101 (1): 100–108.

Harrison P, Chesson H, Kassler W (2000). Sex under the influence: the effect of alcohol policy on sexually transmitted disease rates in the United States. J Law Econom 43: 215–238.

Heath DB (2009). Prohibition, repeal, and historical cycles. *DATA: The Brown University Digest of Addiction Theory & Application*, *28*, 8. Retrieved from http://0-search.ebscohost.com.library.vu.edu.au/login.aspx?direct=true&db=aph&AN=36631344&site=ehost-live.

Hingson R, Heeren T, Winter M (1994). Lower legal blood alcohol limits for young drivers. Public Health Reports109: 738–744. Washington, D.C.: 1974.

Humphreys K, Lembke A (2014). Recovery-oriented policy and care systems in the UK and USA. Drug Alcohol Rev 33: 13–18.

Humphreys K, McLellan AT (2010). Brief intervention, treatment, and recovery support services for Americans who have substance use disorders: an overview of policy in the Obama administration. Psycholog Serv 7: 275–284.

Kaestner R, Yarnoff B (2011). Long-term effects of minimum legal drinking age laws on adult alcohol use and driving fatalities. J Law Econom 54: 325–363.

Kaplan L, Nugent C, Baker M et al. (2010). Introduction: The Recovery Community Services Program. Alcohol Treat Q 28: 244–255.

Kilmer B, Nicosia N, Heaton P et al. (2013). Efficacy of frequent monitoring with swift, certain, and modest sanctions for violations: insights from South Dakota's 24/7 Sobriety Project. Am J Public Health 103: e37–e43.

Latham RTHR (2013). 1918: Brewers Excise and Economic Relief. Act of 2013.

Markowitz S (2000b). An Economic Analysis of Alcohol, Drugs, and Violent Crime in the National Crime Victimization Survey. NBER Working Paper No. 7982. http://www.nber.org/papers/w7982.pdf.

Markowitz S (2000a). The price of alcohol, wife abuse, and husband abuse. South Econom J 67: 279–303.

Markowitz S, Grossman M (2000). The effects of beer taxes on physical child abuse. J Health Econ 19: 271–282.

Markowitz S, Chatterji P, Kaestner R (2003). Estimating the impact of alcohol policies on youth suicides. J Ment Health Policy Econ 6: 37–46.

Marlowe DB, Lambert JB, Thompson RG (1999). Voluntary intoxication and criminal responsibility. Behav Sci Law 17: 195–217.

McCartt AT, Northrup VS (2004). Effects of enhanced sanctions for high-BAC DWI offenders on case dispositions and rates of recidivism. Traffic Inj Prev 5: 270–277.

McMillan GP, Lapham S (2006). Effectiveness of bans and laws in reducing traffic deaths: legalized Sunday packaged alcohol sales and alcohol-related traffic crashes and crash fatalities in New Mexico. Am J Public Health 96: 1944–1948.

Meier PS, Purshouse R, Brennan A (2010). Policy options for alcohol price regulation: the importance of modelling population heterogeneity. Addiction 105: 383–393.

National Highway Traffic Safety Administration (2008). 2008 Data Alcohol-Impaired Driving. NHTSA's National Center for Statistics and Analysis. http://www-nrd.nhtsa.dot.gov/Pubs/811155.pdf.

National Highway Traffic Safety Admininstration (2012). 2011 Motor Vehicle Crashes. Overview 1–5. Washington, DC.

National Highway Traffic Safety Administration (2013). Digest of Impaired Driving and Selected Beverage Control Laws, Retrieved from http://www.nhtsa.gov/Driving+Safety/Impaired+Driving/Digest+of+State+Alcohol+Highway+Safety-Related+Legislation.

NIAAA (2000). Latest approaches to preventing alcohol abuse and alcoholism. Alcohol Res Health 24: 42–51.

Ogundipe KA, Weiss KJ (2009). Drunk driving, implied consent, and self-incrimination. J Am Acad Psychiatry Law 37: 386–391.

Okrent D (2010). Last Call: The Rise and Fall of Prohibition. Scribner, New York, pp. 1–480.

Roth R, Voas R, Marques P (2007). Mandating interlocks for fully revoked offenders: the New Mexico experience. Traffic Inj Prev 8: 20–25.

Schofield TP, Denson TF (2013a). Alcohol outlet business hours and violent crime in New York state. Alcohol Alcohol 48: 363–369.

Schofield TP, Denson TF (2013b). Temporal alcohol availability predicts first-time drunk driving, but not repeat offending. PLoS One 8 (8): e71169.

Sharma A, Vandenberg B, Hollingsworth B (2014). Minimum pricing of alcohol versus volumetric taxation: which policy will reduce heavy consumption without adversely affecting light and moderate consumers? PLoS One 9 (1): e80936.

Shults RA, Elder RW, Sleet DA (2001). Reviews of evidence regarding interventions to reduce alcohol-impaired driving. Am J Prev Med (21(Suppl.)): 66–88.

Shults RA, Elder RW, Nichols JL et al. (2009). Effectiveness of multicomponent programs with community mobilization for reducing alcohol-impaired driving. Am J Prev Med 37: 360–371.

Stout EM, Sloan FA, Liang L et al. (2000). Reducing harmful alcohol-related behaviors: effective regulatory methods. J Stud Alcohol 61: 402–412.

Tippetts AS, Voas RB, Fell JC et al. (2005). A meta-analysis of .08 BAC laws in 19 jurisdictions in the United States. Accid Anal Prev 37: 149–161.

Tunzi M (2001). Can the patient decide? Evaluating patient capacity in practice. Am Fam Physician 64: 299–306.

Van Oorsouw K, Merckelbach H, Ravelli D et al. (2004). Alcoholic blackout for criminally relevant behavior. J Am Acad Psychiatry Law 32: 364–370.

Wagenaar AC, Maldonado-Molina MM (2007). Effects of drivers' license suspension policies on alcohol-related crash involvement: long-term follow-up in forty-six states. Alcohol Clin Exp Res 31: 1399–1406.

Wagenaar AC, Toomey TL (2002). Effects of minimum drinking age laws: review and analyses of the literature from 1960 to 2000. Supplement, (14), J Stud Alcohol: 206–225.

Wagenaar AC, Maldonado-Molina MM, Erickson DJ et al. (2007). General deterrence effects of U.S. statutory DUI fine and jail penalties: long-term follow-up in 32 states. Accid Anal Prev 39: 982–994.

Wagenaar AC, Salois MJ, Komro KA (2009). Effects of beverage alcohol price and tax levels on drinking: a meta-analysis of 1003 estimates from 112 studies. Addiction 104: 179–190.

Watterson RT (1991). Just say no to the charges against you: alcohol intoxication, mental capacity, and criminal responsibility. Bull Am Acad Psychiatry Law 19: 277–290.

Wen H, Cummings JR, Hockenberry JM et al. (2013). State parity laws and access to treatment for substance use disorder in the United States: implications for federal parity legislation. JAMA Psychiatry 70: 1355–1362.

Wilkinson C, Room R (2009). Warnings on alcohol containers and advertisements: international experience and evidence on effects. Drug Alcohol Rev 28: 426–435.

Willis C, Lybrand S, Bellamy N (2004). Alcohol ignition interlock programmes for reducing drink driving recidivism. Cochrane Database Syst Rev 3: 1–27.

Yörük BK (2013). Legalization of Sunday alcohol sales and alcohol consumption in the United States. Addiction 55–61.

Zador PL (1991). Alcohol-related relative risk of fatal driver injuries in relation to driver age and sex. J Stud Alcohol 52: 302–310.

Zador PL, Krawchuk SA, Voas RB (2000). Alcohol-related relative risk of driver fatalities and driver involvement in fatal crashes in relation to driver age and gender: an update using 1996 data. J Stud Alcohol 61: 387–395.

Handbook of Clinical Neurology, Vol. 125 (3rd series)
Alcohol and the Nervous System
E.V. Sullivan and A. Pfefferbaum, Editors

Chapter 39

Clinical management of alcohol use disorders in the neurology clinic

ANNA LEMBKE[1]* AND MARK STANFORD[2]

[1]*Department of Psychiatry and Behavioral Sciences, Stanford University, Stanford, CA, USA*

[2]*Addiction Medicine and Therapy Services, Santa Clara Valley Health and Hospital System, Santa Clara County, CA, USA*

INTRODUCTION

The link between alcohol use disorders (AUDs) and adverse health outcomes is well established. Alcohol affects 20–50% of hospitalized patients, 15–30% of patients seen in primary care settings, and upwards of 60% of patients with psychiatric illnesses. AUDs are a leading cause of early preventable death and are associated with a constellation of medical consequences (Saitz, 2005a).

Patients typically do not directly present their unhealthy alcohol use to their primary or specialty care provider but do present with alcohol-related conditions such as ataxia, cognitive impairment, tremor, gastritis, or psychogenic problems, including anxiety and depression (Aira et al., 2003). Adverse health consequences, particularly the neurologic sequelae, are common in AUDs.

Despite the fact that patients with AUD frequently present in ambulatory care settings, physicians and other healthcare providers seldom screen or intervene for this problem. The reasons are many, including the moral stigma of addiction, lack of physician training in addiction, and a pervasive belief that addiction treatment doesn't work.

The moral stigma associated with addiction continues to affect the care of the addicted patient. CASA Columbia's national survey of the attitudes of US adults vis-à-vis addiction and its treatment found that approximately one-third of Americans believe that addiction is a sign of lack of willpower or self-control, and approximately 50% of physicians believe that addiction is caused by a character defect (CASA Columbia, 2012). These beliefs are at odds with an ever-growing body of evidence showing that addiction is a chronic illness with strong genetic and biologic components. Given these preconceived notions, it is no wonder that physicians do not screen for substance use disorders: Many of them do not actually consider this a medical problem.

Lack of physician training in the area of substance misuse and abuse is another barrier to care. Only 20–30% of primary care physicians feel "very prepared" to detect risky substance use, yet 80% feel "very prepared" to tackle hypertension or diabetes (CASA Columbia, 2012). Instead of targeting addiction, most healthcare professionals address the medical complications of addiction without addressing its core cause.

Many people, including healthcare providers, express a sense of hopelessness about addiction treatment. Yet, the data demonstrate that addiction treatment works at rates on a par with treatment for other chronic illnesses, from depression to asthma (McLellan et al., 2000). In fact, treatments for alcohol addiction reduce alcohol use by 40–60%, as well as decreasing criminal activity, decreasing comorbid physical and mental illnesses, and improving other measures of well-being, such as housing status and employment.

To treat the medical problems related to substance misuse and to target the substance use disorder itself, providers should screen and intervene for substance use disorders within an ambulatory care clinic, even when resources for addiction treatment are limited. An evidence-based algorithm called screening, brief intervention, and referral to treatment (SBIRT) explains how at-risk drinking and AUDs can be identified, and how at-risk drinking can be effectively targeted within ambulatory care settings. In this chapter we review the components of SBIRT and how to implement SBIRT

*Correspondence to: Anna Lembke, M.D., Assistant Professor, Psychiatry Behavioral Sciences, Stanford University, 401 Quarry Road, Stanford, CA 94305, USA. Tel: +1-650-725-9570, E-mail: alembke@stanford.edu

in clinical practice. We also describe clinical presentations of common neurologic disorders associated with acute and chronic alcohol toxicity.

WHAT IS SBIRT?

SBIRT is a method of intervention with individuals with alcohol and other substance use problems. SBIRT is best suited for primary care clinics, hospital emergency rooms, trauma centers, neurology clinics, and community health settings, and is a physician-led prevention approach.

The Substance Abuse Mental Health Services Administration (SAMHSA) recommends the routine use of SBIRT in primary and specialty care settings for the following reasons:

- It is brief. The initial screening is accomplished quickly (modal time about 5 minutes).
- Specific behaviors are targeted. The screening tool addresses a specific behavioral characteristic deemed to be problematic or preconditional to other diagnoses.
- The services occur in a healthcare or other non-substance abuse treatment setting. This may be an emergency department, primary or specialty care physician's office, including the neurology clinic.
- It is comprehensive. SBIRT includes a seamless transition between a quick screening and brief intervention.
- Research and substantial experiential evidence supports the SBIRT model. At a minimum, programmatic outcomes demonstrate a successful approach.

WHAT IS THE EVIDENCE THAT SBIRT WORKS?

Early identification and brief intervention in primary and specialty care settings have been shown to be effective in patients with at-risk alcohol use. Forty percent of patients with risky alcohol use who received screening and brief intervention from their primary or specialty care provider moderated their drinking to safe levels compared with only 20% in control groups. These differences between intervention and control groups were still present 4 years later. There is growing evidence from review studies as well as meta-analyses of randomized clinical trials that, using the SBIRT approach, reduces risky drinking in patients presenting in primary and specialty care settings.

From these and other studies, the World Health Organization, the Institute of Medicine, and other health authorities recommend asking (screening) all patients about their alcohol use patterns and providing brief intervention when necessary. Evaluation of lifestyle health risks should always also include an evaluation of alcohol use. The US Preventative Services Task Force (USPSTF) has recommended that "interventions for risky/harmful alcohol use among adult primary care patients can provide an effective public health prevention approach to reducing problematic drinking" (US Preventative Services Task Force, 2004). The USPSTF also concluded that physician-led brief interventions for risky drinkers should include education and advice to reduce current drinking; feedback about current drinking patterns; and explicit goal setting, usually for moderation and assistance in achieving the goals. The combination of screening all adult primary care patients to identify individuals with unhealthy use and a brief counseling intervention has been proposed as a population-wide, preventive intervention in primary care. Clinical trials support the use of screening and brief intervention in unhealthy alcohol use for adults (Saitz, 2013).

SCREENING

The purpose of alcohol screening is to identify patients with risky alcohol use so that interventions can be provided to reduce drinking and prevent harm (Bradley and Berger, 2013). Screening is not the same as diagnosis, nor should a positive screen be misinterpreted as an AUD. Rather, screening alerts the provider to patients who need follow-up and further assessment of drinking patterns and their association with adverse health effects.

Quantity and frequency of alcohol should be used as the initial method of determining who is engaging in at-risk alcohol use. Evaluation of alcohol consumption patterns should be included during medical consultations or physical examinations, as part of a more general evaluation of lifestyle health risks. It is important to introduce the alcohol discussion with the patient during the medical consultation and always connect that discussion with the presenting medical issues, the diagnosis, and the physical care plan goals.

When screening patients for quantity and frequency of alcohol use, it is important to know what constitutes one standard drink. One standard drink is defined in the alcohol literature as 0.5 fluid ounces of alcohol, which can be found in one 12-oz bottle of beer, one 5-oz glass of wine (there are usually five standard drinks in one bottle of wine), and one 1½-oz "shot glass" of hard liquor (vodka, whiskey, etc.).

An ever-growing body of evidence links quantity and frequency of alcohol consumption to overall morbidity and mortality. Those who drink above a certain level are at increased risk for numerous poor health outcomes, including gastrointestinal problems (Lembke et al., 2011), trauma (Harris et al., 2009), and all-cause mortality (Harris et al., 2010); and one-third are at risk of developing addiction (Saitz, 2005b).

Risky alcohol use for an adult male, based on the evidence above, is defined as a pattern of alcohol use exceeding 14 standard drinks per week or four per occasion, and for an adult female as exceeding seven standard drinks per week or three drinks per occasion. The amounts differ for men and women because of differences in metabolism and vulnerability to alcohol toxicity, with women being more vulnerable to the toxic effects of alcohol than men, even with shorter drinking careers. Amounts below these are considered to be non-risky, or moderate alcohol use, because they do not correlate with adverse health outcomes.

Focusing on these quantifiable anchor points minimizes the problems of subjective report and patient denial. Also, although many individuals, addicted or not, are poor at recalling substance use accurately when asked to retrospectively assess their use over a long time period, such as the past year (even when they are comfortable disclosing it candidly to the clinician), data show that when asking about specific amounts and specific days in the recent past, patients are surprisingly accurate reporters of consumption. Finally, quantity and frequency lead to a number that can be regarded as a "fifth vital sign," easily charted and compared with other numbers past and future, a possible indicator of a problem but not a definitive diagnosis of a problem.

Two brief and easy-to-use screening tools to assess quantity and frequency of alcohol consumption are available and ideal for the neurology clinic, namely, the Timeline-Followback (TLFB) method and the AUDIT-C.

The TLFB method charts the amounts and patterns of substance use in the preceding week, independently of consequences or compulsivity of use (Sobell and Sobell, 1995). By keeping to specifics and a timeline, there is a markedly decreased chance of the patient minimizing consumption, mixing up drinking occasions, or mentally averaging drinking over the period of interest. Ask patients to begin with the day prior to the clinical encounter and move backwards, remembering how much alcohol they consumed on each day in the preceding week. Where appropriate, offer concrete events to stimulate memory, e.g., "Let's start with Monday, that was the first day of the week before mid-term exams." Then calculate the number of standard drinks consumed in a week.

The same principle applies to use of the AUDIT-C, which consists of three questions regarding frequency and quantity of alcohol consumption: (1) "How often do you have a drink containing alcohol?" (never, monthly, 2–3 times per week, etc.); (2) "How many drinks of alcohol do you drink on a typical drinking day?"; and (3) "How often have you had six or more drinks per day?" The last question captures binge drinkers, a well-established drinking pattern which might be missed with the TLFB method, and consists of heavy, condensed periods of consumption interspersed with longer periods of abstinence. Again, starting with quantity and frequency allows the clinician to tackle the issue of alcohol use in a quantitative and non-judgmental way.

If risky drinking is identified based on quantity and frequency, then the healthcare provider is prompted to delve deeper, specifically to address whether the patient meets *Diagnostic and Statistical Manual of Mental Disorders* (DSM-V) criteria for a substance use disorder (American Psychiatric Association, 2013).

The DSM defines a substance use disorder as use of nicotine, alcohol, and/or other drugs with two or more of the following symptoms within a 12-month period: (1) attempting to cut back on substance use without success; (2) consuming more of the substance than planned; (3) spending a lot of time and energy getting, consuming, and recovering from using the substance; (4) experiencing intense desire to consume the substance, often referred to as "craving"; (5) failing to fulfill major life obligations due to substance use; (6) continuing to use the substance despite consequences; (7) giving up or reducing important activities due to substance use; (8) using in dangerous situations; (9) developing tolerance; and (10) experiencing withdrawal.

The new edition of the DSM (DSM-V) has added spectrum qualifiers "mild," "moderate," and "severe," allowing providers to capture the variance in severity that is common with substance use disorders. Severity is rated as follows: If the patient endorses twoor three items on the list, then (s)he has a mild substance use disorder, four or five items and she has a moderate substance use disorder, and six or more items and she has a severe substance use disorder. For example, the patient who describes attempting to cut back without success, spending a lot of time using and recovering from the substance, failing to fulfill major life obligations due to substance use, and experiencing withdrawal has a moderate substance use disorder (American Psychiatric Association, 2013).

The DSM does not rely on frequency and quantity of substance use, not because it assumes these data points are unimportant, but because of the difficulty in standardizing across the wide variety in form (plant, pill, liquid, gas) and potency of addictive substances, not to mention the difficulty in knowing what is in the new synthetic drugs that have been flooding the market over the last decade ("legal highs," "bath salts," etc.). The absence of quantifiable measures in the DSM is an expression of the limitations of a phenomenologic guide rather than a dismissal of its importance. Therefore, using DSM criteria alone, without also assessing quantity and frequency, is inadequate. A patient who drinks a fifth of vodka every night but denies all of the DSM criteria for a substance use disorder nonetheless likely has an AUD and could benefit from an intervention. A quantitative reference point allows clinicians to make the appropriate diagnosis, even in the absence of subjective recognition of addiction ("denial").

Once the patient has screened positive for at-risk alcohol use or an AUD, the algorithm below is useful in determining the "next steps" after an alcohol screen.

SCREENING RESULTS

1. No or minimal problems with drinking: Patient education and no further intervention.
2. Mild to moderate at-risk drinking: Patient education and brief intervention.
3. Moderate to high-risk drinking: Brief treatment.
4. AUD based on DSM criteria: Referral to specialty treatment.

Screening can be repeated at intervals, as needed.

In summary, alcohol use occurs along a continuum from at-risk drinking to AUD, ranging in severity from mild to moderate to severe. Appropriate interventions should be applied based on the severity of the patient's drinking determined through a screening process. The National Institute of Alcoholism and Alcohol Abuse (NIAAA, a division of the National Institutes of Health) focuses on two cohort groups of problem alcohol use: (1) at-risk drinkers (i.e., those who exceed recommended maximum daily levels of drinking but do not necessarily meet criteria for an AUD); and (2) drinkers who meet the DSM V criteria for AUD (American Psychiatric Association, 2013). Brief interventions are not effective for those with AUD who need referral to specialty alcohol treatment, but the process of doing SBIRT can provide a litmus test to identify those patients who have more significant substance use problems (i.e., addiction) and need specialty referral.

BRIEF INTERVENTION

Brief interventions are interactions with patients that are intended to induce a change in a health-related behavior. Brief interventions are generally used as a health management strategy for patients with at-risk alcohol use but who do not necessarily meet criteria for an AUD. Most patients in primary and specialty care medicine are in this at-risk category. The goal of a brief intervention (which usually involves one to five sessions lasting about 5–10 minutes) is to educate patients and increase their motivation to reduce risky behavior and embrace the healthier lifestyle as outlined in their physical care plan.

Research from smoking cessation and problem alcohol use has increased our understanding of the change process, giving new directions for improved health promotion and lasting lifestyle change in patients. Current views depict patients as being in a process of change. When physicians choose a mode of intervention, "one size doesn't fit all." Two important advances include the stages of change model and motivational interviewing. Developed by Stephen Rollnick at Cardiff University School of Medicine in the United Kingdom, motivational interviewing is a collaborative, goal-oriented style of communication with particular attention to the language of change (Miller and Rollnick, 2012). It is designed to strengthen personal motivation for and commitment to a specific goal by eliciting and exploring the person's own reasons for change within an atmosphere of acceptance and compassion. Motivation is multidimensional and not easily assessed. To assist with motivating the behavioral change process, psychologists James Prochaska and Carlo DiClemente identified five primary stages that people move through as they seek to make changes in their lives. Motivational interviewing can assist people with successfully moving through the first three stages of the change process. Through motivational interviewing, the physician can help a patient to consider the possibility of change, contemplate the risks and benefits of change, and prepare for moving forward with making changes in her or his life.

The creators of the stages of change model used factor and cluster analytic methods in retrospective, prospective, and cross-sectional studies of the ways people quit smoking. The "stage-based" approach has been validated and applied to myriad behaviors, including nicotine cessation, exercise, and dietary behavior (Grimley, et al., 1993; Glanz et al., 1994; Prochaska et al., 1994; Hellman, 1997). Simple and effective "stage-based" approaches derived from the stages of change model have demonstrated a widespread utility Additionally, brief interventions (a 5–10-minute exam room consult) are as effective as longer visits and result in significantly longer sustained behavioral change (Calfas et al., 1997).

Behavior change is generally not a discrete, single event. Physicians can sometimes see patients who, after experiencing a medical crisis and being advised to change the contributing behavior, readily comply. More often, physicians see patients who are unable or unwilling to change (Zimmerman et al., 2000). Recently, research has shown that behavioral change is a process of five identifiable stages through which people pass successively and that, for most people, a change in behavior occurs gradually, with the patient moving from being uninterested, unaware, or unwilling to make a change (precontemplation), to considering a change (contemplation), to deciding and preparing to make a change (action). Physicians can facilitate movement through these stages by taking specific actions. Understanding this process provides physicians with additional tools to assist patients, who are often as discouraged as their physicians with their lack of change.

The stages of change model encompasses many concepts from previously developed models. The health belief model, the locus of control model, and behavioral models fit together well within this framework. During the precontemplation stage, patients do not consider change. They may not believe that their behavior is a problem or that it will negatively affect them (health belief model), or they may be resigned to their unhealthy behavior because of previous failed efforts and no longer believe that they have control (external locus of control). During the contemplation stage, patients struggle with ambivalence, weighing the pros and cons of their current behavior and the benefits of and barriers to change (health belief model – Janz and Becker, 1984). Cognitive behavioral models of change (e.g., focusing on coping skills or environmental manipulation) and 12-step programs fit well in the preparation, action, and maintenance stages. The five stages of change are described below:

1. Precontemplation: Patients do not consider changing. Overeaters who are "in denial" may not see that the advice applies to them personally. Patients with hypertension may feel "immune" to the health problems that strike others. Diabetic patients may have tried many times unsuccessfully to lose weight and change their diet habits, but have since given up.
2. Contemplation: Patients are actually ambivalent about changing. Letting go of an enjoyed behavior can result in a feeling of loss and depression or anxiety despite any promised gain. During this stage, patients assess the different and individualized risks and benefits of change.
3. Action: Patients prepare to make a specific change in the near future. They can experiment with small changes as the determination to make more and more change increases. For example, sampling two drinks per day instead of five may be an experimentation with a move towards reducing alcohol intake. Switching to a healthier diet, starting a blood pressure journal, or starting to walk a little for exercise are indicators that they have decided a change is needed. This is the stage that most physicians and other healthcare providers are most eager to get their patients to achieve. However, the provider's eagerness must be tempered and the interventions must be based in the stage where the patient is not in the stage the provider wants the patient in. For example, new year's resolutions broken in February provide evidence that, if the prior stages have been rushed through, the action stage is not enough to sustain.
4. Maintenance: Patients internalize new behavior as part of their lifestyle.
5. Relapse prevention: Discouragement over occasional setbacks may retard the change process and result in the patient giving up. Therefore, encouragement and recognition from the physician of the gains made by the patient are important.

When using motivational interviewing during brief interventions, it is helpful to consider the stage a patient currently is in regarding potential change (Prochaska et al., 1992). Physicians can carefully direct intervention efforts based on consideration of two important factors: (1) the severity of the alcohol problem; and (2) the patient's readiness to change drinking behavior (i.e., the patient's stage of change).

Well-intended advice, a practice all too familiar with physicians, works best with patients who are ready to prepare for change. Using direct persuasion with a patient who is too ambivalent risks generating patient resistance. Patient resistance is seen when the physician moves too fast with the patient in the change process. Note, however, an urgent intervention in patients at any stage when an immediate change is required, e.g., a pregnant woman who is still smoking, or a patient on anticonvulsant meds for traumatic brain injury (TBI) who continues to drink occasionally. In all circumstances, educate patients about the consequences of their substance use: "Smoking affects the developing fetus in this way. . ." The information on smoking, in this example, works best when it addresses the issue that directly concerns the patient. Therefore, using the medical diagnosis and the physical care plan as the basis for motivating the patient to change behaviors is essential. A lecture on the hazards of smoking, drinking, and drug abuse, separate from the medical care plan, is insufficient to create lasting and meaningful change.

Begin the interview with an attempt to understand how the behavior fits into the patient's life. This assessment is the complete intervention in patients who are in precontemplation stage. In the contemplation stage, physicians should discover the patient's ambivalence toward change, including risks and benefits of continued drinking. It is at this point that patients may be more receptive to information about alcohol's effects. In the later readiness stages, physicians can introduce patients to helpful community resources, including Alcoholics Anonymous

(AA) or specialty treatment programs, should the severity of their drinking warrant.

The overall goal to stage-based interventions is to help patients develop their own motivation for change and to guide them toward a healthier lifestyle. This guidance underscores the idea that behavioral change occurs incrementally rather than all at once and the physician can determine his/her patient's stage of change and provide an intervention appropriate to that stage for a best response. This intervention approach is non-directive and therefore removes much of the resistance the patient would otherwise have. Identifying patients' readiness to change through assessing the particular stage of change they are in allows them to do all of the work, including reflecting on the ways alcohol fits into their life, weighing the personal risks and benefits of continued drinking, providing the rationale for change and making the decision to stop drinking. The physician's job is simply to elicit information, encourage patients to reflect, and support their movement toward healthy change.

The wording of questions the physician asks during an intervention requires forethought because wording can bias reactions and responses. When precontemplative patients respond to questions, rather than quickly giving advice, reflect with empathy, instill hope, and empathetically point out discrepancies between patients' stated care plan goals and their statements. Asking patients, "Do you know this will ultimately kill you?" can be viewed as threatening and will result in greater resistance. However, asking patients, "How will you know that it's time to quit?" allows patients to be in charge of their lives and can help foster a meaningful thought process that extends beyond the exam room consult. Good questions from the physician inspirespatients to think about answers that are applied to them and advances them along the change process. Some examples of stage-based questions are shown below.

Precontemplation stage

Goal: To move the patient toward thinking about change

- "What would have to happen for you to know that this is a problem?"
- "What warning signs would let you know that this is a problem?"
- "Have you tried to change this in the past?"

Contemplation stage

Goal: To encourage the patient to examine the benefits and risks of change

- "Why do you want to change at this time?"
- "What were the reasons for not changing?"
- "What would keep you from changing at this time?"
- "What are the barriers today that keep you from change?"
- "What might help you with that aspect?"
- "What things (people, programs, and behaviors) have helped in the past?"
- "What would help you at this time?"

Patients who are in preparation stage begin to experiment with changing a behavior such as cutting down on smoking/drinking or starting to exercise or eat differently, and this is seen as transitioning into more decisive action. While continuing to explore ambivalence, intervention strategies need to shift from motivational to planned action-oriented behavioral skills specific to their stage of change or readiness. During the action and maintenance stages of change, physicians ask patients about their continued successes and challenges with praise, encouragement, and empathy as this can best reinforce the continuation of sustained healthy change.

REFERRAL TO TREATMENT

Patients who meet DSM criteria for an AUD or do not respond to a brief intervention may benefit from referral to an intervention specifically for alcohol use problems. There are many such specialty clinics, residential facilities, and intensive outpatient programs, but sometimes patients are unwilling or unable to participate in such programs often due, for example, to geographic or insurance restrictions.

This chapter focuses on just one of these options, namely AA, simply because it is the least expensive and most available and therefore the most utilized service for AUDs in the world.

AA is an organization founded in the mid-1930s by two confirmed alcoholics who decided to help each other get sober. They found that through mutual support and shared experiences they were able to remain abstinent from alcohol when all other treatments available at that time had failed. Today, AA boasts a membership of more than 4 million people from 400 different societies across the globe.

Despite its popularity, AA is not considered an "evidence-based treatment." This is not to suggest that AA does not work. Rather, AA is not considered "treatment" because it is not administered by professional healthcare providers. Second, it cannot be formally studied like other evidence-based treatments, in which people are divided into case and control groups, because by definition people who go to AA self-refer and cannot be assigned to groups. Nonetheless, more than three decades of accumulated evidence show that AA improves drinking outcomes and other markers of health and well-being for those who actively participate, and the higher the level of participation, the better the outcome. AA involvement is a potent predictor of achieving and maintaining abstinence from alcohol (Emrick et al., 1993, Humphreys, 2004). AA participation also predicts improved psychologic and social functioning (Humphreys, 2004).

For effective referral of patients to AA, healthcare providers need to understand some fundamental facts about the organization, how it works, and which patients are appropriate for referral.

AA is an abstinence-oriented organization, which means that the goal for people in the group is to stop drinking, not "cut back." This approach is distinctly different from patients with alcohol use problems who are working to reduce drinking to healthy levels. The previous section, describing brief interventions for AUD, discusses the importance of establishing mutually agreed-upon goals. If the mutually agreed-upon goal between the healthcare provider and the patient with an AUD is to reduce drinking but not stop, then this patient would not be an appropriate referral to AA. On the other hand, the single criterion for membership in AA is the "desire to stop drinking," which means that patients who are still actively abusing alcohol but would like to stop would be appropriate for AA.

North American research has consistently found no relationship between AA affiliation and demographic variables such as age, social class, race, employment status, and parental socioeconomic status (Trice and Roman, 1970; Emrick et al., 1993). However, AA may work best for the most severely addicted (Trice and Wahl, 1958; Emrick et al., 1993). Therefore, the options for patients with the most refractory and severe alcohol use problems may indeed be AA. Often clinicians assume that a 30-day residential program is mandatory for those severely addicted to alcohol. There is no evidence to support this notion, and the costs of such a program are prohibitive for many people, hovering around $30 000 for a month's stay. AA has the advantage of helping individuals find recovery while still integrating the treatment into their everyday lives and without incurring the burden of high expense. Also, meetings can be found in almost every community center, church basement, or synagogue in almost every city in the United States. The absence of fees and the ubiquity of meetings reduce potential barriers to participation for patients who may not have health insurance or are otherwise limited in their ability to access care.

AA has a developed philosophy and program of change, outlined in a book titled *Alcoholics Anonymous*. The book was published in 1939 and is known universally by its members as *The Big Book*, because of the thick paper on which it was originally printed (Humphreys, 2004). AA views alcoholism as a chronic illness, from which one is never cured but remains in recovery. The core principles of AA are known as the 12 steps, and practicing them is called "working the program." Full sobriety is achieved when the individual is abstinent from alcohol and working the program, which, loosely summarized, consists of living out the ideals of honesty, humility, selflessness, and mindfulness. Although the achievement of abstinence alone is considered a central goal, it is not considered complete recovery in itself (Humphreys, 2004).

AA is not affiliated with any specific religion but does espouse the importance of spirituality and a higher power in the process of transformative healing. The higher power of AA is loosely defined and can consist of anything from a Christ-like deity to the ineffable mystery of the universe to the awe-inspiring support and acceptance of the AA fellowship. The spiritual aspect of AA can be an obstacle to participation for some would-be members. The use of the words "God" and "higher power" throughout the AA literature can be alienating for some persons who do not self-identify as religious. For these individuals in particular, it is essential to point out that God can be defined in any way the individual chooses and therefore need not interfere with membership.

Clinicians who refer patients to AA, or work with patients already active in AA, should monitor their progress in the organization, by asking them for example "What step are you working on now?" or "Are you working with a sponsor?" This communicates to patients that the provider is invested in their well-being outside of clinic time and reinforces a patient's attachment to a therapeutic community. It also helps monitor for relapse, because it has been shown that having a sponsor prevents against relapse (Witbrodt et al., 2012).

NEUROLOGIC MEDICAL COMORBIDITY

Physical examination

The physical examination does not provide much evidence that would suggest unhealthy alcohol use in the early stages of alcohol-related problems (Burge and Schneider, 1999). These patients may have mildly elevated blood pressure and few other abnormal physical findings. Later however, as the consumption and frequency of drinking increase, patients can develop significant signs of alcohol overuse, including gastrointestinal conditions such as an enlarged tender liver; cutaneous issues, including spider angiomata, varicosities, and jaundice; and neurologic signs, including tremor, ataxia, or neuropathies; and cardiac arrhythmias (Mersy, 2003).

Acute presentations

Alcohol intoxication is generally obvious, with the smell of alcohol, nystagmus, dysarthria, skin flushing, hypotension, and ataxia. Presentations depend on the blood alcohol concentration (BAC) but are individual-dependent. This individualized variation can be clinically informative. For example, the ability to maintain a conversation with a blood alcohol of 300 mL/dL indicates severe dependence; these individuals will exhibit withdrawal at alcohol levels associated with intoxication in non-tolerant individuals (i.e., around 100 mg/dL). In practice, treatment should be initiated when withdrawal symptoms occur, rather than waiting for the alcohol level to fall to near zero.

Quite common to emergency departments is the patient with reduced consciousness who reeks of alcohol. While this might be simple inebriation, exclusion of metabolic causes, poisoning, concomitant use of other drugs, or an underlying structural brain lesion needs to be considered. Alcoholics may have cerebral atrophy, predisposing them to subdural hematomas, and disordered coagulation, rendering them liable to intracerebral hemorrhage. These patients need a full examination, blood alcohol testing, and a computed tomography brain scan. Focal neurologic signs are not expected in alcohol intoxication, and a blood alcohol level of <200 mg/dL should not cause coma in the alcohol-tolerant individual. By comparison, serum ethanol levels for non-tolerant individuals, including binge drinkers, are as follows:

- Negative: no alcohol detected
- Lower limit of detection = 10 mg/dL
- >80 mg/dL (>17.4 mmol/L) is considered positive for driving under the influence in most states
- >300–400 mg/dL (65.1–86.8 mmol/L): potentially fatal

Blood alcohol concentration

To convert serum ethanol levels to BAC, move the decimal point three places to the left. For example, a 100 mg/dL serum ethanol level is equivalent to a 0.10 (g/dL) BAC, or 0.10% (weight/volume). This means that one-tenth of a percent of a person's blood volume is alcohol or that a person has 1 part alcohol per 1000 parts blood. Blood alcohol levels are easily quantified using a breathalyzer.

Alcohol withdrawal

Alcohol withdrawal occurs in those physically dependent on alcohol and is the result of the neuroadaptive compensatory changes that occur during prolonged exposure to alcohol's depressant effects (see Chapter 3). This neuroadaptive process represents the neurobiologic foundations of tolerance and includes the downregulation of γ-aminobutyric acid type A receptors and the upregulation of *N*-methyl-D-aspartate (NMDA) receptors (Kosten and O'Conner, 2003, Reoux and Oreskovich, 2006). If a physically dependent individual abruptly stops drinking, the inhibitory effects of alcohol are diminished, while the adaptive changes persist, resulting in NMDA-facilitated hyperactivity that generates alcohol withdrawal symptoms. The most common substance-related cause of seizures is alcohol withdrawal. Typically seizures occur 6–48 hours after the last drink and are generalized tonic-clonic, although partial seizures can also occur. They are usually self-limiting but may give rise to status epilepticus, where alcohol dependence accounts for 9–25% of cases (Daeppen et al., 2002; O'Brien et al., 2007).

Delirium tremens are not common and occur in about 5% of patients hospitalized for alcohol withdrawal. Alcohol-related seizures generally develop 24–72 hours after the last drink. Worsening agitation, distractibility, and delusions generally precede the onset, which is characterized by fluctuating disturbance of consciousness, changes in cognition, exacerbation of autonomic symptoms (sweating, nausea, palpitations, and tremor), and fear or terror (O'Connor, 2007). Persons with a history of hallucinosis (e.g., schizophrenia, bipolar disorders) have a greater susceptibility to delirium tremens than other alcohol-dependent cohort populations (Stanley et al., 2005).

Wernicke–Korsakoff syndrome

Alcohol withdrawal can precipitate Wernicke's encephalopathy. Wernicke's encephalopathy consists of psychomotor slowing or apathy, nystagmus, ataxia, ophthalmoplegia, impaired consciousness, and, if untreated, coma and death. Wernicke's encephalopathy results from inadequate intake or absorption of thiamine plus continued carbohydrate ingestion. Excessive alcohol intake interferes with thiamine absorption from the gastrointestinal tract and hepatic storage of thiamine; the poor nutrition associated with alcoholism often precludes adequate thiamine intake (Martin et al., 2004; Sechi and Serra, 2007). The disorder may remit with treatment, persist, or degenerate into Korsakoff's syndrome (KS), a late complication of persistent Wernicke's encephalopathy, and results in memory deficits, confusion, and behavioral changes. KS is an amnestic syndrome with impaired ability to consolidate new memories and relatively intact intellectual function, perhaps through an interruption of diencephalic–hippocampal circuitry, including between thalamic nuclei and mamillary bodies (see Chapters 12, 12, and 16). KS occurs in 80% of untreated patients with Wernicke's encephalopathy. Wernicke–Korsakoff syndrome, which combines Wernicke encephalopathy and Korsakoff psychosis, occurs in some alcoholics who do not consume foods fortified with thiamine.

The risk factors and early manifestations of Wernicke's must be recognized and treated promptly and adequately with parenteral vitamins, notably, thiamine (B_1). Improvement in confusion usually occurs in 1–2 days, and in ocular abnormalities in days to weeks, while ataxia usually responds within the first week but can take months or much longer to resolve. Treatment consists of thiamine and supportive measures. Loading carbohydrates in patients with thiamine deficiency (i.e., refeeding after starvation or giving intravenous (IV) dextrose-containing solutions to high-risk patients) can trigger Wernicke's encephalopathy. Ensuring that dietary supplies of thiamine are adequate is important regardless of symptoms. Because IV glucose can worsen thiamine deficiency, alcoholics at risk of thiamine deficiency should receive IV thiamine 100 mg before receiving IV glucose solutions (O'Connor, 2013). Patients rarely have truly discrete deficits in forming new memories, generally exhibiting more global deficits along a spectrum of severity. KS is treatable and only about 25% of patients show no recovery.

KS can also present without a definable episode of Wernicke's, developing insidiously or following undiagnosed subacute episodes. This has led to prescribing oral thiamine supplements to chronic alcoholics. Magnesium is a cofactor for many thiamine-dependent enzymes, and deficiency may induce clinical signs of thiamine deficiency (Meier and Daeppen, 2005).

Pellagra

Niacin (vitamin B_3, nicotinic acid) and/or tryptophan deficiency results in skin, gastrointestinal, and mental abnormalities that often progress to memory impairment, delusions, hallucinations, dementia or delirium. Hypertonus and startle myoclonus may be present. Symptoms usually improve with nicotinic acid or nicotinamide (amide of vitamin B_3) replacement therapy (Kumar, 2010; So, 2012).

Blackouts (retrograde amnesia)

Blackouts are periods of amnesia during and after episodes of heavy alcohol consumption. They are similar to episodes of transient global amnesia (which occur in the absence of alcohol consumption) and are attributed to alcohol inhibition of NMDA receptor and impairing long-term potentiation (Eichenbaum, 2002, Hartzler and Fromme, 2003; White et al., 2004).

Central pontine and extrapontine myelinolysis

Hyponatremia is common in alcoholics, especially beer drinkers, due to hyperhydration of fluid intake. It is best treated by restoring normal hydration and diet while abstaining from alcohol. Attempts to correct the electrolyte disturbance with saline (particularly hypertonic saline) may result in demyelination, thought to be triggered by rapid osmotic shifts in the brain causing complement-mediated oligodendrocyte toxicity. This most commonly occurs in the pons (central pontine myelinolysis) but it can also occur elsewhere (extrapontine myelinolysis); the basal ganglia and thalamus are vulnerable (National Institute of Neurological Disorders and Stroke, 2012). If the pons is primarily affected, symptoms include dysarthria, dysphagia, and spastic quadriparesis. Lesions outside the pons have a much more variable presentation, which can include mutism, parkinsonism, dystonia, and catatonia (Kitabayashi et al., 2007; Yoon et al., 2008; Schneider et al., 2012).

Neuropathy

AUDs have been associated with what has been called the "Saturday-night palsy" due to compression of the radial nerve in deep or stuporous sleep (akinesic and mute). Chronic alcoholics can develop a symmetric, bilateral mixed sensory and motor peripheral neuropathy, usually of the lower limbs (Harati and Bosch, 2008). Patients may be asymptomatic or present with pain, numbness, burning feet, and hyperesthesia. This neuropathy is usually attributed to thiamine deficiency, although direct alcohol toxicity (Shy, 2007) may also contribute. Research suggests that, if thiamine deficiency is the predominant cause, the symptoms are motor, whereas direct alcohol toxicity produces more of a sensory-dominant presentation (Koike et al., 2003; National Institute of Neurological Disorders and Stroke, 2012). Recovery occurs with abstinence and thiamine supplements can be used. Chronic cerebellar ataxia affects about one-third of the chronically alcohol-dependent. It is related to a combination of malnutrition and alcohol's direct toxic effects. The ataxia consists of instability of gait and stance with severe lack of coordination of the knee–shin test, and relatively little involvement of the arms. In severe cases, there is evidence of neural degeneration in the anterior and superior portions of the cerebellar vermis with extension into the anterior lobes and flocculi (Shy, 2007).

Cognitive impairment

Alcohol-related brain damage has no single cause. Thiamine deficiency underlies KS on one hand, yet the alcohol-dependent person repeatedly has episodes of drunkenness with subsequent alcohol withdrawal, dietary neglect and malnutrition, severe vitamin deficiencies, TBI, cerebrovascular events, and alcoholic liver disease, all of which can contribute to impairment in brain structure and function. Whether KS or not, abstinence and an enhanced nutritional focus are central to management. Alcohol-related cognitive deficits generally improve with abstinence.

A note about alcohol use as a risk factor for traumatic brain injury or recovery from TBI

Alcohol use and TBI are often closely related. Up to two-thirds of people with TBI have a history of alcohol abuse or risky drinking. Between 30% and 50% of people with TBI were injured while they were drunk and about one-third were under the influence of other drugs. Around half of those who have a TBI cut down on their drinking or stop altogether after injury, but some people with TBI continue to drink heavily, which increases their risk of having negative outcomes (Bjork and Grant, 2009).

In epidemiologic surveys, emergency room patients have reported elevated incidence of recent alcohol use (Cherpitel and Ye, 2008) and have also shown high incidence of actual presence of alcohol or other drugs at the time of injury (Vitale and van de Mheen, 2006). Moreover, alcohol abuse can exacerbate the effects of TBI. Jorge et al. (2005) reported that alcohol abuse or dependence following TBI impaired vocational outcome, possibly by exacerbating the neurologic sequelae of the injury itself. Similarly, Corrigan (1995) reported that substance abuse following TBI portended a poorer trajectory of rehabilitation following injury. In a sample of over 1600 TBI subjects, Horner et al. (2005) reported that a pre-existing substance abuse diagnosis at time of injury was predictive of heavy drinking following the injury. Therefore, screening and brief intervention may be especially important for TBI patients as well as education on TBI recovery for families.

CONCLUSION

In this chapter we reviewed the components of SBIRT as a way of targeting AUD in an ambulatory care setting, specifically the neurology clinic, although the information herein can be integrated into most primary care settings. We emphasize that AUDs exist on a continuum from normative use, to risky use, to a use disorder. Within the subset of pathologic use, the new *Diagnostic and Statistical Manual* further subdivides the spectrum of use disorders into mild, moderate, and severe, based on compulsive use, out-of-control use, and continued use despite consequences. Frequency and quantity of alcohol consumption should be considered the fifth vital sign, and an important starting point for a discussion about alcohol use within the ambulatory care setting. Data show that brief interventions for at-risk or pathologic use can be effectively implemented in 5–10 minutes and should involve exploring the patient's ambivalence about change, understanding the patient's motivation to change, and adjusting the intervention to meet each individual's needs. Patient and provider should agree on a goal and strive toward it, whether that goal is reducing alcohol consumption, reducing risky behaviors associated with alcohol consumption, or abstinence. Healthcare providers can effectively limit medical consequences of chronic heavy alcohol use by screening and intervening for the AUD itself.

References

Aira M, Kauhanen AM, Larivaara P et al. (2003). Factors influencing inquiry about patients' alcohol consumption by primary health care physicians: qualitative semi-structured interview study. Fam Pract 20: 270–275.

American Psychiatric Association (2013). Diagnostic and Statistical Manual of Mental Disorders, American Psychiatric Association, Washington, DC.

Bjork JM, Grant S (2009). Does traumatic brain injury increase risk for substance abuse? J Neurotrauma 26 (7): 1077–1082.

Bradley K, Berger D (2013).In: R Saiz (Ed.), Addressing unhealthy alcohol use in primary care. Springer Science and Business Media, New York, NY.

Burge SK, Schneider FD (1999). Alcohol-related problems: recognition and intervention. Am Fam Physician 15;59 (2): 361–370.

Calfas KJ, Sallis JF, Oldenburg B et al. (1997). Mediators of change in physical activity following an intervention in primary care: PACE. Prev Med 26: 297–304.

CASA Columbia (2012). The National Center on Addiction and Substance Abuse at Columbia University. Annual Report. CASA, New York, NY.

Cherpitel CJ, Ye Y (2008). Drug use and problem drinking associated with primary care and emergency room utilization in the US general population: data from the 2005 national alcohol survey. Drug Alcohol Depend 97: 226–230.

Corrigan JD (1995). Substance abuse as a mediating factor in outcome from traumatic brain injury. Arch Phys Med Rehabil 76: 302–309.

Daeppen JB, Gache P, Landry U et al. (2002). Symptom-triggered vs fixed-schedule doses of benzodiazepine for alcohol withdrawal. Arch Intern Med 162: 1117–1121.

Eichenbaum H (2002). The Cognitive Neuroscience of Memory: An Introduction. Oxford University Press, New York.

Emrick CD, Tonigan JS, Montgomery H et al. (1993). Alcoholics anonymous: What is currently known? In: BS Mccrady, WR Miller (Eds.), Research on Alcoholics Anonymous: Opportunities and alternatives. Rutgers Center for Alcohol Studies, New Brunswick, NJ.

Glanz K, Patterson RE, Kristal AR et al. (1994). Stages of change in adopting healthy diets: fat, fiber, and correlates of nutrient intake. Health Educ Q 21: 499–519.

Grimley DM, Riley GE, Bellis JM et al. (1993). Assessing the stages of change and decision-making for contraceptive use for the prevention of pregnancy, sexually transmitted diseases, and acquired immunodeficiency syndrome. Health Educ Q 29: 455–470.

Harati Y, Bosch EP (2008). Disorders of peripheral nerves. In: WG Bradley, RB Daroff, GM Fenichel et al. (Eds.), Bradley: Neurology in Clinical Practice. 5th ed. Butterworth-Heinemann Elsevier, Philadelphia, Pa, chapter 80.

Harris AHS, Bryson CL, Sun H et al. (2009). Alcohol screening scores predict risk of subsequent fractures. Subst Use Misuse 44: 1055–1069.

Harris A, Bradley KA, Bowe T et al. (2010). Associations between AUDIT-C and mortality vary by age and sex. Popul Health Manag 13: 263–268.

Hartzler B, Fromme K (2003). Fragmentary blackouts: Their etiology and effect on alcohol expectancies. Alcohol Clin Exp Res 27 (4): 628–637.

Hellman EA (1997). Use of the stages of change in exercise adherence model among older adults with a cardiac diagnosis. J Cardiopulm Rehabil 17: 145–155.

Horner MD, Ferguson PL, Selassie AW et al. (2005). Patterns of alcohol use 1 year after traumatic brain injury: a population-based, epidemiological study. J Int Neuropsychol Soc 11: 322–330.

Humphreys K (2004). Circles of Recovery: Self-Help Organizations for Addictions, Cambridge University Press, Cambridge.

Janz NK, Becker MH (1984). The Health Belief Model: a decade later. Health Educ Q 11: 1–47.

Jorge RE, Starkstein SE, Arndt S et al. (2005). Alcohol misuse and mood disorders following traumatic brain injury. Arch Gen Psychiatry 62: 742–749.

Kitabayashi Y, Narumoto J, Shibata K et al. (2007). Neuropsychiatric background of alcohol hallucinosis: A SPECT study. J Neuropsychiatry Clin Neurosci 19: 1,85.

Koike H, Iijima M, Sugiura M et al. (2003). Alcoholic neuropathy is clinicopathologically distinct from thiamine-deficiency neuropathy. Ann Neurol 54 (1): 19–29.

Kosten TR, O'Conner PG (2003). Management of drug and alcohol withdrawal. N Engl J Med 348: 178695.

Kumar N (2010). Neurologic presentation of nutritional deficiencies. Neurol Clin 28: 107–170.

Lembke A, Bradley K, Henderson P et al. (2011). Alcohol screening scores and the risk of new-onset gastrointestinal illness or related hospitalization. J Gen Intern Med 26: 777–782.

Martin PR, Singleton CK, Hiller-Sturmhofel S (2004). The role of thiamine deficiency in alcoholic brain disease. National Institute on Alcohol Abuse and Alcoholism, Bethesda, MD.

Mclellan AT, Lewis DC, O'Brien CP et al. (2000). Drug dependence, a chronic medical illness: implications for treatment, insurance, and outcomes evaluation. JAMA 284: 1689–1695.

Meier S, Daeppen JB (2005). Prevalence, prophylaxis and treatment of Wernicke encephalopathy. Thiamine, how much and how do we give it? http://www.ncbi.nlm.nih.gov/pubmed/16117048.

Mersy DJ (2003). Recognition of alcohol and substance abuse. Am Fam Physician 67: 1529–1532.

Miller WR, Rollnick S (2012). Motivational Interviewing. Helping People Change. Third Edition. Guilford Press, New York.

National Institute of Neurological Disorders and Stroke (2012). Peripheral Neuropathy Fact Sheet. National Institute of Neurological Disorders and Stroke. National Institutes of Health, Bethesda, MD.

O'Brien JM, Lu B, Ali NA et al. (2007). Alcohol dependence is independently associated with sepsis, septic shock, and hospital mortality among adult intensive care unit patients. Crit Care Med 35: 345–350.

O'Connor PG (2007). Alcohol abuse and dependence. In: L Goldman, D Ausiello (Eds.), Cecil Medicine, 23rd ed. Saunders Elsevier, Philadelphia, Pa, chapter 31.

O'Connor PG (2013). In the Merck Manual for Healthcare Professionals. Whitehouse Station, Merck Sharp & Dohme, N.J., U.S.A.

Prochaska JO, DiClemente CC, Norcross JC (1992). In search of how people change. Applications to addictive behaviours. Am Psychol 47 (11): 02.

Prochaska JO, Velicer WF, Rossi JS et al. (1994). Stages of change and decisional balance for 12 problem behaviors. Health Psychol 13: 39–46.

Reoux JP, Oreskovich MR (2006). A comparison of two versions of the clinical institute withdrawal assessment for alcohol: the CIWA-Ar and the CIWA-AD. Am J Addict 15: 85–93.

Saitz R (2005a). Clinical practice. Unhealthy alcohol use. N Engl J Med 352: 596–607.

Saitz R (2005b). Unhealthy alcohol use. N Engl J Med 352: 596–607.

Saitz R (2013). Addressing unhealthy alcohol use in primary care, Springer Science and Business Media, New York.

Schneider P, Nejtek VA, Hurd CL (2012). A case of mistaken identity: alcohol withdrawal, schizophrenia, or central pontine myelinolysis? Neuropsychiatr Dis Treat 8: 49–54.

Sechi G, Serra A (2007). Wernicke's encephalopathy: new clinical settings and recent advances in diagnosis and management. Lancet Neurol 6 (5): 442–455.

Shy ME (2007). Peripheral neuropathies. In: L Goldman, D Ausiello (Eds.), Cecil Medicine. 23rd ed. Philadelphia, Pa, Saunders Elsevier, chapter 446.

Sobell LC, Sobell MB (1995). Alcohol Timeline Followback Users' Manual. Addiction Research Foundation, Toronto, Canada.

Stanley KM, Worrall CL, Lunsford SL et al. (2005). Experience with an adult alcohol withdrawal syndrome. Practice guideline in internal medicine patients. Pharmacotherapy 25 (8): 1073–1083.

Trice HM, Roman PM (1970). Sociopsychological predictors of affiliation with Alcoholics Anonymous: a longitudinal study of "treatment success". Soc Psychiatry 5: 51–59.

Trice HM, Wahl R (1958). A rank order analysis of the symptoms of alcoholism. Q J Stud Alcohol 19: 636–648.

U.S. Preventative Services Task FORCE (2004). Screening and behavioral counseling interventions in primary care to reduce alcohol misuse: recommendation statement. Ann Intern Med 140: 554–556.

Vitale S, van de Mheen D (2006). Illicit drug use and injuries: a review of emergency room studies. Drug Alcohol Depend 82: 1–9.

White E, Signer ML, Kraus CL et al. (2004). Experiential aspects of alcohol-induced blackouts among college students. Am J Drug Alcohol Abuse 30: 205–224.

Witbrodt J, Kaskutas LA, Bond J et al. (2012). Does sponsorship improve outcomes above Alcoholics Anonymous attendance? A latent class growth curve analysis. Addiction 107 (2): 301–311.

Yoon B, Shim YS, Chung SW (2008). Central pontine and extrapontine myelinolysis after alcohol withdrawal. Alcohol Alcohol 43 (6): 647–649.

Zimmerman GL, Olsen CG, Bosworth MF (2000). A ‘Stages of Change’ approach to helping patients change behavior. Mar 1, Am Fam Physician 61 (5): 1409–1416.

Index

NB: Page numbers in *italics* refer to boxes, figures and tables.

D

H

I

O

P

Q

R

FSC
www.fsc.org
MIX
Papier | Fördert gute Waldnutzung
FSC® C083411

Kolibri 360 GmbH
Ferdinand-Jühlke-Straße 7
99095 Erfurt, Deutschland
produktsicherheit@kolibri360.de